Fields Medallists' Lectures

Third Edition

World Scientific Series in 21st Century Mathematics

Published

Vol. 1 Fields Medallists' Lectures
(3rd Edition)
edited by Sir Michael Atiyah, Daniel Iagolnitzer and Chitat Chong

Forthcoming

Vol. 2 Fifty Years of Mathematical Physics
Selected Works of Ludvig Faddeev
edited by Molin Ge and Antti J Niemi

World Scientific Series in 21st Century Mathematics – Vol. 1

Fields Medallists' Lectures

Third Edition

Editors

Sir Michael Atiyah
University of Edinburgh, UK

Daniel Iagolnitzer
CEA-Saclay, France

Chitat Chong
NUS, Singapore

World Scientific

NEW JERSEY • LONDON • SINGAPORE • BEIJING • SHANGHAI • HONG KONG • TAIPEI • CHENNAI • TOKYO

Published by

World Scientific Publishing Co. Pte. Ltd.
5 Toh Tuck Link, Singapore 596224
USA office: 27 Warren Street, Suite 401-402, Hackensack, NJ 07601
UK office: 57 Shelton Street, Covent Garden, London WC2H 9HE

The editors and publisher would like to thank the following organisations and publishers of the various journals and books for their assistance and permission to reproduce the selected reprints found in this volume:

Academia Scientiarum Fennica
American Mathematical Society
Birkhäuser Verlag
Cambridge University Press
Canadian Mathematical Society
Documenta Mathematica
Dunod Editeur
Elsevier Science Publishers
Finnish Academy of Science and Letters
Higher Education Press
Institut Mittag-Leffler
International Mathematical Union
London Mathematical Society
Mathematical Society of Japan
Mir Publishers
Polish Scientific Publishers PWN Ltd.
Springer-Verlag

While every effort has been made to contact the publishers of reprinted papers prior to publication, we have not been successful in some cases. Where we could not contact the publishers, we have acknowledged the source of the material. Proper credit will be accorded to these publishers in future editions of this work after permission is granted.

Fields Medal on cover: photographed by AK Peters.

World Scientific Series in 21st Century Mathematics — Vol. 1
FIELDS MEDALLISTS' LECTURES
3rd Edition

ISBN 978-981-4696-17-3
ISBN 978-981-4696-18-0 (pbk)

Typeset by Stallion Press
Email: enquiries@stallionpress.com

PREFACE TO THE FIRST EDITION

Although the Fields Medal does not have the same public recognition as the Nobel Prizes, they share a similar intellectual standing. It is restricted to one field — that of mathematics — and an age limit of 40 has become an accepted tradition. Mathematics has in the main been interpreted as pure mathematics, and this is not so unreasonable since major contributions in some applied areas can be (and have been) recognized with Nobel Prizes. The restriction to 40 years is of marginal significance, since most mathematicians have made their mark long before this age.

A list of Fields Medallists and their contributions provides a bird's eye view of mathematics over the past 60 years. It highlights the areas in which, at various times, greatest progress has been made. This volume does not pretend to be comprehensive, nor is it a historical document. On the other hand, it presents 22 Fields Medallists and so provides a highly interesting and varied picture.

The contributions themselves represent the choice of the individual Medallists. They are either reproductions of already published works, or are new articles produced for this volume. In some cases they relate directly to the work for which the Fields Medals were awarded. In other cases they relate to more current interests of the Medallists. This indicates that while Fields Medallists must be under 40 at the time of the award, their mathematical development goes well past this age. In fact the age limit of 40 was chosen so that young mathematicians would be encouraged in their future work.

The contribution of each medallist is in most cases preceded by the introductory speech given by another leading mathematician during the prize ceremony, a photograph and an up-to-date biographical notice. The introductory speech outlines the basic works of the medallist at the time of the medal and the reasons why it was awarded.

The Editors

PREFACE TO THE SECOND EDITION

It is a pleasure to present this second edition which includes new contributions, by the 1998 and 2002 medallists.

Most Fields medallists are still living and most of them in full scientific activity. However, in addition to Lars Ahlfors (who was the first medallist with Jesse Douglas in 1936), we were very sad to learn the more recent death of other medallists. Due to the cooperation of his daughter, Mrs Oka, we are pleased to include a work by Kunihiko Kodaira, 1954 medallist who died in 1997 (the year of the first edition of this volume). On the other hand, Laurent Schwartz and Rene Thom also died recently, after the first edition of this book. Both were great scientists who made remarkable contributions to mathematics and other domains of science, and also deeply influenced younger colleagues through their teaching (one of us, D.I., was e.g. a student of Laurent Schwartz at the University of Paris). Both had a warm personality which attracted many friends.

We would like to dedicate this second edition to them.

The Editors

PREFACE TO THE THIRD EDITION

This third edition of the *Fields Medallists's Lectures* adds to the collection lectures of the following Fields medallists: John Milnor (1962), Enrico Bombieri (1974), Gerd Faltings (1986), Andrei Okounkov and Terence Tao (2006), Cédric Villani, Elon Lindenstrauss, Ngô Bảo Châu and Stanislav Smirnov (2010). In addition, it also presents English translations of introductory lectures on the works of four Fields medallists (Michael Atiyah, Alexander Grothendieck, Stephen Smale and Jean-Christophe Yoccoz) originally written in French. This edition brings the collection of available lectures up to the awards of the Fields medals in 2010.

Since the publication of the second edition in 2003, several Fields medallists have passed away. They were Paul Cohen in 2007, Daniel Quillen in 2011, Lars Hörmander and William Thurston in 2012, and Alexander Grothendieck in 2014. Each of them made fundamental contributions to mathematics and transformed the fields in which they worked. We dedicate the third edition of *Fields Medallists' Lectures* to them.

The Editors

RECIPIENTS OF FIELDS MEDALS

1936
Lars V. Ahlfors
Jesse Douglas

1950
Laurent Schwartz
Atle Selberg

1954
Kunihiko Kodaira
Jean-Pierre Serre

1958
Klaus F. Roth
René Thom

1962
Lars Hörmander
John W. Milnor

1966
Michael F. Atiyah
Paul J. Cohen
Alexander Grothendieck
Stephen Smale

1970
Alan Baker
Heisuke Hironaka
Sergei P. Novikov
John G. Thompson

1974
Enrico Bombieri
David B. Mumford

1978
Pierre R. Deligne
Charles L. Fefferman
Gregory A. Margulis
Daniel G. Quillen

1982
Alain Connes
William P. Thurston
Shing-Tung Yau

1986
Simon K. Donaldson
Gerd Faltings
Michael H. Freedman

1990
Vladimir G. Drinfeld
Vaughan F. R. Jones
Shigefumi Mori
Edward Witten

1994
Jean Bourgain
Pierre-Louis Lions
Jean-Christophe Yoccoz
Efim I. Zelmanov

1998
Richard E. Borcherds
William T. Gowers
Maxim Kontsevich
Curtis T. McMullen

2002
Laurent Lafforgne
Vladimir Voevodsky

2006
Andrei Okounkov
Grigori Perelman (declined)
Terence Tao
Wendelin Werner

2010
Elon Lindenstrauss
Ngô Bảo Châu
Stanislav Smirnov
Cédric Villani

CONTENTS

Preface to the First Edition v

Preface to the Second Edition vii

Preface to the Third Edition ix

Recipients of Fields Medals xi

1936 L. V. AHLFORS

Biographical Notice **3**

Quasiconformal Mappings, Teichmüller Spaces, and Kleinian Groups **5**

1950 L. SCHWARTZ

The Work of L. Schwartz by H. Bohr **19**

Biographical Notice **27**

Calcul Infinitesimal Stochastique **29**

1954 K. KODAIRA

Obituary: Kunihiko Kodaira **49**

Biographical Notice **57**

On Kähler Varieties of Restricted Type **59**

1958 K. F. ROTH

The Work of K. F. Roth by H. Davenport **65**

Biographical Notice **71**

Rational Approximations to Algebraic Numbers **73**

R. THOM

The Work of R. Thom by H. Hopf 81

Autobiography 87

1962 L. HÖRMANDER

Hörmander's Work on Linear Differential Operators by L. Gårding 93

Autobiography 99

Looking forward from ICM 1962 103

J. W. MILNOR

The Work of John W. Milnor by H. Whitney 121

Topological Manifolds and Smooth Manifolds 127

1966 M. F. ATIYAH

The Work of Michael F. Atiyah by H. Cartan 135

Biography 141

The Index of Elliptic Operators 143

PAUL J. COHEN

The Continuum Problem by A. Church 157

ALEXANDER GROTHENDIECK

The Works of Alexander Grothendieck by Jean Dieudonné 163

S. SMALE

On the Works of Stephen Smale by R. Thom 169

Biographical Notice 175

A Survey of Some Recent Developments in Differential Topology 181

1970 A. BAKER

The Work of Alan Baker by P. Turán 195

Biography 199

Effective Methods in the Theory of Numbers 201

Effective Methods in Diophantine Problems 211

Effective Methods in Diophantine Problems. II **223**

Effective Methods in the Theory of Numbers/Diophantine Problems **231**

S. NOVIKOV

The Work of Serge Novikov by M. F. Atiyah **235**

Biography **239**

Rôle of Integrable Models in the Development of Mathematics **243**

1974 E. BOMBIERI

The Work of Enrico Bombieri by K. Chandrasekharan **261**

Variational Problems and Elliptic Equations **271**

D. MUMFORD

The Work of David Mumford by J. Tate **283**

Autobiography **289**

Pattern Theory: A Unifying Perspective **291**

1978 G. A. MARGULIS

The Work of Gregory Aleksandrovitch Margulis by J. Tits **327**

Biographical Notice **335**

Oppenheim Conjecture **337**

1982 A. CONNES

The Work of Alain Connes by H. Araki **391**

Biographical Notice **399**

Brisure de Symétrie Spontanée et Géométrie du Point de vue Spectral **401**

W. P. THURSTON

The Work of W. Thurston by C. T. C. Wall **431**

1986 S. K. DONALDSON

The Work of Simon Donaldson by M. F. Atiyah **435**

Biographical Notice **441**

Remarks on Gauge Theory, Complex Geometry and 4-Manifold Topology 443

G. FALTINGS

On Some of the Mathematical Contributions of Gerd Faltings by B. Mazur 463

Neuere Entwicklungen in der Arithmetischen Algebraischen Geometrie 471

M. H. FREEDMAN

The Work of M. H. Freedman by J. Milnor 479

Biographical Notice 485

Betti Number Estimates for Nilpotent Groups 487

1990 V. F. R. JONES

The Work of Vaughan F. R. Jones by J. S. Birman 509

Biographical Notice 521

A Polynomial Invariant for Knots via Von Neumann Algebras 523

Index for Subfactors 535

S. MORI

The Work of Shigefumi Mori by H. Hironaka 563

Biographical Notice 571

Birational Classification of Algebraic Threefolds 573

E. WITTEN

The Work of E. Witten by L. D. Faddeev 587

The Work of Edward Witten by M. F. Atiyah 591

Biographical Notice 597

Geometry and Quantum Field Theory 599

1994 J. BOURGAIN

The Work of Jean Bourgain by L. Caffarelli 613

Biographical Notice 617

Hamiltonian Methods in Nonlinear Evolution Equations 619

P. L. LIONS

The Work of Pierre-Louis Lions by S. R. S. Varadhan 633

Biographical Notice 639

On Some Recent Methods for Nonlinear Partial Differential Equations 643

J. C. YOCCOZ

Presentation of Jean-Christophe Yoccoz by A. Douady 661

Recent Developments in Dynamics 669

E. I. ZELMANOV

The Work of Efim Zelmanov by W. Feit 691

Biographical Notice 701

On the Restricted Burnside Problem 703

1998 R. E. BORCHERDS

The Work of Richard Ewen Borcherds by P. Goddard 713

Biographical Notice 725

What is Moonshine? 727

W. T. GOWERS

The Work of William Timothy Gowers by B. Bollobás 737

Autobiography 747

A New Proof of Szemerédi's Theorem for Arithmetic Progressions of Length Four 753

M. KONTSEVICH

The Work of Maxim Kontsevich by C. H. Taubes 777

Biographical Notice 787

Formal (Non)-Commutative Symplectic Geometry 791

Comments on "Formal (Non)-Commutative Symplectic Geometry" 809

C. T. McMULLEN

The Work of Curtis T. McMullen by J. Milnor 811

Biographical Notice 819

Rigidity and Inflexibility in Conformal Dynamics 821

2002 L. LAFFORGUE

The Work of Laurent Lafforgue by G. Laumon **837**

Biographical Notice **845**

V. VOEVODSKY

The Work of Vladimir Voevodsky by C. Soulé **847**

Biographical Notice **853**

Open Problems in the Motivic Stable Homotopy Theory, I **857**

2006 A. OKOUNKOV

The Work of Andrei Okounkov by G. Felder **887**

Biographical Notice **899**

TERENCE TAO

The Work of Terence Tao by C. Fefferman **901**

Biographical Notice **911**

The Dichotomy between Structure and Randomness, Arithmetic Progressions, and the Primes **913**

2010 NGÔ BAO CHÂU

The Work of Ngô Bao Châu by J. Arthur **943**

Biographical Notice **957**

Endoscopy Theory of Automorphic Forms **959**

E. LINDENSTRAUSS

The Work of Elon Lindenstrauss by H. Furstenberg **987**

Biographical Notice **993**

Equidistribution in Homogeneous Spaces and Number Theory **995**

S. SMIRNOV

The Work of Stanislav Smirnov by H. Kesten **1021**

Biographical Notice **1035**

Discrete Complex Analysis and Probability **1037**

C. VILLANI

The Work of Cédric Villani by H.-T. Yau **1065**

Biographical Notice **1079**

Landau Damping **1081**

Lars V. Ahlfors

LARS VALERIAN AHLFORS

by

J. J. O'CONNOR AND E. F. ROBERTSON

Mathematical Institute

University of St Andrews, Scotland

Lars Ahlfors' father was professor of mechanical engineering at the Polytechnic Institute in Helsingfors. Tragically his mother died in childbirth when he was born. Ahlfors describes his early years in [1]:

> *As a child I was fascinated by mathematics without understanding what it was about, but I was by no means a child prodigy. As a matter of fact I had no access to mathematical literature except in the highest grades. Having seen many prodigies spoilt by ambitious parents, I can only be thankful to my father for his restraint. the high school curriculum did not include any calculus, but I finally managed to learn some on my own, thanks to clandestine visits to my father's engineering library.*

Ahlfors entered Helsingfors University in 1924, and there he was taught by Lindelf and Nevanlinna. He graduated from Helsingfors in 1928. Nevanlinna replaced Weyl in Zurich for the session 1928/29 while Weyl was on leave, and Ahlfors went to Zurich with him. In Zurich Nevanlinna lectured on Denjoy's conjecture on the number of asymptotic values of an entire function. Ahlfors modestly writes in [1]:

> *I had the incredible luck of hitting upon a new approach, based on conformal mappings, which, with very considerable help from Nevanlinna and Pólya, led to a proof of the full conjecture.*

Ahlfors went to Paris with Nevanlinna for three months before returning to Finland. There he was appointed lecturer in mathematics in Turku. He presented his doctoral thesis in 1930, then in the following two years he made a number of visits to Paris and other European centres.

In 1935, Caratheodory, whom Ahlfors had met in Munich during his travels, recommended him for a post at Harvard in the United States. Ahlfors agreed to a three year trial period. In 1936 he was one of the first two recipients of a Fields Medal at the International Congress in Oslo.

In 1938 Ahlfors was offered a chair in mathematics at the University of Helsinki and, being rather homesick, he accepted this rather than remain permanently at Harvard. However a difficult time was approaching with World War II about to begin. The war led to severe problems in Finland and the universities were closed. Ahlfors was unfit for military service so, as he states in [1]:

> *Paradoxically I was myself able to do a lot of work during the war, although without the benefit of accessible libraries.*

Ahlfors' family was evacuated to Sweden during the war and so, when he was offered a chair in Zurich in 1944 it seemed a good chance to be reunited with his family. He met up with his family in Sweden, where Beurling gave them a great deal of help and friendship, but the war made the trip to Switzerland close to impossible. A flight from Stockholm to Prestwick in Scotland was arranged and, in March 1945, they made the trip. From Glasgow they travelled by train to London, then they made the difficult journey across the Channel, across France via Paris to Switzerland. He writes [1]:

> *I cannot honestly say that I was happy in Zurich. The post-war era was not a good time for a stranger to take root in Switzerland. ... My wife and I did not feel welcome outside the circle of our immediate colleagues.*

An offer from Harvard in 1946 was therefore gladly accepted and, on this occasion, he remained there, retiring in 1977.

His books are of lasting importance. Among them are *Complex analysis* (1953),* *Riemann surfaces* (with L Sario) (1960), *Lectures on quasi-conformal mappings* (1966) and Conformal *invariants* (1973). In addition to the topics covered by these texts, Ahlfors did work of major importance on Kleinian groups.

Ahlfors received many honours for his outstanding contributions to mathematics. The award of the first Fields medal, mentioned above, must rank as the most important but another great honour was the award of the Wolf Prize in Mathematics in 1981.

References

[1] O Lehto, *On the life and work of Lars Ahlfors*, The Mathematical Intelligencer 20 (4) (1999), 4–8.

*Allow me [EFR] a personal note on Ahlfors' *Complex analysis.* This was the text recommended to me by Copson who taught me complex analysis and it is indeed a tribute to Ahlfors that Copson, who had himself written a superb book on complex analysis, should recommend Ahlfors' book rather than his own. I found Ahlfors' *Complex analysis* beautifully written, an example of the very highest quality in mathematical texts, combining clarity with an excitement for the topic.

Reprinted from Proc. Int. Congr. Math., 1978

QUASICONFORMAL MAPPINGS, TEICHMÜLLER SPACES, AND KLEINIAN GROUPS

by

LARS V. AHLFORS*

I am extremely grateful to the Committee to select hour speakers for the great honor they have bestowed on me, and above all for this opportunity to address the mathematicians of the whole world from the city of my birth. The city has changed a great deal since my childhood, but I still get a thrill each time I return to this place that holds so many memories for me. I assure you that today is even a more special event for me.

I have interpreted the invitation as a mandate to report on the state of knowledge in the fields most directly dominated by the theory and methods of quasiconformal mappings. I was privileged to speak on the same topic once before, at the Congress in Stockholm 1962, and it has been suggested that I could perhaps limit myself to the developments after that date. But I feel that this talk should be directed to a much wider audience. I shall therefore speak strictly to the non-specialists and let the experts converse among themselves on other occasions.

The whole field has grown so rapidly in the last years that I could not possibly do justice to all recent achievements. A mere list of the results would be very dull and would not convey any sense of perspective. What I shall try to do, in the limited time at my disposal, is to draw your attention to the rather dramatic changes that have taken place in the theory of functions as a direct of the inception and development of quasiconformal mappings. I should also like to make it clear that I am not reporting on my own work; I have done my share in the early stages, and I shall refer to it only when needed for background.

1. Historial remarks. In classical analysis the theory of analytic functions of complex variables, and more particularly functions of one variable, have played a dominant role ever since the middle of the nineteenth century. There was an obvious peak around the turn of the century, centering around names like Poincaré, Klein, Picard, Borel, and Hadamard. Another blossoming took place in the 1920s with the arrival of Nevanlinna's theory. The next decade seemed at the time as a slackening

*This work has been supported by the National Science Foundation of the United States under Grant number MCS77 07782

of the pace, but this deceptive; many of the ideas that were later to be fruitful were conceived at that time.

The war and the first post-war years were of course periods of stagnation. The first areas of mathematics to pick up momentum after the war were topology and functions of several complex variables. Big strides were taken in these fields, and under the leadership of Henri Cartan, Behnke, and many others, the more-dimensional theory of analytic functions and manifolds acquired an almost entirely new structure affiliated with algebra and topology. As a result of this development the gap between the conservative analysts who were still doing conformal mapping and the more radical ones involved with sheaf-theory became even wider, and for some time it looked as if the one-dimensional theory had lost out and was in danger of becoming a rehash of old ideas. The gap is still there, but I shall try to convince you that in the long run the old-fashioned theory has recovered and is doing quite well.

The theory of quasiconformal mappings is almost exactly fifty years old. They were introduced in 1928 by Herbert Grötzsch in order to formulate and prove a generalization of Picard's theorem. More important is his paper of 1932 in which he discusses the most elementary but at the same time most typical cases of extremal quasiconformal mappings, for instance the most nearly conformal mapping of one doubly connected region on another. Grötzsch's contribution is twofold: (1) to have been the first to introduce non-conformal mappings in a discipline that was so exclusively dominated by analytic functions, (2) to have recognized the importance of measuring the degree of quasiconformality by the maximum of the dilatation rather than by some integral mean (this was recently pointed out by Lipman Bers).

Grötzch's papers remained practically unnoticed for a long time. In 1935 essentially the same class of mappings was introduced by M. A. Lavrentiev in the Soviet Union whose work was connected more closely with partial differential equations than with function theory proper. In any case, the theory of quasiconformal mappings, which at that time had also acquired its name, slowly gained recognition, originally as a useful and flexible tool, but inevitably also as an interesting piece of mathematics in its own right.

Nevertheless, quasiconformal mappings might have remained a rather obscure and peripheral object of study if it had not been for Oswald Teichmüller, an exceptionally gifted and intense young mathematician and political fanatic, who suddenly made a fascinating and unexpected discovery. At that time, many special extremal problems in quasiconformal mapping had already been solved, but these were isolated results without a connecting general idea. In 1939 he presented to the Prussian Academy a now famous paper which marks the rebirth of quasiconformal mappings as a new discipline which completely overshadows the rather modest beginnings of the theory. With remarkable intuition he made a synthesis of what was known and proceeded to announce a bold outline of a new program which he presents, rather dramatically, as the result of a sudden revelation that occured to him at night. His main discovery was that the extremal problem of quasiconformal mapping, when applied to Riemann surfaces, leads automatically to an intimate connection with the holomorphic quadratic differentials on the surface. With this connection the whole theory takes on a completely different complexion: A problem concerned with non-conformal mappings turns out to have a solution which is expressed in terms of holomorphic differentials, so that in reality the problem belongs to classical function theory. Even if some of the proofs were only heuristic, it was clear from the start that this paper would have a tremendous impact, although actually its influence was delayed due to the poor communications during the war. In the same paper Teichmüller lays the foundations for what later has become known as the theory of Teichmüller spaces.

2. Beltrami coefficients. It is time to become more specific, and I shall start by recalling the definition and main properties of quasiconformal (q.c.) mappings. To begin with I shall talk only about the two-dimensional case. There is a corresponding theory in several dimensions, necessarily less developed, but full of interesting problems. One of the reasons for considering q.c. mappings, although not the most compelling one, is precisely that the theory does not fall apart when passing to more than two dimensions. I shall return to this at the end of the talk.

Today it can be assumed that even a non-specialist knows roughly what is meant by a q.c. mapping. Intuitively, a homeomorphism is q.c. if small circles are carried into small ellipses with a bounded ratio of the axes; more precisely, it is K-q.c. if the ratio is $\leqslant K$. For a diffeomorphism f this means that the complex derivatives $f_z = \frac{1}{2}(f_x - if_y)$ and $f_{\bar{z}} = \frac{1}{2}(f_x + if_y)$ satisfy $|f_{\bar{z}}| \leqslant k|f_z|$ with $k = (K-1)(K+1)$.

Already at an early stage it became clear that it would not do to consider only diffeomorphisms, for the class of diffeomorphisms lacks compactness. In the beginning rather arbitrary restrictions were introduced, but in time they narrowed down to two conditions, one geometric and one analytic, which eventually were found to be equivalent. The easiest to formulate is the analytic condition which says that f is K-q.c. if it is a weak L^2-solution of a *Beltrami equation*

$$f_{\bar{z}} = \mu f_z \tag{1}$$

where $\mu = \mu_f$, known as a Beltrami coefficient, is a complex-valued measurable function with $\|\mu\|_\infty \leqslant k$.

The equation is classical for smooth μ, but there is in fact a remarkably strong existence and uniqueness theorem without additional conditions. If μ is defined in the whole complex plane, with $|\mu| \leqslant k < 1$ a.e., then (1) has a homeomorphic solution which maps the plane on itself, and the solution is unique up to conformal mappings. Simple uniform estimates, depending only on k, show that the class of K-q.c. mappings is compact.

It must be clear that I am condensing years of research into minutes. The fact is that the post-Teichmüller era of quasiconformal mappings did not start seriously until 1954. In 1957 I. N. Vekua in the Soviet Union proved the existence and uniqueness theorem for the Beltrami equation, and in the same year L. Bers discovered that the theorem had been proved already in 1938 by C. Morrey. The great difference in language and emphasis had obscured the relevance of Morrey's paper for the theory of q.c. mappings. The simplest version of the proof is due to B. V. Boyarski who made it a fairly straightforward application of the Calderón–Zygmund theory of singular integral transforms.

As a consequence of the chain rule the Beltrami coefficients obey a simple composition law:

$$\mu_{g\circ f^{-1}} = \left[\frac{\mu_g - \mu_f}{1 - \bar{\mu}_f \mu_g}(f_z/|f_z|)^2\right] \circ f^{-1}.$$

The interesting thing about this formula is that for any fixed z and f the dependence on $\mu_g(z)$ is complex analytic, and a conformal mapping of the unit disk on itself. This simple fact turns out to be crucial for the study of Teichmüller space.

3. Extremal length. The geometric definition is conceptually even more important than the analytic definition. It makes important use of the theory of extremal length, first developed by A. Beurling for conformal mappings. Let me recall this concept very briefly. If L is a set of locally rectifiable arcs in R^2, then a Borel measurable function $\varrho : R^2 \to R^+$ is said to be *admissible* for L if $\int_\gamma \varrho ds \geqslant 1$ for all $\gamma \in L$. The *module* $M(L)$ is defined as inf $\int \varrho^2 dx$ for all admissible ϱ; its reciprocal is the *extremal length* of L. It is connected with q.c. mappings in the following way: If f is a K-q.c. mapping (according to the analytic definition), then $M(fL) \leqslant KM(L)$. Conversely, this property may be used as a geometric definition of K-q.c. mappings, and it is sufficient that the inequality hold for a rather restrictive class of families L that can be chosen in various ways. This definition has the advantage of having an obvious generalization to several dimensions.

Inasmuch as extremal length was first introduced for conformal mappings, its connection with q.c. mappings, even in more than two dimensions, is another indication of the close relationship between q.c. mappings and classical function theory.

4. Teichmüller's theorem. The problem of extremal q.c. mappings has dominated the subject from the start. Given a family of homeomorphisms, usually defined by some specific geometric or topological conditions, it is required to find a mapping f in the family such that the maximal dilatation, and hence the norm $\|\mu_f\|_\infty$ is a minimum. Because of compactness the existence is usually no problem, but the solution may or may not be unique, and if it is there remains the problem of describing and analyzing the solution.

It is quite obvious that the notion of q.c. mappings generalizes at once to mappings from one Riemann surface to another, each with its own conformal structure, and that the problem of extremal mapping continues to make sense. The Beltrami coefficient becomes a Beltrami differential $\mu(z)d\bar{z}/dz$ of type

$(-1, 1)$. Note that $\mu(z)$ does not depend on the local parameter on the target surface.

Teichmüller considers topological maps $f : S_0 \to S$ from one compact Riemann surface to another. In addition he requires f to belong to a prescribed homotopy class, and he wishes to solve the extremal problem separately for each such class. Teichmüller asserted that there is always an extremal mapping, and that it is unique. Moreover, either there is a unique conformal mapping in the given homotopy class, or there is a constant k, $0 < k < 1$, and a *holomorphic quadratic differential* $\varphi(z)\, dz^2$ on S_0 such that the Beltrami coefficient of the extremal mapping is $\mu_f = k\bar{\varphi}/|\varphi|$. It is thus a mapping with constant dilatation $K = (1 + k)/(1 - k)$. The inverse f^{-1} is simultaneously extremal for the mappings $S \to S_0$, and it determines an associated quadratic differential $\psi(w)\, dw^2$ on S. In local coordinates the mapping can be expressed through

$$\sqrt{\psi(w)}dw = \sqrt{\varphi(z)}\, dz + k\sqrt{\bar{\varphi}}\,(z)\, d\bar{z}.$$

Naturally, there are singularities at the zeros of φ, which are mapped on zeros of ψ of the same order, but these singularities are of a simple explicit nature. The integral curves along which $\sqrt{\varphi}\, dz$ is respectively real or purely imaginary are called horizontal and vertical trajectories, and the extremal mapping maps the horizontal and vertical trajectories on S_0 to corresponding trajectories on S. At each point the stretching is maximal in the direction of the horizontal trajectory and minimal along the vertical trajectory.

This is a beautiful and absolutely fundamental result which, as I have already tried to emphasize, throws a completely new light on the theory of q.c. mappings. In his 1939 paper Teichmüller gives a complete proof of the uniqueness part of his theorem, and it is still essentially the only known proof. His existence proof, which appeared later, is not so transparent, but it was put in good shape by Bers; the result itself was never in doubt. Today, the existence can be proved more quickly than the uniqueness, thanks to a fruitful idea of Hamilton. Unfortunately, times does not permit me to indicate how and why these proofs work, except for saying that the proofs are variational and make strong use of the chain rule for Beltrami coefficients.

5. Teichmüller spaces. Teichmüller goes on to consider the slightly more general case of compact surfaces with a finite number of punctures. Specifically, we say that S is of finite type (p, m) if it is an oriented topological surface of genus p with m points removed. It becomes a Riemann surface by giving it a conformal structure. Following Bers we shall define a conformal structure as a sense-preserving topological mapping σ on a Riemann surface. Two conformal structures σ_1 and σ_2 are equivalent if there is a conformal mapping g of $\sigma_1(S)$ on $\sigma_2(S)$ such that $\sigma_s^{-1} \circ g \circ \sigma_1$ is homotopic to the identity. The equivalence classes $[\sigma]$ are the points of the

Teichmüller space $T(p,m)$, and the distance between $[\sigma_1]$ and $[\sigma_2]$ is defined to be

$$d([\sigma_1],\ [\sigma_2]) = \log\ \inf K(f)$$

where $K(f)$ is the maximal dilatation of f, and f ranges over all mappings homotopic to $\sigma_2 \circ \sigma_1^{-1}$. It is readily seen that the infimum is actually a minimum, and that the extremal mapping from $\sigma_1(S)$ to $\sigma_2(S)$ is as previously described, except that the quadratic differentials are now allowed to have simple poles at the punctures.

With this metric $T(p,m)$ is a complete metric space, and already Teichmüller showed that it is homeomorphic to $R^{6p-6+2m}$ (provided that $2p-2+m>0$).

Let f be a self-mapping of S. It defines an isometry $\tilde{f}$ of $T(p,m)$ which takes $[\sigma]$ to $[\sigma \circ f]$. This isometry depends only on the homotopy class of f and is regarded as an element of the *modular group* $\mathrm{Mod}(p,m)$. It follows from the definition that two Riemann surfaces $\sigma_1(S)$ and $\sigma_2(S)$ are conformally equivalent if and only if $[\sigma_2]$ is the image of $[\sigma_1]$ under an element of the modular group. The quotient space $T(p,m)/\mathrm{Mod}\,(p,m)$ is the Riemann space of algebraic curves or moduli. The Riemann surfaces that allow conformal self-mappings are branch-points of the covering.

6. Fuchsian and quasifuchsian groups. The universal covering of any Riemann surface S, with a few obvious exceptions, is conformally equivalent to the unit disk U. The self-mappings of the covering surface correspond to a group G of fractional linear transformations, also referred to as Möbius transformations, which map U conformally on itself. More generally one can allow coverings with a signature, that is to say regular covering surfaces which are branched to a prescribed order over certain isolated points. In this case G includes elliptic transformations of finite order. It is always discrete.

Any discrete group of Möbius transformations that preserves a disk or a half-plane, for instance U, is called a Fuchsian group. It is a recent theorem, due to Jørgensen, that a nonelementary group which maps U on itself is discrete, and hence Fuchsian, if and only if every elliptic transformation in the group is of finite order. As soon as this condition is fulfilled the quotient U/G is a Riemann surface S, and U appears as a covering of S with a signature determined by the orders of the elliptic transformations. The group acts simultaneously on the exterior U^* of U, and $S^* = U^*/G$ is a mirror image of S. G is determined by S up to conjugation.

A point is a limit point if it is an accumulation point of an orbit. For Fuchsian groups all limit points are on the unit circle; the set of limit points will be referred to as the *limit set* $\Lambda(G)$. Except for some trivial cases there are only two alternatives: either Λ is the whole unit circle, or it is a perfect nowhere dense subset. With an unimaginative, but classical, terminology Fuchsian groups are accordingly classified as being of the first kind or second kind.

If S is of finite type, then G is always of the first kind; what is more, G has a fundamental region with finite noneuclidean area. Consider a q.c. mapping

$f: S_0 \to S$ with corresponding groups G_0 and G. Then f lifts to a mapping $f: U \to U$ (which we continue to denote by the same letter), and if $g_0 \in G_0$ there is a $g \in G$ such that $f \circ g_0 = g \circ f$. This defines an isomorphism $\theta : G_0 \to G$ which is uniquely determined, up to conjugation, by the homotopy class of f. Moreover, f extends to a homeomorphism of the closed disks, and the boundary correspondence is again determined uniquely up to normalization. The Teichmüller problem becomes that of finding f with given boundary correspondence and smallest maximal dilatation. The extremal mapping has a Beltrami coefficient $\mu = k\bar{\varphi}|\varphi|$ where φ is an invariant quadratic differential with respect to G_0.

Incidentally, the problem of extremal q.c. mappings with given boundary values makes sense even when there is no group, but the solution need not be unique. The questions that arise in this connection have been very successfully treated by Hamilton, K. Strebel, and E. Reich.

For a more general situation, let $\mu d\bar{z}/dz$ be any Beltrami differential, defined in the whole plane and invariant under G_0 in the sense that $(\mu \circ g_0)\bar{g}_0'/g_0' = \mu$ a.e. for all $g_0 \in G$. Suppose f is a solution of the Beltrami equation $f_{\bar{z}} = \mu f_z$. It follows from the chain rule that $f \circ g_0$ is another solution of the same equation. Therefore $f \circ g_0 \circ f^{-1}$ is conformal everywhere, and hence a Möbius transformation g. In this way μ determines an isomorphic mapping of G_0 on another group G, but this time G will in general not leave U invariant. For this reason G is a Kleinian group rather than a Fuchsian group. It has two invariant regions $f(U)$ and $f(U^*)$, separated by a Jordan curve $f(\delta U)$. The surfaces $f(U)/G$ and $f(U^*)/G$ are in general not conformal mirror images.

The group $G = fG_0f^{-1}$ is said to be obtained from G_0 by q.c. deformation, and it is called a quasifuchsian group. Evidently, quasifuchsian groups have much the same structure as fuchsian groups, except for the lack of symmetry. The curve that separates the invariant Jordan regions is the image of the unit circle under a q.c. homeomorphism of the whole plane. Such curves are called quasicircles. It follows by a well-known property of q.c. mappings that every quasicircle has zero area, and consequently the limit set $\Lambda(G)$ has zero two-dimensional measure.

Strangely enough, quasicircles have a very simple geometric characterization: A Jordan curve is a quasicircle if and only if for any two points on the curve at least one of the subarcs between them has a diameter at most equal to a fixed multiple of the distance between the points. It means, among other things, that there are no cusps.

7. The Bers representation. There are two special cases of the construction that I have described: (1) If μ satisfies the symmetry condition $\mu(1/\bar{z})\bar{z}^2/z^2 = \bar{\mu}(z)$, then G is again a Fuchsian group and f preserves symmetry with respect to the unit circle. (2) If μ is identically zero in U and arbitrary in U^*, except for being invariant with respect to G_0, then f is conformal in U, and $f(U)/G$ is conformally equivalent to $S = U/G$, while $f(U^*)/G$ is a q.c. mirror image of S.

I shall refer to the second construction as the Bers mapping. Two Beltrami differentials μ_1 and μ_2 will lead to the same group G and to homotopic maps f_1, f_2 if and only if $f_1 = f_2$ on ∂U (up to normalizations). When that is the case we say that μ_1 and μ_2 are equivalent, and that they represent the same point in the Teichmüller space $T(G_0)$ based on the Fuchsian group G_0.

In other words the equivalence classes are determined by the values of f on the unit circle. These values obviously determine $f(U)$, and hence f, at least up to a normalization. One obtains strict uniqueness by passing to the Schwarzian derivative $\varphi = S_f$ defined in U (recall that $S_f = (f''/f')' - \frac{1}{2}(f''/f')^2$). From the properties of the Schwarzian it follows that $\varphi(gz)g'(z)^2 = \varphi(z)$ for all $g \in G_0$. Furthermore, by a theorem of Nehari $|\varphi(z)|(1-|z|^2)^2$ is bounded (actually $\leqslant 6$). Thus φ belongs to the Bers class $B(G_0)$ of bounded quadratic differentials with respect to the group G_0. The Bers map is an injection $T(G_0) \to B(G_0)$.

It is known that the image of $T(G_0)$ under the Bers map is *open*, and as a vector space $B(G_0)$ has a natural complex structure. The mapping identifies $T(G_0)$ with a certain open subset of $B(G_0)$ which in turn endows $T(G_0)$ with its own complex structure. If S is of type (p, m) the complex dimension is $3p - 3 + m$. The nature of the subset that represents $T(p, m)$ in C^{3p-3+m} is not well known. For instance, it seems to be an open problem whether $T(1, 1)$ is a Jordan region in C.

The case where $G = I$, the identity group, is of special interest because it is so closely connected with classical problems in function theory. An analytic function φ, defined on U, will belong to $T(I)$ if and only if it is the Schwarzian S_f of a schlicht (injective) function on U with a q.c. extension to the whole plane. The study of such functions has added new interest to the classical problems of schlicht functions.

To illustrate the point I would like to take a minute to tell about a recent beautiful result due to F. Gehring. Let $\boldsymbol{S}$ denote the space of all $\varphi = Sf$, f analytic and schlicht in U, with the norm $\|\varphi\| = \sup(1 - |z|^2)^2|\varphi(z)|$, and let $\boldsymbol{T} = T(I)$ be the subset for which f has a q.c. extension. Gehring has shown (i) that $\boldsymbol{T} = \text{Int}\, \boldsymbol{S}$, (ii) the closure of $\boldsymbol{T}$ is a proper subset of $\boldsymbol{S}$. To prove the second point, which gives a negative answer to a question raised by Bers maybe a dozen years ago, he constructs, quite explicitly, a region with the property that no small deformation, measured by the norm of the Schwarzian, changes it to a Jordan region, much less to one whose boundary is a quasicircle. I mention this particular result because it is recent and because it is typical for the way q.c. mappings are giving new impulses to the classical theory of conformal mappings.

In the finite dimensional case $T(p, m)$ has a compact boundary in $B(G_0)$. It is an interesting and difficult problem to find out what exactly happens when φ approaches the boundary. The pioneering research was carried out by Bers and Maskit. They showed, first of all, that when φ approaches a boundary point the holomorphic function f will tend to a limit which is still schlicht, and the groups G tend to a limit group which is Kleinian with a single, simply connected invariant region. Such groups were called B-groups (B stands either for Bers or for boundary) in the belief that any such group can be obtained in this manner. It can happen that the invariant simply connected region is the whole set of discontinuity; such groups are said to be *degenerate*. Classically, degenerate groups were not known, but Bers proved that they must exist, and more recently Jørgensen has been able to construct many explicit examples of such groups.

Intuitively, it is clear what should happen when φ goes to the boundary. We are interested to follow the q.c. images $f(U^*)$. In the degenerate case the image disappears completely. In the nondegenerate case the fact that one approaches the boundary must be visible in some way, and the obvious guess is that one or more of the closed geodesics on the surface is being pinched to a point. In the limit $f(S^*)$ would either be of lower genus or would disintegrate to several pieces, and one would end up with a more general configuration consisting of a "surface with nodes", each pinching giving rise to two nodes.

A lot of research has been going on with the intent of making all this completely rigorous, and if I am correctly informed these attempts have been successful, but much remains to be done. This is the general trend of much of the recent investigations of Bers, Maskit, Kra, Marden, Earle, Jørgensen, Abikoff and others; I hope they will understand that I cannot report in any detail on these theories which are still in *status nascendi.*

In a slightly different direction the theory of Teichmüller spaces has been extended to a study of the so-called universal Teichmüller curve, which for every type (p, m) is a fiber-space whose fibers are the Riemann surfaces of that type. A special problem is the existence, or rather non-existence, of holomorphic sections.

The Bers mapping is not concerned with extremal q.c. mappings, and it is rather curious that one again ends up with holomorphic quadratic differentials. The Bers model has a Kählerian structure obtained from an invariant metric, the Petersson–Weil metric, on the space of quadratic differentials. The relation between the Petersson–Weil metric and the Teichmüller metric has not been fully explored and is still rather mystifying.

8. Kleinian groups. I would have preferred to speak about Kleinian groups in a section all by itself, but they are so intimately tied up with Teichmüller spaces that I was forced to introduce Kleinian groups somewhat prematurely. I shall now go back and clear up some of the terminology.

It was Poincaré who made the distinction between Fuchsian and Kleinian groups and who also coined the names, much to the displeasure of Klein. He also pointed out that the action of any Möbius transformation extends to the upper half space, or, equivalently, to the unit ball in three-space. Any discrete group of Möbius transformations is discontinuous on the open ball. Limit points are defined as in the Fuchsian case; they are all on the unit sphere, and the limit set Λ may be regarded either as a set on the Riemann sphere or in the complex plane. The elementary groups with at most two limit points are usually excluded, and in modern terminology a Kleinian group is one whose limit set is nowhere dense and perfect.

A Kleinian group may be looked upon as a Fuchsian group of the second kind in three dimensions. As such it cannot have a fundamental set with finite non-euclidean volume. Therefore, the relatively well developed methods of Lie group theory which require finite Haar measure are mostly not available for Kleinian groups. However, the important method of Poincaré series continues to make sense.

Let G be a Kleinian group, Λ its limit set, and Ω the set of discontinuity, that is to say the complement of Λ in the plane or on the sphere. The quotient manifold Ω/G inherits the complex structure of the plane and is thus a disjoint union of Riemann surfaces. It forms the boundary of a three-dimensional manifold $M(G) = B(1) \cup \Omega/G$.

What is the role of q.c. mappings of Kleinian groups? For one thing one would like to classify all Kleinian groups. It is evident that two groups that are conjugate to each other in the full group of Möbius transformations should be regarded as essentially the same. But as in the case of quasifuchsian groups two groups can also be conjugate in the sense of q.c. mappings, namely if $G' = fGf^{-1}$ for some q.c. mapping of the sphere. In that case G' is a q.c. deformation of G, and such groups should be in the same class.

But this is not enough to explain the sudden blossoming of the theory under the influence of q.c. mappings. As usual, linearization pays off, and it has turned out that infinitesimal q.c. mappings are relatively easy to handle. An infinitesimal q.c. mapping is a solution of $f_{\bar{z}} = \nu$ where the right-hand member is a function of class L^∞. This is a non-homogeneous Cauchy–Riemann equation, and it can be solved quite explicitly by the Pompeiu formula, which is nothing else than a generalized Cauchy integral formula. In order that f induce a deformation of the group ν must be a Beltrami differential, $\nu \in \operatorname{Bel} G$, this time with arbitrary finite bound. There is a subclass N of trivial differentials that induces only a conformal conjugation of G, and the main theorem asserts that the dual space of $\operatorname{Bel} G/N$ can be identified with the space of quadratic differentials on $\Omega(G)/G$ which are of class L^1.

This technique is particularly successful if one looks only at finitely generated groups. In that case the deformation space is finite dimensional, so that there are only a finite number of linearly independent integrable quadratic differentials. This result led me to announce, somewhat prematurely, the so-called *finiteness theorem*: If G is finitely generated, then $S = \Omega(G)/G$ is a finite union of Riemann surface of finite type. I had overlooked that fact that a triply punctured square carries no quadratic differentials. Fortunately, the gap was later filled by L. Greenberg, and again by L. Bers who extended the original method to include differentials of higher order. With this method Bers obtained not only an upper bound for the number of surfaces in terms of the number of generators, but even a bound on the total Poincaré area of S.

It was not unreasonable to expect that finitely generated Kleinian groups would have other simple properties. For instance, since a finitely generated Fuchsian group has a fundamental polygon with a finite number of sides one could hope that every finitely generated Kleinian group would have a finite fundamental polyhedron. All such hopes were shattered when L. Greenberg proved that a degenerate group in the sense of Bers and Maskit can never have a finite fundamental polyhedron. Groups with a finite fundamental polyhedron are called *geometrically finite*, and it has been suggested that one should perhaps be content to study only geometrically finite groups. With his constructive methods that go back to Klein, Maskit has been able to give a complete classification of all geometrically finite groups, and Marden has used three-dimensional topology to study the geometry of the three-manifold. These are very farreaching and complicated results, and it would be impossible for me to try to summarize them even if I had the competence to do so.

9. The zero area problem. An interesting problem that remains unsolved is the following: Is it true that every finitely generated Kleinian group has a limit set with two-dimensional measure zero?

The most immediate reason for raising the question is that it is easy to prove the corresponding property for Fuchsian groups of the second kind, two-dimensional measure being replaced by one-dimensional. How does one prove it? If the limit set of a Fuchsian group has positive measure one can use the Poisson integral to construct a harmonic function on the unit disk with boundary values 1 a.e. on the limit set and 0 elsewhere. If the group is finitely generated the surface must have a finitely generated fundamental group, and it is therefore of finite genus and connectivity. The ideal boundary components are then representable as points or curves. If they are all points the group would be of the first kind, and if there is at least one curve the existence of a nonconstant harmonic function which is zero on the boundary violates the maximum principle. Therefore the limit set must have zero linear measure. The proof is thus quite trivial, but it is trivial only because one has a complete classification of surfaces with finitely generated fundamental groups.

For Kleinian groups it is easy enough to imitate the construction of the harmonic function, which this time has to be harmonic with respect to the hyperbolic metric of the unit ball. If the group is geometrically finite this leads rather easily to a proof of measure zero. For the general case it seems that one would need a better topological classification of three-manifolds with constant negative curvature. It is therefore not suprising that the problem has come to the attention of the topologists, and I am happy to report that at least two leading topologists are actively engaged in research on this problem. I believe that this pooling of resources will be very fruitful, and it would of course not be the first time that analysis topology, and vice versa.

Some time ago W. Thurston became interested in a topological concerning foliations of surfaces, and he proved a theorem which is closely related to Teichmüller theory. I have not seen Thurston's work, but I have seen Bers' interpretation of it as a new extremal problem for self-mappings of a surface. It is fascinating, and I could and perhaps should have talked about it in connection with the Teichmüller extremal problem, but I am a little hesitant to speak about things that are not yet in print, and therefore not quite in the public domain. Nevertheless, since many exciting things have happened quite recently in this particular subject, I am taking upon myself to report very informally on some of the newest developments, including some where I have to rely on faith rather than proofs.

Thurston has now begun to apply his remarkable geometric and topological intuition and skill to the problem of zero measure. I certainly do not want to preempt him in case he is planning to talk about it in his own lecture, and I have seen only glimpses of his reasoning, but it would seem that he can prove zero area for all groups that are limits, in one sense or another, of geometrically finite groups. This would be highly significant, for it would show that all groups on the boundary of Teichmüller space have limit sets with zero measure. It would neither prove nor disprove the original conjecture, but it would be a very big step. Personally, I feel that a definitive solution is almost imminent.

Very recently there was a highly specialized conference on Riemann surfaces in the United States, and there was an air of excitement caused not only by what Thurston had done and was doing, but also by the presence of D. Sullivan who had equally fascinating stories to tell. Sullivan, too, has worked hard on the area problem, and he has come up with a by-product that does not solve the problem, but is extremely interesting in itself. He applies the powerful tool of what has been called topological dynamics. If a transformation group acts on a measure space, the space splits into two parts, a dissipative part with a measurable fundamental set, and a recurrent part whose every measurable subset meets infinitely many of its images in a set of positive measure. This powerful theorem, which goes back to E. Hopf, does not seem to have been familiar to those who have approached Kleinian group from the point of view of q.c. mappings. The dissipative part of a finitely generated group is the set of discontinuity, and nothing more; this is a known theorem. The recurrent part is the limit set, and it is of interest only if it has positive measure. But even if the area conjecture is true Sullivan's work remains significant for groups whose limit set is the whole sphere.

Sullivan has several theorems, but the one that has captured my special interest because I understand it best asserts that there is no invariant vector field supported on the limit set. If the limit set is the whole sphere there is no invariant vector field, period. In an equivalent formulation, the limit set carries no Beltrami differential. It was known before that there are only a finite number of linearly independent Beltrami differentials on the limit set of a finitely generated Kleinian group, but that there are none was a surprise to me, and Sullivan's approach gives results even for groups that are not finitely generated. Sullivan's results, taken as a whole, give a new outlook on the ergodic theory of Kleinian groups. They are related to, but go beyond the results of E. Hopf which were already considered deep and difficult, and as a corollary Sullivan obtains a strenthening of Mostow's rigidity theorem. I cannot explain the proofs beyond saying that they are very clever and show that Sullivan is not only a leading topologist, but also a strong analyst.

10. Several dimensions. In the remaining time I shall speak briefly about the generalizations to more than two dimensions. There are two aspects: q.c. mappings *per se*, and Kleinian groups in several dimensions.

The foundations for q.c. mappings in space are essentially due to Gehring and J. Väisälä, but very important work has also been done in the Soviet Union and Romania. I have already mentioned, in passing, that correct definitions can be based on modules of curve families, and the modules give the only known workable technique. Otherwise, the difficulties are enormous. It is reasonably clear that the Beltrami coefficient should be replaced by a matrix-valued function, but this function is subject to conditions that were already known to H. Weyl, but which are so complicated that nobody has been able to put them to any use. Very little is known about when a region in n-space is q.c. equivalent to a ball, and there is not even an educated guess what Teichmüller's theorem should be replaced by. On the positive side one knows a little bit about boundary correspondence.

In two dimensions there is not much use for mappings that are locally q.c. but not homeomorphic, for by passing to Riemann surfaces they can be replaced by homeomorphisms. In several dimensions the situation is quite different, and there has been rapid growth of the theory of so-called quasiregular mappings from one n-dimensional space to another. It has been developed mostly in the Soviet Union and Finland, and this is perhaps a good opportunity to congratulate the young Finnish mathematicians to their success in this area. In the spirit of Rolf Nevanlinna they have even been able to carry over parts of the value distribution theory to quasiregular functions. In fact, less than a month ago I learned that Rickman has succeeded in proving a generalization of Picard's theorem that I know they have been looking for for a long time. It is so simple that I cannot resist quoting the result: There exists $q = q(n, k)$ such that any K-q.c. mapping $f : R^n \to R^n - \{a_1, \ldots, a_q\}$ is constant. (They believe that the theorem is true with $q = 2$.)

As for Kleinian groups, they generalize trivially to any number of dimensions, and the distinction between Fuchsian and Kleinian groups disappears. Some properties that depend purely on hyperbolic geometry will carry over, but they are not the ones that use q.c. mappings. However, infinitesimal q.c. mappings have an interesting counterpart for several variables. There is a linear differential operator that takes the place of $f_{\bar{z}}$, namely $Sf = \frac{1}{2}(Df + Df') - (1/n)\mathrm{tr}Df \cdot 1_n$ which is symmetric matrix with zero trace. It has the right invariance, and the conditions under which the Beltrami equation $Sf = \nu$ has a solution can be expressed as a linear integral equation. The formal theory is there, but it will take time before it leads to tangible results.

My survey ends here. I regret that there are so many topics that I could not even mention, and that my report has been so conspicuously insufficient as far as research in the Soviet Union is concerned. I know that I have not given a full picture, but I hope that I have given you an idea of the extent to which q.c. mappings have penetrated function theory.

References

The following surveys have been extremely helpful to the author:

Lipman Bers, *Quasiconformal mappings, with applications to differential equations, function theory and topology*, Bull. Amer. Math. Soc. (6) **83** (1977), 1083–1100.

L. Bers and I. Kra (Editors), *A crash course on Kleinian groups*, Lecture Notes in Math., vol. 400, Springer–Verlag, Berlin and New York, 1974.

W. J. Harvey (Editor), *Discrete groups and automorphic functions*, Proc. on Instructional Conf. organized by the London Math. Soc. and the Univ. of Cambridge, Academic Press, London, New York, San Francisco, 1977.

Harvard University
Cambridge, Massachusetts 02138, U.S.A.

Reprinted from Proc. Int. Congr. Math., 1950

THE WORK OF L. SCHWARTZ

by

HARALD BOHR

At a meeting of the organizing committee of the International Congress held in 1924 at Toronto the resolution was adopted that at each international mathematical congress two gold medals should be awarded, and in a memorandum the donor of the fund for the founding of the medals, the late Professor J. C. Fields, expressed the wish that the awards should be open to the whole world and added that, while the awards should be a recognition of work already done, it was at the same time intended to be an encouragement for further mathematical achievements. The funds for the Fields' medals were finally accepted by the International Congress in Zürich in 1932, and two Fields medals were for the first time awarded at the Congress in Oslo 1936 to Professor Ahlfors and Professor Douglas. And now, after a long period of fourteen years, the mathematicians meet again at an international congress, here in Harvard.

In the fall of 1948 Professor Oswald Veblen, as nominee of the American Mathematical Society for the presidency of the Congress in Harvard, together with the chairman of the organizing committee, and with the secretary of the Congress, appointed an international committee to select the two recipients of the Fields medals to be awarded at the Congress in Harvard, the committee consisting of Professors Ahlfors, Borsuk, Fréchet, Hodge, Kolmogoroff, Kosambi, Morse, and myself. With the exception of Professor Kolmogoroff, whose valuable help we were sorry to miss, all the members of the committee have taken an active part in the discussions. As chairman of the committee I now have the honor to inform the Congress of our decisions and to present the gold medals together with an honorarium of $1,500 to each of the two mathematicians selected by the committee.

The members of the committee were, unanimously, of the opinion that the medals, as on the occasion of the first awards in Oslo, should be given to two really young mathematicians, without exactly specifying, however, the notion of being "young". But even with this principal limitation the task was not an easy one, and it was felt to be very encouraging for the expectations we may entertain of the future development of our science that we had to choose among so many young and very talented mathematicians, each of whom should certainly have been worthy of an official appreciation of his work. Our choice fell on Professor *Atle*

Selberg and Professor *Laurent Schwartz*, and I feel sure that all members of the Congress will agree with the committee that these two young mathematicians not only are most promising as to their future work but have already given contributions of the uttermost importance and originality to our science; indeed they have already written their names in the history of mathematics of our century.

I now turn to the work of the slightly older, of the two recipients, the French mathematician *Laurent Schwartz*. Having passed through the old celebrated institution École Normale Supérieure in Paris, he is now Professor at the University of Nancy. He belongs to the group of most promising and closely collaborating young French mathematicians who secure for French mathematics in the years to come a position worthy of its illustrious traditions. Like Selberg, Schwartz can look back on an extensive and varied production, but when comparing the work of these two young mathematicians one gets a strong impression of the richness and variety of the mathematical science and of its many different aspects. While Selberg's work dealt with clear cut problems concerning notions which, as the primes, are, so to say, given *a priori*, one of the greatest merits of Schwartz's work consists on the contrary in his creation of new and most fruitful notions adapted to the general problems the study of which he has undertaken. While these problems themselves are of classical nature, in fact dealing with the very foundation of the old calculus, his way of looking at the problem is intimately connected with the typical modern development of our science with its highly general and often very abstract character. Thus once more we see in Schwartz's work a confirmation of the words of Felix Klein that great progress in our science is often obtained when new methods are applied to old problems. In the short time at my disposal I think I may give the clearest impression of Schwartz's achievements by limiting myself to speak of the very central and most important part of his work, his theory of "distribution". The first publication of his new ideas was given in a paper in the Annales de ℓ'Université de Grenoble, 1948, with the title *Généralisation de la notion de fonction, de dérivation, de transformation de Fourier et applications mathématiques et physiques*, a paper which certainly will stand as one of the classical mathematical papers of our times. As the title indicates it deals with a generalization of the very notion of a function better adapted to the process of differentiation than the ordinary classical one. In trying to explain briefly the new notions of Schwartz and their importance I think I can do no better than start by considering the same example as that used by Schwartz himself, namely, the simple function $f(x)$ which is equal to 0 for $x \leqq 0$ and equal to 1 for $x > 0$. This function has a derivative $f'(x) = 0$ for every $x \neq 0$, but this fact evidently does not tell us anything about the magnitude of the jump of $f(x)$ at the point $x = 0$. In order to overcome this inconvenience the physicist and the technicians had accustomed themselves to say that the function $f(x)$ has as derivative the "Dirac-function" $f'(x) = \delta(x)$ which is 0 for $x \neq 0$ and equal to $+\infty$ for $x = 0$ and moreover has the property that its integral over any interval containing the point $x = 0$ shall be equal to 1. But this is of course not a legitimate way of speaking; from a mathematical point of view — using the idea of a Stieltjes' integral — we naturally would think of the derivative of our function $f(x)$ not as a function but as a mass-distribution, in this case of the particularly simple type with the whole mass 1 placed at the origin $x = 0$. Now, according to a classical theorem of F. Riesz, there is a most intimate connection, in fact a one-to-one correspondence, between an arbitrary mass-distribution μ on the x-axis and a linear continuous functional $\mu(\phi)$ defined in the space of all continuous functions $\phi(x)$, vanishing outside a finite interval, where the topology of the ϕ-space is fixed by the simple claim that convergence of sequence ϕ_n shall mean that the functions $\phi_n(x)$ are all zero outside a fixed finite interval and that the sequence $\phi_n(x)$ shall be uniformly convergent. This correspondence between the mass-distributions μ and the functionals $\mu(\phi)$ is given simply by the relation

$$\mu(\phi) = \int_{-\infty}^{+\infty} \phi(x)\, d\mu.$$

In the Schwartz theory of distributions the new notion, generalizing, or rather replacing, that of a function is nothing else than just such a linear continuous functional, but of a kind essentially different from that above, the underlying ϕ-space and its topology being of a quite different nature. The new notion — once invented — is so easy to explain that I cannot resist the temptation, notwithstanding the general solemn nature of this opening meeting, to go into some detail. Let us consider, then, with Schwartz a quite arbitrary function $f(x)$, assumed only to be integrable in the sense of Lebesgue over any finite interval, and let us try to characterize the function $f(x)$, not as in the classical Dirichlet way by the values it takes for the different values of x, but by what we may call its effect when operating on an arbitrary auxiliary function $\phi(x)$ of which, for the moment, we suppose only, as above, that $\phi(x)$ is continuous and equal to zero outside a finite interval; by the effect of the given function $f(x)$ on the auxiliary function $\phi(x)$ we here mean simply the value of the integral

$$\int_{-\infty}^{+\infty} f(x)\phi(x)\, dx.$$

This integral, obviously, is a linear functional, associated with the function $f(x)$, and we shall denote it $f(\phi)$, using the same letter f, as we wish, so to speak, to identify it with the function $f(x)$ itself. In the special case where the given function $f(x)$ has a continuous derivative $f'(x)$ we may of course, starting with $f'(x)$ instead of with $f(x)$, in the same way build a functional associated with $f'(x)$, i.e., the fuctional

$$f'(\phi) = \int_{-\infty}^{+\infty} f'(x)\phi(x)\, dx.$$

If now — and this is an essential point — we assume also the auxiliary function $\phi(x)$ to have a continuous derivative $\phi'(x)$, we immediately find through partial integration

$$\int_{-\infty}^{+\infty} f'(x)\phi(x)\,dx = -\int_{-\infty}^{+\infty} f(x)\phi'(x)\,dx,$$

i.e., the simple relation

$$f'(\phi) = -f(\phi').$$

In order that the derivative of any function $\phi(x)$ of our space shall also belong to the space, we must obviously assume, with Schwartz, that the auxiliary function $\phi(x)$ to be considered shall possess derivaties not only of the first order but of arbitrarily high order. In the space consisting of all such functions $\phi(x)$, i.e., of all functions $\phi(x)$ zero outside a finite interval and with derivatives of any order, and topologized by the definition that convergence of a sequence ϕ_n shall mean not only as above that the function $\phi_n(x)$ shall all be zero outside a fixed finite interval and that $\phi_n(x)$ shall converge uniformly, but moreover that all the derivated sequences $\phi_n'(x), \phi_n''(x)\ldots$ shall converge uniformly, Schwartz now takes into consideration all continuous linear functionals $J(\phi)$. These functionals $J(\phi)$ are just what Schwartz denotes as "distributions". Among them are in particular the distributions $f(\phi)$ derived in the manner above from an ordinary function $f(x)$, and more generally we have distributions, which we denote by $\mu(\phi)$, which are associated with a mass-distribution μ — evidently the word distribution has been chosen to remind us vaguely of these mass-distributions — but the whole class of Schwartz distributions is far from being exhausted by the special distributions of the type $f(\phi)$ or $\mu(\phi)$. Now — and this is the decisive point in the theory — Schwartz assigns to every one of his distributions $J(\phi)$ another distribution $J'(\phi)$ as the derivative of $J(\phi)$, namely, immediately suggested by the consideration above, the distribution defined by

$$J'(\phi) = -J(\phi').$$

In the special case where the distribution $J(\phi)$ is of the type $f(\phi)$ and moreover is derived from a function $f(x)$ with a continuous derivative, the derivated distribution $J'(\phi) = -J(\phi')$ is, evidently, nothing else than the distribution $f'(\phi)$ associated with the function $f'(x)$. But generally, if $f(x)$ is an arbitrary function with no derivative, the corresponding functional $f(\phi)$ still has a derivative $f'(\phi)$ which, however, is no longer associated with any function, neither, generally, with any mass-distribution, but is just some Schwartz distribution.

And now, one will naturally ask, what has been gained by Schwartz's generalization of a function $f(x)$ to that of a distribution $J(\phi)$. Naturally, the aim of any such generalization of basic notions — as, for instance, the generalization of the notion of a real number to that of a complex number — is, in principle, the same and of a double kind; on the one hand, and this is the primary purpose, one aims at getting simplifications in the treatment of problems concerning the old notions through the greater freedom in carrying out operations, provided by the new notions, and on the other hand, one may hope to meet with new fruitful problems concerning these new notions themselves. In both these respects the theory of Schwartz may be said to be a great success. I think that every reader of his cited paper, like myself, will have left a considerable amount of pleasant excitement, on seeing the wonderful harmony of the whole structure of the calculus to which the theory leads and on understanding how essential an advance its application may mean to many parts of higher analysis, such as spectral theory, potential theory, and indeed the whole theory of linear partial differential equations, where, for instance, the important notion of the "finite part" of a divergent integral, introduced by Hadamard, presents itself in a most natural way when the distributions and not the functions are taken as basic elements. And as to the harmony brought about I shall mention only one single, very simple, but most satisfactory result. Not only has, as we have seen, every distribution $J(\phi)$ a derivative $J'(\phi)$, and hence derivatives of every order, but conversely it also holds that every distribution possesses a primitive distribution, i.e., is the derivative of another distribution which is uniquely determined apart from an additive constant (i.e., of course, a distribution associated with a constant). The simplification obtained, and not least the easy justification of different "symbolic" operations often used in an illegitimate way by the technicians, is of such striking nature that it seems more than a utopian thought that elements of the theory of the Schwartz distributions may find their place even in the more elementary courses of the calculus in universities and technical schools.

Schwartz is now preparing a larger general treatise on the theory of distributions, the first, very rich, volume of which has already appeared. In his introduction to this treatise he emphasizes the fact that ideas similar to those underlying his theory have earlier been applied by different mathematicians to various subjects — here only to mention the methods introduced by Bochner in his studies on Fourier integrals — and that the theory of distibutions is far from being a "nouveauté revolutionaire." Modestly he characterizes his theory as "une synthèse et une simplification". However, as in the case of earlier advances of a general kind — to take only one of the great historic examples, that of Descartes' development of the analytic geometry which, as is well-known, was preceded by several analytic treatments by other mathematicians of special geometric problems — the main merit is justly due to the man who has clearly seen, and been able to shape, the new ideas in their purity and generality.

No wonder that the work of Schwartz has met with very great interest in mathematical circles throughout the world, and that a number of younger mathematicians have taken up investigations in the wide field he has opened for new researches.

And now I have the honor to call upon Professor Selberg and Professor Schwartz to present to them the golden medals and the honorarium.

In the name of the committee, I think I dare say of the whole Congress, I congratulate you most heartily on the awards of the Fields medals. Repeating the wish of Fields himself I may finally express the hope that the great admiration of your achievements of which the medals are a token may also mean an encouragement to you in your future work.

Harvard University
Cambridge, Massachusetts
August 30, 1950

Laurent Schwartz

LAURENT SCHWARTZ

(1915–2002)

Etudes et diplômes

1934–1937	Ecole Normale Supérieure
1937	Agrégation de Mathématíques
1943	Doctorat ès-Sciences, Faculté des Sciences de Strasbourg réfugiée á Clermont Ferrand.

Carrière

1944–1945	Chargé de cours à la Faculté des Sciences de Grenoble
1945–1952	Maître de conférences, puis professeur à la Faculté des Sciences de Nancy
1953–1969	Maître de conférences, puis professeur à la Faculté des Sciences de Paris
1959–1980	Professeur à l'Ecole Polytechnique
1980–1983	Professeur á l'Université Paris VII
1983	Retraite

Prix

1950	Médaille Fields, Congrès international des mathématiciens, Cambridge (Mass.) U.S.A.
1955	Prix Carrière de l'Académie des Sciences, Paris, France
1964	Grand prix de Mathématiques et de Physique, Académie des Scíences, Paris, France
1972	Prix Cognac-Jay, Académie des Sciences, Paris, France (avec J. L. Lions et B. Malgrange).
1956	Professeur honoraire de l'Université d'Amérique, Bogota, Colombie, et professeur honoraire de l'Université de Buenos-Aires, Argentine
1957	Membre corespondant de la Société royale des Sciences de Liège, Belgique
1958	Membre honoraire de l'Union mathématique Argentine, académicien honoraire
1960	Docteur Honoris Causa de l'Université de Humboldt, Berlin, R.D.A.
1962	Docteur Honoris Causa de l'Université libre de Bruxelles, Belgique

1964 Membre correspondant de l'Académie des Sciences du Brésil
1965 Professeur honoraire de l'Université nationale des Ingénieurs du Pérou, Lima, Pérou
1971 Membre honoraire du Tata Institute of Fundamental Research, Bombay, Inde
1972 Membre correspondant, Académie des Sciences, Paris, France
1975 Membre de l'Acad'emie des Sciences
1977 Membre étranger de l'Académie Indienne des Sciences
1981 Docteur Honoris Causa de l'Université de Lund (Suède)
1981 Docteur Honoris Causa de l'Université de Tel-Aviv (Israël)
1985 Docteur Honoris Causa de l'Université de Montréal (Québec)
1988 Membre étranger de l' "Accademia Nazionale dei Lincei", Rome
1991 Membre associé de l'Académie Polonaise des Sciences
1993 Docteur honoris causa de l'Université d'Athènes (Grèce)
1995 Membre honoraire de la Société Mathématique de Moscou (Russie).

Reprinted from Analyse Mathématique et Applications, 1988

CALCUL INFINITESIMAL STOCHASTIQUE

by
L. SCHWARTZ
Centre de Mathématiques, Ecole Polytechnique
Palaiseau, France

Dédié à Jacques–Louis Lions pour son soixantième anniversaire

Cet article ne contient presque pas de résultats nouveaux, mais donne, à partir de rien, et sans démonstrations, de nombreux résultats personnels du calcul infinitésimal stochastique. Je suis très heureux de le dédier à Jacques–Louis Lions.

1. Processus[1]

La donnée initiale est toujours un ensemble Ω (ensemble des échantillons), une tribu $\mathcal{F}$ de parties de Ω, une probabilité $\mathcal{P}$ (mesure ≥ 0 de masse 1) sur $(\Omega, \mathcal{F})$. On se donne en outre une "filtration", $(\mathcal{F}_t)_{t \in \bar{\mathbb{R}}_+}$, de sous tribus de la tribu $\mathcal{P}$-mesurable, $\mathcal{P}$-complète, croissante, continue à droite ($\mathcal{F}_t = \cap_{t'>t} \mathcal{F}_{t'}$, pour $t < +\infty$). On prend ici pour ensemble des temps t l'ensemble $\overline{\mathbb{R}}_+ = \mathbb{R}_+ \cup \{+\infty\}$, compact; d'autres préfèrent prendre $\mathbb{R}_+$; c'est sans importance, puisque nous prendrons plus loin (chapitre 8) des sous-ensembles A de $\overline{\mathbb{R}}_+$ comme ensembles des temps. Nous ajouterons souvent un élément $\overline{+\infty} > +\infty$, isolé, $\mathcal{F}_{\overline{+\infty}} = \mathcal{F}_{+\infty}$.

Une variable aléatoire, par exemple à valeurs dans un espace vectoriel E de dimension finie N, est une application $\mathcal{P}$-mesurable $\Omega \to E$. Un processus X à valeurs dans E est une application $\overline{\mathbb{R}}_+ \times \Omega \to E$; $X_t : \omega \mapsto X(t, \omega)$, l'etat du processus à l'instant t, sera supposé $\mathcal{F}_t$-mesurable; $X(\omega) : t \mapsto X(t, \omega)$ est la trajectoire du processus correspondant à l'échantillon $\omega \in \Omega$. Un processus est donc une trajectoire aléatoire. Puisque X_s est $\mathcal{F}_t$-mesurable pour $s \leq t$, $\mathcal{F}_t$ apparaît comme la tribu du passé (présent inclus !) de l'instant t. Le processus est dit continu, continu à droite, à variation finie, ..., si, pour $\mathcal{P}$-presque tout ω (expression que nous ne répèterons plus), la trajectoire $X(\omega)$ est continue, continue à droite, à variation finie, ...

[1] Il est impossible de donner ici une liste des livres d'initiation aux probabilités. Nous renverrons souvent à Dellacherie–Meyer [1], mais on pourrait aussi souvent renvoyer à Ikeda–Watanabe [1]

2. Processus à variation finie et martingales

Un processus V sera dit *à variation finie* s'il est *continu* à variation finie. Une *martingale* M (voir Dellacherie–Meyer [1], volume 2, chapitre V) est un processus continu intégrable (pour tout t, M_t est intégrable), tel que, si $s \leq t$ et $A \in \mathcal{F}_s$:

$$\int_A M_s d\mathcal{P} = \int_A M_t d\mathcal{P}. \tag{2.1}$$

En probabilités, les intégrales s'écrivent en général E (espérance); (2.1) s'écrit alors

$$E(1_A M_s) = E(1_A M_t). \tag{2.2}$$

On écrit cela aussi en utilisant la notion très féconde d'espérance conditionnelle, mais le lecteur pourra s'en passer pour la suite. L'espérance conditionnelle d'une variable aléatoire intégrable Y, par rapport à une tribu $\mathcal{T}$, P-mesurable, notée $Y^{\mathcal{T}}$ ou $E(Y|\mathcal{T})$, est une variable aléatoire $\mathcal{T}$-mesurable, ayant mêmes intégrales que Y sur les ensembles de la tribu $\mathcal{T}$; elle n'est définie qu'à un ensemble $\mathcal{T}$-mesurable $\mathcal{P}$-négligeable près. Si $\mathcal{T}$ est toute la tribu $\mathcal{P}$-mesurable, $Y^{\mathcal{T}} = Y$; si $\mathcal{T}$ est la tribu triviale, $\mathcal{T} = \{\phi, \Omega\}$, $Y^{\mathcal{T}}$ est la constante égale à l'intégrale $\int_\Omega Y d\mathcal{P} = E(Y)$. Alors M est une martingale si elle est intégrable et si, pour $s \leq t$:

$$M_s = E(M_t|\mathcal{F}_s). \tag{2.3}$$

2'. Martingales locales et temps d'arrêt[2]

On a besoin en pratique de processus un peu plus généraux, les *martingales locales*; local n'a pas ici le sens topologique habituel. Le processus M est une martingale locale s'il est continu, et s'il existe une suite croissante $(T_n)_{n\in\mathbb{N}}$ de temps aléatoires (T_n est une variable aléatoire $\Omega \to [0, \overline{+\infty}]$) tendant *stationnairement* vers $\overline{+\infty}$ (pour presque tout ω, il existe $N(\omega)$ tel que, pour $n \geq N(\omega)$, $T_n(\omega) = \overline{+\infty}$; on écrira $T_n \upuparrows \overline{+\infty}$) et une suite de martingales $(M_n)_{n\in\mathbb{N}}$, tels que, dans $[0, T_n[= \{(t,\omega); t < T_n(\omega)\}$, $M = M_n$; on ne met pas $[0, T_n]$, car cela imposerait que M_o soit intégrable, ce que l'on ne souhaite pas; M_t n'est peut-être intégrable pour aucun t. Mais, par continuité, $M = M_n$ dans $[0, T_n] \cap (\overline{\mathbb{R}}_+ \times \Omega) \cap \{T_n > 0\} = \{(t,\omega); T_n(\omega) > 0,\ 0 \leq t \leq T_n(\omega) \leq +\infty\}$.

Cela s'écrit aussi en utilisant les temps d'arrêt (voir Dellacherie–Meyer [1], volume 1, chapitre IV, 49, page 184). Les temps d'arrêt ont été introduits par Doob. Nous ne les développerons pas ici, malgré leur rôle fondamental, parce que nous ne les utiliserons pas plus loin. Bornons-nous à dire qu'un temps d'arrêt est une variable aléatorie $T : \Omega \to [0, \overline{+\infty}]$, telle que, pour tout $t \leq +\infty$, l'ensemble $\{T \leq t\} = \{\omega; T(\omega) \leq t\}$ soit $\mathcal{F}_t$-mesurable. Un temps d'arrêt sert à arrêter des processus; si X est un processus, le processus arrêté au temps T, noté X^T, est le

[2] Le lecteur pressé pourra passer ce chapitre

processus pour lequel la trajectoire $X^T(\omega)$ coïncide avec $X(\omega)$ aux temps $\leq T(\omega)$, mais reste fixée à $X(T(\omega), \omega)$ aux temps $\geq T(\omega)$; on peut écrire $X_t^T = X_{T \wedge t}$ (voulant dire, avec l'habitude des probabilités de ne jamais écrire ω : $X_t^T(\omega) = X_{T(\omega) \wedge t}(\omega)$). Un théorème de Doob dit qu'un processus arrêté d'une martingale est encore une martingale (voir Dellacherie–Meyer [1], volume 2, chapitre VI, théorème 10). Alors M est une martingale locale s'il existe $(T_n) \upuparrows \overline{+\infty}$ tels que, pour tout n, $1_{\{T_n > 0\}} M^{T_n}$ soit une martingale. *Dans la suite, nous abrègerons martingale locale par martingale.*

3. Semi-martingales

Une semi-martingale (voir Dellacherie–Meyer [1], volume 2, chapitre VII, 2) est un processus continu qui peut s'écrire

$$X = V + M, \tag{3.1}$$

V à variation finie, M martingale (sous-entendu locale). On normalise part M_o (valeur initiale) $= 0$, alors la décomposition est unique, à un ensemble $\mathcal{P}$-négligeable près (ce qui signifie qu'une martingale à variation finie est un processus constant, $M = M_o$). L'unicité de la décomposition est le théorème de Doob–Meyer (voir Dellacherie–Meyer [1], volume 2, chapitre VIII, 45). On écrit habituellement

$$X = \widetilde{X} + X^c; \tag{3.2}$$

$\widetilde{X}$ s'appelle la caractéristique locale de X, X^c sa compensée.

4. Calcul intégral stochastique d'Ito

La tribu optionnelle Opt sur $\overline{\mathbb{R}}_+ \times \Omega$ est (voir Dellacheric–Meyer [1], volume 1, chapitre IV, 61) la tribu engendrée par les processus réels continus à droite et les parties $\mathcal{P}$-négligeables (c.-à-d. dont la projection sur Ω est $\mathcal{P}$-négligeable). (Elle est aussi engendrée par les intervalles stochastiques $[S, T[= \{(t, \omega); S(\omega) \leq t < T(\omega)\}, S, T$ temps d'arrêt, ou simplement les $[S, +\infty] = [S, \overline{+\infty}[$, et les parties $\mathcal{P}$-négligeables).

Si H est une fonction réelle optionnelle bornée sur $\overline{\mathbb{R}}_+ \times \Omega$, on peut définir une intégrale $H.V$,

$$(H.V)_t = \int_{]0,t]} H_s \, dV_s, \quad (H.V)_o = 0, \tag{4.1}$$

où, comme toujours, ω n'est pas marqué; cela veut dire

$$(H.V)(t, \omega) = \int_{]0,t]} H(s, \omega) dV(s, \omega) \quad (d = d_s). \tag{4.2}$$

Elle se calcule individuellement pour tout ω (intégrale de Stieltjes); $H.V$ est encore un processus à variation finie.

Mais Ito a montré que, si H est optionnelle bornée, et si M est une martingale, on peut encore définir une intégrale stochastique $H.M$ (voir Dellacherie–Meyer [1], volume 2, chapitre VIII); mais elle ne se calcule plus individuellement pour tout ω, elle n'a qu'un sens global, comme $\mathcal{P}$-classe de processus (modulo les processus $\mathcal{P}$-presque partout nuls, c.-à-d. nuls pour $\mathcal{P}$-presque tout ω, pour tout t) un peu comme l'intégrale de Fourier d'une fonction de L^2 est seulement une classe de Lebesgue de fonctions de carré intégrable, sans valeur précise en chaque point; la méthode de définition est d'ailleurs la même, c'est un prolongement par continuité; c'est encore une martingale (voulant dire, comme indiqué à la fin du chapitre 2', une martingale locale; même si M est une martingale vraie, et si H est optionnelle bornée, $H.M$ n'est en général qu'une martingale locale).

Si alors X est une semi-martingale, et si H est optionnelle bornée, il y aura encore une intégrale stochastique $H.X$, nulle au temps 0; et $(H.X)^\sim = H.\widetilde{X}, (H.X)^c = H.X^c$. Inversement un théorème de Dellacherie (voir Dellacherie–Meyer [1], volume 2, chapitre VIII, 4) dit que les semi-martingales sont les seuls processus donnant lieu, dans un sens à préciser, à des intégrales stochastiques. De là, on passe facilement à des H optionnelles non seulement bornées, mais X-intégrables. On peut même passer à des H optionnelles qui ne sont plus intégrables, et définir des $H.X$ qui sont seulement des semi-martingales formelles (voir Schwartz [1]), au sens où la dérivée d'une fonction sans dérivée est une distribution, ou fonction formelle; d'ailleurs une semi-martingale formelle peut se définir comme une $H.X$, X semi-martingale, H optionnelle, non nécessairement X-intégrable. On pourra désormais écrire $H.X$, pour H optionnelle arbitraire, X semi-martingale formelle et ce sera encore une semi-martingale formelle. Mais on ne pourra plus parler de sa valeur $X(t, \omega)$, comme une distribution n'a pas de valeur en un point. On a aussi des processus à variation finie formels, et des martingales formelles, et on a toujours:

$$(H.X)^\sim = X.\widetilde{X}, \quad (H.X)^c = H.X^c, \quad H.(K.X) = HK.X. \tag{4.3}$$

Bien entendu, si X est à valeurs dans l'espace vectoriel E, H n'a pas besoin d'être réelle; si H est à valeurs dans $\mathcal{L}(E; F)$, alors $H.X$ est à valeurs dans F. Notons que, si S et T sont des temps d'arrêt, $1_{[S,T[}.X = X^T - X^S$. Nous ne nous occuperons pas des processus formels, mais nous négligerons les problémes d'intégrabilité.

5. Le crochet(3)

Si, dans (2.1), (2.2), (2.3), on remplace $=$ par $\leq$, on a la notion de sous-martingale; de là on passe aux sous-martingales locales, qu'on abrège encore par sous-martingales.

(3) Voir Dellacherie–Meyer [1], volume 2, chapitre VII, 42

Si M est une martingale réelle, son carré M^2 est une sous-martingale ≥ 0; elle est alors une semi-martingale et sa composante à variation finie se note $[M, M]$, qui est un processus croissant. Si X est une semi-martingale quelconque, on pose

$$[X, X] = [X^c, X^c] + X_o^2. \tag{5.1}$$

On peut l'interpréter facilement. En effet, la trajectoire $X(\omega)$ n'est pas en général à variation finie (à cause de la composante martingale X^c), mais elle a une variation quadratique finie, et justement $[X, X]_t(\omega)$ est la variation quadratique de $X(\omega)$ de 0 à t (voir Dellacherie–Meyer [1], volume 2, chapitre VIII, théorème 20):

$$[X, X]_t(\omega) = \lim(X_o^2(\omega) + \sum_i (X_{t_{i+1}\wedge t} - X_{t_i \wedge t})^2(\omega)), \tag{5.2}$$

lorsque le pas de la subdivision tend vers 0 (le pas est mesuré par rapport à une distance sur $\overline{\mathbb{R}}_+$ compatible avec sa topologie compacte); lim. est une limite en probabilité, uniformément en t. Par polarisation, si X, Y sont deux semi-martingales, on définit leur crochet, à variation finie $[X, Y]$. Si X, Y sont à valeurs dans des espaces vectoriels $E, F, [X, Y]$ est à valeurs dans $E \otimes F$; pour $F = E$, on prend en général $[X, Y]$ à valeurs dans le produit tensoriel symétrique $E \odot E$, quotient de $E \otimes E$. On a toujours

$$[H.X, K.Y] = HK.[X, Y]. \tag{5.3}$$

Un théorème, dû à Girsanov (voir Dellacherie–Meyer [1], volume 2, chapitre VII, 45), dit que, si Q est une probabilité équivalente à $\mathcal{P}$ sur $(\Omega, \mathcal{F})$, un processus X est une Q-semi-martingale si et seulement s'il est une $\mathcal{P}$-semi-martingale. Les processus $\widetilde{X}, X^c$, ne sont pas les mêmes pour $\mathcal{P}$ et Q, mais $[X, X]$ est le même d'après (5.2).

$[X, Y]$ est l'unique processus à variation finie tel que $XY - [X, Y]$ soit une martingale nulle au temps 0.

6. Le mouvement brownien

La plus connue des martingales (vraie, pas seulement locale, mais sur $\mathbb{R}_+ \times \Omega$, $\mathbb{R}_+ = [0, +\infty[$) est le mouvement brownien. Dans un espace euclidien E de dimension N, on appelle mouvement brownien normal une martingale B vérifiant:

a) $B_o = 0\,\mathcal{P}$-presque sûrement;
b) Pour $t > s$, $B_t - B_s$ est indépendante de la tribu $\mathcal{F}_s$, et sa loi dans E (la mesure image $(B_t - B_s)(\mathcal{P})$) est la loi de Gauss centrée de paramètre $\sqrt{t-s}$:

$$\left(\frac{1}{\sqrt{2\pi(t-s)}}\right)^N \exp(-|x|^2/2(t-s))dx$$

L'existence d'une telle martingale a été démontrée par N. Wiener; elle n'est évidemment pas unique ($\Omega, \mathcal{P}, \ldots$ peuvent varier), mais elle est unique "en loi". Les plus importants résultats sur le brownien ont été donnés par P. Lévy.

On montre qu'on peut aussi caractériser le brownien par le fait que c'est une martingale dont le crochet $[B, B]$, à valeur dans $E \odot E$. est

$$[B, B]_t = \sum_{k=1}^{N} (e_k \odot e_k)t, \quad \text{ou } [B^i, B^j]_t = \delta^{ij} t,$$

si (e_k) est une base orthonormée quelconque, δ le symbole de Kronecker; ou encore, en omettant toujours ω:

$$t \mapsto B_t \odot B_t - \sum_{k=1}^{N} (e_k \odot e_k)t \text{ est une martingale.}$$

(Voir Dellacherie–Meyer [1], volume 2, chapitre VIII, théorème 5.9). Ce théorème est dû à Paul Lévy.

7. Le calcul différentiel stochastique vectoriel; la formule de changement de variables d'Ito

Soient E, F des espaces vectoriels, U un ouvert de E, Φ une application C^2 : $U \to F, X : \overline{\mathbb{R}}_+ \times \Omega \to E$ une semi-martingale, à valeurs dans U. Alors $\Phi(X)$ est une semi-martingale à valeurs dans F, qui s'écrit (voir Dellacherie–Meyer [1], volume 2, chapitre VIII, théorème 27):

$$\Phi(X) = \Phi(X_o) + \Phi'(X).X + \frac{1}{2}\Phi''(X).[X, X] \tag{7.1}$$

Comme X n'est pas à variation finie, mais a une variation quadratique $[X, X]$, il n'est pas étonnant qu'apparaisse le deuxième terme de la formule de Taylor, avec Φ''. L'interprétation est immédiate: X est à valeurs dans $E, \Phi'(X)$ à valeurs dans $\mathcal{L}(E; F)$, donc $\Phi'(X).X$ à valeurs dans $F; [X, X]$ est à valeurs dans $E \odot E$, $\Phi''(X)$ à valeurs dans l'espace des applications bilinéaires symétriques de $E \times E$ dans F, donc linéaires de $E \odot E$ dans F, et $\Phi''(X).[X, X]$ est bien encore à valeurs dans F. On en déduit:

$$\widetilde{\Phi(X)} = \Phi(X_o) + \Phi'(X).\widetilde{X} + \frac{1}{2}\Phi''(X).[X, X] \tag{7.2}$$

$$(\Phi(X))^c = \Phi'(X).X^c \tag{7.3}$$

$$\frac{1}{2}[\Phi(X), \Phi(X)] = \frac{1}{2}\Phi(X_o) \odot \Phi(X_o) + \frac{1}{2}(\Phi'(X) \odot \Phi'(X)).[X, X]. \tag{7.4}$$

Dans la dernière formule, $\Phi'(X)$ est une application linéaire de E dans F, son carré tensoriel symétrique $\Phi'(X) \odot \Phi'(X)$ est linéaire de $E \odot E$ dans $F \odot F, [X, X]$ est à valeurs dans $E \odot E$, donc $(\Phi'(X) \odot \Phi'(X))[X, X]$ est bien à valeurs dans $F \odot F$.

8. Première extension: formules locales[(4)]

Soit $A \subset \overline{\mathbb{R}}_+ \times \Omega$, optionnel. On dira que A est ouvert si, pour tout ω, la section $A(\omega) = \{t; (t, \omega) \in A\}$ est un ouvert de $\overline{\mathbb{R}}_+$.

On dira qu'un processus X est équivalent à 0 sur A, $X \underset{A}{\sim} 0$, si pour $\mathcal{P}$-presque tout ω, $X(\omega)$ est localement constant sur $A(\omega)$ (localement, cette fois, au sens topologique habituel: $X(\omega)$ est constant sur tout intervalle de $A(\omega)$). On dira que $X \underset{A}{\sim} Y$ si $X - Y \underset{A}{\sim} 0$.

On montre que, pour X, Y, semi-martingales:

$$X \underset{A}{\sim} 0 \Longleftrightarrow \widetilde{X} \underset{A}{\sim} 0\,, \quad X^c \underset{A}{\sim} 0, \tag{8.1}$$

$$X_{\underset{A}{\sim}} \Rightarrow [X, Y] \underset{A}{\sim} 0, \tag{8.2}$$

$$\text{Si } H \text{ est optionnel borné}, \textit{nul} \text{ sur } A, \text{ou si } X \underset{A}{\sim} 0, \text{alors } H.X \underset{A}{\sim} 0. \tag{8.3}$$

Alors

$$X \underset{A}{\sim} X', \quad Y \underset{A}{\sim} Y' \Rightarrow [X, Y] \underset{A}{\sim} [X', Y'] \tag{8.4}$$

(Noter que ce n'est pas vrai pour le produit multiplicatif ordinaire: les équivalences de gauche n'entraînent pas $XY \underset{A}{\sim} X'Y'$).

$$H \underset{A}{=} H', \quad X \underset{A}{\sim} X' \Rightarrow H.X \underset{A}{\sim} H'.X'. \tag{8.5}$$

On peut alors définir des semi-martingales sur des ouverts A de $\overline{\mathbb{R}}_+ \times \Omega$. Un processus $X : A \to E$ est une semi-martingale s'il existe une suite $(A_n)_{n \in \mathbb{N}}$ d'ouverts de réunion A, et de semi-martingales $(X_n)_{n \in \mathbb{N}}$ sur $\overline{\mathbb{R}}_+ \times \Omega$ tel que $X \underset{A_n}{\sim} X_n$; de même pour les processus à variation finie et martingales. Si $A = \overline{\mathbb{R}}_+ \times \Omega$, on retrouve les objets définis antérieurement sur $\overline{\mathbb{R}}_+ \times \Omega$ (à condition d'avoir introduit les martingales locales du chapitre 2'). Si X est une semi-martingale formelle sur A, on peut définir $\widetilde{X}, X^c, [X, X]$, et $H.X$ pour H processus optionnel vrai sur A (restriction à A d'un processus optionnel sur $\overline{\mathbb{R}}_+ \times \Omega$); mais ce ne sont que des processus formels sur A, même si X est une semi-martingale vraie sur A (une égalité $X = \widetilde{X} + X^c$, où X est vraie et $\widetilde{X}, X^c$ seulement formelles, est à comparer à une solution u de l'equation des ondes, où $\frac{\partial u}{\partial t}$ et Δu sont seulement des distributions, pour $\frac{\partial u}{\partial t} - \Delta u = 0$ fonction) et ils ne sont définis qu'à une équivalence près sur A;

[(4)]Voir Schwartz [2], chapitres 2 et 3, et Schwartz [1]

autrement dit ce sont des classes d'équivalence sur A de semi-martingales formelles; et toutes les formules antérieures subsistent dans ce cadre.

La formule (7.1) d'Ito subsiste: si X est une semi-martingale sur A, et si Φ est C^2, $\Phi(X)$ est une semi-martingale sur A, mais les termes du deuxième membre ne sont que des classes d'équivalence sur A de semi-martingales formelles, et (7.2), (7.3), (7.4) subsistent aussi, où même les premiers membres sont des classes d'équivalence sur A de semi-martingales formelles. On conviendra d'appeler *différentielle semi-martingale* une telle classe d'équivalence; si X est une semi-martingale, sa classe se notera dX. Alors $d(H.X) = HdX$; on a $H(K\,dX) = H(d(K.X)) = d(H.(K.X))$, et $(H\,K)dX = d(HK.X)$, donc l'égalité (4.3) donne $H(K\,dX) = (H\,K)dX$, ce qui est une notation cohérente; les différentielles semi-martingales sur A (mais aussi les différentielles à variation finie et les différentielles martingales) forment un module sur l'anneau des fonctions optionnelles sur A, pour l'intégration stochastique.

Par ailleurs il est aussi commode de remplacer $d[X,X]$ par $dX \odot dX$ ou même $dX\ dX$, puisque c'est une variation quadratique (chapitre 5), et alors les formules d'Ito (chapitre 7) si X est une semi-martingale sur A, s'écrivent sous forme différentielle:

$$d(\Phi(X)) = \Phi'(X)dX + \frac{1}{2}\Phi''(X)dX\ dX \tag{8.6}$$

$$d(\widetilde{\Phi(X)}) = \Phi'(X)d\widetilde{X} + \frac{1}{2}\Phi''(X)dX\ dX \tag{8.7}$$

$$d(\Phi(X)^c) = \Phi'(X)dX^c \tag{8.8}$$

$$\frac{1}{2}d(\Phi(X))d(\Phi(X)) = \frac{1}{2}(\Phi'(X) \odot \Phi'(X))dX\ dX. \tag{8.9}$$

9. Deuxième extension: semi-martingales sur des variétés[(5)] de classe C^2

Soit V une telle variété, $X : A \subset \overline{\mathbb{R}}_+ \times \Omega \to V$ un processus. On dira que c'est une V-semi-martingales si, pour toute fonction φ réelle C^2 sur V, $\varphi(X)$ est une semi-martingale réelle. Par Ito, si $V = E$ espace vectoriel, on retrouve la notion antérieure. Si $\Phi : V \to V'$ est une application C^2, $\Phi(X)$ est ene V'-semi-martingale. Si maintement W est une sous-variété (non nécessairement fermée) de V et X une V-semi-martingale prenant ses valeurs dans W, c'est une W-semi-martingale. Bien entendu, on peut parler de V-processus à variation finie, mais pas de V-martingale; $X^c, \widetilde{X}, [X,X]$ n'ont aucun sens, ni la notion d'équivalence sur A, car il n'y a pas d'addition sur V. Mais nous allons donner un sens aux différentielles, en utilisant les espaces vectoriels tangents à V.

(5) Voir Schwartz [2], [3]

D'après ce qui est dit ci-dessus, comme V peut (Whitney) être plongée dans un espace vectoriel E, un V-processus est une V-semi-martingale si et seulement si c'est une E-semi-martingale.

10. Différentielles semi-martingales sections d'un fibré vectoriel optionnel[(6)]

Soient $A \subset \overline{\mathbb{R}}_+ \times \Omega$ ouvert, et G_A un espace fibré optionnel au dessus de A, à fibres vectorielles de dimension finie, de fibre-type G; $G_{(t,\omega)}$ est la fibre au dessus de $(t,\omega) \in A$. Il y a des cartes produits, $G_{A'} \to A' \times G$; la formule de transition d'une carte à une autre est optionnelle, $((t,\omega),g) \mapsto ((t,\omega),\alpha(t,\omega)g)$, où α est optionnelle sur A' à valeurs dans $\mathcal{L}(G;G)$. Alors on appellera *diffèrentielle semi-martingale section de* G_A la donnée, pour chaque carte $G_{A'} \to A' \times G$, d'une G-différentielle semi-martingale dX' sur A', avec la formule de transition $dX' \mapsto \alpha dX'$, pour la structure de module par intégration stochastique définie au chapitre 8. Seule existe ici la différentielle semi-martingale section de G_A, dX, par le système cohérent des dX'; X elle-même n'existe pas; d'abord les différentielles sont des classes d'équivalence de semi-martingales formelles; ensuite, si on peut écrire symboliquement $dX(t,\omega) \in G_{(t,\omega)}, dX(t,\omega)$ est un petit vecteur semi-martingale $\in G_{(t,\omega)}$ (exactement par le même abus de langage par lequel, si T est une section-distribution d'un fibré vectoriel G_V au-dessus d'une variété V, on se perment d'écrire, en l'absence d'étudiants, pour $v \in V$, $T(v) \in G_v$, alors que T n'a de valeur en aucun point), les fibres $G_{(t,\omega)}$ varient avec t,ω, et "l'intégrale" $X_t(\omega) = \int_{]o,t]} dX_s(\omega)$ n'est nulle part!

Mais la structure de module est très riche. Si dX est une G_A-différentielle semi-martingale au dessus de A, β une section optionnelle d'un fibré $\mathcal{L}(G_A;H_A)$ (de fibre $\mathcal{L}(G(t,\omega);H(t,\omega))$ au-dessus de (t,ω)), le produit βdX (intégration stochastique) est une H_A-différentielle semi-martingale. On a de même des différentielles à variation finie et martingales. Toute G_A-différentielle de semi-martingale dX a une décomposition unique $dX = d\widetilde{X} + dX^c, d\widetilde{X}$ G_A-différentielle à variation finie, dX^c G_A-différentielle martingale; et $d[X,X]$ ou $dX\,dX$ est une différentielle à variation finie section de $G_A \odot G_A$.

11. Différents espaces tangents à une variété C^2[(7)]

L'espace m-tangent $T^m(V;v)$ à une variété V de classe C^m, en un point v, est l'espace vectoriel des formes linéaires sur $C^m(V;\mathbb{R})$ (C^m = espace vectoriel des fonctions réelles de classe C^m) qui s'annulent sur les fonctions m-plates en v (une

[(6)] Voir Schwartz [4]
[(7)] Voir Schwartz [3], chapitres 1,2

fonction est m-plate en v si ses dérivées d'ordre $1, 2, \ldots, m$, sont nulles en v; ceci a un sens par des cartes). Un élément de $T^m(V; v)$ est la trace au point v d'un opérateur différentiel d'ordre $\leq m$ sur V, sans terme d'ordre 0; et un tel opérateur différentiel d'ordre $\leq m$ n'est autre qu'un champ de vecteurs m-tangents, ou une section du fibré $T^m(V)$ m-tangent. Nous aurons besoin de $m = 1, 2$; $T^1(V; v) \subset T^2(V; v)$. Le quotient $T^2(V; v)/T^1(V; v)$ est canoniquement isomorphe au produit tensoriel symétrique $T^1(V; v) \odot T^1(V; v)$. Soient $\xi, \eta \in T^1(V; v)$. Ils se prolongent en $\bar{\xi}, \bar{\eta}$, opérateurs différentiels d'ordre 1 sur V; le composé $\bar{\xi}\bar{\eta}$ est un opérateur différentiel d'ordre 2; sa trace $(\bar{\xi}, \bar{\eta})_v$ en v est un élément de $T^2(V; v)$. Il dépend, bien entendu, des prolongements choisis; mais, modulo $T^1(V; v)$, il n'en dépend pas. On définit ainsi une application bilinéaire de $T^1(V; v) \times T^1(V; v)$ dans $T^2(V; v)/T^1(V; v)$. Elle est symétrique, car $[\bar{\xi}, \bar{\eta}]$ (crochet de Lie) est un opérateur différentiel d'ordre 1. Donc elle définit une application linéaire de $T^1(V; v) \odot T^1(V; v)$ dans $T^2(V; v)/T^1(V; v)$, qu'une carte montre être bijective. Si D est un opérateur différentiel d'ordre ≤ 2 sans terme d'ordre 0, c.-à-d. une section du fibré $T^2(V)$, son symbole principal, en théorie des équations aux dérivées partielles, au point v, est son image dans $T^2(V; v)/T^1(V; v)$, donc un élément de $T^1(V; v) \odot T^1(V; v)$, ou un polynôme homogène de degré 2 sur le fibré cotangent $T^{*1}(V; v)$.

Dans une carte de V sur un ouvert d'un vectoriel $E, T^1(V; v) \simeq E, T^2(V; v) \simeq E \oplus (E \odot E)$, le quotient devient facteur direct. On peut d'aileurs identifier $T^m(V; v)$ à l'espace des opérateurs différentiels *à coefficients constants* d'ordre $\leq m$, sans terme d'ordre 0. Si (e_k), $k = 1, 2, \ldots, N$, est une base de E, on peut identifier e_k à la dérivée partielle ∂_k, un élément de $T^1(V; v) \simeq E$ est $\sum_{k=1}^N b^k \partial_k$, et un élément de $T^2(V; v) \simeq E \oplus (E \odot E)$ est

$$\sum_{k=1}^N b^k \partial_k + \frac{1}{2} \sum_{i,j=1}^N a^{i,j} \partial_i \partial_j, \qquad a^{i,j} = a^{j,i};$$

$T^1(V; v)$ est de dimension $N, T^2(V; v)$ de dimension $N + N(N + 1)/2$.

Si on passe d'une carte à une autre, par une application Φ de classe C^2 d'un ouvert de E dans F, et si on représente les vecteure 2-tangents par la différentielle d'ordre 2 de Φ,

$$\begin{pmatrix} \alpha \\ \beta \end{pmatrix} \mapsto \begin{pmatrix} \Phi'(v)\alpha + \Phi''(v)\beta \\ \Phi'(v) \odot \Phi'(v)\beta \end{pmatrix} \in \begin{pmatrix} F \\ \oplus \\ F \odot F \end{pmatrix}, \tag{11.1}$$

ou

$$\begin{pmatrix} \Phi'(v) & \Phi''(v) \\ 0 & \Phi'(v) \odot \Phi'(v) \end{pmatrix} \begin{pmatrix} \alpha \\ \beta \end{pmatrix}.$$

12. Formule d'Ito et espaces tangents

Soit X une V-semi-martingale. Dans une carte sur un ouvert U d'un vectoriel E, où nous l'écrirons encore X, introduisons la semi-martingale complète associée,

$\underline{X}$, $(E \oplus (E \odot E))$-semi-martingale qui n'a de sens que dans cette carte:

$$\underline{X} = \begin{pmatrix} X \\ \frac{1}{2}[X,X] \end{pmatrix}, \qquad \underline{dX} = \begin{pmatrix} dX \\ \frac{1}{2}dX\,dX \end{pmatrix}. \tag{12.1}$$

Dans le changement de cartes $E \to F$, les formules d'Ito (8.6), (8.7), (8.8), (8.9) s'écrivent, si $Y = \Phi(X)$:

$$\underline{dY} = \begin{pmatrix} \Phi'(v) & \Phi''(v) \\ 0 & \Phi'(v) \odot \Phi'(v) \end{pmatrix} \underline{dX}, \tag{12.2}$$

$$\underline{\widetilde{dY}} = \begin{pmatrix} \Phi'(v) & \Phi''(v) \\ & \Phi(v) \odot \Phi'(v) \end{pmatrix} \underline{\widetilde{dX}}, \tag{12.3}$$

$$\underline{dY}^c \text{ qu'on écrit } dY^c = \Phi'(X)dX^c, \tag{12.4}$$

$$\frac{1}{2}dY dY = \Phi'(X) \odot \Phi'(X)\frac{1}{2}dX\,dX. \tag{12.5}$$

En comparant (12.2) et (11.1), on voit que l'on peut considérer $\underline{dX}_t$ comme un petit vecteur semi-martingale $\in T^2(V; X_t)$, $\underline{dX}(t,\omega) \in T^2(V; X(t,\omega))$. Les définitions du chapitre 10 montrent que, si X est une V-semi-martingale sur $A \subset \overline{\mathbb{R}}_+ \times \Omega$, on peut considérer sa différentielle $\underline{dX}$ (voir Schwartz [4], proposition (2.7), p. 710) comme une différentielle semi-martingale section du fibré $T^2(V)$ au dessus de V, le long de X, c.-à-d. une différentielle semi-martingale section du fibré G_A, où G_A est le fibré image réciproque du fibré $T^2(V)$ par l'application $X : A \mapsto V$; la fibre $G_{(t,\omega)}$ est $T^2(V; X(t,\omega))$. De même sa composante à variation finie, $\underline{\widetilde{dX}}$, est une différentielle section du même fibré. Mais la composante martingale dX^c, par la formule (12.4), est une différentielle martingale section du sous-fibré $T^1(V)$ le long de X, tandis que l'image de $\underline{dX}$ dans le quotient $T^2(V)/T^1(V)$ est une différentielle à variation finie section du quotient $T^1(V) \odot T^1(V)$; c'est $\frac{1}{2}dX\,dX$. Ainsi il n'y a pas de décomposition $X = \widetilde{X} + X^c$, ni de crochet $[X,X]$, mais il y a une décomposition $\underline{dX} = \underline{\widetilde{dX}} + dX^c$, $\underline{dX}(t,\omega) \in T^2(V; X(t,\omega))$, $\underline{\widetilde{dX}}(t,\omega) \in T^2(V; X(t,\omega))$, $dX^c(t,\omega) \in T^1(V; X(t,\omega))$, et un produit $\frac{1}{2}dX\,dX(t,\omega) \in T^1(V; X(t,\omega)) \odot T^1(V; X(t,\omega))$, $\frac{1}{2}dX\,dX$ image de $\underline{dX}$ et de $\underline{\widetilde{dX}}$ dans $T^2(V)/T^1(V)$.

13. Equations Différentielles Stochastiques (EDS)

Ici $A = \overline{\mathbb{R}}_+ \times \Omega$. Une EDS sur un ouvert d'un espace vectoriel E de dimension finie N est une équation de la forme

$$dX = \sum_{k=1}^{m} H_k(X)dZ^k, \tag{13.1}$$

où H_k est un champ de vecteurs, Z^k une semi-martingale réelle. Il y a existence et unicité de la solution, pour des H_k localement lipschitziens (avec explosion ou temps de mort), si on se donne la valeur initiale X_o, variable aléatoire $\mathcal{F}_o$-mesurable. Si on fait un changement de cartes, la formule d'Ito montre que les crochets $[Z^i, Z^j]$ interviendront; autant vaut les mettre tout de suite. Nous considérerons donc plutôt une EDS de la forme:

$$dX_t = \sum_{k=1}^{m} H_k(X_t)dZ_t^k + \frac{1}{2}\sum_{i,j=1}^{m} H_{i,j}(X_t)d[Z^i, Z^j]_t, \tag{13.2}$$

où les H_k et les $H_{i,j}$ sont des champs de vecteurs, $H_{j,i} = H_{i,j}$.

Si, dans une carte, les $H_{i,j}$ sont nuls, c'est un accident, ils ne le seront pas dans une autre. En passant à la semi-martingale complète associée, on trouve:

$$\frac{1}{2}dX_t dX_t = \frac{1}{2}\sum_{i,j=1}^{m} H_i(X_t) \odot H_j(X_t)d[Z^i, Z^j]_t \tag{13.3}$$

d'où

$$\underline{dX}_t = \sum_{k=1}^{m} H_k(X_t)dZ_t^k + \frac{1}{2}\sum_{i,j=1}^{m} \underline{H_{i,j}(X_t)}d[Z^i, Z^j]_t, \tag{13.4}$$

où les H_k sont des champs de E-vecteurs, les $H_{i,j}$ des champs de $(E \oplus (E \odot E))$-vecteurs, $\underline{H_{j,i}} = \underline{H_{i,j}}$; la composante de $\underline{H_{i,j}}$ dans $E \odot E$ étant $H_i \odot H_j$.

Mais alors ceci devient indépendant de toute carte. Une EDS sur une variété V (voir Schwartz [3], chapitre 8) sera une équation de la forme (13.4); les Z^k sont des champs de vecteurs 1-tangents donnés; les $\underline{H_{i,j}}$, avec $\underline{H_{j,i}} = \underline{H_{i,j}}$, sont des champs de vecteurs 2-tangents donnés; ces champs doivent être localement lipschitziens, donc V doit être C^2-lipschitz; et on a la condition de compatibilité suivante: la projection de $\underline{H_{i,j}}$ dans le quotient $T^2/T^1 = T^1 \odot T^1$ doit être $H_i \odot H_j$ qui exprime que l'image du deuxième membre dans le quotient $T^2(V; X_t)/T^1(V; X_t) = T^1(V; X_t) \odot T^1(V; X_t)$ est $\frac{1}{2}dX_t dX_t$. Une solution est une V-semi-martingale X, dont la différentielle $\underline{dX}$ au sens du chapitre 12 doit être égale au deuxième membre, différentielle semi-martingale section de $T^2(V)$ le long de X.

On peut donc écrire globalement une EDS sur une variété. Et, pas plus difficilement sur une variété que sur un espace vectoriel, on montre l'existence et l'unicité de la solution pour une condition initiable donnée X_o, $\mathcal{F}_o$-mesurable, avec explosion ou temps de mort.

14. Diffusion brownienne sur une variété

Ici on est sur $\mathbb{R}_+ \times \Omega$, $\mathbb{R}_+ = [0, +\infty[$. Le mouvement brownien sur un espace euclidien E vérifie l'EDS $dX = dB$, donc, compte tenu du chapitre 6:

$$\underline{dX}_t = \begin{pmatrix} dB_t \\ \frac{1}{2}\Sigma_{k=1}^N e_k \odot e_k dt \end{pmatrix};$$

on en déduit

$$\widetilde{\underline{dX}}_t = \begin{pmatrix} 0 \\ \frac{1}{2}\Sigma_{k=1}^N e_k \odot e_k \end{pmatrix} dt. \tag{14.1}$$

L'opérateur différentiel du second ordre correspondant au champ de vecteurs constant $\frac{1}{2}\Sigma_{k=1}^N e_k \odot e_k$ est $\frac{1}{2}\Delta$, qu'on peut écrire comme champ 2-tangent: $v \mapsto \frac{1}{2}\Delta = \frac{1}{2}\Delta(v)$. C'est cela qui traduit que son semi-groupe, de générateur infinitésimal $\frac{1}{2}\Delta$, est le semi-groupe de la chaleur, et que la loi de $X_t = B_t$ est la gaussienne de paramètre $\sqrt{t}$.

Si alors L est un opérateur différentiel du second ordre semi-elliptique ≥ 0 sans terme d'ordre 0, sur une variété V, on appellera *diffusion L-brownienne* une solution de l'EDS (voir Schwartz [5]):

$$\widetilde{\underline{dX}}_t = L(X_t)dt. \tag{14.2}$$

On voit bien que L est un champ de vecteurs 2-tangents, sa valeur en X_t est $L(X_t) \in T^2(V; X_t)$, et $\widetilde{\underline{dX}}_t$ est un petit vecteur à variation finie $\in T^2(V; X_t)$, au sens du chapitre 12, donc (14.2) est bien cohérent. Cette EDS n'est pas mise sous la forme canonique (13.4), assurant l'existence et l'unicité de la solution, pour une condition initiale donnée; on ne donne ici que $\widetilde{\underline{dX}}_t$, il manque la composante martingale dX_t^c, et la condition de compatibilité requise n'est donc pas là non plus. Mais, si L est de range constant (cas strictement elliptique par exemple), la connaissance d'un $\widetilde{\underline{dX}}_t$ proportionnel à dt permet (moyennant un élargissement de $\Omega, \mathcal{P}$), de trouver le terme dX_t^c; ou démontre qu il est nécessairement de la forme

$$dX_t^c = \sum_{l=1}^m \sigma_l(X_t)dB_t^l, \tag{14.3}$$

où $(B^l)_{l=1,2,\dots,m}$ est un brownien normal d'un espace $\mathbb{R}^m$, et où, pour $v \in V$, $\sigma_l(v) \in T^1(V; v)$, et

$$\frac{1}{2}\sum_{l=1}^m \sigma_l(v) \odot \sigma_l(v) = \text{image de } L(v) \text{ dans } T^2(V; v)/T^1(V; v) = T^1(V; v) \odot T^1(V; v).$$

Alors

$$\underline{dX}_t = \sum_{l=1}^m \sigma_l(X_t)dB_t^l + L(X_t)dt, \tag{14.4}$$

donc $\widetilde{\underline{dX}}_t = L(X_t)dt$; c'est une EDS sous la forme canonique (13.4), $\sigma_l(v) \in T^1(V; v)$, $L(v) \in T^2(V; v)$. La condition de compatibilité est vérifiée; l'image dans le quotient $T^2(V; v)/T^1(V; v)$ de $L(v)$ est $\frac{1}{2}\Sigma_{l=1}^m \sigma_l(v) \odot \sigma_l(v)$, et il y a bien une solution unique, pour une condition initiale donnée. Si V est riemannienne, son brownien associé est son L-brownien, $L = \frac{1}{2}\Delta$. Par des méthodes difficiles, on peut montrer, pour L strictement elliptique ≥ 0, l'existence et l'unicité lorsque L a ses coefficients d'ordre 2 continus, et d'ordre 1 boréliens localement bornés: c'est le théorème de Stroock et Varadhan, voir Meyer–Priour–Spitzer [1], article de Priouret, chapitre VI.

15. Mouvement brownien et problème de Dirichlet

Montrons succinctement le rôle que joue le L-brownien pour la résolution d'un problème de Dirichlet pour L. Soit U un ouvert relativement compact de V, de frontière ∂U. Soit X le L-brownien issu de $a \in U$, $X_o = a$. La formule $\underline{\widetilde{dX}}_t = L(X_t)dt$ est une égalité entre petits vecteurs 2-tangents à V (en réalité entre différentielles semi-martingales sections du fibré $T^2(V)$ le long de X). Si φ est une fonction réelle C^2 à support compact sur V, on peut calculer les valeurs de ces petits vecteurs 2-tangents sur φ; la valeur du premier est $d\varphi\widetilde{(X)}_t$, la valeur du deuxième est $L\varphi(X_t)dt$. Donc:

$$d\varphi\widetilde{(X)}_t = L\varphi(X_t)dt. \tag{15.1}$$

Désignons part T le temps d'entrée dans le complémentaire de U (ou dans ∂U):

$$T = \inf\{t \geq 0; X_t \notin U\};$$

c'est un temps d'arrêt (voir chapitre 2'); si on suppose V connexe non compacte, on voit que $T < +\infty$ et que les intégrales qui vont suivre ont un sens. Calculons le $E\int_o^T$ des deux membres de (14.1), en se souvenant que $X_o = a$:

$$E(\varphi(X))_{\widetilde{T}} - \varphi(a) = E\int_o^T (L\varphi)(X_t)dt. \tag{15.2}$$

(si $Y = (\varphi(X))^\sim$ est un processus, Y_T est sa valeur en T, $Y_T(\omega) = Y(T(\omega), \omega)$, c'est une variable aléatoire, à ne pas confondre avec le processus arrêté Y^T).

Mais $\varphi(X)$ et $(L\varphi)(X)$ sont bornés, donc $(\varphi(X))^\sim$ aussi aux temps bornés, donc $(\varphi(X))^c$ aussi, donc c'est une vraie martingale pas seulement locale, et nulle au temps 0; $E(\varphi(X))^c_T = 0$, et $E(\varphi(X))^\sim_t = E(\varphi(X))_T = E\varphi(X_T)$. Donc

$$\varphi(a) = E(\varphi(X_T)) - E\int_o^T (L\varphi)(X_t)dt. \tag{15.3}$$

On définit ainsi deux mesures ≥ 0:

$$\mu_a, \mu_a(\varphi) = E(\varphi(X_T)), \quad \text{de masse} + 1, \quad \text{portée par } \partial U;$$

$$\Gamma_a, \Gamma_a(\psi) = E\int_o^T \psi(X_t)dt, \quad \text{portée par } U.$$

La deuxième est absolument continue par rapport aux mesures de Lebesgue de V. Si en effet A est un ensemble Lebesgue-négligeable,

$$E\int_o^T 1_A(X_t)dt \leq E\int_o^{+\infty} 1_A(X_t)dt = \int_o^{+\infty} \mathcal{P}\{X_t \in A\}dt;$$

mais on démontre que la loi de X_t, loi image $X_t(\mathcal{P})$ de $\mathcal{P}$ par X_t, est toujours absolument continue par rapport aux mesures de Lebesgue, donc $\mathcal{P}\{x_t \in A\} = 0$. Donc A Lebesgue-négligeable est Γ_a-négligeable. Donc $\Gamma_a = G(a,x)dx$, si dx est une mesure de Lebesgue sur V. Alors:

$$\varphi(a) = \mu_a(\varphi) - \int_U G(a,x)L\varphi(x)dx : \tag{15.4}$$

μ_a est la mesure L-harmonique sur ∂U relative à a $\in U$, et $G(a,x)dx$ est le L-noyau de Green. (La formule (15.4) n'est démontrée que si la fonction φ sur $\bar{U}$ est restriction d'une fonction C^2 sur V; il faut que ∂U soit assez régulier pour qu'on puisse l'etendre à des données de Dirichlet plus générales).

16. Le flot d'une EDS (13.4)

Si les coefficients d'une EDS sont C^∞ et en nous bornant au cas où il n'y a pas à considére de temps de mort (cas dit complètement conservatif), on peut appeler $\Phi_t(\omega;x)$ la valeur en (t,ω) de la solution correspondant à la valeur initiale x. Alors un théorème très remarquable dit que l'on peut choisir Φ tel que, pour $\mathcal{P}$-presque tout $\omega, \Phi(\omega;.)$ soit C^∞ en x, à dérivées continues en t,x; Φ est le flot. (Ce qui est remarquable, c'est qu'on n'ait pas seulement: pour tout x, pour presque tout $\omega, \Phi(\omega;x)$ est continue en t, mais: pour presque tout $\omega,\ldots$). Pour une donnée initiale X_o $\mathcal{F}_o$-mesurable, la solution est $\Phi(.,X_o)$. Alors $\Phi_t(\omega;.)$ est un C^∞-difféomorphisme de V sur un ouvert de V, et sur V elle-même moyennant des hypothèses plus restrictives, par exemple si V est compacte, ou si $V = \mathbb{R}^N$ et si les champs H_k de (13.1) sont globalement lipschitziens.

On peut aller plus loin, et résoudre l'équation sans probabilité (voir Schwartz [6], chapitre 4). L'ensemble $(Z^1(\omega), Z^2(\omega), \ldots, Z^m(\omega))$ est une trajectoire $uu \in C([0,+\infty]; \mathbb{R}^m])$; $(H_1(v), H_2(v), \ldots, H_m(v))$ est une application linéaire $H(v)$ de $\mathbb{R}^m$ dans $T^1(V;v), \underline{(H_{i,j}(v))}_{i,j=1,2,\ldots,m}$ une application linéaire de $\mathbb{R}^m \odot \mathbb{R}^m$ dans $T^2(V;v)$. On peut écrire l'équation avec des notations à une variable:

$$\underline{dX}_t = H(X_t)dZ_t + \frac{1}{2}\underline{H}(X_t)d[Z,Z]_t, \tag{16.1}$$

l'image de $\underline{H}(v) \in \mathcal{L}(\mathbb{R}^m \odot \mathbb{R}^m; T^2(V;v))$ dans $\mathcal{L}(\mathbb{R}^m \odot \mathbb{R}^m;\ T^2(V;v)/T^1(V;v))$ étant $H(v) \odot H(v) \in \mathcal{L}(\mathbb{R}^m \odot \mathbb{R}^m; T^1(V;v) \odot T^1(V;v))$.

Ecrivon l'EDS en termes de trajectoires:

$$\begin{gathered}\underline{d\xi}_t = H(\xi_t)dw_t + \frac{1}{2}\underline{H}(\xi_t)d[w,w]_t,\\ w \in C(\overline{\mathbb{R}}_+;\mathbb{R}^m) \text{ donnée}, \xi \in C(\overline{\mathbb{R}}_+;V) \text{ inconnue.}\end{gathered} \tag{16.2}$$

Cette équation a un sens si w est à variation finie, avec $d[w,w] = dwdw = 0$; sinon, pour un w donné, elle n'a pas de sens (pas seulement parce que $[w,w]$ n'a pas de sens, il peut en avoir si w a une variation quadratique, mais parce que $H(\xi_t)dw_t$ n'en a pas). Mais il existe, indépendamment de toute probabilité, une application de $C(\overline{\mathbb{R}}_+;\mathbb{R}^m) \times V$ dans $C(\overline{\mathbb{R}}_+;V)$, appelée flot Φ, ayant la propriété suivante: si Z est une $\mathbb{R}^m$-semi-martingale sur un $(\Omega,\mathcal{F},(\mathcal{F}_t),\mathcal{P}), X_o$ une condition initiale $\mathcal{F}_o$-mesurable, la solution de (15.1) avec la condition initiale X_o est donnée par $X_t(\omega) = \Phi_t(Z(\omega), X_o(\omega))$. L'application unique Φ résout toutes les EDS pour les champs $H, \underline{H}$ donnés, sans probabilité. En particulier, pour $\Omega,\mathcal{F},(\mathcal{F}_t),Z,X_o$ donnés, la solution est la même pour toutes les probabilités $\mathcal{P}$ qui font de Z une semi-martingale. En outre, l'ensemble $\mathcal{W}$ des $w \in C(\overline{\mathbb{R}}_+;\mathbb{R}^m)$ pour lesquels $\Phi(w,\,.)$ est une fonction C^∞ en x, à dérivées continues en t,x, et pour lesquels $\Phi_t(w,\,.)$ est, pour tout t, un C^∞-difféomorphisme de V sur un ouvert de V, est "universellement presque sûr": pour tout $\mathcal{P}$ qui fait de Z une semi-martingale, $Z(\omega) \in \mathcal{W}$ $\mathcal{P}$-presque sûrement.

17. Relèvement d'une semi-martingale par une connexion[(8)]

Soit G_V un fibré sur une variété V, de fibre G_v en $v \in V$. Soit σ une connexion; si $\hat{v} \in G_v, \sigma(\hat{v}) \in \mathcal{L}(T(V;v);T(G_V;\hat{v}))$, $\pi\sigma =$ identité, si π est la projection de G_V sur V. On démontre que cette connexion définit automatiquement des connexions d'ordre supérieur, $\sigma^m, \sigma^m(\hat{v}) \in \mathcal{L}(T^m(V;v);T^m(G_V;\hat{v}))$; σ^2 respecte les structures de sous-espace et quotient du chapitre 11, c.-à-d. σ^2 induit $\sigma^1 = \sigma$ sur $T^1(V;v)$, et définit $\sigma^1 \odot \sigma^1$ sur $T^2(V;v)/T^1(V;v) = T^1(V;v) \odot T^1(V;v)$.

On sait que les connexions définissent des transports parallèles le long de courbes C^1 de V, ou relèvements horizontaux de ces courbes. On peut prolonger ces relèvements aux semi-martingales. Soit X une V-semi-martingale; ses relevées $\hat{X}$ sont des G_V-semi-martingales horizontales, vérifiant la relation différentielle

$$\underline{d\hat{X}}_t = \sigma^2(\hat{X}_t)\underline{dX}_t, \quad dX_t = \pi(\hat{X}_t); \tag{17.1}$$
$$\underline{d\hat{X}}_t \in T^2(V;\hat{X}_t), \quad \sigma^2(\hat{X}_t) \in \mathcal{L}(T^2(V;X_t);T^2(G_V;\hat{X}_t)).$$

Le relèvement s'obtient comme dans le cas d'une courbe C^1. On montre d'abord que toute semi-martingale X sur V est solution globale d'une EDS (13.4) sur V. Mais le relèvement d'une EDS est évident; on posera $\hat{H}(\hat{v}) = \sigma(\hat{v})H(v)$, $\underline{\hat{H}}(\hat{v}) = \sigma^2(\hat{v})\underline{H}(v)$; l'EDS relevée est

$$\underline{d\hat{X}}_t = \hat{H}(\hat{X}_t)dZ_t + \frac{1}{2}\underline{\hat{H}}(\hat{X}_t)d[Z,Z]_t. \tag{17.2}$$

(8)Pour ce chapitre 17, voir Schwartz [3], chapitre 13

Si l'EDS de X satisfait à la condition de compatibilité, celle de $\hat{X}$ aussi; elle a donc une solution unique (avec temps de mort), pour un rèvement initial donné $\hat{X}_o$, au dessus de X_o, $\mathcal{F}_o$-mesurable, et c'est le relèvement de X.

Il y a même un flot, comme au chapitre 16, indépendant de toute probabilité. Ceci, à ma connaissance, n'est démontré nulle part, mais c'est une conséquence facile du chapitre 16. Il existe une application $(w, \hat{x}) \mapsto \hat{w} = \Phi(w, \hat{x})$, de $C(\overline{\mathbb{R}}_+; V) \times G_V$ dans $C(\overline{\mathbb{R}}_+; G_V)$, définie seulement pour $\pi\hat{x} = w_o$, telle que, si X est une $(\Omega, \mathcal{F}, (\mathcal{F}_t), \mathcal{P})$-$V$-semi-martingale, son relèvement, de valeur initiale $\hat{X}_o$ au-dessus de X_o, soit la semi-martingale $\hat{X} = \Phi(X, \hat{X}_o)$, $\hat{X}_t(\omega) = \Phi_t(X(\omega), \hat{X}_o(\omega))$. En particulier, si X est un V-processus sur $(\Omega, \mathcal{F}, (\mathcal{F}_t)_t)$, et si $\hat{X}_o$ est un relèvement $\mathcal{F}_o$-mesurable de $X_o, \hat{X} = \Phi(X, \hat{X}_o)$ est son relèvement correspondant pour toutes les probabilités $\mathcal{P}$ sur Ω qui font de X une semi-martingale.

Supposons que X soit une L-diffusion sur V, suivant le chapitre 14; L est un opérateur différentiel du second ordre ou champ de vecteurs 2-tangents, donc il a un relèvement $\sigma^2 L = \hat{L}$, champe de vecteurs 2-tangents sur G_V ou opérateur différentiel du second ordre, $\hat{L}(\hat{v}) = \sigma^2(\hat{v})L(v)$. Ensuite X a un relèvement horizontal $\hat{X}$ (pour une condition initiale donnée $\hat{X}_o$). De (16.1) on déduit:

$$\widetilde{\underline{d\hat{X}}_t} = \sigma^2(\hat{X}_t)\widetilde{\underline{dX}_t} = \sigma^2(\hat{X}_t)L(X_t)dt = \hat{L}(\hat{X}_t)dt : \tag{17.3}$$

$\hat{X}$ est une $\hat{L}$-diffusion sur G_V. On peut donc construire $\hat{X}$ comme relèvement de X ou comme $\hat{L}$-diffusion, ou encore construire les $\hat{L}$-diffusions directement ou en relevant les L-diffusions. On peut aussi construire une L-diffusion sur V en projetant sur V une $\hat{L}$-diffusion sur G_V.

Il se trouve que, si V est riemannienne, si $L = \frac{1}{2}\Delta$, et si G_V est le fibré des repères orthonormés (fibré principal), muni de la connexion de Levi–Civita, $\hat{L} = \frac{1}{2}\hat{\Delta}$ est facile à construire, parce que le fibré tangent de G_V est trivial! Si $(e_k)_{k=1,2,\ldots,N}$ est la base canonique de $\mathbb{R}^N$, un point $\hat{v}$ de G_v au-dessus de v est un repère $\mathbb{R}^m \to T(V; v); \hat{v}(e_k) \in T(V; v)$ a un relèvement horizontal $\sigma(\hat{v})(\hat{v}(e_k)) \in T(G_V; \hat{v})$, soit $\eta_k(\hat{v})$. Alors chaque η_k est un champ de vecteur 1-tangent ou opérateur différentiel d'ordre 1 sur G_V, appelé champ basique associé à e_k. On peut construire les opérateurs différentiels du deuxième ordre $\eta_k^2 = \eta_k \circ \eta_k$ sur G_V, ou champ de vecteurs 2-tangents. Alors le relèvement $\hat{\Delta}$ de Δ est $\Sigma_{k=1}^N \eta_k^2$. On peut résoudre su G_V l'EDS.

$$d\hat{X}_t = \sum_{k=1}^{m} \eta_k(\hat{X}_t)dB_t^k + \frac{1}{2}\hat{\Delta}(\hat{X}_t)dt, \tag{17.4}$$

qui admet une solution unique pour $\hat{X}_o$ donné; c'est une $\frac{1}{2}\hat{\Delta}$-diffusion; si $X_o = \pi\hat{X}_o$, alors $X = \pi\hat{X}$ sera la $\frac{1}{2}\Delta$-diffusion ou mouvement brownien de valeur initiale X_o d'où une construction du mouvement brownien sur V à partir de son relevé sur le fibré des repères. Avec des notations différentes, et même une conception assez différente, c'est la méthode de Eels–Elworthy (voir Ikeda-Watanabe [1], chapitre V, paragraphe 4). Ce sont ces jeux de relévements par connexion de mouvements browniens qui ont permis à J.M. Bismut de donner une nouvelle démonstration du théorème de l'indice d'Atiyah–Singer pour l'opérateur de Dirac.

18. Semi-martingales à valeurs dans des espaces de dimension infinie[(9)]

Soit $\mathcal{H}$ un Banach. On pourrait appeler $\mathcal{H}$ semi-martingale un processus somme d'un processus à variation finie et d'une martingale. I1 y a cependant deux difficultés. L'une est qu'il y a d'autres processus que les semi-martingales qui donnent lieu à des intégrales stochastiques suivant le chapitre 4, et cela quel que soit $\mathcal{H}$ de dimension infinie. Ce n'est pas trop gênant, mais ce qui l'est plus c'est qu'en général, une semi-martingale ne donne pas lieu à des intégrales stochastiques. C'est toutefois vrai si $\mathcal{H}$ est hilbertien séparable. C'est pourquoi, si G est un espace vectoriel topologique localement convexe séparé, on dira que X, processus à valeurs dans G, est une G-semi-martingale, s'il existe une suite de temps d'arrêt (voir chapitre 2') $(T_n)_{n\in\mathbb{N}} \upuparrows \overline{+\infty}$, et une suite de sous-espaces hilbertiens séparables $\mathcal{H}_n$ de G, tels que $1_{\{T_n>0\}}X^{T_n}$ soit une $\mathcal{H}_n$-semi-martingale. Un cas particulièrement intéressant est celui des espace G de Fréchet nucléaires (Ustunel, voir Schwartz [7]); si X est un G-processus, et si, pour tout $\xi \in G', < \xi, X >$ est une semi-martingale, X est une G-semi-martingale; et même il existe un même $\mathcal{H}$, sous-espace hilbertien séparable de G, tel que X soit une $\mathcal{H}$-semi-martingale. Un cas intéressant est celui de $G = C^\infty(E; F)$, où E et F sont des espaces vectoriels de dimension finie, ou plus généralement $C^\infty(U; F)$, où U est une variété de dimension finie; c'est un Fréchet nucléaire. Si maintenant V est une autre variété, $C^\infty(U; V)$ n'est plus un espace vectoriel, ni même une variété. On peut cependant dire que Φ, $C^\infty(U; V)$-processus, est une $C^\infty(U; V)$-semi-martingale, si, pour toute fonction φ réelle C^∞ sur V, $\varphi \circ \Phi$ est une $C^\infty(U; \mathbb{R})$-semi-martingale. On a alors les trois théorèmes suivants, si U, V, W sont des variétés de dimension finie, Φ une $C^\infty(U; V)$-semi-martingale, Ψ une $C^\infty(V; W)$-semi-martingale, X une U-semi-martingale:

(1) $\Phi(X)$ est une V-semi-martingale;
(2) $\Psi \circ \Phi$ est une $C^\infty(U; W)$-semi-martingale;
(3) si, pour presque tout ω, pour tout t, $\Phi(t, \omega)$ est un difféomorphisme de U sur V, d'inverse $\Phi^{-1}(t, \omega)$, Φ^{-1} est une $C^\infty(V; U)$-semi-martingale.

[(9)]Voir Schwartz [7]

Par exemple, si Φ est le flot d'une EDS sur une variété V (chapitre 16), dans le cas complètement conservatif (par exemple V compacte), Φ est une $C^\infty(V;V)$-semi-martingale.

References

C. Dellacherie, P. A. Meyer

[1] *Probabilités et Potentiels*, Hermann, volume 1 1975, volume 2 1980, volume 3 1983.

N. Ikeda, S. Watanabe

[1] *Stochastic Differential Equations and Diffusion Processes*, North-Holland-Kodansha, 1981.

P. A. Meyer

[1] *Flot d'une équation différentielle stochastique*, in *Séminaire de Probabilités, XV, 1979–80*, Lecture Notes in Mathematics 850, Springer-Verlag, 1981, pp. 103–117.

P. A. Meyer, P. Priouret, F. L. Spitzer

[1] *Ecole d'Eté de Probabilités, St Flour, 1973*, Lecture Notes in Mathematics 390, Springer-Verlag, 1974.

L. Schwartz

[1] *Les semi-martingales formelles*, in *Séminaire de Probabilités, XV, 1979–80*, Lecture Notes in Mathematics 850, Springer-Verlag, 1981, pp. 413–489.

[2] *Semi-martingales sur des Variétés et Martingales Conformes sur des Variétés Analytiques Complexes*, Lecture Notes in Mathematics 780, Springer-Verlag, 1980.

[3] *Géométri Différentielle Stochastique*, in *Séminaire de Probabilités, XVI, 1980–81*, Lecture Notes in Mathematics 921, Springer-Verlag, 1982, Supplément, pp. 1–148.

[4] *Les gros produits tensoriels en analyse et en probabilités*, in *Aspects of Mathematics and its Applications, in honour of L. Nachbin*, Elsevier, 1986.

[5] *Construction Directe du Brownien sur une variété*, in *Séminaire de Probabilités, XIX, 1983–84*, Lecture Notes in Mathematics 1123, Springer-Verlag, 1985, pp. 91–112.

[6] *Calculs Stochastiques Directs sur les Trajectoires et Propriétés de Boréliens Porteurs*, in *Séminaire de Probabilités, XVIII, 1982–83*, Lecture Notes in Mathematics 1059, Springer-Verlag, 1984, pp. 271–326.

[7] *(I) Semi-martingales à valeurs dans des espaces d'applications C^∞ entre espaces vectoriels*, et *(II) Le théorème Φ^{-1}, et les semi-martingales à valeurs dans des espace d'applications C^∞ entre variétés*, C.R. Acad. Sc. Paris 305, série I, 1987, I: pp. 31–35, II: pp. 49–53.

Reprinted from Bull. London Math. Soc., **31**, *1999*

OBITUARY: KUNIHIKO KODAIRA

by

M. F. ATIYAH

University of Edinburgh, Edinburgh EH9 3JZ

Kunihiko Kodaira, who died on 26 July 1997, was the outstanding Japanese mathematician of the post-war period, his fame established by the award of the Fields Medal at the Amsterdam Congress in 1954.

He was born on 16 March 1915, the son of an agricultural scientist who at one time was Vice Minister of Agriculture in the Japanese Government and had also played an active role in agricultural developments in South America. Kodaira studied at Tokyo University, taking degrees in both mathematics and physics. From 1944 to 1951 he was an associate professor of physics at the University. His PhD thesis was published in the *Annals of Mathematics* [**18**], and it immediately attracted international attention. Essentially this filled a significant lacuna in the basic theorem of W. V. D. Hodge on harmonic integrals. Kodaira had worked on this for many years but, because of the war, his research was carried out in isolation from the international community and did not become known until much later.

Hermann Weyl, who had been a keen supporter of Hodge's work, realised the importance of Kodaira's thesis, and arranged for him to come to the Institute for Advanced Study in Princeton in 1949. This was the start of Kodaira's 18-year residence in the United States, a fruitful period which saw the full blossoming of his research, much of it in collaboration with Donald Spencer. Kodaira spent many years at Princeton, divided between the Institute and the University, but the years 1961–67 were more unsettled, seeing him successively at Harvard, Johns Hopkins and finally Stanford. In 1967 he returned to a professorship at the University of Tokyo, where he remained until the normal retiring age. From 1975 to 1985 he worked at Gakushuin University, where retirement restrictions did not apply.

During his time at Princeton, Kodaira continued his involvement with harmonic forms, particularly in their application to algebraic geometry, the area which had also provided the motivation for Hodge's work. The 1950s saw a great flowering of complex algebraic geometry, in which the new methods of sheaf theory, originating in France in the hands of Leray, Cartan and Serre, provided a whole new machinery with which to tackle global problems. Sheaf theory fitted with Hodge theory, so it was natural that Kodaira should have been well placed to exploit the new developments. This he did, in a rapid succession of papers written in collaboration with Donald Spencer. These papers altered the face of algebraic geometry, and provided the framework in which Hirzebruch and others of the younger generation were able to make spectacular progress. Large numbers of problems left unsolved or incomplete by the Italian geometers of the classical school were now disposed of in convincing fashion. In particular, a full understanding of the arithmetic genus in higher dimensions emerged, and the Hirzebruch–Riemann–Roch theorem was the culmination of a long story (and the beginning of a new one).

In addition to the general theorems, which combined sheaf theory and harmonic forms to provide the new foundations, Kodaira made two specific and notable contributions. First, in collaboration with Spencer, he developed the theory of deformations of complex structures. This generalised the classical theory of Riemann surfaces in a very satisfactory manner, although the story of higher dimensions is much richer and more complex.

But perhaps his most striking individual achievement was in the general characterisation of projective algebraic varieties in [38]. A few years earlier, Hodge had drawn attention to what he modestly called 'Kähler manifolds of restricted type', ones where the periods of the Kähler form are integral. Hodge had shown that these possessed all the important properties of algebraic varieties that could be deduced from the theory of harmonic forms. For a short while, André Weil had appropriately christened these 'Hodge manifolds'. Unfortunately for Hodge, this terminology dropped out of use as soon as Kodaira proved in [38] that Hodge manifolds are all projective varieties. This striking theorem of Kodaira generalises the classical results, going back to Riemann and Siegel, characterising complex tori which are algebraic. The proof was typical of Kodaira's work, involving a masterly use of the new analytic methods together with a detailed understanding of the relevant geometry.

In 1955–56, having just completed my PhD, I was a visiting member of the Institute at Princeton, and regularly attended Kodaira's lectures at the University, at which the front rows were filled by the new generation of young geometers: Hirzebruch, Serre, Bott and Singer. The front rows were actually rather crowded, since Kodaira's voice rarely rose above a whisper. Fortunately, he wrote very clearly, very slowly and in very large handwriting, so his lecture notes were impeccable. I remember attending his Fields Medal Lecture at Amsterdam in 1954, where he alternated slowly between the microphone at the dais and the blackboard behind him. Not much information was conveyed by this process, but the audience did not mind, because his manner was an attractive combination of extreme shyness, genuine humility and repressed humour. There was a twinkle in his eye which relieved the embarrassment.

The Kodaira–Spencer collaboration was more than just a working relationship. The two men had very different personalities, which were complementary. Kodaira's shyness and reticence were balanced by Spencer's dynamism. In the world of university politics, Spencer was able to exercise his talents on Kodaira's behalf, providing a protective environment in which Kodaira's mathematical talents could flourish. At home, Kodaira was equally protected by his wife Seiko, the sister of the distinguished mathematician S. Iyanaga, and a much more worldly figure than her husband. She even mowed the grass of their Princeton garden.

After Kodaira's return to Japan, he gave lectures and ran seminars which attracted many able students. Kodaira's influence was so pronounced that one could say that he established a new school of Japanese algebraic geometers. It is noteworthy that the other two Japanese mathematicians to receive Fields Medals (Hironaka and Mori) are also algebraic geometers. In Princeton his one outstanding student was W. L. Baily Jr, who became a close friend and eventually learned Japanese.

Kodaira's contributions were widely recognised. The work for which he received the Fields Medal was reported on, in his inimitable style, by Hermann Weyl, and was published in the Proceedings of the Amsterdam Congress. In his home country, Kodaira received the Japan Academy Prize and the Cultural Medal, the highest level of recognition for cultural achievement. In 1988 he was awarded the Wolf Prize. He was a member of the Japan Academy and a foreign associate of the National Academy of Sciences, and he was elected an Honorary Member of the London Mathematical Society on 15 June 1979. In the last decade of his life he suffered from ill health, and this prevented him from putting in an appearance at the International Congress in Kyoto in 1990, even though he was chairman of the organising committee.

Everyone who came into contact with Kodaira realised what a unique and charming individual he was: a deep thinker, a hard-working mathematician, but incredibly quiet and bashful. Even in private conversation you had to listen carefully, and occasionally there would be some subtle humour with a flicker of a smile.

I am grateful to Donald Spencer for letting me see in advance the article he wrote on Kodaira for the *Notices of the American Mathematical Society* (Volume 45, Number 3, March 1998).

References

[1] 'Die Kuratowskische Abbildung und der Hopfsche Erweiterungssatz', *Compositio Math.* **7** (1939) 177–184.

[2] 'On some fundamental theorems in the theory of operators in Hilbert space', *Proc. Imp. Acad. Tokyo* **15** (1939) 207–210.

[3] 'On the theory of almost periodic functions in a group', *Proc. Imp. Acad. Tokyo* **16** (1940) 136–140.

[4] 'Über die Differenzierbarkeit der einparametrigen Untergruppe Liescher Gruppen', *Proc. Imp. Acad. Tokyo* **16** (1940) 165–166.

[5] (with M. Abe) 'Über zusammenhängende kompakte abelsche Gruppen', *Proc. Imp. Acad. Tokyo* **16** (1940) 167–172.

[6] 'Über die Gruppe der messbaren Abbildungen', *Proc. Imp. Acad. Tokyo* **17** (1941) 18–23.
[7] 'Über die Beziehung zwischen den Massen und den Topologien in einer Gruppe', *Proc. Phys.-Math. Soc. Japan* **23** (3) (1941) 67–119.
[8] (with S. Kakutani) 'Normed ring of a locally compact Abelian group', *Proc. Imp. Acad. Tokyo* **19** (1943) 360–365.
[9] 'Über die Harmonischen Tensorfelder in Riemannschen Mannigfaltigkeiten I', *Proc. Imp. Acad. Tokyo* **20** (1944) 186–198.
[10] 'Über die Harmonischen Tensorfelder in Riemannschen Mannigfaltigkeiten II', *Proc. Imp. Acad Tokyo* **20** (1944) 257–261.
[11] 'Über die Rand- und Eigenwertprobleme der linearen elliptischen Differentialgleichungen zweiter Ordnung', *Proc. Imp. Acad. Tokyo* **20** (1944) 262–268.
[12] 'Über die Harmonischen Tensorfelder in Riemannschen Mannigfaltigkeiten III', *Proc. Imp. Acad. Tokyo* **20** (1944) 353–358.
[13] (with S. Kakutani) 'Über das Haarsche Mass in der lokal bikompakten Gruppe', *Proc. Imp. Acad. Tokyo* **20** (1944) 444–450.
[14] 'Relations between harmonic fields in Riemannian manifolds', *Math. Japan* **1** (1948) 6–23.
[15] 'On singular solutions of second order differential operators I, General theory', *Sūgaku (Math.)* **1** (1948) 177–191.
[16] 'On the existence of analytic functions on closed analytic surfaces', *Kodai Math. Sem. Rep.* **2** (1949) 21–26.
[17] 'On singular solutions of second order differential equations II, Applications to special problems', *Sūgaku (Math.)* **2** (1949) 113–139.
[18] 'Harmonic fields in Riemannian manifolds (generalized potential theory)', *Ann. Math.* **50** (2) (1949) 587–665.
[19] 'The eigenvalue problem for ordinary differential equations of the second order and Heisenberg's theory of S-matrices', *Amer. J. Math.* **71** (1949) 921–945.
[20] 'On ordinary differential equations of any even order and the corresponding eigenfunction expansions', *Amer. J. Math.* **72** (1950) 502–544.
[21] (with S. Kakutani) 'A non-separable translation invariant extension of the Lebesgue measure space', *Ann. Math.* **52** (2) (1950) 574–579.
[22] (with G. de Rham) *Harmonic Integrals* (Institute for Advanced Study, Princeton, 1950).
[23] 'Green's forms and meromorphic functions on compact analytic varieties', *Canad. J. Math.* **3** (1951) 108–128.
[24] 'The theorem of Riemann–Roch on compact analytic surfaces', *Amer. J. Math.* **73** (1951) 813–875.
[25] (with W.-L. Chow) 'On analytic surfaces with two independent meromorphic functions', *Proc. Nat. Acad. Sci. U.S.A.* **38** (1952) 319–325.
[26] 'On the theorem of Riemann–Roch for adjoint systems on Kählerian varieties', *Proc. Nat. Acad. Sci. U.S.A.* **38** (1952) 522–527.
[27] 'Arithmetic genera of algebraic varieties', *Proc. Nat. Acad. Sci. U.S.A.* **38** (1952) 527–533.
[28] 'The theorem of Riemann–Roch for adjoint systems on 3-dimensional algebraic varieties', *Ann. Math.* **56** (2) (1952) 298–342.

[29] 'The theorem of Riemann–Roch for adjoint systems on Kählerian varieties', *Contributions to the Theory of Riemann Surfaces*, Ann. Math. Stud. **30** (Princeton University Press, Princeton, NJ, 1953) 247–264.

[30] (with D. C. Spencer) 'On arithmetic genera of algebraic varieties', *Proc. Nat. Acad. Sci. U.S.A.* **39** (1953) 641–649.

[31] 'On cohomology groups of compact analytic varieties with coefficients in some analytic faisceaux', *Proc. Nat. Acad. Sci. U.S.A.* **39** (1953) 865–868.

[32] (with D. C. Spencer) 'Groups of complex line bundles over compact Kähler varieties', *Proc. Nat. Acad. Sci. U.S.A.* **39** (1953) 868–872.

[33] (with D. C. Spencer) 'Divisor class groups on algebraic varieties', *Proc. Nat. Acad. Sci. U.S.A.* **39** (1953) 872–877.

[34] 'On a differential-geometric method in the theory of analytic stacks' *Proc. Nat. Acad. Sci. U.S.A.* **39** (1953) 1268–1273.

[35] (with D. C. Spencer) 'On a theorem of Lefschetz and the lemma of Enriques–Severi–Zariski', *Proc. Nat. Acad. Sci. U.S.A.* **39** (1953) 1273–1278.

[36] 'Some results in the transcendental theory of algebraic varieties', *Ann. Math.* **59** (2) (1954) 86–134.

[37] 'On Kähler varieties of restricted type', *Proc. Nat. Acad. Sci. U.S.A.* **40** (1954) 313–316.

[38] 'On Kähler varieties of restricted type (an intrinsic characterization of algebraic varieties)', *Ann. Math.* **60** (2) (1954) 28–48.

[39] 'Some results in the transcendental theory of algebraic varieties', *Proc. Int. Congr. Math. 1954*, Vol. III (North-Holland, Amsterdam, 1956) 474–480.

[40] 'Characteristic linear systems of complete continuous systems', *Amer. J. Math.* **78** (1956) 716–744.

[41] (with D. C. Spencer) 'On the variation of almost-complex structure', *Algebraic Geometry and Topology. A Symposium in Honor of S. Lefschetz* (Princeton University Press, Princeton, NJ, 1957) 139–150.

[42] (with F. Hirzebruch) 'On the complex projective spaces', *J. Math. Pures Appl.* **36** (9) (1957) 201–216.

[43] (with D. C. Spencer) 'On deformations of complex analytic structures I, II', *Ann. Math.* **67** (2) (1958) 328–466.

[44] (with D. C. Spencer) 'A theorem of completeness for complex analytic fibre spaces', *Acta Math.* **100** (1958) 281–294.

[45] (with L. Nirenberg and D. C. Spencer) 'On the existence of deformations of complex analytic structures', *Ann. Math.* **68** (2) (1958) 450–459.

[46] (with D. C. Spencer) 'On a theorem of completeness of characteristic systems of complete continuous systems', *Amer. J. Math.* **81** (1959) 477–500.

[47] (with D. C. Spencer) 'Existence of complex structure on a differentiable family of deformations of compact manifolds', *Ann. Math.* **70** (2) (1959) 145–166.

[48] 'Differential geometry of a certain class of complex pseudo group structures', *Sūgaku* **11** (1959/1960) 183–187.

[49] (with D. C. Spencer) 'On deformations of complex analytic structures III, Stability theorems for complex structures', *Ann. Math.* **71** (2) (1960) 43–76.

[50] 'On compact complex analytic surfaces I', *Ann. Math.* **71** (2) (1960) 111–152.

[51] 'On deformations of some complex pseudo-group structures', *Ann. Math.* **71**(2) (1960) 224–302.
[52] 'On compact analytic surfaces', *Analytic Functions* (Princeton University Press, Princeton, NJ, 1960) 121–135.
[53] (with D. C. Spencer) 'Multifoliate structures', *Ann. Math.* **74** (2) (1961) 52–100.
[54] 'A theorem of completeness for analytic systems of surfaces with ordinary singularities', *Ann. Math.* **74** (2) (1961) 591–627.
[55] 'A theorem of completeness of characteristic systems for analytic families of compact submanifolds of complex manifolds', *Ann. Math.* **75** (2) (1962) 146–162.
[56] 'On stability of compact submanifolds of complex manifolds', *Amer. J. Math.* **85** (1963) 79–94.
[57] 'On the structure of compact complex analytic surfaces I, II', *Proc. Nat. Acad. Sci. U.S.A.* **50** (1963) 218–221; **51** (1963) 1100–1104.
[58] 'On compact analytic surfaces II, III', *Ann. Math.* **77** (2) (1963) 563–626; **78** (1963) 1–40.
[59] 'On the structure of compact complex analytic surfaces I', *Amer. J. Math.* **86** (1964) 751–798.
[60] 'On characteristic systems of families of surfaces with ordinary singularities in a projective space', *Amer. J. Math.* **87** (1965) 227–256.
[61] 'Complex structures on $S^1 \times S^3$', *Proc. Nat. Acad. Sci. U.S.A.* **55** (1966) 240–243.
[62] 'On the structure of compact complex analytic surfaces II', *Amer. J. Math.* **88** (1966) 682–721.
[63] 'A certain type of irregular algebraic surfaces', *J. Anal. Math.* **19** (1967) 207–215.
[64] 'Pluricanonical systems on algebraic surfaces of general type', *Proc. Nat. Acad. Sci. U.S.A.* **58** (1967) 911–915.
[65] 'Pluricanonical systems on algebraic surfaces of general type', *J. Math. Soc. Japan* **20** (1968) 170–192.
[66] 'On the structure of compact complex analytic surfaces III', *Amer. J. Math.* **90** (1968) 55–83.
[67] 'On the structure of complex analytic surfaces IV', *Amer. J. Math.* **90** (1968) 1048–1066.
[68] 'On homotopy K3 surfaces', *Essays on Topology and Related Topics* (*Mémoires dédiés à Georges de Rham*) (Springer, New York, 1970) 58–69.
[69] (with J. Morrow) *Complex Manifolds* (Holt, Rinehart and Winston, New York, 1971).
[70] 'Holomorphic mappings of polydisks into compact complex manifolds', *J. Differential Geom.* **6** (1971/1972) 33–46.
[71] *Kunihiko Kodaira: Collected Works*, Vols I, II, III (ed. W. L. Baily, Iwami Shoten Publishers, Tokyo; Princeton University Press, Princeton, NJ, 1975).
[72] *Theory of Complex Manifolds*, Vols I, II, III (Iwami Shoten Publishers, Tokyo, 1983).
[73] *Introduction to Complex Analysis* (transl. A. Sevenster, Cambridge University Press, 1984).
[74] *Complex Manifolds and Deformation of Complex Structures*, Grundlehren Math. Wiss. **283** (transl. K. Akao, Springer, New York, 1986).

Kunihiko Kodaira

BIOGRAPHY

Kunihiko KODAIRA: March 16, 1915–July 26, 1997

Place of Birth: Tokyo, Japan

Married: Sei Iyanaga (sister of Shokichi Iyanaga)

Education:

Dai-Ichi High School	1932–1935
Mathematics Institute, University of Tokyo	1935–1938
Physics Institute, University of Tokyo	1938–1941

Academic Career:

Research Fellow, Physics Institute, University of Tokyo	1941
Associate Professor, Mathematical Institute, Tokyo Bun-Rika University	1942
Associate Professor, Physics Institute, University of Tokyo	1942–1955
Member, School of Mathematics, Institute for Advanced Study, NJ	1949–1961
Visiting Professor, Johns Hopkins University	1950–1951
Associate Professor, Princeton University	1952–1955
Professor, Princeton University	1955–1961
Professor, Harvard University	1961–1962
Professor, Johns Hopkins University	1962–1965
Professor, Stanford University	1965–1967
Professor, University of Tokyo	1967–1975
Professor, Gakushuin University	1975–1985
Dean, Faculty of Science, University of Tokyo	1971–1973

Member:

Japan Academy of Science	1965
Gottingen Akademie	1974
National Academy of Sciences	1975

Honorary Member:

National Academy of Liberal Arts	1978
London Mathematical Society	1979

Medals:

Fields Medal	1954
Japan Academy Prize	1957
The Order of Culture Medal	1957
Fujiwara Prize	1975
Wolf Prize	1984

Reprinted from Proc. Nat. Acad. Sci. U.S.A., **40**, *1954*

ON KÄHLER VARIETIES OF RESTRICTED TYPE*

by
K. KODAIRA
Department of Mathematics, Princeton University

1. Introduction

This is a preliminary report of a paper concerning Kähler varieties of restricted type. A compact complex analytic variety V of complex dimension $n \geqq 2$ is called a Kähler variety of *restricted type* or a *Hodge variety* if V carries a Kähler metric $ds^2 = 2\sum g_{\alpha\bar{\beta}}(dz^\alpha d\bar{z}^\beta)$ such that the associated exterior form $\omega = i\sum g_{\alpha\bar{\beta}}dz^\alpha d\bar{z}^\beta$ belongs to the cohomology class of an *integral* 2-cocycle on V. In what follows, such a metric will be called a *Hodge metric.* It is well known that *every nonsingular algebraic variety in a projective space carries a Hodge metric.* Our main theorem asserts that the converse of this proposition holds. Namely, we have:

Main Theorem. *A compact complex analytic variety is (bi-regularly equivalent to) a nonsingular algebraic variety imbedded in a projective space if it carries a Hodge metric.*

2. Proof of the Main Theorem

We give here a rief outline of the proof of our main theorem. Let V be a Hodge variety of complex dimension $n \geqq 2$ and let $\omega = i\sum g_{\alpha\bar{\beta}}dz^\alpha d\bar{z}^\beta$ be the closed 2-form associated with the Hodge metric on V. Then there exists a complex line bundle F_1 over V whose characteristic class $c(F_1)$ contains ω. Letting $F = hF_1$ where h is a *sufficiently large positive integer*, we associate with F a "meromorphic" mapping Φ_F of V into a projective space $\mathfrak{S}$ in the following manner: Take a base $\{\varphi_0, \varphi_1, \ldots, \varphi_\lambda, \ldots, \varphi_d\}$ of the linear space $\Gamma(F)$ consisting of all holomorphic sections of the bundle F, and, letting $\{U_j\}$ be a finite convering of V, denote by $\varphi_{\lambda j}(z)$ the fiber co-ordinate of $\varphi_\lambda(z)$ over U_j, where z is a point in U_j. Then, considering $(\varphi_{0j}(z), \ldots, \varphi_{\lambda j}(z), \ldots, \varphi_{dj}(z))$ as the homogeneous co-ordinates of a

*This work was supported by a research project at Princeton University sponsored by the Office of Ordnance Research, U.S. Army.

point in the projective space $\mathfrak{S}$, we have

$$(\varphi_{0j}(z), \ldots, \varphi_{dj}(z)) = (\varphi_{0k}(z), \ldots, \varphi_{dk}(z))$$

for

$$z \in U_j \cap U_k.$$

Hence we can define a mapping Φ_F of V into $\mathfrak{S}$ by

$$z \to \Phi_F(z) = (\varphi_{0j}(z), \varphi_{1j}(z), \ldots, \varphi_{dj}(z)).$$

In view of a theorem of Chow, every compact analytic subvariety of $\mathfrak{S}$ is an a lgebraic variety. Therefore, it is sufficient to prove that Φ_F is a bi-regular mapping of V into $\mathfrak{S}$.

In order to show that the mapping Φ_F is locally bi-regular at each point p on V, we construct the quadratic transform $\tilde{V} = Q_p(V)$ of V with respect to the center p. The quadratic transformation Q_p maps p into a subvariety $S = Q_p(p)$ of V which is (bi-regularly equivalent to) an $(n-1)$-dimensional projective space. Moreover, the complex line bundle $\{S\}$ over $\tilde{V}$ satisfies the relation[a]

$$\{S\}_S = -\{e\}, \tag{1}$$

where $\{S\}_s$ denotes the restriction of $\{S\}$ to S and $\{e\}$ is the complex line bundle over S determined by the hyperplane e on S. We note that the mapping Q_p is bi-regular between $V - p$ and $\tilde{V} - S$. Now, let $\tilde{F} = Q_p F$ be the complex line bundle over $\tilde{V}$ induced by F in a canonical manner. Since the bundle $\tilde{F}$ is trivial over S, Q_p induces *the isomorphism*

$$\Gamma(F) \cong \Gamma(\tilde{F}). \tag{2}$$

Denoting by K the canonical bundle over V, we infer that the canonical bundle $K(\tilde{V})$ over $\tilde{V}$ is given by

$$K(\tilde{V}) = \tilde{K} + (n-1)\{S\}, \tag{3}$$

where $\tilde{K} = Q_p K$. Letting

$$E_m = \tilde{F} - m\{S\}, \quad m = 0, 1, 2,$$

we therefore have

$$E_m - K(\tilde{V}) = \tilde{F} - \tilde{K} - (n+m-1)\{S\}. \tag{4}$$

By means of (1) it can be shown that the characteristic class $c(-\{S\})$ of $-\{S\}$ contains a closed real (1, 1)-form σ whose restriction σ_S to S is equal to the fundamental form on the projective space S. With the help of this form, σ, we infer

[a]This relation is due to F. Hirzebruch.

from (4) that $c(E_m - K(\tilde{V}))$ contains a closed positive (1, 1)-form for $m = 1$, 2. Consequently, we have [7, Theorem 3]

$$H^1(\tilde{V}, \Omega(E_m)) = 0, \quad \text{for } m = 1, 2, \tag{5}$$

where $\Omega(E_m)$ denotes the sheaf over $\tilde{V}$ of germs of holomorphic sections of E_m. Since the restriction $\tilde{F}_S$ is a trivial bundle over S, we infer from (1) that $(E_m)_S = \{me\}$. Hence we have the exact sequences,

$$0 \to \Omega(E_{m+1}) \to \Omega(E_m) \xrightarrow{r} \Omega(\{me\}) \to 0,$$

where r denotes the restriction map to S. Combined with (5), these sequences yield

$$\begin{aligned} &0 \to \Gamma(E_1) \to \Gamma(\tilde{F}) \xrightarrow{r^*} C_S \to 0, \\ &0 \to \Gamma(E_2) \to \Gamma(E_1) \xrightarrow{r^*} \Gamma(\{e\}) \to 0, \end{aligned} \tag{6}$$

where C_S is the space of constants on S. Now it is easy to derive from (2) and (6) that the mapping Φ_F is bi-regular in a neighborhood U_p of p, while p is an arbitrary point on V. Consequently, *Φ_F is a locally bi-regular mapping of V into $\mathfrak{S}$.*

To prove that $\Phi_F(p) \neq \Phi_F(q)$ for any pair of points p, q, $p \neq q$, on V, we form the quadratic transform $\tilde{V} = Q_p Q_q(V)$ and let

$$\begin{aligned} E' &= \tilde{F} - \{S\}, \\ E'' &= \tilde{F} - \{S\} - \{T\}, \end{aligned}$$

where $\tilde{F} = Q_p Q_q F$, $S = Q_p Q_q(p)$, and $T = Q_p Q_q(q)$. Then we get, as above, the exact sequences

$$\begin{aligned} &0 \to \Gamma(E') \to \Gamma(\tilde{F}) \xrightarrow{r_S^*} C_S \to 0, \\ &0 \to \Gamma(E'') \to \Gamma(E') \xrightarrow{r_T^*} C_T \to 0, \end{aligned} \tag{7}$$

where r_S^* and r_T^* are the restriction maps to S and T, respectively. On the other hand, we have the isomorphism $\Gamma(\tilde{F}) \cong \Gamma(\tilde{F})$. From these results it follows that $\Phi_F(p) \neq \Phi_F(q)$. Thus we conclude that *$z \to \Phi_F(z)$ is a one-to-one bi-regular mapping of V into $\mathfrak{S}$.* This completes the proof of our main theorem.

3. Applications

Now we mention several applications of our main theorem. First, let V be a compact complex analytic manifold with a Hermitian metric $\sum g_{\alpha\bar{\beta}}(dz^\alpha d\bar{z}^\beta)$. Letting

$$R_{\alpha\bar{\beta}} = -\frac{\partial^2 \log g}{\partial z^\alpha \partial \bar{z}^\beta}, \quad \text{where } g = \det(g_{\alpha\bar{\beta}}),$$

be the Ricci curvature, we infer readily that the metric

$$ds^2 = -[\text{or}+] \sum \left(\frac{R_{\alpha\bar{\beta}}}{\pi}\right)(dz^\alpha d\bar{z}^\beta)$$

is a Hodge metric if it is positive definite. Hence we obtain the following theorem:

Theorem 1. *Every compact Hermitian manifold with negative (or positive) definite Ricci curvature is a nonsingular algebraic variety imbedded in a projective space.*

Second, let $\mathfrak{B}$ be a bounded domain on the space of n complex variables and let Δ be a discontinuous group of analytic automorphisms of $\mathfrak{B}$ such that $\mathfrak{B}$ is compact with respect to Δ. Moreover, assume that no element of Δ except the identity has a fixed point in $\mathfrak{B}$. Then the factor space $V = \mathfrak{B}/\Delta$ is a compact analytic manifold. Now it can be shown that the Bergmann metric,

$$ds^2 = \frac{1}{\pi}\sum \frac{\partial^2 \log K(z,\bar{z})}{\partial z^\alpha \partial \bar{z}^\beta}(dz^\alpha d\bar{z}^\beta),$$

of $\mathfrak{B}$ induces on $V = \mathfrak{B}/\Delta$ a Hodge metric, where $K(z,\bar{z})$ denotes Bergmanns kernel function for the domain $\mathfrak{B}$. Consequently, we obtain the following theorem:

Theorem 2.[b] *The compact analytic manifold $V = \mathfrak{B}/\Delta$ is a nonsingular algebraic variety imbedded in a projective space.*

This theorem implies the existence of "sufficiently many" automorphic functions on $\mathfrak{B}$ with respect to Δ.

Finally, we consider projective bundles over algebraic varieties. Let B be an analytic fibre bundle over a Hodge variety V whose fibre is a complex projective space and whose structure group is the group of projective transformations. Then it can be shown that B is also a Hodge variety. Hence we obtain the following theorem:

Theorem 3.[c] *Every projective bundle over a nonsingular algebraic variety in a projective space is also a nonsingular algebraic variety imbedded in a projective space.*

References

[1] K. Kodaira, On Kähler Varieties of Restricted Type (an instrinsic characterization of algebraic uarieties), *Ann. Math.* **60** (1954) 28–48.

[2] W. V. D. Hodge, A Special Type of Kähler Manifolds, *Proc. London Math. Soc.* **1** (1951) 104–117.

[3] A. Weil, On Picard Varieties, *Am. J. Math.* **74** (1952) 865–894.

[4] K. Kodaira and D. C. Spencer, Groups of Complex Line Bundles over Compact Kähler Varieties, *Proc. Nat. Acad. Sci. U.S.A.* **39** (1953) 868–872.

[5] W. L. Chow, On Compact Complex Analytic Varieties, *Am. J. Math.* **71** (1949) 893–914.

[b] A. Borel has communicated to the author the fact that a similar result has been obtained recently by J.-P. Serre.

[c] The author wishes to express his sincere thanks to A. Borel for valuable suggestions concerning these applications.

[6] See F. Hirzebruch, Über vierdimensionale Riemannsche Flächen mehrdeutiger analytischer Funktionen von zwei komplexen Veränderlichen, *Math. Ann.* **126** (1953) 1–22.

[7] K. Kodaira, On a Differential-geometric Method in the Theory of Analytic Stacks, *Proc. Nat. Acad. Sci. U.S.A.* **39** (1953) 1268–1273.

Reprinted from Proc. Int. Congr. Math., 1958

THE WORK OF K. F. ROTH

by
H. DAVENPORT

On the three previous occasions on which Fields Medals have been presented, the addresses on the achievements of the recipients have been given either by the Chairman or by a member of the awarding Committee. On this occasion, Professor Siegel was to have spoken about the work of Dr Roth, but as he is unfortunately unable to be present the duty has devolved on me. It is a pleasant duty, in that it requires me to pay tribute to the work of a colleague and friend.

Dr Roth's greatest achievement is by now well known to mathematicians generally; it is his solution, in 1955, of the principal problem concerning approximation to algebraic numbers by rational numbers.

If α is any irrational number, whether algebraic or not, there are infinitely many rational numbers p/q such that

$$\left|\frac{p}{q} - \alpha\right| < \frac{1}{q^2};$$

for example the convergents to the continued fraction for α. It is therefore natural to attempt to characterize irrational numbers in terms of the exponents μ for which there are infinitely many approximations satisfying

$$\left|\frac{p}{q} - \alpha\right| < \frac{1}{q^\mu}.$$

For convenience, I denote by $\bar{\mu} = \bar{\mu}(\alpha)$ the upper bound of such exponents μ. Obviously $\overline{\mu}(\alpha) \geqslant 2$.

The problem is: what can be said about the value of $\overline{\mu}(\alpha)$ when α is *algebraic*? In 1844 Liouville showed, in a very simple manner, that $\overline{\mu}(\alpha) \leqslant n$ if α is an algebraic number of degree n. In fact, if α is a root of the irreducible equation $f(x) = 0$, where $f(x)$ has integral coefficients (not all 0), then on the one hand

$$\left|f\left(\frac{p}{q}\right)\right| \geqslant \frac{1}{q^n},$$

and on the other hand it is easily seen that

$$\left|f\left(\frac{p}{q}\right)\right| = \left|f\left(\frac{p}{q} - \alpha + \alpha\right)\right| < c\left|\frac{p}{q} - \alpha\right|,$$

where c depends only on α. Comparison of these inequalities leads to the result. If $n = 2$ we get $\bar{\mu}(\alpha) = 2$; thus quadratic irrationals are about as badly approximable as any irrational number can be.

There are simple considerations which suggest that Liouville's result is far from being best possible. But it was not until 1908 that this was proved; in that year the Norwegian mathematician Axel Thue showed that $\bar{\mu}(\alpha) \leqslant \frac{1}{2}n + 1$. In 1921 Siegel made further very substantial progress, and obtained $\bar{\mu}(\alpha) < 2\sqrt{n}$ approximately, the precise result being a little better than this. In 1947 Dyson improved Siegel's inequality to $\bar{\mu}(\alpha) \leqslant \sqrt{(2n)}$.

In all this work, extending over a period of 40 years, the basic idea was the use of polynomials in two variables. Suppose $f(x_1, x_2)$ is a polynomial with integral coefficients, of degree r_1 in x_1 and r_2 in x_2, and suppose p_1/q_1 and p_2/q_2 are two rational approximations to α. Then

$$\left|f\left(\frac{p_1}{q_1}, \frac{p_2}{q_2}\right)\right| \geqslant \frac{1}{q_1^{r_1} q_2^{r_2}},$$

provided of course that

$$f\left(\frac{p_1}{q_1}, \frac{p_2}{q_2}\right) \neq 0.$$

Suppose further that $f(\alpha, \alpha) = 0$ and that the Taylor expansion of $f(x_1, x_2)$ in powers of $x_1 - \alpha$ and $x_2 - \alpha$ has all its 'early' coefficients zero, a condition which can be made precise in various ways. Then one can obtain an upper bound for

$$\left|f\left(\frac{p_1}{q_1}, \frac{p_2}{q_2}\right)\right|$$

in terms of

$$\left|\frac{p_1}{q_1} - \alpha\right| \quad \text{and} \quad \left|\frac{p_2}{q_2} - \alpha\right|,$$

and the principle is to combine this with the previous lower bound in such a way as to establish that p_1/q_1 and p_2/q_2 cannot both be very good approximations to α. Finally, p_1/q_1 and p_2/q_2 are chosen in a suitable way from the infinite sequence of approximations.

The proof of the existence of a polynomial $f(x_1, x_2)$ with all the desired properties is a difficult matter, and the condition that

$$f\left(\frac{p_1}{q_1}, \frac{p_2}{q_2}\right) \neq 0$$

is particularly troublesome. No explicit construction for such a polynomial has yet been found. During the course of their work, the four mathematicians I have mentioned developed methods of great subtlety and interest, and other important ideas which are relevant to the problem were contributed by Gelfond, Mahler and Schneider.

In 1955 Roth finally solved the problem: he proved that $\bar{\mu}(\alpha) = 2$ for any algebraic number α. The achievement is one that speaks for itself; it closes a chapter, and a new chapter will now be opened. Roth's theorem settles a question which is both of a fundamental nature and of extreme difficulty. It will stand as a landmark in mathematics for as long as mathematics is cultivated.

It is not my intention to describe or analyse Dr Roth's proof, particularly as he will be speaking about it himself. My own impression of his proof is that it is a structure, inevitably of some complexity, every part of which fits into its proper place and carries its proper share of the total load. As you have probably anticipated from my description of previous work, it uses polynomials in an arbitrarily large number of variables, instead of in two variables. It had indeed long been realized that this would be necessary, but the difficulties in the way had appeared to be quite insuperable.

I turn now to another achievement of Dr Roth, which seems to me to be also of the first magnitude, though the problem to which it relates is perhaps of less universal interest. Let

$$n_1, n_2, n_3, \ldots$$

be a sequence of natural numbers, and suppose that no three of the numbers are in arithmetic progression; in other words,

$$n_i + n_j \neq 2n_k$$

unless $i = j = k$. It was conjectured by Erdős and Turán in 1935 (though the conjecture is believed to be older) that such a sequence must have zero density, that is, the number $N(x)$ of terms not exceeding x must satisfy

$$\frac{N(x)}{x} \to 0 \quad \text{as} \quad x \to \infty.$$

This problem was the subject of several interesting and ingenious papers, but it resisted all attempts at solution for a long time. The conjecture was proved by Dr Roth in 1952, and his proof is one of great interest and originality. He first considers a set of numbers $n_1, n_2, \ldots, n_\tau \leqslant x$ with the property in question, for which τ is a maximum, and proves that such a set, if dense, would have to have considerable regularity of distribution. This regularity is of two kinds: regularity of distribution in position and regularity of distribution among the residue-classes to any modulus. Then he applies the analytic method developed by Hardy and Littlewood for problems of an additive character, and the features of regularity prove to be just sufficient to give the estimates necessary for the method to succeed. The final conclusion is that $N(x) < cx/(\log \log x)$, where c is an absolute constant. I can recall no other instance, of comparable importance, in which the Hardy–Littlewood method has been used to elucidate the additive properties of an unknown sequence, instead of a special sequence such as the kth powers or the primes.

There are other achievements of Dr Roth which stand out by their originality and novelty, and it is with reluctance that I pass over them. Those I have outlined already will, I am sure, satisfy you that the recognition which has come to him is well merited.

The Duchess, in *Alice in Wonderland*, said that there is a moral in everything if only you can find it. It is not difficult to find a moral in Dr Roth's work. It is that the great unsolved problems of mathematics may still yield to direct attack, however difficult and forbidding they appear to be, and however much effort has already been spent on them.

Klaus F. Roth

KLAUS F. ROTH

Date of birth: 29 October 1925

Place of birth: Breslau

Son of: Late Dr. Franz Roth and Mathilde Roth (née Liebrecht)

Married: Melek Khairy, 1955

Education:

St. Paul's School (1939–43)
Peterhouse, Cambridge (B.A., 1945)
University College London (M.Sc., 1948; Ph.D., 1950)

Academic career:

Assistant Master, Gordonstoun School, 1945–46;
Member of Dept. of Mathematics, University College London, 1948–66 (Assistant Lecturer, Lecturer 1948–56, Reader 1956–61, Professor 1961–66);
Professor of Pure Mathematics (Theory of Numbers), Imperial College, London, 1966–88;
Hon. Research Fellow in the Dept. of Mathematics, University College London, 1996–present.

Visiting appointments:

Visiting Lecturer, Mass. Inst. of Techn., 1956–57;
Visiting Professor, Mass. Inst. of Techn., 1965–66;
Visiting Professor, University of Colorado, Spring Semester 1969;
Visiting Professor, Imperial College, London, 1988–96.

Medals:

Fields Medal, 1958;
De Morgan Medal (London Math. Soc., 1983);
Sylvester Medal (Royal Soc., 1991).

Memberships:

Fellow Royal Soc., 1960;
Foreign Hon. Mem. Amer. Acad. of Arts and Scis., 1966;
Fellow Univ. College London, 1979;

Hon. Fellow Peterhouse, Cambridge, 1989;
Hon. Fellow Royal Soc. of Edinburgh, 1993.

Address:
24 Burnsall Street, London SW3 3ST; and "Colbost", 16A Drummond Road, Inverness, IV2 4NB, Scotland.

Reprinted from Proc. Int. Congr. Math., 1958

RATIONAL APPROXIMATIONS TO ALGEBRAIC NUMBERS

by

K. F. ROTH

1. Let α be any algebraic irrational number and suppose there are infinitely many rational approximations h/q to α such that

$$\left|\alpha - \frac{h}{q}\right| < \frac{1}{q^\kappa}. \tag{1}$$

I proved in 1955 that this implies $\kappa \leqslant 2$. In this talk I shall try to outline the proof, and to say a few words about some of the possible extensions, and about the limitations of the method.

It is easily seen that there is no loss of generality in supposing that α is an algebraic integer. Accordingly, we shall suppose that α is a root of the polynomial

$$f(x) = x^n + a_1 x^{n-1} + \cdots + a_n \tag{2}$$

with integral coefficients and highest coefficient 1.

2. Previous work on the problem entailed the use of polynomials in two variables. It has long been recognized that further progress would demand the use of polynomials in more than two variables, and that polynomials in a large number of variables would have to be used to obtain the full result. It is not difficult to formulate properties which a polynomial should have in order to be useful for our purpose.

Suppose that $h_1/q_1, \ldots, h_m/q_m$ are rational approximations to α, all satisfying (1). Let $Q(x_1, \ldots, x_m)$ denote a polynomial with integral coefficients, of degree at most r_j in x_j for each j. Then

$$\left|Q\left(\frac{h_1}{q_1}, \ldots, \frac{h_m}{q_m}\right)\right| \geqslant \frac{1}{P}, \tag{3}$$

where $P = q_1^{r_1} \ldots q_m^{r_m}$, provided, of course, that

$$Q\left(\frac{h_1}{q_1}, \ldots, \frac{h_m}{q_m}\right) \neq 0.$$

Let the Taylor expansion of $Q(h_1/q_1, \ldots, h_m/q_m)$ in powers of $(h_1/q_1) - \alpha, \ldots,$ $(h_m/q_m) - \alpha$ be

$$\sum_{i_1} \cdots \sum_{i_m} Q_{i_1, \ldots, i_m} \left(\frac{h_1}{q_1} - \alpha\right)^{i_1} \cdots \left(\frac{h_m}{q_m} - \alpha\right)^{i_m}.$$

Suppose now that Q has the properties:

A:
$$\sum_{i_1} \cdots \sum_{i_m} |Q_{i_1}, \ldots, i_m| < P^{\Delta},$$

where Δ is small;

B:
$$Q_{i_1,\ldots,i_m} = 0 \text{ for all } i_1, \ldots, i_m \text{ satisfying}$$

$$q_1^{i_1} \cdots q_m^{i_m} \leqslant P^{\phi} \quad (\phi > 0).$$

Then each term in this Taylor expansion with a non-zero coefficient has

$$\left|\frac{h_1}{q_1} - \alpha\right|^{i_1} \cdots \left|\frac{h_m}{q_m} - \alpha\right|^{i_m} < \frac{1}{(q_1^{i_1} \cdots q_m^{i_m})^{\kappa}} < \frac{1}{P^{\kappa\phi}}.$$

Hence
$$\left|Q\left(\tfrac{h_1}{q_1}, \ldots, \tfrac{h_m}{q_m}\right)\right| < \tfrac{1}{P^{\kappa\phi-\Delta}}. \tag{4}$$

Comparison of (3) and (4) yields

$$\kappa < \frac{1+\Delta}{\phi}. \tag{5}$$

We must bear in mind that to obtain (5) we used, in addition to A and B, the condition

C:
$$Q\left(\tfrac{h_1}{q_1}, \ldots, \tfrac{h_m}{q_m}\right) \neq 0.$$

To prove our theorem we shall establish the existence of a polynomial Q satisfying the conditions A, B, C with ϕ near to $\frac{1}{2}$. For this purpose it will be necessary to take m large and to choose the approximations $h_1/q_1, \ldots, h_m/q_m$ suitably. Only after choosing these approximations do we choose the polynomial Q, and the latter choice depends on the former.

[With $m = 2$ it is only possible to satisfy condition B with ϕ of the order of $n^{-\frac{1}{2}}$ (where n is the degree of α), and this leads to an estimate of the type $\kappa < cn^{\frac{1}{2}}$.]

3. The logical structure of the proof is as follows. We suppose $\kappa > 2$; m is chosen sufficiently large and is fixed throughout. A small positive number $\delta (< 1/m)$ will be fixed until the end of the proof, when we let $\delta \to 0$. We denote by Δ any function of δ and m such that $\Delta \to 0$ as $\delta \to 0$ for fixed m.

We begin by choosing $h_1/q_1, \ldots, h_m/q_m$ [with $(h_j, q_j) = 1$] from the assumed infinite sequence of approximations to α (satisfying (1)) by first taking q_1 sufficiently large (in terms of m, δ, α), then taking q_2 sufficiently large in terms of q_1, and so on. It will in fact suffice if

$$\frac{\log q_j}{\log q_{j-1}} > \delta^{-1} \quad (j = 2, \ldots, m).$$

Then we choose integers $r_1, \ldots, r_m$ which are sufficiently large in relation to $q_1, \ldots, q_m$ and which satisfy

$$q_1^{r_1} \leqslant q_j^{r_j} < q_1^{r_1(1+\frac{1}{10}\delta)}. \tag{6}$$

This presents no difficulty. We note that (6) implies

$$q_1^{mr_1} \leqslant P < q_1^{mr_1(1+\Delta)}. \tag{7}$$

Condition B now takes the form that the Taylor coefficients $Q_{i_1,\ldots,i_m}$ vanish for all $i_1, \ldots, i_m$ satisfying

$$\frac{i_1}{r_1} + \cdots + \frac{i_m}{r_m} < m\phi + \Delta. \tag{8}$$

4. We shall first outline a proof of the existence of a polynomial Q^* satisfying conditions A and B only (with ϕ near to $\frac{1}{2}$). This proof, due essentially to Siegel, is based on the use of Dirichlet's compartment principle. The question of satisfying C as well, which gives rise to the principal difficulty, is deferred until later.

We put $B_1 = q_1^{\delta r_1}$ and consider all polynomials $W(x_1, \ldots, x_m)$ of degree at most r_j in x_j, having positive integral coefficients, each less than B_1. We try to find two such polynomials W', W'' such that their derivatives of order $i_1, \ldots, i_m$ are equal when $x_1 = \ldots = x_m = \alpha$, for all $i_1, \ldots, i_m$ satisfying (8). Since any such derivative is of the form

$$A_0 + A_1\alpha + \cdots + A_{n-1}\alpha^{n-1},$$

where $A_0, \ldots, A_{n-1}$ are integers, one can estimate the number of possibilities for a derivative for given $i_1, \ldots, i_m$. This number can be shown to be less than $B_1^{n(1+3\delta)}$. The number of polynomials W is about B_1^r, where $r = (r_1 + 1) \ldots (r_m + 1)$. Thus the number of polynomials W will exceed the number of possible distinct sets of derivatives provided that the number of sets of $i_1, \ldots, i_m$ satisfying (8), and with no i exceeding the corresponding r, is less than about $r/\{n(1 + 3\delta)\}$. The number of integer points $(i_1, \ldots, i_m)$ in the region defined by the above conditions can be shown to be less than $\frac{2}{3}r/n$ if ϕ is chosen so that

$$m\phi + \Delta = \frac{1}{2}m - 3nm^{\frac{1}{2}}. \tag{9}$$

The polynomial $Q^* = W' - W''$ satisfies B, by its definition, and can be shown by a process of simple estimation to satisfy A. Furthermore, on letting $\delta \to 0$ we would obtain

$$\phi = \frac{1}{2} - 3nm^{-\frac{1}{2}},$$

so that ϕ could be assumed to be sufficiently near to $\frac{1}{2}$, since m is large. [It is in order to be able to choose ϕ near to $\frac{1}{2}$, that we must work with polynomials in many variables.]

5. To find a polynomial Q which satisfies A, B and C as well, we seek a derivative

$$Q = \frac{1}{j_1!}\left(\frac{\partial}{\partial x_1}\right)^{j_1} \cdots \frac{1}{j_m!}\left(\frac{\partial}{\partial x_m}\right)^{j_m} Q^*,$$

of not too high an order, of the polynomial Q^* just considered. We want Q not to vanish at $(h_1/q_1, \ldots, h_m/q_m)$. The 'order' is measured by $j_1/r_1 + \cdots + j_m/r_m$. The replacement of Q^* by Q will involve a weakening of condition B, but provided the order in question is a Δ, this will make no difference on letting $\delta \to 0$. There will also be an effect on condition A, but this turns out to be insignificant. Condition C is the essential requirement now.

The existence of such a derivative, whose order is a Δ, is not easy to establish. One would in fact expect this to cause difficulty, as the choice of Q^* was designed to make Q^* very small at $(h_1/q_1, \ldots, h_m/q_m)$.

At this stage it is convenient to introduce the notion of the index of a polynomial at a point. We define the index of a polynomial at the point $(\alpha_1, \ldots, \alpha_m)$ relative to positive parameters $r_1, \ldots, r_m$ to be the minimal order of derivative (measured as above) which does not vanish at the point $(\alpha_1, \ldots, \alpha_m)$. In this language, we need to show that the index of Q^* at $(h_1/q_1, \ldots, h_m/q_m)$ is a Δ.

For polynomials in two variables, two quite different lines of reasoning have been used to obtain upper bounds for the index of Q^* at $(h_1/q_1, \ldots, h_m/q_m)$. The first, due to Siegel, is algebraic in nature. It is based on the principle that, under certain conditions, the sum of the indices of a polynomial at a finite number of points (not restricted to be rational) is bounded in terms of its degrees in the various variables. Since Q^* satisfies condition B (with an appropriate ϕ), it has an almost maximal index (in a certain sense) at the point $(\alpha, \ldots, \alpha)$ and at the points obtained by replacing α by its conjugates; and it can be deduced that the index of Q^* is small at any other point. I have been unable to extend this method to polynomials in more than 2 variables.

The second method, due to Schneider, is arithmetic in nature. It is based on the principle that, under certain conditions, the index of a polynomial at a rational point is bounded in terms of the magnitude of its coefficients. Since the coefficients of Q^* are not too large, this leads to a result of the desired kind.

My treatment is based on Schneider's approach and enables me to prove the following lemma.

Principal lemma. Let $0 < \delta < m^{-1}$, *and let* $r_1, \ldots, r_m$ *be positive integers satisfying*

$$r_m > 10\delta^{-1}, \quad \frac{r_{j-1}}{r_j} > \delta^{-1} \quad (j = 2, \ldots, m).$$

Let $q_1, \ldots, q_m$ *be positive integers satisfying*

$$q_1 > c = c(m, \delta), \quad q_j^{r_j} \geqslant q_1^{r_1}.$$

Consider any polynomial R, not identically zero, with integral coefficients of absolute value at most $q_1^{\delta r_1}$ and of degree at most r_j in x_j. Then

$$\text{index } R < 10^m \delta^{(\frac{1}{2})^m},$$

where the index is taken at a point $(h_1/q_1, \ldots, h_m/q_m)$ relative to $r_1, \ldots, r_m$; the h's being integers relatively prime to the corresponding q's.

This suffices for the purpose of finding Q; the hypotheses of the lemma are satisfied when $R = Q^*$, and the lemma shows that the index of Q^* at $(h_1/q_1, \ldots, h_m/q_m)$ is a Δ, as required.

6. The proof of the lemma is self-contained, as indeed it must be, for it uses induction on the number of variables, whereas in the main proof the number m is fixed. Furthermore, the lemma has to be generalized before the induction can be set up.

We consider the class of all polynomials $R(x_1, \ldots, x_m)$ with integral coefficients, each coefficient being numerically at most B, say, and of degree at most r_j in x_j. We obtain, under certain conditions, an upper bound for the indices of polynomials of this class at a point $(h_1/q_1, \ldots, h_m/q_m)$ relative to $r_1, \ldots, r_m$. During the course of the proof, which is by induction on m, it is necessary to consider various different sets of values of the parameters involved. The final estimate is of the type required to establish the lemma.

The case $m = 1$ is simple. Suppose the coefficients of a polynomial $R(x_1)$ are numerically less than B. If θ_1 is the index of R at h_1/q_1 relative to r_1, the polynomial $R(x_1)$ is divisible by

$$\left(x_1 - \frac{h_1}{q_1}\right)^{\theta_1 r_1}.$$

It follows from Gauss's theorem on the factorization of polynomials with integral coefficients into polynomials with rational coefficients, that

$$R(x_1) = (q_1 x_1 - h_1)^{\theta_1 r_1} R^*(x_1),$$

where $R^*(x_1)$ is a polynomial with integral coefficients. Hence the coefficient of the highest term in R^* is an integral multiple of $q_1^{\theta_1 r_1}$, so that

$$q_1^{\theta_1 r_1} \leqslant B, \quad \theta_1 \leqslant \frac{\log B}{r_1 \log q_1}.$$

This gives an upper bound of the required type for $m = 1$.

Now suppose that upper bounds of this kind have been obtained for $m = 1, 2, \ldots, p-1$, where $p \geqslant 2$. We wish to deduce an upper bound for the indices for classes of polynomials in p variables.

For any given polynomial $R(x_1, \ldots, x_p)$ we consider all representations of the form

$$R = \phi_0(x_p)\psi_0(x_1, \ldots, x_{p-1}) + \cdots + \phi_{l-1}(x_p)\psi_{l-1}(x_1, \ldots, x_{p-1}), \tag{10}$$

where the ϕ_ν and ψ_ν are polynomials with rational coefficients, subject to the condition that the ϕ_ν and ψ_ν are of degree at most r_j in x_j. Such a representation is possible, e.g. with $l-1=r_p$ and $\phi_\nu(x_p)=x_p^\nu$. From all such representations we select one for which l is least.

In this representation the polynomials ϕ form a linearly independent set, and so do the ψ's. Thus the Wronskian $W(x_p)$ of the ϕ's is not identically zero, and the same is true of a certain generalized Wronskian $G(x_1,\ldots,x_{p-1})$ of the ψ's. From (10) and the rule for multiplication of determinants by rows, it follows that

$$G(x_1,\ldots,x_{p-1})\,W(x_p)=F(x_1,\ldots,x_p) \tag{11}$$

is a certain determinant whose elements are all of the form

$$R_{j_1,\ldots,j_p}(x_1,\ldots,x_p).$$

Since G and W have rational coefficients, there is an equivalent factorization of F in the form

$$F(x_1,\ldots,x_p)=U(x_1,\ldots,x_{p-1})\,V(x_p), \tag{12}$$

where U and V have integral coefficients.

If the coefficients of R are assumed to be numerically less than B, this will imply an upper bound for the coefficients of F; and this, in turn, will imply upper bounds for the coefficients of U and V. The induction hypothesis then gives us upper bounds for the indices of U and V at the points $(h_1/q_1,\ldots,h_{p-1}/q_{p-1})$ and h_p/q_p respectively; and by a multiplicative property of indices, (12) then yields an upper bound for the index of F at $(h_1/q_1,\ldots,h_p/q_p)$.

On the other hand, F is obtained from R by the operations of differentiation, addition and multiplication; and by using some simple properties of indices relating to these operations, one obtains a lower bound for the index of F in terms of the index of R. Thus the upper bound for the index of F leads to an upper bound for the index of R.

In this way it is possible to set up the induction on m, although the details are somewhat more complicated than they are made to appear above.

This concludes the outline of the proof of our theorem. We note that the proof of the existence of a polynomial Q satisfying conditions A, B, C of §2 is very indirect, and it would be of considerable interest if such a polynomial could be obtained by a direct construction.

7. The theorem can be generalized and extended in various ways. For example, instead of considering rational approximations to the algebraic number α, we may consider approximations to α by algebraic numbers β (a) lying in a fixed algebraic field, or (b) of fixed degree. In each case the accuracy of the approximation is measured in terms of $H(\beta)$, the maximum absolute value of the rational integral coefficients in the primitive irreducible equation satisfied by β.

The results already found by Siegel can be improved in both cases. In case (a) the best possible result has been obtained.[†] In case (b), Siegel's result is significant only if the degree of β is not too large compared to the degree of α, and I do not know how to obtain an improvement which does not suffer from a similar limitation.

The theorem can also be extended to p-adic and g-adic number fields, and this has been done by Ridout and Mahler respectively.

Various deductions can be made from the theorem. For example, for a given α and $\kappa > 2$, it is possible to estimate the number of solutions of (1), as has been done by Davenport and myself. This leads to estimates for the number of solutions of certain Diophantine equations.

The method is subject to a severe limitation, however, due to the role played by the selected approximations $h_1/q_1, \ldots, h_m/q_m$. One cannot answer questions of the following type:

(i) Can one give, in terms of α and κ (if $\kappa > 2$), an upper bound for the greatest denominator q among the finite number of solutions h/q of (1)?

(ii) Can one prove that

$$\left|\alpha - \frac{h}{q}\right| < q^{-(2+f(q))}$$

has only a finite number of solutions h/q for some explicit function $f(q)$ such that $f(q) \to 0$ as $q \to \infty$?

Liouville's result

$$\left|\alpha - \frac{h}{q}\right| > c(\alpha)q^{-n}$$

remains the only known result of its type for which an explicit value of the constant can be given.

Our method can only throw light on such questions if some assumption is made concerning the 'gaps' between the convergents to α. It would appear that a completely new idea is needed to obtain any information concerning problems of the above type.

One outstanding problem is to obtain a theorem analogous to ours concerning simultaneous approximations to two or more algebraic numbers by rationals of the same denominator. In the case of such simultaneous approximation to two algebraic numbers α_1, α_2 (subject to a suitable independence condition), one would expect the inequalities

$$\left|\alpha_1 - \frac{h_1}{q}\right| < q^{-\kappa}, \quad \left|\alpha_2 - \frac{h_2}{q}\right| < q^{-\kappa}$$

to have at most a finite number of solutions for any $\kappa > \frac{3}{2}$. But practically nothing is known in this connection.

A complete solution of the problem of simultaneous approximations could lead to the complete solution of many others, such as, for example, case (b) of the first problem mentioned in this section.

[†]See W. J. LeVeque, *Topics in Number Theory*, Addison Wesley, 1956.

Reprinted from Proc. Int. Congr. Math., 1958

THE WORK OF R. THOM

by
H. HOPF[1]

René Thom was born in Montbéliard in 1923. He studied at the Ecole Normale Supérieure in Paris from 1943 to 1946 and then went to the University of Strasbourg where he is now "Professeur sans Chaire". In Strasbourg, he prepared his thesis "Espaces fibrés en sphères et carrés de Steenrod". He presented it in Paris in 1951 in order to get the degree of Docteur ès Sciences. In 1954 his paper "Quelques propriétés globales des variétés différentiables" appeared in Comm. Math. Helvet. 28. It was prepared from his thesis. There, one finds the foundations of the theory of cobordism. Today, four years later, one can say that, for a long time, only few events have so strongly influenced Topology and, through topology, other branches of mathematics as the advent of this work.

I would like to try to briefly describe here the basic idea and the grounds of the theory of cobordism.

One considers k-dimensional compact, oriented manifolds without boundary $A, B, \ldots$; it is not necessarily supposed that they are connected (a 1-dimensional A is a finite union of mutually disjoint closed lines, a 2-dimensional A is a finite union of closed surfaces which do not intersect, etc,...) $A + B$ denotes the disjoint union of copies of A and B, $-A$ denotes the manifold which is homeomorphic to A but has the opposite orientation, $A - B$ denotes the sum of A and $(-B)$. $A.B$ is, for manifolds of arbitrary dimensions, the Cartesian product. Besides, one considers, for each k, $(k+1)$-dimensional compact oriented manifolds U, V with boundary whose boundaries are k-dimensional manifolds. If, for a Ω k-dimensional A, there exists a $(k+1)$-dimensional U whose boundary is a copy of A, one says "A is bordant" and one writes: $A \sim 0$; when $A - B$ is bordant, then one says "A and B are cobordant" and one writes: $A \sim B$. The relation $\sim$ defines the "cobordism classes" of the manifolds $A, B, \ldots$ This classification is compatible with addition; the classes of each dimension k form an abelian group Ω with respect to addition; the 0-Element

[1]English translation from German by Stefan Weinzierl, with revisions by C. Kopper and V. Rivasseau.

is the class of the bordant A (i.e. which are boundaries). This classification is also compatible with the Cartesian multiplication; so the groups Ω^k with $k = 0, 1, 2, \ldots$ are fused to become the Ring

$$\Omega = \Omega^0 + \Omega^1 + \cdots + \Omega^k + \cdots$$

This is the Thom algebra.

Here one may ask: "Is all this not totally trivial? Can such a primitive definition be the origin of new and interesting insights?" But the same question was raised in 1955 when Hurewicz defined the Homotopy groups, and when the consequences of that primitive definition were realized many a mathematician had to admit "that is something that I could not have discovered. It would have been too easy for me". Indeed it required an ingenious mathematician like Hurewicz, not to be afraid of the apparently simple, but to see how deeply these simple concepts intrude upon the heart of the problem. I think Hurewicz was one of the greatest geometers of our time — we feel painfully for his early death during this conference — and personally I can hardly make a greater compliment to a mathematician than to place him near Hurewicz.

Then, Thom's discovery of cobordism and the algebra Ω reminds me again and again — in spite of the different contexts — of the discovery of the Homotopy groups by Hurewicz.

One of the, by no means trivial, insights which Thom had obviously from the beginning was that the notion of cobordism is particularly suited for the study of differentiable manifolds. We will restrict from now on to those. A differentiable manifolds can always be equipped with a Riemannian Metric, that is a metric that is euclidean at the infinitesimal level; this is how the orthogonal groups enter the game. Thom connects the orthogonal groups $\mathrm{SO}(n)$ with certain topological complexes $M(\mathrm{SO}(n))$; with the help of some highly non-trivial tools, developed only in recent years (Eilenberg, Maclane, Cartan, Serre) he shows that the Groups Ω^k are isomorphic to certain homotopy Groups of the spaces $M(\mathrm{SO}(n))$ and he succeeds in many cases to calculate these Groups or at least to make rather precise statements about them. The main results are the following:

$$\Omega^o = \mathbb{Z},\ \Omega^1 = \Omega^2 = \Omega^3 = 0,\ \Omega^4 = \mathbb{Z},\ \Omega^5 = \mathbb{Z}_2,\ \Omega^6 = \Omega^7 = 0\,,$$

where $\mathbb{Z}$ denote the infinite cyclic group, $\mathbb{Z}_2$ denote the Group of order 2. For all k, which are $\not\equiv 0 \bmod 4$, Ω^k is finite. For $k = 4m$, Ω^k is a direct sum of $\pi(n)$ Groups $\mathbb{Z}$, where $\pi(m)$ is the number of partitions (decompositions into positive summands) of m, and of a finite Group. The finite Groups just mentioned are very hard to determine; however one can get rid of these groups by replacing Ω by the "weak" Thom algebra Ω', which is obtained from Ω by admitting rational numbers as coefficients of the elements of Ω (in such a way that Ω' is the tensor product of Ω with the field of rational numbers); now one can describe exactly not only the additive but also the multiplicative structure of the algebra: Ω' is the algebra of the polynomials with rational coefficients in the variables $P^0, P^2, \dots, P^{2m}$, where P^{2m} is represented by the Complex Projective Space of $2m$ complex dimensions (that means $4m$ real dimensions). It follows that for each manifold M there exists an entire multiple nM, which is cobordant with well-defined integer linear combination of cartesian products of complex projective spaces of even dimension.

All this serves primarily to study those properties of manifolds which are invariants of the cobordism classes. In the first place, these are the properties which are connected with the "Pontrjagine numbers" of the manifold whose study is *a priori* very difficult. The theorem which was just mentioned shows that it is essential, on the one hand, to represent the considered manifold in the algebra Ω and, on the other hand, to determine the Pontrjagine numbers of the spaces P^m; the rule, which these spaces, stemming from Classical Algebraic Geometry, are playing for the topology of arbitrary (differentiable) manifolds, is very remarkable.

With this framework, we find the so far most important consequence of the cobordism theory; it constitutes one of the pillars on which F. Hirzebruch's theory of the Riemann–Roch theorem is founded, a theory which in turn is considered to be one of the most important mathematical achievements in recent years (F. Hirzebruch, Neue topologische Methoden in der algebraischen Geometrie, Springer–Verlag, 1956).

A different application of the theory of cobordism was given by J. Milnor; e gave a proof of the spectacular fact that the sphere S^7 can have several distinct differentiable structures; this shows for the first time that there are two differentiable manifolds which can be mapped onto each other homeomorphically and differentiably; on the top of Milnor's proof, there is the theorem of Thom, that $\Omega^7 = 0$, i.e. that every 7-dimensional (closed, orientable) manifold is a boundary (bordant) (Ann. Math. (2), 64, 1956).

I have no time here to present more of the many interesting results of Thom, neither from the papers mentioned above, nor from those I have not touched. But I have one remark of principal nature.

Topology is today, like many other branches of mathematics, in a process of powerful and consequent algebraization; this process has had remarkable success in clarification simplification and unification, and it has led to some unexpected new results; it is not simply that algebra only serves as a tool for the treatment of topological problems; rather one finds that most of the problems have an aspect of particularly algebraic character themselves. But the great successes of this development give rise, in my opinion, to a certain danger, the danger of a perturbation of the mathematical balance: there is a tendency to neglect totally the geometrical content of the topological problems and situations; but this neglect would imply an impoverishment of mathematics. Just in view of this danger, I feel that the achievements of Thom have something extraordinarily encouraging and enjoyable: of course Thom also masters and uses modern algebraic methods and sees the algebraic aspects of his problems, but his basic ideas, the grand simplicity of which I already talked of are of a very geometric and intuitive nature. These ideas have significantly enriched mathematics, and everything seems to indicate that the impact of Thom's ideas — whether they find their expression in the already known or in forthcoming works — is not exhausted by far.

René Thom

AUTOBIOGRAPHY OF RENÉ THOM

When I was asked what I had gained from receiving the Fields medal (in Edinburgh 1958), I thought it would be simpler to come up with a short autobiography, adding a few personal comments whenever they might seem necessary.

Childhood: 1923–1940

I was born on September 2nd 1923 in Montbéliard (Doubs). In those days, Montbéliard looked like a medieval town. It lies on an affluent part of the Doubs, the Allan, which, flowing from the Swiss Jura, bathes it on the southern side, whilst to the north, the river is fed by another affluent part from the southern Vosges, the Luzine. At that time the Luzine was not covered over; nor were the two little watercourses coming from the east. The canals drained water from the Savoureuse and were built by the princes of Montbéliard in order to strengthen the fortifications of the medieval castle, a very well preserved castle set upon a rock betwixt Allan and Luzine. Our house was at Number 21, rue de Belfort, right at the southern foot of the castle rock, from which it was separated by a space 10 or 11 meters wide. What remained of the old rampart walk which, after the fall of the princely house, had been "privatised" into little gardens. This detail played an important part in my childhood, for I often went to play in this space called "the garden". For years we grew flowers and vegetables and kept a chicken-run. Our house was very old indeed; to reach the upper floors we had to climb a spiral stone staircase (called a "yorbe" in the local patois). The date 1631 could still be seen on a stone doorway at the foot of the stairs. I had a relatively happy childhood. My father and mother got on well. They had a small grocery-cum-drugstore at the same address, 21 rue de Belfort. My father was a tall, well-built man, with a good secondary education. He enjoyed reciting in Latin and, though he was a fervent agnostic, singing the Latin mass. My mother was a loving if sometimes rather forceful woman. I had an elder brother, (Robert) with whom I (René) didn't hit it off too well, because, at that age, he used to take unfair advantage of the strength of his three extra years. I must mention how well I remember my grandmothers. I had two of them, Papa's mother, whose maiden name was Blazer, and Mama's, born a Ramel. Both lived with us. I even had (up to the age of four) a great grandmother, née Beucler. These old ladies took a great interest in the children of the family and made up the framework of my life up to the time when I was sent to the Primary School, Market Building, in 1931. I did well in school from the start and passed the scholarship exam (the first notable event leading to my university cursus). It must be said that although they were shopkeepers, my parents were not particularly good at commerce and, financially speaking, the situation was shaky. It stayed much the same, with no major incidents, until 1940, the year war broke out.

1940 — The War

At the start of the war, our parents insisted on our leaving for the south, as so many people were doing at the time. They themselves stayed at home, but my brother Robert and I took to our bikes and the southbound roads. On the plateau where the southern Doubs makes a loop, near Maiche, we came upon the last skirmishes of a routed army, forced to cross the border into Switzerland. We joined their convoy and so passed into Switzerland. The surprising warmth with which we were welcomed there, all those people offering food and drink at the roadside, still fills me with emotion. Thus we made our way to Bienne, where we were given a place to sleep in the straw. Later we were allowed to stay for some time in Switzerland, haymaking in the Gruyère valley near Romont. But towards the end of September a convoy was organized to bring us back to France by train, to Lyons in the free zone. Luckily my mother had a friend in Lyons, related to one of our uncles, who was able to take care of us. I had already passed the scientific *Baccalauréat* (*Math. Elem.*) in 1940 in Besançon, so I now made the most of our stay in Lyons and passed the *Baccalauréat* in Philosophy in June 1941. But we couldn't stay there forever and started thinking of going home, which we did at the end of June 1941. We took the train to a village near the Loue, a frontier river flowing into the Doubs. Guides helped us cross the river by night with our bikes. From there we cycled to Besançon where we boarded a train to Montbéliard. Needless to say, our parents were overjoyed to have us back.

This period was short-lived, for very soon it was time to think of my future studies. Thanks to the good advice of G. Becker, my teacher in *Premiére* (a man of letters!) I was directed towards the *Lycée Saint-Louis* in Paris. After a first interview, I was accepted there without further discussion. I tried for the *École Normale Supérieure* (Paris) in 1942 but failed and had to try again in 1943. This time I was successful (though not brilliantly so!). All this brings us to October 1943, when I entered the ENS, rue d'Ulm.

The periodic trips home to see my parents had become difficult, for Montbéliard was now in a "forbidden zone". Nevertheless, thanks to the complicity of railway workers on the eastern network, it was still possible to cross from one zone to another by stratagem, and I did so several times. The commitment and resourcefulness of railway workers was admirable (for example, we could use the dog compartment of the wagon and cross over by Vitrey). Unfortunately my brother Robert had to leave Montbéliard for the STO. He was working in a factory at Mödling, southern suburb of Vienna, right up to the end of the war three years later. In Montbéliard itself the food situation was pretty bleak. I was able to go back there for good in 1945, when the French were drawing near from the south. Of my studies at the Lycée St Louis, I have retained a special admiration for the teachers we had, both in *Hypotaupe* and in *Taupe*. I entered ENS in October 1943 (with a modest rank) and started on my "higher" education.

ENS, 1943–1946

I remember those years at the ENS under the Occupation as a horribly tense period. It is true that the school authorities did all they could to make our stay easier. The political antagonism between the Director Carcopino and the Deputy Director Bruhat obviously did not make the task of administration any easier. But I have to say that, until G. Bruhat was arrested, there were no major disturbances (at least in so far as I was concerned). The yellow star, of course, soon made its appearance to set its mark on the Jews. Only the last year, after the "victory", was a year of opening, bringing with it the impression of once more living life to the full. Of this rebirth I recall a sensation of freedom that I found it hard to control. But the evenings spent in students' rooms with friends like Duvert, Pechmajou, Pecker, and others, will stay forever in my memory.

Henri Cartan was our master there from day one, teaching us the rudiments of Bourbakism, spheres and balls in all dimensions. Later I had a glimpse of Malraux speechifying in a corridor. It was too late to plunge further into literary country, although I had a feeling that a whole universe lay there to explore. But the evident superiority of the arts students was so crushing that we did not contest it.

From the mathematical point of view this was the time when we made the acquaintance of Bourbaki. And at the end of 1946 I decided to go to Strasbourg, to follow Henri Cartan.

Strasbourg. Entering the CNRS

Towards November 1946 I moved into a rented room in Strasbourg; at first I was way off on the road to Lyons near Graffenstaden, but later found a room in town at the Krutenau. It was then that I got to know mathematicians of my own age, or near enough, some of whom had already given proof of their talent. I should mention Koszul who drew my attention to the new Steenrod squares, Wu Wen Tsün, a discreet and kindly Chinese whose calculations were legendary in their reliability, Reeb, creator of the theory of foliation, and Charles Ehresmann, another expert in Topology, who was to stay there a few years, and whose theories struck me profoundly. For all that, the formation of fundamental notions in Algebraic Topology was not self-evident and a great deal was said (in Paris, not in Strasbourg) about the difference between a "shell" theory and a "blanket" theory. The years 1940–1950, though they came in with a mighty war, were astonishingly fertile for the foundations of Algebraic Topology. They were happy years for me. I met the girl who was to become my wife and bear me three children. I came to know some amazing minds, some very general theories, including the "categories" developed by Ehresmann. My first piece of work in Mathematics was a C.R.A.S. Note on

Morse theory, seen from the then novel point of view of a cellular subdivision of the manifold, a point of view which later helped considerably in Smale's resolution of Milnor's conjecture. Then, in 1951, I defended my thesis under the direction of Henri Cartan: "Fibre Spaces in Spheres and Steenrod Squares".

It was also in 1951 that I obtained a fellowship for the United States. So there I went, leaving the children, Françoise and her little sister Elisabeth, with my wife and mother-in-law. This gave me the opportunity to learn some English and to see such famous people as Einstein and Hermann Weyl (whose last lectures I listened to). In Princeton, I met N. Steenrod, sadly already suffering from the spinal disorder of which he was later to die. The English I learned during that stay has stood me in good stead, for it is that little part that has not since been forgotten. I was also able to attend the seminars of Calabi and Kodaira.

Back in France in 1953, I was reunited with my family and was given a chair in Strasbourg made vacant by the departure of Chabauty for Grenoble. This period in Strasbourg was very fertile for me, although complicated for the first few months by the necessity of commuting between Strasbourg and Grenoble. It was around that time that I began to be interested in the theory of differentiable maps. A certain conversation comes to mind in which de Rham drew my attention to *Sard's Theorem*, according to which the set of singular values $F(\Sigma)$ of the critical set Σ of a smooth map $F : X \to M$ is nil in M. This theorem was to contribute in an essential way to the rise of differential topology by restriction to generic situations. It was the basic tool for my work on smooth manifolds.

Thus I obtained results that were to be rewarded by the Fields Medal in 1958: the construction of the theory of cobordism. It must be said that Serre and Milnor played an essential part in this progress. Milnor, through his discovery of exotic spheres, showed that intuition is not always enough... : there exist spheres which are not the boundaries of balls (in C^∞)!

After 1958

I have little to say about the attribution of the Fields Medal in itself. It was an impressive ceremony, at which I bowed respectfully before the Lord Mayor of Edinburgh. But at bottom this success left me with a slightly bitter taste, for I have the impression that work was done just a very little while later that was greater in depth and sagacity than mine and whose authors were quite as deserving, if not more so, of the medal (such as my co-medallist Roth). I am thinking too of Barry Mazur's demonstration of the *Schönfliess conjecture*: Every shpere S^{n-1} in R^n with regular boundary is the boundary of an n-ball. Not to mention the discovery by Milnor of exotic shperes. In this sense, the Fields medal meant for me a certain fragility which the future was to make even more visible. If there was any consequence of my receiving the Fields medal, it was the invitation to come to the newly created Institut des Hautes Études Scientifiques (IHES). Finally, three or four years later, I took it up and moved with my family to Bures-sur-Yvette.

I had been invited by the founding director, Léon Motchane, with whom I had many friendly conversations. Relations with my colleague Grothendieck were less agreeable for me. His technical superiority was crushing. His seminar attracted the whole of Parisian mathematics, whereas I had nothing new to offer. That made me leave the strictly mathematical world and tackle more general notions, like the theory of morphogenesis, a subject which interested me more and led me towards a very general form of "philosophical" biology. The director, L. Motchane, had no objection (or if he did he kept it to himself). And so I ran what was known as a "crazy" seminar, that lasted the best part of my first year at the Institute. Three or four years passed, during which I returned to Aristotle and classical Greek. My first book, "Stabilité structurelle et Morphogenèse" (Structural Stability and Morphogenesis) had difficulty in getting to press with my editor (Pergamon Press). It was only with the help of Christopher Zeeman (then professor at the University of Warwick) that it was eventually published, and that the notion of "catastrophe" appeared in the literature (and in the press).

Catastrophe theory has as earlier origin. From the Strasbourg days I had been interested in the (classical) theory of envelopes; I looked for the structure of singular sets of envelopes. It was then that the notion of "generic singularity" came to me, starting with the case of envelopes of curves in the plane. I recalled H. Whitney's result (October 1947) on the generic singularities of plane to plane mappings with the notion of "cusp".

From there on I began to look at the structure of smooth maps of Euclidean spaces (and/or manifolds) with respect to each other. In fact, in Strasbourg, I had persuaded a physicist colleague (Ph. Pluvinage) to set up an optical apparatus allowing me to study the structure of generic caustics. It was the fruit of an old curiosity which had led me to write a paper "On envelope theory".

I have no clear memory of exactly when I launched "catastrophe theory". It was already there in 1972 when the first edition of "Stabilité structurelle et Morphogenèse" came out. But it was the fruit of a long semi-conscious genesis. Catastrophe terminology has often been held against me. It was certainly not introduced just to draw attention. It was meant to mark the presence of a pregnant discrete detail in the midst of an undifferentiated homogeneous continuum, and that is the fundamental characteristic of the *eidos* within the *genos*, the most evident manifestation of phenomenal existence.

It is a fact that catastrophe theory (CT) is dead. But one could say that it died of its own success. It was brought down by the extension from analytical (or algebraic) models to models that were only smooth. For as soon as it became clear that the theory did not permit quantitative prediction, all good minds, following Rutherford's phrase, decided that it was of no value. When it comes down to it, this extension resulted from B. Malgrange's extension of the preparation theorem to the class C^∞. Thus CT fell victim to its own progress. As for understanding the value of strictly qualitative theory, that takes a certain breadth of vision. Zeeman's models required that intelligence, and perhaps it is an intuitive quality vital to the global perception of a conflict.

The sociological history of CT is not without interest. It was marked by enormous popular success, appearing after the IMU congress of Vancouver (1972), where E. C. Zeeman was the perfect propagandist. The theory was world renowned until it was ruined by the success of B. Malgrange. Some notions remain, introduced by algebraists, such as the concept of versal unfolding of a singularity, now undoubtedly classic.

However that may be, the wish to escape from the evident superiority of excellent mathematicians, led me to move into other areas of research, including those connected with biology. I can claim no success there, at least not in society. Gangway for molecules! even if it takes millions of computer operations to calculate the interaction between two macromolecules. My retreat towards the global perception has led me to a few (perhaps obvious) discoveries in linguistics and biology. Maybe they will be resurrected in days to come...

To conclude, I would like to say just this. The Fields medal brought me the freedom to choose what research I wanted to do, and that is essential. When we were driving from France to Scotland, Suzanne, my wife, felt a little sick during the long journey, a "specific" sickness that lasted right up to the ceremony in Edinburgh. Then she discovered her pregnant state. Back in Strasbourg, a son, Christian, our third child, was born to us on April 17th 1959. This fact anticipates by some twenty years the theory of "pregnance" expounded later in "Esquisse d'une Sémiophysique". One might say — and it is hardly surprising — that the Fields medal confers on its recipient a certain "pregnance", both in the realm of mathematics and elsewhere. It remains up to him to stay worthy of it when the inevitable falling-off of old age sidles in.

René Thom
(trad. Vendla Meyer)

Reprinted from Proc. Int. Congr. Math., 1962

HÖRMANDER'S WORK ON LINEAR DIFFERENTIAL OPERATORS

by
L. GÅRDING

Lars Hörmander was born 1931, studied at Lund University and is now professor of mathematics at Stockholm University. He has been given a Fields Medal for his outstanding work in the theory of partial differential equations. Before trying to describe some of his results I shall have to lead you through a somewhat lengthy introduction.

Let me first acquaint you with a few notations. Coordinates and derivatives in real n-space will be denoted by x_k and $D_k = i^{-1}\partial/\partial x_k$ respectively. Higher derivatives will be written as powers $D^\alpha = D_1^{\alpha_1} \dots D_n^{\alpha_n}$; we put $|\alpha| = \alpha_1 + \dots + \alpha_n$. A linear differential operator of order m is a sum

$$P = P(x, D) = \sum_{|\alpha| \leq m} a_\alpha(x) D^\alpha$$

with principal part

$$P_m(x, D) = \sum_{|\alpha| = m} a_\alpha(x) D^\alpha.$$

We assume that the coefficients are smooth complex functions. Substituting complex numbers ξ_j for the derivatives we get the characteristic polynomials

$$P(x, \xi) = \sum_{|\alpha| \leq m} a_\alpha(x) \xi^\alpha$$

and $P_m(x, \xi)$. The Fourier transform will convert a differential operator with constant coefficients $P(D)$ into multiplication by $P(\xi)$. The adjoint and the conjugate of P are defined by

$$P^* = \sum D^\alpha \bar{a}_\alpha(x)$$

and

$$\overline{P} = \sum \bar{a}_\alpha(x) D^\alpha.$$

Linear differential equations

$$Pu = v \tag{1}$$

and systems of such equations occur in many mathematical models of physics, e.g. hydrodynamics, elasticity, thermodynamics, electricity and magnetism, and magnetohydrodynamics. We can think of u as a function describing some aspect of the state of the physical system and v as some kind of exterior force. The manifold of solutions u gives us the possible states under the action v. Additional restraints mostly given as boundary conditions give us a solution describing a particular situation. In practice it is the boundary conditions that account for the variety while P is given by the model itself.

It was pointed out very emphatically by Hadamard that it is not natural to consider only analytic solutions and source functions even if P has analytic coefficients. This reduces the interest of the Cauchy–Kowalevski theorem which says that (1) has locally analytic solutions if P and v are analytic. The Cauchy–Kowalevski theorem does not distinguish between classes of differential operators which have, in fact, very different properties.

If

$$P = D_1^2 + \cdots + D_n^2$$

is Laplace's operator, we are dealing with potential theory and if

$$P = D_1^2 - D_2^2 - \cdots - D_n^2$$

is the wave operator, we are studying wave propagation. The solutions of the homogeneous equation $Pu = 0$ behave quite differently. In the first case they are harmonic and hence real analytic, in the second case the example $u =$ arbitrary function of $x_1 \pm x_2$ shows that they may have very complicated singularities.

A natural reaction to all this is the question: what is it in P that makes this difference? More generally we can ask for properties of P or rather its characteristic polynomial which are intrinsic in the sense that they are more or less equivalent to properties of the solutions. Questions of this nature have no physical background but a very solid motivation: mathematical curiosity. They lead Hadamard to the fruitful notion of correctly set boundary problems. The first complete results, however, are due to Petrovsky who proved among other things that

$$Pu = 0 \Rightarrow u \text{ analytic} \tag{2}$$

and

$$\xi \text{ real } \neq 0 \Rightarrow P_m(\xi) \neq 0 \tag{3}$$

are equivalent properties for operators with constant coefficients. The second condition has an immediate extension to operators with variable coefficients. They are said to be elliptic. A simple sample is Laplace's operator. Actually Petrovsky's main achievement on that occasion is a proof that all solutions of suitably defined elliptic nonlinear analytic systems are analytic. The problem goes back to one of Hilbert's problems from the Paris congress in 1900.

Petrovsky felt that his results were just a beginning and in a lecture in 1945 he explicitly asked for a general theory of linear differential operators including those which do not appear in the mathematical models of physics. At the same time the theory of distributions appeared as a new tool in analysis. In his book on distributions Laurent Schwartz pointed out that the equivalence of (2) and (3) holds if the first u in (2) is assumed to be a distribution. He also stated a number of problems about differential operators. Since then a rather comprehensive theory has been worked out. Many people have contributed but the deepest and most significant results are due to Hörmander. I can describe only a few of them.

One of Schwartz's questions was the following one: what becomes of (3) if we replace (2) by the weaker statement

$$Pu = 0 \Rightarrow u \in C^\infty ? \tag{4}$$

The answer, namely

$$\text{Im}\,\xi \text{ bounded}, \quad \xi \to \infty \Rightarrow P(\xi) \to \infty,$$

was obtained by Hörmander in his thesis as a byproduct of a study of operators with constant coefficients, which I shall not go into. The corresponding operators are called hypoelliptic. A characteristic point of the proof is that the theorem of the closed graph is used to replace (4) by an equivalent inequality which establishes the problem in a suitable analytical form. The last part of the thesis deals with variable coefficients and establishes an important inequality which can be described as follows.

Let Ω be an open part of R^n and let H_0^k be the closure of $C_0 = C_0^\infty(\Omega)$ in the norm

$$|u|_k^2 = \int \sum_{|\alpha| \le k} |D^\alpha u(x)|^2 dx.$$

It is a Hilbert space and we can identify its dual H^{-k} with the space of distributions u in Ω for which

$$|u|_{-k} = \sup |(u, v)|/|v|_k < \infty,$$

where $(u, v) = \int u(x)\overline{v(x)}dx$ and v runs over C_0. We observe that $H^{-0} = H_0^0$ and that H^{-k} increases with k. A simple result from the theory of linear operators tells us that if A and B are Banach spaces and L is a densely defined linear operator such that

$$A \xrightarrow{L} B \text{ has a continuous inverse}, \tag{5}$$

then the adjoint map

$$B^* \xrightarrow{L^*} A^*$$

is onto. By a piece of ingenious and powerful analysis Hörmander established (5) in the form of an inequality

$$|u|_{m-1} \le C|Pu|_0 \quad (u \in C_0)$$

and hence also the solvability of

$$P^* u = v \in H^{1-m}. \tag{6}$$

It is assumed that

$$\Omega \text{ is sufficiently small,} \tag{7}$$

$$P_m \text{ is real,} \tag{8}$$

$$\xi \text{ real } \neq 0 \Rightarrow \sum_1^n \left| \frac{\partial P_m(x,\xi)}{\partial \xi_k} \right| > 0. \tag{9}$$

The solvability of (6) was a great breakthrough in the theory and the conditions (7) and (9) but perhaps not (8) of reasonable generality. There were, however, surprises to come. Let us weaken (6) to

$$Pu = v \text{ has a solution (possibly a distribution) for every } v \in C_0. \tag{10}$$

By an ingenious but very special reasoning Hans Lewy proved in 1957 that the operator

$$D_1 + iD_2 + i(x_1 + ix_2)D_3$$

does not have this property. An almost complete investigation of the situation was given by Hörmander in 1960. Rephrasing (10) in the form of an inequality he got the following result. Put

$$C(x,D) = (P(x,D)\overline{P}(x,D) - \overline{P}(x,D)P(x,D))_{2m-1}.$$

This operator only depends on P_m and vanishes if P_m is real. Then one has

$$(10) \Rightarrow (P_m(x,\xi) = 0, \quad \xi \text{ real} \Rightarrow C(x,\xi) = 0). \tag{11}$$

Furthermore, this statement almost has a converse. We say that P is principally normal if there exists a polynomial $Q_{m-1}(x,\xi)$ of degree $m-1$ in ξ such that

$$C(x,\xi) = P_m(x,\xi)\bar{Q}_{m-1}(x,\xi) + \overline{P}_m(x,\xi)Q_{m-1}(x,\xi)\,.$$

This property implies the right side of (11) and Hörmander proved that

$$P \text{ principally normal} \tag{12}$$

together with (7) and (9) implies the solvability of (6). The discovery of these results and their proofs is a first-class achievement. Recently he has given a global version of the theory.

I will finish by touching upon the problem of unique continuation which is the following. Consider a regular surface $S : s(x) = 0$ in R^n and a solution u of $Pu = 0$ which vanishes on one side of S. Does it vanish on the other side? If the coefficients of P are analytic the answer is yes provided S in non-characteristic for P, i.e.

$$P_m(x, \text{grad } s(x)) \neq 0. \tag{13}$$

This is an old result by Holmgren. The same problem for non-analytic coefficients is much harder and there are counterexamples to show that (13) is not enough.

The first positive results are due to Carleman for two variables and to Calderón in the general case. Hörmander has clarified the situation considerably by inventing a sufficient convexity condition bearing on S and P which also comes close to being necessary.

I hope that this sketch has given you an idea of the power and the drive that characterizes Hörmander's work.

References

L. Hörmander, *Linear Partial Differential Operators*, Springer 1963.

Lars Hörmander

AUTOBIOGRAPHY OF LARS HÖRMANDER

I was born on January 24, 1931, in a small fishing village on the southern coast of Sweden where my father was a teacher. After elementary school there and "realskola" in a nearby town which could be reached daily by train I went to Lund to attend "gymnasium", as my older brothers and sisters had done before me. I was more fortunate than they, for the principal was just starting an experiment which meant that three years were decreased to two with only three hours daily in school. This meant that I could mainly work on my own, with much greater freedom than the universities in Sweden offer today, and that suited me very well. I was also lucky to get an excellent and enthusiastic mathematics teacher who was a docent at the University of Lund. He encouraged me to start reading mathematics at the university level, and it was natural to follow his advice and go on to study mathematics at the University of Lund when I finished "gymnasium" in 1948.

In 1950 I got a masters degree and started as a graduate student. Marcel Riesz was my advisor, as he had been for my "gymnasium" mathematics teacher. Riesz was close to his retirement in 1952, and his lectures which I had actually attended since 1948 were not devoted to partial differential equations where he had recently made major contributions but rather to his earlier interests in classical function theory and harmonic analysis. My first mathematical attempts were therefore in that area. Although they did not amount to much this turned out to be an excellent preparation for working in the theory of partial differential equations. That became natural when Marcel Riesz retired and left for the United States while the two new professors Lars Gårding and Åke Pleijel appointed in Lund were both working on partial differential equations.

After a year's absence for military service 1953–1954, spent largely in defense research which gave ample opportunity to read mathematics, I finished my thesis in 1955 on the theory of linear partial differential operators. It was to a large extent inspired by the thesis of B. Malgrange which was announced in 1954, combined with techniques developed for hyperbolic differential operators by J. Leray and L. Gårding. Soon after that I was ready for my first visit to the United States, where in 1956 I spent the Winter and Spring quarters at the University of Chicago, the summer at the Universities of Kansas and Minnesota, and the fall in New York at what is now called the Courant Institute. (At the time R. Courant was still the director and it was called the Institute of Mathematical Sciences.) In Chicago there was no activity at all in my field, but the Zygmund seminar, conducted in his absence by E. M. Stein and G. Weiss, gave a useful addition to my background in harmonic analysis. At the other places I visited there was much to learn in my proper field.

At the end of this stay I was appointed to a full professorship at the University of Stockholm (called Stockholms Högskola then), which I had applied for before leaving for the United States. I took up my duties there in January 1957 and remained as professor until 1964. However, already during the academic year 1960–61 I was back in the United States as a member of the Institute for Advanced Study. During the summers of 1960 and 1961 I lectured at Stanford University and wrote a major part of my first book on partial differential equations. It was published by Springer Verlag in the Grundlehren Series in 1963 after the manuscript had been completed and polished back in Stockholm during the academic year 1961–62.

The 1962 International Congress of Mathematicians was held in Stockholm. In view of the small number of professors in Sweden at the time it was inevitable that I should be rather heavily involved in the preparations but it came as a complete surprise to me when I was informed that I would receive one of the Fields medals at the congress.

Some time after the two summers at Stanford I received an offer of a part time appointment as professor at Stanford University. I had declared that I did not want to leave Sweden, so the idea was that I should spend the Spring and Summer quarters at Stanford but remain in Stockholm most of the academic year there, from September through March. A corresponding partial leave of absence was granted by the ministry of education in Sweden and the arrangement became effective in 1963. However, I had barely arrived at Stanford when I received an offer to come to the Institute for Advanced Study as permanent member and professor. Although I had previously been determined not to leave Sweden, the opportunity to do research full time in a mathematically very active environment was hard to resist. After an attempt to create a research professorship for me in Sweden had failed, I finally decided in the fall of 1963 to accept the offer from the Institute and resign from the universities of Stockholm and Stanford to take up the new position in Princeton in the fall of 1964.

At that time the focus of interest in Princeton was definitely not in analysis which was felt both as a challenge and as a great opportunity to broaden my mathematical outlook. However, it turned out that I found it hard to stand the demands on excellence which inevitably accompany the privilage of being an Institute professor. After two years of very hard work I felt that my results were not up to the level which could be expected. Doubting that I would be able to stand a lifetime of this pressure I started to toy with the idea of returning to Sweden when a regular professorship became vacant. An opportunity arose in 1967, and I decided to take it and return as professor in Lund from the fall term 1968. After the decision had been taken I felt much more relaxed, and my best work at the Institute was done during the remaining year.

So in 1968 I had completed a full circle and was back in Lund where I had started as an undergraduate in 1948. I have remained there since then, with interruptions for some visits mainly to the United States. During the Fall term of 1970 I was visiting professor at the Courant Institute in New York, during the Spring term 1971 I was a member of the Institute for Advanced Study, and during the Summer quarter 1971 I was back at Stanford University, where I also lectured during the Summer quarters of 1977 and 1982. During the academic year 1977–1978, I was again a member of the Institute for Advanced Study which had a special year in microlocal analysis then, and during the Winter quarter 1990 I was visiting professor at the University of California in San Diego.

After five years devoted to writing a four volume monograph on linear partial differential operators I spent the academic years 1984–1986 as director of the Mittag–Leffler Institute in Stockholm. I had only accepted a two year appointment with a leave of absence from Lund since I suspected that the many administrative duties there would not agree very well with me. The hunch was right, and since 1986 I have been in Lund where I became professor emeritus in January 1996.

In my contribution to this volume I have summed up my work in partial differential equations and microlocal analysis which can be considered as the continuation of the work which gave me a Fields medal. Another major interest has been the theory of functions of several complex variables and its applications to the theory of partial differential equations. Lecture notes on this subject written at Stanford during the summer of 1964 became a book published in 1966, and extended editions were published in 1973 and 1990. A book on convexity theory published in 1994 is also to a large extent devoted to this field.

LOOKING FORWARD FROM ICM 1962

by

LARS HÖRMANDER

Department of Mathematics, University of Lund

1. Introduction

When I received the invitation to contribute to a volume entitled *Fields Medallists' Lectures*, intended to be similar to *Nobel Lectures in Physics*, my first reaction was quite negative. The Nobel prizes and the Fields medals are so very different in character; while Nobel prizes are supposed to be given for work of already recognized importance, often the work of a lifetime, the Fields medals are given "in recognition of work already done, and *as an encouragement for further achievement on the part of the recipients*". However, since this may imply an obligation to account for the expected further achievements, I have decided to contribute to this volume a brief survey of the later development of the work for which I assume that I received a Fields medal in 1962. It will not be a complete survey even of the development of these topics in the theory of linear partial differential equations since 1962, for I shall concentrate on my own work and only mention work by others interacting with it.

At the ICM in 1962 I gave a half hour lecture [H1] with the following table of contents:

1. Notations
2. Equations without solutions
3. Existence theorems
4. Hypoelliptic operators
5. Holmgren's uniqueness theorem
6. Carleman estimates
7. Uniqueness of the Cauchy problem
8. Unique continuation of singularities

The paper [H1] was written just after the manuscript of my first book [H2] had been completed, and all the topics 2–8 are presented in detail there. When discussing the later development I shall often compare it with the new version [H3] published in four volumes about 20 years later.

2. Pseudodifferential Operators and The Wave Front Set

In [H2] the main tool for proving *a priori* estimates for differential operators was the Fourier transformation. When the coefficients were variable their arguments were first "frozen" at a point, sometimes after a preliminary integration by parts as in [H2, Chap. VIII]. For the success of the procedure it was of course necessary that the error committed in freezing the coefficients was in some sense small compared to the quantities to be estimated. This worked well also for the proof of uniqueness theorems originally proved by Calderón [Ca] using singular integral operators; in fact, it was possible to reduce his regularity assumptions. Singular integral operators were not included in [H2], for they appeared to have many drawbacks: they required an artificial reduction to operators of order 0, only principal symbols were handled, and the role of the Fourier transformation was so suppressed that the calculations involved seemed artificial. These objections were removed by the introduction of pseudodifferential operators by Kohn and Nirenberg [KN]. They considered operators of the form

$$\begin{aligned} a(x, D)u(x) &= (2\pi)^{-n} \int a(x, \xi) e^{i\langle x, \xi\rangle} \hat{u}(\xi) d\xi, \\ \hat{u}(\xi) &= \int e^{-i\langle x, \xi\rangle} u(x) dx, \quad u \in C_0^\infty(\mathbf{R}^n), \end{aligned} \tag{2.1}$$

where a is asymptotically a sum of terms which are homogeneous of integer order, which means that this algebra of operators is essentially generated by singular integral operators, differential operators and standard potential operators. However, the important point was that the calculus formulas for the symbols a were essentially the same as the familiar formulas for differential operators, not only on a principal symbol level. It was then easy and natural to introduce more general symbols which were useful in the further development of the theory of linear differential operators. The properties of parametrices of hypoelliptic operators with constant coefficients led me to introduce in [H4] the symbol class $S^m_{\varrho,\delta}$ of C^∞ functions in $\mathbf{R}^n \times \mathbf{R}^n$ such that

$$\left|\partial_x^\beta \partial_\xi^\alpha a(x, \xi)\right| \le C_{\alpha,\beta}(1 + |\xi|)^{m - \varrho|\alpha| + \delta|\beta|}. \tag{2.2}$$

When $0 \le \delta < \varrho \le 1$ operators with such symbols give a calculus with very good properties, and if $\delta \ge 1 - \varrho$, hence $\varrho > \frac{1}{2}$, it is invariant under changes of variables so it can be transplanted to smooth manifolds. (Such operators had already been used in [H5] to construct left parametrices of hypoelliptic differential operators, but I had not understood that an algebra of operators could be obtained in this way.)

Pseudodifferential operators were used at once to localize the study of singularities of solutions of partial differential equations. However, it took a few years before this was codified in the notion of wave front set: if $u \in \mathcal{D}'(\mathbf{R}^n)$ then the wave front set $WF(u) \subset T^*(\mathbf{R}^n)$ was defined in [H6] as the intersection of the characteristic sets of all pseudodifferential operators A such that $Au \in C^\infty(\mathbf{R}^n)$. By standard elliptic regularity theory it follows that the projection of $WF(u)$ in $\mathbf{R}^n$ is equal to the singular support of u, and the definition guarantees that $WF(Au) \subset WF(u)$ if A is a pseudodifferential operator. The wave front set describes both the location of the singularities and the directions of the frequencies which cause the singularities. A similar resolution of analytic singularities was given independently by Sato [Sa] but is technically more difficult to explain. For the original definitions we refer to [SKK], and various alternative definitions can be found in [H3, Chap. VIII, IX].

The proof of estimates is often reduced to positivity of an operator in L^2. The classical Gårding inequality states that if $a(x, D)$ is a differential operator with $a \in S^m_{1,0}$, $m > 0$, and $a(x,\xi) \geq c(1 + |\xi|)^m$ for some $m > 0$ and $c > 0$, then

$$\mathrm{Re}\ (a(x, D)u, u) \geq -C(u, u), \quad u \in C_0^\infty(\mathbf{R}^n), \tag{2.3}$$

where $(\cdot\,,\cdot)$ is the scalar product in L^2. This remains true for pseudodifferential operators with symbol in $S^m_{\varrho,\delta}$, where $0 \leq \delta < \varrho \leq 1$. The proof is quite trivial in the pseudodifferential framework since $a(x, D) + a(x, D)^* = b(x, D)^* b(x, D) + c(x, D)$ for some $b \in S^{m/2}_{\varrho,\delta}$ and $c \in S^0_{\varrho,\delta}$. A stronger result, called the "sharp Gårding inequality" was proved in [H7]; an extended version given in [H3, Theorem 18.6.7] states that if $a \in S^{\varrho-\delta}_{\varrho,\delta}$ and $a \geq 0$ then

$$\mathrm{Re}\ (a(x, D)u, u) \geq -\,C(u, u)\,, \quad u \in C_0^\infty(\mathbf{R}^n). \tag{2.4}$$

An important refinement of the right-hand side was given by Melin [Mel] (see also [H3, Theorem 22.3.3]), and Fefferman–Phong [FP] (see also [H3, Theorem 18.6.8]) proved that (2.4) remains valid when $a \in S^{2(\varrho-\delta)}_{\varrho,\delta}$ and $a \geq 0$. This is often a very significant technical improvement.

The study of solvability of (pseudo)differential equations led Beals–Fefferman [BF1] to introduce much more general symbol classes than the classes $S^m_{\varrho,\delta}$ above, which could be tailored to the equation being studied. A further extension was made in [H8] (see [H3, Chap. XVIII]) using a variant of the definition (2.1) proposed already by Weyl [W] in his work on quantum mechnanics. It is the symplectic invariance of the approach of Weyl which allows a greater generality. Even more general symbols have been studied by Bony and Lerner [BL].

3. Fourier Intergral Operators

Since pseudodifferential operators cannot increase the wave front set, a (pseudo)differential operator cannot have a pseudodifferential fundamental solution (or para-metrix) unless it is hypoelliptic. Parametrices of a different kind were constructed already in 1957 by Lax [La] for certain hyperbolic differential operators, as linear combinations of what is now called Fourier integral operators. In the simplest case these are of the form (2.1) with a modified exponent,

$$Au(x) = (2\pi)^{-n} \int a(x,\eta) e^{i\varphi(x,\eta)} \hat{u}(\eta)\, d\eta,$$
$$\hat{u}(\eta) = \int e^{-i\langle y,\eta\rangle} u(y)\, dy, \quad u \in C_0^\infty(\mathbf{R}^n), \tag{3.1}$$

where $\varphi(x,\eta)$ is positively homogeneous of degree 1 in η and $\det \partial^2\varphi/\partial x \partial\eta \neq 0$. (A reference to this construction in [La] was given in [H2, p. 230] but it could not be studied with the techniques used in [H2].) With an operator of the form (3.1) there is associated a canonical transformation χ with the generating function φ,

$$\chi : (\partial\varphi(x,\eta)/\partial\eta, \eta) \mapsto (x, \partial\varphi(x,\eta)/\partial x); \tag{3.2}$$

it has the important property that

$$WF(Au) \subset \chi WF(u), \quad u \in \mathcal{D}'. \tag{3.3}$$

A systematic study of such operators was initiated in [H9] in connection with a study of spectral asymptotics (see Section 9 below), and a thorough, more general and global theory was presented in [H10]. Fourier integral operators permit simplifications of (pseudo)differential operators, for as observed by Egorov [Eg1] conjugation by an invertible Fourier integral operator changes the principal symbol by composition with the canonical transformation. This was exploited in [DH], a sequel to [H10] written jointly with J. J. Duistermaat. A somewhat different exposition of these matters is given in [H3, Chap. XXV]. Some early papers in the area have been collected and provided with an introduction and bibliography by Brüning and Guillemin [BG].

4. Hypoellipticity

A (pseudo)differential operator P in an open set $X \subset \mathbf{R}^n$ (or a manifold X) is called hypoelliptic if

$$\text{sing supp } u = \text{sing supp } Pu, \quad u \in \mathcal{D}'(X). \tag{4.1}$$

Hypoelliptic differential operators with constant coefficients in $\mathbf{R}^n$ were characterised in my thesis (see [H2, Chap. IV]). The simplest class of hypoelliptic differential operators with variable coefficients is the class of elliptic operators: A differential operator P of order m is elliptic if the principal symbol $p(x,\xi)$, which is homogeneous of degree m, never vanishes when $\xi \in \mathbf{R}^n \setminus \{0\}$. If $H_{(m)}^{\text{loc}} = \{u; D^\alpha u \in L^2_{\text{loc}}, |\alpha| \leq m\}$, then P is elliptic if and only if $Pu \in H_{(0)}^{\text{loc}}$ implies $u \in H_{(0)}^{\text{loc}}$. The definition of the Sobolev spaces $H_{(s)}^{\text{loc}}$ can be extended to all $s \in \mathbf{R}$ in a unique way such that for all elliptic pseudodifferential operators of order $m \in \mathbf{R}$ we have $Pu \in H_{(0)}^{\text{loc}} \iff u \in H_{(m)}^{\text{loc}}$, and then we have more generally $Pu \in H_{(s)}^{\text{loc}} \iff u \in H_{(s+m)}^{\text{loc}}$.

Elliptic operators (with smooth coefficients) have been understood for a very long time, for they can easily be treated as mild perturbations of elliptic operators with constant coefficients. A class of hypoelliptic differential operators which can be studied similarly starting from hypoelliptic differential operators with constant coefficients was investigated by B. Malgrange and myself; the results were included in [H2, Chap. VII]. At the time they seemed quite general, but hypoellptic operators not covered by these results were soon found by Treves [T1]. In [H5] his ideas were developed as a very primitive and incomplete version of pseudodifferential operator theory already mentioned. In the more mature form of pseudodifferential operators in [H4] it was a simple consequence of the calculus, that if P is a pseudodifferential operator with symbol $p \in S^m_{\varrho,\delta}$, where $0 \le \delta < \varrho \le 1$, and if

$$|p(x,\xi)^{-1}\partial_x^\beta \partial_\xi^\alpha p(x,\xi)| \le C_{\alpha,\beta}|\xi|^{-\varrho|\alpha|+\delta|\beta|}, \qquad |\xi| > C,$$

$$|p(x,\xi)^{-1}| \le C|\xi|^{m'}, \qquad |\xi| > C,$$

then P is hypoelliptic. (In [H4, Section 4] there is actually a more general result for systems.) When $1 - \varrho \le \delta$ this gives a class of hypoelliptic operators which is invariant under changes of variables. However, that requires $\varrho > \frac{1}{2}$, and it is easy to see that second order differential operators satisfying this condition must in fact be elliptic. This led to the detailed study in [H11] of second order hypoelliptic operators.

A classical model equation, the Kolmogorov equation, was known in the theory of Brownian motion but was more familiar to probabilists than to experts in partial differential equations. The Kolmogorov operator is

$$P = (\partial/\partial x)^2 + x\partial/\partial y - \partial/\partial t$$

in $\mathbf{R}^3$ with coordinates (x, y, t). Kolmogorov [Ko] constructed an explicit fundamental solution, singular only on the diagonal, which implies hypoellipticity. Freezing the coefficients would give an operator with constant coefficients acting only along a two dimensional subspace, so the operator is obviously not covered by the results mentioned so far. However, the vector fields $\partial/\partial x$ and $x\partial/\partial y - \partial/\partial t$ do not satisfy the Frobenius integrability condition so the operator does not act only along submanifolds. The importance of this fact was established in [H11] where it was proved more generally that if

$$P = \sum_1^r X_j^2 + X_0 + c, \tag{4.2}$$

where $X_0, \dots, X_r$ are smooth real vector fields in a manifold M of dimension n, and if among the operators $X_j, [X_j, X_k], [X_j, [X_k, X_l]], \dots$ obtained by taking repeated commutators it is possible to find a basis for the tangent space at any point in M, the P is hypoelliptic. The condition on the vector fields is necessary in the sense that if the rank is smaller than n in an open subset then the operator only acts in less than n local coordinates, if they are suitably chosen, so it cannot be hypoelliptic.

However, the condition can be relaxed at smaller subsets. (See e.g. [OR], [KS], [BM].) Simplified proofs of these results (with slightly less precise regularity) due to J. J. Kohn can be found in [H3, Section 22.2]. There is now a very extensive literature on operators of the form (4.2) (see e.g. [RS], [HM], [NSW], [SC]), partly motivated by the importance in probability theory.

If P is of the form (4.2), then the principal symbol of P is ≤ 0. Many of the results concerning operators of the form (4.2) have been extended to general pseudodifferential operators such that the principal symbol (microlocally) takes its values in an angle $\subset \mathbf{C}$ with opening $< \pi$. We refer to [H3, Chap. XXII] for such results and references to their origins.

A pseudodifferential operator of order m and type 1, 0 is called *subelliptic* with loss of δ derivaties, if $0 < \delta < 1$ and

$$u \in \mathcal{D}'(\Omega), \quad Pu \in H_{(s)}^{\text{loc}} \Longrightarrow u \in H_{(s+m-\delta)}^{\text{loc}}. \tag{4.3}$$

For $\delta = 0$ this condition would be equivalent to ellipticity, and the assumption $\delta < 1$ implies that (4.3) depends only on the principal symbol $p(x, \xi)$ of P, which we assume to be homogeneous. In [H7] it was proved that $\delta \geq \frac{1}{2}$ when (4.3) is valid and P is not elliptic, and that (4.3) is valid for $\delta = \frac{1}{2}$ if and only if

$$\{\operatorname{Re} p(x,\xi), \operatorname{Im} p(x,\xi)\} > 0, \quad \text{when } p(x,\xi) = 0. \tag{4.4}$$

(The term subellipticity was introducted in [H3] just for the case $\delta = \frac{1}{2}$. There are pseudodifferential operators with this property but no differential operators since the Poisson bracket is then an odd function of ξ.) In [H7] a very implicit condition for subellipticity with loss of $\frac{1}{2}$ derivatives was also given for systems. The condition was made somewhat more explicit in [H12] but there are no really satisfactory results on subellipticity for systems with loss of δ derivatives even for $\delta = \frac{1}{2}$, and very little is known for $\delta \in (\frac{1}{2}, 1)$. However, for the scalar case Egorov [Eg2] found necessary and sufficient conditions for subellipticity with loss of δ derivatives; the proof of sufficiency was completed in [H13]. The results prove that the best δ is always of the form $k/(k+1)$ where k is a positive integer, and the conditions replacing (4.4) then involve the Poisson brackets of $\operatorname{Re} p$ and $\operatorname{Im} p$ of order $\leq k$, at the characteristics. A slight modification of the presentation in [H13] is given in [H3, Chap. XXVII], but it is still very complicated technically. Another approach which also covers systems operating on scalars has been given by Nourrigat [No] (see also the book [HN] by Helffer and Nourrigat), but it is also far from simple so the study of subelliptic operators may not yet be in a final form.

The hypoelliptic operators discussed here are all microlocally hypoelliptic in the sense that (4.1) can be strengthened to

$$WF(u) = WF(Pu), \quad u \in \mathcal{D}'(X). \tag{4.1$'$}$$

This remains valid in an open conic subset of the cotangent bundle if the conditions above are only satisfied there.

5. Solvability and Propagation of Singularities

The discovery by Lewy [Lew] that the equation

$$\partial u/\partial x_1 + i\partial u/\partial x_2 + 2i(x_1 + ix_2)\partial u/\partial x_3 = f$$

for most $f \in C^\infty(\mathbf{R}^3)$ (in the sense of category) has no solution in any open subset of $\mathbf{R}^3$ led to a systematic study of local solvability of differential equations in [H14], [H15] and [H2, Chap. VI, VIII]. It was proved there that if $p(x,\xi)$ is the principal symbol of a differential operator P of order m in $\mathbf{R}^n$ then the equation $Pu = f$ has no distribution solution in any neighborhood of x^0 for most $f \in C^\infty$ unless

$$p(x,\xi) = 0 \Longrightarrow \{\operatorname{Re} p(x,\xi), \operatorname{Im} p(x,\xi)\} = 0, \tag{5.1}$$

when x is in a neighborhood of x^0. On the other hand, if

$$\{\operatorname{Re} p(x,\xi), \operatorname{Im} p(x,\xi)\} = \operatorname{Re}(p(x,\xi)a(x,\xi)), \tag{5.2}$$

in a neighborhood of x^0 for some polynomial a in ξ with C^1 coefficients, and

$$p'_\xi(x,\xi) \neq 0 \quad \text{for } \xi \in \mathbf{R}^n \setminus \{0\}, \tag{5.3}$$

it was proved that the equation $Pu = f$ has a solution $u \in H^{\text{loc}}_{(s+m-1)}$ in a neighborhood of x^0 for every $f \in H^{\text{loc}}_{(s)}$. There is of course a substantial gap between the conditions (5.1) and (5.2). For first order differential equations it was filled to a large extent by Nirenberg–Treves [NT1]. The study of boundary problems for elliptic operators led to the extension of the solvability problem from differential to pseudodifferential operators in [H7]. There it was proved that the results of [H14], [H15] remain valid for pseudodifferential operators if (5.1), (5.2) are modified to

$$p(x,\xi) = 0 \Longrightarrow \{\operatorname{Re} p(x,\xi), \operatorname{Im} p(x,\xi)\} \leq 0, \tag{5.1$'$}$$

$$\{\operatorname{Re} p(x,\xi), \operatorname{Im} p(x,\xi)\} \leq \operatorname{Re}(p(x,\xi)a(x,\xi)). \tag{5.2$'$}$$

Thus (5.1)$'$ forbids $\operatorname{Im} p$ to change sign from $-$ to $+$ at a simple zero in the forward direction on a bicharacteristic of $\operatorname{Re} p$. Nirenberg and Treves [NT2] proved that such sign changes must not occur at any zeros of finite order either, and later work based on an idea of Moyer [Mo] has shown that local solvability implies that there are no such sign changes at all. (See [H3, Theorem 26.4.7].) This is called the condition (Ψ). For differential operators we have $p(x,-\xi) = (-1)^m p(x,\xi)$ and it follows that (Ψ) implies that there are no sign changes at all, which is called condition (P). It was also proved in [NT2] that local solvability follows from condition (P) and (5.3) provided that p is real analytic, a condition which was later removed by Beals and Fefferman [BF2]. However, it is still not known whether there is local solvability under condition (Ψ); there are many positive and negative results by Lerner [Ler1], [Ler2], [Ler3] and others. For these matters we refer to a recent survey paper [H16] and to [H3, Chap. XXVI].

The existence theorems just mentioned, first established in [H17], are not only local. They are valid for arbitrary compact sets which do not trap any complete bicharacteristics for the operator (see [H3, Theorem 26.11.1]), and they also give C^∞ solutions for C^∞ data. The key to such semiglobal results is the study of propagation of singularities of solutions of pseudodifferential equations. In [H1] there was just a very primitive result of this kind, Theorem 8.1, stating that if the principal symbol p of a differential operator P is real and satisfies (5.3), then a distribution u such that $Pu \in C^\infty$ is in C^∞ in a neighborhood of a point x^0 if there is a C^2 hypersurface through x^0 which has positive curvature at x^0 with respect to tangential bicharacteristics such that $u \in C^\infty$ outside the surface. A first step toward more precise results was taken by Grushin [Gr] who proved in the constant coefficient case that if $x^0 \in$ sing supp u then sing supp u contains a bicharacteristic line through x^0. An extension of this result to operators with variable coefficients was announced in [H18, p. 39]. However, the conclusion was only local which was a great weakness, for if one has concluded that an interval from x^0 to x^1 on a bicharacteristic Γ_0 is contained in sing supp u, it follows that there is an interval around x^1 on a bicharacteristic Γ_1 contained in sing supp u, but there is no guarantee that Γ_1 should be a continuation of Γ_0. The proof announced in [H18] depended on an early local version of the theory of Fourier integral operators. The global theory presented in [H10] was developed precisely to remove this flaw so that it could be proved directly that a complete bicharacteristic must be contained in sing supp u. However, just as this machinery was completed the idea of the wave front set presented in [H6] was conceived. The right statement on the propagation of singularities is that if $Pu \in C^\infty$ and $(x^0, \xi^0) \in WF(u)$, then the bicharacteristic strip starting at (x^0, ξ^0) is contained in $WF(u)$, provided that the principal symbol p is real valued. (If one removes the hypothesis $Pu \in C^\infty$ the conclusion is that the bicharacteristic strip remains in $WF(u)$ until it encounters $WF(Pu)$.) A bicharacteristic curve is the base projection of a bicharacteristic strip, so it is not determined by its starting point whereas a bicharacteristic strip is. This means that for the improved microlocal version of the theorem on propagation of singularities the local result implies the global one. A simple proof was outlined in [H6]. In [H19] the propagation theorem was extended to symbols with $\text{Im}\, p \geq 0$, and in [DH] an analogue for the case of characteristics where $d\,\text{Re}\, p$ and $d\,\text{Im}\, p$ are linearly independent was established under condition (P). The main point in [H17] was a fairly complete discussion of propagation of singularities for arbitrary pseudodifferential operators satisfying condition (P). The results on propagation of singularities were completed by Dencker [D], replacing for some bicharacteristic a weaker result in [H17] which gave the same existence theorems though. Some of the results on propagation of singularities remain valid for operators satisfying only condition (Ψ), but there are no satisfactory general results in that case.

6. Holmgren's Uniqueness Theorem

The classical uniqueness theorem of Holmgren states that a classical solution of a linear differential equation $P(D)u = 0$ vanishing on one side of a C^1 surface must vanish in a neighborhood of every noncharacteristic point. This was extended to distribution solutions in [H2], and by a simple geometric argument it was concluded (see [H1, Theorem 5.1]) that if the surface is in C^2 the assertion remains valid at characteristic points where the curvature of the surface is positive with respect to the corresponding tangential bicharacteristic.

It had been observed already by John [Jo] that the proof of Holmgren's uniqueness theorem could be modified to proving analyticity theorems such as the analyticity of solutions of elliptic differential equations with analytic coefficients. This observation could be reversed when the analytic singularities had been microlocalized to a set similar to the wave front set. The definition of Sato [Sa] (see also [SKK]) mentioned above works for hyperfunctions, whereas a definition of such a set $WF_A(u)$ modelled on the definition of $WF(u)$ introduced in [H20] only works for distributions. Since the equivalence with the definition of Sato in this case has been verified by Bony [Bo] we shall use the notation $WF_A(u)$ for arbitrary hyperfunctions u. (A proof of the equivalence of these definitions and another definition due to Bros and Iagolnitzer [BI] in the case of distributions is also given in [H3, Chap. VIII, IX].) The base projection of $WF_A(u)$ is of course the analytic singular support sing $\text{supp}_A u$, the complement of the largest open set where u is a real analytic function.

The connection with Holmgren's uniqueness theorem is given by two facts:

(i) If P is a differential operator with real analytic coefficients and $Pu = 0$, then $WF_A(u)$ is contained in the characteristic set of P.

(ii) If u is a hyperfunction vanishing on one side of a C^2 surface passing through $x^0 \in \text{supp}\, u$, then $WF_A(u)$ contains (x^0, ν) if ν is conormal to the surface at x^0.

Part (i) is a microlocal version of the standard analytic regularity theorem for elliptic differential equations, and part (ii) follows from a part of the arguments used originally to prove Holmgren's uniqueness theorem. If the principal part p of P is real valued then a theorem on propagation of analytic singularities combined with (i) and (ii) proves that if $Pu = 0$ and u vanishes on one side of a C^1 hypersurface through $x^0 \in \text{supp}\, u$ then the surface is characteristic at x^0 and the full bicharacteristic strip through a conormal must remain in $WF_A(u)$ so the base projection stays in supp u. This is a great improvement of [H1, Theorem 5.1] first obtained independently by Kawai in [Ka] and myself in [H20]. Many other improvements of Holmgren's uniqueness theorem have been obtained through a deeper understanding of (i) and (ii); a recent survey is given in [H21].

7. Analytic Hypoellipticity and Propagation of Analytic Singularities

Having made no contributions to this area beyond the first steps taken in [H20] we shall content ourselves here with some references to results which are relevant in connection with the improvements of Holmgren's uniqueness theorem discussed in the preceding section. A differential operator with real analytic coefficients is called analytic hypoelliptic if

$$\text{sing supp}_A\, u = \text{sing supp}_A\, Pu, \quad u \in \mathcal{D}'(X), \tag{7.1}$$

and P is said to be microlocally analytically hypoelliptic if

$$WF_A(u) = WF_A(Pu), \quad u \in \mathcal{D}'(X). \tag{7.2}$$

(One sometimes insists on these equalities for all hyperfunctions.) If P is subelliptic in the C^∞ sense then it was proved by Treves [T2] that P is analytically hypoellptic, and by Trépreau [Tre] in even greater generality that P is also microlocally analytically hypoelliptic.

However, operators of the form (4.2) with analytic coefficients satisfying the commutator condition which implies hypoellipticity are not always analytically hypoelliptic even if $X_0 = 0$. (Lower order terms are not expected to affect analytic hypoellipticity.) Simple examples are given in e.g. Christ [Ch]. However, if the characteristic set is a symplectic manifold then microlocal analytic hypoellipticity has been proved, also for some classes of operators with higher order multiplicities for the characteristics (see [Tar], [T3], [Met], [Sj1], and also [DT] for recent results and additional references).

Concerning propagation of analytic singularities we shall content ourselves with referring to the very general results by Kawai and Kashiwara [KK] and Grigis, Schapira and Sjöstrand (see [Sj2]), although these are in no way the last words on the subject.

8. Carleman Estimates

Section 6 in [H1] and Chap. VIII in [H2] were devoted to Carleman estimates of the form

$$\tau \int |D^{m-1}u|^2 e^{2\tau\varphi} dx \le C_1 \int |Pu|^2 e^{2\tau\varphi}\, dx$$
$$+ C_2 \sum_0^{m-2} \tau^{2(m-j)-1} \int |D^j u|^2 e^{2\tau\varphi}\, dx\,, \quad u \in C_0^\infty(K) \tag{8.1}$$

where P is a differential operator of order m in a neighborhood of the compact set $K \subset \mathbf{R}^n$. Such estimates with $C_2 \ne 0$ were applied to the proof of existence theorems and some very weak results on propagation of singularities, and these results have been made obsolete by those discussed in Section 5. The estimates with $C_2 = 0$,

$$\tau \int |D^{m-1}u|^2 e^{2\tau\varphi}\, dx \le C_1 \int |Pu|^2 e^{2\tau\varphi}\, dx, \quad u \in C_0^\infty(K), \tag{8.1$'$}$$

are still of interest though. It was proved in [H2] that (8.1)$'$ implies that if p is the principal symbol of P then

$$|\xi + i\tau\varphi'(x)|^{2(m-1)} \le C_1\{\overline{p(x,\xi+i\tau\varphi'(x))}, p(x,\xi+i\tau\varphi'(x))\}/2i\tau, \tag{8.2}$$

$$\text{if } \xi \in \mathbf{R}^n,\ \tau > 0, p(x,\xi+i\tau\varphi'(x)) = 0. \tag{8.3}$$

When (5.2) is fulfilled it is easy to conclude as a limiting case when $\tau \to 0$ that

$$|\xi|^{2(m-1)} \le C_1 \mathrm{Re}\{\overline{p(x,\xi)}, \{p(x,\xi), \varphi(x)\}\}, \tag{8.2$'$}$$

$$\text{if } \xi \in \mathbf{R}^n,\ p(x,\xi) = 0. \tag{8.3$'$}$$

It was proved in [H2] that conversely the conditions (8.2),(8.3) imply (8.1)$'$ (with a larger constant C_1) when (5.2) is valid with $a(x,\xi)$ in C^1 and polynomial in ξ. It was remarked in [H1, p. 343] that (8.2), (8.3) imply (5.1) but that it might be possible to eliminate the extraassumption (5.2) to a large extent. This was done in [H3, Chap. XXVIII] where (5.2) was weakened to

$$|\{\overline{p(x,\xi)}, p(x,\xi)\}| \le C_3 |p(x,\xi)||\xi|^{m-1}, \tag{8.4}$$

In the application of the estimate (8.1)$'$ to the proof of uniqueness theorems only the level sets of φ are important. Replacing φ by an increasing convex function of φ such as $e^{\lambda\varphi}$ with some large positive λ will add a positive term in the right-hand side of (8.2), (8.2)$'$ unless

$$\langle p'_\xi(x,\xi+i\tau\varphi'(x))\rangle\varphi'(x) = \{p(x,\xi+i\tau\varphi'(x)), \varphi(x)\} = 0. \tag{8.5}$$

If (8.2), (8.2)$'$ are only assumed to be valid when (8.5) is added to the hypotheses (8.3), (8.3)$'$ (with $\tau = 0$ in (8.5) in the second case), one can modify φ in this way without changing the level sets so that (8.2), (8.2)$'$ become valid with another constant C_1 under the hypotheses (8.3), (8.3)$'$ only. By the standard Carleman argument it follows then that if $Pu = 0$, $u \in H_{m-1}^{\mathrm{loc}}$ in a neighborhood of x^0 and $u = 0$ when $\varphi(x) > \varphi(x^0)$, then $u = 0$ in a neighborhood of x^0. (The proof uses also that the convexity conditions (8.2),(8.2)$'$ are stable under small perturbations.)

Already in [H2, Section 8.9] some examples based on constructions by A. Pliš and P. Cohen were given which proved that the convexity assumption in this uniqueness result could not in general be relaxed. A more systematic study of such examples was given in [H22]. However, these constructions relied on perturbations of P by terms of lower but positive order. Much better examples were constructed by Alinhac [Al] who proved that uniqueness fails after addition to P of a suitable C^∞ term of order 0 if (5.1) is not valid, or p has real coefficients and $\varphi(x) \le \varphi(x^0)$ on a bicharacteristic curvethrough x^0, or the right-hand side of (8.2) vanishes for a suitable family of zeros satisfying (8.3) and (8.5). For first order differential operators it was proved by Strauss and Treves [ST] that there is uniqueness for all non-characteristic surfaces if condition (P) is fulfilled, but recently Colombini and Del Santo [CD] proved that the condition (8.4) cannot be replaced by condition (P) in the uniqueness theorem above; in their example there are no non-trivial solutions of (8.3), (8.3)′ satisfying (8.5).

A surprising new discovery concerning the uniqueness of the Cauchy problem was made a few years ago by Robbiano [Ro] who proved that for a differential operator in $\mathbf{R}^{n+1}$ of the form

$$P = D_t^2 - A(x, D_x)$$

which is hyperbolic with respect to t, there is uniqueness for the Cauchy problem on every surface which is cylindrical in the t direction. This would be false in general for lower order terms with coefficients depending on t. A quantitatively more precise form of this result given in [H23] states that there is unique continuation of solutions of the equation $Pu = 0$ across a timelike surface with conormal (τ, ξ) at a point (t, x) provided that

$$27\tau^2/23 - a(x, \xi) < 0;$$

it is of course classical that there is unique continuation across space like surfaces, that is, surfaces with $\tau^2 - a(x, \xi) > 0$. Very recently Tataru [Tat] has proved that there is unique continuation across all non-characteristic surfaces. This is a special case of a general result stating that if the principal symbolis translation invariant along a linear subspace $\mathcal{A}$ and all coefficients are real analytic in the direction of $\mathcal{A}$, then the uniqueness theorem above is valid if the convexity conditions (8.2), (8.2)′ are satisfied when to the conditions (8.3), (8.3)′ and (8.5) is added that ξ is a conormal of $\mathcal{A}$. In [H24] it is proved that it suffices to assume that the restriction of the principal symbol to the conormal bundle of $\mathcal{A}$ and its parallel spaces is translation invariant, and there are similar improvements of the other results of [Tat] as well.

9. Spectral Asymptotics

When I was a graduate student in Lund the two professors, Lars Gårding and Åke Pleijel, were both working on asymptotic properties of eigenvalues and eigenfunctions of elliptic differential operators P, and so were most of the graduate students. I chose a different direction for my thesis but became of course aware of the state of the field.

Since the work of Carleman in the 1930's the main approach to such questions has been to study the kernel of some function of P, such as the resolvent or the Laplace transform. When pseudo differential operators had appeared and been recognized as a powerful tool for the construction of parametrices, it was natural to try their strength in this field.

Let $P(x, D)$ be an elliptic differential operator of order m with C^∞ coefficients in an open set $\Omega \subset \mathbf{R}^n$ with a self-adjoint extension $\mathcal{P}$ in $L^2(\Omega)$ which is bounded below. Let (E_λ) be the spectral resolution of $\mathcal{P}$ and let $e(x, y, \lambda)$ be the kernel of E_λ. Gårding [Gå1] proved using the simplest asymptotic properties of the resolvent of $\mathcal{P}$ that

$$R(x, \lambda) = \lambda^{-n/m} e(x, x, \lambda) - (2\pi)^{-n} \int_{p(x,\xi)<1} d\xi \tag{9.1}$$

converges to 0 as $\lambda \to +\infty$. Here p is the principal symbol of P. For operators with constant coefficients he proved in [Gå2] that $R(x, \lambda) = O(\lambda^{-1/m})$. For second order operators with variable coefficients this was proved by Avakumovič [Av] and in part by Lewitan [Lewi]. By a fairly straightforward application of pseudo differential calculus to the construction of the resolvent $(\mathcal{P} - z)^{-1}$ it was proved in [H25] that $R(x, \lambda) = O(\lambda^{-\theta/m})$ for every $\theta < \frac{1}{2}$ (every $\theta < 1$ if the coefficients of the principal part are constant). The same result was obtained independently with different methods by Agmon and Kannai [AK]; in fact, Agmon had been the first to prove such bounds with a positive θ. It was also proved in [H25] that the value of θ has decisive importance for the summability properties of the eigenfunction expansion with respect to $\mathcal{P}$. The reason why Avakumovič obtained the optimal value $\theta = 1$ was that he used the Hadamard construction of parametrices for second order differential equations which takes advantage of geodesic coordinate systems. In [H9] it was proved that $R(x, \lambda) = O(\lambda^{-1/m})$ in general by applying Fourier integral operators to construct a parametrix for the hyperbolic pseudodifferential operator

$$i\partial/\partial t - \mathcal{P}^{1/m}$$

after a reduction to a compact manifold and a positive operator $\mathcal{P}$ which makes $\mathcal{P}^{1/m}$ a well defined pseudodifferential operator. This construction, which takes into account the geometrical optics description of propagation of singularities, was the starting point for the work on Fourier integral operators described in Section 3. What is required is only an understanding of the operator $e^{-it\mathcal{P}^{1/m}}$ for small values of $|t|$. Later work by many mathematicians where this unitary group is studied also for large values of t has led to much deeper understanding of the eigenvalues of elliptic differential operators, in particular the connection between clustering of eigenvalues and closed bicharacteristics. Some of this work is covered by [H3, Chap. XXIX]. (In the first edition there is an error in Theorem 29.1.4 corrected in the second edition.) However, it would carry too far to give a survey of this work here.

References

[AK] S. Agmon and Y. Kannai, On the asymptotic behavior of spectral functions and resolvent kernels of elliptic operators. *Israel J. Math.* **5** (1967), 1–30.

[Al] S. Alinhac, Non-unicité du problème de Cauchy. *Ann. of Math.* **117** (1983), 77–108.

[Av] V. G. Avakumovič, Über die Eigenfunktionenauf geschlossenen Riemannscghen Mannigfaltigkeiten. *Math. Z.* **65** (1956), 327–344.

[BF1] R. Beals and C. Fefferman, Spatially inhomogeneous pseudo-differential operators I. *Comm. Pure Appl. Math.* **27** (1974), 1–24.

[BF2] ———, On local solvability of linear partial differential equations. *Ann. of Math.* **97** (1973), 482–498.

[BM] D. Bell and S.-E. Mohammed, An extension of Hörmander's theorem for infinitely degenerate second-order operators. *Duke Math. J.* **78** (1995), 453–475.

[Bo] J.-M. Bony, Équivalence des diverses notions de spectre singulier analytique. Sém. Goulaouic-Schwartz 1976–77, Exposé no. III.

[BL] J.-M. Bony and N. Lerner, Quantification asymptotique et microlocalisations d'ordre supérieur I. *Ann. Sci. École Norm. Sup.* **22** (1989), 377–433.

[BI] J. Bros and D. Iagolnitzer, Tuboïdes et structure analytique des distributions. Sém. Goulaouic-Schwartz 1974–75, Exposés XVI et XVIII.

[BG] J. Brüning and V. Guillemin, *Mathematics past and present. Fourier integral operators.* Springer Verlag, 1991.

[Ca] A. P. Calderón, Uniqueness in the Cauchy problem for partial differential equations. *Amer. J. Math.* **80** (1958), 16–36.

[Ch] M. Christ, Certain sums of squares of vector fields fail to be analytic hypoelliptic. *Comm. Partial Differential Equations* **16** (1991), 1695–1707.

[CD] F. Colombini and D. Del Santo, Condition (P) is not sufficient for uniqueness in the Cauchy problem. *Comm. Partial Differential Equations* **20** (1995), 2113–2128.

[D] N. Dencker, On the propagation of singularities for pseudo-differential operators of principal type. *Ark. Mat.* **20** (1982), 23–60.

[DH] J. J. Duistermaat and L. Hörmander, Fourier integral operators II. *Acta Math.* **128** (1972), 183–269.

[DT] M. Derridj and D. S. Tartakoff, Microlocal analyticity for the canonical solution to $\bar{\partial}_b$ on strictly pseudoconvex CR manifolds of real dimension three. *Comm. Partial Differential Equations* **20** (1995), 1871–1926.

[Eg1] Ju. V. Egorov, The canonical transformations of pseudo-differential operators. *Uspehi Mat. Nauk* **24** (1969), no 5:235–236.

[Eg2] ———, Subelliptic operators. *Uspehi Mat. Nauk* **30** (1975), no 2:59–118, no 3:55–105.

[FP] C. Fefferman and D. H. Phong, On positivity of pseudo-differential operators. *Proc. Nat. Acad. Sci.* **75** (1978), 4673–4674.

[Gå1] L. Gårding, On the asymptotic distribution of the eigenvalues and eigenfunctions of elliptic differential operators. *Math. Scand.* **1** (1953), 237–255.

[Gå2] ———, On the asymptotic properties of the spectral function belonging to a self-adjoint semi-bounded extension of an elliptic differential operator. *Kungl. Fysiogr. Sällsk. i Lund Förh.* **24:21** (1954), 1–18.

[Gr] V. V. Grushin, On the solutions of partial differential equations. *Doklady Akad. Nauk SSSR* **139** (1961), 17–19, (in Russian).

[HN] B. Helffer and J. Nourrigat, *Hypoellipticitémaximale pour des opérateurs polynômes de champs de vecteurs.* Birkhäuser, 1985, Progress in Mathematics **58**.

[H1] L. Hörmander, Existence, uniqueness and regularity of solutions of linear differential equations. in *Proc. Int. Congr. Math. 1962*, pp. 339–346.

[H2] ———, *Linear partial differential operators.* Springer Verlag, 1963.

[H3] ———, *The analysis of linear partial differential operators I–IV.* Springer Verlag, 1983–85.

[H4] ———, Pseudo-differential operators and hypo elliptic equations, in *Amer. Math. Soc. Symp. on Singular Integrals*, pp. 138–183, 1966.

[H5] ———, Hypoelliptic differential operators. *Ann. Inst. Fourier Grenoble* **11** (1961), 477–492.

[H6] ———, Linear differential operators, in *ActesCongr. Int. Math. Nice* **1** (1970), 121–133.

[H7] ———, Pseudo-differential operators and non-elliptic boundary problems. *Ann. of Math.* **83** (1966), 129–209.

[H8] ———, The Weyl calculus of pseudo-differential operators. *Comm. Pure Appl. Math.* **32** (1979), 359–443.

[H9] ———, The spectral function of an elliptic operator. *Acta Math.* **121** (1968), 193–218.

[H10] ———, Fourier integral operators I. *Acta Math.* **127** (1971), 79–183.

[H11] ———, Hypoelliptic second order differential equations. *Acta Math.* **119** (1967), 147–171.

[H12] ———, On the subelliptic test estimates. *Comm. Pure Appl. Math.* **33** (1980), 339–363.

[H13] ———, Subelliptic operators. in *Seminaron singularities of solutions of differential equations*, pp. 127–208. Princeton University Press, 1979.

[H14] ———, Differential operators of principal type. *Math. Ann.* **140** (1960), 124–146.

[H15] ———, Differential equations without solutions. *Math. Ann.* **140** (1960), 169–173.

[H16] ———, On the solvability of pseudodifferential equations in *Structure of solutions of differential equations*, ed. M. Morimoto and T. Kawai, World Scientific, 1996, pp. 183–213.

[H17] ———, Propagation of singularities and semi-global existence theorems for (pseudo-)differential operators of principal type. *Ann. of Math.* **108** (1978), 569–609.

[H18] ———, On the singularities of solutions of partial differential equations, in *Proc. Int. Conf. on Funct. Anal. and Related topics, Tokyo, April* 1969, pp. 31–40.

[H19] ———, On the existence and the regularity of solutions of linear pseudo-differential equations. *Ens. Math.* **17** (1971), 99–163.

[H20] ———, Uniqueness theorems and wave front sets for solutions of linear differential equations with analytic coefficients. *Comm. Pure Appl. Math.* **24** (1971), 671–704.

[H21] ———, Remarks on Holmgren's uniqueness theorem. *Ann. Inst. Fourier Grenoble* **43** (1993), 1223–1251.

[H22] ———, Non-uniqueness for the Cauchy problem. *Springer Lecture Notes in Math.* **459** (1974), 36–72.

[H23] ———, A uniqueness theorem for second order hyperbolic differential equations. *Comm. Partial Differential Equations* **17** (1992), 699–714.

[H24] ———, On the uniqueness of the Cauchy problem under partial analyticity assumptions in *Geometrical Optics and Related Topics*, F. Columbini and N. Lerner, eds. Birkhäuser, Boston, 1997.

[H25] ———, On the Riesz means of spectral functions and eigenfunction expansions for elliptic differential operators, in *The Belfer Graduate School Science Conference Nov. 1966*, pp. 155–202, 1969.

[HM] L. Hörmander and A. Melin, Free systems of vector fields. *Ark. Mat.* **16** (1978), 83–88.

[Jo] F. John, On linear differential equations with analytic coefficients. Unique continuation of data. *Comm. Pure Appl. Math.* **2** (1949), 209–253.

[KK] M. Kashiwara and T. Kawai, Microhyperbolic pseudodifferential operators I. *J. Math. Soc. Japan* **27** (1975), 359–404.

[Ka] T. Kawai, On the theory of Fourier hyperfunctions and its application to partial differential equations with constant coefficients. *J. Fac. Sci. Tokyo* **17** (1970), 467–517.

[KN] J. J. Kohn and L. Nirenberg, On the algebra of pseudo-differential operators. *Comm. Pure Appl. Math.* **18** (1965), 269–305.

[Ko] A. N. Kolmogorov, Zufällige Bewegungen. *Ann. of Math.* **35** (1934), 117–117.

[KS] S. Kusuoka and D. Strook, Applications of the Malliavin calculus, Part II. *J. Fac. Sci. Univ. Tokyo.*

[La] P. D. Lax, Asymptotic solutions of oscillatory initial value problems. *Duke Math. J.* **24** (1957), 627–646.

[Ler1] N. Lerner, Sufficiency of condition (ψ) for local solvability in two dimensions. *Ann. of Math.* **128**(1988), 243–258.

[Ler2] ———, Nonsolvability in L^2 for a first order operator satisfying condition (ψ). *Ann. of Math.* **139** (1994), 363–393.

[Ler3] ———, An iff solvability condition for the oblique derivative problem. Séminaire EDP 90–91, École Polytechnique, Exposé 18.

[Lewi] B. M. Lewitan, On the asymptotic behavior of the spectral function of a self-adjoint differential equation of the second order. *Izv. Akad. Nauk SSSR* **16** (1952), 325–352; II. **16** (1955), 33–58.

[Lew] H. Lewy, An example of a smooth linear partial differential equation without solution. *Ann. of Math.* **66**(1957), 155–158.

[Mel] A. Melin, Lower bounds for pseudo-differential operators. *Ark. Mat.* **9** (1971), 117–140.

[Met] G. Métivier, Analytic hypoellipticity for operators with multiple characteristics. *Comm. Partial Differential Equations* **6** (1981), 1–90.

[Mo] R. D. Moyer, Local solvability in two dimensions: Necessary conditions for the principle-type case. Mimeographed manuscript, University of Kansas (1978).

[NSW] A. Nagel, E. M. Stein and S. Wainger, Balls and metrics defined by vector fields I: Basic properties. *ActaMath.* **155** (1985), 103–147.

[NT1] L. Nirenberg and F. Treves, Solvability of a firstorder linear partial differential equation. *Comm. Pure Appl. Math.* **16** (1963), 331–351.

[NT2] ———, On local solvability of linear partial differential equations. Part I: Necessary conditions. *Comm. Pure Appl. Math.* **23** (1970), 1–38; Part II: Sufficient conditions. *Comm. Pure Appl. Math.* **23** (1970), 459–509; Correction. *Comm. Pure Appl. Math.* **24** (1971), 279–288.

[No] J. Nourrigat, Subelliptic systems. *Comm. in Partial Differential Equations* **15** (1990), 341–405.

[OR] O. A. Olejnik and E. V. Radkevič, Second order equations with non-negative characteristic form, in *Matem. Anal.* 1969 (R. V. Gamkrelidze, ed.) Moscow, 1971, (Russian). English translation Plenum Press, New York, London (1973).

[Ro] L. Robbiano, Théorème d'unicité adapté aucontrôle des solutions des problèmes hyperboliques. *Comm. Partial Differential Equations* **16** (1991), 789–800.

[RS] L. P. Rothschild and E. M. Stein, Hypoelliptic differential operators and nilpotent groups. *Acta Math.* **137** (1976), 247–320.

[SC] A. Sánchez-Calle, Fundamental solutions and geometry of the sum of squares of vector fields. *Invent. Math.* **78** (1984), 143–160.

[Sa] M. Sato, Regularity of hyperfunction solutions of partial differential equations, in *Actes Congr. Int. Math. Nice* **2** (1970), 785–794.

[SKK] M. Sato, T. Kawai and M. Kashiwara, Hyperfunctions and pseudodifferential equations. *Lecture Notes in Mathematics* **287** (1973), 265–529.

[Sj1] J. Sjöstrand, Analytic wavefront set and operators with multiple characteristics. *Hokkaido Math. J.* **12** (1983), 392–433.

[Sj2] ———, Singularités analytiques microlocales. *Astérisque* **95** (1982), 1–166.

[ST] M. Strauss and F. Treves, First order linear P.D.E.'s and uniqueness of the Cauchy problem, *J. Differential Equations* **38** (1980), 374–392.

[Tar] D. Tartakoff, The local real analyticity of solutions to $\Box_b$ and the $\bar{\partial}$-Neumann problem. *Acta Math.* **145** (1980), 177–204.

[Tat] D. Tataru, Unique continuation for solutions to PDE's; between Hörmander's theorem and Holmgren's theorem. *Comm. Partial Differential Equations* **20** (1995), 855–884.

[Tre] J.-M. Trépreau, Sur l'hypoellipticité analytique microlocale des opérateurs de type principal. *Comm. Partial Differential Equations* **9** (1984), 1119–1146.

[T1] F. Treves, Opérateurs différentiels hypoelliptiques. *Ann. Inst. Fourier Grenoble* **9** (1959), 1–73.

[T2] ———, Analytic hypo-elliptic PDE's of principal type. *Comm. Pure Appl. Math.* **24** (1971), 537–570.

[T3] ———, Analytic hypoellipticity of a class of pseudodifferential operators with double characteristics. *Comm. Partial Differential Equations* **3** (1978), 85–116.

[W] H. Weyl, Quantenmechanik und Gruppentheorie. *Zeitschrift für Physik* **46** (1927), 1–47, Collected works, Vol. III, pp. 90–135.

Box 118, S-221 00 Lund, Sweden
E-mail address: ivhmaths.lth.se

THE WORK OF JOHN W. MILNOR

by
HASSLER WHITNEY

To aid in understanding the significance of Milnor's work, I will first say a few words about the recent history of algebraic topology. In the early thirties, the subject seemed to have reached a certain level of completeness in its basic methods and results; the famous textbooks of Lefschetz, of Seifert and Threlfall, and of Alexandroff and Hopf, gave a very good picture of the field. The term "algebraic" had not yet been applied. Then in 1935, the sudden explosion of "cohomology theory", with its various kinds of applications, provided a great impetus to research. In developing these applications to new heights, various algebraic problems arose; one could often write out a formula which described a situation, but could not understand the geometric meaning of the formula. In the forties, powerful and general algebraic machinery came into being; this enabled problems, which had before seemed hopelessly complex, to be answered in relatively simple algebraic terms. The subject of "algebraic topology" was in full swing.

The new algebraic methods now began to grow by themselves; new concepts, such as exact sequences, sheaves, homological algebra, spread out not only through topology but through neighboring domains of algebra and analysis, where they have had extraordinary success. As pointed out by H. Hopf, in presenting the work of the Fields Medalist R. Thom at the last Congress at Edinburgh, the geometric point of view tended to be swamped in the algebraic machinery. Papers in algebraic topology had commonly the appearance of being pure algebra. Then in the mid fifties, some great discoveries of a more geometric nature took place, which brought on a resurgence of the geometric point of view. The definition of cobordism and its basic properties and applications by Thom in 1954 was one such discovery; for this he was awarded a Fields Medal in 1958, as mentioned above. In 1956, the mathematical world was astounded by Milnor's proof that the 7-dimensional sphere S^7 was capable of several differential structures. This at first might have seemed like an isolated fact; but through Milnor and others it has led to a vast field of work, which has now acquired a name: *differential topology*.

To explain this work, I must give some definitions. A *differential manifold* is a topological manifold with given local coordinate systems, such that the transformation from one coordinate system to an overlapping one is differentiable. I (and some others) have purposely altered the standard term, "differentiable manifold", precisely because Milnor's result shows that one must tell, not merely whether a manifold is "differentiable", but which differential structure one has in mind. (Recently M. Kervaire has shown that there are topological manifolds which admit no differential structure at all.) A homeomorphism between two topological manifolds is a one-one correspondence between them which is continuous in both directions; a *diffeomorphism* is a homeomorphism that is differentiable in both directions.

We may now state Milnor's result: There are differential manifolds which are homeomorphic to the 7-sphere S^7, but not diffeomorphic to it.

A first reaction to such a theorem might be: What is the use in showing that S^7 can be distorted to such a degree that the original differential structure is lost, but a new one is taken on, distinct from the first? The fact is, the new manifolds are constructed not by pathological means, but by simple and well known methods; the actual differential manifolds had certainly been considered, but one had not suspected that they were homeomorphic to the sphere, and yet had different differential structures.

The manifolds are defined as follows. Start with the Cartesian product $S^4 \times S^3$. Cut S^4 apart at its equator S^3, thus cutting the product manifold apart. Now glue the two resulting bounded manifolds together again at their boundaries, identifying (u, v) with (u, v'), where $v' = u^h v u^j$; the multiplication here is quaternion multiplication in $S^3 \subset R^4$. Taking $h + j = l, h - j = k$, for any odd integer k, gives a manifold M_k^7; it is the total space of a sphere bundle over S^4, and has an obvious differential structure. It is easy to show that there is a differentiable real valued function on M_k^7 which admits just two critical points, these being nondegenerate; from this it follows that M_k^7 is homeomorphic to S^7.

An invariant λ for closed oriented differential manifolds M^7 is now defined, as follows: As part of Thom's cobordism theory, M^7 is the boundary of a differential manifold B^8. (For the M_k^7, one may take B_k^8 to be the total space of the associated disk bundle.) Now $\lambda(M^7)$ is defined as a certain function of the "index" and the first Pontrjagin class of B^8; it is an integer mod 7, independent of the choice of B^8. Moreover, $\lambda(M_k^7) \equiv k^2 - 1 \bmod 7$, and $\lambda(S^7) \equiv 0$; hence M_3^7 for instance is homeomorphic but not diffeomorphic to S^7, and it thus may be considered as defining a new differential structure on the topological manifold S^7.

Milnor later extended this work to a study of spheres of dimension $4m - 1$, with $m > 2$. The 31-sphere, for example, was shown to admit more than sixteen million distinct differential structures.

The next job in building up a theory of these structures was to form them into some sort of mathematical system. Milnor showed that an abelian group θ^n could be constructed from them, for each dimension n. We give a brief description of θ^n. The *connected sum* $M_1 \# M_2$ of two differential n-manifolds M_1 and M_2 is formed by removing a small n-cell from each, and then gluing together the resulting boundaries, with proper regard to orientation. Two closed differential manifolds M and M' are *h – cobordant* (homotopy cobordant) if the disjoint sum $M + (-M')$ is the boundary of some differential manifold W, such that both M and M' are deformation retracts of W; this is an equivalence relation. Now the group θ^n is the abelian group of h-cobordism classes of differential manifolds which are homotopy n-spheres (that is, have the same homotopy type as S^n), under the connected sum operation.

Through some deep work of S. Smale and others, it is known that the elements of θ^n correspond precisely to the distinct diffeomorphism classes of S^n. Hence, for instance, the order of θ^n is the number of distinct differential structures on S^n. These groups have wide importance; for instance, they serve as coefficient groups in the study of obstructions to the construction of differential structures on a manifold. They play a role, relative to differentiability, similar to the role of the homotopy groups in more purely topological questions.

Kervaire and Milnor, in a joint study, have shown that the groups θ^n are always finite, except that the structure of θ^3 is unknown. (If the Poincaré conjecture were proved, it would follow that θ^3 is trivial.) They have found the orders of many of these groups; for instance:

n	1	2	3	4	5	6	7	8	9	10	11	12	13	14	15
order of θ^n	1	1	?	1	1	1	28	2	8	6	992	1	3	2	16256

The order of θ^{4m-1} is given in terms of the order of the stable homotopy group $\pi_{k+4m-1}(S^k)$ (k large) and the Bernoulli number B_m.

We must mention the recent solution by Milnor of two outstanding problems of topology; in each case the solution was made possible by deep work of someone else. First, when R. Bott proved a theorem about the divisibility of the highest Pontrjagin class of the $4m$-sphere, Milnor saw that, combining this with results of Wu Wen-Tsün, it followed that the n-sphere is parallelizable only if $n = 1$, 3 or 7. (Kervaire and Hirzebruch also discovered this independently.) A further consequence of this is that there exists a division algebra of rank n over the real numbers if and only if $n = 1$, 2, 4 or 8. (In these cases, examples are: the real numbers, the complex numbers, the quarternions, and the Cayley numbers.)

Second, the so-called "Hauptvermutung" for complexes, that homeomorphic polyhedra have isomorphic cell subdivisions, was formulated in a paper by H. Tietze in 1908 (it was conjectured earlier). In this paper, he studied the "lens spaces" $L(p, q)$. These are manifolds, defined as follows. Set $\omega = e^{2\pi i/p}$. Taking the unit sphere S^3 in the space of pairs (z_1, z_2) of complex numbers and identifying any two points $(z_1, z_2), (\omega z_1, \omega^q z_2)$ of S^3 defines $L(p, q)$. Tietze showed that p is a topological invariant. With the help of a very recent theorem of B. Mazur on the existence of a diffeomorphism between certain product spaces, Milnor shows that $L(p, q) \times R^n$ is diffeomorphic to $L(p, q') \times R^n$ provided that $\pm qq'$ is a quadratic residue mod p, and $n > 3$. Now construct the finite complex X_q from the product $L(7, q) \times \sigma^n$, σ^n being an n-simplex, by adjoining a cone over the boundary $L(7, q) \times \partial\sigma^n$. Then for $n \geqslant 3$, X_1 is homeomorphic with X_2, as can be proved from the result stated above. Now the Hauptvermutung is *disproved* by showing that X_1 and X_2 have no isomorphic cell subdvisions. This is accomplished by slightly generalizing the torsion invariant of Reidemeister, Franz and de Rham, and following some methods developed by J. H. C. Whitehead.

One possible peculiarity of differential structures appears as a corollary of the work just mentioned: The bounded manifolds $L(7,1) \times D^5$ and $L(7,2) \times D^5$ (where D^5 is the unit 5-disk) are not diffeomorphic, but their interiors are.

To look ahead, Milnor is now developing a theory of "microbundles", which injects a new and basic tool into the subjects discussed above. You will hear about this from him in his address at this Congress.

In conclusion, it is apparent that differential topology is a strong young field, with a bright present and a great outlook, and that its vitality is largely due to the fine achievements of Milnor. It gives me especial pleasure to report on this; in the thirties I was much concerned with the relations between differential and topological properties, and I felt that my own contributions were merely some beginnings of what might come at some time in the future. The future has arrived!

John Milnor

TOPOLOGICAL MANIFOLDS AND SMOOTH[1] MANIFOLDS

by

J. MILNOR

Suppose that one is given a topological manifold M (i.e. a Hausdorff space with a countable basis where each point has a neighborhood homeomorphic to some euclidean space). Then one can ask the following two questions:

Problem 1. Can M be given the structure of a smooth manifold? In more intuitive terms: can M be imbedded in a high dimensional euclidean space so as to have a continuously turning tangent plane?

Problem 2. If such a smoothness structure exists, is it essentially unique? More precisely, given two such structures on M, does there exist a homeomorphism of M onto itself which carries one structure to the other?[2]

The first problem was answered negatively when M. Kervaire gave an example of a compact triangulable 10-dimensional manifold which is not smoothable. (Thus if Kervaire's manifold is imbedded in some euclidean space, its image must have "angles" or "corners" or worse singularities.) Other such examples, in other dimensions, have been given by Smale, Tamura, Wall and by Eells and Kuiper (references [5], [6], [14], [18], [21]).

The second problem was answered negatively when the author showed that the 7-dimensional sphere possesses several essentially distinct smoothness structures (see [8], [9], [13], [17]).

Thus the two problems are non-trivial. They lead naturally to the following.

Problem 3. Given a topological manifold M, can one make a classification of all possible smoothness structures on M?

The answer must surely depend on a detailed knowledge of the topology of M.

Quite a bit of progress on these questions has been made during the last few years. Suppose for example that M is the topological sphere S^n. Define two smoothness structures on S^n to be *equivalent* if there exists an orientation preserving homeomorphism of S^n to itself which carries one smoothness structure to the other. For $n \neq 4$ it is known that the set of such equivalence classes can be made into an abelian group, which is denoted by Γ_n. The structure of this group for many small

[1]The word *smooth* will be used as a synonym for "differentiable of class C^∞".

[2]This is equivalent to the question as to whether the two resulting smooth manifolds are diffeomorphic to each other.

values of n has been determined by Kervaire and Milnor, making use of work by Smale (references [7], [10], [14], [15]). (For the cases $n < 4$ see [11], [22].) For example one has:

$$\Gamma_1 = \Gamma_2 = \Gamma_3 = \Gamma_5 = \Gamma_6 = 0, \quad \Gamma_7 = Z_{28}, \quad \Gamma_8 = Z_2.$$

The groups $\Gamma_n, n \neq 4$, are all finite.(3)

Now let M be an arbitrary triangulated manifold. J. Munkres, in reference [12], has defined a sequence of obstructions, whose vanishing implies that M can be given a smoothness structure (see also Thom [19], [20]). These obstructions are homology classes of M with coefficients in the groups Γ_i Similarly, if one is given two different smooth manifolds with the same underlying complex, Munkres [11] has defined a sequence of obstruction classes whose vanishing implies that the two manifolds are diffeomorphic. Again the groups Γ_i occur as coefficient groups.

One interesting application of Munkres' results has been made by J. Stallings. In reference [16], Stallings shows that the euclidean space $R^n, n \neq 4$, has an essentially unique smoothness structure.

In the remainder of this lecture, I would like to introduce a quite different tool, which I hope will be used in the future to attack these problems; namely the theory of microbundles.

A "microbundle" is an object something like a fibre bundle having the euclidean space R^n as fibre. However the fibre in a microbundle is not an honest topological space, but is only a "germ" of a topological space. This can be made precise as follows.

Definition. An R^n-*microbundle* over B is a commutative diagram

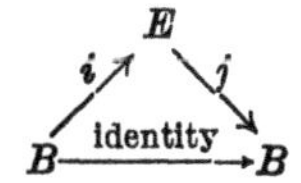

where B, E are topological spaces, and i, j are continuous maps; such that the following "local triviality" condition is satisfied:

Requirement. For each $b \in B$ there should exist neighborhoods U of b and V of $i(b)$, with

$$i(U) \subset V, \; j(V) = U$$

so that V is homeomorphic to $U \times R^n$ under a homomorphism which makes the following diagram commutative

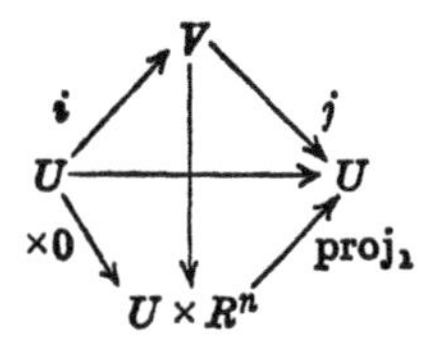

[3]Here Z_k denotes the cyclic group of order k. For $n = 4$ the group Γ_4 must be defined somewhat differently. Nothing is known about the structure of Γ_4.

Here R^n denotes the n-dimensional euclidean space, $\times 0$ denotes the mapping $u \to (u, 0)$, and proj_1 denotes the projection to the first factor: $\text{proj}_1\ (u, x) = u$.

Such a microbundle will be denoted by a single German letter, such as $\mathfrak{x}$. The spaces E and B will be called the *total space* and the *base space* respectively. The maps i, j will be called the *injection* and the *projection* maps of $\mathfrak{x}$.

Note that this condition of local triviality depends only on that portion of E which lies in a arbitrarily small neighborhood of $i(B)$. If E_0 is any neighborhood of $i(B)$ in E then we will take the point of view that the new microbundle

$$B \xrightarrow{\ i\ } E_0 \xrightarrow{\ j|E_0\ } B, \qquad B \xrightarrow{\text{identity}} B$$

can be identified with the original one. More precisely, and more generally:

Definition. A second microbundle $\mathfrak{x}'$ over B with diagram

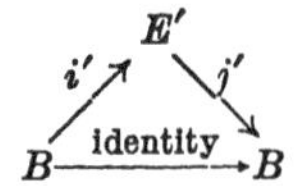

is *isomorphic* to $\mathfrak{x}$ if there exist neighborhoods E_0 of $i(B)$ in E and E_0' of $i'(B)$ in E', and a homeomorphism from E_0 to E_0' which makes the following diagram commutative.

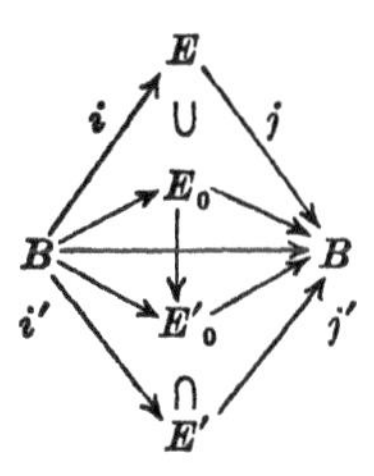

Here are some examples of microbundles.

Example 1. For any topological space B and any integer $n \geqslant 0$ one has the *trivial* microbundle $\mathfrak{e}^n$ with diagram

$$B \xrightarrow{\ \times 0\ } B \times R^n \xrightarrow{\ \text{proj}_1\ } B, \qquad B \xrightarrow{\text{identity}} B$$

More generally any microbundle isomorphic to $\mathfrak{e}^n$ is called a trivial micro-bundle.

Example 2. Let ξ be a vector bundle over B with total space $E(\xi)$ and projection map $p : E(\xi) \to B$. There is a standard cross-section

$$z: B \to E(\xi)$$

which assigns to each $b \in B$ the zero vector in the vector space $p^{-1}(b)$. The *underlying microbundle* $|\xi|$ of ξ is defined to be the microbundle

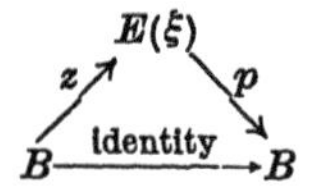

Example 3. Let M be a topological manifold. Then the *tangent microbundle* ‡ of M is defined to be the microbundle

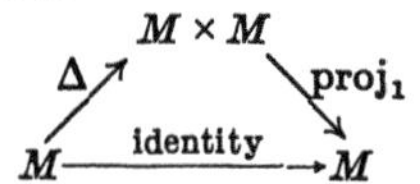

where Δ denotes the diagonal map. Thus the "fibre" over a point $x_0 \in M$ is the set of all pairs (x_0, y) where y ranges over an arbitrary neighborhood of x_0 in M. The local triviality condition can be verified as follows. Given $x_0 \in M$ let U be a neighborhood homeomorphic to R^n under a homeomorphism h, and let $V = U \times U$. Then V is homeomorphic to $U \times R^n$ under the homeomorphism,

$$f(u_1, u_2) - (u_1, h(u_2) - h(u_1)),$$

which makes the following diagram commutative.

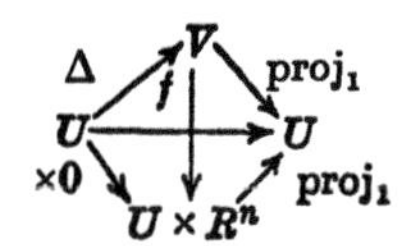

Now suppose that M can be made into a smooth manifold. Then, using the smoothness structure, one can also define the *tangent vector bundle* τ of M. The following result is fundamental.

THEOREM 1. *In this situation the underlying microbundle* $|\tau|$ *is isomorphic to the tangent microbundle* ‡ of M.

The proof can be outlined as follows. Choose a Riemannian metric on M. Then for any tangent vector $v \in E(\tau)$ which is not too long, there exists a geodesic segment

$$\gamma_v : [0, 1] \to M,$$

whose velocity vector at 0 is the given vector v. Now the correspondence

$$v \to (\gamma_v(0), \gamma_v(1))$$

defines the required homeomorphism between a neighborhood of $z(M)$ in $E(t)$ and a neighborhood of the diagonal in $M \times M$.

COROLLARY. *If M can be smoothed then the tangent microbundle ‡ is isomorphic to $|\xi|$ for some vector bundle ξ over M.*

A fundamental conjecture would be the converse proposition:

Problem. If ‡ is isomorphic to $|\xi|$ for some ξ, does it follow that M can be given a smoothness structure?

The following partial result can be proved.

THEOREM 2. *If the tangent microbundle of M is isomorphic to $|\xi|$ for some vector bundle ξ, then the Cartesian product $M \times R^{4n+1}$ can be given a smoothness structure.*

I will not try to describe the proof, which is based on a method due to M. Curtis and R. Lashof [4].

Many standard constructions for vector bundles carry over immediately to microbundles. For example given a microbundle $\mathfrak{x}$ over B, and given a map $\dagger : B' \to B$, one can construct the *induced microbundle*$\dagger^*\mathfrak{x}$ over B'.

THEOREM 3. (Homotopy theorem.) *If B' is paracompact, and if $g : B' \to B$ is homotopic to f, then the induced bundle $g^*\mathfrak{x}$ is isomorphic to $\dagger^*\mathfrak{x}$.*

The proof is similar to the usual proof for vector bundles.

Given two microbundles $\mathfrak{x}$ and $\mathfrak{y}$ over the same base space B, one can construct the *Whitney sum* $\mathfrak{x} \oplus \mathfrak{y}$, a new vector bundle over B. By definition, $\mathfrak{x} \oplus \mathfrak{y}$, is equal to $\Delta^*(\mathfrak{x} \times \mathfrak{y})$, where $\mathfrak{x} \times \mathfrak{y}$ denotes the Cartesian product microbundle

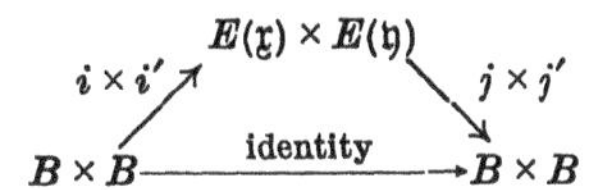

and where $\Delta : B \to B \times B$ denotes the diagonal map.

Definition. Two microbundles $\mathfrak{x}$ and $\mathfrak{x}'$ over B belong to the same s-class if there exist integers m, m' so that the Whitney sum $\mathfrak{x} \oplus \mathfrak{e}^m$ is isomorphic to $\mathfrak{x}' \oplus \mathfrak{e}^{m\prime}$. (Here $\mathfrak{e}^{m'}$ denotes the trivial R^m-microbundle over B.)

THEOREM 4. *Let B be a finite dimensional complex. Then the s-classes of microbundles over B form an abelian group with respect to the Whitney sum operation.*

The proof is more difficult than the corresponding proof for vector bundles. The key step, showing that for each $\mathfrak{x}$ there exists a $\mathfrak{y}$ with $\mathfrak{x} \oplus \mathfrak{y}$ trivial, is proved by induction on the dimension of B.

This group of s-classes of microbundles will be denoted by $k_{\text{Top}}B$. The analogous group whose elements are s-classes of vector bundles over B will be denoted by $k_{\text{Orthog}}B$. Note that the correspondence $\xi \to |\xi|$ gives rise to a natural homomorphism

$$k_{\text{Orthog}}B \to k_{\text{Top}}B.$$

Note also that the groups $k_{\text{Top}}\ B$ (or $k_{\text{Orthog}}\ B$) behave somewhat like cohomology groups. Thus any map $f : B' \to B$ induces a homomorphism

$$f^* : k_{\text{Top}} B \to k_{\text{Top}} B'.$$

If f is a homotopy equivalence, then f^* is an isomorphism.

The groups $k_{\text{Orthog}} B$ well known through the work of Atiyah, Hirze-bruch, Adams and others (see [1], [2], [3]). Unfortunately very little is known about $k_{\text{Top}} B$. For example it is not known whether the groups $k_{\text{Top}} S^n$ are finite, countably infinite, or uncountably infinite. Even the group $k_{\text{Top}} S^1$ seems forbiddingly difficult to compute.

The following qualitative result can be obtained.

THEOREM 5. *There exists a finite complex* X_1 *for which the canonical homomorphism*

$$k_{\text{Orthog}} X_1 \to k_{\text{Top}} X_1$$

has a non-trivial kernel. Furthermore there exists a finite complex X_2 *so that the canonical homomorphism*

$$k_{\text{Orthog}} X_2 \to k_{\text{Top}} X_2$$

is not onto.

Thus the theory of microbundles is essentially distinct from the theory of vector bundles. The proof of Theorem 5 is quite difficult. It is based on joint research with M. Kervaire [7].

(Actually the proof of Theorem 5 gives a specific example of such a complex X_1: namely a 7-sphere with an 8-cell attached by a map of degree 7. For X_2 the proof shows only that one of two possibilities will work. If $k_{\text{Top}} S^8$ is infinite then the 8-sphere itself will serve as a complex X_2. If $k_{\text{Top}} S^8$ is finite, then the 8-sphere with a 9-cell attached by a map of degree 3 will serve.)

Each half of Theorem 5 has an interesting consequence.

COROLLARY 1. *The tangent vector bundle of a certain smooth manifold* M_1 *is not a topological invariant.*

Proof. Choose an open set U_1 in some euclidean space R^n which has the same homotopy type as X_1. Then there exists a vector bundle ξ over U_1 whose s-class is non-trivial, and belongs to the kernel of the homomorphism $k_{\text{Orthog}} U_1 \to k_{\text{Top}} U_1$. Thus the underlying microbundle $|\xi|$ is s-trivial. Without loss of generality we may assume that $|\xi|$ itself is trivial.

Let ε^p denote the trivial vector bundle, with total space $U_1 \times R^p$, where p is the fibre dimension of ξ. Since $|\varepsilon^p|$ is isomorphic to $|\xi|$ it follows that some neighborhood M_1 of $U_1 \times 0$ in $U_1 \times R^p$ is homeomorphic to some neighborhood M_1' of the zero cross-section in $E(\mathfrak{x})$.

But each of the bundles ε^p and ξ can be given the structure of a smooth vector bundle. Hence the open sets $M_1 \subset E(\varepsilon^p)$ and $M_1' \subset E(\xi)$ can be considered as smooth manifolds. Clearly the manifold M_1 is parallelizable. However the tangent vector bundle of M_1, restricted to U_1 is isomorphic to

$$(\text{tangent bundle of } U_1) \oplus \xi \cong \varepsilon^n \oplus \xi.$$

Thus M_1' is not parallelizable. This completes the proof of Corollary 1.

COROLLARY 2. *There exists a topological manifold M_2 such that no Cartesian product $M_2 \times M'$ can be given a smoothness structure.*

Sketch of proof. Let U_2 be an open subset of some euclidean space R^n having the homotopy type of X_2. Then there exists a microbundle $\mathfrak{x}$ over U_2 whose s-class does not belong to the image of the homomorphism

$$k_{\text{Orthog}} U_2 \to k_{\text{Top}} U_2.$$

Let M_2 be the total space of this microbundle. We may assume that M_2 is a manifold.

It can be shown that the tangent microbundle of M_2, restricted to U_2, is isomorphic to the Whitney sum

$$(\text{tangent microbundle of } U_2) \oplus \mathfrak{x} \cong \mathfrak{e}^n \oplus \mathfrak{x}.$$

Since this is not isomorphic to $|\xi|$ for any vector bundle ξ over U_2, it follows from the Corollary to Theorem 1 that M_2 is not smoothable.

Given any positive integer p, a similar argument shows that the product $M_2 \times R^p$ is not smoothable. But this implies that no Cartesian product $M_2 \times M'$ can be smoothable; and proves Corollary 2.

References

[1] ADAMS, J. F., Vector fields on spheres. *Ann. Math.*, 75 (1962), 603–632.

[2] ATIYAH, M. & HIRZEBRUCH, F., Riemann-Roch theorems for differentiable manifolds. *Bull. Amer. Math. Soc.*, 65 (1959), 276–281.

[3] ——— Vector bundles and homogeneous spaces. *Proc. Symp. Pure Math.*, 3. Differential Geometry, *Amer. Math. Soc.*, (1961), 7–38.

[4] CURTIS, M. & LASHOF, R., On product and bundle neighborhoods. (To appear.)

[5] EELLS, J. & KUIPER, N., Manifolds which are like projective planes. *Publ. Math. (I.H.E.S.), Paris*, 1962.

[6] KERVAIRE, M., A manifold which does not admit any differentiable structure. *Comment. Math. Helv.*, 34 (1960), 257–270.

[7] KERVAIRE, M. & MILNOR, J., Groups of homotopy spheres. (To appear.)

[8] MILNOR, J., On manifolds homeomorphic to the 7-sphere. *Ann. Math.*, 64 (1956), 399–405.

[9] ——— Differentiable structures on spheres. *Amer. J. Math.*, 81 (1959), 962–972.

[10] ——— Sommes de variétés différentiables et structures différentiables des sphéres. *Bull. Soc. Math. France*, 87 (1959), 439–444.

[11] MUNKRES, J., Obstructions to the smoothing of piecewise-differentiable homeomorphisms. *Ann. Math.*, 72 (1960), 521–554.

[12] ——— Obstructions to imposing differentiable structures. (To appear.)

[13] SHIMADA, N., Differentiable structures on the 15-sphere and Pontrjagin classes of certain manifolds. *Nagoya Math. J.*, 12 (1957), 59–69.

[14] SMALE, S., Generalized Poincare's conjecture in dimensions greater than four. *Ann. Math.*, 74 (1961), 391–406.

[15] ——— On the structure of manifolds. *Amer. J. Math.*, 84 (1962), 387–399.

[16] STALLINGS, J., The piecewise-linear structure of Euclidean space. *Proc. Cambridge Philos. Soc.*, 58 (1962), 481–488.

[17] TAMURA, I., Homeomorphy classification of total spaces of sphere bundles over spheres. *J. Math. Soc. Japan*, 10 (1958), 29–43.

[18] ——— 8-manifolds admitting no differentiable structure. *J. Math. Soc. Japan*, 13 (1961), 377–382.

[19] THOM, R., Les structures différentiables des boules et des sphéres. *Colloque Géom. Diff. Globale.* Bruxelles, 1958, 27–35. Centre Belge Rech. Math., Louvain, 1959.

[20] ——— Des variétés triangulées aux variétés différentiables. *Proc. Internat. Congress Math. Edinburgh*, 1958, 248–255. Cambridge Univ. Press, 1960.

[21] WALL, C. T. C., Classification of $(n-1)$-connected $2n$-manifolds. *Ann. Math.*, 75 (1962), 163–189.

[22] WHITEHEAD, J. H. C., Manifolds with transverse fields in euclidean space. *Ann. Math.*, 73 (1961), 154–212.

THE WORK OF MICHAEL F. ATIYAH*

by
HENRI CARTAN

I will discuss very briefly the works of Atiyah in three areas, for that matter closely interconnected: K-theory, the index formula, and the "Lefschetz formula". I will leave aside his other contributions, in Algebraic Geometry or the theory of cobordism, although they are very interesting; and I will also pass over in silence the very recent results, still unpublished, of which the author himself will be speaking in his lecture during this Congress.

1. *K*-theory. Most of the works of Atiyah in K-theory have been done in collaboration with F. Hirzebruch. It was in 1956 that the fundamental book by Hirzebruch ("Neue topologische Methoden in der algebraischen Geometrie"†) appeared, of which the ultimate goal was the famous theorem that now bears the name "Riemann-Roch-Hirzebruch theorem". It was about algebraic geometry over the complex field. Shortly afterwards, Grothendieck sought and obtained a purely algebraic proof (valid over any algebraically closed base field, of any characteristic) of a more general theorem [1], since instead of considering an algebraic variety X he studied a morphism $X \to Y$ (the case treated by Hirzebruch being that where the algebraic variety Y is reduced to a point). It is on this occasion that Grothendieck introduced a contravariant functor which, to each algebraic variety X, associates a *ring* constructed using the isomorphism classes of algebraic vector bundles with base X. Atiyah and Hirzebruch [2] had the idea of doing the same for a *compact* topological space and for the classes of *complex* vector bundles with base X (that is to say, locally trivial topological bundles). One thus defines a ring $K(X)$ for any compact space X, whence the name of K-theory. There is also a KO-theory for *real* vector bundles, and a KSp-theory for quaternionic vector bundles.

We confine ourselves, for simplicity, to K-theory. One defines the relative groups $K(X,Y)$ (for Y a closed subspace of X), and then, by suspension, the groups $K^n(X,Y)$ for n an integer $\leqslant 0$, with $K^0(X,Y) = K(X,Y)$. One then has an exact sequence

$$\cdots \to K^n(X,Y) \to K^n(X) \to K^n(Y) \to K^{n+1}(X,Y) \to \cdots$$

*Translated from French to English by Chee Whye Chin.
† "Topological methods in algebraic geometry".

analogous to the exact sequence of cohomology. On the other hand, Atiyah observes that the famous Bott periodicity theorem [which concerns the homotopy groups of the infinite unitary group $U = \varinjlim_{m} U(m)$] can be expressed by an explicit isomorphism

$$K^n(X) \approx K^{n+2}(X),$$

which allows one to define the functor K^n also for n an integer > 0. In this way, one obtains a "cohomological theory" in the sense of Eilenberg-Steenrod, except that one of the Eilenberg-Steenrod axioms (the axiom "of dimension") is not satisfied. This theory was first dubbed "extraordinary cohomology".

If one wants to compare extraordinary cohomology with ordinary cohomology, one can roughly say this: instead of considering, as in ordinary cohomology, the homotopy classes of maps from a space X to the Eilenberg-MacLane space $K(\pi, n)$, one considers, in K-theory, the infinite unitary group U (or, what amounts to be the same, the infinite complex linear group), and its classifying space BU; it is these which serve as spaces for comparison. The existing relationships between the two cohomological theories (ordinary and extraordinary) are expressed by a spectral sequence, and the "Chern character"

$$ch \; : \; K^*(X, Y) \to H^*(X, Y; \mathbf{Q})$$

is a multiplicative homomorphism from one theory to the other.

The importance of "extraordinary cohomology" was quickly highlighted by the applications that Atiyah and Hirzebruch made, in Algebraic Topology and elsewhere [3]. Let us mention several examples which illustrate these applications of K-theory:

- a theorem of "Riemann-Roch-Grothendieck" type, this time valid for differentiable manifolds [4];
- the computation of $K(X)$ for certain homogeneous spaces, and the relation of this question to the representation theory of compact Lie groups [2];
- non-embedding theorems [5]: for example, complex projective space $P_n(\mathbf{C})$ cannot be differentiably embedded into affine space $\mathbf{R}^{4n-2\alpha(n)}$, where $\alpha(n)$ denotes the number of digits 1 in the dyadic expansion of the integer n;
- criteria for recognizing whether a cohomology class of a compact complex analytic variety can be represented by an analytic sub-variety [6].

All these applications are due to the collaboration of Atiyah with Hirzebruch. There are others; for example, it is through K-theory and the introduction of certain functors $K \to K$ (the idea of which came essentially from Grothendieck) that J.F. Adams [7] was able to completely solve a classical problem, which had remained unanswered for a long time: the exact determination, as a function of the integer n, of the maximum number of linearly independent vector fields on the sphere S^n (see the lecture of Adams at the Stockholm Congress in 1962).

2. The index theorem. But the most beautiful application of K-theory was to be made by Atiyah himself: I am speaking of the *index theorem* (1963), proved in collaboration with I. Singer [8].

Let D be an elliptic operator on a compact differentiable manifold X (assumed without boundary), mapping from the vector space $\Gamma(E)$ of differentiable sections of a complex vector bundle E to the space $\Gamma(F)$ of differentiable sections of a complex vector bundle F. One knows that the kernel and cokernel of the linear map $D : \Gamma(E) \to \Gamma(F)$ are finite dimensional; the *index* $i(D)$ is the integer defined by

$$i(D) = \dim(\operatorname{Ker} D) - \dim(\operatorname{Coker} D).$$

The works of several Soviet mathematicians had made obvious the fact that $i(D)$ does not change when D varies in a continuous manner, and I.M. Gelfand, in 1960 [9], had conjectured that $i(D)$ should therefore be computable in terms of purely topological invariants related to the given X and D. This is the problem which Atiyah and Singer have completely solved. The homogeneous terms of the highest degree of the operator D define a "symbol" $\sigma(D)$ which first of all allows one to define the ellipticity of D, and then, by involving K-theory, Chern character and the Todd class of the cotangent bundle of X (which has an almost complex structure), to finally define a cohomology class, element of $H^*(X;\mathbf{Q})$. Its component of degree $n = \dim X$ is an element of $H^n(X;\mathbf{Q}) \approx \mathbf{Q}$ (one supposes that X is orientable, for simplicity). Hence a *rational number* $i_t(D)$ is attached to D (and X), and is defined up to a sign; one can call it the "topological index" of D. The theorem of Atiyah-Singer then says that *the topological index* $i_t(D)$ *is equal to the index* $i(D)$ (with suitable conventions on the orientation). This theorem thus establishes a bridge between two vast areas of mathematics: analysis of partial differential equations on one hand, algebraic topology on the other.

Let us observe that, by definition, $i(D)$ is an integer. It follows that the rational number $i_t(D)$ coming from Algebraic Topology is, in fact, an integer. In this way one obtains, in a natural manner, by the choice of appropriate elliptic operators, all the "integrality theorems" related to characteristic classes of manifolds (integrality of the L-genus, of the Todd genus, of the $\hat{A}$-genus). Conversely, any information given by Algebraic Topology gives a result that is of interest in Analysis; for example, one sees easily that the topological index $i_t(D)$ is zero if the manifold X is of odd dimension.

The proof of the theorem $i(D) = i_t(D)$ is laborious, but very interesting, because it leads to the introduction of operators more general than differential operators, namely the singular integral operators of Calderon-Zygmund and Seeley. The proof also relies on a theory of cobordism which constitutes a generalization (relatively easy) of that due to Thom. In fact there is a new, more recent, proof of the index formula, which avoids the use of cobordism.

Instead of considering a single elliptic operator, one can consider a sequence of differential operators

$$\text{(D)} \qquad \Gamma(E_0) \to \Gamma(E_1) \to \cdots \to \Gamma(E_k)$$

forming a "complex" (i.e.: the composite of two consecutive operators is zero). One defines the "ellipticity" of such a complex. To each elliptic complex, one again attaches a rational number $i_t(D)$. On the other hand, the homology groups of the elliptical complex (D) are finite dimensional vector spaces; let $\chi(D)$ denote the alternating sum of their dimensions (it is a kind of Euler-Poincaré invariant). Then one has the theorem:

$$\chi(D) = i_t(D).$$

This more general form of the index theorem is very useful in applications. For example, if one applies it to a compact complex analytic variety X, and to the "complex" defined by the d'' differential operator (also written as $\overline{\partial}$) of differential forms, one recovers exactly the statement of the Riemann-Roch-Hirzebruch theorem. The latter was proven previously only for algebraic varieties without singularity; it is henceforth valid for any compact analytic variety.

I am leaving aside the index theorem for manifolds with boundaries [10]; it requires a new definition of ellipticity taking into account the "conditions at the limits". The question has been completely solved by Atiyah in collaboration with Bott and Singer.

3. Fixed point formulas. The index theorem is, in reality, only an extreme case of a situation of which another extreme case is, when the elliptic complex is that defined by the exterior differentiation operator on differential forms, the Lefschetz formula for fixed points (assumed isolated) for a transformation of a compact manifold X to itself. There are many intermediate cases, of which investigation is in progress. The results already obtained are due to the collaboration of Atiyah and Bott [11]. Let us explain with an example what it is about: let X be a compact, complex analytic variety, and let $\mathfrak{f} : X \to X$ be a holomorphic map; one knows that the cohomology vector spaces $H^q(X, \mathcal{O})$ with coefficients in the sheaf $\mathcal{O}$ of holomorphic functions are finite dimensional; let $L(\mathfrak{f})$ denote the alternating sum of traces

$$(-1)^q \operatorname{Tr}\left(\mathfrak{f}|_{H^q(X,\mathcal{O})}\right).$$

This is an integer; in the case where $\mathfrak{f}$ is the identity, this integer is none other than the first term of the Riemann-Roch-Hirzebruch equation. In the general case, one would like to express this integer in terms of the topological properties of $\mathfrak{f}$ in the neighbourhood of the set of fixed points of $\mathfrak{f}$. If $\mathfrak{f}$ is the identity, we can consider that the formula of Hirzebruch (proved by Atiyah-Singer) solves the problem. Suppose on the contrary that $\mathfrak{f}$ has only a finite number of fixed points P, and that the differential $d\mathfrak{f}_P$ does not admit the eigenvalue 1 (as is the case when $\mathfrak{f}$ is a transformation of finite order). Then the determinant

$$\det_C(1 - d\mathfrak{f}_P)$$

is a complex number $\neq 0$; the theorem proved by Atiyah and Bott says that, under these hypotheses, the integer $L(\mathfrak{f})$ is equal to the sum of the reciprocals of these complex numbers.

Limiting ourselves to this example, let us simply add that the results already obtained furnish a proof "without calculation" of the formula of H. Weyl giving the character of a representation of a semisimple group, and they also allow one to solve the Conner-Floyd problems on compact manifolds on which a finite group acts. Let us also note that, according to Hirzebruch [12], one can deduce from that a formula of Langlands giving the dimension of vector spaces of automorphic forms for a discrete group with compact quotient.

In conclusion, we are indebted to Michael Atiyah for several major contributions which put Topology and Analysis in close relation. Each of them was achieved in collaboration; without diminishing in any way the part which came from collaborators as eminent as Hirzebruch, Singer or Bott, there is no doubt that in each case the personal involvement of Atiyah was decisive. He gives us the example of a mathematician for whom clarity of ideas and vision of all phenomena are combined harmoniously with creative imagination, and also with perseverance that leads to major achievements.

References

[1] Borel A., Serre J. P., Le théorème de Riemann–Roch, *Bull. Soc. Math. France*, **86** (1958), 97–136.
[2] Atiyah M. F., Hirzebruch F., Vector bundles and homogeneous spaces, Symp. Pure Math., no. 3, A. M. S., 1961.
[3] Atiyah M. F., The Grothendieck ring in Geometry and Topology, Proc. Int. Congress Math., Stockholm (1962), 442–446.
[4] Atiyah M. F., Hirzebruch F., Riemann–Roch theorems for differentiable manifolds, *Bull. A. M. S.*, **65** (1959), 276–281.
[5] Atiyah M. F., Hirzebruch F., Quelques théorèmes de non plongement pour les variétés différentiables, *Bull. Soc. Math. France*, **89** (1959), 383–396.
[6] Atiyah M. F., Hirzebruch F., Analytic cycles on complex manifolds, *Topology*, **1** (1962), 25–46.
[7] Adams J. F., Vector fields on spheres, *Ann. Math.*, **75** (1962), 603–632.
[8] Atiyah M. F., Singer I., The index of elliptic operators on compact manifolds, *Bull. A. M. S.*, **69** (1963), 422–433. See also two seminars held in 1963–64, one by R. Palais (Annals of Math. Studies, no. 57, Princeton Univ. Press, 1965), another by H. Cartan and L. Schwartz (Secrétariat Math., Inst. H. Poincaré, Paris 1965).
[9] Gelfand I. M., On elliptic equations, Uspehi Mat. Nauk, **15**, No. 3 (1960), 121–132. English translation: Russian Math. Surveys, **15**, No. 3 (1960), 113.
[10] Atiyah M. F., Bott R., The index problem for manifolds with boundary, Differential Analysis, Bombay, 1964.
[11] Atiyah M. F., Bott R., Report on the Woods Hole Fixed Point Theorem, Seminar (1964).
[12] Hirzebruch F., Elliptische Differentialoperatoren auf Mannigfaltigkeiten, Festschrift Weierstrass, Westdeutsche Verlag Köln u. Opladen, 1965, 583–608.

Michael F. Atiyah

MICHAEL F. ATIYAH

Michael F. Atiyah was born in 1929. His father was a distinguished Lebanese and his mother came from a Scottish background. He was educated at Victoria College, Cairo, and Manchester Grammar School. After National Service, he went to Trinity College, Cambridge, where he obtained his BA and PhD degrees and continued with further research, finally as a University lecturer and Fellow of Pembroke College. In 1961 he moved to Oxford, initially appointed to a Readership, and later to the Savilian Professorship of Geometry. From 1969 he was Professor of Mathematics at the Institute for Advanced Study in Princeton, USA (where he had held a Commonwealth Fund Fellowship in 1955) until 1972 when he returned to Oxford as a Royal Society Research Professor and Fellow of St Catherine's College. He held this post until 1990 when he became Master of Trinity College, Cambridge, and Director of the new Isaac Newton Institute for Mathematical Sciences.

Michael Atiyah has contributed to a wide range of topics in Mathematics centering around the interaction between geometry and analysis. His first major contribution (in collaboration with F. Hirzebruch) was the development of a new and powerful technique in topology (K-theory) which led to the solution of many outstanding difficult problems. Subsequently (in collaboration with I. M. Singer) he established an important theorem dealing with the number of solutions of elliptic differential equations. This "index theorem" had antecedents in algebraic geometry and led to important new links between differential geometry, topology and analysis. Combined with considerations of symmetry it led (jointly with R. Bott) to a new and refined "fixed-point theorem" with wide applicability.

All these ideas were subsequently found to be directly relevant to gauge theories of elementary particle physics. The index theorem could be interpreted in terms of quantum theory and has proved a useful tool for theoretical physicists.

Beyond these linear problems, gauge theories involved deep and interesting non-linear differential equations. In particular, the Yang–Mills equations have turned out to be particularly fruitful for mathematicians. Atiyah initiated much of the early work in this field and his student Simon Donaldson went on to make spectacular use of these ideas in 4-dimensional geometry.

Most recently Atiyah has been influential in stressing the role of topology in quantum field theory and in bringing the work of theoretical physicists, notably E. Witten, to the attention of the mathematical community.

In the past few years, he has taken on major responsibilities in the educational and scientific arena. As President of the Royal Society from 1990–95 he was much involved with science policy, both nationally and internationally. In Cambridge, as Master of Trinity College and first Director of the Isaac Newton Institute for Mathematical Sciences, he has a substantial involvement both in education and in research.

He has received numerous honours, including the Fields Medal awarded to him in Moscow in 1966. He was knighted in 1983 and made a member of the Order of Merit in 1992.

Distributed in conjunction with the Colloquium Lectures given at the Fairmont Hotel, Dallas, Texas, January 25–28, 1973. Seventy-ninth annual meeting of the Amer. Math. Soc.

THE INDEX OF ELLIPTIC OPERATORS

by

MICHAEL F. ATIYAH

Introduction

The index theorem is an outgrowth of the Riemann–Roch theorem in algebraic geometry and in these lectures I shall follow its historical development, starting from the theory of algebraic curves and gradually leading up to the modern developments. Since the Riemann–Roch theorem has been a central theorem in algebraic geometry the history of the theorem is to a great extent a history of algebraic geometry. My purpose therefore is really to use the theorem as a focus for a general historical survey. For convenience I shall divide up the four lectures roughly according to the following periods:

(1) The classical era pre 1939, from Riemann to Hodge.
(2) The post war period up to 1954, culminating in Hirzebruch's proof of the Riemann–Roch theorem.
(3) 1955–62, notable for the work of Grothendieck and the introduction of K-theory.
(4) 1963–: the move from algebraic geometry to general elliptic operators.

I should emphasize that this is not meant to be a balanced account of the whole subject either in terms of time-scale or in terms of content. It is simply a personal point of view emphasizing those aspects which have been my own particular specialty. In particular my interest lies mainly in the transcendental theory connecting algebraic geometry (over the complex field) with holomorphic functions of several complex variables and the theory of harmonic functions. I am concerned with the Laplace operator and other operators of the same type, namely the elliptic ones. I shall therefore essentially ignore the purely algebraic aspects of the subject which have of course been the center of great interest in recent years. However, most important ideas straddle the border between the algebraic and transcendental areas and it is impossible to make a strict separation.

Finally, I should like to make it clear that these notes are only an approximate indication of the lectures: they are a complement and not a substitute for the spoken word.

1. The Classical Period

1.1. *Algebraic curves*

Classical algebraic geometry starts from the study of algebraic curves given by a polynomial equation $f(x, y) = 0$ or, to include points "at infinity," a homogeneous equation $f(x, y, z) = 0$ where (x, y, z) are homogeneous coordinates for a point in the projective plane. The coefficients of f and the values of (x, y, z) are complex numbers. Assuming that f is irreducible and non-singular (i.e., the partial derivatives $\frac{\partial f}{\partial x}, \frac{\partial f}{\partial y}, \frac{\partial f}{\partial z}$ do not vanish simultaneously for $(x, y, z) \neq 0$) the curve is a compact subspace of the projective plane which can be locally parametrized by one complex variable (one of the six ratios x/y, y/z ... will do). In other words the curve is a compact Riemann surface or in modern parlance a compact one-dimensional complex analytic manifold. In fact, the theories of

(a) algebraic curves
(b) compact Riemann surfaces
(c) algebraic function fields of one variable

essentially coincide, being simply the geometric, analytic and algebraic aspects of the same mathematical entity. One also gets different points of view according as to how much emphasis one places on the choice of coordinates in the plane. In the early days one considered the (inhomogeneous) equation $f(x, y) = 0$ as defining y as a (multi-valued) algebraic function of x. Then as x wandered over the complex line (including ∞) y described a multiple-sheeted covering with branch points. The Riemann surface of the curve was thus viewed as always sitting over the 2-sphere, and it was not until much later that the idea of an abstract Riemann surface evolved.

On a compact Riemann surface the basic objects of study are naturally the holomorphic and meromorphic functions. Globally any holomorphic function must be a constant (Liouville's theorem) and, more generally, any meromorphic function is a rational function of (x, y) (when we consider the algebraic point of view). This is the basis for the use of global transcendental methods in the study of algebraic functions. From a quantitative point of view the simplest question we can ask is: how many linearly independent global meromorphic functions are there with a prescribed set of poles (of given multiplicities)? Thus if $P_1, \ldots, P_r$ are distinct points on the Riemann surface and $n_1, \ldots, n_r$ are positive integers we consider all meromorphic functions φ having at each P_i a pole of order $\leq n_i$ (and no other poles). Putting formally $D = \sum_{i=1}^{r} n_i P_i$ and denoting by $H(D)$ the space of all φ as above we get, by Liouville's theorem

$$h(D) = \dim\ H(D) \leq 1 + d$$

where $d = \deg\ D = \sum_{i=1}^{r} n_i$. If the Riemann surface is the 2-sphere, so that we are dealing just with rational functions of one variable, then we have equality. But if the Riemann surface has *genus* $g \geq 1$ we get instead the *Riemann–Roch formula* $h(D) = 1 + d - (g - i(D))$. Here $i(D)$, the "index of specialty" of D, denotes the number of (independent) holomorphic differentials which vanish on D (i.e., have zeros at P_i of order $\geq n_i$): thus $0 \leq i(D) \leq g$. Now the total number of zeros of a holomorphic differential is always $2g - 2$ and so $i(D) = 0$ if deg $D > 2g - 2$. Thus if the number of poles is large the Riemann–Roch formula gives $h(D) = 1 + d - g$. Note also that we can allow the integers n_i to be negative provided we interpret a pole of negative order as a zero in the obvious way.

Remark. For a meromorphic differential ω Cauchy's theorem implies that $\sum_P \mathrm{Res}_P \omega = 0$ (since the Riemann surface is closed and so has no boundary). This gives restrictions on the principal parts of φ and Riemann–Roch asserts that these are the only restrictions.

The only numerical invariant of a Riemann surface is its genus and it appears in many guises. Thus

(1) $g =$ number of "handles" $= \frac{1}{2}$ first Betti number
(2) $g =$ number of independent holomorphic differentials
(3) $2g - 2 =$ number of zeros - poles of a meromorphic differential
(4) $1 - g =$ constant term in the (linear) polynomial in m given by $h(mD)$ for m large: this is called the *arithmetic genus*.

1.2. *Algebraic surfaces*

For a non-singular algebraic surface we have more invariants, and the genus of a curve generalizes in several different directions. Thus we have

(1) Betti numbers: $B_1 = B_3$ and B_2
(2) Number of independent holomorphic 1-forms: g_1
Number of independent holomorphic 2-forms: g_2
(3) The divisor class of zeros - poles of a meromorphic 2-form: $-C_1$.

Note that $-C_1$ is a class of curves and so, by taking its self-intersection, we obtain a numerical invariant denoted by C_1^2. More generally, if D is a given curve we can define an intersection number $C_1 \cdot D$.

Suppose now $D = \sum n_i P_i$ is a divisor, i.e., a formal linear combination of irreducible curves P_i, and denote by $H(D)$ the space of meromorphic functions φ with poles along P_i of order $\leq n_i$. Then the Italian algebraic geometers proved a weak form of Riemann–Roch, namely

$$h(D) = \dim\ H(D) \geq \frac{D^2}{2} + \frac{C_1 D}{2} + \frac{C_1^2 + C_2}{12} - i(D)$$

where D^2 is the self-intersection number of D, $C_2 = 2 - 2B_1 + B_2$ is the Euler–Poincaré characteristic, and the index of specialty $i(D)$ is the number of holomorphic

2-forms vanishing on D. The difference between the left and right hand sides of this inequality was termed the *super-abundance* of D and remained a mystery for a long time. Moreover, if D was say a plane section (assuming the surface in P_3) then $i(mD)$ and the super-abundance of mD vanished for large m, $h(mD)$ was then a polynomial (of degree 2) in m with constant term $\frac{C_1^2+C_2}{12}$. Generalizing property (4) of the genus of a curve it was known that this "arithmetic genus" was given by

$$1 - g_1 + g_2 = \frac{C_1^2 + C_2}{12}.$$

1.3. *Higher dimensions*

For algebraic varieties of higher dimension the first problem is to provide the appropriate generalization of the known invariants for curves and surfaces. One main step was taken by J. A. Todd [12] in the mid-thirties (and independently by Eger) by showing that one could define for each i a canonical subvariety C_i of (complex) co-dimension i (unique up to an appropriate equivalence) generalizing C_1 and C_2 for surfaces. It was claimed by Severi that the number $\sum(-1)^i g_i$ should be expressible in terms of the C_i and Todd found explicitly the polynomials that were needed for the first few dimensions. Thus

$$\sum_{i=0}^{n}(-1)^i g_i = T_n(C_1, \ldots, C_n)$$

where T_n was a polynomial of weight n in the C_i. These polynomials now known as the Todd polynomials begin as follows:

$$T_1 = \tfrac{1}{2}C_1, \quad T_2 = \tfrac{C_1^2+C_2}{12}, \quad T_3 = \tfrac{C_1C_2}{24},$$

$$T_4 = \tfrac{1}{720}(-C_4 + C_3C_1 + 3C_2^2 + 4C_2C_1^2 - C_1^4).$$

Moreover, it was also conjectured by Severi that $\sum(-1)^i g_i$ should again coincide with the arithmetic genus, i.e., the constant term in $h(mD)$ for m large and D a hyperplane section.

The other major step in the thirties was the development by Hodge [11] of the theory of harmonic forms. In particular Hodge introduced numerical invariants $h^{p,q}$ which refined the Betti numbers and were related to holomorphic differentials. Precisely

$$B_i = \sum_{p+q=i} h^{p,q} \qquad g_p = h^{p,0}$$

and $h^{p,q} = h^{q,p} = h^{n-p,\, n-q}$. An important formula due to Hodge was the *signature formula*. If the complex dimension n is even (e.g. for a surface) the middle homology group has a quadratic form given by intersection: the signature τ is defined as $p - q$ where p, q are the number of $+$ and $-$ signs in a diagonalization of the quadratic form. Hodge's signature theorem asserts that

$$\tau = \sum_{p,q} (-1)^q h^{p,q}.$$

The formal similarity between this alternating sum, that for the arithmetic genus and the Euler characteristic

$$C_n = \sum_{p,q} (-1)^{p+q} h^{p,q}$$

is very striking. The full implications of this similarity were not however understood until much later.

So much for the new numerical invariants in higher dimensions. The Riemann–Roch theorem however did not seem to generalize even as an inequality for dimensions ≥ 3, although one still knew (from the work of Hilbert) that $h(mD)$ was a polynomial in m of degree n provided D was a hyperplane section and m was large.

2. Sheaf Theory

In the post war years powerful new methods were introduced into algebraic geometry. The theory of sheaves developed by J. Leray and applied by H. Cartan and J -P. Serre to complex analysis led to rapid developments, notably the work of Kodaira–Spencer and of Hirzebruch [10]. In addition to sheaf theory many other techniques and ideas were taken over from algebraic topology. One of the simplest and most useful concepts was that of a *vector bundle.* A vector bundle over a space X is roughly speaking a continuous family of vector spaces parametrized by the points of X. If X is a differentiable manifold we can consider differentiable bundles, where the family varies differentiably while if X is a complex manifold we can consider holomorphic bundles where the family varies holomorphically. In differential geometry vector bundles arise naturally by taking for example the family of tangent spaces or more generally the tensor spaces of various types. A section S of a vector bundle E is a function $x \longmapsto s(x) \in E_x$ which is continuous, differentiable or holomorphic as the case may be. For the tangent vector bundle a section is a tangent vector field, for the bundle of skew-symmetric covariant p-tensors it is an exterior differential p-form.

Chern showed that a complex vector bundle E over a space X had certain invariantly associated cohomology classes $C_i(E) \in H^{2i}(X)$. If X is an algebraic variety and E its tangent bundle these are dual to the classes defined by Todd. These classes are now called the *Chern classes* of E. For real vector bundles similar invariants had earlier been introduced by Whitney and Pontrjagin.

For a holomorphic vector bundle E over a compact complex manifold X the space $H^0(X, E)$ of holomorphic sections is *finite-dimensional.* For example, if E is the bundle Ω^p of p-forms then dim $H^0(X, \Omega^p) = g_p$ in our previous notation.

Vector bundles of dimension one — also called line-bundles — turn up in algebraic geometry in quite another direction. Over the projective space $P_{n-1}(\mathbb{C})$ we have a natural line-bundle obtained by associating to each point $x \in P_{n-1}(\mathbb{C})$ the line L_x in $\mathbb{C}^n$ defined by x. In fact, it is better to work with the dual L_x^* and these give a bundle H over $P_{n-1}(\mathbb{C})$. The important point is that the holomorphic sections of H are precisely the linear forms on $\mathbb{C}^n$. More generally the homogeneous polynomials on $\mathbb{C}^n$ of degree k are the holomorphic sections of the line-bundle $H^k = H \otimes H \otimes \cdots \otimes H (k \text{ times})$. If now $X \subset P_{n-1}$ is an algebraic subvariety we get a line-bundle H_X by restriction and if D is a hyperplane section then

$$H(D) \cong H^0(X, H_X).$$

More generally, given any divisor D we can associate to it a line-bundle L so that

$$H(D) \cong H^0(X, L)$$

and moreover $C_1(L)$ is the class dual to the $(2n-2)$-cycle defined by D.

We are now ready to import sheaf cohomology. Not only can we define the groups $H^0(X, L)$ but there are also *cohomology groups* $H^q(X, L)$ for $q = 0, 1, \ldots, n$, and the same holds for holomorphic vector bundles of any dimension. For example, taking the bundle Ω^p of p-forms it turns out that the Hodge numbers can be identified as

$$h^{p,q} = \dim\ H^q(X, \Omega^p).$$

In view of the symmetry $h^{p,q} = h^{q,p}$ we see that

$$\sum_{i=0}^{n} (-1)^i g_i = \sum_{i=0}^{n} (-1)^i \dim\ H^i(X, \Omega^0) = \chi(X, \Omega^0)$$

is a holomorphic Euler characteristic (Ω^0 denotes here the trivial line-bundle — whose sections are just functions on X). Euler characteristics are well-known to have better properties than dimensions of individual cohomology groups so it is quite reasonable to expect a formula for $\chi(X, L)$ in terms of $C_1(L)$ and the Chern classes† $C_i(X)$. For a curve X it is clear from the classical Riemann–Roch that we have such a formula provided

$$\dim\ H^1(X, L) = i(D) = \dim\ H^0(X, \Omega^1 \otimes L^*).$$

This is, in fact, a special case of the Serre duality theorem which asserts that

$$H^q(X, E) \text{ and } H^{n-q}(X, \Omega^n \otimes E^*)$$

are canonically dual for any holomorphic vector bundle E.

For a surface we then see that $H^1(X, L)$ should be the mysterious superabundance. For dimension ≥ 3 it is also clear why even an inequality is not available in classical terms because both H^1 and H^2 enter and have opposite signs.

†The standard convention is to write $C_i(X)$ for the Chern classes of the tangent bundle of X.

With these preliminaries we can now formulate the Hirzebruch form of the Riemann–Roch theorem for line-bundles:

$$\chi(X,L) = \frac{C_1(L)^n}{n!} + \frac{T_1 \cdot C_1(L)^{n-1}}{(n-1)!} + \frac{T_2 \cdot C_1(L)^{n-2}}{(n-2)!} + \cdots + T_n$$

where the T_i are the Todd polynomials in the Chern classes of X. Moreover, Hirzebruch gave a generating function for the Todd polynomials as follows: we put

$$\prod_{i=1}^{n} \frac{x_i}{1-e^{-x_i}} = 1 + T_1 + T_2 + \cdots + T_n$$

where the T_i are polynomials in the elementary symmetric functions C_j of $x_1, \ldots, x_n$.

Not only does this give a complete generalization of the classical Riemann–Roch but it also disposes of the Severi–Todd conjectures concerning the arithmetic genus. In fact, taking L to be the trivial bundle, $C_1(L) = 0$ and we get

$$\sum (-1)^i g_i = \chi(X, \Omega^0) = T_n.$$

Moreover, if $L = H_X^m$ then Kodaira (and also Cartan–Serre) proved that $H^q(X,L) = 0$ for $q \geq 1$ and m large. On the other hand

$$\chi(X, H_X^m) = \sum_{j=0}^{n} \frac{T_j C^{n-j}}{j!} \cdot m^{n-j}$$

where $C = C_1(H_X)$ is the class of a hyperplane section. This identifies the Hilbert polynomial in m and shows that the constant term is in fact T_n.

More generally still Hirzebruch [10] established a formula for $\chi(X,E)$ for any holomorphic vector bundle E. This may be written

$$\chi(X,E) = \sum_{j=0}^{n} \frac{\mathrm{ch}_{n-j}(E) \cdot T_j}{j!} \tag{HRR}$$

where $\mathrm{ch}\, E = \sum \mathrm{ch}_k(E)$ (the Chern character) is a polynomial in the Chern classes of E defined by the identities

$$\mathrm{ch}\, E = \sum_{i=1}^{q} e^{x_i} \quad \sum_{j=0}^{q} C_j(E) = \prod_{i=1}^{q} (1 + x_i).$$

Applying this formula to the vector bundles Ω^p, summing and using the Hodge signature theorem we end up with a formula for the signature (when $n = 2k$)

$$\tau = L_k(p_1, \ldots, p_k)$$

where the p_i are related to the Chern classes C_j of X by

$$\sum p_i = \prod (1 + x_j^2) \quad \sum C_i = \prod (1 + x_j)$$

and the polynomials L_k are given by

$$\prod_{i=1}^{n} \frac{x_i}{\tanh x_i} = 1 + L_1 + \cdots + L_k$$

(that is, we express the L_i in terms of the elementary symmetric functions p_i of the x_j^2).

This Hirzebruch signature formula now makes sense for any oriented differentiable manifold of dimension $4k$ — not necessarily a complex manifold. In fact, Hirzebruch established this formula first using Thom's cobordism theory and used it as a step to the Riemann–Roch theorem.

One of the interesting consequences of HRR is that the Todd polynomial $T_n(C_1, \ldots, C_n)$ gives an integer even though it has large denominators. For example when $n = 2$, $T_2 = \frac{C_1^2 + C_2}{12}$ and so we deduce that $C_1^2 + C_2 \equiv 0 \bmod 12$. Many interesting integrality theorems of this kind can be deduced from HRR and these have been very important in differential topology. For example, the exotic differentiable structures on spheres discovered by Milnor were detected using such integrality theorems and the relationship is quite deep. Although much mystery remains the topological significance of HRR became clearer later on as we shall see.

3. K-Theory

It might seem that HRR was the last word on the subject, but around 1957 Grothendieck introduced some revolutionary new ideas (see [7]). Part of Grothen dieck's aim was to give a purely algebraic proof of HRR without transcendental methods. However his ideas go far beyond this and have had a very significant impact in the transcendental and topological domain.

Grothendieck starts by observing that $\chi(X, E)$ is an additive invariant of E. That is,

$$\chi(X, E) = \chi(X, E') + \chi(X, E'')$$

whenever E' is a sub-vector bundle of E and E'' is the quotient bundle. The right-hand side of HRR is also such an additive invariant. Grothendieck therefore conceived the idea of studying all such additive invariants, by introducing an abelian group $K(X)$ with one generator $[E]$ for every vector bundle on X and one relation $[E] = [E'] + [E'']$ for every exact sequence $0 \to E' \to E \to E'' \to 0$ (i.e., E' a sub-bundle and $E'' \cong E/E'$). Clearly the additive invariants of bundles with values in some abelian group A are just given by homomorphisms $K(X) \to A$. For example the Chern character $\mathrm{ch}\, E$ defined earlier is additive and so defines a homomorphism $\mathrm{ch} : K(X) \to H^*(X, \mathbf{Q})$. In fact, the tensor product of bundles turns $K(X)$ into a commutative ring and ch is even a ring homomorphism. Moreover, Grothendieck showed that $K(X)$ had many of the formal properties of cohomology and he was able to compute it for certain important spaces such as projective space.

For a (holomorphic) map $f : Y \to X$ of non-singular algebraic varieties Grothendieck was able to define a homomorphism $f_! : K(Y) \to K(X)$ which generalized the Euler characteristic χ. More precisely, if X is a point then $K(X) \cong \mathbb{Z}$ and $f_![E] = \chi(Y, E)$. His version of RR was then that the diagram

$$\begin{array}{ccc} K(Y) & \xrightarrow{\text{ch}} & H^*(Y, \mathbf{Q}) \\ \downarrow f_! & & \downarrow f_* \\ K(X) & \xrightarrow{\text{ch}} & H^*(X, \mathbf{Q}) \end{array}$$

commutes up to multipliers $\mathcal{T}_Y, \mathcal{T}_X$ (the total Todd classes $1+T_1+T_2+\cdots$ of Y and X respectively), where f_* denotes the map on cohomology given by using Poincaré duality. Thus for a vector bundle E on Y we have

$$\mathcal{T}_X \cdot \text{ch } f_![E] = f_*\{\mathcal{T}_Y \cdot \text{ch } E\}. \qquad \text{(GRR)}$$

Clearly this reduces to HRR when X is a point.

The greater generality of GRR is not only more appealing but it also leads to a simpler and more natural proof. Thus to prove (HRR) we factor the map $Y \to$ point into $Y \xrightarrow{i} P \xrightarrow{\pi}$ point where P is projective space and i is some embedding: we then prove GRR for i and π separately and this implies it for the composition $\pi \circ i$. For π it is easy because $K(P)$ is completely known so the main step in the proof is the proof for the embedding i.

Almost at the same time there were startling new developments on the topological front. Bott discovered his famous periodicity theorems concerning the stable homotopy groups of the classical groups.† For $\text{GL}(N, C)$ these assert

$$\pi_n(\text{GL}(N, C)) \cong \pi_{n+2}(\text{GL}(N, C)) \text{ for large } N$$

from which one deduces that $\pi_n \cong \mathbb{Z}$ for n odd and 0 for n even. Since a map $S^{2n-1} \to \text{GL}(N, C)$ defines a complex N-dimensional vector bundle over S^{2n} Bott's result classifies vector bundles of large dimension over even spheres.

Shortly after all this I became interested in a topological problem‡ concerning complex projective spaces and I found it convenient to combine the work of Grothendieck and Bott. The results were so successful that it soon became apparent that a powerful topological tool was in the making. Hirzebruch and I therefore systematically transposed the formalism of Grothendieck to the topological context (see[2]). Thus we defined a new $K(X)$ using continuous vector bundles over any (compact) space X. It turned out that Bott's periodicity theorem was the basic building stone of the new theory and a substitute for the corresponding theorems of Grothendieck. For example, in both contexts, a basic result asserts that

$$K(X \times P) \cong K(X) \otimes K(P)$$

where P is a complex projective space.

†For a historical discussion see Bott's Colloquium Lectures published in [8].
‡See [4].

Topological K-theory has by now justified itself by numerous important applications. Interestingly enough one of the most subtle and striking was the solution by J. F. Adams [1] of the vector-field problem on spheres. This is in essence the real counterpart of the problem which started the whole thing off — though I must point out that the real case is considerably more delicate and used much more of the formal structure of K-theory.

In view of its mixed parentage it should surprise no one to learn that the integrality theorems derived from HRR were now explained in satisfactory terms by K-theory. Moreover, similar results, following from GRR, could now be proved in the topological context. All the same there was still more to follow as we shall see in the next lecture.

As a preparation for later on it is useful here to recall how the generator of the group $\pi_{2n-1}(\mathrm{GL}(N, C))$ can be constructed. Assume that we have $2n$ complex $N \times N$ matrices $A_1, \ldots, A_{2n}$ satisfying the Clifford identities

$$A_i^2 = -1, \quad A_i A_j = -A_j A_i \quad \text{for } i \neq j$$

so that $(\sum x_i A_i)^2 = -\sum x_1^2$. Thus for a unit vector $x = (x_1, \ldots, x_{2n})$ the matrix $A(x) = \sum x_i A_i$ is invertible and so $x \longmapsto A(x)$ defines a map $S^{2n-1} \to \mathrm{GL}(N, C)$. The Bott generator is given by taking N as small as possible: in fact, this value is 2^{n-1} and the matrices A_i are essentially unique. For $n = 1$, putting $J = A_1^{-1} A_2$ we recognize the usual map $x \longmapsto x_1 + J x_2$ sending S^1 into the non-zero complex numbers. For $n = 2$ we get the unit quaternions mapped in $\mathrm{GL}(2, C)$.

Note that the Clifford matrices arose in the work of Dirac in which he sought to express the Laplace operator[†] as the square of a first order system:

$$-\sum \frac{\partial^2}{\partial x_i^2} = \left(\sum A_i \frac{\partial}{\partial x_i}\right)^2.$$

4. The Index Theorem

In attempting to understand further one of the integrality theorems of Hirzebruch, Singer and I were led (in 1961) to rediscover the Dirac operator and its curved analogue on a Riemannian manifold. We then conjectured that the Hirzebruch integer in question should be expressible in terms of *harmonic spinors* in much the same way as the arithmetic genus was expressed in terms of holomorphic forms. While pondering this question our attention was drawn to some recent papers of Gel′fand [9] and his associates. In these papers the index problem was posed for general elliptic differential operators and some first steps were taken towards its solution.

[†]Of course Dirac was concerned with the indefinite Lorentz metric but the algebra is the same.

A partial differential operator[†] $P = \sum_{|\alpha|\leq k} a_\alpha(x)\frac{\partial}{\partial x^\alpha}$ (with smooth coefficients $a_\alpha(x)$) is said to be elliptic (or order k) if $\sum_{|\alpha|=k} a_\alpha(x)\xi^\alpha \neq 0$ for $0 \neq \xi \in \mathbb{R}^n$. If P is acting on vector-valued functions so that the $a_\alpha(x)$ are square matrices the condition is that the symbol

$$\sigma_P(x,\xi) = \sum_{|\alpha|=k} a_\alpha(x)\xi^\alpha$$

has non-zero determinant for $\xi \neq 0$. These definitions make sense on a manifold X and when vector-valued functions are generalized to sections of vector bundles. Thus in general P is a linear operator $C^\infty(X,E) \to C^\infty(X,F)$ (where C^∞ denotes the C^∞ sections). Elliptic operators share the main qualitative properties of the Laplace operator. In particular if X is compact the space of solutions of $Pu = 0$ has finite-dimension (say α) and the equation $Put = v$ can be solved provided v satisfies a number (say β) of linear relations. The index of P is the number $\alpha - \beta$. By introducing metrics and passing to the adjoint P^* this can be expressed

$$\text{index } P = \dim(\text{Ker } P) - \dim(\text{Ker } P^*).$$

The most significant property of this index is that it is stable under perturbation and hence in particular it depends only on the highest order terms. It is therefore reasonable to expect a topological formula for it in terms of the geometrical data given by the symbol $\sigma_P(x,\xi)$. To interpret σ_P geometrically we must view ξ as a vector in the cotangent space T_x^*, and then $\sigma_P(x,\xi)$ is a linear map $E_x \to F_x$. Since σ is an isomorphism for $\xi \neq 0$ we can construct a vector bundle $V(\sigma)$ on the double $\sum X$ of the unit ball bundle BX of T^*X (using a Riemannian metric). Namely we lift E to one copy of BX, F to the other and identify E with F along the boundary SX using σ. The index of P depends only on $V(\sigma)$ and in fact the index extends to give a homomorphism

$$K(\Sigma X) \longrightarrow \mathbb{Z}.$$

This shows that K-theory is just the right tool to study the general index problem.

It is not hard to reformulate Riemann–Roch so that it appears as an index problem. For Riemann surfaces this is quite clear, we simply take the $\overline{\partial}$-operator. Its kernel consists of the holomorphic functions (or more generally of holomorphic sections) and the kernel of its adjoint can be identified with holomorphic differentials. Thus $h(D) - i(D)$ is just the index of the $\overline{\partial}$-operator of the line-bundle corresponding to the divisor D. In higher dimensions we have to replace $\overline{\partial}$ by $\overline{\partial}+\overline{\partial}^*$ (where $\overline{\partial}^*$ is the adjoint of $\overline{\partial}$) acting from forms of type (0, even) to forms of type (0, odd).

[†]We use the usual abbreviated notation $\alpha = (\alpha_1,\ldots,\alpha_n), |\alpha| = \sum \alpha_i, \frac{\partial}{\partial x^\alpha} = \left(\frac{\partial}{\partial x_1}\right)^{\alpha_1}\cdots (\partial x_n)^{\alpha_n}$ and $\xi^\alpha = \xi_1^{\alpha_1}\cdots\xi_n^{\alpha_n}$.

The Hirzebruch signature formula for an oriented (real) $4k$-manifold can also be viewed as an index formula for a suitable first-order system. Finally the Dirac operator leads to an index problem and this was precisely what Singer and I were studying.

We see, therefore, that there are numerous classical examples of first-order elliptic systems which give interesting index problems. Now it might seem that the general index problem for higher-order operators would be vastly more difficult than for the classical first-order ones. In fact, this is not so: *a solution of the index problem for all classical operators implies a solution of all index problems.* This surprising and striking fact is essentially a consequence of Bott's periodicity theorem which shows that the generating bundle on S^{2n} is in fact given by the symbol of the Dirac operator as we saw in the last lecture. More precisely one shows that the group $K(\sum X)$ is (essentially) generated by the symbols of classical operators.[†]

With the whole machinery of K-theory at our disposal and with the formulae of Hirzebruch for various classical operators it was not difficult for Singer and me to guess what the general index formula ought to be. It can be written as follows:

$$\text{index } P = \sum_{k=0}^{n} T_k(X)\text{ch}_{n-k}(V(\sigma_P))$$

where $T_k(X)$ denote the Todd polynomials of the complexified tangent bundle of X and n is the (real) dimension of X.

As I have explained it is only necessary to establish this formula for all classical operators. In our first proof [5] we followed Hirzebruch's procedure of using cobordism. The main point was to show there was some formula for index P involving characteristic classes of X and $V(\sigma)$ — the precise formula could then be found by computing various special cases. Very recently an alternative approach has been found [3] in which Riemannian geometry replaces cobordism.

There is also another quite different proof [6] which involves embedding X in R^n and transferring the given index problem on X to one on R^N (or rather on the N-sphere $R^N \cup \infty$). This is in the spirit of the proof of GRR, though of

Thus K-theory which was introduced to study a very special index problem, namely Riemann–Roch, turned out to be precisely the right tool for the general case. As I have mentioned this depends on Bott's periodicity theorem, itself closely related to Grothendieck's work on K-theory. In trying to unravel this story Bott and I were led to various new proofs of the periodicity theorem (see [8]) in which the index of certain operators played a key role. In fact, the deeper one digs the more one finds that K-theory and index theory are one and the same subject!

The various proofs of the index theorem have different merits and lead to generalizations in different directions. I cannot pursue these here but suffice it to say that there is plenty of life in the subject yet.

[†]For X even-dimensional and orientable.

References

This is a very small and somewhat arbitrary selection from the vast literature. For more references consult [10].

1. J. F. Adams, *Vector fields on spheres*, Ann. of Math. **75** (1962), 603–632.
2. M. F. Atiyah, *K-theory*, Benjamin, 1967.
3. ______, R. Bott and V. K. Patodi, *On the heat equation and the index theorem*, Inven. Math. **19** (1973), 279–330 and **28** (1975), 277–280.
4. ______, and J. A. Todd, *On complex Stiefel manifolds*, Proc. Camb. Phil. Soc. **56** (1960), 342–353.
5. ______, and I. M. Singer, *The index of elliptic operators on compact manifolds*, Bull. Amer. Math. Soc. **69** (1963), 422–433.
6. ______, *The index of elliptic operators: I*, Ann. of Math. **83** (1968), 484–530.
7. A. Borel and J -P. Serre, Le Théorème de Riemann–Roch (d'après Grothendieck), *Bull. Soc. Math. France* **86** (1958), 97–136.
8. R. Bott, *The periodicity theorem for the classical groups and some of its applications*, Adv. Math. **4** (1970), 353–411.
9. I. M. Gel′fand, *On elliptic equations*, Russian Math. Surveys **15** (1960), no. 3, 113.
10. F. Hirzebruch, *Topological Methods in Algebraic Geometry*, 3rd edition, Springer, 1966.
11. W. V. D. Hodge, *The Theory and Application of Harmonic Integrals*, Cambridge, 1941.
12. J. A. Todd, *The arithmetical invariants of algebraic loci*, Proc. London Math. Soc. **43** (1937), 190–225.

Postscript (1996)

Developments related to index theory since 1973 are very extensive. A short list of (inter-linked) major topics includes:

1. C^*-algebras and associated K-theory as developed by Kasparov and others.
2. Non-commutative differential geometry developed by Connes.
3. Local versions in the context of Riemannian Geometry, developed in particular by Bismut.
4. Super-symmetry formalism, borrowed from theoretical physics, by Quillen and others.
5. Infinite-dimensional versions as exploited by physicists, chiefly Witten.
6. Elliptic cohomology and modular forms.

PAUL J. COHEN AND THE CONTINUUM PROBLEM

by

ALONZO CHURCH

On the occasion of the award of a prize to Paul Cohen, and in spite of significant contributions by him to analysis, to topological groups, and to the theory of differential equations, I believe that the audience will agree that it is appropriate to devote the entire time allowed for exposition to the continuum problem.

For here is another case, of the sort which arises from time to time in the history of mathematics, in which a mathematician who has done important work in other fields turns to a field not properly his own to solve an outstanding problem that has baffled the specialists. As a consequence of the tremendous growth of mathematics the universal mathematician of other days is no longer a possibility — David Hilbert was certainly the last of them. The next best thing is that abler men should not confine themselves too closely to one field or be afraid to turn to an area in which they may not have all the expert knowledge of those who have concentrated their work in it. Certainly Paul Cohen's results have been and will be greatly extended, and the method of his proof greatly improved, by the specialists in set theory. But we are concerned today with the initial break-through.

Number one of the Hilbert problems, placed even before the problem of the consistency of arithmetic which occupied so much of Hilbert's own attention in the latter part of his life, is "Cantor's problem of the cardinal number of the continuum." So it is titled in the contemporary English translation of Hilbert's famous paper. Hilbert himself in German uses Cantor's original term "Mächtigkeit," which has no good English translation. Hilbert does not say that the order in which the problems are numbered gauges their relative importance, and it is not meant to suggest that he intended this. But he does mention the arithmetical formulation of the concept of the continuum and the discovery of non-Euclidean geometry as being the outstanding mathematical achievements of the preceding century, and gives this as a reason for putting problems in these areas first.

Already in 1878 Cantor stated the continuum hypothesis as a conjecture. But there is a sense in which the continuum problem dates from Cantor's statement at the end of a paper which appeared in the *Mathematische Annalen* in 1884. Here it is proved that a closed infinite subset of the (linear) continuum must have the cardinal number either of the natural numbers or of the whole continuum. Then it is said that the result can be extended to subsets which are not closed, and a proof will be provided. The paper closes with the words "Fortsetzung folgt." But the promised Fortsetzung never did folgen, and it seems clear that the proof Cantor believed he had broken down.

Cantor passes at once from the proposition that there is no cardinal number between that of the natural numbers and that of the continuum to the second form of the continuum hypothesis, that the cardinal number of the continuum is $\aleph_1$. Of course this depends on a tacit assumption of the axiom of choice, in particular as Sierpiński's result of 1947 deriving the axiom of choice from the generalized continuum hypothesis was not then available (or the background that made this result possible).[1] Hilbert is more cautious and states as a separate problem, subsidiary to the continuum problem, the question whether the continuum can be well ordered. Zermelo's paper which explicitly states the axiom of choice for the first time (in the strong form which Zermelo later called "Prinzip der Auswahl"), and shows as a consequence of it that every set can be well ordered, followed Hilbert's paper on mathematical problems by only four years.

It was Cantor's original point of view that the transfinite cardinal and ordinal numbers are two different kinds of generalizations of the natural numbers and are to serve the same purposes for transfinite sets which the natural numbers do for finite sets. If this program is to be fulfilled, one evidently must be able to answer at least the simplest and most immediate questions that arise about the cardinal numbers of the most commonly used mathematical sets, among them the continuum. This is clearly the reason why the frustrating difficulties of the continuum problem acquired the importance that they did for the Cantor theory. Surely neither Cantor nor Hilbert could have surmised that the ultimate solution would take the negative form that it has. Yet Hilbert is quite explicit that in general it may happen that the solution of a problem must be in the form of an impossibility proof.

The antinomies of set theory, which first came to the attention of the mathematical public through Burali-Forti's paper of 1897, played an important role in the progress towards the ultimate solution, as it was the antinomies that forced the transition from the older naive and "genetic" use of sets in mathematics to an axiomatic basis for set theory. And it is of course only by the axiomatic method that a proof of the impossibility of a proof becomes possible.

Within axiomatic set theory there was a proof of the independence of the axiom of choice by Fraenkel as early as 1922. But this was unsatisfactory in that it referred to axioms of set theory so formulated as to admit a domain of Urelemente, or non-sets, of unspecified structure, and the possibility remained open that the axiom of choice would lose its independence upon adding axioms specifying the structure of the domain of Urelemente (or most simply, upon adding as an axiom that there are no Urelemente). Extensions of Fraenkel's result and improvements of his method, by Fraenkel himself, Lindenbaum, Mostowski, and more recently Shoenfield and

[1]Sierpiński points out that this result had been announced by Lindenbaum and Tarski in 1926. Their proof was never published.

Mendelson, either did not remove this objection or only mitigated it (in the sense that independence from quite the full usual system of axioms for set theory was not yet proved).

A much more important step — which constitutes in fact the first half of the solution of continuum problem, and on which subsequent work heavily depends — was taken by Kurt Gödel in 1938–40. Abstracts of Gödel's methods and results appeared in 1938 and 1939, and the monograph containing the full proofs, in 1940. Gödel's method is to set up what has since been called an inner model of set theory. I.e., set theory without axiom of choice is used to set up a model of set theory in which both the axiom of choice and the generalized continuum hypothesis hold. (The generalized continuum hypothesis is the proposition that the power set of a set of cardinal number $\aleph_\alpha$ has the cardinal number $\aleph_{\alpha+1}$, Cantor's original continuum hypothesis, in its second form, being the special case of this in which $\alpha = 0$.) The result of Gödel's procedure, setting up an inner model, is a relative consistency proof for the axiom of choice and for the generalized continuum hypothesis: If set theory without axiom of choice is consistent, it remains so upon introducing both the axiom of choice and the generalized continuum hypothesis as additional axioms.

After the (relative) consistency proof, the second half of the negative solution of the continuum problem is of course independence. A partial step in this direction was taken by Gödel, who in 1942 found a proof of the independence of the axiom of constructibility in type theory. According to his own statement (in a private communication), he believed that this could be extended to an independence proof of the axiom of choice; but due to a shifting of his interests toward philosophy, he soon afterwards ceased to work in this area, without having settled its main problems. The partial result mentioned was never worked out in full detail or put into form for publication.

These climactic results, the independence in set theory of the axiom of choice (even the weak form of the axiom of choice which concerns a countable set of pairs) and of the continuum hypothesis from the axiom of choice, remained for Paul Cohen in 1963–64. It is no part of our present purpose to describe the details of his method. Let it only be said that, besides the now well-known notion of *forcing*, it depends on an adaptation of Gödel's method of 1940 for setting up models of set theory, on a modification of the earlier methods of Fraenkel, Mostowski, and others in connection with the independence of the axiom of choice, and on the result of Skolem that there exists a countable model of set theory (a model having the cardinal number of the natural numbers).

The feeling that there is an absolute realm of sets, somehow determined in spite of the non-existence of a complete axiomatic characterization, receives more of a blow from the solution (better, the unsolving)[2] of the continuum problem than from the famous Gödel incompleteness theorems. It is not a question of realism (miscalled

[2]I borrow this whimsical term from W. W. Boone.

"Platonism") versus either conceptualism or nominalism, but if one chooses realism, whether there can be a "genetic" realism without axiomatic specification. The Gödel-Cohen results and subsequent extensions of them have the consequence that there is not one set theory but many, with the difference arising in connection with a problem which intuition still seems to tell us must "really" have only one true solution.

I know of mathematicians who hold that the axiom of choice has the same character of intuitive self-evidence that belongs to the most elementary laws of logic on which mathematics depends. It has never seemed so to me. But how shall one argue matters of intuition? The point is, I know of no one who maintains such self-evidence for the continuum hypothesis.

The realist will expect that the reality independent of the human mind which he maintains must have many ramifications, and will take what has now become the classical mathematical view, dating from the nineteenth century discussions of non-Euclidean geometry, that all the ramifications equally demand exploration. The same view is possible also for him who takes the intermediate position between radical realism and conceptualism by holding that mathematical and physical objects alike, not excluding such basic logico-mathematical objects as sets, have their reality only relative to and within a certain theory. And if a choice must in some sense be made among the rival set theories, rather than merely and neutrally to develop the mathematical consequences of the alternative theories, it seems that the only basis for it can be the same informal criterion of simplicity that governs the choice among rival physical theories when both or all of them equally explain the experimental facts.

References

[1] Cantor G., Ein Beitrag zur Mannigfaltigkeitslehre, *Journal fü die reine und angewandte Mathematik*, **84** (1878), 242–258. [See p. 257.]

[2] Cantor G., Ueber unendliche, lineare Punktmannigfaltigkeiten, *Mathematische Annalen*, **23**, No. 6 (1884), 453–488.

[3] Burali-Forti Cesare, Una questione sui numeri transfiniti, *Rendiconti del Circolo Matematico di Palermo*, **11** (1897), 154–164.

[4] Hilbert David, Mathematische Probleme, *Nachricshten von der K. Gesellschaft der Wissenschaften zu Göttingen*, Math.-Phys. K1., 1900, pp. 253–297. Reprinted with additions in *Archiv der Mathematik und Physik*, ser. 3, vol. 1 (1901), 44–63, 213–237. English translation in *Bulletin of the American Mathematical Society*, **8** (1901–2), 437–479.

[5] Zermelo Ernst, Beweis, daβ jede Menge wohlgeordnet werden kann, *Mathematische Annalen*, **59** (1904), 514–516.

[6] Fraenkel A., Der Begriff ≪defint≫ und die Unabhängigkeit des Auswahlaxioms, *Sitzungsberichie der Preussischen Akademie der Wissenschaften*, Phys.-Math. K1., 1922, pp. 253–257.

[7] Skolem Thoralf, Einige Bermerkungen zur axiomatischen Begründung der Mengenlehre, Wissenschaftliche Vorträge gehalten auf dem Fünften Kongress der Skandinavischen Mathematiker in Helsingfors vom 4, bis 7, Juli 1922, Helsingfors, 1923, 217–232.

[8] Lindenbaum A., Tarski A., Communication sur les recherches de la théorie des ensembles, *Comptes Rendus des Séances de la Société des Sciences et des Lettres de Varsovie*, Classe III, **19** (1926), see p. 314.

[9] Cödel Kurt, Über formal unentscheidbare Sätze der Principia Mathematica und verwandter Systeme I, *Monatshefte für Mathematik und Physik*, **38** (1931), 173–198.

[10] Fraenkel A., Ueber eine abgeschwaechte Fassung des Auswahlaxioms, *The Journal of Symbolic Logic*, **2** (1937), 1–25.

[11] Lindenbaum Adolf, Mostowski Andrzej, Über die Unabhängigkeit des Auswahlaxioms und einiger seiner Folgerungén, *Comptes Rendus des Séances de la Société des Sciences et des Lettres de Varsovie*, Classe III, **31** (1938), 27–32.

[12] Gödel Kurt, The consistency of the axiom of choice and of the generalized continuum-hypothesis, *Proceedings of the National Academy of Sciences*, **24** (1938), 556–557.

[13] Gödel Kurt, Consistency-proof for the generalized continuum-hypothesis, *Proceedings of the National Academy of Sciences*, **25** (1939), 220–224.

[14] Mostowski Andrezej, Über die Unabhängigkeit des Wohlordnungssatzes vom Ordnungsprinzip, *Fundamenta Mathematicae*, **32** (1939), 201–252.

[15] Gödel Kurt, The consistency of the axiom of choice and of the generalized continuum-hypothesis with the axioms of set theory, Princeton, 1940, 66 pp.

[16] Sierpiński Waclaw, L'hypothèse généralisée du continu et l'axiome du choix, *Fundamenta Mathematicae*, **34** (1947), 1–5.

[17] Shoenfield J. R., The independence of the axiom of choice, Abstract, *The Journal of Symbolic Logic*, **20** (1955), 202.

[18] Mendelson Elliott, The independence of a weak axiom of choice, *The Journal of Symbolic Logic*, **21** (1956), 350–366.

[19] Mendelson Elliott, The axiom of Fundierung and the axiom of choice, *Archiv für mathematische Logik und Grundlagenforschung*, **4** (1958), 65–70.

[20] Cohen Paul J., A minimal model for set theory, *Bulletin of the American Mathematical Society*, **69** (1963), 537–540.

[21] Cohen Paul J., The independecne of the continuum hypothesis, *Proceedings of the National Academy of Sciences*, **50** (1963), 1143–1148, and **51** (1964), 105–110.

[22] Cohen Paul J., Independence results in set theory. The theory of models, Proceedings of the 1963 International Symposium at Berkeley, Amsterdam (1965), 39–54.

Paul J. Cohen

THE WORKS OF ALEXANDER GROTHENDIECK*

by
JEAN DIEUDONNÉ

Alexander Grothendieck is not yet 40 years old, and already the magnitude of his work and the extent of his influence on contemporary mathematics are so great that it is not possible to give anything other than a very distorted idea in such a short presentation.

Everyone knows that Grothendieck is the principal architect of the renovation in Algebraic Geometry that is being accomplished before our eyes. Of course, this renovation has been prepared on one hand by the works of Weil-Zariski, who laid the foundation of "abstract" Algebraic Geometry over an arbitrary field, and on the other hand by those of Serre, who introduced into theory the powerful tools of sheaves and homological algebra. But Grothendieck knew how to extend these ideas to their full scope by developing them in their general form, free from parasite restrictions which hindered their use; and he has added many completely original ideas.

In its "affine" form, modern Algebraic Geometry coincides with commutative algebra; already in classical Algebraic Geometry, an affine variety was associated with the ring of "regular" functions on the variety. Conversely, an *arbitrary* commutative ring A (with a unit element) is now put in bijective correspondence with a geometric object, the "affine scheme of the ring A", which is the set of prime ideals of A, given with a certain topology and a sheaf whose fibers are the local rings at the prime ideals of A. The benefits of this formulation are: 1° that it provides a geometric intuition which is a very valuable guide (such as suggesting analogies with differentiable manifolds or analytic varieties); 2° that it goes beyond the "affine" point of view and leads to the idea of general "schemes" (generalizing the "abstract varieties" of Weil) by the simple topological process of "gluing" topological spaces equipped with sheaves (an idea due to Serre).

This framework is complemented by two new ideas: 1° the focus on the notion of *morphism*, which, in the affine case, corresponds to that of a homomorphism of rings respecting the unit element, and which allows one to "relativize" all the concepts from the classical theory; 2° the general notion of "*base change*": given a morphism $X \to S$, for every "base change" morphism $S' \to S$, one forms in a canonical way a new scheme $X' = X_{(S')}$ and a new morphism $X' \to S'$ by a process which, in the

*Translated from French to English by Chee Whye Chin.

affine case, corresponds to the tensor product of rings, and which encompasses the classical "extension of the base field" from the Weil-Zariski period.

These notions, as well as that of *flatness* (due to Serre, but of which Grothendieck has considerably expanded in usage) are the basis of a technique of remarkable power and flexibility. Among the many tools thus forged, let us mention especially:

I) The passage to the *projective limit* for schemes, which in many cases allows one to reduce problems to the case where the rings being considered are finitely generated algebras over **Z** (thus making concrete the famous thesis of Kronecker).

II) The theory of *excellent* rings, which systematizes and completes the deep results of Zariski-Nagata on noetherian local rings, and which can be used in general problems thanks to the passage to the projective limit, the finitely generated **Z**-algebras being excellent rings.

III) The theory of *relative cohomology* for schemes and its relations with the notion of depth (due to Auslander-Buchsbaum and Serre).

IV) The theory of *formal schemes*: these are *inductive limits* of schemes, an operation which, in the affine case, corresponds to the completion of local rings, but having a more general scope; for example, in certain questions, it allows one to reduce a problem in characteristic $p > 0$ to a problem in characteristic 0; it is also in this framework that the theory of "holomorphic functions" of Zariski is formulated (and furthermore put in a much more general cohomological form).

V) The use of the notion of a *representable functor*, replacing the more restrictive "universal problem"; being given a set E attached to a morphism $X \to S$, in a way "compatible" with base change, it amounts to knowing whether there exists a scheme Z over S such that E is identified in a natural way with the set of all *sections* of the morphism $Z \to S$.

VI) The theory of *descent*. Many problems are simplified when one makes an appropriate "base change" (for example, in classical Algebraic Geometry, when one passes from the base field to an algebraic closure of that field). The question is to go back to the initial situation and deduce consequences there from what happens after the base change; that is the goal of the theory of "descent", which provides criteria allowing this to be done in cases sufficiently general for many applications.

VII) Pushing this idea further, Grothendieck generalized in a very original way the notion of topology and the cohomology of sheaves on a topological space (the theory of "sites" and "topos"): the role played by the open sets of a topological space is taken by morphisms $X' \to X$ of a special type, usually the *etale* morphisms (analogues of the "unramified coverings" of an open set in an analytic variety).

Before going further, it should be noted that one does not do justice to the above theories by describing them, rather contemptuously, as "technical results"; some of them are indeed easy extensions of classical methods, but many require totally new methods of attack, relying on subtle considerations in commutative algebra or homological algebra, and which by themselves already constitute an imposing work of which there are few examples.

However, it is indeed true that in the minds of Grothendieck, all these methods were not developed for their own sake, but with a view towards tackling some fundamental problems in Algebraic Geometry. Among those where he has made significant progress (partly with the collaboration of his students), we must mention:

1. The determination, in characteristic $p > 0$, of the prime-to-p part of the fundamental group of an algebraic curve.
2. The definitions of the formal "moduli" scheme (isomorphism classes of schemes), of the Picard scheme (divisor classes), of the Hilbert scheme (set of all subvarieties of a given variety, whose structure as a scheme is destined to replace the classical "Chow coordinates").
3. (In collaboration with M. Demazure). A vast theory of "group schemes", generalizing the theory of algebraic groups of Chevalley.
4. (In collaboration with M. Artin and J. Verdier). The definition of the "etale cohomology" of schemes, where, thanks to the theory of "sites", one already has in any characteristic methods and results analogous to those provided by algebraic topology for algebraic varieties over the field of complex numbers (theorems of finiteness and duality, Lefschetz formula, comparison with the "topological" cohomology in the classical case); thanks to these results, Grothendieck was able to show a part of the famous "Weil conjectures": rationality of the L-functions attached to varieties over a finite field, and their expression in terms of homological invariants.
5. Finally, the earliest of Grothendieck's works in Algebraic Geometry, the generalization of the theorem of Riemann-Roch-Hirzebruch and its purely algebraic proof in any characteristic. It is on this occasion that Grothendieck introduced the first notions of "K-theory" (or, as we now say, "Grothendieck groups (or rings)"). This idea was striking to many, especially topologists and algebraists, who have drawn brilliant well-known applications in multiple areas.

It is appropriate to also mention the works of Grothendieck in homological algebra, a little earlier than his proof of the Riemann-Roch theorem, and which have expanded and made more flexible the results of Cartan-Eilenberg, especially in giving a "good" definition of the cohomology of sheaves on an arbitrary space.

Finally, I have said nothing of the first papers of Grothendieck on topological vector spaces (1950–55), in part because they are very well known and are being increasingly used in functional analysis, especially the theory of nuclear spaces, which "explains" the phenomena encountered in the theory of distributions. I have personally had the privilege of witnessing up close, at that time, the blossoming talent of this extraordinary "beginner" who at the age of 20 was already a master; and, with 10 years of hindsight, I still consider that the work of Grothendieck in this period remains, with that of Banach, one that has left the strongest mark on this part of mathematics.

If we have to look for a spiritual kinship of Grothendieck, it seems to me that it is Hilbert whom we can best compare him with: like Hilbert, his motto could be: "simplify by generalizing", by seeking the deep springs of mathematical phenomena; but, like Hilbert also, when this in-depth analysis has led to a point where only a frontal attack remains possible, he almost always finds in his rich imagination the battering ram which breaks the obstacle. The comparison is perhaps heavy to carry, but Grothendieck is of a size not to be overwhelmed.

A. Grothendieck

ON THE WORKS OF STEPHEN SMALE*

by
RENÉ THOM

The first scientific work of S. Smale is his Ph.D. thesis defended in 1956 at the University of Michigan (Ann Arbor). Written under the direction of Raoul Bott, it already reflects a glowing mastery. The essential result, now well known, is the theorem for lifting homotopies of the immersions of a manifold modulo a sub-manifold. Established by sophisticated geometrical constructions, this result reflected its author's first rate capacities for intuition. Thanks to him, one could establish a conjecture — then about ten years old — of C. Ehresmann on the classification of the immersions from one manifold into another; it followed that it was possible, by a regular deformation (that is to say without leaving the immersions) to transform the canonical embedding of the "2-sphere" in Euclidean space R^3 of three dimensions into an antipodal embedding; this result did not go without arousing the curiosity of topologists, many of whom contrived to clarify this deformation. But the thesis of Smale gave more than this curiosity, it opened an avenue of attack throughout a domain of questions which were previously out of reach, and a whole chapter of Differential Topology, the study of immersions and embeddings of a differentiable manifold into another, marked by the works of M. Hirsh, Haefliger, etc., came out from it more or less directly.

With the great 1960 works on the Poincaré conjecture, we come to the best known part of the work of Smale, that which doubtlessly earned us his presence here. One already knew, by Morse theory, that any compact manifold can be decomposed into cells from a gradient flow, and that, if one is given at each critical level the attaching map of the cell from the corresponding gradient flow, one is able to reconstruct the manifold; one has already started — as a result of the works of Kervaire, Milnor, Wallace — to practice "surgery" of manifolds, that is to say the technique of transforming to a simpler manifold by the resection of a pair of dual handles.

The huge merit of Smale, in this question, is to have dared to undertake what any other mathematician of the time would have regarded as hopeless: being given a Morse function on a homotopy sphere, simplify the presentation of this manifold by eliminating by surgery the pairs of super-critical points of index $k, k+1$. One knew too well the difficulties in low dimensions (three or four) to hope that this was possible; Smale dared, and succeeded. He understood that the difficulties encountered were a phenomenon special to low dimensions: by confining

*Translated from French to English by Chee Whye Chin.

to dimension five or higher, one can work more comfortably, and surgery can be effected more easily; by very ingenious constructions, Smale then overcame the last difficulties: he eliminates on a homotopy sphere all pairs of critical points of index k different from zero and n; he thus obtains a manifold on which there exists a function having only two critical points (minimum and maximum). By a theorem of Reeb, this manifold is a topological sphere (but not necessarily diffeomorphic to the usual sphere, as was shown by the examples due to Milnor).

This extremely brilliant result was completed soon afterwards by the so-called h-cobordism theorem, which generalizes it. If two compact manifolds M_1 and M_2 form the boundary of a common bounding manifold W of which they are deformation retracts, and if they are simply connected, then M_1 and M_2 are diffeomorphic. This theorem allows one, using a later result of Novikov and Browder, to reduce the problem of the classification of differentiable manifolds to a problem purely of homotopy, and in fact to a problem of (admittedly difficult) algebra. The techniques used in the proof of the h-cobordism theorem have probably not yielded all their fruits, and more recent works, such as those of J. Cerf, have expanded the scope of their application.

With the contemporary result of B. Mazur on the conjecture of Schönflies, the works of Smale has turned a page in Algebraic Topology. One can say that the topology of "spaces", of differentiable manifolds, is henceforth quasi-completed. There certainly remain many unresolved questions: the algebraic structures defined by the classification are not clear — in general —; but will they be someday? There remains little beyond the theory — which although has made much progress recently — of such things, still somewhat pathological, as the piecewise-linear manifolds and the purely topological manifolds. Under these conditions, if Topology wants to renew itself, and not to be confined to difficult problems of vain technicality, it must be concerned with renewing its materials and tackling new problems. With the geometrical objects associated with differentiable structures: differential forms, tensors, laminations, differential operators, a huge field is open to the topologist. We have also seen that one of our laureates has been rewarded for a result along this route.

Aside from essentially linear, classical analysis, there is the practically unexplored domain of non-linear analysis; there, the topologist can again hope for better use of his methods, and perhaps of his essential quality, namely the intrinsic vision of things. This is what Smale understood very quickly; in an article in collaboration with R. Palais, he defined the best possible conditions for the application of Morse theory to the calculus of variations; from these, he then deduced theorems relating to the existence of solutions for non-linear elliptic problems. But very soon, Smale turns to a theory — then much neglected —: the qualitative theory of differential systems on a manifold. Being almost alone, Smale reads Poincaré and Birkhoff; before the inextricability of the problem, he understands very quickly the benefits of an essential notion, that of "structural stability" introduced by Andronov and Pontryagin, this concept aims to characterize, among the vector fields on a manifold, those which enjoy a propriety of qualitative stability, in the following sense: any field (Z) sufficiently close to a given field (X) (with respect to the C^1-topology) gives rise to a field of trajectories homeomorphic to the field defined by (X). The central problem is thus the problem of approximation: can any vector field be approached by a structurally stable field? This problem, solved in the affirmative by Peixoto for compact manifolds of dimension at most two, was posed for the higher dimensions. Smale then constructed a compact manifold M of dimension four, and a field (X) on M, such that no field (Z) sufficiently close to (X) is structurally stable. The general problem of the stability of differential systems is thus solved in the negative. However, the very notion of structural stability is far from having lost all its benefits: first, because there exists, in the functional space of vector fields on a manifold M, a "relatively important" open set of structurally stable fields: those formed by the vector fields of generic gradient type (without recurrence) and, probably, a class of fields defined by Smale (the so-called Morse-Smale fields), which have recurrence (with closed trajectories) but in a benign and severely controlled form. But, by the theory of configurations of trajectories associated to the homoclinic points of Poincaré, Smale was soon convinced that other fields, with complex and rigid topology, are structurally stable. He returned to the brilliant works of the Soviet school (with Sinai, Arnold, Anosov) for establishing the existence of an extended class of structurally stable fields, of the type of geodesic flow on a riemannian manifold with negative curvature. These works exerted a great influence on Smale, and inflected his research in the current direction, namely the obvious setting of "piecewise structural stability", each "piece" being linked to a rigid configuration of recurrent (non-wandering) trajectories in the sense of Birkhoff. These investigations are ongoing and seem very promising.

I would be remiss if I did not emphasize one last point: if the works of Smale do not perhaps possess the formal perfection of a definitive work, it is because Smale is a pioneer who takes his risks with quiet courage; in a completely unexplored area, in a geometric jungle of inextricable wealth, he is the first to have traced the route and paved the way. And one can predict that his work will take on a fundamental importance in future, comparable to that of the great forerunners, Poincaré and Birkhoff. After all many classical problems, such as the problem of Fermat or the conjecture of Riemann, may still have to wait a few years for their solution; but if science wants to use the differential tool to describe natural phenomena, it cannot afford to ignore much longer the topological structure of the attractors of a structurally stable dynamical system, because any "physical state" having a certain stability, a certain permanence, is necessarily represented by such an attractor. According to certain views of Smale, these attractors would be homogeneous spaces of Lie groups of a special type. If these views could be extended to Hamiltonian systems, one could perhaps explain the appearance — so far not understood — of Lie groups in the physics of elementary particles. In this sense, the problem of Smale is — in my eyes — of essential epistemological importance.

Stephen Smale

STEPHEN SMALE

1930	Born July 15 at Flint, Michigan
1952	B.S., University of Michigan
1953	M.S., University of Michigan
1957	Ph.D., University of Michigan
1956–68	Instructor, University of Chicago
1958–60	Member, Institute for Advanced Study, Princeton
1962	Visiting Professor, College de France, Paris (Spring)
1960–61	Associate Professor of Mathematics, University of California, Berkeley
1961–64	Professor of Mathematics, Columbia University
1964–present	Professor of Mathematics, University of California, Berkeley
1960–62	Alfred P. Sloan Research Fellow
1966	Member, Institute for Advanced Study, Princeton (Fall)
1967–68	Research Professor, Miller Institute for Basic Research in Science, Berkeley
1969–70	Visiting Member, Institut des Hautes Études Scientifiques (Fall)
1972–73	Visiting Member, Institut des Hautes Études Scientifiques (Fall)
1972–73	Visiting Professor, University of Paris, Orsay (Fall)
1974	Visiting Professor, Yale University (Fall)
1976	Visiting Professor, Instituto de Matematica Pura e Aplicada, Rio de Janeiro (April)
1976	Visiting Professor, Institut des Hautes Études Scientifiques (May, June)
1976–present	Professor without stipend, Department of Economics, University of California, Berkeley
1979–80	Research Professor, Miller Institute for Basic Research in Science, Berkeley
1987	Visiting Scientist, IBM Corporation, Yorktown Heights (Fall)
1987	Visiting Professor, Columbia University (Fall)

Honors

1960–62	Alfred P. Sloan Research Fellowship
1964–present	Foreign Membership, Brazilian Academy of Science
1965	Veblen Prize for Geometry, American Mathematical Society
1966	Fields Medal, International Mathematical Union

1967	University of Michigan Sesquicentennial Award
1967–present	Membership, American Academy of Arts and Sciences
1967–68	Research Professorship, Miller Institute for Basic Research in Science, Berkeley
1970–present	Membership, National Academy of Sciences
1972	Colloquium Lecturership, American Mathematical Society
1974	Honorary D.Sc. University of Warwick, England
1979–80	Research Professorship, Miller Institute for Basic Research in Science, Berkeley
1983–present	Election as Fellow, The Econometric Society
1983	Faculty Research Lecturership, University of California, Berkeley
1987	Honorary D.Sc., Queens University, Kingston, Ontario
1988	Chauvenet Prize, Mathematical Association of America
1989	Von Neumann Award, Society for Industrial and Applied Mathematics
1990	Honorary Membership in Instituto de Mathematica, Pura e Aplicada (IMPA), Rio de Janeiro
1990	Research Professorship, Miller Institute for basic Research in Science, Berkeley
1991–92	Trinity Mathematical Society, Dublin, Honorary Member Bishop Berkeley Lecture, Trinity College Dublin Distinguished Lecture, Fields Institute

Bibliography

1. A note on open maps, *Proceedings of the AMS* **8** (1957), 391–393.
2. A Vietoris mapping theorem for homotopy, *Proceedings of the AMS* **8** (1957), 604–610.
3. Regular curves on Riemannian manifolds, *Transactions of the AMS* **87** (1958), 492–512.
4. On the immersions of manifolds in Euclidean space (with R. K. Lashof), *Annals of Mathematics* **68** (1958), 562–583.
5. Self-intersections of immersed manifolds (with R. K. Lashof), *Journal of Mathematics and Mechanics* **8** (1959), 143–157.
6. A classification of immersions of the two-sphere, *Transactions of the AMS* **90** (1959), 281–290.
7. The classification of immersions of spheres in Euclidean space, *Annals of Mathematics* **69** (1959), 327–344.
8. Diffeomorphisms of the two-sphere, *Proceedings of the AMS* **10** (1959), 621–626.
9. On involutions of the 3-sphere (with Morris Hirsch), *American Journal of Mathematics* **81** (1959), 893–900.
10. Morse inequalities for a dynamical system, *Bulletin of the AMS* **66** (1960), 43–49.
11. The generalized Poincaré conjecture in higher dimensions, *Bulletin of the AMS* **66** (1960), 373–375.
12. On dynamical systems, *Boletin de la Sociedad Mathematica Mexicana* (1960), 195–198.
13. On gradient dynamical systems, *Annals of Mathematics* **74** (1961), 199–206.
14. Generalized Poincaré conjecture in dimensions greater than 4, *Annals of Mathematics* **74** (1961), 391–406.
15. Differentiable and combinatorial structures on manifolds, *Annals of Mathematics* **74** (1961), 498–502.
16. On the structure of 5–manifolds, *Annals of Mathematics* **75** (1962), 38–46.
17. On the structure of manifolds, *American Journal of Mathematics* **84** (1962), 387–399.
18. Dynamical systems and the topological conjugacy problem for diffeomorphisms, *Proceedings of the International Congress of Mathematicians*, Stockholm, 1962. Almquist and Wiksells Uppsala 1963.
19. A survey of some recent developments in differential topology, *Bulletin of the AMS* **69** (1963), 131–146.
20. Stable manifolds of difeomorphisms and differential equations, *Annali della Scuola Normali Superioro di Pisa. Serie III* **XVII** (1963), 97–116.
21. A generalized Morse theory (with R. Palais), *Bulletin of the AMS* **70** (1964), 165–172.
22. Morse theory and non-linear generalization of the Dirichlet problem, *Annals of Mathematics* **80** (1964), 382–396.
23. Diffeomorphisms with many periodic point, *Differential and Combinatorial Topology* (*A symposium in honor of Marston Morse*), Princeton University Press, Princeton, NJ, 1965, pp. 63–80.
24. An infinite dimensional version of Sard's theorem, *American Journal of Mathematic* **87** (1965), 861–866.

25. On the Morse index theorem, *Journal of Mathematics and Mechanics* **14** (1965), 1049–1056.
26. A structurally stable differentiable homeomorphism with an infinite number of periodic points, *Report on the Symposium on Non-Linear Oscillations*, Kiev Mathematics Institute, Kiev, 1963, pp. 365–366.
27. On the calculus of variations, *Differential Analysis*, Bombay, Tata Institute 1964, pp. 187–189.
28. Structurally stable systems are not dense, *American Journal of Mathematics* **88** (1966), 491–496.
29. Dynamical systems on n-dimensional manifolds, *Differential Equations and Dynamical Systems* (1967), pp. 483–486. Academic Press N. Y. 1967.
30. Differentiable dynamical systems, *Bulletin of the AMS* **73** (1967), 747–817.
31. What is global analysis? *American Mathematics Monthly* Vol. 76, (1969), 4–9.
32. Nongenericity of Ω-stability (with R. Abraham), *Global Analysis. Proc. of Symposia in Pure Mathematics*, Vol. 14, American Mathematical Society, Providence, RI, 1970, pp. 5–8.
33. Structural stability theorems (with J. Palis), *Global Analysis. Proc. of Symposia in Pure Mathematics*, Vol. 14, American Mathematical Society, Providence, RI, 1970, pp. 223–231.
34. Notes on differential dynamical systems, *Global Analysis. Proc. of Symposia in Pure Mathematics*, Vol. 14, American Mathematical Society, Providence, RI, 1970, pp. 277–287.
35. The Ω-stability theorem, *Global Analysis. Proc. of Symposia in Pure Mathematics*, Vol. 14, American Mathematical Society, Providence, RI, 1970, pp. 289–297.
36. Topology and mechanics, *I. Inventiones Mathematicae* **10** (1970), 305–331.
37. Topology and mechanics, *II. Inventiones Mathematicae* **11** (1970), 45–64.
38. Stability and genericity in dynamical systems, Seminaire Bourbaki (1960–70) pp. 177–186, Springer, Berlin.
39. Problems on the nature of relative equilibria in celestial mechanics, *Proc. of Conference on Manifolds, Amsterdam*, Springer Verlag, Berlin, 1970, pp. 194–198.
40. On the mathematical foundations of electrical circuit theory, *Journal of Differential Geometry* **7** (1972), 193–210.
41. Beyond hyperbolicity (with M. Shub), *Annals of Mathematics* **96** (1972), 591–597.
42. Personal perspectives on mathematics and mechanics, *Statistical Mechanics: New Concepts, New Problems, New Applications*, edited by Stuart A. Rice, Karl F. Freed, and John C. Light, University of Chicago Press, Chicago 1972, pp. 3–12.
43. Stability and isotopy in discrete dynamical systems, *Dynamical Systems*, edited by M. M. Peixoto, Academic Press, New York, 1973.
44. Global analysis and economics I, *Pareto Optimum and a Generalization of Morse Theory, Dynamical Systems*, edited by M. M. Peixoto, Academic Press, New York, 1973.
45. Global analysis and economics IIA, *Journal of Mathematical Economics*, April **1** (1974), pp. 1–14.
46. Global analysis and economics III, Pareto optima and price equilibria, *Journal of Mathematical Economics* **I** (1974), 107–117.

47. Global analysis and economics IV, Finiteness and stability of equilibria with general consumption sets and production, *Journal of Mathematical Economics* **I** (1974), 119–127.
48. Global analysis and economics V, Pareto theory with constraints, *Journal of Mathematical Economics* **I** (1974), 213–221.
49. Optimizing several functions, *Manifolds — Tokyo 1973, Proc. of the International Conference on Manifolds and Related Topics in Topology, Tokyo*, 1973, University of Tokyo Press, Tokyo 1975, pp. 69–75.
50. A mathematical model of two cells via Turing's equation, *Lectures on Mathematics in the Life Sciences* **6** (1974), 17–26.
51. Sufficient conditions for an optimum, *Warwick Dynamical Systems 1974*, Lecture Notes in Mathematics, Springer–Verlag, Berlin, 1975, pp. 287–292.
52. Differential equations (with I. M. Singer), *Encyclopedia Britannica Fifteenth Edition*, 1974, pp. 736–767. Helen Hemingway Berton Publisher.
53. Dynamics in general equilibrium the theory, *American Economic Review* **66** (1976), 288–294.
54. Global analysis and economics VI: Geometric analysis of Parieto Optima and price equilibria under classical hypotheses, *Journal of Mathematical Economics* **3** (1976), 1–14.
55. The qualitative analysis of a difference equation of population growth (with R. F. Williams), *Journal of Mathematical Biology* **3** (1976), 1–4.
56. On the differential equations of species in competition, *Journal of Mathematical Biology* **3** (1976), 5–7.
57. A convergent process of price adjustment, *Journal of Mathematical Economics* **3** (1976), 107–120.
58. Exchange processes with price adjustment, *Journal of Mathematical Economics* **3** (1976), 211–226.
59. Global stability questions in dynamical systems, *Proceedings of Conference on Qualitative Analysis, Washington D. C.*
60. Dynamical systems and turbulence, *Turbulence Seminar: Berkeley 1976/77*, Lecture Notes in Mathematics edited by P. Bernard and T. Ratiu, No. 615, Springer–Verlag, Berlin, 1977, pp. 48–71.
61. Some dynamical questions in mathematical economics, *Colloques Internationaux du Centre National de la Recherche Scientifique*, No. 259: Systems Dynamiques et Modeles Economiques, (1977).
62. An approach to the analysis of dynamic processes in economic systems, *Equilibrium and disequilibrium in Economic Theory*, Reidel, Dordrecht, 1977, pp. 363–367.
63. On comparative statics and bifurcation in economic equilibrium theory, *Annals of the New York Academy of Sciences* **316** (1979), 545–548.
64. On algorithms for solving $f(x) = 0$ (with Morris W. Hirsch), *Communication on Pure and Applied Mathematics* **55** (1980), 1–12.
65. Smooth solutions of the heat and wave equations, *Commentarii Mathematici Helvetici* **55** (1980), 1–12.
66. The fundamental theorem of algebra and complexity theory, *Bulletin of the American Mathematical Society* **4** (1981), 1–36.
67. The prisoner's dilemma and dynamical systems associated to non-cooperative games, *Econometrica* **48** (1980), 1617–1634.

68. Global analysis and economics, *Handbook of Mathematical Economics*, Vol. 1, edited by K. J. Arrow and M. D. Intrilligator, North-Holland, Amsterdam, 1981, 331–370.
69. *The Mathematics of Time: Essays on Dynamical Systems, Economic Processes and Related Topics*, Springer–Verlag, Berlin, 1980.
70. The problem of the average speed of the simplex method, *Proceedings of the XIth International Symposium on Mathematical Programming*, edited by A. Bachem, M. Grotschel, and B. Korte, Springer–Verlag, Berlin, 1983, pp. 530–539.
71. On the average number of steps of the simplex method of linear programming, *Mathematical Programming* **27** (1983), 241–262.
72. On the steps of Moscow University, *Mathematical Intelligencer*, **6**, No. 2 (1984), 21–27.
73. Gerard Debreu wins the Nobel Prize, *Mathematical Intelligencer*, **6**, No. 2 (1984), 61–62.
74. Scientists and the arms race (in German translation), *Natur-Wissenschafter Gegen Atomrustung*, edited by Hans–Peter Durr, Hans–Peter Harjes, Matthias Krech, and Peter Starlinge, Spiegel–Buch, Hamburg, 1983, pp. 327–334.
75. Computational complexity: On the geometry of polynomials and a theory of cost Part I (with Mike Shub), *Annales Scientifiques de l'Ecole Normale Superieure* **18** (1985), 107–142.
76. On the efficiency of algorithms of analysis, *Bulletin of the American Mathematical Society* **13** (1985), 87–121.
77. Computational complexity: On the geometry of polynomials and a theory of cost: Part II (with M. Shub), *SIAM Journal of Computing* **15** (1986), 145–161.
78. On the existence of generally convergent algorithms (with M. Shub), *Journal of Complexity* **2** (1986), 2–11.
79. Newton's method estimates from data at one point, *The Merging of Disciplines: New Directions in Pure, Applied, and Computational Mathematics*, edited by Richard E. Ewing, Kenneth I. Gross, and Clyde F. Martin, Springer–Verlag, New York, 1986, pp. 185–196.
80. On the topology of algorithms, **I**, *Journal of Complexity* **3** (1987), 81–89.
81. Algorithms for solving equations, *Proceedings of the International Congress of Mathematicians, August 3–11, 1986, Berkeley, California USA*, American Mathematical Society, Providence, RI, 1987, pp. 172–195.
82. Global analysis in economic theory, *The New Palgrave: A Dictionary of Economics*, Vol. 2, edited by John Eatwell, Murray Milgate, and Peter Newman, Macmillan, London, 1987, pp. 532–534.
83. The Newtonian contribution to our understanding of the computer, *Queen's Quarterly* **95** (1988), 90–95.
84. On a theory of computation and complexity over the real numbers: NP-completeness, recursive functions and universal machines (with Lenore Blum and Mike Shub), *Bulletin of the American Mathematical Society (New Series)* **21** (1989), 1–46.
85. Some remarks on the foundations of numerical analysis, *SIAM Review* **32**(2) (1990), 221–270.
86. The story of the higher dimensional Poincaré conjecture, *Mathematical Intelligencer* **12**, No. 2 (1990), 40–51.
87. Dynamics retrospective: Great problems, attempts that failed, *Physica D* **51** (1991), 267–273.

Reprinted from Bull. Amer. Math. Soc. **69** *(2) (1963), 133–145*

A SURVEY OF SOME RECENT DEVELOPMENTS IN DIFFERENTIAL TOPOLOGY

by

S. SMALE

1. We consider differential topology to be the study of differentiable manifolds and differentiable maps. Then, naturally, manifolds are considered equivalent if they are diffeomorphic, i.e., there exists a differentiable map from one to the other with a differentiable inverse. For convenience, differentiable means C^∞; in the problems we consider, C' would serve as well.

The notions of differentiable manifold and diffeomorphism go back to Poincaré at least. In his well-known paper, *Analysis situs* [**27**] (see pp. 196–198), topology or analysis situs for Poincaré was the study of differentiable manifolds under the equivalence relation of diffeomorphism. Poincaré used the word homeomorphism to mean what is called today a diffeomorphism (of class C'). Thus differential topology is just topology as Poincaré originally understood it.

Of course, the subject has developed considerably since Poincaré; Whitney and Pontrjagin making some of the major contributions prior to the last decade.

Slightly after Poincaré's definition of differentiable manifold, the study of manifolds from the combinatorial point of view was also initiated by Poincaré, and again this subject has been developing up to the present. Contributions here were made by Newman, Alexander, Lefschetz and J. H. C. Whitehead, among others.

What started these subjects? First, it is clear that differential geometry, analysis and physics prompted the early development of differential topology (it is this that explains our admitted bias toward differential topology, that it lies close to the main stream of mathematics). On the other hand, the combinatorial approach to manifolds was started because it was believed that these means would afford a useful attack on the differentiable case. For example, Lefschetz wrote [**13**, p. 361], that Poincaré tried to develop the subject on strictly "analytical" lines and after his *Analysis situs*, turned to combinatorial methods because this approach failed for example in his duality theorem.

An address delivered before the Stillwater meeting of the Society on August 31, 1961, by invitation of the Committee to Select Hour Speakers for Western Sectional Meetings; received by the editors November 28, 1962.

Naturally enough, mathematicians have been trying to relate these two viewpoints that have developed side by side. S. S. Cairns is an example of such.

In the last decade, the three domains, differential topology, combinatorial study of manifolds, and the relations between the two, have all advanced enormously. Of course, these developments are not isolated from each other. However, we would like to make the following important point.

It has turned out that the main theorems in differential topology did not depend on developments in combinatorial topology. In fact, the contrary is the case; the main theorems in differential topology inspired corresponding ones in combinatorial topology, or else have no combinatorial counterpart as yet (but there are also combinatorial theorems whose differentiable analogues are false).

Certainly, the problems of combinatorial manifolds and the relationships between combinatorial and differentiable manifolds are legitimate problems in their own right. An example is the question of existence and uniqueness of differentiable structures on a combinatorial manifold. However, we don't believe such problems are the goal of differential topology itself. This view seems justified by the fact that today one can substantially develop differential topology most simply without any reference to the combinatorial manifolds.

We have not mentioned the large branch of topology called homotopy theory until now. Homotopy theory originated as an attack on the homeomorphism or diffeomorphism problem, witness the "Poincaré Conjecture" that the homotopy groups characterize the homeomorphism type of the 3-sphere, and the Hurewicz conjecture that the homeomorphism type of a closed manifold is determined by the homotopy type. One could attack the homotopy problem more easily than the homeomorphism one and, for many years, most of the progress in topology centered around the homotopy problem.

The Hurewicz conjecture turned out not to be true, but amazingly enough, as we shall see, the last few years have brought about a reduction of a large part of differential topology to homotopy theory. These problems do not belong so much to the realm of pure homotopy theory as to a special kind of homotopy theory connected with vector space bundles and the like, as exemplified by work around the Bott periodicity theorems.

Of course, there are a number of important problems left in differential topology that do not reduce in any sense to homotopy theory and topologists can never rest until these are settled. But, on the other hand, it seems that differential topology has reached such a satisfactory stage that, for it to continue its exciting pace, it must look toward the problems of analysis, the sources that led Poincaré to its early development.

We here survey some developments of the last decade in differential topology itself. Certainly, we make no claims for completeness. A notable omission is the work of Thom, on cobordism, and the study of differentiable maps. The reader is referred to expositions of Milnor [**21**] and H. Levine [**14**] for accounts of part of this work.

2. We now discuss what must be considered a fundamental problem of differential topology, namely, the diffeomorphism classification of manifolds.

The classification of closed orientable 2-manifolds goes back to Riemann's time. The next progress on this problem was the development of numerical and algebraic invariants which were able to distinguish many nondiffeomorphic manifolds. These invariants include, among others, the dimension, betti numbers, homology and homotopy groups and characteristic classes.

For dimension greater than two, there was still (at the beginning of 1960) no known case where the existing numerical and algebraic invariants determined the diffeomorphism class of the manifold. The simplest case of this problem (or so it appeared) was that posed by Poincaré: Is a 3-manifold which is closed and simply connected, homeomorphic (equivalently diffeomorphic) to the 3-sphere? This has never been answered.

The surprising thing is, however, that without resolving this problem, the author showed that in many cases, the known numerical and algebraic invariants were sufficient to characterize the diffeomorphism class of a manifold. Generally speaking in fact, considerable information on the structure of manifolds was found. We will now give an account of this.

To see how manifolds can be constructed, one defines the notion of *attaching a handle*. Let M^n be a compact manifold with boundary ∂M (we remind the reader that everything is considered from the C^∞ point of view, manifolds, imbeddings, etc.) and let D^s be the s-disk (i.e., the unit disk of Euclidean s-space E^0). Suppose $f : (\partial D^s) \times D^{n-s} \to M$ is an imbedding. Then $X(M; f; s)$, "M with a handle attached by f_0^n is defined by identifying points under f and imposing a differentiable structure on $M \cup_f D^s \times D^{n-s}$ by a process called "straightening the angle." Similarly if $f_i : (\partial D_1^s) \times D_1^{n-s} \to M$, $i = 1, \ldots, k$, are imbeddings with disjoint images one can define $X(M; f_1, \ldots, f_k; s)$. If M itself is a disk, then $X(M; f_1, \ldots, f_k; s)$ is called a *handlebody*.

(2.1) Theorem. *Let f be a C^∞ function on a closed (i.e. compact with empty boundary) manifold with no critical points on $f^{-1}[-\epsilon, \epsilon]$ except k nondegenerate ones on $f^{-1}(0)$, all of index s. Then $f^{-1}[-\infty, \epsilon]$ is diffeomorphic to $X(f^{-1}[-\infty, -\epsilon]; f_1, \cdots, f_k; s)$ (for suitable f_i).*

(For a reference to the notion of nondegenerate critical point and its index, see e.g. [**1**].) This might well be regarded as the basic theorem of finite dimensional Morse theory. Morse [**23**] was concerned with the homology version of this theorem, Bott [**1**], the homotopy version of 2.1. The proof of 2.1 itself is based on the ideas of the proofs of the weaker statements.

(2.2) Theorem. (Morse–Thom). *On every closed manifold W, there exists a C^∞ function with nondegenerate critical points.*

For a proof see [**41**].

By combining 2.1 and 2.2 we see that every manifold can be obtained by attaching handles successively to a disk (we have been restricting ourselves to the compact, empty boundary case only for simplicity).

The main idea of the following theory is to remove superfluous handles (or equivalently critical points) without changing the diffeomorphism type of the manifold. For this one starts (after 2.2) with a "nice function," a function on M given by 2.2 with the additional property that the handles are attached in order according to their dimension (the s in $D^s \times D^{n-s}$).

(2.3) Theorem. *On every closed manifold there exists a function f with nondegenerate critical points such that at each critical point, the value of f is the index.*

This was proved by A. Wallace [**44**] and the author [**38**] independently by different methods. (For a general references to this section see [**34**; **37**].)

The actual removing of the extra handles is the main part and for this one needs extra assumptions. The next is the central theorem (or its generalization "2.4′" to include manifold with boundary).

(2.4) Theorem. *Let M be a simply connected closed manifold of dimension > 5. Then on M there is a nondegenerate (and nice) function with the minimal number of critical points consistent with the homology structure of M.*

We make more explicit the conclusion of 2.4. Let $\sigma_{i1}, \cdots, \sigma_{ip(i)}, \tau_{i1}, \cdots, \tau_{ig(i)}$, $0 \leqq i \leqq n$, be a set of generators for a direct sum decomposition of $H_i(M)$, σ_{ij} free, τ_{ij} of finite order. Then one can obtain the function of 2.4 with type numbers M_i (the number of critical points of index i) satisfying $M_i = p(i) + g(i) + g(i-1)$.

The first special case of 2.4 is

(2.5) Theorem. *Let M be a simply connected closed manifold of dimension greater than 5 with no torsion in the homology of M. Then there exists a nondegenerate function on M with type numbers equal to the betti numbers of M.*

We should emphasize that 2.4 and 2.5 should be interpreted from the point of view of 2.1. One may apply 2.5 to the case of a "homotopy sphere" (using 2.1 of course).

(2.6) Theorem. *Let M^n be a simply connected closed manifold with the homology groups of a sphere, $n > 5$. Then M can be obtained by "gluing" two copies of the n-disk by a diffeomorphism from the boundary of one to the boundary of the other.*

For $n = 5$, the theorem is true and can be proved by an additional argument.

Theorem 2.6 implies the weaker statement ("the generalized Poincaré conjecture in higher dimensions") that a homotopy sphere of dimension $\geqq 5$ is homeomorphic to S^n, see [**33**]. Subsequently, Stallings [**39**] and Zeeman [**50**] found a proof of this last statement.

Theorem 2.4 was developed through the papers [**34; 35**] and appears in the above form in [**37**]. Rather than try to give an idea of the proof of 2.4, we refer the reader to these papers. One may also refer to [**2; 3**] and [**15**].

The analogue of 2.4, say 2.4′ is also proved for manifolds with boundary [**37**] and this analogue implies the h-cobordism theorem stated below. The simplest case of 2.4′ is

(2.7) Disk Theorem. *Let M^n be a contractible compact manifold, $n > 5, \partial M$ connected and simply connected. Then M^n is diffeomorphic to the disk D^n.*

We now discuss another aspect of the preceding theory, the relationship between diffeomorphism and h-cobordism. Two oriented, closed manifolds M_1, M_2 are *cobordant* if there exists a compact oriented manifold W with $\partial W = M_1 - M_2$ (taking into account orientations). Thom in [**40**] reduced the cobordism classification of manifolds to a problem in homotopy theory which has since been solved.

Two closed oriented manifolds M_1, M_2 are h-cobordant (following Milnor [**18**] if one can choose W as above so that the inclusions $M_i \to W, i = 1, 2$ are homotopy equivalences. The idea of h-cobordism (homotopy-cobordism) is to replace diffeomorphism by a notion of equivalence, *a priori* weaker than diffeomorphism and amenable to study using homotopy and cobordism theory.

The following theorem was prove in [**37**], with special cases in [**34; 35**]. It is also a consequence of 2.4′.

(2.8) The h-Cobordism Theorem. *Let M_1^n, M_2^n be closed oriented simply connected manifolds, $n > 4$, which are h-cobordant. The M_1 and M_2 are diffeomorphic.*

Milnor [**20**] has shown that 2.7 is false for nonsimply connected manifolds. On the other hand Mazur [**16**] has generalized 2.7 to a theorem which includes the nonsimply connected manifolds.

Theorem 2.8 reduces the diffeomorphism problem for a large class of manifolds to the h-cobordism problem. This h-cobordism problem has been put into quite good shape for homotopy spheres by Milnor [**19**], Kervaire and Milnor [**12**], and recently Novikov [**24**] has found a general theorem.

Kervaire and Milnor show that homotopy spheres of dimension n, with equivalence defined by h-cobordism form an abelian group $\mathcal{H}^n$. Their main theorem is the following.

(2.9) Theorem. *$\mathcal{H}^n$ is finite, $n \neq 3$.*

Kervaire and Milnor go on to find much information about the structure of this finite abelian group, in particular, to find its order for $4 \leqq n \leqq 18$. The main technique in the proof of 2.9 is what is called spherical modification or surgery, see [**44; 22**].

Putting together 2.8 and 2.9 one obtains a classification of the simplest type of closed manifolds, the homotopy spheres, of dimension n for $5 \leqq n \leqq 18$ and for general n the finiteness theorem. This also gives theorems on differentiable structures on spheres. See [**35**] or [**12**].

Most recently, Novikov has found a very general theorem on the h-cobordism structure of manifolds (and hence by 2.8, the diffeomorphism structure). We refer the reader to [**24**] for a brief account of this.

Lastly we mention some specific results on the manifold classifaction problem [**36; 37**].

(2.10) Theorem. (a) *There is a* **1–1** *correspondence between simply-connected closed 5-manifolds with vanishing 2nd Stiefel–Whitney class and finitely generated abelian groups, the correspondence given by $M \to$ Free part $H_2(M) + \frac{1}{2}$ Torsion part $H_2(M)$.*

(b) *Every 2-connected 6-manifold is diffeomorphic to S^6 or a sum (in the sense of* [**18**]) *of r copies of $S^3 \times S^3$, r a positive integer.*

C. T. C. Wall has very general theorems which extend the above results to $(n-1)$-connected $2n$-manifolds [**43**].

3. We give a substantial account of immersion theory because the main problem here has been completely reduced to homotopy theory.

An immersion of one mainifold M^k in a second X^n is a C^∞ map $f : M \to X$ with the property that for each $p \in M$; in some coordinate systems (and hence all) about p and $f(p)$, the Jacobian matrix of f has rank k. A regular homotopy is a homotopy $f_t : M \to X$, $0 \leqq t \leqq 1$ which for each t is an immersion and which has the additional property that the induced map $F_t : T_M \to T_X$ (the derivative) on the tangent spaces is continuous (on $T_M \times I$). One could obtain an equivalent theory by requiring in place of the last property that f_t be a differentiable map of $M \times I$ into X.

The fundamental problem of immersion theory is: given manifolds M and X, find the equivalence classes of immersions of M in X, equivalent under regular homotopy. This includes in particular the problem of whether M can be immersed in X at all. This general problem is in good shape. The complete answer has been given recently in terms of homotopy theory as we shall see.

The first theorem of this type was based on "general position" arguments and proved by Whitney [**46**] in 1936.

(3.1) Theorem. *Given manifolds M^k, X^n, any two immersions $f, g: M \to X$ which are homotopic are regularly homotopic if $n \geqq 2k+2$. If $n \geqq 2k$ there exists an immersion of M in X.*

Recent proofs of the second part of this theorem can be found in [**17**] and [**28**].

The first statement of 3.1 is equally true with $2k+2$ replaced by $2k+1$. Most of the theorems in this survey on the existence of immersions and imbeddings can be strengthened with an approximation property of some sort. Although these are important, for simplicity we omit them.

The first immersion theorem for which arguments transcending general position are needed was the Whitney–Graustein theorem [**45**] (proved in a paper by Whitney who gives much credit to Graustein). For an immersion $f: S^1 \to E^2$, S^1, E^2 oriented, the induced map on the tangent vectors yields a map of S^1 into S^1; the degree (an integer) of this map times 2π is called by Whitney the rotation number.

(3.2) Theorem (Whitney–Graustein). *Two immersions of S^1 in E^2 are regularly homotopic if and only if they have the same rotation number. There exists an immersion of S^1 in E^2 with prescribed rotation number of the form an integer times 2π.*

The next step in the theory of immersions was again taken by Whitney [**48**] in 1944 with the following theorem.

(3.3) Theorem. *Every k-manifold can be immersed in E^{2k-1} for $k > 1$.*

The proof of this is quite difficult and involved a careful analysis of the critical points of a differentiable mapping of M^k into E^{2k-1}. In this dimension, these critical points are isolated in a suitable approximation of a given map, but have to be removed to obtain an immersion. The difficulties in the study of singularities of differentiable maps have limited this method, although, very recently, Haefliger [**4**] has used effectively Whitney's ideas in the above proof, both in studying imbeddings and immersions. We shall say more about this later.

Some of the recent progress in imbeddings and immersions can be measured by Whitney's statement in the above paper. "It is a highly difficult problem to see if the imbedding and immersion theorems of the preceding and the present one can be improved upon." He goes on to ask if every open or orientable M^4 may be imbedded in E^7 and immersed in E^6. The complex projective plane cannot be immersed in E^6, but every open M^4 can be imbedded in E^7. See Hirsch [**8; 9**]. Whitney finally asks if every M^3 can be imbedded in E^5. Hirsch has proved this is so if M^3 is orientable [**10**].

The next progress in the subject of immersions occured in papers [**30; 31**] and [**32**] of the author in 1957–1959. The first generalized the Whitney–Graustein theorem for circles immersed in the plane to circles immersed in an arbitrary manifold, and here methods were introduced which soon led to the solution of the general problem mentioned previously.

Consider "based" immersions of $S^1 = \{0 \leqq \theta \leqq 2\pi\}$ in a manifold X, that is those which map $\theta = 0$ into a fixed point x_0 of X and the positive unit tangent at $\theta = 0$ into a fixed tangent vector of X at x_0. To each based immersion of S^1 in X, the differential of the immersion associates an element of the fundamental group of T'_X, the unit tangent bundle of X.

(3.4) Theorem. *The above is a* **1–1** *correspondence between (based) regular homotopy classes of based immersions of* S^1 *in* X *and* $\pi_1(T'_X)$.

In the next paper [**31**] corresponding theorems are proved with S^1 replaced by S^2. A noteworthy special case of the theorem proved there is that any two immersions of S^2 in E^3 are regularly homotopic. It is a good mental exercise to check this for a reflection through a plane of S^2 in E^3 and the standard S^2 in E^3. This check has been carried through independently by A. Shapiro and N. H. Kuiper, unpublished.

In [**32**], the classification, under regular homotopy, of immersions of S^k in E^n is given (any k, n). This can be stated as follows.

If $f, g : S^k \to E^n$ are based immersions (i.e., at a fixed point x_0 of S^k, $f(x_0) = g(x_0)$ is prescribed and the derivatives of f and g at x_0 are prescribed and equal) one can define an invariant $\Omega(f, g) \in \pi_k(V_{n,k})$ where $\pi_k(V_{n,k})$ is the kth homotopy group of the Stiefel manifold of k-frames in E^n.

(3.5) Theorem. *Based immersions* $f, g : S^k \to E^n$ *are (based) regularly homotopic if and only if* $\Omega(f, g) = 0$. *Furthermore, given a based immersion* $f : S^k \to E^n$ *and* $\Omega_0 \in \pi_k(V_{n,k})$, *there is a based immersion* $g : S^k \to E^n$ *such that* $\Omega(f, g) = \Omega_0$.

The content of Theorem 3.2 is that the homotopy group $\pi_k(V_{n,k})$ classifies immersions of S^k in E^n. Information on the groups can be found in [**25**]. An application of this theorem is that immersions of S^k in E^{2k} are classified by the integers if k is even, the correspondence given by the intersection number.

R. Thom in [**42**] has given a rough exposition of the proof of the previous theorem, which contributes to the theory of conceptualizing part of the proof.

M. Hirsch in his thesis [**8**], using the results of [**32**], has generalized 3.5 to the case of immersions of an arbitrary manifold in an arbitrary manifold. If M^k and X^n are manifolds T_M, T_X their tangent bundles, a monomorphism $\phi : T_M \to T_X$ is fiber preserving map which is a vector space monomorphism on each fiber. For each immersion $f : M \to X$ the derivative is a monomorphism $\phi_f : T_M \to T_X$.

(3.6) Theorem. *If* $n > k$, *the map* $f \to \phi_f$ *induces a* **1–1** *correspondence between regular homotopy classes of immersions of* M *in* X *and (monomorphism) homotopy classes of monomorphisms of* T_M *into* T_X.

In this theorem one can replace homotopy classes of monomorphisms of T_M into T_X by equivariant homotopy classes (equivariant with respect to the action of $GL(k)$) of the associated k-frame bundles of T_M and T_X respectively. Still another interpretation is that the regular homotopy classes of immersions of M in X are in a **1–1** correspondence with homotopy classes of cross-sections of the bundle associated to the bundle of k-frames of M whose fiber is the bundle of k-frames of X. Recently Hirsch (unpublished) has established the theorem for the case $n = k$ provided M is not closed.

Theorem 3.6 includes as special cases all the previous theorems mentioned here on immersions and has the following consequences as well.

(3.7) Theorem. *If $n > k$ and M^k is immersible in E^{n+r} with a normal r-field, then it is immersible in E^n. Conversely, if M^k is immersible in E^n then (trivially) it is immersible in E^{n+r} with a normal r-field.*

(3.8) Theorem. *If M^k is parallelizable (admits k independent continuous tangent vector fields), it can be immersed in E^{k+1}. Every closed 3-manifold can be immersed in E^4; every closed 5-manifold can be immersed in E^8.*

Theorem 3.6 is the fundamental theorem of immersion theory. It reduces all questions pertaining to the existence or classification of immersions to a homotopy problem. The homotopy problem, though far from being solved, has been studied enough to yield much information on immersions through Theorem 3.6 as can be seen for example in Theorem 3.8. Most further work on the existence and classification of immersions would thus seem to lie outside of differential topology proper and in the corresponding homotopy problems.

We note that Haefliger [**4**] has very recently given another very different proof of Theorem 3.6 under the additional assumption $n > 3(k+1)/2$. See also [**7**].

We return now to discuss very briefly some of the methods used to prove the theorems of the previous section. The first step is to introduce function spaces of immersions. If M^k, X^n are manifolds, let $\mathrm{Im}^r(M, X)$ be the space of all immersions of M in X endowed with the C^r topology, $1 \leqq r \leqq \infty$. This means roughly that two immersions are close if they are pointwise close and their first r derivatives are close. Of course $\mathrm{Im}^r(M, X)$ might be empty! A point in $\mathrm{Im}^r(M, X)$ is an immersion of M in X and an arc in $\mathrm{Im}^1(M, X)$ is a regular homotopy, so the main problem amounts to finding the arc-components of $\mathrm{Im}^1(M, X)$ or $\pi_0(\mathrm{Im}^1(M, X))$. One now generalizes the problem to finding not only $\pi_0(\mathrm{Im}^1(M, X))$, but all the homotopy groups of $\mathrm{Im}^1(M, X)$. The homotopy groups of $\mathrm{Im}^r(M, X)$ do not depend on r and we sometimes omit it. To find these homotopy groups one uses the exact homotopy sequence of a fiber space and one of the main problems becomes, to show certain maps are fiber maps.

The following in fact is perhaps the most difficult part of [**32**].

(3.9) Theorem. *Define a map* $\pi : \mathrm{Im}^2(D^k, E^n) \to \mathrm{Im}^2(\partial D^k, E^n)$ *by restricting an immersion of* D^k *to the boundary. If* $n < k+1$, π *has the covering homotopy property.*

Actually, one uses an extension of this theorem to the case where boundary conditions involving first order derivates are incorporated into the range space of π. Since it was first proved, Theorem 3.9 has been generalized and strengthened. The general version, due to Hirsch and Palais [**11**] is as follows.

(3.10) Theorem. *Let* V *be a submanifold of a manifold* M, X *another manifold and* $\pi : \mathrm{Im}(M, X) \to \mathrm{Im}(V, X)$ *defined by restriction. Then* π *is a fiber map in the sense of Hurewicz (and hence has the covering homotopy property).*

A version of 3.10 is also proved with the boundary conditions mentioned above.

An idea not present in the author's original proof of 3.9, but introduced by Thom in [**42**], was to prove theorems of types 3.9 and 3.10 by first explicitly proving the coresponding theorem for spaces of imbeddings, this theorem being much easier and quite useful itself. The final version of this intermediate result is due to Palais [**26**].

(3.11) Theorem. *Let* M *be a compact manifold,* V *a submanifold, and* X *any manifold. Let* $\mathcal{E}(M, X)$, $\mathcal{E}(V, X)$ *be the respective spaces of imbeddings with the* C^r *topology,* $1 \leqq r \leqq \infty$, *and* $\pi : \mathcal{E}(M, X) \to \mathcal{E}(V, X)$ *defined by restriction. Then* π *is a locally trivial fiber map.*

Using Theorems 3.9, 3.10 and an induction basically derived from the fact that the dimension of the boundary of a manifold is one less than the manifold itself, one obtains weak homotopy equivalence theorems. The most general one is due to Hirsch and Palais [**11**]. Given manifolds M, X let $K(M, X)$ be the space of monomorphisms of T_M into T_X with the compact open topology. Then as described in Sec. 2, there is a map

$$\alpha : \ \mathrm{Im}(M, X) \to K(M, X).$$

(3.12) Theorem. *The map* α *induces an isomorphism on all the homotopy groups (is a weak homotopy equivalence) if* $\dim X > \dim M$.

Theorem 3.12 applied to the zeroth homotopy groups or arc-components of $\mathrm{Im}(M, X)$ and $K(M, X)$ yields Theorem 3.6. Theorem 3.12 was first proved for $\mathrm{Im}(S^k, E^n)$ in [**32**].

4. An imbedding (or differentiable imbedding) is an immersion which is also a homeomorphism onto its image. A regular (or differentiable) isotopy is a regular homotopy which at each stage is an imbedding. The fundamental problem of imbedding theory is; given manifolds M^k, X^n, classify the imbeddings of M in

X under equivalence by regular isotopy. This includes the problem: does there exist an imbedding of M in X? Our discussion of imbedding theory is limited to work on this problem. The difficulty of the general problem is indicated by the special case of imbeddings of S^1 in E^3. This problem of classifying "classical" knots is far from being settled (and of course we omit any discussion of this special case although it could well be considered within the scope of differential topology).

Again the first theorems are due to Whitney in 1936 and are proved by general position arguments [**46**].

(4.1) Theorem. *A manifold M^k can always be imbedded in E^{2k+1}. Any two homotopic imbeddings of M in X^{2k+3} are regularly isotopic.*

One can replace X^{2k+3} by X^{2k+2} here.

See [**17**] or [**28**] for recent proofs of the first statement of 6.1.

In 1944, Whitney proved the much harder theorem [**47**].

(4.2) Theorem. *Every k-manifold can be imbedded in E^{2k}.*

The methods used in this paper have been important in subsequent developments in imbedding theory. A. Shapiro, in fact, has considerably developed Whitney's ideas in the framework of obstruction theory. Only the first stage of Shapiro's work is in print [**29**]. Besides being mostly unpublished, the theory has the further disadvantage from our point of view that it is a theory of imbedding for complexes and does not directly apply to give imbeddings (differentiable) of manifolds. On the other hand, Shapiro's work has in part inspired the important theorems of Haefliger that we will come to shortly.

Wu Wen Tsun in a number of papers, see e.g. [**49**], has a theory of imbedding and isotopy of complexes which overlaps with Shapiro's work. Shapiro's (unpublished) theorems on the existence of imbeddings of complexes in Euclidean space seem much stronger than those of Wu Wen Tsun. On the other hand, Shapiro works with spaces derived from the two-fold product of a space, while Wu studies stronger invariants derived from the p-fold products. Also Wu Wen Tsun not only considers existence of imbeddings but isotopy problems as well, including the following one for the differentiable case [**49**]. The proof is based on Whitney's paper [**47**].

(4.3) Theorem. *Any two imbeddings of a connected manifold M^k in E^{2k+1} are regularly isotopic.*

Haefliger [**4**] has taken a big step forward in the theory of imbeddings with the following theorems, proved by strong extensions of the work of Whitney, Shapiro and Wu Wen Tsun. Haefliger's main theorem can be expressed as follows.

As imbedding $f : M \to E^n$ induces a map $\phi_f : M \times M - M \to S^{n-1}$($M \times M - M$ is the product with the diagonal deleted) by $\phi_f(x,y) = (f(x) - f(y)/\|f(y) - f(x)\|$.

Then clearly ϕ_f is equivariant with respect to the involution on $M \times M - M$ which interchanges factors and the antipodal map of S^{n-1}.

(4.4) Theorem. *If* $n > 3(k+1)/2$, *the map* $f \rightarrow \phi_f$ *induces a* **1–1** *correspondence between regular isotopy classes of* M *in* E^n *and equivariant homotopy classes of* $M \times M - M$ *into* S^{n-1}.

The equivariant homotopy classes are in a **1–1** correspondence with homotopy classes of cross-sections of the following bundle E. Let M^* be the quotient space of $M \times M - M$ under the above involution. The two involutions described above define an action of the cyclic group of order two on $(M \times M - M) \times S^{n-1}$. The orbit space of this action is our bundle E with base M^* and fiber S^{n-1}.

Haefliger actually proves 4.4 with E^n replaced by an arbitrary manifold X^n.

Another of Haefliger's theorems is the following.

(4.5) Theorem. *If* M^k *and* X^n *are manifolds which are respectively* $(r-1)$-*connected,* r-*connected and* $n \geqq 2k - r + 1$ *then*

(a) *if* $2r < n$, *any continuous map of* M *in* X *is homotopic to an imbedding;*
(b) *if* $2r < n+1$, *two homotopic imbeddings of* M *in* X *are regularly isotopic. Thus if* $n > 3(k+1)/2$, *any two imbeddings of* S^k *in* E^n *are regularly isotopic.*

Hirsch has proved some theorems on the existences of imbeddings of manifolds in Euclidean space. Perhaps the most interesting is the following [**10**].

(4.6) Theorem. *Every orientable* 3-*manifold can be imbedded in* E^5.

We do not discuss here, in general, the highly unstable problem of imbeddings of M^k in X^n where $n \leqq k+2$ except to mention that 2.7 is relevant to the differentiable Schonflies problem.

Since this section was first written Haefliger has obtained several further important results on imbeddings; see [**5**].

References

[1] R. Bott, *The stable homotopy of the classical groups*, Ann. of Math. (2) **70** (1959), 313–337.
[2] H. Cartan, Seminar, 1961–1962, Paris.
[3] J. Cerf, *Travaux de Smale sur la structure des variétés*, Seminaire Bourbaki, 1961–1962, No. 230, Paris.
[4] A. Haefliger, *Differentiable imbeddings*, Bull. Amer. Math. Soc. **67** (1961), 109–112.
[5] ———, *Knotted* $(4k-1)$-*spheres in* $6k$-*space*, Ann. of Math. (2) **75** (1962), 452–466.
[6] ———, *Plongements differentiables de variété dans variétés*, Comment. Math. Helv. **36** (1962).

[7] A. Haefliger and M. Hirsch, *Immersions in the stable range*, Ann. of Math. (2) **75** (1962), 231–241.

[8] M. Hirsch, *Immersions of manifolds*, Trans. Amer. Math. Soc. **93** (1959), 242–276.

[9] ______, *On imbedding differentiable manifolds in Euclidean space*, Ann. of Math. (2) **73** (1961), 566–571.

[10] ______, *The imbedding of bounding manifolds in Euclidean space*, Ann. of Math. (2) **74** (1961) 494–497.

[11] M. Hirsch and R. Palais, unpublished.

[12] M. Kervaire and J. Milnor, *Groups of homotopy spheres*. I, Ann. of Math. (2) (1963) (to appear).

[13] S. Lefschetz, *Topology*, Amer. Math. Soc. Colloq. Publ. Vol. 12, Amer. Math. Soc., Providence, R. I., 1930.

[14] H. Levine, *Singularities of differentiable mappings*. I, Mathematisches Institut der Universität, Bonn, 1959.

[15] B. Mazur, *The theory of neighborhoods*, Harvard University.

[16] ______, *Simple neighborhoods*, Bull Amer. Math. Soc. **68** (1962), 87–92.

[17] J. Milnor, *Differential topology*, Princeton, 1959.

[18] ______, *Sommes de variétés différentiables et structures différentiables des sphères*, Bull. Soc. Math. France **87** (1959), 439–444.

[19] ______, *Differentiable manifolds which are homotopy spheres*, Princeton, 1959.

[20] ______, *Two complexes which are homeomorphic but combinatorial distinct*, Ann. of Math. (2) **74** (1961), 575–590.

[21] ______, *On cobordism, a survey of cobordism theory*, Enseignement Math. **111** (1962), 16–23.

[22] ______, *A procedure for killing homotopy groups of differentiable manifolds*, Proc. Sympos. Pure Math. Vol. 3, Amer. Math. Soc., Providence, R. I., 1961.

[23] M. Morse, *The calculus of variations in the large*, Amer. Math. Soc. Colloq. Publ. Vol. 18, Amer. Math. Soc., Providence, R. I., 1934.

[24] S. P. Novikov, *Diffeomorphisms of simply connected manifolds*, Dokl. Akad. Nauk SSSR **143** (1962), 1046–1049 = Soviet Math. Dokl. **3** (1962), 540–543.

[25] G. F. Paechter, *The groups* $\Pi_r(V_{n,m})$. I, Quart J. Math. Oxford Ser. (2) **7** (1956), 249–268.

[26] R. Palais, *Local triviality of the restriction map for embeddings*, Comment. Math. Helv. **34** (1960), 305–312.

[27] H. Poincaré, *Analysis situs*, Collected Works, Vol. 6 Gauthier–Villars, Paris, 1953.

[28] L. S. Pontrjagin, *Smooth manifolds and their applications in homotopy theory*, Trudy Mat. Inst. Steklov **45** (1955) = Amer. Math. Soc. Transl. (2) **11** (1955).

[29] A. Shapiro, *Obstructions to the imbedding of a complex in a Euclidean space*. I. *The first obstruction*, Ann. of Math. (2) **66** (1957), 256–269.

[30] S. Smale, *Regular curves on Riemannian manifolds*, Trans. Amer. Math. Soc. **81** (1958), 492–512.

[31] ______, *A classification of immersions of the two-sphere*, Trans. Amer. Math. Soc. **90** (1959), 281–290.

[32] ______, *The classification of immersions of spheres in Euclidean spaces*, Ann. of Math. (2) **69** (1959), 327–344.

[33] ———, *The generalized Poincaré conjecture in higher dimensions*, Bull. Amer. Math. Soc. **66** (1960), 373–375.
[34] ———, *The generalized Poincaré conjecture in dimensions greater than four*, Ann. of Math. (2) **74** (1961), 391–406.
[35] ———, *Differentiable and combinatorial structures on manifolds*, Ann. of Math. (2) **74** (1961), 498–502.
[36] ———, *On the structure of* 5-*manifolds*, Ann. of Math. (2) **75** (1962), 38–46.
[37] ———, *On the structure of manifolds*, Amer. J. Math. (to appear).
[38] ———, *On gradient dynamical systems*, Ann. of Math. (2) **74** (1961), 199–206.
[39] J. Stallings, *Polyhedral homotopy-spheres*, Bull. Amer. Math. Soc. **66** (1960), 485–488.
[40] R. Thom, *Quelques propriétés globales des variétés differentiables*, Comment. Math. Helv. **29** (1954), 17–85.
[41] ———, *Les singularités des applications différentiables*, Ann. Inst. Fourier (Grenoble) **6** (1956), 43–87.
[42] ———, *La classification des immersions d'après Smale*, Séminaire Bourbaki, December 1957, Paris.
[43] C. T. C. Wall, *Classification of* $(n-1)$-*connected* $2n$-*manifolds*, Ann. of Math. (2) **75** (1962), 163–189.
[44] A. H. Wallace, *Modifications and cobounding manifolds*, Canad. J. Math. **12** (1960), 503–528.
[45] H. Whitney, *On regular closed curves in the plane*, Compositio Math. **4** (1937), 276–284.
[46] ———, *Differentiable manifolds*, Ann. of Math. (2) **37** (1936), 645–680.
[47] ———, *The self-intersections of a smooth* n-*manifold in* $2n$-*space*, Ann. of Math. (2) **45** (1949), 220–246.
[48] ———, *The singularities of a smooth* n-*manifold in* $(2n-1)$-*space*, Ann. of Math. (2) **45** (1949), 247–293.
[49] Wu Wen Tsun, *On the isotopy of* C^r *manifolds of dimension* n *in Euclidean* $(2n+1)$-*space*, Science Records (N. S.) **2** (1958), 271–275.
[50] E. C. Zeeman, *The generalized Poincaré conjecture*, Bull. Amer. Soc. **67** (1961), 270.

Columbia University

Reprinted from Actes, Congrès Int. Math., 1970. Tome 1, p. 3 à 5

THE WORK OF ALAN BAKER

by
PAUL TURÁN

The theory of transcendental numbers, initiated by Liouville in 1844, has been enriched greatly in recent years. Among the relevant profound contributions are those of A. Baker, W. M. Schmidt and V. A. Sprindzuk. Their work moves in important directions which contrast with the traditional concentration on the deep problem of finding significant classes of functions assuming transcendental values for all non-zero algebraic values of the independent variable. Among these, Baker's have had the heaviest impact on other problems of mathematics. Perhaps the most significant of these impacts has been the application to diophantine equations. This theory, carrying a history of more than thousand years, was, until the early years of this century, little more than a collection of isolated problems subjected to ingenious *ad hoc* methods. It was A. Thue who made the breakthrough to general results by proving in 1909 that all diophantine equations of the form $f(x,y) = m$, where m is an integer and f is an irreducible homogeneous binary form of degree at least three, with integer coefficients, have at most finitely many solutions in integers. This theorem was extended by C. L. Siegel and K. F. Roth (himself a Fields medallist) to much more general classes of algebraic diophantine equations in two variables of degree at least three. They even succeeded in establishing general upper bounds on the number of such solutions. A *complete* resolution of such problems however, requiring a knowledge of *all* solutions, is basically beyond the reach of these methods, which are what are called "ineffective". Here Baker made a brilliant advance. Considering the equation $f(x,y) = m$, where m is a positive integer, $f(x,y)$ an irreducible binary form of degree $n \geqq 3$, with integer coefficients, he succeeded in determining an *effective* bound B, depending only on n and on the coefficients of f, so that

$$\max(|x_0|, |y_0|) \leqq B$$

for any solution (x_0, y_0). Thus, although B is rather large in most cases, Baker has provided, in principle at least, and for the first time, the possibility of determining *all* the solutions explicitly (or the nonexistence of solutions) for a large *class* of equations. This is an essential step towards the positive aspects of Hilbert's tenth problem the interest of which is largely increased by the recent negative solution of the general problem by Ju. V. Matyaszevics. The significance of his theorem is also enhanced by the fact that the so-called elliptic and hyperelliptic equations fall, after appropriate transformation, under its scope and again he gave explicit upper bounds on the totality of their solutions.

Joint work of Baker with J. Coates made effective for curves of genus 1 Siegel's classical theorem. Elaborating these methods and results Coates found among others the first *explicit* lower bound tending to infinity with n for the maximal primefactor of $|f(n)|$ where $f(x)$ stands for an arbitrary polynomial with integer coefficients apart from a trivial exception. The more fact that the maximal primefactor of $|f(n)|$ tends to infinity with n (conjectured for polynomials of second degree by Gauss) was established by K. Mahler several decades ago as well as an *explicit* lower bound for $n = 2$ by him and S. Chowla.

In collaboration with H. Davenport, Baker has shown by some examples how the upper bounds thus obtained permit actually the determination of *all* solutions.

As another consequence of his results he gave an *effective* lower bound for the approximability of algebraic numbers by rationals, the first one which is better than Liouville's.

As mentioned before, these results are all consequences of his main results on transcendental numbers. As is well known, the seventh problem of Hilbert asking whether or not α^β is transcendental whenever α and β are algebraic, certain obvious cases aside, was solved independently by A. O. Gelfond and T. Schneider in 1934. Shortly afterwards Gelfond found a stronger result by obtaining an *explicit* lower bound for $|\beta_1 \log \alpha_1 + \beta_2 \log \alpha_2|$ in terms of $\alpha_\nu' s$'s and of the degrees and heights of the $\beta_\nu' s$'s when the $\log \alpha_\nu' s$ are linearly independent. After Gelfond realised in 1948, in collaboration with Ju. V. Linnik, the significance of an *effective* lower bound for the three-term sum, he and N. I. Feldman soon discovered an *ineffective* lower bound for it. The transition from this important first step to *effective* bound for the three-term sum, and more generally for the k-term sum, resisted all efforts until Baker's success in 1966. This success enabled Baker to obtain a vast generalization of Gelfond–Schneider's theorem by showing that if $\alpha_1, \alpha_2, \ldots, \alpha_k$ ($\neq 0, 1$) are algebraic, $\beta_1, \beta_2, \ldots, \beta_k$ linearly independent, algebraic and irrational, then $\alpha_1^{\beta_1} \alpha_2^{\beta_2} \cdots \alpha_k^{\beta_k}$ is transcendental. Some further appreciation of the depth of this result can be gained by recalling Hilbert's prediction that the Riemann conjecture would be settled long before the transcendentality of α^β. The analytic prowess displayed by Baker could hardly receive a higher testimonial. On the other hand, his brilliant achievement shows, after Gelfond–Schneider once more, that mathematics offers no scope for a doctrine of papal infallibility concerning its future. Among his other results generalizing transcendentality theorems of Siegel and Schneider I shall mention only one special case, in itself sufficiently remarkable, according to which the sum of the circumferences of two ellipses, whose axes have algebraic lengths, is transcendental.

His pathbreaking role is not diminished but perhaps even emphasized by the fact that in 1968 Feldman found another important lower estimate for the k-term sum which is stronger in its dependence upon the maximal height of the β_ν coefficients; it is weaker in its dependence upon the maximal height of the α's which is relevant in most applications at present. It is reasonable to expect also new applications depending more on the former.

The 1948 discovery of Gelfond and Linnik, mentioned above, revealed an unexpected connection between such lower bounds for the three-term sum and a classical class-number problem. This has as its goal the determination of all algebraic extensions $R(\theta)$ of the rational field with class number 1. In its full generality this seems hopelessly out of reach at present. Restricting themselves to the imaginary quadratic case $R(\sqrt{-d})$, $d > 0$, H. Heilbronn and E. Linfoot showed in 1934 that at most ten such "good" fields can exist. Nine of these were found explicitly. Concerning the tenth it was known that its d would have to exceed $\exp(10^7)$. Hence, if it can be shown that there exists an upper bound $d_0 < \exp(10^7)$ for all "good" d's then the tenth possible field cannot exist. Now the Gelfond–Linnik discovery was that the afore mentioned *effective* lower bound for the three-term sum could furnish such an *effective* d_0. Baker found that one of his general results implies an upper bound $d_0 = 10^{500}$ enough by far for this purpose. This outcome provides a striking new example, illustrating once more how effectivity can play a *decisive* role in essential problems. Again, the value of this approach is of course not diminished by H. M. Stark's outstanding achievement in showing the non-existence of the tenth field, simultaneously and independently, by quite different methods.

To illustrate further the many-sided applicability of Baker's work, I mention that it could be employed to make effective some ineffective results of Linnik on the coefficients of a complete reduced set of binary quadratic forms belonging to a fixed negative discriminant (Linnik had used ideas from ergodic theory).

As one can guess, obtaining such long-sought solutions was a very complicated task. It is very difficult to attempt even a sketch of the underlying ideas in the short time at my disposal beyond the remark that they are of hard-analysis type. Fortunately, you will have the opportunity of hearing about them in some detail from Baker himself in his address to this Congress. To conclude, I remark that his work exemplifies two things very convincingly. Firstly, that beside the worthy tendency to start a theory in order to solve a problem it pays also to attack specific difficult problems directly. Particularly is this the case with such problems where rather singular circumstances do not make it probable that a solution would fall out as an easy consequence of a general theory. Secondly, it shows that a direct solution of a deep problem develops itself quite naturally into a healthy theory and gets into early and fruitful contact with other significant problems of mathematics. So, let the two different ways of doing mathematics live in peaceful coexistence for the benefit of our science.

P. Turán
Mathematical Institute
of the Hungarian
Academy of Sciences,
Budapest, Hungary

Alan Baker
Trinity College
Cambridge
(Grande-Bretagne)

Alan Baker

ALAN BAKER

Professor Alan Baker obtained his first degree from University College, University of London and M.A., Ph.D. from Trinity College, Cambridge Universyt. He has been Professor of Pure Mathematics at Cambridge University since 1974.

He was awarded a Fields Medal in 1970 at the International Congress of Mathematicians. In addition, he has a distinguished career with many awarded honours which include Fellow of Trinity College, Cambridge (1964), Adams Prize, Cambridge (1972), Fellow of the Royal Society (1973), First Turán Lecturer, Hungary (1978), Fellow, UCL (1979), Foreign Honorary Fellow of the Indian National Science Academy (1980).

He was Director of Studies in Mathematics, Trinity College, Cambridge (1968–74), Member of the Institute for Advanced Study, Princeton (1970), co-chairman of a program at MSRI, Berkeley (1993), and he has held visiting professorships at many universities including Stanford University (1974), University of Hong Kong (1988) and ETH, Zürich (1989).

Among his many important publications are "Transcendental Number Theory" (1975), "Transcendence Theory: Advances and Applications" (1977), (ed jtly) "A Concise Introduction to the Theory of Numbers" (1984), and "New Advances in Transcendence Theory" (1988), (ed).

Professor Alan Baker is well known internationally for his work in several areas of Number Theory. In particular, he succeeded in obtaining a vast generalization of the Gelfond–Schneider Theorem which is the solution to Hilbert's seventh problem. From this work he generated a large category of transcendental numbers not previously identified and showed how the underlying theory could be used to solve a wide range of Diophantine problems.

Because of his profound and significant contributions to Number Theory, Professor Alan Baker was awarded the Fields Medal. This is awarded every four years and is the most highly regarded international medal for outstanding discoveries in mathematics. Indeed, the Fields Medal is generally considered the equivalent of the Nobel Prize. In the Ceremony for the Award at the International Congress of Mathematicians, Professor P. Turán reported on Baker's work and said

> "... Some further appreciation of the depth of this (Baker's) result can be gained by recalling Hilbert's prediction that the Riemann conjecture would be settled long before the transcendentality of α^{β}. The analytic prowess displayed by Baker could hardly receive a higher testimonial ..."

This biographical sketch was prepared by the University of Hong Kong in connection with a Public Lecture given on 31 March 1995.

Reprinted from Proc. Int. Congr. Math., 1970

EFFECTIVE METHODS IN THE THEORY OF NUMBERS

by

A. BAKER

1. Problems concerning the determination of the totality of integers possessing certain prescribed properties such as, for instance, solutions of systems of Diophantine equations or inequalities, have captured man's imagination since antiquity, and a wide variety of different techniques have been employed through the centuries to resolve a diverse multitude of problems in this field. Most of the early work tended to be of an *ad hoc* character, the arguments involved being specifically related to the particular numerical example under consideration, but gradually the emphasis has altered and the trend in recent times has been increasingly towards the development of general coherent theories. Two particular advances stand out in this connexion. First, investigations of Thue [39] in 1909 and Siegel [33] in 1929 led to the discovery of a simple necessary and sufficient condition for any Diophantine equation $F(x, y) = 0$, where F denotes a polynomial with integer coefficients, to possess only a finite number of solutions in integers; this occurs, namely, if and (reading "ganzartige" for "integer") only if the curve has genus at least 1 or genus 0 and at least three infinite valuations. The proof depends upon, amongst other things, Weil's well-known generalization [40] of Mordell's finite basis theorem and the earlier pioneering work of Thue and Siegel [32] concerning rational approximations to algebraic numbers. Secondly, in answer to a question raised by Gauss in his famous Disquisitiones Arithmeticae, Hecke, Mordell, Deuring and Heilbronn [29] showed in 1934 that there could exist only finitely many imaginary quadratic fields with any given class number, a result later to be incorporated in the celebrated Siegel–Brauer formula. These theorems and all their many ramifications, though of major importance in the evolution of much of modern number theory, nevertheless suffer from one basic limitation that of their non-effectiveness. The arguments depend on an assumption, made at the outset, that the relevant aggregates possess one or more elements that are, in a certain sense, large, and they provide no way of deciding whether or not these hypothetical elements exist. Thus the work leads merely to an estimate for the number of elements in question and throws no light on the fundamental problem of determining their totality.

Some special effective results in the context of the Thue–Siegel theory were obtained in 1964 by means of certain properties peculiar to Gauss' hypergeometric function, in particular, the classic fact, certainly known to Padé, that quotients of such functions serve to represent the convergents to rational powers of $1 - x$ (see [1, 2, 3]), but the first effective results applicable in a general context came in 1966 from a completely different source. One of Hilbert's famous list of problems raised at the International Congress held in Paris in 1900 asked whether an irrational quotient of logarithms of algebraic numbers is transcendental. An affirmative answer was obtained independently by Gelfond [26] and Schneider [30] in 1934, and shortly afterwards Gelfond established an important refinement giving a positive lower bound for a linear form in two logarithms (cf. [27]). It was natural to conjecture that an analogous result would hold for linear forms in arbitrarily many logarithms of algebraic numbers and a theorem of this nature was proved in 1994 [4]. The techniques devised for the demonstration form the basis of the principal effective methods in number theory known to data. I shall first describe briefly the main arguments and shall then proceed to discuss some of their applications(*).

2. The key result, which serves to illustrate most of the principal ideas, states that if $\alpha_1, \ldots, \alpha_n$ are non-zero algebraic numbers such that $\log \alpha_1, \ldots, \log \alpha_n$ are linearly independent over the rationals then $1, \log \alpha_1, \ldots, \log \alpha_n$ are linearly independent over the field of all algebraic numbers. This implies, in particular, that $e^{\beta_0}\alpha_1^{\beta_1} \cdots \alpha_n^{\beta_n}$ is transcendental for all non-zero algebraic numbers $\alpha_1, \ldots, \alpha_n$, $\beta_0, \ldots, \beta_n$. It will suffice to sketch here the proof of a somewhat weaker result namely, if $\alpha_1, \ldots, \alpha_n$, $\beta_1, \ldots, \beta_{n-1}$ are non-zero algebraic numbers such that $\alpha_1, \ldots, \alpha_n$ are multiplicatively independent, then the equation $\alpha_1^{\beta_1} \cdots \alpha_{n-1}^{\beta_{n-1}} = \alpha_n$ is untenable; it is under these conditions that our arguments assume their simplest form. We suppose the opposite and derive a contradiction. The proof depends on the construction of an auxiliary function of several complex variables which generalizes the function of a single variable employed originally by Gelfond. Functions of many variables were utilized by Schneider [31] in his studies concerning Abelian integrals but, for reasons that will shortly be explained, there seemed to be severe limitations to their serviceability in wider settings. The function that proved to be decisive in the present context is given by

$$\Phi(z_1, \ldots, z_{n-1}) = \sum_{\lambda_1=0}^{L} \cdots \sum_{\lambda_n=0}^{L} p(\lambda_1, \ldots, \lambda_n)\alpha_1^{(\lambda_1+\lambda_n\beta_1)z_1} \cdots \alpha_{n-1}^{(\lambda_{n-1}+\lambda_n\beta_{n-1})z_{n-1}},$$

where L is a large parameter and the $p(\lambda_1, \ldots, \lambda_n)$ denote rational integers not all 0. By virtue of the initial assumption we see at once that

$$\Phi(z, \ldots, z) = \sum_{\lambda_1=0}^{L} \cdots \sum_{\lambda_n=0}^{L} p(\lambda_1, \ldots, \lambda_n)\alpha_1^{\lambda_1 z} \cdots \alpha_n^{\lambda_n z}$$

(*)For a fuller survey of the applications see [13].

and so, for any positive integer l, the value of Φ at $z_1 = \cdots = z_{n-1} = l$ is an algebraic number in a fixed field. Moreover, apart from a multiplicative factor given by products of powers of the logarithms of the α's, the same holds for any derivative

$$\Phi_{m_1,\ldots,m_{n-1}} = (\partial/\partial z_1)^{m_1} \cdots (\partial/\partial z_{n-1})^{m_{n-1}}\Phi.$$

It follows from a well-known lemma on linear equations that, for any integers h, k, with hk^{n-1} a little less than L^n, one can choose the $p(\lambda_1, \ldots, \lambda_n)$ such that

$$\Phi_{m_1,\ldots,m_{n-1}}(l, \ldots, l) = 0 \quad (1 \leqslant l \leqslant h, m_1 + \cdots + m_{n-1} \leqslant k)$$

and, furthermore, an explicit bound for $|p(\lambda_1, \ldots, \lambda_n)|$ can be given in terms of h, k and L.

The real essence of the argument is an extrapolation procedure which shows that the above equation remains valid over a much longer range of values for l, provided that one admits a small diminution in the range of values for $m_1 + \cdots + m_{n-1}$. Although interpolation arguments have long been a familiar feature of transcendental number theory, work in this connexion has hitherto always involved an extension in the order of the derivatives while leaving the points of interpolation fixed; when dealing with functions of many variables, however, this type of argument requires that the points in question admit a representation as a Cartesian product and, as far as I can see, the condition can be satisfied only with respect to special multiply-periodic functions. Our algorithm proceeds by induction and it will suffice to illustrate the first step. We suppose that $m_1 + \cdots + m_{n-1} \leqslant \frac{1}{2}k$ and we prove that then

$$f(z) = \Phi_{m_1,\ldots,m_{n-1}}(z, \ldots, z)$$

vanishes at $z = l$, where $1 \leqslant l \leqslant h^2$. Now the condition $hk^{n-1} \leqslant L^n$ allows one to take $L \leqslant k^{1-\varepsilon}$ for some $\varepsilon > 0$ and h about $k^{\frac{1}{4}\varepsilon}$. This "saving" by an amount ε is crucial for it leads to a sharp bound for $|f(z)|$ on a circle centre the origin and radius slightly larger than h^2, thus including all the points l as above. Further, apart from a trivial multiplicative factor, $f(l)$ represents an algebraic integer in a fixed field and a similar bound obtains for each of the conjugates. But, by construction, we have

$$f_m(r) = 0 \left(0 \leqslant m \leqslant \frac{1}{2}k, 1 \leqslant r \leqslant h\right),$$

and the maximum-modulus principle applied to the function $f(z)/F(z)$, where

$$F(z) = \{(z-1) \cdots (z-h)\}^{[\frac{1}{2}k]},$$

now shows that $|f(l)|$ is sufficiently small to ensure that the norm of the algebraic integer is less than 1. Hence $f(l) = 0$ as required. The argument is repeated inductively and after a finite number of steps we conclude that

$$\Phi(l, \ldots, l) = 0 \quad (1 \leqslant l \leqslant (L+1)^n).$$

But these represent linear equations in the $p(\lambda_1, \ldots, \lambda_n)$. The determinant of coefficients is of Vandermonde type and since, by hypothesis, $\alpha_1, \ldots, \alpha_n$ are multiplicatively independent, it does not vanish. The contradiction establishes our result.

3. The argument just described is capable of considerable refinement and generalization. In particular several other auxiliary functions can be taken in place of Φ, the points of extrapolation can be varied and greater use can be made in the latter part of the exposition of our information regarding the partial derivatives. Thus, for instance, results in the context of elliptic functions have been derived and, in particular, the transcendence has been established of any non-vanishing linear combination with algebraic coefficients of periods and quasi-periods associated with a Weierstrass $\wp$-function with algebraic invariants [10, 11, 12]. More relevant to the main theme of this talk, however, are refinements giving quantitative lower bounds for linear forms in logarithms. The main change in the preceding discussion required to obtain results of this nature is the replacement of the maximum-modulus principle by the Hermite interpolation formula. With this device one can show that

$$|\beta_0 + \beta_1 \log \alpha_1 + \cdots + \beta_n \log \alpha_n| > Ce^{-(\log H)^\kappa},$$

where $\alpha_1, \ldots, \alpha_n$ denote non-zero algebraic numbers such that $\log \alpha_1, \ldots, \log \alpha_n$ are linearly independent over the rationals, $\beta_0, \ldots, \beta_n$ denote algebraic numbers, not all 0, with degrees and heights at most d and H respectively, $\kappa > n + 1$ and $C > 0$ depends only on n, $\log \alpha_1, \ldots, \log \alpha_n$, κ and d; by the height of an algebraic number we mean the maximum of the absolute values of the relatively prime integer coefficients in its minimal defining polynomial [5]. With more complicated adaptations the number on the right can be strengthened to $CH^{-\kappa}$, where $\kappa > 0$ is specified like C above; this was shown by Feldman [22, 23]. In applications it frequently suffices to have simply a lower bound of the form $e^{-\delta H}$, valid for any $\delta > 0$ and all $H > C$, where C now depends on δ, and interest then attaches to the exact expression for C. Some explicit forms have been calculated (cf. [5, 6, 23, 24]) but there is certainly scope for improvement here and, indeed, the general efficacy of our methods seems to be closely linked to our progress in this connexion.

4. We now discuss some applications of our results in the theory of Diophantine equations. To begin with, they can be utilized to obtain a complete resolution of the equation originally considered by Thue, namely $f(x, y) = m$, where f denotes an irreducible binary form with integer coefficients and degree at least 3 [6]. Indeed our arguments enable us to find more generally all algebraic integers x, y in a given field K satisfying any equation $\beta_1 \cdots \beta_n = m$ where $\beta_j = x - \alpha_j y$, $n \geqslant 3$ and $\alpha_1, \ldots, \alpha_n$, m denote algebraic integers in K subject only to the condition that the α's are all distinct (cf. [15]). For denoting by $\theta^{(1)}, \ldots, \theta^{(d)}$ the field conjugates of any element θ of K and by $\eta_1, \ldots, \eta_r$ a fundamental system of units in K, it is readily seen that an associate

$$\gamma_i = \beta_i \eta_1^{b_{i1}} \cdots \eta_r^{b_{ir}}$$

of β_i can be determined such that

$$|\log|\gamma_i^{(j)}|| \leqslant C_1 \quad (1 \leqslant j \leqslant d),$$

where $C_1, C_2, \ldots,$ can be effectively computed in terms of f and m. Writing

$$H_i = \max|b_{ij}| \quad \text{and} \quad H_l = \max\ H_i$$

we have $|\beta_l^{(h)}| \leqslant C_2 e^{-H_l/C_3}$ for some h; and without loss of generality we can suppose that $\beta_l^{(h)} = \beta_l$. From the initial equation we see that $|\beta_k| \geqslant C_4^{-1}$ for some $k \neq l$ and if now j is any suffix other than k or l, the identity

$$(\alpha_k - \alpha_l)\beta_j - (\alpha_j - \alpha_l)\beta_k = (\alpha_k - \alpha_j)\beta_l$$

gives

$$\eta_1^{b_1} \cdots \eta_r^{b_r} - \alpha_{r+1} = \omega,$$

where

$$b_s = b_{ks} - b_{js}, \quad 0 < |\omega| < C_5 e^{-H_l/C_6}$$

and α_{r+1} is an element of K with degree and height $\leqslant C_7$. Now $|b_s| \leqslant 2H_l$ and hence the work of §3 can be applied to obtain a bound for H_l, whence also for all the conjugates of the β's and, finally, for the conjugates of x and y.

The last result enables one to solve many other Diophantine equations in two unknowns. In particular, one can now effectively determine all rational integers x, y satisfying $y^m = f(x)$, where m is any integer $\geqslant 2$ and f is a polynomial with integer coefficients possessing at least three simple zeros [8]. This includes the celebrated Mordell equation $y^2 = x^3 + k$, the hyperelliptic equation and the Catalan equation $x^n - y^m = 1$ with prescribed m, n. The demonstration involves ideal factorizations in algebraic number fields similar to those appearing in the first part of the proof of the Mordell–Weil theorem; in special cases one has readier arguments and, in particular, the elliptic equation has been efficiently treated by means of Hermite's classical theory of the reduction of binary quartic forms [7]. There is, moreover, little difficulty in carrying out the work more generally when the coefficients and variables represent algebraic integers in a fixed field, and, indeed, Coates and I have used this extension to give a new and effective proof Siegel's theorem on $F(x, y) = 0$ (see §1) in the case of curves of genus 1 [15, 21]. Here the equation of the curve is reduced to canonical form by means of a birational transformation similar to that described by Chevalley, the rational functions defining the transformation being constructed to possess poles only at infinity and thus be integral over a polynomial ring. Explicit upper bounds have been established in each instance for the size of all the solutions [6, 7, 8, 15]. The bounds tend to be large, with repeated exponentials, and current research in this field is centred on techniques for reducing their magnitude. In particular, Siegel [34] has recently given some improved estimates for units in algebraic number fields which should prove useful for this purpose, and, furthermore, devices have been obtained which, for a wide range of numerical examples, would seem to render the problem of determining the complete list of solutions in question accessible to practical computation (cf. [16]).

5. Finally we mention some further results that have been obtained as a consequence of these researches. One of the first applications was to establish an effective algorithm for resolving the old conjecture that there are only nine imaginary quadratic fields with class number 1 [4, 18]. The connexion between this problem and inequalities involving the logarithms of algebraic numbers was demonstrated by Gelfond and Linnik [28] in 1949 by way of an expression for a product of L-functions analogous to the well-known Kronecker limit formula. By a remarkable coincidence, Stark [38] established the conjecture at about the same time by an entirely different method with its origins in a paper of Heegner. Attention has subsequently focussed on the problem of determining all imaginary quadratic fields with class-number 2, and I am happy to report that an algorithm for this purpose was obtained very recently by means of a new result relating to linear forms in three logarithms [9, 14](*). It seems likely that this latest development will lead to advances in other spheres.

Among the original motivations of our studies was the search for an effective improvement on Liouville's inequality of 1844 relating to the approximation of algebraic numbers by rationals; from the work described in §4 we have now

$$|\alpha - p/q| > cq^{-n}e^{(\log\ q)^{1/\kappa}}$$

for all algebraic numbers α with degree $n \geqslant 3$ and all rationals $p/q(q > 0)$ where $\kappa > n$ and $c = c(\alpha, \kappa) > 0$ is effectively computable [6, 25]. For some particular α, such as the cube roots of 2 and 17, sharper results in this direction have been obtained from the work on the hypergeometric function mentioned in §1. Further, in the special case when p, q are comprised solely of powers of fixed sets of primes, a much stronger result can be obtained directly from the inequalities referred to in §3; indeed we have then

$$|\alpha - p/q| > c(\log\ q)^{-\kappa}$$

where $c > 0$, $\kappa > 0$ are effectively computable in terms of the primes and α, and this in fact furnishes an improvement on Ridout's generalization of Roth's theorem.

Analogues of the arguments of §3 and §4 in the p-adic realm have been given by Coates [19, 20]; his work leads, in particular, to an effective determination of all rational solutions of the equations discussed earlier with denominators comprised solely of powers of fixed sets of primes and so, more especially, provides a means for finding all elliptic curves with a given conductor (see also [35, 36, 37]). Furthermore, Brumer obtained in 1967 a natural p-adic analogue of the main theorem on logarithms which, in conjunction with work of Ax, resolved a well-known problem of Leopoldt on the non-vanishing of the p-adic regulator of an Abelian number field [17].

(*)See also Stark's address to this Congress.

6. And now I must conclude my survey. It will be appreciated that I have been able to touch upon only a few of the diverse results that have been established with the aid of the new techniques, and, certainly, many avenues of investigation await to be explored. The work has demonstrated, in particular, a surprising connexion between the apparently unrelated seventh and tenth problems of Hilbert, as well as throwing an effective light on both of the fundamental topics referred to at the beginning concerning Diophantine equations and class numbers. Though the strength of this illumination has been steadily growing, and indeed the respective regions of shadow in these contexts have been receding at a remarkably similar rate, it would appear nevertheless that several further ideas will be required before our theories can be regarded as, in any sense, complete. The main feature to emerge is, I think, that the principal passage to effective methods in number theory lies, at present, deep in the domain of transcendence, and it is to be hoped that the territory so far gained in this connexion will be much extended in the coming years.

References

[1] A. Baker. Rational approximations to certain algebraic numbers, *Proc. London Math. Soc.* (3), 4 (1964), pp. 385–398.

[2] ______. Rational approximations to $\sqrt[3]{2}$ and other algebraic numbers, *Quart. J. Math. Oxford Ser.* (2), 15 (1964), pp. 375–383.

[3] ______. Simultaneous rational approximations to certain algebraic numbers, *Proc. Cambridge Philos. Soc.*, 63 (1967), pp. 693–702.

[4] ______. Linear forms in the logarithms of algebraic numbers. *Mathematika*, 13 (1966), pp. 204–216.

[5] ______. Linear forms in the logarithms of algebraic numbers II, III, IV, *Mathematika*, 14 (1967), pp. 102–107, 220–228; 15 (1968), pp. 204–216.

[6] ______. Contributions to the theory of Diophantine equations. I: On the representation of integers by binary forms. II: The Diophantine equation $y^2 = x^3 + k$, *Philos. Trans. Roy. Soc. London*, A 263 (1968), pp. 173–208.

[7] ______. The Diophantine equation $y^2 = ax^3 + bx^2 + cx + d$, *J. London Math. Soc.*, 43 (1968), pp. 1–9.

[8] ______. Bounds for the solutions of the hyperelliptic equation, *Proc. Cambridge Philos. Soc.*, 65 (1969), pp. 439–444.

[9] ______. A remark on the class number of quadratic fields, *Bull. London Math. Soc.*, 1 (1969), pp. 98–102.

[10] ______. On the periods of the Weierstrass $\wp$-function, *Symposia Mathematica*, vol. IV (Bologna, 1970), pp. 155–174.

[11] ______. On the quasi-periods of the Weierstrass ζ-function, *Göttingen Nachrichten* (1969), No. 16, pp. 145–157.

[12] ______. An estimate for the $\wp$-function at an algebraic point, *American J. Math.*, 92 (1970), pp. 619–622.

[13] ———. Effective methods in Diophantine problems, *Proc. Symposia Pure Maths (American Math. Soc.)*, vol. 20 (1971) pp. 195–205.

[14] ———. Imaginary quadratic fields with class number 2, *Annals of Math.*, **94** (1971), 139–152.

[15] ———. and J. Coates. Integer points on curves of genus 1, *Proc. Cambridge Philos. Soc.*, 67 (1970), pp. 595–602.

[16] ———. and H. Davenport. The equations $3x^2 - 2 = y^2$ and $8x^2 - 7 = z^2$. *Quart. J. Math., Oxford Ser.* (2), 20 (1969), pp. 129–137.

[17] Armand Brumer. On the units of algebraic number fields, *Mathematika*, 14 (1967), pp. 121–124.

[18] P. Bundschuh and A. Hock. Bestimming aller imaginär-quadratischen Zahlkorper der Klassenzahl Eins mit Hilfe eines Satzes von Baker, *Math. Z.*, 111 (1969), pp. 191–204.

[19] J. Coates. An effective p-adic analogue of a theorem of Thue, *Acta Arith.*, 15 (1969), pp. 279–305.

[20] ———. An effective p-adic analogue of a theorem of Thue. II: On the greatest prime factor of a binary form. III: The Diophantine equation $y^2 = x^3 + k$. *Acta Arith.*, 16 (1970), pp. 399–412, 425–435.

[21] ———. Construction of rational functions on a curve, *Proc. Cambridge Philos. Soc.*, 68 (1970), pp. 105–123.

[22] N. I. Feldman. Estimate for a linear form of logarithms of algebraic numbers, *Mat. Sb.*, 76 (118) (1968), pp. 304–319; *Math. USSR Sb.*, 5 (1968), pp. 291–307.

[23] ———. An improvement of the estimate for a linear form in the logarithms of algebraic numbers, *Mat. Sb.*, 77 (119) (1968), pp. 423–436; *Math. USSR Sb.*, 6 (1968), pp. 393–406.

[24] ———. An inequality for a linear form in the logarithms of algebraic numbers, *Mat. Zametki*, 6 (1969), pp. 681–689.

[25] ———. Refinements of two effective inequalities of A. Baker, *Mat. Zametki*, 6 (1969), pp. 767–769.

[26] A. O. Gelfond. On Hilbert's seventh problem, *Doklady Akad. Nauk SSSR*, 2 (1934), pp. 1–6; *Izv. Akad. Nauk SSSR*, 7 (1934), pp. 623–634.

[27] ———. *Transcendental and algebraic numbers* (New York, 1960).

[28] ———. and Yu V. Linnik. On Thue's method and the problem of effectiveness in quadratic fields, *Doklady Akad. Nauk SSSR*, 61 (1948), pp. 773–776.

[29] H. Heilbronn. On the class-number in imaginary quadratic fields, *Quart. J. Math. Oxford Ser.*, 5 (1934), pp. 150–160.

[30] Th. Schneider. Transzendenzuntersuchungen periodischer Funktionen I. Transzendenz von Potenzen. *J. Reine angew. Math.*, 172 (1934), pp. 65–69.

[31] ———. Zur Theorie der Abelschen Funktionen und Integrale, *J. reine angew. Math.*, 183 (1941), pp. 110–128.

[32] C. L. Siegel. Approximation algebraischer Zahlen, *Math. Z.*, 10 (1921), pp. 173–213.

[33] ———. Über einige Anwendungen diophantischer Approximationen, *Abh. Preuss. Akad. Wiss.* (1929), No. 1.

[34] ———. Abschätzung von Einheiten, *Göttingen Nachrichten*, 9 (1969), pp. 71–86.

[35] V. G. Sprindzuk. Concerning Baker's theorem on linear forms in logarithms, *Dokl. Akad. Nauk BSSR*, 11 (1967), pp. 767–769.

[36] ———. Effectivization in certain problems of Diophantine approximation theory, *Dokl. Akad. Nauk BSSR*, 12 (1968), pp. 293–297.

[37] and A. I. Vinogradov. The representation of numbers by binary forms, *Mat. Zametki*, 3 (1968), pp. 369–376.

[38] H. M. Stark. A complete determination of the complex quadratic fields of class-number one, *Michigan Math. J.*, 14 (1967), pp. 1–27.

[39] A. Thue. Über Annäherungswerte algebraischer Zahlen, *J. reine angew. Math.*, 135 (1909), pp. 284–305.

[40] A. Weil. L'arithmétique sur les courbes algébriques, *Acta Math.*, 53 (1928), pp. 281–315.

Trinity College
Cambridge
England

Reprinted from Inst. on Number Theory

EFFECTIVE METHODS IN DIOPHANTINE PROBLEMS

by
A. BAKER

1. Introduction

These notes are intended to serve as a guide to the various effective results in the theory of numbers that have been obtained as a consequence of recent researches. Most of the theorems derive in some way from the author's papers on the logarithms of algebraic numbers and we shall begin with an account of this work. The reader will be referred to the original memoirs for proofs.

2. On the Logarithms of Algebraic Numbers

At the International Congress of mathematicians held in Paris 1900, Hilbert raised, as the seventh of his famous list of 23 problems, the question whether an irrational logarithm of an algebraic number to an algebraic base is transcendental. The question is capable of various alternative formulations; thus one can ask whether an irrational quotient of natural logarithms of algebraic numbers is transcendental, or whether α^β is transcendental for any algebraic number $\alpha \neq 0, 1$ and any algebraic irrational β. A special case relating to logarithms of rational numbers had been posed by Euler more than a century before but no apparent progress had been made towards its solution. Indeed Hilbert expressed the opinion that the resolution of the problem lay farther in the future than a proof of the Riemann hypothesis or Fermat's last theorem.

The first significant advance was made by Gelfond in 1929. Employing interpolation techniques of the kind that he had utilized previously in researches on integral integer-valued functions, Gelfond showed that the logarithm of an algebraic number to an algebraic base cannot be an imaginary quadratic irrational, that is, α^β is transcendental for any algebraic number $\alpha \neq 0, 1$ and any imaginary quadratic irrational β; in particular, one sees that $e^\pi = (-1)^{-i}$ is transcendental. The result was extended to real quadratic irrationals β by Kuzmin in 1930. But it was clear that direct appeal to an interpolation series for $e^{\beta z}$, on which the Gelfond–Kuzmin work was based, was not appropriate for more general β, and further progress awaited a new idea. The search for the latter was concluded successfully by Gelfond and Schneider, independently, in 1934. The arguments they discovered were applicable for any irrational β and, though differing in detail, both depended on the construction of an auxiliary function that vanished at certain selected points. A similar technique had been used a few years earlier by Siegel in the course of his investigations on the Bessel functions. Herewith, Hilbert's seventh problem was finally solved.

The Gelfond–Schneider theorem shows that for any nonzero algebraic numbers α_1, α_2, β_1, β_2 with log α_1, log α_2 linearly independent over the rationals we have

$$\beta_1 \log \alpha_1 + \beta_2 \log \alpha_2 \neq 0.$$

It was natural to conjecture that an analogous theorem would hold for arbitrarily many logarithms of algebraic numbers and moreover it was soon realized that generalizations of this kind would have important consequences in number theory. But, for some thirty years, the problem of extension seemed resistant to attack. It was finally settled in 1966 [6], and the techniques devised for its solution have been the main instruments in establishing the various results described herein. The original theorem has been extended slightly to include also the case in which an additional nonzero algebraic number is present on the left and now reads as follows:

Theorem 1 [7]. *If $\alpha_1, \ldots, \alpha_n$ denote nonzero algebraic numbers such that $\log \alpha_1, \ldots, \log \alpha_n$ are linearly independent over the rationals then $1, \log \alpha_1, \ldots, \log \alpha_n$ are linearly independent over the field of all algebraic numbers.*

Here $\log \alpha_1, \ldots, \log \alpha_n$ are any fixed determinations of the logarithms. The proof of the theorem depends on the construction of an auxiliary function of several complex variables which generalizes the original function of a single variable employed by Gelfond. The subsequent arguments, however, involve an extrapolation procedure that is special to the present context and for which there is no precise earlier counterpart. Quantitative extensions of Theorem 1 will be discussed in the next section and applications of the results to various branches of number theory will be the theme of §§4 to 6.

We record now a few immediate corollaries of Theorem 1.

Theorem 2. *Any nonvanishing linear combination of logarithms of algebraic numbers with algebraic coefficients is transcendental.*

Theorem 3. *$e^{\beta_0}\alpha_1^{\beta_1} \cdots \alpha_n^{\beta_n}$ is transcendental for any nonzero algebraic numbers $\alpha_1, \ldots, \alpha_n, \beta_1, \ldots, \beta_n$.*

Theorem 4. *$\alpha_1^{\beta_1} \cdots \alpha_n^{\beta_n}$ is transcendental for any algebraic numbers $\alpha_1, \ldots, \alpha_n$, other than 0 or 1, and any algebraic number $\beta_1, \ldots, \beta_n$ with 1, $\beta_1, \ldots, \beta_n$ linearly independent over the rationals.*

Particular cases of the above theorems show that $\pi + \log \alpha$ is transcendental for any algebraic number $\alpha \neq 0$, and $e^{\alpha\pi+\beta}$ is transcendental for any algebraic numbers α, β with $\beta \neq 0$. One might also mention an analogy with Lindemann's classical theorem; this asserts that

$$\beta_1 \exp \alpha_1 + \cdots + \beta_n \exp \alpha_n \neq 0$$

for any distinct algebraic numbers $\alpha_1, \ldots, \alpha_n$ and any nonzero algebraic numbers $\beta_1, \ldots, \beta_n$; Theorem 1 shows that the same holds with "exp" replaced by "log" provided that the logarithms are linearly independent over the rationals.

3. Lower Bounds for Linear Forms

By the *height* of an algebraic number we shall mean the maximum of the absolute values of the relatively prime integer coefficients in its minimal defining polynomial. Soon after obtaining his solution to the seventh problem of Hilbert, Gelfond established an important refinement expressing a positive lower bound for a linear form in two logarithms; he proved that for any nonzero algebraic numbers α_1, α_2, β_1, β_2 with $\log \alpha_1$, $\log \alpha_2$ linearly independent over the rationals and any $\kappa > 5$ we have

$$|\beta_1 \log \alpha_1 + \beta_2 \log \alpha_2| > Ce^{-(\log H)^\kappa},$$

where H denotes the maximum of the heights of β_1, β_2, and $C > 0$ denotes a computable number depending only on $\log \alpha_1$, $\log \alpha_2$, and the degrees of β_1, β_2. Gelfond later improved the condition $\kappa > 5$ to $\kappa > 2$. Also he showed that, as a corollary to the Thue-Siegel theorem, about which we shall speak in §5, an inequality of the form

$$|b_1 \log \alpha_1 + \cdots + b_n \log \alpha_n| > Ce^{-\delta H}$$

holds, where $\delta > 0$ and $b_1, \ldots, b_n$ are rational integers, not all 0, with absolute values at most H; but here $C > 0$ could not be effectively computed.

After the demonstration of Theorem 1 it proved relatively easy to obtain extensions of Gelfond's inequalities relating to arbitrarily many logarithms of algebraic numbers, and indeed the following theorem was established [7]

Theorem 5. *Let $\alpha_1, \ldots, \alpha_n$ denote nonzero algebraic numbers with $\log \alpha_1, \ldots, \log \alpha_n$ linearly independent over the rationals, and let $\beta_0, \ldots, \beta_n$ denote algebraic numbers, not all 0, with degrees and heights at most d and H respectively. Then, for any $\kappa > n+1$, we have*

$$|\beta_0 + \beta_1 \log \alpha_1 + \cdots + \beta_n \log \alpha_n| > Ce^{-(\log H)^\kappa}$$

where $C > 0$ denotes an effectively computable number depending only on n, $\log \alpha_1, \ldots, \log \alpha_n$, and d.

Very recently the number on the right of the above inequality has been improved by Feldman to $C^{-1}H^{-\kappa}$, where $C > 0$, $\kappa > 0$ depend only on n, $\log\alpha_1, \ldots, \log\alpha_n$ and d [23], [24]. Estimates for C and κ have been explicitly calculated, but their values are large; the estimate for C takes the form $C' \exp[(\log A)^{\kappa'}]$ where A denotes the maximum of the heights of $\alpha_1, \ldots, \alpha_n$, κ' depends only on n, C' depends only on n, d and the degrees of $\alpha_1, \ldots, \alpha_n$. The value of C and, in particular, its dependence on A is of importance in applications; a more special result, but giving sharper estimates with respect to the parameters other than H, was recently established by the author:

Theorem 6 [7]. *Suppose that $\alpha_1, \ldots, \alpha_n$ are $n \geq 2$ nonzero algebraic numbers and that the heights and degrees of $\alpha_1, \ldots, \alpha_n$ do not exceed integers A, d respectively, where $A \geq 4$, $d \geq 4$. Suppose further that $0 < \delta \leq 1$ and let $\log\alpha_1, \ldots, \log\alpha_n$ denote the principal values of the logarithms. If rational integers $b_1, \ldots, b_n$ exist with absolute values at most H such that*

$$0 < |b_1 \log\alpha_1 + \cdots + b_n \log\alpha_n| < e^{-\delta H}$$

then

$$H < (4^{n^2}\delta^{-1}d^{2n}\log A)^{(2n+1)^2}.$$

Theorem 6 will be the only transcendence result to which we shall refer directly in the sequel. It is useful for application to a wide class of Diophantine problems and yields estimates that will be found, in many cases, to be accessible to practical computation [17].

4. On Imaginary Quadratic Fields with Class Number 1

In 1966 Stark [32] and the author [6], [12] showed independently how one could resolve the long-standing conjecture that the only imaginary quadratic fields $Q(\sqrt{-d})$ with class number 1 are those given by $d = 1, 2, 3, 7, 11, 19, 43, 67, 163$. Heilbronn and Linfoot had proved in 1934 that there could be at most ten such fields, and calculations had shown that the tenth field, if it existed, would satisfy $d > \exp(2{\cdot}2 \times 10^7)$. The work of Stark was motivated by an earlier paper of Heegner, and the work of the author was based on an idea of Gelfond and Linnik. We shall sketch briefly the latter method.

Let $-d < 0$ and $k > 0$ denote the discriminants of the quadratic fields $Q(\sqrt{-d})$ and $Q(\sqrt{k})$ respectively, and suppose that the corresponding class numbers are given by $h(d)$ and $h(k)$. Suppose also that $(d, k) = 1$ and let

$$\chi(n) = (k/n), \quad \chi'(n) = (d/n)$$

denote the usual Kronecker symbols. Further let

$$f = f(x, y) = ax^2 + bxy + cy^2$$

run through a complete set of inequivalent quadratic forms with discriminant $-d$. From a well-known formula for $L(s,\chi)L(s,\chi\chi')$ with $s>1$ we obtain, on taking limits as $s\to 1$ and applying classical results of Dirichlet,

$$h(k)h(kd)\log\epsilon = \frac{1}{12}\pi k\sqrt{d}\left(\sum_{f}\chi(a)a^{-1}\right)\prod_{p|k}(1-p^{-2})$$

$$+B_0+\sum_{f}\sum_{r=-\infty;r\neq 0}^{\infty} B_r e^{\pi irb/(ka)},$$

where ϵ denotes the fundamental unit in the field $Q(\sqrt{k})$, $h(kd)$ denotes the class number of the field $Q(\sqrt{(-kd)})$, $B_0=-\log p\sum_f \chi(a)$ if k is the power of a prime p, $B_0=0$ otherwise and, for $r\neq 0$, we have

$$|B_r|\leq k|r|e^{-\pi|r|\sqrt{d}/(ak)}.$$

On choosing $k=21$ or 33, so that $h(k)=1$ and the units ϵ are given by $\alpha_1=\frac{1}{2}(5+\sqrt{21})$ and $\alpha_2=23+4\sqrt{33}$ respectively, we see that, if $h(d)=1$,

$$|h(21d)\log\alpha_1-\frac{32}{21}\pi\sqrt{d}|<e^{-\pi\sqrt{d}/100},\quad |h(33d)\log\alpha_2-\frac{80}{33}\pi\sqrt{d}|<e^{-\pi\sqrt{d}/100}.$$

Since, in particular, $h(21d)<4\sqrt{d}$, $h(33d)<4\sqrt{d}$, it follows, on writing

$$\delta^{-1}=14\times 10^3,\quad H=140\sqrt{d},\quad b_1=35h(21d),\quad b_2=-22h(33d),$$

that the hypotheses of Theorem 6 are satisfied. We conclude that $H<10^{250}$ whence $d<10^{500}$. Note that, instead of Theorem 6, it would in principle have sufficed to have appealed to the earlier work of Gelfond and moreover, since $\pi=-2i\log i$, to have referred to the above formulae for just one value of k. A similar argument leads to the bound $d<10^{500}$ for all imaginary quadratic fields $Q(\sqrt{-d})$ with class number 2, where d denotes a square-free positive integer with $d\not\equiv 3 \pmod 8$ [12].[1]

5. On the Representation of Integers by Binary Forms

We come now to the fundamental work begun by Thue in 1909 and subsequently developed by Siegel, Roth and others. Thue obtained a nontrivial inequality expressing a limit to the accuracy with which any algebraic number (not itself rational) can be approximated by rationals and thereby showed, in particular, that the Diophantine equation $f(x,y)=m$, where f denotes an irreducible binary form with integer coefficients and degree at least 3, possesses only a finite number of solutions in integers x, y. The work was much extended by Siegel first in 1921

[1] For another proof of the class number 1 result see P. Bundschuh and A. Hock [19]. An interesting application of Theorem 5 in a related context has been given by E. A. Anfert'eva and N. G. Čudakov [2].

when he strengthened the basic approximation inequality and then in 1929 when he applied his result, together with the Mordell–Weil theorem, to give a simple necessary and sufficient condition for any equation of the form $f(x, y) = 0$, where f denotes a polynomial with integer coefficients, to possess only a finite number of integer solutions. Siegel's work gave rise to many further developments; in particular, Mahler obtained far-reaching p-adic generalizations of the original theorems, and Roth succeeded in further improving the approximation inequality, establishing a result that is essentially best possible.

All the work which I have just described, however, suffers from one basic limitation, that of its noneffectiveness. The proofs depend on an assumption, made at the outset, that the Diophantine inequalities, or the corresponding equations, possess at least one solution in integers with large absolute values and the arguments provide no way of deciding whether or not such a solution exists. Thus although the Thue-Siegel theory supplies information on the number of solutions of the equation $f(x, y) = 0$, it does not enable one to determine whether or not a particular equation of this type is soluble; of course, such a determination would amount to a solution of Hilbert's tenth problem for polynomials in two unknowns. For the special equation $f(x, y) = m$, where $f(x, 1)$ has at least one complex zero, another proof of the finiteness of the number of solutions was given by Skolem in 1935 by means of a p-adic argument very different from the original, but here the work depends on the compactness property of the p-adic integers and so is again noneffective.

As a consequence of Theorem 6 one can now give a new and effective proof of Thue's result on the representation of integers by binary forms.

Theorem 7 [8]. *If f denotes an irreducible binary form with degree $n \geq 3$ and with integer coefficients then, for any positive integer m, the equation $f(x, y) = m$ has only a finite number of solutions in integers x, y, and these can be effectively determined.*

The method of proof leads in fact to an explicit bound for the size of all the solutions; assuming that $\mathcal{H}$ is some number exceeding the maximum of the absolute values of the coefficients of f we have

$$\max(|x|, |y|) < \exp\left\{ n\mathcal{H})^{(10n)^5} + (\log m)^{2n+2} \right\}.$$

In view of the mean-value theorem, this corresponds to an effective result on the approximation of algebraic numbers by rationals; indeed one can show that for any algebraic number α with degree $n \geq 3$ and for any $\kappa > n$ there exists an effectively computable number $c = c(\alpha, \kappa) > 0$ such that

$$|\alpha - p/q| > cq^{-n} \exp\left[(\log q)^{1/\kappa}\right]$$

for all integers p, q ($q > 0$). Some slightly stronger quantitative results in this direction have been established for certain fractional powers of rationals [3], [4], [5] but here the work depends on particular properties of Gauss' hypergeometric function and is therefore of a special nature.

As regards the proof of Theorem 7, we assume, without loss of generality, that the coefficient of x^n in $f(x, y)$ is ± 1 and we denote the zeros of $f(x, 1)$ by $\alpha_1, \ldots, \alpha_n$, where it is assumed that $\alpha^{(1)}, \ldots, \alpha^{(s)}$ only are real and $\alpha^{(s+1)}, \ldots, \alpha^{(s+t)}$ are the complex conjugates of $\alpha^{(s+t+1)}, \ldots, \alpha^{(n)}$; thus it is implied that $n = s + 2t$. The algebraic number field generated by $\alpha = \alpha^{(1)}$ over the rationals will be denoted by K, and $\theta^{(1)}, \ldots, \theta^{(n)}$ will represent the conjugates of any elements θ of K corresponding to the conjugates $\alpha^{(1)}, \ldots, \alpha^{(n)}$ of α. C_1, $C_2, \ldots$ will denote numbers greater than 1 which can be specified explicitly in terms of m, n and the coefficients of f. Finally we denote by $\eta_1, \ldots, \eta_r$ a set of $r = s + t - 1$ units in K such that

$$\left| \log |\eta_i^{(j)}| \right| < C_1 \ (1 \leq i, j \leq r)$$

and such that also the determinant Δ of order r with $\log |\eta_i^{(j)}|$ in the ith row and jth column satisfies $|\Delta| > C_2^{-1}$.

Suppose now that x, y are rational integers satisfying $f(x, y) = m$ and put $\beta = x - \alpha y$. Clearly β is an algebraic integer in K, and we have $|\beta^{(1)} \cdots \beta^{(n)}| = m$. Further it is easily seen that an associate γ of β can be determined such that

$$\left| \log |\gamma^{(j)}| \right| < C_3 \quad (1 \leq j \leq n).$$

Writing $\gamma = \beta \eta_1^{b_1} \cdots \eta_r^{b_r}$ and $H = \max |b_j|$ we deduce from the equations

$$\log |\gamma^{(j)}/\beta^{(j)}| = b_1 \log |\eta_1^{(j)}| + \cdots + b_r \log |\eta_r^{(j)}| \quad (1 \leq j \leq r)$$

that the maximum of the absolute values of the numbers on the left must exceed $C_4^{-1} H$, whence

$$\log |\beta^{(l)}| \leq -(C_4^{-1} H - C_3)/(n - 1)$$

for some l. In particular we have $|\beta^{(l)}| \leq C_5$ and so $|\beta^{(k)}| \geq C_6^{-1}$ for some $k \neq l$. Since $n \geq 3$, there exists a superscript $j \neq k$, l, and we have the identity

$$(\alpha^{(k)} - \alpha^{(l)})\beta^{(j)} - (\alpha^{(j)} - \alpha^{(l)})\beta^{(k)} = (\alpha^{(k)} - \alpha^{(j)})\beta^{(l)}.$$

This gives $\alpha_1^{b_1} \cdots \alpha_r^{b_r} - \alpha_{r+1} = \omega$, where

$$\alpha_s = \eta_s^{(k)}/\eta_s^{(j)} \qquad (1 \leq s \leq r),$$

$$\alpha_{r+1} = \frac{(\alpha^{(j)} - \alpha^{(l)})\gamma^{(k)}}{(\alpha^{(k)} - \alpha^{(l)})\gamma^{(j)}} \quad \text{and} \quad \omega = \frac{(\alpha^{(k)} - \alpha^{(j)})\beta^{(l)}\gamma^{(k)}}{(\alpha^{(k)} - \alpha^{(l)})\beta^{(k)}\gamma^{(j)}}.$$

Now by the choice of k and l we see that $0 < |\omega| < C_7 \exp(-H/C_8)$. Further, the degrees and heights of $\alpha_1, \ldots, \alpha_{r+1}$ are bounded above by numbers depending only on f and m. Noting that $|e^z - 1| < \frac{1}{4}$ implies that $|z - ik\pi| < 4|e^z - 1|$ for some rational integer k, we easily obtain an inequality of the type considered in Theorem 6. Hence we conclude that $H < C_9$; this gives $|\beta^{(j)}| < C_{10}$ for each j and thus

$$\max(|x|, |y|) < C_{11} \max(|\beta^{(1)}|, |\beta^{(2)}|) < C_{12}.$$

6. The Elliptic and Hyperelliptic Equations

Theorem 7 and its natural generalization to algebraic number fields can be used to solve effectively many other Diophantine equations in two unknowns. In particular it enables one to treat $y^2 = x^3 + k$ for any $k \neq 0$, an equation with a long and famous history in the theory of numbers [9], [27]. The work rests on the classical theory of the reduction of binary cubic forms, due mainly to Hermite, and the techniques used by Mordell in establishing the finiteness of the number of solutions of the equation. More generally, by means of the theory of the reduction of binary quartic forms one can prove:

Theorem 8 [10]. *Let a $(\neq 0)$, b, c, d denote rational integers with absolute values at most $\mathcal{H}$, and suppose that the cubic on the right of the equation*

$$y^2 = ax^3 + bx^2 + cx + d$$

has distinct zeros. Then all solutions in integers x, y satisfy

$$\max (|x|, |y|) < \exp \{10^6 \mathcal{H}^{10^6}\}.$$

Still more generally one can give bounds for all the solutions in integers x, y of equations of the form $y^m = f(x)$, where $m \geq 2$ and f denotes a polynomial with integer coefficients [11]. The work here, however, is based on a paper of Siegel and involves the theory of factorization in algebraic number fields; the bounds are therefore much larger than that specified in Theorem 8. As immediate consequences of the results one obtains inequalities of the type

$$|x^m - y^n| > c\,(\log \log x)^{1/n^2} \quad (m, n \geq 3),$$

where $c = c(m, n) > 0$; in particular one can effectively solve the Catalan equation $x^m - y^n = 1$ for any given m, n.

We referred earlier to the celebrated Theorem of Siegel on the equation $f(x, y) = 0$. By means of appropriate extensions of the results described above, a new and effective proof of Siegel's theorem in the case of curves of genus 1 has recently been obtained.

Theorem 9 [16]. *Let $F(x, y)$ be an absolutely irreducible polynomial with degree n and with integer coefficients having absolute values at most $\mathcal{H}$ such that the curve $F(x, y) = 0$ has genus 1. Then all integer solutions x, y of $F(x, y) = 0$ satisfy*

$$\max (|x|, |y|) < \exp \exp \exp \left\{ (2\mathcal{H})^{10^{n^{10}}} \right\}.$$

The proof of the theorem involves a combination of some work of J. Coates [22] on the construction of rational functions on curves with prescribed poles, together with the techniques just mentioned for treating the equation $Y^2 = f(X)$. More precisely it is shown by means of the Riemann–Roch theorem that the integer solutions of $F(x, y) = 0$ can be related by a birational transformation to the solutions of an equation $Y^2 = f(X)$ as above, where now f denotes a cubic in X and where the coefficients and variables denote algebraic integers in a fixed field. From bounds for X, Y and their conjugates we immediately obtain the desired bounds for x, y.

In a recent series of papers [20], [21], Coates has generalized many of the theorems described above by employing analysis in the p-adic domain. In particular he has obtained explicit upper bounds for all integer solutions x, y, $j_1, \ldots, j_s$ of equations of the type

$$f(x,y) = mp_1^{j_1} \cdots p_s^{j_s} \quad \text{and} \quad y^2 = x^3 + kp_1^{j_1} \cdots p_s^{j_s},$$

where $p_1, \ldots, p_s$ denote fixed primes. The work involves, amongst other things, utilization of the Schnirelman line integral and the theory of S-units in algebraic number fields. As particular applications of his results, one can now give explicit lower bounds of the type $c(\log\log x)^{1/4}$ for the greatest prime factor of a binary form $f(x,y)$, and one can determine effectively all elliptic curves with a given conductor.

Several other extensions, applications and refinements of the theorems discussed here have been obtained by N. I. Feldman [25], [26], V. G. Sprindžuk [29], [30] and A. I. Vinogradov [31]. Recently Siegel [28] established some improved estimates for units in algebraic number fields which are likely to be of value in reducing the size of bounds. And, in 1967, Brumer [18] derived a natural p-adic analogue of Theorem 1 which, in conjunction with work Ax [1], resolved a well-known problem of Leopoldt on the nonvanishing of the p-adic regulator of an abelian number field.

7. On the Weierstrass Elliptic Functions

Let $\mathcal{P}(z)$ denote a Weierstrass $\mathcal{P}$-function, let g_2, g_3 denote the usual invariants occurring in the equation

$$(\mathcal{P}'(z))^2 = 4(\mathcal{P}(z))^3 - g_2\mathcal{P}(z) - g_3$$

and let ω, ω' denote any pair of fundamental periods of $\mathcal{P}(z)$. Seigel proved in 1932 that if g_2, g_3 are algebraic then at least one of ω, ω' is transcendental; hence both are transcendental if $\mathcal{P}(z)$ admits complex multiplication. Seigel's work was much improved by Schneider in 1937; Schneider showed that if g_2, g_3 are algebraic then any period of $\mathcal{P}(z)$ is transcendental and moreover, the quotient ω/ω' is transcendental except in the case of complex multiplication. Furthermore Schneider proved that if $\zeta(z)$ is the corresponding Weierstrass ζ-function, given by $\mathcal{P}(z) = -\zeta'(z)$, and if $\eta = 2\zeta(\frac{1}{2}\omega)$ then any linear combination of ω, η with algebraic coefficients, not both 0, is transcendental.

By means of techniques similar to those used in the proof of Theorem 1 these results can now be generalized as follows. Let $\mathcal{P}_1(z)$, $\mathcal{P}_2(z)$ be Weierstrass $\mathcal{P}$-functions (possibly with $\mathcal{P}_1 = \mathcal{P}_2$) for which the invariants g_2, g_3 are algebraic and let $\zeta_1(z)$, $\zeta_2(z)$ be the associated Weierstrass ζ-functions. Further let ω_1, ω_1'

and ω_2, ω_2' be any pairs of fundamental periods of $\mathcal{P}_1(z)$, $\mathcal{P}_2(z)$ respectively, and put $\eta_1 = 2\zeta(\frac{1}{2}\omega_1)$, $\eta_2 = 2\zeta(\frac{1}{2}\omega_2)$. We have

Theorem 10 [13], [14]. *Any nonvanishing linear combination of ω_1, ω_2, η_1, η_2 with algebraic coefficients is transcendental.*

It will be recalled that ω_1, ω_2 and η_1, η_2 can be expressed as elliptic integrals of the first and second kinds respectively and so one sees, for instance, that the theorem establishes the transcendence of the sum of the circumferences of two ellipses with algebraic axes-lengths. Also, by an appropriate refinement of Theorem 10, one can obtain an upper estimate for the values assumed by a $\mathcal{P}$-function with algebraic invariants at an algebraic point. In particular, for any positive integer n, we have $|\mathcal{P}(n)| < C\exp[(\log n)^\kappa]$ for some absolute constant $\kappa > 0$ and some $C > 0$ depending only on g_2, g_3.[2] The proof of Theorem 10 utilizes results on the division values of the elliptic functions.

8. Concluding Remarks

The three main problems left open by the work discussed here are

(i) To determine effectively all imaginary quadratic fields with a given class number ≥ 2.
(ii) To find an effective algorithm for determining all the integer points on any curve of genus ≥ 2.
(iii) To establish, under suitable conditions, the algebraic independence of the logarithms of algebraic numbers.

The resolution of these problems would represent a considerable advance in our knowledge.

References

1. James Ax, *On the units of an algebraic number field*, Illinois J. Math. **9** (1965), 584–589. MR **31** #5858.
2. E. A. Anfert'eva and N. G. Čudakov, *The minima of a normed function in imaginary quadratic fields*, Dokl. Akad. Nauk SSSR **183** (1968), 255–256 = Soviet Math. Dokl. **9** (1968), 1342–1344; erratum, ibid. **187** (1969). MR **39** #5472.
3. A. Baker, *Rational approximations to certain algebraic numbers*, Proc. London Math. Soc. (3) **4** (1964), 385–398. MR **28** #5029.

[2]An account of this work is given in a paper submitted to the American J. Math [15]; it is easily seen that $|\mathcal{P}(n)| > Cn$ for some $C > 0$ and infinitely many n.

4. ______, *Rational approximations to* $\sqrt[3]{2}$ *and other algebraic numbers*, Quart. J. Math. Oxford Ser. (2) **15** (1964), 375–383. MR **30** #1977.
5. ______, *Simultaneous rational approximations to certain algebraic numbers, Proc. Cambridge Philos. Soc.* **63** (1967), 693–702. MR **35** #4167.
6. ______, *Linear forms in the logarithms of algebraic numbers*, Mathematika **13** (1966), 204–216. MR **36** #3732.
7. ______, *Linear forms in the logarithms of algebraic numbers.* II, III, IV, Mathematika **14** (1967), 102–107, 220–228; ibid. **15** (1968), 204–216. MR **36** #3732.
8. ______, *Contributions to the theory of Diophantine equations.* I: *On the representation of integers by binary forms*, Philos. Trans. Roy. Soc. London A **263** (1967/68), 173–291. MR **37** #4005.
9. ______, *Contributions to the theory of Diophantine equations.* II: *The Diophantine equation* $y^2 = x^3 + k$, Philos. Trans. Roy. Soc. London Ser. A **263** (1967/68), 193–208. MR **37** #4006.
10. ______, *The Diophantine equation* $y^2 = ax^3 + bx^2 + cx + d$, J. London Math. Soc. **43** (1968), 1–9. MR **38** #111.
11. ______, *Bounds for the solutions of the hyperelliptic equation*, Proc. Cambridge Philos. Soc. **65** (1969), 439–444. MR **38** #3226.
12. ______, *A remark on the class number of quadratic fields*, Bull. London Math. Soc. **1** (1969), 98–102. MR **39** #2723.
13. ______, *On the quasi-periods of the Weierstrass* ζ*-function*, Nachr. Akad. Wiss. Göttingen Math.-Phys. K1. II **1969**, 145–157.
14. ______, *On the periods of the Weierstrass* $\mathcal{P}$*-function*, Proc. Sympos. Math. (Rome), vol. IV, 1968 (Academic Press, London 1976) pp. 155–174.
15. ______, *An estimate for the* $\mathcal{P}$*-function at an algebraic point*, Amer. J. Math., **92** (1970), 619–622.
16. A. Baker and J. Coates, *Integer points on curves of genus* 1, Proc. Cambridge Philos. Soc. **67** (1970), 595–602.
17. A. Baker and H. Davenport, *The equations* $3x^2 - 2 = y^2$ and $8x^2 - 7 = z^2$, Quart. J. Math. Oxford Ser. (2) **20** (1969), 129–137.
18. Armand Brumer, *On the units of algebraic number fields*, Mathematika **14** (1967), 121–124. MR **36** #3746.
19. P. Bundschuh and A. Hock, *Bestimmung aller imaginärquadratischen Zahl-körper der Klassenzahl Eins mit Hilfe eines Satzes von Baker*, Math. Z. **111** (1969), 191–204.
20. J. Coates, *An effective p-adic analogue of a theorem of Thue*, Acta Arith. **15** (1968/69), 279–305. MR **39** #4095.
21. ______, *An effective p-adic analogue of a theorem of Thue.* II: *On the greatest prime factor of a binary form.* III: *The Diophantine equation* $y^2 = x^3 + k$, Acta Arith. **16** (1970), 399–412, 425–435.
22. ______, *Construction of rational functions on a curve*, Proc. Cambridge Philos. Soc. **68** (1970), 105–123.
23. N. I. Feld'man, *Estimate for a linear form of logarithms of algebraic numbers*, Mat. Sb. **76** (118) (1968), 304–319 = Math. USSR Sb. **5** (1968), 291–307. MR **37** #4025.
24. ______, *An improvement of the estimate of a linear form in the logarithms of algebraic numbers*, Mat. Sb. **77** (119) (1968), 423–436 = Math. USSR Sb. **6** (1968), 393–406. MR **38** #1059.

25. ———, *An inequality for a linear form in the logarithms of algebraic numbers*, Mat. Zametki **6** (1969), 681–689.
26. ———, *Refinement of two effective inequalities of A. Baker*, Mat. Zametki **6** (1969), 767–769.
27. L. J. Mordell, *A chapter in the theory of numbers*, Cambridge Univ. Press, Cambridge; Macmillan, New York, 1947. MR **8**, 502.
28. C. L. Siegel, *Abschätzung von Einheiten*, Nachr. Akad. Wiss. Göttingen Math.-Phys. K1. II **1969**, 71–86.
29. V. G. Sprindžuk, *Concerning Baker's theorem on linear forms in logarithms*, Dokl. Akad. Nauk BSSR **11** (1967), 767–769. MR **36** #1396.
30. ———, *Effectivization in certain problems of Diophantine approximation theory*, Dokl. Akad. Nauk BSSR **12** (1968), 293–297. MR **37** #6247.
31. V. G. Sprindžuk and A. I. Vinogradov, *The representation of numbers by binary forms*, Mat. Zametki **3** (1968), 293–297. MR **37** #151.
32. H. M. Stark, *A complete determination of the complex quadratic fields of class-number one*, Michigan Math. J. **14** (1967), 1–27. MR **36** #5102.

Additional Comment (added to first edition). The first of the three problems was solved by Goldfeld, Gross and Zagier via the theory of elliptic curves; see *Bull. Amer. Math. Soc.* **13** (1985), 23–37, *Inventiones Math.* **84** (1986), 225–320. The other two problems remain open.

Trinity College
Cambridge
England

Reprinted from Proc. Symp. Pure. Math.

EFFECTIVE METHODS IN DIOPHANTINE PROBLEMS. II

by
A. BAKER

1. Introduction

Three years ago, at a conference held in Stony Brook, I surveyed the theories which had then recently been developed for the effective resolution of a diverse collection of Diophantine problems [1] (see also [2]). Since that time, several of the topics have been considerably expanded and I should like to use the opportunity provided by the present Symposium to bring the account up to date.

2. Lower Bounds for Linear Forms

One of the most active fields of research has been concerned with improved bounds for linear forms in the logarithms of algebraic numbers. In particular, much study has been made of the special situation, of considerable importance in applications, when one of the algebraic numbers has a large height relative to the remainder. The primary result obtained in this connexion reads as follows.

Theorem 1. *Let $\alpha_1, \ldots, \alpha_n$, $\beta_1, \ldots, \beta_n$ be nonzero algebraic numbers with degrees at most d, let $\alpha_1, \ldots, \alpha_{n-1}$ have heights at most A' and let α_n and $\beta_1, \ldots, \beta_n$ have heights at most A and B respectively. If $\varepsilon > 0, \delta > 0$ and*

$$0 < |\beta_1 \log \alpha_1 + \cdots + \beta_n \log \alpha_n| < e^{-\delta H}$$

for some $H > \exp((\log B)^{1/2})$, then $H < C(\log A)^{1+\varepsilon}$, where $C = C(n, d, \varepsilon, \delta, A')$ is effectively computable.

Special cases of the theorem were proved by Stark [21] and myself [3] in connexion with certain class number problems (see §4) and the full result was obtained by a combination of our methods [9]. Previous work, as described in [1], had led to a

AMS 1970 *subject classifications.* Primary 10–02, 10F35; Secondary 10B45, 10F25, 10H10, 12A25, 12A50.

similar theorem but with $1+\varepsilon$ replaced by a number greater than $n-1$. The condition $H > \exp((\log B)^{1/2})$ can be relaxed to $H > (\log B)^{cn^2/\varepsilon}$ for a sufficiently large absolute constant c, provided that $\varepsilon < 1$, and, furthermore, one can replace $(\log A)^{\varepsilon}$ by some power of $\log\log A$, though this power is usually large. Very recently, by further developments of the arguments, it has been shown that, in the case when $\beta_1, \ldots, \beta_{n-1}$ are rational integers and $\beta_n = -1$, conditions frequently satisfied in applications, the exponent $1+\varepsilon$ can be replaced by 1, which is best possible.

Theorem 2 [5], [6]. *Let $\alpha_1, \ldots, \alpha_n$ be nonzero algebraic numbers with degrees at most d and let the heights of $\alpha_1, \ldots, \alpha_{n-1}$ and α_n be at most A' and A ($\geqq 2$) respectively. If, for some $\varepsilon > 0$, there exist rational integers $b_1, \ldots, b_{n-1}$ with absolute values at most B such that*

$$0 < |b_1 \log \alpha_1 + \cdots + b_{n-1} \log \alpha_{n-1} - \log \alpha_n| < e^{-\varepsilon B},$$

then $B < C \log A$ for some effectively computable number C depending only on n, d, A' and ε.

Theorem 2 is, in fact, an immediate consequence of another theorem [6] to the effect that there exists $C = C(n, d, A')$ such that, for any δ with $0 < \delta < \frac{1}{2}$, the inequalities

$$0 < |b_1 \log \alpha_1 + \cdots + b_n \log \alpha_n| < (\delta/B')^{C \log A} e^{-\delta B}$$

have no solution in rational integers $b_1, \ldots, b_{n-1}$ and $b_n (\neq 0)$ with absolute values at most B and B' respectively. Clearly, on taking $\delta = 1/B$ and assuming $B' \leqq B$, the number on the right becomes at least $C^{-\log A \log B}$ for some effectively computable C, and this bound is best possible with respect to A when B is fixed and with respect to B when A is fixed. Corollaries relating to the theory of Diophantine equations will be discussed in §3.

The proofs of the above theorems depend upon several new developments in the earlier works. In particular, the underlying auxiliary functions are now considerably more involved; the argument leading to Theorem 2, for instance, utilizes a function of the form

$$\sum_{\lambda_{-1}=0}^{L_{-1}} \cdots \sum_{\lambda_n=0}^{L_n} p(\lambda_{-1}, \ldots, \lambda_n)\, \Lambda(z_0)\, \alpha_1^{\gamma_1 z_1} \cdots \alpha_{n-1}^{\gamma_{n-1} z_{n-1}},$$

where the $p(\lambda_{-1}, \ldots, \lambda_n)$ are, as usual, rational integers, $\gamma_r = \lambda_r + b_r \lambda_n$ and

$$\Lambda(z) = \Delta(z + \lambda_{-1}; h, \lambda_0 + 1, m_0) \prod_{r=1}^{n-1} \Delta(\gamma_r; m_r),$$

$$\Delta(z; k) = \frac{1}{k!}(z+1) \cdots (z+k), \quad \Delta(z; k, l, m) = \frac{1}{m!} \frac{d^m}{dz^m} (\Delta(z; k))^l.$$

Further, the inductive nature of the expositions is substantially modified and the ultimate contradiction is obtained now by an appeal to certain algebraic lemmas relating to Kummer theory, quite different from the techniques employed previously. The reader is referred to the original memoirs for details.

3. Diophantine Equations

In my earlier survey, I discussed the fundamental theorem of Thue on $f(x,y)=m$, where f denotes an irreducible binary form with integer coefficients and degree $n \geqq 3$. More especially, I described how the theorem could be made effective, and indeed how one could establish an upper bound

$$\max(|x|,|y|) < C\exp\{(\log m)^{\kappa}\},$$

applicable for all integer solutions x, y, where $\kappa > n$ and C is computable in terms of κ and the coefficients of f. In view of Theorem 2, one can now strengthen the number on the right to Cm^c, where c can be computed like C, and this gives at once

Theorem 3. *For any algebraic number α with degree $n \geqq 3$ there exist positive effectively computable numbers c, κ depending only on α, with $\kappa < n$, such that*

$$|\alpha - p/q| > cq^{-\kappa}$$

for all rationals p/q $(q > 0)$.

Feldman [13] first obtained this result from a special case of Theorem 2, involving certain restrictions on α_n, and his arguments rested on rather different adaptations in the basic theory of linear forms in logarithms; yet another approach, employing p-adic analysis, was described by Sprindžuk [19], [20] at about the same time.

Several other new theorems on rational approximations to algebraic numbers follow from the general result cited after the enunciation of Theorem 2. Thus, for instance, it shows that

$$|\alpha - pp'/qq'| > Q^{-\kappa \log \ \log Q'},$$

where p', q' are comprised solely of powers of fixed sets of primes and Q, Q' are the maxima of the absolute values of p, q and p', q' respectively; this furnishes a further improvement on Ridout's generalization of Roth's theorem (cf. the recent survey of Schmidt [18]). Furthermore one sees that

$$|\alpha^{1/m} - p/q| > cq^{-\kappa \log m}$$

for any algebraic number α, where c, κ are positive numbers effectively computable in terms of α, and this is sharper than the Thue-Siegel inequality when the integer m is large. The latter theorem recalls to mind the very first effective results in this context, derived by means of special properties of Gauss's hypergeometric function (see [1]). When applicable, this method gives surprisingly strong estimates for the solutions of Diophantine equations, and it has not been dormant. In particular,

Feldman [12] and Osgood [15], [16] have widely applied ideas of this nature to study effectively certain equations of norm form in several variables.

4. Class Numbers

I described at Stony Brook the transcendental method for determining all the imaginary quadratic fields with class number 1, and I remarked also that the same techniques could be used to treat the analogous problem for class number 2 when the discriminants of the fields are even. Since then, a complete resolution of the class number 2 problem has been obtained, and I should like to indicate the main new idea very briefly. A fuller account is provided by the text of the lecture I delivered a year or so ago in Washington [4].

If $Q((-d)^{1/2})$ has class number 2 and odd discriminant $-d < -15$, then $d = pq$, where p, q are primes congruent to 1 and 3 (mod 4) respectively. Denoting by $\chi'(n)$ one of the generic characters associated with forms of discriminant $-d$ and writing

$$\chi_{pq}(n) = \left(\frac{-pq}{n}\right), \quad \chi_p(n) = \left(\frac{p}{n}\right), \quad \chi_q(n) = \left(\frac{-q}{n}\right), \quad \chi(n) = \left(\frac{k}{n}\right),$$

where k is an integer $\equiv 1 \pmod 4$ and $(k, pq) = 1$, we obtain

$$L(1,\chi)L(1,\chi\chi_{pq}) + L(1,\chi\chi_p)L(1,\chi\chi_q) = \sum \chi(f)/f,$$

where $f = f(x, y)$ denotes the principal form with discriminant $-d$, and the sum is over all integers x, y not both 0. Now if k is not a prime power, for instance if $k = 21$, then the sum on the right approximates to a rational multiple of π^2, and on substituting for the L-functions on the left from Dirichlet's formulae we obtain an inequality of the type considered in Theorem 1; this leads at once to the desired effective bound for d.

By somewhat similar techniques, Schinzel and I [8] have recently shown that every genus of primitive binary quadratic forms with discriminant D represents a positive integer $\leqq c(\varepsilon)|D|^{3/8+\varepsilon}$ for any $\varepsilon > 0$, where $c(\varepsilon)$ depends only on ε. Our proof involves Siegel's theorem on L-functions and so does not enable $c(\varepsilon)$ to be effectively computed when $\varepsilon < \frac{1}{8}$; on the other hand, an effective estimate would, as we show, yield a complete determination of all the "numeri idonei" of Euler, and, of course, this would include the class number 1 and 2 results to which I have just referred.

5. Elliptic Functions

The main result on elliptic functions cited in [1] has been extended recently by Coates [11].

Theorem 4. *Any nonvanishing linear combination of ω_1, ω_2, η_1, η_2 and $2\pi i$ with algebraic coefficients is transcendental.*

Here ω_1, ω_2 denote a pair of fundamental periods of a Weierstrass $\mathcal{P}$-function with algebraic invariants g_2, g_3 and $\eta_1 = 2\zeta(\frac{1}{2}\omega_1)$, $\eta_2 = 2\zeta(\frac{1}{2}\omega_2)$, where $\zeta(z)$ denotes the associated Weierstrass ζ-function. The new feature in Theorem 4 is the inclusion of $2\pi i$, this extension having been gained, however, at the cost of some restriction in the hypotheses. The result is of particular interest in view of the Legendre relation $\omega_1\eta_2 - \omega_2\eta_1 = 2\pi i$, showing that the five numbers in question are algebraically dependent. Furthermore, one sees that the theorem includes the transcendence of such numbers as $\pi + \omega$ and $\pi + \eta$ for any period ω of $\mathcal{P}(z)$ and quasi-period η of $\zeta(z)$.

Some quantitative estimates in connexion with Theorem 4 have recently been derived by a student of mine, D. W. Masser; in particular, he has proved [14]:

Theorem 5. *For any positive integer n and any $\varepsilon > 0$, we have*

$$|\mathcal{P}(n)| < Cn^{(\log\ \log n)^{7+\varepsilon}}$$

where C depends only on g_2, g_3 and ε.

Moreover he has shown that a similar estimate obtains for $\mathcal{P}(\pi + n)$ and indeed for $\mathcal{P}(\alpha)$, where α is any nonzero algebraic number. Theorem 5 compares well with the lower bound $|\mathcal{P}(n)| > Cn$ valid for some $C > 0$ and infinitely many n, and it improves upon the result mentioned in [1], where an unspecified power of log n occurred in place of log log n. It seems likely that this general area of study will be considerably developed in the next few years (cf. [10]).

6. Further Results and Problems

In a lecture at the same conference in Stony Brook to which I referred at the beginning, Chowla raised the problem whether there exists a rational-valued function $f(n)$, periodic with prime period p, such that $\sum f(n)/n = 0$. He proved some twenty years ago that this could not hold for odd functions f if $\frac{1}{2}(p-1)$ is prime, a condition subsequently removed by Siegel, and recently he showed that the same is true for even functions f if $f(0) = 0$. In a forthcoming paper [7] by Birch, Wirsing and myself, it is shown that there is in fact no function f with these properties. The arguments involve an appeal to the basic result on the linear independence of the logarithms of algebraic numbers, but otherwise the proof runs on classical lines. Our work enables us to treat more generally functions f that take algebraic values and are periodic with any modulus q, and we prove thereby

Theorem 6. *If $(q, \phi(q)) = 1$ and χ runs through all nonprincipal characters* mod q *then the $L(1, \chi)$ are linearly independent over the rationals.*

Theorem 6 plainly generalizes Dirichlet's famous result on the nonvanishing of $L(1,\chi)$; it does not, however, give a new proof of this result, for the latter is, in fact, utilized in the demonstration. It would be of much interest to know whether the theorem is valid when $(q,\phi(q))>1$.

Finally, I should like to discuss some possible future avenues of investigation. First, one would like to have a theorem of the nature of Theorem 1 in which A denotes the height of all the α's and not just α_n; some work in this direction has been carried out by Ramachandra [17] and his pupil T. N. Shorey, and they have applied their results to certain questions in prime number theory. But, at the moment, the theorems are rather special and one would hope for considerable improvements here. Secondly, it is almost certain that Theorems 1 and 2 have natural p-adic analogues, and these would enable many of the Diophantine results obtained earlier to be strengthened. In particular, they would give an inequality of the form $||(3/2)^n|| > 2^{-\delta n}$, valid for all $n > n_0$, where n_0 is effectively computable, δ is an absolute constant with $0<\delta<1$ and $||x||$ denotes the distance of x from the nearest integer. If, moreover, the value of δ were such that $2^{-\delta} > \frac{3}{4}$ then this would settle an outstanding question in connexion with Waring's problem. But, of course, it may be difficult to obtain such a precise value of δ from the present analysis. Thirdly, one would like to obtain a value of κ in Theorem 3 depending only on n and indeed of the same order of magnitude as the Siegel exponent; this would naturally lead to an effective determination of all the integer points on a curve of arbitrary genus, that is, to a complete solution to the first problem mentioned at the end of [1]. Since the magnitude of κ depends on the value of C in Theorem 2, this again reflects on the basic theory of linear forms in logarithms. And lastly, one would like an extension of Theorem 2 in which $b_1,\ldots,b_{n-1}$ denote arbitrary algebraic numbers and not merely rational integers; this too seems difficult to obtain with our present techniques.

References

1. A. Baker, *Effective methods in Diophantine problems*, Proc. Sympos. Pure Math., vol. 20, Amer. Math. Soc., Providence, R.I., 1971, pp. 195–205.
2. ———, *Effective methods in the theory of numbers*, Proc. Internat. Congress Math. (Nice, 1970), vol. 1, Gauthier-Villars, Paris, 1971, pp. 19–26.
3. ———, *Imaginary quadratic fields with class number* 2, Ann. of Math. (2) **94** (1971), 139–152.
4. ———, *On the class number of imaginary quadratic fields*, Bull. Amer. Math. Soc. **77** (1971), 678–684.
5. ———, *A sharpening of the bounds for linear forms in logarithms*, Acta Arith. **21** (1972), 117–129.
6. ———, *A sharpening of the bounds for linear forms in logarithms.* II, Acta Arith., **24** (1973), 33–36.

7. A. Baker, B. J. Birch and E. A. Wirsing, *On a problem of Chowla*, J. Number Theory, **5** (1973), 224–236.
8. A. Baker and A. Schinzel, *On the least integers represented by the genera of binary quadratic forms*, Acta Arith. **18** (1971), 137–144.
9. A. Baker and H. M. Stark, *On a fundamental inequality in number theory*, Ann. of Math. (2) **94** (1971), 190–199.
10. J. Coates, *An application of the division theory of elliptic functions to Diophantine approximation*, Invent. Math. **11** (1970), 167–182.
11. ———, *The transcendence of linear forms in* ω_1, ω_2, η_1, η_2, $2\pi i$, Amer. J. Math. **93** (1971), 385–397.
12. N. I. Fel'dman, *Effective bounds for the number of solutions of certain Dio-phantine equations*, Mat. Zametki **8** (1970), 361–371. (Russian) MR **42** #7590.
13. ———, *An effective sharpening of the exponent in Liouville's theorem*, Izv. Akad. Nauk SSSR Ser. Mat. **35** (1971), 973–990 = Math. USSR Izv. **5** (1971), 985–1002.
14. D. W. Masser, *On the periods of the exponential and elliptic functions*, Proc. Cambridge Philos. Soc., **73** (1973), 339–350.
15. C. F. Osgood, *The simultaneous Diophantine approximation of certain kth roots*, Proc. Cambridge Philos. Soc. **67** (1970), 75–86. MR **40** #2612.
16. ———, *On the simultaneous Diophantine approximation of values of certain algebraic functions*, Acta Arith. **19** (1971), 343–386.
17. K. Ramachandra, *A note on numbers with a large prime factor*. III, Acta Arith. **19** (1971), 49–62.
18. W. M. Schmidt, *Approximation to algebraic numbers*, Enseignement Math. **17** (1971), 187–253.
19. V. G. Sprindžuk, *A new application of p-adic analysis to representations of numbers by binary forms*, Izv. Akad. Nauk SSSR Ser. Mat. **34** (1970), 1038–1063 = Math. USSR Izv. **4** (1970), 1043–1069. MR **42** #5910.
20. ———, *On rational approximations to algebraic numbers*, Izv. Akad. Nauk SSSR Ser. Mat. **35** (1971), 991–1007 = Math. USSR Izv. **5** (1971), 1003–1019.
21. H. M. Stark, *A transcendence theorem for class number problems*, Ann. of Math. (2) **94** (1971), 153–173.

Trinity College
Cambridge
England

EFFECTIVE METHODS IN THE THEORY OF NUMBERS/DIOPHANTINE PROBLEMS

by
A. BAKER

Although the theory described in the preceding papers has progressed greatly in the intervening 25 years or so since they were published and modern surveys of the field would look very different, it became apparent on reading through them that I could not possibly reflect again the novelty of the results and the excitement of their discovery that is evident in these works. They have therefore been left as they stand and I give now a short note to indicate some of the main developments.

The basic theory of logarithmic forms has been much refined. The best result to date, at least as far as the fundamental rational case is concerned, is proved in Baker and Wüstholz [3]. It is shown that if

$$\Lambda = b_1 \log \alpha_1 + \cdots + b_n \log \alpha_n \neq 0,$$

in the now familiar notation, then

$$\log |\Lambda| > -(16nd)^{2(n+2)} \log A_1 \cdots \log A_n \log B,$$

where the A's signify the respective classical heights of the α's and B denotes a bound for the absolute values of the b's; further, as usual, d signifies the degree of the number field generated by the α's over the rationals. A still stronger result is given in [3] in terms of the logarithmic Weil height; this has several nice properties, for instance, it is semi-multiplicative, and it has become the standard height employed in the field. The proof of the inequality depends on many new ingredients, among them Kummer theory, Δ-functions, successive minima, Blaschke products and multiplicity estimates on group varieties.

The latter estimates have themselves been the outcome of a remarkable series of discoveries. They originate from techniques involving commutative algebra introduced by Nesterenko [17] in connection with studies on E-functions; they were developed by Brownawell and Masser [4], Masser and Wüstholz [11], Philippon [18] and especially by Wüstholz [29]. He obtained, in 1983, the critical result extending a zero estimate on group varieties appertaining to a single differential operator to arbitrarily many such operators. The work has found widespread application; in particular it has yielded an abelian analogue of the famous Lindemann theorem and it has solved a long-standing problem concerning periods of elliptic integrals (see [26], [27], [28]). The key result is now the analytic subgroup theorem established in Wüstholz [30]; this furnishes a generalisation in terms of algebraic groups of the basic qualitative theorem on logarithmic independence that customarily bears my name. Among the many sequels, Masser and Wüstholz have used this sphere of ideas to yield an isogeny result which makes effective one of the principal components in Faltings' well known proof of the Mordell conjecture (see [12] to [16]).

Another extensive area of application of logarithmic forms has been in connection with the solution of Diophantine equations. Indeed it now provides the standard method for effectively solving algebraic equations in two integer variables (see [1]). Wider aspects, appertaining for instance to norm form, discriminant form and index form equations, have been brought to light, most notably by Győry (see the abundant literature cited in [2]). My original work in this field proceeded by way of the S-unit equation and used linear forms in complex logarithms as above; this would still seem to be the most universal method. Recently, however, a number of writers have developed an alternative approach that, when applicable, is more direct (see [8], [23]); it is based on linear forms in elliptic logarithms. Here the best result to date is due to Hirata–Kohno [9] and the explicit version needed for determining the complete list of points on the relevant Diophantine curve has been worked out by S. David [5]. In either case, one meets the old problem of computing all solutions of a Diophantine inequality below a rather large bound; Davenport and I described a method of dealing with the problem based on a simple lemma in Diophantine approximation and two further computational techniques have been described subsequently, one due to Grinstead and Pinch [19], based on recurrence sequences, and the other due to Tzanakis and de Weger [24] based on an algorithm of A. K. Lenstra, H. W. Lenstra and L. Lovász. Now that the constants in our estimates for $\log|\Lambda|$ are much reduced and computers generally have become more powerful, these methods are seen to be very efficient — specific and fully worked instances are given e.g. in Gaál *et al.* [6], [7] and Tzanakis and de Weger [25]. It should be remarked that the p-adic theory of linear forms in logarithms also features in the practical solution of Diophantine equations; here the most precise results to date are due to Yu Kunrui [31]. Moreover, the fact that our lower bound for $\log|\Lambda|$ depends now on each of the $\log A$'s only to the first power, which is best possible, has led to the effective analysis of the remarkable class of exponential Diophantine equations; see the tract by Shorey and Tijdeman [21] and, for the Catalan equation in particular, the book by Ribenboim [20].

The theory has been applied in other diverse areas. They include p-adic L-functions (Ax and Brumer), Knot Theory (Riley), Modular Forms (Odoni), Ramanujan Functions (Murty) and Primitive Divisors (Schinzel and Stewart). Further, there is a close connection with the so-called abc-conjecture of Oesterlé and Masser (see [22]); in fact, before the conjecture itself was formulated, studies on logarithmic forms had already led Mason [10] to a statement and proof of its analogue for function fields. Indeed it has become evident that there is now considerable interplay between transcendence theory and many aspects of arithmetical algebraic geometry and this would seem to be a major trend for the future.

References

[1] A. Baker, *Transcendental Number Theory* (3rd ed., Cambridge Math. Library series, 1990).
[2] A. Baker (ed.), *New Advances in Transcendence Theory* (Cambridge, 1988).
[3] A. Baker and G. Wüstholz, Logarithmic forms and group varieties, *J. reine angew. Math.* **442** (1993), 19–62.
[4] W. D. Brownawell and D. W. Masser, Multiplicity estimates for analytic functions I, II, *J. reine angew. Math.* **314** (1980), 200–216; *Duke Math. J.* **47** (1980), 273–295.
[5] S. David, Minorations de formes linéaires de logarithmes elliptiques, *Publ. Math. Univ. Pierre et Marie Curie* **106**, Problèmes diophantiens 1991–2, exposé no. 3.
[6] I. Gaál, On the resolution of inhomogeneous norm form equations in two dominating variables, *Math. Computation* **51** (1988), 359–373.
[7] I. Gaál, A. Pethö and M. Pohst, On the resolution of index form equations in biquadratic number fields III: The bicyclic biquadratic case, *J. Number Theory* **53** (1995), 100–114.
[8] J. Gebel, A. Pethö and H. G. Zimmer, Computing integral points on elliptic curves, *Acta Arith.* **68** (1994), 171–192.
[9] N. Hirata-Kohno, Formes linéaires de logarithmes de points algébriques sur les groupes algébriques, *Invent. Math.* **104** (1991), 401–433.
[10] R. C. Mason, *Diophantine Equations over Function Fields* (LMS Lecture Note Series 96, Cambridge, 1984).
[11] D. W. Masser and G. Wüstholz, Zero estimates on group varieties I, II, *Invent. Math.* **64** (1981), 489–516; **80** (1985), 233–267.
[12] D. W. Masser and G. Wüstholz, Estimating isogenies on elliptic curves, *Invent. Math.* **100** (1990), 1–24.
[13] D. W. Masser and G. Wüstholz, Galois properties of division fields of elliptic curves, *Bull. London Math. Soc.* **25** (1993), 247–254.
[14] D. W. Masser and G. Wüstholz, Periods and minimal abelian subvarieties, *Ann. Math.* **137** (1993), 407–458.
[15] D. W. Masser and G. Wüstholz, Isogeny estimates for abelian varieties and finiteness theorems, *Ann. Math.* **137** (1993), 459–472.
[16] D. W. Masser and G. Wüstholz, Endomorphism estimates for abelian varieties, *Math. Zeit.* **215** (1994), 641–653.
[17] Yu. V. Nesterenko, Bounds on the orders of the zeros of a class of functions and their application to the theory of transcendental numbers, *Math. USSR Izvest.* **11** (1977), 253–284.
[18] P. Philippon, Lemmes de zéros dans les groupes algébriques commutatifs, *Bull. Soc. Math. France* **114** (1986), 355–383; **115** (1987), 397–398.
[19] R. G. E. Pinch, Simultaneous Pellian equations, *Math. Proc. Camb. Phil. Soc.* **103** (1988), 35–64.
[20] P. Ribenboim, *Catalan's Conjecture* (Academic Press, 1994).
[21] T. N. Shorey and R. Tijdeman, *Exponential Diophantine Equations* (Cambridge, 1986).

[22] C. L. Stewart and Kunrui Yu, On the *abc*-conjecture, *Math. Ann.* **291** (1991), 225–230.
[23] R. J. Stroeker and N. Tzanakis, Solving elliptic diophantine equations by estimating linear forms in elliptic logarithms, *Acta Arith.*, **67** (1994), 177–196.
[24] N. Tzanakis and B. M. M. de Weger, On the practical solution of the Thue equation, *J. Number Theory* **31** (1989), 99–132.
[25] N. Tzanakis and B. M. M. de Weger, Solving a specific Thue-Mahler equation, *Math. Comp.* **57** (1991), 799–815.
[26] G. Wüstholz, Uber das abelsche Analogon des Lindemannsche Satzes I, *Invent. Math.* **72** (1983), 363–388.
[27] G. Wüstholz, Zum Periodenproblem, *Invent. Math.* **78** (1984), 381–391.
[28] G. Wüstholz, Transzendenzeigenschaften von Perioden elliptischer Integrale, *J. reine angew. Math.* **354** (1984), 164–174.
[29] G. Wüstholz, Multiplicity estimates on group varieties, *Ann. Math.* **129** (1989), 471–500.
[30] G. Wüstholz, Algebraische Punkte auf analytischen Untergruppen algebraischer Gruppen, *Ann. Math.* **129** (1989), 501–517.
[31] Yu Kunrui, Linear forms in p-adic logarithms I, II, III, *Acta Arith.* **53** (1989), 107–186; *Compositio Math.* **74** (1990), 15–133; **91** (1994), 241–276.

Additional Comment (added to second edition). Three striking recent achievements in this field have been the solution to a problem of Wolfart in hypergeometric theory through studies of Cohen and Wüstholz and subsequent work of Edixhoven and Yafaev on the André-Oort conjecture (see *A Panorama in Number Theory or the View from Baker's Garden*, Cambridge 2002, pp. 89–106), the solution to a century-old problem on primitive divisors of Lucas and Lehmer numbers by Bilu, Hanrot and Voutier (see *J. reine angew. Math.* **539** (2001), 75–112) and the remarkable solution by Mihailescu to the Catalan Conjecture (as reported by Bilu, see *Séminaire Bourbaki*, Exposé 909, 2002).

Reprinted from Actes du Congr. Int. des Math., 1970. Tome 1, p. 11 à 13

THE WORK OF SERGE NOVIKOV

by

M. F. ATIYAH

It gives me great pleasure to report on the work of Serge Novikov. For many years he has been generally acknowleged as one of the most outstanding workers in the fields of Geometric and Algebraic Topology. In this rapidly developing area, which has attracted many brilliant young mathematicians, Novikov is perhaps unique in demonstrating great originality and very powerful technique both in its geometric and algebraic aspects.

Novikov made his first impact, as a very young man, by his calculation of the unitary cobordism ring of Thom (independently of similar work by Milnor). Essentially Thom had reduced a geometrical problem of classification of manifolds to a difficult problem of homotopy theory. Despite the great interest aroused by the work of Thom this problem had to wait several years before its successful solution by Milnor and Novikov. Many years later Novikov returned to this area and, combining cobordism with homotopy theory, he developed some very powerful algebraic machinery which gives one of the most refined tools at present available in Algebraic Topology. In his early work it was a question of applying homotopy to solve the geometric problem of cobordism; in this later work it was the reverse, cobordism was used to attack general homotopy theory.

On the purely geometric side I would like to single out a very beautiful and striking theorem of Novikov about foliations on the 3-dimensional sphere. Perhaps I should remind you that a foliation of a manifold is (roughly speaking) a decomposition into manifolds (of some smaller dimension) called the leaves of the foliation: one leaf passing through each point of the big manifold. If the leaves have dimension one then we are dealing with the trajectories (or integral curves) of a vector field, and closed trajectories are of course particularly interesting. In the general case a basic question therefore concerns the existence of *closed leaves*. Very little was known about this problem. Thus even in the simplest case of a foliation of the 3-sphere into 2-dimensional leaves the answer was not known until Novikov, in 1964, proved that every foliation in this case does indeed have a closed leaf (which is then necessarily a torus). Novikov's proof is very direct and involves many delicate geometric arguments. Nothing better has been proved since in this direction.

Undoubtedly the most important single result of Novikov, and one which combines in a remarkable degree both algebraic and geometric methods, is his famous proof of the topological invariance of the Pontrjagin classes of a differentiable manifold. In order to explain this result and its significance I must try in a few minutes to summarize the history of manifold theory over the past 20 years. Fortunately, during this Congress you will be able to hear many more detailed and comprehensive surveys.

There are 3 different kinds or categories of manifold: differentiable, piece-wise linear (or combinatorial) and topological. For each category the main problem is to understand the structure or to give some kind of classification. There was no clear idea about the distinction between these 3 categories until Milnor produced his famous example of 2 different differentiable structures on the 7-sphere. After that the subject developed rapidly with important contributions from many people, including Novikov, so that in a few years the distinction between differentiable and piece-wise linear manifolds, and their classification, was very understood. However, there were still no real indications about the status of topological manifolds. Were they essentially similar to piece-wise linear manifolds or were they quite different? Nobody knew. In fact, there were no known invariants of topological manifolds except homotopy invariants. On the other hand, there were many invariants known for differentiable or piece-wise linear manifolds which were finer than homotopy invariants. Notable among these were the Pontrjagin classes. For a differentiable manifold these are cohomology classes which measure, in some sense, the amount of global twisting in the tangent spaces. For a manifold with a global parallelism like a torus they are zero. In the context of Riemannian geometry there is a generalized Gauss–Bonnet theorem which expresses them in terms of the curvature. In any case their definition relies heavily on differentiability. Around 1957 it was shown by Thom, Rohlin and Svarc, using important earlier work of Hirzebruch, that the Pontrjagin classes are actually piece-wise linear invariants (provided we use rational or real coefficients). When Novikov, in 1965, proved their topological invariance this was the first real indication that topological manifolds might be essentially similar to piece-wise linear ones. It was a big break-through and was quickly followed by very rapid progress which, in the past few years, through the work of many mathematicians — notably Kirby and Siebenmann — has resulted in fairly complete information about the topological piece-wise linear situation. Thus we now know that nearly all topological manifolds can be triangulated and essentially in a unique way. You will undoubtedly hear about this in the Congress lectures.

Perhaps you will understand Novikov's result more easily if I mention a purely geometrical theorem (not involving Pontrjagin classes) which lies at the heart of Novikov's proof. This is as follows:

Theorem (*). — *If a differentiable manifold X is homeomorphic to a product $M \times R^n$ (where M is compact, simply-connected and has dimension ≥ 5) then X is diffeomorphic to a product $M' \times R^n$.*

Here both M, M' are differentiable manifolds. The theorem thus asserts that a topological factorization implies a differentiable factorization: it is clearly a deep result. Combined with the earlier Thom–Hirzebruch work it leads easily to the invariance of the Pontrjagin classes.

I hope I have now indicated the importance of this result of Novikov's and its place in the general development of manifold theory. I would like also to stress the remarkable nature of the proof which combines very ingenious geometric ideas with considerable algebraic virtuosity. One aspect of the geometry is particularly worth mentioning. As is well-known many topological problems are very much easier if one is dealing with simply-connected spaces. Topologists are very happy when they can get rid of the fundamental group and its algebraic complications. Not so Novikov! Although the theorem above involves only simply-connected spaces, a key step in his proof consists in perversely introducing a fundamental group, rather in the way that (on a much more elementary level) puncturing the plane makes it non-simply-connected. This bold move has the effect of simplifying the geometry at the expense of complicating the algebra, but the complication is just manageable and the trick works beautifully. It is a real master stroke and completely unprecedented. Since then a somewhat analogous device has proved crucial in the important work of Kirby mentioned earlier.

I hope this brief report has given some idea of the real individuality of Novikov's work, its variety and its importance, all of which fully justifies the award of the Fields Medal. It is all the more remarkable when we remember that he worked in relative isolation from the main body of mathematicians in his particular field. We offer him our heartiest congratulations in the full confidence that he will continue, for many years to come, to produce mathematics of the highest order.

Michael Atiyah
Institute for Advance Study
Department of Mathematics,
Princeton, New Jersey 08540
(U. S. A.)

Serge Novikov
Steklov Mathematical Institute
ul Vavilova 42,
Moscow V 333
(U. R. S. S.)

*This formulation is due to L. Siebenmann.

Sergei Novikov

SERGEI NOVIKOV

Novikov was born on March 20, 1938 in Gorki (Nizni Novgorod), into a family of outstanding mathematicians. His father, Petr Sergeevich Novikov (1901–1975), was an academician, an outstanding expert in mathematical logic, algebra, set theory, and function theory; his mother, Lyudmila Vsevolodovna Keldysh (1904–1976), was a professor, a well-known expert in geometric topology and set theory. Novikov received his mathematical education in the Faculty of Mathematics and Mechanics of Moscow University (1955–1960), and he has worked there since 1964 in the Department of Differential Geometry; since 1983 he has been head of the Department of Higher Geometry and Topology of Moscow University.

Novikov married Eleanor Tsoi (who also received her mathematical education at Moscow State University, 1955–1960). They have three children: two daughters, Irina (1964) and Maria (1965), and a son, Peter (1973).

In 1960 Novikov enrolled as a research student at the Steklov Institute of Mathematics, where his supervisor was M. M. Postnikov; since 1963 he has been on the staff there. He was awarded the degree of Ph.D. there in 1964, and that of Doctor of Science in 1965. In 1966 he was elected Corresponding Member of the Academy of Sciences of the USSR, and in 1981 a full member. Since 1984 he has been head of the Department of Geometry and Topology of the Mathematical Institute of the Academy of Sciences and in charge of the problem committee of *Geometriya I topologiya* (Geometry and Topology) at the Mathematics Division of the Academy of Sciences of the USSR. He has been head of the Mathematics Division at the L. D. Landau Institute of Theoretical Physics of the Academy of Sciences since 1971, where he works closely with the physicists. During the period 1985–1996 Novikov served as President of the Moscow Mathematical Society, and during 1986–1990 he was also a Vice-President of the International Association in Mathematical Physics.

In the Gorbachev–Elzyn era, Novikov started to visit western countries more actively. Before that it was very difficult for Soviet scientists to attend scientific conferences abroad. In 1970 Novikov was not permitted to attend a ceremony at the International Congress of Mathematicians in Nice where he was awarded the Fields Medal for his work in topology because he had signed some letters defending dissidents. In 1991 he was able to work for the first time in his life for a half year in Paris at the Laboratory of Theoretical Physics, Ec. Norm. Superior de Paris. Beginning in 1992 he regularly worked at the University of Maryland at College Park as a Visiting Professor. In September 1996 he became a full time professor at the University of Maryland at College Park in the Department of Mathematics and the Institute for Physical Science and Technology (IPST). He also continues to work in Moscow for a period during the winter and summer.

Since 1971 his scientific work has played an important part in building a "bridge" between modern mathematics the theoretical physics. Some of Novikov's papers can be divided as follows:

Papers before 1971

Methods of calculating stable homotopy groups, Complex cobordism theory.

The classification of smooth simply-connected manifolds of dimension $n \geq 5$ with respect to diffeomorphisms. Topological Invariance of rational Pontryagyin classes, higher signatures.

The qualitative theory of foliations of codimension 1 on three-dimensional manifolds.

Papers after 1971

Methods of qualitative theory of dynamical systems in the theory of homogeneous cosmological models (of spatially homogeneous solutions to Einstein equations).

Periodic problems in the theory of solitons (non-linear waves) and in the spectral theory of linear operators, Riemann surfaces and Θ-functions in mathematical physics.

The Hamiltonian formalism of completely integrable systems, Hamiltonian hydrodynamic type systems, and applications of Riemannian geometry.

Ground states of a two-dimensional non-relativistic particle with spin $1/2$ in a doubly-periodic topologically non-trivial magnetic field. Topological invariants for generic operators; Laplace transformations and exactly solvable two-dimensional Schrödinger operators in magnetic fields, and a discrete analogue of this theory.

Multi-valued functional in mechanics and quantum field theory. Analogue of the Morse theory for the closed 1-forms; foliations given by the closed 1-forms; the special case of 3-torus; and applications in the quantum theory of normal metal, observable topological numbers.

Analogues of the Fourier–Laurent series on Riemann surfaces, Virasoro algebras, operator construction of string theory.

Novikov's main area of current scientific interests: Geometry, Topology and Mathematical Physics.

Awards and Honors

1966–1981	Corresponding member of the Academy of Sciences of the USSR
1967	Lenin Prize
1970	Fields Medal of the International Mathematical Union
1981	Lobachevskii International Prize of the Academy of Sciences of the USSR
1981	Full Member of the Academy of Sciences of the USSR
1987	An Honorary Member of the London Math. Society
1988	Honorary Member of the Serbian Academy of Art and Sciences
1988	Honorary Doctor of the University of Athens
1991	Foreign Member of the "Academia de Lincei", Italy
1992	Member of Academia Europea
1994	Foreign Member of the National Academy of Sciences of the USA
1996	Member of Pontifical Academy of Sciences (Vatican)

Students of Sergei Novikov

More than 30 of Novikov's students have been awarded the Candidate Degree (equivalent to Ph.D.), and of these V. M. Buchstaber, A. S. Mishchenko, O. I. Bogoyavalenskii, I. M. Krichever, B. A. Dubrovin, G. G. Kasparov, F. A. Bogomolov, S. P. Tsarev, I. A. Taimanov, A. P. Veselov, M. A. Brodskii, V. V. Vedenyapin, R. Nadiradze, V. L. Golo, S. M. Gusein–Zade have been awarded the degree of Doctor of Science (Scientific Degree, equivalent to the level of full professor in the former USSR and in Russia).

In addition to those mentioned above, other pupils of Novikov with the Candidate Degree (corresponding to Ph.D. level in the West) include I. A. Volodin, N. V. Panov, A. L. Brakhman, P. G. Grinevich, O. I. Mokhov, A. V. Zorich, F. A. Voronov, G. S. D. Grigoryan, A. S. Lyskova, E. Potemin, M. Pavlov, L. Alania, D. Millionshikov, V. Peresetski, I. Dynnikov, A. Maltsev, V. Sadov, Le Tu Thang, S. Piunikhin and A. Lazarev.

Reprinted from Symposium Current State and Prospects of Mathematics, Barcelona, June 1991

RÔLE OF INTEGRABLE MODELS IN THE DEVELOPMENT OF MATHEMATICS

by
SERGEI NOVIKOV

The history of mathematics and theoretical physics shows that the starting ideas of the best mathematical methods were discovered in the process of solving integrable models. Mathematical discoveries of the last twenty years will be especially discussed as by-products of the famous integrable systems of the soliton and quantum theories.

Starting as a Topologist

Before discussing certain models of mathematical physics and explaining their rôle in the development of modern mathematics, I want to say a few words about my own experiences as a mathematician. Let me start by reconstructing my career, not from the point when I was awarded the Fields Medal, but much earlier.

I started my mathematical life working in algebraic topology, and continued in this area for more than ten years; in fact, I still consider myself first as an algebraic topologist. When I started doing mathematics, in the mid-fifties, Russia was a very dark country, living behind the iron curtain. However, we had a very large and powerful mathematical school, whose leading person at that time was Andrei Kolmogorov in Moscow. He was the greatest mathematician, I think, after Poincaré, Hilbert and Hermann Weyl. A lot of famous mathematicians were his former students: Gel′fand, Arnol′d and many others (not me).

There was a common point of view in the Moscow mathematical school, concerning what was important and what was not important. The "important" areas of science were set theory, logic, functional analysis, and partial differential equations (not in the sense of solving models, but in the sense of proving rigorous theorems and establishing foundations). In Russian mathematics of that period — as in French mathematics — the main goal was to construct some kind of axiomatization, and the leading mathematicians were pursuing that. Topology was not existent in Russia in that period; there were only some remains of Pontrjagin's scientific school.

At the end of 1956, I was a second-level undergraduate student and had to choose one area — at least for some time — in order to be able to participate in seminars. I was attracted by an announcement posted in the Faculty of Mathematics and Mechanics of Moscow University. It was signed by Postnikov, Boltyanskiĭ, and Albert Shvarts. (The latter was a graduate student, but he was not considered as a "young mathematician"; in Russia, people aged 25 were not "young" in that period.) It was written in the announcement that there was a very new and exciting science, namely modern algebraic topology, opposite the nonsense of point-set topology (maybe my translation is not very exact). These people were punished after that announcement; especially Shvarts (Postnikov and Boltyanskiĭ were professors, so life was somehow easier for them). Paul Alexandrov, the famous topologist — who just continued a science which was thirty years old at that time — was terrified. So there was no place for Shvarts to continue his job in Moscow University, and he had to leave. Then he started learning about Fredholm operators; later, he moved to quantum physics and participated in the discovery of instantons, in cooperation with Polyakov. He was the first to discover nontrivial topological quantum field theory, ten years ago. In some sense, he followed the same way in science as me, but he was the most active during that period.

My friends were Arnol′d and Sinaĭ, who were children of Kolmogorov's seminar, and Anosov, a child of Pontrjagin's seminar in control theory. Some people asked me why I was trying to learn such a strange science, which was "completely useless", instead of studying important sciences like probability or partial differential equations. Thus topology was completely outside of the interest of our community in Russia. Postnikov told me that there were no prospects in topology, yet I could perhaps find something in cohomological algebra. Only Shvarts was enthusiastic about topology; however, he left Moscow very soon after he finished his thesis.

I published my first paper when I was 21. I was not "young" at that period because people like, for example, Arnol′d, wrote their first papers at age 18 or 19. This was completely normal. I come from a mathematical family, and my mother complained that "Everybody has published scientific papers, except my son".

I first worked in homotopy theory. Postnikov and Dyn′kin had made a very good translation of a collection of famous papers, mainly by French mathematicians: Serre, Cartan, Thom, ... We learnt them in our seminar. I was impressed by the excellent papers of the leading person in homotopy theory at that time: Frank Adams (who died recently). He started as an extremely brilliant scientist, solving famous problems.

This was a very interesting period. For example, nontrivial Hopf algebras — which are now very popular in the framework of quantum groups — were discovered during that time, shortly after the axiomatization of the work of Hopf by Armand Borel. The first persons who wrote papers about Hopf algebras were Adams and Milnor; before them, Hopf algebras were just cohomology rings of H-spaces and Lie groups. My first papers were dedicated to applications of Hopf algebras to the computation of homotopy groups of spheres and Thom complexes, which are important in cobordism theory. After that, I moved to the theory of differentiable manifolds, under the influence of several people who started visiting Russia at that period. John Milnor, Fritz Hirzebruch and Steve Smale helped me to start differential topology. Dynamical systems also started, in connection with topology and new kinds of algebra, after Milnor and Smale. When I write my memoirs, I will write something about all this, because I have found inaccuracies in many historical articles and books. For example, Smale's influence on the crucial point of the theory of dynamical systems in structural stability started in Russia. This relevant fact is missing in the historical literature, yet I was a direct witness of it.

Anosov and myself, together with Arnol′d, Sinaĭ, Shafarevich and Manin, organized a group of people who learnt different branches of mathematics from each other. Later Gel′fand's group joined us. People from partial differential equations started to interact with us after the discovery of the index of operators in the early sixties by Wolpert a strange person from Bielorussia who appeared in Moscow at that time. This was done before the Atiyah–Singer paper; in fact, Atiyah and Singer wrote their paper after the publicity that Gel′fand made of Wolpert's achievements.

Topology started to be recognized as something serious more or less after 1961. The main question that I wanted to answer was "For what are we working?". As I said, I had good connections. I consulted friends like Arnol′d and Anosov; as a result, I got involved with foliations, in connection with problems of dynamical systems which were originated by Smale. Other friends helped me in connection with index problems from Gel′fand's school, by teaching me about partial differential equations — I also wrote something in that area — and learning topology in their turn. People working in algebraic geometry were also extremely useful; they helped me in the use of certain algebraic concepts in topology. However, I found out very soon that, no matter how far I was moving into mathematics, I was not able to answer my basic question, concerning the goal of what we were doing. I found that the theory of partial differential equations was as abstract as topology, and probability even more (I never worked in probability, but my fried Sinaĭ explained this to me; he moved from that area to dynamical systems himself). Dynamical systems was a much more beautiful and newer area; however, it played no rôle in the real world either, because nobody knew enough about it; it was too hard for people working in natural sciences in that period.

The Split between Mathematics and Physics

Arnol′d taught me, in his seminar, analytical mechanics and elements of hydrodynamics, in the framework of classical mechanics (not of quantum mechanics). Indeed, in the mid-twenties, after the creation of quantum mechanics, there was a very serious split between mathematics and physics in Russia (and not only in Russia). The best mathematicians of Kolmogorov's period — with a very few exceptions — never knew even the mathematical language of theoretical physics. The new language of theoretical physics started to be constructed more or less in 1925. According to physicists of that period, the crucial point in the divergence between mathematics and physics was not the creation of relativity — the rôle of relativity was realized later — but the creation of quantum theory. In Moscow,

Gel′fand was probably the only one who learnt the new physics. Sometimes physicists participated in Gel′fand's seminar; however, in the late fifties Gel′fand stopped this job completely and physicists disappeared from that seminar for as long as twenty years.

I know a lot of childish tales from mathematicians about physicists, and from physicists about mathematicians, normally based on a lack of information. I have continually heard them for the last thirty years, even from great mathematicians or from the best physicists. I heard the latest yesterday: René Thom — one of my great teachers — spoke in his talk about the weakness of the "Landau theory of turbulence," which was pointed out to him by Arnol′d (who is also my friend, and with whom I have a lot of family connections). This was typically caused by a lack of information; one should not worry too much about these things happening. The story is that Landau never had any theory of turbulence. Landau had been interested in hydrodynamics since 1940, for twenty years at least, starting from his famous papers in superfluidity. For twenty-five years he was the only person who claimed that turbulence was purely a dynamical effect, a result of some global dynamics. It is a stupid idea, he said, to think that Navier–Stokes in nondeterministic: If one looks at a turbulent flow locally along time, one immediately observes that it is a very well defined flow; thus it is not reasonable to say that it is "something nondeterministic." Landau produced the idea that it is the result of something global, as a dynamical system. The weak point of his ideas was that nothing serious was known in the physics community about new complicated examples of dynamical systems (even in pure mathematics they appeared relatively late). He said that these were perhaps some complicated infinite-dimensional tori embedded in a functional space. This may be the origin of the tale. (I should add that I have no interest in criticizing bad mathematicians. The criticism is only interesting if it is addressed to a good scientist. I would be happy if there is a revenge with the same weapons.)

Arnol′d, who is in some sense my teacher in questions of mechanics, started coming to my lectures in the early sixties, when I was 23 years old (I remember that he was one of the three people who attended my lectures). He was shocked by the idea of *transversality* and *generic position*. Transversality was a completely new concept, even for people who were famous in the theory of functions of real variables in 1961. Thus Arnol′d also learned something from me, while I learnt mechanics from him.

He told me that Kolmogorov proposed him to improve the result which is now called the *Kolmogorov–Arnol′d–Moser theorem*. Kolmogorov found the basic ideas and invited Arnol′d to continue them and furnish a rigorous proof. Kolmogorov also asked him to learn mechanics. Thus he read a lot of books, starting from Appell and some Russian books written by people in classical mechanics; however, as he said, he could not understand what mechanics really was. Then he found the book of Landau and Lifshits (which was not yet famous at that time among the mathematical community). He told me that, after reading this book, he finally understood what mechanics was, and, after that, he understood how bad the book was. Arnol′d himself wrote a brilliant book on mathematical understanding of classical mechanics. I would honestly say that I do not like that book, because he completely reconstructed the ideology. The book of Landau and his school was just a starting point to develop a great science; it contained many initial points allowing further progress. In Arnol′d's reconstruction, the mathematics is, of course, much better — it is a very good book for pure mathematicians —, but starting points for future research areas are missing. People who read Arnol′d's book arrive at an endpoint.

Learning Physics

My friend Manin had the same views as me about those books. We both independently decided, at the same time, to start learning quantum physics (my brother* used to tell me that mathematicians should know everything about quantum theory). I first tried to learn quantum field theory as mathematicians normally do, and found out that this task was completely impossible. It might even be stupid to do so. Instead, I very much like the style of Einstein's lectures or the best lectures of Landau. I understood that *naturality* was the base of that science, exactly as I had earlier realized in topological books. The topologists of that period, like Jean–Pierre Serre, René Thom or John Milnor, sometimes omitted definitions in their lectures; they just said "This definition is natural." I recognized this style of "naturality" in the best physicists; in the lectures of Einstein, in the best books of Landau. (Not all books of Landau are equally good, but the collection of all of them is very valuable. Our students who want to do theoretical physics must know all these books at the age of 22. It is their common starting point.)

We discussed with Manin some paradoxes and unclear features of quantum theory, about which we shared a common point of view. I remember Manin telling me that every mathematician would find unclear points, but it would be a mistake to stop at those points and stand on criticism against them. Many mathematicians, including my students, have important difficulties in learning theoretical physics. They want to learn it as if it were mathematics: If they find something that they do not understand, they stop. I may definitely say that physicists also find a lot of nonunderstandable things; however, one must go ahead and think about such things only after having done a lot of exercises and reached a certain level.

It is very difficult to carry out Hilbert's program and to write theoretical physics in an axiomatic style. Hilbert did important work after Einstein's discovery of general relativity. He realized that the Einstein equation was an Euler–Lagrange equation for some functional. Thus he confirmed, in the case of the Einstein

*Leonid Keldysh, a leading quantum solid state physicist in Russia.

equation, that the axioms for any fundamental physics theory have to be started from a Lagrangian principle. Hilbert's program was useful for Hilbert himself, because he used it in that way. However, I do not like the experiences of some of my friends — extremely good mathematical physicists — who work in Hilbert's program, trying to make physics rigorous. This is, I think, impossible. One may prove a good theorem here, a good theorem there, about some physical situations. However, I think that Richard Feynman is completely right when he claims that it cannot be done globally (perhaps it is sometimes possible locally). The development of physics is more rapid than the flux of theorems which try to axiomatize it. The percentage of things which may be done rigorously is going to zero; the number of good theorems is increasing, but the ratio is going down very rapidly.

I spent at least five years, between 1965 and 1970, just learning physics. Sometimes, half of my working time was dedicated to learning it as a student, from the earliest books of Landau, Lifshits, Einstein and others. In 1970, I wanted to make contact with physicists and people from Landau's school (they worked in the newborn Landau Institute, which was created in 1965). There were increasing rumours among physicists — even among engineers — that something very interesting was being developed, namely algebraic topology (of course, things like dynamical systems were also "topology" for them). Isaac Khalatnikov, who was director of the Institute, informed me that people at Landau Institute had some very good problems in general relativity. They needed a topologist, so I joined them. We started working together at the end of 1970. At that time, I knew general relativity, which I had learnt earlier as a part of differential geometry. (But not in mathematical books; definitely, the best books on general relativity are not mathematical. It is better to read books by physicists, even in order to understand what is the best mathematics therein. Einstein's lectures are suitable to start with. Also the books of Landau–Lifshits, Misner, Thorne, are extremely good.)

First Contributions to the Domain of Physics

The first paper which I was able to write was a joint paper with my collaborator Bogoyavlenskiĭ. He was my student at that time. We worked together in general relativity and just applied our knowledge about dynamical systems. My acquaintance with people in dynamical systems, which started ten years before, became fruitful in that period.

In this situation, I worked in the theory of homogeneous cosmological models, studying the space of homogeneous solutions of the Einstein equation, which leads to complicated dynamical systems, especially near cosmological singularities. There is an extremely unusual feature in this area, even from the point of view of the theory of dynamical systems. It is completely concrete; the ideas of genericity cannot be applied.

Physicists normally make the following criticism about such remarkable papers as the one by Ruelle and Takens. It is very good to construct abstractly nice examples of complicated dynamical systems. (A lot of such results were given before Ruelle and Takens, starting from Marston Morse in the thirties. Smale and our people — like Anosov, Arnol′d and Sinaĭ — also found a lot of remarkable examples of this kind.) But when dealing with real Navier–Stokes equations or with real Einstein equations, the question is "Is this situation realizable or not?" I may definitely say that it has never been realized up to now in hydrodynamics.

Even remarkable specialists in ergodic theory, like Sinaĭ, were not able to understand, for ten years, how such Hamiltonian systems — like Einstein equations of homogeneous cosmological models — could lead to nontrivial ergodic properties, i.e., to nontrivial strange attractors. As far as I know, this is a unique example of a strange attractor which can be investigated analytically.

There is a particularly interesting geometry in the phase space describing the dynamical system of general relativity. In fact, some people from the Landau school had discovered these anomalous regimes before us; but nobody believed them, not even in the community of theoretical physicists. (I am not speaking about mathematics now; that job was totally done within the physics community.)

Bogoyavlenskiĭ and myself developed a certain technique of dynamical systems. As a special case of our technique, we were able to compare these strange anomalous regimes with strange attractors. They were computed analytically — not numerically — because of this very special strange geometry. But they were not *generic* attractors. In concrete systems, generic ideology does not work, because fundamental systems always have some important hidden symmetry.

We worked rigorously. This does not mean that we proved any theorem rigorously describing attractors. Rather, our skeleton was constructed in such a way that there were points with zero Lyapunov eigenvalues. We worked rigorously in the sense that our computations were done without arithmetical mistakes, in the process of solving those models.

At that time, my former student Bogoyavlenskiĭ presented his second dissertation in order to become full professor. It was the continuation of those methods on dynamical systems (later, he published a book in English about qualitative methods in gas dynamics and general relativity). The famous physicist, Zel′dovich, was the person who evaluated Bogoyavlenskiĭ's papers. He said that remarkable results had been obtained, not only in relativity, but also in gas dynamics. Moreover, after the official speech, Zel′dovich expressed to me unofficially how much he liked that technique, and how complicated it is for astrophysicists: nobody will use it for a long period! I have to say that, at that time, even the elementary Poincaré two-dimensional plane qualitative theory was considered as a high-level theory in the physics community. Only the best people were able to use it. The level of nonlinear science in the community of the best theoretical physicists was very low.

Modern Developments in Topology

My own work at the Landau Institute has been divided into two parts. One part was dedicated mainly to topology (I was paid to work as a topologist, and people just consulted me about "modern mathematics"). A lot of topology appeared in this period, also related to instantons. I remember my fried Polyakov — who is six years younger than me — visiting me and asking "Did you hear anything about characteristic classes?" I told him that this was a very trivial concept from differential geometry. After I displayed the elementary formulae for dimension 4 (using tensor language), he said "It is quadratic!" Next day, he told me that he had found what had become the famous self-duality equation. Characteristic classes were quadratic, so it was possible to combine them with the Yang–Mills functional in that way!

I also interacted as a consultant with other people, leading to a famous discovery about the rôle of homotopy theory in liquid crystals and in anomalous kinds of superconductors in the early seventies. Fifty years before that, people had made interesting observations in optical experiments, like singularities in liquid crystals such as cholesterine (this is something that you can buy in a pharmacy and freeze up to -150 degrees in order to observe phases which display singularities). Nobody was able to realize before the early seventies that this phenomenon can be, in fact, explained by elementary homotopy theory. There were two groups working independently in that direction around 1974: one was in France; the other was at the Landau Institute.

I was not able to find anything substantially new and interesting in topology for myself until 1980. Before that, topology worked as it had been built. The new era began when topology started to apply discoveries of physicists inside topology itself. There was a huge noise in the physics community about topology during the seventies; also among applied physicsts, working in low-temperatures physics, liquid crystals, and related fields. It was a common opinion of physicists that the main new thing which physics borrowed from mathematics in the last ten years was topology. (You may find this assertion, for example, in an article written in the early 80's by Anderson, the famous applied physicist in superconductivity, and Nobel Prize winner). A lot of new things indeed appeared during the eighties, as I will next explain.

Thus I started producing new work in topology in the past decade only. Before that I sold my knowledge about topology to physicists. In fact, dynamical systems was more my own way of solving models. The new direction in my work started with the discovery of soliton theory, which was done in the late sixties. It was an extremely interesting finding, which led, among other things, to modern integrable systems, conformal field theory, and quantum group theory (as a late by-product).

Integrable Models in Classical Mathematics and Mechanics

Everyone knows the rôle of the famous two-body problem, solved by Newton, in the development of the mathematical methods of physics. For a long period after that, people used the method of the exact analytic solution for some differential equations as a principal tool in mathematical physics. They simplified their problem if it was (or looked) too difficult, and after that tried to find the exact solution.

A lot of work has been done in the process of searching for special "integrable cases" of famous problems, like the motion of the top, for example. All mathematical methods — like power and trigonometric series, Fourier–Laplace (and other) integral transformations, complex analysis and symmetry arguments — were discovered and developed for that in the nineteenth century. These methods led sometimes to remarkable negative results, i.e., to proofs that certain models are not solvable in principle.

Some strange integrable cases which do not admit any obvious symmetry were discovered in the nineteenth century: the integrability of geodesics on 2-dimensional ellipsoids in Euclidean 3-space (Jacobi), the motion of the top with special parameters for constant gravity (Kovalevskaya), and some others. Riemann surfaces and θ-functions of genus 2 played the leading rôle in their integrability. What kind of hidden symmetry can be found behind this? It was not completely clear until the discoveries of soliton theory.

We have to add to these discoveries a new understanding of the deep hidden algebraic symmetry in the two-body problem, based on the so-called Laplace–Runge–Lentz vector of integrals. The "hidden" group generated by it acts on the energy levels in the phase space; this group is isomorphic to SO(4) for the negative energy levels corresponding to the closed elliptic orbits, and isomorphic to SO(3, 1) for the positive levels corresponding to the hyperbolic noncompact orbits.

Generic spherically symmetric forces lead to the existence of nonperiodic orbits in an arbitrarily small neighborhood of any periodic orbit in the phase space (which is 6-dimensional for a particle in 3-space). The Kepler two-body problem is exceptional; all orbits are closed for any negative energy. As a consequence, any small spherically symmetric perturbation of the gravity forces leads to the famous displacement of the perihelium. This displacement was extremely important for the astronomical testing of celestial mechanics and general relativity. The group SO(3) is not enough for the periodicity of orbits! A very large hidden symmetry is needed here.

We shall discuss later the fundamental rôle of the same symmetry for applications of quantum mechanics to the structure of atoms, discovered in the 1920's by Pauli.

Integrable Models in Quantum and Statistical Mechanics (1925–1965)

The integrable models of classical mechanics were forgotten by almost everyone except a few very narrow experts in these very classical problems. A new era of qualitative methods started with Poincaré. But the very new important branches of modern physics, like special and general relativity and quantum mechanics, were discovered in the first quarter of the twentieth century. Some initial problems for the Einstein and Schrödinger equations were also explicitly solved in the classical style.

For two particles with opposite charges, the nonrelativistic electric force is mathematically the same as the gravity force. The exact solution of this quantum nonrelativistic "two-body" problem preserves the same type of hidden symmetry as in classical mechanics; there is the quantized Laplace–Runge–Lentz vector commuting with the Hamiltonian (energy operator) and generating the Lie algebra of the group SO(3, 1) for positive energy and SO(4) for negative energy. As a consequence, the spectrum of this quantum system (the Balmer spectrum) is more degenerate than the spherical symmetry requires (the group SO(3)). The stationary localized states have negative energy and can be considered as the quantized periodic orbits for the classical Kepler-type problem for electric forces. Together with the famous Pauli principle (two electrons cannot occupy the same state in any family of all possible states which present the orthogonal basis), this spectrum leads to an approximate explanation of the Mendeleev classification of elements.

One may observe that hidden symmetry of this sort is an exceptional property of the τ^{-2}-type forces (and also of the linear forces) but these exceptional cases appear in the most fundamental problems of classical and quantum theory as a first good approximation. There is a common belief in the community of theoretical physicists that *the most fundamental mathematical principles of physics should be described (at least in some good first approximation) by objects containing enormously large hidden symmetry.*

It is very difficult to classify the wide collection of concrete problems solved by physicists in the process of developing quantum theory between 1925 and 1965. We want to point out the well-known results of Bethe and Onsager, who solved some multiparticle one-dimensional quantum models and the models of statistical mechanics. The famous *Bethe Ansatz* for the construction of eigenfunctions was discovered and the two-dimensional Ising model was solved. But the influence of these discoveries on the mathematical methods of physics was fully recognized only much later, after the discovery of soliton theory in the process of quantization of its methods.

Soliton Theory

In the early sixties, people working in the theory of plasma observed that the Korteweg–de Vries (KdV) equation appears as the universal first approximation for the propagation of waves in many nonlinear media, combining nonlinearity and dispersion (if viscosity can be neglected). Before that, KdV was known for many years only as a very special system for waves in shallow water. People started to investigate the KdV equation in the sixties, and the most important discovery was made in the papers of Kruskal–Zabusky (1965), Gardner–Greene–Kruskal–Miura (1967), Lax (1968). This highly nontrivial system is in a sense exactly integrable by a very strange and new procedure (*inverse scattering transform*).

Let us be more precise now. The KdV system has a form

$$u_t = 6uu_x - u_{xxx}.$$

Its simplest solutions, known since the nineteenth century, are the *solitons*

$$u(x - vt) = -\frac{3a}{\operatorname{ch}^2 (3a)^{\frac{1}{2}} (x + 12\,at)}$$

and the *knoidal waves*

$$u(x - vt) = 2\wp(x - vt) + \text{constant}.$$

The soliton is localized and the knoidal wave is periodic in x. Here $\wp$ is the doubly-periodic Weierstrass elliptic function, whose degeneration is exactly the soliton.

Consider the Sturm–Liouville operator and the third-order operator

$$L = -\partial_x^2 + u(x,t), \quad A = -3\partial_x^3 + 4u\partial_x + 2u_x.$$

Their commutator

$$[L, A] = 6uu_x - u_{xxx} = Q_1$$

is multiplication by the function Q_1. It means that the *Lax equation*

$$\frac{\partial L}{\partial t} = [L, A]$$

is formally equivalent to the nonlinear KdV equation.

Inverse Scattering Transform

The famous GGKM procedure may be immediately deduced from the Lax equation. Suppose all functions $u(x,t)$ are *rapidly decreasing* for $x \to \pm\infty$. Consider the special solutions of the linear Sturm–Liouville equation with exponential asymptotics

$$L\psi = \lambda\psi, \quad L\varphi = \lambda\varphi, \quad k^2 = \lambda,$$
$$\psi_\pm \to e^{\pm ikx}, \quad x \to -\infty,$$
$$\varphi_\pm \to e^{\pm ikx}, \quad x \to +\infty.$$

There is a unimodular transformation from the basis ψ to φ:

$$\varphi_+ = a\psi_+ + b\psi_-$$
$$\varphi_- = \bar{b}\psi_+ + \bar{a}\psi_-$$
$$|a|^2 - |b|^2 = 1.$$

The coordinates $a(k)$, $b(k)$ determine the so-called *scattering data* for the Schrödinger operator L with localized potential. The potential may be completely reconstructed by the *inverse scattering transform* from the function $[a(k),\ b(k)]$ with the proper analytical properties, plus a finite number of "discrete data," as was known long ago in the fifties.

The GGKM theorem states that:

(a) $\dfrac{da(k)}{dt} = 0, \quad \dfrac{db(k)}{dt} = (ik)^3 b(k).$

(b) The discrete eigenvalues are the integrals of motion.

(c) Local densities of the integrals of motion for KdV may be constructed in the following way. Consider the formal solution for the Riccati equation

$$\chi_x + \chi^2 = u - \lambda, \quad \chi = k + \sum_{n\geq 1} \frac{P_n(u, u_x, \ldots)}{(2k)^n}.$$

Then the integrals

$$I_m = \int P_{2m+3}\, dx$$

are the local conservative quantities for KdV.

(d) Exact (multisoliton) solutions will be obtained from the *reflexionless potentials* $b(k) \equiv 0$.

KdV Hierarchy. Hamiltonian Properties. Generalization

There is an infinite number of operators

$$A_0 = \partial_x\,, \quad A_1 = A, \ldots A_n = \partial_x^{2n+1} + \cdots, \ldots$$

such that the commutator $[L, A_n]$ is multiplication by a certain polynomial

$$Q_n(u, u_x, u_{xx}, \ldots, u_{2n+1}).$$

The *KdV hierarchy* is the collection of nonlinear systems

$$\frac{\partial u}{\partial t_n} = Q_n,$$

equivalent to the collection of Lax-type equations

$$\frac{\partial L}{\partial t_n} = [L, A_n].$$

All these flows commute with each other; we may find a common solution

$$u(x, t_1, t_2, t_3, \ldots), \quad t_0 = x.$$

The detailed study of the polynomials Q_n was done in the late sixties by Gardner. In particular they have the following form (*Gardner form*)

$$Q_n = \frac{\partial}{\partial x}\left(\frac{\delta I_{n+1}}{\delta u(x)}\right)$$

$$I_{-1} = \int u\,dx, \quad I_0 = \int u^2\,dx, \quad I_1 = \int\left(\frac{u_x^2}{2} + u^3\right)dx, \ldots$$

As observed by Gardner, Zakharov and Faddeev (1971), this is the form of the Hamiltonian equation corresponding to the GZF–Poisson bracket

$$\{u(x), u(y)\} = \delta'(x - y)$$

and to the Hamiltonian

$$H_n = I_{n+1}.$$

The inverse scattering transform may be treated as a functional analog of the transformation from $u(x)$ to the action-angle variables as in analytical mechanics, which were useful for the semiclassical quantization (Zakharov and Faddeev).

The quantization program was started after 1975 by Faddeev, Takhtadzhyan, Sklyanin and others in Leningrad. Different groups starting from 1971 discovered many new interesting nonlinear systems integrable by the Lax equation and inverse scattering transform. Some famous systems — well-known before — were solved by that method (for example, nonlinear Schrödinger, sine–Gordon, discrete Volterra system or discrete KdV, Toda lattice and many others, including some special two-dimensional systems which proved to be very important later).

Starting from 1976, different groups found other interesting properties and generalizations of the Hamiltonian formalism of KdV theory. Interesting new Poisson brackets with large hidden algebraic symmetry were discovered; for example, the Lenart–Magri second bracket for KdV, and the Gel′fand–Dikiĭ brackets for the generalizations of KdV hierarchy to the scalar operators of higher order.

The Periodic and Quasiperiodic Problems. Riemann Surface and θ-Functions

The appropriate approach to solving the periodic problem for KdV was found by the present author (1974). After that, it was completely solved by Novikov and Dubrovin (1974), Lax (1975), It.s and Matveev (1975), Mckean and Van Moerbeke (1975) and by Krichever (1976) for the (2 + 1)-system of KP-type (Kadomtsev–Petviashvili). The basis of this approach is the KdV-invariant class of "finite-gap" Schrödinger operators with periodic and quasiperiodic potentials, whose spectrum on the line **R** has a finite number of gaps only. These potentials satisfy the stationary KdV and higher KdV equations

$$\left[L, \sum_{j=0}^{N} c_j A_j\right] = 0,$$

where N is the number of finite gaps.

This system is a completely integrable finite-dimensional Hamiltonian system. Its solution was found explicitly in terms of the θ-functions of a Riemann surface whose genus is equal to N:

$$u(x,t) = \text{constant} - 2\partial_x^2 \log \theta\,(U_x + V_y + U_0),$$

where U, V are certain N-vectors.

Finite-gap potentials form a dense family in the space of continous periodic functions (Marchenko–Ostrovskiĭ, 1977) and even in the space of quasiperiodic functions. For the periodic potential $u(x+T) = u(x)$ the Schrödinger operator L commutes with the shift $\hat{T} : x \to x+T$. Therefore, there is a common eigenfunction for the operators L and $\hat{T}$ (which is familiar in solid state physics)

$$L\Psi = \lambda\Psi$$

$$\hat{T}\Psi = e^{ipT}\Psi.$$

Here $p = p(\lambda)$ is some multivalued function of λ and the spectrum is exactly the set of all λ such that $p(\lambda) \in \mathbf{R}$.

There are exactly two Bloch eigenfunctions $\Psi_\pm$ for each complex value of λ (except for a countable number of branching points). The function Ψ is meromorphic on some hyperelliptic Riemann surface Γ (two-covering of the λ-plane), generically of infinite genus. For the real potential $u(x)$ on the line **R** the branching points are real; they exactly coincide with the endpoints of the spectrum. This is the *spectral curve* Γ.

The generic real nonsingular solution of the commutativity equation

$$\left[L, \sum_{j=0}^{N} c_j A_j\right] = 0$$

is quasiperiodic (it contains a dense family of periodic solutions with different periods). All of them have the spectral curve Γ of finite genus N and vice-versa. The potential $u(x)$ and the eigenfunction Ψ may be expressed in terms of θ-functions corresponding to the surface Γ. There was a very interesting formal algebraic investigation of the commuting ordinary differential operators in the twenties (Burchnall and Chaundy); a Riemann surface was discovered based on the polynomial relation between L and A, which they found for any commuting pair, with no formal connection between periodicity and our surface. There is a theorem stating that, for operators of relatively prime orders with periodic coefficients, these two surfaces in fact coincide. It may be not so in more complicated cases.

The general elementary idea between the appearance of Riemann surfaces in the theory of finite-dimensional integrable systems can now be explained. It is very useful to relate the original Lax representation for KdV-type systems to the compatibility condition of the two linear 2×2 systems whose coefficients depend on λ (*zero curvature equation*):

(a) LAX EQUATION (1968)

$$\frac{\partial L}{\partial t} = [L, A].$$

(b) ZERO CURVATURE EQUATION (1974, for KdV hierarchy and sine–Gordon)

$$\frac{\partial \Psi}{\partial t} = \Lambda(\lambda)\Psi$$

$$\frac{\partial \Psi}{\partial x} = Q(\lambda)\Psi$$

$$\frac{\partial \Lambda}{\partial x} - \frac{\partial Q}{\partial t} = [\Lambda, Q].$$

For the ordinary KdV we have:

$$Q = \begin{pmatrix} 0 & 1 \\ u - \lambda & 0 \end{pmatrix}$$

$$\Lambda_1 = \Lambda = \begin{pmatrix} -u_x & 2u + 4\lambda \\ -4\lambda^2 + 2\lambda u - u_{xx} + 2u^2 & u_x \end{pmatrix}.$$

For higher KdV we have the same Q; the corresponding matrices Λ_n are polynomial in λ. For the stationary equation ($\partial_t = 0$) we have " Lax-type" representations for the finite-dimensional system

$$\frac{\partial \Lambda}{\partial x} = [Q, \Lambda].$$

The Riemann surface Γ appears here. It has genus one for the ordinary stationary KdV, and the Weierstrass elliptic function appears:

$$\det(\Lambda - \mu I) = P(\lambda, \mu) = 0.$$

The function $\psi(x, P)$ is meromorphic on the surface Γ. Its coefficients are the integrals of the system (they do not depend on x),

For KdV and higher KdV, all matrices Λ_n have zero trace and the Riemann surfaces are hyperelliptic:

$$\mu^2 = R_{2n+1}(\lambda),$$

where n is the genus of Γ.

The function ψ was explicitly found from the "algebro-geometrical data": The Riemann surface Γ and the poles of Ψ (Ψ has exactly n poles, which do not depend on x after the proper normalization). This approach works for all known nontrivial integrable systems. As is now know, this mechanism is valid also for the classical systems mentioned above (Jacobi, Clebsch, Kovalevskaya, ...). An important discovery was made by Krichever (1976), who improved this approach and found the generalizations to the (2 + 1)-systems like KP. *All Riemann surfaces appear in the theory of KP.* This last property is very important for applications of these ideas to different problems (like Schottky-type problems in the theory of θ-functions, analogs of Fourier–Laurent bases on Riemann surfaces — which are useful in string theory —, the formalism of τ-functions, etc). *The classical functional constructions were greatly extended after the periodic theory of solitons.*

Conclusion

To conclude this long discussion, we now present the collection of the different branches of mathematics and theoretical physics which were involved in the integrable soliton models. Important new connections between them were discovered.

(1) Nonlinear waves in continuum media (including plasma and nonlinear optics).
(2) Quantum theory; scattering theory and periodic crystals.
(3) Hamiltonian dynamics.
(4) Algebraic geometry of Riemann surfaces and Abelian varieties (θ-functions).

Quantization of the soliton methods led to the discovery of the *quantum inverse transform*, started by Faddeev, Sklyanin, Takhtadzhyan, Zamolodchikov, Belavin and others. An important new object, the so-called *Yang–Baxter equation*, started to play the leading rôle. Later (Sklyanin, Drinfel′d, Jimbo) Hopf algebras appeared here with the special *universal R-matrix* of Yang–Baxter, which led to the now very popular *quantum groups*.

Recall that from the self-dual Yang–Mills equation (Polyakov and others) the theory of instantons appeared, with remarkable applications in four-dimensional topology (Donaldson and other people in Atiyah's school). The theory of Yang–Baxter equations led to the Jones polynomials in the theory of knots. The famous 2-dimensional conformal field theories also gave us the remarkable collection of integrable models which are also the most beautiful new algebraic objects. They have deep connections with soliton theory and quantum groups.

As a result, I may formulate the following thesis: *A large part of the most important discoveries in mathematics and in the mathematical methods of physics was done in the process of developing the theory of integrable models.*

*Sergei Novikov**
Landau Institute for Theoretical Physics
Academy of Sciences of the USSR
GSP-1 117940 Kosygina Str. 2
Moscow V-334
Russia

Prepared from the author's text and the videotape of the talk, by Regina Martinez.

*Univ. of Maryland at College Park, IPST and Department of Math, College Park, MD, 20742-2431, USA, e-mail: novikov@ipst.umd.edu tel:(301) 4054836(0).

Proceedings of the International Congress of Mathematicians
Vancouver, 1974

THE WORK OF ENRICO BOMBIERI

by
K. CHANDRASEKHARAN

Bombieri's work ranges over many fields: number theory, univalent functions, several complex variables, partial differential equations, algebraic geometry. I do not seek to describe it all. I shall not touch upon his work in algebraic geometry, nor shall I anticipate his article in these PROCEEDINGS on partial differential equations. I shall speak only about three of his contributions. They should give some idea of the variety and depth of his work.

1. The distribution of primes. First among Bombieri's achievements is his remarkable theorem on the distribution of primes in arithmetical progressions, which he obtained by an application of the method of the *large sieve* (Mathematika **12** (1965), 201–225).

The prime number theorem for the arithmetical progression $a + mq$, where a and q are integers, $q > 0$, $(a, q) = 1$, $0 \leqq a < q$, and $m = 1, 2, \ldots$, is equivalent to the assertion that

$$\psi(x; q, a) \sim x/\varphi(q),$$

as $x \to \infty$, where φ stands for Euler's function, and

$$\psi(x; q, a) = \sum_{n \leqq x:\ n \equiv a (\mathrm{mod} q)} \Lambda(n),$$

where $\Lambda(n) = \log p$ if is n a power of a prime p, and $\Lambda(n) = 0$ otherwise.

Bombieri's theorem is concerned with an estimate of the error term $E(x; q, a) = \psi(x; q, a) - x/\varphi(q)$, *not* for an individual q, but *on the average* over q, up to a certain bound. It states that given a positive constant A, there exists a positive constant $B = B(A)$, such that

$$\sum_{q \leqq Q} \max_{y \leqq x} \max_{a, (a,q)=1} |E(y; q, a)| \ll x(\log x)^{-A}, \tag{1}$$

if $Q = x^{1/2}(\log x)^{-B}$.

A slightly weaker result, which is less widely applicable, was obtained independently by A. I. Vinogradov (also in 1965) by a different method.

The significance of Bombieri's theorem becomes clear, if we note that, for any *fixed* q (that is, fixed relative to x), the best result so far known is that

$$E(x; q, a) = O(x \exp(-c(\log x)^{\theta})),$$

with $c > 0$, $\frac{1}{2} \leqq \theta < 1$. If q is a function of x, the main term $x/\varphi(q)$ in the asymptotic formula for ψ decreases as q increases. Therefore estimates uniform in q are required. But an estimate which is *uniform* in q requires a strong restriction on the range of q (in the present state of knowledge). Such an estimate was first deduced by Arnold Walfisz (1936) from a theorem of C. L. Siegel (1935) on the location of the real zeros of Dirichlet's L-functions with real, nonprincipal, characters. It is as follows:

$$E(x; q, a) = O(x \exp(-c_0(\log x)^{1/2}))$$

where c_0 is a positive constant, *uniformly* for $q \leqq \log^{\alpha} x$, where α is a given positive number however large.

If, on the other hand, one *assumes* the "extended Riemann hypothesis", that not only the Riemann zeta-function but all the L-functions, modulo q, of Dirichlet, have all their zeros in the critical strip *on* the critical line, one would get the much stronger estimate: $E(x; q, a) = O(x^{1/2} \log^2 x)$, *uniformly* for $q \leqq x$. This would, if used on the left-hand side of (1), give a result comparable to Bombieri's, with $B = A + 2$, but even this, it is to be noted, is *not* significant if q *exceeds* $x^{1/2}$.

Bombieri's theorem may therefore, and *does* sometimes, serve as a substitute for the assumption of the extended Riemann hypothesis, which has far-reaching implications in number theory. His proof is as remarkable as his result. To explain it one might perhaps cast a glance backwards.

A *sieve*, in simple terms, is a combination of (i) a finite sequence $\mathscr{N}$ of integers, (ii) a finite set of distinct primes $\mathscr{P}$, and (iii) corresponding to each prime $p \in \mathscr{P}$, a subset Ω_p of residue classes modulo p. If one *sieves out*, or deletes, from the given sequence $\mathscr{N}$, all those integers whose residue class modulo p belongs to Ω_p for some $p \in \mathscr{P}$, the problem is to estimate, from above and from below, the number of integers left over in $\mathscr{N}$ after the *sieving* (or deletion).

A sieve is called *large* or *small*, according as $|\Omega_p|$, the number of residue classes in Ω_p, is, on the average, large or small.

If we take $\mathscr{N}$ to be the sequence of consecutive integers $1, 2, \ldots, N$; $\mathscr{P}$ to be the set of *all* primes $p \leqq N^{1/2}$; and Ω_p to consist of the single residue class 0 (mod p) for each $p \leqq N^{1/2}$, we get the (ancient) sieve of Eratosthenes, which is obviously a *small* sieve. The elements left over in $\mathscr{N}$ after the sieving are the integer 1, together with all primes p, such that $N^{1/2} < p \leqq N$.

Viggo Brun was the first to introduce, in 1920, an ingenious sieve method to prove that every sufficiently large even integer is a sum of two integers, each of which has not more than nine prime factors. Improvements of Brun's method were made in later years by H. Rademacher, T. Estermann, G. Ricci, and A. A. Buchstab; until Atle Selberg, during the years 1946–1951, developed a sieve method more general and more powerful than Brun's and its improved versions.

We are here concerned, however, with the method of the *large sieve*, which is different from the small sieves of Brun and of Selberg, and which, when *combined* with analytical arguments, yields results that are beyond the reach of the other sieves.

The idea of the *large sieve* originated with Yu. V. Linnik in 1941, in his attempt to tackle I. M. Vinogradov's hypothesis (which is yet to be proved or disproved) on $h_2(p)$, the least quadratic nonresidue modulo p. The hypothesis is that given $\varepsilon > 0$, there exists a constant $c = c(\varepsilon)$ such that $h_2(p) < cp^{\varepsilon}$. Linnik sought to estimate the number of primes $p \leqq x$, say, for which $h_2(p) > p^{\varepsilon}$ for any given $\varepsilon > 0$.

Let $n_1, n_2, \ldots, n_Z$ be Z integers, such that $M + 1 \leqq n_1 < n_2 < \cdots < n_Z \leqq M + N$. Let the prime p be called *exceptional*, if the number of residue classes *not* represented by the numbers $(n_j), j = 1, 2, \ldots, Z$, *greater than* τp, where τ is a fixed number such that $0 < \tau < 1$. Linnik proved that for *any* such sequence (n_j), the number of exceptional primes $p \leqq N^{1/2}$ does not exceed $c_1 N/\tau^2 Z$, where c_1 is an *absolute* constant. As an application, he proved the striking theorem that the number of primes $p \leqq N$ for which the least quadratic nonresidue is greater than N^{ε}, for a fixed number $\varepsilon > 0$, is bounded. It follows that the number of primes $p \leqq X$ for which the least quadratic nonresidue is greater than p^{ε} is $\ll \log\log X$.

In the context of the definition of a sieve, the sequence $(n_j), j = 1, 2, \ldots, Z$, may be looked upon as the sequence of elements *left over* in the interval $[M+1, M+N]$, *after* a sieving has been effected (*on* the sequence of all integers in that interval, for example), with a *sieving set* $\{\Omega_p\}$ of residue classes modulo $p, p \leqq N^{1/2}$, which has the property that for each *exceptional* $p \leqq N^{1/2}$, the corresponding Ω_p has *more than* τp elements. Hence the name large sieve.

The next important step was taken by A. Rényi. If $Z(p, a)$ denotes the number of elements in the given sequence (n_j) such that $n_j = a \pmod p$, Linnik's result takes the form: The number of primes $p \leqq N^{1/2}$ such that $Z(p, a) = 0$ for *at least* τp values of a, where $0 < \tau < 1$, does *not exceed*

$$c_1 N/\tau^2 Z.$$

Rényi considered instead the sum

$$S_X = \sum_{p \leqq X} p \sum_{\alpha=0}^{p-1} (Z(p, a) - Z/p)^2,$$

and proved in 1950 that

$$S_x \leqq 2NZ, \quad \text{for } X = (N/12)^{1/3}. \tag{2}$$

Again, in the context of the definition of a sieve, we have $Z(p, a) = 0$ if $a \in \Omega_p$, so that

$$S_X \geqq \sum_{p \leqq X} \frac{|\Omega_p|}{p} Z^2,$$

which, when combined with (2), gives an upper bound for Z–and also Linnik's result *provided that* $X = N^{1/2}$.

As an application of his inequality, Rényi proved the striking theorem that every sufficiently large *even* integer is the sum of a prime and an almost prime (that is, an integer which is the product of a bounded number of prime factors).

Though Rényi's inequality yields more precise information than Linnik's result for the range of primes $p \ll N^{1/3}$, it does not work for the wider range $p \ll N^{1/2}$ of Linnik, which is more appropriate in the context of arithmetical applications. This defect was sought to be repaired by many mathematicians. It was not until 1965, however, that important further progress was made by K. F. Roth (Mathematika **12** (1965), 1–9) and, independently, by Bombieri (Mathematika **12** (1965), 201–225). Roth proved that Rényi's inequality (2) holds for $X = (N/\log N)^{1/2}$, and Bombieri that it holds for $X = N^{1/2}$ (with $\ll$ in place of $\leqq$).

Bombieri proceeded to place Rényi's inequality in a more general setting and proved, by a simple and ingenious argument, an inequality for trigonometrical double sums, which is as follows: Let $x_1, x_2, \ldots, x_R$ be real numbers which are δ-well-spaced, in the sense that $||x_k - x_l|| \geqq \delta > 0$ for $k \neq l$ (where $||\theta||$, for any real θ, denotes the distance of θ from the nearest integer). Let $T(x) = \sum_{n=M+1}^{M+N} a_n e^{2\pi i n x}$, where the (a_n) are complex numbers. Then

$$\sum_{k=1}^{R} |T(x_k)|^2 \leqq \left(N + \frac{2}{\delta}\right) \sum_{n=M+1}^{M+N} |a_n|^2 \tag{3}$$

(Acta Arith. **18** (1971), 401–404; Proc. Internat. Conf. Number Theory, Moscow, 1971). This corresponds, as Bombieri has shown, to something like Bessel's inequality in a Hilbert space.

If we take x_k to be rational, $x_k = a/q$, say, where $(a,q) = 1$, $q \leqq Q$, with $a_n = 1$ for $n = n_j$ and $a_n = 0$ for $n \neq n_j$, we get (more than) Rényi's inequality (2) in case q is a prime, and something similar to the inequality given by Selberg's upper-bound sieve, in case q is composite.

Thus many results previously obtained by Selberg's method can now be proved by using (3).

Bombieri then considered the analogue of his *large-sieve inequality* (3) for sums of Dirichlet characters $\mathcal{X}$ modulo q instead of trigonometrical sums. The connecting link is the Gaussian sum

$$G(\mathcal{X}) = \sum_{a=1}^{q} \mathcal{X}(a) \exp(2\pi i a/q),$$

since

$$\begin{aligned} \sum_{\mathcal{X}} |G(\mathcal{X})|^2 \mathcal{X}(m)\overline{\mathcal{X}(n)} &= \varphi(q) S_{m-n,q}, \quad &&\text{if } (mn,q) = 0, \\ &= 0, &&\text{if } (mn,q) > 1, \end{aligned}$$

where

$$S_{m,q} = \sum_{a=1:(a,q)=1}^{q} \exp(2\pi i a m/q)$$

is the well-known Ramanujan sum (*not* to be confused with S_X in (2)).

Vital for Bombieri's proof of his theorem on arithmetical progressions is the following inequality: Let Q be *any* finite set of positive integers, (a_n) *any* complex numbers. Then

$$\sum_{q\in Q}\frac{1}{\varphi(q)}\sum_{\mathcal{X}}|G(\mathcal{X})|^2.\left|\sum_{X<n\leqq Y}\mathcal{X}(n)a_n\right|^2$$

$$\leqq 7D\max(Y-X,M^2)\sum_{X<n\leqq Y} d(n)a_n|^2. \tag{4}$$

Here $\sum_{\mathcal{X}}$ denotes summation over *all* characters χ modulo $q, d(n)$ denotes the divisor function, $D=D(q)=\max_{q\in Q} d(q)$, $M=M(Q)=\max_{q\in Q} q$.

By skilful and *repeated* application of this inequality, with different choices of X, Y, and a_n, Bombieri deduced a *new* type of *density theorem* for the zeros of L-functions. The theorem gives an estimate for the sum

$$\sum_{q\in Q}\frac{1}{\varphi(q)}\sum_{\mathcal{X}}|G(\mathcal{X})|^2 N(\alpha,T;\mathcal{X}),$$

which is *uniform* with respect to Q, for $\frac{1}{2}\leqq\alpha\leqq 1$, $T\geqq 2$. Here $N(\alpha,T;\mathcal{X})$ denotes the number of zeros of Dirichlet's function $L(s,\mathcal{X})$ in the rectangle $\alpha\leqq \operatorname{Re} s\leqq 1$, $\frac{1}{2}\leqq\alpha\leqq 1$, $|\operatorname{Im} s|\leqq T$, in the complex s-plane.

From his density theorem Bombieri deduced his theorem on primes in arithmetical progressions, by an appeal to classical arguments in the theory of L-functions, combined with an application of the Siegel-Walfisz theorem (stated at the beginning).

Bombieri's work has given rise to a general method for treating problems that were previously solved either on the assumption of the extended Riemann hypothesis, or by Linnik's 'dispersion method', or by highly complicated, ad hoc methods. It thus furnishes a new approach to such important results as I. M. Vinogradov's theorem (1937) that every sufficiently large *odd* integer is a sum of three primes, or Linnik's theorem (1961) that *every* sufficiently large integer is a sum of a prime and two squares, or Chen's result (1967) that every sufficiently large *even* integer is a sum of a prime and an integer with at most two prime factors. Bombieri's theorem represents a deep synthesis of the most important modern methods in prime number theory. It has not put an end to any one question; rather it has led to many new ones.

His inequality for sums of Dirichlet characters has been extended to general multiplicative characters of the form $\mathcal{X}(n)n^{it}$ which are "δ-well-spaced". In consequence, the *best* bounds *so far* known have been obtained for $N(\alpha,T;\mathcal{X})$, yielding as special cases such results as the following: The difference between the consecutive primes p_{n+1}, p_n has the estimate $p_{n+1}-p_n \ll p_n^{7/12+\varepsilon}$, for every $\varepsilon>0$. (It is known that the Riemann hypothesis implies this with the exponent $1/2$ in place of $7/12$.) The "density hypothesis" $N(\alpha,T;\mathcal{X}0)\ll T^{2(1-\alpha)+\varepsilon}$ holds for $\alpha>13/16$.

(Here $\mathcal{X}0$ is the principal character, so that the zeros are those of Riemann's zeta-function.) Bombieri's method has also been generalized to algebraic number fields. Many mathematicians have played a part in the development of his method — H. Davenport, H. Halberstam, P. X. Gallagher, H. L. Montgomery, G. Halász, M. N. Huxley, M. Jutila and, more recently, M. Forti and C. Viola, to mention but a few. There is little doubt that Bombieri's theorems have inspired that development.

2. Univalent functions and the local Bieberbach conjecture. Bombieri's work on the local validity of the Bieberbach conjecture is an impressive achievement in an altogether different branch of mathematics. It shows his power and ingenuity in attacking problems of 'hard analysis'.

Let $\mathscr{S}$ denote the family of functions $f(z) = z + a_2z^2 + z_3z^3 + \cdots$ which are (normalized) holomorphic and univalent in the unit disc $|z| < 1$. Bieberbach's conjecture is that if $f(z) \in \mathscr{S}$, then Re $a_n \leqq n$, with the equality holding only if $f(z) = z/(1 - pz)^2$, and $p^{n-1} = 1$. The conjecture has so far been proved for $2 \leqq n \leqq 6$ on the one hand, and for a large number of subfamilies of $\mathscr{S}$ on the other.

In 1965 P. R. Garabedian and M. Schiffer raised the question of the *local* validity of that conjecture, that is: If $2 - \mathrm{Re}\, a_2$ is small enough, is it true that $n - \mathrm{Re}\, a_n$ is non-negative? They answered it in the affirmative if n is *even*. They proved the existence of a positive constant ε_{2m} say, such that if $|2 - a_2| < \varepsilon_{2m}$, then Re $a_{2m} < 2m$, with the equality holding if and only if $f(z) = z(1 - z)^{-2} = \sum_{n=1}^{\infty} nz^n$, the Koebe function.

Bombieri proved this in 1967 for *all* n, odd as well as even, the case of n odd being the more difficult (Invent. Math. **4** (1967), 26–67). To be precise, he proved that

$$\liminf_{a_2 \to 2} \frac{n - \mathrm{Re}\, a_n}{2 - \mathrm{Re}\, s_2} > 0, \quad \text{if } n \text{ is even,}$$

and

$$\liminf_{a_3 \to 3} \frac{n - \mathrm{Re}\, a_n}{3 - \mathrm{Re}\, a_3} > 0, \quad \text{if } n \text{ is odd,}$$

where the 'lim inf' is taken over all functions of the family $\mathscr{S}$.

An independent, though less direct, proof of this has since been published by Garabedian and Schiffer (Arch. Rational Mech. Anal. **26** (1967), 1–32).

Bombieri's proof is based on an ingenious combination of K. Löwner's 'parametric method' with the theory of the 'second variation' developed by P. L. Duren and M. Schiffer. He uses the results of A. C. Schaefer and D. C. Spencer on Löwner curves, as well as an earlier result of his own concerning a set of quadratic forms (Q_n), in an infinite number of variables, which had been encountered by Duren and Schiffer in their theory of the second variation. These quadratic forms Q_n have the property that: (a) if Q_n is an indefinite form, then Bieberbach's conjecture is false for that n; (b) if Q_n is positive definite, then every analytic variation of the Koebe function decreases Re a_n. Duren and Schiffer proved (1962/63) that Q_n is positive definite for $n = 2, 3, \ldots, 9$, and the same was checked with a computer for all $n \leqq 100$. Bombieri proved that Q_n is positive definite for *all* n (Boll. Un. Mat. Ital. (3) **22** (1967), 25–32).

3. Several complex variables. Bombieri's theorem concerning algebraic values of meromorphic maps (Invent. Math. **10** (1970), 267–287; **11** (1970), 163–166), motivated though it is by the theory of transcendental numbers, is an incursion of geometric integration theory in the analysis of functions of several complex variables. The theorem is as follows:

THEOREM. *Let K be an algebraic number field. Let $f_1, f_2, \ldots, f_N$ be meromorphic functions of finite order in C^n. Suppose that at least $n+1$ of them are algebraically independent oyer K. and that for any j with $1 \leqq j \leqq N$ and ν with $1 \leqq \nu \leqq n$, the partial derivative $\partial f_i / \partial z_\nu$ is a polynomial in $f_1, f_2, \ldots, f_N$ with coefficients in K. Then, the set of points in C^n at which all the f_j are defined, and have values in K, is contained in an algebraic hypersurface in C^n. (If the given functions are of order $\leqq \rho$, then the degree of the hypersurface $\leqq n(n+1)\rho[K : Q] + n$.)*

The case $n = 1$ was proved by S. Lang after previous work by Th. Schneider; It unifies divers results due to A. O. Gelfond and to Schneider, and contains, in particular, the transcendency of e^α for $\alpha \neq 0$, α algebraic, and of α^β for $\alpha \neq 0, 1, \alpha$ and β algebraic and β irrational. While Bombieri's extension does not seem immediately to lead to new theorems on transcendency, variants of it are applicable to the study of n-parameter subgroups of algebraic groups (Invent. Math. **11** (1970), 1–14).

But the real interest of the paper, once again, arises from the proof which contains an existence theorem and a structure theorem. The existence theorem, which generalizes previous work of L. Hörmander and of A. Martineau, states that for any pluri-subharmonic function p on $C^n, p \not\equiv -\infty$, there exists a nonzero entire function f on C^n, with

$$\int_{C^n} |f(z)|^2 e^{-p(z)} (1 + |z|^2)^{-3n} \, dz < +\infty.$$

The structure theorem, on the other hand, gives a sufficient condition for a current of degree (1,1) to be integration on an analytic set of codimension 1. Several authors had previously attempted, without success, to produce workable conditions of that type. Bombieri's result has since been used by F. Reese Harvey and James King (Invent, Math. **15** (1972), 47–52) to characterize those currents of degree $(k, k), k \geqq 1$, on a complex manifold that correspond to integration over (linear combinations of) complex subvarieties, thus settling a conjecture of P. Lelong which had been open for several years that those are precisely the positive currents that are d-closed and whose densities (or Lelong numbers) are locally bounded away from zero. For his proof, Bombieri makes use of Hörmander's work on L^2-estimates and existence theorems for solutions of the $\bar{\partial}$-Neumann problem, besides ideas from H. Federer's work in geometric measure theory. It bears the hallmark of a highly original analyst.

4. I have not spoken about Bombieri's contributions to the theory of partial differential equations and minimal surfaces — in particular, to the solution of Bernstein's problem in higher dimensions. Nor have I spoken about the fact that he was among the first to give effective applications of Dwork's method in the p-adic approach to Andrdé Weil's zeta-function. But I hope I have said enough to show that Bombieri's *versatility* and *strength* have combined to create many original patterns of ideas which are both rich and inspiring. It is in recognition of these qualities that he has been awarded a Fields Medal. To him mathematics is a private garden; may it bring forth many new blooms.

Eidg. Technische Hochschule
Zürich, Switzerland

Enrico Bombieri

Proceedings of the International Congress of Mathematicians
Vancouver, 1974

VARIATIONAL PROBLEMS AND ELLIPTIC EQUATIONS

by
ENRICO BOMBIERI

I. Variational problems. In this expository article I will be concerned with second-order, nonlinear, elliptic equations arising from variational problems. Perhaps the simplest example is the

Dirichlet problem. Find a function $u(x)$ harmonic in a given bounded open set Ω and taking given boundary values on $\partial\Omega$.

The variational formulation of Dirichlet's problem is expressed through the

Dirichlet principle. The function $u(x)$ is the unique solution of the variational problem

$$\int_\Omega |Du|^2 dx = \min, \quad u = f \text{ on } \partial\Omega,$$

where Du denotes the gradient of u.

The approach to the Dirichlet problem through the Dirichlet principle was soon criticized because the existence of a minimum for the Dirichlet integral was not obvious; in particular, some conditions are needed in order to have a finite Dirichlet integral. This is not unnatural to assume a priori, since, for example, in physical models the Dirichlet integral represents the energy of a system, which should be finite to start with. Once these limitations of the variational approach were understood, its usefulness became clear and the Dirichlet principle became again a respectable tool in mathematics.

More generally, one may ask to minimize the functional

$$J[u] = \int_\Omega f(x, u, Du)\, dx$$

under appropriate boundary conditions for the competing functions u. Actually $u(x)$ may be a vector-valued function. If $J[u] = \min$, then $J[u] \leqq J[u+\varepsilon v]$ for every v with compact support in Ω and, expanding $J[u+\varepsilon v]$ in a Taylor series in ε,

$$J[u+\varepsilon v] = J[u] + \varepsilon\delta J[u] + \varepsilon^2\delta^2 J[u] + \cdots,$$

we see that we need $\delta J[u] = 0$ and $\delta^2 J[u] \geqq 0$ for all such v, i.e, (writing $p = (p_1, \ldots, p_n)$ for Du),

$$\delta J[u] = \int_\Omega \left(\sum_i \frac{\partial f}{\partial p_i} \frac{\partial v}{\partial x_i} + \frac{\partial f}{\partial u} v \right) dx$$

$$= \int_\Omega \left(-\sum_i \frac{\partial}{\partial x_i} \left(\frac{\partial f}{\partial p_i} \right) + \frac{\partial f}{\partial u} \right) v \, dx = 0,$$

and we obtain the well-known Euler equation

$$\sum_i \frac{\partial}{\partial x_i} \left(\frac{\partial f}{\partial p_i} \right) = \frac{\partial f}{\partial u}.$$

A simple condition, which implies $\delta^2 J \geqq 0$, is

$$\sum_{ij} \frac{\partial^2 f}{\partial p_i \partial p_j} \xi_i \xi_j > 0, \quad \xi \in \boldsymbol{R}^n, \ \xi \neq 0,$$

which expresses a kind of convexity condition for the functional $J[u]$. If this condition is satisfied, one says that the integrand $f(x, u, p)$ is regular elliptic. In case one considers vector-valued solutions $u = u^1, \ldots, u^\lambda, \ldots, u^N)$, the regularity condition imposed on $f = f(x, u^\lambda, p^\lambda)$ is

$$\sum_{ij} \sum_{\lambda\mu} \frac{\partial^2 f}{\partial p_i^\lambda \partial p_j^\mu} \eta^\lambda \eta^\mu \xi_i \xi_j > 0$$

at every point $(x, u^\lambda, p^\lambda)$ and all $\eta \in \boldsymbol{R}^N$, $\xi \in \boldsymbol{R}^n$, $\eta, \xi \neq 0$.

In his 19th problem of his address at the International Congress of Mathematicians in 1900, Hilbert raised the question whether solutions of regular elliptic, analytic variational problems are necessarily analytic. This problem of regularity, together with the problem of existence of solutions, form two central questions in the theory of variational problems.

II. Elliptic equations: the early work. In his celebrated thesis of 1904, S. Bernstein proved the remarkable result that C^3 solutions of a single elliptic, nonlinear, analytic equation in two variables are necessarily analytic; this was considered at the time a solution to Hilbert's 19th problem. Having thus attacked the problem of regularity, he went on with the existence problem in an important series of papers, between 1906 and 1912. We owe to him the basic idea (and the name) of an "a priori estimate", which still has a central role in the theory: If we have the right majorizations for all solutions (and their derivatives) of an elliptic equation, then existence and regularity of solutions of the Dirichlet problem will follow. Since in obtaining these estimates we assume "a priori" that we are dealing with smooth solutions, we have the name "a priori estimates". Bernstein himself showed how to prove such estimates in some important cases, using the maximum principle and what is known today as the method of barriers.

Bernstein's work was rather involved and relied heavily on analyticity, and was later improved and generalized to several variables and elliptic systems by the work of several authors, among which are Gevrey, Giraud, Lichtenstein, H. Lewy, E. Hopf, T. Rado, I. Petrowsky and Bernstein himself. However, one had to wait until the years between 1932 and 1937 before the basic reasons for the importance of the "a priori estimates" in the existence problem were fully understood and clarified through the work of Schauder, Leray and Caccioppoli and in particular the classical paper of Leray and Schauder of 1934.

Consider for example a quasi-linear equation

$$\sum a_{ij}(x, u, Du) D_i D_j u = 0 \quad \text{in } \Omega,$$

$$u = f \quad \text{on } \partial\Omega.$$

We denote by T the operator which to a function u associates the unique solution v of the linear Dirichlet problem

$$\sum a_{ij}(x, u, Du) D_i D_j v = 0 \quad \text{in } \Omega,$$

$$v = f \quad \text{on } \partial\Omega.$$

Since the latter problem is linear, it is much easier to solve, and the question is reduced to finding a fixed point $u = Tu$ for the operator T. The main point is that very general fixed point theorems are available if we have the right "a priori estimates" for the solutions of the original equation and of the linearized equation. The advantage of this procedure over an iteration scheme $u_{n+1} = Tu_n$ (used by Bernstein) is obvious: If uniqueness is not satisfied, the iteration need not converge.

The fundamental "a priori estimates" for the linearized equation were found by Schauder; the search for such estimates in the nonlinear case is still today more of an art than of a method.

III. Direct methods and weak solutions. Another approach to the existence problem in the variational case is provided by the so-called "direct methods in the calculus of variations". Roughly speaking, one wants to show

(A) the integrand $J[u]$ is lower semicontinuous and bounded below, with respect to a suitable notion of convergence in some admissible class of competing functions u;

(B) a minimizing sequence $\{u_n\}$, i.e., $J[u_n] \to \operatorname{Inf} J[u]$ converges to an admissible u; hence $J[u] = \min$ by (A).

This idea was used perhaps for the first time by Zaremba and also by Hilbert in his investigations on the Dirichlet principle. It became a standard approach to variational problems in the hands of Lebesgue, Courant, Fréchet and especially Tonelli, If the integrand $J[u]$ satisfies an inequality $f(x, u, p) \geqq m_1|p|^r - m_2$, $m_1 > 0$, with $1 \leqq r < +\infty$, then Tonelli's method, using absolutely continuous functions and uniform convergence, works provided $r \geqq n = \dim \Omega$, which is a too strong condition if $n \geqq 3$. A notable success of this method was however Haar's work of 1927 on functionals of the type $J[u] = \int_\Omega f(Du)dx$, for the case of $n = 2$ variables. Here one assumes that Ω is a smooth convex domain, and the boundary values are also smooth, satisfying a certain "three-point condition". The class of competing functions used by Haar is a class of functions satisfying a uniformly bounded Lipschitz condition.

The deep reason for the limitation of Tonelli's approach was found only later, through the fundamental work of Sobolev and Morrey in 1938. The Sobolev spaces $H^{k,p}(\Omega)$ are the Banach spaces of functions on Ω whose derivatives of order $\leqq k$ are in L^p. Sobolev discovered the fundamental embedding theorems for these spaces, the simplest being (one assumes Ω bounded and $\partial\Omega$ smooth):

(i) if $f \in H^{1,p}(\Omega)$, $1 \leqq p < n$, then $f \in L^s(\Omega)$ with $s = np/(n-p)$, and

$$||f||_{L^s} \leqq C(\Omega)||f||H^{1,p};$$

(ii) if $f \in H^{1,p}(\Omega)$, $p > n$, then f satisfies a Hölder condition in Ω.

The new approach to the existence problem could now be summarized as follows:

(A) the integrand $J[u]$ determines naturally a function space $\mathscr{F}$ (usually a Sobolev space), in which the lower semicontinuity becomes a natural statement;

(B) by means of "a priori estimates" one shows that there exists a convergent minimizing sequence (here the Sobolev embedding theorems are often crucial).

From (A) and (B) one deduces the existence of a solution in the function space $\mathscr{F}$. However, one expects the solution so obtained to be very smooth. In some cases, e.g., those in which Tonelli's method works, the smoothness of solutions is automatic (compare (ii) of Sobolev's embedding theorem); in general, there remains the difficult problem of "regularization":

(C) the "weak solutions" so obtained are in fact differentiable solutions.

The necessary results about lower semicontinuity have been obtained by Serrin; stages (B) and (C) require an extensive use of "a priori estimates", the regularization part being often difficult if not intractable.

This approach led to remarkable results especially in two cases: nonlinear second-order equations in $n = 2$ variables, where one could also use tools from quasi-conformal mapping (Morrey, Bers, Nirenberg), and linear equations and systems with smooth coefficients (we may mention the work of Ladyzenskaya and Caccioppoli of 1951 for second-order equations, and the general theory of Friedrichs, F. John, Agmon-Douglis-Nirenberg of 1959, who also considered higher-order systems and the problem of boundary regularity).

The first breakthrough in the nonlinear case came in 1957–1958 when De Giorgi and independently Nash for parabolic equations succeeded in proving Hölder continuity of weak solutions of uniformly elliptic equations

$$\sum D_i(a_{ij}(x)D_j u) = 0$$

with measurable coefficients a_{ij} and with the ellipticity condition

$$m|\xi|^2 \leqq \sum a_{ij}(x)\xi_i\xi_j \leqq M|\xi|^2,$$

where m, M are positive constants independent of x.

This result has some striking applications to nonlinear problems. De Giorgi himself showed how his theorem implied that weak extremals of uniformly elliptic analytic integrands of the type $\int_\Omega f(Du)dx = \min$ are indeed analytic in Ω. Stampacchia and Gilbarg found another application, namely the extension of Haar's theorem to the case of $n > 2$ variables; further important applications and generalizations have been given by Morrey, Ladyzenskaya and Uraltseva, Oleinik and many others, in particular to the study of second-order quasi-linear equations which are quadratic in the first-order derivatives.

Of great importance was also a new proof of De Giorgi's theorem, found by Moser in 1960, using the Sobolev inequalities rather than the isoperimetric inequalities of De Giorgi. This also led to a proof of the Harnack inequality; If $\Omega' \Subset \Omega$ and if u is a positive solution in Ω of a uniformly elliptic equation $\sum D_i(a_{ij}(x)D_j u) = 0$, then $\max_{\Omega'} u \leqq C\ \min_{\Omega'}\ u$, where C depends only on Ω', Ω and the ellipticity constant $L = M/m$. Hence one obtains a Liouville theorem: A bounded solution over $\boldsymbol{R}^n$ of such an equation is necessarily a constant.

IV. Weak solutions of elliptic systems. The problem of the extension of De Giorgi's regularization to systems of equations or to higher-order equations remained outstanding for awhile, until in 1968 De Giorgi found an example of a uniformly elliptic linear system of variational type with bounded measurable coefficients, with the discontinuous solution $x/|x|$. By adapting De Giorgi's example, in 1969 Giusti and Miranda showed that if $n > 2$ the integrand

$$\int |D\boldsymbol{u}|^2 + \left[\sum_{ij}\left(\delta_{ij} + \frac{4}{n-2}\frac{u^i u^j}{1+|\boldsymbol{u}|^2}\right) D_j u^i\right]^2 dx$$

with $\boldsymbol{u} = (u^1, \ldots, u^n)$ is a regular uniformly elliptic analytic integrand, while $u = x/|x|$ is an extremal which is not real analytic at $x = 0$. These examples pointed out the great importance of the results obtained by Morrey in 1968 on the regularity problem for systems in $n > 2$ variables.

Here the breakthrough came with the introduction of new powerful compactness methods, originally introduced by De Giorgi and especially Almgren in 1960–1966, in the study of minimal surfaces.

In rather crude terms, the idea behind the use of compactness methods may be described as follows. Suppose we want to prove an "a priori estimate" of local nature for solutions of a class of variational problems which is invariant by linear changes of the coordinates. If the estimate we want fails in every neighborhood of a point x_0, this means that we can find a sequence of elliptic equations or systems over a fixed domain Ω, and a sequence of solutions, for which the desired estimate fails in smaller and smaller neighborhoods of x_0. By performing a linear change of coordinates, we can expand these neighborhoods to a fixed neighborhood of x_0, and in doing so we have to replace our equations by new equations still in the same class and defined over larger and larger domains. Using the appropriate compactness theorems then one shows that this sequence of equations and solutions converges in some sense to a limiting equation, now defined over $\boldsymbol{R}^n$, and to a limiting solution for which the desired "a priori estimate" still fails. The main point however is that, in doing so, we have replaced an elliptic operator by its "tangent operator" at x_0, and thus the limiting equation is often of a very simple type, for example linear with constant coefficients, and, for it, it may be easy to check that the "a priori estimate" we want does in fact hold. This gives a contradiction and establishes the local estimate we were looking for. In the nonlinear case, convergence to a limiting equation is usually obtained by assuming certain mild conditions about the local behaviour of solutions at a point. If these conditions are valid almost everywhere, which is often the case because of measure theoretic arguments, one ends up with estimates which are valid only near almost every point, and in turn one establishes only regularity almost everywhere.

In this way it was proved by Morrey in 1968 that weak solutions of a large class of nonuniformly elliptic analytic variational problems of the type

$$\int f(x, D\boldsymbol{u})dx = \min,$$

and also of uniformly elliptic analytic variational problems of the type

$$\int f(x, \boldsymbol{u}, D\boldsymbol{u})dx = \min$$

are in fact analytic almost everywhere. Giusti and Miranda, in 1970–1972, extended and substantially simplified this work, and they have also been able to obtain good estimates for the Hausdorff dimension of the exceptional set in which the solutions are not analytic.

The outstanding problem here is to determine the structure of the singular set; for example, is it semi-analytic? In special cases, one can even prove that solutions are everywhere analytic, and it is an interesting open question to find good conditions which imply regularity everywhere.

V. The minimal surface equation. A well-known variational problem is the *Problem of Plateau.* Find a surface of least area among all surfaces having a prescribed boundary.

This is not a regular variational problem, if taken in this generality, and it is not possible for me to explain in this article all the new fundamental results obtained between 1960 and 1974 by Federer, Fleming, Reifenberg, De Giorgi, Almgren, Allard and many others. I will restrict my attention instead to the case of minimal graphs (the nonparametric Plateau problem) and to some special questions about the parametric Plateau problem in codimension one.

If the graph $y = u(x)$ of a function $u(x), x \in \Omega \subset \boldsymbol{R}^n$, is a solution of Plateau's problem, then it minimizes the area functional $\int_\Omega (1 + |Du|^2)^{1/2} dx$, and the associated Euler equation is

$$\sum D_i(D_i u/W) = 0, \quad W = (1 + |Du|^2)^{1/2},$$

which expresses the fact that the graph has mean curvature 0 at every point.

The strong nonlinearity of this equation gives rise to unexpected phenomena, which have no counterpart in the theory of linear equations. For $n = 2$ variables:

(i) the Dirichlet boundary value problem is soluble for arbitrary continuous data if and only if Ω is convex (Bernstein, Finn);

(ii) a solution defined over a disk minus the centre extends to a solution over the disk, i.e., isolated singularities are removable (Bers);

(iii) if $u > 0$ is a solution over $|x| < R$ then

$$(1 + |Du(0)|^2)^{1/2} \leqq \exp(\pi u(0)/2R)$$

and this estimate is sharp (Finn, Serrin);

(iv) a solution defined over $\boldsymbol{R}^2$ is linear (Bernstein).

The solution of the analogous problems for $n > 2$ variables has been obtained only recently. We have:

(i) the Dirichlet boundary value problem is soluble for arbitrary continuous data if and only if $\partial\Omega$ has positive mean curvature at every point (Serrin, Bombieri, De Giorgi and Miranda, 1968);

(ii) a solution defined over Ω minus K, where K is a compact subset of Ω with $(n-1)$-dimensional Hausdorff measure 0, extends to the whole of Ω (De Giorgi and Stampacchia, 1965);

(iii) if $u > 0$ is a solution over $|x| < R$ then

$$|Du(0)| < c_1 \exp(c_2 u(0)/R)$$

(Bombieri, De Giorgi and Miranda, 1968);

(iv) if $n \leqq 7$, a solution defined over $\boldsymbol{R}^n$ is linear (Fleming's new proof in 1962 for the case $n = 2$, De Giorgi for $n = 3$ in 1964, Almgren for $n = 4$ in 1966, Simons for $n \leqq 7$ in 1968); on the other hand, if $n \geqq 8$, there are solutions defined over $\boldsymbol{R}^n$ which are not linear (Bombieri, De Giorgi and Giusti, 1969).

What about the methods of proof? In his talk at the International Congress of Mathematicians in 1962, L. Nirenberg made the statement that "most results for nonlinear problems are still obtained via linear ones, i.e. despite the fact that the problems are nonlinear not because of it". The minimal surface equation is no exception to this statement; but since the linearization procedure is rather unusual, it is worthwhile to describe it.

Let us define a vector ν with components

$$\nu_i = -(D_i u)/W, \quad i = 1, \ldots, n,$$

$$\nu_{n+1} = 1/W,$$

and differential operators

$$\delta_i = D_i - \nu_i \sum_{j=1}^{n+1} \nu_j D_j, \quad i = 1, \ldots, n+1,$$

in $\boldsymbol{R}^{n+1}$.

If we denote by S the graph of $x_{n+1} = u(x)$ in $\boldsymbol{R}^{n+1}$, then the vector ν is the normal unit vector to S at the point $P = (x, u(x))$ and the operators are δ_i the projections of the operators D_i on the tangent space to S at the point P. The "Laplacian" $\mathscr{D} = \sum \delta_i \delta_i$ is actually the Laplace-Beltrami operator on S, and the fact that S has mean curvature 0 at every point is nicely expressed by the fact that the coordinate functions x_i are harmonic on S for the Laplace-Beltrami operator. Moreover it can be shown that the normal vector ν satisfies the nonlinear elliptic system $\mathscr{D}\nu + c^2(x)\nu = 0$ on S where $c^2(x) = \sum_{ij}(\delta_i \nu_j)^2$ is the sum of the squares of the principal curvatures of S at P. In particular since $\nu_{n+1} > 0$ it follows that $\mathscr{D}\nu_{n+1} \leqq 0$, i.e., ν_{n+1} is superharmonic on S.

Now we have two main facts (Miranda, 1967):

(a) if f has compact support and S is minimal, then

$$\int \delta_i f d||S|| = 0, \quad \text{all } i,$$

or in other words the operators δ_i can be integrated by parts on the surface S;

(b) if f has compact support, S is minimal and $1 \leqq p < n$ then

$$\left(\int |f|^{np/(n-p)} d||S|| \right)^{(n-p)/n} \leqq c(p,n) \int |\delta f|^p d||S||,$$

or in other words we have a uniform Sobolev inequality on S for the differential operators δ_i.

We can use (a) and (b) together with De Giorgi's regularization technique (which is highly nonlinear) to investigate the differential inequality $\mathscr{D}\nu_{n+1} \leqq 0$, and eventually one arrives at the "a priori estimate" (iii). The solubility of the Dirichlet problem, and also the analyticity of weak solutions, depends on this "a priori estimate".

More generally, one may investigate uniformly elliptic equations of the type $\sum \delta_i(a_{ij}(x)\delta_j u) = 0$ on an absolutely minimizing surface S of codimension one (Bombieri and Giusti, 1972). Thus one obtains the extension of the Moser-Harnack theorem to these equations, and as an application one gets that if u is a positive harmonic function on a minimal surface in $\boldsymbol{R}^{n+1}$ without boundary, then u is constant. Since the coordinate functions x_i are harmonic on S, one gets as a corollary a theorem of Miranda that a minimal surface without boundary contained in a half-space is a hyperplane. Also, a minimal surface without boundary is connected (Bombieri and Giusti, 1972).

The extension of Bernstein's theorem up to dimension 7, and the construction of a counterexample in dimension $n \geqq 8$, depends on different ideas. It was Fleming in 1962 who used compactness techniques to show that the failure of Bernstein's theorem in dimension n implied the existence of a singular minimal cone in $\boldsymbol{R}^{n+1}$. De Giorgi later proved that in fact one would get the existence of such a cone in $\boldsymbol{R}^n$, and in this way extended Bernstein's theorem through dimension $n = 3$. Then the question centered about the existence of minimal cones, and eventually Simons succeeded in proving the nonexistence of singular minimal cones in $\boldsymbol{R}^n$, $n \leqq 7$. Moreover, Simons proved that the cone in $\boldsymbol{R}^8$ given by $x_1^2 + x_2^2 + x_3^2 + x_4^2 = x_5^2 + x_6^2 + x_7^2 + x_8^2$ was at least a locally minimal cone, i.e., area would increase with every sufficiently small deformation. Making use of the invariance of this cone by $\mathrm{SO}(3) \times \mathrm{SO}(3)$, Bombieri, De Giorgi and Giusti proved that this cone was in fact minimal in the large, by reducing the problem to a question about a system of first-order ordinary differential equations. It was natural to see whether this cone was associated with the failure of Bernstein's theorem in dimension 8, and this was obtained by constructing explicitly a subsolution u^-, and a supersolution u^+, of the minimal surface equation in $\boldsymbol{R}^8$, with the property that $u^- \leqq u^+$ everywhere and that no function between u^- and u^+ could be linear. Now an application of the maximum principle and also of the "a priori estimate" for the gradient obtained before showed the existence of a solution u defined everywhere comprised between u^- and u^+. It should be noted that the choice of u^- and u^+ was in fact suggested by the results obtained in the investigation of Simons' cone.

VI. Further results. I will end this article by mentioning some results and directions of research which I could not treat more explicitly, but which seem to me of great importance.

First of all, the facts which I have stated about the minimal surface equation are not limited to that special case. A whole class of elliptic equations can be treated with similar methods, among which are the equations of surfaces with prescribed mean curvature, the equation of capillarity phenomena, and many others. Here much recent work has been done by Ladyzenskaya and Uraltseva, Bombieri and Giusti, Trudinger, Finn, Serrin and many others.

Second, and more importantly, I have limited myself in this article to variational problems of a nonparametric nature. The parametric point of view, in which one considers functionals on geometrical objects rather than on functions, has led to the modern geometric measure theory, the theory of integral currents and varifolds and of parametric elliptic integrands. Here the work of Federer, Fleming and especially Almgren is outstanding. Also, among more recent developments, I may mention the work of Allard on the first variation of a varifold and that of Jean Taylor on the structure of the singular set of soap films and soap bubbles.

Another fruitful idea which I could not treat in this article is that of variational problems in which the solutions have to satisfy additional constraints. Here one may ask for solutions satisfying inequalities, thus obtaining classical problems with obstacles, or asking for solutions with gradient not exceeding certain bounds (an example is the potential equation for a subsonic gas flow), or one may impose convexity, as for the Monge-Ampére equations, and so on. Here the theory of variational inequalities begins to give a general foundation for many problems of this type. Fortunately many of these questions will receive special attention in these PROCEEDINGS, and I have to refer to the more specialized articles for further illustrations of the directions in which the theory of variational problems and of elliptic equations is moving.

References

Section II

S. Bernstein, *Sur la nature analytique des solutions des équations aux deriveés partielles du second ordre.* Math. Ann. **59** (1904), 20–76.

J. Leray and J. Schauder, *Topologie et équations fonctionnelles*, Ann. Sci. Ecol. Norm. Sup. (3), **51** (1934), 45–78.

Section III

C. B. Morrey, Jr., *Multiple integrals in the calculus of variations*, Die Grundlehren der math. Wissenschaften, Band 130, Springer-Verlag, New York, 1966. MR **34** #2380.

L. Nirenberg, *Some aspects of linear and nonlinear partial differential equations*, Proc. Internat. Congress Math. (Stockholm, 1962), Inst. Mittag-Leffler, Djursholm, 1963, pp. 147–162. MR **31** #471.

O. A. Ladyženskaya and N. N. Uraltseva, *Linear and quasi-linear equations of elliptic type*, "Nauka", Moscow, 1964; English transl., Academic Press, New York, 1968. MR **35** #1955; **39** #5941.

S. Agmon, A. Douglis and L. Nirenberg, *Estimates near the boundary for solutions of elliptic partial differential equations satisfying general boundary conditions.* I, II, Comm. Pure Appl. Math. **12** (1959), 623–727; ibid. **17** (1964), 35–92. MR **23** #A2610; **28** #5252.

E. De Giorgi, *Sulla differenziabilitá e l'analiticitá delle estremali degli integrali multipli regolari*, Mem. Accad. Sci. Torino CI. Sci. Fis. Mat. Nat. (3) **3** (1957), 25–43. MR **20** # 172.

J. Moser, *On Harnack's theorem for elliptic differential equations*, Comm. Pure Appl. Math. **14** (1961), 577–591. MR **28** #2356.

Section IV

E. De Giorgi, *Un esempio di estremali discontinue per un problema variazionale di tipo ellittico*, Boll. Un. Mat. Ital. (4) **1** (1968), 135–147. MR **37** #3411.

E. Giusti and M. Miranda, *Un esempio di soluzioni discontinue per un problema di minima relativo ad un integrale regolare del calcolo delle variazioni*, Boll, Un, Mat. Ital. (4) **1** (1968), 219–226. MR 38 #591.

C. B. Morrey, Jr., *Partial regularity results for non-linear elliptic systems*, J. Math. Mech. **17** (1967/68), 649–670. MR **38** #6224.

E. Giusti, *Regolaritá parziale delle soluzioni di sistemi ellittici quasi-lineari di ordine arbitrario*, Ann. Scuola Norm. Sup. Pisa (3) **23** (1969), 115–141. MR **40** #527.

Section V

R. Finn, *On equations of minimal surface type*, Ann. of Math. (2) **60** (1954), 397–416. MR **16**, 592.

S. Bernstein, *Über eine geometrisches Theorem und seine Anwendung auf die partiellen Differentialgleichungen vom elliptischen Typus*, Math. Z. **26** (1927), 551–558.

J. Serrin, *The problem of Dirichlet for quasilinear elliptic differential equations with many independent variables*, Philos. Trans. Roy. Soc. London Ser. A **264** (1969), 413–496. MR **43** #7772.

E. Bombieri, E. De Giorgi and M. Miranda, *Una maggiorazione a priori relativa alle ipersuperfici minimali nonparametriche*, Arch. Rational Mech. Anal. **32** (1969), 255–267. MR **40** #1898.

J. Simons, *Minimal varieties in Riemannian manifolds*, Ann. of Math. (2) **88** (1968), 62–105. MR **38** #1617.

E. De Giorgi and G. Stampacchiá, *Sulle singolárita eliminabili delle ipersuperficie minimali*, Atti Accad. Naz. Lincei Rend. CI. Sci. Fis. Mat. Natur. (8) **38** (1965), 352–357. MR **32** #4612.

E. Bombieri, E. De Giorgi and E. Giusti, *Minimal cones and the Bernstein problem*, Invent. Math. **7** (1969), 243–268. MR **40** #3445.

E. Bombieri and E. Giusti, *Harnack's inequality for elliptic differential equations on minimal surfaces*, Invent. Math. **15** (1972), 24–46. MR **46** #8057.

Section VI

O. A. Ladyženskaya, and N. N. Uraltseva, *Local estimates for gradients of solutions of nonuniformly elliptic and parabolic equations*, Comm. Pure Appl. Math. **23** (1970), 677–703. MR **42** #654.

E. Bombieri and E. Giusti, *Local estimates for the gradient of non-parametric surfaces of prescribed mean curvature*, Comm. Pure Appl. Math. **26** (1973), 381–394.

N. Trudinger, *Gradient estimates and mean curvature* (to appear).

R. Finn, *Capillarity phenomena*, Uspehi Mat. Nauk (to appear).

H. Federer, *Geometric measure theory*, Die Grundlehren der math. Wissenschaften, Band 153, Springer-Verlag, New York, 1969. MR **41** #1976.

F. J. Almgren, Jr., *Existence and regularity almost everywhere of solutions to elliptic variational problems among surfaces of varying topological type and singularity structure*, Ann. of Math, (2) **87** (1968), 321–391, MR **37** #837.

W. K. Allard, *On the first variation of a varifold*, Ann. of Math. (2) **95** (1972), 417–491, MR **46** #6136.

J. E. Taylor, *Regularity of the singular sets of two-dimensional area-minimizing flat chains modulo* 3 *in* R^3 Invent. Math, **22** (1973), 119–160.

F. J. Almgren, Jr., *Existence and regularity almost everywhere of solutions to elliptic variational problems with constraints* (to appear).

Scuola Normale Superiore
Pisa, Italy

Reprinted from Proc. Int. Congr. Math., 1974

THE WORK OF DAVID MUMFORD

by
J. TATE

It is a great pleasure for me to report on Mumford's work. However I feel there are many people more qualified than I to do this. I have consulted with some of them and would like to thank them all for their help, especially Oscar Zariski.

Mumford's major work has been a tremendously successful multi-pronged attack on problems of the existence and structure of varieties of moduli, that is, varieties whose points parametrize isomorphism classes of some type of geometric object. Besides this he has made several important contributions to the theory of algebraic surfaces. I shall begin by mentioning briefly some of the latter and then will devote most of this talk to a discussion of his work on moduli.

Mumford has carried forward, after Zariski, the project of making algebraic and rigorous the work of the Italian school on algebraic surfaces. He has done much to extend Enriques' theory of classification to characteristic $p > 0$, where many new difficulties appear. This work is impossible to describe in a few words and I shall say no more about it except to remark that our other Field's Medallist, Bombieri, has also made important contributions in this area, and that he and Mumford have recently been continuing their work in collaboration.

We have a good understanding of divisors on an algebraic variety, but our knowledge about algebraic cycles of codimension > 1 is still very meager. The first case is that of 0-cycles on an algebraic surface. In particular, what is the structure of the group of 0-cycles of degree 0 modulo the subgroup of cycles rationally equivalent to zero, i.e., which can be deformed to 0 by a deformation which is parametrized by a line. This group maps onto the Albanese variety of the surface, but what about the kernel of this map? Is it "finite-dimensional"? Severi thought so; but Mumford proved it is not, if the geometric genus of the surface is $\geqq 1$. Mumford's proof uses methods of Severi, and he remarks that in this case the techniques of the classical Italian algebraic geometers seem superior to their vaunted intuition. However, in other cases Mumford has used modern techniques to justify Italian intuition, as in the construction by him and M. Artin of examples of unirational varieties X which are not rational, based on 2-torsion in $H^3(X, \boldsymbol{Z})$.

Probably Mumford's most famous result on surfaces is his topological characterization of nonsingularity. Let P be a normal point on an algebraic surface V in a complex projective space. Mumford showed that if V is topologically a manifold at P, then it is algebraically nonsingular there. Indeed, consider the intersection K of V with a small sphere about P. This intersection K is 3-dimensional and if V is a manifold at P, then K is a sphere and its fundamental group is trivial. Mumford showed how to compute this fundamental group $\pi_1(K)$ in terms of the diagram of the resolution of the singularity of V at P, and then he showed that $\pi_1(K)$ is not trivial unless the diagram is, i.e., unless V is nonsingular at P. A by-product of this proof is the fact that the Poincaré conjecture holds for the 3-manifolds which occur as K's. Mumford's paper was a critical step between the early work on singularities of branches of plane curves (where K is a torus knot) and fascinating later developments. Brieskorn showed that the analogs of Mumford's results are false in general for V of higher dimension. Consideration of the corresponding problem there led to the discovery of some beautiful relations between algebraic geometry and differential topology, including simple explicit equations for exotic spheres.

Let me now turn to Mumford's main interest, the theory of varieties of moduli. This is a central topic in algebraic geometry having its origins in the theory of elliptic integrals. The development of the algebraic and global aspects of this subject in recent years is due mainly to Mumford, who attacked it with a brilliant combination of classical, almost computational, methods and Grothendieck's new scheme-theoretic techniques.

Mumford's first approach was by the 19th century theory of invariants. In fact, he revived this moribund theory by considering its geometric significance. In pursuing an idea of Hilbert, Mumford was led to the crucial notion of "stable" objects in a moduli problem. The abstract setting behind this notion is the following: Suppose G is a reductive algebraic group acting on a variety V in projective space $\boldsymbol{P}_N$ by projective transformations. Then the action of G is induced by a linear and unimodular representation of some finite covering G^* of G on the affine cone $\boldsymbol{A}^{N+1}$ over the ambient $\boldsymbol{P}_N$. Mumford defines a point $x \in V$ to be *stable* for the action of G on V, relative to the embedding $V \subset \boldsymbol{P}_N$, if for one (and hence every) point $x^* \in \boldsymbol{A}^{N+1}$ over x, the orbit of x^* under G^* is closed in $\boldsymbol{A}^{N+1}$, and the stabilizer of x^* is a finite subgroup of G^*. His fundamental theorem is then that the set of stable points is an open set V_s in V, and V_s/G is a quasi-projective variety.

For example, suppose $V = (\boldsymbol{P}_n)^m$ is the variety of ordered m-tuples of points in projective n-space and G is PGL_n acting diagonally on V via the Segre embedding. Then a point $x = (x_1, x_2, \ldots, x_m) \in V$ is stable if and only if for each proper linear subspace $L \subset \boldsymbol{P}_n$, the number of points $x_i \in L$ is strictly less than $m(\dim L + 1)$ $/(n+1)$. In case $n = 1$, for example, this means that an m-tuple of point on the projective line is unstable if more than half the points coalesce. The reason such m-tuples must be excluded is the following: Let $P_t = (tx_1, tx_2, \ldots, tx_r, x_{r+1}, \ldots, x_m)$ and $Q_t = (x_1, \ldots, x_r, t^{-1}x_{r+1}, \ldots, t^{-1}x_m)$, where the x_i are pairwise distinct. Then P_t is in the same orbit as Q_t, for $t \neq 0, \infty$, but $P_0 = (0, \ldots, 0, x_{r+1}, \ldots, x_m)$ is not in the same orbit as $Q_0 = (x_1, \ldots, x_r, \infty, \ldots, \infty)$ unless $m = 2r$, and even then is not in general. Thus if we want a separated orbit space in which $\lim_{t\to 0}$ (Orbit P_t) is unique, we must exclude P_0 or Q_0; and it is natural to exclude the one with more than half its components equal.

Using the existence of the orbit spaces V_s/G, Mumford was able to construct a moduli scheme over the ring of integers for polarized abelian varieties, relative Picard schemes (following a suggestion of Grothendieck), and also moduli varieties for "stable" vector bundles on a curve in characteristic 0. The meaning of stability for a vector bundle is that all proper sub-bundles are less ample than the bundle itself, if we measure the ampleness of a bundle by the ratio of its degree to its rank. In the special example $V = (\boldsymbol{P}_n)^m$ mentioned above, the results can be proved by explicit computations which work in any characteristic and even over the ring of integers. But in its general abstract form Mumford's theory was limited to characteristic 0 because his proofs used the semisimplicity of linear representations. He conjectures that in characteristic p, linear representations of the classical semisimple groups have the property that complementary to a stable line in such a representation there is always a stable hypersurface (though not necessarily a stable hyperplane which would exist if the representation were semisimple). If this conjecture is true[1] then Mumford's treatment of geometric invariant theory would work in characteristic p. Seshadri has proved the conjecture in case of SL_2. He has also shown recently that the conjecture can be circumvented, by giving different more complicated proofs for the main results of the theory which work in any characteristic.

For moduli of abelian varieties and curves, Mumford has given more refined constructions than those furnished by geometric invariant theory. In three long papers in *Inventiones Mathematicae* he has developed an algebraic theory of theta functions. Classically, over the complex numbers, a theta function for an abelian variety A can be thought of as a complex function on the universal covering space $H_1(A, \boldsymbol{R})$ which transforms in a certain way under the action of $H_1(A, \boldsymbol{Z})$. For Mumford, over any algebraically closed field k, a theta function is a k-valued function on $\Pi_{l\in S} H_1(A, \boldsymbol{Q}_l)$ (étale homology) which transforms in a certain way under $\Pi_{l\in S} H_1(A, \boldsymbol{Z}_l)$. Here S is any finite set of primes $l \neq$ char (k), though in treating some of the deeper aspects of the theory Mumford assumed $2 \in S$. In order to get an idea of what these theta functions accomplish let us consider a classical special case. Let A be an elliptic curve over $\boldsymbol{C}$ with its points of order 4 marked. Then we get a canonical embedding $A \hookrightarrow \boldsymbol{P}^3$ via the theta functions $\theta\begin{bmatrix} a \\ b \end{bmatrix}$; $a, b = 0, 1$. Let 0_A be the origin on A, whose coordinates in $\boldsymbol{P}_3$ are the "theta Nullwerte". Then A is the intersection of all quadric surfaces in $\boldsymbol{P}_3$ which

[1](ADDED DURING CORRECTION OF PROOFS). The conjecture is true; shortly after the Congress, it was proved by W. Haboush in general and by E. Formanek and C. Procesi for $\mathrm{GL}(n)$ and $\mathrm{SL}(n)$.

pass through the orbit of 0_A under a certain action of $(\boldsymbol{Z}/4\boldsymbol{Z}) \times (\boldsymbol{Z}/4\boldsymbol{Z})$ on $\boldsymbol{P}_3$. Thus 0_A determines A and can be viewed as a "modulus". Moreover, 0_A lies on the quartic curve $\theta \begin{bmatrix} 0 \\ 0 \end{bmatrix}^4 = \theta \begin{bmatrix} 0 \\ 1 \end{bmatrix}^4 + \theta \begin{bmatrix} 1 \\ 0 \end{bmatrix}^4$ in the plane $\theta \begin{bmatrix} 1 \\ 1 \end{bmatrix} = 0$, and that curve minus a finite set of points is a variety of moduli for elliptic curves with their points of order 4 marked. Mumford's theory generalizes this construction to abelian varieties of any dimension, with points of any order $\geqq 3$ marked, in any characteristic $\neq 2$. The moduli varieties so obtained have a natural projective embedding, and their closure in that embedding is, essentially, an algebraic version of Satake's topological compactification of Siegel's moduli spaces. Besides these applications to moduli, the theory gives new tools for the study of a single abelian variety by furnishing canonical bases for all linear systems on it.

Next I want to mention briefly p-adic uniformization. Motivated by the study of the boundary of moduli varieties for curves, i.e., of how nonsingular curves can degenerate, Mumford was led to introduce p-adic Schottky groups, and to show how one can obtain certain p-adic curves of genus $\geqq 2$ transcendentally as the quotient by such groups of the p-adic projective line minus a Cantor set. The corresponding theory for genus 1 was discovered by the author, but the generalization to higher genus was far from obvious. Besides its significance for moduli, Mumford's construction is of interest in itself as a highly nontrivial example of "rigid" p-adic analysis.

The theta functions and p-adic uniformization give some insight into what happens on the boundary of the varieties of moduli of curves and abelian varieties, but a much more detailed picture can now be obtained by Mumford's theory of toroidal embeddings. This theory, which unifies ideas that had appeared earlier in the works of several investigators, reduces the study of certain types of varieties and singularities to combinatorial problems in a space of "exponents". The local model for a toroidal embedding is called a torus embedding. This is a compactification $\overline{V}$ of a torus V such that the action of V on itself by translation extends to an action of V on $\overline{V}$. The coordinate ring of V is linearly spanned by the monomials $x^a = x_1^{a_1} x_2^{a_2} \cdots x_n^{a_n}$, $n = \dim V$, with positive or negative integer exponents a_i. Viewing the exponent vectors a as integral points in Euclidean n-space, define a *rational cone* in that space to be a set consisting of r's such that $(r, a) \geqq 0$ for $a \in S$, where S is some finite set of exponent vectors. For each rational cone σ, the monomials x^a such that $(r, a) \geqq 0$ for all $r \in \sigma$ span the coordinate ring of an affine variety $V(\sigma)$ which contains V as an open dense subvariety, if σ contains no nonzero linear subspace of $\boldsymbol{R}^n$. Now if we decompose $\boldsymbol{R}^n$ into the union of a finite number of rational cones σ_α in such a way that each intersection $\sigma_\alpha \cap \sigma_\beta$ is a face of σ_α and σ_β, then the union of the $V(\sigma_\alpha)$ is a compactification of V of the type desired. All such compactifications $\overline{V}$ of V can be obtained in this way and the invariant sheaves of ideals on them can be described in terms of the decomposition into cones. One can also read off whether $\overline{V}$ is nonsingular, and if it is not one can desingularize it by suitably subdividing the decomposition. In short, there is a whole dictionary for translating questions about the algebraic geometry of V and $\overline{V}$ into combinatorial questions about decompositions of $\boldsymbol{R}^n$ into rational cones.

Mumford with the help of his coworkers has used these techniques to prove the following semistable reduction theorem. If a family of varieties X_t over $\boldsymbol{C}$, in general nonsingular, is parametrized by a parameter t on a curve C, and if X_{t_0} is singular, then one can pull back the family to a ramified covering of C in a neighborhood of t_0 and blow it up over t_0 in such a way that the new singular fiber is of the stablest possible kind, i.e., is a divisor whose components have multiplicities 1 and cross transversally. The corresponding problem in characteristic p is open. For curves in characteristic p the result was proved by Mumford and Deligne and was a crucial step in their proof of the irreducibility of the moduli variety for curves of given genus.

Toroidal embeddings can also be used to construct explicit resolutions of the singularities of the projective varieties $\overline{D/\Gamma}$, where D is a bounded symmetric domain, Γ is an arithmetic group, and the bar denotes the "minimal" compactification of Baily and Borel. The construction of these resolutions is a big step forward. With them one has a powerful tool to analyse the behavior of functions at the "boundary", compute numerical invariants, and, generally, study the finer structure of these varieties.

I hope this report, incomplete as it is, gives some idea of Mumford's achievements and their importance. I heartily congratulate him on them and wish him well for the future!

Harvard University
Cambridge
Massachusetts 02138
U.S.A.

David Mumford

AUTOBIOGRAPHY OF DAVID MUMFORD

I was born in 1937 in an old English farm house. My father was British, a visionary with an international perspective, who started an experimental school in Tanzania based on the idea of appropriate technology and worked during my childhood for the newly created United Nations. He imbued me with old testament ideas of your obligation to fully use your skills and I learned science with greed. My mother came from a well-to-do New York family. I grew up on Long Island Sound, and went to Exeter and Harvard.

At Harvard, a classmate said "Come with me to hear Professor Zariski's first lecture, even though we won't understand a word" and Oscar Zariski bewitched me. When he spoke the words 'algebraic variety', there was a certain resonance in his voice that said distinctly that he was looking into a secret garden. I immediately wanted to be able to do this too. It led me to 25 years of struggling to make this world tangible and visible. Especially, I became obsessed with a kind of passion flower in this garden, the moduli spaces of Riemann. I was always trying to find new angles from which I could see them better. There were other amazing people in this garden, the acerbic and daunting Andre Weil, Alexander Grothendieck who truly seemed to be another species, and my fellow travelers, Artin, Griffiths and Hironaka, who made it seem progress was possible. I stayed at Harvard from my freshman year through most of my career as this was the world center for algebraic geometry.

At Radcliffe I met Erika Jentsch. She rescued me from what might have been an all too private world and we had a beautiful family of four children, Stephen, Peter, Jeremy and Suchitra. We spent summers in Maine, in the woods and on the sea sailing along the coast. She became a poet, kept winning prizes and I learned to go places as 'the spouse'. My life changed in the 80's. Erika passed away. I turned from algebraic geometry to an old love — is there a mathematical approach to understanding thought and the brain? This is applied mathematics and I have to say I don't think theorems are very important here. I met remarkable people who showed me the crucial role played by statistics, Grenander, Geman and Diaconis. My mother always said, as you get older, your horizons expand. With my second wife, Jenifer Gordon, we share seven children, now grown up with dreams and families of their own, and I took a new job at Brown, which is the world center for this approach to intelligence. The article reproduced here is a survey of this theory, meant to convince my mathematical colleagues I hadn't gone mad.

Oct. 17, 1996

PATTERN THEORY: A UNIFYING PERSPECTIVE

by
DAVID MUMFORD*

Department of Mathematics, Harvard University

1.1. Introduction

The term "Pattern Theory" was introduced by Ulf Grenander in the 70's as a name for a field of applied mathematics which gave a theoretical setting for a large number of related ideas, techniques and results from fields such as computer vision, speech recognition, image and acoustic signal processing, pattern recognition and its statistical side, neural nets and parts of artificial intelligence (see Grenander 1976–81). When I first began to study computer vision about ten years ago, I read parts of this book but did not really understand his insight. However, as I worked in the field, every time I felt I saw what was going on in a broader perspective or saw some theme which seemed to pull together the field as a whole, it turned out that this theme was part of what Grenander called pattern theory. It seems to me now that this is the right framework for these areas, and, as these fields have been growing explosively, the time is ripe for making an attempt to reexamine recent progress and try to make the ideas behind this unification better known. This article presents pattern theory from my point of view, which may be somewhat narrower than Grenander's, updated with recent examples involving interesting new mathematics.

The problem that Pattern Theory aims to solve, which I would like to call the 'Pattern Understanding Problem', may be described as follows:

the analysis of the patterns generated by the world in any modality, with all their naturally occurring complexity and ambiguity, with the goal of reconstructing the processes, objects and events that produced them and of predicting these patterns when they reoccur.

These patterns may occur in the signals generated by one of the basic animal senses. For example *vision* usually refers to the analysis of patterns detected in the electromagnetic signals of wavelengths 400–700 nm incident at a point in space from different directions. *Hearing* refers to the analysis of the patterns present in

*Supported in part by NSF Grant DMS 91-21266 and by the Geometry Center, University of Minnesota, a STC funded by NSF, DOE and Minnesota Technology Inc.

the oscillations of 60–20,000 hertz in air pressure at a point in space as a function of time, both with and without human language. On the other hand, these patterns may occur in the data presented to a higher processing stage, in so-called cognitive thought. As an example, *medical expert systems* are concerned with the analysis of the patterns in the symptoms, history and tests presented by a patient: this is a higher level modality, but still one in which the world generates confusing but structured data from which a doctor seeks to infer hidden processes and events. Finally, these patterns may be temporal patterns occurring in a sequence of motor actions and the resulting sensations, e.g. a sequence of complex signals 'move-feel-move-feel- etc.' involving motor commands and the resulting tactile sensation in either an animal or a robot.

Pattern Theory is one approach to solving this pattern understanding problem. It contains three essential components which we shall detail in Sections 1.2, 1.3 and 1.4 below. These concern firstly the abstract mathematical setup in which we frame the problem; secondly a hypothesis about the specific mathematical objects that arise in natural pattern understanding problems; and thirdly a general architecture for computing the solution.

Before launching into this description of pattern theory, we want to give two examples of simple sensory signals and the patterns that they exhibit. This will serve to motivate and make more concrete the theory which follows. The first example involves speech: Figure 1.1 shows the graph of the pressure $p(t)$ while the word "SKI" is being pronounced. Note how the signal shows four distinct wave forms: something close to white noise during the pronunciation of the sibilant 'S', then silence followed by a burst which conveys the plosive 'K', then an extended nearly musical note for the vowel 'I'. The latter has a fundamental frequency corresponding to the vibration of the vocal cords, with many harmonics whose power peeks around three higher frequencies, the formants. Finally, the amplitude of the whole is modulated during the pronunciation of the word. In this example, the goal of auditory signal processing is to identify these four wave forms, characterize each in terms of its frequency power spectrum, its frequency and amplitude modulation, and then, drawing on a memory of speech sounds, identify each wave form as being produced by the corresponding configurations of the speaker's vocal tract, and finally label each with its identity as an english phoneme. In addition, one would like to describe explicitly the stress, pitch and the quality of the speaker's voice, using this later to help disambiguate the identity of the speaker and the intent of the utterance.

Figure 1.2a shows the graph of the light intensity $I(x, y)$ of a picture of a human eye: it would be hard to recognize this as an eye, but the black and white image defined by the same function is shown in Figure 1.2b. Note again how the domain of the signal is naturally decomposed into regions where I has different values or different spatial frequency behavior: the pupil, the iris, the whites of the eyes, the lashes, eyebrows and skin. These six regions are the six types of visible surface in this image, characterized in real world terms by distinct albedo, texture and geometry.

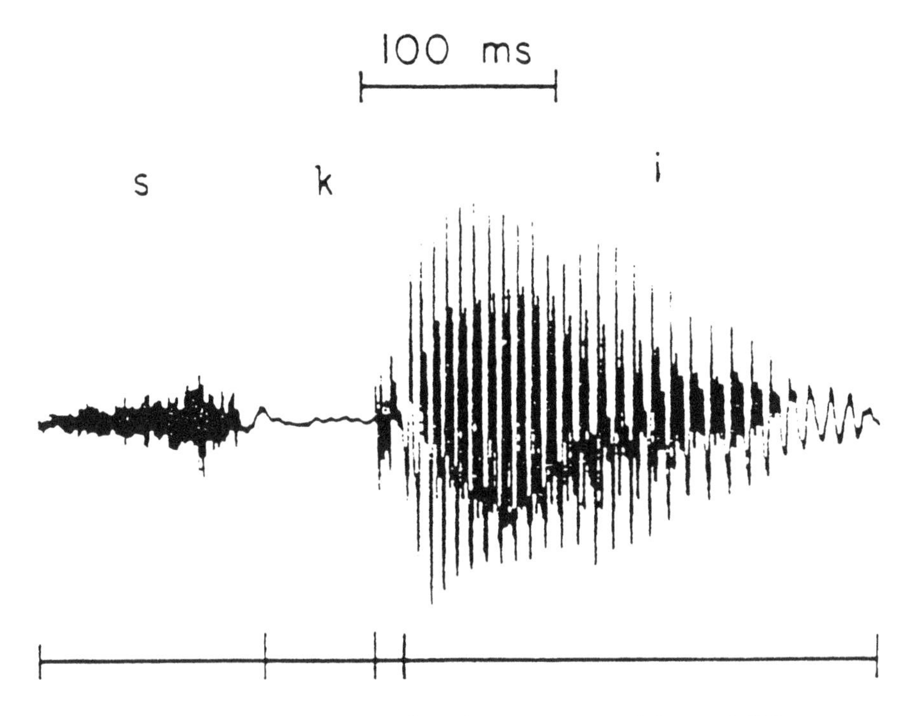

Figure 1.1. Acoustic waveform for an utterance of the word *SKI*.

The goal of visual signal processing is again to describe this function of two variables as being built up from simpler signals on subdomains, on which it either varies slowly or is statistically regular, i.e. approximately stationary and hence to describe the distinct 3D visible surface patches. In other words, each surface patch is assumed to have an identifiable 'texture', which may be the result of its particular spatial power spectrum or may result from it being composed of some elementary units, called *textons*, which are repeated with various modifications. These modifications in particular include spatial distortion, contrast modulation and interaction with larger scale structures. This description of the signal may be computed either prior to or concurrent with a comparison of the signal with remembered eye shapes, which include a description of the expected range of variation of eyes, specific descriptions of the eyes of well-known people, etc.

Note that in both of these examples, something rather remarkable happened. The simplest description of the signal as an abstract function, e.g. having subdomains on which it is relatively homogeneous, leads naturally to a description of the true processes and objects that produced the signal. We will discuss this further below, in connection with the idea of 'minimum description length'.

In order to understand what the field of pattern theory is all about, it is necessary to begin by addressing a major misconception — namely that the whole problem is essentially trivial. The history of computer speech and image recognition projects, like the history of AI, is along one of ambitious projects which attained their goals with carefully tailored artificial input but which failed as soon as more of the complexities of real world data were present. The source of this misconception, I believe, is our subjective impression of perceiving instantaneously and effortlessly the significance of the patterns in a signal, e.g. the word being spoken or which face is being seen. Many psychological experiments however have shown that what we perceive is not the true sensory signal, but a rational reconstruction of what the signal should be. This means that the messy ambiguous raw signal never makes it to our consciousness but gets overlaid with a clearly and precisely patterned version whose computation demands extensive use of memories, expectations and logic. An example of how misleading our impressions are is shown in Figure 1.3: the reader will instantly recognize that it is an image of an old man sitting on a park bench. But ask yourself — how did you know that? His face is almost totally obscure, with his hand merging with his nose; the most distinct shape is that of his hat, which by itself could be almost anything; even his jacket merges in many places with the background because of its creases and the way light strikes them,

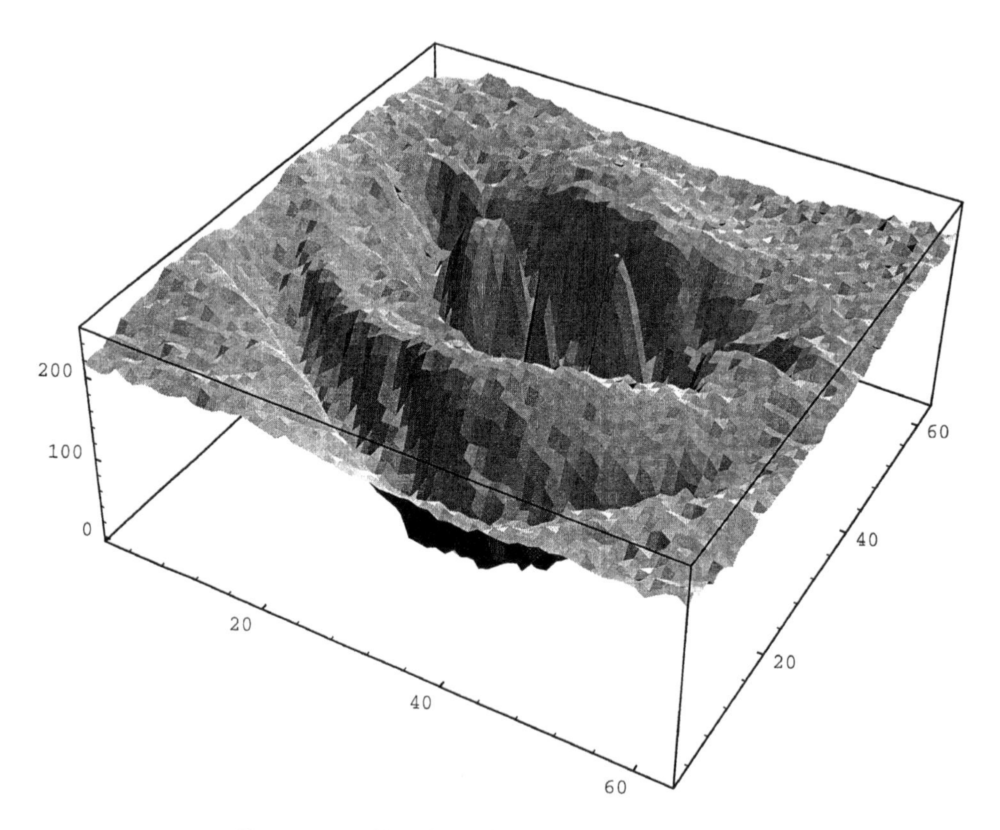

Figure 1.2a. Visual waveform for an image of an eye.

Figure 1.2b. Identical waveform presented using variable intensity.

Figure 1.3. A challenge image for computers to recognize.

so no simple algorithm is going to trace its contour without getting lost. However, when you glance at the picture, in your mind's eye, you 'see' the face and its parts distinctly, the man's jacket is a perfectly clear coherent shape, whose creases in fact contribute to your perception of its 3-dimensional structure instead of confusing you. The ambiguities which must, in fact, have been present in the early stages of your processing of this image never become conscious because you have found an explanation of every peculiarity of the image, a match with remembered shapes and lighting effects. In fact, the problems of pattern theory are hard, and although major progress has been made in both speech and vision in the last 5 years, a robust solution has not been achieved!

1.2. Mathematical Formulation of the Problem

To make the field of pattern theory precise, we need to formulate it mathematically. There are three parts to this which all appear in Grenander's work: the first is the description of the players in the field, the fundamental mathematical objects which appear in the pattern understanding problem. The second is to restrict the possible generality of these objects by using something about the nature of the world. This gives us a more circumscribed, more focussed set of problems to study. Finally, the goal of the field being the reconstruction of hidden facts about the world, we aim primarily for algorithms, not theorems, and the last part is the general framework for these algorithms. In this section, we look at the first part, the basic mathematical objects of pattern theory. This formulation of the pattern understanding problem is not unique to pattern theory. It has been introduced many times, e.g. in the hidden Markov models of speech understanding, in work on expert systems, in connectionist analyses of neural nets, etc.

There are two essentially equivalent formulations, one using Bayesian statistics and one using information theory. The statistical approach (see for instance D. Geman, 1991) is this: consider all possible signals $f(\mathbf{x})$ which may be perceived. These may be considered as elements of a space Ω_{obs} of functions $f : \mathcal{D} \to \mathcal{V}$. For instance, speech is defined by pressure $P : [t_1, t_2] \to \mathbf{R}_+$ as a function of time, color vision is defined by intensity I on a domain $\mathcal{D} \subset S^2$ of visible rays with values in the convex cone of colors $\mathcal{V}_{\text{RGB}} \subset \mathbf{R}^3$, or these functions may be sampled on a discrete subset of the above domains, or their values may be approximated to finite precision, etc. The first basic assumption of the statistical approach is that nature determines a probability p_{obs} on a suitable σ-algebra of subsets of Ω_{obs}, and that, in life, one observes random samples from this distribution. These signals, however, are highly structured as a result of their being produced by a world containing many processes, objects and events which don't appear explicitly in the signal. This means many more random variables are needed to describe the state of the world. The second assumption is that the possible states w of the world form a second probability space Ω_{wld} and that there is a big probability distribution $p_{o,w}$ on $\Omega_{\text{obs}} \times \Omega_{\text{wld}}$ describing the probability of both observing f and the world being in state w. Then p_{obs} should be the marginal distribution of $p_{o,w}$ on Ω_{obs}. The mathematical problem, in this setup, is to infer the state of the world w, given the measurement f, and for this we may use Bayes's rule:

$$p(w|f) = \frac{p(f|w) \cdot p(w)}{p(f)} \tag{1.1}$$

leading to the *maximum a posteriori reconstruction of the state of the world*†:

$$\text{MAP estimate of } w = \arg\max_w [p(f|w) \cdot p(w)] \tag{1.2}$$

To use the statistical approach, therefore, we must construct the probability space $(\Omega_{\text{obs}} \times \Omega_{\text{wld}}, p_{o,w})$ and finding algorithms to compute the MAP-estimate.

The information theoretic approach has its roots in work of Barlow (1961) (see also Rissanen, 1989). Assume D and V, hence Ω_{obs} are finite. The idea is that instead of writing out any particular perceptual signal f in raw form as a table of values, we seek a method of encoding f which minimizes its expected length in bits: i.e. we take advantage of the patterns possessed by most f to encode them in a compressed form. We consider coding schemes which involve choosing various auxiliary variables w and then encoding the particular f using these w (e.g. w might determine a specific typical signal f_w and we then need only to encode the deviation $(f - f_w)$). We write this:

$$\text{length}(\text{code}(f, w)) = \text{length}(\text{code}(w)) + \text{length}(\text{code}(f \text{ using } w)). \tag{1.3}$$

The mathematical problem, in the information theoretic setup, is, for a given f, to find the w leading to the shortest encoding of f, and moreover, to find the encoding *scheme* leading to the shortest expected coding of all f's. This optimal choice of w is called the *minimum description length* or MDL estimate of w:

$$\text{MDL est. of } w = \arg\min_w [\text{len}(\text{code}(w)) + \text{len}(\text{code}(f \text{ using } w))]. \tag{1.4}$$

Finding the optimal encoding scheme for all the signals of the world is obviously impractical and perhaps even contradictory (recall the problem of the finding the smallest integer not definable in less than twenty words!). What is really meant by MDL is that we seek some approximation to the optimal encoding, using coding schemes constrained in various ways†. Pattern theory proposes that there are quite specific kinds of encoding schemes which very often give good results and which may, therefore, be built into our thinking: we shall discuss these in the next section.

†Other statistical procedures can be used for estimating the state of the world: e.g. taking the mean of various world variables in the posterior distribution, or minimum risk estimates, etc.

†One could argue that the hypothesis of vengeful sky gods with human emotions was an early MDL hypothesis for the deeper processes of the world, and that modern science only bettered its description length when people sought to describe the quantitative signals of the world with more bits and with much larger samples including those outliers which we call 'experiments'.

There is a close link between the Bayesian and the information-theoretic approaches which comes from Shannon's optimal coding theorem. This theorem states that given a class of signals f, the coding scheme for such signals for which a random signal has the smallest expected length satisfies:

$$\text{len}(\text{code}(f)) = -\log_2 p(f) \tag{1.5}$$

(where fractional bit lengths are achieved by actually coding several f's at once, and doing this the LHS gets asymptotically close to the RHS when longer and longer sequences of signals are encoded at once). We may apply Shannon's theorem both to encoding w and to encoding f, given w. For these encodings $\text{len}(\text{code}(w) = -\log_2 p(w)$ and $\text{len}(\text{code}(f \text{ using } w) = -\log_2 p(f|w)$. Therefore, taking $\log_2$ of equation (1.2), we get equation (1.4) and find that the MAP estimate of w is the same as the MDL estimate.

There is an additional wrinkle in the link between the two approaches. Why shouldn't the search for optimal encodings of world signals lead you to odd combinatorial coding schemes which have nothing to do with what is actually happening in the world? In the previous paragraph, we have assumed that the encoding scheme for f was based on encoding the true world variables w and using them to encode f. But one can imagine discovering some strange numerology (like Kepler's hypothesis for spacing the orbits of the planets via the inscribed and circumscribed spheres of the platonic polyhedra) which gave a concise description of some class of signals or measurements but had nothing to do with the true objects or processes of the world. This does not seem to happen in real life! We encountered an example of this in Section 1.1 where we noticed that finding time intervals in which a speech signal had a nearly constant spectrogram, hence was concisely codable, also gave us the time intervals in which the mouth of the speaker was in a particular position, i.e. a specific phoneme was being articulated. Similarly, finding parts of images where the texture was nearly constant usually gives coherent 3D surface patches. In fact, it seems that the search for short encodings leads you *automatically, without prior knowledge of the world* to the same hidden variables on which the Bayesian theory is based. Insofar as this can be relied on, the information-theoretic approach has the great advantage over the Bayesian approach that it does not require that you have *a prior* knowledge of the physics, chemistry, biology, sociology, etc. of the world, but gives you a way of discovering these facts.

A very simple example may be useful here. Suppose five different bird songs are heard regularly in your back yard. You can assign a short distinctive code to each such song, so that instead of having to remember the whole song from scratch each time, you just say to yourself something like "Aha, song #3 again". Note that in doing so, you have automatically learned a world variable at the same time: the number or code you use for each song is, in effect, a name for a species, and you have rediscovered part of Linnaean biology. Moreover, if one bird is the most frequent singer, you will probably use the shortest code, e.g. "song #1", for that bird. In this way, you are also learning the probability of different values for the variable "song #x", as in Shannon's fundamental theorem. In Section 1.5.4, we will give a more extended example of how this works.

1.3. Four Universal Transformations of Perceptual Signals

The above formulation of the pattern understanding problem provides a framework in which to analyze signals, but it says nothing about the nature of the patterns which are to be expected, what distortions, complexities and ambiguities are to be expected, hence what sorts of probability spaces Ω_{obs} are we likely to encounter, how shall we encode them, etc. What gives the theory its characteristic flavor is the hypothesis that the world does not have an infinite repertoire of different tricks which it uses to disguise what is going on. Consider the coding schemes used by engineers for the transmission of electrical signals: they use a small number of well-defined transformations such as AM and FM encoding, pulse coding, etc. to convert information into a signal which can be efficiently communicated. Analogous to this, the world produces sound to be heard, light to be seen, surfaces to be felt, etc. which are all, in various ways, reflections of its structure. We may think of these signals as the productions of a particularly perverse engineer, who sets us the problem of decoding this message, e.g. of recognizing a friend's face or estimating the trajectory of oncoming traffic, etc. The second contention of pattern theory is that such signals are derived from the world by *four types of transformations or deformations*, which occur again and again in different guises. In the terminology of Grenander (1976) simple unambiguous signals from the world are referred to as *pure images* and the transformations on them are called *deformations*, which produce the actually observed perceptual signals which he called *deformed images*. The bad news is that these four transformations produce much more complex effects than the coding schemes of engineers, hence the difficulty of decoding them by the standard tricks of electrical engineering. The good news is that these transformations are not arbitrary recursive operations which produce unlearnable complexity.

A very similar situation occurs in the study of the syntax of languages. In the formal study of the learnability of the syntax of language, Gold's theorem gives very strong restrictions on what languages can be learned if their syntax is at all general (see Osherson & Weinstein (1984) for an excellent exposition). In contrast, Chomsky (1981) has suggested that all languages have essentially the same syntax, with individual languages differing only by the setting of a small number of parameters. In fact, transformational grammar has a very similar structure to pattern theory: each sentence has an underlying deep structure, analogous to Grenander's pure images. It is subjected to a restricted set of transformations, analogous to Grenander's deformations. And finally one observes the spoken surface form of the sentence, analogous to Grenander's deformed image.

The study of perceptual signals suggests a small number of special transformations, or deformations, that the languages in which perceptual signals speak are of very special types. The exact set of these is not completely clear at this point and my choices are not exactly the same as Grenander's. But to make progress, we must make some hypothesis and so I give here a set of four transformations which seem to me to suffice. These are:

(i) *Noise and blur.* These effects are the bread and butter of standard signal processing, caused for instance by sampling error, background noise and imperfections in your measuring instrument such as imperfect lenses, veins in front of the retina, dust and rust. A typical form of this transformation is given by the formula:

$$I \to (I * \sigma)(x_i) + n_i \tag{1.6}$$

where σ is a blurring kernel, x_i are the points where thesignal is sampled and n_i is some kind of additive noise, e.g. Gaussian, but of course much more complex formulae are possible. Especially significant is that Gaussian noise is usually a poor model of the noise effects, for example when the noise is caused by finer detail which is not being resolved. Rosenfeld calls such an n *clutter*, which conveys what it often represents. The key feature of noise, in whatever guise, is that it has no significant remaining patterns, hence cannot be compressed significantly by recoding. These transformations are part of what Grenander calls *changes in contrast.* When they are present, the unblurred noiseless I should be one of the variables w as getting rid of noise and blur reveals the hidden processes of the world more clearly.

(ii) *Superposition.* Signals typically can be decomposed into simpler components. Fourier analysis is the best studied example of this, in which a signal is written as a linear combination of sinusoidal functions. But the whole development of wavelets has shown that there are many other such expansions, appropriate for particular classes of signals. Most such superpositions have the property that the various components have different scales, but some may combine several on the same scale (e.g. one can superimpose 2D Gabor functions with different orientations; for another example, faces with arbitrary illumination can be approximated by the superposition of about half a dozen 'eigenfaces'). Most such superpositions are linear, but some may be more complex (e.g. in amplitude modulation, a low frequency signal plus a large enough positive constant is multiplied by a high frequency 'carrier'). In images, local properties such as sharp edges and texture details may be constructed by adding Gabor functions or model step edges with small support, while global properties like slowly varying shading or large shapes may be obtained by adding slowly varying functions with large support. In speech, information is conveyed by the highest frequency formants, by the lower frequency vibration of the vocal cords and the even slower modulation of stress. A typical form of this transformation is given by the formula:

$$I_1, \ldots, I_n \longrightarrow (I_1 + \cdots + I_n) \quad \text{or} \quad \sigma(I_1, \ldots, I_n) \tag{1.7}$$

where $I_1, \ldots, I_n$ represent component signals often in disjoint frequency bands, which can be combined either additively or by some more complex rule σ. The individual components I_k of I should be included in the variables w.

(iii) *Domain warping.* Two signals generated by the same object or event in different contexts typically differ because of expansions or contractions of their domains, possibly at varying rates: phonemes may be pronounced faster or slower, the image of a face is distorted by varying expression and viewing angle. In speech, this is called 'time warping' and in vision, this is modeled by 'flexible templates'. In both cases, a diffeomorphism of the domain of the signal brings the variants much closer to each other, so that this transformation is given by:

$$I \longrightarrow (I \circ \psi) \tag{1.8}$$

where ψ represents a diffeomorphism of the domain of I. These transformations are what Grenander calls *background deformations.* The diffeomorphism ψ should be one of the variables w.

(iv) *Interruptions.* Natural signals are usually analyzed best after being broken up into pieces consisting of their restrictions to subdomains. This is because the world itself is made up of many objects and events and different parts of the signal are caused by different objects or events. For instance, an image may show different objects partially occluding each other at their edges, as in Figure 1.3 where the old man is an object which occludes part of the park bench or as in a tiger seen behind a fragmented foreground of occluding leaves. In speech, the phonemes naturally break up the signal and, on a larger scale, one speaker or unexpected sound may interrupt another. Such a transformation is given by a formula like:

$$I_1, I_2 \longrightarrow (I_1|_{D'}, I_2|_{D-D'}) \tag{1.9}$$

where I_2 represents the background signal which is interrupted by the signal I_1 on a part D' of its domain D, (or I_2 may only be defined on $D - D'$). This type of deformation is called *incomplete observations* by Grenander. The components I_k and the domain D' should be included in the variables w.

What makes pattern theory hard is not that any of the above transformations are that hard to detect and decode in isolation, but rather that all four of them tend to coexist, and then the decoding becomes hard.

Requires Pattern Synthesis

Taking the Bayesian definition of the objects of pattern theory, we note that the probability distribution $(\Omega_{\text{obs}} \times \Omega_{\text{wld}}, p_{o,w})$ allows you to do two things. On the one hand, we can use it to define the MAP-estimate of the state of the world; but we can also sample from it, possibly fixing some of the world variables w, using this distribution to construct sample signals f generated by various classes of objects or events. A good test of whether your prior has captured all the patterns in some class of signals is to see if these samples are good imitations of life. For this reason, Grenander's idea was that the analysis of the patterns in a signal and the synthesis of these signals are inseparable problems, using a common probabilistic model: computer vision should not be separated from computer graphics, nor speech recognition from speech generation.

Many of the early algorithms in pattern recognition were purely *bottom-up*. For example, one class of algorithms started with a signal, computed a vector of 'features', numerical quantities thought to be the essential attributes of the signal, and then compared these feature vectors with those expected for signals in various categories. This was used to classify images of alpha-numeric characters or phonemes for instance. Such algorithms give no way of reversing the process, of generating typical signals. The problem these algorithms encountered was that they had no way of dealing with anything unexpected, such as a smudge on the paper partially obscuring a character, or a cough in the middle of speech. These algorithms did not say what signals were expected, only what distinguished typical signals in each category.

In contrast, a second class of algorithms works by actively reconstructing the signal being analyzed. In addition to the bottom-up stage, there is a *top-down* stage in which a signal with the detected properties is synthesized and compared to the present input signal. What needs to be checked is whether the input signal agrees with the synthesized signal to within normal tolerances, or whether the residual is so great that the input has not been correctly or fully analyzed. This architecture is especially important for dealing with the fourth type of transformation: interruptions. When these are present, the features of the two parts of the signal get confused. Only when the obscuring signal is explicitly labelled and removed, can the features of the background signal be computed. We may describe this top-down stage as 'pattern reconstruction' in distinction to the bottom-up purely pattern recognition stage. A flow chart for such algorithms is shown in Figure 1.4.

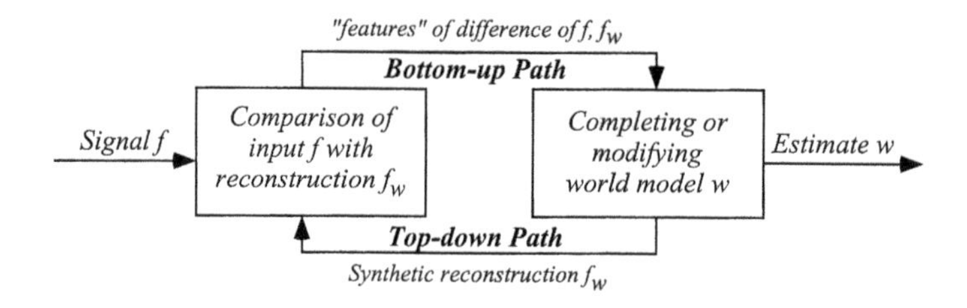

Figure 1.4. The fundamental architecture of pattern theory.

The question of whether the correct interpretation of real world signals can be solved by a purely bottom-up algorithm, or whether it requires an iterative bottom-up/top-down architecture has been widely debated for a long time. The first person, to my knowledge, to introduce the above type of iterative architecture, was Donald MacKay (1956). On the other hand, Marr was a strong believer in the purely feedforward architecture, claiming that one needed merely to develop better algorithms to deal with things like interruptions in a purely feedforward way. A strong argument for the necessity of a top-down stage for the recognition of heavily degraded signals, such as faces in deep shadow, is given in Cavanagh (1991). Neural net theory has gone in both directions: while 'back propagation' nets categorize in an exclusively bottom-up manner (only using feedback in their learning phase), the 'attractive neural nets' with symmetric connections (Hopfield, 1982; D. Amit, 1989) seek not merely to categorize but also to construct the prototype ideal version of the category by a kind of pattern completion which they call 'associative memory'. What these nets do not do is to go back and attempt to compare this reconstruction with the actual input to see if the full input has been 'explained'. One demonstration system that does this is Grossberg and Carpenter's 'adaptive resonance theory' (Carpenter & Grossberg, 1987). A proposal for the neuroanatomical substrate for such bottom-up/top-down loops in mammalian cortex is put forth in Mumford (1991–92). One reason Marr rejected the complex top-down feedback architecture is that it seemed to take too long to converge to be biologically plausible. This argument, however, ignores the fact that a large proportion of the time, we can anticipate the next stimulus, either by extrapolating from the preceding stream of stimuli or by drawing on memories of shapes and sounds. In this situation, the top-down pathway may actively synthesize a guess for the next stimulus even before it arrives, and convergence is fast unless something totally unexpected happens.

The third part of our definition of pattern theory is the hypothesis that no practical feedforward algorithm exists for computing the most likely values of the world variables w from signals f. But that if your algorithm explicitly models the generation process, starting with a guess for w (or a set of guesses), then generating an f_w, then deforming w, combining and extending these guesses, you can solve the problem of computing the most likely w in a reasonable time.

1.5. Examples

In this section, we want to present several examples of interesting mathematics which have come out of pattern theory, in attempting to come to grips with one or another of the above universal transformations. These examples are from vision because this is the field I know best, but many of these ideas are used in speech recognition too.

1.5.1. *Pyramids and wavelets*

The problem of analyzing functions that convey information on more than one scale, has arisen in many disciplines. The classical method of separating additively combined scales is, of course, Fourier analysis. But what is usually required is to analyze a function locally simultaneously in its original domain *and* in the domain of its Fourier transform, and Fourier analysis does not do this. In vision, moving closer or farther from an object by a factor σ changes the image $I(x,y)$ of the object into the new image $I(\sigma x, \sigma y)$, thus any feature which occurs on one scale in one image is equally likely to occur at any other scale in a second image. In computer vision, at least as far back as the early 70's, this problem led to the idea of analyzing an image by means of a 'pyramid', e.g. Uhr, 1972; Rosenfeld & Thurston, 1971. In its original incarnation, the main idea was to compute a series of progressively coarser resolution images by blurring and resampling, e.g. a set of $(2^n \times 2^n)$-pixel images, for $n = 10, 9, \ldots, 1$. Putting these together, the resulting data structure looks like an exponentially tapering pyramid. Instead of running algorithms that took time proportional to the width of the image, one ran the algorithms up and down the pyramid, possibly in parallel at different pixels, in time proportional to *log*(width). Typical algorithms that were studied at this time are morphological ones, involving for instance linking and marking extended contours, which have nothing to do with filtering or linear expansions. The bottom layer of this so-called *Gaussian pyramid* held the original image, with both high and low frequency components, although it was used only to add local or high-frequency information.

In the early 80's, the idea of using the pyramid to separate band pass components of a signal and thus to expand that signal arose both in computer vision (Burt & Adelson, 1983) (where they *subtracted* successive layers of the Gaussian pyramid, producing what they called the *Laplacian pyramid*) and in petroleum geology (Grossman & Morlet, 1984). Figure 1.5 shows this Laplacian pyramid for a face: note that the high-frequency differences show textures and sharp edges, while the low frequency differences show large shapes. This work led directly to the idea of wavelets and wavelet expansions which now seem to be the most natural way to analyze a signal locally in both space and frequency. Mathematically, the idea is simply to expand an arbitrary function $f(\mathbf{x})$ of n variables as a sum:

$$f(\mathbf{x}) = \sum_{\text{scale } k \in Z} \left[\sum_{\vec{n} \in \text{lattice} L} \ \sum_{\text{fin.\# of } \alpha} a_{k,\vec{n},\alpha} \psi_\alpha(\lambda^k \mathbf{x} + \vec{n}) \right] \tag{1.10}$$

where the ψ_α are suitable functions, called wavelets, with mean 0. Usually $\lambda = 2$, and, at least in dimension 1, there is a single α and wavelet ψ_α. The original expansions of Burt and Adelson, which are not quite of this form, have been reinvestigated from a more mathematical point of view recently in Mallat (1989). The basic link between the expansion in (1.10) and pyramids is this: define a space V_m to be the set of f's whose expansions involve only terms with $k \leq m$. This

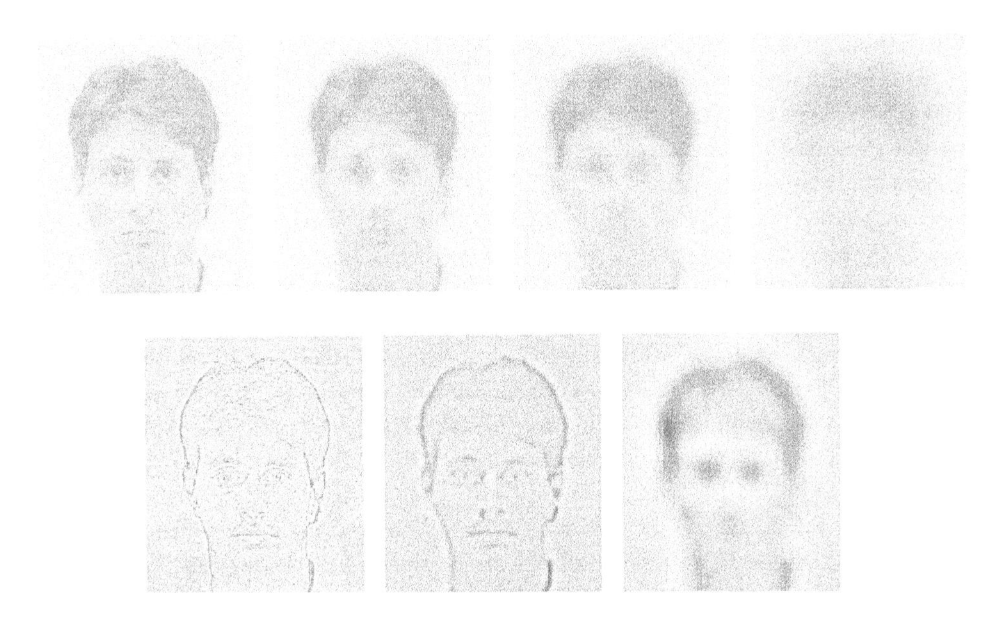

Figure 1.5. The Gaussian and Laplacian pyramids for a face.

defines a 'multi-resolution ladder' of subspaces of functions with more and more detail:

$$\cdots \subset V_{-1} \subset V_0 \subset V_1 \subset \cdots \subset L^2(\mathbf{R}^n) \tag{1.11}$$

such that $f(x) \longmapsto f(2x)$ maps V_m isomorphically onto V_{m+1}. Then one may think of V_m as functions which have been blurred and sampled at a spacing 2^{-m}: i.e. the level of the pyramid of $(2^m \times 2^m)$-pixel images. The mathematical development of the theory of these expansions is due especially to Meyer and Daubechies (see Meyer, 1986; Daubechies, 1988; Daubechies, 1990), who showed that (i) with *very* careful choice of ψ, this expansion is even an orthogonal one, (ii) for many more ψ, the functions on the right form an unconditional but not orthogonal basis of $L^2(\mathbf{R}^n)$ and (iii) for even more ψ, the functions on the right form a 'frame', a set of functions that spans $L^2(\mathbf{R}^n)$ and gives a canonical though non-unique expansion of every f.

From the perspective of pattern theory, we want to make two comments on the theory of wavelets. The first is that they fit in very naturally with the idea of minimum description length. Looked at from the point of view of optimal linear encoding of visual and speech signals (i.e. encoding by linear combinations of the function values), the idea of wavelet expansions is very appealing. This was pointed out early on by Burt and Adelson and data compression has been one of the main applications of wavelet theory ever since. Moreover, its further development leads beyond the classical idea of expanding a function in terms of a fixed basis to the idea of using a much larger spanning set which *oversamples* a function space and using suitably chosen subsets of this set in terms of which to expand or approximate the given function (see Coifman, Meyer & Wickerhouser, 1990, where *wavelet libraries* are introduced). Even though the data needed to describe this expansion or approximation is now both the particular subset chosen and the coefficients, this may be a more efficient code. If so, it should lead us to the correct variables w for describing the world (cf. Section 1.2): for example, expanding a speech signal using wavelet libraries, different bases would naturally be used in the time domains during which different phonemes were being pronounced — thus the break up of the signal into phonemes is discovered as a consequence of the search for efficient coding! It also appears that nature uses wavelet type encoding: there are severe size restrictions on the optic nerve connecting the retina with the higher parts of the brain and the visual signal is indeed transmitted using something like a Burt–Adelson wavelet expansion (Dowling, 1987).

The second point is that wavelets, even in their oversampled form, are still just the linear side of pyramid multi-scale analysis. In our description of multi-scale transformations of signals in Section 1.3, we pointed out that the two scales can be combined by multiplication or a more general function σ as well as by addition. To decode such a transformation, we need to perform some local non-linear step, such as rectification or auto-correlation, at each level of the pyramid before blurring and resampling. An even more challenging and non-linear extension is to a *multi-scale description of shapes*: e.g. subsets $S \subset \mathbf{R}^2$ with smooth boundary. The analog of blurring a signal is to let the boundary of S evolve by diffusion proportional to its curvature (see Gage & Hamilton, 1986; Grayson, 1987; Kimia, Tannenbaum & Zucker, 1993). Although there is no theory of this at present, one should certainly have a multi-scale description of S starting from its coarse diffused form — which is nearly round — and adding detailed features at each scale. In yet another direction, face recognition algorithms have been based on matching a crude blurry face template at a low resolution, and then refining this match, especially at key parts of the face like the eyes. This is the kind of general pyramid algorithm that Rosenfeld proposed many years ago (Rosenfeld & Thurston, 1971), many of which have been successfully implemented by Peter Burt and his group at the Sarnoff Laboratory (Burt & Adelson, 1983).

1.5.2 *Segmentation as a free-boundary value problem*

A quite different mathematical theory has arisen out of the search for algorithms to detect transformations of the 4th kind, interruptions. Evidence for an interruption or a discontinuity in a perceptual signal comes from two sources: the relative homogeneity of the signal on either side of the boundary and the presence of a large change in the signal across the boundary. One approach is to model this as a variational problem: assuming that a blurred and noisy signal f is defined on a domain $D \subset \mathbf{R}^n$, one seeks a set $\Gamma \subset D$ and a smoothed version g of f which is allowed to be discontinuous on Γ such that:

- g is as close as possible to f,
- g has the smallest possible gradient on $D - \Gamma$,
- Γ has the smallest possible $(n-1)$-volume.

These conditions define a variational problem, namely to minimize the functional

$$E(g, \Gamma) = \mu^2 \underset{D}{\int \cdots \int} (f-g)^2 + \underset{D-\Gamma}{\int \cdots \int} \|\nabla g\|^2 + \nu |\Gamma| \tag{1.12}$$

where μ and ν are suitable constants weighting the three terms and $|\Gamma|$ is the $(n-1)$-volume of Γ. The g minimizing E may be understood as the optimal piecewise smooth approximation to the quite general function f. In Grenander's terms, the function g is the pure image and f is the deformed image; I like to call g a *cartoon* for the signal f. The Γ minimizing E is a candidate for the boundaries of parts of the domain D of f where different objects or events are detected. Segmenting the domain of perceptual signals by such variational problems was proposed independently by S. and D. Geman and by A. Blake and A. Zisserman (see Geman & Geman, 1984, and Blake & Zisserman, 1987) for functions on discrete lattices, and was extended by Mumford & Shah (1989) to the continuous case. In the case of visual signals, the domain D is 2-dimensional and we want to decompose D into the parts on which different objects in the world are projected. When you reach the edge of an object as seen from the image plane, the signal f typically will be more or less discontinuous (depending on noise and blur and the lighting effects caused by the grazing rays emitted by the surface as it curves away from the viewer). An example of the solution of this variational problem is shown in Figure 1.6: Figure 1.6a is the original image of the eye, 1.6b shows cartoon g and 1.6c shows the boundaries Γ. This is a case where the algorithm succeeds in finding the 'correct' segmentation, but it doesn't always work so well. Figure 1.7 gives the same treatment as Figure 1.6, to the 'oldman' image. Note that the algorithm fails to find the perceptually correct segmentation in several ways: the man's face is connected to his black coat and the black bar of the bench and the highlights on the back of his coat are treated as separate objects. One reason is that the surfaces of objects are often textured, hence the signal they emit is only statistically homogeneous. More sophisticated variational problems are needed to segment textured objects (see below).

From a mathematical standpoint, it is important to know if this variational problem is well-posed. It has been proven that E has a minimum if Γ is allowed to be a closed rectifiable set of finite Hausdorff $(n-1)$-dimensional measure and g is taken in a certain space SBV ('special bounded variation', which means that the distributional derivative of g is the sum of an L^2-vector field plus a totally singular distribution supported on Γ) (see DeGiorgi, Carriero & Leaci, 1988; Ambrosio &

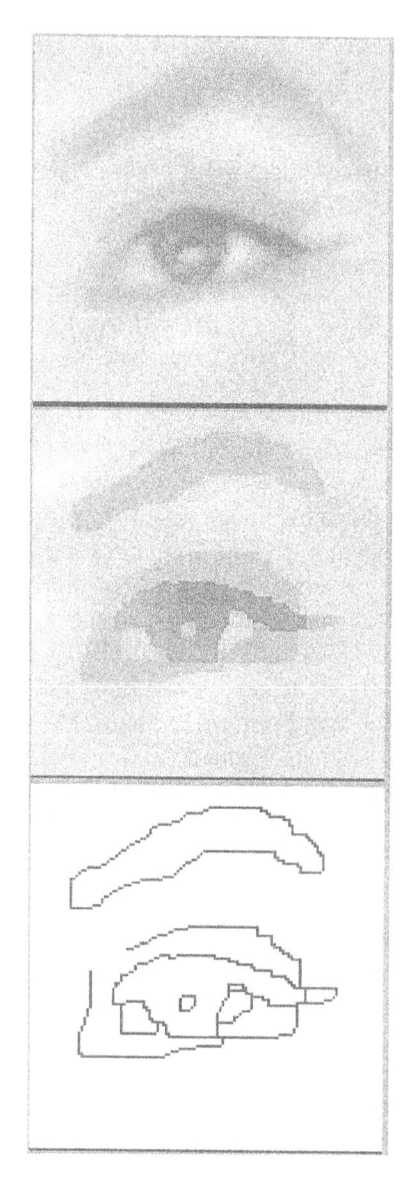

Figure 1.6. Segmentation of the eye-image via optimal piecewise smooth approximation.

Tortorelli, 1989; Dal Maso, Morel & Solimini, 1989). Unfortunately, it seems hard to check whether these minima are 'nice' when f is, e.g. whether, when $n = 2, \Gamma$ is made up of a finite number of differentiable arcs, though Shah and I have conjectured that this is true. Of course, if the signal is replaced by a sampled version, the problem is finite dimensional and certainly well-posed.

This variational problem fits very nicely into both the Bayesian framework and the information theoretic one. Geman and Geman introduced it, for discrete

Figure 1.7. Segmentation of the oldman-image via optimal piecewise smooth approximation.

domains, in the Bayesian setting. The basic idea is to define probability spaces by Gibbs fields. Let $D = \{x_\alpha\}$ be the domain, $\{f_\alpha\}$ and $\{g_\alpha\}$ the values of f and g at x_α. To describe Γ, for each pair of 'adjacent pixels' α and β, let $\ell_{\alpha,\beta} = 1$ or 0 depending on whether or not Γ separates the pixels α and β: these random variables are called the *line process*. Then we define an initial probability distribution on the random variables $\{\ell_{\alpha,\beta}\}$ by the formula:

$$p(\{\ell_{\alpha,\beta}\}) = \frac{e^{-\nu(\sum \ell_{\alpha,\beta})}}{Z_1} \tag{1.13}$$

where Z_1 is the usual normalizing constant. This just means that boundaries Γ get less and less probable, the bigger they are. Next, we put a joint probability distribution on $\{g_\alpha\}$ and on the line process by the formula:

$$p(\{g_\alpha\}, \{\ell_{\alpha,\beta}\}) = \frac{e^{-\sum_{\alpha,\beta \text{ adj}}(1-\ell_{\alpha,\beta})\cdot(g_\alpha - g_\beta)^2}}{Z_2}. \tag{1.14}$$

This is a discrete form of the previous E: if adjacent pixels α and β are *not* separated by Γ, then $\ell_{\alpha,\beta} = 0$ and the probability of $\{g_\alpha\}$ goes down as $|g_\alpha - g_\beta|$ gets larger, but if they *are* separated, then $\ell_{\alpha,\beta} = 1$ and g_α and g_β are independent. Together, the last two equations define an intuitive prior on $\{g_\alpha, \ell_{\alpha,\beta}\}$ enforcing the idea that g is smooth except across the boundary Γ. The data term in the Bayesian approach makes the observations $\{f_\alpha\}$ equal to the model $\{g_\alpha\}$ plus Gaussian noise, i.e. it defines the conditional probability by the formula:

$$p(\{f_\alpha\}|\{g_\alpha, \ell_{\alpha,\beta}\}) = \frac{e^{-\mu^2\cdot\sum_\alpha (f_\alpha - g_\alpha)^2}}{Z_3}. \tag{1.15}$$

Multiplying (1.12), (1.13) and (1.14) defines a probability space $(\Omega_{\text{obs}} \times \Omega_{\text{wld}}, p_{o,w})$ as in Section 1.2 and taking $-log$ of this probability, we get back a discrete version of E up to a constant. Thus the MAP-estimate of the world variables $\{g_\alpha, \ell_{\alpha,\beta}\}$ is essentially the minimum of the functional E.

This probability space is closely analogous to that introduced in physics in the Ising model. In terms of this analogy, the discontinuities Γ of the signal are exactly the interfaces between different phases of some material in statistical mechanics (specifically in the Ising model of where the magnetic field of the iron atoms are oriented up or down).

From the information-theoretic perspective, we want to interpret E as the bit length of a suitable encoding of the image $\{f_\alpha\}$. These ideas have not been fully developed, but for the simplified model in which Γ is assumed to divide up the domain into pieces $\{D_k\}$ on which the image is approximately a constant $\{g_k\}$, this interpretation was pointed out by Leclerc (1989). We imagine encoding the image by starting with a 'chain code' for Γ: the length of this code will be proportional to its length $|\Gamma|$. Then we encode the constants $\{g_k\}$ up to some accuracy by a constant times the number of these pieces k. Finally, we encode the deviation of the image from these constants by Shannon's optimal encoding based on the assumption that $f_\alpha = g_k +$ Gaussian noise n_α. The length of this encoding will be a constant times the first term in E. (If g is not locally constant, we may go on to interpret the second term in E as follows: consider the Neumann boundary value problem for the laplacian Δ acting on the domain $D - \Gamma$. We may expand g in terms of its eigenfunctions, and encode g by Shannon's optimal encoding assuming these coefficients are independently normally distributed with variances going down with the corresponding eigenvalues. The length will be this second term.)

Many variants of this Gibbs field or 'energy functional' approach to perceptual signal processing have been investigated. Some of these seek to incorporate texture segmentation, e.g. Geman *et al.* (1990) and Lee *et al.* (1992) (which proposes an algorithm that should also segment most phonemes in speech) and others to deal with the asymmetry of boundaries caused by the 3D-world: at a boundary, one side is in front, the other in back (Nitzberg & Mumford, 1990). The 'Hidden Markov Models' used in speech recognition are Gibbs fields are of this type. To clarify the relationship, recall that HMM's are based on modelling speech by a Markov chain whose underlying graph is made up of subgraphs, one for each phoneme and whose states predict the power spectrum of the speech signal in local time intervals. Assuming a specific speech signal f is being modelled, HMM-theory computes the MAP sample of this chain conditional on the observed power spectra. Note that any sample of the chain defines a segmentation of time by the set $\Gamma = \{t_k\}$ of times at which the sample moves from the subchain for one phoneme to another, and each interval $t_k \leq t \leq t_{k+1}$ is associated to a specific phoneme a_k. Let A be the string $\{a_1 a_2, \ldots, a_N\}$. Taking $-log$ of the probability, the MAP estimate of the chain is the pair $\{\Gamma, A\}$ minimizing an energy E of the form:

$$E(A, \Gamma) = \sum_k \text{dist.}(f\big|_{t_k}^{t_{k+1}}, \text{phoneme } a_k) + \nu|\Gamma| \tag{1.16}$$

which is clearly analogous to the E's defined above.

Finally, some physiological theories have been proposed in which various areas of cortex (e.g. V1 and V2) compute the segmentation of images by an algorithm analogous to minimizing (1.12) (Grossberg & Mignolla, 1985). It has also been used in computing depth from stereo (see Belhumeur, this volume; Geiger *et al.*, 1992), computing the so-called optical flow field, the vector field of moving objects across the focal plane (Yuille & Grzywacz, 1989; Hildreth, 1984) and many other applications.

We have not mentioned the problem of actually computing or approximating the minimum of energy functionals like E. Four methods have been proposed: in case $n = 1$, we can use *dynamic programming* to find the global minimum fast and efficiently. This applies to the speech applications and is one reason why speech recognition is considerably ahead of image analysis. For any n, Geman & Geman (1984) applied a Monte Carlo algorithm due to Kirkpatrick *et al.* (1983) known as *simulated annealing*. Making this work is something of a black art, as the theoretical bounds on its correctness are astronomical; still it is always an easy thing to try as a first step. A third method, which seems the most reliable at this point, is the *graduated non-convexity* method introduced in Blake & Zisserman (1987). It is based on putting the functional E in a family E_t such that $E = E_0$ and E_1 is a convex functional, hence has a unique local minimum. One then starts with the minimum of E_1 and follows it as $t \to 0$. The final idea is related to the third and that is to use *mean field theory* as in statistical physics: this often leads to approximations to the Gibbs field which allow us to put E in a family becoming convex in the limit (see Geiger & Yuille, 1989).

1.5.3 *Random diffeomorphisms and template matching*

The third example concerns the identification of objects in an image, putting them in categories such as 'the letter A', 'a hammer' or 'my Grandmother's face'. One of the biggest obstacles in these problems is the variability of the shapes which belong to such categories. This variability is caused, for example, by changes in the orientation of the object and the viewpoint of the camera, changes in individual objects such as varying expressions on a face and differences between objects of the same category such as different fonts for characters, different brands of hammer, etc. If the shapes were not too variable, one could hope to introduce average examples of each letter, of each tool, of the faces of everyone you know — 'templates' for each of these objects — and recognize each such object as it is perceived by comparing it to the various templates stored in memory. Unfortunately, the variations are usually too large for this to work, and, worse than that, some variations occur commonly, while others do not (e.g. faces get wrinkled but never become wavy like water). What we need to do is to explicitly model the common variations and use our knowledge to see if a suitably varied template fits! A large part of this variation can be modelled by domain warping, the third of the transformations introduced in Section 1.3 and this leads one to study *deformable templates*, templates whose parts can be changed in size and orientation and shifted relative to each other. These were first introduced in computer vision by Fischler & Elschlager (1973) who called them 'templates with springs' but the idea is well-known in biology, e.g. in the famous and beautiful book by Thompson (1917) (see Figure 1.8a, showing the deformations between three primate skulls).

Mathematically, we can describe flexible templates as follows. We must construct four things: (i) a standard image I_T on a domain D_T which can be a set of pixels or can also be reduced to a graph of 'parts' of the object, (ii) a space of allowable maps $\psi : D_T \to D$ or $(D \cup \{\text{missing}\})$, (iii) a measure $E(\psi)$ of the degree of deformation in the map ψ, the 'stretching of the springs' and (iv) a measure of the difference d between the standard image I_T and the deformation $\psi^*(I)$ of the observed image I. Here ψ is typically a diffeomorphism, 'missing' is an extra element in the range of ψ to allow certain parts of the standard image to be missing in the observed image and $\psi^*(I)$ is a 'pull-back' of I which may be just the composition of I and ψ if D_T is a set of pixels, or may be some set of local 'features' of I when D_T is a graph of parts. The basic problem is then to compute:

$$\arg\min_{\psi}[d(\psi^*(I), I_T) + E(\psi)], \tag{1.17}$$

in order to find the optimal match of the template with the observed image.

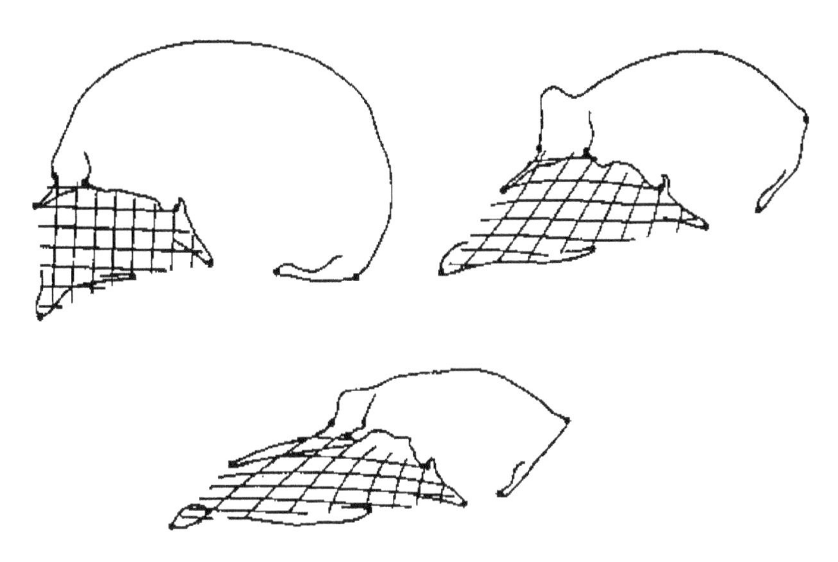

Figure 1.8a. Diffeomorphism between primate skulls.

Figures 1.8b, 1.8c and 1.8d show three examples of such matches. Figure 1.8b from Yamamoto & Rosenfeld (1982) applies these ideas to the recognition of chinese characters or kanji. In this application D_T is a 1-dimensional polygonal skeleton of the outline of the character, and ψ is a piecewise linear embedding of D_T in the domain D of the character image. Figure 1.8c from Y. Amit (1991) applies these ideas to tracing a hand in an X-ray by comparing it with a standard hand. Here ψ is a small deformation of the identity defined by a wavelet expansion of its (x, y)-coordinates and the prior $E(\psi)$ is a weighted L^2-norm of the expansion coefficients. Finally Figure 1.8d from Yuille, Hallinan & Cohen (1992) applies these ideas to the recognition of eyes. Here D_T has two parts, a pair of parabolas representing the outline of the eye and a black circle on a white ground representing the iris/pupil on the eyeball. ψ is linear on each parabola and on the circle, but the range of the first may *occlude* the range on the second to allow the iris/pupil to be partially hidden. This is incorporated in a careful definition of d.

An interesting mathematical side of this theory is the need for a careful definition and comparative study of various priors on the spaces of diffeomorphisms ψ. One can, for instance, define various measures $E(\psi)$ based (i) on the square norm of the Jacobian, as in harmonic map theory, (ii) on the area of the graph, as in geometric measure theory, (iii) on the stress of the map as in elasticity theory or (iv) on second derivatives of ψ, which give more control over the minima. Mumford (1991) discusses some of these measures, but the best approach is unclear and the restriction of maps to be diffeomorphisms is not always natural. An interesting neurophysiological aside is that the anatomy of the cortex of mammals seems well equipped to perform domain warping. The circuitry of the cortex is based on two types of connections: local connections within disjoint subsets of the cortex known as *cortical areas*, and global connections, called *pathways*, between the two

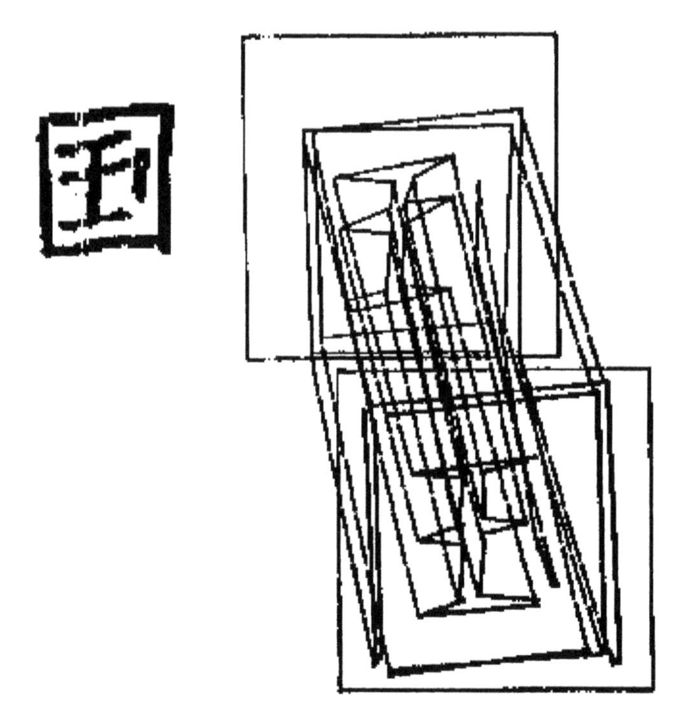

Figure 1.8b. Diffeomorphism between kanji.

distinct areas. The pathways occur in pairs, setting up maps which are crudely homeomorphisms between the cortical surfaces of the two areas which are inverse to each other. These pathways are not exactly point to point maps, however, because of the multiple synapses of their axons, hence the pair of inverse pathways may be able to shift a pattern of excitation by small amounts in any direction.

1.5.4 *The Stereo correspondence problem via minimum description*

As described in Section 1.2, there are two approaches to the problems of pattern theory: the first is to use all the geometry, physics, chemistry, biology and sociology that we know about the world and try to define from this high-level knowledge an appropriate probabilistic model $(\Omega_{\text{obs}} \times \Omega_{\text{wld}}, p_{o,w})$ of the world and our observations. The second involves *learning this model* using only the patterns and the internal structure of the signals that are presented to us. Almost all research to date in computer vision falls in the first category, while the standard approach to speech recognition starts with the first but significantly improves on it using the 'EM-algorithm', a learning algorithm in the second category.

However, newborn animals seem to rely as strongly on learning their environment as on a genetically transmitted knowledge of it: it is not hard to imagine that a baby growing up in a virtual reality governed by quite unusual physics would learn

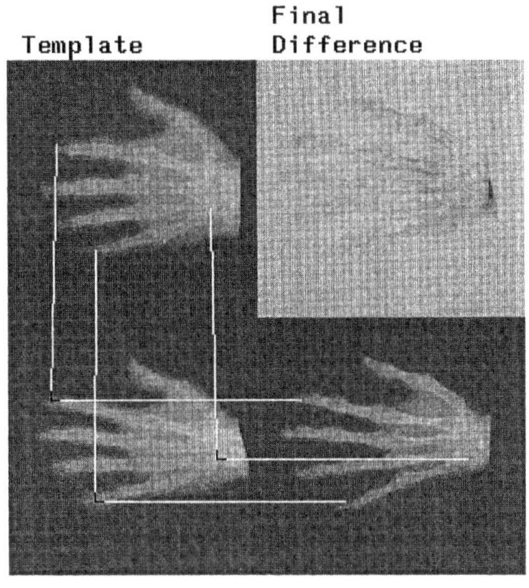

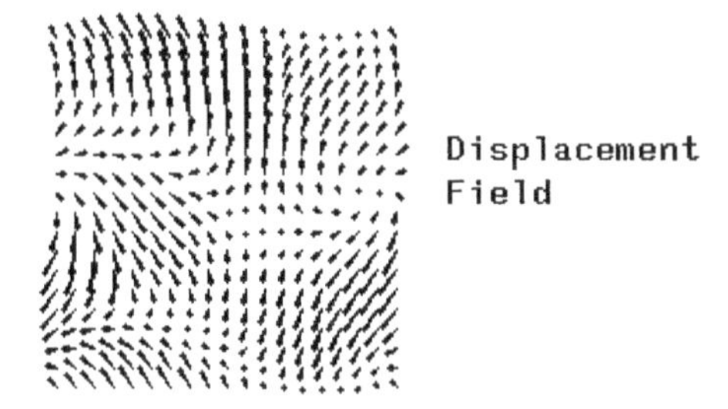

Figure 1.8c. Diffeomorphism between X-rays of hands.

these just as rapidly as the physics of its ancestral world. Humans can read scanning electron microscope images, which are produced by totally different reflectance rules from normal images. All of this suggests the possibility of discovering universal pattern analysis algorithms which learn patterns from scratch. One of the great appeals of the idea of pattern theory is the hope that the structure of the world can be discovered in this way. It is in this spirit that we present the final example. It is not an extensive theory like the previous three, but illustrates how the minimum description length principle can lead one to uncover the hidden structure of the world in a remarkably direct way. Closely related ideas can be found in Becker & Hinton (1992).

We are concerned with the problem of stereo vision. If we view the world with two eyes or with two cameras separated by a known distance, and either identically oriented or with a known difference of orientations, then we can use trigonometry to infer the 3-dimensional structure of the world: see Figure 1.9. More precisely, the

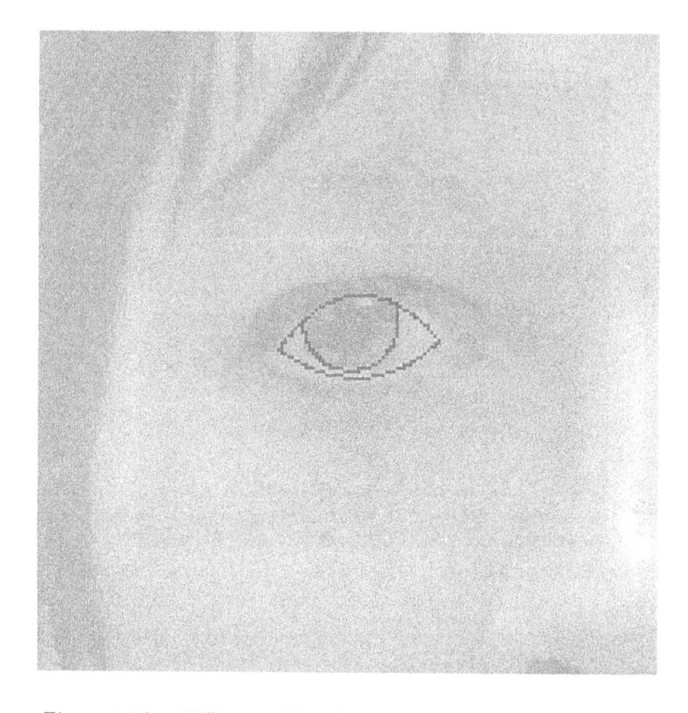

Figure 1.8d. Diffeomorphism from a cartoon eye to a real eye.

two imaging systems produce two images, I_L and I_R (the left and right images). Suppose a point A in the world visible in both images appears as $A_L \in D_L$ and $A_R \in D_R$ in the domains of the two images. The coordinates of A_L and A_R plus the known geometry of the imaging system give the 3-dimensional coordinates of A. However, to use this, we need to first find the pair of corresponding points A_L and A_R: finding these is called the *correspondence problem*. Notice from Figure 1.9 that the geometry of the imaging system gives us one simplification: all points A in a fixed 3-dimensional plane π, through the centers of the two lenses, are seen as points $A_L \in \ell_L$ and $A_R \in \ell_R$, where ℓ_L and ℓ_R are the intersections of π with the two focal planes, and are called *epipolar lines*. Moreover, when we are looking at a single relatively smooth surface S in the 3-dimensional world, S is visible from the left and right eye as subdomains $S_L \subset D_L$ and $S_R \subset D_R$ and the corresponding points on these subdomains define a diffeomorphism $\psi : S_L \to S_R$ carrying each epipolar line in the left domain to the corresponding epipolar line in the right. This leads us to a problem like that in the last section. But there is a further twist: at the edges of objects, each of the two eyes can typically see a little further around one edge, producing pixels in one domain D_L or D_R with no corresponding pixel in the other domain. In this way, the domain is often segmented into subdomains corresponding to distinct objects. (See Belhumeur (1993)).

My claim is that the minimum description length principle alone leads you naturally to discover all this structure, without any prior knowledge of 3-dimensions.

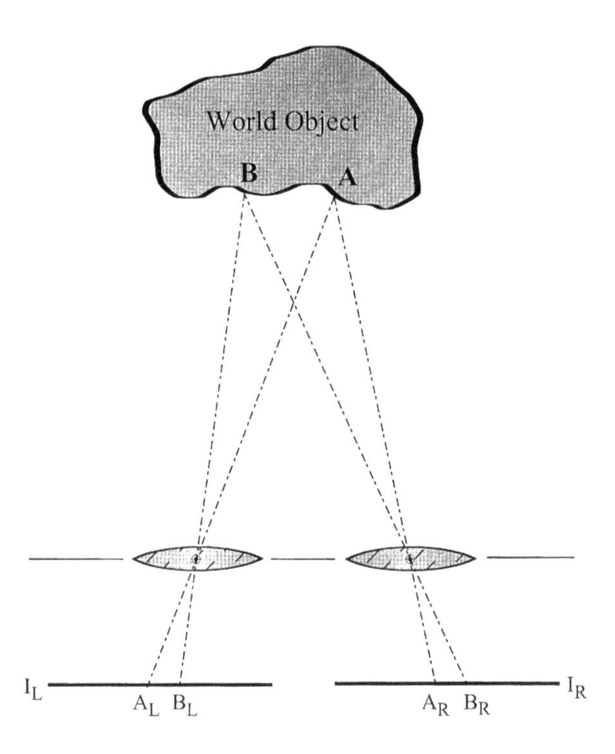

Figure 1.9. The geometry of stereo vision, in a plane through the centers of the two lenses.

The argument is summarized in Figure 1.10. In this figure, I have represented a series of increasingly complex stereo images in diagrammatic form. Firstly, in order to represent the essentials concisely, I have used only a single pair of epipolar lines ℓ_L and ℓ_R instead of the full domains D_L and D_R. Secondly, instead of graphing the complex intensity function, we have used small symbols (squares, circles, triangles, stars, etc.) to denote local intensity functions with various characteristics. Thus a square on both lines represents local intensities which are similar functions. On the left, at each stage in Figure 1.10, we see the plane π in the world, with the visible surface points, and the left and right eyes. In the middle, we see the left and right images I_L and I_R which this scene produces, as well as dotted lines connecting corresponding points A_L and A_R. On the right we give a method of encoding the image data.

Stage 0 represents a simple flat object seen from the front: it produces images I_L and I_R, but we assume that our pattern analysis begins with naively encoding the images independently. At stage 1, the same scene is seen, but now the analysis uses the much more concise method of encoding only I_L, the fixed translation d by which corresponding points differ and a possible small residual $\Delta I(x) = I_R(x) - I_L(x + d)$. Clearly this is more concise. At stage 2, the scene is more complex: a surface of varying distance is seen, hence the displacement between corresponding points (called the *disparity*) is not constant. To adapt the previous encoding to this

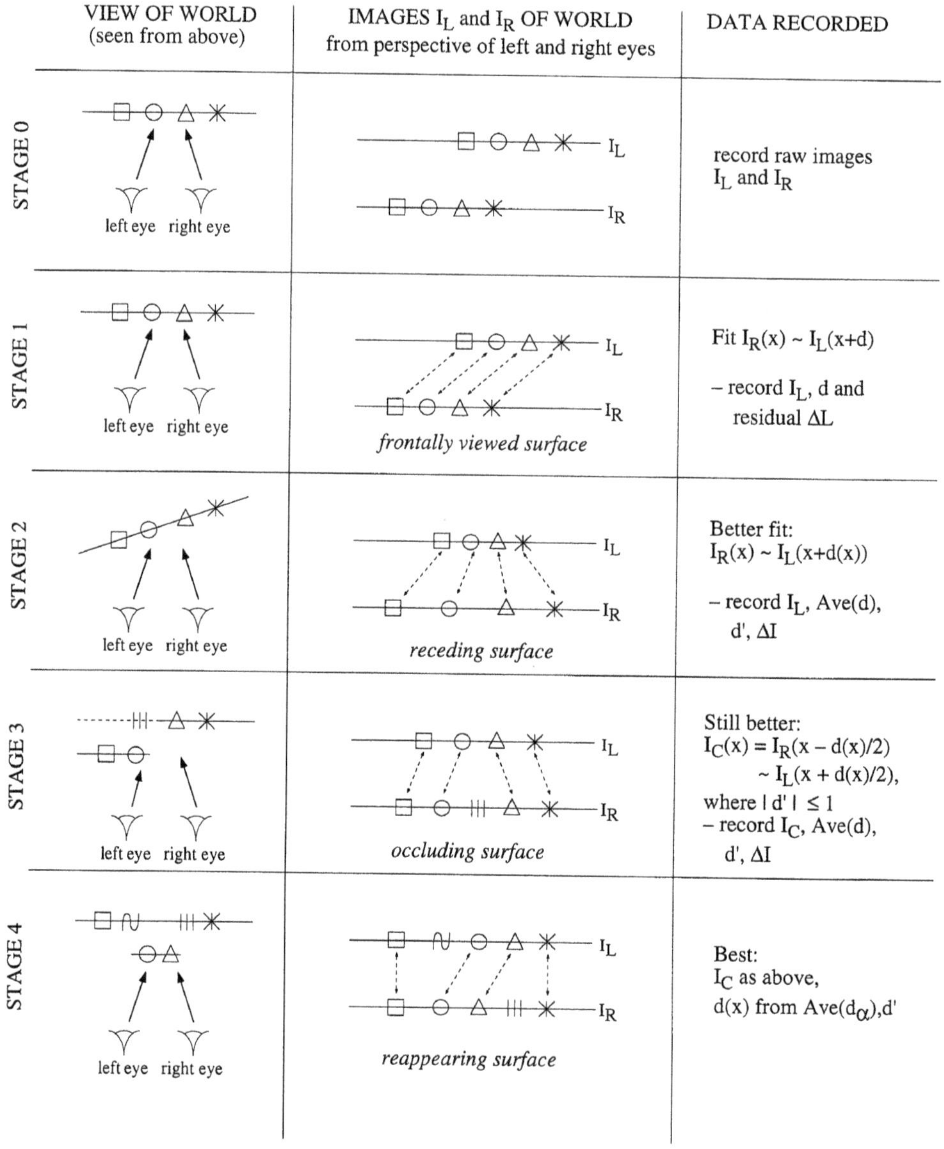

	VIEW OF WORLD (seen from above)	IMAGES I_L and I_R OF WORLD from perspective of left and right eyes	DATA RECORDED
STAGE 0	left eye right eye	I_L I_R	record raw images I_L and I_R
STAGE 1	left eye right eye	I_L I_R *frontally viewed surface*	Fit $I_R(x) \sim I_L(x+d)$ – record I_L, d and residual ΔL
STAGE 2	left eye right eye	I_L I_R *receding surface*	Better fit: $I_R(x) \sim I_L(x+d(x))$ – record I_L, Ave(d), d', ΔI
STAGE 3	left eye right eye	I_L I_R *occluding surface*	Still better: $I_C(x) = I_R(x - d(x)/2) \sim I_L(x + d(x)/2)$, where $\lvert d' \rvert \leq 1$ – record I_C, Ave(d), d', ΔI
STAGE 4	left eye right eye	I_L I_R *reappearing surface*	Best: I_C as above, d(x) from Ave(d_α),d'

Figure 1.10. Discovering the world via MDL.

situation, one could take a mean value of d and have a bigger residual ΔI. But this residual could be quite big and a better scheme is replace the fixed d by a function $d(x)$ and encode I_L, the mean and derivative $(\overline{d}, d')$ of d and the residual ΔI. Now in stage 3, we encounter a new wrinkle: the scene consists in two surfaces, one occluding the other. Notice that a little bit of the back surface is visible to one eye only. To include this complexity, we go over to a more symmetrical treatment of the two eyes and encode a combined *cyclopean* image $I_C(x)$, where

$$I_C(x) = I_R\left(x - \frac{d(x)}{2}\right), I_L\left(x + \frac{d(x)}{2}\right) \text{ or their average} \tag{1.18}$$

depending on whether the point is visible only to the right eye, only to the left eye or to both eyes. To make this representation unique, it is easy to see that we must require that $|d'(x)| \leq 1$. Then we encode the scene via $(I_C, \overline{d}, d', \Delta I)$. In the final stage 4, we introduce the possibility of a surface disappearing behind another *and then reappearing*. The point is that each surface has its own average disparity, and it now becomes more efficient to record d by several means $\overline{d}_\alpha$, one for each surface, and the derivative d'. Thus we see how the search for minimum length encoding leads us naturally, first to the third coordinate of world points, then to smooth descriptions of surfaces in terms of their tangent planes and finally to explicit labelling of distinct surfaces in the visible field.

Although this approach might seem very abstract and impossible to implement biologically, G. Hinton (unpublished) has developed neural net theories incorporating both MDL and feed-back. These might be able to learn stereo exactly as outlined in this section.

1.6. Pattern Theory and Cognitive Information Processing

The examples of the last section all concern pattern theory as a theory for analyzing sensory input. The examples come from vision, but most of the ideas could apply to hearing or touch too. The purpose of this section is to ask the question: to what extent is pattern theory relevant to all cognitive information processing, both 'higher level' thinking as studied in cognitive psychology and AI, and the output stages of an intelligent agent, motor control and action planning. I believe that in many ways the same ideas are applicable on a theoretical level and that there is physiological evidence that the same algorithms are applied throughout the cortex.

In the introduction, we gave medical expert systems as another example of pattern theory. In this extension, we considered the data available to a physician — symptoms, test results and the patient's history — as an encoded version of the full state of the world, a 'deformed image' in Grenander's terminology. The full state of the world, the 'pure image', in this case means the diseases and processes present in the patient. Inferring these hidden random variables can and has been studied as a problem in Bayesian statistics, exactly as in Section 1.2: see, for instance, Pearl (1988) and Lauritzen & Spiegelhalter (1988). In particular, describing the probability distribution on all the random variables as a Gibbs field, as in Section 1.5, has been shown to be a powerful technique for introducing realistic models of the probability distribution in the real world. Figure 1.11, from the article Lauritzen & Spiegelhalter (1988), shows a simplified set of such random variables and the graph on which a Gibbs distribution can be based. Whether

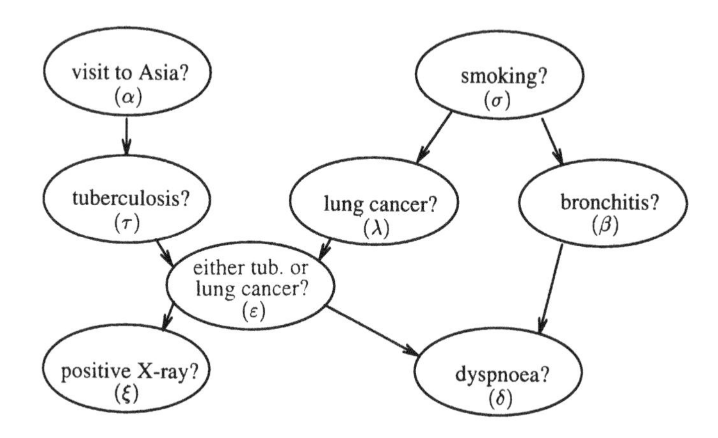

Figure 1.11. Causal graph in a toy medical expert system.

or not pattern theory extends in an essential way to these types of problems hinges on whether the transformations described in Section 1.3 generate the kind of probability distributions encountered with higher level variables. To answer this, it is essential to look at test cases which are not too artificially simplified (as is done all too often in AI), but which incorporates the typical sorts of complexities and complications of the real world. While I do not think this question can be definitively answered at present, I want to make a case that the four types of transformations of Section 1.3 are indeed encoding mechanisms encountered at all levels of cognitive information processing.

The first class of transformations, noise and blur, certainly occur at all levels of thought. In the medical example, errors in tests, the inadequacies of language in conveying the nature of a pain or symptom, etc. all belong to this category. Uncertainty over facts, misinterpretations and confusing factors are within this class. The simplest model leads to multi-dimensional normal distributions on a vector P of 'features' being analyzed.

The fourth category of transformation, 'interruptions', are also obviously universal. In any cognitive sphere, the problem of separating the relevant factors for a specific event or situation being analyzed from the extraneous factors involved with everything else in the world, is clearly central. The world is a complex place with many, many things happening simultaneously, and highlighting the 'figure' against the 'ground' is not just a sensory problem, but one encountered at every level. Another way this problem crops up is that a complex of symptoms may result from one underlying cause or from several, and, if several causes are present, their effects have to be teased apart in the process of pattern analysis. As proposed in Section 1.4, pattern synthesis — actively comparing the results of one cause with the presenting symptoms P followed by analysis of the residual, the unexplained symptoms, is a universal algorithmic approach to these problems.

The second of the transformations, 'multi-scale superposition', can be applied to higher level variables as follows: philosophers, psychologists and AI researchers have all proposed systematizing the study of concepts and categories by organizing them in hierarchies. Thus psychologists (see Rosch, 1978) propose distinguishing *superordinate categories, basic level categories and subordinate categories*: for instance, a particular pet might belong to the superordinate category 'animal', the basic-level category 'dog' and the subordinate category 'terrier'. In AI, this leads to graphical structures called *semantic nets* for codifying the relationships between categories (see Findler, 1979). These nets always include ordered links between categories, called *isa* links, meaning that one category is a special case of another. I want to propose that cognitive multi-scale superposition is precisely the fact that to analyze a specific situation or thing, some aspects result from the situation belonging to very general categories, others from very specific facts about the situation that put it in very precise categories. Thus sensory thinking requires we deal with large shapes with various overall properties, supplemented with details about their various parts, precise data on location, proportions, etc.; cognitive thinking requires we deal with large ideas with various general properties, supplemented with details about specific aspects, precise facts about occurrence, relationships, etc.

Finally, how about 'domain warping'? Consider a specific example first. Associated to a cold is a variety of several dozen related symptoms. A person may, however, be described as having a sore throat, a chest cold, flu, etc.: in each case the profile of their symptoms shifts. This may be modelled by a map from symptom to symptom, carrying for instance the modal symptom of soreness of throat to that of coughing. The more general cognitive process captured by domain warping is that of making an *analogy*. In an analogy, one situation with a set of participants in a specific relationship to each other is mapped to another situation with new participants in the same relationship. This map is the ψ in Section 1.5, and the constraints on ψ, such as being a diffeomorphism, are now that it preserve the appropriate relationships. The idea of domain warping applying to cognitive concepts seems to suggest that higher level concepts should form some kind of geometric space. At first this sounds crazy, but it should be remembered that the entire cortex, high and low level areas alike, has the structure of a 2-dimensional sheet. This 2-dimensional structure is used in a multitude of ways to organize sensory and motor processes efficiently: in some cases, sensory maps (like the retinal response and patterns of tactile responses) are laid out geometrically. In other cases, interleaved stripes carry intra-hemispheric and inter-hemispheric connections. In still other cases, there are 'blobs' in which related responses cluster. But, in all cases, adjacency in this 2D sheet allows a larger degree of cross-talk and interaction than with non-adjacent areas and this seems to be used to develop responses to related patterns. My suggestion is: is this spatial adjacency used to structure abstract thought too[†]?

[†]I have argued elsewhere that the remarkable anatomical uniformity of the neo-cortex suggests that common mechanisms, such as the 4 universal transformations of pattern theory, are used

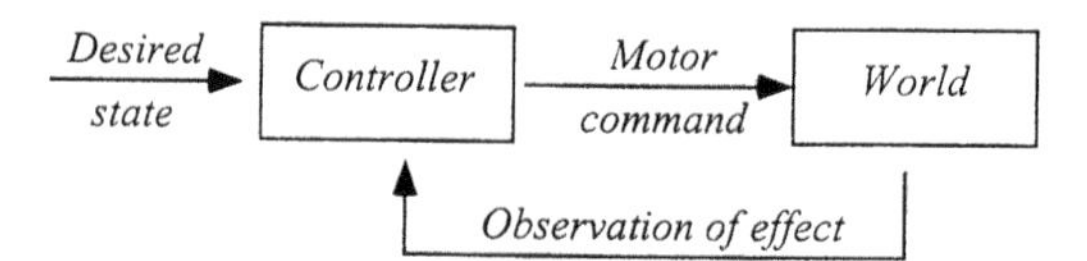

Figure 1.12a. The flow chart of control theory.

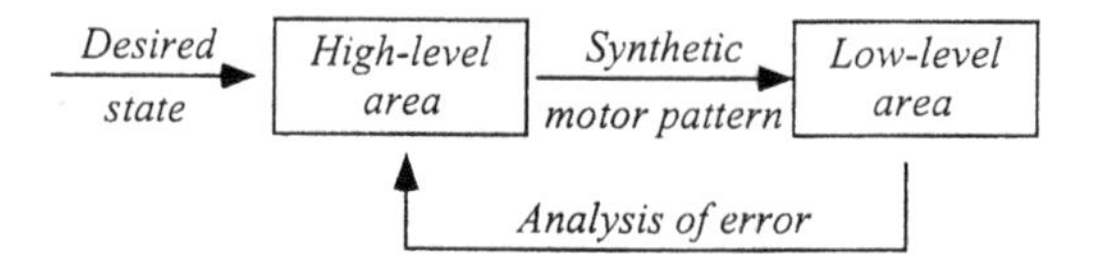

Figure 1.12b. A motor task via pattern theory.

To conclude, we want to discuss briefly how pattern theory helps the analysis of motor control and action planning, the output stage of a robot. Control theory has long been recognized as the major mathematical tool for analyzing these problems but it is not, in fact, all that different from pattern theory. In Figure 1.12a, we give the customary diagram of what control theory does. The controller is a black box which compares the current observation of the state of the world with the desired state and issues an updated motor command, which in turn affects the black box called the world. This diagram is very similar to Figure 1.4, which described how pattern analysis and pattern synthesis formed a loop used in the algorithm for reconstructing the hidden world variables from the observed sensory ones. Figure 1.12b presents the modification of Figure 1.4 to a motor task. Here a high-level area or 'black box' is in a loop with a low-level area: the high-level area compares the desired state with an analysis of the error and generates an updated motor command sequence by pattern synthesis. The low-level area, either by actually carrying out an action and observing its consequences, or by internal simulation, finds that it falls short in various ways, and send its pattern analysis of this error back up. Notice that the four transformations of Section 1.3 will occur or should be used in the top-down pattern synthesis step. Noise and blur are the inevitable consequences of the inability to control muscles perfectly, or eliminate external uncontrollable interference. Domain warping is the bread-and-butter of control theory — speeding up or slowing down an action by modifying the forces in order that it optimizes performance. Multi-scale superposition is what hierarchical

throughout the cortex (Mumford, 1991–92; 1993). The referree has pointed out that 'the uniformity of structure may reflect common machinery at a lower level. For example, different computers may have similar basic mechanisms at the level of registers, buses, etc., which is a low level of data handling. Similarly in the brain, the apparent uniformity of structure may be at the level of common lower-level mechanisms rather than the level of dealing with universal transformations.' This is a certainly an alternative possibility, quite opposite to my conjectural link between the high-level analysis of pattern theory and the circuitry of the neo-cortex.

control is all about: building up an action first in large steps, then refining these steps in their parts, eventually leading to detailed motor commands. Finally, interruptions are the terminations of specific control programs, either by success or by unexpected events, where quite new programs take over. In general, we seek to anticipate these and set up successor programs beforehand, hence we need to actively synthesize these as much as possible.

In summary, my belief is that pattern theory contains the germs of a universal theory of thought, one which stands in opposition to the accepted analysis of thought in terms of logic. The extraordinary similarity of the structure of all parts of the human cortex to each other and of human cortex with the cortex of the most primitive mammals suggests that a relatively simple universal principal governs its operation, even in complex processes like language (see Mumford, 1991–92, 1993) where these physiological links are developed: pattern theory is a proposal for what these principles may be.

References

Ambrosio, L. & Tortorelli, V. (1991). Approximations of functionals depending on jumps by elliptic functionals via gamma-convergence. *Comm. Pure & Applied Math.*, **43**, 999–1036.

Amit, D. (1989). *Modelling Brain Function*, Camb. Univ. Press.

Amit, Y. (1991). A non-linear variational problem for image matching. Preprint, Brown Dept. of Applied Math.

Barlow, H. (1961). Possible principles underlying the transformation of sensory messages. In *Sensory Communication*, ed. W. Rosenblith. MIT Press.

Becker, S. & Hinton, G. (1992). Self-organizing neural network that discovers surfaces in random-dot stereograms, *Nature*, **235**, 161–163.

Belhumeur, P., (1993). A Binocular Stereo Algorithm for Reconstructing Surfaces in the Presence of Half-Occlusion, *Proc. Int. Conf. Comp. Vision*, 431–438.

Blake, A. & Zisserman, A. (1987). *Visual Reconstruction*. Cambridge, MA: MIT Press.

Burt, P. & Adelson, E. (1983). The Laplacian pyramid as a compact image code. *IEEE Trans. on Comm.*, **31**, 532–540.

Carpenter, G. & Grossberg, S. (1987). A massively parallel architecture for a self-organizing neural pattern recognition machine. *Comp. Vision, Graphics & Image Proc.*, **37**, 54–115.

Cavanagh, P. (1991). What's up in top-down processing. *Proc. 13th ECVP.*

Chomsky, N. (1981). *Lectures on Government and Binding*, Dordrecht:Foris.

Coifman, R., Meyer, Y. & Wickerhouser, V. (1990). Wavelet analysis and signal processing. Preprint, Yale Univ. Math. Dept.

Dal Maso, G., Morel, J-M. & Solimini, S. (1992). A variational method in image segmentation, *Acta Math.*, to appear.

Daubechies, I. (1988). Orthonormal bases of compactly supported wavelets. *Comm. Pure & Applied Math.*, **49**, 909–996.

Daubechies, I. (1990). The wavelet transform, time-frequency localization and signal analysis. *IEEE Trans. Inf. Theory*, 961–1005.

DeGiorgi, E., Carriero, M. & Leaci, A. (1989). Existence theorem for a minimum problem with free discontinuity set. *Arch. Rat. Mech. Anal.*, **108**, 195–218.

Dowling, J. (1987). *The Retina.* Harvard Univ. Press.

Findler, N. ed. (1979). *Associative Networks.* Academic Press.

Fischler, M. & Elschlager, R. (1973). The representation and matching of pictorial structures. *IEEE Trans. on Computers*, **22**, 67–92.

Gage, M. & Hamilton, R. (1986). The heat equation shrinking convex plane curves. *J. Diff. Geom.*, **23**, 69–96.

Geiger, D. & Yuille, A. (1989). A common framework for image segmentations. Harvard Robotics Lab. Tech Report, 1989.

Geiger, D., Ladendorf, B. & Yuille, A. (1992). Occlusions and binocularstereo. *Proc. European Conf. Comp. Vision*, Springer Lecture Notes in Comp. Sci., **588**.

Geman, D. (1991). Random fields and inverse problems in imaging. Springer Lecture Notes in Mathematics, **1427**.

Geman, S. & Geman, D. (1984). Stochastic relaxation, Gibbs distribution and Bayesian restoration of images. *IEEE Trans. Pattern Anal. Mach. Intell.*, **6**, 721–741.

Geman, S., Geman, D., Graffigne, C. & Dong, P. (1990). Boundary detection by Constrained Optimization. *IEEE Trans. Pattern Anal. and Mach. Int.*, **12**.

Grayson, M. (1987). The heat equation shrinks embedded plane curves to round points. *J. Diff. Geom.*, **26**, 285–314.

Grenander, U. (1976–1981). *Lectures in Pattern Theory I, II and III: Pattern Analysis, Pattern Synthesis and Regular Structures*, Springer–Verlag.

Grossberg, S. & Mingolla, E. (1985). Neural dynamics of form perception: Boundary completion, illusory figures and neon color spreading. *Psych. Rev.*, **92**, 173–211.

Grossman, A. & Morlet, J. (1984). Decomposition of Hardy functions into square integrable wavelets of constant shape. *SIAM J. Math. Anal.*, **15**, 723–736.

Hertz, J., Krogh, A. & Palmer, R. (1991) *Introduction to the Theory of Neural Computation.* Addison–Wesley.

Hildreth, E. (1984). *The Measurement of Visual Motion.* Cambridge, MA: MIT Press.

Hopfield, J. (1982). Neural networks and physical systems with emergent collective computational abilities. *Proc. Nat. Acad. Sci.*, **79**, 2554–2558.

Kass, M., Witkin, A. & Terzopoulos, D. (1987). Snakes: Active contour models. *IEEE Proc. 1st Int. Conf. Computer Vision*, 259–268.

Kimia, B., Tannenbaum, A. & Zucker, S. (1993). Shapes, shocks and deformations I. *Int. J. Comp. Vision.*

Kirkpatrick, S., Geloti, C. & Vecchi, M. (1983), Optimization by simulated annealing. *Science*, **220**, 671–680.

Lauritzen, S. & Spiegelhalter, D. (1988). Local computations with probabilities on graphical structures and their applications to expert systems, *J. Royal Stat. Soc. B*, **50**, 157–224.

Leclerc, Y. (1989), Constructing simple stable descriptions for image partitioning. *Int. J. Comp. Vision*, **3**, 73–102.

Lee, T. S., Mumford, D. & Yuille, A. (1992). Texture segmentation by minimi- zing vector-valued energy functionals. *Proc. Eur. Conf. Comp. Vision*, Springer Lecture Notes in Comp. Sci., **1427**.

MacKay, Donald (1956). The epistemological problem for automata. In *Automata Studies*, ed. Shannon & McCarthy, pp. 2235–2251. Princeton Univ. Press.

Mallat, S. (1989). A theory of multi-resolution signal decomposition: The wavelet representation. *IEEE Trans. PAMI*, **11**, 674–693.

Meyer, Y. (1986). Principe d'incertitude, bases hilbertiennes et algèbres d'opéra- teurs. *Séminaire Bourbaki*, Springer Lecture Notes in Mathematics.

Mumford, D. (1991). Mathematical theories of shape: Do they model perception? *Proc. SPIE*, **1570**, 2–10.

Mumford, D. (1991–92). On the computational architecture of the neocortex I and II. *Biol. Cybernetics*, **65**, 135–145 & **66**, 241–251.

Mumford, D. (1993). Neuronal architectures for pattern-theoretic problems. *Proc. Idyllwild conference on large scale neuronal theories of the brain.* To appear, MIT press.

Mumford, D. & Shah, J. (1989). Optional approximation by piecewise smooth functions and associated variational problems. *Comm. Pure & Applied Math.*, **42**, 577–685.

Nitzberg, M. & Mumford, D. (1990). The 2.1*D* sketch. *Proc. 3rd IEEE Int. Conf. Comp. Vision*, 138–144.

Osherson, D. & Weinstein, S. (1984). Formal learning theory. In *Handbook of Cognitive Neuroscience*, ed. M. Gazzaniga, pp. 275–292. Plenum Press.

Pearl, J. (1988). *Probabilistic Reasoning in Intelligent Systems.* Morgan–Kaufman.

Perona, P. & Malik, J. (1987). Scale-space and edge detection using anisotropic diffusion. *IEEE Workshop on Computer Vision*, Miami.

Rissanen, J. (1989). *Stochastic Complexity in Statistical Inquiry.* World Scientific.

Rosch, E. (1978). Principles of Categorization. In *Cognition and Categorization*, eds. E. Rosch and B. Lloyd, L. Erlbaum.

Rosenfeld, A. & Thurston, M. (1971). Edge and curve detection for visual scene analysis. *IEEE Trans. on Computers*, **C-20**, 562–569.

Thompson, D'Arcy (1917). *On Growth and Form.* Camb. Univ. Press.

Uhr, L., (1972). Layered "recognition cone" networks that preprocess, classify and describe. *IEEE Trans. on Computers*, **C-21**, 758–768.

Yamamoto, K. & Rosenfeld, A. (1982). Recognition of handprinted Kanji characters by a relaxation method. *Proc. 6th Int. Conf. Pattern Recognition*, 395–398.

Yuille, A. & Grzywacz, N. (1989). A mathematical analysis of the motion coherence theory. *Int. J. Comp. Vision*, **3**, 155–175.

Yuille, A., Hallinan, P., & Cohen, D. (1992). Feature extraction from faces using deformable templates. *Int. J. Comp. Vision*, **6**.

Reprinted from Proc. Int. Congr. Math., 1978

THE WORK OF GREGORY ALEKSANDROVITCH MARGULIS

by
J. TITS

The work of Margulis belongs to combinatorics, differential geometry, ergodic theory, the theory of dynamical systems and the theory of discrete subgroups of real and p-adic Lie groups. In this report, I shall concentrate on the last aspect which covers his main results.

1. Discrete subgroups of Lie groups. The origin. Discrete subgroups of Lie groups were first considered by Poincaré, Fricke and Klein in their work on Riemann surfaces: if M is a Riemann surface of genus $\geqslant 2$, its universal covering is the Lobatchevski plane (or Poincaré half-plane), therefore the fundamental group of M can be identified with a discrete subgroup Γ of $\mathrm{PSL}_2(\boldsymbol{R})$; the problem of uniformization and the theory of differentials on M lead to the study of automorphic forms relative to Γ.

Other discrete subgroups of Lie groups, such as $\mathrm{SL}_n(\boldsymbol{Z})$ (in $\mathrm{SL}_n(\boldsymbol{R})$) and the group of "units" of a rational quadratic form (in the corresponding orthogonal group) play an essential role in the theory of quadratic forms (reduction theory) developed by Hermite, Minkowski, Siegel and others. In constructing a space of moduli for abelian varieties, Siegel was led to consider the "modular group" $\mathrm{Sp}_{2n}(\boldsymbol{Z})$, a discrete subgroup of $\mathrm{Sp}_{2n}(\boldsymbol{R})$.

The group $\mathrm{SL}_n(\boldsymbol{Z})$, the group of units of a rational quadratic form and the modular group are special instances of "arithmetic groups", as defind by A. Borel and Harish-Chandra. A well-known theorem of those authors, generalizing classical results of Fricke, Klein, Siegel and others, asserts that if Γ is an arithmetic subgroup of a semi-simple Lie group G, then the volume of G/Γ (for any G-invariant measure) is finite; we say that Γ has *finite covolume* in G. The same holds for $G{=}\mathrm{PSL}_2(\boldsymbol{R})$ if Γ is the fundamental group of a Riemann surface "with null boundary" (for instance, a compact surface minus a finite subset).

Already Poincaré wondered about the possibility of describing all discrete subgroups of finite covolume in a Lie group G. The profusion of such subgroups in $G{=}\mathrm{PSL}_2(\boldsymbol{R})$ makes one at first doubt of any such possibility. However, $\mathrm{PSL}_2(\boldsymbol{R})$ was for a long time the only simple Lie group which was known to contain non-arithmetic discrete subgroups of finite covolume, and further examples discovered in 1965 by Makarov and Vinberg involved only few other Lie groups, thus adding credit to conjectures of Selberg and Pyatetski–Shapiro to the effect that "for most semisimple Lie groups" discrete subgroups of finite covolume are necessarily arithmetic. Margulis' most spectacular achievement has been the complete solution of that problem and, in particular, the proof of the conjectures in question.

2. The noncocompact case. Selberg's conjecture. Let G be a semisimple Lie group. To avoid inessential technicalities, we assume that G is the group of real points of a real simply connected algebraic group $\mathcal{G}$ which we suppose embedded in some $\mathrm{GL}_n(\boldsymbol{R})$, and that G has no compact factor. Let Γ be a discrete subgroup of G with finite covolume and irreducible in the sense that its projection in any nontrivial proper direct factor of G is nondiscrete. Suppose that the real rank of G is $\geqslant 2$ (this means that G is not a covering group of the group of motions of a real, complex or quaternionic hyperbolic space or of an "octonionic" hyperbolic plane) and that G/Γ *is not compact.* Then, Selberg's conjecture asserts that Γ is arithmetic which, in this case, means the following: there is a base in $\boldsymbol{R}^n$ with respect to which $\mathcal{G}$ is defined by polynomial equations with rational coefficients and such that Γ is commensurable with $\mathcal{G}(\boldsymbol{Z}) = G \cap GL_n(\boldsymbol{Z})$ (i.e. $\Gamma \cap \mathcal{G}(\boldsymbol{Z})$ has finite index in both Γ and $\mathcal{G}(\boldsymbol{Z})$). Selberg himself proved that result in the special case where G is a direct product of (at least two) copies of $\mathrm{SL}_2(\boldsymbol{R})$.

A first important step toward the understanding of noncompact discrete subgroups of finite covolume was the proof by Každan and Margulis [2] of a related, more special conjecture of Selberg: *under the above assumptions* (except that no hypothesis is made on $\mathrm{rk}_{\boldsymbol{R}} G$), Γ *contains nontrivial unipotent elements of* G (i.e. elements all of whose eigenvalues are 1). This was a vast generalization of results already known for $\mathrm{SL}_2(\boldsymbol{R})$ and products of copies of $\mathrm{SL}_2(\boldsymbol{R})$ (Selberg); in view of a fundamental theorem of Borel and Harish–Chandra ("Godement's conjecture"), it had to be true if Γ was to be arithmetic. Let us also note in passing another remarkable byproduct of Každan–Margulis' method: *given* G, *there exists a neighborhood* W *of the identity in* G *such that for every* Γ (*cocompact or not*), *some conjugate of* Γ *intersects* W *only at the identity*; in particular, *the volume of* G/Γ *cannot be arbitrarily small* (for a given Haar measure in G). For $G{=}\mathrm{SL}_2(\boldsymbol{R})$, the last assertion had been proved by Siegel, who had also given the exact lower bound of vol (G/Γ) in that case. A. Borel reported on those results of Každan and Margulis at Bourbaki Seminar [26].

The existence of unipotent elements in Γ was giving a hold on its structure. In [6], Margulis announced, among others, the following result which was soon recognized by the experts as a crucial step for the proof of Selberg's conjecture:

in the space of lattices in $\boldsymbol{R}^n$, *the orbits of a one-parameter unipotent subsemigroup of* $\mathrm{GL}_n(\boldsymbol{R})$ *"do not tend to infinity"* (*in other words, a closed orbit is periodic*).

For a couple of years, Margulis' proof remained unpublished and every attempt by other specialists to supply it failed. When it finally appeared in [9], the proof came as a great surprise, both for being rather short and using no sophisticated technique: it can be read without any special knowledge and gives a good idea of the extraordinary inventiveness shown by Margulis throughout his work.

Using unipotent element, it is relatively easy to show that, G and Γ being as above, there is a $\boldsymbol{Q}$-structure $\mathcal{G}$ on G such that $\Gamma \subset \mathcal{G}(\boldsymbol{Q})$. The main point of Selberg's conjecture is then to show that the matrix coefficients of the elements of Γ have bounded denominators. In [15], Margulis announced a complete proof of the conjecture and gave the details under the additional assumption that the $\boldsymbol{Q}$-rank of $\mathcal{G}$ is at least 2. Another proof under the same restriction was given independently by M. S. Raghunathan. The much more difficult case of a $\boldsymbol{Q}$-rank one group is treated by Margulis in [19], by means of a very subtle and delicate analysis of the set of unipotent elements contained in Γ. The main techniques used in [15] and [19] are those of algebraic group theory and p-adic approximation.

3. The cocompact case. Rigidity. Margulis was invited to give an address at the Vancouver Congress, no doubt with the idea that he would expose his solution of Selberg's conjecture. Instead, prevented (at this time) from attending the Congress, he sent a report on completely new and totally unexpected results on the cocompact case [18].

That case, about which nothing was known before, presented two great additional difficulties which nobody knew how to handle. On the one hand, if G/Γ is compact, Γ contains no unipotent element, so that the main technique used in the other case is not available. But there is another basic difficulty in the very notion of arithmetic group: let G, Γ be as in Sec. 2 except that G/Γ is no longer assumed to be noncompact; then Γ is said to be arithmetic if there exist an algebraic linear semi-simple simply connected group $\mathcal{H}$ defined over $\boldsymbol{Q}$ and a homomorphism $\alpha : \mathcal{H}(\boldsymbol{R}) \to G$ with compact kernel such that Γ is commensurable with $\alpha(\mathcal{H}(\boldsymbol{Z}))$. The point is that in the noncocompact case, α is necessarily an isomorphism. In the general case, there is *a priori* no way of knowing what $\mathcal{H}$ will be (in fact, for a given $G, \mathcal{H}$ tan have an arbitrarily large dimension). A conjecture, more or less formulated by Pyatetski–Shapiro at the 1966 Congress in Moscow, to the effect that also in the cocompact case, assuming again $\mathrm{rk}_R G \geqslant 2$, Γ had to be arithmetic, was certainly more daring at the time and seemed completely out of reach. It was the proof of that conjecture that Margulis sent, without warning, to the Vancouver Congress.

Arithmetic subgroups of Lie groups are in some sense "rigid"; intuitively, this follows from the impossibility to alter an algebraic number continuously without destroying the algebraicity. On the other hand, theorems of Selberg, Weil and Mostow showed that in semi-simple Lie groups different from $\mathrm{SL}_2(\boldsymbol{R})$ (up to local isomorphism) cocompact discrete subgroups are rigid, and Selberg had observed that rigidity implies a "certain amount of arithmeticity": in fact, it is readily seen to imply that Γ is contained in $\mathcal{G}(\boldsymbol{K})$ for some algebraic group $\mathcal{G}$ and some number field $\boldsymbol{K}$. As before, the crux of the matter is the proof that the matrix coefficients of the elements of Γ have bounded denominators. This is achieved by Margulis through a "superrigidity" theorem which, for groups of real rank at least 2, is a vast generalization of Weil's and Mostow's rigidity theorems:

Assume $\mathrm{rk}_R G \geqslant 2$, *let* F *be a locally compact nondiscrete field and let* $\varrho : \Gamma \to \mathrm{GL}_n(F)$ *be a linear representation such that* $\varrho(\Gamma)$ *is not relatively compact and that its Zariski closure is connected; then* $F = \boldsymbol{R}$ *or* C *and* ϱ *extends to a rational representation of* $\mathcal{G}$.

The proof of this theorem is relatively short (considering the power of the result), but is a succession of extraordinarily ingenious arguments using a great variety of very strong techniques belonging to ergodic theory (the "multiplicative ergodic theorem"), the theory of unitary representations, the theory of functional spaces (spaces of measurable maps), algebraic geometry, the structure theory of semi-simple algebraic groups, etc. In 1975–1976, I devoted my course at the Collège de France to those results of Margulis; I believe that I learned more mathematics during that year than in any other year of my life. A summary of the main ideas of that beautiful piece of work is given in [27].

Another, quite different proof of the superrigidity theorem and its application to arithmeticity (both in the cocompact and the noncocompact case) — using the work of H. Furstenberg — can be found in [20].

4. Other results.

4.1. *S-arithmetic groups.* Let K be a number field, S a finite set of places of K including all places at infinity, $\mathfrak{O}$ the ring of elements of K which are integral at all finite places not belonging to S, $\mathcal{H} \subset \mathcal{GL}_n$ a simply connected semisimple linear algebraic group defined over K and $\mathcal{H}(\mathfrak{O}) = \mathcal{H} \cap \mathrm{GL}_n(\mathfrak{O})$. Then, $\mathcal{H}(\mathfrak{O})$ injects as a discrete subgroup of finite covolume in the product $H = \Pi_{v \in S} \mathcal{H}(K_v)$, where K_v denotes the completion of K at v.

(Example: if $\mathfrak{O}$ is the ring of rational numbers whose denominator is a power of 2, $\mathrm{SL}_n(\mathfrak{O})$ is a discrete subgroup of finite covolume of $\mathrm{SL}_n(\boldsymbol{R}) \times \mathrm{SL}_n(\boldsymbol{Q}_2)$). Now let G be a direct product of simply connected semi-simple real or p-adic Lie groups. A discrete subgroup Γ of G is called S-arithmetic if there exist $K, S, \mathcal{H}$ as above and a homomorphism $\alpha : H \to G$ with compact kernel such that Γ and $\alpha(\mathcal{H}(\mathfrak{O}))$ are commensurable. All results of [18], stated above for ordinary Lie groups and arithmetic groups, are in fact proved by Margulis in the more general framework described here. In particular, he show that *if* G *is as above, if the rank of* G *(i.e. the sum of the relative ranks of its factors) is at least 2 and if* Γ *is a discrete subgroup of finite covolume in* G, *which is irreducible (as defined in n° 2), then* Γ *is* S*-arithmetic.*

4.2. *"Abstract" isomorphisms.* The very general and powerful superrigidity theorem (in the framework of 4.1) has far-reaching consequences besides the arithmeticity. For instance, it enables Margulis to solve almost completely the problem of "abstract isomorphisms" between groups of points of algebraic simple groups over number fields or arithmetic subrings of such fields; his result embodies, in the arithmetic case, all those obtained before on that problem by Dieudonné, O'Meara and his school, A. Borel and me, and goes considerably further.

4.3. *Normal subgroups.* Let G be as in 4.1 and let Γ be an irreducible discrete subgroup of G of finite covolume. Margulis was able to show (cf. [27]) that *if* rk $G \geqslant 2$, *then every noncentral normal subgroup of* Γ *has finite index.* (In fact, the conditions of Margulis' theorem are more general: under suitable hypotheses, G is allowed to have factors defined over locally compact local fields of finite characteristic.) So far, the only results known in that direction — results of Mennicke, Bass, Milnor, Serre, Raghunathan — were connected with the congruence subgroup problem and valid only in the cases where that problem has a positive solution.

4.4. *Action on trees.* In a paper which appeared in the Springer Lecture Notes, no 372, Serre showed that the group of integral points of a simple Chevalley group-scheme of rank $\geqslant 2$ cannot act without fixed point on a tree; this also means that such a group is not an amalgam in a nontrivial way. Serre points out that his method of proof does not extend to congruence subgroups and asks whether the result generalizes to such subgroups or to other arithmetic groups. With his own methods, Margulis was able to solve at once the problem in its widest generality: *if G is as in* 4.1, *of rank at least 2, and if Γ is an irreducible discrete subgroup of finite covolume in G, then Γ cannot act without fixed point on a tree.*

5. Conclusion. Margulis has completely or almost completely solved a number of important problems in the theory of discrete subgroups of Lie groups, problems whose roots lie deep in the past and whose relevance goes far beyond that theory itself. It is not exaggerated to say that, on several occasions, he has bewildered the experts by solving questions which appeared to be completely out of reach at the time. He managed that through his mastery of a great variety of techniques used with extraordinary resources of skill and ingenuity. The new and most powerful methods he has invented have already had other important applications besides those for which they were created and, considering their generality, I have no doubt that they will have many more in the future.

I wish to conclude this report by a nonmathematical comment. This is probably neither the time nor the place to start a polemic. However, I cannot but express my deep disappointment — no doubt shared by many people here — in the absence of Margulis from this ceremony. In view of the symbolic meaning of this city of

Helsinki,[1] I had indeed grounds to hope that I would have a chance at last to meet a mathematician whom I know only through his work and for whom I have the greatest respect and admiration.

References

Published work of G. A. Margulis

1. *Positive harmonic functions on nilpotent groups*, Dokl. Akad. Nauk S.S.S.R. **166** (1966), 1054–1057.
2. (with D. A. Každan) *Proof of a conjecture of Selberg*, Mat. Sb. **75** (1968), 163–168.
3. *Appendix to "Certain smooth ergodic systems" by D. V. Anosov and Ya. G. Sinaĭ*, Uspehi Mat. Nauk **22** (1967), 169–171.
4. *Discrete subgroups of real semi-simple Lie groups*, Mat. Sb. **80** (1960), 600–615.
5. *Certain applications of ergodic theory to the investigation of manifolds of negative curvature*, Funkcional. Anal. i Priložen, **3** (4) (1969), 89–90.
6. *On the problem of arithmeticity of discrete groups*, Dokl. Akad. Nauk S.S.S.R. **187** (1969), 518–520.
7. *Certain measures that are connected with U-flows*, Funkcional. Anal. i Priložen **4** (1) (1970), 62–76.
8. *Isometry of closed manifolds of constant negative curvature having the same fundamental group*, Dokl. Akad. Nauk S.S.S.R. **192** (1970), 736–737.
9. *On the action of unipotent groups in the space of lattices*, Mat. Sb. **86** (1971), 552–556.
10. *Nonuniform lattices in semi-simple algebraic groups, Lie groups and their representations*, Summer school of the Bolyai Janos Math. Soc. (Budapest, 1971), Akademiai Kiado, Budapest, 1975.
11. *Correspondence of boundaries in biholomorphic mappings*, Tezis dokladov Vsesoyuznoi konferencii po teorii funkcii kompleksnogo peremennogo (1971).
12. *Metric questions in the theory of U-systems*, in Devyataya Letnyaya Mat. Škola, Kiev. Inst. Mat. Akad. Nauk S.S.S.R., 1972, pp. 342–348.
13. *Arithmeticity of nonuniform lattices*, Funkcional. Anal. i Priložen **7** (3) (1973), 88–89.
14. *Explicit constructions of dilatators*, Probl. Peredači Informacii **9** (4) (1973), 71–80.
15. *Arithmetic properties of discrete groups*, Uspehi Mat. Nauk S.S.S.R. **29** (1974), 49–98.
16. *Arithmeticity and finite dimensional representations of uniform lattices*, Funkcional. Anal. Priložen **8** (3) (1974), 77–78.
17. *Probabilistic characteristics of graphs with high connectivity*, Probl. Peredači Informacii **10** (2) (1974), 101–108.
18. *Discrete groups of motions of manifolds with nonpositive curvature*, Proc. Internat. Congr. Math. (Vancouver, 1974), vol. 2, 1975, pp. 21–34.

[1]The address was delivered in Finlandia Hall, where the 1975 Helsinki Agreements were concluded.

19. *Arithmeticity of nonuniform lattices in weakly noncompact groups*, Funkcional. Anal. Priložen **9** (1) (1975), 35–44.
20. *Arithmeticity of irreducible lattices in semisimple groups of rank greater than 1*, Appendix to the Russian translation of M. Raghunathan's book: Discrete Subgroups of Lie Groups, Moscow, 1977, pp. 277–313.
21. (with G. A. Soifer), *Criteria for the existence of maximal subgroups of infinite index in finitely generated linear groups*, Dokl. Akad. Nauk S.S.S.R. **234** (1977), 1261–1264.
22. *Cobounded subgroups in algebraic groups over local fields*, Funkcional. Anal. i Priložen **11** (1977), 45–57.
23. *Factor groups of discrete subgroups and measure theory*, Funkcional. Anal. i Priložen **12** (4) (1978), 64–76.
24. *Factor groups of discrete subgroups*, Dokl. Akad. Nauk S.S.S.R. **242** (1978) 533–536.
25. *Finiteness of factor groups of discrete subgroups*, Funkcional. Anal. i Priložen (to appear).

Other references

26. A. Borel, *Sous-groupes discrets de groups semi-simples* (d'après D. A. Kajdan et G. A. Margoulis), Sém. Bourbaki, 1969, Lecture Notes in Math., vol. 179, Springer-Verlag, Berlin, pp. 199–215.
27. J. Tits, *Travaux de Margulis sur les sous-groupes discrets de groupes de Lie*, Sém. Bourbaki, 1976, Lecture Notes in Math., vol. 567, Springer-Verlag, Berlin, pp. 174–190.

Collège De France
75231 Paris Cedex 05,
France

Gregory A. Margulis

GREGORY A. MARGULIS

Date of birth: 24 February 1946
Place of birth: Moscow, USSR (Russia)

Education:

Moscow High School N 721	(1962)
Moscow University:	
(a) undergraduate student	(1962–67)
(b) graduate student	(1967–70)

Scientific degrees:

Candidate of Science (∼ Ph.D.),	
Moscow University	(1970)
Doctor of Science, Minsk	(1983)

Academic career:

Institute for Problems in Information Transmission:	
Junior scientific worker	(1970–74)
Senior scientific worker	(1974–86)
Leading scientific worker	(1986–91)
Yale University, Professor	(1991–present)

Visiting positions held:

1. University of Bonn — 1979 (3 months)
2. Max-Planck Institute, Bonn — 1988 (4 months)
3. IHES, Bures-sur-Yvette — 1989 (2 months)
4. College de France — 1990 (1 months) visiting professor
5. Harvard University — 1990–91 (fall semester) visiting professor
6. Institute for Advanced Study, Princeton — 1991 (4 months)
7. Columbia University — 1994 (spring semester), Eilenberg Chair
8. Newton Institute for Mathematical Sciences, Cambridge — 2000 (4 months)

Honors and awards:

Prize of Moscow Mathematical Society for young mathematicians	(1968)
Fields Medal	(1978)
Medal of College de France (in connection with the visiting professorship)	(1991)
Foreign Honorary Member of the American Academy of Arts and Sciences	(1991)
Humboldt Prize	(1995)
Honorary Fellow of Tata Institute of Fundamental Research	(1996)
Lobachevsky Prize	(1997)
Honorary Doctoral Degree fro the University of Bielefeld	(1999)
Member of the National Academy of Science	(2001)

OPPENHEIM CONJECTURE

by
G. A. MARGULIS*

Department of Mathematics, Yale University
P.O. Box 208283
New Haven, CT 06520-8283, USA
E-mail: margulismath.yale.edu

The main purpose of this paper is to give a survey of results and methods connected with a conjecture on values of indefinite irrational quadratic forms at integer points stated by Oppenheim in 1929 and proved by the author in 1986. The different approaches to this and related conjectures (and theorems) involve analytic number theory, the theory of Lie groups and algebraic groups, ergodic theory, representation theory, reduction theory, geometry of numbers and some other topics. A comprehensive survey of the methods related to the Oppenheim conjecture is thus a long story.

Let B be a real nondegenerate indefinite quadratic form in n variables. We will say that B is *rational* if it is a multiple of a form with rational coefficients and *irrational* otherwise. According to Meyer's theorem (see [Ca3]) if B is rational and $n \geq 5$ then B represents zero over $\mathbb{Z}$ nontrivially, i.e. there exist integers $x_1, \ldots, x_n$ not all equal to 0 such that $B(x_1, \ldots, x_n) = 0$. Let us set

$$m(B) = \inf\{|B(x)| \mid x \in \mathbb{Z}^n, x \neq 0\}.$$

Then Meyer's theorem is equivalent to the statement that if B is rational and $n \geq 5$ then $m(B) = 0$. In 1929 A. Oppenheim [Op1, 2] conjectured that if $n \geq 5$ then $m(B) = 0$ also for irrational B. Later it was realized that $m(B)$ should be equal to 0 under a weaker condition $n \geq 3$ (for diagonal forms it was stated as a conjecture in [DavH]; let us remind the reader that a quadratic form Q is called *diagonal* if $Q(x_1, \ldots, x_n) = \lambda_1 x_1^2 + \cdots + \lambda_n x_n^2$). Of course if n is 3 or 4 one has to assume that B is irrational because there exist rational nondegenerate indefinite quadratic forms in 3 and 4 variables which do not represent zero over $\mathbb{Z}$ nontrivially. Thus the Oppenheim conjecture with the generalization for $n = 3$ and $n = 4$ states that if $n \geq 3$ and B is not proportional to a form with rational coefficients, then for any $\varepsilon > 0$ there exist integers $x_1, \ldots, x_n$ not all equal to 0 such that $|B(x_1, \ldots, x_n)| < \varepsilon$. (Let us note that the condition "$n \geq 3$" cannot be replaced by the condition "$n \geq 2$".

*Partially supported by NSF grant DMS-9424613.

To see this, consider the form $x_1^2 - \lambda x_2^2$ where λ is an irrational positive number such that $\sqrt{\lambda}$ has a continued fraction development with bounded partial quotients; for example $\lambda = (1+\sqrt{3})^2 = 4+2\sqrt{3}$.) This conjecture was proved by studying orbits of the orthogonal group $SO(2,1)$ on the space $\Omega \cong SL(3,\mathbb{R})/SL(3,\mathbb{Z})$ of unimodular lattices in $\mathbb{R}^3$ (see [Marg3–6]).

Before the Oppenheim conjecture was proved it was extensively studied mostly using analytic number theory methods. We will describe this earlier development in Sec. 1. In Sec. 2, we begin with a 1955 paper of Cassels and Swinnerton-Dyer and proceed to describe results related to the Oppenheim conjecture up to the present. In Sec. 3 we present a proof of a strengthened Oppenheim conjecture based largely on a consideration of unipotent flows on homogeneous spaces. We point out other phenomena related to unipotent flows.

Let us now make the following remark. If B is a real irrational nondegenerate indefinite quadratic form in n variables and $m < n$ then $\mathbb{R}^n$ contains a rational subspace L of dimension m such that the restriction of B to L is irrational nondegenerate and indefinite (this is a standard fact; the proof can be found in Sec. 5 of [DanM1]). Hence if the Oppenheim conjecture is proved for some n_0, then it is proved for all $n \geq n_0$. In particular, it is enough to prove the conjecture for $n = 3$.

The author would like to thank S. G. Dani, A. Eskin and G. D. Mostow for their comments on a preliminary version of the paper.

1. Analytic Number Theory Methods

1.1. Chowla's result and the distribution of values of positive definite quadratic forms at integer points. The first result on the Oppenheim conjecture was obtained in 1934 by Chowla [Ch] who proved the conjecture for indefinite diagonal forms

$$B(x_1, \dots, x_n) = \lambda_1 x_1^2 + \cdots + \lambda_r x_r^2 + \lambda_{r+1} x_{r+1}^2 + \cdots + \lambda_n x_n^2$$

such that $n \geq 9$ and all the ratios $\lambda_i/\lambda_j (i \neq j)$ are irrational. His proof used a theorem of Jarnik and Walfisz on the number of integer points in a large ellipsoid. Indeed Chowla's proof can be applied also in the following cases: (a) $r \geq 5$, $\lambda_i > 0$, $1 \leq i \leq r$, and at least one ratio $\lambda_i/\lambda_j, 1 \leq i < j \leq r$ is irrational; (b) $n - r \geq 5$, $\lambda_i < 0$, $r+1 \leq i \leq n$, and at least one ratio λ_i/λ_j, $r+1 \leq i < j \leq n$, is irrational.

Let us describe now the result of Jarnik and Walfisz. Let Q be a positive definite real quadratic form in n variables and let $A_Q(X)$ denote the number of integer solutions of the inequality $Q(x) \leq X$. One can interpret $A_Q(X)$ as the number of points from the lattice $\mathbb{Z}^n$ in the n-dimensional ellipsoid $\{x \in \mathbb{R}^n \mid Q(x) \leq X\}$. Let $\mathcal{T}_Q(X)$ denote the volume of this ellipsoid. We have

$$\mathcal{T}_Q(X) = \mathcal{C}_Q X^{n/2}, \quad \mathcal{C}_Q = \frac{\pi^{n/2}}{\sqrt{D}\Gamma\left(\frac{n}{2}+1\right)} \tag{1}$$

where D is the determinant of Q. In 1930 Jarnik and Walfisz [JW] proved that if $n \geq 5$ and Q is diagonal and irrational then

$$A_Q(X) = \mathcal{T}_Q(X) + o(X^{n/2-1}). \tag{2}$$

They showed also that "o " cannot be replaced by any specific function. It means that for any positive function $\varphi(X), X > 0$, with $\lim_{X\to\infty} \varphi(X) = 0$, there exists a diagonal irrational form Q such that

$$\limsup_{X\to\infty} \frac{|A_Q(X) - \mathcal{T}_Q(X)|}{X^{n/2-1}\varphi(X)} = \infty.$$

In fact the set of all such Q is of second category in the Baire sense (as a subset of the space of positive definite diagonal forms). This can be easily proved by a standard topological argument using the fact that for any rational Q

$$\limsup_{X\to\infty} \frac{|A_Q(X) - \mathcal{T}_Q(X)|}{X^{n/2-1}} > 0.$$

Indeed the only case $n = 5$ was considered in [JW], because for $n \geq 6$ the estimate (2) had been proved in an earlier paper by Jarnik. The proof of (2) in this case $n = 5$ given in [JW] uses a modification of Jarnik's method.

It follows immediately from (1) and (2) that for any fixed $\varepsilon > 0$

$$A_Q(X+\varepsilon) - A_Q(X) \sim \varepsilon \mathcal{C}_Q \frac{n}{2} X^{n/2-1} \text{ as } X \to \infty.$$

In particular, the gaps between successive values of the quadratic form Q must tend to 0 or, equivalently,

$$A_Q(X+\varepsilon) - A_Q(X) > 0 \tag{3}$$

for every positive ε and all $X \geq X_0(\varepsilon)$.

Now let $B, \lambda_1, \ldots, \lambda_n$ be the same as at the beginning of this subsection. We assume that $n \geq 9$ and all the ratios $\lambda_i/\lambda_j (i \neq j)$ are irrational. The Chowla's argument is the following one. Of nine positive or negative numbers, at least five must have the same sign. Hence we may assume that $\lambda_1, \ldots, \lambda_5$ are positive and that at least one of the numbers $\lambda_6, \ldots, \lambda_n$ is negative. Now applying (3) to the quadratic form

$$Q(x_1, \ldots, x_5) = \lambda_1 x_1^2 + \cdots + \lambda_5 x_5^2$$

we get that there exist $y \geq X_0(\varepsilon), y > 0$, and integers $m_1, \ldots, m_n$ such that

$$\sum_{s=6}^{n} \lambda_s m_s^2 = -y$$

and

$$y < \lambda_1 m_1^2 + \cdots + \lambda_5 m_5^2 \leq y + \varepsilon.$$

Then $|B(m_1, \dots, m_n)| < \varepsilon$ where the m's are not all zero; this is the desired statement.

It was conjectured in [DavL] that (2) and (3) are true for a general positive definite irrational quadratic form in $n \geq 5$ variables. These conjectures have not yet been proved even under the assumption that n is very large. However, in 1972 Davenport and Lewis, in the same paper [DavL], partially resolved the question regarding gaps (that is, the generalization of (3)). They proved the following:

1.1.1. Theorem. *There exists an integer n_0 (absolute) with the following property:*

Let $Q(x) = Q(x_1, \dots, x_n)$ be a positive definite quadratic form with real coefficients and suppose that $n \geq n_0$. Then, if $x_1^*, \dots, x_n^*$ are integers with $\max_i |x_i^*|$ sufficiently large, there exist integers $x_1, \dots, x_n$, not all zero, such that

$$|Q(x + x^*) - Q(x^*)| < 1. \tag{4}$$

Clearly, in this theorem one can replace 1 by any positive ε. As it is noticed in [DavL] and [Lew] the result of Davenport and Lewis is imperfect in two ways. (a) One could have $Q(x + x^*) = Q(x^*)$ even if Q is irrational, and indeed this may well happen if Q represents a rational form in 4 variables (i.e. if there exists a rational 4-dimensional subspace L such that the restriction of Q to L is proportional to a form with integral coefficients). (b) Even if (a) does not occur, no deduction can be made about the gaps since the result does not prevent that the elements of $Q(\mathbb{Z}^n)$ occur in clumps with large gaps between the clumps.

1.1.2. The proof of Theorem 1.1.1 uses the Hardy–Littlewood circle method and representation properties of positive definite diagonal forms with integral coefficients. In 1986 Cook and Raghavan [CoR2] gave an estimate for the size of n_0 and also gave a lower bound for the number of solutions of the inequality (4). Their proof is similar to the proof given in [DavL]. It is also noticed in [CoR2] that Davenport and Lewis overlooked the (trivial) solution $x = -2x^*$ for (4). Let us state the theorem of Cook and Raghavan.

1.1.2. Theorem [CoR2, Theorem 2]. *There exists an integer $n_0 \leq 995$ and a constant $\tau > 0$ with the following property:*

Let $Q(x) = Q(x_1, \dots, x_n)$ be a positive definite quadratic form with real coefficients and suppose that $n \geq n_0$. Then, if $x_1^*, \dots, x_n^*$ are integers with $\max |x_i^*|$ sufficiently large, then there exist at least $[\|x^*\|^\tau]$ integer points $x \in \mathbb{Z}^n$ such that

$$|Q(x + x^*) - Q(x)| < 1.$$

1.2. Diagonal forms in $n \geq 5$ variables. In 1946, Davenport and Heilbronn proved the Oppenheim conjecture for diagonal forms in $n \geq 5$ variables. Strictly speaking, they considered only forms in five variables. But the case $n \geq 5$ immediately reduces to the case $n = 5$ by a trivial argument. In fact, Davenport and Heilbronn proved the following stronger theorem which gives lower bounds for the number of integral solutions of some quadratic inequalities in some domains.

1.2.1. Theorem [DavH]. *Let $\lambda_1, \dots, \lambda_5$ be real numbers, not all of the same sign, and none of them zero, such that at least one of the ratios λ_r/λ_s is irrational. Let*

$$B(x_1, \dots, x_5) = \lambda_1 x_1^2 + \dots + \lambda_5 x_5^2.$$

Then there exist arbitrarily large integers P such that the inequalities

$$1 \le x_1 \le P, \dots, 1 \le x_5 \le P \ , \ |B(x_1, \dots, x_5)| < 1$$

have more than γP^3 integral solutions, where $\gamma = \gamma(\lambda_1, \dots, \lambda_5) > 0$.

1.2.2. Remarks. (a) There is a footnote in [DavH] on page 186 that the same method, together with the use of an inequality due to Hua (Quarterly J. of Math., 9 (1938), 199–202), proves the corresponding theorem for $\lambda_1 x_1^k + \dots + \lambda_s x_s^k$, where $s = 2^k + 1$.

(b) Clearly the inequality $|B(x)| < \varepsilon$ is equivalent to the inequality $|B_\varepsilon(x)| < 1$ where $B_\varepsilon = B/\varepsilon$. Therefore Theorem 1.4 implies that for any $\varepsilon > 0$ there exist arbitrarily large integers P such that the inequalities

$$1 \le x_1 \le P, \dots, 1 \le x_5 \le P \ , \ |B(x_1, \dots, x_5)| < \varepsilon$$

have more than $\gamma_\varepsilon P^3$ integral solutions where $\gamma_\varepsilon = \gamma(\varepsilon^{-1}\lambda_1, \dots, \varepsilon^{-1}\lambda_5) > 0$.

1.2.3. The Davenport–Heilbronn proof uses a modification of the Hardy–Littlewood method. We will describe now this proof following the original paper [DavH] and Lewis's survey [Lew].

We assume, without loss of generality, that λ_1/λ_2 is irrational. The constants implied by the symbol O depend only on $\varepsilon, \lambda_1, \dots, \lambda_5$.

Let $e(x) = e^{2\pi i x}$. For any positive integer P we define

$$S(\alpha) = S_P(\alpha) = \sum_{x=1}^{P} e(\alpha x^2), I(\alpha) = I_P(\alpha) = \int_0^P e(\alpha x^2) dx.$$

Direct computations or standard facts about Fourier transform show that for any $t \in \mathbb{R}$

$$\int_{-\infty}^{\infty} e(\alpha t) \left(\frac{\sin \pi\alpha}{\pi\alpha} \right)^2 d\alpha = \max(0, 1 - |t|). \tag{5}$$

It is clear that

$$e(B(x_1, \dots, x_5)) = e(\lambda_1 x_1^2 + \dots + \lambda_5 x_5^2) = e(\lambda_1 x_1^2) \cdot \ldots \cdot e(\lambda_5 x_5^2).$$

Therefore the equality (5) implies the following two equalities:

$$\sum_{x_1=1}^{P} \cdots \sum_{x_5=1}^{P} \max(0, 1 - |B(x_1, \dots, x_5)|)$$

$$= \int_{-\infty}^{\infty} S(\lambda_1\alpha) \dots S(\lambda_5\alpha) \left(\frac{\sin \pi\alpha}{\pi\alpha} \right)^2 d\alpha, \tag{6}$$

$$I = \int_0^P \cdots \int_0^P \max(0, 1 - |B(x_1, \ldots, x_5)|) dx_1 \ldots dx_5$$

$$= \int_{-\infty}^{\infty} I(\lambda_1 \alpha) \ldots I(\lambda_5 \alpha) \left(\frac{\sin \pi\alpha}{\pi\alpha}\right)^2 d\alpha \tag{7}$$

A straightforward integration shows that

$$I > \gamma_1 P^3 \tag{8}$$

where $\gamma_1 = \gamma_1(\lambda_1, \ldots, \lambda_5) > 0$ is a positive constant independent of P [DavH, Lemma 10]. In fact, it is not difficult to prove the following asymptotic formula,

$$I \sim \beta P^3 \tag{9}$$

where

$$\beta = \int_{L \cap \Omega} \frac{dA}{\|\nabla B\|},$$

L is the lightcone $B = 0$, $\Omega = \{(x_1, \ldots, x_5) \mid 0 \leq x_1 \leq 1, \ldots, 0 \leq x_5 \leq 1\}$ and dA is the area element on L.

In view of (6), (7) and (8), the theorem will be proved if we show that for some $\delta > 0$ there exist infinitely large numbers P such that the difference between the right-hand sides of (6) and (7) is $O(P^{3-\delta})$. To do this we divide the remaining part of the proof into three steps.

Step 1. If $\alpha = O(P^{-\frac{3}{2}})$ then $S(\alpha) - I(\alpha) = O(1)$ [DavH, Lemma 5]. But it is trivial that $I(\alpha) \leq P$ for any α. Hence

$$\prod_{r=1}^{5} S(\lambda_r \alpha) - \prod_{r=1}^{5} I(\lambda_r \alpha) = O(P^4) \text{ on } |\alpha| < P^{-\frac{3}{2}}. \tag{10}$$

Further,

$$S(\alpha) = O(|\alpha|^{-\frac{1}{2}-\varepsilon}) \text{ if } \alpha = O(P^{-1}), \alpha \neq 0, \tag{11}$$

for any $\varepsilon > 0$ [DavH, Lemma 7] and

$$I(\alpha) = \frac{1}{2|\alpha|^{\frac{1}{2}}} \int_0^{P^2|\alpha|} e(\pm x) x^{-\frac{1}{2}} dx = O(|\alpha|^{-\frac{1}{2}}) \tag{12}$$

for any $\alpha \neq 0$. It follows easily from the estimates (10), (11) and (12) that

$$\int_{-P^{-1}}^{P^{-1}} S(\lambda_1 \alpha) \ldots S(\lambda_5 \alpha) \left(\frac{\sin \pi\alpha}{\pi\alpha}\right)^2 d\alpha$$

$$= \int_{-\infty}^{\infty} I(\lambda_1 \alpha) \ldots I(\lambda_5 \alpha) \left(\frac{\sin \pi\alpha}{\pi\alpha}\right)^2 + O(P^{\frac{5}{2}}). \tag{13}$$

Step 2. Let $r(t)$ denote the number of representations of an integer t as the sum of two integral squares. It is well known that $r(t) = O(t^\varepsilon)$ for any $\varepsilon > 0$ [DavH, Lemma 2]. Hence for any $m \in \mathbb{R}$ and any $\varepsilon > 0$

$$\begin{aligned}\int_m^{m+1} |S(\alpha)|^4 d\alpha &= \int_0^1 |S(\alpha)|^4 d\alpha \\ &= \int_0^1 |\sum_{x=1}^{P}\sum_{y=1}^{P} e(\alpha(x^2+y^2))|^2 d\alpha \\ &\le \sum_{t=1}^{2P^2} r^2(t) = O(P^{2+\varepsilon}). \qquad (14)\end{aligned}$$

It follows easily from (14) [DavH, Lemma 8] that for $\mu \ge 0$ and $1 \le r \le 5$

$$\int_\mu^\infty |S(\lambda_r\alpha)|^4 \left(\frac{\sin \pi\alpha}{\pi\alpha}\right)^2 d\alpha = O\left(\frac{P^{2+\varepsilon}}{\mu+1}\right). \qquad (15)$$

Using the trivial estimate $|S(\alpha)| \le P$ and the standard inequality comparing arithmetic and geometric means we get from (15) the following two estimates:

$$\int_{|\alpha|>\mu} S(\lambda_1\alpha)\ldots S(\lambda_5\alpha)\left(\frac{\sin \pi\alpha}{\pi\alpha}\right)^2 d\alpha = O\left(\frac{P^{3+\varepsilon}}{\mu+1}\right), \qquad (16)$$

$$\begin{aligned}&\int_{\mu_1<|\alpha|<\mu_2} S(\lambda_1\alpha)\ldots S(\lambda_5\alpha)\left(\frac{\sin \pi\alpha}{\pi\alpha}\right)^2 d\alpha \\ &= \sup\{\min(|S(\lambda_1\alpha)|, |S(\lambda_2\alpha)|)\ |\mu_1 < |\alpha| < \mu_2\}(\mu_2-\mu_1)O(P^{2+\varepsilon}) \qquad (17)\end{aligned}$$

for any $\mu \ge 0$ and $0 \le \mu_1 < \mu_2$.

Step 3. In previous two steps, P was an arbitrary positive integer. We now restrict P to a sequence of squares of the denominators of the partial fraction convergents to the irrational number λ_1/λ_2. Then there exist integers a, q such that the greatest common divisor (a, q) is 1,

$$|\lambda_1/\lambda_2 - a/q| < \frac{1}{q^2} \text{ and } P = q^2. \qquad (18)$$

Let ρ be an absolute constant with $0 < \rho < \frac{1}{10}$. Assume that we can show

$$\min(|S(\lambda_1\alpha)|, |S(\lambda_2\alpha)|) = O(P^{1-\rho+\varepsilon}) \text{ on } P^{-1} \le |\alpha| \le P^\rho. \qquad (19)$$

Then using (6), (7), (13), (16) and (17), we get the desired estimate:

$$\sum_{x_1=1}^{P} \cdots \sum_{x_5=1}^{P} \max(0, 1 - |B(x_1, \dots, x_5)|)$$

$$- \int_0^P \cdots \int_0^P \max(0, 1 - |B(x_1, \dots, x_5))dx_1 \dots dx_5$$

$$= \int_{-\infty}^{\infty} S(\lambda_1\alpha)\dots S(\lambda_5\alpha) \left(\frac{\sin \pi\alpha}{\pi\alpha}\right)^2 d\alpha$$

$$- \int_{-\infty}^{\infty} I(\lambda_1\alpha)\dots I(\lambda_5\alpha) \left(\frac{\sin \pi\alpha}{\pi\alpha}\right)^2 d\alpha$$

$$= O(P^{3-\rho+\varepsilon}) \tag{20}$$

(we use (16) for $\mu = P^\rho$ and (17) for $\mu_1 = P^{-1}$ and $\mu_2 = P^\rho$).

To prove (19) we take α with $P^{-1} \leq |\alpha| \leq P^\delta$ and consider rational approximations $a_1/q_1, a_2/q_2$ to $\lambda_1\alpha$ and $\lambda_2\alpha$, respectively, such that

$$\begin{cases} (a_1, q_1) = 1, q_1 \leq P, |\lambda_1\alpha - a_1/q_1| \leq P^{-1}q_1^{-1}, \\ (a_2, q_2) = 1, q_2 \leq P, |\lambda_2\alpha - a_2/q_2| \leq P^{-1}q_2^{-1} \end{cases}. \tag{21}$$

If $a_1 = 0$ or $a_2 = 0$ then (19) follows from (11). If $q_1 \geq P^{2\rho}$ or $q_2 \geq P^{2\rho}$ then (19) follows from Weyl's inequality for exponential sums [DavH, Lemma 6]. We may therefore suppose that

$$a_1 \neq 0, a_2 \neq 0, q_1 < P^{2\rho}, q_2 < P^{2\rho}. \tag{22}$$

We can rewrite the inequalities (21) as

$$\lambda_1\alpha = \frac{a_1}{q_1}(1 + b_1), \lambda_2\alpha = \frac{a_2}{q_2}(1 + b_2), |b_1| \leq P^{-1}, |b_2| \leq P^{-1}. \tag{23}$$

It follows from (18) and (23) that

$$\frac{a_1q_2}{a_2q_1} - \frac{a}{q} = O(P^{-1}). \tag{24}$$

Since $|\alpha| < P^\rho$, (21) implies that $a_2/q_2 = O(P^\rho)$. But $0 < \rho < 1/10$ and $q = P^{1/2}$. Hence it follows from (22) that $a_2q_1q = O(P^{5\rho+\frac{1}{2}}) = o(P) = o(q^2)$. It shows that (24) is impossible for sufficiently large P because $(a, q) = 1$. This completes the proof of Theorem 1.2.1.

1.3. A modification of the theorem of Davenport and Heilbronn. The kernel $(\frac{\sin \pi\alpha}{\pi\alpha})^2$ is the Fourier transform of the function $q(t) = \max(0, 1 - |t|)$. We can replace q by another function. Let $\varphi(t)$ be a continuous function on $\mathbb{R}$ with a compact support, let $\hat{\varphi}(\alpha)$ denote the Fourier transform of φ and let

$$A_\varphi = \sup\{(1 + \alpha^2)|\hat{\varphi}(\alpha)| \mid -\infty < \alpha < \infty\}.$$

We assume that $A_\varphi < \infty$. Then the estimates (13), (15), (16) and (17) will be still true if we replace $\left(\frac{\sin \pi\alpha}{\pi\alpha}\right)^2$ by $\hat{\varphi}(\alpha)$ and multiply the constants implied by the symbol O by A_φ. Therefore the following analogue of (20) is true:

$$\begin{aligned} &\sum_{x_1=1}^{P} \cdots \sum_{x_5=1}^{P} \varphi(B(x_1,\ldots,x_5) \\ &\quad - \int_0^P \cdots \int_0^P \varphi(B(x_1,\ldots,x_5))dx_1\ldots dx_5 \\ &\quad = \int_{-\infty}^{\infty} S(\lambda_1\alpha)\ldots S(\lambda_5\alpha)\hat{\varphi}(\alpha)d\alpha \\ &\quad - \int_{-\infty}^{\infty} I(\lambda_1\alpha)\ldots I(\lambda_5\alpha)\hat{\varphi}(\alpha)d\alpha \\ &\quad = A_\varphi \cdot O(P^{3-\rho+\varepsilon}). \end{aligned} \tag{25}$$

Let us fix real numbers $a < b$. Let us denote $P^{-\frac{\varrho}{2}}$ by θ and consider the following two functions:

$$\varphi_1(t) = \begin{cases} 0 & \text{if } t \le a \text{ or } t \ge b \\ 1 & \text{if } a+\theta \le t \le b-\theta \\ \frac{t-a}{\theta} & \text{if } a \le t \le a+\theta \\ \frac{b-t}{\theta} & \text{if } b-\theta \le t \le b, \end{cases}.$$

$$\varphi_2(t) = \begin{cases} 0 & \text{if } t \le a-\theta \text{ or } t \ge b+\theta \\ 1 & \text{if } a \le t \le b \\ \frac{a-t}{\theta} & \text{if } a-\theta \le t \le a \\ \frac{t-b}{\theta} & \text{if } b \le t \le b+\theta \end{cases}.$$

Thus φ_1, φ_2 are piecewise linear functions and $\varphi_1 \le \chi_{[a,b]} \le \varphi_2$ where $\chi_{[a,b]}$ denotes the characteristic function of the segment $[a,b]$. We also have that $A_{\varphi_1} = O(\theta^{-1}) = O(P^{\frac{\varrho}{2}})$, $A_{\varphi_2} = O(\theta^{-1}) = O(P^{\frac{\varrho}{2}})$ and

$$\int_0^P \cdots \int_0^P (\varphi_1 - \varphi_2)(B(x_1,\ldots,x_5))dx_1\ldots dx_5 = O(\theta P^3) = O(P^{3-\frac{\varrho}{2}})$$

(the last estimate can be obtained by a straightforward integration). Let us introduce the following notation:

$$\Omega = \{(x_1,\ldots,x_5) \mid 0 < x_1 \le 1,\ldots,0 < x_5 \le 1\}$$

and

$$V^B_{(a,b)} = \{v = (x_1, \dots, x_5) \in \mathbb{R}^5 \mid a < B(v) < b\}.$$

Then using the above observations we get from (25) that

$$\begin{aligned} &|V^B_{(a,b)} \cap P\Omega \cap \mathbb{Z}^n| - \mathrm{Vol}(V^B_{(a,b)} \cap P\Omega) \\ &= \sum_{x_1=1}^{P} \cdots \sum_{x_5=1}^{P} \chi_{[\mathrm{ah},\mathrm{bh}]}(B(x_1, \dots, x_5)) \\ &- \int_0^P \cdots \int_0^P \chi_{[\mathrm{ah},\mathrm{bh}]}(B(x_1, \dots, x_5))dx_1 \dots dx_5 = O(P^{3-\frac{\varrho}{2}+\varepsilon}). \end{aligned} \tag{26}$$

Let us recall that ρ is an arbitrary constant with $0 < \rho < \frac{1}{10}$ and ε is an arbitrary positive number. Thus we obtained the following modification of the theorem of Davenport and Heilbronn.

1.3.1. Theorem. *Let $\lambda_1, \dots, \lambda_5$ be real numbers, not all of the same sign, and none of them zero, such that λ_1/λ_2 is irrational. Let*

$$B(x_1, \dots, x_5) = \lambda_1 x_1^2 + \dots + \lambda_5 x_5^2$$

and let P belong to a sequence of squares of the denominators of the partial convergents to the irrational number λ_1/λ_2. Let us fix real numbers $a < b$. Then the number of the integral solutions of the inequalities

$$1 \le x_1 \le P, \dots, 1 \le x_5 \le P, \ a < B(x_1, \dots, x_5) < b$$

is

$$\begin{aligned} &\mathrm{Vol}(V^B_{(a,b)} \cap P\Omega) + O(P^{3-\frac{\varrho}{2}}) \\ &= \mathrm{Vol}\{v = (x_1, \dots, x_5) \in \mathbb{R}^5 \mid 1 \le x_1 \le P, \dots, 1 \le x_5 \le P, a < B(v) < b\} \\ &\quad + O(P^{3-\frac{\varrho}{2}}) \end{aligned}$$

where ρ is an arbitrary constant with $0 < \rho < 1/10$.

1.3.2. Remarks. (a) It is not difficult to show that

$$\mathrm{Vol}(V^B_{(a,b)} \cap P\Omega) = \beta P^3 + O(P^2)$$

where $\beta > 0$ is the same as in (9).

(b) Theorem 1.3.1 (without the remainder term) was proved in [Br].

1.4. Results for general quadratic forms. Watson, in a 1953 paper [Wa2], extended the result of Davenport and Heilbronn to forms which include a single cross-product term (say $\lambda_6 x_4 x_5$). As in [DavH], the proof in [Wa2] is based on the estimates of trigonometric sums and it seems that one can obtain an analogue of Theorem 1.3.1 for forms considered in [Wa2] (for some positive ρ). In another 1953 paper [Wa1], Watson proved the Oppenheim conjecture for the following two types of diagonal quadratic forms in three and four variables:

(A) $x^2 - a\theta y^2 - (a\theta + 1)z^2$ where a is a positive integer and θ is the positive root of the equation $\theta^2 = a\theta + 1$;
(B) $x^2 + dy^2 - \theta^2(z^2 + dw^2)$ where d is a positive integer, θ is any number in the quadratic field $\mathbb{Q}(\sqrt{N})$ and N is a non-square integer of the form $u^2 + dv^2$ with integral u, v.

In fact a more precise result is proved in [Wa1] for the forms of type (A). It is shown that there exists a number $\mathcal{C} = \mathcal{C}(a)$ such that the inequalities

$$0 < x \leq X, 0 < y \leq X, 0 < z \leq X, \ |x^2 - a\theta y^2 - (a\theta + 1)z^2| < \mathcal{C}X^{-2}$$

are soluble in integers x, y, z for any positive integer X. The approach in [Wa1] is based on some elementary properties of continued fractions.

In 1956–57 Davenport [Dav1, 3] proved the conjecture for quadratic forms of signature $(r, n - r)$ provided that either

$$r \geq 16 \quad \text{and} \quad n - r \geq 16, \text{ or}$$

$$13 \leq r \leq 15 \quad \text{and} \quad n > (9r - 20)/(r - 12).$$

As in [DavH] the proof in [Dav1, 3] uses a modification of the Hardy–Littlewood method with the replacement of the kernel $(\sin \pi\alpha/\pi\alpha)^2$ by the kernel

$$K(\alpha) = \frac{4}{3}\frac{\sin 4\pi\alpha/3}{4\pi\alpha/3}\left(\frac{\sin 2\pi\alpha/3m}{2\pi\alpha/3m}\right)^m.$$

More precisely, this modification leads to the conclusion that either the desired result holds, or the restriction of the form αQ (for a certain real α) to a certain 5-dimensional sublattice of $\mathbb{Z}^n$ is an indefinite form with almost integral coefficients. In the latter case, Davenport uses the estimate of Cassels for the magnitude of the least solutions of homogeneous quadratic equations (see Theorem 1.4.1 below). A similar approach was used in a 1959 paper [DavR] by Davenport and Ridout where the Oppenheim conjecture was proved under the weaker hypothesis that

$$n \geq 21 \quad \text{and} \quad \min(r, n - r) \geq 6.$$

A different method was used by Birch and Davenport in [BiD3] where they proved the conjecture under the hypothesis that either

$$1 \leq \min(r, n - r) \leq 4 \quad \text{and} \quad n \geq 21, \text{ or}$$

$$\min(r, n - r) > 4 \quad \text{and} \quad n \geq 17 + \min(r, n - r).$$

The idea of [BiD3] is to prove that the restriction of B to a certain 5-dimensional sublattice of $\mathbb{Z}^n$ is an indefinite form which is almost diagonal. To the almost diagonal form in 5 variables Birch and Davenport apply a quantitative improvement of Theorem 1.2.1 (see Theorem 1.4.2 below). Their method was very soon modified by Ridout [Ri] who settled the case

$$n \geq 21 \quad \text{and} \quad \min(r, n-r) = 5.$$

The combination of the above-mentioned results from [DavR], [BiD1] and [Ri] proves the Oppenheim conjecture for $n \geq 21$. For a rather detailed description of the papers [Dav1, 3], [DavR], [BiD1] and [Ri] see the above-mentioned survey [Lew].

In 1975 using the methods of linear and half-dimensional sieves, Iwaniec [Iw] proved the conjecture for forms in 4 variables of the type

$$x_1^2 + x_2^2 - \theta(x_3^2 + x_4^2)$$

where θ is a real positive irrational number [Iw]. In 1986 R. Baker and Schlickewei [Bas] settled the following cases:

(a) $n = 18, r = 9$;
(b) $n = 19, 8 \leq r \leq 11$;
(c) $n = 20, 7 \leq r \leq 13$.

As it is noticed in [Bas], their method is an elaboration of that of Davenport and Ridout [DavR] and the new weapon is Schlickewei's extension of the work [BiD1] of Birch and Davenport on quadratic equations in several variables.

It seems that the methods of analytic number theory are not sufficient to prove the Oppenheim conjecture for general quadratic forms in a small number of variables. In connection with this let us quote a remark from the just mentioned paper [Bas] that the Oppenheim conjecture "does not seem likely to be settled soon at the present rate of progress".

Let us state now the above-mentioned results of Cassels on the least solutions of quadratic equations and of Birch and Davenport on a quantitative improvement of Theorem 1.2.1 of Davenport and Heilbronn.

1.4.1 Theorem [Ca1], [Dav2]). *Let*

$$f(x_1, \ldots, x_n) = \sum_{i=1}^{n} \sum_{j=1}^{n} f_{ij} x_i x_j$$

be a quadratic form in n variables ($n \geq 2$) with integral coefficients. If the equation $f = 0$ has a non-trivial solution in integers, then it has a solution $(x_1, \ldots, x_n) \in \mathbb{Z}^n$ satisfying

$$0 < \sum_i x_i^2 \leq \gamma_{n-1}^{n-1} \left(2 \sum_{i,j} f_{ij}^2 \right)^{\frac{1}{2}(n-1)},$$

where γ_{n-1} is Hermite's constant, defined as the upper bound of the minima of positive definite quadratic forms in $n-1$ variables of determinant 1.

This formulation is taken from [Dav2]. An example given by Kneser [Ca1] shows that the exponent $\frac{1}{2}(n-1)$ cannot be improved for any n. Watson [Wa3] proved that for quadratic forms of signature (r, s), the exponent $\frac{1}{2}(n-1)$ can be replaced by

$$\theta = \max\left(2, \frac{1}{2}r, \frac{1}{2}s\right),$$

but cannot be replaced by any number smaller than $\frac{1}{2}h$ where an integer h is defined by

$$h \min(r, s) \leq \max(r, s) < (h+1)\min(r, s).$$

1.4.2. Theorem [BiD2]. *For $\delta > 0$ there exists C_δ with the following property. For any real $\lambda_1, \ldots, \lambda_5$ not all of the same sign and all of absolute value 1 at least, there exist integers $x_1, \ldots, x_n$ which satisfy both inequalities*

$$|\lambda_1 x_1^2 + \cdots + \lambda_5 x_5^2| < 1 \quad \text{and} \quad 0 < |\lambda_1| x_1^2 + \cdots + |\lambda_5| x_5^2 < C_\delta |\lambda_1 \ldots \lambda_5|^{1+\delta}.$$

An important part in the proof of this theorem is a modification of Theorem 1.4.1 given in [BiD2].

1.4.3. As before, let B be a real nondegenerate indefinite quadratic form in n variables. As it is noticed in [Lew] Birch, Davenport and Ridout actually showed in the series of papers [Dav1, 3], [DavR], [BiD3] and [Ri] that $B(\mathbb{Z}^n)$ contains 0 as a nonisolated accumulation point if B is irrational and $n \geq 21$. On the other hand, Oppenheim [Op4] proved in 1952 that if $n \geq 3$ and $\{x \in \mathbb{Z}^n \mid 0 < |B(x)| < \varepsilon\} \neq \emptyset$ for any $\varepsilon > 0$ then $B(\mathbb{Z}^n)$ is dense in $\mathbb{R}$. Combining these two results we get that $B(\mathbb{Z}^n)$ is dense in $\mathbb{R}$ if B is irrational and $n \geq 21$.

Let us mention another result of Oppenheim which he also obtained in 1952. He proved that if B is irrational, $n \geq 4$ and $B(x) = 0$ for some nonzero $x \in \mathbb{Z}^n$ then $B(\mathbb{Z}^n)$ is dense in $\mathbb{R}$ (this result follows from Theorem 1 in [Op4], Theorems 1 and 2 in [Op5], and Theorem 2 in [Op6] for $n \geq 4$).

2. Oppenheim Conjecture And Subgroup Actions On Homogeneous Spaces

2.1 Paper [CaS] of Cassels and Swinnerton-Dyer. Let G be a Lie group, and Γ a discrete subgroup of G. Consider the action of a subgroup H of G on G/Γ by left translations. If H is one-parameter or cyclic then the study of this action becomes the usual theory of flows on homogeneous spaces. However, for our purposes we have to consider multidimensional H. In particular, it turns out that the Oppenheim conjecture is equivalent to the statement that any relatively compact orbit of $SO(2,1)$ in $SL(3,\mathbb{R})/SL(3,\mathbb{Z})$ is compact (see Proposition 2.2.4 below). While preparing this survey the author realized that in implicit form this equivalence appears already in Sec. 10 of an old paper [CaS] of Cassels and Swinnerton-Dyer.

Though the language of the theory of dynamical systems is not used there, the paper [CaS] seems to be one of the first papers (maybe even the first one) on dynamical properties actions of multidimensional subgroups on homogeneous spaces. Let us first reproduce the abstract of [CaS].

"Isolation theorems for the minima of factorizable homogeneous ternary cubic forms and of indefinite ternary quadratic forms of a new strong type are proved. The problems whether there exist such forms with positive minima other than multiples of forms with integer coefficients are shown to be equivalent to problems in the geometry numbers of a superficially different type. A contribution is made to the study of the problem whether there exist real φ, ψ such that $x|\varphi x - y||\psi x - z|$ has a positive lower bound for all integers $x > 0, y, z$. The methods used have wide validity."

As we can see from this abstract two types of forms are considered in [CaS]. The first type consists of products of three linear forms in 3 variables, and the second one consists of indefinite ternary quadratic forms. Let us state first some of the results from [CaS] about products of linear forms. For any function f on $\mathbb{R}^n$ we will denote $\inf\{|f((x)||x \in \mathbb{Z}^n, x \neq 0\}$ by $m(f)$.

2.1.1. Theorem [CaS, Theorem 2]. *Let $f(x, y, z) = L_1L_2L_3$ be the product of three real linear forms which represent zero (over $\mathbb{Q}$)) only trivially, and suppose that f has integer coefficients. Let (δ_1, δ_2) be any open interval however small. Then there is a neighbourhood of f (in the space of products of three linear forms) such that all forms f^* in the neighbourhood which are not multiples of f itself take some value (on $\mathbb{Z}^3$) in the interval (δ_1, δ_2). In particular, to any given $\delta > 0$ we can choose a neighbourhood in which $m(f^*) < \delta$.*

Let us sketch the proof of this theorem. Since $f = L_1L_2L_3$ has integer coefficients and $f(x) \neq 0$ for any $x \in \mathbb{Z}^3 - \{0\}$, there exist $S, T \in SL(3, \mathbb{Z})$ and two independent units η_1, ζ_1 of a totally real cubic field K, with conjugates η_2, ζ_2, η_3, ζ_3 such that

$$S^mT^nL_j = \eta_j^m\zeta_j^nL_j \quad (j = 1, 2, 3)$$

for each pair of integers m, n. Let us write

$$\begin{aligned} f^* &= (1 + \varepsilon_0)L_1^*L_2^*L_3^*, \\ L_1^* &= L_1 + \varepsilon_{12}L_2 + \varepsilon_{13}L_3, \\ L_2^* &= \varepsilon_{21}L_1 + L_2 + \varepsilon_{23}L_3, \\ L_3^* &= \varepsilon_{31}L_1 + \varepsilon_{32}L_2 + L_3. \end{aligned}$$

We may suppose without loss of generality that $\varepsilon_0 = 0$ and that

$$\varepsilon_{31} = \max|\varepsilon_{ij}| > 0.$$

Replacing if necessary S by S^mT^n for suitable m, n, we may also suppose that

$$\eta_1 > \eta_2 > \eta_3.$$

Then, using the independence of units η_1 and ζ_1, it is not difficult to show that the upper topological limit of the sequence of sets

$$\{S^m T^n f_r^* \mid m > 0, n \in \mathbb{Z}\}$$

contains $f + \lambda L_1^2 L_2$ for any $\lambda > 0$ when f_r^* is a form of the same type as f^* and $f_r^* \to f$. But $(S^m T^n f_r^*)(\mathbb{Z}^3) = f_r^*(\mathbb{Z}^3)$. Hence if $f_r^*(\mathbb{Z}^3) \cap (\delta_1, \delta_2) = \emptyset$ for every r, we have that

$$(f + \lambda L_1^2 L_2)(\mathbb{Z}^3) \cap (\delta_1, \delta_2) = \emptyset$$

for every $\lambda > 0$. This contradicts the existence of $w \in \mathbb{Z}^3$ such that

$$f(w) < \delta_1, L_1(w) \neq 0, \text{ and } L_2(w) > 0.$$

2.1.2. Theorem [CaS, Theorem 5]. *Assume that $m(L_1 L_2 L_3) = 0$ for any real linear forms L_1, L_2, L_3 in x, y, z such that $L_1 L_2 L_3$ is not a multiple of a form with integer coefficients. Then for any D_0 however large, there are only a finite number of inequivalent sets of forms L_1, L_2, L_3 with determinant $\leq D_0$ such that $m(L_1 L_2 L_3) = 1$.*

Here two sets of forms are considered *equivalent* if the corresponding products $L_1 L_2 L_3$ can be transformed one into the other by an integral unimodular transformation on x, y, z. Let us reproduce the argument from [CaS] which deduces Theorem 2.1.2 from Theorem 2.1.1.

If there are infinitely many inequivalent sets of forms L_1, L_2, L_3 with determinant $\leq D_0$ such that $m(L_1 L_2 L_3) = 1$, then there are infinitely many lattices F in $\mathbb{R}^3$ of determinant at most D_0 such that $X_1 X_2 X_3 \geq 1$ for any $(X_1, X_2, X_3) \in F - \{0\}$; and none of these lattices is obtainable from another by a diagonal transformation $X_j \to \lambda_j X_j$. By Mahler's compactness criterion the set of these lattices must contain a convergent subsequence $\{\Delta_r\}$. (Mahler's compactness criterion states that the set A of lattices in $\mathbb{R}^n$ is relatively compact in the space of lattices if and only if there exist $D > 0$ and a neighborhood U of 0 in $\mathbb{R}^n$ such that, for any lattice $\Lambda \in A$, the determinant of Λ is not greater than D and $\Lambda \cap U = \{0\}$.) Let us choose a basis of the limit lattice Δ and write X_1, X_2, X_3 in this basis. Then we get a set of forms L_1, L_2, L_3 with determinant $\leq D_0$ such that $m(L_1 L_2 L_3) \geq 1$. By the assumption in the formulation of Theorem 2.1.2, $L_1 L_2 L_3$ is a multiple of a form with integer coefficients. But this is in contradiction with Theorem 2.1.1 and the choice of the subsequence $\{\Delta_r\}$.

2.1.3. Theorem [CaS, Theorem 6]. *As in Theorem 2.1.2, assume that $m(L_1 L_2 L_3) = 0$ for any real linear forms L_1, L_2, L_3 in x, y, z such that $L_1 L_2 L_3$ is not a multiple of a form with integer coefficients. Then*

$$(*) \qquad \liminf_{n \to \infty} n \|n\alpha\| \|n\beta\| = 0$$

for any real α and β where $\|x\|$ denotes the distance from x to the closest integer.

To deduce Theorem 2.1.3 from Theorem 2.1.1 we apply Mahler's compactness criterion to the sequence $\{A^n\Lambda \mid n \in \mathbb{N}^+\}$ where Λ is the lattice in (X_1, X_2, X_3)-space with points

$$\begin{bmatrix} X_1 = x \\ X_2 = \alpha x - y \\ X_3 = \beta x - z \end{bmatrix} \quad (x, y, z \quad \text{integers})$$

and $A(X_1, X_2, X_3) = (\frac{1}{4}X_1, 2X_2, 2X_3)$.

Remark. The statement (*) is the famous Littlewood conjecture still not settled.

2.1.4. Theorem [CaS, Theorem 4]. *The following two statements are equivalent:*

(a) *There exist real linear forms L_1, L_2, L_3 in x, y, z such that $L_1L_2L_3$ is not a multiple of a form with integer coefficients and $m(L_1L_2L_3) = 1$.*
(b) *There exist real linear forms M_1, M_2, M_3 in x, y, z such that*

$$\min\{m(M_1M_2M_3), m(M_1M_2(M_2 + M_3))\} = 1.$$

The proof of this theorem is analogous to the proof of the corresponding Theorem 2.1.8 for ternary quadratic forms and uses the following:

2.1.5. Lemma [CaS, Lemma 6]. *Let L_1, L_2, L_3 be three real linear forms in x_1, x_2, x_3 of non-zero determinant, each of which represents zero (over $\mathbb{Q}$) only trivially. Suppose there exist a transformation $T \in SL_3(\mathbb{Z})$ (other than the identity) and constants c_1, c_2, c_3 such that*

$$c_j > 0, c_1c_2c_3 = 1$$

and

$$TL_j = c_jL_j\ ,\ 1 \leq j \leq 3.$$

Then there is a multiple of $L_1L_2L_3$ with integer coefficients.

2.1.6. Let us state now theorems from [CaS] about indefinite ternary quadratic forms.

Theorem [CaS, Theorem 8]. *Let $B(x, y, z)$ be a non-singular indefinite ternary quadratic form with integer coefficients, and let (δ_1, δ_2) be any open interval however small. Then there is a neighborhood of B such that all (quadratic) forms B^* in the neighborhood which are not multiples of B itself take some value in the interval (δ_1, δ_2).*

The proof of this theorem is similar to the proof of Theorem 2.1.1. The only difference is that instead of commuting matrices $S, T \in SL(3, \mathbb{Z})$ one has to consider $S, T \in SL(3, \mathbb{Z})$ such that $SB = TB = B$, S (resp. T) has an eigenvalue $\lambda > 1$ (resp. $\mu > 1$), and λ and μ are elements of distinct quadratic fields. (Recall that if $SB = B$ then S has three eigenvalues $\lambda, \pm 1, \pm\lambda^{-1}$.)

2.1.7. The following theorem is deduced from Theorem 2.1.6 in the same way as Theorem 2.1.2 is deduced from Theorem 2.1.1. As before we say that a quadratic form is rational if it is a multiple of a form with rational coefficients and irrational otherwise. Two forms B_1 and B_2 are considered *equivalent* if B_1 can be transformed into $aB_2, a \neq 0$, by an integral unimodular transformation of x, y, z.

Theorem [CaS, Theorem 10]. *Assume that $m(B) = 0$ for any irrational indefinite ternary quadratic form B. Then for any D_0 however large there are only a finite number of inequivalent indefinite ternary quadratic forms B with determinant at most D_0 such that $m(B) = 1$.*

2.1.8. Another result is the following:

Theorem [CaS, Theorem 9]. *The following two statements are equivalent:*

(a) *There is an irrational indefinite ternary quadratic form B such that $m(B) > 0$.*
(b) *There are ternary linear forms M_1, M_2, M_3 such that*

$$\min\{m(M_2^2 - M_1M_3), m(M_2^2 - M_3(M_1 + M_3))\} = 1.$$

The implication (b) $\Rightarrow$ (a) is straightforward. We will explain how (a) implies (b) in the next subsection 2.2.

2.2. Paper [CaS] (continuation). For any real quadratic form B in n variables, let us denote the special orthogonal group

$$SO(B) = \{g \in SL(n, \mathbb{R}) \mid gB = B\} \text{ by } H_B,$$

and

$$\inf\{|B(x)||x \in \Lambda, x \neq 0\} \text{ by } m(B, \Lambda).$$

It is clear that $m(B, h\Lambda) = m(B, \Lambda)$ for any $h \in H_B$. This remark and Mahler's compactness criterion imply the following lemma (which was not explicitly stated in [CaS] but was implicitly used several times in Sec. 10 of [CaS]).

2.2.1. Lemma. *Let B be a real quadratic form in n variables, and Λ a lattice in $\mathbb{R}^n$ (resp. Ψ a set of unimodular lattices). Then the orbit $H_B\Lambda$ (resp. the set $H_B\Psi$) is relatively compact in the space Ω_n of lattices if and only if $m(B, \Lambda) > 0$ (resp. $\inf\{m(B, \Lambda) \mid \Lambda \in \Psi\} > 0$). In particular, $H_B\mathbb{Z}^n$ is relatively compact if and only if $m(B) > 0$.*

2.2.2. Lemma. *Let $B(x_1, x_2, x_3)$ be an indefinite quadratic form not representing 0 (over $\mathbb{Q}$).*

(a) [CaS, Lemma 14]. *Assume that $H_B \cap SL(3, \mathbb{Z})$ contains two non-commuting linear transformations S and T each of which has three distinct eigenvalues. Then the form B is rational.*
(b) *Assume that $H_B/H_B \cap SL(3, \mathbb{Z})$ is compact. Then the form B is rational.*

The connected component of identity of the group H_B is isomorphic to $PSL(2,\mathbb{R}) = SL(2,\mathbb{R})/\{\pm E\}$, and under this isomorphism hyperbolic elements of $PSL(2,\mathbb{R})$ correspond to elements with three positive distinct eigenvalues. On the other hand, any cocompact discrete subgroup of $PSL(2,\mathbb{R})$ contains non-commuting hyperbolic elements. Therefore (b) follows from (a). Let us sketch the proof of (a) given in [CaS].

Let $\lambda_1 = \lambda, \lambda_2 = 1, \lambda_3 = \lambda^{-1}$ be the eigenvalues of S. Then there exist linear forms ξ_1, ξ_2, ξ_3 such that

$$S\xi_j = \lambda_j \xi_j,\ 1 \le j \le 3,$$

ξ_2 has rational coefficients and the coefficients of ξ_1, ξ_3 lie in a quadratic field and are conjugates. On the other hand, $SB = B$ and the linear subspace of S-invariant quadratic forms on $\mathbb{R}^3$ has dimension 2. Hence

$$B(x_1, x_2, x_3) = \rho\xi_2^2 + \sigma\xi_1\xi_3,$$

where ρ, σ are real numbers and ξ_2^2 and $\xi_1\xi_3$ have rational coefficients. Applying the same argument for T instead of S, we get that there exist $\lambda_1^* = \lambda^*, \lambda_2^* = 1, \lambda_3^* = \lambda^{*-1}$ and linear forms $\xi_1^*, \xi_2^*, \xi_3^*$ such that

$$T\xi_j^* = \lambda_j^* \xi_j^*\ ,\ 1 \le j \le 3,$$

$$B(x_1, x_2, x_3) = \rho^*\xi_2^{*2} + \sigma^*\xi_1^*\xi_3^*$$

where ρ^*, σ^* are real numbers and ξ_2^* and $\xi_1^*\xi_3^*$ have rational coefficients. Since S and T are non-commuting elements of H_B with three distinct eigenvalues, we have that ξ_2^* is not a multiple of ξ_2. Hence after a suitable coordinate change we may assume that

$$B(x_1, x_2, x_3) = \rho_1 x_1^2 + \sigma\xi_1\xi_3 = \rho_1^* x_2^2 + \sigma^*\xi_1^*\xi_3^*$$

where $\rho_1, \rho_1^*, \sigma, \sigma^*$ are real non-zero numbers and $\xi_1\xi_3,\ \xi_1^*\xi_3^*$ are quadratic forms in x_1, x_2, x_3 with rational coefficients. Now, comparing the coefficients on both sides and using the non-singularity of B, we see that $\rho_1/\sigma^*, \rho_1^*/\sigma$ and σ/σ^* are rational numbers. This implies that B is rational.

2.2.3. If B_1 and B_2 are two indefinite ternary quadratic forms then there exist $g \in SL(3,\mathbb{R})$ and $\lambda \neq 0$ such that $gB_1 = \lambda B_2$. Thus, instead of studying $B(\mathbb{Z}^3)$ and $H_B\mathbb{Z}^3$ for a general ternary quadratic form B, we can fix one such form B_0 and consider $B_0(\Lambda)$ and $H_{B_0}\Lambda$ for a general unimodular lattice $\Lambda \in \Omega_3 = SL(3,\mathbb{R})/SL(3,\mathbb{Z})$. Now, using Lemmas 2.2.1 and 2.2.2 we get the following:

2.2.4. Proposition. *Let us fix an indefinite ternary quadratic form B_0, and let us denote H_{B_0} by H and the stabilizer $\{h \in H \mid Hz = z\}$, $z \in \Omega_3 = SL(3,\mathbb{R})/SL(3,\mathbb{Z})$, by H_z. Then the following two statements are equivalent:*

(a) *If B is an irrational indefinite ternary quadratic form then $m(B) = 0$ (i.e. the Oppenheim conjecture is true).*

(b) *If $z \in \Omega_3$ and the orbit Hz is relatively compact in Ω_3, then the quotient space H/H_z is compact.*

Remarks. (I) To prove the implication (a) $\Rightarrow$ (b) we have to use the classical fact that $H_B/H_B \cap SL(3,\mathbb{Z})$ is compact if a form B is rational and does not represent 0.

(II) Let G be a second countable locally compact group, Γ a discrete subgroup of G, F a closed subgroup of G, and $z \in G/\Gamma$. It is well known that the orbit Fz is closed if and only if the natural map $F/F_z \to Fz$ is proper. Therefore (b) is equivalent to the statement:

(b$'$) Any relatively compact orbit of H on Ω_3 is compact.

2.2.5. Let us write

$$w = \begin{pmatrix} 1 & 0 & 1 \\ 0 & 1 & 0 \\ 0 & 0 & 1 \end{pmatrix}.$$

The set of values of the form $x_2^2 - x_3(x_1 + x_3)$ on a lattice $L \subset \mathbb{R}^3$ coincides with the set of values of the form $x_2^2 - x_1x_3$ on wL. Therefore the statement (b) in the formulation of Theorem 2.1.8 is equivalent to the following statement:

(c) There is a lattice L in $\mathbb{R}^3$ such that

$$\min(m(B_0, L), m(B_0, wL)) = 1$$

where $B_0(x_1, x_2, x_3) = x_2^2 - x_1x_3$.

Combining this observation with Lemma 2.2.1 and Proposition 2.2.4 (together with Remark (II)) we get that the implication (a) $\Rightarrow$ (b) in Theorem 2.1.8 is a consequence of the following assertion:

2.2.6. Proposition. *Let $B_0(x_1, x_2, x_3) = x_2^2 - x_1x_3$, $H = H_{B_0}$ and $z \in \Omega_3 = SL(3,\mathbb{R})/SL(3,\mathbb{Z})$. Assume that the orbit Hz is relatively compact in Ω_3 but not closed. Then there is $y \in \overline{Hz}$ such that $wy \in \overline{Hz}$ (as usual $\bar{A}$ denotes the closure of A).*

This proposition is easily derived from the following two lemmas.

2.2.7. Lemma. *Let G be a second countable locally compact group, Γ a discrete subgroup of G, and F a closed subgroup of G.*

(a) *If $Y \subset G/\Gamma$, $FY{=}Y$ and Y is not closed, then the closure of the set $\{g \in G - F \mid gY \cap Y \neq \emptyset\}$ contains e. In particular if $z \in G/\Gamma$ and the natural map $F/F_z \to Fz$ is not proper, then the closure of the set $\{g \in G - F \mid gFz \cap Fz \neq \emptyset\}$ contains e.*

(b) *Let Y and Y' be closed F-invariant subsets of G/Γ, and let $M \subset G$. Suppose that Y is compact and $mY \cap Y' \neq \emptyset$ for any $m \in M$. Then $gY \cap Y' \neq \emptyset$ for any $g \in \overline{FMF}$.*

To prove (a) it is enough to notice that if $\{g_i\} \subset G$, $\lim_{i\to\infty} g_i = e$ and $g_i g_j^{-1} \in F$ for all i and j, then $g_i \in F$ for any i. To prove (b), consider the set $S = \{g \in G \mid gY \cap Y' \neq \emptyset\}$. Since Y is compact and Y' is closed, the set S is closed. On the other hand $M \subset S$ and (since Y and Y' are F-invariant) $FSF = S$. Hence $S \supset \overline{FMF}$.

2.2.8. Lemma. *Let $B_0(x_1, x_2, x_3) = x_2^2 - x_1x_3$, $H = H_{B_0}$, and $M \subset SL(3, \mathbb{R}) - H$. Suppose that $e \in \bar{M}$. Then $w \in \overline{HMH}$.*

Let $\{g_n\}$ be a sequence of elements of M such that $g_n \to e$ as $n \to \infty$. It is easy to check that there exist $h_n, h_n' \in H$ such that $\|h_n\| < c$, $h_n' \to e$ as $n \to \infty$, and

$$\|h_n g h_n' h_n^{-1} - e\| \leq c|(h_n g h_n' h_n^{-1})_{13}|$$

where c is an absolute constant. Therefore for every $\varepsilon > 0$, we can find $g = (g_{ij}) \in HMH$ such that

$$\|g - e\| < \varepsilon \text{ and } \|g - e\| < c|g_{13}|.$$

Consider an element

$$g^* = \begin{pmatrix} |g_{13}|^{-\frac{1}{2}} & 0 & 0 \\ 0 & 1 & 0 \\ 0 & 0 & |g_{13}|^{\frac{1}{2}} \end{pmatrix} g \begin{pmatrix} |g_{13}|^{\frac{1}{2}} \mathrm{sgn} g_{13} & 0 & 0 \\ 0 & 1 & 0 \\ 0 & 0 & |g_{13}|^{-\frac{1}{2}} \mathrm{sgn} g_{13} \end{pmatrix}.$$

The first and last factors are elements of H. Hence $g^* \in HMH$. Clearly

$$\|g_{ii}^* - 1\| \leq c\varepsilon \ , \ g_{13}^* = 1,$$

$$\|g_{ij}^*\| \leq c\varepsilon^{\frac{1}{2}} \qquad \text{otherwise.}$$

It implies that $g^* \to w$ when $\varepsilon \to 0$. This proves the lemma.

2.2.9. Remarks. (a) In Proposition 2.2.6 and Lemma 2.2.8, B_0 can be replaced by any indefinite ternary quadratic form of the type

$$ax_1x_3 + bx_2^2 + cx_2x_3 + dx_3^2.$$

Also, w can be replaced by

$$w(t) = \begin{pmatrix} 1 & 0 & t \\ 0 & 1 & 0 \\ 0 & 0 & 1 \end{pmatrix}.$$

(b) The above proof of Theorem 2.1.8 is essentially a translation of the proof from [CaS], though these two proofs superficially look quite different.

2.3. Closures of orbits of orthogonal groups and integer solutions of quadratic inequalities. Proposition 2.2.4 establishes the equivalence between the Oppenheim conjecture and the statement that any relatively compact orbit of $SO(2, 1)$ in $SL(3, \mathbb{R})/SL(3, \mathbb{Z})$ is compact. As it is noticed in 2.1, in implicit form this equivalence appears already in Sec. 10 of [CaS] (and, in fact, is used to prove some of the results of the paper [CaS]). However the language of the theory of dynamical systems is not used in [CaS]. Because of that, for a long time the relation between the Oppenheim conjecture and problems in dynamical systems remained unknown to most (and probably all) people studying subgroup actions on homogeneous spaces. This changed when, in the mid-seventies, M.S. Raghunathan rediscovered the above-mentioned equivalence and noticed that the Oppenheim conjecture would follow from a conjecture about closures of orbits of unipotent subgroups (see [Dan3]). The Raghunathan conjecture states that if G is a connected Lie groups, Γ a lattice in G (that is, Γ is a discrete subgroup such that G/Γ carries a G-invariant probability measure), and U an Ad-unipotent subgroup of G (that is, Adu is a unipotent linear transformation for any $u \in U$), then for any $x \in G/\Gamma$ there exists a closed connected subgroup $L = L(x)$ containing U such that the closure of the orbit Ux coincides with Lx. In the literature it was first stated in [Dan3] and in a more general form in [Marg5] (when the subgroup U is not necessarily unipotent but generated by unipotent elements). Raghunathan's conjecture and some other related conjectures will be discussed in Sec. 3.8.

Let us briefly describe the argument of Raghunathan reducing the Oppenheim conjecture to his conjecture.

Let B be a real indefinite quadratic form in $n \geq 3$ variables and let $H = SO(B)$. If U is a unipotent subgroup of H and the closure of $U\mathbb{Z}^n$ in the space of lattices coincides with the orbit $L_U\mathbb{Z}^n$ where L_U is a closed connected subgroup of $SL(n, \mathbb{R})$, then using Borel's density theorem [Bo1] it is not difficult to prove that L_U is an algebraic subgroup defined over $\mathbb{Q}$. Thus the number of all possible L_U is at most countable. This countability quite easily implies that there exists a unipotent subgroup $U_0 \subset SL(n, \mathbb{R})$ such that $L_{U_0} \supset L_U$ for any other unipotent subgroup U. But the connected component H^0 of identity in H is generated by unipotent elements (because $n \geq 3$). Hence $L_{U_0} \supset H^0$. On the other hand $H^0 = SO(B)^0$ is a maximal proper connected subgroup in $SL(n, \mathbb{R})$. Thus either $L_{U_0} = SL(n, \mathbb{R})$ or $L_{U_0} = H^0$. In the first case $m(B) = 0$ (according to Lemma 2.2.1) and in the second case B is rational (because L_{U_0} is defined over $\mathbb{Q}$).

Inspired by Raghunathan's observations the author started to work on the homogeneous space approach to the Oppenheim conjecture and eventually established the following theorem which proves the conjecture.

2.3.1. Theorem (see [Marg5, Theorem 1] or one of the papers [Marg3,4,6]). *Let B be a real irrational indefinite quadratic form in $n \geq 3$ variables. Then for any $\varepsilon > 0$ there exist integers $x_1, \ldots, x_n$ not all equal to 0 such that $|B(x_1, \ldots, x_n)| < \varepsilon$.*

2.3.2. The equivalence given by Proposition 2.2.4 is used in [Marg3–6], and indeed Theorem 2.3.1 is deduced there from the following reformulation of the statement (b) in Proposition 2.2.4.

Theorem [Marg5, Theorem 2]. *Let $G = SL(3, \mathbb{R})$ and $\Gamma = SL(3, \mathbb{Z})$. Let us denote by H the group of elements of G preserving the form $2x_1x_3 - x_2^2$ and by $\Omega_3 = G/\Gamma$ the space of lattices in $\mathbb{R}^3$ having determininant 1. Let G_y denote the stabilizer*

$\{g \in G \mid gy = y\}$ of $y \in \Omega_3$. If $z \in \Omega = G/\Gamma$ and the orbit Hz is relatively compact in Ω then the quotient space $H/H \cap G_z$ is compact.

The proof of this theorem given in [Marg5] depends to a large extent on studying the closures of orbits of unipotent subgroups but it does not use the argument of Raghunathan mentioned above.

2.3.3. In the 1952 paper [Op4] Oppenheim modified his conjecture replacing the inequality $|B(x)| < \varepsilon$ by a slightly stronger inequality $0 < |B(x)| < \varepsilon$. When informed by A. Borel of this fact, the author showed that the Oppenheim conjecture in this modified form can also be deduced from Theorem 2.3.2 (see [Marg5, Theorem 1′]). On the other hand, according to the result of Oppenheim proved in the same paper [Op4] and mentioned above in 1.5, if $n \geq 3$ and $\{x \in \mathbb{Z}^n \mid 0 < |B(x)| < \varepsilon\} \neq \emptyset$ for any $\varepsilon > 0$ then $B(\mathbb{Z}^n)$ is dense in $\mathbb{R}$. Thus we have the following:

Theorem. *If B is a real irrational indefinite quadratic form in $n \geq 3$ variables, then $B(\mathbb{Z}^n)$ is dense in $\mathbb{R}$ or, in other words, for any $a < b$ there exists integers $x_1, \ldots, x_n$ such that*

$$a < B(x_1, \ldots, x_n) < b.$$

2.3.4. An integer vector $x \in \mathbb{Z}^n$ is called *primitive* if $x \neq ky$ for any $y \in \mathbb{Z}^n$ and $k \in \mathbb{Z}$ with $|k| \geq 2$. The set of all primitive vectors in $\mathbb{Z}^n$ will be denoted by $\mathcal{P}(\mathbb{Z}^n)$. As in [Bo2] let us say that a subset $(x_1, \ldots, x_n)$ of $\mathbb{Z}^n$ $(m \leq n)$ is *primitive* if it is a part of a basis of $\mathbb{Z}^n$. If $m < n$, this condition is equivalent to the existence of $g \in SL_n(\mathbb{Z})$ such that $ge_i = x_i (i = 1, \ldots, m)$. We can state now a strengthening of Theorem 2.3.3.

2.3.4. Theorem. *Let B be a real irrational indefinite quadratic form in $n \geq 3$ variables, and let B_2 be the corresponding bilinear form defined by $B_2(v, w) = \frac{1}{4}\{B(v + w) - B(v - w)\}$ for all $v, w \in \mathbb{R}^n$.*

(a) [DanM1, Theorem 1]. *The set $\{B(x) \mid x \in \mathcal{P}(\mathbb{Z}^n)\}$ is dense in $\mathbb{R}$.*
(b) [BoP, Corollary 7.8] (see also [DanM1, Theorem 1] for $m \leq 2$). *Let $m < n$ and $y_1, \ldots, y_m$ be elements of $\mathbb{R}^n$. Then there exists a sequence $(x_{j,1}, \ldots, x_{j,m})$ $(j = 1, \ldots)$ of primitive subsets of $\mathbb{R}^n$ such that*

$$B_2(y_a, y_b) = \lim_{j \to \infty} B_2(x_{j,a}, x_{j,b}) \ (1 \leq a, b \leq m).$$

(c) (see [BoP, Corollary 7.9] or [Bo2, Theorem 2]). *Let $c_i \in \mathbb{R}$ $(i = 1, \ldots, n - 1)$. Then there exists a sequence $(x_{j,1}, \ldots, x_{j,n-1})$ $(j = 1, 2, \ldots)$ of primitive subsets of $\mathbb{Z}^n$ such that*

$$\lim_{j \to \infty} B(x_{j,i}) = c_i \quad (i = 1, \ldots, n - 1).$$

This is deduced from the following Theorem 2.3.5 by an extension of the argument reducing Theorems 2.3.1 and 2.3.3 to Theorem 2.3.2. Theorem 2.3.5 is in fact an easy consequence of a general result of M. Ratner on orbit closures (see Theorem 3.8.4 below). However, for $n = 3$ this theorem had been earlier proved in [DanM1]. The proof given in [DanM1] uses the technique which involves, as in [Marg5], finding orbits of larger subgroups inside closed sets invariant under unipotent subgroups.

2.3.5. Theorem. *Let B be a real indefinite quadratic form in $n \geq 3$ variables. Let us denote by H the special orthogonal group $SO(B)$ and by $\Omega_n = SL(n,\mathbb{R})/SL(n,\mathbb{Z})$ the space of lattices in $\mathbb{R}^n$ with discriminant 1. Then any orbit of H in Ω_n either is closed and carries a H-invariant probability measure or is dense.*

2.3.6. Remarks. (a) Let $B(x_1, x_2, x_3) = 2x_1x_3 - x_2^2$, $H = SO(B)$, and let

$$w(t) = \begin{pmatrix} 1 & 0 & t \\ 0 & 1 & 0 \\ 0 & 0 & 1 \end{pmatrix},$$

$W^+ = \{w(t) \mid t \geq 0\}$, $W^- = \{w(t) \mid t \leq 0\}$. It is easy to see that if L is a lattice in $\mathbb{R}^3$ then there exist $w_1 \in W^+, w_2 \in W^-$ and primitive vectors $x_1, x_2 \in \mathcal{P}(L)$ such that $B(w_1x_1) = B(w_2x_2) = 0$. Because of that, in order to prove Theorems 2.3.1, 2.3.3 and 2.3.4 (a), it is enough to prove that if $x \in \Omega_3$ then either Hx is closed and carries a H-invariant probability measure or there exist a $y \in \overline{Hx}$ such that W^+y or W^-y is contained in $\overline{Hx}$. Using this observation it is possible to simplify the proofs of Theorems 2.3.1, 2.3.3 and 2.3.4 (a). This was done in [DanM3] and [Dan8] for Theorem 2.3.4 (a) (see also Sec. 3.7 below) and in [Marg7] and [Si] for Theorems 2.3.1 and 2.3.3. It should also be mentioned that the proof given in [Dan8] does not depend on the axiom of choice (other proofs involve existence of minimal invariant subsets for various actions, which depends on Zorn's lemma and, in turn, on the axiom of choice).

(b) Let Q be a nondegenerate quadratic form in $n \geq 2$ variables. Borel and Prasad noted (see [Bo2, Proposition 4]) that Q is irrational if and only if, given $\varepsilon > 0$, there exist $x, y \in \mathbb{Z}^n$ such that $0 < |Q(x) - Q(y)| < \varepsilon$. This is related to the conjecture mentioned in 1.1 about the gaps between successive values of Q.

(c) Borel and Prasad generalized Theorems 2.3.1, 2.3.3 and 2.3.4 for a family $\{B_s\}$ where $s \in S$, S is a finite set of places of a number field k containing the set S_∞ of archimedean places, B_s is a quadratic form on k_s^n, and k_s is the completion of k at s (see [BoP] and [Bo2]). To prove these generalizations they use S-arithmetic analogs of Theorems 2.3.2 and 2.3.5.

2.4. Quantitative results. In connection with Theorem 2.3.3, it is natural to study the following problem. Let B be a real irrational indefinite quadratic form in $n \geq 3$ variables, let $a < b$, and let $N^B_{(a,b)}(T)$ denote the number of integral points v in a ball of radius T with $a < B(v) < b$. According to Theorem 2.3.3, $N^B_{(a,b)}(T) \to \infty$

when $T \to \infty$. It is well known that as $T \to \infty$

$$|\{v \in \mathbb{R}^n \mid \|v\| < T\}| \sim \operatorname{Vol}\{v \in \mathbb{Z}^n \mid \|v\| < T\}.$$

Therefore one might expect that

$$\lim_{T\to\infty} \frac{N^B_{(a,b)}(T)}{\operatorname{Vol}(\{v \in \mathbb{R}^n \mid a < Q(v) < b, \|v\| < T\})} = 1. \tag{1}$$

It was shown by Dani and the author that (1) is true but only if lim is replaced by lim inf. Rather surprisingly the asymptotic formula (1) holds if the signature of B is not (2,1) or (2,2) and does not hold for general irrational forms of these two signatures (it was proved recently by Eskin, Mozes and the author). Before we give the precise formulations, let us introduce some notations and make some preliminary remarks.

Let ν be a continuous positive function on the sphere $\{v \in \mathbb{R}^n \mid \|v\| = 1\}$ and let $\Omega = \{v \in \mathbb{R}^n \mid \|v\| < \nu(v/\|v\|)\}$. We denote by $T\Omega$ the dilate of Ω by T. Define the following sets:

$$V_{(a,b)}(\mathbb{R}) = V^B_{(a,b)}(\mathbb{R}) = \{x \in \mathbb{R}^n \mid a < B(x) < b\} \text{ and}$$

$$V_{(a,b)}(\mathbb{Z}) = V^B_{(a,b)}(\mathbb{Z}) = \{x \in \mathbb{Z}^n \mid a < B(x) < b\}.$$

One can verify that as $T \to \infty$

$$\operatorname{Vol}(V_{(a,b)}(\mathbb{R}) \cap T\Omega) \sim \lambda_{B,\Omega}(b-a)T^{n-2} \tag{2}$$

where

$$\lambda_{B,\Omega} = \int_{L\cap\Omega} \frac{dA}{\|\nabla B\|}, \tag{3}$$

L is the light cone $B = 0$ and dA is the area element on L.

Let $p \geq 2, q \geq 1, p \geq q$, let $\mathcal{O}(p,q)$ denote the space of quadratic forms of signature (p,q) and discriminant ± 1, let $n = p+q$, (a,b) be an interval, and let $\mathcal{K}$ be a compact subset of $\mathcal{O}(p,q)$.

2.4.1. Theorem [DanM5, Corollary 5]. *(I) For any $\theta > 0$ there exists a finite subset $\mathcal{L} = \mathcal{L}(\theta, a, b)$ of $\mathcal{K}$ such that each form $B \in \mathcal{L}$ is rational and for any compact subset $\mathcal{F}$ of $\mathcal{K} - \mathcal{L}$ there exists $T_0 = T_0(\theta, a, b, \mathcal{F}) \geq 0$ such that for all B in $\mathcal{F}$ and $T \geq T_0$*

$$|V_{(a,b)}(\mathbb{Z}) \cap T\Omega| \geq (1-\theta)\operatorname{Vol}(V_{(a,b)}(\mathbb{R}) \cap T\Omega). \tag{4}$$

(II) If $n \geq 5$, then for any $\varepsilon > 0$ there exist $c = c(\varepsilon, \mathcal{K}) > 0$ and $T_0 = T_0(\varepsilon, \mathcal{K}) > 0$ such that for all $B \in \mathcal{K}$ and $T \geq T_0$

$$|V_{(-\varepsilon,\varepsilon)}(\mathbb{Z}) \cap T\Omega| \geq c \operatorname{Vol}(V_{(-\varepsilon,\varepsilon)}(\mathbb{R}) \cap T\Omega). \tag{5}$$

For a single B a similar estimate, but only with a positive constant rather than one arbitrarily close to 1, had been obtained earlier by Dani and Mozes, and also independently by Ratner (both unpublished).

2.4.2. Theorem (see [EMM1, Theorem 1] or [EMM2, Theorem 2.1]). *If $p \geq 3$ then as $T \to \infty$*

$$|V_{(a,b)}(\mathbb{Z}) \cap T\Omega| \sim \lambda_{B,\Omega}(b-a)T^{n-2} \tag{6}$$

for any irrational form $B \in \mathcal{O}(p,q)$ where $\lambda_{B,\Omega}$ is as in (3).

If the signature B is (2,1) or (2,2) then no universal formula like (6) holds. In fact, we have the following theorem:

2.4.3. Theorem (see [EMM1, Theorem 2] or [EMM2, Theorem 2.2]). *Let Ω_0 be the unit ball, and let $q = 1$ or 2. Then for every $\varepsilon > 0$ and every interval (a,b) there exists a quadratic irrational form B of signature $(2,q)$, and a constant $c > 0$ such that for an infinite sequence $T_j \to \infty$*

$$|V_{(a,b)}(\mathbb{Z}) \cap T\Omega_0| > cT_j^q(\log T_j)^{1-\varepsilon}. \tag{7}$$

The case $q = 1, b \leq 0$ of this theorem was noticed by Sarnak and worked out in detail in [Br]. The quadratic forms constructed are of the type $x_1^2 + x_2^2 - \alpha x_3^2$, or $x_1^2 + x_2^2 - \alpha(x_3^2 + x_4^2)$, where α is extremely well approximated by squares of rational numbers (in fact we can take α which belong to a set of second category in the Baire sense). Let us also note that in the statement of Theorem 2.4.2, $(\log T)^{1-\varepsilon}$ can be replaced by $\log T/\nu(T)$ where $\nu(T)$ is any unbounded increasing function.

However, in the (2,1) and (2,2) cases, there is an upper bound of the form $cT^q \log T$. This upper bound is effective, and is uniform over compact subsets of $\mathcal{O}(p,q)$. There is also an effective uniform upper bound for the case $p \geq 3$.

2.4.4. Theorem (see [EMM1, Theorem 3] or [EMM2, Theorem 2.3]). *If $p \geq 3$ there exists a constant $c = c(\mathcal{K}, a, b, \Omega)$ such that for any $B \in \mathcal{K}$ and all $T > 1$*

$$|V_{(a,b)}(\mathbb{Z}) \cap T\Omega| < cT^{n-2}. \tag{8}$$

If $p = 2$ and $q = 1$ or $q = 2$, then there exists a constant $c = c(\mathcal{K}, a, b, \Omega)$ such that for any $B \in \mathcal{K}$ and $T > 2$

$$|V_{(a,b)}(\mathbb{Z}) \cap T\Omega| < cT^{n-2}\log T. \tag{9}$$

Also, for the (2,1) and (2,2) case, the following "almost everywhere" result is true:

2.4.5. Theorem (see [EMM1, Theorem 4] or [EMM2, Theorem 2.5]). *The asymptotic formula* (6) *holds for almost all quadratic forms of signature $(p,q) = (2,1)$ or $(2,2)$.*

Sarnak [Sar] recently proved that (6) holds for almost all forms within the following two-parameter family of quadratic forms of signature (2,2).

$$(x_1^2 + 2bx_1x_2 + cx_2^2) - (x_3^2 + 2bx_3x_4 + cx_4^2),$$

$b, c \in \mathbb{R}$, $c - b^2 > 0$. This family arises in problems related to quantum chaos.

Let us also state a "uniform" version of Theorem 2.4.2.

2.4.6. Theorem (see [EMM, Theorem 5] or [EMM2, Theorem 2.5]). *For any $\theta > 0$ there exists a finite subset $\mathcal{P} = \mathcal{P}(\theta, a, b)$ of $\mathcal{K}$ such that each $B \in \mathcal{P}$ is rational and for any compact subset $\mathcal{F}$ of $\mathcal{K} - \mathcal{P}$ there exists $T_0 = T_0(\theta, a, b, \mathcal{P}) \geq 0$ such that for all B in $\mathcal{F}$ and $T \geq T_0$*

$$(10) \quad (1-\theta)\lambda_{B,\Omega}(b-a)T^{n-2} \leq |V_{(a,b)}(\mathbb{Z}) \cap T\Omega| \leq (1+\theta)\lambda_{B,\Omega}(b-a)T^{n-2}$$

where $\lambda_{B,\Omega}$ is as in (3).

2.4.7. Remark. Let $\mathcal{P}(\mathbb{Z}^n)$ denote as before the set of primitive vectors in $\mathbb{Z}^n$. If we consider $|V_{(a,b)}(\mathbb{R}) \cap T\Omega \cap \mathcal{P}(\mathbb{Z}^n)|$ instead of $|V_{(a,b)}(\mathbb{Z}) \cap T\Omega|$ then Theorems 2.4.1, 2.4.2, 2.4.3, 2.4.5, and 2.4.6 hold provided one replaces $1-\theta$ by $(1-\theta)/\zeta(n)$ in Theorem 2.4.1 and $\lambda_{B,\Omega}$ by $\lambda'_{B,\Omega} = \lambda_{B,\Omega}/\zeta(n)$ in Theorems 2.4.2, 2.4.3, 2.4.5, and 2.4.6, where ζ is the Riemann zeta function.

2.4.8. Let $\Gamma = SL(n,\mathbb{Z}), G = SL(n,\mathbb{R})$, $\Omega_n = G/\Gamma = \{$the space of unimodular matrices in $\mathbb{R}^n$ with discriminant 1$\}$. One can associate to an integrable function ψ on $\mathbb{R}^n$ a function $\tilde{\psi}$ on $\Omega_n = G/\Gamma$ by setting

$$(11) \qquad \tilde{\psi}(g\Gamma) = \sum_{v \in g\mathbb{Z}^n, v \neq 0} \tilde{\psi}(v), \quad g \in G.$$

According to a theorem of Siegel

$$(12) \qquad \int_{\mathbb{R}^n} \psi dm^n = \int_{G/\Gamma} \tilde{\psi} d\mu$$

where m^n is the Lebesgue measure on $\mathbb{R}^n$ and μ is the G-invariant probability measure on G/Γ.

In [DanM5], the proof of Theorem 2.4.1 is based on the following identity which is immediate from the definitions:

$$(13) \qquad \int_T^{\ell T} \int_F \sum_{v \in g\mathbb{Z}^n} \psi(u_t k v) d\sigma(k) dt = \int_T^{\ell T} \int_F \tilde{\psi}(u_t k g\Gamma) d\sigma(k) dt$$

where $\{u_t\}$ is a certain one-parameter unipotent subgroup of $SO(p,q)$, F is a Borel subset of the maximal compact subgroup K of $SO(p,q), \sigma$ is the normalized Haar measure on K, and ψ is a continuous function on $\mathbb{R}^n - \{0\}$ with compact support. The number $|V_{(a,b)}(\mathbb{Z}) \cap T\Omega|$ can be approximated by the sum over m of the integrals on the left-hand side of (13) for an appropriate choice of $g, \psi = \psi_i, F = F_i, 1 \leq i \leq m$. The right-hand side of (13) can be estimated, uniformly when $g\Gamma$ belongs to certain compact subsets of $SL(n,\mathbb{R})/\Gamma$, using (12) and a refined version of Ratner's uniform distribution theorem (see Theorem 3.9.3 below). To prove the assertion (II) of Theorem 2.4.1 we have to use the following fact which is essentially equivalent to Meyer's theorem: if $n \geq 5$ then any closed orbit of $SO(p,q)$ in $SL(n,\mathbb{R})/\Gamma$ is unbounded.

Let us state the just-mentioned version of the uniform distribution theorem.

2.4.9. Theorem [DanM5, Theorem 3]. *Let G be a connected Lie group, Γ a lattice in G, and μ the G-invariant probability measure on G/Γ. For any closed subgroups H*

and W of G let $X(H,W)=\{g\in G \mid Wg\subset gH\}$. Let $U=\{u_t\}$ be an Ad-unipotent one-parameter subgroup of G and let φ be a bounded continuous function on G/Γ. Let $\mathcal{D}$ be a compact subset of G/Γ and let $\varepsilon>0$ be given. Then there exist finitely many proper closed subgroups $H_1=H_1(\varphi,\mathcal{D},\varepsilon),\dots,H_k=H_k(\varphi,\mathcal{D},\varepsilon)$ such that $H_i\cap\Gamma$ is a lattice in H_i for all i, and compact subsets $C_1=C_1(\varphi,\mathcal{D},\varepsilon),\dots,C_k=C_k(\varphi,\mathcal{D},\varepsilon)$ of $X(H_1,U),\dots,X(H_k,U)$ respectively, for which the following holds. For any compact subset F of $\mathcal{D}-\bigcup_{1\le i\le k} C_i\Gamma/\Gamma$ there exist a $T_0\ge 0$ such that for all $x\in F$ and $T>T_0$

$$\left|\frac{1}{T}\int_0^T \varphi(u_t x)dt-\int_{G/\Gamma}\varphi d\mu\right|<\varepsilon.$$

2.4.10. The function $\tilde{\psi}$ defined by (12) is unbounded for any continuous nonzero function ψ on $\mathbb{R}^n$. Therefore we cannot use (13) and Theorem 2.4.9 to get the asymptotic of $|V_{(a,b)}(\mathbb{Z})\cap T\Omega|$, because this theorem is proved (and in general true) only for bounded continuous functions. On the other hand, as it was done in [DanM5] one can get lower bounds by considering bounded continuous function $f\le\tilde{\psi}$ and applying Theorem 2.4.9 to f.

If $\delta>0$ and h is a nonnegative function then for any $\varepsilon>0$

$$h=h_\varepsilon+\varepsilon h^{1+\delta}$$

where h_ε is bounded. Therefore it would be possible to obtain the asymptotic of $|V_{(a,b)}(\mathbb{Z})\cap T\Omega|$ from (13) and Theorem 2.4.9 if we knew that for some $\delta>0$ and any continuous nonnegative function ψ on $\mathbb{R}^n$ with compact support,

$$\limsup_{T\to\infty}\frac{1}{T}\int_0^T\int_K \tilde{\psi}(u_t k g\Gamma)^{1+\delta}d\sigma(k)dt<\infty. \tag{14}$$

Using methods of the papers [Dan2], [Dan6] and [Marg2] it is possible to prove Theorem 3.3.4 below which implies that for $\theta=1/n^3$

$$\limsup_{T\to\infty}\left\{\sup_{g\in L}\frac{1}{T}\int_0^T\tilde{\psi}(u_t g\Gamma)^\theta dt\right\}<\infty$$

for any compact subset $L\subset SL(n,\mathbb{R})$ and any unipotent subgroup $\{u_t\}\subset SL(n,\mathbb{R})$. But these methods are certainly not enough to prove (14). In [EMM1, 2] we present a different argument and essentially prove (14) for all $0\le\delta<1$ in the case where $p\ge 3$ and $q\ge 1$. If $(p,q)=(2,2)$ or $(p,q)=(2,1)$ then (14) is in general not true even for $\delta=0$ but it becomes true for $\delta=0$ if we replace in (14) the factor $1/T$ by $1/(T\log T)$. Let us note that for technical reasons we consider in [EMM1, 2] not the integrals from (14) but the integrals of the form

$$\int\tilde{\psi}(a_t k)^{1+\delta}d\sigma(k) \tag{15}$$

where $\{a_t\}$ is a certain diagonalizable subgroup of $SO(p,q)$. Because of that we do not use Theorem 2.4.9 directly. Instead of that we first deduce the following result from Theorem 2.4.9. This relies on the well-known fact that in the hyperbolic plane large circles are well approximated by horocycles and that $SO(p,q)$ is a maximal proper subgroup in $SL(n,\mathbb{R})$.

2.4.11. Theorem [EMM2, Theorem 4.5]. *Let $\{a_t \mid t \in \mathbb{R}\}$ be a self-adjoint one-parameter subgroup of $SO(2,1)$. Let $p \geq 2$ and $q \geq 1$. Denote $p+q$ by n, $SO(p,q)$ by H, $SL(n,\mathbb{R})$ by G, and $SO(p) \times SO(q)$ by K. Let Γ be a lattice in G and let φ be any continuous function on G/Γ vanishing outside a compact set. Then for any $\varepsilon > 0$ and any bounded measurable function r on K and every compact subset $\mathcal{D}$ of G/Γ there exist finitely many points $x_1, \ldots, x_\ell \in G/\Gamma$ such that*

(i) *the orbits $Hx_1, \ldots, Hx_\ell$ are closed and have finite H-invariant measure;*
(ii) *for any compact subset F of $\mathcal{D} - \bigcup_{1\leq i\leq \ell} Hx_i$ there exists $T_0 > 0$ such that for all $x \in F$ and $t > T_0$*

$$\left|\int_K \varphi(a_t kx)r(k)d\sigma(k) - \int_{G/\Gamma} \varphi d\mu \int_K r d\sigma\right| \leq \varepsilon.$$

2.4.12. We will describe now the argument from [EMM2] how to obtain upper bounds for the integrals (15).

Let Δ be a lattice in $\mathbb{R}^n$. We say that a subspace L of $\mathbb{R}^n$ is Δ-*rational* if $L \cap \Delta$ is a lattice in L. For any Δ-*rational* subspace L, we denote by $d_\Delta(L)$ or simply by $d(L)$ the volume of $L/(L\cap\Delta)$. Let us note that $d(L)$ is equal to the norm of $e_1 \wedge \cdots \wedge e_\ell$ in the exterior power $\wedge^\ell(\mathbb{R}^n)$ where $\ell = \dim L$ and $(e_1, \ldots, e_\ell)$ is a basis over $\mathbb{Z}$ of $L \cap \Delta$. If $L = \{0\}$ we write $d(L) = 1$. It is clear that $d_\Delta(\mathbb{R}^n) = 1$ if and only if $\Delta \in \Omega_n = SL(n,\mathbb{R})/SL(n,\mathbb{Z})$.

Let us introduce the following notation:

$$\alpha_i(\Delta) = \sup\left\{\frac{1}{d(L)} \mid L \text{ is a } \Delta\text{-rational subspace of dimension } i\right\}, 0 \leq i \leq n,$$

$$\alpha(\Delta) = \max_{0\leq i\leq n} \alpha_i(\Delta).$$

It is a standard fact from the geometry of numbers that for any bounded function f on $\mathbb{R}^n$ vanishing outside a compact subset, there exists a positive constant $c = c(f)$ such that

$$\tilde{f}(\Delta) < c\alpha(\Delta) \tag{16}$$

for any lattice Δ in $\mathbb{R}^n$.

Let $\{e_1, \ldots, e_n\}$ be the standard basis of $\mathbb{R}^n$. Let B_0 be the quadratic form defined by

$$B_0\left(\sum_{i=1}^n v_i e_i\right) = 2v_1 v_n + \sum_{i=2}^p v_i^2 - \sum_{i=p+1}^{n-1} v_i^2 \tag{17}$$

for all $v_1, \ldots, v_n \in \mathbb{R}$. The form B_0 has signature (p, q). Let $H = SO(B_0)$. For $t \in \mathbb{R}$, let a_t be the linear map so that $a_t e_1 = e^{-t} e_1, a_t e_n = e^t e_n$, and $a_t e_i = e_i$, $2 \leq i \leq n-1$. Then the one-parameter subgroup $\{a_t\}$ is contained in H. Let $\hat{K}$ be the subgroup of $SL(n, \mathbb{R})$ consisting of orthogonal matrices, and let $K = H \cap \hat{K}$. It is easy to check that K is a maximal compact subgroup of H, and consists of all $h \in H$ leaving invariant the subspace spanned by $\{e_1 + e_n, e_2, \ldots, e_p\}$.

Necessary upper bounds for the integrals (15) are deduced from (16) and the following main integrability estimates:

2.4.13. Theorem [EMM2, Theorems 3.2 and 3.3].

(a) *If $p \geq 3, q \geq 1$ and $0 < s < 2$, or if $p = 2$, $q \geq 1$ and $0 < s < 1$, then for any lattice Δ in $\mathbb{R}^n$*

$$\sup_{t>0} \int_K \alpha(a_t k \Delta)^s d\sigma(k) < \infty. \tag{18}$$

(b) *If $p = 2$ and $q = 2$, or if $p = 2$ and $q = 1$, then for any lattice Δ in $\mathbb{R}^n$,*

$$\sup_{t>1} \frac{1}{t} \int_K \alpha(a_t k \Delta) d\sigma(k) < \infty. \tag{19}$$

The upper bounds (18) *and* (19) *are uniform as Δ varies over compact sets the space of lattices.*

Sketch of the proof. For any $t \geq 0$ and any continuous action of the group H on a topological space X, we consider the averaging operator A_t:

$$(A_t f)(x) = \int_K f(a_t k x) d\sigma(k), x \in X.$$

If $X = H$, f is left K-invariant and $K \backslash H$ is $(n-1)$-dimensional hyperbolic space (or, equivalently, $q = 1$), then $(A_t f)(h)$ can be interpreted as the average of f over the sphere of radius $2t$ in $K \backslash H$, centered at Kh.

The main idea of the proof is to show that the α_i^s satisfy certain systems of integral inequalities which imply the desired bounds. If $0 < s < 2, p \geq 3$ and $0 < i < n$ or $p = 2, q = 2$ and $i = 1$ or 3, or if $0 < s < 1$, we prove that for any $c > 0$ there exist $t = t(s, c) > 0$ and $w = w(s, c) > 1$ such that

$$A_t \alpha_i^s \leq \frac{c}{2} \alpha_i^s + w^2 \max_{0 < j \leq \min(n-i, i)} \sqrt{\alpha_{i+j}^s \alpha_{i-j}^s}. \tag{20}$$

The following inequalities also hold:

$$A_t \alpha_i^* \leq \alpha_i^* + w^2 \sqrt{\alpha_{3-i}^*}, 1 \leq i \leq 2, \tag{21}$$

for $(p, q) = (2, 1)$ and

$$A_t \alpha_2^\# \leq \alpha_2^\# + w^2 \sqrt{\alpha_1 \alpha_3} \tag{22}$$

for $(p,q)=(2,2)$ and $i=2$, where α_1^*, α_2^* and $\alpha_2^\#$ are suitably modified functions α_1 and α_2 (the ratios α_1/α_1^*, α_1^*/α_1, α_2/α_2^*, α_2^*/α_2, $\alpha_2/\alpha_2^\#$ and $\alpha_2^\#/\alpha_2$ are bounded).

Let $f_i(h) = \alpha_i(h\Delta), h \in H$. It follows from its definition that each α_i is K-invariant. Hence

$$f_i(Kh) = f_i(h), h \in H, 0 \leq i \leq n. \tag{23}$$

In view of (20) we have

$$A_t f_i^s \leq \frac{c}{2} f_i^s + w^2 \max_{0<j\leq \min(n-i,i)} \sqrt{f_{i+j}^s f_{i-j}^s}. \tag{24}$$

Let us denote $q(i) = i(n-i)$. Then by direct computations $2q(i) - q(i+j) - q(i-j) = -2j^2$. Therefore we get from (24) that for any $i, 0 < i < n$, and any positive $\varepsilon < 1$

$$\begin{aligned} A_t(\varepsilon^{q(i)} f_i^s) &\leq \frac{c}{2}\varepsilon^{q(i)} f_i^s \\ &\quad + w^2 \max_{0<j\leq \min(n-i,i)} \varepsilon^{q(i)-\frac{q(i+j)+q(i-j)}{2}} \sqrt{\varepsilon^{q(i+j)} f_{i+j}^s \varepsilon^{q(i-j)} f_{i-j}^s} \\ &\leq \frac{c}{2}\varepsilon^{q(i)} f_i^s + \varepsilon w^2 \max_{0<j\leq \min(n-i,i)} \sqrt{\varepsilon^{q(i+j)} f_{i+j}^s \varepsilon^{q(i-j)} f_{i-j}^s}. \end{aligned} \tag{25}$$

Consider the linear combination

$$f_{\varepsilon,s}(h) = \sum_{0\leq i\leq n} \varepsilon^{q(i)} f_i^s(h) = \sum_{0\leq i\leq n} \varepsilon^{q(i)} \alpha_i(h\Delta)^s. \tag{26}$$

Since $\varepsilon^{q(i)} f_i^s < f_{\varepsilon,s}, f_0 = 1$ and $f_n = 1/d(\Delta)$, the inequalities (25) imply the following inequality:

$$A_t f_{\varepsilon,s} < 1 + d(\Delta)^{-s} + \frac{c}{2} f_{\varepsilon,s} + n\varepsilon w^2 f_{\varepsilon,s}. \tag{27}$$

Taking $\varepsilon = \frac{c}{2nw^2}$ we see that

$$A_{t(s,c)} f_{\varepsilon,s} < c f_{\varepsilon,s} + b \tag{28}$$

where $b = 1 + d(\Delta)^{-s}$.

It is well known and easy to show that for every neighborhood V of e in H there exists a neighborhood U of e in K such that

$$a_t U a_r \subset K V a_t a_r K$$

for any $t \geq 0$ and $r \geq 0$. On the other hand the positive function $f_{\varepsilon,s}$ is left K-invariant (because of (23)) and the logarithm $\log f_{\varepsilon,s}$ is equicontinuous with respect to a left-invariant uniform structure on H. Using these facts and the inequality (28) for sufficiently small c, it is not difficult to show that

$$\sup_{t>0} (A_t f_{\varepsilon,s})(e) < \infty. \tag{29}$$

Combining (26) and (29) with the definition of A_t and α we get (18) (the assertion (a)). The proof of (b) is similar but we have to use (21) and (22), in addition to (20) (we first use (20) for $s = 1/2$ and after that use (21), (22) and (20) for $s = 1$).

To obtain (20) we have to apply the following two lemmas.

Lemma 1. *Let*

$$F(i) = \{x_1 \wedge \cdots \wedge x_i \mid x_1, \ldots, x_i \in \mathbb{R}^{p+q}\} \subset \wedge^i(\mathbb{R}^{p+q}).$$

If $0 < s < 2, p \geq 3$ *and* $0 < i < n$ *or* $p = 2, q = 2$ *and* $i = 1$ *or 3, or if* $0 < s < 1$, *then*

$$\lim_{t\to\infty} \sup_{v\in F(i), \|v\|=1} \int_K \frac{d\sigma(k)}{\|a_t kv\|^s} = 0. \tag{30}$$

Lemma 2. *For any two* Δ*-rational subspaces* L *and* M

$$d(L)d(M) \geq d(L \cap M)d(L + M). \tag{31}$$

Lemma 1 is deduced from some standard facts about Lie groups and their finite dimensional representations. Lemma 2 easily follows from the inequality $\|x \wedge y\| \leq \|x\|\|y\|$, $x \in \wedge^i(\mathbb{R}^n), y \in \wedge^j(\mathbb{R}^n)$.

It follows from (30) that for any $c > 0$ there exist $t = t(s, c) > 0$ such that

$$\int_K \frac{d\sigma(k)}{d_{a_t k\Lambda}(a_t kL)^s} < \frac{c}{2}\frac{1}{d_\Lambda(L)^s} \tag{32}$$

for every lattice $\Lambda \subset \mathbb{R}^n$ and every Λ-rational subspace L of dimension i. Let $L_i = L_{i,\Lambda}$ be a Λ-rational subspace of dimension i such that

$$\frac{1}{d_\Lambda(L_i)} = \alpha_i(\Lambda). \tag{33}$$

Let $w = e^t$. We have that

$$w^{-1} \leq \frac{\|a_t v\|}{\|v\|} \leq w, v \in \wedge^j(\mathbb{R}^n), 0 < j < n. \tag{34}$$

Let us denote the set of Λ-rational subspaces L of dimension i with $d_\Lambda(L) < w^2 d_\Lambda(L_i)$ by Ψ_i. We get from (34) that for a Λ-rational i-dimensional subspace $L \notin \Psi_i$

$$d_{a_t k\Lambda}(a_t kL) > d_{a_t k\Lambda}(a_t kL_i), k \in K. \tag{35}$$

It follows from (32), (33), (35) and the definition of α_i and A_t that

$$(A_t\alpha_i^s)(\Lambda) < \frac{c}{2}\alpha_i^s(\Lambda) \quad \text{if} \quad \Psi_i = \{L_i\}. \tag{36}$$

Assume now that $\Psi_i \neq \{L_i\}$. Let $M \in \Psi_i, M \neq L_i$. Then $\dim(M+L_i) = i+j, j > 0$. Now using (33), (34) and Lemma 2 we get that for any $k \in K$

$$\alpha_i(a_t k\Lambda) < w\alpha_i(\Lambda) = \frac{w}{d_\Lambda(L_i)} < \frac{w^2}{\sqrt{d_\Lambda(L_i)d_\Lambda(M)}}$$

$$\leq \frac{w^2}{\sqrt{d_\Lambda(L \cap M)d_\Lambda(L+M)}} \leq w^2\sqrt{\alpha_{i+j}(\Lambda)\alpha_{i-j}(\Lambda)}. \tag{37}$$

Hence

$$(A_t\alpha_i^s)(\Lambda) \leq w^2 \max_{0<j\leq \min(n-i,i)} \sqrt{\alpha_{i+j}^s \alpha_{i-j}^s} \quad \text{if} \quad \Psi_i \neq L_i. \tag{38}$$

Combining (36) and (38) we get (20).

2.4.14. Remark. Theorem 1.4.2 of Birch and Davenport implies that if B is a real indefinite form in five variables then for any $\delta > 0$ there exists $C = C(\delta, B)$ with the following property. For any $\varepsilon > 0$ there exist $x \in \mathbb{Z}^5, x \neq 0$, such that $|B(x)| < \varepsilon$ and $\|x\| < C\varepsilon^{-5-\delta}$. It would be interesting to prove a similar result for nondiagonal forms (maybe with the replacement of $\varepsilon^{-5-\delta}$ by $\varepsilon^{-n-\delta}$ for some n). It would also be interesting to obtain an analog of Theorem 1.3.1 for nondiagonal forms. At present it is not clear how the homogeneous space approach can be applied to solve these two problems.

2.5. Markov spectrum. Let $\mathcal{O}_n$ denote the set of all nondegenerate indefinite quadratic form in n variables, let $d(B)$ denote the discriminant of a form $B \in \mathcal{O}_n$, and let, as before, $m(B) = \inf\{B(x) \mid x \in \mathbb{Z}^n, x \neq 0\}$. Let us set $\mu(B) = m(B)^n/|d(B)|$. It is clear that $\mu(B) = \mu(B')$ if B and B' are equivalent (in the sense of 2.1.7). Let M_n denote $\mu(\mathcal{O}_n)$. The set M_n is called the *n-dimensional Markov spectrum.* It easily follows from Mahler's compactness criterion that M_n is bounded and closed.

In 1880, Markov [Mark1] described the intersection of M_2 with the segment $(4/9, \infty)$ and described corresponding quadratic forms. It follows from this description that $M_2 \cap (4/9, \infty)$ is a discrete countable subset of $(4/9, 4/5]$ and that, for any $a > 4/9$, there are only finitely many equivalence classes of forms $B \in \mathcal{O}_2$ with $\mu(B) > a$. On the other hand, the intersection $M_2 \cap [0, 4/9]$ is not countable and moreover has quite a complicated topological structure (see [Ca2], [CuF], [Po]).

It follows from Meyer's theorem and Theorem 2.3.1 that $M_n = \{0\}$ if $n \geq 5$. Theorems 2.1.7 and 2.3.1 imply that if n is 3 or 4 and $\varepsilon > 0$ then the set $M_n \cap (\varepsilon, \infty)$ is finite and moreover there are only finitely many equivalence classes of forms $B \in \Phi_n$ with $\mu(B) > \varepsilon$. For rational forms B, this result had been proved earlier by Vulakh [Vu2]. As noted in [Vu1], Vulakh obtained the complete description of spectra of nonzero minima of rational Hermitian forms.

For $n = 3$, Markov [Mark2] determined the first 4 values of $\mu(B)^{-1}$ and Venkov [Ven] determined the next 7 values. For $n = 4$ and forms of signature

(2, 2), Oppenheim [Op3] determined the first 7 values of $\mu(B)^{-1}$. For $n = 4$ and forms of signature (3, 1) he [Op1] determined also the first value of $\mu(B)^{-1}$ which is 7/4.

Recently, using a computer, Grunewald and Martini gave complete lists of representatives of 290 equivalence classes of $B \in \Phi_3$ for which $\mu(B) > 3/46$ and of representatives of 257 equivalence classes of $B \in \Phi_4$ for which $\mu(B) > 1/36$. On the basis of these computations, they suggested the following conjectural asymptotic formulas:

$$\Phi_3(x) \approx 1,2x^2 \quad \text{and} \quad \Phi_4(x) \approx 1,2x^{3/2} \text{ as } x \to \infty,$$

where

$$\Phi_n(x) = |\{B \in \mathcal{O}_n \mid \mu(B) \leq x^{-1}\}|.$$

3. Unipotent Flows On Homogeneous Spaces

3.1 Growth properties of polynomials. It was observed a long time ago (already in [Marg2] and probably even earlier) that certain growth properties of polynomials of bounded degrees play an important role in the study of unipotent flows on homogeneous spaces. We now state some of these properties in the form of several elementary lemmas which easily follow from Lagrange interpolation formula.

Let $\mathcal{P}_m$ (resp. $\mathcal{P}_m^+$) denote the space of all (resp. all non-negative) polynomials on $\mathbb{R}$ of degree at most m.

3.1.1. Lemma. *For any $k > 1$ and $m \in \mathbb{N}$ there exist $0 < \varepsilon_1(k,m) < \varepsilon_2(k,m)$ such that the following holds: Let $c > 0$ and $t_1 \leq t_2$. If $P \in \mathcal{P}_m^+$ is such that $P(t) \leq c$ for all $t \in [t_1, t_2]$ and $P(t_2) = c$ then the values of P at all points of one of the intervals*

$$[t_1 + k(t_2 - t_1), t_1 + k^2(t_2 - t_1)],$$
$$[t_1 + k^3(t_2 - t_1), t_1 + k^4(t_2 - t_1)], \ldots, [t_1 + k^{2m+1}(t_2 - t_1), t_1 + k^{2m+2}(t_2 - t_1)]$$

are greater than $c\varepsilon_1(k,m)$ and smaller than $c\varepsilon_2(k,m)$.

3.1.2. Lemma. *For any $k > 1$ and $m \in \mathbb{N}$ there exists $\bar{\varepsilon}(k,m) > 0$ such that the following holds: Let $c > 0$ and $t_1 \leq t_2$. If $P \in \mathcal{P}_m^+$ is such that $P(t) = c$ for some $t \in [t_1, t_2]$, $P(t_2) < c\bar{\varepsilon}(k,m)$ then there exists $t \in [t_2, t_1 + k(t_2 - t_1)]$ such that $P(t) = c\bar{\varepsilon}(k,m)$.*

3.1.3. Lemma. *For any $m \in \mathbb{N}$ and $\varepsilon > 0$ there exists $\alpha = \alpha(\varepsilon, m) > 0$ such that the following holds: Let $c > 0, t_1 \leq t_2$, and $P \in \mathcal{P}_m$ be such that $|P(t)| < c$ for all $t \in [t_1, t_2]$. Then $|P(t) - P(t')| < \varepsilon c$ if $t_1 \leq t < t' \leq t_2$ and $t' - t < \alpha(t_2 - t_1)$.*

3.1.4. Lemma 3.1.3 can be viewed as a weak version of the following:

Lemma. *For any $m \in \mathbb{N}^+$ there exists $\beta = \beta(m) > 0$ such that the following holds: Let $c > 0$ and $t_1 \leq t_2$. If $P \in \mathcal{P}_m$ is such that $|P(t)| < c$ for all $t \in [t_1, t_2]$ then*

$$|P^{(i)}(t)| < \frac{c\beta^i}{(t_2 - t_1)^i}$$

for any $t \in [t_1, t_2]$ and $i, 1 \leq i \leq n$, where $P^{(i)}$ denotes the i-th derivative of P.

3.1.5. Lemma. *For any $m \in \mathbb{N}^+$ and $\varepsilon > 0$ there exists $\delta = \delta(\varepsilon, m) > 0$ such that the following holds: Let $c > 0$ and $t_1 \leq t_2$. If $P \in \mathcal{P}_m^+$ is such that $P(t) = c$ for some $t \in [t_1, t_2]$ then*

$$\ell(\{t \in [t_1, t_2] \mid P(t) < \delta c\}) < \varepsilon(t_2 - t_1)$$

where ℓ denotes the Lebesgue measure on $\mathbb{R}$.

3.1.6. There is a slightly stronger version of Lemma 3.1.5 which can be easily deduced from Lemmas 3.1.3 and 3.1.5.

Lemma. *For any $m \in \mathbb{N}^+, \varepsilon > 0$ and $k > 1$ there exists $\delta' = \delta'(\varepsilon, m, k) > 0$ such that the following holds: Let $c > 0$ and $t_1 < t_2$. If $P \in \mathcal{P}_m^+$ is such that $P(t) = c$ for some $t \in [t_1, t_2]$ then*

$$\begin{aligned} &\ell(\{t \in [t_1, t_2] \mid P(t) < \delta' c\}) \\ &\leq \varepsilon \ell(\{t \in [t_1, t_2] \mid ck^{-1} < P(t) < c\}). \end{aligned}$$

3.1.7. Remark. Lemmas 3.1.1–3.1.6. are standard and simple and, as it was noticed, they easily follow from Lagrange interpolation formula. There are slightly stronger versions of Lemmas 3.1.3 [Rat5, Proposition 1.1] and 3.1.5 [DanM5, Lemma 4.1]. These versions can also be easily deduced from Lagrange interpolation formula. To prove Lemmas 3.1.1–3.1.6, one can also apply another standard argument which is based on the linear rescaling of variables and on the compactness of the unit ball in the space of polynomials on [0,1] of degree at most m (see for example [Marg2] and [Dan2] for Lemmas 3.1.1. and 3.1.2. and [Rat6, Lemma 3.1] for Lemma 3.1.5).

3.2. Unipotent groups of linear transformations. Let $\{u(t) \mid t \in \mathbb{R}\}$ be a unipotent one-parameter subgroup of $SL(n, \mathbb{R})$. Matrix coefficients of $u(t)$ are polynomials in t of degree at most $n - 1$, and $\|u(t)v\|^2$ is a polynomial in t of degree at most $2n - 2$ for any $v \in \mathbb{R}^n$.

3.2.1. Applying Lemma 3.1.6 to the polynomials $P_v(t) = \|u(t)v\|^2$ we get the following:

Lemma. *For any $\varepsilon > 0$ and $k > 1$ there exists $\tilde{\delta} = \tilde{\delta}(\varepsilon, k, n) > 0$ such that the following holds: Let $c > 0, t_1 < t_2$ and $v \in \mathbb{R}^n$. If $\|u(t)v\| = c$ for some $t \in [t_1, t_2]$*

then

$$\ell(\{t \in [t_1, t_2] \mid \|u(t)v\| < \tilde{\delta}c\})$$
$$\leq \varepsilon\ell(\{t \in [t_1, t_2] \mid ck^{-1} < \|u(t)v\| < c\}).$$

3.2.2. To put Lemma 3.2.1 in a more general context, let us consider a one-parameter group $\{T^t\}$ of homeomorphisms of a locally compact space X and a subset A of X (not necessarily closed). We say that A is *avoidable with respect to* $\{T^t\}$ if for any compact subset C of $X - A$ and any $\varepsilon > 0$ there exists a neighbourhood Ψ of A in X such that for any $x \in C$ and any $t_1 \leq 0 \leq t_2$, we have

$$\ell(\{t \in [t_1, t_2] \mid T^t x \in \Psi\}) \leq \varepsilon(t_2 - t_1).$$

Lemma 3.2.1 essentially means that $\{0\}$ is avoidable with respect to $\{u(t)\}$. Let us note that $\{0\}$ is not avoidable with respect to any non-quasiunipotent one-parameter subgroup $\{a_t\}$ of $GL(n, \mathbb{R})$ (an element g of $GL(n, \mathbb{R})$ is called *quasiunipotent* if all eigenvalues of g have absolute value 1). Lemma 3.2.1 can be considered as a special case ($A = \{0\}$) of the following assertion which can be easily deduced from Proposition 4.2 in [DanM5].

Proposition. *Any algebraic subvariety A of $\mathbb{R}^n$ is avoidable with respect to $\{u(t)\}$.*

3.2.3. Lemma. *For any $\varepsilon > 0$ there exists $\tilde{\alpha} = \tilde{\alpha}(\varepsilon, n) > 0$ such that the following holds: Let $c > 0, t_1 \leq t_2$, and $v \in \mathbb{R}^n$ be such that $\|u(t)v\| < c$ for all $t \in [t_1, t_2]$. Then $\|u(t)v - u(t')v\| < \varepsilon c$ if $t_1 \leq t < t' \leq t_2$ and $t' - t < \tilde{\alpha}(t_2 - t_1)$.*

To prove this lemma, it is enough to notice that the coordinates of $u(t)v$ are polynomials in t of degree at most n and after that apply Lemma 3.1.3.

3.2.4. Let $V_0 = \{v \in \mathbb{R}^n \mid u(t)v = v \text{ for all } t \in \mathbb{R}\}$. Using an easy compactness argument one can deduce from Lemma 3.2.3 the following:

Lemma (cf. [Wi, Proposition 6.3]). *For any $\varepsilon > 0$ there exists $\theta = \theta(\varepsilon, n, \{u_t\}) > 0$ such that the following holds: Let $t_1 \leq 0 \leq t_2$, and $v \in \mathbb{R}^n$ be such that $\|v\| < \theta$ and $\|u(t)v\| < 1$ for all $t \in [t_1, t_2]$. Then for any $t \in [t_1, t_2]$ one can find $q_t \in V_0$ such that $\|u(t)v - q_t\| < \varepsilon$.*

3.2.5. We already defined V_0. Let $V_i = \{v \in \mathbb{R}^n \mid u(t)v - v \in V_{i-1} \text{ for all } t \in \mathbb{R}\}, 1 \leq i \leq n-1$. Then $\dim V_i/V_{i-1} \leq \dim V_{i-1}/V_{i-2}$ and $V_{n-1} = \mathbb{R}^n$. We can represent $\mathbb{R}^n$ as $V_0 \oplus V_1/V_0 \oplus \cdots \oplus V_{n-1}/V_{n-2}$. By Jordan decomposition, after the identification of V_i/V_{i-1} with a certain subspace $V_i' \subset V_{i-1}'$ of $V_0, 1 \leq i \leq n-1$, the following holds: for any $v \in \mathbb{R}^n$ there exists a polynomial function $\varphi_v : \mathbb{R} \to V_0$ of degree at most $n - 1$ such that

$$u(t)v = (\varphi_v(t), \varphi_v'(t), \ldots, \varphi_v^{(i)}(t), \ldots, \varphi_v^{(n-1)}(t))$$
$$\in \mathbb{R}^n = V_0 \oplus V_1/V_0 \oplus \cdots \oplus V_i/V_{i-1} \oplus \cdots \oplus V_{n-1}/V_{n-2},$$

Now applying Lemma 3.1.4 we get a stronger version of Lemma 3.2.4.

Lemma. *There exist* $\tilde{\beta} = \tilde{\beta}(n) > 0$ *and, for every* $v \in \mathbb{R}^n$, *a polynomial function* $\varphi_v : \mathbb{R} \to V_0$ *of degree at most* $n-1$ *with the following property: Let* $c > 0$, $t_1 \le 0 \le t_2, t_1 \ne t_2$, *and* $v \in \mathbb{R}^n$ *be such that* $\|u(t)v\| < c$ *for all* $t \in [t_1, t_2]$. *Then*

$$\|u(t)v - (\varphi_v(t), \dots, \varphi_v^{(i-1)}(t), 0, \dots, 0)\| < \frac{c\tilde{\beta}^i}{(t_2 - t_1)^i}$$

for any $t \in [t_1, t_2]$ *and* $1 \le i \le n-1$.

This lemma is similar to Corollary 3.1 in [Rat2]. The proof given in [Rat2] also uses Jordan decomposition.

3.2.6. Remark. Let $G = SL(2, \mathbb{R}), h(t) = \begin{pmatrix} 1 & t \\ 0 & 1 \end{pmatrix}$, $H = \{h(t) \mid t \in \mathbb{R}\}$, and let d_G denote a left invariant metric on G. Lemma 3.2.4 (or Lemma 3.2.5) applied to the one-parameter unipotent group $\{\mathrm{Ad}h(t)\}$ implies a property of horocycle flows which Ratner terms the H-property (see Definition 1 in [Ra4]). This property in the form given in [Rat10] states that given $0 < \varepsilon < 1$ and $p, N > 0$ there exist $\delta(\varepsilon, p, N)$, $\alpha(\varepsilon) \in (0,1)$ such that if $d_G(x, e) < \delta(\varepsilon, p, N)$ for some $x \in G - H$ then there are $L, T > 0$ with $N < L < T$, $L \ge \alpha(\varepsilon)T$ such that either $d_G(xh(t), h(t+p)) \le p\varepsilon$ for all $t \in [T-L, T]$ or $d_G(xh(t), h(t-p)) \le p\varepsilon$ for all $t \in [T-L, T]$. The H-property was generalized by Witte in [Wi].

3.2.7. It is possible to state and prove generalizations of Lemmas 3.2.1, 3.2.3–3.2.5 for multidimensional unipotent groups U. But for noncomutative U it involves a technical construction of subsets of U replacing intervals in the one-dimensional case. Let us state only a lemma which for one-parameter U easily follows from Lemma 3.2.5. In the general case this lemma can be proved by induction on n using the Lie–Kolchin theorem (see for example [Marg7, Sec. 5]).

Lemma. *Let* U *be a connected group of unipotent linear transformations of* $\mathbb{R}^n$, *and let* $Y \subset \mathbb{R}^n$. *Let* $L = \{v \in \mathbb{R}^n \mid Uv = v\}$ *and* $p \in L \cap \bar{Y}$. *Suppose that* $L \cap Y = \emptyset$. *Then* $\overline{UY} \cap L$ *contains the image of a nonconstant polynomial map* $\varphi : \mathbb{R} \to L$ *such that* $\varphi(0) = p$.

3.3. Recurrence to compact sets. Let $\Omega_n = SL(n, \mathbb{R})/SL(n, \mathbb{Z})$ denote the space of unimodular lattices in $\mathbb{R}^n$. For any lattice $\Delta \in \Omega_n$ we define

$$\beta(\Delta) = \inf\{\|v\| \mid v \in \Lambda, v \ne 0\}.$$

According to Mahler's compactness criterion, a subset $K \subset \Omega_n$ is relatively compact in Ω_n if and only if $\inf\{\beta(\Delta) \mid \Delta \in K\} > 0$.

3.3.1. Theorem [Marg2]. *Let* $\{u(t)\}$ *be a one-parameter unipotent subgroup of* $SL(n, \mathbb{R})$. *For any lattice* $\Delta \in \Omega_n$ *there exists* $\varepsilon(\Delta) > 0$ *such that the set* $\{t \ge 0 \mid \beta(u(t)\Delta) > \varepsilon(\Delta)\}$ *is unbounded. Equivalently, for any* $x \in \Omega_n$ *the "positive semi-orbit"* $\{u(t)x \mid t \ge 0\}$ *does not tend to infinity. That is, there exists a compact subset* $K = K(x) \subset \Omega_n$ *such that* $\{t \ge 0 \mid u(t)x \in K\}$ *is unbounded.*

For $n = 2$ the proof of this theorem is simple. Indeed, if $n = 2$, $\Delta \in \Omega_n$, v_1 and v_2 are nonzero primitive vectors in Δ, $v_1 \neq \pm v_2$, then the sets $\{t \in \mathbb{R} \mid \|u(t)v_1\| < 1\}$ and $\{t \in \mathbb{R} \mid \|u(t)v_2\| < 1\}$ are disjoint. After that it remains to apply Lemma 3.2.1. However for $n \geq 3$ the proof is much harder. It is based on a rather complicated induction argument and uses Lemmas 3.1.1 and 3.1.2 applied to the polynomials $P_L(t) = d^2_{u(t)\Delta}(u(t)L)$ where L is a Δ-rational subspace of $\mathbb{R}^n$ and $d_{u(t)\Delta}(u(t)L)$ is defined in 2.4.12.

3.3.2. Theorem [DanM5, Theorem 6.1]. *Let G be a connected Lie group, Γ a lattice in G, F a compact subset of G/Γ and $\varepsilon > 0$. Then there exists a compact subset K of G/Γ such that for any Ad-unipotent one-parameter subgroup $\{u(t)\}$ of G, any $x \in F$, and $T \geq 0$,*

$$\ell(\{t \in [0, T] \mid u(t)x \in K\} \geq (1 - \varepsilon)T.$$

This theorem is essentially due to Dani. He proved it in [Dan5] for semisimple groups G of $\mathbb{R}$-rank 1 and in [Dan6] for arithmetic lattices. The general case can be easily reduced to these two cases using the arithmeticity theorem. In the case of arithmetic lattices the theorem can be considered as the quantitative version of Theorem 3.3.1 and the proof given in [Dan6] is similar in principle to the proof of Theorem 3.3.1 in [Marg2].

3.3.3. The following theorem is a strengthening of Theorem 3.3.1 and of Theorem 3.3.2 for arithmetic Γ.

Theorem (see [DanM1, Theorem 1.1] or [Dan6, Proposition 2.7]). *Given $\varepsilon > 0$ and $\theta > 0$ there exists $\delta > 0$ such that for any lattice $\Delta \in \Omega_n$, any unipotent one-parameter subgroup $\{u(t)\}$ of $SL(n, \mathbb{R})$ and $T \geq 0$ at least one of the following holds*:

(i) $\ell(\{t \in [0, T] \mid \beta(u(t)\Delta) \geq \delta\} \geq (1 - \varepsilon)T$;
(ii) *there exists a Δ-rational subspace L of $\mathbb{R}^n$ such that $P_L(t) < \theta$ for all $t \in [0, T]$, where $P_L(t) = d^2_{u(t)\Delta}(u(t)L)$ denotes the same as in 3.3.1.*

Remark. About different generalizations of Theorems 3.3.1–3.3.3 see [Sha2] and [EMS2].

3.3.4. Using a modification of the approach developed in [Marg2] and [Dan6] it is possible to prove the following improvement of Theorem 3.3.3. It is stated for polynomial maps. Let us recall that any one-parameter unipotent subgroup $\{u(t)\} \subset \mathbb{R}^n$ can be considered as a polynomial map from $\mathbb{R}$ into $SL(n, \mathbb{R})$ of degree at most $n - 1$. The proof will appear elsewhere.

Theorem. *Let $t_1 \leq t_2$, $0 < \varepsilon < q \leq 1$, and let $h : \mathbb{R} \to GL_n(\mathbb{R})$ be a polynomial map of degree not greater than r. Then at least one of the following holds:*

(i) $\ell(\{t \in [t_1, t_2] \mid \beta^2(h(t)\mathbb{Z}^n) < \varepsilon\} \leq \varepsilon^{\frac{1}{2nr}} q^{-\frac{1}{2nr}} 4^{n+2} 2nr(t_2 - t_1)$;

(ii) *there exists a $\mathbb{Z}^n$-rational subspace L of $\mathbb{R}^n$ such that $d^2_{h(t)\mathbb{Z}^n}(h(t)L) < q$ for all $t \in [0, T]$.*

3.3.5. Using Theorem 3.3.3 it is not difficult to prove the following:

Corollary [DanM1, Corollary 1.3]. *Let $G = SL(n, \mathbb{R})$ and $\Gamma = SL(n, \mathbb{Z})$. There exists a compact subset C of $\Omega_n = G/\Gamma$ such that the following holds: Let U be a connected unipotent subgroup of G. Let $\mathcal{N}_G(U)$ be the normalizer of U in G and let $\{f(t)\}_{t \geq 0}$ be a curve in $\mathcal{N}_G(U)$ such that if L is a proper nonzero U-invariant subspace then L is invariant under $f(t)$ for all t and $|\det f(t)|_L| \to \infty$ as $t \to \infty$ ($\det f(t)|_L$ denotes the determinant of the restriction of $f(t)$ to L). Then for all $g \in G$, $C \cap Uf(t)g\Gamma/\Gamma$ is nonempty for all large t. If F is the subgroup generated by U and $\{f(t) \mid t \geq 0\}$, then every nonempty closed F-invariant subset contains a minimal closed F-invariant subset.*

3.3.6. Using an improvement of Theorem 3.3.2 obtained in [Dan6], Dani proved the following result.

Theorem (see [Dan6, Theorem 3.8] or [DanM1, Theorem 1.4]). *Let G be a connected Lie group and let Γ be a lattice in G. Then there exists a compact subset C of G/Γ such that for any closed connected Ad-unipotent subgroup U of G and any $g \in G$ either $C \cap Ug\Gamma/\Gamma$ is non-empty or $g^{-1}Ug$ is contained in a proper closed subgroup F such that $F\Gamma$ is closed and $F \cap \Gamma$ is a lattice in F.*

3.3.7. Corollary [DanM1, Corollary 1.5]. *Let G be a connected Lie group, Γ a lattice in G, and U a connected Ad-unipotent subgroup of G. Then any nonempty closed U-invariant subset of G/Γ contains a minimal closed U-invariant subset.*

3.3.8. Using Theorem 3.3.1 and its analog for non-arithmetic lattices, the author proved the following:

Theorem [Marg9, Theorem 1]. *Let G, Γ and U be the same as in Corollary 3.3.7. Then every minimal closed U-invariant subset of G/Γ is compact.*

The proof of this theorem also uses a general result about actions of nilpotent Lie groups on locally compact spaces.

3.4. Finiteness of ergodic measures. Suppose that G is a locally compact group acting continuously on a locally compact space X. We say that a subgroup H of G *has property* (D) *with respect to* X if for every H-invariant locally finite Borel measure μ on X, there exist Borel subsets $X_i \subset X$, $i \in \mathbb{N}^+$, such that $\mu(X_i) < \infty$, $HX_i = X_i$ and $X = \cup X_i$. If H has property (D) with respect to X then every H-ergodic H-invariant locally finite Borel measure on X is finite. (When G and X are separable and metrizable, the converse is also true.)

3.4.1. Theorem (see [Dan2, Theorem 3.3] for arithmetic lattices and [Dan5, Theorem 4.3] in the general case). *Let G be a connected Lie group and let Γ be a lattice in G. Then any unipotent subgroup U of G has property (D) with respect to G/Γ.*

This theorem is deduced from (a weak version of) Theorem 3.3.2 and the individual ergodic theorem.

3.4.2. We say that a subgroup H of a locally compact group G has *Mautner property* with respect to a subgroup F of G if, for any continuous unitary representation ρ of G on a Hilbert space W and any $w \in W$ such that $\rho(F)w = w$, we have that $\rho(H)w = w$. It is easy to see that if H has Mautner property with respect to F and F has property (D) with respect to X, then H has property (D) with respect to X. On the other hand, in view of a general result of Moore [Mo], if G is a connected Lie group and $H \subset G$ is a connected subgroup generated by its Ad-unipotent elements then H has Mautner property with respect to its maximal Ad-unipotent subgroup. Using these facts, one can deduce from Theorem 3.4.1 the following.

Theorem (cf. [Marg5, Theorem 7]). *Let G be a connected Lie group, Γ a lattice in G, and H a connected subgroup of G. Suppose that the quotient of H by its unipotent radical is semisimple. Then H has property (D) with respect to G/Γ.*

(By the unipotent radical of H we mean the maximal connected normal Ad-unipotent subgroup of H.)

3.4.3. The following is a special case of Theorem 3.4.2.

Theorem (cf. [Marg5, Theorem 8]). *Let G, Γ, and H be the same as in Theorem 3.4.2 and let $x \in G/\Gamma$. Suppose that the orbit Hx is closed in G/Γ. Then $H \cap G_x$ is a lattice in H, where $G_x = \{g \in G \mid gx = x\}$ is the stabilizer of x.*

3.4.4. Applying Theorem 3.4.3 to the case where $G = SL(n,\mathbb{R}), \Gamma = SL(n,\mathbb{Z})$, $x = e\Gamma$, and $H \subset G$ is the set of $\mathbb{R}$-rational points of a $\mathbb{Q}$-subgroup of G, we obtain the following special but important case of the theorem of Borel and Harish–Chandra on the finiteness of the volume of the quotient by an arithmetic subgroup.

Theorem. *Let H be a connected $\mathbb{Q}$-group. Suppose that the quotient of H by its unipotent radical is semisimple. Then $H(\mathbb{Z})$ is a lattice in $H(\mathbb{R})$.*

3.4.5. Combining Theorem 3.4.3 with Borel's density theorem [Bo1], it is easy to prove the following:

Proposition (cf. [BoP, Proposition 1.1]). *Let H be a connected subgroup of $SL(n,\mathbb{R})$ generated by its unipotent elements and such that $H\mathbb{Z}^n$ is closed in Ω_n. Then H is the connected component of the identity of the set of $\mathbb{R}$-rational points of a $\mathbb{Q}$-subgroup of $SL(n)$.*

3.4.6. Let us state now a result which can be considered as a generalization of Lemma 2.2.2 (b). It can be easily deduced from Proposition 3.4.5 (by a modification of the argument from Sec. 1.3 of [Marg5]).

Proposition. *Let B be a real indefinite quadratic form in $n \geq 3$ variables. Let us denote by H the special orthogonal group $SO(B)$. If the orbit $H\mathbb{Z}^n$ is closed in Ω_n then B is rational.*

3.5. General properties of minimal invariant sets. As before, in this and following subsections we denote by $\bar{A}$ the closure of a subset A of a topological space, by $\mathcal{N}_G(F)$ the normalizer of a subgroup F in a group G, and by e the identity element of G.

Let G be an arbitrary second countable locally compact group, and Ω a homogeneous space of G. The stabilizer $\{g \in G \mid gy = y\}$ of $y \in \Omega$ is denoted by G_y.

3.5.1. Lemma. *Let F be a closed subgroup of G, and Y a minimal closed F-invariant subset of Ω. Then for every nonempty open in Y subset U of Y and every compact subset K of Y, there exists a compact subset $L = L(U, K)$ of F such that $Ly \cap U \neq \emptyset$ for any $y \in K$.*

This lemma follows from the density of Fy in Y for every $y \in Y$.

3.5.2. Lemma. *Let F and H, $F \subset H$, be closed subgroups of G. Let Y be a minimal compact F-invariant subset of Ω, and let $y \in Y$. Suppose that there exists a sequence $\{f_n\} \subset F$ such that $\{f_n y\}$ is relatively compact in Ω_n but $\{f_n H_y\}$ is not relatively compact in H/H_y where $H_y = H \cap G_y$. Then the closure of $\{g \in G - H \mid gy \in Fy\}$ contains e.*

Proof. We can assume that the sequence $\{f_n\}$ tends to infinity in H/H_y. Let $W \subset G$ be a relatively compact neighborhood of e. By Lemma 3.5.1 there exist $\ell_n \in F$ and $g_n \in W$ such that $\ell_n f_n y = g_n y$ and the sequence $\{\ell_n\}$ is relatively compact in F. Then $\ell_n f_n H_y$ tends to infinity in H/H_y as $n \to \infty$. But $g_n^{-1} \ell_n f_n y = y$ and $\{g_n^{-1}\}$ is relatively compact in G. Therefore $g_n \notin H$ for almost all n. Thus $W \cap \{g \in G - H \mid gy \in Fy\} \neq \emptyset$. This proves the lemma.

3.5.3. We will need another simple result.

Lemma (see [Marg5, Lemmas 1 and 2] or [DanM1, Lemma 2.1]). *Let F, P and Q be closed subgroups of G such that $F \subset P \cap Q$. Let X and Y be closed subsets of Ω invariant under the actions of P and Q respectively. Suppose also that X is a compact minimal F-invariant subset. Let M be a subset of G such that $gX \cap Y$ is nonempty for all $g \in M$. Then $hX \subset Y$ for all $h \in \overline{QMP} \cap \mathcal{N}_G(F)$. If further $X = Y$ and $P = Q$ then X is invariant under the closed subgroup generated by $\overline{QMP} \cap \mathcal{N}_G(F)$.*

3.6. Topological limits of double cosets in algebraic groups. We define *a real algebraic group* to be a group of $\mathbb{R}$-rational points of an $\mathbb{R}$-group or, in other words, an algebraic subgroup of $GL(m, \mathbb{R})$.

3.6.1. Lemma (cf. [BoP, Proposition 2.3] and [Marg5, Lemma 5]). *Let G be a real algebraic group, U a connected unipotent subgroup of G and $M \subset G$. Suppose that*

$e \in \bar{M} - M$ and $M \subset G - \mathcal{N}_G(U)$. Then there is a rational map $\psi : U \to \mathcal{N}_G(U)$ such that $\psi(e) = e$, $\psi(U) \subset \overline{UMU}$ and $\psi(U) \not\subset KU$ for any compact subset K of $\mathcal{N}_G(U)$.

Sketch of the proof. The connected unipotent subgroup U is algebraic and has no (nontrivial) rational characters. So, in view of Chevalley's theorem, there exists $n \in \mathbb{N}^+$, a faithful rational representation $\alpha : G \to GL(n, \mathbb{R})$ and $x_0 \in \mathbb{R}^n$ such that $U = \{g \in G \mid \alpha(g)x_0 = x_0\}$. Let $L = \{v \in \mathbb{R}^n \mid Uv = v\}$. It is easy to show that $L \cap \alpha(G)x_0 = \alpha(\mathcal{N}_G(U))x_0$. It remains to apply Lemma 3.2.7 to $\alpha(U)$, $Y = \alpha(M)x_0$ and $p = x_0$ and to notice that, since U is unipotent, there exists a rational section $s : G/U \to G$.

Remark. For $G = SL(2, \mathbb{R})$ the lemma had been known for a long time. In particular, this case of the lemma was used by A. Kirillov approximately in 1960 to prove that $SL(2, \mathbb{R})$ has Mautner property with respect to any nontrivial unipotent subgroup of $SL(2, \mathbb{R})$.

3.6.2. Combining Lemmas 3.5.3 and 3.6.1 we get the following:

Proposition. *Let G be a real algebraic group, U a connected unipotent subgroup of G, and Γ a discrete subgroup of G. Let X be a compact minimal U-invariant subset of G/Γ, and Y a closed U-invariant subset of G/Γ. Suppose that e belongs to the closure of the subset $\{g \in G - \mathcal{N}_G(U) \mid gX \cap Y \neq \emptyset\}$. Then there is a rational map $\psi : U \to \mathcal{N}_G(U)$ such that $\psi(e) = e, \psi(U) \not\subset KU$ for any compact subset K of $\mathcal{N}_G(U)$, and $hX \subset Y$ for any $h \in \psi(U)$. If further $X = Y$ then X is invariant under the closed subgroup generated by $\psi(U)$.*

3.6.3. Proposition. *Let G, U and Γ be the same as in Proposition 3.6. Let μ be a finite Borel U-ergodic U-invariant measure on G/Γ. Suppose that for any closed subset $X \subset G/\Gamma$ of μ-positive measure, the closure of the subset $\{g \in G - \mathcal{N}_G(U) \mid gX \cap X \neq \emptyset\}$ contains e. Then there is a rational map $\psi : U \to \mathcal{N}_G(U)$ such that $\psi(e) = e, \psi(U) \not\subset KU$ for any compact subset K of $\mathcal{N}_G(U)$, and the measure μ is invariant under $\psi(U)$.*

This proposition can be considered as the measure theoretical analog of Proposition 3.6.2 (for $X = Y$). It is a weak form of Proposition 8.2 in [MargT2], and it also immediately follows from Ratner's measure classification theorem (see Theorem 3.9.1 below). The proof given in [MargT2] follows the same strategy as in the proof of Proposition 3.6.2 but it is technically much more complicated and involves a multidimensional generalization of the Birkhoff individual ergodic theorem. This generalization is in fact the replacement of (trivial) Lemma 3.5.1 which we implicitly used in the proof of Proposition 3.6.2.

Remark. There are certain similarities between the proof in [MargT2] of the above proposition and the proofs of [Rat2, Theorem 3.1, Lemma 3.3, Lemma 4.2] and [Rat3, Lemma 3.1] but, as far as the author can see, the proofs are prompted by different lines of thought.

3.6.4. Let B_0 be an indefinite quadratic form in n variables of the type

$$B_0(x, y, z) = 2xz - Q(y) = 2xz - \langle Ay, y\rangle, x, z \in \mathbb{R}, y \in \mathbb{R}^{n-2},$$

where Q is a form in $n-2$ variables, A is the symmetric matrix corresponding to Q and $\langle\, ,\rangle$ is the standard inner product on $\mathbb{R}^{n-2}$. Let 1_i denote the unit $i \times i$ matrix and let, for $t \in \mathbb{R}$ and $c \in \mathbb{R}^{n-2}$.

$$d(t) = \begin{pmatrix} e^t & 0 & 0 \\ 0 & 1_{n-2} & 0 \\ 0 & 0 & e^{-t} \end{pmatrix},$$

$$v_1(c) = \begin{pmatrix} 1 & Ac & \langle Ac, c\rangle/2 \\ 0 & 1_{n-2} & c \\ 0 & 0 & 1 \end{pmatrix},$$

$$v_2(t) = \begin{pmatrix} 1 & 0 & t \\ 0 & 1_{n-2} & 0 \\ 0 & 0 & 1 \end{pmatrix}.$$

Let $H = SO(B_0)$,

$$D = \{d(t) \mid t \in \mathbb{R}\}, \quad V_1 = \{v_1(c) \mid c \in \mathbb{R}^{n-2}\},$$
$$V_2 = \{v_2(t) \mid t \in \mathbb{R}\},$$
$$V_2^+ = \{v_2(t) \mid t \geq 0\}, \quad V_2^- = \{v_2(t) \mid t \leq 0\},$$

We note that D normalizes V_1 and V_2, and the subgroups D and V_1 are contained in H.

Lemma. *Let M be a subset of $G - HV_2$ such that $e \in \bar{M}$. Then $\overline{HMV_1}$ contains either V_2^+ or V_2^-.*

Proof. Let Φ denote the linear space of real quadratic forms in n variables. The group $G = SL(n, \mathbb{R})$ naturally acts on Φ. The stabilizer of B_0 under this action is H. Hence we can identify $H\backslash G$ and GB_0. Under this identification, Hg corresponds to $g^{-1}B_0$ and, in particular, $Hv_2(t)$ corresponds to $B_0 + tB_1$ where $B_1(x, y, z) = 2z^2$. Therefore it is enough to prove that if $Y \subset GB$, $B \in \bar{Y}$ and $Y \cap (B_0 + \mathbb{R}B_1) = \emptyset$ then $\overline{V_2 Y}$ contains either $\{B_0 + tB_1 \mid t \geq 0\}$ or $\{B_0 + tB_1 \mid t \leq 0\}$. But $GB \cap (\mathbb{R}B_0 + \mathbb{R}B_1) \subset (\pm B_0 + \mathbb{R}B_1)$ (because forms from GB have the same discriminant as B). Thus in view of Lemma 3.2.7, it remains to show that

$$\mathbb{R}B_0 + \mathbb{R}B_1 = \{Q \in \Phi \mid V_1 Q = Q\}. \tag{1}$$

It is not difficult to prove (1) by direct calculations. These caluclations can be simplified if we notice that the right-hand side in (1) is invariant under D, and hence it is spanned by eigenvectors of D. We omit the details.

3.7. Proof of Theorem 2.3.4 (a). We will use the notation introduced in 3.6.4. Let $G = SL(n, \mathbb{R})$, $\Gamma = SL(n, \mathbb{Z})$ and $\Omega_n = G/\Gamma$. According to Remark 2.3.6 (a), in order to prove Theorem 2.3.4 (a), it is enough to prove (for $n = 3$) the following:

3.7.1. Proposition. *If $x \in \Omega_n$ then either Hx is closed and carries a H-invariant probability measure or there exist a $y \in \overline{Hx}$ such that $\overline{Hx} \supset V_2 y$.*

This proposition will be proved in 3.7.3. Before that we need to state Proposition 3.7.2.

3.7.2. Proposition. *Let $x \in \Omega_n$. Suppose that Hx is not closed. Then $\overline{Hx}$ contains a minimal compact V_1-invariant subset Z such that the closure of the set*

$$\{g \in G - HV_2 \mid gz \in \overline{Hx}\}$$

contains e for every $z \in Z$.

This proposition will be proved in 3.7.8 after we prove, in 3.7.4–3.7.7, some auxillary lemmas.

3.7.3. Proof of Proposition 3.7.1. Suppose that Hx is not closed. Let Z be a subset defined in Proposition 3.7.2. Then in view of Lemmas 3.5.3 and 3.6.4, $\overline{Hx}$ contains either $V_2^+ Z$ or $V_2^- Z$ (we apply Lemma 3.5.3 to $P = F = V_1, Q = H, X = Z$ and $Y = \overline{Hx}$). On the other hand, it easily follows from Theorem 3.3.1 that the closure of any semi-orbit $V_2^+ x$ or $V_2^- x, x \in \Omega_n$, contains a V_2-orbit. Hence $\overline{Hx} \supset V_2 y$ for some $y \in \Omega_n$. Now it remains to note that, according to Theorem 3.4.3, any closed H-orbit carries a H-invariant probability measure.

3.7.4. Lemma. (a) *There exists a compact subset C of Ω_n such that $DV_1 C = \Omega_n$ and moreover for every $x \in \Omega_n, C \cap V_1 d(t)x$ is nonempty for all large t.*

(b) *Let S be a subgroup of G which contains DV_1. Then every closed S-invariant subset of Ω_n contains a minimal closed S-invariant subset.*

Proof. It is easy to show that any proper nonzero V_1-invariant subspace L of $\mathbb{R}^n$ is contained in $\{(0, y, z) \mid y \in \mathbb{R}^{n-2}, z \in \mathbb{R}\}$ and contains $\{(0, 0, z) \mid z \in \mathbb{R}\}$. Hence L is D-invariant and $\det d(t) \mid_L \to \infty$ as $t \to +\infty$. Now the assertion (a) follows from Corollary 3.3.5, and (b) follows from (a) and Zorn's lemma.

3.7.5. Let

$$L = \left\{ \begin{pmatrix} 1 & 0 & 0 \\ 0 & C & 0 \\ 0 & 0 & 1 \end{pmatrix} \mid C \in SO(Q) \right\} \subset H = SO(B_0).$$

We note that L normalizes V_1 and V_2, and that DLV_1 is a closed normal subgroup of index 2 in $H \cap N_G(V_2)$. Let $F = DLV_1V_2$.

Lemma. *Let $g \in G$ and $\Lambda = F \cap g\Gamma g^{-1}$. Then at least one of the following holds:*

(i) *there exists a one-parameter subgroup $\{u(t)\}$ of V_1 such that $u(t)\Lambda$ tends to infinity in F/Λ as $t \to \infty$;*

(ii) *for every compact subset K of F/Λ, there exists $T = T(K)$ such that $K \cap V_1 d(t)\Lambda$ is empty for all $t > T$.*

Proof. Let us denote the Zariski closure of Λ by Φ. Then $g^{-1}\Phi g$ is the Zariski closure of the subgroup $g^{-1}\Lambda g$ of $\Gamma = SL(n, \mathbb{Z})$. Hence $g^{-1}\Phi g$ is is defined over $\mathbb{Q}$ and has no $\mathbb{Q}$-rational characters with infinite image. So, in view of Chevaley's theorem, there exist $m \in \mathbb{N}^+$, a faithful rational representation $\alpha : G \to GL(m, \mathbb{R})$ and $x_0 \in \mathbb{R}^m$ such that $\Phi = \{g \in G \mid \alpha(g)x_0 = x_0\}$. If $U = \{u(t)\}$ is a one-parameter unipotent subgroup of G then either $\alpha(U)x_0 = x_0$ or $\alpha(u(t))x_0$ tends to infinity in $\mathbb{R}^m$ as $t \to \infty$. This implies that if (i) does not hold then

$$\alpha(V_1)x_0 = x_0, \text{ and hence } V_1 \subset \Phi. \tag{2}$$

Thus we can assume that (2) is true. Then

$$\alpha(V_1 d(t)\Lambda)x_0 = \alpha(d(t)V_1\Lambda)x_0 = \alpha(d(t))x_0. \tag{3}$$

All eigenvalues of $\alpha(d(t))$ are real. Hence if $\alpha(D)x_0 \neq x_0$ then $\alpha(d(t))x_0$ tends either to infinity or to 0 as $t \to \infty$. This and (3) imply that if (ii) does not hold then $DV_1 \subset \Phi$. But any containing DV_1 closed subgroup of $F = DLV_1V_2$ is not unimodular (because D centralizes DL and $V_1V_2 \subset W = U_{d(-1)}$). Thus Φ and $g^{-1}\Phi g$ are not unimodular. This contradicts the fact that $g^{-1}\Phi g$ is defined over $\mathbb{Q}$ and has no $\mathbb{Q}$-rational characters with infinite image.

3.7.6. Lemma. *Let Y be a minimal closed DLV_1-invariant subset of Ω_n, and let $y \in Y$. Then the closure of $\{g \in G - F \mid gy \in DLV_1 y\}$ contains e.*

Proof. We write $y = g_0\Gamma$. Then the stabilizer $F_y = F \cap G_y$ coincides with $F \cap g_0 \Gamma g^{-1}$. In view of Theorem 3.3.1 and Lemma 3.7.4 (a), the following two assertions hold:

(i) for any one-parameter subgroup $\{u(t)\}$ of $V_1, u(t)y$ does not tend to infinity in Ω_n;
(ii) there exists a compact subset C of Ω_n such that $C \cap V_1 d(t)y$ is nonempty for all $t > 0$.

From these assertions and Lemma 3.7.5 we deduce the existence of a sequence $\{f_n\} \subset DV_1$ such that $\{f_n y\}$ is relatively compact in Ω_n but $\{f_n F_y\}$ is not relatively compact in F/F_y. Now the lemma follows from Lemma 3.5.2.

3.7.7. Lemma. *Let $h_1, h_2, h \in H$ and $t_1, t_2, t \in \mathbb{R}$. If $t_1 \neq 0$ and*

$$h_1 v_2(t_1) h_2 v_2(t_2) = h v_2(t), \tag{4}$$

then $h_2 \in N_G(V_2)$.

Proof. As in the proof of Lemma 3.6.4, let $B_1(x, y, z) = 2z^2$. A direct computation shows that $v_2(s)B_1 = B_0 - sB_1, s \in \mathbb{R}$. But H preserves B_0 and (4) is equivalent to the equality $v_2(t_1)h_2 = h_1^{-1}hv_2(t-t_2)$. Hence

$$(5) \qquad B_0 - t_1B_1 = v_2(t_1)h_2B_0 = h_1^{-1}hv_2(t-t_2)B_0$$
$$= h_1^{-1}h(B_0 - (t-t_2)B_1) = B_0 - (t-t_2)h_1^{-1}hB_1.$$

Thus $t_1B_1 = (t-t_2)h_1^{-1}hB_1$. On the other hand, it is easy to check that if $h_0 \in H$ and h_0B_1 is a multiple of B_1 then $h_0 \in N_G(V_2)$. Therefore $h_2^{-1}h \in N_G(V_2)$. It implies that

$$h_2 = v_2(-t_1)h_1^{-1}hv_2(t-t_2) \in V_2N_G(V_2)V_2 = N_G(V_2).$$

3.7.8. Proof of Proposition 3.7.2. Let us denote by X the set of $y \in \overline{Hx}$ such that the closure of the set $\{g \in G - H \mid gy \in \overline{Hx}\}$ contains e. Since Hx is not closed, $X \neq \emptyset$ (see Lemma 2.2.7 (a)). Then by Lemma 3.7.4 X contains a minimal closed DLV_1-invariant subset Y, and by Corollary 3.3.7 and Theorem 3.3.8 Y contains a minimal compact V_1-invariant subset Z. Take $z \in Z \subset X$ and assume that the closure of $\{g \in G - HV_2 \mid gz \in \overline{Hx}\}$ does not contain e. Then there exists a sequence $\{g_n\} \subset HV_2 - H$ such that $g_nz \in \overline{Hx}$ and $g_n \to e$ as $n \to \infty$. In view of Lemma 3.7.6, there exist $a_n \in DLV_1$ and $s_n \in G - F$ such that

$$s_nz = a_nz \quad \text{and} \quad s_n \to e \text{ as } n \to \infty.$$

Replacing $\{g_n\}$ and $\{s_n\}$ by subsequences if necessary, we can assume that $a_ng_na_n^{-1} \to e$ as $n \to \infty$ and that $s_n \in HV_2$. Since $g_n \in HV_2 - H$, $a_n \in DLV_1 \subset H$, and DLV_1 normalizes V_2, we have that $a_ng_na_n^{-1} \in HV_2 - H$. But $s_n \in HV_2 - N_G(V_2)$ for all large n (because DLV_1 is open in $H \cap N_G(V_2)$, and therefore $F = DLV_1V_2$ is open in $HV_2 \cap N_G(V_2)$). Hence by Lemma 3.7.7 $a_ng_na_n^{-1}s_n \notin HV_2$ for all large n. On the other hand $a_ng_na_n^{-1}s_nz = a_ng_nz \in \overline{Hx}$ and $a_ng_na_n^{-1}s_n \to e$ as $n \to \infty$. We have a contradiction with our previous assumption that the closure of $\{g \in G - HV_2 \mid gz \in \overline{Hz}\}$ does not contain e.

3.8. Conjectures of Dani and Raghunathan. Uniform distribution. In 1936 Hedlund [He] proved that if $G = SL(2,\mathbb{R})$, Γ is a lattice in G, and U is a unipotent one-parameter subgroup of G, then every U-orbit on G/Γ is either periodic or dense in G/Γ. In particular, if G/Γ is compact then all U-orbits are dense. In 1972 Furstenberg [Fu] proved that for the same G, Γ and U, the action of U on the compact space G/Γ is uniquely ergodic, i.e. the G-invariant probability measure is the only U-invariant probability measure on G/Γ.

Furstenberg's theorem was generalized by Bowen [Bow], Veech [Vee] and Ellis and Perrizo [EP]. In particular they showed that if G is a connected simple Lie group and Γ is a uniform lattice on G (that is, Γ is a lattice such that G/Γ is compact), then the action on G/Γ by left translations of any nontrivial horosperical subgroup U of G is uniquely ergodic. (By the *horospherical subgroup* corresponding to an element g of a Lie group F we mean the subgroup

$$U_g = \{u \in F \mid g^i u g^{-i} \to e \text{ as } j \to +\infty\}.$$

It is well known that U_g is a connected closed Ad-unipotent subgroup and g normalizes U_g. Any nontrivial connected unipotent subgroup of $SL(2,\mathbb{R})$ is horospherical.)

In a 1981 paper [Dan3] Dani formulated two conjectures. One is Raghunathan's conjecture which we mentioned in 2.3. The second conjecture is due to Dani himself and may be stated as follows (in a slightly stronger form than in [Dan3]). If G is a connected Lie group, Γ is a lattice in G, U is an Ad-unipotent subgroup of G, and μ is a finite Borel U-ergodic U-invariant measure on G/Γ, then there exists a closed subgroup F of G such that μ is F-invariant and $\text{supp}\mu = Fx$ for some $x \in G/\Gamma$ (a measure for which this condition holds is called *algebraic*). In the same paper [Dan3] Dani proved his conjecture when G is reductive and U is a maximal horospherical subgroup of G. In another paper [Dan7], he proved Raghunathan's conjecture in the case when G is reductive and U is an arbitrary horospherical subgroup of G. Starkov [Sta1, 2] proved Raghunathan's conjecture for solvable G. We remark that the proof given in [Dan7] is restricted to horospherical U and the proof given in [Sta1, 2] cannot be applied for nonsolvable G.

The first result on Raghunathan's conjecture for nonhorospherical subgroups of semisimple groups was obtained by Dani and the author in [DanM2]. We proved the conjecture in the case when $G = SL(3,\mathbb{R})$ and $U = \{u(t)\}$ is a one-parameter unipotent subgroup of G such that $u(t) - 1$ has rank 2 for all $t \neq 0$. Though this is only a very special case, the proof in [DanM2] together with the methods developed in [Marg5] and [DanM1] suggests an approach for proving the Raghunathan conjecture in general (cf. [Sha3]). This approach is based on the technique which involves (as in [Marg5] and [Dan1]) finding orbits of larger subgroups inside closed sets univariant under unipotent subgroups by studying the minimal invariant sets, and the limits of orbits of sequences of points tending to a minimal invariant set.

Let G be a Lie group, Γ a lattice in G, and $\{u_t \mid t \in \mathbb{R}\}$ a one-parameter subgroup of G. We say that the orbit $\{u_t x\}$ of $x \in G/\Gamma$ is *uniformly distributed with respect to a probability measure* μ *on* G/Γ if for any bounded continuous function f on G/Γ

$$\frac{1}{T}\int_0^T f(u_t x)dt \to \int_{G/\Gamma} f d\mu \text{ as } T \to \infty.$$

The above-mentioned result of Furstenberg implies that if $G = SL(2,\mathbb{R})$, Γ is a uniform lattice in G and $\{u_t\}$ is a one-parameter unipotent subgroup of G then every orbit of $\{u_t\}$ on G/Γ is uniformly distributed with respect to the G-invariant probability measure. For nonuniform lattices Γ in $G = SL(2,\mathbb{R})$ Dani and Smillie [DanS] proved that every nonperiodic orbit of $\{u_t\}$ on G/Γ is uniformly distributed with respect to the G-invariant probability measure. For $\Gamma = SL(2,\mathbb{Z})$ this result had been proved earlier in [Dan4]. In the paper [Dan4], Dani also stated another conjecture that, for any one-parameter unipotent subgroup $\{u_t\}$ of $SL(n,\mathbb{R})$ and any point $x \in G/\Gamma$, the orbit $\{u_t x\}$ is uniformly distributed with respect to a probability measure μ_x on $SL(n,\mathbb{R})/SL(n,\mathbb{Z})$. This conjecture was reproduced in [Marg4] where it was stated for an arbitrary lattice in connected Lie group. For nilpotent Lie groups it was proved by Lesigne [Les1, 2].

3.9. Ratner's results on unipotent flows. Conjectures of Dani and Raghunathan were eventually proved in full generality by Ratner. In fact, she proved the results for a larger class of actions. Let us first state Ratner's measure classification theorem for actions of connected Lie subgroups H. This theorem proved to be an important and useful tool for the study of (unipotent) flows on homogeneous spaces.

3.9.1. Theorem [Rat2–4]. *Let G be a connected Lie group and Γ a discrete subgroup of G (not necessarily a lattice). Let H be a Lie subgroup of G that is generated by the Ad-unipotent one-parameter subgroups contained in it. Then any finite H-ergodic H-invariant measure μ on G/Γ is algebraic.*

3.9.2. Remarks. (a) It immediately follows from the result of Moore on Mautner property (mentioned in 3.4.2) that μ is U-ergodic for any maximal unipotent subgroup U of H. Therefore it is enough to prove Theorem 3.9.1 for unipotent H.

(b) Some ingredients of Ratner's original proof of Theorem 3.9.1 are described for $SL(2,\mathbb{R})$ in [Rat6], and in the general case it is sketched in [Rat8]. We refer the reader also to a shorter proof given by Tomanov and the author (see [MargT1, 2] for the crucial case where G is algebraic and [MargT3] for a simple reduction to that case). The proof in [MargT2] is mostly based on some versions of Proposition 3.6.3. This proof bears a strong influence of Ratner's arguments but is substantially different in approach and methods.

3.9.3. Using Theorem 3.9.1 together with Theorem 3.3.2 and a simple result about the countability of a certain (depending on Γ) set of subgroups of G, Ratner proves in [Rat5] the following uniform distribution theorem which can be considered as the quantitative strengthening of Raghunathan's conjecture for one-parameter unipotent subgroups. (Ratner also proves in [Rat5] an analogous statement for cyclic unipotent subgroups.)

Theorem. *If G is a connected Lie group, Γ is a lattice in G, $\{u_t\}$ is a one-parameter Ad-unipotent subgroup of G and $x \in G/\Gamma$, then the orbit $\{u_t x\}$ is uniformly distributed with respect to an algebraic probability measure μ_x on G/Γ.*

This theorem proves the above-mentioned conjecture from [Dan4] and [Marg4]. An alternative approach to the proof of the uniform distribution theorem for the case of a "regular unipotent subgroup", also using Theorems 3.9.1 and 3.3.2, is given in [Sha1]. In another paper [Sha2] Shah proved an analogue of the uniform distribution theorem for multidimensional unipotent subgroups.

Let us recall that a refined version of Ratner's uniform distribution theorem is stated in 2.4.9. This version is due to Dani and the author. The proof which we give in [DanM5] uses, as in [Rat5] and [Sha2], Theorems 3.9.1 and 3.3.2. The reduction to these theorems is essentially based on a variation of the following assertion: Let H be a connected closed subgroup of G such that $H \cap \Gamma$ is a lattice in G, and let $X(H, U) = \{g \in G \mid Ug \subset gH\}$. Then the subset $X(H, U)\Gamma$ of G/Γ is avoidable with respect to $\{u_t\}$ (in the sense of 3.2.2). In the proof of (the variation of) this assertion we use Proposition 3.2.2.

3.9.4. The uniform distribution theorem, in combination with the countability result mentioned in 3.9.3, rather easily implies the following theorem. This theorem proves a generalized version of Raghunathan's conjecture mentioned in 2.3.

Theorem [Rat5]. *Let G and H be as in Theorem 3.9.1. Let Γ be a lattice in H. Then for any $x \in G/\Gamma$, there exists a closed connected subgroup $L = L(x)$ containing H such that $\overline{Hx} = Lx$ and there is an L-invariant probability measure supported on Lx.*

3.10. Remarks. (a) Let G be a connected semisimple Li group without compact factors, Γ an irreducible lattice in G, and g a semisimple element of G such that the horospherical subgroup $U = U_g$ corresponding to g is nontrivial. We are given $f \in L^2(U)$ with compact support and a uniformly continuous $\psi \in L^2(G/\Gamma)$. Then according to [KM, Proposition 2.2.1], for any compact subset L of G/Γ and any $\varepsilon > 0$ there exists $T > 0$ such that

$$(*) \qquad \left| \int_U f(u)\psi(g^{-n}ux)dm(u) - \int_U f dm \int_{G/\Gamma} \psi d\mu \right| \leq \varepsilon$$

for all $x \in L$ and $t \geq T$, where m is Haar measure on U and μ is the G-invariant probability measure on G/Γ. Using this assertion in combination with Theorem 3.3.6, it is not very difficult to prove for horospherical subgroups of semisimple Lie groups the conjectures of Dani and Raghunathan and a suitable version of the uniform distribution theorem. This is one of the reasons why the horospherical subgroup case is much simpler than the general case. The proof of (*) uses mixing properties of the action of g on G/Γ and the fact that any neighborhood V of e in G contains another neighborhood W such that $g^{-n}WUg^n \subset VU$ for all positive n (so called "banana argument"). Let us note that similar methods were used in [Bow], [EMc], [EP], [Marc], [Marg1] and [Rat2] (chronologically the first reference in this list is [Marg1] where an analog of (*) for Anosov flows was proved using a version of the above argument).

(b) Recently Raghunathan informed the author that his interest in orbit closures was triggered by some early papers of J. S. Dani and S. G. Dani (in particular [DanJ] and [Dan1]). After S. G. Dani proved some of his results about actions of horospherical subgroups, Raghunathan began wondering about arbitrary unipotent subgroups. Finally, Dani's theorems on horospherical flows, the result from [Marg2] that unipotent one-parameter subgroups cannot have non-compact orbits in the space of lattices (Theorem 3.3.1 of the present paper) and some observations about the Oppenheim conjecture led Raghunathan to his conjecture. His hope was that one could use a downward induction on dimension starting with a horosphere (in some sense, this is the strategy in [DanM2]). He also hoped that unipotent flows are likely to have "manageable behavior" because of the slow divergence of orbits of unipotent one-parameter subgroups (in contrast to the exponential divergence of orbits of diagonalizable subgroups).

(c) Analogs of most of the above results from 2.3, 2.4.1 and 3.2–3.9 hold in p-adic and S-arithmetic cases (see [Bo2], [BoP], [MargT1–3], [Rat7–10]).

(d) We refer to surveys [Bo2], [Dan9], [Dan10], [Marg8] and [Rat10] for further information about the topics which we considered in Secs. 2 and 3.

References

[Bas] R. C. Baker and H. P. Schlickewey, *Indefinite quadratic forms*, Proc. London Math. Soc. **54** (1987), 383–411.

[BiD1] B. J. Birch and H. Davenport, *Quadratic equations in several variables*, Proc. Cambridge Philos. Soc. **54** (1958), 135–138.

[BiD2] ———, *On a theorem of Davenport and Heilbronn*, Acta Math. **100** (1958), 259–279.

[BiD3] ———, *Indefinite quadratic forms in many variables*, Mathematika **5** (1958), 8–12.

[Bo1] A. Borel, *Density properties of certain subgroups of semisimple groups*, Ann. Math. **72** (1960), 179–188.

[Bo2] ———, *Values of indefinite quadratic forms at integral points and flows on spaces of lattices*, Bull. Amer. Math. Soc. **32** (1995), 184 204.

[BoP] A. Borel and G. Prasad, *Values of isotropic quadratic forms at S-integral points*, Compositio Mathematica **83** (1992), 347–372.

[Bow] R. Bowen, *Weak mixing and unique ergodicity on homogeneous spaces*, Israel J. Math. **23** (1976), 267–273.

[Br] T. Brennan, *Distribution of values of diagonal quadratic forms at integer points*, Princeton University undergraduate thesis, 1994.

[Ca1] J. W. S. Cassels, *Bounds for the least solutions of homogeneous quadratic equations*, Proc. Cambridge Philos. Soc. **51** (1955), 262–264; Addendum, ibid. **52** (1956), 602.

[Ca2] ———, *An Introduction to Diophantine Approximation*, Cambridge Univ. Press, Cambridge, 1957.

[Ca3] ———, *Rational Quadratic Forms*, Academic Press, London, New York, 1978.

[CaS] J. W. S. Cassels and H. P. F. Swinnerton-Dyer, *On the product of three homogeneous forms and indefinite ternary quadratic forms*, Philos. Trans. Roy. Soc. London **248, Ser. A** (1955), 73–96.

[Ch] S. Chowla, *A theorem on irrational indefinite quadratic forms*, J. London Math. Soc. **9** (1934), 162–163.

[CoR1] R. J. Cook and S. Raghavan, *Indefinite quadratic polynomials of small signature*, Monatsh. Math. **97** (1984), 169–176.

[CoR2] ———, *On positive definite quadratic polynomials*, Acta Arith. **45** (1986), 319–328.

[CuF] T. Cusick and M. Flahive, *The Markov and Lagrange spectra*, Math. Surveys, vol. **30** AMS, Providence 1989.

[DanJ] J. S. Dani, *Density properties of orbits under discrete groups*, J. Ind. Math. Soc. **39** (1975), 189–218.

[Dan1] S. G. Dani, *Bernoullian translations and minimal horospheres on homogeneous spaces*, J. Ind. Math. Soc. **40** (1976), 245–284.

[Dan2] ———, *On invariant measures, minimal sets and a lemma of Margulis*, Invent. Math. **51** (1979), 239–260.

[Dan3] ———, *Invariant measures and minimal sets of horospherical flows*, Invent. Math. **64** (1981), 357–385.

[Dan4] ———, *On uniformly distributed orbits of certain horocycle flows*, Ergod. Theor. Dynam. Syst. **6** (1982), 139–158.

[Dan5] ———, *On orbits of unipotent flows on homogeneous spaces*, Ergod. Theor. Dynam. Syst. **4** (1984), 25–34.

[Dan6] ———, *On orbits of unipotent flows on homogeneous spaces II*, Ergod. Theor. Synam. Syst. **6** (1986), 167–182.

[Dan7] ———, *Orbits of horospherical flows*, Duke Math. J. **53** (1986), 177–188.

[Dan8] ———, *A proof of Margulis' theorem on values of quadratic forms, independent of the axiom of choice*, L'enseig. Math. **40** (1994), 49–58.

[Dan9] ———, *Flows on homogeneous spaces and Diophantine approximation*, Proc. ICM 94, 780–789.

[Dan10] ———, *Flows on homogeneous spaces: A review*, Proc. of the Warwick Symposium "Ergodic Theory of $\mathbb{Z}^d$-action", London Math. Soc., Lect. Notes Series 228.

[DanM1] S. G. Dani and G. A. Margulis, *Values of quadratic forms at primitive integral points*, Invent. Math. **98** (1989), 405–424.

[DanM2] ———, *Orbit closures of generic unipotent flows on homogeneous spaces of* $SL(3,\mathbb{R})$, Math. Ann. **286** (1990), 101–128.

[DanM3] ———, *Values of quadratic forms at integral points: An elementary approach*, L'enseig. Math. (2) **36** (1990), 143–174.

[DanM4] ———, *Asymptotic behaviour of trajectories of unipotent flows on homogeneous spaces*, Proc. Indian Acad. Sci. **101** (1991), 1–17.

[DanM5] ———, *Limit distributions of orbits of unipotent flows and values of quadratic forms*, Advances in Soviet Math. **16** (1993), 91–137.

[DanS] S. G. Dani and J. Smillie, *Uniform distribution of horocycle flows for Fuchsian groups*, Duke Math. J. **51** (1984), 185–194.

[Dav1] H. Davenport, *Indefinite quadratic forms in many variables*, Mathematika **3** (1956), 81–101.

[Dav2] ———, *Note on a theorem of Cassels*, Proc. Cambridge Philos. Soc. **53** (1957), 539–540.

[Dav3] ———, *Indefinite quadratic forms in many variables* (II), Proc. London Math. Soc. (3) **8** (1958), 109–126.

[DavH] H. Davenport and H. Heilbronn, *On indefinite quadratic forms in five variables*, J. London Math. Soc. **21** (1946), 185–193.

[DavL] H. Davenport and D. J. Lewis, *Gaps between values of positive definite quadratic forms*, Acta Arith. **22** (1972), 87–105.

[DavR] H. Davenport and D. Ridout, *Indefinite quadratic forms*, Proc. London Math. Soc. (3) **9** (1959), 544–555.

[EMc] A. Eskin and C. McMullen, *Mixing, counting and equidistribution in Lie groups*, Duke Math. J. **71** (1993), 181–209.

[EMM1] A. Eskin, G. Margulis and S. Mozes, *On a quantitative version of the Oppenheim conjecture*, ERA, Amer. Math. Soc. **01** (1995), 124–130.

[EMM2] ———, *Upper bounds and asymptotics in a quantitative version of the Oppenheim conjecture*, Preprint.

[EMS1] A. Eskin, S. Mozes and N. Shah, *Unipotent flows and counting lattice points on homogeneous varieties*, Ann. Math. **143** (1996), 253–299.

[EMS2] ———, *Non-divergence of translates of certain algebraic measures*, Geom. Functional Anal., to appear.

[EP] R. Ellis and W. Perrizo, *Unique ergodicity of flows on homogeneous spaces*, Israel J. Math. **29** (1978), 176–284.

[Fu] H. Furstenberg, *The unique ergodicity of the horocycle flow*, Recent Advances in Topological Dynamics, Lectures Notes in Math., 95–115; vol. 318, Springer, New York, 1972.

[Gh] E. Ghys, *Dynamique des flots unipotents sur les espaces homogènes*, Astérisque 206 (Sém. Bourbaki 1992–93, no. 747) (1992), 93–136.

[He] G. Hedlund, *Fuchsian groups and transitive horocycles*, Duke Math. J. **2** (1936), 530–542.

[Iw] H. Iwaniec, *On indefinite quadratic forms in four variables*, Acta Arith. **33** (1977), 209–229.

[JW] V. Jarnik and A. Walfisz, *Über Gitterpunkte in mehrdimensionalen Ellipsoiden*, Math. Zeitschrift **32** (1930), 152–160.

[KM] D. Kleinbock and G. Margulis, *Bounded orbits of nonquasiunipotent flows on homogeneous spaces*, Amer. Math. Soc. Transl. (2) **171** (1996), 141–172.

[Les1] E. Lesigne, *Théorèmes ergodiques pour une translation sur nilvariété*, Ergod. Theor. Dynam. Syst. **9** (1989), 115–126.

[Les2] ———, *Sur une nilvariété, les parties minimales associées à une translation son uniquement ergodiques*, Ergod. Theor. Dynam. Syst. **11** (1991), 379–391.

[Lew] D. J. Lewis, *The distribution of values of real quadratic forms at integer points*, Proc. Sympos. Pure Math., vol. XXIV, Amer. Math. Soc. Providence, RI, 1973, 159–174.

[Marc] B. Marcus, *Unique ergodicity of the horocycle flow: variable negative curvature case*, Israel J. Math. **21** (1975), 133–144.

[Marg1] G. A. Margulis, *On some problems in the theory of U-systems* (*in Russian*), Thesis, Moscow University, 1970.

[Marg2] ———, *On the action of unipotent groups in the space of lattices*, Proc. of the summer school on group representations (Bolyai Janos Math. Soc., Budapest, 1971), 365–370; Akadémiai Kiado, Budapest 1975.

[Marg3] ———, *Formes quadratiques indéfinies et flots unipotents sur les spaces homogénes*, C. R. Acad. Sci. Paris Ser 1 **304** (1987), 247–253.

[Marg4] ———, *Lie groups and ergodic theory*, Algebra Some Current Trends (L. L. Avramov, ed.)(Proc. Varna 1986, Lecture Notes in Math., vol. 1352, Springer, New York, 130–146.

[Marg5] ———, *Discrete subgroups and ergodic theory*, Proc. of the conference "Number theory, trace formulas and discrete groups" in honour of A. Selberg (Oslo, 1987), 377–398; Academic Press, Boston, MA 1989.

[Marg6] ———, *Indefinite quadratic forms and unipotent flows on homogeneous spaces*, Dynamical Systems and Ergodic Theory, vol. 23, Banach Center Publ., PWN-Polish Scientific Publ., Warsaw, 1989, 399–409.

[Marg7] ———, *Orbits of group actions and values of quadratic forms at integral points*, Festschrift in honour of I. I. Piatetski-Shapiro (Isr. Math. Conf. Proc., vol. 3, pp. 127–151) Jerusalem: The Weitzmann Science Press of Israel, 1990.

[Marg8] ———, *Dynamical and ergodic properties of subgroup actions on homogeneous spaces with applications to number theory*, Proc. of ICM (Kyoto, 1990), Math. Soc. of Japan and Springer, 1991, 193–215.

[Marg9] ———, *Compactness of minimal closed invariant sets of actions of unipotent groups*, Geometriae Dedicata **37** (1991), 1–7.

[MargT1] G. A. Margulis and G. M. Tomanov, *Measure rigidity for algebraic groups over local fields*, C. R. Acad. Sci., Paris 315 (1992), 1221–1226.

[MargT2] ———, *Invariant measures for actions of unipotent groups over local fields on homogeneous spaces*, Invent. Math. **116** (1994), 347–392.

[MargT3] ———, *Measure rigidity for almost linear groups and its applications*, J. d'Analyse Math. **69** (1996), 25–54.

[Mark1] A. A. Markov, *On binary quadratic forms of positive determinant*, SPb, 1880; Usp. Mat. Nauk 3 (5) (in Russian) (1948), 7–51.

[Mark2] ———, *Sur les formes quadratiques ternaires indéfinies*, Math. Ann. **56** (1903), 233–251.

[Mo] C. C. Moore, *The Mautner phenomenon for general unitary representations*, Pacific J. Math. **86** (1980), 155–169.

[Op1] A. Oppenheim, *The minima of indefinite quaternary quadratic forms*, Proc. Nat. Acad. Sci. USA **15** (1929), 724–727.

[Op2] ———, *The minima of indefinite quaternary quadratic forms*, Ann. Math. **32** (1931), 271–298.

[Op3] ———, *The minima of quaternary quadratic forms of signature zero*, Proc. London Math. Soc. **37** (1934), 63–81.

[Op4] ———, *Values of quadratic forms I*, Quart. J. Math. Oxford Ser. (2) **4** (1953), 54–59.

[Op5] ———, *Values of quadratic forms II*, Quart. J. Math. Oxford Ser. (2) **4** (1953), 60–66.

[Op6] ———, *Values of quadratic forms III*, Monatsh. Math. **57** (1953), 97–101.

[Po] A. Pollington, *Number theory with an emphasis on Markov spectrum*, Marcel Dekker, New York, 1993.

[Rat1] M. Ratner, *Horocycle flows, joinings and rigidity of products*, Ann. Math. **118** (1983), 277–313.

[Rat2] ———, *Strict measure rigidity for unipotent subgroups of solvable groups*, Invent. Math. **101** (1990), 449–482.

[Rat3] ———, *On measure rigidity of unipotent subgroups of semisimple groups*, Acta Math. **165** (1990), 229–309.

[Rat4] ———, *On Raghunathan's measure conjecture*, Ann. of Math. **134** (1991), 545–607.

[Rat5] ———, *Raghunathan's topological conjecture and distribution of unipotent flows*, Duke Math. J. **63** (1991), 235–280.

[Rat6] ———, *Raghunathan's conjecture for* $SL_2(\mathbb{R})$, Israel J. Math. **80** (1992), 1–31.

[Rat7] ———, *Raghunathan's conjectures for p-adic Lie groups*, Int. Math. Res. Notices, no. 5 (1993), 141–146.

[Rat8] ———, *Invariant measures and orbit closures for unipotent actions on homogeneous spaces*, Geom. Functional Anal. **4** (1994), 236–256.

[Rat9] ———, *Raghunathan's conjectures for cartesian products of real and p-adic groups*, Duke Math. J. **77** (1995), 275–382.

[Rat10] ———, *Interactions between ergodic theory, Lie groups and number theory*, Proc. of ICM (Zürich, 1994), Birkhäuser, 1995 (to appear) **94**, 157–182.

[Ri] D. Ridout, *Indefinite quadratic forms*, Mathematika **5** (1968), 122–124.

[Sar] P. Sarnak, *Values at integers of binary quadratic forms*, In preparation.

[Sha1] N. A. Shah, *Uniformly distributed orbits of certain flows on homogeneous spaces*, Math. Ann. **289** (1991), 315–334.

[Sha2] ———, *Limit distributions of polynomial trajectories on homogeneous spaces*, Duke Math. J. **75** (1994), 711–732.

[Sha3] ———, *Ph. D. Thesis*, Tata Institute for Fundamental Research.

[Si] J. -C. Sikorav, *Valeurs des formes quadratiques indéfinies irrationneles*, in Séminaire de Theorie des Nombres, Paris 1987–88, pp. 307–315, Progress in Math., **81**, Birkhauser 1990.

[Sta1] A. N. Starkov, *Solvable homogeneous flows*, Math. Sbornik (in Russian) **176** (1987), 242–259.

[Sta2] ———, *The ergodic decomposition of flows on homogeneous spaces of finite volume*, Math. Sbornik (in Russian) **180** (1989), 1614–1633.

[Vee] W. A. Veech, *Unique ergodicity of horospherical flows*, Amer. J. Mth. **99** (1977), 827–859.

[Ven] B. A. Venkov, *On Markov's extremal problem for ternary quadratic forms*, Izv. Akad. Nauk SSSR Ser. Mat. 9 (in Russian) (1945), 429–434.

[Vu1] L. Ya. Vulakh, *On minima of rational indefinite Hermitian forms*, Ann. N. Y. Acad. Sci. **70** (1983), 99–106.

[Vu2] ———, *On minima of rational indefinite quadratic forms*, J. Number Theory **21** (1985), 275–285.

[Wa1] G. L. Watson, *On indefinite quadratic forms in three and four variables*, J. London Math. Soc. **28** (1953), 239–242.

[Wa2] ———, *On indefinite quadratic forms in five variables*, Proc. London Math. Soc. (30 **3** (1953), 170–181.

[Wa3] ———, *Least solutions of homogeneous quadratic equations*, Proc. Cambridge Philos. Soc. **53** (1957), 541–543.

[Wi] D. Witte, *Rigidity of some translations on homogeneous spaces*, Invent. Math. **85** (1985), 1–27.

Reprinted from Proc. Int. Congr. Math., 1983

THE WORK OF ALAIN CONNES

by
HUZIHIRO ARAKI

The theory of operator algebras, after being quietly nourished in somewhat isolation for 30 years or so, started a revolutionary development around the late 1960's. Alain Connes came into this field just when the smokes of the first stage of the revolution were settling down. He immediately led the field to breathtaking achievements beyond the expectation of experts.

His most remarkable contributions are: (1) general classification and a structure theorem for factors of type III, obtained in his thesis [12], (2) classification of automorphisms of the hyperfinite factor [29], which served as a preparation for the next contribution, (3) classification of injective factors [31], and (4) applications of the theory of C^*-algebras to foliations and differential geometry in general [44, 50] — a subject currently attracting a lot of attention.

In this report, I shall mostly concentrate on the first three aspects which form a well-established and most spectacular part of the theory of von Neumann algebras.

1. Classification of Type III Factors

In the latter half of 1930's, Murray and von Neumann initiated the study of what are now called von Neumann algebras (i.e. weakly closed *-sub-algebra of the *-algebra $L(H)$ of all bounded linear operators on a Hilbert space H) and classified the factors (i.e. von Neumann algebras with trivial centers) into the types I_n, $n = 1, 2, \ldots$, and I_∞ (isomorphic to $L(H)$ with $\dim H = n$ and ∞), II_1, II_∞ and III. (In the following we restrict our attention to von Neumann algebras M on separable Hilbert space H.) Only three type III factors (and only three type II_1 factors) had been known to be mutually non-isomorphic till 1967, when Powers showed the existence of a continuous family of mutually non-isomorphic type III factors.

Traces provided a tool for a systematic analysis of type II factors at an earlier stage, while the non-existence of traces made type III factors remain untractable till the late 1960's, when the Tomita–Takesaki theory was created and furnished a powerful tool for type III. To introduce notation, let M be a von Neumann algebra on a separable Hilbert space H and let $\Psi \in H$ be cyclic (i.e. $M\Psi$ be dense in H) and separating (i.e. such that $x\Psi = 0$ for $x \in M$ implies $x = 0$). The conjugate linear operator $S_\Psi x\Psi \equiv x^*\Psi$, $x \in M$, has a closure $\bar{S}_\Psi$ and defines the positive selfadjoint operator $\Delta_\Psi \equiv S_\Psi^* \bar{S}_\Psi$, called the *modular operator*. The Tomita–Takesaki theory says that $x \in M$ implies $\sigma_t^\varphi(x) \equiv \Delta_\Psi^{it} x \Delta_\Psi^{-it} \in M$. The one-parameter group of (*-)automorphisms σ_t^ψ of M depends only on the positive linear functional $\psi(x) \equiv (x\Psi, \Psi)$ and is called the *group of modular automorphisms.*

Connes [12] has shown that the modular automorphisms for different ψ's are mutually related by inner automorphisms (in other words, the independence of σ_t^Ψ from ψ in the quotient Out M of the group Aut M of all automorphisms modulo the subgroup Int M of all inner automorphisms) and introduced the following two isomorphism invariants for M:

$$S(M) = \bigcap_\Psi \mathrm{Sp}\Delta_\Psi \qquad \text{(Sp denotes the spectrum)},$$

$$T(M) = \{t \in \boldsymbol{R} : \sigma_t^\psi \in \mathrm{Int}\boldsymbol{M}\}.$$

(In a more general case, $S(M) = \cap\ \mathrm{Sp}\Delta_\psi$, where the intersection is taken over faithful normal semifinite weights ψ.) It turns out that $S(M)\backslash\{0\}$ is a closed multiplicative subgroup of $\boldsymbol{R}_+^*$, and this leads to the classification of type III factors into the types III$_\lambda$, $0 \leq \lambda \leq 1$:

$$S(M) = \{\lambda^n : n \in \boldsymbol{Z}\} \cup \{0\} \qquad \text{if } 0 < \lambda < 1,$$

$$S(M) = \boldsymbol{R}_+^* \qquad \text{if } \lambda = 1, \qquad S(M) = \{0, 1\} \qquad \text{if } \lambda = 0.$$

The Powers examples $R_\lambda, 0 < \lambda < 1$, due to Powers, are of types III$_\lambda$ and hence mutually non-isomorphic. The two invariants $r_\infty(M)$ and $\varrho(M)$ of Araki and Woods, introduced for a systematic classification of infinite tensor products of type I factors (including R_λ), are shown to be equivalent to $S(M)$ and $T(M)$ for them [7].

2. Structure Analysis of Type III Factors

Connes [12] went on and succeeded in analysis of the structures of type III$_\lambda$ factors M, $0 \leq \lambda < 1$, in terms of a type II von Neumann algebra N (with a non-trivial center) and an automorphism θ of N, such that M is the so-called *crossed product* $N \times_\theta \boldsymbol{Z}$ of N by θ.

For $0 < \lambda < 1$, θ should scale a trace τ of a type II$_\infty$ factor N in the sense that $\tau(\theta x) = \lambda\tau(x)$ for all $x \in N_+$ and $N_i \times_{\theta_i} \boldsymbol{Z}$, $i = 1, 2$, are isomorphic if and only if there exists an isomorphism π of N_1 onto N_2 such that $\pi^{-1}\theta_2\pi\theta_1^{-1}$ is inner, or equivalently (in view of a later result of Connes and Takesaki), $\pi\theta_1\pi^{-1} = \theta_2$. This means that the pair (N, θ) is uniquely determined by M and the classification of M is reduced to that of the pair (N, θ).

For $\lambda = 0$, θ should scale down a trace τ of a type II$_\infty$ von Neumann algebra N in the sense that $\tau(\theta x) \leq \varrho\tau(x)$ for all $x \in N_+$ for some $\varrho < 1$ and, again, there is a somewhat more complicated uniqueness result for the pair (N, θ).

Motivated by the above results of Connes, a general structure theorem including the type III$_1$ has been obtained by Takesaki in terms of a one-parameter group θ_t of trace-scaling automorphisms of a type II$_\infty$ von Neumann algebra N.

In the process of developing the above classification and structure theory, Connes introduced two important technical tools, namely the unitary Radon–Nikodym cocycle (equivalently, relative modular operators) useful in application to quantum statistical mechanics, non-commutative L_p theory, etc., and the Connes spectrum useful in the analysis of C^* dynamical systems.

3. Classification of Automorphisms of the Hyperfinite Factor

A von Neumann algebra, containing an ascending sequence of finite-dimensional subalgebras with a dense union, is called *approximately finite-dimensional* (AFD). AFD factors of type II_1, as shown by Murray and von Neumann, are all isomorphic to what is called the *hyperfinite factor*, denoted by R in the following. Connes [29] has given a complete classification of automorphisms of R modulo inner automorphisms (i.e. the conjugacy class of Out R). Namely, a complete set of isomorphism invariants in Out R for an $\alpha \in \mathrm{Aut}\, R$ is given by the pair of the outer period p $(= 2, 3, \ldots)$, which is the smallest $p > 0$ such that α^p is inner, defined to be 0 for outer aperiodic α, and the obstruction γ which is the p-th root of 1 (1 for $p = 0$) such that $\alpha^p = \mathrm{Ad}U$, $\alpha(U) = \gamma U$, where $(\mathrm{Ad}U)(x) = UxU^*$. Although the result is about a specific R, this factor R is in the bottom of all known AFD factors and the result that outer aperiodic automorphisms of R are all conjugate up to inner automorphisms is essential for the results described in the next section.

As a by-product, Connes [23] solved negatively one of the old problems on von Neumann algebras by exhibiting, for each $0 < \lambda < 1$, factors of type III_λ, not anti-isomorphic to themselves.

4. Classification of AFD Factors

A complete classification of AFD factors of type III_λ, $\lambda \neq 1$, is what I consider the most distinguished work of Connes. It turns out that an AFD factor of type III_λ is unique and is isomorphic to R_λ for each $0 < \lambda < 1$, while AFD factors of type III_0 are isomorphic to Krieger's factors associated with single non-singular ergodic transformations of the Lebesgue measure space, their isomorphism classes being in one-to-one correspondence with the metric equivalence classes of non-singular non-transitive ergodic flows on the Lebesgue measure space.

One of the most important technical ingredients of the proof is the equivalence of various concepts about a Neumann algebra which arose over years in the theory of von Neumann algebras. Murray and von Neumann found a factor N of type II_1 non-isomorphic to R (distinguished by Property Γ). In 1962, Schwartz distinguished N, R and $N \otimes R$ by the following property P: A von Neumann algebra M on a Hilbert space H has the property P iff for any $T \in L(H)$, the norm closed convex hull of the uTu^* with u varying over all unitary operators in M intersects M'. Any AFD factors posseses Property P.

The Property P for M implies the existence of a projection of norm 1 from $L(H)$ to M'. This is the Hakeda–Tomiyama extension property for M', called Property

E by Connes, and is stable under taking the intersection of a decreasing family, the weak closure of the union of an ascending family, the commutant, tensor products and crossed product by an amenable group. Thus the Property P implies Property E for M.

Any projection E of norm 1 from a C^*-algebra $\mathfrak{A}_1$ to its subalgebra $\mathfrak{A}_2$ is shown by Tomiyama to be a completely positive map satisfying the property of the conditional expectation: $E(axb) = aE(x)b$ for any $a, b \in \mathfrak{A}_2$ and $x \in \mathfrak{A}_1$. A C^*-algebra $\mathfrak{A}$ with unit is called *injective* if any completely positive unit preserving linear map θ from $\mathfrak{A}$ into another C^*-algebra $\mathfrak{B}$ with unit has an extension $\bar{\theta}$ to any C^*-algebra $\mathfrak{A}_1$ containing $\mathfrak{A}$ as a completely positive unit preserving linear map from $\mathfrak{A}_1$ into $\mathfrak{B}$. A von Neumann algebra M is injective if and only if it has Property E.

Effros and Lance called a von Neumann algebra M *semidiscrete* if the identity map from M into M is a weak pointwise limit of completely positive maps of finite rank and proved that M is injective if it is semidiscrete.

Connes [31] unified all these concepts by showing that they are all equivalent. The core result is the isomorphism of all injective factors of type II_1 to the unique hyperfinite factor R; it is established by a highly involved and technical proof, utilizing a theorem on tensor products of C^*-algebras, the property Γ of a factor which Murray and von Neumann introduced to distinguish some factors, properties of Aut N and Int N of a factor N, the ultra product R^ω for a free ultrafilter ω, an argument analogous to Day–Namioka proof of Fϕlner's characterization of amenable groups, etc.

The uniqueness of injective factors of type II_1 then implies the uniqueness of injective factors of type II_∞. Together with an earlier uniqueness result for trace-scaling automorphisms of $R \otimes B(H)$ (exhibiting the unique injective factor of type II_∞), it also implies the uniquencess of injective factors of type III_λ, $0 < \lambda < 1$. With the help of an earlier result of Krieger, injective factor of type III_0 are also completely classified by the isomorphism class of the so-called flow of weight. Thus Connes succeeded in a complete classification of AFD factors (which is as much as saying injective factors) except for the case of type III_1, which still remains open.

The work of Connes also shows that any continuous representation of a separable locally compact group G generates an injective von Neumann algebra if G/G_0 is amenable, where G_0 is the connected component of the identity (in particular, if G is connected or amenable).

5. Other Works

After his success in the almost complete classification of injective factors, Connes turned his attention to application of operator algebras to differential geometry. Connes developed a non-commutative integration theory, which provides a method of integration over a family of ergodic orbits or over the set of leaves of a foliation. One significant outcome of this theory is an index theorem of foliation. I am sure that this subject will rapidly develop much further. For a survey of the present status, we refer to [44], [50].

The works on positive cones [13] provide a geometric characterization of von Neumann algebras through the associated natural positive cone in the Hilbert space and lead to some applications.

A work connected with Kazdan's property T [42] provides a simple example of continuously many non-isomorphic factors of type II, and answers a question of Murray and von Neumann about the fundamental group of a factor of type II_1.

I hope that I have conveyed to you some feeling about the incredible power of Alain Connes and the richness of his contributions.

References

Published works of Alain Connes

[1] Sur une généralisation de la notion de corps ordonné, *C. R. Acad. Sci. Paris Sér A-B* **269** (1969), pp. A337–A340.

[2] Ordres faibles et localisation des zéros de polynômes, *C. R. Acad. Sci. Paris Sér. A-B* **269** (1969), pp. A373–A376.

[3] Ordres faibles et localization de zéros de polynômes, *Séminaire Choquet: 1968/69, Initiation à l'Analyse*, Exp. 5, 27 pp. Secrétariat mathématique, Paris, 1969.

[4] Détermination de modèles minimaux en analyse non standard et application, *C. R. Acad. Sci. Paris Sér. A-B* **271** (1970), pp. A969–A971.

[5] Ultrapuissances et applications dans le cadre de l'analyse non standard, *Séminaire Choquet: 1969/70, Initiation à l'Analyse*, Fasc. 1, Exp. 8, 25 pp. Secrétariat mathématique, Paris, 1970.

[6] Un nouvel invariant pour les algèbres de von Neumann, *C. R. Acad. Sci. Paris Sér. A* **273** (1971), pp. 900–903.

[7] Calcul des deux invariants d'Araki–Woods par la théorie de Tomita–Takesaki, *C. R. Acad. Sci. Paris Sér. A* **274** (1972), pp. 175–177.

[8] États presque périodiques sur une algèbre de von Neumann, *C. R. Acad. Sci. Paris Sér. A* **274** (1972), pp. 1402–1405.

[9] Groupe modulaire d'une algèbre de von Neumann, *C. R. Acad. Sci. Paris Sér. A* **274** (1972), pp. 1923–1926.

[10] Une classification des facteurs de type III, *C. R. Acad. Sci. Paris Sér. A* **275** (1972), pp. 523–525.

[11] (Co-author: A. Van Daele) The Group Property of Invariant S of von Neumann Algebras, *Math. Scand.* **32** (1973), pp. 187–192.

[12] Une classification des fracteurs de type III, *Ann. Sci. Ecole Norm. Sup. 4ème Sér.* **6** (1973), pp. 133–252.

[13] Caractérisation des algèbres de von Neumann comme espaces vectoriels ordonnés, *Ann. Inst. Fourier* (Grenoble) **26** (1974), pp. 121–155.

[14] Almost Periodic States and Factors of type III_1, *J. Functional Analysis* **16** (1974), pp. 415–445.

[15] (Co-author: E. J. Woods) Existence de facteurs infinis asymptotiquement abéliens, *C. R. Acad. Sci. Paris Sér. A* **279** (1974), pp. 189–191.

[16] Sur le théorème de Radon–Nikodym pour les poids normaux fidèles semi-finis, *Bull. Sci. Math. 2ème Sér.* **97** (1973), pp. 253–258.
[17] On Hyperfinite Factors of Type III_0 and Krieger's Factors, *J. Functional Analysis* **18** (1975), pp. 318–327.
[18] (Co-author: E. Stφrmer) Entropy for Automorphisms of Finite von Neumann Algebras, *Acta Math.* **134** (1975), pp. 289–306.
[19] On the Hierarchy of W. Krieger, *Illinois J. Math.* **19** (1975), pp. 428–432.
[20] (Co-author: M. Takesaki) The Flow of Weights on Factors of Type III, *Tōhoku Math. J.* **29** (1977), pp. 437–575.
[21] (Co-author: M. Takesaki) Flot des poids sur les facteurs de type III, *C. R. Acad. Sci. Paris Sér. A* **278** (1974), pp. 945–948.
[22] A Factor not Antiisomorphic to Itself, *Bull. London Math. Soc.* **7** (1975) pp. 171–174.
[23] A Factor not Antiisomorphic to Itself, *Ann. Math.* (2) **101** (1975), pp. 536–554.
[24] Periodic Automorphisms of the Hyperfinite Factor of Type II_1, *Acta Sci. Math.* (Szeged) **39** (1977), pp. 39–66.
[25] The Tomita–Takesaki Theory and Classification of Type III Factors. In: D. Kastler (ed.), *Proceedings of the International School of Physics E. Fermi, Course* **60**, pp. 29–46.
[26] (Co-authors: P. Ghez, R. Lima, D. Testard and E. J. Woods) *Review of Crossed Product of von Neumann Algebras.* Mimeographed notes (un-published).
[27] Structure Theory for Type III Factors. In: *Proceedings of the International Congress of Mathematicians*, vol. 2 (1975), pp. 87–91.
[28] Classification of Automorphisms of Hyperfinite Factors of Type II_1 and II_∞ and Application to Type III Factors, *Bull. Amer. Math. Soc.* **81** (1975), pp. 1090–1092.
[29] Outer Conjugacy Classes of Automorphisms of Factors, *Ann. Sci. École Norm. Sup. 4ème Sér.* **8** (1975), pp. 383–419.
[30] Sur la classification des facteurs de type II, *C. R. Acad. Sci. Paris Sér. A* **281** (1975), pp. 13–15.
[31] Classification of Injective Factors, *Ann. Math.* **104** (1976), pp. 73–115.
[32] (Co-author: E. Stφrmer) Homogeneity of the State Space of Factors of Type III_1, *J. Functional Analysis*, **28** (1978), pp. 187–196.
[33] On the Cohomology of Operator Algebras, *J. Functional Analysis* **28** (1978), pp. 248–253.
[34] (Co-author: W. Krieger) Measure Space Automorphisms, the Normalizers of Their Full Groups and Approximate Finiteness, *J. Functional Analysis* **24** (1977), pp. 336–352.
[35] (Co-author: W. Krieger) *Outer Conjugacy of Non-Commutative Bernoulli Shifts.* Mimeographed notes (unpublished).
[36] The Tomita–Takesaki Theory and Classification of Type III Factors. In: D. Kastler (ed.), *C^*-algebras and Their Applications to Statistical Mechanics and Quantum Field Theory*, Proc. Int. School of Physics "Enrico Fermi", Varenna, 1973, Ital. Phys. Soc., 1976, pp. 29–46.
[37] The von Neumann Algebra of a Foliation. In: *Mathematical Problems in Theoretical Physics*, Lecture Notes in Phys. 80, Springer Verlag, 1978, pp. 145–151.
[38] Outer Conjugacy of Automorphisms of Factors, *Symposia Math.* Vol. 20, Academic Press, 1976, pp. 149–159.
[39] On the Equivalence Between Injectivity and Semi-Discreteness for Operator Algebra. In: *Colloq. Int. CNRS* no. 274 (Marseille, 1977), 1979, pp. 107–112.

[40] Von Neumann Algebras. In: O. Lehto (ed.), *Proceedings of the International Congress of Mathematicians, Helsinki 1978*, Acad. Sci. Fennica, 1980, pp. 97–109.
[41] (Co-author: E. J. Woods) A Construction of Approximately Finite-Dimensional non-ITPFI factors, *Canad. Math. Bull.* **23** (1980), pp. 227–230.
[42] A factor of Type II_1 with Countable Fundamental Group, *J. Operator Theory* **4** (1980), pp. 151–153.
[43] On the Spatial Theory of von Neumann Algebras, *J. Functional Analysis* **35** (1980), pp. 153–164.
[44] Sur la théorie non commutative de l'integration. In: P. de la Harpe (ed.), *Algèbres d'Operateurs*, Lecture Notes in Math. 725, Springer, 1979, pp. 19–143.
[45] Feuilletages et algèbres d'operateurs, *Séminaire Bourbaki 1979/80*, n° 551, Lecture Notes in Math. 842 (1981), pp. 139–155.
[46] C^*-algèbres et géométrie differentielle, *C. R. Acad. Sci. Paris* **290** A (1980), pp. 599–604.
[47] An Analogue of the Thom Isomorphism for Cross Products of a C^*-algebra by an Action of $\boldsymbol{R}$, *Adv. in Math.* **39** (1981), pp. 31–55.
[48] (Co-author: H. Moscovici) The L^2-Index Theorem for Homogeneous Spaces of Lie Groups, *Ann. of Math.* **115** (1982), pp. 291–330.
[49] (Co-authors: J. Feldman and B. Weiss) Amenable Equivalence Relations Are Generated by a Single Transformation, *Ergodic Theory Dynamical Systems* **1** (1981), pp. 431–450.
[50] A Survey of Foliations and Operator Algebras, *Proceedings of Symposia in Pure Mathematics* (AMS), **38** (1982), Part 1, pp. 521–628.
[51] (Co-author: G. Skandalis) Theorème de l'indice pour les feuiletages, *C. R. Acad. Sci. Paris, Sér.* I, **292** (1981), pp. 871–876.
[52] (Co-author: V. Jones) A II_1 Factor with Two Nonconjugate Cartan Subalgebras, *Bull. Amer. Math. Soc.* **6** (1982), pp. 211–212.
[53] (Co-author: B. Weiss) Property T and Asymptotically Invariant Subspaces, *Israel J. Math.* **37** (1980), pp. 209–210.
[54] Spectral Sequence and Homology of Currents for Operator Algebras, *Faβungsbericht 42/81*, Oberwolfach (1981), pp. 4–5.
[55] (Co-author: G. Skandalis) *The Longitudinal Index Theorem for Foliations*, IHES Preprint M/82/24 (1982).
[56] Classification des facteurs. In: *Proceedings of Symposia in Pure Mathematics* (AMS), **38** (1982), Part 2, pp. 43–109.
[57] (Co-author: H. Moscovici) L^2-Index Theory on Homogeneous Spaces and Discrete Series Representations, In: *Proceedings of Symposia in Pure Mathematics* (AMS), **38** (1982), Part 1, pp. 419–433.

Research Institute for Mathematical Sciences
Kyoto University
Kyoto 606, Japan

Alain Connes

ALAIN CONNES

Professor at *Collège de France*, Paris, and at *Institut des Hautes Études Scientifiques*, Bures-sur-Yvette

Addresses: College de France, 3 Rue d'Ulm, Paris 75005
I.H.E.S, Le Bois-Marie, 35, Route de Chartres, F-91440 Bures-sur-Yvette
Tel. +33 (0) 60 92 66 00
Fax +33 (0) 60 92 66 09

Born in Draguignan (Var), France, on April 1st 1947

Home address: 5, impasse Carrière
F-91310 LONGPONT Tel and Fax: 33 1 64 49 02 89

Studied at the École Normale Supérieure, Paris	1966–70
Researcher with the *Centre National de Recherche Scientifique*	1970–74
Lecturer, then professor at the University of Paris VI	1976–80
Director of research CNRS	1981–89
Long-term professor at the IHES	1979–present
Professor at the *Collège de France*	1984–present

Invited one-hour lecturer, ICM Helsinki	1978
Prix Ampère, Académie des Sciences, Paris	1980
Foreign member of the Danish Academy of Sciences	1980
Member of the French Academy of Sciences	1982
Fields Medal	1982
Foreign member of the Norwegian Academy of Sciences	1993
Foreign member of the Canadian Academy of Sciences	1996
Docteur Honoris Causa de l'Université de Rome Tor Vergata	1997
Foreign Assoc. Member of National Academy of Sciences USA	1997

BRISURE DE SYMÉTRIE SPONTANÉE ET GÉOMÉTRIE DU POINT DE VUE SPECTRAL

par
ALAIN CONNES

1. Généralités

La géométrie de Riemann admet pour données préalables une *variété* M dont les points $x \in M$ sont localement paramétrés par un nombre fini de coordonnées réelles x^μ, et la *métrique* donnée par l'élément de longueur infinitésimal

$$ds^2 = g_{\mu\nu}\, dx^\mu\, dx^\nu. \tag{1}$$

La distance entre deux points $x, y \in M$ est donnée par

$$d(x,y) = \text{Inf Longueur}\ \gamma, \tag{2}$$

où γ varie parmi les arcs joignant x à y, et

$$\text{Longueur}\ \gamma = \int_x^y ds. \tag{3}$$

La théorie de Riemann est à la fois assez souple pour fournir (au prix d'un changement de signe) un bon modèle de l'espace temps de la relativité générale et assez restrictive pour mériter le nom de géométrie. Le point essentiel est que le calcul différentiel et intégral permet de passer du local au global et que les notions simples de la géométrie euclidienne telle celle de droite continuent à garder un sens. L'équation des géodésiques:

$$\frac{d^2\, x^\mu}{dt^2} = -\Gamma^\mu_{\nu\rho} \frac{dx^\nu}{dt}\ \frac{dx^\rho}{dt} \tag{4}$$

(où $\Gamma^\mu_{\nu\rho} = \frac{1}{2}\, g^{\mu\alpha}(g_{\alpha\nu,\rho} + g_{\alpha\rho,\nu} - g_{\nu\rho,\alpha})$) pour la métrique $dx^2 + dy^2 + dz^2 - (1 - 2V(x,y,z))dt^2$ donne l'équation de Newton dans le potentiel V (cf. [W] pour un énoncé plus précis). Les données expérimentales récentes sur les pulsars binaires confirment [DT] la relativité générale et l'adéquation de la géométrie de Riemann comme modèle de l'espace temps à des échelles suffisamment grandes. La question ([R]) de l'adéquation de cette géométrie comme modèle de l'espace temps à très courte échelle est controversée mais la longueur de Planck

$$\ell_p = (G\hbar/c^3)^{1/2} \sim 10^{-33}\text{cm} \tag{5}$$

est considérée comme la limite naturelle sur la détermination précise des coordonnées d'espace temps d'un évènement. (Voir par exemple [F] ou [DFR] pour l'argument physique, utilisant la mécanique quantique, qui établit cette limite.)

Dans cet exposé nous présentons une nouvelle notion d'espace géométrique qui en abandonnant le rôle central joué par les *points* de l'espace permet une plus grande liberté dans la description de l'espace temps à courte échelle. Le cadre proposé est suffisamment général pour traiter les espaces discrets, les espaces riemanniens, les espaces de configurations de la théorie quantique des champs et les duaux des groupes discrets non nécessairement commutatifs. Le problème principal est d'adapter à ce cadre général les notions essentielles de la géométrie et en particulier le calcul infinitésimal. Le formalisme opératoriel de la mécanique quantique joint à l'analyse des divergences logarithmiques de la trace des opérateurs donnent la généralisation cherchée du calcul différentiel et intégral (Section II). Nous donnons quelques applications directes de ce calcul (Théorèmes 1, 2, 4).

La donnée d'un espace géométrique est celle d'un *triplet spectral*:

$$(\mathcal{A}, \mathcal{H}, D) \tag{6}$$

où $\mathcal{A}$ est une algèbre involutive d'opérateurs dans l'espace de Hilbert $\mathcal{H}$ et D un opérateur autoadjoint non borné dans $\mathcal{H}$. L'algèbre involutive $\mathcal{A}$ correspond à la donnée de l'espace M selon la dualité Espace $\leftrightarrow$ Algèbre classique en géométrie algébrique. L'opérateur $D^{-1} = ds$ correspond à l'élément de longueur infinitésimal de la géométrie de Riemann.

Il y a deux différences évidentes entre cette *géométrie spectrale* et la géométrie riemannienne. La première est que nous ne supposerons pas en général la commutativité de l'algèbre $\mathcal{A}$. La deuxième est que ds, étant un opérateur, ne commute pas avec les éléments de $\mathcal{A}$, même quand $\mathcal{A}$ est commutative.

Comme nous le verrons, des relations de commutation très simples entre ds et l'algèbre $\mathcal{A}$, jointes à la dualité de Poincaré caractérisent les triplets spectraux (6) qui proviennent de variétés riemanniennes (Théorème 6). Quand l'algèbre $\mathcal{A}$ est commutative sa fermeture normique dans $\mathcal{H}$ est l'algèbre des fonctions continues sur un espace compact M. Un point de M est un caractère de $\bar{\mathcal{A}}$, i.e. un homomorphisme de $\bar{\mathcal{A}}$ dans $\mathbb{C}$

$$\begin{aligned} &\chi : \bar{\mathcal{A}} \to \mathbb{C}, \chi(a+b) = \chi(a) + \chi(b), \chi(\lambda a) = \lambda\chi(a), \chi(ab) = \chi(a)\chi(b), \\ &\forall a, b \in \bar{\mathcal{A}}, \forall \lambda \in \mathbb{C}. \end{aligned} \tag{7}$$

Soit par exemple $\mathcal{A}$ l'algèbre $\mathbb{C}\Gamma$ d'un groupe discret Γ agissant dans l'espace de Hilbert $\mathcal{H} = \ell^2(\Gamma)$ de la représentation régulière (gauche) de Γ. Quand le groupe Γ et donc l'algèbre $\mathcal{A}$ sont commutatifs les caractères de $\bar{\mathcal{A}}$ sont les éléments du dual de Pontrjagin de Γ

$$\hat{\Gamma} = \{\chi : \Gamma \to U(1); \chi(g_1 g_2) = \chi(g_1)\chi(g_2) \qquad \forall g_1, g_2 \in \Gamma\}. \tag{8}$$

Les notions élémentaires de la géométrie différentielle pour l'espace $\hat{\Gamma}$ continuent à garder un sens dans le cas général où Γ n'est plus commutatif grâce au dictionnaire suivant dont la colonne de droite n'utilise pas la commutativité de l'algèbre $\mathcal{A}$:

Espace X	Algèbre $\mathcal{A}$
Fibré vectoriel	Module projectif de type fini
Forme différentielle de degré k	Cycle de Hochschild de dimension k
Courant de de Rham de dimension k	Cocycle de Hochschild de dimension k
Homologie de de Rham	Cohomologie cyclique de $\mathcal{A}$

L'intérêt de la généralisation ci-dessus au cas non commutatif est illustré par exemple par la preuve de la conjecture de Novikov pour les groupes Γ qui sont hyperboliques [CM1].

Dans le cas général la notion de point, donnée par (7) est de peu d'intérêt, par contre celle de mesure de probabilité garde tout son sens. Une telle mesure φ est une forme linéaire positive sur $\mathcal{A}$ telle que $\varphi(1) = 1$

$$\varphi : \bar{\mathcal{A}} \to \mathbb{C}, \varphi(a^*a) \geq 0, \qquad \forall\, a \in \bar{\mathcal{A}}, \varphi(1) = 1. \tag{9}$$

Au lieu de mesurer les distances entre les points de l'espace par la formule (2) nous mesurons les distances entre états φ, ψ sur $\bar{\mathcal{A}}$ par une formule duale qui implique un *sup* au lieu d'un *inf* et n'utilise pas les arcs tracés dans l'espace

$$d(\varphi, \psi) = \mathrm{Sup}\{|\varphi(a) - \psi(a)|;\ a \in \mathcal{A}, \|a \in \mathcal{A}, \|[D, a]\| \leq 1\}. \tag{10}$$

Vérifions que cette formule redonne la distance géodésique dans le cas riemannien. Soit M une variété riemannienne compacte munie d'une K-orientation, i.e. d'une structure spinorielle. Le triplet spectral $(\mathcal{A}, \mathcal{H}, D)$ associé est donné par la représentation

$$(f\,\xi)(x) = f(x)\,\xi(x) \qquad \forall\, x \in M,\ f \in \mathcal{A},\ \xi \in \mathcal{H} \tag{11}$$

de l'algèbre des fonctions sur M dans l'espace de Hilbert

$$\mathcal{H} = L^2(M, S) \tag{12}$$

des sections de carré intégrable du fibré des spineurs.

L'opérateur D est l'opérateur de Dirac (cf. [L-M]). On vérifie immédiatement que le commutateur $[D, f]$, $f \in \mathcal{A}$ est l'opérateur de multiplication de Clifford par le gradient ∇f de f et que sa norme hilbertienne est:

$$\|[D, f]\| = \operatorname*{Sup}_{x \in M} \|\nabla f\| = \text{Norme lipschitzienne de } f. \tag{13}$$

Soient $x, y \in M$ et φ, ψ les caractères correspondants: $\varphi(f) = f(x)$, $\psi(f) = f(y)$. $\forall\, f \in \mathcal{A}$ la formule (10) donne le même résultat que la formule (2), i.e. donne la distance géodésique entre x et y.

Contrairement à (2) la formule duale (10) garde un sens en général et en particulier pour les espaces discrets ou totalement discontinus.

La notion usuelle de *dimension* d'un espace est remplacée par un *spectre de dimension* qui est un sous-ensemble de $\mathbb{C}$ dont la partie réelle est bornée supérieurement par $\alpha > 0$ si

$$\lambda_n^{-1} = O(n^{-\alpha}) \tag{14}$$

où λ_n est la n-ième valeur propre de $|D|$.

La relation entre le local et le global est donnée par la formule locale de l'indice (Théorème 4) ([CM2]).

La propriété caractéristique des *variétés différentiables* qui est transposée au cas non commutatif est la *dualité de Poincaré*. La dualité de Poincaré en homologie ordinaire est insuffisante pour caractériser le type d'homotopie des variétés ([Mi-S]) mais les résultats de D. Sullivan ([S2]) montrent (dans le cas PL, simplement connexe, de dimension ≥ 5 et en ignorant le nombre premier 2) qu'il suffit de remplacer l'homologie ordinaire par la KO-homologie.

De plus la K-homologie admet grâce aux résultats de Brown Douglas Fillmore, Atiyah et Kasparov une traduction algébrique très simple, donnée par

Espace X	Algèbre $\mathcal{A}$
$K_1(X)$	Classe d'homotopie stable de triplet spectral $(\mathcal{A}, \mathcal{H}, D)$
$K_0(X)$	Classe d'homotopie stable de triplet spectral $\mathbb{Z}/2$ gradué

(i.e. pour K_0 on suppose que $\mathcal{H}$ est $\mathbb{Z}/2$ gradué par γ, $\gamma = \gamma^*$, $\gamma^2 = 1$ et que $\gamma a = a\gamma \; \forall\, a \in \mathcal{A}$, $\gamma D = -D\gamma$).

Cette description suffit pour la K-homologie complexe qui est périodique de période 2.

Dans le cas non commutatif la *classe fondamentale* d'un espace est une classe μ de KR-homologie pour l'algèbre $\mathcal{A} \otimes \mathcal{A}^0$ munie de l'involution

$$\tau(x \otimes y^0) = y^* \otimes (x^*)^0 \qquad \forall\, x, y \in \mathcal{A} \tag{15}$$

où $\mathcal{A}^0$ désigne l'algèbre opposée de $\mathcal{A}$. Le produit intersection de Kasparov [K] permet de formuler la dualité de Poincaré par l'invertibilité de μ. La KR-homologie est périodique de période 8 et la dimension modulo 8 est spécifiée par les règles de commutation suivantes, où J est une isométrie antilinéaire dans $\mathcal{H}$ qui implémente l'involution τ

$$JxJ^{-1} = \tau(x) \qquad \forall\, x \in \mathcal{A} \otimes \mathcal{A}^0. \tag{16}$$

On a $J^2 = \varepsilon$, $JD = \varepsilon' DJ$, $J\gamma = \varepsilon'' \gamma J$ où $\varepsilon, \varepsilon', \varepsilon'' \in \{-1, 1\}$ et si n désigne la dimension modulo 8

$n =$	0	1	2	3	4	5	6	7
ε	1	1	-1	-1	-1	-1	1	1
ε'	1	-1	1	1	1	-1	1	1
ε''	1		-1		1		-1	

L'isométrie antilinéaire J est donnée dans le cas riemannien par l'opérateur de conjugaison de charge et dans le cas non commutatif par l'opérateur de Tomita [Ta] qui, dans le cas où une algèbre d'opérateurs $\mathcal{A}$ admet un vecteur cyclique qui est cyclique pour le commutant $\mathcal{A}'$, établit un antiisomorphisme

$$a \in \mathcal{A}'' \to Ja^*J^{-1} \in \mathcal{A}'. \tag{17}$$

La donnée de μ ne spécifie que la classe d'homotopie stable du triplet spectral $(\mathcal{A}, \mathcal{H}, D)$ muni de l'isométrie J (et de la $\mathbb{Z}/2$ graduation γ si n est pair). La non trivialité de cette classe d'homotopie est visible dans la forme d'intersection

$$K(\mathcal{A}) \times K(\mathcal{A}) \to \mathbb{Z} \tag{18}$$

donnée par l'indice de Fredholm de D à coefficient dans $K(\mathcal{A} \otimes \mathcal{A}^0)$.

Pour comparer les triplets spectraux dans la classe μ, nous utiliserons la fonctionnelle spectrale suivante

$$\mathrm{Trace}\,(\overline{\omega}(D)) \tag{19}$$

où $\overline{\omega} : \mathbb{R} \to \mathbb{R}_+$ est une fonction positive convenable.

L'algèbre $\mathcal{A}$ une fois fixée, une géométrie spectrale est déterminée par la classe d'équivalence unitaire du triplet spectral $(\mathcal{A}, \mathcal{H}, D)$ avec l'isométrie J. Si l'on note π la représentation de $\mathcal{A}$ dans $\mathcal{H}$ l'équivalence unitaire entre $(\pi_1, \mathcal{H}_1, D_1, J_1)$ et $(\pi_2, \mathcal{H}_2, D_2, J_2)$ signifie qu'il existe un unitaire $U : \mathcal{H}_1 \to \mathcal{H}_2$ tel que

$$U\pi_1(a)U^* = \pi_2(a) \qquad \forall\, a \in \mathcal{A},\ UD_1U^* = D_2,\ UJ_1U^* = J_2 \tag{20}$$

(et $U\gamma_1 U^* = \gamma_2$ dans le cas où n est pair).

Le groupe $\mathrm{Aut}(\mathcal{A})$ des automorphismes de l'algèbre involutive $\mathcal{A}$ agit sur l'ensemble des géométries spectrales par composition

$$\pi'(a) = \pi(\alpha^{-1}(a)) \qquad \forall\, a \in \mathcal{A},\ \alpha \in \mathrm{Aut}(\mathcal{A}). \tag{21}$$

Le sous-groupe $\mathrm{Aut}^+(\mathcal{A})$ des automorphismes qui préservent la classe μ agit sur la classe d'homotopie stable déterminée par μ et préserve par construction la fonctionnelle d'action (19). En général ce groupe est non compact, et il coïncide par exemple dans le cas riemannien avec le groupe $\mathrm{Diff}^+(M)$ des difféomorphismes qui préservent la K-orientation, i.e. la structure spinorielle de M. A l'inverse le groupe d'isotropie d'une géométrie donnée, est automatiquement *compact* (pour $\mathcal{A}$ unifère). Ceci montre que la fonctionnelle d'action (19) donne automatiquement naissance au phénomène de brisure de symétrie spontanée (Figure 1).

Nous montrerons que pour un choix convenable de l'algèbre $\mathcal{A}$ la fonctionnelle d'action (19) ajoutée au terme $\langle \xi, D\xi \rangle$, $\xi \in \mathcal{H}$ donne le modèle standard de Glashow–Weinberg–Salam couplé à la gravitation. L'algèbre $\mathcal{A}$ est le produit tensoriel de l'algèbre des fonctions sur un espace riemannien M par une algèbre non commutative de dimension finie dont les données phénoménologiques spécifient la géométrie spectrale.

2. Un Calcul Infinitésimal

Nous montrons comment le formalisme opératoriel de la mécanique quantique permet de donner un sens précis à la notion de variable infinitésimale. La notion d'infinitésimal est sensée avoir un sens intuitif évident. Elle résiste cependant fort bien aux essais de formalisation donnés par exemple par l'analyse non standard. Ainsi, pour prendre un exemple précis ([B-W]), soit $dp(x)$ la probabilité pour qu'une fléchette lancée au hasard sur la cible Ω termine sa course au point $x \in \Omega$ (Figure 2). Il est clair que $dp(x) < \varepsilon \forall \varepsilon > 0$ et que néanmoins la réponse $dp(x) = 0$ n'est

Figure 1.

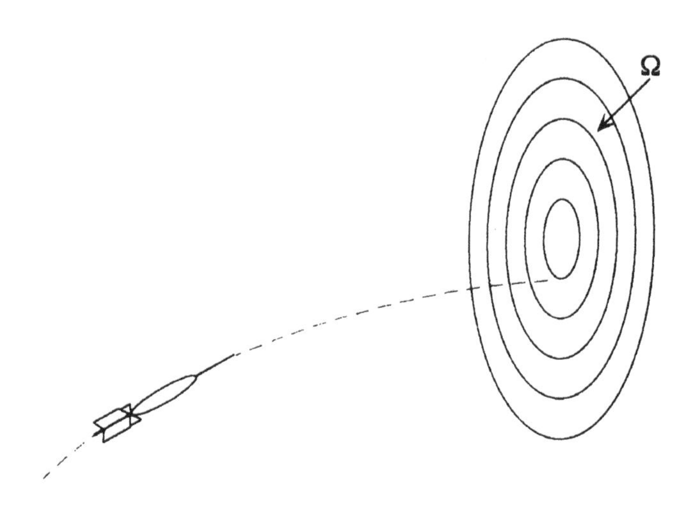

Figure 2.

pas satisfaisante. Le formalisme usuel de la théorie de la mesure ou des formes différentielles contourne le problème en donnant un sens à l'expression

$$\int f(x)\,dp(x) \qquad f:\Omega\to\mathbb{C} \tag{1}$$

mais est insuffisant pour donner un sens par exemple à $e^{-\frac{1}{dp(x)}}$. La réponse, à savoir un réel non standard, fournie par l'analyse non standard, est également décevante: tout réel non standard détermine canoniquement un sous-ensemble non Lebesgue mesurable de l'intervalle $[0,1]$ de sorte qu'il est impossible ([Ste]) d'en exhiber un seul. Le formalisme que nous proposons donnera une réponse substantielle et calculable à cette question.

Le cadre est fixé par un espace de Hilbert séparable $\mathcal{H}$ décomposé comme somme directe de deux sous-espaces de dimension infinie. Donner cette décomposition revient à donner l'opérateur linéaire F dans $\mathcal{H}$ qui est l'identité, $F\xi = \xi$, sur le premier sous-espace et moins l'identité, $F\xi = -\xi$ sur le second; on a

$$F = F^*, \; F^2 = 1. \tag{2}$$

Le cadre ainsi déterminé est unique à isomorphisme près. Le début du dictionnaire qui traduit les notions classiques en language opératoriel est le suivant:

Classique	Quantique
Variable complexe	Opérateur dans $\mathcal{H}$
Variable réelle	Opérateur autoadjoint
Infinitésimal	Opérateur compact
Infinitésimal d'ordre α	Opérateur compact dont les valeurs caractéristiques μ_n vérifient $\mu_n = O(n^{-\alpha}),\ n \to \infty$
Différentielle d'une variable réelle ou complexe	$đf = [F, f] = Ff - fF$
Intégrale d'un infinitésimal d'ordre 1	$⨍ T =$ Coefficient de la divergence logarithmique dans la trace de T.

Les deux premières lignes du dictionnaire sont familières en mécanique quantique. L'ensemble des valeurs d'une variable complexe correspond au *spectre* d'un opérateur. Le calcul fonctionnel holomorphe donne un sens à $f(T)$ pour toute fonction holomorphe f sur le spectre de T. Les fonctions holomorphes sont les seules à opérer dans cette généralité ce qui reflète la différence entre l'analyse complexe et l'analyse réelle où les fonctions boréliennes arbitraires opèrent. Quand $T = T^*$ est autoadjoint $f(T)$ a un sens pour toute fonction borélienne f. Notons que toute variable aléatoire usuelle X sur un espace de probabilité, (Ω, P) peut être trivialement considérée comme un opérateur autoadjoint. On prend $\mathcal{H} = L^2(\Omega, P)$ et

$$(T\xi)(p) = X(p)\xi(p) \qquad \forall p \in \Omega, \xi \in \mathcal{H}. \tag{3}$$

La mesure spectrale de T redonne la probabilité P.

Passons à la troisième ligne du dictionnaire. Nous cherchons des "variables infinitésimales", i.e. des opérateurs T dans $\mathcal{H}$ tels que

$$\|T\| < \varepsilon \qquad \forall \varepsilon > 0, \tag{4}$$

où $\|T\| = \mathrm{Sup}\,\{\|T\xi\| \; ; \; \|\xi\| = 1\}$ est la norme d'opérateur. Bien entendu si l'on prend (4) au pied de la lettre on obtient $\|T\| = 0$ et $T = 0$ est la seule solution.

Mais on peut affaiblir (4) de la manière suivante

(5) $\forall\,\varepsilon > 0$, $\exists$ sous-espace de dimension finie $E \subset \mathcal{H}$ tel que $\|T/E^\perp\| < \varepsilon$

où $E^\perp$ désigne l'orthogonal de E dans $\mathcal{H}$

$$E^\perp = \{\xi \in \mathcal{H}\ ;\ \langle \xi, \eta\rangle = 0 \quad \forall\,\eta \in E\} \tag{6}$$

qui est un sous-espace de codimension finie de $\mathcal{H}$. Le symbole $T/E^\perp$ désigne la restriction de T à ce sous-espace

$$T/E^\perp : E^\perp \to \mathcal{H}. \tag{7}$$

Les opérateurs qui satisfont la condition (5) sont les *opérateurs compacts*, i.e. sont caractérisés par la compacité normique de l'image de la boule unité de $\mathcal{H}$. L'opérateur T est compact ssi $|T| = \sqrt{T^*T}$ est compact, et ceci a lieu ssi le spectre de $|T|$ est une suite de valeurs propres $\mu_0 \geq \mu_1 \geq \mu_2 \ldots,\ \mu_n \downarrow 0$.

Ces valeurs propres sont les valeurs caractéristiques de T et on a

$$\mu_n(T) = \text{Inf}\,\{\|T - R\|\ ;\ R \text{ opérateur de rang } \leq n\} \tag{8}$$

$$\mu_n(T) = \text{Inf}\,\{\|T/E^\perp\|\ ;\ \dim E \leq n\}. \tag{9}$$

Les opérateurs compacts forment un idéal bilatère $\mathcal{K}$ dans l'algèbre $\mathcal{L}(\mathcal{H})$ des opérateurs bornés dans $\mathcal{H}$ de sorte que les règles algébriques élémentaires du calcul infinitésimal sont vérifiées.

La taille d'un infinitésimal $T \in \mathcal{K}$ est gouvernée par l'ordre de décroissance de la suite $\mu_n = \mu_n(T)$, quand $n \to \infty$. En particulier pour tout réel positif α la condition

$$\mu_n(T) = O(n^{-\alpha}) \qquad \text{quand } n \to \infty \tag{10}$$

(i.e. il existe $C > 0$ tel que $\mu_n(T) \leq Cn^{-\alpha}\forall\, n \geq 1$) définit les infinitésimaux d'ordre α. Ils forment de même un idéal bilatère, comme on le voit en utilisant (8), (cf. [Co]) et de plus

$$T_j \text{ d'ordre } \alpha_j \Rightarrow T_1T_2 \text{ d'ordre } \alpha_1 + \alpha_2. \tag{11}$$

(Pour $\alpha < 1$ l'idéal correspondant est un idéal normé obtenu par interpolation réelle entre l'idéal $\mathcal{L}^1$ des opérateurs traçables et l'idéal $\mathcal{K}$ ([Co]).) Ainsi, hormis la commutativité, les propriétés intuitives du calcul infinitésimal sont vérifiées.

Comme la taille d'un infinitésimal est mesurée par une suite $\mu_n \to 0$ il pourrait sembler inutile d'utiliser le formalisme opératoriel. Il suffirait de remplacer l'idéal $\mathcal{K}$ de $\mathcal{L}(\mathcal{H})$ par l'idéal $c_0(\mathbb{N})$ des suites convergeant vers 0 dans l'algèbre $\ell^\infty(\mathbb{N})$ des suites bornées. Cette version commutative ne convient pas car tout élément de $\ell^\infty(\mathbb{N})$ a un spectre ponctuel et une mesure spectrale discrète. Ce n'est que la *non commutativité* de $\mathcal{L}(\mathcal{H})$ qui permet la coexistence de variables ayant un spectre de Lebesgue avec des variables infinitésimales.

En fait la ligne suivante du dictionnaire utilise de manière cruciale la non commutativité de $\mathcal{L}(\mathcal{H})$. La différentielle df d'une variable réelle ou complexe

$$df = \Sigma \frac{\partial f}{\partial x^{\mu}} dx^{\mu} \tag{12}$$

est remplacée par le commutateur

$$đ f = [F, f]. \tag{13}$$

Le passage de (11) à (12) est semblable à la transition du crochet de Poisson $\{f, g\}$ de deux observables f, g de la mécanique classique, au commutateur $[f, g] = fg - gf$ d'observables quantiques.

Etant donnée une algèbre $\mathcal{A}$ d'opérateurs dans $\mathcal{H}$ la *dimension* de l'espace correspondant (au sens du dictionnaire 1) est gouvernée par la taille des différentielles $đ f$, $f \in \mathcal{A}$. En dimension p on a

$$đ f \text{ d'ordre } \frac{1}{p}, \forall f \in \mathcal{A}. \tag{14}$$

Nous verrons très vite des exemples concrets où p est la dimension de Hausdorff d'un ensemble de Julia. Des manipulations algébriques très simples sur la fonctionnelle

$$\tau(f^0, \ldots, f^n) = \text{Trace}\,(f^0\, đ f^1 \ldots đ f^n) \qquad n \text{ impair}, \; n > p \tag{15}$$

montrent que τ est un cocycle cyclique et permettent de transposer les idées de la topologie différentielle en exploitant *l'intégralité* du cocycle τ, i.e. $\langle \tau, K_1(\mathcal{A}) \rangle \subset \mathbb{Z}$.

Si le dictionnaire s'arrêtait là, il nous manquerait un outil vital du calcul infinitésimal, la *localité*, i.e. la possibilité de négliger les infinitésimaux d'ordre > 1 dans un calcul. Dans notre cadre les infinitésimaux d'ordre > 1 sont contenus dans l'idéal bilatère suivant

$$\left\{ T \in \mathcal{K} \; ; \; \mu_n(T) = o\left(\frac{1}{n}\right) \right\} \tag{17}$$

où le petit o à la signification usuelle.

Ainsi, si nous utilisons la trace comme dans (16) pour intégrer nous rencontrons deux problèmes:

(a) les infinitésimaux d'ordre 1 ne sont pas dans le domaine de la trace,

(b) la trace des infinitésimaux d'ordre > 1 n'est pas nulle.

Le domaine naturel de la trace est l'idéal bilatère $\mathcal{L}^1(\mathcal{H})$ des opérateurs traçables

$$\mathcal{L}^1 = \left\{ T \in \mathcal{K} \; ; \; \sum_{o}^{\infty} \mu_n(T) < \infty \right\}. \tag{18}$$

La trace d'un opérateur $T \in \mathcal{L}^1(\mathcal{H})$ est donnée par la somme

$$\text{Trace}\,(T) = \sum \langle T\xi_i, \xi_i \rangle \tag{19}$$

indépendamment du choix de la base orthonormale (ξ_i) de $\mathcal{H}$. On a

$$\text{Trace}\,(T) = \sum_{o}^{\infty} \mu_n(T) \qquad \forall\, T \geq 0. \tag{20}$$

Soit $T \geq 0$ un infinitésimal d'ordre 1, le seul contrôle sur $\mu_n(T)$ est

$$\mu_n(T) = O\left(\frac{1}{n}\right) \tag{21}$$

ce qui ne suffit pas pour assurer la finitude de (20). Ceci précise la nature du problème (a) et de même pour (b) puisque la trace ne s'annule pas pour le plus petit idéal de $\mathcal{L}(\mathcal{H})$, l'idéal $\mathcal{R}$ des opérateurs de rang fini.

Ces deux problèmes sont résolus par la trace de Dixmier [Dx], i.e. par l'analyse suivante de la divergence logarithmique des traces partielles

$$\text{Trace}_N(T) = \sum_o^{N-1} \mu_n(T), \; T \geq 0. \tag{22}$$

Il est utile de définir $\text{Trace}_\Lambda(T)$ pour tout $\Lambda > 0$ par la formule d'interpolation

$$\text{Trace}_\Lambda(T) = \text{Inf}\,\{\|A\|_1 + \Lambda\|B\| \; ; \; A + B = T\} \tag{23}$$

où $\|A\|_1$ est la norme $\mathcal{L}^1$ de A, $\|A\|_1 = \text{Trace}\,|A|$. Cette formule coïncide avec (22) pour Λ entier et donne l'interpolation affine par morceaux. On a de plus ([Co])

$$\text{Trace}_\Lambda(T_1 + T_2) \leq \text{Trace}_\Lambda(T_1) + \text{Trace}_\Lambda(T_2) \qquad \forall\,\Lambda \tag{24}$$

$$\text{Trace}_{\Lambda_1+\Lambda_2}(T_1 + T_2) \geq \text{Trace}_{\Lambda_1}(T_1) + \text{Trace}_{\Lambda_2}(T_2) \qquad \forall\,\Lambda_1, \Lambda_2 \tag{25}$$

où T_1, T_2 sont *positifs* pour (25).

Soit $T > 0$ infinitésimal d'ordre 1 on a alors

$$\text{Trace}_\Lambda(T) \leq C\,\log\Lambda \tag{26}$$

et la propriété remarquable *d'additivité asymptotique* du coefficient de la divergence logarithmique (26) est la suivante: $(T_j \geq 0)$,

$$|\tau_\Lambda(T_1 + T_2) - \tau_\Lambda(T_1) - \tau_\Lambda(T_2)| \leq 3C\,\frac{\log(\log\Lambda)}{\log\Lambda} \tag{27}$$

où pour tout $T \geq 0$ on pose

$$\tau_\Lambda(T) = \frac{1}{\log\Lambda}\int_e^\Lambda \frac{\text{Trace}_\mu(T)}{\log\mu}\,\frac{d\mu}{\mu} \tag{28}$$

qui est la moyenne de Cesaro sur le groupe $\mathbb{R}_+^*$ des échelles, de la fonction $\frac{\text{Trace}_\mu(T)}{\log\mu}$.

Il résulte facilement de (27) que toute limite simple τ des fonctionnelles non linéaires τ_Λ définit une trace positive et linéaire sur l'idéal bilatère des infinitésimaux d'ordre 1

$$\tau(\lambda_1\,T_1 + \lambda_2\,T_2) = \lambda_1\;\tau(T_1) + \lambda_2\;\tau(T_2) \qquad \forall\,\lambda_j \in \mathbb{C}$$

$$\tau(T) \geq 0 \qquad \forall\,T \geq 0 \tag{29}$$

$$\tau(ST) = \tau(TS) \qquad \forall\,S \text{ borné}$$

$$\tau(T) = 0 \text{ si } \mu_n(T) = o\left(\frac{1}{n}\right).$$

En pratique le choix du point limite τ est sans importance car dans tous les exemples importants (et en particulier comme corollaire des axiomes dans le cadre général, cf. Section IV) la condition suivante de *mesurabilité* est satisfaite:

$$\tau_\Lambda(T) \text{ est convergent quand } \Lambda \to \infty. \tag{30}$$

Pour les opérateurs mesurables la valeur de $\tau(T)$ est indépendante de τ et est notée

$$⨍ T. \tag{31}$$

Le premier exemple intéressant est celui des opérateurs pseudodifférentiels T sur une variété différentiable M. Quand T est d'ordre 1 (au sens de (21)) il est mesurable et $⨍ T$ est le résidu non commutatif de T ([Wo], [Ka]). Ce résidu a une expression locale très simple en terme du noyau distribution $k(x,y)$, $x,y \in M$. Quand T est d'ordre 1 (au sens de (21)) le noyau $k(x,y)$ admet une divergence logarithmique au voisinage de la diagonale

$$k(x,y) = a(x) \log|x-y| + 0(1) \tag{32}$$

où $|x-y|$ est la distance riemannienne dont le choix n'affecte pas la 1-densité $a(x)$. On a alors (à normalisation près)

$$⨍ T = \int_M a(x) \tag{33}$$

et le terme de droite de cette formule se prolonge de manière quasi évidente à tous les opérateurs pseudodifférentiels (cf. [Wo]) si l'on note que le noyau d'un tel opérateur admet un développement asymptotique de la forme

$$k(x,y) = \sum a_k(x, x-y) + a(x) \log|x-y| + 0(1) \tag{34}$$

où $a_k(x,\xi)$ est homogène de degré $-k$ en la variable ξ, et où la 1-densité $a(x)$ est définie de manière intrinsèque.

En fait le même principe de prolongement de $⨍$ à des infinitésimaux d'ordre < 1 s'applique aux opérateurs hypoelliptiques et plus généralement (cf. Théorème 4) aux triplets spectraux dont le spectre de dimension est simple.

Après cette description passons à des exemples. La variable infinitésimale $dp(x)$ qui donne la probabilité dans le jeu de fléchettes (Figure 2) est donnée par l'opérateur

$$dp = \Delta^{-1} \tag{35}$$

où Δ est le Laplacien de Dirichlet pour le domaine Ω. Il agit dans l'espace de Hilbert $L^2(\Omega)$ ainsi que l'algèbre des fonctions $f(x_1,x_2)$, $f : \Omega \to \mathbb{C}$, qui agissent par opérateurs de multiplication (cf. (3)). Le théorème de H. Weyl montre immédiatement que dp est d'ordre 1, que $f\,dp$ est mesurable et que

$$⨍ f\,dp = \int_\Omega f(x_1,x_2)\,dx_1 \wedge dx_2 \tag{36}$$

donne la probabilité usuelle.

Montrons maintenant comment utiliser notre calcul infinitésimal pour donner un sens à des expressions telles que l'aire d'une variété de dimension 4, qui sont dépourvues de sens dans le calcul usuel.

Il y a, à équivalence unitaire et multiplicité près, une seule quantification du calcul infinitésimal sur $\mathbb{R}$ qui soit invariante par translations et dilatations. Elle est donnée par la représentation de l'algèbre des fonctions f sur $\mathbb{R}$ comme opérateurs de multiplication dans $L^2(\mathbb{R})$ (cf. (3)), alors que l'opérateur F dans $\mathcal{H} = L^2(\mathbb{R})$ est la transformation de Hilbert ([St])

$$(37)\quad (f\xi)(s) = f(s)\,\xi(s) \qquad \forall\, s \in \mathbb{R},\ \xi \in L^2(\mathbb{R}),\ (F\xi)(t) = \frac{1}{\pi i}\int \frac{\xi(s)}{s-t}\,ds.$$

On a une description unitairement équivalente pour $S^1 = P_1(\mathbb{R})$ avec $\mathcal{H} = L^2(S^1)$ et

$$(38)\quad F\,e_n = \mathrm{Sign}\,(n)\,e_n,\ \ e_n(\tau) = \exp\,(in\tau) \qquad \forall\,\tau \in S^1,\ (\mathrm{Sign}\,(0) = 1).$$

L'opérateur $\text{đ} f = [F, f]$, pour $f \in L^\infty(\mathbb{R})$, est représenté par le noyau $\frac{1}{\pi i}\,k(s,t)$, avec

$$(39)\qquad k(s,t) = \frac{f(s) - f(t)}{s-t}.$$

Comme f et F sont des opérateurs bornés il en est de même de $\text{đ} f = [F, f]$ pour toute f mesurable bornée sur S^1, ce qui donne un sens à $|\text{đ} f|^p$ pour tout $p > 0$. Soient par exemple $c \in \mathbb{C}$ et J l'ensemble de Julia associé à l'itération de la transformation

$$(40)\qquad \varphi(z) = z^2 + c,\ \ J = \partial B,\ \ B = \{z \in \mathbb{C}\ ;\ \sup_{n\in\mathbb{N}} |\varphi^n(z)| < \infty\}.$$

Pour c petit J est une courbe de Jordan et B la composante bornée de son complément. Soit $Z : S^1 \to J$ la restriction à $S^1 = \partial D$, $D = \{z \in \mathbb{C},\, |z| < 1\}$ d'une équivalence conforme $D \sim B$. Comme (par un résultat de D. Sullivan) la dimension de Hausdorff p de J est > 1 (pour $c \neq 0$) la fonction Z n'est nulle part à variation bornée et la valeur absolue $|Z'|$ de la dérivée de Z au sens des distributions n'a pas de sens. Cependant $|\text{đ} Z|$ est bien défini et on a:

Théorème 1. — (a) $|\text{đ} Z|$ *est un infinitésimal d'ordre* $\frac{1}{p}$.

(b) *Pour toute fonction continue* h *sur* J, *l'opérateur* $h(Z)\,|\text{đ} Z|^p$ *est mesurable.*

(c) $\exists\,\lambda > 0$:

$$⨍ h(Z)\,|\text{đ} Z|^p = \lambda \int h\,d\Lambda_p \qquad \forall\, h \in C(J),$$

où $d\Lambda_p$ *désigne la mesure de Hausdorff sur* J.

L'énoncé (a) utilise un résultat de V. V. Peller qui caractérise les fonctions f pour lesquelles $\mathrm{Trace}\,(|\text{đ} f|^\alpha) < \infty$. La constante λ gouverne le développement asymptotique de la distance dans $L^\infty(S^1)$ entre Z et les fonctions rationnelles ayant au plus n-pôles hors du disque unité. Cette constante est de l'ordre de $\sqrt{p-1}$ et s'annule pour $p = 1$. Cela tient à une propriété spécifique de la dimension 1, à savoir que pour $f \in C^\infty(S^1)$ $\text{đ} f$ n'est pas seulement d'ordre $(\dim S^1)^{-1} = 1$ mais est traçable, avec

$$(41)\qquad \mathrm{Trace}\,(f^0\,\text{đ} f^1) = \frac{1}{\pi i}\int_{S_1} f^0\,df^1 \qquad \forall\, f^0, f^1 \in C^\infty(S^1).$$

En fait par un résultat classique de Kronecker $đ f$ est de rang fini ssi f est une fraction rationnelle (cf. [P]).

Le calcul différentiel quantique s'applique de la même manière à l'espace projectif $P_1(K)$ sur un corps local arbitraire K (i.e. un corps localement compact non discret) et est invariant par le groupe des transformations projectives. Les cas spéciaux $K = \mathbb{C}$ et $K = \mathbb{H}$ (corps des quaternions) sont des cas particuliers du calcul sur les variétés compactes conformes orientées de dimension paire, $M = M_{2n}$, qui se définit ainsi:

$$(42) \quad \mathcal{H} = L^2(M, \Lambda^n T^*), \; (f\xi)(p) = f(p)\,\xi(p) \qquad \forall\, f \in L^\infty(M), \; F = 2P - 1$$

où le produit scalaire sur l'espace de Hilbert des formes différentielles de degré $n = \frac{1}{2} \dim M$ est donné par $\langle \omega_1, \omega_2 \rangle = \int \omega_1 \wedge * \omega_2$ et ne dépend que de la structure conforme de M. L'opérateur P est le projecteur orthogonal sur le sous-espace des formes exactes.

Prenons d'abord $n = 1$. Un calcul immédiat donne

$$(43) \qquad ⨍\, đ f \, đ g = -\frac{1}{\pi} \int df \wedge * dg \qquad \forall\, f, g \in C^\infty(M).$$

Soit alors X une application (C^∞) de M dans l'espace $\mathbb{R}^N$ muni de la métrique riemannienne $g_{\mu\nu}\, dx^\mu\, dx^\nu$; on a

$$(44) \qquad ⨍\, g_{\mu\nu}\, đ X^\mu\, đ X^\nu = -\frac{1}{\pi} \int_M g_{\mu\nu}\, dX^\mu \wedge * dX^\nu$$

où le terme de droite est l'action de Polyakov de la théorie des cordes. Pour $n = 4$ l'égalité (44) n'a pas lieu, l'action définie par le terme de droite n'est pas intéressante car elle n'est pas invariante conforme. Le terme de gauche est parfaitement défini par le calcul quantique et est invariant conforme: on a

Théorème 2. — *Soit X une application C^∞ de M_4 dans $(\mathbb{R}^N, g_{\mu\nu}\, dx^\mu\, dx^\nu)$,*

$$⨍\, g_{\mu\nu}(X)\, đ X^\mu\, đ X^\nu = (16\pi^2)^{-1} \int_M g_{\mu\nu}(X)$$

$$\left\{ \frac{1}{3}\, r \langle dX^\mu, dX^\nu \rangle - \Delta \langle dX^\mu, dX^\nu \rangle + \langle \nabla dX^\mu, \nabla dX^\nu \rangle - \frac{1}{2}\, (\Delta X^\mu)\, (\Delta X^\nu) \right\} dv$$

où pour écrire le terme de droite on utilise sur M une structure riemannienne η quelconque compatible avec la structure conforme. Ainsi la courbure scalaire r, le Laplacien Δ et la connection de Levi Civita ∇ se réfèrent à η, mais le résultat n'en dépend pas.

Le Théorème 2 est à rapprocher de la formule suivante qui exprime l'action de Hilbert–Einstein comme l'aire d'une variété de dimension 4 (cf. [Kas] [K-W])

$$(45) \qquad ⨍\, ds^2 = \frac{-1}{96\pi^2} \int_{M_4} r \sqrt{g}\; d^4 x$$

($dv = \sqrt{g}\; d^4x$ est la forme volume et $ds = D^{-1}$ est l'élément de longueur, i.e. l'inverse de l'opérateur de Dirac).

Quand la métrique $g_{\mu\nu}\, dx^\mu\, dx^\nu$ sur $\mathbb{R}^N$ est invariante par translations, la fonctionnelle d'action du Théorème 2 est donnée par l'opérateur de Paneitz sur M. C'est un opérateur d'ordre 4 qui joue le rôle du Laplacien en géométrie conforme ([B-O]). Son anomalie conforme a été calculée par T. Branson [B].

Reprenons le cas $n = 2$ et modifions la structure conforme de M par une différentielle de Beltrami $\mu(z,\bar z)\, d\bar z/dz$, $|\mu(z,\bar z)| < 1$ en utilisant pour définir les angles en $z \in M$

$$X \in T_z(M) \to \langle X, dz + \mu(z,\bar z)\, d\bar z\rangle \in \mathbb{C} \tag{46}$$

au lieu de $\langle X, dz\rangle$. Le calcul quantique sur M associé à la nouvelle structure conforme s'obtient simplement en remplaçant l'opérateur F par l'opérateur F',

$$F' = (\alpha F + \beta)(\beta F + \alpha)^{-1}, \ \alpha = (1 - m^2)^{-1/2}, \ \beta = m(1 - m^2)^{-1/2} \tag{47}$$

où m est l'opérateur dans $\mathcal{H} = L^2(M, \Lambda^1 T^*)$ donné par l'endomorphisme du fibré vectoriel $\Lambda^1 T^* = \Lambda^{(1,0)} \oplus \Lambda^{(0,1)}$ de matrice,

$$m(z,\bar z) = \begin{bmatrix} 0 & \bar\mu(z,\bar z)\, d\bar z/dz \\ \mu(z,\bar z)\, dz/d\bar z & 0 \end{bmatrix}. \tag{48}$$

Les propriétés cruciales de l'opérateur $m \in \mathcal{L}(\mathcal{H})$ sont

$$\|m\| < 1, \ m = m^*, \ m\,f = f\,m \qquad \forall\, f \in \mathcal{A} = C^\infty(M) \tag{49}$$

et la déformation (47) de F est un cas particulier de la:

Proposition 3. — *Soient $\mathcal{A}$ une algèbre involutive d'opérateurs dans $\mathcal{H}$ et $N = \mathcal{A}' = \{T \in \mathcal{L}(\mathcal{H})\,;\, Ta = aT \quad \forall\, a \in \mathcal{A}\}$ l'algèbre de von Neumann commutant de $\mathcal{A}$.*

(a) *L'égalité suivante définit une action du groupe $G = GL_1(N)$ des éléments inversibles de N sur les opérateurs F, $F = F^*$, $F^2 = 1$*

$$g(F) = (\alpha F + \beta)\,(\beta F + \alpha)^{-1} \qquad \forall\, g \in G$$

où $\alpha = \frac{1}{2}(g - (g^{-1})^*)$, $\beta = \frac{1}{2}(g + (g^{-1})^*)$.

(b) *On a $[g(F), a] = Y[F,a]\,Y^* \quad \forall\, a \in \mathcal{A}$, où $Y = (\beta F + \alpha)^{*-1}$.*

L'égalité (b) montre que pour tout idéal bilatère $J \subset \mathcal{L}(\mathcal{H})$ la condition

$$[F, a] \in J \tag{50}$$

est préservée par la déformation $F \to g(F)$. Comme seule la *mesurabilité* de la différentielle de Beltrami μ est requise pour que m vérifie (49), seule la mesurabilité de la structure conforme sur M est requise pour que le calcul quantique associé soit défini. De plus (b) montre que la condition de régularité sur $a \in L^\infty(M)$ définie par (50) ne dépend que de la structure quasiconforme de la variété M ([CST]). Un homéomorphisme local φ de $\mathbb{R}^n$ est *quasiconforme* ssi il existe $K < \infty$ tel que

$$H_\varphi(x) = \operatorname*{Lim\,sup}_{r\to 0} \frac{\max|\varphi(x) - \varphi(y)|\,;\, |x-y| = r}{\min|\varphi(x) - \varphi(y)|\,;\, |x-y| = r} \le K, \ \forall\, x \in \text{Domaine}\,\varphi. \tag{51}$$

Une structure quasiconforme sur une variété topologique M_n est donnée par un atlas quasiconforme. La discussion ci-dessus s'applique au cas général (n pair) ([CST]) et montre que le calcul quantique est bien défini pour toute variété quasiconforme. Le résultat de D. Sullivan [S] basé sur [Ki] montre que toute variété topologique M_n, $n \neq 4$ admet une structure quasiconforme. En utilisant le calcul quantique et la cohomologie cyclique à la place du calcul différentiel et de la théorie de Chern–Weil on obtient ([CST]) une formule locale pour les classes de Pontrjagin topologiques de M_n.

3. La Formule De L'indice Locale Et La Classe Fondamentale Transverse

Nous montrons dans cette section que le calcul infinitésimal ci-dessus permet le passage du local au global dans le cadre général des triplets spectraux $(\mathcal{A}, \mathcal{H}, D)$. Nous appliquons ensuite le résultat général au produit croisé d'une variété par le groupe des difféomorphismes.

Nous ferons l'hypothèse de régularité suivante sur $(\mathcal{A}, \mathcal{H}, D)$

$$a \text{ et } [D,a] \in \cap \operatorname{Dom} \delta^k, \ \forall a \in \mathcal{A} \tag{1}$$

où δ est la dérivation $\delta(T) = [|D|, T]$.

Nous désignerons par $\mathcal{B}$ l'algèbre engendrée par les $\delta^k(a)$, $\delta^k([D,a])$. La *dimension* d'un triplet spectral est bornée supérieurement par $p > 0$ ssi $a(D+i)^{-1}$ est un infinitésimal d'ordre $\frac{1}{p}$ pour tout $a \in \mathcal{A}$. Quand $\mathcal{A}$ est unifère cela ne dépend que du spectre de D.

La notion précise de dimension est définie comme le sous-ensemble $\Sigma \subset \mathbb{C}$ des singularités des fonctions analytiques

$$\zeta_b(z) = \operatorname{Trace}(b|D|^{-z}) \qquad \operatorname{Re} z > p, \ b \in \mathcal{B}. \tag{2}$$

Nous supposerons que Σ est discret et simple, i.e. que les ζ_b se prolongent à $\mathbb{C}/\Sigma$ avec des pôles simples en Σ.

Nous renvoyons à [CM2] pour le cas de spectre multiple.

L'indice de Fredholm de l'opérateur D détermine une application additive, $K_1(\mathcal{A}) \xrightarrow{\varphi} \mathbb{Z}$ donnée par l'égalité

$$\varphi([u]) = \operatorname{Indice}(PuP), \quad u \in GL_1(\mathcal{A}) \tag{3}$$

où P est le projecteur $P = \frac{1+F}{2}$, $F = \operatorname{Signe}(D)$.

Cette application est calculée par l'accouplement entre $K_1(\mathcal{A})$ et la classe de cohomologie du cocycle cyclique suivant

$$\tau(a^0, \ldots, a^n) = \operatorname{Trace}(a^0[F,a^1] \ldots [F,a^n]) \qquad \forall a^j \in \mathcal{A} \tag{4}$$

où $F = \operatorname{Signe} D$ et où n est un entier impair $n \geq p$.

Le problème est que τ est difficile à déterminer en général car la formule (4) implique la trace ordinaire au lieu de la trace locale $⨍$.

Ce problème est résolu par la formule suivante:

Théorème 4. ([CM2]) — *Soit $(\mathcal{A}, \mathcal{H}, D)$ un triplet spectral vérifiant les hypothèses* (1) *et* (2).

(a) *L'égalité* $⨍ P = \mathrm{Res}_{z=0} \operatorname{Trace}(P|D|^{-z})$ *définit une trace sur l'algèbre engendrée par* $\mathcal{A}$, $[D, \mathcal{A}]$ *et* $|D|^z$, $z \in \mathbb{C}$.

(b) *La formule suivante n'a qu'un nombre fini de termes non nuls et définit les composantes* $(\varphi_n)_{n=1,3,\ldots}$ *d'un cocycle dans le bicomplexe* (b, B) *de* $\mathcal{A}$

$$\varphi_n(a^0, \ldots, a^n) = \sum_k c_{n,k} ⨍ a^0 [D, a^1]^{(k_1)} \ldots [D, a^n]^{(k_n)} \, |D|^{-n-2|k|} \quad \forall a^j \in \mathcal{A}$$

où l'on note $T^{(k)} = \nabla^k(T)$ *et* $\nabla(T) = D^2 T - T D^2$, *et où* k *est un multiindice,* $c_{n,k} = (-1)^{|k|} \sqrt{2i} (k_1! \cdots k_n!)^{-1} \left((k_1+1) \ldots (k_1 + k_2 + \cdots + k_n + n)\right)^{-1} \Gamma\left(|k| + \frac{n}{2}\right)$, $|k| = k_1 + \cdots + k_n$.

(c) *L'accouplement de la classe de cohomologie cyclique* $(\varphi_n) \in HC^*(\mathcal{A})$ *avec* $K_1(\mathcal{A})$ *donne l'indice de Fredholm de* D *à coefficient dans* $K_1(\mathcal{A})$.

Rappelons que le bicomplexe (b, B) est donné par les opérateurs suivants agissant sur les formes multilinéaires sur l'algèbre $\mathcal{A}$

$$(b\varphi)(a^0, \ldots, a^{n+1}) =$$

$$\sum_0^n (-1)^j \, \varphi(a^0, \ldots, a^j a^{j+1}, \ldots, a^{n+1}) + (-1)^{n+1} \, \varphi(a^{n+1} a^0, a^1, \ldots, a^n) \tag{5}$$

$$B = AB_0, B_0 \, \varphi(a^0, \ldots, a^{n-1}) = \varphi(1, a^0, \ldots, a^{n-1}) - (-1)^n \, \varphi(a^0, \ldots, a^{n-1}, 1)$$

$$(A\psi)(a^0, \ldots, a^{n-1}) = \sum_0^{n-1} (-1)^{(n-1)j} \, \psi(a^j, a^{j+1}, \ldots, a^{j-1}). \tag{6}$$

Nous renvoyons à [Co] pour la normalisation de l'accouplement entre HC^* et $K(\mathcal{A})$.

Remarques. — (a) L'énoncé du Théorème 4 reste valable si l'on remplace dans toutes les formules l'opérateur D par $D|D|^\alpha$, $\alpha \geq 0$.

(b) Dans le cas pair, c'est-à-dire si l'on suppose que $\mathcal{H}$ est $\mathbb{Z}/2$ gradué par γ, $\gamma = \gamma^*$, $\gamma^2 = 1$, $\gamma a = a \gamma \;\; \forall a \in \mathcal{A}$, $\gamma D = -D\gamma$, on a une formule analogue pour un cocycle (φ_n), n pair qui donne l'indice de Fredholm de D à coefficient dans K_0. Cependant la composante φ_0 ne s'exprime pas en terme du résidu $⨍$ car elle est non locale pour $\mathcal{H}$ de dimension finie (cf. [CM2]).

(c) Quand le spectre de dimension Σ a de la multiplicité on a une formule analogue mais qui implique un nombre fini de termes correctifs, dont le nombre est borné indépendamment de la multiplicté (cf. [CM2]).

Le spectre de dimensions d'une variété V est $\{0, 1, \ldots, n\}$, $n = \dim V$, et est simple. La multiplicité apparait pour les variétés singulières et les ensembles de Cantor donnent des exemples de points complexes, $z \notin \mathbb{R}$ dans ce spectre. Nous discutons maintenant une construction géométrique générale pour laquelle les hypothèses (1) et (2) sont vérifiées. Il s'agit de construire la classe fondamentale en K-homologie d'une variété K-orientée M sans briser la symétrie du groupe $\mathrm{Diff}^+(M)$ des difféomorphismes de M qui préservent la K-orientation. De manière plus précise nous cherchons un triplet spectral, $(C^\infty(M), \mathcal{H}, D)$ de la même classe de K-homologie que l'opérateur de Dirac associé à une métrique riemannienne (cf. I

(11) et (12)) mais qui soit équivariant par rapport au groupe $\mathrm{Diff}^+(M)$ au sens de [K]. Cela signifie que l'on a une représentation unitaire $\varphi \to U(\varphi)$ de $\mathrm{Diff}^+(M)$ dans $\mathcal{H}$ telle que

$$U(\varphi)\, f\, U(\varphi)^{-1} = f \circ \varphi^{-1} \qquad \forall\, f \in C^\infty(M),\ \varphi \in \mathrm{Diff}^+(M) \tag{7}$$

et que

$$U(\varphi)\, D\, U(\varphi)^{-1} - D \text{ est borné pour tout } \varphi \in \mathrm{Diff}^+(M). \tag{8}$$

Lorsque D est l'opérateur de Dirac associé à une structure riemannienne le symbole principal de D détermine cette métrique et les seuls difféomorphismes qui vérifient (8) sont les isométries.

La solution de ce problème est essentielle pour définir la géométrie transverse des feuilletages et elle est effectuée en 2 étapes. La première est l'utilisation ([Co1]) de la métrique de courbure négative de l'espace $GL(n)/O(n)$ et de l'opérateur "dual Dirac" de Miscenko et Kasparov pour se ramener à l'action de $\mathrm{Diff}^+(M)$ sur l'espace total P du fibré des métriques de M. La deuxième, dont l'idée est due à Hilsum et Skandalis ([HS]) est l'utilisation des opérateurs hypoelliptiques pour construire l'opérateur D sur P.

On notera qu'alors que la géométrie équivariante obtenue pour P est de dimension finie et vérifie les hypothèses (1) (2) la géométrie obtenue sur M en utilisant le produit intersection avec le "dual Dirac" est de dimension infinie et θ-sommable

$$\mathrm{Trace}\,(e^{-\beta D^2}) < \infty \qquad \forall\, \beta > 0. \tag{9}$$

Par construction, le fibré $P \xrightarrow{\pi} M$ est le quotient $F/O(n)$ du $GL(n)$ fibré principal F des repères sur M par l'action du groupe orthogonal $O(n) \subset GL(n)$. L'espace P admet la structure canonique suivante: le feuilletage vertical $V \subset TP$, $V = \mathrm{Ker}\,\pi_*$ et les structures euclidiennes suivantes sur les fibrés V et $N = (TP)/V$. Le choix d'une métrique riemannienne $GL(n)$-invariante sur $GL(n)/O(n)$ détermine la métrique sur V et celle de N est la métrique tautologique: $p \in P$ détermine une métrique sur $T_{\pi(p)}(M)$ qui grâce à π_* est isomorphe à N_p.

Cette construction est fonctorielle pour les difféomorphismes de M.

Le calcul hypoelliptique adapté à cette structure est un cas particulier du calcul pseudodifférentiel sur les variétés de Heisenberg ([BG]). Il modifie simplement l'homogénéité des symboles $\sigma(p,\xi)$ en utilisant les homothéties:

$$\lambda \cdot \xi = (\lambda \xi_v, \lambda^2 \xi_n),\ \forall\, \lambda \in \mathbb{R}_+^* \tag{10}$$

où ξ_v, ξ_n sont les composantes verticales et normales du covecteur ξ. La formule (10) dépend de coordonnées locales (x_v, x_n) adaptées au feuilletage vertical mais le calcul pseudodifférentiel correspondant n'en dépend pas. Le symbole principal d'un opérateur hypoelliptique d'ordre k est une fonction, homogène de degré k pour (10), sur le fibré $V^* \oplus N^*$. Le noyau distribution $k(x,y)$ d'un opérateur pseudodifférentiel

T dans le calcul hypoelliptique admet un développement au voisinage de la diagonale de la forme

$$k(x,y) \sim \sum a_j(x, x-y) + a(x) \log |x-y|' + 0(1) \tag{11}$$

où a_j est homogène de degré $-j$ en $x-y$ pour (10) et où la métrique $|x-y|'$ est localement de la forme

$$|x-y|' = ((x_v - y_v)^4 + (x_n - y_n)^2)^{1/4}. \tag{12}$$

Comme dans le calcul pseudodifférentiel ordinaire, le résidu se prolonge aux opérateurs de tout degré et est donné par l'égalité

$$⨍ T = \frac{1}{v+2m} \int a(x) \tag{13}$$

où la 1-densité $a(x)$ ne dépend pas du choix de la métrique $|\ |'$ et où $v = \dim V$, $m = \dim N$ de sorte que $v + 2m$ est la dimension de Hausdorff de l'espace métrique $(P, |\ |')$.

L'opérateur D est défini par l'équation $D|D| = Q$ où Q est l'opérateur différentiel hypoelliptique de degré 2 obtenu en combinant (quand v est pair) l'opérateur $d_V d_V^* - d_V^* d_V$ de signature où d_V est la différentiation verticale, avec l'opérateur de Dirac transverse. (On utilise le revêtement métaplectique $M\ell(n)$ de $GL(n)$ pour définir la structure spinorielle sur M.) La formule explicite de Q utilise une connection affine sur M mais le choix de cette connection n'affecte pas le *symbole principal hypoelliptique* de Q et donc de D ce qui assure l'invariance (8) de D par rapport aux difféomorphismes de M.

Donnons la formule explicite de Q dans le cas $n = 1$, i.e. pour $M = S^1$. On remplace P par la suspension $SP = \mathbb{R} \times P$ pour se ramener au cas où la dimension verticale v est paire. Un point de $SP = \mathbb{R} \times P$ est paramétré par 3 coordonnées $\alpha \in \mathbb{R}$ et $p = (s, \tau)$ où $\tau \in S^1$ et où $s \in \mathbb{R}$ définit la métrique $e^{2s}(d\tau)^2$ en $\tau \in S^1$.

On munit SP de la mesure $\nu = d\alpha\, ds\, d\tau$ et l'on représente l'algèbre $C_c^\infty(SP)$ par opérateurs de multiplication dans $\mathcal{H} = L^2(SP, \nu) \otimes \mathbb{C}^2$. La fonctorialité de la construction ci-dessus donne la repésentation unitaire suivante du groupe $\mathrm{Diff}^+(S^1)$

$$(U(\varphi)^{-1}\xi)(\alpha, s, \tau) = \varphi'(\tau)^{1/2}\, \xi(\alpha, s - \log \varphi'(\tau), \varphi(\tau)). \tag{14}$$

Enfin l'opérateur Q est donné par la formule

$$Q = -2\partial_\alpha \partial_s \sigma_1 + \frac{1}{i}\, e^{-s} \partial_\tau \sigma_2 + \left(\partial_s^2 - \partial_\alpha^2 - \frac{1}{4}\right) \sigma_3 \tag{15}$$

où $\sigma_1, \sigma_2, \sigma_3 \in M_2(\mathbb{C})$ sont les 3 matrices de Pauli.

L'opérateur ∂_τ est de *degré* 2 dans le calcul hypoelliptique et l'on vérifie que Q est hypoelliptique.

Un long calcul donne le résultat suivant ([CM3]):

Théorème 5. — *Soit $\mathcal{A}$ l'algèbre produit croisé de $C_c^\infty(SP)$ par* $\mathrm{Diff}^+(S^1)$.

(a) *Le triplet spectral $(\mathcal{A}, \mathcal{H}, D)$ (où $\mathcal{A}$ agit dans $\mathcal{H}$ par* (14) *et $D|D| = Q$) satisfait les hypothèses* (1) *et* (2) *et son spectre de dimension est $\Sigma = \{0, 1, 2, 3, 4\}$.*

(b) *La seule composante non nulle du cocycle associé (Théorème 4) est φ_3 et elle est cohomologue à 2ψ où ψ est le 3-cocycle cyclique classe fondamentale transverse du produit croisé.*

L'intégralité de 2ψ, i.e. de l'accouplement $\langle 2\psi, K_1(\mathcal{A})\rangle$ résulte alors du Théorème 4. Le 3-cocycle ψ est donné par (cf. [Co])

$$\psi(f^0U(\varphi_0),\; f^1U(\varphi_1), f^2U(\varphi_2), f^3U(\varphi_3)) =$$

$$\int h^0\, dh^1 \wedge dh^2 \wedge dh^3 \text{ si } \varphi_0\varphi_1\varphi_2\varphi_3 = 1 \tag{16}$$

$$\text{et } = 0 = \text{si } \varphi_0\varphi_1\varphi_2\varphi_3 \neq 1$$

$$\text{avec } h^0 = f^0, h^1 = (f^1)^{\varphi_0}, h^2 = (f^2)^{\varphi_0\varphi_1}, h^3 = (f^3)^{\varphi_0\varphi_1\varphi_2}.$$

L'homologie entre φ_3 et 2ψ met en évidence l'action sur l'algèbre $\mathcal{A}$ de l'algèbre de Hopf engendrée par les transformations linéaires suivantes (pour la relation de δ_3 avec l'invariant de Godbillon Vey, voir [Co]) de $\mathcal{A}$

$$\delta_1(fU(\varphi)) = (\partial_\alpha f)\,U(\varphi),\; \delta_2(fU(\varphi)) = (\partial_s f)\,U(\varphi),$$

$$\delta_3(fU(\varphi)) = f\,e^{-s}\,\partial_\tau \log(\varphi^{-1})'\,U(\varphi),\; X(fU(\varphi)) = e^{-s}(\partial_\tau\, f)\,U(\varphi) \tag{17}$$

dont la compatibilité avec la multiplication de $\mathcal{A}$ est régie par le coproduit

$$\Delta\delta_j = \delta_j \otimes 1 + 1 \otimes \delta_j \qquad j = 1,2,3 \tag{18}$$

(i.e. les δ_j sont des dérivations de $\mathcal{A}$)

$$\Delta X = X \otimes 1 + 1 \otimes X - \delta_3 \otimes \delta_2 \tag{19}$$

où (19) montre que X est de degré 2.

4. La Notion De Variété Et Les Axiomes De La Géométrie

Commençons par spécifier la place de la géométrie riemannienne dans notre cadre en caractérisant (Théorème 6) les triplets spectraux correspondants. Soit $n \in \mathbb{N}$ la dimension, le triplet $(\mathcal{A}, \mathcal{H}, D)$ est supposé $\mathbb{Z}/2$ gradué par γ, $\gamma = \gamma^*$, $\gamma^2 = 1$ quand n est pair.

Les axiomes *commutatifs* sont les suivants:

(1) (Dimension) $ds = D^{-1}$ est infinitésimal d'ordre $\frac{1}{n}$.

(2) (Ordre un) $[[D, f], g] = 0 \quad \forall\, f, g \in \mathcal{A}$.

(3) (Régularité) Pour tout $f \in \mathcal{A}$, f et $[D, f]$ appartiennent à $\bigcap_k$ Domaine δ^k, où δ est la dérivation $\delta(T) = [|D|, T]$.

(4) (Orientabilité) Il existe un cycle de Hochschild $c \in Z_n(\mathcal{A}, \mathcal{A})$ tel que $\pi(c) = 1$ (n impair) ou $\pi(c) = \gamma$ (n pair), où π: $\mathcal{A}^{\otimes(n+1)} \to \mathcal{L}(\mathcal{H})$ est l'unique application linéaire telle que $\pi(a^0 \otimes a^1 \otimes \ldots \otimes a^n) = a^0[D, a^1] \ldots [D, a^n] \quad \forall\, a^j \in \mathcal{A}$.

(5) (Finitude) Le $\mathcal{A}$-module $\mathcal{E} = \bigcap_k$ Domaine D^k est projectif de type fini et l'égalité suivante définit une structure hermitienne sur $\mathcal{E}$

$$\langle a\xi, \eta\rangle = \int\!\!\!\!\!- a(\xi,\eta)\, ds^n \qquad \forall\, \xi,\eta \in \mathcal{E},\ a \in \mathcal{A}.$$

(6) (Dualité de Poincaré) La forme d'intersection $K_*(\mathcal{A}) \times K_*(\mathcal{A}) \to \mathbb{Z}$ donnée par la composition de l'indice de Fredholm de D avec la diagonale, $m_* : K_*(\mathcal{A}) \times K_*(\mathcal{A}) \to K_*(\mathcal{A} \otimes \mathcal{A}) \to K_*(\mathcal{A})$, est *inversible.*

(7) (Réalité) Il existe une isométrie antilinéaire J sur $\mathcal{H}$ telle que $Ja^*J^{-1} = a \quad \forall a \in \mathcal{A}$ et $J^2 = \varepsilon$, $JD = \varepsilon' DJ$, $J\gamma = \varepsilon''\gamma J$ où la table des valeurs de $\varepsilon, \varepsilon', \varepsilon'' \in \{-1, 1\}$ en fonction de n modulo 8 est donnée en (I.16).

Les axiomes (2) et (4) donnent la présentation de l'algèbre abstraite notée $(\mathcal{A}, ds)$ engendrée par $\mathcal{A}$ et $ds = D^{-1}$.

Théorème 6. — *Soit $\mathcal{A} = C^\infty(M)$ où M est une variété compacte de classe C^∞.*

(a) *Soit π une représentation unitaire de $(\mathcal{A}, ds)$ satisfaisant les conditions* (1) *à* (7). *Il existe alors une unique structure riemannienne g sur M telle que la distance géodésique soit donnée par :*

$$d(x,y) = \operatorname{Sup}\{|a(x) - a(y)|\ ;\ a \in \mathcal{A},\ \|[D,a]\| \leq 1\}.$$

(b) *La métrique $g = g(\pi)$ ne dépend que de la classe d'équivalence unitaire de π et les fibres de l'application $\{$classe d'équivalence unitaire$\} \to g(\pi)$ forment un nombre fini d'espaces affines $\mathcal{A}_\sigma$ paramétrés par les structures spinorielles σ de M.*

(c) *La fonctionnelle $\int\!\!\!\!- ds^{n-2}$ est quadratique et positive sur chaque $\mathcal{A}_\sigma$ où elle admet un unique minimum π_σ.*

(d) *π_σ est la représentation de $(\mathcal{A}, ds)$ dans $L^2(M, S_\sigma)$ donnée par les opérateurs de multiplication et l'opérateur de Dirac associé à la connection de Levi Civita de la métrique g.*

(e) *La valeur de $\int\!\!\!\!- ds^{n-2}$ en π_σ est l'action de Hilbert–Einstein de la métrique g:*

$$\int\!\!\!\!\!- ds^{n-2} = -c_n \int r\sqrt{g}\, d^n x,\ c_n = \tfrac{n-2}{12}\,(4\pi)^{-n/2}\, 2^{[n/2]}\,\Gamma\left(\tfrac{n}{2}+1\right)^{-1}.$$

L'exemple le plus simple pour comprendre la signification du théorème et de vérifier que la géométrie du cercle S^1 de longueur 2π est entièrement spécifiée par la présentation:

$$U^{-1}[D,U] = 1, \text{où } UU^* = U^*U = 1. \tag{1}$$

L'algèbre $\mathcal{A}$ étant celle des fonctions C^∞ de l'opérateur unitaire U, on a $S^1 =$ Spectre $(\mathcal{A})$ et l'égalité (1) est le cas le plus simple de l'axiome 4.

Remarques. — (a) L'hypothèse $\mathcal{A} = C^\infty(M)$ devrait résulter des axiomes (1) – (7) (et de la commutativité de $\mathcal{A}$). Il résulte de (3) et (5) que si $\mathcal{A}''$ est l'algèbre de von

Neumann engendrée par $\mathcal{A}$ on a:

$$\mathcal{A} = \left\{ T \in \mathcal{A}'' \; ; \; T \in \bigcap_{k>0} \operatorname{Dom} \delta^k \right\}, \tag{2}$$

ce qui montre que $\mathcal{A}$ est uniquement spécifiée dans $\mathcal{A}''$ par la donnée de D. Cela montre que $\mathcal{A}$ est stable par calcul fonctionnel C^∞ dans sa fermeture normique $A = \bar{\mathcal{A}}$ et en particulier que

$$\operatorname{Spectre} \mathcal{A} = \operatorname{Spectre} A. \tag{3}$$

Soit X cet espace compact; on devrait déduire des axiomes que l'application de X dans $\mathbb{R}^N$ donnée par les $a_i^j \in \mathcal{A}$ qui interviennent dans le cycle de Hochschild c de (4) est un plongement de X comme sous-variété C^∞ de $\mathbb{R}^N$ (cf. Proposition 15, p. 312 de [Co]).

(b) Rappelons qu'un cycle de Hochschild $c \in Z_n(\mathcal{A}, \mathcal{A})$ est un élément de $\mathcal{A}^{\otimes(n+1)}$, $c = \sum a_i^0 \otimes a_i^1 \ldots \otimes a_i^n$ tel que $bc = 0$, où b est l'application linéaire $b : \mathcal{A}^{\otimes n+1} \to \mathcal{A}^{\otimes n}$ telle que:

$$b(a^0 \otimes \ldots \otimes a^n) =$$

$$\sum_0^{n-1} (-1)^j \, a^0 \otimes \ldots \otimes a^j a^{j+1} \otimes \ldots \otimes a^n + (-1)^n \, a^n a^0 \otimes a^1 \otimes \ldots \otimes a^{n+1}.$$

La classe de Hochschild du cycle c détermine la *forme volume*.

(c) Nous utilisons la convention selon laquelle la courbure scalaire r est positive pour la sphère S^n, en particulier le signe de l'action $⨍ ds^{n-2}$ est le bon pour la formulation euclidienne de la gravitation. Par exemple, pour $n = 4$, l'action de Hilbert–Einstein $-\frac{1}{16\pi G} \int r \sqrt{g} \, d^4x$ coïncide avec l'aire $\frac{1}{\ell_p^2} ⨍ ds^2$ en unité de Planck.

(d) Quand M est une variété spinorielle l'application $\pi \to g(\pi)$ du théorème est surjective et si l'on fixe le cycle $c \in Z_n(\mathcal{A}, \mathcal{A})$ son image est l'ensemble des métriques dont la forme volume est fixée — (b).

(e) Si l'on supprime l'axiome 7 on a un résultat analogue au théorème en remplaçant les structures spinorielles par les structures spin^c ([LM]), mais l'on n'a plus unicité dans (c) à cause de la liberté dans le choix de la connection spinorielle.

(f) Il résulte de l'axiome 4 et de ([Co], Théorème 8, p. 309) que les opérateurs $a \, ds^n$, $a \in \mathcal{A}$ sont automatiquement mesurables de sorte que le symbole $⨍$ qui apparait dans 5 est bien défini.

Passons au cas général non commutatif. Étant donnée une algèbre involutive $\mathcal{A}$ d'opérateurs dans l'espace de Hilbert $\mathcal{H}$ la théorie de Tomita [Ta] associe à tout vecteur $\xi \in \mathcal{H}$ cyclique pour $\mathcal{A}$ et pour son commutant $\mathcal{A}'$,

$$\overline{\mathcal{A}\xi} = \mathcal{H}, \; \overline{\mathcal{A}'\xi} = \mathcal{H} \tag{4}$$

une involution antilinéaire isométrique $J : \mathcal{H} \to \mathcal{H}$ obtenue à partir de la décomposition polaire de l'opérateur

$$S \, a\xi = a^* \xi \qquad \forall \, a \in \mathcal{A} \tag{5}$$

et qui vérifie la propriété de commutativité suivante:

$$JA''J^{-1} = A'. \tag{6}$$

On a donc en particulier $[a, b^0] = 0 \quad \forall\, a, b \in \mathcal{A}$ où

$$b^0 = Jb^*J^{-1} \qquad \forall b \in \mathcal{A} \tag{7}$$

de sorte que $\mathcal{H}$ devient un $\mathcal{A}$-bimodule en utilisant la représentation de l'algèbre opposée $\mathcal{A}^0$ donnée par (7). Dans le cas commutatif on a $a^0 = a \quad \forall\, a \in \mathcal{A}$ de sorte que l'on ne perçoit pas la nuance entre module et bimodule.

Le théorème de Tomita est l'outil nécessaire pour assurer la substance des axiomes dans le cas général. Les axiomes (1) (3) et (5) sont inchangés, dans l'axiome de réalité (7) on remplace l'égalité $Ja^*J^{-1} = a \quad \forall\, a \in \mathcal{A}$ par

$$[a, b^0] = 0 \qquad \forall\, a, b \in \mathcal{A} \text{ où } b^0 = Jb^*J^{-1} \tag{7'}$$

et l'axiome (2) (ordre un) se formule ainsi

$$[[D, a], b^0] = 0 \qquad \forall\, a, b \in \mathcal{A}. \tag{2'}$$

(On notera que comme a et b^0 commutent $2'$ équivaut à $[[D, a^0], b] = 0 \quad \forall\, a, b \in \mathcal{A}$.)

L'axiome $(7')$ fait de $\mathcal{H}$ un $\mathcal{A}$-bimodule et donne une classe μ de KR^n-homologie pour l'algèbre $\mathcal{A} \otimes \mathcal{A}^0$ munie de l'automorphisme antilinéaire τ

$$\tau(x \otimes y^0) = y^* \otimes x^{*0}.$$

Le produit intersection de Kasparov [K] permet alors de formuler la dualité de Poincaré, comme l'invertibilité de μ

$$\exists\, \beta \in KR_n(\mathcal{A}^0 \otimes \mathcal{A}),\ \beta \otimes_{\mathcal{A}} \mu = \mathrm{id}_{\mathcal{A}^0},\ \mu \otimes_{\mathcal{A}^0} \beta = \mathrm{id}_{\mathcal{A}}. \tag{6'}$$

Ceci implique l'isomorphisme $K_*(\mathcal{A}) \xrightarrow{\cap \mu} K^*(\mathcal{A})$. La forme d'intersection

$$K_*(\mathcal{A}) \times K_*(\mathcal{A}) \to \mathbb{Z}$$

est obtenue à partir de l'indice de Fredholm de D à coefficient dans $K_*(\mathcal{A} \otimes \mathcal{A}^0)$ et n'utilise plus l'application diagonale $m : \mathcal{A} \otimes \mathcal{A} \to \mathcal{A}$ qui n'est un homomorphisme que dans le cas commutatif. Cette forme d'intersection est quadratique ou symplectique selon la valeur de n modulo 8.

L'homologie de Hochschild à coefficient dans un bimodule garde tout sons sens dans le cas général et l'axiome (4) prend la forme suivante

$(4')$ Il existe un cycle de Hochschild $c \in Z_n(\mathcal{A}, \mathcal{A} \otimes \mathcal{A}^0)$ tel que $\pi(c) = 1$ (n impair) ou $\pi(c) = \gamma$ (n pair).

(Où $\mathcal{A} \otimes \mathcal{A}^0$ est le $\mathcal{A}$ bimodule obtenu par restriction à la sous-algèbre $\mathcal{A} \otimes 1 \subset \mathcal{A} \otimes \mathcal{A}^0$ de la structure de $\mathcal{A} \otimes \mathcal{A}^0$ bimodule de $\mathcal{A} \otimes \mathcal{A}^0$, i.e.

$$a(b \otimes c^0)d = abd \otimes c^0 \qquad \forall\, a, b, c, d \in \mathcal{A}.)$$

Les axiomes (1), (3) et (5) sont inchangés dans le cas non commutatif et la démonstration de la mesurabilité des opérateurs $a(ds)^n$, $a \in \mathcal{A}$ reste valable en général.

Nous adopterons les axiomes (1), (2′), (3), (4′), (5), (6′) et (7′) dans le cas général comme définition d'une *variété spectrale* de dimension n. L'algèbre $\mathcal{A}$ étant fixée nous parlerons de géométrie spectrale sur $\mathcal{A}$ comme dans I.20 et I.21. On démontre que l'algèbre de von Neumann $\mathcal{A}''$ engendrée par $\mathcal{A}$ dans $\mathcal{H}$ est automatiquement finie et hyperfinie et on a la liste complète de ces algèbres à isomorphisme près [Co]. L'algèbre $\mathcal{A}$ est stable par calcul fonctionnel C^∞ dans sa fermeture normique $A = \bar{\mathcal{A}}$ de sorte que $K_j(\mathcal{A}) \simeq K_j(A)$, i.e. $K_j(\mathcal{A})$ ne dépend que de la topologie sous-jacente (définie par la C^* algèbre A). L'entier $\chi = \langle \mu, \beta \rangle \in \mathbb{Z}$ donne la caractéristique d'Euler sous la forme

$$\chi = \operatorname{Rang} K_0(\mathcal{A}) - \operatorname{Rang} K_1(\mathcal{A})$$

et le Théorème 4 en donne une formule locale.

Le groupe $\operatorname{Aut}(\mathcal{A})$ des automorphismes de l'algèbre involutive $\mathcal{A}$ joue en général le rôle du groupe $\operatorname{Diff}(M)$ des difféomorphismes d'une variété M. (On a un isomorphisme canonique $\operatorname{Diff}(M) \xrightarrow{\alpha} \operatorname{Aut}(C^\infty(M))$ donné par

$$\alpha_\varphi(f) = f \circ \varphi^{-1} \qquad \forall\, f \in C^\infty(M),\ \varphi \in \operatorname{Diff}(M).)$$

Dans le cas général non commutatif, parallèlement au sous-groupe normal $\operatorname{Int}\mathcal{A} \subset \operatorname{Aut}\mathcal{A}$ des automorphismes intérieurs de $\mathcal{A}$,

$$\alpha(f) = ufu^* \qquad \forall\, f \in \mathcal{A} \tag{8}$$

où u est un élément unitaire de $\mathcal{A}$ (i.e. $uu^* = u^*u = 1$), il existe un feuilletage naturel de l'espace des géométries spectrales sur $\mathcal{A}$ en classes d'équivalences formées des *déformations intérieures* d'une géométrie donnée. Une telle déformation est obtenue sans modifier ni la représentation de $\mathcal{A}$ dans $\mathcal{H}$ ni l'isométrie antilinéaire J par la formule

$$D \to D + A + JAJ^{-1} \tag{9}$$

où $A = A^*$ est un opérateur autoadjoint arbitraire de la forme

$$A = \Sigma\, a_i[D, b_i],\ a_i, b_i \in \mathcal{A}. \tag{10}$$

Le nouveau triplet spectral obtenu continue à vérifier les axiomes (1) — (7′).

L'action du groupe $\operatorname{Int}(\mathcal{A})$ sur les géométries spectrales (cf. I.21) se réduit à une transformation de jauge sur A, donnée par la formule

$$\gamma_u(A) = u[D, u^*] + uAu^*. \tag{11}$$

L'équivalence unitaire est implémentée par la représentation suivante du groupe unitaire de $\mathcal{A}$ dans $\mathcal{H}$,

$$u \to uJuJ^{-1} = u(u^*)^0. \tag{12}$$

La transformation (9) se réduit à l'identité dans le cas riemannien usuel. Pour obtenir un exemple non trivial, il suffit d'en faire le produit par l'unique géométrie spectrale sur l'algèbre de dimension finie $\mathcal{A}_F = M_N(\mathbb{C})$ des matrices $N \times N$ sur $\mathbb{C}$, $N \geq 2$. On a alors $\mathcal{A} = C^\infty(M) \otimes \mathcal{A}_F$, $\operatorname{Int}(\mathcal{A}) = C^\infty(M, PSU(N))$ et les déformations intérieures de la géométrie sont paramétrées par les potentiels de jauge pour une théorie de jauge de groupe $SU(N)$. L'espace $P(\mathcal{A})$ des états purs de l'algèbre $\mathcal{A}$ est le produit $P = M \times P_{N-1}(\mathbb{C})$ et la métrique sur $P(\mathcal{A})$ déterminée par la formule I.10 dépend du potentiel de jauge A. Elle coïncide avec la métrique de Carnot [G] sur P définie par la distribution horizontale de la connection associée à A (cf. [Co3]). Le groupe $\mathrm{Aut}(\mathcal{A})$ des automorphismes de $\mathcal{A}$ est le produit semi direct

$$\mathrm{Aut}(\mathcal{A}) = \mathcal{U} \rtimes \mathrm{Diff}(M) \tag{13}$$

du groupe $\mathrm{Int}(\mathcal{A})$ des transformations de jauges locales par le groupe des difféomorphismes. En dimension $n = 4$, les fonctionnelles d'action de Hilbert–Einstein pour la métrique riemannienne et de Yang–Mills pour le potentiel vecteur A apparaissent simplement, et avec les bons signes, dans le développement asymptotique en $\frac{1}{\Lambda}$ du nombre $N(\Lambda)$ de valeurs propres de D qui sont $\leq \Lambda$. On régularise cette expression en la remplaçant par

$$\mathrm{Trace}\ \overline{\omega}\left(\frac{D}{\Lambda}\right) \tag{14}$$

où $\overline{\omega} \in C_c^\infty(\mathbb{R})$ est une fonction paire qui vaut 1 sur l'intervalle $[-1,1]$, (cf. [CC]). Les seuls autres termes non nuls du développement asymptotique sont un terme cosmologique, un terme de gravité de Weyl et un terme topologique.

Un exemple plus élaboré de variété spectrale est le tore non commutatif $\mathbb{T}^2_\theta$. Le paramètre $\theta \in \mathbb{R}/\mathbb{Z}$ définit la déformation suivante de l'algèbre des fonctions C^∞ sur le tore $\mathbb{T}^2$, de générateurs U, V. Les relations

$$VU = \exp 2\pi i\theta\, UV \quad \text{et} \quad UU^* = U^*U = 1,\ VV^* = V^*V = 1 \tag{15}$$

définissent la structure d'algèbre involutive de $\mathcal{A}_\theta = \{\Sigma\, a_{n,m} U^n V^n\ ;\ a = (a_{n,m}) \in \mathcal{S}(\mathbb{Z}^2)\}$ où $S(\mathbb{Z}^2)$ est l'espace de Schwartz des suites à décroissance rapide. Comme pour les courbes elliptiques on utilise comme paramètre pour définir la géométrie de $\mathbb{T}^2_\theta$ un nombre complexe τ de partie imaginaire positive et, à isométrie près, cette géométrie ne dépend que de l'orbite de τ pour $PSL(2,\mathbb{Z})$ [Co]. Le phénomène nouveau qui apparaît est *l'équivalence de Morita* qui relie entre elles les algèbres $\mathcal{A}_{\theta_1}, \mathcal{A}_{\theta_2}$ lorsque θ_1 et θ_2 sont dans la même orbite de l'action de $PSL(2,\mathbb{Z})$ sur $\mathbb{R}$ [Ri].

Étant données une variété spectrale $(\mathcal{A}, \mathcal{H}, D)$ et une équivalence de Morita entre $\mathcal{A}$ et une algèbre $\mathcal{B}$ donnée par

$$\mathcal{B} = \mathrm{End}_{\mathcal{A}}(\mathcal{E}) \tag{16}$$

où $\mathcal{E}$ est une $\mathcal{A}$-module à droite, projectif de type fini et hermitien, on obtient une géométrie spectrale sur $\mathcal{B}$ par le choix d'une *connection hermitienne* sur $\mathcal{E}$. Une telle connection ∇ est une application linéaire $\nabla : \mathcal{E} \otimes_{\mathcal{A}} \Omega^1_D$ vérifiant les règles ([Co])

$$\nabla(\xi a) = (\nabla\xi)a + \xi \otimes da \qquad \forall\, \xi \in \mathcal{E},\ a \in \mathcal{A} \tag{17}$$

$$(\xi, \nabla\eta) - (\nabla\xi, \eta) = d(\xi, \eta) \qquad \forall\, \xi, \eta \in \mathcal{E} \tag{18}$$

où $da = [D, a]$ et où $\Omega^1_D \subset \mathcal{L}(\mathcal{H})$ est le $\mathcal{A}$-bimodule formé par les opérateurs de la forme (10).

Toute algèbre $\mathcal{A}$ est Morita équivalente à elle-même (avec $\mathcal{E} = \mathcal{A}$) et quand on applique la construction ci-dessus on obtient les déformations intérieures de la géométrie spectrale.

5. La Géométrie Spectrale De L'espace Temps

L'information expérimentale et théorique dont on dispose sur la structure de l'espace temps est résumée par la fonctionnelle d'action suivante, $\mathcal{L} = \mathcal{L}_E + \mathcal{L}_G + \mathcal{L}_{G\varphi} + \mathcal{L}_\varphi + \mathcal{L}_{\varphi f} + \mathcal{L}_f$, où $\mathcal{L}_E = -\frac{1}{16\pi G}\int r\sqrt{g}\,d^4x$ est l'action de Hilbert–Einstein et les 5 autres termes constituent le modèle standard de la physique des particules, couplé de manière minimale à la gravitation. Outre la métrique $g_{\mu\nu}$ ce Lagrangien implique plusieurs champs de bosons et de fermions. Les bosons de spin 1 sont le photon γ, les bosons médiateurs $W^\pm$ et Z et les huit gluons. Les bosons de spin 0 sont les champs de Higgs φ qui sont introduits pour briser la parité et pour que le mécanisme de brisure de symétrie spontanée confère une masse aux diverses particules sans contredire la renormalisabilité des champs de jauge non abéliens. Tous les fermions sont de spin $\frac{1}{2}$ et forment 3 familles de quarks et leptons.

Les champs impliqués dans le modèle standard ont a priori un statut très différent de celui de la métrique $g_{\mu\nu}$. Le groupe de symétrie de ces champs, à savoir le groupe des transformations de jauge locales:

$$\mathcal{U} = C^\infty(M, U(1) \times SU(2) \times SU(3)) \tag{1}$$

est a priori très différent du groupe $\mathrm{Diff}(M)$ de symétries de $\mathcal{L}_E$. Le groupe de symétrie naturel de $\mathcal{L}$ est le produit semidirect $\mathcal{U} \rtimes \mathrm{Diff}(M) = G$. La première question à résoudre si l'on veut donner une signification purement géométrique à $\mathcal{L}$ est de trouver un espace géométrique X tel que $G = \mathrm{Diff}(X)$. Ceci détermine, en tenant compte du relèvement des difféomorphismes aux spineurs, l'algèbre $\mathcal{A}$:

$$\mathcal{A} = C^\infty(M) \otimes \mathcal{A}_F, \ \ \mathcal{A}_F = \mathbb{C} \oplus \mathbb{H} \oplus M_3(\mathbb{C}), \tag{2}$$

où l'algèbre involutive $\mathcal{A}_F$ est la somme directe des algèbres $\mathbb{C}, \mathbb{H}$ des quaternions et $M_3(\mathbb{C})$ des matrices 3×3 complexes.

L'algèbre $\mathcal{A}_F$ correspond à un espace *fini* dont les fermions du modèle standard et les paramètres de Yukawa (masses des fermions et matrice de mélange de Kobayashi Maskawa) déterminent la géométrie spectrale de la manière suivante. L'espace de Hilbert $\mathcal{H}_F$ est de dimension finie et admet pour base la liste des fermions élémentaires. Par exemple pour la 1ère génération de leptons cette liste est

$$e_L, e_R, \nu_L, \bar{e}_L, \bar{e}_R, \bar{\nu}_L. \tag{3}$$

L'algèbre $\mathcal{A}_F$ admet une représentation naturelle dans $\mathcal{H}_F$ (cf. [Co3]) et en désignant par J_F l'unique involution antilinéaire qui échange f et $\bar{f}$ pour tout vecteur de la base, on a la commutation

$$[a, Jb^*J^{-1}] = 0 \qquad \forall\, a, b \in \mathcal{A}_F. \tag{4}$$

L'opérateur D_F est simplement donné par la matrice $\begin{bmatrix} Y & 0 \\ 0 & \bar{Y} \end{bmatrix}$ où Y est la matrice de couplage de Yukawa. De plus les propriétés particulières de Y assurent la commutation

$$[[D_F, a], b^0] = 0 \qquad \forall\, a, b \in \mathcal{A}_F. \tag{5}$$

La $\mathbb{Z}/2$ graduation naturelle de $\mathcal{H}_F$ vaut 1 pour les fermions gauches $(e_L, \nu_L \ldots)$ et -1 pour les fermions droits; on a

$$\gamma_F = \varepsilon\, \varepsilon^0 \text{ où } \varepsilon = (1, -1, 1) \in \mathcal{A}_F. \tag{6}$$

Nous renvoyons à [Co3] pour les vérifications des axiomes (1) — (7′). Le seul défaut est que le nombre de générations introduit une multiplicité dans la forme d'intersection, $K_0(\mathcal{A}) \times K_0(\mathcal{A}) \to \mathbb{Z}$, donnée par un multiple entier de la matrice 3×3

$$\begin{bmatrix} -1 & 1 & -1 \\ 1 & 0 & 1 \\ -1 & 1 & 0 \end{bmatrix}. \tag{7}$$

Nous reviendrons à la fin de cet exposé sur la signification de la géométrie spectrale $(\mathcal{A}_F, \mathcal{H}_F, D_F) = F$.

Le pas suivant consiste à calculer les déformations intérieures (formule III.9) de la géométrie produit $M \times F$ où M est une variété riemannienne spinorielle de dimension 4. Le calcul donne les bosons de jauge du modèle standard, $\gamma, W^{\pm}, Z$, les huits gluons et les champs de Higgs φ avec les bons nombres quantiques et montre que

$$\mathcal{L}_{\varphi f} + \mathcal{L}_f = \langle \psi, D\psi \rangle \tag{8}$$

où $D = D_0 + A + JAJ^{-1}$ est la déformation intérieure de la géométrie produit (donnée par l'opérateur $D_0 = \not{\partial} \otimes 1 + \gamma_5 \otimes D_F$).

La structure de produit de $M \times F$ donne une bigraduation de Ω_D^1 et une décomposition $A = A^{(1,0)} + A^{(0,1)}$ de A qui correspond à la décomposition (8). Le terme $A^{(1,0)}$ rassemble tous les bosons de spin 1 et le terme $A^{(0,1)}$ les bosons de Higgs qui apparaissent comme des termes de différence finie sur l'espace F. Cette bigraduation existe sur l'analogue Ω_D^2 des 2-formes ([Co]) et décompose la courbure $\theta = dA + A^2$ en trois termes $\theta = \theta^{(2,0)} + \theta^{(1,1)} + \theta^{(0,2)}$ 2 à 2 orthogonaux pour le produit scalaire

$$\langle \omega_1, \omega_2 \rangle = \int\!\!\!\!\!\!- \; \omega_1\, \omega_2^*\, ds^4. \tag{9}$$

Ainsi l'action de Yang–Mills, $\langle \theta, \theta \rangle = ⨍ \theta^2\, ds^4$ se décompose comme somme de 3 termes et on démontre que ces termes sont respectivement $\mathcal{L}_G$, $\mathcal{L}_{G\varphi}$ et $\mathcal{L}_\varphi$ pour $(2,0)$, $(1,1)$ et $(0,2)$ respectivement [Co].

L'action de Yang-Mills $⨍ \theta^2\, ds^4$ utilise la décomposition $D = D_0 + A + JAJ^{-1}$ et n'est donc pas, a priori, une fonction ne dépendant que de la géométrie définie par D. Nous avons vu en III.14 que, dans un cas plus simple, la combinaison $\mathcal{L}_E + \mathcal{L}_G$ apparaît directement dans le développement asymptotique du nombre de valeurs propres inférieures à Λ. Le même principe (cf. [CC]) s'applique au modèle standard et conduit à la fonctionnelle suivante

$$\text{Trace}\left(\varpi\left(\frac{D}{\Lambda}\right)\right) + \langle\psi, D\psi\rangle \tag{10}$$

dont le développement asymptotique ([CC]) donne $\mathcal{L}+$ un terme de gravité de Weyl et un terme en $r\varphi^2$ qui est le seul terme que l'on peut rajouter à $\mathcal{L}$ sans altérer le modèle standard. Nous renvoyons à [CC] pour l'interprétation physique de ces résultats.

La géométrie finie F ci-dessus était dictée par les résultats expérimentaux et il reste à en comprendre la signification conceptuelle à partir de l'analogue des groupes de Lie en géométrie non commutative, i.e. la théorie des groupes quantiques. Le fait simple (cf. [M]) est que le revêtement spinoriel Spin(4) de $SO(4)$ n'est pas un revêtement maximal parmi les groupes quantiques. On a Spin(4) $= SU(2) \times SU(2)$ et même le groupe $SU(2)$ admet grâce aux résultats de Lusztig des revêtements finis de la forme (Frobenius à l'∞):

$$1 \to H \to SU(2)_q \to SU(2) \to 1, \tag{11}$$

où q est une racine de l'unité, $q^m = 1$, m impair. Le cas le plus simple est $m = 3$, $q = \exp\left(\frac{2\pi i}{3}\right)$. Le groupe quantique fini H a une algèbre de Hopf de dimension finie très voisine de $\mathcal{A}_F$, et la représentation spinorielle de H définit un bimodule sur cette algèbre de Hopf de structure très voisine du bimodule $\mathcal{H}_F$ sur $\mathcal{A}_F$. Cela suggère d'étendre la géométrie spinorielle ([LM]) aux revêtements quantiques du groupe spinoriel, ce qui nécessite même pour parler de G-fibré principal, d'introduire un minimum de non commutativité (du style $C^\infty(M) \otimes \mathcal{A}_F$) dans l'algèbre des fonctions.

Mentionnons enfin que nous avons négligé dans cet exposé la nuance importante entre les signatures riemanniennes et lorentziennes.

Références

[At] M. F. Atiyah — *K-theory and reality*, Quart. J. Math. Oxford (2), **17** (1966), 367–386.

[B-G] R. Beals and P. Greiner — *Calculus on Heisenberg manifolds*, Annals of Math. Studies **119**, Princeton Univ. Press, Princeton, N.J., 1988.

[B] T. P. Branson — *An anomaly associated with 4-dimensional quantum gravity*, to appear in C.M.P.

[B-O] T. P. Branson and B. Ørsted — *Explicit functional determinants in four dimensions*, Proc. Amer. Math. Soc., **113** (1991), 669–682.

[B-W] A. R. Bernstein and F. Wattenberg — *Non standard measure theory*, **in** Applications of model theory to algebra analysis and probability, Edited by W. A. J. Luxenburg Halt, Rinehart and Winstin (1969).

[C-J-G] A. H. Chamseddine, J. Fröhlich and O. Grandjean — *The gravitational sector in the Connes–Lott formulation of the Standard model*, J. Math. Phys., **36** n° 11 (1995).

[C-C] A. Chamseddine and A. Connes — *The spectral action principle*, to appear.

[Co] A. Connes — *Noncommutative geometry*, Academic Press (1994).

[Co1] A. Connes — *Cyclic cohomology and the transverse fundamental class of a foliation*, Geometric methods in operator algebras, (Kyoto, 1983), pp. 52–144, Pitman Res. Notes in Math. **123** Longman, Harlow (1986).

[Co2] A. Connes — *Noncommutative geometry and reality*, Journal of Math. Physics **36** n°11 (1995).

[Co3] A. Connes — *Gravity coupled with matter and the foundation of noncommutative geometry*, to appear in C.M.P.

[Co-L] A. Connes and J. Lott — *Particle models and noncommutative geometry*, Nuclear Phys. B, **18B** (1990), suppl. 29–47 (1991).

[C-M1] A. Connes and H. Moscovici — *Cyclic cohomology, the Novikov conjecture and hyperbolic groups*, Topology **29** (1990), 345–388.

[C-M2] A. Connes and H. Moscovici — *The local index formula in noncommutative geometry*, GAFA, **5** (1995), 174–243.

[C-M3] A. Connes and H. Moscovici — *Hypoelliptic Dirac operator, diffeomorphisms and the transverse fundamental class.*

[Co-S] A. Connes and G. Skandalis — *The longitudinal index theorem for foliations*, Publ. Res. Inst. Math. Sci. Kyoto **20** (1984), 1139–1183.

[C-S-T] A. Connes, D. Sullivan and N. Teleman — *Quasiconformal mappings, operators on Hilbert space, and local formulae for characteristic classes*, Topology, Vol. 33, n°4 (1994), 663–681.

[D-T] T. Damour and J. H. Taylor — *Strong field tests of relativistic gravity and binary pulsars*, Physical Review D, Vol. 45 n°6 (1992), 1840–1868.

[Dx] J. Dixmier — *Existence de traces non normales*, C. R. Acad. Sci. Paris, Ser. A-B **262** (1966).

[D-F-R] S. Doplicher, K. Fredenhagen and J. E. Roberts — *Quantum structure of space time at the Planck scale and Quantum fields*, to appear in CMP.

[F] J. Fröhlich — *The noncommutative geometry of two dimensional supersymmetric conformal field theory*, Preprint ETH (1994).

[G-F] K. Gawedzki and J. Fröhlich — *Conformal Field theory and Geometry of Strings*, CRM Proceedings and Lecture Notes, Vol. 7 (1994), 57–97.

[Gh] E. Ghys — *L'invariant de Godbillon–Vey*, Sém. Bourbaki 1988/89, exposé 706, Astérisque **177–178**, S.M.F. (1989), 155–181.

[Gi] P. Gilkey — *Invariance theory, the heat equation and the Atiyah–Singer index theorem*, Math. Lecture Ser. **11**, Publish or Perish, Wilmington, Del., 1984.

[G] M. Gromov — *Carnot–Caratheodory spaces seen from within*, Preprint IHES/M/94/6.

[G-K-P] H. Grosse, C. Klimcik and P. Presnajder — *On finite 4-dimensional quantum field theory in noncommutative geometry*, CERN Preprint TH/96 — 51 Net Hepth/9602115.

[H-S] M. Hilsum, G. Skandalis — *Morphismes K-orientés d'espaces de feuilles et fonctorialité en théorie de Kasparov*, Ann. Sci. Ecole Norm. Sup. (4) **20** (1987), 325–390.

[I-K-S] B. Iochum, D. Kastler and T. Schücker — *Fuzzy mass relations for the Higgs*, J. Math. Phys. **36** n° 11 (1995).

[K-W] W. Kalau and M. Walze — *Gravity, noncommutative geometry and the Wodzicki residue*, J. of Geom. and Phys. **16** (1995), 327–344.

[K] G. Kasparov — *The operator K-functor and extensions of C^*-algebras*, Izv. Akad. Nauk. SSSR Ser. Mat., **44** (1980), 571–636.

[Ka] C. Kassel — *Le résidu non commutatif*, Séminaire Bourbaki 1988/89, exposé 708, Astérisque **177–178**, S.M.F. (1989) Vol. 199–229.

[Kas] D. Kastler — *The Dirac operator and gravitation*, Commun. Math. Phys. **166** (1995), 633–643.

[Ki] R. C. Kirby — *Stable homeomorphisms and the annulus conjecture*, Ann. Math. **89** (1969), 575–582.

[L-M] B. Lawson and M. L. Michelson — *Spin Geometry*, Princeton 1989.

[M] Y. Manin — *Quantum groups and noncommutative geometry*, Centre Recherche Math. Univ. Montréal (1988).

[Mi-S] J. Milnor and D. Stasheff — *Characteristic classes*, Ann. of Math. Stud., **76** Princeton University Press, Princeton, N.J. (1974).

[P] S. Power — *Hankel operators on Hilbert space*, Res. Notes in Math. **64**, Pitman, Boston, Mass. (1982).

[Ri1] M.A. Rieffel — *Morita equivalence for C^*-algebras and W^*-algebras*, J. Pure Appl. Algebra **5** (1974), 51–96.

[Ri2] M.A. Rieffel — *C^*-algebras associated with irrational rotations*, Pacific J. Math. **93** (1981), 415–429; MR 83b:46087.

[R] B. Riemann — *Mathematical Werke*, Dover, New York (1953).

[St] E. Stein — *Singular integrals and differentiability properties of functions*, Princeton Univ. Press, Princeton, N.J. (1970).

[Ste] J. Stern — *Le problème de la mesure*, Sém. Bourbaki 1983/84, exposé 632, Astérisque **121–122** (1985), 325–346.

[S1] D. Sullivan — *Hyperbolic geometry and homeomorphisms* **in** Geometric Topology, Proceed. Georgia Topology Conf. Athens, Georgia (1977), 543–555.

[S2] D. Sullivan — *Geometric periodicity and the invariants of manifolds*, Lecture Notes in Math. **197**, Springer (1971).

[Ta] M. Takesaki — *Tomita's theory of modular Hilbert algebras and its applications*, Lecture Notes in Math. **128**, Springer (1970).

[W] Weinberg — *Gravitation and Cosmology*, John Wiley and Sons, New York London (1972).

[Wo] M. Wodzicki — *Noncommutative residue, Part I. Fundamentals K-theory, arithmetic and geometry*, Lecture Notes in Math., 1289, Springer–Berlin (1987).

Alain CONNES
Collège de France
3, rue d'Ulm
75005 PARIS
et
I.H.E.S.
35, route de Chartres
91440 BURES-sur-YVETTE

Reprinted from Proc. Int. Congr. Math., 1983

THE WORK OF W. THURSTON

by

C. T. C. WALL

Thurston has fantastic geometric insight and vision; his ideas have completely revolutionized the study of topology in 2 and 3 dimensions, and brought about a new and fruitful interplay between analysis, topology and geometry.

The central new idea is that a very large class of closed 3-manifolds should carry a hyperbolic structure — be the quotient of hyperbolic space by a discrete group of isometries, or equivalently, carry a metric of constant negative curvature. Although this is a natural analogue of the situation for 2-manifolds, where such a result is given by Riemann's uniformization theorem, it is much less plausible — even counter-intuitive — in the 3-dimensional situation. The case of a manifold fibred over a circle with fibre a surface of genus exceeding 1 seems particularly implausible, and this was the case Thurston examined first. The fibration is determined by a homeomorphism h of the surface, and in seeking to put h (and hence its iterates) into normal form, he was led to consider the images of curves under high iterates of h: these may eventually become dense in some regions, leading to measured foliations. In general, he was led to consider a lamination, which is a disjoint union of injectively immersed curves, which may be dense in some regions and not in others. These ideas gave rise to a geometric model of Teichmüller space and its compactification, which revolutionized thinking in this already highly developed subject.

In this, Thurston was able to draw on his previous work on foliation theory. He swept through this subject producing startling new examples (the Godbillon-Vey invariant takes on uncountably many values), extending the Haefliger foliation theory to closed manifolds by an entirely novel geometric technique, calculating homology of classifying spaces of foliations and relating it to homology of diffeomorphism groups, etc. One dramatic example: any closed manifold of Euler characteristic zero admits a codimension 1 foliation.

The analysis of the diffeomorphisms of a surface concludes with a partition of the surface into well defined pieces, on each of which h has a structure of particularly simple form: the "generic" case is that in which h is Anosov. This partition gives a partition of the total space of the fibration mentioned above, and again we have a geometric structure on each piece. This leads to a reformulation of the project, which occurred at a timely moment in the independent development of 3-dimensional topology.

In the late 1950's, Papakyriakopoulos obtained fundamental new results on embeddings of discs and spheres into 3-manifolds, one consequence of which was a unique decomposition of M by spheres into irreducible pieces. It seemed that these were largely determined by their fundamental groups, and much work went into studying properties of these, though even a basic question like "are the fundamental groups of knots residually finite?" remained unanswered. A method was developed by Haken and Waldhausen using successive decompositions of M by "essential" embedded surfaces to answer such questions: this gave excellent results in the cases to which it applied. These are the manifolds M formerly called "sufficiently large" but now, following Thurston, "Haken 3-manifolds". The condition is that there exists an embedded surface of positive genus whose fundamental group maps injectively to $\pi_1(M)$. If M has a boundary component of positive genus, or if the first Betti number is non-zero, such a surface can be constructed. Yet more particular are the examples given by Seifert fibre spaces. A close analysis of these by Jaco and Shalen and (independently) Johannsen led to a decomposition of M into a "Seifert" piece and an "atoroidal" piece: in the latter, every embedded torus is parallel to the boundary.

Thurston was now able to conjecture that every irreducible, atoroidal 3-manifold has a hyperbolic structure, and to prove it in the case of Haken manifolds. This includes, for example, the complement of any knot in S^3 (other than torus knots and companion knots) — which allows one to prove the residual finiteness mentioned above. It also led to the solution of the Smith Conjecture — that the fixed point set of a periodic homeomorphism of S^3 is always unknotted — a problem which had attracted a great deal of attention over a forty year period. In fact, Thurston formulates the general conjecture in more attractive terms: every compact 3-manifold has a canonical decomposition into pieces each of which has a geometric structure.

Here, a type of geometric structure is defined by a (simply-connected) model manifold X, with a Riemannian metric, and its group G_x of isometrics. To say that M has a structure of type X means that there are local coordinate charts from M to X, with transformations between different charts given on their overlaps by elements of G_X. From this, Thurston constructs a developing map φ: $\tilde{M} \to X$ from the universal covering of M: the structure on M is complete if φ is a homeomorphism. In order to be of interest, X must satisfy some conditions (e.g. G_X is transitive): such X he calls *geometries*. Thurston then shows that there are just 8 three-dimensional geometries: in addition to the sphere, Euclidean and hyperbolic space, and products of two-dimensional models with a line, there are 3 further cases, in each of which X is a Lie group. Although the hyperbolic structures are by far the most profound, this general theory of geometric structures has clarified and synthesized much previous work in the other cases also: the list of 8 three-dimensional geometries has not been previously obtained.

The main theorem proved to date is that every compact Haken 3-manifold admits a geometrical decomposition as above. However, hyperbolic structures have also been obtained for numerous non-Haken manifolds. There is also an extended conjecture: that if the manifold has a finite group of automorphisms, then a decomposition, and geometric structures on the pieces, can be found so that the group respects these. This is now known in most cases where a decomposition exists. Of particular difficulty are finite group actions on the 3-sphere. Thurston has shown that most of these are equivalent to orthogonal actions, but the fixed-point free case still eludes the method. Results on these problems have been obtained by other authors using minimal surface theory. Thurston's method involves using his main theorem to obtain a hyperbolic structure on the subset where the group acts freely.

As might be expected, the proof of the general result is long and involves many new ideas: it is not yet all available in detail. All I can do here is to mention a few of the ingredients.

Thurston showed that one can pass from any point in Teichmüller space to any other point by a unique "left earthquake". An example of an earthquake is an incomplete Dehn twist: cut a Riemann surface along a simple closed geodesic and then identify the banks after moving each point on one by the same distance. In a general earthquake the simple closed geodesic is replaced by a lamination. Thurston's earthquake theorem was used by his student Kerckhoff to solve the Nielsen realization problem (every finite subgroup of the Teichmüller modular group has a fixed point).

Discrete isometry groups of hyperbolic 3-space were first studied by Poincaré, who dubbed them "Kleinian groups". These have been much studied by analysts, and Ahlfors' finiteness theorem obtained in the 1960's was a fundamental result. Thurston has studied deformations of such groups (in order to patch together hyperbolic structures defined on two pieces of a manifold): this involves a deep study of limit sets. He has shown that a quasiconformal map on S^3 which conjugates one Kleinian group to another extends to H^3 as a quasiconformal volume preserving map with the same property. Typically, one of these groups is Fuchsian, with limit set S^1, but the image Jordan curve is fractal, with Hausdorff dimensional $d > 1$. It can be constructed as the boundary of a disc obtained by bending the standard D^2 along all the curves of a lamination. Thurston has also shown that for a large class of Kleinian groups (including "degenerate" ones), the limit set has measure zero: thus proving another conjecture which had resisted repeated earlier attempts.

Thurston's work has had an enormous influence on 3-dimensional topology. This area has a strong tradition of "bare hands" techniques and relatively little interaction with other subjects. Direct arguments remain essential, but 3-dimensional topology has now firmly rejoined the main stream of mathematics.

Reprinted from Proc. Int. Congr. Math., 1986

THE WORK OF SIMON DONALDSON

by

MICHAEL F. ATIYAH

In 1982, when he was a second-year graduate student, Simon Donaldson proved a result [**1**] that stunned the mathematical world. Together with the important work of Michael Freedman (described by John Milnor), Donaldson's result implied that there are "exotic" 4-spaces, i.e., 4-dimensional differentiable manifolds which are topologically but not differentiably equivalent to the standard Euclidean 4-space R^4. What makes this result so surprising is that $n = 4$ is the only value for which such exotic n-spaces exist. These exotic 4-spaces have the remarkable property that (unlike R^4) they contain compact sets which cannot be contained inside any differentiably embedded 3-sphere!

To put this into historial perspective, let me remind you that in 1958 Milnor discovered exotic 7-spheres, and that in the 1960s the structure of differentiable manifolds of dimension ≥ 5 was actively developed by Milnor, Smale (both Fields Medallists), and others, to give a very satisfactory theory. Dimension 2 (Riemann surfaces) was classical, so this left dimensions 3 and 4 to be explored. At the last Congress, in Warsaw, Thurston received a Fields Medal for his remarkable results on 3-manifolds, and now at this Congress we reach 4-manifolds. I should emphasize that the stories in dimensions 3, 4 and $n \geq 5$ are totally different, with the low-dimensional cases being much more subtle and intricate.

Although I have highlighted the exotic 4-space as a spectacular corollary of the Freedman/Donaldson results, this is a by-product; their work is actually devoted to studying *closed* 4-manifolds. To such a 4-manifold, one associates standard topological invariants. In particular, for an *oriented* manifold, one gets a symmetric integer matrix of determinant ± 1 defined by the intersection properties of the 2-cycles (and depending on a choice of basis). Freedman showed that all such matrices can occur for topological 4-manifolds. Donaldson's result was that, among positive definite matrices, only those equivalent to the unit matrix can occur for differentiable 4-manifolds.[1] This is a severe restriction and shows that the differentiable and topological situations are totally different.

The surprise produced by Donaldson's result was accentuated by the fact that his methods were completely new and were borrowed from theoretical physics, in the form of the Yang–Mills equations. These equations are essentially a nonlinear generalization of Maxwell's equations for electro-magnetism, and they are the variational equations associated with a natural geometric functional. Differential geometers study connections and curvature in fibre bundles, and the Yang–Mills functional is just the L^2-norm of the curvature. If the group of the fibre bundle is the circle, we get back the linear Maxwell theory, but for nonabelian Lie groups, we get a nonlinear theory. Donaldson uses only the simplest nonabelian group, namely SU(2), although in principle other groups can and will perhaps be used.

Physicists are interested in these equations over Minkowski space-time, where they are hyperbolic, and also over Euclidean 4-space, where they are elliptic. In the Euclidean case, solutions giving the absolute minimum (for given boundary conditions at ∞) are of special interest, and they are called *instantons*.

Several mathematicians (including myself) worked on instantons and felt very pleased that they were able to assist physics in this way. Donaldson, on the other hand, conceived the daring idea of reversing this process and of using instantons on a general 4-manifold as a new geometrical tool. In this he has been brilliantly successful: he has unearthed totally new phenomena and simultaneously demonstrated that the Yang–Mills equations are beautifully adapted to studying and exploring this whole new field.

Of course, the use of differential equations in geometry is not new; the study of geodesics or minimal surfaces are classical examples. However, in these cases a solution of the differential equation (e.g., a minimal surface) is used as a geometrical object. Donaldson's use of instantons is quite different. I should explain that instantons as solutions of a minimization problem are not unique but typically depend on a finite number of continuous parameters, and it is the nonlinear space of these instanton parameters that Donaldson uses as a geometrical tool. The closest prior example of such an approach is the (linear) Hodge theory of harmonic forms. In fact, Hodge was directly motivated by Maxwell's equations, and instantons are a natural nonlinear generalization of harmonic forms. In the linear case the parameter space is of course linear and determined by its dimension, but in the nonlinear case there is much more information embodied in the parameter space, which is a topologically interesting manifold.

The success of Donaldson's program depends on having a thorough understanding of the analysis of the Yang–Mills equation. On needs existence, regularity, and convergence theorems, all of which are quite delicate, involving both local and global

[1] Actually, in [**1**] Donaldson restricted himself to simply connected manifolds, but more recently he has succeeded in removing this restriction.

aspects. Fortunately, C. H. Taubes [**6, 7**] and K. Uhlenbeck [**8, 9**] have provided these analytical foundations, and so one can proceed to use instantons as an effective geometric tool. However, instantons cannot be bought off the shelf: to use them one has to understand and become involved with the full details of the analysis, and Donaldson has had to do this in order to put them to geometric use.

The Yang–Mills equations depend on fixing a background metric on the 4-manifold and, as in Hodge theory, Donaldson has to study the effect of varying the metric in order to derive results which depend only on the underlying manifold. Because of the nonlinearity, this is a more serious problem than in Hodge theory and great care is needed.

In fact, the Yang–Mills equations depend only on the conformal class of the metric and this conformal invariance is fundamental in physics where it implies the absence of a basic length scale. Analytically it is a source of difficulty making the equations a delicate border-line case where certain compactness arguments just fail, so that a sequence of instanton solutions can pick up Dirac delta functions in the limit. It is, however, just this delicate failure that Donaldson exploits geometrically: instead of the delta functions being regarded as undesirable singularities, they provide the key link between the 4-manifold and the instanton parameter space. One might say that the physicist's ambivalence to particles and fields is the essence of Donaldson's theory.

When Donaldson proved his first result it was by no means clear if this was some isolated case or whether instantons could be used more generally. Since then, however, Donaldson has, with great insight and skill, developed and exploited instantons with remarkable success. He has extended his results to the case of indefinite intersection matrices, providing further constraints on the topology of differentiable 4-manifolds. He has also, in the other direction, produced new invariants of 4-manifolds which can be used to distinguish smooth manifolds which are topologically equivalent. In particular, he has shown that complex algebraic surfaces (of complex dimension 2 and so of real dimension 4) appear to play a key role. In a very elegant paper [**2**] he proved an existence theorem which showed that, on an algebraic surface, instantons (or rather their parameter spaces) have a purely algebraic description, coinciding with what algebraic gometers call stable vector bundles. His new invariants can then be calculated algebraically and he used this [**3**] to exhibit two algebraic surfaces which are homeomorphic but not diffeomorphic. One of these surfaces is rational and his results strongly suggest that the rationality of an algebraic surface may be a differentiable property (it is *not* topological).

I indicated earlier that mathematicians has been working on the original physicists' problem of explicitly finding all instantons on Euclidean 4-space. In a short but decisive paper [**4**] Donaldson linked this problem with algebraic vector bundles on the complex projective plane (viewed as a compactification of $R^4 = C^2$). He also applied similar ideas [**5**] to solve a related but more difficult physical problem, that of magnetic monopoles. He proved the remarkably simple result that the parameter space of monopoles of magnetic charge k can be identified with the space of rational functions of a complex variable of degree k.

When Donaldson produced his first few results on 4-manifolds, the ideas were so new and foreign to geometers and topologists that they merely gazed in bewildered admiration. Slowly the message has gotten across and now Donaldson's ideas are beginning to be used by others in a variety of ways.

From what I have said you can see that Donaldson has opened up an entirely new area; unexpected and mysterious phenomena about the geometry of 4-dimensions have been discovered. Moreover, the methods are new and extremely subtle, using difficult nonlinear partial differential equations. On the other hand, this theory is firmly in the mainstream of mathematics, having intimate links with the past, incorporating ideas from theoretical physics, and tying in beautifully with algebraic geometry. It is remarkable and encouraging that such a young mathematician can understand and harness such a wide range of ideas and techniques in so short a time and put them to such brilliant use. It is an indication that mathematics has not lost its unity, or its vitality.

References

[1] S. K. Donaldson, *Self-dual connections and the topology of smooth 4-manifolds*, Bull. Amer. Math. Soc. **8** (1983), 81–83.

[2] ———, *Anti-self-dual connections over complex algebraic surfaces and stable vector bundles*, Proc. London Math. Soc. **50** (1985), 1–26.

[3] ———, *La topologic differentielle des surfaces complexes*, C. R. Acad. Sci. Paris **301** (1985), 317–320.

[4] ———, Instantons and geometric invariant theory, Comm. Math. Phys. **93** (1984), 453–460.

[5] ———, *Nahm's equations and the classification of monopoles*, Comm. Math. Phys. **96** (1984), 387–407.

[6] ———, C. H. Taubes, *Self-dual connections on non-self-dual 4-manifolds*, J. Differential Geom. **17** (1982), 139–170.

[7] ———, *Self-dual connections on manifolds with indefinite intersection matrix*, J. Differential Geom. **19** (1984), 517–560.

[8] K. K. Uhlenbeck, *Connections with L^p bounds on curvature*, Comm. Math. Phys. **83** (1982), 11–30.

[9] ———, *Removable singularities in Yang–Mills fields*, Comm. Math. Phys. **83** (1982), 31–42.

University of Oxford
Mathematical Institute
Oxford OX1 3LB, England

Simon K. Donaldson

SIMON K. DONALDSON

Full Name: Simon Kirwan Donaldson

Date of birth: 20th August 1957

Place of birth: Cambridge, England

Marital status: Married (1986), three children

Education: Sevenoaks School, Kent (1970–75)

Pembroke College, Cambridge University (1976–80)

Worcester College, Oxford University (1980–83)

Degrees: B.A., Cambridge (1979)

D. Phil, Oxford (1983), M.A. (1984)

Positions held: Junior Research Fellow,

All Souls College, Oxford (1983–85)

Visiting member, Institute for Advanced Study,

Princeton (1983–84)

Post still held: Wallis Professor of Mathematics,

Oxford University (1985–present)

Prizes, etc.: London Mathematical Society, Junior Whitehead Prize, 1985.

Fellow of the Royal Society 1986; Royal Medal, 1992.

Field Medal of the International Mathematical Union, 1986.

Sir William Hopkins Prize of the Cambridge Philosophical

Society, 1991.

Crafoord Prize of Royal Swedish Academy

of Sciences, 1994. (with S-T Yau)

Reprinted from Proc. Int. Congr. Math., 1983

REMARKS ON GAUGE THEORY, COMPLEX GEOMETRY AND 4–MANIFOLD TOPOLOGY

by

S. K. DONALDSON

The Mathematical Institute, Oxford

This article is made up of a number of rather different parts. Nearly all my research work has fallen under two headings:

(1) Differential geometry of holomorphic vector bundles,
(2) Applications of gauge theory to 4-manifold topology.

These two areas intertwine, but have also developed into separate, fairly well-defined fields. In the last part of the article, which is the only part which contains any significant technical content, I will explain how — motivated by recent work of Tian — one can extend some of the ideas which are well known under heading (1) above to problems in Kahler–Einstein geometry. The second and third parts of the article deal with heading (2), and I will give a very rapid survey of the development of this field, mainly through reference to other survey articles, and describe a result of Fintushel and Stern in order to illustrate some of the problems of current research interest. In the first part of the article, as a piece of self-indulgence, I will reminisce briefly about my own early encounters with these two fields.

1. Reminiscences, 1980–83

When I arrived in Oxford in 1980 as a beginning doctoral student I was fortunate to be presented with a problem by my research supervisor, Nigel Hitchin, which turned out to be outstandingly fruitful and prescient. The problem concerned Yang–Mills connections on holomorphic vector bundles over Kahler manifolds. If X is a Kahler manifold and E is a holomorphic vector bundle over X then the curvature $F(A)$ of a compatible connection A on E is a bundle-valued form of type $(1, 1)$. The Kahler metric defines a contraction $\Lambda : \Omega^{1,1} \to \Omega^0$ so one gets an endomorphism $\Lambda F(A)$ of E. The equation $\Lambda F(A) = 0$ is a special form of the Yang–Mills equations, which co-incides with the instanton equation in the case of complex dimension 2, and Hitchin suggested that the condition that a holomorphic bundle admit such a connection should be related to the algebro-geometric condition of *stability*. (Similar conjectures were made independently at about the same time by Kobayashi). One of the main pieces of the evidence for the conjecture was the old (1963) result of Narasimhan and Seshadri, dealing with the case of Riemann surfaces, which had been cast in "Yang–Mills" form, and studied in great depth, by Atiyah and Bott shortly before [1]. I soon began to tinker with analytical approaches to Hitchin's conjecture, perhaps motivated by the renowned work of Yau, a few years before, on the Calabi conjecture. I also spent many hours with the paper of Eells and Sampson [15] on harmonic maps, since it seemed a good idea to try to find the desired connections as a limit of solutions to a nonlinear "heat" equation, in the manner pioneered by Eells and Sampson. In one way or another I put together various things in the year 1980–81 which seemed to me to be heading towards an analytical proof of Hitchin's conjecture, but in the autumn of 1981 this work was interrupted by the realisation that Yang–Mills theory, in the shape of the instanton equations, could be applied to problems in 4-manifold topology.

This topological turn came about as an accidental by-product of the project described above. In retrospect, one can certainly see that this development was bound to occur sooner or later. The decisive shift was really one of attitude: the main focus of the work by mathematicians in the late 1970's on the instanton equation involved the interaction with twistor geometry, culminating in the famous ADHM construction of all instantons over 4-sphere, and this discussion was tied to manifolds with special geometric structures, rather than general Riemannian 4-manifolds. For example large parts of the foundational paper [2] of Atiyah, Hitchin and Singer carry over to general 4-manifolds without any change, although the whole paper restricts attention to the class of "self-dual" manifolds. Perhaps the first work in which the general case was considered was that of Taubes [30], which proved an existence theorem for instantons using Taubes' celebrated "grafting" technique. In fact this change in emphasis was not completely clear-cut in Taubes' work: the first version of [30] restricted attention to a manifolds admitting a metric with a certain positive curvature property (which was exploited in a Weitzenboch formula). The revised, published, version relaxed this hypothesis to consider 4-manifolds with no self-dual harmonic forms. It is a very simple consequence of the Hodge theory, although this was not pointed out explicitly by Taubes in [30], that this condition is a purely topological one — it is the same as saying that the intersection form of the 4-manifold is negative definite.

I studied Taubes' paper in detail in 1980–81: it fitted in with my thinking about Hitchin's problem in the following way. In studying the nonlinear heat equation mentioned above the essential thing was to obtain analytical compactness theorems which would allow one to get some kind of limit. This is closely related to understanding the compactness of instanton moduli spaces: now the algebro-geometric literature contained various examples of moduli spaces of stable bundles, and one can observe in these examples that the moduli spaces have natural compactifications in which one adjoins points "at infinity" made up of configurations of point in the underlying complex surface. It was therefore natural to make the hypothesis, assuming the conjectured relation between instantons and holomorphic bundles, that instanton moduli spaces over general 4-manifolds should be compactified by adjoining configurations of points — this is the now well-known "bubbling" phenomenon, the points correspond to sequences of connections whose curvature densities converge to delta-functions on the manifold. Taubes' grafting construction then appears as the inverse procedure, constructing connections "near to infinity" in the moduli space.

In fact the papers of Uhlenback [32, 33] which appeared about that time contained essentially all the analysis required to put this picture on a firm footing. The papers do not discuss "bubbling" explicitly — perhaps the arguments were supposed to be obvious to experts by analogy with the work of Sacks and Uhlenbeck [27] in the harmonic maps case — so it took some months before I really absorbed the material, and in that time it seemed natural to test the compactification hypothesis by contemplating various examples. How about the 4-manifold $\overline{\mathbf{CP}}^2$ (the complex projective plane with reversed orientation)? This has negative definite form, so Taubes' work applies. The deformation theory of [2] says that the moduli space M (in the simplest, charge 1, case) is 5-dimensional and it is natural to guess that Taubes' construction produces a collar $\overline{\mathbf{CP}}^2 \times (0,1) \subset M$ and the complement should be compact, according to the ideas above. But I had learnt about cobordism theory, so something funny must happen inside M, since the complex projective plane is not a boundary. The solution of course is that there is a reducible connection in the moduli space, where the bundle reduces to $S^1 \subset \mathrm{SU}(2)$, and this is a singular point. In fact, in view of the symmetry group $\mathrm{SU}(3)$ of the projective plane which must act on M, about the only thing that could happen is that M is a *cone* over $\overline{\mathbf{CP}}^2$, with the reducible connection as vertex. (This was later confirmed by explicit constructions of Buchdahl [6] and myself [8].) From this it was a short step to consider a general 4-manifold with definite intersection form, obtaining the picture described in the accompanying article, reproduced from the Proceedings of the 1983 International Congress, and leading to the conclusion that the intersection form must be standard. The point I wish to make is that the chain of reasoning was to a large extent a product of naivete; the initial impetus being the desire to test the compactification hypothesis for instanton moduli spaces. Moreover, it was not immediately clear to me what use the argument should be put to: I did not even know that there were any non-standard quadratic forms! At that time I shared an office in the Mathematical Institute with Mike Hopkins, and he put me straight on this point. Then one wondered if it was known for elementary reasons that 4-manifolds could not have non-standard intersection forms — I had no particular acquaintance with 4-manifold theory at that time. Another possibility seemed to lie in the orientation of the moduli space: maybe there were subtleties of index theory involved here; but this idea did not last long since Atiyah (who was supervising my work by this time) quickly furnished a proof that the moduli spaces were orientable. All in all, within a few months, it became clear that if this chain of reasoning could be made solid then one would get important new results about smooth 4-manifolds, in sharp contrast with those of Freedman (which appeared, by coincidence, at much the same time) in the topological case.

I would like to mention one other ingredient in this chain of ideas which raises certain wider issues. This concerns the local structures of the instanton moduli spaces. Atiyah, Hitchin and Singer's deformation theory for instantons was adapted from the work of Kuranishi and others, stretching back to Riemann, on the deformations of complex structures on manifolds — the main ideas apply to a wide variety of structures. In these theories one has cohomology groups H^0, H^1, H^2, and the moduli space of structures is locally a manifold, modelled on H^1, if H^0 (which is the Lie algebra of the automorphism group) and H^2 (the obstruction space) vanish. Now the classical problems to which this theory is applied are typically very "rigid" — there is just *one* moduli space of complex structures associated to a particular smooth manifold; take it or leave it. The obstruction spaces H^2 (for example in the case of complex surfaces) are often very large, and it is hard to say much about the moduli spaces on general grounds. The instanton problem, at least once one moves away from 4-manifolds with special structures, is rather different since one has the infinite dimensional space of metrics on the underlying 4-manifold which appear as parameters: each metric gives a moduli space. One can go further and consider arbitrary deformations of the instanton equation of various kinds: provided the deformation is reasonably controlled these give moduli spaces of solutions which serve just as well for the topological arguments one wants to make. Thus the situation is very "flexible". With this shift in perspective one is studying the instanton moduli spaces essentially in the framework of "Fredholm differential topology" and there are general transversality arguments that mean one can reduce to the case when the obstruction spaces vanish. The arguments are not difficult, and go back at least to Smale [29]: the issue is more a matter of change of view-point than anything else. One can carry this "flexible" point of view over to other moduli problems: contemplating what the structure of a moduli space should be after some kind of generic deformation. One area in which this has been done is in the moduli spaces of complex curves, where the generic deformations arise naturally from the consideration of almost-complex structures. But in other problems it is not so clear what significance such generic deformations should have. Another issue is that in higher dimensions one encounters over-determined equations, reflected in further cohomology groups $H^i, i \geq 3$, and it is not so clear what the right differential topological framework should be. For example, in theory of the Casson invariant, which is closely allied to the instanton theory, one is familiar with the idea that the space of irreducible SU(2)-representations of a 3-manifold fundamental group $\pi_1(M^3)$ "should be" a finite set of points; in the sense that it becomes so after a generic perturbation, although the actual representation space could be a variety of higher dimension. It is not clear whether there is any useful extension of this notion to representations of the fundamental group of higher dimensional manifolds.

I will now close this reminisence with a few concluding remarks. First, for completeness, I will go back to the conjecture of Hitchin and Kobayashi. I put together a proof of the conjecture for the case of complex surfaces in mid-1983 [7]: this proof was really a hybrid of two different lines of attack, one of which was written up in [14] and the other, dealing with the higher dimensional case but using a substantial input from algebraic geometry, in [10]. Meanwhile Uhlenbeck and Yau, deploying more powerful analytical techniques, gave what is probably the most natural proof in [34]. Turning back to 4-manifolds: a striking development, which came out in late 1983 was in the work of Fintushel and Stern who found an approach which gave the main results on definite forms, also using instanton moduli spaces but for connections with structure group SO(3). Their approach avoided the analysis of the boundary of the moduli space, which was precisely the most difficult technical part of the original argument and as I have described above, the route by which the argument was discovered! Finally, I would like to mention the beginnings of the development of 4-manifold *invariants* defined by instanton moduli spaces. In their applications, these are complementary to the first results: rather than showing that some intersection forms do not occur for any smooth 4-manifold, one wants to distinguish manifolds with the same intersection form. The general scheme by which one could try to obtain invariants from instanton moduli spaces was fairly clear to me by the middle of 1983: that is, one considers the pairings

$$\langle\, \mu(\alpha_1) \cup \cdots \cup \mu(\alpha_r), [M]\,\rangle, \tag{1}$$

where $[M]$ is the fundamental homology class of a moduli space, interpreted via one's understanding of its compactification, and the $\mu(\alpha_i)$ are cohomology classes over the moduli space which one can construct by a general procedure (that I had learnt from the paper of Atiyah and Bott). The problem for some time was how to calculate any examples, and using these to obtain new information about 4-manifolds. But what seemed to be clear was the existence of the following dichotomy: *either* these pairings would give new information, which would be good, *or* they could be expressed in terms of some known data, in which case it would be a good problem to find formulae for them.

2. Gauge Theory and 4–Manifolds: Survey of Surveys

Following the reminiscence above, I will now pass rapidly over the ensuing decade. An enormous amount of work was done in this period, by many mathematicians, on the interaction between gauge theory (in the shape of the instanton equations) and 4-manifold topology; and there exist a number of surveys of these developments. My contribution [9] to the Proceedings of the 1986 International Congress was written when the foundations of the theory, and the first applications, were complete. This is also roughly the material covered in the book [14]. In the late 1980's there was a great volume of work, much of it devoted to calculations of particular invariants. Perhaps the first breakthrough in the way of calculations was the work of Friedman and Morgan on elliptic surfaces [18]. One can also refer to the article [17], which contains many conjectures which set the direction for much of this research. My contribution [11] to the Proceedings of the 1992 European Congress surveys this period of very diverse developments. A particularly notable development in the late 1980's came through the work of Floer, leading to the Floer homology groups of 3-manifolds and the development of "cutting and pasting" formulae for instanton invariants. This is a topic which is not very well served, at present, in the literature, but there are various expositions of the ideas, for example [12] and the articles in the volume [19].

In a sense the survey article [11] was written at the "wrong" time, coming just before the dramatic discoveries by Kronheimer and Mrowka [23] of fundamental structural formulae for the instanton invariants (although this timing may perhaps add to its interest as a historial record). Let us recall that the simplest set of invariants of a 4-manifold X are polynomial functions on $H_2(X)$:

$$f_d(\alpha) = \langle\, \mu(\alpha)^d, M_X^{2d}\, \rangle, \tag{2}$$

where M_X^{2d} is a moduli space of instantons over X of dimension $2d$. Kronheimer and Mrowka's work applies to simply connected 4-manifolds X with $b^+(X)$, the dimension of a maximal positive subspace for the intersection form, at least 3, and satisfying a further technical condition of being of "simple type". In this case they show that the generating function

$$f(\alpha) = \sum_d \frac{1}{d!}\, f_d(\alpha)$$

can be written as

$$f(\alpha) = e^{\alpha.\alpha/2} \sum_i a_i e^{\kappa_i.\alpha}, \tag{3}$$

where the κ_i run over a finite collection of "basic classes" in $H^2(X)$, and the a_i are numerical co-efficients. Kronheimer and Mrowka's proof was long and difficult, but their result lead to a radical advance in the field: it became possible to find all the invariants for many 4-manifolds whereas a few years before it was a major task to compute a single one.

Hard on the heels of this structural theorem came the renowned work of Seiberg and Witten (1994) which more-or-less brings our story up to date. All the work we have discussed so far falls within the realms of geometry, topology and global analysis. While some of the technicalities may be complicated, the foundations are perfectly secure. By contrast, Seiberg and Witten's work depended essentially on Quantum Field Theory. Witten had shown in 1989 [35] that the cohomology pairings could be obtained from functional integrals of the shape

$$\iint e^{-\|F(A)\|^2} \mathcal{F}(A, \phi) \mathcal{D}A \mathcal{D}\phi, \tag{4}$$

where the integral runs over the space of connections A and certain auxiliary fields ϕ. In 1994, Seibergy and Witten carried through a sophisticated analysis of this Quantum Field Theory invoking a new "S-duality" principle. A prominent feature of their theory is a complex parameter u, related to the auxiliary fields appearing in the functional integral, and S-duality goes over to modular properties with respect to the parameter u. Geometrically, u parametrises a family of elliptic curves, which degenerate when $u = \pm 1$. The analysis of the theory at the points ± 1 lead Seiberg and Witten to write down their celebrated equations; these are classical equations for a connection a on a $U(1)$ bundle $L \to X$ and a spinor field ψ over a 4-manifold, coupled to the line bundle L. The Seiberg–Witten equations have the shape:

$$D_a \psi = 0, \qquad F_a^+ = \psi^* \psi.$$

(The equations really only involve a Spin^c structure, so one does not need to restrict to Spin 4-manifolds.) Seiberg and Witten predicted that the basic classes κ_i of Kronheimer and Mrowka, and the co-efficients a_i, can be obtained in a more direct way from the moduli spaces of solutions to the Seiberg–Witten equations. But in any event, as Witten [36] and others quickly showed, the new equations could be used to obtain essentially all the results of the old theory in a much simpler way, and many new results besides. These developments, at least up to the spring of 1995, are surveyed in the article [13].

It is amusing to speculate about the other ways in which this story might have run. It would have certainly been quite possible for geometers to have written down the Seiberg–Witten equations, and discovered the invariants they lead to, at any time in the last decade — they fit in well with a large body of work on the differential geometry of Yang–Mills–Higgs theories, albeit that this has mainly emphasised the case of Riemann surfaces. Indeed one can imagine a story in which the Seiberg–Witten equations were discovered before the instanton theory in which case, going back to the dichotomy at the end of the previous section, it would have been very natural to search for formulae relating the two.

3. Elliptic Functions and Modular Forms

The structural formula (3) of Kronheimer and Mrowka pretty much completes the theory of the instanton invariants for a large class of 4-manifolds. But when one moves outside this class, for example to manifolds such as the complex projective plane with $b^+ < 3$, many interesting questions remain. At the time of writing great progress is being made on these questions by Gottsche and other [20, 21], the striking thing being the appearance of complicated formulae involving modular forms. The first formula of this kind was found in 1993 by Fintushel and Stern [16]. This deals with the relation between the invariants of a 4-manifold X and its "blow-up" $X' = X\#\overline{\mathbf{CP}}^2$. One can analyse a moduli space $M_{X'}$ of instantons over X' by considering a family of Riemannian metrics on X' in which the "neck" of the connected sum is pulled out into a long tube. This is the standard technique in the Floer theory and, roughly speaking, it allows instantons over X' to be analysed in terms of solutions over the two summands. It is quite straightforward to obtain formulae in this way for low-dimensional moduli spaces $M_{X'}$, or more precisely when the "contribution" to the dimension from the $\overline{\mathbf{CP}}^2$ side is small. If we write e for the generator of $H_2(\overline{\mathbf{CP}}^2) \subset H_2(X')$ then one wants to compute

$$\langle\, \mu(e)^d \phi', M_{X'}\, \rangle, \tag{5}$$

where ϕ' is a cohomology class over the moduli space associated to homology classes in X. For general reasons one expects this to be expressed in the form

$$\left\langle\, \sum b_{d,i} v^i \phi, M_X \,\right\rangle, \tag{6}$$

where ϕ is the cohomology class over M_X defined in the same way as $\phi' \in H^*(M_{X'})$, $v \in H^4(M_X)$ is the 4-dimensional class $\mu(\text{point})$ and the $b_{d,i}$ are universal numerical co-efficients. The problem becomes harder as the exponent d increases: the difficulty is that the points at infinity in the compactifications of the moduli spaces play an essential role, and a direct attack — as in [25] — requires a detailed and laborious description of the ends of the moduli spaces. Fintushel and Stern found a trick which by-passes these difficulties. They considered the double blow-up $X'' = X\#\overline{\mathbf{CP}}^2\#\overline{\mathbf{CP}}^2$, with two cohomology classes $e_1, e_2, \in H_2(X'')$ corresponding to the two summands. The class e_1+e_2 is represented by a sphere of self-intersection -2, so X'' can also be regarded as a "generalised connected sum"

$$X'' = Y\#_{P^3} N, \tag{7}$$

where P^3 is a copy of the real projective 3-space embedded in X'', decomposing this 4-manifold into a neighbourhood N of the -2 sphere and its complement Y. Now the same neck-stretching analysis can be applied to this description of N, much as for connected sums (since P^3 is not very different from a 3-sphere). In particular we can analyse a pairing of the form

$$\langle\, \mu(e_1 + e_2)^4 \psi, M_{X''}\, \rangle,$$

where $e_1 + e_2$ is the generator of $H_2(N)$ and ψ is a cohomology class associated to the other piece Y. This is similar to the analysis of the problem (5) on X' with the particular value $d = 4$, which is small enough to be tractable, using the arguments like those of [18]. One gets a formula relating the pairings for three different moduli spaces: $M_{X''}, M^*_{X''}, M^{**}_{X''}$ say, over X'':

$$\langle\, \mu(e_1 + e_2)^4 \psi, M_{X''}\, \rangle = -4\langle\, \mu(e_1 + e_2)^2 v\psi, M^*_{X''}\, \rangle - 4\langle\, \psi, M^{**}_{X''}\, \rangle,$$

where v is the 4-dimensional class, as before. In particular, Fintushel and Stern observed that this applies when ϕ is a power $\mu(e_1 - e_2)^n$, since the class $e_1 - e_2$ is supported in Y, so

$$\langle \mu(e_1 + e_2)^4 \mu(e_1 - e_2)^n, M_{X''} \rangle = -4\langle \mu(e_1 + e_2)^2 \mu(e_1 - e_2)^n v, M^*_{X''} \rangle$$
$$-4\langle \mu(e_1 - e_2)^n, M^{**}_{X''} \rangle.$$

On the other hand, the different terms can be expanded out using the blow-up formula (6) for the two summands, and the formula translates into a recursion relation for the co-efficients $b_{d,i}$. The key thing is that the exponent n is arbitrary — increasing n does not make the stretching analysis any harder in the decomposition (7) — and this gives information about the high values of d which appear intractable from the point of view of the original connected sum decomposition. (Actually, Fintushel and Stern's argument does rather more than we have said, since it also gives the existence of the general shape (6) of the formula, without having to assume this *a priori*.) In turn, by elementary manipulation, the recursion relation for the co-efficient $b_{d,i}$ can be expressed as a differential equation for the generating function

$$B(x,t) = \sum b_{d,i} x^i t^d.$$

The differential equation is

$$B^{(4)}B - 4B'''B' + 3(B'')^2 + 4x(B''B - (B')^2) + 2B^2 = 0,$$

where all derivatives are with respect to t, the variable x being thought of as a parameter. This can be integrated three times, using information about some low order terms to fix the constants of integration, to show that the function $y = -(\log B)'' - \frac{x}{3}$ satisfies the familiar equation

$$(y')^2 = 4y^3 - g_2 y - g_3,$$

defining the Weierstrasse $\wp$-function, where $g_2 = 4\left(\frac{x^2}{3} - 1\right)$, $g_3 = \frac{8x^3 - 36x}{27}$, and Finstushel and Stern deduce finally that

$$B(x,t) = e^{-t^2 x/6} \sigma_x(t),$$

where $\sigma_x(t)$ is a particular Weierstrasse σ-function (a solution of $\left(\frac{\sigma'}{\sigma}\right)' = -\wp$) corresponding to these values of g_2, g_3.

While the geometric input into the argument of Fintushel and Stern is comparatively straightforward, the argument is indirect and one gets little insight into why the answer should turn out to involve these particular special functions. The work of Gottsche has a similar flavour. Thus what is lacking at the moment is a conceptual proof of these formulae. Elliptic curves and the associated functions have a central place in the Seiberg–Witten Quantum Field Theory analysis, so it does seem very likely that there are conceptual derivation from that point of view. On the other hand the Quantum Field Theory arguments of Seiberg and Witten lie at present a long way outside the boundary of standard mathematics: both with regard to the techniques and language, and probably also with regard to the underlying foundations (because of the notorious problems of defining functional integrals rigorously). So an exciting problem at the moment is to throw some kind of bridge across this gap. As well as the conceptual attraction of this problem, there are many concrete open questions that one would like to answer. For example one would like to blend this blow-up formula of Fintushel and Stern into the general Floer theory, which applies to other kinds of decompositions of 4-manifolds — the equivariant Floer theory of Austin and Braam [3] is limited at present by precisely the same kind of dimensional restrictions which obstructed a direct attack on the blow-up formula.

The questions we have outlined in the previous paragraph, concluding our brief survey of this line of work on 4-manifolds and gauge theory, can perhaps be best thought of in a wider context. Over the last decade ideas from Quantum Field Theory have cut a wide swathe through many different fields of Geometry and Topology, often predicting results that are, at best, only proved later by more laborious, less perspicuous, techniques. So it is natural to expect that the line of work we have been discussing will flow into this larger stream, which probably makes up the dominant and most exciting line of research in geometry at the moment.

4. Stability and Kahler–Einstein Geometry

In this section we return to some of the themes from complex differential geometry, sketched in Section 1. The Hitchin–Kobayashi conjecture involves the solution of a nonlinear partial differential equation, and a certain amount of detailed analysis is inevitably involved. However the problem fits into a general framework, in large part going back to Atiyah and Bott [1], involving the geometry of infinite dimensional groups and spaces. Apart from anything else, this framework is useful as a guide to the analysis, and the differential-geometric calculations that arise. Let us briefly recall this "standard picture".

Suppose we have the following data:

(1) a complex manifold $\mathcal{C}$ with Kahler metric, giving a symplectic form Ω;
(2) a Lie group $\mathcal{G}$ which acts on $\mathcal{C}$ by holomorphic isometries;
(3) an equivariant *moment map* $\mu : \mathcal{C} \to \mathrm{Lie}(\mathcal{G})^*$;
(4) a complexification $\mathcal{G}^c$ of $\mathcal{G}$, and an extension of the action to a holomorphic action of $\mathcal{G}^c$ on $\mathcal{C}$

Then the general principle is that one has an identification of "symplectic" and complex quotients:

$$\mathcal{C}_s/\mathcal{G}^c = \mu^{-1}(0)/\mathcal{G}, \tag{8}$$

where $\mathcal{C}_s \subset \mathcal{C}$ is an open, $\mathcal{G}^c$-invariant, subset of "stable points". The important thing is that the notion of stability should depend only on the holomorphic geometry of the situation, while the symplectic quotient depends on the Hermitian geometry. To define the stable points it is useful to suppose that we have the additional data:

(5) a $\mathcal{G}^c$-equivariant holomorphic line bundle $\mathcal{L} \to \mathcal{C}$, with a $\mathcal{G}$-invariant unitary connection having curvature $i\Omega$ and with the action defined by the moment map μ. (This means that the infinitesimal action $\xi \in \mathrm{Lie}(\mathcal{G})$ on a section s of $\mathcal{L}$ is $\nabla_{v(\xi)} s + i\mu(\xi)s$, where $v(\xi)$ is the vector field on $\mathcal{C}$ defined by ξ.)

Then to test the stability of a point $x \in \mathcal{C}$ we consider the $\mathcal{G}^c$-orbit $\mathcal{O} \subset \mathcal{L}$ of any non-zero $\hat{x} \in \mathcal{L}$ lying over x. The point x is stable if the orbit $\mathcal{O}$ is a *closed* subspace of the total space of the line bundle. The point x satisfies $\mu(x) = 0$ if and only if $\hat{x}$ minimises the norm within its orbit $\mathcal{O}$: the content of the main principle (8) is that the norm has a minimum, unique up to the action of $\mathcal{G}$, in each closed orbit.

In finite dimensions the main example of this set-up is when $\mathcal{G}$ is a compact Lie group, $\mathcal{C}$ is a complex projective space and $\mathcal{L}$ is the tautological bundle. Thus we are dealing with linear actions of $\mathcal{G}$ and its complexification on $\mathbf{C}^n$, and the orbits in the tautological bundle can be identified with the orbits in $\mathbf{C}^n$. This is the setting for the Hilbert–Mumford Geometric Invariant Theory, which is the original source of these ideas [24].

There are a number of general features of this set-up which give interesting results in particular cases. For example, it is easy to prove in the general picture that if $x \in \mathcal{C}$ is a point with $\mu(x) = 0$ then the stabiliser of x in $\mathcal{G}^c$ is the complexification of the stabiliser in $\mathcal{G}$. Now, following an idea which apparently goes back to Cartan (see [22], page 196), and which was carried through by Richardson [26], consider a finite-dimensional complex vector space V with a fixed non-degenerate symmetric form and the action of $\mathcal{G}^c = O(V)$ on the subspace

$$\mathcal{C} \subset \mathbf{P}(\mathrm{Hom}(V \otimes V, V)),$$

representing Lie brackets (i.e. solutions of the Jacobi identity), whose Killing form is the given form, and so make V into a semi-simple Lie algebra. The subgroup $\mathcal{G}$ is the compact subgroup of maps which preserve in addition a fixed Hermitian form on V. Then, just as in the first part of [26], one shows that all the points in $\mathcal{C}$ are stable. Then one can apply the general principles above to find a norm-minimising representative $\hat{x}$ for any isomorphism class of semi-simple Lie algebra structure on V. The stabiliser of $\hat{x}$ in $\mathcal{G}^c$ is the Lie group corresponding to this Lie algebra, and one obtains the result that this group is the complexification of a compact subgroup — the stabiliser in $\mathcal{G}$. This point of view allows one to simplify the calculations in [26], as well as fitting them into a general setting.

The Hitchin–Kobayashi problem leads to a well-known infinite-dimensional example of this standard picture. Here one considers a C^∞ vector bundle E over a Kahler manifold X with a fixed Hermitian metric on E. The space $\mathcal{C}$ is the space of $\overline{\partial}$-operators on E, which can also be regarded as the space of unitary connections. The group $\mathcal{G}$ is the group of unitary automorphisms of E, which acts on $\mathcal{C}$ in the ordinary way by gauge transformations. The complexification $\mathcal{G}^c$ is the group of complex linear automorphisms of E which acts on $\mathcal{C}$ by conjugation of the $\overline{\partial}$-operators. There is a $\mathcal{G}^c$-invariant subspace $\mathcal{C}^{\text{int}}$ of integrable $\overline{\partial}$-operators and the orbits in this subspace precisely parametrise the unitary connections compatible with a given holomorphic structure. The moment map μ can be identified with the contraction ΛF of the curvature of a connection. Provided the underlying manifold X is algebraic, there is also a line bundle $\mathcal{L}$ over $\mathcal{C}$, which can be identified with a certain "determinant line bundle" [10]. In this setting the Kobayashi–Hitchin conjecture can be regarded as two statements:

(1) the identification $\mathcal{C}_s/\mathcal{G}^c = \mu^{-1}(0)/\mathcal{G}$ holds in this infinite-dimensional case,
(2) the notion of stability in the infinite-dimensional picture matches up with the notion of stability found earlier by algebraic-geometers (by embedding the bundle-classification problem in a finite-dimensional Geometric Invariant Theory picture of the above kind.)

Now we turn to the recent work of Tian [31]. This concerns the question: when does a compact complex manifold V admit a Kahler–Einstein metric? More precisely one considers some initial Kahler metric ω_0 and the set of metrices in the same cohomology class, which can be represented in the form $\omega = \omega_0 + i\overline{\partial}\partial\phi$, for a real-valued Kahler potential ϕ. One seeks a metric which satisfies the Einstein equation: $\text{Ric} = \lambda g$. The cases when $\lambda \leq 0$ are well-understood through the renowned work of Calabi, Yau and others. When $\lambda > 0$ the problem is still not completely solved. It has been known for a long time that the there are obstructions to the existence of a solution coming from the holomorphic vector fields on the manifold: Tian finds a new kind of obstruction involving Geometric Invariant theory. An obvious condition for a solution with $\lambda > 0$ to exist is that the complex manifold V be a *Fano manifold* i.e. that the anticanonical bundle K_V^* be ample. Then for large s the sections of K_V^{-s} give a projective embedding $V \to \mathbf{P}^n$, and hence an orbit $[V]$ of the natural $SL(n+1)$ action on the Hilbert scheme parametrising subvarieties of $\mathbf{P}^n$. Tian proves that if V admits a Kahler–Einstein metric this orbit must be stable, and conjectures that the converse is true.

This conjecture of Tian, whose solution would represent a complete solution of the existence problem for Kahler–Einstein metrics, clearly falls into the same general pattern as the Kobayashi–Hitchin conjecture, so it is natural to ask whether the problem can be set up in the standard picture described above. We will now show that, with one reservation, this can indeed be done. (This builds on unpublished work of Atiyah and Quillen in the case of one complex dimension.)

Consider a compact symplectic manifold (M^{2n}, ω), and suppose for simplicity that $H^1(M) = 0$. Let $\mathcal{J}$ be the space of almost-complex structures on M which are compatible with ω. This is the space of sections of a bundle over M with fibre the Siegel upper half space $Sp(2n)/U(n)$, which has a $Sp(2n)$-invariant Kahler metric. This fibre-metric, together with the volume form on M, endows $\mathcal{J}$ with the structure of an infinite-dimensional Kahler manifold. More explicity, given one almost complex structure $J \in \mathcal{J}$ a variation of almost complex structure can be represented by a tensor $\mu \in \Omega^{0,1}_J(T)$, where we use the usual notation from complex geometry. The $(1, 0)$ forms for the new almost-complex structure J' have the shape $\alpha + \mu(\alpha)$, where α is in $\Omega^{1,0}_J$. Using the hermitian metric defined by J and ω we can identify $\overline{T}^*$ with T, and so $\Omega^{0,1}(T)$ with the sections of $T \otimes T$. Then, to first order in μ, the condition that J' be compatible with ω is that μ is a section of $s^2(T) \subset T \otimes T$. So the tangent space of $\mathcal{J}$ at J is the space of sections of the complex vector bundle $s^2(T)$, and this has a Hermitian metric defined by integration over M, in the usual way. It is not hard to see that this hermitian structure on $\mathcal{J}$ is Kahler — by Darboux's theorem one can essentially reduce to the case of $\mathrm{Maps}(M, Sp(2n)/U(n))$. We may also consider an infinite dimensional complex subvariety $\overset{\mathrm{int}}{\mathcal{J}} \subset \mathcal{J}$ representing integrable almost complex structures.

Now let $\mathcal{G}$ be the identity component of the group of symplectomorphisms of (M, ω). The Lie algebra of $\mathcal{G}$ can be identified with the space C^∞_0 of functions on M with integral 0: a function H defines the usual Hamiltonian vector field X_H on M with

$$i_{X_H}(\omega) = dH.$$

We want to identify a moment map for the action of $\mathcal{G}$ on $\mathcal{J}$. To do this we review some (almost) complex differrential geometry. Suppose E is any complex vector bundle over M and ∇ is a covariant derivative on E. Then given an almost-complex structure $J \in \mathcal{J}$ we may decompose $\nabla = \nabla'_J + \nabla''_J$, where $\nabla'_J : \Omega^0(E) \to \Omega^{1,0}_J(E)$ and $\nabla''_J : \Omega^0(E) \to \Omega^{0,1}_J(E)$. We observe that if E has a hermitian metric there is a unique way to reconstruct a compatible covariant derivative ∇ from its $(0, 1)$ component ∇''_J — the proof is just the same as in the familar, integrable, case. If E is the trivial bundle we write $\partial_J, \overline{\partial}_J$ for the $(1, 0)$ and $(0, 1)$ components of the derivative d. We may prolong these to the other differential forms, getting

$$\partial_J : \Omega^{p,q}_J \to \Omega^{p+1,q}_J,$$
$$\overline{\partial}_J : \Omega^{p,q}_J \to \Omega^{p.q+1}_J,$$

but it is not in general true that $d = \partial + \overline{\partial}$. This is where the *Nijenhius tensor* N — the obstruction to integrability — enters. The tensor N lies in $T \otimes \Lambda^{0,2}$ and one has

$$d = \partial + \overline{\partial} + i_N + i_{\overline{N}},$$

where $i_N : \Omega^{p,q}_J \to \Omega^{p-1,q+2}_J, i_{\overline{N}} : \Omega^{p,q}_J \to \Omega^{p+2,q-1}_J$ are algebraic operators defined by contraction and wedge product with N.

For our immediate purposes, we just need to consider the operator $\overline{\partial}_J : \Omega^{1,0}_J \to \Omega^{1,1}_J$. This can be regarded as a ∇''-operator on the bundle T^*M, viewed as a complex vector bundle over M using the structure J. By the observation above there is therefore a preferred unitary connection ∇ on T^*M and hence a connection on the "canonical line bundle" $K_M = \Lambda^n_C T^*M$. Let ρ be $-i$ times the curvature form of this connection on K_M, so ρ is a real 2-form on M. Finally define a function S on M by

$$S\omega^n = n!\,\rho \wedge \omega^{n-1}.$$

We might call S the *Hermitian scalar curvature*: in the integrable case the connection above coincides with the Levi–Civita connection and, up to a factor, S is the usual Riemannian scalar curvature. We can now state the result we have been driving at:

Propostion 9. *The map $J \mapsto S(J)$ is an equivariant moment map for the $\mathcal{G}$-action on $\mathfrak{J}$, under the natural pairing:*

$$(S, H) \mapsto \int_M SH\frac{\omega^n}{n!}.$$

We proceed now with the proof of Proposition 9. Fix an almost-complex structure $J \in \mathfrak{J}$; let

$$P : C_0^\infty \to \Gamma(S^2(T)) \subset \Omega_J^{0,1}$$

be the operator representing the infinitesimal action of $\mathcal{G}$ on $\mathfrak{J}$ and

$$Q : \Gamma(s^2(T)) \to C_0^\infty$$

be the operator which represents the derivative of $J \mapsto S(J)$. By the definition of a moment map what we need to show is that, for all H and μ, $(P(H), J\mu) = (H, Q(\mu))$, where $(\ ,\ )$ denotes the usual (real) L^2 inner product. Expressed in terms of the formal adjoints this is $Q = P^* \circ J$. The operator P can be factored as $P = P_2P_1$, where $P_1 : C_0^\infty \to \Gamma(T)$ maps H to the Hamiltonian vector field X_H and P_2 maps a vector field X to the infinitesimal variation in complex structure given by the Lie derivative L_XJ.

Lemma 10. $L_XJ = \nabla_J'' X - \overline{X}.(N)$, *where ∇_J'' is the operator on the complex tangent bundle induced from $\overline{\partial}J$ on the cotangent bundle, and the pairing in the second term is the contraction*

$$\overline{T} \otimes (\Lambda^{0,2} \otimes T) \to \Lambda^{0,1} \otimes T.$$

If the variation in almost-complex structure are identified with $\Omega^{0,1}(T)$, as above, then the action of L_XJ on a $(1,0)$ form α is just the (0,1) part of $L_X\alpha$. But $L_X\alpha = (di_X - i_Xd)\alpha$. The (0, 1) component of this is made up of the three terms: $\overline{\partial}(i_X\alpha) - i_X\overline{\partial}\alpha - i_{\overline{X}}(i_N\alpha)$. The first two terms give $(\nabla_J''X)\alpha$ by the definition of ∇'' on T and hence we get the formula stated in the Lemma.

We now turn to the other operator Q. This can also be written as a composite $Q = Q_1Q_2$, factoring through the infinitesimal change in the connection on K_M, but in doing this we need to keep in mind that the line bundle K_M itself depends upon the almost complex structure. If μ is a deformation of almost complex structure to a new structure J' then the map $\alpha \mapsto \alpha + \mu(\alpha)$ gives an isomorphism between $\Omega_J^{1,0}$ and $\Omega_{J'}^{1,0}$ which is a Hermitian isometry to first order in μ, and with this identification fixed we can regard the connections $\nabla_J, \nabla_{J'}$ as two unitary connections on the same bundle over M, and similarly for the connections on the exterior power K_M. The two connections on K_M differ by an imaginary 1-form $i\psi$ say over M, then we define

$$Q_2 : \Gamma(s^2(T)) \to \Gamma(T^*M), \qquad Q_1 : \Gamma(T^*M) \to C^\infty(M)$$

by $Q_2(\mu) = \psi$ and $Q_1(\psi)\frac{\omega^n}{n!} = d\psi \wedge \omega^{n-1}$.

The main task remaining is to identify the infinitesimal change in the connection.

Lemma 11. *Under a change in almost complex structure μ, and using the bundle identification above, the change in the connection on T^*M is, to first order in μ,*

$$\partial\mu - N.\overline{\mu} \in \Omega^{1,1}(T),$$

where we regard $\Omega^{1,1}(T)$ as identified with the 1-forms on M with values in the skew-adjoint endomorphisms of TM, by the map $\sigma \mapsto \sigma - \sigma^*$.

The proof of this Lemma is slightly confusing. To set the scene consider the case of a Hermitian bundle E over M. If we *fix* an almost complex structure J on M and vary the ∇''-operator on E by $\sigma \in \Omega^{0,1}(EndE)$ then the change in the corresponding connections is by $\sigma - \sigma^*$. On the other hand, suppose we fix a connection ∇ on E and vary the almost complex structure by μ then, under the identifications we have made, the new ∇'' operator is

$$\nabla''_{J'} s = \nabla'' s + \mu \nabla' s, \tag{12}$$

where μ acts on the $\Omega^{1,0}$ factor of $\nabla'(s) \in \Omega^{1,0}(E)$. Our case is complicated because both the connection and the almost complex structure are varying. We need to compute

$$\nabla_{J'} : \Omega^{1,0}_{J'} \to \Omega^{1,1}_{J'}.$$

We have identified $\Omega^{1,0}_J$ with $\Omega^{1,0}_{J'}$ by $\alpha \mapsto \alpha + \mu(\alpha)$, and similarly we can identify $\Omega^{1,1}_J$ with $\Omega^{1,1}_{J'}$ by, to first order in μ,

$$\theta \mapsto \theta + \mu(\theta) + \overline{\mu}(\theta).$$

This means that if χ is a complex 2-form on M with components $\chi = \chi^{2,0} + \chi^{1,1} + \chi^{0,2}$ with respect to the J decomposition then, to first order in μ, the (1, 1) component of χ with respect to the J' decomposition is represented by

$$\chi^{1,1} - \mu(\chi^{2,0}) - \overline{\mu}(\chi^{0,2}).$$

Putting all this together, the operator $\overline{\partial}_{J'}$ maps a (1,0) form α to the $\Omega^{1,1}_{J'}$ component of $d(\alpha + \mu(\alpha))$ which, to first order in μ, is

$$\overline{\partial}_J(\alpha) + \partial_J(\mu(\alpha)) - \mu(\partial\alpha) - \overline{\mu}(i_N(\alpha)). \tag{13}$$

Now let use temporarily write $E = T^*M$, so $\nabla'_J : E \to \Omega^{1,0}(E)$. We can write:

$$\partial_J(\mu(\alpha)) = \nabla'_J(\mu)\alpha + \mu.(\nabla'_J\alpha),$$

where in the second term μ acts on the "bundle" factor E in $\Omega^{1,0}(E)$. This is not the same as the action in the second term of (12), where μ acts on the "cotangent" factor $\Omega^{1,0}$. (The confusing thing here is that E is the cotangent bundle!) The two differ by the action of μ on the antisymmetrisation of $\nabla'_J\alpha \in \Gamma(T^* \otimes T^*)$. This antisymmetrisation is just the tensor $\partial_J\alpha$, since the torsion of our connection is given by the Nijenhius tensor N, and has no component mapping T^* to $\Lambda^{2,0}(T^*)$. So the difference is precisely the other term $\mu(\partial\alpha)$, in (13). Thus we get

$$\nabla''_{J'}\alpha = (\nabla''_J + \mu\nabla'_J)\alpha + \sigma\alpha,$$

where

$$\sigma\alpha = (\nabla'\mu)\alpha - \overline{\mu}(i_N\alpha).$$

The term $\nabla''_J + \mu\nabla'_J$ just represents the way that the ∇''-operator would vary with a fixed connection, so the change in connection is given by $\sigma - \sigma^*$ as required.

Using the Lemma above, we have

$$Q_2(\mu) = \text{Re}\,(\text{Tr}(\nabla'_J\mu - \overline{\mu}.N)). \tag{14}$$

Here the trace is taken using our identification of $\Omega^{1,1}(T)$ with the (0, 1)-forms with values in the skew-adjoint endomorphisms of the cotangent bundle. At this point we bring in the fact that μ lies in the *symmetric* part of $T \otimes T \cong \Omega^{0,1}(T)$. Thus the term $\nabla_J\mu$ can be regarded as a tensor in $T^* \otimes T \otimes T$ which is symmetric in the last two factors. This means that the trace of this term in (14) can be obtained equally via the contraction

$$\Lambda : \Omega^{1,1}(T) \to \Omega^0(T),$$

and the standard identification of the tangent and cotangent bundles. Now the usual Kahler identity

$$\overline{\partial}^* = -i[\Lambda, \partial]$$

holds also in the almost-complex case, and in sum we get

$$Q_2(\mu) = (\nabla''_J)^*\mu - \text{Re}\,(\text{Tr}(\overline{\mu}.N)), \tag{15}$$

and we note that the first term is linear in μ while the second is anti-linear.

The proof of (9) is now just a matter of matching up the various pieces. (To fix the signs one needs to write things out a little more explicitly, since our notation is somewhat compressed.) Using the standard (metric) identification of tangent and cotangent bundles, we have

$$P_1^* = Q_1 \circ J$$

$$P_2^* = -J \circ Q_2 \circ J,$$

so $P^* = Q_* \circ J$ as required.

We will now discuss the relevance of this moment-map calculation to the Kahler–Einstein problem. First observe that for any J in $\mathcal{J}$ the integral of $S(J)$ over the manifold is a topological invariant,

$$\int_M S\,\frac{\omega^n}{n!} = \int_M \rho\wedge\omega^{n-1} = 2\pi c_1(M)\cup[\omega]^{n-1}[M].$$

We fix d such that the integral of $S-d$ is zero, then $S-d$ is also a moment map and the relevant moment map equation is $S-d=0$, i.e. constant Hermitian scalar curvature. In the integrable case one has the following standard argument. The Bianchi identity give $\overline{\partial}S=\overline{\partial}^*\rho$, where ρ is essentially the Ricci tensor, regarded as a $(1,1)$ form. So if the scalar curvature is constant then the Ricci tensor is harmonic, and if $c_1(M)$ is a multiple of the Kahler class considered, we can conclude from the uniqueness of the harmonic representative that $\rho=\lambda\omega$. Thus, given the appropriate cohomological setting, our moment map equation does yield the Kahler–Einstein solutions.

The feature which prevents us from immediately fitting this set-up into the standard picture at the beginning of this section is the fact that the complexification of the group $\mathcal{G}$ does not exist. However we may certainly complexify the Lie algebra and this automatically acts on $\mathcal{J}$, since the structure on $\mathcal{J}$ is integrable. At each point $J\in\mathcal{J}$ we get a subspace of $T\mathcal{J}_J$ spanned by this complexified action and these subspaces form an integrable, holomorphic distribution on $\mathcal{J}$. Thus we get a foliation of $\mathcal{J}$ which plays the role of the "orbits" of the mythical group $\mathcal{G}^c$. (All of these remarks are to be taken rather formally, since we do not want to get involved here with the infinite-dimensional differential-topological aspects of the set-up). Moreover the subspace $\mathcal{J}^{\text{int}}$ is invariant under this foliation, so we can restrict attention to this if we prefer. In this subspace, the setting for the classical Kahler-Einstein problem, the foliation has a straightforward geometric meaning. To see this we consider the infinitesimal action of a function iH, in the imaginary part of the complexified Lie algebra. By definition this is just the variation $JP(H)$ which is

$$\mu = J(\overline{\partial}_J X_H + \overline{X}_H.N) = \overline{\partial}_J(JX_H) + J\overline{X}_H.N.$$

In the integrable case, when N is zero, we have $\mu=\overline{\partial}_J(JX_H)=L_{JX_H}(J)$, which is just the natural action of the vector field JX_H on the complex structure. Thus the geometric effect of applying iH is the same as keeping the complex structure fixed and varying the symplectic form to $\omega'=\omega-L_{JX_H}\omega=\omega-dJdH=\omega-2i\overline{\partial}\partial H$. This is just the familar parametrisation of Kahler forms by Kahler potentials. Conversely, suppose Y is a vector field on M such that the corresponding variation in complex structure $L_Y J=\overline{\partial}Y$ is tangent to $\mathcal{J}$, i.e. lies in s^2T. Let $\gamma\in\Omega^{0,1}$ correspond to Y under the isomorphism $T=\overline{T}^*$ given by the metric. Then one checks that the condition that $\overline{\partial}Y$ lies in s^2T is the same as saying that $\overline{\partial}\gamma\in\Omega^{0,2}$ vanishes. The Dolbeault cohomology group $H^{0,1}$ is zero since $H^1(M)=0$, so we can write

$\gamma = \overline{\partial} f$ for some complex-valued function $f = H_1 + iH_2$ on M. Then this means that $Y = X_{H_1} + iX_{H_2}$. So we conclude that each leaf of the foliation of $\mathfrak{J}^{\text{int}}$ corresponds, up to the diffeomorphism of M, to the set of Kahler metrics, in the given cohomology class, for a fixed complex structure on M. In sum, we have fitted Tian's conjecture — which bears on the case when $c_1 = \lambda[\omega]$ — into the familar picture, to wit: there should be a notion of a stable "orbit" which on the one hand is equivalent to the algebraic geometers notion of stability via the Hilbert scheme and on the other hand admits the standard identification:

$$\mathfrak{J}_s^{\text{int}}/\mathcal{G}^c = \mu^{-1}(0)/\mathcal{G}.$$

Another facet of the theory that would be worth studying is the geometry of the equivariant line bundle $\mathcal{L}$ over $\mathcal{J}$: it is natural to expect that this can be identified with a determinant line bundle furnished with a metric via zeta-function regularisation, making contact with the theory developed by Quillen, Bismut and others.

As well as giving additional motivation for Tian's conjecture, the work above throws light on a number of known results in Kahler geometry. For example:

(1) Matsushima's Theorem that the holomorphic automorphism group of a compact Kahler–Einstein manifold is reductive: the complexification of its isometry group;
(2) the Mabuchi "K-energy" [5], which is just the "norm functional" in this context;
(3) Bando and Mabuchi's result [6] on the uniqueness of Kahler–Einstein metrics, which can be seen as a facet of general convexity property of the norm functional;
(4) the existence of a natural Kahler metric (the Weil–Peterson metric) on the moduli space of Kahler–Einstein structures.

In large part, these follow without further calculation, if one is familiar with the corresponding results in the standard picture. In addition, our set-up suggests that it may be worthwhile to study various extensions of Tian's conjecture. On the one hand one might expect to extend the ideas to constant scalar curvature Kahler metrics. In the case of surfaces there has been substantial work on these by Le Brun and others. On the other hand, we have seen that the integrability of the almost complex structure is not particularly relevant, at least to the formal picture, so one might seek extensions of the results in the realm of almost-complex geometry. In this case the geometrical meaning of the leaves of the foliation is less clear: one gets an equivalence relation on almost-Kahler structures generated by a modification of the notion of Kahler potential — two structures $(J_0, \omega_0), (J_1, \omega_1)$ are equivalent if they can be joined by a path (J_t, ω_t) for which there are functions H_t with

$$\frac{d}{dt}\omega_t = d(J_t(dH_t)), \qquad \frac{d}{dt}J_t = N_t.\overline{X}_t,$$

where X_t is the Hamiltonian vector field defined by H_t and the sympletic form ω_t.

References

[1] M. F. Atiyah and R. Bott, *The Yang–Mills equations over Riemann surfaces*, Phil. Trans. Roy. Soc. Lond. A **308** (1982), 523–615.

[2] M. F. Atiyah, N. J. Hitchin and I. M. Singer, *Self-duality in four-dimensional Riemannian geometry*, Proc. Roy. Soc London, Ser. A **308** (1978), 425–461.

[3] D. Austin and P. Braam, *Equivariant Floer theory and gluing Donaldson polynomials*, Topology **35** (1996), 167–201.

[4] S. Bando, *The K-energy map, almost Einstein–Kahler metrics and an inequality of the Miyaoka–Yau type*, Tohuku Math. J. **39** (1987), 231–235.

[5] S. Bando and T. Mabuchi, *Uniqueness of Einstein–Kahler metrics modulo connected group actions*, Algebraic Geometry: Advanced Studies in Pure Math, Vol. 10, 1987.

[6] N. Buchdahl, *Instantons on* $\mathbf{CP}^2$, J. Differential Geometry **24** (1986), 19–52.

[7] S. K. Donaldson, *Anti-self-dual Yang–Mills connections over complex algebraic surfaces and stable vector bundles*, Proc. Lond. Math. Soc. **50** (1985), 1–26.

[8] ———, *Vector bundles on the flag manifold and the Ward correspondence*, Geometry Today (ed. Arbarello *et al.*), Birkhauser.

[9] ———, *The geometry of 4–manifolds*, Proc. Int. Congr. Math. I, Berkeley 1986, pp. 43–61.

[10] ———, *Infinite determinants, stable bundles and curvature*, Duke Math. J. **54** (1987), 231–247.

[11] ———, *Gauge Theory and Four–Manifold Topology*, Proc. European Cong. Math. I, Paris 1992, Birkhauser.

[12] ———, *On the work of Andreas Floer*, Jber. d. Deutsche Math.–Verein **95** (1993), 103–120.

[13] ———, *The Seiberg–Witten equations and 4-manifold topology*, Bull. Amer. Math. Soc. **33** (1996), 45–70.

[14] S. K. Donaldson and P. B. Kronheimer, *The geometry of four-manifolds*, Oxford U. P., 1990.

[15] J. Eells and J. Sampson, *Harmonic mappings into Riemannian manifolds*, Amer. J. Maths. **86** (1964), 109–160.

[16] R. Fintushel and R. Stern, *The blow-up formula for Donaldson invariants*, Annals Math. **143** (1996), 529–546.

[17] R. Friedman and J. Morgan, *Algebraic surfaces and 4-manifolds: some conjectures and speculations*, Bull. Amer. Math. Soc. **18** (1988), 1–19.

[18] ———, *Smooth Four-manifolds and complex surfaces*, Springer, 1994.

[19] *The Floer Memorial Volume* (*Ed. Hofer, Taubes, Weinstein, Zehnder*), Birkhauser, 1995.

[20] L. Göttsche, *Modular forms and Donaldson invariants for manifolds with* $b^+ = 1$, Preprint (1995).

[21] L. Göttsche and D. Zagier, *Jacobi forms and the structure of Donaldson invariants for manifolds with* b^+1, preprint (1996).

[22] S. Helgason, *Differential geometry, Lie groups and Symmetric Spaces*, Academic Press, 1978.

[23] P. B. Kronheimer and T. S. Mrowka, *Recurrence relations and asymptotics for four-manifold invariants*, Bull. Amer. Math. Soc. **30** (1994), 215–221.

[24] D. Mumford, J. Fogarty and F. Kirwan, *Geometric Invariant Theory* (*3rd. ed.*), Springer, 1994.

[25] P. Orzvath, *Some blow-up formulae for* SU(2) *Donaldson polynomials*, J. Differential Geometry **40** (1994), 411–447.

[26] R. Richardson, *Compact real forms of a complex semisimple Lie algebra*, J. Differential Geometry **2** (1968), 411–420.

[27] J. Sacks and K. Uhlenback, *The existence of minimal* 2-*spheres*, Annals. Math. **113** (1981), 1–24.

[28] N. Seiberg and E. Witten, *Monopoles, duality and chiral symmetry breaking in* $N = 2$ *super-symmetric QCD*, Nucl. Phys. B **431** (1994), 581–640.

[29] S. Smale, *An infinite-dimensional version of Sard's theorem*, Amer. J. Math. **87** (1964), 861–866.

[30] C. Taubes, *Self-dual connections over non-self dual* 4-*manifolds*, J. Differential Geometry **17** (1982), 139–170.

[31] G. Tian, *Kahler–Einstein metrics with positive scalar curvature*, MIT preprint (1996).

[32] K. Uhlenbeck, *Removable singularities in Yang–Mills fields*, Commun. Math. Phys. **83** (1982), 11–29.

[33] ———, *Connections with* L^p *bounds on curvature*, Commun. Math. Phys. **83** (1982), 31–42.

[34] K. K. Uhlenbeck and S.-T. Yau, *On the existence of Heritian–Yang–Mills connections on stable vector bundles*, Commun. Pure Appl. Math. **39** (1986), 5257–5293.

[35] E. Witten, *Topological Quantum Field Theory*, Commun. Math. Phys. **117** (1988), 353–386.

[36] ———, *Monopoles and* 4-*manifolds*, Math. Res. Letters **1** (1994), 769–796.

Proceedings of the International Congress of Mathematicians
Berkeley, California, USA, 1986

ON SOME OF THE MATHEMATICAL CONTRIBUTIONS OF GERD FALTINGS

by
B. MAZUR

One of the recent great moments in mathematics was when Gerd Faltings revealed the circle of ideas which led him to a proof of the conjecture of Mordell ([**1**]; see also [**2, 3**]).

The conjecture, marvelous in the simplicity of its statement, had stood as a goad and an elusive temptation for over half a century: it is even older than the Fields Medal! In modern language it takes the following form:

If K is any number field and X is any curve of genus > 1 defined over K, then X has only a finite number of K-rational points.

To get a feeling for our level of ignorance in the face of such questions, consider that, before Faltings, there was not a single curve X (of genus > 1) for which we knew this statement to be true for all number fields K over which X is defined!

Already in the twenties, Weil and Siegel made serious attempts to attack the problem. Siegel, influenced by Weil's thesis, used methods of diophantine approximation, to prove that the number of *integral* solutions to a polynomial equation $f(X, Y) = 0$ (i.e., solutions in the ring of integers of a number field K) is finite, provided that f defines a curve over K of genus > 0, or a curve of genus 0 with at least three points at infinity.

In his thesis, Weil generalized Mordell's theorem on the finite generation of the group of rational points on an elliptic curve, to abelian varieties of any dimension. Weil then hoped to use this finite generation result for the rational points on the jacobian of a curve to go on to show that when a curve of genus > 1 is imbedded in its jacobian, only a finite number of the rational points of the jacobian can lie on the curve. Not finding a way to do this, he decided to call his proof of finite generation (the "theorem of Mordell-Weil") a thesis, despite Hadamard's advice not to be satisfied with half a result!

After this work of Weil and Siegel there was little progress for thirty years. It was in the sixties and early seventies that several new developments occurred in algebraic geometry and number theory which were to influence Faltings (work of Grothendieck, Serre, Mumford, Lang, Néron, Tate, Manin, Shafarevich, Parsin, Arakelov, Zarhin, Raynaud, and others).

These developments, which enter in an essential way in the work of Faltings, encompass three grand mathematical themes, and Faltings proved the conjecture of Mordell, by first establishing the truth of some other outstanding conjectures — fundamental to arithmetic and to arithmetic algebraic geometry. In the next few minutes I should like to try to convey some sense of the sweep of Faltings's accomplishment by touching on those themes, and those conjectures.

1. The arithmetization of geometry and the geometrization of arithmetic. Nowadays the analogy between number fields and fields of rational functions on algebraic curves (over finite fields) is so well imprinted upon our view of both number theory and the theory of algebraic curves that it is hard to imagine how we might deal with either theory, if deprived of that analogy. A firm understanding of the power of such analogies was present already in the work of Kronecker, and of Dedekind and Weber at the end of the last century. This understanding was deepened by the development of algebraic number theory, in the hands of Artin and Chevalley, of algebraic geometry, via the foundations developed by Zariski, Weil, and Serre, and more recently, of arithmetic algebraic geometry, whose foundations are given by the language of schemes, of Grothendieck.

In the language of schemes, a smooth curve over a finite field and the ring of integers in a number field are not merely analogous: they are two instances of the same notion (*regular schemes of dimension one, of finite type over* $\mathbf{Z}$). Similarly, a family of curves over a "base" curve over a finite field and a curve over the ring of integers in a number field are companion instances of the same notion.

This is not to say that this analogy is thoroughly understood! Why, it has only been relatively recently, thanks to the groundbreaking work of Arakelov, that we have begun to see a format for bringing the *archimedean places* of a number field into the geometric picture.

Moreover, this "synthetic view," extraordinarily efficacious for carrying problems and conjectures from the realm of function fields to the realm of number fields and back again, is far less satisfactory when it comes to carrying the *proofs* of those conjectures from one realm to the other.

For example, the analogue of Mordell's conjecture in the function field case was first settled by Manin back in 1963. A different proof was given by Grauert in 1965. Arakelov (using a beautiful idea of Parsin) found another proof in 1971. Yet another proof, also using Parsin, was given by Zarhin in 1974.

Even when we were armed with these approaches to the function field case, the number field case seemed intractable for almost a decade until Faltings discovered a method, analogous to that of Parsin-Zarhin, which established the classical conjecture of Mordell over number fields. To this day we lack number field analogues of the other approaches to the problem — say, of Manin's original proof, or of Arakelov's (Do they exist?). Judging from Faltings's published contributions to

Arakelov's theory [4] one might imagine that he himself had been simultaneously pursuing two approaches to the Mordell conjecture in the number field case: one suggested by Arakelov's work, and the other by Zarhin's. The problem yielded, in 1983, to the second approach.

Faltings's method in the number field case, and Zarhin's in the function field case, was to settle Mordell's conjecture by first answering a more "geometric" question:

Kodaira had raised the problem of studying, or perhaps "classifying," all (truly varying) families of smooth curves of a given genus over a fixed (not necessarily complete) base curve. It was Shafarevich, at the 1962 International Congress, who first brought attention to the analogue of Kodaira's problem in arithmetic, and to its significance. One version of this analogue, now known as *Shafarevich's Conjecture for Curves*, states that

There are only a finite number of nonisomorphic curves of a given genus > 1 defined over a fixed number field and possessing good reduction outside a fixed finite set of primes in the ring of integers of that number field.

One way of paraphrasing Kodaira's original problem is by formulating the "function field analogue" of Shafarevich's conjecture, where the ring of integers in a number field is replaced by a base curve over a finite field. By an ingenious argument which happily worked as well in the number field case as in the function field case, Parsin had shown in 1968 that Mordell's conjecture follows from Shafarevich's conjecture.

Very roughly, Parsin's idea is as follows: Fix $g > 1$. Given a curve X of genus g and a rational point P on X over a number field K, Parsin produced a convering Y of X which is ramified only above P, and whose number field of definition and set of bad primes are "uniformly bounded" in terms of the data: g, the number field of definition of X and P, and the set of bad primes of X. Since Y determines the pair (X, P) up to finite ambiguity, it follows that if there are only finitely many such Y's (*Shafarevich's conjecture*) then there only finitely many such P's (*Mordell's conjecture*).

In their respective contexts, both Zarhin's and Faltings's attack on Mordell's conjecture is to prove Shafarevich's conjecture for curves (over function fields, and over number fields, respectively), and then to appeal to Parsin's idea.

2. Curves and abelian varieties. It was Weil, in his proof of the "Riemann hypothesis for curves over finite fields," who first made essential use of the passage from curves to abelian varieties to derive important consequences for the arithmetic of curves.

The "geometric" insight, that in pursuing questions about curves it sometimes pays to appeal to their jacobians, goes further back. Indeed, the fact that we see such a close relationship between curves and their jacobians is one of our many legacies from the Italian school of algebraic geometry.

With the trend towards the arithmetization of geometry, it was natural to study closely models for the jacobians of curves, or more generally for abelian varieties, over the rings of integers of number fields. Néron, in 1964, discovered the remarkable, and remarkably useful, fact that any abelian variety over a number field has a "best" model over the ring of integers of the number field — "best," from the point of view of niceness of the reduction of the model modulo prime ideals in the ring. Although Néron, at the time he did his work, was not aware of Kodaira's contributions in the complex analytic case, one may view Néron's theory as a far-reaching amplification and arithmetization of a program initiated by Kodaira.

Néron models now play an important role in any close arithmetic study of abelian varieties, and in particular, they play a role in the detailed analysis of compactifications of moduli spaces for abelian varieties. The systematic *arithmetic* study of moduli spaces and their compactifications — a study initiated by the magnificent work of Mumford — in turn plays a key role in Faltings's approach. Compactification of moduli spaces of abelian varieties over **Z**, incidentally, is a subject to which Faltings has returned more recently: By refining, in certain respects, the work of Ching-Li Chai, Faltings has clarified some questions of arithmetic compactifications, and has thereby significantly simplified, and rendered more natural, the logical structure of his proof of Mordell's conjecture. In his initial proof [**1**] Faltings took a somewhat more circuitous route, using moduli spaces of curves rather than of abelian varieties (and the published account of the technical issues was supremely succinct, requiring a certain expertise on the part of the reader).

Thanks to the close relationship between curves and their jacobians (the classical theorem of Torelli plus a finiteness result concerning polarizations), Shafarevich's conjecture for curves (of genus > 1) reduces to a similar conjecture (also called *Shafarevich's conjecture*) for abelian varieties:

There are only a finite number of abelian varieties of fixed dimension over a fixed number field whose Néron models possess good reduction outside a fixed finite set of primes of the number field.

This conjecture was also settled affirmatively by the work of Faltings. It was settled in tandem with another basic arithmetic question:

3. Abelian varieties and Galois representations. It was by considering number-theoretic analogies of the classical conjectures of Hodge (concerning algebraic cycles) and geometric analogues of the conjecture of Birch and Swinnerton-Dyer that Tate in 1963 formulated the following conjecture, which links the problem of classifying abelian varieties (up to isogeny) to that of classifying the Galois representations to which they give rise:

Let l be a prime number. Let K be a number field, and $\overline{K}$ an algebraic closure of K. An abelian variety A over K is determined up to isogeny (over K) by the natural *representation of Gal($\overline{K}/K$) on the Q_l-vector space*

$$V_l(A) = \mathrm{Hom}(Q_l, A_l(K)),$$

where $A_l(K)$ denotes the group of $\overline{K}$-valued points of A, of l-power order.

This clearly basic 'link' between abelian varieties (denizens of algebraic geometry) and Galois representations (ostensibly more 'elementary' creatures) was proved by Tate himself in 1967 for a finite field K, but over number fields it was not even known to be the case for elliptic curves, before the work of Faltings.

By a beautiful argument (which makes crucial use of two theories upon which we shall comment in a moment, and) which shuttles back and forth between the conjecture of Shafarevich for abelian varieties and the conjecture of Tate, Faltings showed that both of these conjectures are true. The phrase "shuttles back and forth" is quite inadequate to characterize Faltings's mode of argument, which captures in its weave all the mathematical themes upon which I have touched.

The "two theories" referred to are the *Theory of Heights* and the *Theory of p-Divisible Groups* (and, more generally, of *Group Schemes of Exponent p*).

The *Theory of Heights* was initiated by Weil in 1928 as a technique for "counting" rational points on abelian varieties and was used in an essential manner in his proof of the theorem "of Mordell-Weil." This theory was further developed by Néron, by Tate, and more recently was given a new twist in the work of Arakelov.

The *Theory of p-Divisible Groups* was invented by Serre and Tate, and, independently, by Barsotti in the mid-sixties to provide a technique to analyze the way in which p-power torsion points on abelian varieties "degenerate" when specialized to characteristic p. In 1966 Tate proved the analogue of his conjecture on abelian varieties for p-divisible groups over a local field of characteristic 0. Faltings uses this theorem, and, moreover, makes essential use of an important refinement of it (covering the case of group schemes of exponent p) due to Raynaud.

Even the above recitation does not completely exhaust the list of longstanding conjectures established by Faltings in the course of his work on the Mordell conjecture. For example, an important adjunct to the conjecture of Tate concerning the representations of the Galois group Gal($\overline{K}/K$) on l-power torsion points of an abelian variety A defined over a number field K is the assertion of *semisimplicity* of the representation of Gal($\overline{K}/K$) on $V_l(A)$, This *Semisimplicity Conjecture* has also been proved by Faltings (as its "function field analogue" had been proved by Zarhin in the course of his work). Moreover the proof of semisimplicity plays a structural role in the proof of the other conjectures.

The above *Semisimplicity Conjecture* had been formulated by Grothendieck as the "dimension one case" of a more general question (semisimplicity of Galois representations acting on the d-dimensional l-adic cohomology of irreducible smooth projective varieties over number fields, for any d). One could find support for Grothendieck's conjecture, at the time he made it, in the work of Serre concerning the richness of the action of Galois on the torsion points of elliptic curves defined over number fields. Thanks to Faltings, we now know the *Semisimplicity Conjecture of Grothendieck* to be true for $d = 1$. This result of Faltings, incidentally, together with a technique coming from Faltings's proof, has very recently been used by Serre in a deep study of the action of Galois on torsion points of abelian varieties of arbitrary dimension g, defined over number fields.

The general case of Grothendieck's conjecture (i.e., for $d > 1$) is still open.

We have been discussing only Gerd Faltings's approach to the conjecture of Mordell, but his other mathematical contributions, whether they be concerned with moduli spaces of abelian varieties, the Riemann-Roch theorem for arithmetic surfaces, or p-adic Hodge theory, all immediately impress one as the work of a marvelously original mind from which we may expect similarly wonderful things in the future.

References

1. G. Faltings, *Endlichkeitssätze für abelsche Varietäten über Zahlkörpern,* Invent. Math. **73** (1983), 349–366.
2. G. Faltings, G. Wustholtz, *et al.*, *Rational points* (Seminar Bonn/Wuppertal, 1983/84), Aspects of Math., vol. E6, Vieweg, Braunschweig-Wiesbaden, 1984.
3. L. Szpiro, *Séminaire sur les pinceaux arithmétiques: La conjecture de Mordell,* Astérisque, no. 127, Soc. Math. France, Paris, 1985.
4. G. Faltings, *Calculus on arithmetic surfaces*, Ann. of Math. (2) **119** (1984), 387–424.

Harvard University, Cambridge, Massachusetts 02138, USA

Gerd Faltings

Proceedings of the International Congress of Mathematicians
Berkeley, California, USA, 1986

NEUERE ENTWICKLUNGEN IN DER ARITHMETISCHEN ALGEBRAISCHEN GEOMETRIE

by
GERD FALTINGS

Ich möchte einige Höhepunkte in der Entwicklung der arithmetischen algebraischen Geometrie in den letzten yier Jahren darstellen darstellen. Naturgemäß ist die Auswahl rein subjektiv, und die Darstellung erhebt keinen Anspruch auf Vollständigkeit. Wir konzentrieren uns auf die folgenden vier Themen:

(a) Die Formel von Gross-Zagier,
(b) Die Vermutungen von Tate, Schafarevitsch, und Mordell,
(c) Hodge-Tate Strukturen auf der p-adischen Kohomologie,
(d) Die Beilinson-Vermutung.

Zu zweien von den obigen Themenkreisen habe ich persönlich Beiträge geleistet, und ich darf daher vielleicht mit einigem Recht versuchen, die Ergebnisse vor diesem großen Kreis von Zuhörern darzustellen. Bei den beiden anderen Gebieten muß mich darauf beschränken, das wiederzugeben, was ich in letzter Zeit aus verschiedenen Quellen gelernt habe. Meine Entschuldigung für diese sicherlich unvollkommene Darstellung besteht darin, daß die Plenar-Vorträge doch für einen weiteren Zuhörer-Kreis als die Sektions-Vorträge bestimmt sind. Für eine kompetentere Darstellung kann ich aber nur auf diese verweisen.

1. Die Formel von Gross-Zagier. Als Hintergrund dienen zwei Schwierigkeiten, die oft in der diophantischen Geometrie auftreten. Die eine ist die Bestimmung der Nullstellen-Ordnung von L-Reihen. Dies sind Verallgemeinerungen der klassischen L-Reihen. Es handelt sich um Dirichlet-Reihen $\sum a_n \cdot n^{-s}$, welche in einer Halbebene $\mathrm{Re}(s) \geq s_0$ konvergieren. Die L-Reihe gehört zu einer arithmetischen Varietät $\mathbf{X}$, und die Koeffizienten a_n berechnen sich aus dem Reduktions-Verhalten von $\mathbf{X}$ in den endlichen Stellen. Es wird vermutet, daß die L-Reihen sich zu meromorphen Funktionen auf der ganzen komplexen Ebene fortsetzen, und sogar eine Funktional-Gleichung erfüllen. Dies ist gezeigt worden für eine Reihe von arithmetischen Varietäten $\mathbf{X}$, vor allem für gewisse Shimura-Varietäten. Weiter gibt es Vermutungen, daß die Ordnungen der Nullstellen beziehungsweise Pole der L-Reihe an ganzzahligen Stellen mit geometrischen Invarianten zusammenhängen.

Es handelt sich um die Vermutung von Birch und Swinnerton-Dyer und allgemeiner um die Tate-Vermutung.

Da numerische Berechnungen stets mit einem Fehler behaftet sind, kann man mit ihnen nie direkt zeigen, daß eine analytische Funktion an einer bestimmten Stelle verschwindet, sondern höchstens das Gegenteil. Dies macht es schwierig, die oben angeführten Vermutungen zu testen. Dies andere Schwierigkeit betrifft rationale Punkte auf abelschen Varietäten, speziell auf elliptischen Kurven. Man beweist relativ einfach, daß es sich um Torsionspunkte handelt, doch das Gegenteil ist etwas schwieriger. Dabei möchte man oft gerne zeigen, daß die Mordell-Weil Gruppe positiven Rang besitzt, zum Beispiel falls dies von der Vermutung von Birch und Swinnerton-Dyer vorausgesagt wird. Die Formel von B. Gross und D. Zagier erlaubt es nun in einigen Fällen, beide Probleme zu lösen. Sie liefert nämlich die Gleichheit zwischen der Néron-Tate Höhe gewisser rationaler Punkte auf speziellen elliptischen Kurven (Heegner-Punkte auf Weil-Kurven), und der Ableitung einer zugehörigen L-Reihe an der Stelle $s = 1$. Dies kann in beiden Richtungen ausgenutzt werden: Ist der rationale Punkt ein Torsionspunkt (leicht zu zeigen), so verschwindet die Ableitung der L-Reihe (schwierig). Umgekehrt: Ist die Ableitung der L-Reihe verschieden von Null (leicht), so haben wir einen Punkt unendlicher Ordnung gefunden (schwierig).

Die bekannteste Anwendung bis jetzt ist die Bestimmung effektiver unterer Schranken für die Klassenzahlen imaginär quadratischer Zahlkörper. Dies folgt aus einer älteren Arbeit von D. Goldfeld, vorausgesetzt man findet eine L-Reihe wie oben mit einer Nullstelle der Ordnung mindestens drei. Dieses aber leistet gerade die Formel von Gross und Zagier.

Alles in allem handelt es sich um eine schöne Entdeckung, welche wir aber leider noch nicht "erklären" können: Warum ist sie richtig?

Literatur: [**G1**, **G2**, **O**, **Zg**].

2. Die Mordell-Vermutung. Kommen wir zum Themenkreis der Mordell-Vermutung. Diese besagt, daß auf einer Kurve vom Geschlecht gröer als eins über einem Zahlkörper nur endlich viele rationale Punkte liegen. Dies wurde 1922 von L. Mordell vermutet, in der Arbeit in welcher er bewies, daß die rationalen Punkte auf einer elliptischen Kurve ein endlich erzeugte abelsche Gruppe bilden. Allerdings schreibt Mordell selbst, daß er keinen Beweisansatz kennt, und nicht einmal plausibel machen kann warum dies so sein sollte. Die ersten Fortschritte erzielten A. Weil and C. L. Siegel: Der erste verallgemeinerte Mordell's Satz auf abelsche Varietäten, während Siegel mit Hilfe der diophantischen Approximation und A. Weil's Satz die Vermutung für ganze Punkte zeigen konnte. Allerdings führten diese Ansätze im Weiteren nicht mehr zu so groß en Fortschritten, und interessanterweise benutzt der endgültige Beweis ganz andere Methoden.

Diese wurden entwickelt zum Beweis der Mordell-Vermutung über Funktionenkörpern. Dies gelang zuerst J. Manin und H. Grauert, und die für uns wichtige Methode geht zurück auf A. N. Parshin, der die Behauptung reduzierte auf die

Schafarevitsch-Vermutung: Über einem Zahlkörper gibt es bis auf Isomorphie nur endlich viele abelsche Varietäten vorgegebener Dimension, welche gute Reduktion außerhalb einer festen endlichen Menge von Stellen des Körpers haben. Die Verbindung zwischen beiden Vermutungen geschieht, indem man zu einem rationalen Punkt auf einer Kurve zunächst eine Überlagerung konstruiert, welche genau über diesem Punkt verzweigt. Auf diese Weise erhält man eine neue Kurve, die schlechte Reduktion höchstens an den Stellen haben kann, wo dies der Fall ist für die ursprüngliche Kurve, plus eventuell einige weitere Stellen (zum Beispiel alle Stellen der Charakteristik zwei). Die Isomorphieklasse dieser Überlagerung bestimmt den rationalen Punkt, und man benötigt nur noch den Satz von Torelli sowie die Vermutung von Schafarevitsch für die Jacobi-Varietät der Überlagerungskurve. Diese Methode liefert übrigens auch den Satz von Siegel über ganze Punkte auf affinen elliptischen Kurven.

Wie aber zeigt man nun die Vermutung von Schafarevitsch? Sie erschien zunächst recht unwahrscheinlich, da über Funktionenkörpern wohl die Mordell-Vermutung gilt, aber nicht diese stärkere Aussage. Trotzdem kommt man bei Zahlkörpern zum Ziel, da man die Tate-Vermutung benutzen kann: Diese reduziert das Problem auf eine Frage über l-adische Darstellungen, und diese kann man einfacher lösen als das ursprüngliche Problem.

Für alles dies benötigt man etwas Information über den Modulraum der prinzipal polarisierten abelschen Varietäten, insbesondere über seine Kompaktifizierungen. Genauer gesagt geht es um die Bestimmung der Höhe einer abelschen Varietät, das heiß t des zugehörigen rationalen Punktes im Modulraum. Dazu benötigt man eine Kompaktifizierung des Modulraums, ein amples Geradenbündel darauf, sowie eine hermite'sche Metrik für das Geradenbündel. Dies erforderte zunächst einige Tricks, doch mittlerweile besitzen wir eine befriedigende Theorie: Zunächst konstruiert man die toroidale Kompaktifizierung über den ganzen Zahlen. Über dieser dehnt sich die universelle abelsche Varietät aus zu einer semiabelschen Varietät, und man erhält ein Geradenbündel ω, das Determinantenbündel der relativen Differentiale dieser semiabelschen Varietät. Die globalen Schnitte der Potenzen von ω sind gerade die Siegel'schen Modulformen mit ganzzahligen Fourierkoeffizienten. Mit Hilfe von Theta-Reihen ergibt sich, daß eine Potenz von ω global erzeugt ist, und das Bild der korrespondierenden Abbildung von der toroidalen Kompaktifizierung in den projektiven Raum ist die arithmetische Version der minimalen (oder Satake-, oder Baily-Borel-) Kompaktifizierung. Nach Konstruktion gibt es darauf ein kanonisches amples Geradenbündel, und es bleibt die Definition einer guten Metrik. Wieder gibt es nur eine kanonische Wahl, nämlich Quadratintegration von Differentialen. Leider hat diese Metrik Singularitäten im Unendlichen, doch sind diese so mild, daß sie den Gang der Dinge nicht behindern.

Auf diese Weise erhält man eine einfache Definition der Höhe einer abelschen Varietät, mit der sich arbeiten läßt. Es handelt sich im Wesentlichen um das Quadrat-Integral eines Erzeugenden der ganzzahligen Differentiale. Dies liefert eine numerische Invariante der abelschen Varietät, so daß es bis auf Isomorphie nur endlich viele abelsche Varietäten gibt, für die diese Invariante nicht größer als ein fester vorgegebener Wert ist. Erwähnen wir noch daß die arithmetische Kompaktifizierung des Modulraums auch andere Anwendungen besitzt. Insbesondere ist sie sehr nützlich für eine arithmetische Theorie der Siegelschen Modulformen.

Bleibt schließ lich der Beweis der Tate-Vermutung. Genauer gesagt handelt es sich nur um einen Spezialfall der allgemeinen Vermutung, betreffend Endomorphismen abelscher Varietäten. Der Beweis geht zurück auf J. Tate und J. G. Zarhin, und man braucht als neue Ingredienz nur noch die Kompaktifizierung des Modulraums der abelschen Varietäten sowie einige Resultate von J. Tate und M. Raynaud über p-divisible Gruppen. Genauer gesagt betrachtet man eine l-divisible Untergruppe der abelschen Varietät, und dividiert sukzessiv durch die verschieden Stufen dieser Untergruppe. Dies liefert eine Folge von abelschen Varietäten, und man muß zeigen, daß unendlich viele dieser isomorph sind. Dazu betrachtet man die zugehörige numerische Invariante, wie weiter oben definiert. Es reicht wenn die Folge dieser reellen Zahlen stationär wird. Es geht also darum, wie sich die Höhe einer abelschen Varietät (so wie oben definiert) bei Isogenien ändert. Man zeigt daß die Variation der Höhe in einer Isogenieklasse beschränkt ist, sogar mit einer ganz effektiv berechenbaren Schranke.

Alles in allem zeigen diese Resultate aufs neue die ungeheure Kraft der Grothendieck'schen Methoden in der algebraischen Geometrie. Man hat aber das Gefühl daß diese Sätze gewissermaß en den vorläufigen Abschluß einer Entwicklung darstellen, und daß wir neue Methoden benötigen, etwa um rationale Punkte auf Varietäten höherer Dimension zu studieren. Dieses wird sicher einiges Experimentieren erfordern, genau wie ehedem die Verhältnisse bei Kurven erarbeitet werden muß ten.

Literatur: [**D, F2, F3, F4, FW, SI, S2, Zl, Z2**].

3. *P*-adische Darstellungen. Es handelt sich um das Studium der p-adischen étalen Kohomologie einer algebraischen Mannigfaltigkeit über einem p-adischen Körper. Man findet Beziehungen zur kohärenten Kohomologie, und erhält so eine Art von p-adischer Hodge-Theorie. Auß erdem gibt es Zusammenhänge mit der kristallinen Kohomologie. Es geht dabei um A. Grothendieck's "mysterious functor."

Versuchen wir zu erklären, worum es sich handelt. Sei K ein p-adischer Körper, etwa eine endliche Erweiterung der p-adischen Zahlen $\mathbb{Q}_p$. Mit $\mathbb{C}_p$ bezeichnen wir die Komplettierung des algebraischen Abschlusses $\overline{K}$ von K, Darauf operiert stetig die Galois-Gruppe $\mathcal{G} = \mathrm{Gal}(\overline{K}/K)$, und $\mathbb{C}_p(n)$ bezeichnet den Twist mit der n-ten Potenz des zyklotomischen Charakters $\mathcal{G} \to \mathbb{Z}_p^*$. Als erstes bemerken wir, daß der ungeheuer groß e Körper $\mathbb{C}_p$ einer Behandlung zugänglich ist: Er enthält den Zwischenkörper, der durch die Einheitswurzeln von p-Potenzordnung erzeugt wird, und ist fast unverzweigt über diesem. Deshalb kann man viele Probleme auf den kleineren Körper zurückführen, woman sie durch direkte Rechnung löst. Tate und Raynaud haben gezeigt, daß für eine eigentliche glatte K-Varietät $\mathbf{X}$ gilt:

$$H^1(\mathbf{X}, \mathbb{Z}_p) \otimes \mathbb{C}_p \cong H^1(\mathbf{X}, \mathcal{O}_{\mathbf{x}}) \otimes \mathbb{C}_p \oplus H^0(\mathbf{X}, \mathcal{O}_{\mathbf{x}}) \otimes \mathbb{C}_p(-1) \quad (\text{als } \mathcal{G} - \text{Moduln}).$$

Dazu sei bemerkt daß die verschiedenen Twists $\mathbb{C}_p(m)$ nicht isomorph sind, und daß ein $\mathbb{C}_p$-Vektorraum mit semilinearer $\mathcal{G}$-Aktion einen größ ten Unterraum besitzt, welcher isomorph zu einer direkte Summe von $\mathbb{C}_p(m)$'s ist.

Es wurde auch vermutet, und kann inzwischen beweisen werden, daß allgemeiner gilt:

$$H^n(\mathbf{X}, \mathbb{Z}_p) \otimes \mathbb{C}_p \cong \sum_{a+b=n} H^a(\mathbf{X}, \Omega^b_{\mathbf{X}}) \otimes \mathbb{C}_p(-b).$$

Insbesondere kann man damit einen rein algebraischen Beweis für das Degenerieren der Hodge Spektral-Sequenz geben. Es folgt auch die Symmetric der Hodge-Zahlen, zumindest für projektive Varietäten. Bekanntlich reicht dies schon, um viele Resultate zu beweisen, für die man üblicherweise analytische Methoden benutzt. Ein Beispiel ist des Verschwindungssatz von Kodaira.

Die Beweismethode benutzt eine Art von intermediärer Kohomologie, in die sich die Kohomologien auf beiden Seiten der obigen Abbildung natürlich abbilden. Dies reicht, da beide Seiten alle üblichen Axiome erfüllen, wie etwa Künneth-Formel oder Poincaré-Dualität. Die Details sind etwas kompliziert, doch im Wesentlichen kann man die intermediäre Theorie wie folgt beschreiben:

Sei R ein affiner Ring der Varietät, $\overline{R}$ die p-adische Komplettierung der maximalen unverzweigten Erweiterung von R. Dann betrachte man die Galois-Kohomologie von $\overline{R}$. Die Galois-Kohomologie von $\mathbb{Z}_p$ bildet sich da hinein ab, und dies ergibt die eine natürliche Transformation. Die andere erhält man durch Betrachtung der Differentiale. Einige Details: Wir nehmen an daß R genügend viele Einheiten besitzt. Durch Adjungieren der p-Potenzwurzeln dieser Einheiten erhalten wir eine gut zu kontrollierende Erweiterung $\tilde{R}$ von R, welche unverzweigt ist in Charakteristik 0. Über R ist $\overline{R}$ fast unverzweigt in Kodimension ≤ 2, und man kann mit einiger Mühe Resultate von $\tilde{R}$ auf $\overline{R}$ übertragen. Sei zum Beispiel $\overline{V}$ die Komplettierung des ganzen Abschlusses des diskreten Bewertungsringes V von K. Die exakte $\mathcal{G}$-lineare Sequenz

$$0 \to \Omega_{\overline{V}/V} \otimes_{\overline{V}} \overline{R} \to \Omega_{\overline{R}/R} \to \Omega_{\overline{R}/R\overline{V}} \to 0,$$

zusammen mit der Bestimmung des ersten Terms durch J. Fontaine, führt zu der gesuchten zweiten Abbildung, das heiß t zu einer Beziehung zwischen Differentialen und Galois-Kohomologie.

In einigen speziellen Fällen kann man noch mehr aussagen: Man findet eine Beziehung zwischen der p-adischen étalen Kohomologie und der kristallinen Kohomologie, so daß die eine die andere eindeutig bestimmt. Dies ist ein Fall des "mysterious functor," nach dem schon A. Grothendieck gesucht hat. Der Beweis von J. Fontaine und W. Messing benutzt die "syntomic topology," und wird sicher in einem der Spezialvorträge ausführlicher dargestellt werden.

Auß erdem sei erwähnt, daß Fontaine kürzlich eine weitere Vermutung von Schafarevitsch bewiesen hat: Es gibt keine abelsche Varietät über $\mathbb{Q}$, welche überall gute Reduktion hat. Der Beweis benutzt die Theorie der endlichen Gruppenschemata, und natürlich p-adische Darstellungen.

Literatur: [**F5, F6, Fo**].

4. Arithmetische Schnitt-Theorie. Beim Studium der Zahlkörper hat es sich gezeigt, daß man am best en versucht, die unendlichen Stellen und die endlichen ähnlich zu behandeln. Wenn man nun Varietäten über Zahlkörpern betrachtet, so ist es klar was man an den endlichen Stellen zu tun hat: Man benötigt gute Modelle über den entsprechenden diskreten Bewertungsringen. Was aber entspricht dem im Unendlichen? Die Erfahrung zeigt, daß man die algebraischen Objekte mit Metriken versehen muß. Wenn man zum Beispiel Geradenbündel auf einer algebraischen Kurve (über dem Zahlkörper) betrachtet, so benötigt man zunächst eine Ausdehnung auf ein Modell über den ganzen Zahlen. Dies erledigt die endlichen Stellen. Fürs Unendliche versieht man die Geradenbündel mit Metriken. Auf diese Weise hat S. Arakelov eine Schnitt-Theorie für arithmetische Flächen entwickelt, und man kann zeigen, daß ein Teil der Theorie der algebraischen Flächen sich auf diesen Fall überträgt. Dies gilt etwa für den Satz von Riemann-Roch oder für den Indexsatz von Hodge.

Man macht dabei die folgenden Überlegungen: Zunächst definiert man Volumenformen auf der Kohomologie eines metrisierten Geradenbündels. Dies ist ein Resultat über Riemann'sche Flächen, also rein lokal an den unendlichen Stellen. Die Definition des Volumens ist so gemacht, daß der übliche Beweis des Satzes von Riemann-Roch für algebraische Flächen übertragen werden kann. Aus dem Riemann-Roch leitet man eine Beziehung her zwischen Schnitt-Theorie und Néron-Tate Höhen, und der Satz von Hodge folgt. Auß erdem gelten einige weitere Resultate, insbesondere ist bei Kurven vom Geschlecht größ er als Eins das Quadrat des dualisierenden Geradenbündels ω kleiner oder gleich Null. Was aber etwa fehlt, ist ein Analogon zur Hodge-Theorie oder zum Frobenius.

Es stellt sich nun die Frage, was man bei höheren Dimensionen tun soil. Schon an den endlichen Stellen treten neue Schwierigkeiten auf, da die Theorie der minimalen Modelle fehlt, Trotzdem haben A. Beilinson, H. Gillet, und C. Soulé eine Schnitt-Theorie entwickelt. Es handelt sich um eine sehr schöne Mischung aus Analysis und Algebra, und meiner Ansicht nach liegt hier ein reiches Feld für künftige Forschungen.

Ein verwandtes Feld sind die Beilinson-Vermutungen über spezielle Werte von L-Reihen. Dabei scheint mir ein sehr wichtiger Punkt die Frage nach der Konstruktion von Motiven. Ihre Existenz wurde von A. Grothendieck vermutet, und diente als Leitmotiv für viele kohomologische Untersuchungen. Sie sollten eine Art von universeller Kohomologie-Theorie bilden. Zur Zeit kennen wir eigentlich nur 1-Motive, und dies sind im Wesentlichen semiabelsche algebraische Gruppen.

Beilinson hat mit Hilfe der algebraischen K-Theorie einige Kandidaten für Motive höherer Dimension gefunden.

Literatur: [**Bl, B2, B3, Fl, GS**].

Literature

[Bl] A. Beilinson, *Higher regulators and values of L-functions,* Modern Problems in Mathematics, VINITI Series **24** (1984), 181–238.

[B2] ———, *Notes on absolute Hodge cohomology*, preprint, 1984.

[B3] ———, *Height pairings between algebraic cycles*, preprint, 1985.

[D] P. Deligne, Seminaire Bourbaki **616** (1984).

[Fl] G. Faltings, *Calculus on arithmetic surfaces*, Ann. of Math. **119** (1984), 387–424.

[F2] ———, *Endlichkeitssätze für abelsche Varietäten über Zahlkörpern*, Invent. Math. **73** (1983), 349–366.

[F3] ———, *Die Vermutungen von Tate und Mordell*, Jahresber. Deutsch. Math.-Verein. **86** (1984), 1–13.

[F4] ———, *Arithmetische Kompaktifizierung des Modulraums der abelschen Varietäten,* Lecture Notes in Math., vol. 1111, Springer-Verlag, 1985, pp. 321–383.

[F5] ———, *Hodge-Tate structures and modular forms*, submitted to Math. Ann.

[F6] ———, *P-adic Hodge-theory*, Manuscript, Princeton University, Princeton, N.J., 1985.

[Fo] J. Fontaine, *II n'y a pas de variété abelienne sur* $\mathbb{Z}$, Invent. Math. **81** (1985), 515–538.

[FW] G. Faltings und G. Wüstholz, *Rational points*, Vieweg, Braunschweig, 1984.

[Gl] D. Goldfeld, *The class number of quadratic fields and the conjecture of Birch and Swinnerton-Dyer*, Ann. Scuola Norm. Sup. Pisa **3** (1976), 623–663.

[G2] ———, *Gauss' class number problem for imaginary quadratic fields*, Bull. Amer. Math. Soc. (N.S.) **13** (1985), 23–37.

[GS] H. Gillet und C. Soulé, *Intersection sur les variétés d'Arakelov*, C. R. Acad. Sci. Paris **299** (1984), 563–566.

[GZ] B. Gross und D. Zagier, *Points de Heegner et dérivés de fonctions L*, C. R. Acad. Sci. Paris Sér. I Math. **297** (1983), 85–87.

[O] J. Osterté, Seminaire Bourbaki **631** (1984).

[S1] L. Szpiro, Seminaire Bourbaki **619** (1984).

[S2] ———, *Séminaire sur les pinceaux arithmétiques. La conjecture de Mordell*, Astérisque **127** (1985).

[Zl] J. G. Zarhin, *Isogenics of abelian varieties over fields of finite characteristic*, Math. USSR-Sb. **24** (1974), 451–461.

[Z2] ———, *Endomorphisms of abelian varieties over fields of finite characteristic*, Math. USSR Izv. **9** (1975), 255–260.

[Zg] D. Zagier, *Modular points, modular curves, modular surfaces and modular forms*, Lecture Notes in Math., vol. 1111, Springer-Verlag, pp. 225–248.

PRINCETON UNIVERSITY, PRINCETON, NEW JERSEY 08544, USA

Reprinted from Proc. Int. Congr. Math., 1986

THE WORK OF M. H. FREEDMAN

by
JOHN MILNOR

Michael Freedman has not only proved the Poincaré hypothesis for 4-dimensional topological manifolds, thus characterizing the sphere S^4, but has also given us classification theorems, easy to state and to use but difficult to prove, for much more general 4-manifolds. The simple nature of his results in the topological case must be contrasted with the extreme complications which are now known to occur in the study of differentiable and piecewise linear 4-manifolds.

The "n-dimensional Poincaré hypothesis" is the conjecture that every topological n-manifold which has the same homology and the same fundamental group as an n-dimensional sphere must actually be homeomorphic to the n-dimensional sphere. The cases $n = 1, 2$ were known in the nineteenth century, while the cases $n \geq 5$ were proved by Smale, and independently by Stallings and Zeeman and by Wallace, in 1960–61. (The original proofs needed an extra hypothesis of differentiability or piecewise linearity, which was removed by Newman a few years later.) The 3- and 4-dimensional cases are much more difficult.

Freedman's 1982 proof of the 4-dimensional Poincaré hypothesis was an extraordinary tour de force. His methods were so sharp as to actually provide a complete classification of all compact simply connected topological 4-manifolds, yielding many previously unknown examples of such manifolds, and many previously unknown homeomorphisms between known manifolds. He showed that a compact simply connected 4-manifold M is characterized, up to homeomorphism, by two simple invariants. The first is the 2-dimensional homology group

$$H_2 = H_2(M;\ Z) \cong Z \oplus \cdots \oplus Z,$$

together with the symmetric bilinear intersection pairing

$$\omega : H_2 \otimes H_2 \to Z.$$

This pairing, which is defined as soon as we choose an orientation for M, must have determinant ± 1 Poincaré duality. The second is the Kirby–Siebenmann obstruction class, an element

$$\sigma \in H^4(M;\ Z/2) \cong Z/2$$

that vanishes if and only if M is stably smoothable. In other words, σ is zero if and only if the product $M \times R$ can be given a differentiable structure, or equivalently a piecewise linear structure. These two invariants ω and σ can be prescribed arbitrarily, except for a relation in one special case. If the form ω happens to be even, that is if $\omega(x,x) \equiv 0 \pmod 2$ for every $x \in H_2$, then the Kirby–Siebenmann obstruction must be equal to the Rohlin invariant:

$$\sigma \equiv \text{signature } (\omega)/8 \pmod 2.$$

(Freedman's original proof that these two invariants characterize M up to homeomorphism required an extra hypothesis of "almost-differentiability," which was later removed by Quinn.)

If the intersection form $\omega \neq 0$ is indefinite or has rank at most eleven, then it follows from known results about quadratic forms that M can be built up (nonuniquely) as a connected sum of copies of four simple building blocks, each of which may be given either the standard or the reversed orientation. One needs the product $S^2 \times S^2$, the complex projective plane CP^2, and two exotic manifolds which were first constructed by Freedman. One of these is a nondifferentiable analogue of the complex projective plane, and the other is the unique manifold whose intersection form ω is positive definite and even of rank eight. (This ω can be identified with the lattice generated by the root vectors of the Lie group E_8. As noted by Rohlin in 1952, a 4-manifold with such an intersection form can never be differentiable.) By way of contrast, if we allow positive definite intersection forms, then the number of distinct simply connected manifolds grows more than exponentially with increasing middle Betti number.

Freedman's methods extend also to noncompact 4-manifolds. For example, he showed that the product $S^3 \times R$ can be given an exotic differentiable structure, which contains a smoothly embedded Poincaré homology 3-sphere and hence cannot be smoothly embedded in euclidean 4-sphere [**11**, **16**]. His methods apply also to many manifolds which are not simply connected [**22**]. For example, a "flat" 2-sphere in 4-space is unknotted if and only if its complement has free cyclic fundamental group; and a flat 1-sphere in S^3 has trivial Alexander polynomial if and only if it bounds a flat 2-disk in the unit 4-disk whose complement has free cyclic fundamental group.

The proofs of these results are extremely difficult. The basic idea, which had been used in low dimensions by Moebius and Poincaré, and in high dimensions by Smale and Wallace, is to build the given 4-manifold up inductively, starting with a 4-dimensional disk, by successively adding handles. The essential difficulty, which does not arise in higher dimensions, occurs when we try to control the fundamental group by inserting 2-dimensional handles, since a 2-dimensional disk immersed in a 4-manifold will usually have self-intersections. This problem was first attacked by Casson, who showed how to construct a generalized kind of 2-handle with prescribed boundary within a given 4-manifold. Freedman's major technical tool is a theorem which asserts that every Casson handle is actually homeomorphic to the standard open handle, (closed 2-disk) $\times$ (open 2-disk). The proof involves a delicately controlled infinite repetition argument in the spirit of the Bing school of topology, and is nondifferentiable in an essential way.

Papers of M. H. Freedman

1. *Automorphisms of circle bundles over surfaces*, Geometric Topology, Lecture Notes in Math., vol. 438, Springer-Verlag, 1974, pp. 212–214.
2. *On the classification of taut submanifolds*, Bull. Amer. Math. Soc. **81** (1975), 1067–1068.
3. *Uniqueness theorems for taut submanifolds*, Pacific J. Math. **62** (1976), 379–387.
4. *Surgery on codimension* 2 *submanifolds*, Mem. Amer. Math. Soc. No. 119 (1977).
5. *Une obstruction élémentaire à l'existence d'une action continue de groupe dans une variété* (with W. Meeks), C. R. Acad. Sci. Paris Ser. **A 286** (1978), 195–198.
6. Λ-*splitting* 4-*manifolds* (with L. Taylor), Topology **16** (1977), 181–184.
7. *A geometric proof of Rochlin's theorem* (with R. Kirby), Proc. Sympos. Pure Math., vol. 32, part 2, Amer. Math. Soc., Providence, R.I., 1978, pp. 85–98.
8. *Remarks on the solution of first degree equations in groups*, Algebraic and Geometric Topology, Lecture Notes in Math., vol. 664, Springer-Verlag, 1978, pp. 87–93.
9. *Quadruple points of* 3-*manifolds in* S^4, Comment. Math. Helv. **53** (1978), 385–394.
10. *A converse to (Milnor–Kervaire theorem)* $\times R$ *etc....*, Pacific J. Math. **82** (1979), 357–369.
11. *A fake* $S^3 \times R$, Ann. of Math. **110** (1979), 177–201.
12. *Cancelling* 1-*handles and some topological imbeddings*, Pacific J. Math. **80** (1979), 127–130.
13. *A quick proof of stable surgery* (with F. Quinn), Comment. Math. Helv. **55** (1980), 668–671.
14. *Planes triply tangent to curves with nonvanishing torsion*, Topology **19** (1980), 1–8.
15. *Slightly singular* 4-*manifolds* (with F. Quinn), Topology **20** (1981), 161–173.
16. *The topology of* 4-*manifolds*, J. Differential Geom. **17** (1982), 357–454 (see also F. Quinn, ibid, p. 503).
17. *A surgery sequence in dimension four; the relations with knot concordance*, Invent. Math. **68** (1982), 195–226.
18. *Closed geodesics on surfaces* (with J. Hass, P. Scott), Bull. London Math. Soc. **14** (1982), 385–391.
19. *A conservative Dehn's Lemma*, Low Dimensional Topology, Contemp. Math., vol. 20, Amer. Math. Soc., Providence, R.I., 1983, pp. 121–130.
20. *Imbedding least area incompressible surfaces* (with J. Hass, P. Scott), Invent. Math. **71** (1983), 609–642.
21. *Homotopically trivial symmetries of Haken manifolds are toral* (with S.-T. Yau), Topology **22** (1983), 179–189.
22. *The disk theorem for 4-dimensional manifolds*, Proc. Internat. Congr. Math. Warsaw 1983, vol. 1, Pol. Sci. Publ., 1984, pp. 647–663.

23. *There is no room to spare in four-dimensional space*, Notices Amer. Math. Soc. **31** (1984), 3–6.
24. *Atomic surgery problems* (with A. Casson), Four-Manifold Theory, Contemp. Math., vol. 35, Amer. Math. Soc., Providence, R.I., 1984, pp. 181–199.

Other References

25. H. Poincaré, *Cinquième complément á l'Analysis situs* (1904), Oeuvres 6, Paris, 1953, pp. 435–498.
26. L. Siebenmann, *Amorces de la chirurgie en dimension* 4, *un* $S^3 \times R$ *exotique* (d'apres A. Casson et M. H. Freedman), Séminaire Bourbaki (1978–79), Exp. No. 536, Lecture Notes in Math., vol. 770, Springer-Verlag, 1980, pp. 183–207.
27. ———, *La conjecture de Poincaré topologique en dimension* 4 (d'apres M. H. Freedman), Seminaire Bourbaki (1981–82), Exp. No. 588, Astérisque, no. 92–93, Soc. Math. France, Paris, 1982, pp. 219–248.

Institute for Advanced Study
Princeton
New Jersey 08540, USA

Michael H. Freedman

MICHAEL H. FREEDMAN

Personal Data

Date of birth:	April 21, 1951
Place of birth:	Los Angeles, California
Citizenship:	USA
Office Address:	Department of Mathematics, 0112 University of California, San Diego 9500 Gilman Drive La Jolla, California 92093-0112
Office Phone:	(619) 534-2647
E-mail:	mfreedman@ucsd.edu

Education

1968–69	— Studied at University of California, Berkeley, California
1969–73	Ph.D., Princeton University, Princeton, New Jersey

Membership

American Mathematical Society
National Academy of Sciences
American Academy of Arts and Sciences
New York Academy of Sciences
Associate Editor for: Annals of Mathematics (1984–91)
Journal of American Mathematical Society
Journal of Differential Geometry
Mathematical Research Letters, and Topology

Honors and Awards

1980–83	Alfred P. Sloan Fellow
1982	Invited Lecturer, Int'l Congress of Mathematicians, Warsaw, Poland
1984	California Scientist of the Year
1984–89	MacArthur Foundation Fellow
1984	Elected to National Academy of Sciences
1985	Elected to American Academy of Arts and Sciences
1985	Charles Lee Powell Endowed Chair in Mathematics, UCSD
1986	American Mathematical Society's Veblen Prize
1986	Fields Medal, Int'l Congress of Mathematicians, Berkeley, CA

1987	National Medal of Science (White House, June 1987)
1988	Humboldt Award, Germany
1989	Elected to New York Academy of Sciences
1994	Guggenheim Fellowship Award

Professional Experience

1973–75	Lecturer, Dept. of Mathematics, Univ. of California, Berkeley, CA
1975–76	Member, Institute for Advanced Study, Princeton, New Jersey
1976–79	Assistant Professor, Dept. of Mathematics, Univ. of California, San Diego (USCD), La Jolla, CA
1979–80	Associate Professor, Dept. of Mathematics, UCSD
1980	Member, Institute for Advanced Study, Princeton, New Jersey
1981–82	Associate Professor, Dept. of Mathematics, UCSD
1982–present	Professor, Dept. of Mathematics, UCSD
1985–present	Charles Lee Powell Chair Professor, Dept. of Mathematics, UCSD

BETTI NUMBER ESTIMATES FOR NILPOTENT GROUPS

by

MICHAEL FREEDMAN
University of California in San Diego, La Jolla, CA, 92093-0112
E-mail address: freedman@euclid.ucsd.edu

RICHARD HAIN
Department of Mathematics, Duke University, Durham, NC 27708-0320
E-mail address: hain@duke.math.edu

PETER TEICHNER
University of California in San Diego, La Jolla, CA, 92093-0112
E-mail address: teichner@euclid.ucsd.edu

ABSTRACT. We prove an extension of the following result of Lubotzky and Magid on the rational cohomology of a nilpotent group G: if $b_1 < \infty$ and $G \otimes \mathbb{Q} \neq 0, \mathbb{Q}, \mathbb{Q}^2$ then $b_2 > b_1^2/4$. Here the b_i are the rational Betti numbers of G and $G \otimes \mathbb{Q}$ denotes the Malcev-completion of G. In the extension, the bound is improved when we know that all relations of G all have at least a certain commutator length. As an application of the refined inequality, we show that each closed oriented 3-manifold falls into exactly one of the following classes: it is a rational homology 3-sphere, or it is a rational homology $S^1 \times S^2$, or it has the rational homology of one of the oriented circle bundles over the torus (which are indexed by an Euler number $n \in \mathbb{Z}$, e.g. $n = 0$ corresponds to the 3-torus) or it is of *general type* by which we mean that the rational lower central series of the fundamental group does not stabilize. In particular, any 3-manifold group which allows a maximal torsion-free nilpotent quotient admits a rational homology isomorphism to a torsion-free nilpotent group.

1. The Main Results

We analyze the rational lower central series of 3-manifold groups by an extension of the following theorem of Lubotzky and Magid [15, (3.9)].

The first author was supported in part by an NSF grant and the Guggenheim Foundation.
The second author was supported by the IHES and grants from the NSF.
The third author was supported by the IHES and the Humboldt foundation.

Theorem 1. *If G is a nilpotent group with $b_1(G) < \infty$ and $G \otimes \mathbb{Q} \neq \{0\}, \mathbb{Q}, \mathbb{Q}^2$, then*

$$b_2(G) > \frac{1}{4}\, b_1(G)^2.$$

Here $b_i(G)$ denotes the ith rational Betti number of G and $G \otimes \mathbb{Q}$ the Malcev-completion of G. (We refer the reader to Appendix A, where we have collected the group theoretic definitions.)

This Theorem is an analogue of the Golod–Shafarevich Theorem (see [9] or [14, p. 186]), which states that if G is a finite p-group, then the inequality $r > d^2/4$ holds, where d is the minimal number of generators in a presentation of G, and r is the number of relations in any presentation. It can be used to derive a result for finitely generated nilpotent groups similar to the one above, but with $b_2(G)$ replaced by the minimal number of relations in a presentation for G, see [6, p. 121].

We shall need a refined version of this result in the case where the relations of G are known to have a certain commutator length. In order to make this precise, let

$$H_2(G;\mathbb{Q}) =: \Phi_2^{\mathbb{Q}}(G) \supseteq \Phi_3^{\mathbb{Q}}(G) \supseteq \Phi_4^{\mathbb{Q}}(G) \supseteq \cdots$$

be the rational Dwyer filtration of $H_2(G;\mathbb{Q})$ (defined in Appendix A).

Theorem 2. *If G is a nilpotent group with $b_1(G) < \infty$, $H_2(G;\mathbb{Q}) = \Phi_r^{\mathbb{Q}}(G)$ and $G \otimes \mathbb{Q} \neq \{0\}, \mathbb{Q}, \mathbb{Q}^2$, then*

$$b_2(G) > \frac{(r-1)^{(r-1)}}{r^r}\, b_1(G)^r.$$

The statement that $H_2(G;\mathbb{Q}) = \Phi_r^{\mathbb{Q}}(G)$ is equivalent to the statement that in a minimal presentation of the Lie algebra of $G \otimes \mathbb{Q}$, all relations lie in the rth term of the lower central series of the free Lie algebra on the minimal generating set $H_1(G;\mathbb{Q})$. We know of no analogue of this result for p-groups, although there should be one. The result is a corollary of a more technical and stronger result, Proposition 5. Note that Theorem 1 is the special case of Theorem 2 with $r = 2$.

The intuitive idea behind these results is that $H_1(G)$ corresponds to generators of G and $H_2(G)$ to its relations. For example, if G is abelian, then the number of (primitive) relations ($\approx b_2$) grows quadratically in the number of generators ($\approx b_1$) because the commutator of a pair of generators has to be a consequence of the primitive relations. Similarly, if G is nilpotent, then the relations have to imply that for some r, all r-fold commutators in the generators vanish. If $H_2(G) = \Phi_r(G)$, then all relations are in fact r-fold commutators because, by Dwyer's Theorem (see Appendix A), this condition is equivalent to G/Γ_r being isomorphic to F/Γ_r where F is the free group on $b_1(G)$ generators. Thus b_2 should grow as the r-th power of b_1. But if some of the relations are shorter commutators, then they can imply a vast number of relations among the r-fold commutators. Therefore, one can only expect a lower order estimate in this case.

Example. Let $x_1, \ldots, x_4$ be generators of $H^1(\mathbb{Z}^4; \mathbb{Z})$. Then the central extension

$$1 \longrightarrow \mathbb{Z} \longrightarrow G \longrightarrow \mathbb{Z}^4 \longrightarrow 1$$

classified by the class $x_1x_2 + x_3x_4 \in H^2(\mathbb{Z}^4; \mathbb{Z})$ satisfies $b_1(G) = 4, b_2(G) = 5$. This shows that the lower bound for b_2 in Theorem 1 cannot be of the form $b_1(b_1 - 1)/2$ with equality in the abelian case.

This paper is organized as follows. In Section 2 we give applications of the above results to 3-dimensional manifolds. Section 3 explains how to derive the nilpotent classification of 3-manifolds from Theorem 2. In Section 4 we give further examples and in Section 5 we prove Theorem 2 modulo two key lemmas. These are proven in Sections 6 and 7. Appendices A and B contain background information from group theory respectively rational homotopy theory.

Acknowledgement: All three authors would like to thank the IHES for its hospitality and support as well as for bringing us together.

2. Applications to 3-Manifolds

We will look at a 3-manifold through nilpotent eyes, observing only the tower of nilpotent quotients of the fundamental group, but never the group itself. This point of view has a long history in the study of link complements and it arises naturally if one studies 3- and 4-dimensional manifolds together. For example, Stallings proved that for a link L in S^3 the nilpotent quotients $\pi_1(S^3 \setminus L)/\Gamma_r$ are invariants of the topological concordance class of the link L. These quotients contain the same information as Milnor's μ-invariants which are generalized linking numbers. For precise references about this area of research and the most recent applications to 4-manifolds see [7] and [8].

Let M be a closed oriented 3-manifold and $\{\Gamma_r^{\mathbb{Q}} \mid r \geq 1\}$ the rational lower central series (see Appendix A) of $\pi_1(M)$. Similarly to Stallings' result, the quotients $(\pi_1(M)/\Gamma_r^{\mathbb{Q}}) \otimes \mathbb{Q}$ are invariants of rational homology H-cobordism between such 3-manifolds.

Definition. A 3-manifold M is of *general type* if

$$\pi_1(M) = \Gamma_1^{\mathbb{Q}} \supsetneqq \Gamma_2^{\mathbb{Q}} \supsetneqq \Gamma_3^{\mathbb{Q}} \supsetneqq \cdots$$

and *special* if, for some $r > 0$, $\Gamma_r^{\mathbb{Q}} = \Gamma_{r+1}^{\mathbb{Q}}$. (Following the terminology used in group theory, the fundamental group of a special 3-manifold is called *rationally prenilpotent*, compare the Appendix A.)

Our nilpotent classification result reads as follows.

Theorem 3. *If a closed oriented* 3*-manifold* M *is special, then the maximal torsion-free nilpotent quotient of its fundamental group is isomorphic to exactly one of the groups*

$$\{1\}, \quad \mathbb{Z} \quad \text{or} \quad H_n.$$

In particular, this quotient is a torsion-free nilpotent 3*-manifold group of nilpotency class* < 3.

Here the groups $H_n, n \geq 0$, are the central extensions

$$1 \longrightarrow \mathbb{Z} \longrightarrow H_n \longrightarrow \mathbb{Z}^2 \longrightarrow 1$$

classified by the Euler class $n \in H^2(\mathbb{Z}^2; \mathbb{Z}) \cong \mathbb{Z}$. This explains the last sentence in our theorem because H_n occurs as the fundamental group of a circle bundle over the 2-torus with Euler class n.

Since the groups above have nilpotency class < 3, it is very easy to recognize the class to which a given 3-manifold belongs; one simply has to compute $\pi_1(M)/\Gamma_3^{\mathbb{Q}}$. Note, in particular, that a 3-manifold M is automatically of general type if its first rational Betti number satisfies $b_1 M > 3$.

If $b_1 M = 0$, then M is a rational homology sphere, if $b_1 M = 1$, then it is a rational homology $S^1 \times S^2$. In the case $b_1 M = 2$, the cup-product between the two 1-dimensional cohomology classes vanishes and one can compute a Massey triple-product to obtain the integer $n \in \mathbb{Z}$ (note that $H^1(M; \mathbb{Z}) \cong \mathbb{Z}^2$ and thus one can do the computation integrally). It determines whether M is of general type ($n = 0$) or whether it belongs to one of the groups H_n, $n > 0$. Finally, if $b_1 M = 3$, then M is of general type if and only if the triple cup-product between the three 1-dimensional cohomology classes vanishes, otherwise it is equivalent to the 3-torus, i.e., to H_0.

One should compare the above result with the list of nilpotent 3-manifold groups from [22]:

finite	$\mathbb{Z}/k$	$Q_{2^n} \times \mathbb{Z}/(2k+1)$
infinite	$\mathbb{Z}$	H_k

In the case of finite groups, $k \geq 0$ is the order of the cyclic group $\mathbb{Z}/k$ and the quaternion group Q_k. In the infinite case, it is the Euler number that determines H_k. In Section 3 we will outline why no other nilpotent groups occur.

The reason why we used the rational version of the lower central series is that our Betti number estimates only give rational information. The integral version of Theorem 3 was in the meantime proven by the third author [21] using completely different methods.

The following result is the rational version of Turaev's Theorem 2 in [23].

Theorem 4. *A finitely generated nilpotent group G satifies*

$$G \otimes \mathbb{Q} \cong (\pi_1(M)/\Gamma_r) \otimes \mathbb{Q}$$

for a closed oriented 3-manifold M if and only if there exists a class $m \in H_3(G; \mathbb{Q})$ such that the composition

$$H^1(G; \mathbb{Q}) \xrightarrow{\cap m} H_2(G; \mathbb{Q}) \twoheadrightarrow H_2(G; \mathbb{Q})/\Phi_r^{\mathbb{Q}}(G)$$

is an epimorphism.

Here $\Phi_r^{\mathbb{Q}}(G)$ is the r-th term in the rational Dwyer filtration of $H_2(G; \mathbb{Q})$ (defined in the Appendix A).

3. Low-dimensional Surgery

In this section we first show how Theorems 4 and 1 imply Theorem 3. Then we recall the proof of Theorem 4 and finally we give a short discussion of nilpotent 3-manifold groups.

First note that by definition $\Phi_r^{\mathbb{Q}} = 0$ if $\Gamma_{r-1}^{\mathbb{Q}} = \{0\}$. So by Theorem 4, in order for a group G to be the maximal nilpotent quotient of a rationally prenilpotent 3-manifold group, it must possess the following property $\cap$ (here rational coefficients are to be understood):

$$\cap : \qquad \begin{cases} \text{There exits } m \in H_3(G) \text{ such that} \\ \cap m : H^1(G) \longrightarrow H_2(G) \text{ is an epimorphism.} \end{cases}$$

Theorem 3 now follows immediately from the following:

Proposition 1. *A finitely generated torsion-free nilpotent group satisfying property $\cap$ is isomorphic to exactly one of the groups:*

$$\{1\}, \quad \mathbb{Z} \quad \text{or} \quad H_n.$$

Observe that $H_0 \otimes \mathbb{Q} = \mathbb{Z}^3 \otimes \mathbb{Q} = \mathbb{Q}^3$. When $n > 0$ each of the groups $H_n \otimes \mathbb{Q}$ is isomorphic to the $\mathbb{Q}$ points of the Heisenberg group $H_{\mathbb{Q}}$. On the other hand, the groups $\mathbb{Z}^k$ are the only finitely generated torsion-free nilpotent groups whose Malcev completion is $\mathbb{Q}^k$ and, for $n > 0$, the groups H_n are the only finitely generated torsion-free nilpotent groups with Malcev completion $H_{\mathbb{Q}}$. Therefore, Proposition 1 follows from the following result and the fact that the Malcev completion of a finitely generated nilpotent group is a uniquely divisible nilpotent group with $b_1 < \infty$.

Proposition 2. *A uniquely divisible nilpotent group with $b_1 < \infty$ satisfying property $\cap$ is isomorphic to exactly one of the groups:*

$$\{0\}, \quad \mathbb{Q}, \quad \mathbb{Q}^3 \quad \text{or} \quad H_{\mathbb{Q}}.$$

Property $\cap$ implies $b_1 \geq b_2$. This, combined with the inequality $b_2 > b_1^2/4$ from Theorem 1, implies that Proposition 2 follows from the result below. The proof will be given in Section 4.

Proposition 3. *Suppose that G is a uniquely divisible nilpotent group with $b_1(G) < \infty$ satisfying property $\cap$. If $b_1 = b_2 = 2$, then G is isomorphic to $H_{\mathbb{Q}}$. If $b_1 = b_2 = 3$, then G is isomorphic to $\mathbb{Q}^3$.*

This finishes the outline of the proof of Theorem 3.

Proof of Theorem 4. Given a closed oriented 3-manifold M, we may take a classifying map $M \to K(\pi_1(M), 1)$ of the universal covering and compose with the projection $\pi_1(M) \twoheadrightarrow \pi_1(M)/\Gamma_r$ to get a map $u : M \to K(\pi_1(M)/\Gamma_r, 1)$ and a commutative diagram

$$\begin{array}{ccc} H^1(M) & \xrightarrow[\cong]{\cap [M]} & H_2(M) \\ {\scriptstyle u^*}\uparrow & & \downarrow{\scriptstyle u_*} \\ H^1(\pi_1(M)/\Gamma_r) & \xrightarrow{\cap u_*[M]} & H_2(\pi_1(M)/\Gamma_r) \end{array}$$

Clearly u induces an isomorphism on H_1, and therefore u^* is an isomorphism. Thus the "only if" part of our theorem follows directly from Dwyer's theorem (see the Appendix A). Here we could have worked integrally or rationally.

To prove the "if" part, we restrict to rational coefficients. Let $u : M \to K(G, 1)$ be a map from a closed oriented 3-manifold with $u_*[M] = m$, the given class in $H_3(G)$. Such a map exists because rationally oriented bordism maps onto homology by the classical result of Thom. Now observe that we are done by Dwyer's theorem (and the above commutative diagram) if the map u induces an isomorphism on H_1. If this is not the case, we will change the map u (and the 3-manifold M) by surgeries until it is an isomorphism on H_1: We describe the surgeries as attaching 4-dimensional handles to the upper boundary of $M \times I$. Then M is the lower boundary of this 4-manifold and the upper boundary is denoted by M'. If the map u extends to the 4-manifold, then the image of the fundamental class of the new 3-manifold (M', u') is still the given class $m \in H_3(G)$.

First add 1-handles $D^1 \times D^3$ to $M \times I$ and extend u to map the new circles to the (finitely many) generators of G. This makes u' an epimorphism (on H_1). Then we want to add 2-handles $D^2 \times D^2$ to $M' \times I$ to kill the kernel K of

$$u'_* : H_1(M') \longrightarrow H_1(G).$$

We can extend the map u' to the 4-manifold

$$W := M' \times I \cup \text{ 2-handles}$$

if we attach the handles to curves in the kernel of $u : \pi_1(M) \to G$.

Now observe that the new upper boundary M'' of W still maps onto $H_1(G)$ because one can obtain M' from M'' by attaching 2-handles $(2 + 2 = 4)$. But the problem is that $\mathrm{Ker}\{u''_* : H_1(M'') \to H_1(G)\}$ may contain new elements which are meridians to the circles c we are trying to kill. If c has a dual, i.e. if there is a surface in M' intersecting c in a point, then these meridians are null-homologous. But since we only work rationally we may assume all the classes in the kernel to have a dual and we are done.

The more involved integral case is explained in detail in [23] but we do not need it here. □

We finish this section by explaining briefly the table of nilpotent 3-manifold groups given in Section 2. The finite groups G in the table have representations into $SO(4)$ such that the corresponding action of G on the 3-sphere is free. Thus the corresponding 3-manifolds are homogenous spaces S^3/G.

To explain why only cyclic and quaternion groups can occur, first recall that a finite group is nilpotent if and only if it is the direct product of its p-Sylow subgroups [13]. Now it is well known [2] that the only p-groups with periodic cohomology are the cyclic groups and, for $p = 2$, the quaternion groups. The only fact about the group we use is that it acts freely on a homotopy 3-sphere, and thus has 4-periodic cohomology.

To understand why only the groups $\mathbb{Z}$ and H_n occur as infinite nilpotent 3-manifold groups, first notice that a nilpotent group is never a nontrivial free product. Except in the case $\pi_1(M) = \mathbb{Z}$, the sphere theorem implies that the universal cover of the corresponding 3-manifold must be contractible. In particular, the nilpotent fundamental groups $(\neq \mathbb{Z})$ of a 3-manifolds must have homological dimension 3. It is then easy to see that the groups H_n are the only such groups. For more details see [22] or [21].

4. Examples

In this section we prove Proposition 3 and give an example concerning the difference between integral and rational prenilpotence of 3-manifold groups.

Proof of Proposition 3. The key to both cases is the following commutative diagram (later applied with $r = 2, 3, 4$). Here all homology groups, as well as the groups Γ_r and Φ_r, have rational coefficients, i.e., we suppress the letter $\mathbb{Q}$ from the notation.

$$\begin{array}{ccccccc}
\Phi_{r+1}(G) & \longrightarrow & H_2(G) & \xrightarrow{p_*} & H_2(G/\Gamma_r) & \longrightarrow & \Gamma_r(G)/\Gamma_{r+1}(G) \\
 & & {\scriptstyle \cap m}\big\uparrow{\scriptstyle \cong} & & {\scriptstyle \cap p_*(m)}\big\uparrow & & \\
 & & H^1(G) & \underset{\cong}{\xleftarrow{p^*}} & H^1(G/\Gamma_r) & &
\end{array}$$

The upper line is *short* exact by the 5-term exact sequence and the definition of Φ_{r+1}. Since we consider cases with $b_1 = b_2$, the map $\cap m$ is actually an isomorphism. Consider first the case $b_1(G) = b_2(G) = 3$ and $r = 2$ in the above diagram. Since $H_3(G/\Gamma_2) \cong H_3(\mathbb{Q}^3) \cong \mathbb{Q}$, there are only two cases to consider:

(i) $p_*(m) \neq 0$: Then, by Poincaré duality for $\mathbb{Q}^3$, the map $\cap p_*(m)$ is an isomorphism and thus $p_* : H_2(G) \to H_2(G/\Gamma_2)$ is also an isomorphism. By Stallings' Theorem this implies that $G \cong G/\Gamma_2 \cong \mathbb{Q}^3$.
(ii) $p_*(m) = 0$: Then $p_* : H_2(G) \to H_2(G/\Gamma_2)$ is the zero map and thus $\Phi_3(G) = \mathbb{Q}^3$. This contradicts Theorem 2 with $r = 3$, since $3 \not> 4$.

Now consider the case $b_1(G) = b_2(G) = 2$. If $p_* : H_2(G) \to H_2(G/\Gamma_2) \cong \mathbb{Q}$ is onto, then by Stallings' Theorem, G would be abelian and thus have $b_2(G) = 1$, a contradiction. Therefore, $H_2(G) = \Phi_3(G)$ and G/Γ_3 is the rational Heisenberg group $H_\mathbb{Q}$. This follows from Dwyer's Theorem by comparing G to the free group F on 2 generators and noting that $H_\mathbb{Q} \cong F/\Gamma_3 \otimes \mathbb{Q}$.

Now consider the above commutative diagram for $r = 3$. Since $H_3(H_\mathbb{Q}) \cong \mathbb{Q}$ there are again only two cases to consider:

(i) $p_*(m) \neq 0$: Then by Poincaré duality for $H_\mathbb{Q}$, the map $\cap p_*(m)$ is an isomorphism and thus $p_* : H_2(G) \to H_2(G/\Gamma_2)$ is also an isomorphism. By Stallings' Theorem this implies that $G \cong G/\Gamma_3 \cong H_\mathbb{Q}$.
(ii) $p_*(m) = 0$: Then $p_* : H_2(G) \to H_2(G/\Gamma_2)$ is the zero map and thus $H_2(G) = \Phi_4(G)$.

Unfortunately, this does not contradict Theorem 2 with $r = 4$ (since $2 > 27/16$) and we have to go one step further. Again by Dwyer's Theorem G/Γ_4 is isomorphic to $K := F/\Gamma_4 \otimes \mathbb{Q}$. One easily computes that the cap-product map

$$\cap : H_3(K) \otimes H^1(K) \longrightarrow H_2(K)$$

is identically zero and thus the above commutative diagram for $r = 4$ shows that $H_2(G) = \Phi_5(G)$. Now Theorem 2 does lead to the contradiction $2 > 8192/3125$. □

We believe that the above proof is unnecessarily complicated because Theorem 2 does not give the best possible estimate. In fact, we conjecture that there is no nilpotent group with $b_1 = b_2 = 2$ and $\Phi_4 \cong \mathbb{Q}^2$, and that a nilpotent group with $b_1 = b_2 = 3$ is always rationally $\mathbb{Q}^3$ (without assuming property $\cap_\mathbb{Q}$). However, we found the following:

Example. Besides the Heisenberg group $H_\mathbb{Q}$, there are other nilpotent groups with $b_1 = b_2 = 2$. For example, take a nontrivial central extension

$$0 \longrightarrow \mathbb{Q} \longrightarrow G \longrightarrow H_\mathbb{Q} \longrightarrow 0.$$

By Proposition 3, G cannot satisfy property $\cap_\mathbb{Q}$, which can be checked directly.

We finish this section by discussing the 2-torus bundle over the circle with holonomy given by

$$(z_1, z_2) \mapsto (\bar{z}_1, \bar{z}_2), \ \ z_i \in S^1 \subset \mathbb{C}.$$

The fundamental group G of this 3-manifold is the semidirect product of $\mathbb{Z}$ with $\mathbb{Z}^2$, where a generator of the cyclic group $\mathbb{Z}$ acts as minus the identity on the normal subgroup $\mathbb{Z}^2$. One computes that

$$\Gamma_r(G) = 2^{r-1} \cdot \mathbb{Z}^2$$

and thus G is not prenilpotent. However, one sees that $\Gamma_r^{\mathbb{Q}}(G) = \mathbb{Z}^2$ for all r — i.e., G is rationally prenilpotent with maximal torsion-free nilpotent quotient $\mathbb{Z}$.

This example illustrates three phenomena. Firstly, the rational lower central series stabilizes more often than the integral one, even for 3-manifold groups. Secondly, going to the maximal torsion-free nilpotent quotient kills many details of the 3-manifold group which can be still seen in the nilpotent quotients. Finally, unlike in the geometric theory, even a 2-fold covering can alter the class (Theorem 3) to which a 3-manifold belongs.

5. The Proof of Theorem 2

The proof of the estimate in Theorem 2 consists of three steps which are parallel to those in Roquette's proof of the Golod–Shafarevich Theorem, as presented in [14]. We begin by recalling the structure of the proof:

1. The $\mathbb{F}_p$-homology of a p-group G can be calculated using a free $\mathbb{F}_p[G]$- resolution of the trivial module $\mathbb{F}_p$ in place of a free $\mathbb{Z}[G]$-resolution of $\mathbb{Z}$. This puts us in the realm of linear algebra.
2. There is a *minimal resolution* of $\mathbb{F}_p$ over $\mathbb{F}_p[G]$. This is an exact sequence

$$\cdots \longrightarrow C_n \longrightarrow C_{n-1} \longrightarrow \cdots \longrightarrow C_0 \longrightarrow \mathbb{F}_p \longrightarrow 0$$

of free $\mathbb{F}_p[G]$-modules C_i such that, for all $i \geq 0$, one has

$$\operatorname{rank}_{\mathbb{F}_p[G]} C_i = \dim_{\mathbb{F}_p} H_i(G; \mathbb{F}_p).$$

3. Let I be the kernel of the augmentation $\mathbb{F}_p[G] \to \mathbb{F}_p$. Define a generating function $d(t)$ by

$$d(t) := \sum_{k \geq 0} \dim_{\mathbb{F}_p}(I^k / I^{k+1})\, t^k.$$

This function is a polynomial which is positive on the unit interval. The proof is completed by filtering the first two pieces of a minimal resolution by powers of I. This leads to an inequality

$$(\text{a quadratic polynomial}) d(t) \geq 1$$

for $t \in [0, 1]$ from which one deduces the result by looking at the discriminant of the quadratic.

We now consider the analogues of these for a finitely generated torsion free nilpotent group G. We reduce to the case of a uniquely divisible group with $b_1 < \infty$ by replacing G by $G \otimes \mathbb{Q}$.

Ad(1) The $\mathbb{Q}$-homology of G can be clearly calculated using a free resolution of $\mathbb{Q}$ over the group ring $\mathbb{Q}[G]$ rather than over $\mathbb{Z}[G]$. More importantly, we can also resolve $\mathbb{Q}$ as a module over the I-adic completion of $\mathbb{Q}[G]$, where I is again the augmentation ideal. This step involves an Artin–Rees Lemma proved in this setting in [3]. Let A denote the completion $\mathbb{Q}[G]\hat{}$.

Ad(2) There is a minimal resolution of $\mathbb{Q}$ over A in the same sense as above. More precisely, there is minimal resolution which takes into account the Dwyer filtration of $H_2(G)$. The construction of such a resolution will be given in Section 6. It is crucial that one uses the completed group ring A rather than $\mathbb{Q}[G]$.

Ad(3) Define a generating function $d(t)$ by

$$d(t) := \sum_{k \geq 0} \dim_{\mathbb{Q}}(I^k/I^{k+1})\, t^k.$$

The proof of the following result is at the end of this section.

Proposition 4. *With notation as above,*

$$d(t) = \prod_{r \geq 1} \frac{1}{(1-t^r)^{g_r}}$$

where g_r is the dimension of $\Gamma_r(G)/\Gamma_{r+1}(G)$. In particular, $d(t)$ is a rational function.

Define p_r to be the dimension of $\Phi_r^{\mathbb{Q}}(G)/\Phi_{r+1}^{\mathbb{Q}}(G)$. Recall that, roughly speaking, p_r is the number of relations which are r-fold commutators but not $(r+1)$-fold commutators. Let b_i be the ith rational Betti numbers of G. Set

$$p(t) := \sum_{r \geq 2} p_r\, t^r - b_1\, t + 1.$$

The following result will be proven in Section 7 using the resolution constructed in Section 6.

Proposition 5. *For all $0 < t < 1$ one has the inequality $p(t)\, d(t) \geq 1$.*

Now it is easy to prove Theorem 2: First note that when $0 < t < 1$, the generating function $d(t)$ is positive, and thus $p(t)$ is also positive in this interval. Now assume that $H_2(G; \mathbb{Q}) = \Phi_r^{\mathbb{Q}}(G)$ for some $r \geq 2$ and that $G \neq \{0\}, \mathbb{Q}$. Then $b_1 > 1, b_2 > 0$ and $p(t) \leq q(t)$ for each t in $0 < t < 1$, where

$$q(t) := b_2\, t^r - b_1\, t + 1.$$

The polynomial $q(t)$ has a minimum at

$$t_0 := \sqrt[r-1]{\frac{b_1}{r\,b_2}}.$$

The desired inequality follows from the inequality $q(t_0) > 0$ after a little algebraic manipulation. But in order to do this we need $t_0 < 1$, which is equivalent to the condition $b_1 < r\,b_2$.

We know that $0 \le q(1) = b_2 - b_1 + 1$, and thus $b_2 \ge b_1 - 1$. Therefore, the only case where $t_0 < 1$ is not satisfied is $(b_1, b_2) = (2, 1)$ (and $r = 2$). We claim that $(b_1, b_2) = (2, 1)$ implies $G \cong \mathbb{Q}^2$.

Consider the projection $G \twoheadrightarrow G/\Gamma_2 \cong \mathbb{Q}^2$. If the induced map on $H_2(\ ;\mathbb{Q})$ is onto, then we are done by Stallings' Theorem. If not then we know that $H_2(G;\mathbb{Q}) = \Phi_3(G)$, and are in a case where $t_0 < 1$ since $r = 3$. But then $q(t_0) > 0$ leads to the contradiction $b_2 = 1 > 32/27$. □

Proof of Proposition 4. Set

$$\mathrm{Gr}(G) = \bigoplus_{r\ge 1} \Gamma_r(G)/\Gamma_{r+1}(G).$$

This is a graded Lie algebra. Observe that $\mathrm{Gr}(A) \cong \mathrm{Gr}(\mathbb{Q}[G])$. Then, by a theorem of Quillen [18], we have an isomorphism $\mathrm{Gr}(\mathbb{Q}[G]) \cong U(\mathrm{Gr}(G))$ of graded Hopf algebras. Here $U(\mathrm{Gr}(G))$ denotes the universal enveloping algebra of $\mathrm{Gr}(G)$. We shall write $\mathcal{G}$ for $\mathrm{Gr}(G)$ and $\mathcal{G}_r$ for $\Gamma_r(G)/\Gamma_{r+1}(G)$.

By the Poincaré–Birkhoff–Witt Theorem, there is a graded coalgebra isomorphism of $U\mathcal{G}$ with the symmetric coalgebra $S\mathcal{G}$ on $\mathcal{G}$. Note that we have the isomorphism

$$S\mathcal{G} \cong \bigotimes_{r\ge 1} S\mathcal{G}_r,$$

of graded vector spaces, where the tensor product on the right is finite as $\mathcal{G}$ is nilpotent. Since $\mathcal{G}_r$ has degree r, the generating function of the symmetric coalgebra $S\mathcal{G}_r$ is $1/(1-t^r)^{g_r}$. The result follows. □

6. The Minimal Resolution

In this section we use techniques from rational homotopy theory to prove the existence of the minimal resolutions needed in the proof of Theorem 2. We obtain the minimal resolution using Chen's method of formal power series connections, which provides a minimal associative algebra model of the loop space of a space. Chen's theory is briefly reviewed in Appendix B.

The precise statement of the main result is:

Theorem 5. *If G is a nilpotent group with $b_1(G) < \infty$, then there is a free $\mathbb{Q}[G]^\frown$ resolution*

$$\cdots \xrightarrow{\delta} C_2 \xrightarrow{\delta} C_1 \xrightarrow{\delta} C_0 \longrightarrow \mathbb{Q} \longrightarrow 0$$

of the trivial module $\mathbb{Q}$ with the properties:

1. *C_k is isomorphic to $H_k(G) \otimes \mathbb{Q}[G]^\frown$;*
2. *$\delta(\Phi_m^{\mathbb{Q}}) \subseteq H_1(G) \otimes I^{m-1}$, where I denotes the augmentation ideal of $\mathbb{Q}[G]^\frown$.*

Denote the nilpotent Lie algebra associated to G by $\mathfrak{g}$. This is a finite dimensional Lie algebra over $\mathbb{Q}$ as G is nilpotent and the $\mathbb{Q}$ Betti numbers of G are finite. Denote the standard (i.e., the Chevelley–Eilenberg) complex of cochains of $\mathfrak{g}$ by $\mathcal{C}^\bullet(\mathfrak{g})$. By Sullivan's theory of minimal models, this is the minimal model of the classifying space BG of G.

Let (ω, ∂) be a formal power series connection associated to $\mathcal{C}^\bullet(\mathfrak{g})$.[1] Note that there is a natural isomorphism between $\widetilde{H}_\bullet(\mathcal{C}^\bullet(\mathfrak{g}))$ and the reduced homology groups $\widetilde{H}_\bullet(G, \mathbb{Q})$. Set

$$A_\bullet(G) = (T(\widetilde{H}_\bullet(G)[1])^\frown, \partial).$$

Since $A_\bullet(G)$ is complete, we may assume, after taking a suitable change of coordinates of the form

$$Y \mapsto Y + \text{ higher order terms}, \quad Y \in \widetilde{H}_\bullet(G)[1],$$

that

$$\partial Y \in I^k \iff k = \max\{m \geq 1 : \partial(Y + \text{ higher order terms}) \in I^m\}. \tag{1}$$

Since $\mathcal{C}^\bullet(\mathfrak{g})$ is a minimal algebra in the sense of Sullivan, we can apply the standard fact [1, (2.30)] (see also, [11, 2.6.2]) to deduce that $H^k(B(\mathcal{C}(\mathfrak{g})))$ vanishes when $k > 0$. It follows from Theorem B.2 that $H_k(A_\bullet(G))$ also vanishes when $k > 0$.

When $k = 0$, Chen's Theorem B.1 yields a complete Hopf algebra isomorphism

$$H_0(A_\bullet(G)) \cong U^\frown\mathfrak{g} \cong \mathbb{Q}[G]^\frown.$$

These two facts are key ingredients in the construction of the resolution. Another important ingredient is the Adams–Hilton construction:

Proposition 6. *There is a continuous differential δ of degree -1 on the free $A_\bullet(G)$-module*

$$W := H_\bullet(G) \otimes A_\bullet(G)$$

[1] The definition and notation can be found in Appendix B.

such that:

1. *the restriction of δ to $A_\bullet = 1 \otimes A_\bullet$ is ∂;*
2. *δ is a graded derivation with respect to the right $A_\bullet(X)$ action;*
3. *(W, δ) is acyclic.*

Proof. To construct the differential, it suffices to define δ on each $Y \otimes 1$, where $Y \in H_\bullet(G)$. To show that the tensor product is acyclic, we will construct a graded map $s : W \to W$ of degree 1, such that $s^2 = 0$ and

$$\delta s + s\delta = \mathrm{id} - \epsilon. \tag{2}$$

Here ϵ is the tensor product of the augmentations of $H_\bullet(G)$ and $A_\bullet(G)$. Define $s(1) := 0 =: \delta(1)$. For $Y \in \widetilde{H}_\bullet(G)$, we shall denote the corresponding element of $\widetilde{H}_\bullet(G)[1]$ by $\overline{Y}$. For $Y, Y_1, Y_2, \ldots, Y_k \in \widetilde{H}_\bullet(G)$, define

$$s(1 \otimes \overline{Y}_1\overline{Y}_2 \cdots \overline{Y}_k) := Y_1 \otimes \overline{Y}_2 \cdots \overline{Y}_k \text{ and } s(Y \otimes \overline{Y}_1\overline{Y}_2 \cdots \overline{Y}_k) := 0.$$

We shall simply write Y for $Y \otimes 1$ and $\overline{Y}$ for $1 \otimes \overline{Y}$. Since δ is a derivation, it suffices to define it on the $A_\bullet(G)$-module generators $1, Y, \overline{Y}$ of W. We already know how to define δ on the $\overline{Y}$'s. In order that (2) holds when applied to Y, we have to define

$$\delta Y := \overline{Y} - s\delta\overline{Y}.$$

Then (2) also holds when applied to $\overline{Y}$ and this implies $\delta^2 Y = 0$. One can now verify (2) by induction on degree using the fact that s is right $A_\bullet(G)$-linear. □

The way to think of this result is that $A_\bullet(G)$ is a homological model of the loop space $\Omega_x BG$ of the classifying space of G. One can take $H_\bullet(G)$ with the trivial differential to be a homological model for BG. The complex (W, δ) is a homological model of the path-loops fibration whose total space is contractible. This picture should help motivate the next step, the construction of the minimal resolution.

Filter the complex (W, δ) by degree in $H_\bullet(G)$. This leads to a spectral sequence

$$E^1_{p,q} = H_p(G) \otimes H_q(A_\bullet(G)) \implies H_{p+q}(W, \delta).$$

(This is the algebraic analogue of the Serre spectral sequence for the path-loops fibration.) But since $H_q(A_\bullet(G))$ vanishes when $q > 0$, the spectral sequence collapses at the E^1-term to a complex

$$\cdots \to H_2(G) \otimes H_0(A_\bullet(G)) \to H_1(G) \otimes H_0(A_\bullet(G)) \to H_0(A_\bullet(G)) \to \mathbb{Q} \to 0$$

which is acyclic as (W, δ) is. The existence of the resolution now follows from the fact that

$$H_0(A_\bullet(G)) \cong \mathbb{Q}[G]^\wedge.$$

The $\mathbb{Q}[G]^\wedge$ linearity of the differential follows directly from the second assertion of Proposition 6.

It remains to verify the condition satisfied by the differential. Observe that $H_0(A_\bullet(G))$ is the quotient of $T(H_1(G))\hat{}$ by the closed ideal generated by the image of

$$\partial : H_2(G) \to T(H_1(G))\hat{}.$$

Define a filtration

$$H_2(G;\mathbb{Q}) = L_2 \supseteq L_3 \supseteq \cdots$$

by $L_k = \partial^{-1} I^k$. It is not difficult to verify that in the complex $H_\bullet(G) \otimes H_0(A_\bullet(G))$ we have

$$\delta(L_k \otimes \mathbb{Q}) \subseteq H_1(G) \otimes I^{k-1}.$$

So, to complete the proof, we have to show that $\Phi_k^{\mathbb{Q}} = L_k$.

Consider the spectral sequence dual to the one in the proof of Theorem B.2. It has the property that $E^r_{-1,2}$ is a subspace of $H_2(G;\mathbb{Q})$ and $E^r_{-p,p}$ is a quotient of $H_1(G;\mathbb{Q})^{\otimes p}$. We thus have a filtration

$$H_2(G;\mathbb{Q}) = E^1_{-1,2} \supseteq E^2_{-1,2} \supseteq E^3_{-1,2} \supseteq \cdots$$

of $H_2(G;\mathbb{Q})$. From the standard description of the terms of the spectral sequence associated to a filtered complex, it is clear from condition (1) that when $r \geq 2$

$$L_r = E^{r-1}_{-1,2}.$$

We know from [4, (2.6.1)] that

$$E^p_{-p,p} = E^\infty_{-p,p} \cong I^p/I^{p+1},$$

where I denotes the augmentation ideal of $\mathbb{Q}[\pi_1(G)]$. Consequently, a group homomorphism $G \to H$ induces an isomorphism

$$\mathbb{Q}[G]/I^k \to \mathbb{Q}[H]/I^k$$

if and only if the induced map $H_2(G;\mathbb{Q})/L_k \to H_2(H;\mathbb{Q})/L_k$ is an isomorphism. The proof is completed using the fact (cf. [11, (2.5.3)]) that a group homomorphism $G \to H$ induces an isomorphism $(G \otimes \mathbb{Q})/\Gamma_k \to (H \otimes \mathbb{Q})/\Gamma_k$ if and only if the induced map $\mathbb{Q}[G]/I^k \to \mathbb{Q}[H]/I^k$ is an isomorphism, [17].

7. The Inequality

In this section, we prove Proposition 5. We suppose that G is a nilpotent group with $b_1(G) < \infty$. Denote $\mathbb{Q}[G]\hat{}$ by A and its augmentation ideal by I. Recall that

$$d(t) = \sum_{l \geq 0} d_l\, t^l,$$

where $d_l = \dim I^l/I^{l+1}$, and that this is a rational function whose poles lie on the unit circle.

Consider the part

$$H_2(G) \otimes A \xrightarrow{\delta} H_1(G) \otimes A \to A \to \mathbb{Q} \to 0$$

of the resolution constructed in Section 6. Define a filtration

$$E^0 \supseteq E^1 \supseteq E^2 \supseteq \cdots$$

of $H_2(G) \otimes A$ by

$$E^l := \delta^{-1}\left(H_1(G) \otimes I^l\right).$$

Then we have an exact sequence

$$0 \to E^l/E^{l+1} \to H_1(G) \otimes I^l/I^{l+1} \to I^{l+1}/I^{l+2} \to 0 \tag{3}$$

for all $l \geq 0$. Set $e_l := \dim E^l/E^{l+1}$ and

$$e(t) := \sum_{l \geq 0} e_l\, t^l \in \mathbb{R}[[t]].$$

Proposition 7. *The series for $e(t)$ converges to the rational function*

$$\frac{b_1(G)\, td(t) - d(t) + 1}{t}$$

all of whose poles lie on the unit circle.

Proof. Because the sequence (3) is exact, we have $e_l = b_1(G)\, d_l - d_{l+1}$. This implies that

$$te(t) = b_1(G)\, td(t) - d(t) + 1.$$

from which the result follows. □

Our final task is to bound the e_l. Our resolution has the property that

$$\delta(\Phi_m \otimes I^{l-m+2}) \subseteq H_1(G) \otimes I^{l+1}.$$

This implies that

$$\sum_{m \geq 2} \Phi_m \otimes I^{l-m+2} \subseteq E^{l+1},$$

so that the linear map

$$H_2(G) \otimes A \Big/ \left(\sum_{m \geq 2} \Phi_m \otimes I^{l-m+2}\right) \to H_2(G) \otimes A/E^{l+1}$$

is surjective. This implies that

$$\dim\left(H_2(G) \otimes A/(\sum_{m \geq 2} \Phi_m \otimes I^{l-m+2})\right) \geq e_0 + e_1 + \cdots + e_l. \tag{4}$$

To compute the dimension of the left hand side, we apply the following elementary fact from linear algebra.

Proposition 8. *Suppose that*

$$B = B^0 \supseteq B^1 \supseteq B^2 \supseteq \cdots$$

and

$$C = C^0 \supseteq C^1 \supseteq C^2 \supseteq \cdots$$

are two filtered vector spaces. Define a filtration F of $B \otimes C$ by

$$F^k := \sum_{i+j=k} B^i \otimes C^j.$$

Then there is a canonical isomorphism

$$F^k/F^{k+1} \cong \bigoplus_{i+j=k} A^i/A^{i+1} \otimes B^j/B^{j+1}.$$

□

Applying this with $B = H_2(G)$ with the filtration $\Phi_\bullet$, and $C = A$ with the filtration $I^\bullet$, we deduce that

$$\dim\left(H_2(G) \otimes A/\left(\sum_{m\geq 2} \Phi_m \otimes I^{l-m+2}\right)\right)$$
$$= \sum_{m+k\leq l+1} (\dim \Phi_m/\Phi_{m+1})(\dim I^k/I^{k+1}) = \sum_{m+k\leq l+1} p_m d_k$$

where $p_m = \dim \Phi_m$. Combined with (4), this implies that

$$\sum_{m+k\leq l+1} p_m d_k \geq e_0 + \cdots + e_l.$$

Using geometric series, this can be assembled into the following inequality which holds when $0 < t < 1$:

$$\frac{(p_2\, t + p_3\, t^2 + \cdots + p_{l+1}\, t^l)\, d(t)}{1-t} \geq \frac{e(t)}{1-t}.$$

Plugging in the formula for $e(t)$ given in Proposition 7, we deduce the desired inequality

$$d(t)\left(\sum_{m\geq 2} p_m\, t^m - b_1(G)\, t + 1\right) \geq 1$$

when $0 < t < 1$.

Appendix A

Here we collect the necessary group theoretic definitions. Let G be a group.

- $\Gamma_r(G)$ denotes the r-th term of the *lower central series* of G which is the subgroup of G generated by all r-fold commutators. We abbreviate $G/\Gamma_r(G)$ by G/Γ_r. We say G is *nilpotent* if $\Gamma_r(G) = \{1\}$ for some r and it is then said

to have *nilpotency class* $< r$. For example, abelian groups have $\Gamma_2 = \{1\}$ and are of nilpotency class 1.
- G is *prenilpotent* if it's lower central series stabilizes at some term Γ_r, i.e., $\Gamma_r(G) = \Gamma_{r+1}(G)$. This happens if and only if G has a maximal nilpotent quotient (which is then isomorphic to G/Γ_r).

The main homological tool in dealing with nilpotent groups is the following result of Stallings:

Theorem A.1. [20] *If $f : G \to H$ is a homomorphism of groups inducing an isomorphism on H_1 and an epimorphism on H_2, then the induced maps $G/\Gamma_r \to H/\Gamma_r$ are isomorphisms for all $r \geq 1$.*

- The *Dwyer filtration* of $H_2(G;\mathbb{Z})$
$$H_2(G;\mathbb{Z}) =: \Phi_2(G) \supseteq \Phi_3(G) \supseteq \Phi_4(G) \supseteq \cdots$$
is defined by (see [8] for a geometric definition of Φ_r using gropes)
$$\Phi_r(G) := \mathrm{Ker}(H_2(G;\mathbb{Z}) \longrightarrow H_2(G/\Gamma_{r-1};\mathbb{Z})).$$

This filtration is used in Dwyer's extension of Stallings' Theorem:

Theorem A.2. [5] *If $f : G \longrightarrow H$ induces an isomorphism on $H_1(\ ;\mathbb{Z})$, then for $r \geq 2$ the following three conditions are equivalent:*

1. *f induces an epimorphism $H_2(G;\mathbb{Z})/\Phi_r(G) \to H_2(H;\mathbb{Z})/\Phi_r(H)$;*
2. *f induces an isomorphism $G/\Gamma_r \to H/\Gamma_r$;*
3. *f induces an isomorphism $H_2(G;\mathbb{Z})/\Phi_r(G) \to H_2(H;\mathbb{Z})/\Phi_r(H)$, and an injection $H_2(G;\mathbb{Z})/\Phi_{r+1}(G) \to H_2(H;\mathbb{Z})/\Phi_{r+1}(H)$.*

There are rational versions of the above definitions.

- $\Gamma_r^{\mathbb{Q}}$ denotes the r-th term of the *rational lower central series* of G, which is defined by
$$\Gamma_r^{\mathbb{Q}}(G) := \mathrm{Rad}(\Gamma_r(G)) := \{g \in G \mid g^n \in \Gamma_r \text{ for some } n \in \mathbb{Z}\}.$$
It has the (defining) property that $G/\Gamma_r^{\mathbb{Q}} = (G/\Gamma_r)/\,\mathrm{Torsion}$. (Note that the torsion elements in a nilpotent group form a subgroup.) G is *rationally nilpotent* if $\Gamma_r^{\mathbb{Q}}(G) = \{1\}$ for some r.
- G is *rationally prenilpotent* if $\Gamma_r^{\mathbb{Q}}(G) = \Gamma_{r+1}^{\mathbb{Q}}(G)$ for some r. This happens if and only if G has a maximal torsion-free nilpotent quotient (which is then isomorphic to $G/\Gamma_r^{\mathbb{Q}}$).

- The *rational Dwyer filtration* of $H_2(G;\mathbb{Q})$

$$H_2(G;\mathbb{Q}) =: \Phi_2^{\mathbb{Q}}(G) \supseteq \Phi_3^{\mathbb{Q}}(G) \supseteq \Phi_4^{\mathbb{Q}}(G) \supseteq \dots$$

is defined by

$$\Phi_r^{\mathbb{Q}}(G) := \Phi_r(G) \otimes \mathbb{Q} = \mathrm{Ker}(H_2(G;\mathbb{Q}) \longrightarrow H_2(G/\Gamma_{r-1}^{\mathbb{Q}};\mathbb{Q})).$$

- The *Malcev completion* [16] $G \otimes \mathbb{Q}$ of a nilpotent group G may be defined inductively through the central extensions determining G as follows: If G is abelian then one takes the usual tensor product of abelian groups to define $G \otimes \mathbb{Q}$. It comes with a homomorphism $\epsilon : G \to G \otimes \mathbb{Q}$ which induces an isomorphism on rational cohomology. Using the fact that the cohomology group H^2 classifies central extensions, one can then define the Malcev completion for a group which is a central extension of an abelian group. It comes again with a map ϵ as above. The Serre spectral sequence then shows that ϵ induces an isomorphism on rational cohomology. Therefore, one can repeat the last step to define $(G \otimes \mathbb{Q}, \epsilon)$ for an arbitrary nilpotent group G.
The map $\epsilon : G \to G \otimes \mathbb{Q}$ is universal for maps of G into uniquely divisible nilpotent groups and it is characterized by the following properties:
  1. $G \otimes \mathbb{Q}$ is a uniquely divisible nilpotent group.
  2. The kernel of ϵ is the torsion subgroup of G.
  3. For every element $x \in G \otimes \mathbb{Q}$ there is a number $n \in \mathbb{N}$ such that x^n is in the image of ϵ.

A version of Stallings' Theorem holds in the rational setting:

Theorem A.3. [20] *If $f : G \to H$ is a homomorphism of groups inducing an isomorphism on $H_1(\ ;\mathbb{Q})$ and an epimorphism on $H_2(\ ;\mathbb{Q})$, then the induced maps $(G/\Gamma_r) \otimes \mathbb{Q} \to (H/\Gamma_r) \otimes \mathbb{Q}$ are isomorphisms for all $r \geq 1$.*

A good example to keep in mind when trying to understand this theorem is the inclusion of the Heisenberg group H_1 into the Heisenberg group H_n. This induces an isomorphism on rational homology. Both groups are torsion free nilpotent of class 2.

There is also a rational analogue of Dwyer's theorem.

Theorem A.4. [5] *If $f : G \longrightarrow H$ induces an isomorphism on $H_1(\ ;\mathbb{Q})$, then for $r \geq 2$ the following three conditions are equivalent:*

1. *f induces an epimorphism $H_2(G;\mathbb{Q})/\Phi_r^{\mathbb{Q}}(G) \to H_2(H;\mathbb{Q})/\Phi_r^{\mathbb{Q}}(H)$;*
2. *f induces an isomorphism $(G/\Gamma_r) \otimes \mathbb{Q} \to (H/\Gamma_r) \otimes \mathbb{Q}$;*
3. *f induces an isomorphism $H_2(G;\mathbb{Q})/\Phi_r^{\mathbb{Q}}(G) \to H_2(H;\mathbb{Q})/\Phi_r^{\mathbb{Q}}(H)$, and an injection $H_2(G;\mathbb{Q})/\Phi_{r+1}^{\mathbb{Q}}(G) \to H_2(H;\mathbb{Q})/\Phi_{r+1}^{\mathbb{Q}}(H)$.*

Appendix B

In this appendix, we give a brief review of Chen's method of formal power series connections. Two relevant references are [4] and [10]. There is also an informal discussion of the ideas behind Chen's work in [12].

Fix a field F of characteristic zero. Suppose that $\mathcal{A}^\bullet$ is an augmented commutative d.g. algebra over F. Suppose in addition that $\mathcal{A}^\bullet$ is positively graded (i.e., $\mathcal{A}^k = 0$ when $k < 0$) and that $H^\bullet(\mathcal{A}^\bullet)$ is connected (i.e., $H^0(\mathcal{A}^\bullet) = F$). For simplicity, we suppose that each $H^k(\mathcal{A}^\bullet)$ is finite dimensional. Typical examples of such $\mathcal{A}^\bullet$ in this theory are the F (=$\mathbb{Q}$, $\mathbb{R}$ or $\mathbb{C}$) de Rham complex of a connected space with finite rational betti numbers, or the Sullivan minimal model of such an algebra.

Set

$$H_k(\mathcal{A}^\bullet) = \mathrm{Hom}_F(H^k(\mathcal{A}^\bullet), F)$$

and

$$\widetilde{H}_\bullet(\mathcal{A}^\bullet) = \sum_{k>0} H_k(\mathcal{A}^\bullet).$$

Denote by $\widetilde{H}_\bullet(\mathcal{A}^\bullet)[1]$ the graded vector space obtained by taking $\widetilde{H}_\bullet(\mathcal{A}^\bullet)$ and lowering the degree of each of its elements by one.[2]

Denote the free associative algebra generated by the graded vector space V by $T(V)$. Assume that V_n is non-zero only when $n \geq 0$. Denote the ideal generated by V_0 by I_0. Let $T(V)\hat{}$ denote the I_0-adic completion

$$\varprojlim T(V)/I_0^m.$$

By defining each element of V to be primitive, we give $T(V)\hat{}$ the structure of a complete graded Hopf algebra. Recall that the set of primitive elements $PT(V)$ of $T(V)$ is the free Lie algebra $\mathbb{L}(V)$, and the set of primitive elements of $T(V)\hat{}$ is the closure $\mathbb{L}(V)\hat{}$ of $\mathbb{L}(V)$ in $T(V)\hat{}$.

Set $A_\bullet = T(\widetilde{H}_\bullet(\mathcal{A}^\bullet))\hat{}$. Choose a basis $\{X_i\}$ of $\widetilde{H}_\bullet(\mathcal{A}^\bullet)$. Then $A_\bullet$ is the subalgebra of the non-commutative power series algebra generated by the X_i consisting of those power series with the property that each of its terms has a fixed degree. For each multi-index $I = (i_1, \dots, i_s)$, we set

$$X_I = X_{i_1} X_{i_2} \cdots X_{i_s}.$$

If $X_i \in \widetilde{H}_{p_i}(\mathcal{A}^\bullet)$, we set $|X_I| = -s + p_1 + \cdots + p_s$.

[2] Here we are using the algebraic geometers' notation that if V is a graded vectorspace, then $V[n]$ is the graded vector space with $V[n]_m = V_{m+n}$.

A *formal power series connection* on $\mathcal{A}^\bullet$ is a pair (ω, ∂). The first part

$$\omega = \sum_I w_I X_I$$

is an element of the completed tensor product $\mathcal{A}^\bullet \widehat{\otimes} PA_\bullet$, where w_I is an element of $\mathcal{A}^\bullet$ of degree $1 + |X_I|$. The second is a graded derivation

$$\partial : A_\bullet \to A_\bullet$$

of degree -1 with square 0 and which satisfies the minimality condition $\partial(IA_\bullet) \subseteq I^2 A_\bullet$. Here $I^k A_\bullet$ denotes the kth power of the augmentation ideal of $A_\bullet$. These are required to satisfy two conditions. The first is the "integrability condition"

$$\partial\omega + d\omega + \frac{1}{2}[J\omega, \omega] = 0.$$

Here $J : \mathcal{A}^\bullet \to \mathcal{A}^\bullet$ is the linear map $a \mapsto (-1)^{\deg a} a$. The value of the operators d, ∂ and J on ω are obtained by applying the operators to the appropriate coefficients of ω:

$$d\omega = \sum dw_I X_I, \quad \partial\omega = \sum w_I \, \partial X_I, \quad J\omega = \sum Jw_I X_I.$$

The second is that if

$$\sum_i w_i X_i \tag{5}$$

is the reduction of ω mod $I^2 A_\bullet$, then each w_i is closed and the $[w_i]$ form a basis of $H^{>0}(\mathcal{A}^\bullet)$ dual to the basis $\{X_i\}$ of $\widetilde{H}_\bullet(\mathcal{A}_\bullet)$.

Such formal connections always exist in the situation we are describing. To justify the definition, we recall one of Chen's main theorems. It is the analogue of Sullivan's main theorem about minimal models.

Theorem B.1. *Suppose that X is a smooth manifold with finite betti numbers, x a fixed point of X, and suppose that $\mathcal{A}^\bullet$ is the F-de Rham complex of X, with the augmentation induced by x. If X is simply connected, the connection gives a natural Lie algebra isomorphism*

$$\pi_\bullet(X, x)[1] \otimes F \cong H_\bullet(PA_\bullet, \partial).$$

If X is not simply connected, then the connection gives a Lie algebra isomorphism

$$\mathfrak{g}(X, x) \cong H_0(PA_\bullet, \partial)$$

and complete Hopf algebra isomorphisms

$$\mathbb{Q}[\pi_1(X, x)]^\wedge \cong U\widehat{\mathfrak{g}}(X, x) \cong H_0(A_\bullet, \partial).$$

Here $\mathfrak{g}(X,x)$ denotes the F form of the Malcev Lie algebra associated to $\pi_1(X,x)$ and $U\widehat{\mathfrak{g}}(X,x)$ is the completion of its enveloping algebra with respect to the powers of its augmentation ideal. □

We can apply the bar construction to the augmented algebra $\mathcal{A}^\bullet$ to obtain a commutative d.g. Hopf algebra $B(\mathcal{A}^\bullet)$. (We use the definition in [4].) One can define the *formal transport map* of such a formal connection. It is defined to be the element

$$T = 1 + \sum [w_I]\, X_I + \sum [w_I|w_J]\, X_I X_J + \sum [w_I|w_J|w_K]\, X_I X_J X_K + \cdots$$

of $B(\mathcal{A}^\bullet)\widehat{\otimes} A_\bullet$. It induces a linear map

$$\Theta : \mathrm{Hom}_F^{\mathrm{cts}}(A_\bullet, F) \to B(\mathcal{A}^\bullet).$$

Here $\mathrm{Hom}_F^{\mathrm{cts}}(A_\bullet, F)$ denotes the continuous dual

$$\varinjlim \mathrm{Hom}_F(A_\bullet/I^k A_\bullet, F)$$

of $A_\bullet$. It is a commutative Hopf algebra. The map Θ takes the continuous functional ϕ to the result of applying it to the coefficients of T:

$$\Theta(\phi) =$$
$$1 + \sum [w_I]\phi(X_I) + \sum [w_I|w_J]\phi(X_I X_J) + \sum [w_I|w_J|w_K]\phi(X_I X_J X_K) + \cdots$$

(Note that this is a finite sum as ϕ is continuous.) The properties of the formal connection imply that Θ is a d.g. Hopf algebra homomorphism (cf. [10, (6.17)]). The basic result we need is:

Theorem B.2. *The map Θ induces an isomorphism on homology.*

We conclude by giving a brief sketch of the proof. One can filter $A_\bullet$ by the powers of its augmentation ideal. This gives a dual filtration of $\mathrm{Hom}_F^{\mathrm{cts}}(A_\bullet, F)$. The corresponding spectral sequence has E_1 term

$$E_1^{-s,t} = [\widetilde{H}^\bullet(\mathcal{A}^\bullet)^{\otimes s}]^t.$$

On the other hand, one can filter $B(\mathcal{A}^\bullet)$ by the "bar filtration" to obtain a spectral sequence, also with this E_1 term. It is easy to check that Φ preserves the filtrations and therefore induces a map of spectral sequences. The condition on the w_i in (5) implies that the map on E_1 is an isomorphism. The result follows.

References

[1] S. Bloch, I. Kriz: *Mixed Tate Motives.*, Ann. of Math. 140 (1994), 557–605.

[2] K. S. Brown: *Cohomology of Groups.* Graduate texts in Math. 87, Springer-Verlag, 1982.

[3] K. S. Brown and E. Dror: *The Artin-Rees property and homology.* Israel J. Math. 22 (1975).

[4] K. T. Chen: *Iterated path integrals.* Bull. Amer. Math. Soc., 83 (1977), 831–879.
[5] W. G. Dwyer: *Homology, Massey products and maps between groups.* J. of Pure and Applied Algebra 6, no. 2, (1975), 177–190.
[6] J. D. Dixon, M. P. F. du Sautoy, A. Mann and D. Segal: *Analytic Pro-p Groups.* Lecture Notes in Math. 57, Cambridge Univ. Press, 1991.
[7] M. H. Freedman and P. Teichner: 4-*Manifold Topology I: Subexponential Groups.* Inventiones Mathematicae, 122 (1995), 509–529.
[8] M. H. Freedman and P. Teichner: 4-*Manifold Topology II: The Dwyer Filtration.* Inventiones Mathematicae, 122 (1995), 531–557.
[9] E. S. Golod and I. R. Shafarevich: *On the class field tower* (*in Russian*). Izvest. Akad. Nauk SSSR 28 (1964), 261–272.
[10] R. Hain: *Iterated integrals and homotopy periods.* Mem. Amer. Math. Soc. 291 (1984).
[11] R. Hain: *The de Rham homotopy theory of complex algebraic varieties I.* K-Theory 1 (1987), 271–324.
[12] R. Hain, Ph. Tondeur: *The life and work of Kuo–Tsai Chen.* Ill. J. Math. 34 (1990), 175–190.
[13] B. Huppert: *Endliche Gruppen I.* Springer-Verlag 1983.
[14] D. L. Johnson *Presentation of Groups.* London Math. Soc. LNS 22, Cambridge Univ. Press, 1976.
[15] A. Lubotzky, A. Magid: *Cohomology, Poincaré series, and group algebras of unipotent groups*, Amer. J. Math. 107 (1985), 531–553.
[16] A. L. Malcev: *Nilpotent groups without Torsion.* Izvest. Akad. Nauk SSSR, ser. Mat. 13 (1949), 201–212.
[17] D. Passman: *The Algebraic Structure of Group Rings.* John Wiley and Sons, New York, 1977.
[18] D. G. Quillen: *On the associated graded ring of a group ring.* J. of Algebra 10 (1968), 411–418.
[19] D. G. Quillen: *Rational Homotopy Theory.* Ann. of Math. 90 (1969), 205–295.
[20] J. Stallings: *Homology and central series of groups.* J. of Algebra 2 (1965), 1970–1981.
[21] P. Teichner: *Maximal nilpotent quotients of* 3-*manifold groups.* MSRI Preprint 1996.
[22] C. Thomas: *Nilpotent groups and compact* 3-*manifolds.* Proc. Cambridge Phil. Soc. 64 (1968), 303–306.
[23] V. G. Turaev: *Nilpotent homotopy types of closed* 3-*manifolds.* Proceedings of a Topology conference in Leningrad 1982, Springer LNM 1060.

Reprinted from Proc. Int. Congr. Math., 1990

THE WORK OF VAUGHAN F. R. JONES

by
JOAN S. BIRMAN
Department of Mathematics, Columbia University, New York, NY 10027, USA

It gives me great pleasure that I have been asked to describe to you some of the very beautiful mathematics which resulted in the awarding of the Fields Medal to Vaughan F. R. Jones at ICM '90.

In 1984 Jones discovered an astonishing relationship between von Neumann algebras and geometric topology. As a result, he found a new polynomial invariant for knots and links in 3-space. His invariant had been missed completely by topologists, in spite of intense activity in closely related areas during the preceding 60 years, and it was a complete surprise. As time went on, it became clear that his discovery had to do in a bewildering variety of ways with widely separated areas of mathematics and physics, some of which are indicated in Figure 1. These included (in addition to knots and links) that part of statistical mechanics having to do with exactly solvable models, the very new area of quantum groups, and also Dynkin diagrams and the representation theory of simple Lie algebras. The central connecting link in all this mathematics was a tower of nested algebras which Jones had discovered some years earlier in the course of proving a theorem which is known as the "Index Theorem".

My plan is to begin by discussing the Index Theorem, and the tower of algebras which Jones constructed in the course of his proof. After that, I plan to return to the chart in Figure 1 in order to indicate how this tower of algebras served as a bridge between the diverse areas of mathematics which are shown on the chart. I will restrict my attention throughout to one very special example of the tower construction, and so also to one special example of the associated link invariants, in order to make it possible to survey a great deal of mathematics in a very short time. Even with the restriction to a single example, this is a very ambitious plan. On the other hand, it only begins to touch on Vaughan Jones' scholarly contributions.

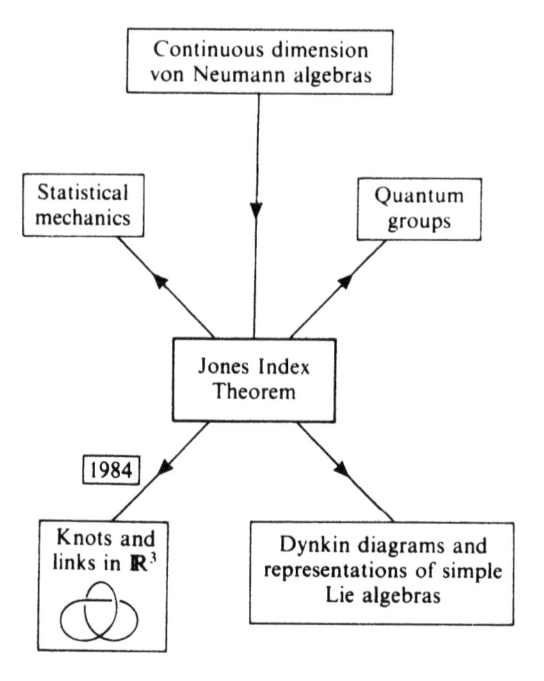

Fig. 1. The Jones Index Theorem

1. The Index Theorem

Let $\mathbf{M}$ denote a von Neumann algebra. Thus $\mathbf{M}$ is an algebra of bounded operators acting on a Hilbert space $\mathcal{H}$. The algebra $\mathbf{M}$ is called a *factor* if its center consists only of scalar multiples of the identity. The factor is *type* II_1 if it admits a linear functional, called a trace, $\mathrm{tr} : \mathbf{M} \to \mathbb{C}$, which satisfies the following three conditions:

$$\mathrm{tr}(\mathbf{xy}) = \mathrm{tr}(\mathbf{yx}) \text{ for all } \mathbf{x}, \mathbf{y} \in \mathbf{M}$$

$$\mathrm{tr}(\mathbf{1}) = 1$$

$$\mathrm{tr}(\mathbf{xx}^*) > 0 \text{ for all } \mathbf{x} \in \mathbf{M}, \text{ where } \mathbf{x}^* \text{ is the adjoint of } \mathbf{x}$$

In this situation it is known that the trace is unique, in the sense that it is the only linear form satisfying the first two conditions. An old discovery of Murray and von Neumann was that factors of type II_1 provide a type of "scale" by which one can measure the dimension $\dim_{\mathbf{M}}(\mathcal{H})$ of $\mathcal{H}$. The notion of dimension which occurs here generalizes the familiar notion of integer-valued dimensions, because for appropriate $\mathbf{M}$ and $\mathcal{H}$ it can be any non-negative real number or ∞.

The starting point of Jones' work was the following question: if $\mathbf{M}_1$ is a type II_1 factor and if $\mathbf{M}_0 \subset \mathbf{M}_1$ is a *subfactor*, is there any restriction on the real numbers which occur as the ratio

$$\lambda = \dim_{\mathbf{M}_0}(\mathcal{H}) / \dim_{\mathbf{M}_1}(\mathcal{H}) \ ?$$

The question has the flavor of questions one studies in Galois theory. On the face of it, there was no reason to think that λ could not take on any value in $[1, \infty]$ so Jones' answer came as a complete surprise. He called λ the *index* $[\mathbf{M}_1 : \mathbf{M}_0]$ of $\mathbf{M}_0$ in $\mathbf{M}_1$, and proved a type of rigidity theorem about type II_1 factors and their subfactors.

The Jones Index Theorem. *If* $\mathbf{M}_1$ *is a* II_1 *factor and* $\mathbf{M}_0$ *a subfactor, then the possible values of the index* $[\mathbf{M}_1 : \mathbf{M}_0]$ *are restricted to:*

$$[4, \infty] \cup \{4\cos^2(\pi/p), \quad \textit{where } p \in \mathbb{N},\, p \geq 3\}.$$

Moreover, each real number in the continuous part of the spectrum $[4, \infty]$ *and also in the discrete part* $\{4\cos^2(\pi/p),\, p \in \mathbb{N},\, p \geq 3\}$ *is realized.*

We now sketch the idea of the proof, which is to be found in [Jo1]. Jones begins with the type II_1 factor $\mathbf{M}_1$ and the subfactor $\mathbf{M}_0$. There is also a tiny bit of additional structure: In this setting there exists a map $\mathbf{e}_1 : \mathbf{M}_1 \to \mathbf{M}_0$, known as the *conditional expectation* of $\mathbf{M}_1$ on $\mathbf{M}_0$. The map $\mathbf{e}_1$ is a *projection*, i.e. $(\mathbf{e}_1)^2 = \mathbf{e}_1$.

His first step is to prove that the ratio λ is independent of the choice of the Hilbert space $\mathcal{H}$. This allows him to choose an appropriate $\mathcal{H}$ so that the algebra $\mathbf{M}_2 = \langle \mathbf{M}_1, \mathbf{e}_1 \rangle$ generated by $\mathbf{M}_1$ and $\mathbf{e}_1$ makes sense. He then investigates $\mathbf{M}_2$ and proves that it is another type II_1 factor, which contains $\mathbf{M}_1$ as a subfactor, moreover the index $|\mathbf{M}_2 : \mathbf{M}_1|$ is equal to the index $|\mathbf{M}_1 : \mathbf{M}_0|$, i.e. to λ. Having in hand another II_1 factor $\mathbf{M}_2$ and its subfactor $\mathbf{M}_1$, there is also a trace on $\mathbf{M}_2$ which (by the uniqueness of the trace) coincides with the trace on $\mathbf{M}_1$ when it is restricted to $\mathbf{M}_1$, and another conditional expectation $\mathbf{e}_2 : \mathbf{M}_2 \to \mathbf{M}_1$. This allows Jones to iterate the construction, to build algebras $\mathbf{M}_1, \mathbf{M}_2 \ldots$ and from them a family of algebras:

$$\mathbf{J}_n = \{\mathbf{1}, \mathbf{e}_1, \ldots, \mathbf{e}_{n-1}\} \subset \mathbf{M}_n, \quad n = 1, 2, 3, \ldots.$$

Rewriting history a little bit in order to make the subsequent connection with knots a little more transparent, we now replace the $\mathbf{e}_k$'s by a new set of generators which are units, defining:

$$\mathbf{g}_k = q\mathbf{e}_k - (1 - \mathbf{e}_k),$$

where

$$(1 - q)(1 - q^{-1}) = 1/\lambda.$$

The $\mathbf{g}_k$'s generate $\mathbf{J}_n$, because the $\mathbf{e}_k$'s do, and we can solve for the $\mathbf{e}_k$'s in terms of the $\mathbf{g}_k$'s. So

$$\mathbf{J}_n = \mathbf{J}_n(q) = \{\mathbf{1}, \mathbf{g}_1, \ldots, \mathbf{g}_{n-1}\},$$

and we have a *tower of algebras*, ordered by inclusion:

$$\mathbf{J}_1(q) \subset \mathbf{J}_2(q) \subset \mathbf{J}_3(q) \subset \ldots.$$

The parameter q, which replaces the index λ, is the quantity now under investigation.

The parameter q is woven into the construction of the tower. First, defining relations in $\mathbf{J}_n(q)$ depend upon q:

$$\mathbf{g}_i\mathbf{g}_k = \mathbf{g}_k\mathbf{g}_i \quad \text{if } |i-k| \geq 2, \tag{1}$$

$$\mathbf{g}_i\mathbf{g}_{i+1}\mathbf{g}_i = \mathbf{g}_{i+1}\mathbf{g}_i\mathbf{g}_{i+1}, \tag{2}$$

$$\mathbf{g}_i^2 = (q-1)\mathbf{g}_i + q, \tag{3q}$$

$$\mathbf{1} + \mathbf{g}_i + \mathbf{g}_{i+1} + \mathbf{g}_i\mathbf{g}_{i+1} + \mathbf{g}_{i+1}\mathbf{g}_i + \quad + \mathbf{g}_i\mathbf{g}_{i+1}\mathbf{g}_i = 0. \tag{4}$$

A second way in which q enters into the structure involves the trace. Recall that since $\mathbf{M}_n$ is type II_1 it supports a unique trace, and since $\mathbf{J}_n$ is a subalgebra it does too, by restriction. This trace is known as a *Markov trace*, i.e. it satisfies the important property:

$$\operatorname{tr}(\mathbf{w}\mathbf{g}_n) = \tau(q)\operatorname{tr}(\mathbf{w}) \quad \text{if } \mathbf{w} \in \mathbf{J}_n, \tag{5q}$$

where $\tau(q)$ is a fixed function of q. Thus, for each fixed value of q the trace is multiplied by a fixed scalar when one passes from one stage of the tower to the next, by multiplying an arbitrary element of $\mathbf{J}_n$ by the new generator $\mathbf{g}_n$ of $\mathbf{J}_{n+1}$.

Relations (1) and (2) above have an interesting geometric meaning, familiar to topologists. They are defining relations for the *n-string braid group*, $\mathbf{B}_n$, discovered by Emil Artin [Ar] in a foundational paper written in 1923. We pause to discuss braids.

An n-braid may be visualized by a weaving pattern of strings in 3-space which join n points on a horizontal plane to n corresponding points on a parallel plane, as illustrated in the example in Figure 2, where $n = 4$. In the case $n = 3$, the familiar braid in a person's hair gives another example. The strings are allowed to be stretched and deformed, the key features being that strings cannot pass through one-another and always proceed directly downward in their travels from the upper plane to the lower one. The equivalence class of weaving patterns under such deformations is an *n-braid*. One multiplies braids by concatenation and erasure of the middle plane. This multiplication makes them into a group, the n-string braid group $\mathbf{B}_n$. The identity is a braid which, when pulled taut, goes over to n straight lines. Generators are the $n-1$ elementary braids which (by an abuse of notation) we continue to call $\mathbf{g}_1, \ldots, \mathbf{g}_{n-1}$. The pictures in Figure 3 show that relations (1) and (2) hold between the generators of $\mathbf{B}_n$. In fact, Artin proved they are *defining relations* for $\mathbf{B}_n$. Thus for each n there is a homomorphism from the n-string braid group $\mathbf{B}_n$ into the Jones algebra $\mathbf{J}_n(q)$, and from the group algebra $\mathbb{C}\mathbf{B}_n$ onto $\mathbf{J}_n(q)$.

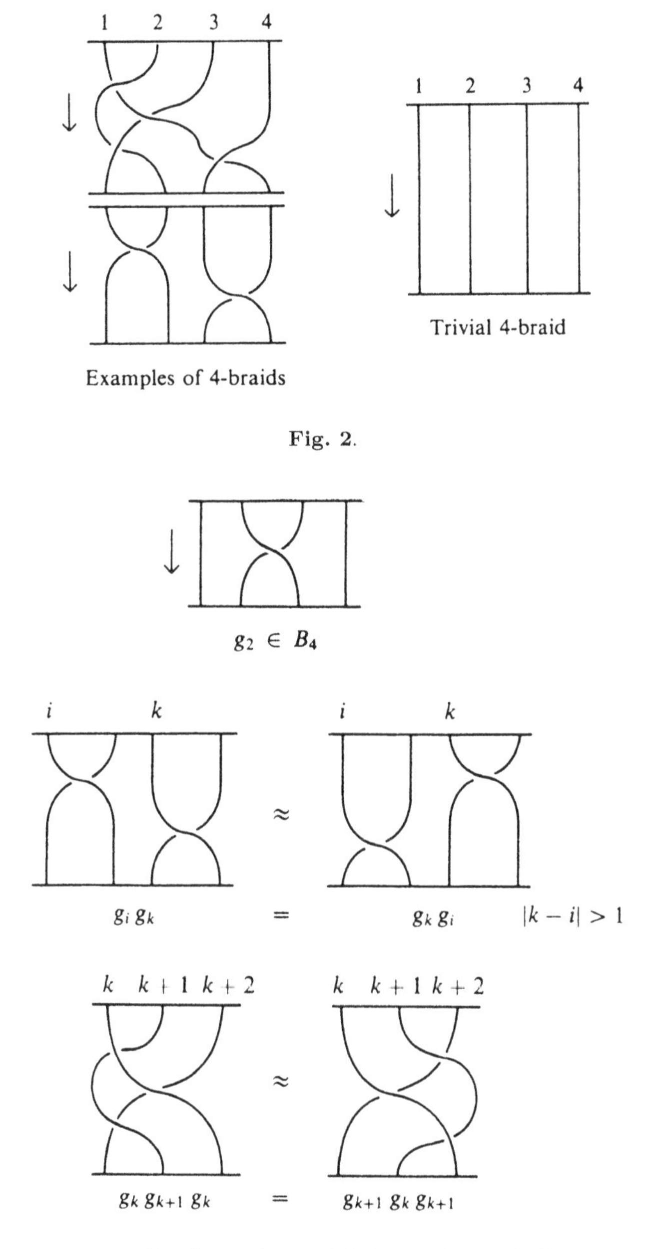

Fig. 2.

Fig. 3. Generators and defining relations in $\mathbf{B}_n$

Returning to the business at hand, i.e. the proof of the Index Theorem, Jones next shows that properties (1), (2), (3_q) and (5_q) suffice for the calculation of the trace of an arbitrary element $\mathbf{x} \in \mathbf{J}_n(q)$. It turns out that trace($\mathbf{x}$) is an *integer polynomial* in $(\sqrt{q})^{\pm 1}$. (We will meet it again in a few moments as the Jones polynomial associated to $\mathbf{x}$.) Jones proof of the Index Theorem is concluded when he shows that the infinite sequence of algebras $\mathbf{J}_n(q)$, with the given trace, could not exist if q did not satisfy the restrictions of the Index Theorem.

2. Knots and Links

We have already seen hints of topological meaning in $\mathbf{J}_n(q)$ via braids. There is more to come. Knots and links are obtained from braids by identifying the initial points and end points of a braid in a circle, as illustrated in Figure 4. It was proved by J. W. Alexander in 1928 that every knot or link arises in this way. Earlier we described an equivalence relation on weaving patterns which yields braids, and there is a similar (but less restrictive) equivalence relation on knots, i.e. a knot or link *type* is its equivalence class under isotopy in 3-space. Note that isotopy in 3-space which takes one closed braid representative of a link to another closed braid representative will pass through a sequence of representatives which are not closed braids in an obvious way. For example see the 2-component link which is illustrated in Figure 4. The left picture is an obvious closed braid representative, whereas the right is not.

Let $\mathbf{B}_\infty$ denote the disjoint union of all of the braid groups $\mathbf{B}_n$, $n = 1, 2, 3, \ldots$. In 1935 the mathematician A. A. Markov proposed the equivalence relation on $\mathbf{B}_\infty$ which corresponds to link equivalence [M]. Remarkably, the properties of the trace, or more particularly the facts that $\text{tr}(\mathbf{xy})=\text{tr}(\mathbf{yx})$ together with property (5_q), were exactly what was needed to make the trace polynomial into an invariant on Markov's equivalence classes! Using Markov's proposed equivalence relation (which was proved to be the correct one in 1972 [Bi]), Jones proved, with almost no additional work beyond results already established in [Jo1], the following theorem:

Theorem [Jo3]. *If* $\mathbf{w} \in \mathbf{B}_\infty$, *then (after multiplication by an appropriate scalar, which depends upon the braid index* n*) the trace of the image of* $\mathbf{w}$ *in* $\mathbf{J}_n(q)$ *is a*

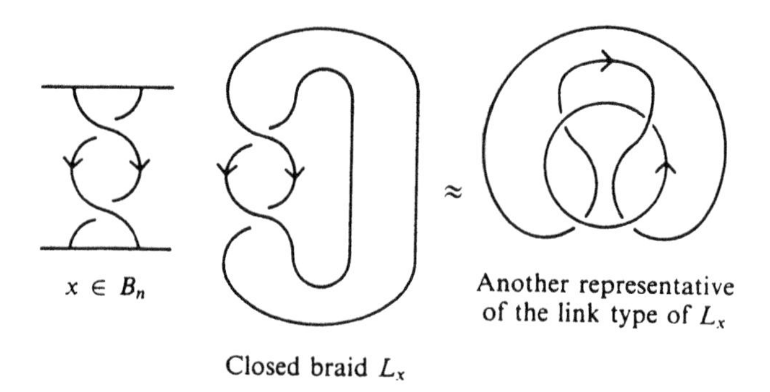

Fig. 4. Braids determine links

polynomial in $(\sqrt{q})^{\pm 1}$ *which is an invariant of the link type defined by the closed braid* $\mathbf{L_w}$.

The invariant of Jones' theorem is the one-variable *Jones polynomial* $\mathbf{V}_x(q)$. Notice that the independent "variable" in this polynomial is essentially the index of a type II_1 subfactor in a type II_1 factor! It's discovery opened a new chapter in knot and link theory.

3. Statistical Mechanics

We promised to discuss other ways in which the work of Jones was related to yet other areas of mathematics and physics, and begin to do so now. As it turned out, when Jones did his work the family of algebras $\mathbf{J}_n(q)$ were already known to physicists who were concerned with *exactly solvable models* in Statistical Mechanics. (For an excellent introduction to this topic, see R. Baxter's article in these Proceedings.) One of the simplest examples in this area is known as the *Potts model*. In that model one considers an array of "atoms" arranged at the vertices of a planar lattice with m rows and n columns as in Figure 5. Each "atom" in the system has various possible spins associated to it, and in the simplest case, known as the Ising model, there are two choices, "+" for spin up and "−" for spin down. We have indicated one of the 2^{nm} choices in Figure 5, determining a *state* of the system. The goal is to compute the free energy of the system, averaged over all possible states.

Letting σ_i denote the spin at site i, we note that an edge e contributes an *energy* $E_e(\sigma_i, \sigma_j)$, where σ_i and σ_j are the states of the endpoints of e. Let E be the collection of lattice edges. Let β be a parameter which depends upon the

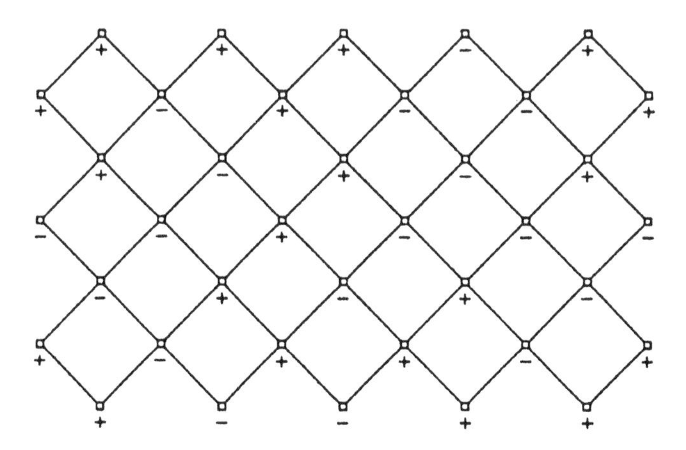

Fig. 5.

temperature. Then the Gibbs *partition function* $\mathbf{Z}$ is given by the formula:

$$\mathbf{Z} = \sum_{\sigma_1,\ldots,\sigma_{mn}} \prod_{e\in E} \exp(-\beta E_e(\sigma_i,\sigma_j)).$$

All of this is microscopic, nevertheless the major macroscopic thermodynamic quantities are functions of the partition function. In particular, the free energy $\mathbf{F}$, the object of interest to us at this time, is given by $\mathbf{Z} = \exp(-\beta\mathbf{F})$.

To compute the manner in which the atoms in one row of the lattice interact with atoms in the next, physicists set up the *transfer matrix* $\mathbf{T}$, which expresses the row-to-row interactions. It turned out that, in order to be able to calculate the free energy, the transfer matrices must satisfy conditions known as the *Yang-Baxter equations* and (to the great surprise of everyone) they turned out to be the braid relations (1) and (2) in disguise! (Remark: before Jones' work, to the best of our knowledge, it was not known that the Yang-Baxter equation was related to braids or knots.) Even more, the algebra which the transfer matrices generate in the Ising model, known to physicists as the *Temperley-Lieb algebra*, is our algebra $\mathbf{J}_n(q)$. The partition function $\mathbf{Z}$ is related in a very simple way to the transfer matrix:

$$\mathbf{Z} = \text{trace}(\mathbf{T})^m.$$

In fact, it is closely related to the Jones trace.

The initial discovery of a relationship between the Potts model and links was reported on in [Jo3]. It opened a new chapter in the flow of ideas between mathematics and physics. We give an explicit example of a way in which the relationship of Jones' work to physics led to new insight into mathematics. Learning that the partition function was a sum over states of the system, Louis Kauffman was led to seek a decomposition of the Jones polynomial into a related sum over "states" of knot diagram, and arrived in [K1] at an elegant "states model" for the Jones polynomial. The Jones polynomial, and Kauffman's states model for it, were later seen to generalize to other polynomial invariants with associated states models, for links in S^3 and eventually into invariants for 3-manifolds M^3 and links in 3-manifolds. The full story is not known at this writing, however we refer the reader to V. Turaev's article in these *Proceedings* for an excellent account of it, as of August 1990.

4. Quantum Groups and Representations of Lie Algebras

We begin by explaining the structure of the algebra $\mathbf{J}_n(q)$. It will be convenient to begin with another algebra $\mathbf{H}_n(q)$, which is generated by symbols $\mathbf{g}_1,\ldots,\mathbf{g}_{n-1}$ which now have a third meaning), with defining relations (1), (2) and (3_q). The algebra $\mathbf{H}_n(q)$ is very well-known to mathematicians. It's the *Iwahori-Hecke algebra*, also known as the *Hecke algebra of the symmetric group* [Bo]. Its relationship with the symmetric group is simple to describe and beautiful. Notice that when $q = 1$, relation (3_q) simplifies to $(\mathbf{g}_k)^2 = 1$. One recognizes (1), (2) and (3_1) as defining relations for the group algebra $\mathbb{C}\mathbf{S}_n$ of the symmetric group $\mathbf{S}_n$. Here $\mathbf{g}_k$ is to be re-interpreted as a transposition which exchanges the symbols $\mathbf{k}$ and $\mathbf{k+1}$. In this way we may view $\mathbf{H}_n(q)$ as a "q-deformation" of the complex group algebra $\mathbb{C}\mathbf{S}_n = \mathbf{H}_n(1)$.

The algebra $\mathbb{C}\mathbf{S}_n$ is *rigid*, that is if one deforms it in this way its irreducible summands continue to be irreducible summands of the same dimension, in fact $\mathbf{H}_n(q)$ is actually algebra-isomorphic to $\mathbb{C}\mathbf{S}_n$ for generic q. Thus $\mathbf{H}_n(q)$ is direct sum of finite dimensional matrix algebras, its irreducible summands being in one-to-one correspondence with the irreducible representations of the symmetric group $\mathbf{S}_n$. In this setting, Jones showed in [Jo2] that for generic q the algebra $\mathbf{J}_n(q)$ may be interpreted as the algebra associtated to the q-deformations of those irreducible representations of $\mathbf{S}_n$ which have Young diagrams with at most two rows.

We now explain how $\mathbf{H}_n(q)$ is related to quantum groups. It will be helpful to recall the classical picture. The fundamental representation of the Lie group GL_n acts on $\mathbb{C}^n$, and so its k-fold tensor product acts naturally on $(\mathbb{C}^n)^{\otimes k}$. The symmetric group $\mathbf{S}_k$ also acts naturally on $(\mathbb{C}^n)^{\otimes k}$, permuting factors. (Remark: In this latter action, the representations of $\mathbf{S}_k$ which are relevant are those whose Young diagrams have $\leq n$ rows.) As is well known, the actions of GL_n and $\mathbf{S}_k$ are each other's commutants in the full group of linear transformations of $(\mathbb{C}^n)^{\otimes k}$. If one replaces GL_n and $\mathbb{C}\mathbf{S}_k$ by the quantum group $U_q(GL_n)$ and the Hecke algebra $\mathbf{H}_k(q)$ respectively, then the remarkable fact is that $U_q(GL_n)$ and $\mathbf{H}_k(q)$ are still each other's commutants [Ji]. The corresponding picture for $\mathbf{J}_n(q)$ is obtained by restricting to GL_2 and to representations of $\mathbf{S}_k$ having Young diagrams with at most 2 rows.

We remark that these are not isolated instances of algebraic accidents, but rather special cases of a phenomenon which relates a large part of the mathematics of quantum groups to finite dimensional matrix representations of the group algebra $\mathbb{C}\mathbf{B}_n$ which support a Markov trace (e.g. see [BW]).

5. Dynkin Diagrams

Dynkin diagrams arise in the tower construction which we described in § 1 via the inclusions of the algebras $\mathbf{J}_n(q)$ in the Jones tower. The inclusions for the Jones tower are very simple, and correspond to the Dynkin diagram of type A_n. However, other, more complicated towers may be obtained by replacing the II_1 factor $\mathbf{M}_1$ in the tower construction of § 1 above by $\mathbf{M}_1 \cap (\mathbf{M}_0)'$, where $(\mathbf{M}_0)'$ is the commutant of $\mathbf{M}_0$ in $\mathbf{M}_1$. We refer the reader to [GHJ] for an introduction to this topic and a discussion of the "derived tower" and the Dynkin diagrams which occur. The connections with the representations of simple Lie algebras can be guessed at from our discussion in § 4 above.

6. Concluding Remarks

I hope I have succeeded in showing you some of the ways in which Jones' work created bridges between the areas of mathematics which were illustrated in Figure 1. To conclude, I want to indicate very briefly some of the ways in which those bridges have changed the mathematics which many of us are doing.

There is another link polynomial in the picture, the famous *Alexander polynomial*. It was discovered in 1928, and was of fundamental importance to knot theory, both in the classical case of knots in S^3 and in higher dimensional knotting. Shortly after Jones' 1984 discovery, it was learned that in fact both the Alexander and Jones polynomial were specializations of the 2-variable Jones polynomial. That discovery was made simultaneously by five separate groups of authors: Freyd and Yetter, Hoste, Lickorish and Millett, Ocneanu, and Przytycki and Traczyk, a simple version of the proof of Ocneanu being given in [Jo3]. One of the techniques used in the proof was the combinatorics of link diagrams, and that technique led to the discovery of yet another polynomial, by Louis Kauffman [K2].

From the point of view of algebra, the Jones polynomial comes from a trace function on $\mathbf{J}_n(q)$, and the 2-variable Jones polynomial from a similar trace on the full Hecke algebra $\mathbf{H}_n(q)$. Beyond that, there is another algebra, the so-called Birman–Wenzl algebra [BW], and Kauffman's polynomial is a trace on it. Even more, physicists who had worked with solutions to the Yang Baxter equation, realized that they knew of still other Markov traces, so they began to grind out still other polynomials, in initially bewildering confusion. That picture is fairly well understood at this moment, however the work of Witten [W] indicates there are still other, related, link invariants. The generalizations are vast, with much work to be done.

There is also a different and very direct way in which Jones had had equal influence. His style of working is informal, and one which encourages the free and open interchange of ideas. During the past few years Jones wrote letters to various people which described his important new discoveries at an early stage, when he did not yet feel ready to submit them for journal publication because he had much more work to do. He nevertheless asked that his letters be shared, and so they were widely circulated. It was not surprising that they then served as a rich source of ideas for the work of others. As it has turned out, there has been more than enough credit to go around. His openness and generosity in this regard have been in the best tradition and spirit of mathematics.

References

[Ar] Artin, E.: Theorie der Zöpfe. Hamburg Abh. **4** (1925) 47–72.

[Bi] Birman, J.: Braids, links and mapping class groups. Ann. Math. Stud. Princeton Univ. Press, Princeton 1974, pp. 37–69.

[Bo] Bourbaki, N.: Groupes et algebres de Lie. Hermann, Paris 1968, Chapitre IV.

[BW] Birman, J., Wenzl, H.: Braids, link polynomials and a new algebra. Trans. AMS **313** (1989) 249–273.

[GHJ] Goodman, F., de la Harpe, P., Jones, V.: Coxeter graphs and towers of algebras. MSRI Publications, vol. 15. Springer, Berlin Heidelberg New York 1989.

[Jo1] Jones, V. F. R.: Index for subfactors. Invent. math. **72** (1983) 1–25.

[Jo2] Jones, V. F. R.: Braid groups, Hecke algebras and subfactors. In: Geometric methods in operator algebras. Pitman Research Notes in Mathematics, vol. 123, 1986, pp. 242–273.

[Jo3] Jones, V. F. R.: Hecke algebra representations of braid groups and link polynomials. Ann. Math. **126** (1987) 335–388.

[Ji] Jimbo, M.: A q-analogue of $U(gl(N+1))$, Hecke algebra, and the Yang Baxter equation. Letters in Math. Physics **II** (1986) 247–252.

[K1] Kauffman, L.: A states model for the Jones polynomial. Topology **26** (1987) 385–407.

[K2] Kauffman, L.: An invariant of regular isotopy. Trans. AMS (to appear).

[M] Markov, A. A.: Über die freie Äquivalenz geschlossener Zöpfe. Recueil Mathematique Moscou **1** (1935) 73–78.

[W] Witten, E.: Quantum field theory and the Jones polynomial. In: Braid group, knot theory and statistical mechanics, eds. Yang and Ge. World Scientific Press, 1989.

Vaughan F. R. Jones

VAUGHAN F. R. JONES

Date and Place of Birth: 31 December 1952, Gisborne, New Zealand
Married (Martha Weare Jones [nee Myers], April 7, 1979 in New Jersey);
3 children
(Bethany Martha Jones; Ian Randal Jones; Alice Collins Jones)

Schooling:

Preparatory School (1961–65), St. Peter's School, Cambridge, NZ
Secondary School (1966–69), Auckland Grammar School, Auckland, NZ
Undergrad Studies (1970–73), Univ. of Auckland, Auckland, NZ
Graduate Studies (1974–76), Ecole de Physique, Geneva, Switzerland
(1976–80), Ecole de Mathematiques, Geneva, Switzerland

Degrees: B.Sc., 1972, Auckland Univ
M.Sc., 1973, Auckland Univ. (First Class Honors)
Docteur es Sciences [Math], 1979, Geneva.
(Supervisor: A. Haefliger)

Employment:

1974	Junior Lecturer, Univ. of Auckland
1975–80	Assistant, Univ. de Geneve, Switzerland
1980–81	E. R. Hedrick Asst. Prof., UCLA, Los Angeles, California
1981–82	Visiting Lecturer, Univ. of Pennsylvania, Philadelphia
1981–84	Assistant Professor, Univ. of Pennsylvania
1984–85	Associate Professor, Univ. of Pennsylvania
1985–present	Professor of Mathematics, Univ. of California, Berkeley

Special Awards:

1969	Universites Entrance Scholarship
1969	Gillies Scholarship (for study at Auckland Univ.)
1969	Phillips Industries Bursary
1973	Swiss Government Scholarship (for study in Switzerland)
1973	F. W. W. Rhodes Memorial Scholarship
1980	Vacheron Constantin Prize (for thesis, Univ. de Geneve)
1983	Alfred P. Sloan Research Fellowship
1986	Guggenheim Fellowship
1990	Fellow of the Royal Society
1990	Fields Medal
1991	New Zealand Government Science Medal

1992	Honorary Vice President for life, International Guild of Knot Tyers
1992	Honorary D.Sc., University of Auckland
1993	Elected to American Academy of Arts and Sciences
1993	Honorary D.Sc. University of Wales

Reprinted from Bull. Amer. Math. Soc., **12** *(1), 1985*

A POLYNOMIAL INVARIANT FOR KNOTS VIA VON NEUMANN ALGEBRAS[1]

by

VAUGHAN F. R. JONES[2]

A theorem of J. Alexander [**1**] asserts that any tame oriented link in 3-space may be represented by a pair (b, n), where b is an element of the n-string braid group B_n. The link L is obtained by closing b, i.e., tying the top end of each string to the same position on the bottom of the braid as shown in Figure 1. The closed braid will be denoted $b^\wedge$.

Thus, the trivial link with n components is represented by the pair $(1, n)$, and the unknot is represented by $(s_1, s_2 \cdots s_{n-1}, n)$ for any n, where $s_1, s_2, \ldots, s_{n-1}$ are the usual generators for B_n.

The second example shows that the correspondence of (b, n) with $b^\wedge$ is many-to-one, and a theorem of A. Markov [**15**] answers, in theory, the question of when two braids represent the same link. A Markov move of type 1 is the replacement of (b, n) by (gbg^{-1}, n) for any element g in B_n, and a Markov move of type 2 is the replacement of (b, n) by $(bs_n^{\pm 1}, n+1)$. Markov's theorem asserts that (b, n) and (c, m) represent the same closed braid (up to link isotopy) if and only if they are equivalent for the equivalence relation generated by Markov moves of types 1 and 2 on the *disjoint* union of the braid groups. Unforunately, although the conjugacy problem has been solved by F. Garside [**8**] within each braid group, there is no known algorithm to decide when (b, n) and (c, m) are equivalent. For a proof of Markov's theorem see J. Birman's book [**4**].

The difficulty of applying Markov's theorem has made it difficult to use braids to study links. The main evidence that they might be useful was the existence of a representation of dimension $n-1$ of B_n discovered by W. Burau in [**5**]. The representation has a parameter t, and it turns out that the determinant of 1-(Burau matrix) gives the Alexander polynomial of the closed braid. Even so, the Alexander polynomial occurs with a normalization which seemed difficult to predict.

1980 *Mathematics Subject Classification.* Primary 57M25; Secondary 46L10.

[1]Research partially supported by NSF grant no. MCS-8311687.

[2]The author is a Sloan foundation fellow.

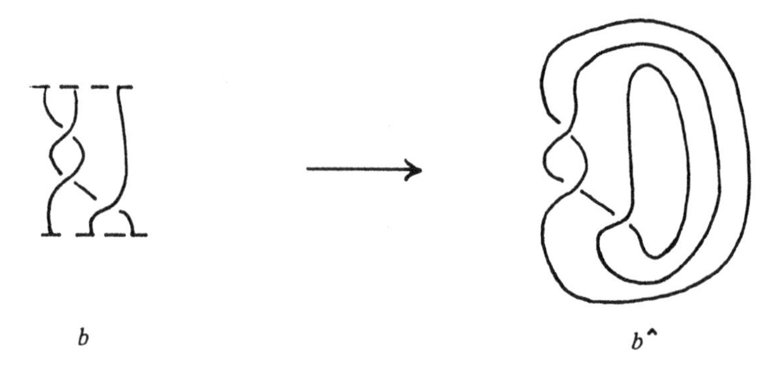

Figure 1.

In this note we introduce a polynomial invariant for tame oriented links via certain representations of the braid group. That the invariant depends only on the closed braid is a direct consequence of Markov's theorem and a certain trace formula, which was discovered because of the uniqueness of the trace on certain von Neumann algebras called type II_1 factors.

Notation. In this paper the Alexander polynomial Δ will always be normalized so that it is symmetric in t and t^{-1} and satisfies $\Delta(1) = 1$ as in Conway's tables in [**6**].

While investigating the index of a subfactor of a type II_1 factor, the author was led to analyze certain *finite-dimensional* von Neumann algebras A_n generated by an identity 1 and n projections, which we shall call $e_1, e_2, \ldots, e_n$. They satisfy the relations

(I) $e_i^2 = e_i, e_i^* = e_i$,
(II) $e_i e_{i\pm 1} e_i = t/(1+t)^2 e_i$,
(III) $e_i e_j = e_j e_i$ if $|i-j| \geq 2$.

Here t is a complex number. It has been shown by H. Wenzl [**24**] that an arbitrarily large family of such projections can only exist if t is either real and positive or $e^{\pm 2\pi i/k}$ for some $k = 3, 4, 5, \ldots$. When t is one of these numbers, there exists such an algebra for all n possessing a trace tr: $A_n \to \mathbf{C}$ completely determined by the normalization tr(1)=1 and

(IV) $\mathrm{tr}(ab) = \mathrm{tr}(ba)$,
(V) $\mathrm{tr}(we_{n+1}) = t/(1+t)^2\ \mathrm{tr}(w)$ if w is in A_n,
(VI) $\mathrm{tr}(a^*a) > 0$ if $a \neq 0$

(note $A_0 = \mathbf{C}$).

Conditions (I)–(VI) determine the structure of A up to $*$-isomorphism. This fact was proved in [**9**], and a more detailed description appears in [**10**]. Remember that a finite-dimensional von Neumann algebra is just a product of matrix algebras, the $*$ operation being conjugate-transpose.

For real t, D. Evans pointed out that an explicit representation of A_n on $\mathbf{C}^{2^n+2}$ was discovered by H. Temperley and E. Lieb [**23**], who used it to show the equivalence of the Potts and ice-type models of statistical mechanics. A readable account of this can be found in R. Baxter's book [**2**]. This representation was rediscovered in the von Neumann algebra context by M. Pimsner and S. Popa [**18**], who also found that the trace tr is given by the restriction of the Powers state with $t = \lambda$ (see [**18**]).

For the roots of unity the algebras A_n are intimately connected with Coxeter groups in a way that is far from understood.

The similarity between relations (II) and (III) and Artin's presentation of the n-string braid group,

$$\{s_1, s_2, \ldots, s_n \,:\, s_i s_{i+1} s_i = s_{i+1} s_i s_{i+1},\, s_i s_j = s_j s_i \text{ if } |i-j| \geq 2\},$$

was first pointed out by D. Hatt and P. de la Harpe. It transpires that if one defines $g_i = \sqrt{t}(te_i - (1-e_i))$, the g_i satisfy the correct relations, and one obtains representations r_t of B_n by sending s_i to g_i.

Theorem 1. *The number* $(-(t+1)/\sqrt{t})^{n-1}\ \mathrm{tr}(r_t(b))$ *for* b *in* B_n *depends only on the isotopy class of the closed braid* $b^\wedge$.

Definition. If L is a tame oriented classical link, the *trace invariant* $V_L(t)$ is defined by

$$V_L(t) = (-(t+1)/\sqrt{t})^{n-1}\ \mathrm{tr}(r_t(b))$$

for any (b, n) such that $b^\wedge = L$.

The Hecke algebra approach shows the following.

Theorem 2. *If the link* L *has an odd number of components,* $V_L(t)$ *is a Laurent polynomial over the integers. If the number of components is even,* $V_L(t)$ *is* $\sqrt{t}$ *times a Laurent polynomial.*

The reader may have observed that the von Neumann algebra structure (i.e., the $*$ operation) and condition (VI) are redundant for the definition of $V_L(t)$. This explains why V_L can be extended to all values of t except 0. However, it must be pointed out that for positive t and the relevant roots of unity, the presence of *positivity* gives a powerful method of proof.

The trace invariant depends on the oriented link but not on the chosen orientation. Let $L^\sim$ denote the mirror image link of L.

Theorem 3. $V_{L^\sim}(t) = V_L(1/t)$.

Thus, the trace invariant can be used to detect a lack of amphicheirality. It seems to be very good at this. A glance at Table 1 shows that it distinguishes the trefoil knot from its mirror image and hence, via Theorem 6, it distinguishes the two granny knots and the square knot.

Conjecture 4. If L is not amphicheiral, $V_{L^\sim} \neq V_L$.

There is some evidence for this conjecture, but only $10 hangs on it. In this direction we have the following result, where b is in B_n, b_+ is the sum of the positive exponents of b, and b_- is the (unsigned) sum of the negative ones in some expression for b as a word on the usual generators.

Theorem 5. *If $b_+ - 3b_- - n + 1$ is positive, then $b^\wedge$ is not amphicheiral.*

For $b_- = 0$, i.e., positive braids, this follows from a recent result of L. Rudolf [**21**]. Also, if the condition of the theorem holds, we conclude that $b^\wedge$ is not the unknot. This is similar in kind to a recent result of D. Bennequin [**3**].

The connected sum of two links can be handled in the braid group provided one pays proper attention to the components being joined. Let us ignore the subtleties and state the following (where # denotes the connected sum).

Theorem 6. $V_{L_1 \# L_2} = V_{L_1} V_{L_2}$.

As evidence for the power of the trace invariant, let us answer two questions posed in [**4**]. Both proofs are motivated by the fact, shown in [**10**], that $r_t(B_n)$ is sometimes finite.

Theorem 7. *For every n there are infinitely many words in B_{n+1} which give close braids inequivalent to closed braids coming from elements of the form $Us_n^{-1}Vs_n$, where U and V are in B_n.*

Explicit examples are easy to find; e.g., all but a finite number of powers of $s_n^{-1}s_2s_3$ will do.

Theorem 8 (see [**4** p. 217, q. 8]). *If b is in B_n and there is an integer k greater than 3 for which $b \in \ker r_t$, $t = e^{2\pi i/k}$, then $b^\wedge$ has braid index n.*

Here the braid index of a link L is the smallest n for which there is a pair (b, n) with $b^\wedge = L$. The kernel of r_t is not hard to get into for these values of t.

Corollary 9. *If the greatest common divisor of the exponents of $b \in B_n$ is more than 1, then the braid index of $b^\wedge$ is n.*

More interesting examples can be obtained by using generators and relations for certain finite groups; e.g., the finite simple group of order 25,920 (see [**10**, **7**]).

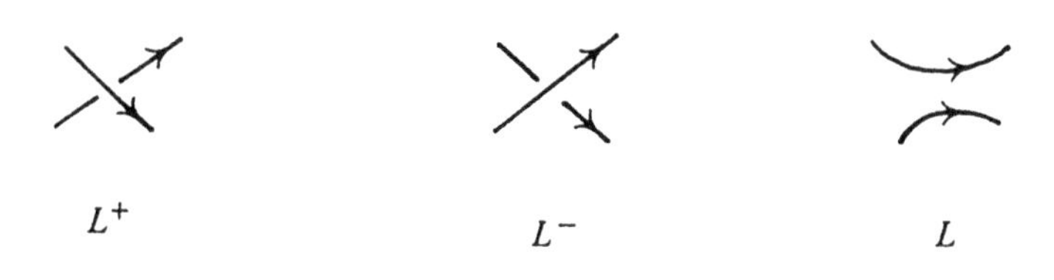

Figure 2.

In general, the trace invariant can probably be used to determine the braid index in a great many cases.

Note also that the trace invariant detects the kernel of r_t.

Theorem 10. *For* $t = e^{2\pi i/k}, k = 3, 4, 5, \ldots,$ $V_{b^\wedge}(t) = (-2\cos\pi/k)^{n-1}$ *if and only if* $b \in \ker r_t$ *(for* $b \in B_n$*).*

Corollary 11. *For transcendental* $t, b \in \ker r_t$ *if and only if* $V_{b^\wedge}(t) = (-(t+1)/\sqrt{t})^{n-1}$.

For transcendental t, r_t is very likely to be faithful.

There is an alternate way to calculate V_L without first converting L into a closed braid. In [**6**] Conway describes a method for rapidly computing the Alexander polynomials of links inductively. In fact, his first identity suffices in principle — see [**11**]. This identity is as follows.

Let L^+, L^-, and L be links related as in Figure 2, the rest of the links being identical. Then $\Delta_{L^+} - \Delta_{L^-} = (\sqrt{t} - 1/\sqrt{t})\Delta_L$.

For the trace invariant we have

Theorem 12. $1/tV_{L^+} - tV_{L^-} = (\sqrt{t} - 1/\sqrt{t})V_L$.

Corollary 13. *For any link* $L, V_L(-1) = \Delta_L(-1)$.

That the trace invariant may always be calculated by using Theorem 12 follows from the proof of the same thing for the Alexander polynomial. We urge the reader to try this method on, say, the trefoil knot.

The special nature of the algebras A_n when t is a relevant root of unity can be exploited to give information about V_L at these values.

Theorem 14. *If* K *is a knot then* $V_K(e^{2\pi i/3}) = 1$.

Theorem 15. $V_L(1) = (-2)^{p-1}$, *where* p *is the number of components of* L.

A more subtle analysis at $t = 1$ via the Temperley–Lieb–Pimsner–Popa representation gives the next result.

Theorem 16. *If K is a knot then $d/dtV_K(1) = 0$.*

It is thus sensible to simplify the trace invariant for knots as follows.

Definition 17. If K is a knot, define W_K to be the Laurent polynomial

$$W_K(t) = (1 - V_K(t))/(1 - t^3)(1 - t).$$

Amphicheirality is less obvious for W. In fact, $W_{K^\sim}(t) = 1/t^4 W_K(1/t)$. It is amusing that for W the unknot is 0 and the trefoil is 1. The connected sum is also less easy to see in the W picture. For the record the formula is

$$W_{K_1\#K_2} = W_{K_1} + W_{K_2} - (1-t)(1-t^3)W_{K_1}W_{K_2}.$$

Corollary 18. $\Delta_K(-1) \equiv 1$ *or* $5 \pmod 8$.

When $t = i$ the algebras A_n are the complex Clifford algebras. This together with a recent result of J. Lannes [**13**] allows one to show the following.

Theorem 19. *If K is a knot the Arf invariant of K is $W_K(i)$.*

Corollary 20. $\Delta_K(-1) = 1$ *or* $5 \pmod 8$ *when the Arf invariant is* 0 *or* 1, *respectively.*

This is an alternate proof of a result in Levine [**14**]; also see [**11**, p. 155]. Note also that Corollary 20 allows one to define an Arf invariant for links as $V(i)$. It may be zero and is always plus or minus a power of two otherwise.

The values of V at $e^{\pi i/3}$ are also of considerable interest, as the algebra A_n is then related to a kind of cubic Clifford algebra. Also, in this case, $r_t(B_n)$ is always a finite group, so one can obtain a rapid method for calculating $V(t)$ without knowing V completely. We have included this value of V in the tables. Note that it is always in $1 + 2\mathbf{Z}(e^{i\pi/3})$.

There is yet a third way to calculate the trace invariant. The decomposition of A_n as a direct sum of matrix algebras is known [**10**], and H. Wenzl has explicit formulae for the (irreducible) representations of the braid group in each direct summand. So in principle this method could always be used. This brings in the Burau representation as a direct summand of r_t. For 3 and 4 braids this allows one to deduce some powerful relations with the Alexander polynomial. An application of Theorem 16 allows one to determine the normalization of the Alexander polynomial in the Burau matrix for proper knots, and one has the following formulae.

Theorem 21. *If b in B_3 has exponent sum e, and $b^\wedge$ is a knot, then*

$$V_{b^\wedge}(t) = t^{e/2}(1 + t^e + t + 1/t - t^{e/2-1}(1 + t + t^2)\Delta_{b^\wedge}(t)).$$

Theorem 22. *If b in B_4 has exponent sum e, and $b^\wedge$ is a knot, then*

$$t^{-e}V(t)+t^eV(1/t)=(t^{-3/2}+t^{-1/2}+t^{1/2}+t^{3/2})(t^{e/2}+t^{-e/2})$$
$$-(t^{-2}+t^{-1}+2+t+t^2)\Delta(t)$$

(where $V=V_{b^\wedge}$ and $\Delta=\Delta_{b^\wedge}$).

These formulas have many interesting consequences. They show that, except in special cases, e is a knot invariant. They also give many obstructions to being closed 3 and 4 braids.

Corollary 23. *If K is a knot and $|\Delta_K(i)|>3$, then K cannot be represented as a closed 3 braid.*

Of the 59 knots with 9 crossings or less which are known not to be closed 3 braids, this simple criterion establishes the result for 43 of them, at a glance.

Corollary 24. *If K is a knot and $\Delta_K(e^{2\pi i/5})>6.5$, then K cannot be represented as a closed 4 braid.*

For $n>4$ there should be no simple relation with the Alexander polynomial, since the other direct summands of r_t look less and less like Burau representations.

In conclusion, we would like to point out that the q-state Potts model could be solved if one understood enough about the trace invariant for braids resembling certain braids discovered by sailors and known variously as the "French sinnet" (sennit) or the "tresse anglaise", depending on the nationality of the sailor. See [**21**, p. 90].

The author would like to thank Joan Birman. It was because of a long discussion with her that the relation between condition (V) and Markov's theorem became clear.

Tables. A single example should serve to explain how to read the tables. The knot 8_8 has trace invariant

$$t^{-3}(-1+2t-3t^2+5t^3-4t^4+4t^5-3t^6+2t^7-t^8).$$

Its W invariant is

$$t^{-3}(1-t+2t^2-t^3+t^4).$$

A braid representation for it is

$$s_1^{-1}s_2s_1^2s_3^{-1}s_2^2s_3^{-2}\text{ in }B_4.$$

Also note that $w=e^{\pi i/3}$.

Table 1. The trace invariant for prime knots to 8 crossings.

knot	braid rep.	p_0	pol (v)	v(w)	p_0	pol (w)
0_1	1	0	1	1	0	0
3_1	1^3	1	101 − 1	$i\sqrt{3}$	0	1
4_1	$12^{-1}12^{-1}$	− 2	1 − 11 − 11	− 1	− 2	− 1
5_1	1^5	2	101 − 11 − 1	− 1	0	1101
5_2	$2^2 1^{-1} 21^2$	1	1 − 12 − 11 − 1	− 1	0	101
6_1	$12^{-1}13^{-1}23^{-1}2^{-1}$	− 4	1 − 11 − 22 − 11	$i\sqrt{3}$	− 4	− 10 − 1
6_2	$1^{-1}21^{-1}2^3$	− 1	1 − 12 − 22 − 21	1	− 1	− 11 − 1
6_3	$1^{-1}2^2 1^{-2} 2$	− 3	− 12 − 23 − 22 − 1	1	− 3	1 − 11
7_1	1^7	3	101 − 11 − 11 − 1	− 1	0	1111101
7_2	$1^{-1}3^3 21^2 3^{-1}2$	1	1 − 12 − 22 − 11 − 1	1	0	10101
7_3	$2^5 12^2 1^{-2}$	2	1 − 12 − 23 − 21 − 1	1	0	110201
7_4	$3^2 1^{-1}23^{-1}21^2 2$	1	1 − 23 − 23 − 21 − 1	− 1	0	10201
7_5	$2^3 12^4 1^{-2}$	2	1 − 13 − 33 − 32 − 1	− 1	0	1102 − 11
7_6	$3^{-1}1^{-1}2113^{-1}2^{-3}$	− 6	− 12 − 34 − 33 − 21	− 1	− 6	1 − 12 − 1
7_7	$13^{-1}23^{-1}21^{-1}23^{-1}2$	− 3	− 13 − 34 − 43 − 21	$-i\sqrt{3}$	− 3	1 − 21 − 1
8_1	$12^{-1}3^{-1}214^{-2}\cdot$ $3^{-1}2^{-1}4$	− 6	1 − 11 − 22 − 22 − 11	1	− 6	−10−10−1
8_2	$2^5 1^{-1}21^{-1}$	0	1 − 12 − 23 − 32 − 21	− 1	1	1 − 11 − 1
8_3	$123^{-1}4^{-1}3^{-1}2\cdot$ $1^{-1}3^2 243^{-1}2^{-2}$	− 4	1 − 12 − 33 − 32 − 11	− 3	− 4	−10−20−1
8_4	$11132^{-1}3^{-2}12^{-1}$	− 3	1 − 12 − 33 − 33 − 21	− 1	− 3	−10−21−1
8_5	$1112^{-1}1112^{-1}$	0	1 − 13 − 33 − 43 − 21	$i\sqrt{3}$	1	1 − 21 − 1
8_6	$3^{-2}12^{-1}132^{-3}$	− 7	1 − 23 − 44 − 43 − 11	1	− 7	−11−21−1
8_7	$2^{-2}12^{-1}1^4$	− 2	− 12 − 24 − 44 − 32 − 1	1	− 2	1−12−11
8_8	$1^{-1}2113^{-1}223^{-2}$	− 3	− 12 − 35 − 44 − 32 − 1	1	− 3	1 − 12 − 11
8_9	$2^3 1^{-1}21^{-3}$	− 4	1 − 23 − 45 − 43 − 21	1	− 4	−11−21−1
8_{10}	$221^{-2}2^3 1^{-1}$	− 2	− 12 − 35 − 45 − 42 − 1	$i\sqrt{3}$	− 2	1 − 13 − 11
8_{11}	$12^{-2}32^{-1}3^{-2}12^{-1}$	− 7	1 − 23 − 55 − 44 − 21	$-i\sqrt{3}$	− 7	−11−22−1
8_{12}	$12^{-1}34^{-1}34^{-1}\cdot$ $213^{-1}2^{-1}$	− 4	1 − 24 − 55 − 54 − 21	− 1	− 4	−11−31−1
8_{13}	$1123^{-1}21^{-1}3^{-2}2$	− 3	13 − 45 − 55 − 32 − 1	− 1	− 3	1 − 22 − 11
8_{14}	$11221^{-1}3^{-1}23^{-1}2$	− 1	1 − 24 − 56 − 54 − 31	− 1	− 1	−12−22−1
8_{15}	$111231^{-1}2^3 32^{-1}$	2	1 − 25 − 56 − 64 − 31	$-i\sqrt{3}$	0	1103−22−1
8_{16}	$112^{-1}112^{-1}12^{-1}$	− 2	− 13 − 46 − 66 − 53 − 1	1	− 2	1 − 23 − 21
8_{17}	$21^{-1}21^{-1}221^{-2}$	− 4	1 − 35 − 67 − 65 − 31	1	− 4	−12−32−1
8_{18}	$(12^{-1})^3$	− 4	1 − 46 − 79 − 76 − 41	3	− 4	−13−33−1
8_{19}	$121212^2 1$	3	10100 − 1	$-i\sqrt{3}$	0	11111
8_{20}	$21^3 21^{-3}$	− 1	− 12 − 12 − 11 − 1	$i\sqrt{3}$	− 1	101
8_{21}	$2^3 12^2 1^{-2}21^{-1}$	1	2 − 23 − 32 − 21	$i\sqrt{3}$	0	1 − 11 − 1

Table 2. The trace invariant for some divers knots and links.

link	po	pol(V)	v(w)	po	pol(w)	braid rep.
10_{141} (1)	−2	1 − 23 − 34 − 32 − 21	$i\sqrt{3}$	−2	− 11 − 11 − 1	$2^{-4}112221$
KT(2)	−4	− 12 − 22001 − 22 − 21	1	−4	1 − 110 − 11 − 1 $1^{-1}3^{-1}2^{-1}$	$1113323^{-1}1^{-2}2.$
C(3)	−4	− 12 − 22001 − 22 − 21	1	−4	1 − 110 − 11 − 1	$22213^{-1}2^{-2}1.$ $2^{-1}13^{-1}$
2_1^2	1/2	− 10 − 1	− i			1^2
4_1^2 (4)	3/2	− 10 − 11 − 1	− i			1^4
4_1^2 (4)	1/2	− 11 − 10 − 1	i			$12^{-1}122$
5_1^2	− 7/2	1 − 21 − 21 − 1	i			$12^{-1}12^{-2}$
6_1^2	5/2	− 10 − 11 − 11 − 1	$\sqrt{3}$			1^6
6_2^2	3/2	− 11 − 22 − 21 − 1	− i			2221121^{-1}
6_3^2	− 3/2	− 12 − 22 − 31 − 1	$\sqrt{3}$			$21^{-1}23^{-1}2123^{-1}$
H_1 (5)	1/2	− 11 − 10 − 1	i			$12^{-1}122$
H_2 (5)	− 3/2	1 − 10 − 10 − 1	− i			$12^{-3}122$
W(6)	− 3/2	− 11 − 21 − 21	− i			$11221^{-1}2^{-2}$
6_1^3	− 1	1 − 13 − 13 − 21	$i\sqrt{3}$			$221^{-1}221^{-1}$
6_2^3	− 3	− 13 − 24 − 23 − 1	1			$12^{-1}12^{-1}12^{-1}$
6_3^3	2	10102	1			122122
A(7)	5/2	− 10 − 32 − 34 − 22 − 1	3 i			11222333
B(7)	5/2	− 10 − 32 − 34 − 22 − 1	3 i			11122333

Table Notes

(1) Compare 8_5 which has the same Alexander polynomial.
(2) The Kinoshita–Terasaka knot with 11 crossings. See [**12**].
(3) This is Conway's knot with trivia Alexander polynomial. See [**20**].
(4) Same link, different orientation.
(5) These links have homeomorphic complements.
(6) The Whitehead link.
(7) Two composite links with the same trace invariant.

Added in Proof. The similarity between the relation of Theorem 12 and Conway's relation has led several authors to a two-variable generalization of V_L. This has been done (independently) by Lickorish and Millett, Ocneanu, Freyd and Yetter, and Hoste.

References

1. J. W. Alexander, *A lemma on systems of knotted curves*, Proc. Nat. Acad. **9** (1923), 93–95.
2. R. J. Baxter, *Exactly solved models in statistical mechanics*, Academic Press, London, 1982.
3. D. Bennequin, *Entrelacements et structures de contact*, These, Paris, 1982.
4. J. Birman, *Braids, links and mapping class groups*, Ann. Math. Stud. **82** (1974).
5. W. Burau, *Uber Zopfgruppen und gleichsinning verdrillte Verkettunger*, Abh. Math. Sem. Hanischen Univ. **11** (1936), 171–178.
6. J. H. Conway, *An enumeration of knots and links*, Computational Problems in Abstract Algebra, Pergamon, New York, 1970, pp. 329–358.
7. H. Coxeter, *Regular complex polytopes*, Cambridge Univ. Press, 1974.
8. F. Garside, *The braid group and other groups*. Quart J. Math. Oxford Ser. **20** (1969), 235–254.
9. V. F. R. Jones, *Index for subfactors*, Invent. Math. **72** (1983), 1–25.
10. ———, *Braid groups, Hecke algebras and type* II_1 *factors*, Japan–U.S. Conf. Proc. 1983.
11. L. H. Kaufman, *Formal knot theory*, Math. Notes, Princeton Univ. Press, 1983.
12. S. Kinoshita and H. Terasaka, *On unions of knots*, Osaka Math. J. **9** (1959), 131–153.
13. J. Lannes, *Sur l'invariant de Kervaire pour les noeuds classiques*, École Polytechnique, Palaiseau, 1984 (preprint).
14. J. Levine, *Polynomial invariants of knots of codimension two*, Ann. of Math. (2) **84** (1966), 534–554.
15. A. A. Markov, *Uber die freie Aquivalenz geschlossener Zopfe*, Mat. Sb. **1** (1935), 73–78.
16. K. Murasugi, *On closed* 3*-braids*. Mem. Amer. Math. Soc.
17. K. A. Perko, *On the classification of knots*, Proc. Amer. Math. Soc. **45** (1974), 262–266.
18. M. Pimsner and S. Popa, *Entropy and index for subfactors*, INCREST, Bucharest, 1983 (preprint).
19. R. T. Powers, *Representations of uniformly hyperfinite algebras and the associated von Neumann algebras*, Ann. of Math. (2) **86** (1967), 138–171.
20. D. Rolfsen, *Knots and links*, Publish or Perish Math. Lecture Ser., 1976.
21. L. Rudolph, *Nontrivial positive braids have positive signature*, Topology **21** (1982), 325–327.
22. S. Svensson, *Handbook of Seaman's ropework*, Dodd, Mead, New York, 1971.

23. H. N. V. Temperley and E. H. Lieb, *Relations between the percolation and colouring problem and other graph-theoretical problems associated with regular planar lattices: some exact results for the percolation problem*, Proc. Roy. Soc. (London) (1971), 251–280.
24. H. Wenzl, *On sequences of projections*, Univ. of Pennsylvania, 1984 (preprint).

Department of Mathematics,
University of Pennsylvania,
Philadelphia, Pennsylvania 19104

Mathematical Sciences Research Institute,
2223 Fulton Street, Room 3603,
Berkeley, California 94720

Reprinted from Invent. Math. **72** *(1983), 1–25*

INDEX FOR SUBFACTORS

by

V. F. R. JONES

Department of Mathematics, University of Pennsylvania,
Philadelphia, PA 19104, USA

1. Introduction

One of the first things Murray and von Neumann did with their theory of continuous dimension for subspaces affiliated with a type II_1 factor was to define an invariant for its action on a Hilbert space. This invariant has come to be known as the coupling constant and it measures the relative mobility of the factor and its commutant. To be precise, if M is a II_1 factor on $\mathcal{H}$ and M' is its commutant, the coupling constant C_M is infinite if M' is an infinite factor and if M' is finite, one takes any non-zero vector ξ in $\mathcal{H}$ and one considers the closed subspaces $\overline{M\xi}$ and $\overline{M'\xi}$ affiliated with M' and M, respectively. It is a nontrivial result of Murray and von Neumann that the ratio $C_M = \dim_{M'}(\overline{M\xi})/\dim_M(\overline{M'\xi})$ does not depend on ξ. This real number between 0 and ∞ is called the coupling constant.

From a more modern point of view, the coupling constant measures the imension of a Hilbert space on which M acts and may be defined in terms of intertwining maps in the category of normal M-modules. This leads to the notation $C_M = \dim_M(\mathcal{H})$. This notation is more natural and has been useful in the study of the von Neumann algebra of a foliation where the dimensions of certain geometric Hilbert spaces are measured by a von Neumann algebra (see [8]). We will use this more suggestive notation while keeping the term "coupling constant". Two normal representations of M are unitarily equivalent iff they have the same coupling constant.

It is an easy observation that the coupling constant can be used to define a conjugacy invariant for subfactors of II_1 factors. We call this invariant the index since if the subfactor comes from a subgroup in the group constructions of II_1 factors, the conjugacy invariant is the index of the subgroup. The index is defined in general as $\dim_N(L^2(M, \mathrm{tr}))$ where N is the subfactor and tr is the trace on M. This definition was probably noticed by Murray and von Neumann and appears more or less explicitly in works by Goldman [14], Suzuki [26] and others. In fact Goldman proves that if a subfactor has index 2 then the whole factor may be

Supported in part by NSF Grant # MCS 79–03041

expressed as the crossed product of the subfactor by a $\mathbb{Z}_2$ action. This is analogous to the fact that any subgroup of index 2 of a group is normal. It may be combined with Connes' classification of periodic automorphisms of the hyperfinite II_1 factor R ([6]) to yield the pleasing positive result that there is, up to conjugacy, only one subfactor of index 2 of R.

The question immediately arises: what possible values can the index take? The difficulty in answering this question is that there is very little to play with given an arbitrary subfactor of a II_1 factor. The only general result available is the existence of a conditional expectation onto the subfactor shown by Umegaki in [30]. In fact this tool will allow us to determine completely the possible values of the index for subfactors of R. Experience with dimension in II_1 factors suggests that the index will take on a continuum of values. This is indeed true but surprisingly one cannot turn on this effect until index 4 and even then it involves the fundamental group. It seems entirely plausible that for II_1 factors without full fundamental group, whose existence was shown by Connes in [9], the index may take only countably many values.

What, then, are the possible values of the index for subfactors of R? The answer is $\{4\cos^2 \pi/n | n = 3, 4, \ldots\} \cup \{r \in \mathbb{R} | r \geqq 4\} \cup \{\infty\}$. The value ∞ and the real values $\geqq 4$ are easily obtained (2.1.19 and 2.2.5). The situation between 1 and 4 is more difficult. The basic idea of the analysis for index < 4 is as follows: Let $N \subseteq M$ be II_1 factors. One represents M on $L^2(M, \text{tr})$ and considers the extension e_N to $L^2(M, \text{tr})$ of the conditional expectation onto N. One defines $\langle M, e_N \rangle$ to be the II_1 factor generated by M and e_N on $L^2(M, \text{tr})$. (This construction appears also in [5], [24]).) The crucial observation at this point is that the index of M in $\langle M, e_N \rangle$ is the same as that of N in M. Thus one may iterate this extension process and one obtains a sequence of II_1 factors, each one obtained from the previous one by adding a projection. The inductive limit gives a II_1 factor and if the projections in the construction are numbered $e_1, e_2, \ldots$, then they satisfy $e_i e_{i\pm1} e_i = \tau e_i$, $e_i e_j = e_j e_i$ if $|i - j| \geqq 2$ and $\text{tr}(e_{i_1} e_{i_2} \ldots e_{i_n}) = \tau^n$ if $|i_j - i_k| \geqq 2$ for $j \neq k$. Here τ is the reciprocal of the index of N in M and tr denotes the trace on the inductive limit. Analysis of the algebra generated by the e_i's yields that if $\tau > 1/4$ it can only be $\frac{1}{4}\sec^2\pi/n$ for $n = 3, 4, \ldots$. A large bonus of the analysis is that it shows how to construct subfactors with these values as τ.

It seem strange that the set of values contains a discrete and a continuous part, but this may yet be understood by the fact that, if the index is less than 4, the relative commutant is trivial while the constructions available for the continuous part all have nontrivial relative commutant. We have very little information on what happens to the values of the index if the relative commutant is required to be trivial, but note that a result of Popa in [23], together with Connes' result on injective factors [7] shows that any II_1 factor has a subfactor of infinite index with trivial relative commutant.

Before closing the introduction, I would like to pose 4 problems which I hope will lead to a more profound understanding of subfactors.

Problem 1 (due to Connes). What are the possible values of the index for subfactors of R with trivial relative commutant? Or, what is $\mathcal{C}_{\mathcal{R}}$?

Problem 2. For each $n = 3, 4, 5, \ldots$ are there only finitely many subfactors of R up to conjugacy with index $4\cos^2 \pi/n$?

Problem 3. If N is a subfactor of R, is N conjugate to $N \otimes R$ in a decomposition $R \cong R \otimes R$? ($[R : N] < \infty$)

Problem 4. If N is a subfactor of M which is regular and has trivial relative commutant, is M the crossed product of N by a group action? (True if $M = R$ by a result of Ocneanu [22], see [17].)

It would be impossible to thank everyone who helped me with this paper. I have tried to mention individual contributions in the text, but let me also thank especially B. Baker, A. Connes, F. Goodman, R. Powers, M. Takesaki, and A. Wassermann for many fruitful conversations. This paper is an extended version of the Comptes Rendus note [18].

2. Generalities

2.1. *The global index*

If M is a finite factor acting on a Hilbert space $\mathcal{H}$ with finite commutant M', the coupling constant $\dim_M(\mathcal{H})$ of M is defined as $\mathrm{tr}_M(E_\xi^{M'})/\mathrm{tr}_{M'}(E_\xi^M)$ where ξ is a non-zero vector in $\mathcal{H}$, tr_A denotes the normalized trace and E_ξ^A is the projection onto the closure of the subspace $A\xi$. This definition, due to Murray and von Neumann in [20], is independent of ξ.

We recall some rules of calculation associated with $\dim_M$. (See [10, p. 263].)

$$\dim_M(\mathcal{H}) > 0 \tag{2.1.1}$$

$$\dim_M(\mathcal{H}) = (\dim_{M'}(\mathcal{H}))^{-1} \tag{2.1.2}$$

$$\text{If } e \text{ is a projection in } M',\ \dim_{M_e}(e\mathcal{H}) = \mathrm{tr}_{M'}(\mathrm{e})\ \dim_M(\mathcal{H}) \tag{2.1.3}$$

$$\text{If } E \text{ is a projection in } M,\ \dim_{M_e}(e\mathcal{H}) = (\mathrm{tr}_M(e))^{-1} \dim_M(\mathcal{H}) \tag{2.1.4}$$

$$\text{If } M \otimes 1 \text{ is the amplification of } M \text{ on } \mathcal{H} \otimes \mathcal{K},$$

$$\dim_M(\mathcal{H} \otimes \mathcal{K}) = \dim_{\mathbb{C}}(\mathcal{K}) \dim_{\mathcal{M}}(\mathcal{H}) \tag{2.1.5}$$

$$\dim_M(\mathcal{H}) = 1 \quad \text{iff } M \text{ is standard on } \mathcal{H}, \text{ i.e. there is a cyclic trace vector for } M. \tag{2.1.6}$$

Agree to put $\dim_M(\mathcal{H}) = \infty$ if M' is infinite.

Proposition 2.1.7. *Let M be as above and N be a subfactor. The number $\dim_N(\mathcal{H})/\dim_M(\mathcal{H})$ is independent of $\mathcal{H}$ provided M' is finite.*

Proof. Any two such representations differ by an amplification and an induction. By (2.1.3) and (2.1.5), both $\dim_M$ and $\dim_N$ are multiplied by the same constants in this process. □

Definition. If N is a subfactor of M, the number $\dim_N(\mathcal{H})/\dim_M(\mathcal{H})$ defined in 2.1.7 is called the (global) *index* of N in M and written $[M : N]$. Note that $[M : N] = \infty$ means that N' is infinite for any normal representation of M.

By (2.1.7) and (2.1.6), $[M : N] = \dim_N(L^2(M, \mathrm{tr}))$. Thus $[M : N]$ is a conjugacy invariant for N as a subfactor of M.

The rules of calculation for $\dim_M$ give some rules for $[M : N]$.

Proposition 2.1.8. *If $P \subseteq Q \subseteq M$ are II_1 factors then*

$$[M : M] = 1 \tag{2.1.9}$$

$$[M : P] \geqq 1 \tag{2.1.10}$$

$$[M : P] = [M : Q]\,[Q : P] \tag{2.1.11}$$

$$[M : P] \geqq [M : Q] \tag{2.1.12}$$

$$[M : P] = [M : Q] \quad \textit{implies } P = Q \tag{2.1.13}$$

$$[M : P] = [P' : M'] \quad \textit{if } P' \textit{ is finite.} \tag{2.1.14}$$

Proof. The only nontrivial property is (2.1.13). To prove it first note that by (2.1.11) and (2.1.9) we may suppose that $M = Q$ and that M acts on $L^2(M, \mathrm{tr})$ with cyclic trace vector ξ. Then $P\xi$ is dense in $\mathcal{H}$ by hypothesis. But then for any $a \in M$ there is a net b_n of elements of P with $b_n\xi \to a\xi$ in $\mathcal{H}$, i.e. $\|b_n - a_n\|_2 \to 0$. By [10, Lemma 1, p. 270], this implies $a \in P$. □

We next examine how the index behaves under tensor products.

Proposition 2.1.15. *Let N_1 and N_2 be subfactors of the finite factors M_1 and M_2, respectively. Then $N_1 \otimes N_2$ is a subfactor of $M_1 \otimes M_2$ and $[M_1 \otimes M_2 : N_1 \otimes N_2] = [M_1 : N_1]\,[M_2 : N_2]$.*

Proof. Let M_1 and M_2 act with cyclic trace vectors $\xi_1 \in \mathcal{H}_1$ and $\xi_2 \in \mathcal{H}_2$ respectively. Then if e_1 and e_2 are the projections onto $\overline{N_1\xi_1}$ and $\overline{N_2\xi_2}$, $e_1 \otimes e_2$ is the projection onto $\overline{N_1 \otimes N_2(\xi_1 \otimes \xi_2)}$. Moreover $M_1 \otimes M_2$ is standard on $\mathcal{H}_1 \otimes \mathcal{H}_2$ and $(N_1 \otimes N_2)' = N_1' \otimes N_2'$. Thus $\mathrm{tr}_{(N_1 \otimes N_2)'}(e_1 \otimes e_2) = \mathrm{tr}_{N_1'}(e_1)\mathrm{tr}_{N_2'}(e_2)$ which gives the desired result. □

Proposition 2.1.16. *Let N_i, M_i be as above and suppose $N_i' \cap M_i = \mathbb{C}$, $i = 1, 2$. Then $(N_1 \otimes N_2)' \cap (M_1 \otimes M_2) = \mathbb{C}$.*

Proof. $(N_1 \otimes N_2)' \cap (M_1 \otimes M_2) = (N_1' \otimes N_2') \cap M_1 \otimes M_2$ (see [28, p. 227]) and $(N_1' \otimes N_2') \cap (M_1 \otimes M_2) = (N_1' \cap M_1) \otimes (N_2' \cap M_2)$. □

We now introduce two isomorphism invariants for II_1 factors.

Definition. If M is a finite factor let

$$\begin{aligned} \mathcal{I}_M = \{r \in \mathbb{R} \cup \{\infty\} | &\text{ there is a } \mathrm{II}_1 \text{ subfactor } N \text{ of } M \\ &\text{ with } [M : N] = r\} \\ \mathcal{C}_M = \{r \in \mathbb{R} \cup \{\infty\} | &\text{ there is a } \mathrm{II}_1 \text{ subfactor } N \text{ of } M \\ &\text{ with } [M : N] = r \text{ and } N' \cap M = \mathbb{C}\}. \end{aligned}$$

The determination of $\mathcal{I}_M$ and $\mathcal{C}_M$ is in general rather difficult. We shall gather some immediate results about them.

Proposition 2.1.17. *If $M \cong M \otimes M$ then both $\mathcal{I}_M$ and $\mathcal{C}_M$ are subsemigroups (with 1) of $\{r \in \mathbb{R} | r \geqq 1\} \cup \{\infty\}$ under multiplication.*

Proof. This follows from 2.1.15 and 2.1.16. □

Lemma 2.1.18. *If N is a hyperfinite subfactor of M then $[M : N] < \infty$ implies that M is hyperfinite.*

Proof. If M acts in such a way that N$'$ is finite, then N$'$ is hyperfinite (e.g. [29]) and by [7], M' and so M is hyperfinite. □

Corollary 2.1.19. *For any II_1 factor M, $\infty \in \mathcal{I}_M$.*

Proof. If $M \cong R$ then $R \otimes 1 \subseteq R \otimes R$ is of infinite index. Otherwise by [21] there is a hyperfinite subfactor of M which is of infinite index by 2.1.18. □

Proposition 2.1.20. *For any separable II_1 factor M, $\infty \in \mathcal{C}_M$.*

Proof. A result of Popa in [23] says that we can find a maximal abelian subalgebra A and a unitary u with $u\,A\,u^* = A$ such that u and A generate a subfactor N of M, isomorphic to R. Since N contains a maximal abelian subalgebra, $N' \cap M = \mathbb{C}$.

So if M is not hyperfinite, combining this result with 2.1.18 gives the required subfactor. If $M = R$, see [17] or Sec. 2.3. □

In this paper we will show that

$$\mathcal{I}_M \cap [1, 4) = \mathcal{C}_M \cap [1, 4) \subseteq \{4 \cos^2 \pi/n | n = 3, 4, \ldots\}$$

and that

$$\mathcal{I}_R = \{4\cos^2 \pi/n | n = 3, 4, \ldots\} \cup \{r \in \mathbb{R} | r \geqq 4\} \cup \{\infty\}.$$

2.2. *The local index*

If the global index of a subfactor is finite, one may define a finer invariant, which I call the local index, obtained by restricting the trace on N' to $N' \cap M$. I would like to thank A. Connes for suggesting this approach.

Definition. Let $N \subseteq M$ be II_1 factors and let $p \in N' \cap M$ be a projection. The *index of N at p* will be $[M_p : N_p] = [M : N]_p$.

Lemma 2.2.1. *The index at p and the global index are related by the formula*

$$[M : N]_p = [M : N]\, \mathrm{tr}_M(p)\, \mathrm{tr}_{N'}(p).$$

Proof. If M begins in standard form on $\mathcal{H}$ then by 2.1.4, $\dim_{M_p}(p\mathcal{H}) = \mathrm{tr}_M(p)^{-1}$. Also $\dim_N(\mathcal{H}) = [M : N]$ so by 2.1.3, $\dim_{N_p}(p\mathcal{H}) = [M : N]\, \mathrm{tr}_{N'}(p)$. Thus $[M : N]_p = \dim_{N_p}(p\mathcal{H}) / \dim_{M_p}(p\mathcal{H}) = [M : N]\, \mathrm{tr}_M(p)\, \mathrm{tr}_{N'}(p)$. □

Lemma 2.2.2. *If $\{p_i\}$ is a partition of unity in $N' \cap M$ then*

$$[M : N] = \sum_i \mathrm{tr}_M(p_i)^{-1} [M : N]_{p_i}.$$

Proof. For each i, $[M : N]\, \mathrm{tr}_{N'}(p_i) = \mathrm{tr}_M(p_i)^{-1} [M : N]_{p_i}$ and summing over i gives the result. □

Corollary 2.2.3. *If $[M : N] < \infty$ then $N' \cap M$ is finite dimensional.*

Proof. If $N' \cap M$ were infinite dimensional we could find arbitrarily large partitions of unity and by 2.2.2 $[M : N] = \infty$. □

In fact with a little more care one may obtain the bound $[M : N] \geqq \dim_{\mathbb{C}}(N' \cap M)$.

Corollary 2.2.4. *If $[M : N] < 4$, $N' \cap M = \mathbb{C}$.*

Proof. If $N' \cap M \neq \mathbb{C}$, then it contains two mutually orthogonal non-zero projections and since $[M : N]_{p_i} \geqq 1$, $[M : N] \geqq \mathrm{tr}(p_1)^{-1} + \mathrm{tr}(p_2)^{-1} \geqq 4$. □

Corollary 2.2.5. *If M has fundamental group $= \mathbb{R}$, $\mathcal{I}_M$ contains $\{r \in \mathbb{R} | r \geqq 4\}$.*

Proof. We must exhibit for any $r \geqq 4$ a subfactor of index r. Choose $d \in (0, 1)$ with $1/d + 1/(1-d) = r$ and choose a projection p with $\mathrm{tr}_M(p) = d$. Then M_p and

M_{1-p} are isomorphic so choose some isomorphism $\theta: M_p \to M_{1-p}$ and let N be the subfactor $\{x + \theta(x) | x \in M_p\}$. Then $N_p = M_p$ and $N_{1-p} = M_{1-p}$ so by 2.2.2, $[M : N] = 1/d + 1/(1-d) = r$. □

2.3. *Examples*

We give three examples of subfactors and their indices.

Example 2.3.1. Suppose $M = N \otimes P$ where P is a type I_n factor. Then $[M : N \otimes 1] = n^2$. This follows from 2.1.5.

Thus for any II_1 factor M, $\mathcal{I}_M$ always contains $\{n^2 | n \in \mathbb{Z} - \{0\}\} \cup \{\infty\}$. It is not inconceivable that there are II_1 factors for which $\mathcal{I}_M$ is no larger than this set.

Example 2.3.2. Let A be a von Neumann algebra and G a countable discrete group of automorphisms for which the crossed product $A \rtimes G$ is a finite factor. If H is a subgroup of G such that $A \rtimes H$ is a finite factor then $[A \rtimes G : A \rtimes H] = [G : H]$.

Proof. Write G as a disjoint union of cosets, $G = \coprod_{i \in I} Hg_i$. Then let $V_i = \overline{(A \rtimes H)\, u_{g_i}}$. The V_i are subspaces of $L^2(A \rtimes G, \mathrm{tr})$ affiliated with $(A \rtimes H)'$ which are mutually orthogonal, where tr is the extension to $A \rtimes G$ of a faithful normal G-invariant trace on A and the u_g's are the implementing unitaries of the crossed product. Moreover if p_i is the projection onto V_i then $\sum_{i \in I} p_i = 1$ and the p_i are mutually equivalent in $(A \rtimes H)'$ since $V_i = Ju_{g_i}JV_0$ where $V_0 = \overline{A \rtimes H}$ and J is the involution on $L^2(A \rtimes G, \mathrm{tr})$. This is because $Ju_{g_i}J \in (A \rtimes G)' \subseteq (A \rtimes H)'$.

Thus if $[G : H] = \infty$, $(A \rtimes H)'$ is infinite so $[A \rtimes G : A \rtimes H] = \infty$. If $[G : H] < \infty$, $(A \rtimes H)'$ is finite since reduction by p_0 puts $A \rtimes H$ in standard form. So $[G : H]\, \mathrm{tr}_{(A \rtimes H)'}(p_0) = 1$ which establishes that $[A \rtimes G : A \rtimes H] = [G : H]$. □

This example is the justification for the name "index".

Example 2.3.3. If M is a II_1 factor and G is a finite group of outer automorphisms of M with fixed point algebra M^G, $[M : M^G] = |G|$.

Proof. This result could be established using 2.3.2 but I shall give a different proof which brings in the basic construction of Chapter 3.

Let M act on $L^2(M, \mathrm{tr})$ and let u_g be the unitaries extending the action of G on M. Then the u_g's act also on M' and it is established in [1] that $(M^G)'$ is isomorphic in the obvious way to $M' \rtimes G$. The projection onto $\overline{M^G}$ is $|G|^{-1} \sum_{g \in G} u_g$ and by the isomorphism with the crossed product its trace is $|G|^{-1}$. Thus $[M : M^G] = |G|$. □

3. The Basic Construction

3.1. *Extending finite von Neumann algebras by subalgebras*

Let M be a finite von Neumann algebra with faithful normal normalized trace tr and let N be a von Neumann subalgebra. By [30] there is a conditional expectation $E_N : M \to N$ defined by the relation $\mathrm{tr}(E_N(x)y) = \mathrm{tr}(xy)$ for $x \in M,\ y \in N$. The map E_N is normal and has the following properties:

$$E_N(axb) = aE_N(x)b \quad \text{for } x \in M,\ a, b \in N \text{ (the bimodule property)} \tag{3.1.1}$$

$$E_N(x^*) = E_N(x)^* \quad \text{for all } x \in M \tag{3.1.2}$$

$$E_N(x^*)E_N(x) \leqq E_N(x^*x) \quad \text{and} \quad E_N(x^*x) = 0 \quad \text{implies } x = 0. \tag{3.1.3}$$

Let ξ be the canonical cyclic trace vector in $L^2(M, \mathrm{tr})$. Identify M with the algebra of left multiplication operators on $L^2(M, \mathrm{tr})$. The conditional expectation E_N extends to a projection e_N on $\mathcal{H}$ via $e_N(x\xi) = E_N(x)\xi$. Let J be the involution $x\xi \mapsto x^*\xi$.

Proposition 3.1.4.

(i) *For* $x \in M,\ e_N x e_N = E_N(x)e_N$.
(ii) *If* $x \in M$ *then* $x \in N$ *iff* $e_N x = x e_N$.
(iii) $N' = \{M' \cup \{e_N\}\}''$.
(iv) J *commutes with* e_N.

Proof. (i) If y is an arbitrary element of M then $e_N x e_N(y\xi) = e_N x(E_N(y)\xi) = E_N(x)E_N(y)\xi$ by 3.1.1, and $E_N(x)e_N(y\xi) = E_N(x)E_N(y)\xi$. But the vectors $y\xi$ are dense in $L^2(M, \mathrm{tr})$.

(ii) Relation 3.1.1 shows that e_N commutes with N as in (i). Moreover if $x \in M$ and $e_N x = x e_N$ then $(e_N x)\xi = E_N(x)\xi = (xe)\xi = x\xi$. Since ξ is separating, $x = E_N(x)$.
(iii) It suffices to show that $\{M' \cup \{e_N\}\}' = N$. This follows from (i).
(iv) This follows from 3.1.2. □

These calculations lead to the following.

Definition. Let $\langle M, e_N\rangle$ be the von Neumann algebra on $L^2(M, \mathrm{tr})$ generated by M and e_N. This is the basic construction.

Proposition 3.1.5. (i) $\langle M, e_N\rangle = JN'J$.

(ii) *Operators of the form* $a_0 + \sum_{i=1}^{n} a_i e_N b_i$ *with* $a_i,\ b_i \in M$, *give a dense* $*$-*subalgebra of* $\langle M, e_N\rangle$.
(iii) $x \mapsto x e_N$ *is an isomorphism of* N *onto* $e_N\langle M, e_N\rangle e_N$.

(iv) *The central support of* e_N *in* $\langle M, e_N\rangle$ *is* 1.
(v) $\langle M, e_N\rangle$ *is a factor iff* N *is.*
(vi) $\langle M, e_N\rangle$ *is a finite iff* N' *is.*

Proof. (i) and (ii) follow immediately from 3.1.4. For (iii), to show that $e_N\langle M, e_N\rangle e_N \subseteq Ne_N$ it suffices by (ii) to show that $e_N(ae_Nb)e_N \in Ne_N$. This follows from (i) of 3.1.4. Moreover if $xe_N = 0$ then $xe_N\xi = x\xi = 0$ and ξ is separating. Thus $x \mapsto xe_N$ is injective. Affirmations (iv), (v) and (vi) are now easy. □

We want to consider special traces on $\langle M, e_N\rangle$.

Definition. If P is a subalgebra of $\langle M, e_N\rangle$, a trace Tr on $\langle M, e_N\rangle$ is called a (τ, P) trace if Tr extends tr and $\mathrm{Tr}(e_N x) = \tau\,\mathrm{tr}(x)$ for $x \in P$.

Lemma 3.1.6. *A* (τ, N) *trace is a* (τ, M) *trace.*

Proof. If $x \in M$, $\mathrm{Tr}(xe_N) = \mathrm{Tr}(e_N x e_N) = \mathrm{Tr}(E_N(x)e_N) = \tau\,\mathrm{tr}(E_N(x)) = \tau\,\mathrm{tr}(x)$. □

We shall now concentrate on the case where M and N are factors.

Proposition 3.1.7. *If* M *and* N *are factors then* $[M : N] < \infty$ *iff* $\langle M, e_N\rangle$ *is finite and in this case the canonical trace* Tr *on* $\langle M, e_N\rangle$ *is a* (τ, M) *trace where* $\tau = [M : N]^{-1}$. *In particular* $\mathrm{Tr}(e_N) = [M : N]^{-1}$. *Also* $[\langle M, e_N\rangle : M] = [M : N]$.

Proof. By 3.1.6 it suffices to show that Tr is a (τ, N) trace. But consider the map $y \to \mathrm{Tr}(e_N y)$ defined on N. This is a trace by (ii) of 3.1.4 so since N is a factor there is a constant K such that $\mathrm{Tr}(e_N y) = K\,\mathrm{tr}(y)$. Moreover the trace of e_N in N' is by definition τ so by (iv) of 3.1.4, (i) of 3.1.5 and uniqueness of the trace, this is the same as $\mathrm{Tr}(e_N)$. Thus $K = \tau$, and Tr is a (τ, N) trace.

To prove this last assertion note that

$$\begin{aligned}[\langle M, e_N\rangle : M] &= \dim_{M'}(L^2(M, \mathrm{tr}))/\dim_{N'}(L^2(M, \mathrm{tr})) = [\dim_{N'}(L^2(M, \mathrm{tr}))]^{-1}\\ &= \dim_N(L^2(M, \mathrm{tr})) = [M : N].\end{aligned}$$

□

For some general results about projections onto finite subalgebras see Skau's paper [24].

Finally in this section I show that the basic construction is generic for subfactors of finite index in the sense that all such subfactors are of the form $M \subseteq \langle M, e_N\rangle$, although not canonically.

Lemma 3.1.8. *Let* N *be a* II_1 *factor acting on* $L^2(N, \mathrm{tr})$ *and let* M *be a* II_1 *factor containing* N. *Then* $[M : N]$ *is finite and there is a subfactor* P *of* N *such that* $M = \langle N, e_P\rangle$.

Proof. Since M' is a II_1 factor, $[M:N]<\infty$. Let $P=JM'J$. Then $[N:P] = [M:P]$ as in 3.1.7 and $\langle N,e_p\rangle$ is a subfactor of M with $[M:\langle N,e_p\rangle]=1$. Thus by 2.1.13, $M=\langle N,e_p\rangle$. □

Corollary 3.1.9. *Let $N\subseteq M$ be II_1 factors with $[M:N]<\infty$ then there is a subfactor P of N and an isomorphism $\theta: M\to\langle N,e_P\rangle$ with $\theta|_N=\text{id}$.*

Proof. Represent M on $L^2(M,\text{tr})$. Choose a projection $p\in M'$ with $\text{tr}(p) = [M:N]^{-1}$. Then by [20], on $pL^2(M,\text{tr})$, M and N act with N in standard form. By 3.1.8 we are through. □

3.2. *Inclusions of complex semisimple algebras*

Let $N\subseteq M$ be finite dimensional complex semisimple algebras and $N=\bigoplus_{i=1}^{n}N_i$, $M=\bigoplus_{j=1}^{m}M_j$ be their canonical decompositions as direct sums of simple algebras, $N_i\cong M_{n_i}(C)$, $M_j\cong M_{m_j}(C)$. The inclusion of N in M is specified up to conjugacy by an $n\times m$ matrix $A_N^M=(a_{ij})$ where (a_{ij}) is the number of simple components of a simple M_j module viewed as an N_i module. This may be zero or any positive integer. If $\{p_i\}$ are the central idempotents for the N_i and q_j those for the M_j, $a_{ij}=0$ iff $p_iq_j=0$. We will call the matrix A_N^M the inclusion matrix.

The inclusion can also be described diagramatically as follows:

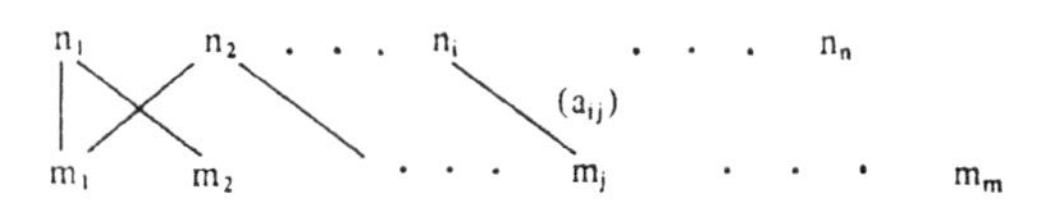

Here there are a_{ij} lines between n_i and m_j. This diagram will be called the Bratteli diagram after [4].

If the identity of M is the same as that of N we have the obvious relation $m_j=\sum_{i=1}^{n}a_{ij}n_i$ which we shall write as

$$\vec{m}=\vec{n}\,A_N^M. \tag{3.2.1}$$

A concrete example is

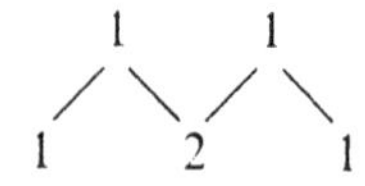

which is the diagram for the inclusion of $\mathbb{C}S_2$ in $\mathbb{C}S_3$.

If $N \subseteq M \subseteq P$, note the formula

$$A_N^P = A_N^M A_M^P. \tag{3.2.2}$$

If V is a faithful M-module, the centres of M and M'' are identical so that (if 3.2.1 holds) in the decompositions of M' and N' as simple algebras we may write $N' = \bigoplus_{i=1}^{n} N_i'$ and $M' = \bigoplus_{j=1}^{m} M_j'$. The following formula is in [3, Sec. 5, ex. 17].

$$A_{M'}^{N'} = (A_N^M)^T. \tag{3.2.3}$$

Since there is only one normalized trace on $M_n(\mathbb{C})$, a trace on $M = \bigoplus_j M_j$ may be specified by a column vector $\vec{t} = \begin{pmatrix} t_1 \\ t_2 \\ \vdots \\ t_m \end{pmatrix}$ where t_j is the trace of a minimal idempotent in M_j. The trace of the identity is the product $\vec{m} \cdot \vec{t}$, where $\vec{m} = (m_1, m_2, \ldots, m_m)$.

If $\vec{t}$ defines a trace on M whose restriction to N is defined by the vector $\vec{s}$ then the following relation is immediate.

$$\vec{s} = A_N^M \vec{t}. \tag{3.2.4}$$

Conversely if $\vec{s}$ and $\vec{t}$ define traces on N and M, respectively, then they agree on N if 3.2.4 holds.

3.3. *Finite dimensional C^*-algebras*

A finite dimensional C^*-algebra is of course semisimple so the discussion and notation of Sec. 3.2 applies. The presence of the $*$-operation allows us to perform the basic construction of Sec. 3.1. The main question to be answered in this section is: given a faithful trace tr on the finite dimensional C^*-algebra M, when does there exist a (faithful) (τ, M) trace on $\langle M, e_N \rangle$? All traces in this section will be positive, i.e. $\vec{t}$ is a positive vector in the natural ordering of $\mathbb{R}^n$.

We begin with a lemma giving more information on the basic construction in finite dimensions.

Lemma 3.3.1. *Let $N \subseteq M$ be finite dimensional C^*-algebras and let* tr *be a faithful positive normalized trace on M. Let e_N and $\langle M, e_N \rangle$ be as in Sec. 3.1. Suppose $\{p_i | i = 1, 2, \ldots, n\}$ are the minimal central projections of N. Then*

(i) *Jp_iJ are the minimal central projections of $\langle M, e_N \rangle$*
(ii) *$A_M^{\langle M, e_N \rangle} = (A_N^M)^T$ (with the obvious identification of the indices, $p_i \leftrightarrow Jp_iJ$)*
(iii) *$e_N Jp_iJ = e_N p_i$*
(iv) *$x \to e_N x Jp_iJ$ is an isomorphism from p_iN onto $(e_N Jp_iJ)\langle M, e_N \rangle (e_N Jp_iJ)$.*

Proof. (i) The p_i are the minimal central projections of N' and $\langle M, e_N \rangle = JN'J$.

(ii) This follows from 3.2.3. and (i).

(iii) If $x \in M$, $(e_N J p_i J)(x\xi) = e_N(xp_i\xi) = E_N(x)p_i\xi$, and

$$(e_N p_i)(x\xi) = e_N(p_i x\xi) = p_i E_N(x)\xi = E_N(x)p_i\xi.$$

(iv) Injectivity follows from (iii), and (iii) of 3.1.5. If $y \in p_i e_N \langle M, e_N \rangle p_i e_N$ then $y = e_N z$ for $z \in N$ and $p_i z = z$ by (iii) of 3.1.5 so $y = e_N p_i z$. □

Theorem 3.3.2. *Let M, N and* tr *be as above and let* tr *be given on M by the vector $\vec{t}$ and on N by the vector $\vec{s}$. Then there is a (τ, M) trace* Tr *on $\langle M, e_N \rangle$ iff*

(i) $A^T A \vec{t} = (1/\tau) \vec{t}$
(ii) $AA^T \vec{s} = (1/\tau) \vec{s}$

where $A = A_N^M$.

Proof. ($\Rightarrow$) Let Tr be given on $\langle M, e_N \rangle$ by $\vec{r}$, r_i being the trace of a minimal projection in $J_{P_i} J \langle M, e_N \rangle$. By (iv) of 3.3.1, such a minimal projection may be chosen of the form $e_N q$, q being a minimal projection in $p_i N$. Since Tr is a (τ, M) trace, $r_i = \tau s_i$ so $\vec{r} = \tau \vec{s}$. But by (ii) of 3.3.1, 3.2.2 and 3.2.4, $\vec{s} = AA^T \vec{r} = \tau AA^T \vec{s}$, i.e. $AA^T \vec{s} = (1/\tau) \vec{s}$. Also $\vec{t} = A^T \vec{r} = \tau A^T AA^T \vec{r} = \tau A^T A \vec{t}$.

($\Leftarrow$) Define a trace Tr on $\langle M, e_N \rangle$ by the vector $\vec{r} = \tau \vec{s}$. It is a faithful positive trace since tr is. By 3.1.6 it suffices to show that it extends tr and that it is a (τ, N) trace. For the former, by 3.2.4 we need $A^T \vec{r} = \vec{t}$. But $A \vec{t} = \vec{s}$ and by (i), $A^T \vec{s} = (1/\tau) \vec{t}$ so $A^T \vec{r} = \vec{t}$. For the latter let q be a minimal projection in $p_i N$. Then $e_N q$ is a minimal projection in $J_{P_i} J \langle M, e_N \rangle$ and by definition $\mathrm{Tr}(e_N q) = \tau\, \mathrm{tr}(q)$. The map $x \to \mathrm{Tr}(e_N x)$ is a trace on $p_i N$ so by uniqueness and linearity, $\mathrm{Tr}(e_N x) = \tau\, \mathrm{tr}(x)$ for all $x \in N$. □

3.4. *Two extensions; Goldman's theorem*

Let N be a proper von Neumann subalgebra of the finite von Neumann algebra M with faithful normal normalized trace tr. Suppose there is a faithful normal (τ, M) trace Tr on $\langle M, e_N \rangle$. Then we may form the extension $\langle \langle M, e_N \rangle, e_M \rangle$.

Proposition 3.4.1.

(i) $e_M e_N e_M = \tau e_M$
(ii) $e_N e_M e_N = \tau e_N$
(iii) $e_M \wedge e_N = e_M \wedge e_N^{\perp} = e_M^{\perp} \wedge e_N = 0.$

Proof. (i) By (i) of 3.1.4, $e_M e_N e_M = E_M(e_N) e_M$ and since Tr is a (τ, M) trace, $E_M(e_N) = \tau$.

(ii) By (ii) of 3.1.5 it suffices to verify the relation on vectors of the form $a\xi$ and $ae_N b\xi$ where $a, b \in M$. But

$$e_N e_M e_N(a\xi) = e_N(E_M(e_N a)\xi) = \tau e_N(a\xi),$$

and

$$e_N e_M e_N(ae_N b\xi) = e_N e_M(E_N(a)e_N b\xi) = e_N(\tau E_N(a)b\xi)$$

while $\tau e_N(ae_N b\xi) = \tau e_N E_N(a)b\xi$.

(iii) $e_N \wedge e_M = s - \lim_{n\to\infty}(e_N e_M e_N)^n = 0$ since $\tau < 1$.

The other relations follow from $(1-\tau) < 1$. □

Now suppose there is a faithful $(\tau, \langle M, e_N\rangle)$ trace Tr on $\langle\langle M, e_N\rangle, e_M\rangle$.

Corollary 3.4.2. *If $\tau \neq 1/2$, the von Neumann algebra generated by e_N and e_M is isomorphic to $M_2(\mathbb{C}) \oplus \mathbb{C}$. If $\tau = 1/2$ it is isomorphic to $M_2(\mathbb{C})$ and we have the relation $e_M + e_N - e_M e_N - e_N e_M = 1/2$.*

Proof. The relations of 3.4.1 show immediately that the von Neumann algebra generated by e_N and e_M has dimension at most 5 and is not abelian. But $\mathrm{Tr}(e_M \vee e_N) = 2\tau - \mathrm{Tr}(e_M \wedge e_N) = 2\tau$. This is enough to prove the affirmations about the structure of the algebra. An easy calculation shows that $p = (1-\tau)^{-1}(e_M + e_N - e_M e_N - e_N e_M)$ is a projection with $pe_N = e_N$ and $pe_M = e_M$. Thus $p = e_1 \vee e_2$ and if $\tau = 1/2$, $p = 1$. □

Corollary 3.4.3. (Goldman's theorem [14]). *Let N be a subfactor of the II_1 factor M with $[M : N] = 2$. Then M decomposes as the crossed product of N by an outer action of $\mathbb{Z}_2$.*

Proof. By 3.1.9 we know that M is of the form $\langle N, e_p\rangle$ for a subfactor of index 2 of N. Thus N is generated by N and $u = 2e_p - 1$. Moreover since tr is a $(1/2, N)$ trace on M, $\mathrm{tr}(ux) = 0$ for $x \in N$, and the relationship of 3.4.2 implies that, if $v = 2e_N - 1$, $uv = -vu$. Thus for $x \in N$, $vuxu^* = uxu^*v$. So uxu^* commutes with e_N and by (ii) of 3.1.4, $uxu^* \in N$. Thus by [25], M is the crossed product of N by a $\mathbb{Z}_2$ action which is necessarily outer since M is a factor. □

4. Possible Values of the Index

4.1. *Certain algebras generated by projections*

Chapter 4 will be largely devoted to proving the following result.

Theorem 4.1.1. *Let M be a von Neumann algebra with faithful normal normalized trace* tr. *Let $\{e_i | i = 1, 2, \ldots\}$ be projections in M satisfying*

(a) $e_ie_{i\pm 1}e_i = \tau e_i$ *for some* $\tau \leqq 1$
(b) $e_ie_j = e_je_i$ *for* $|i-j| \geqq 2$
(c) $\mathrm{tr}(we_i) = \tau\, \mathrm{tr}(w)$ *if* w *is a word on* $1, e_1, e_2, \ldots, e_{i-1}$

Then if P *denotes the von Neumann algebra generated by the* e_i's,

(i) $P \cong R$ (*the hyperfinite* II_1 *factor*)
(ii) $P_\tau = \{e_2, e_3, \ldots\}''$ *is a subfactor of* P *with* $[P : P_\tau] = \tau^{-1}$.
(iii) $\tau \leqq 1/4$ *or* $\tau = \frac{1}{4}\sec^2 \pi/n$, $n = 3, 4, \ldots$.

In this section we will prove (i) and (ii). We begin the proof with some notation.

Definition. Let $A_{m,n}$ be the $*$-algebra generated by 1, e_m, $e_{m+1}, \ldots$, e_n for $1 \leqq m \leqq n \leqq \infty$. Let A_n be $A_{1,n}$, $A_0 = \mathbb{C}$. Thus $A_\infty = A_{1,\infty}$ is the $*$-algebra generated by 1 and the e_i's.

Next some combinatorial results. If w is an (associative) word on the e_i's, call it *reduced* if it is of minimal length for the grammatical rules $e_ie_{i\pm1}e_i \leftrightarrow e_i$, $e_ie_j \leftrightarrow e_je_i$ for $|i-j| \geqq 2$, $e_i^2 \leftrightarrow e_i$.

Lemma 4.1.2. *Let* $e_{i_1}e_{i_2}\ldots e_{i_k}$ *be a reduced word. Then if* $m = \max\{i_1, i_2, \ldots, i_k\}$, m *occurs only once in the list* $i_1, i_2, \ldots, i_k$.

Proof. By induction on the length of a reduced word. It is trivial for words of length $\leqq 1$. Suppose true for words of length $\leqq n$ and let w be a reduced word of length $n+1$. Suppose $w = w_1e_mw_2e_mw_3$ where m is the maximum index and w_2 does not contain e_m. Then there are 2 possibilities.

(a) w_2 does not contain e_{m-1}. In this case e_m commutes with all the e_i's in w_2 so the length of w may be shortened using $e_m^2 \to e_m$.
(b) w_2 contains e_{m-1}. Then since w is reduced, so is w_2 and by induction $w_2 = v_1e_{m-1}v_2$ where v_1 and v_2 are words on e_1, e_2, $\ldots$, e_{m-2}. But then e_m commutes with v_1 and v_2 so that the length of w may be reduced using $e_me_{m-1}e_m \to e_m$. □

It is clear that in the algebra A, any word on the e_i's is proportional to a reduced word.

Corollary 4.1.3. (i) A_n *is finite dimensional.*

(ii) *For* $x \in A_n$, $e_{n+1}xe_{n+1} = E_{A_{n-1}}(x)e_{n+1}$ (*here* $E_{A_{n-1}} : A_n \to A_{n-1}$ *is with respect to the restriction of* tr).
(iii) $x \mapsto xe_{n+1}$ *is an isomorphism of* A_{n-1} *onto* $e_{n+1}\, A_{n+1}\, e_{n+1}$.

Proof. (i) If there are only finitely many reduced words, A_n is finite dimensional. But this follows immediately by induction from 4.1.2.

(ii) First of all $\mathrm{tr}(xe_{n+1}) = \tau\ \mathrm{tr}(x)$ for $x \in A_n$ follows immediately from 4.1.1(c) so that $E_{A_{n-1}}(e_n) = \tau$. By 4.1.2 and linearity it suffices to consider x of the form we_nw' with w and w' in A_{n-1}. But then $e_{n+1}xe_{n+1} = \tau ww'e_{n+1}$ by 4.1.1 (a) and (b), and $E_{A_{n-1}}(x)e_{n+1} = \tau ww'e_{n+1}$ by the bimodule property of $E_{A_{n-1}}$.

(iii) If w is a reduced word on $e_1, e_2, \ldots, e_{n+1}$ write $w = xe_{n+1}y$ with $x, y \in A_n$. Then by (ii), $e_{n+1}\, we_{n+1} = te_{n+1}$ for $t \in A_{n-1}$. Thus $e_{n+1}A_{n+1}e_{n+1} \subseteq A_{n-1}e_{n+1}$. To show that the map is an isomorphism, suppose $xe_{n+1} = 0$, $x \in A_{n-1}$. Then $xx^*e_{n+1} = 0$ so $\mathrm{tr}(xx^*) = 0$, i.e. $x = 0$. □

Aside 4.1.4. In fact it is possible to uniquely order reduced words by pushing $e_{\max}$ to the right as far as possible. It is easy to show that such an ordered reduced word is of the form

$$(e_{j_1}e_{j_1-1}\ldots e_{k_1})(e_{j_2}e_{j_2-1}\ldots e_{k_2})\ldots(e_{j_p}e_{j_p-1}\ldots e_{k_p})$$

where j_p is the maximum index, $j_i \geqq k_i$ and $j_{i+1} > j_i$, $k_{i+1} > k_i$. To each such word we may associate an increasing path on the integer lattice between (0,0) and $(n+1,\, n+1)$, which does not cross the diagonal. For instance $(e_3e_2e_1)(e_4e_3)(e_5e_4)$ in A_5 would correspond to the path (I owe this observation to H. Wilf):

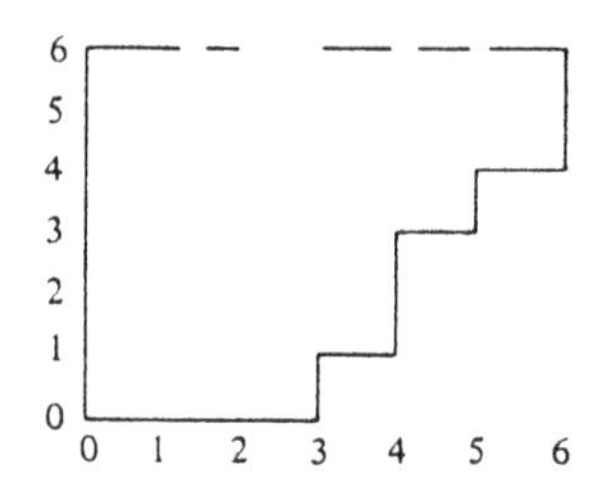

It is well known that such paths are counted by the Catalan numbers $1/(n+2)\binom{2(n+1)}{n+1}$ so we obtain $\dim\ A_n \leqq 1/(n+2)\binom{2(n+1)}{n+1}$. Uniqueness and linear independence of ordered reduced words would follow from $\dim A_n = 1/(n+2)\binom{2(n+1)}{n+1}$ which we will prove in Sec. 5.1 for $\tau \leqq 1/4$. See Aside 5.1.1.

We now want to study traces on A_∞. For this define a *totally reduced word* to be a reduced word on the e_i's where we also allow cyclic permutations. The following result is obvious.

Remark 4.1.5. Any trace on A_n is determined by its effect on totally reduced words.

Lemma 4.1.6. Any totally reduced word is of the form $w = e_{i_1}e_{i_2}\ldots e_{i_n}$ with $|i_j - i_k| \geqq 2$, $j \neq k$, and $\mathrm{tr}(w) = \tau^n$.

Proof. The last assertion is immediate from 4.1.1(c). We prove the first assertion by induction on the length of a totally reduced word. It suffices to prove that if $m = \max\{i_1, i_2, \ldots, i_n\}$ in a totally reduced word $e_{i_1}e_{i_2}\ldots e_{i_n}$, then e_{m-1} does not occur. For this note that by a cyclic permutation we may suppose the word is of the form $e_m w$ and that e_{m-1} occurs at most once in w (since a totally reduced word is reduced). But then we may proceed to $e_m w e_m$ using $e_m^2 = e_m$ and then eliminate e_{m-1} from w using (a) and (b) of 4.1.1. □

We shall now show that any normal normalized trace on P is equal to tr. For this it suffices to show that it equals tr on completely reduced words. For each subset $I \subseteq N$ with $|i-j| \geqq 2$ whenever $i,\ j \in I,\ i \neq j$, define A_I to be the algebra generated by $\{e_i | i \in I\}$.

Lemma 4.1.7. *For any finite permutation σ of I, there is a unitary $u \in A_\infty$ such that $ue_k u^* = e_{\sigma(k)}$ for all $k \in I$.*

Proof. It suffices to show that any transposition $e_i \leftrightarrow e_j (j \geqq i)$ can be effected by a unitary. We may even suppose that no k strictly between i and j is in I. If we can find a unitary $u \in A_{i,j}$ with $ue_i u^* = e_j$ and $ue_j u^* = e_i$ then this u will do since it commutes with all the other e_k's, $k \in I$. But for this it suffices to show that e_i and e_j are equivalent in $A_{i,j}$. And $e_i e_{i+1} \ldots e_j$ is a multiple of a partial isometry v with $vv^* = e_i,\ v^*v = e_j$. □

Corollary 4.1.8. *Any normal normalized trace on P is equal to* tr *on A_I''.*

Proof. Since all the e_i's are independent for tr, A_I'' may be identified with an infinite tensor product Bernoulli shift algebra. By 4.1.7 the normalizer induces the obvious action of S_∞ on A_I''. This action is well known to be ergodic. Hence any invariant measure which is absolutely continuous with respect to tr is proportional to tr. □

Corollary 4.1.9. *P is a* II_1 *factor isomorphic to R.*

Proof. By 4.1.4, 4.1.5 and 4.1.8 there is only one normal normalized trace on P. Thus P is a factor. By [21] and (a) of 4.1.3, $P \cong R$. □

Note that our proof that P is a factor follows similar lines to the scheme laid out in [27].

Corollary 4.1.10. *P_τ is a subfactor of P.*

Proof. Writing $f_i = e_{i+1}$, the f_i satisfy the same relations as the e_i. □

Lemma 4.1.11. *For each n, the map $e_i \to e_{n-i}$ extends to a* tr*-preserving $*$-automorphism σ_n of A_n.*

Proof. The map obviously extends to a $*$-automorphism of the free involutive monoid on the self-adjoint e_i. It thus suffices to know that if w is a word on $e_1, e_2, \ldots, e_n$ then $\text{tr}(\sigma_n(w)) = \text{tr}(w)$. For then if $x = \sum_i c_i w_i$ for $c_i \in \mathbb{C}$,

$$\text{tr}(xx^*) = \text{tr}\left(\sum_{i,j} c_i \bar{c}_j w_i w_j^*\right) = \text{tr}\left(\sum_{i,j} c_i \bar{c}_j \sigma(w_i)\sigma(w_j^*)\right)$$

so that $\sum_i c_i w_i \to \sum_i c_i \sigma(w_i)$ is a well defined isometry for the definite hermitian scalar product defined by tr. But the trace of w is determined by 4.1.1 (a) and (b) and the formula of 4.1.6, all of which are invariant under the interchange $e_i \leftrightarrow e_{n-1}$. □

Corollary 4.1.12.

(i) $E_{A_{2,n}}(e_1) = \tau$
(ii) *for* $x \in A_{2,n}$, $e_1 x e_1 = E_{A_{3,n}}(x) e_1$
(iii) $E_{P_\tau}(e_1) = \tau$, $e_1 x e_1 = E_{A''_{3,\infty}}(x) e_1$ *for* $x \in P_\tau$.

Proof. (i) and (ii) follow from σ_n applied to (ii) of 4.1.3, and (iii) is just the limit as $n \to \infty$ of (i) and (ii). □

Proof of (ii) of 4.1.1 (calculation of $[P : P_\tau]$). Do the basic construction to obtain $\langle P, e_{P_\tau}\rangle$. Then $e_{P_\tau} e_1 e_{P_\tau} = \tau e_{P_\tau}$ follows from (i) of 4.1.12. We further claim that $e_1 e_{P_\tau} e_1 = \tau e_1$. By (iii) of 4.1.12, elements of the form $a_0 + \sum_{i=1}^{n} a_i e_1 b_i$ with a_i, $b_i \in P_\tau$ are dense in P so it suffices to verify $e_1 e_{P_\tau} e_1 = \tau e_1$ on $x\xi$ and $x e_1 y \xi$ with $x, y \in P_\tau$. But $e_1 e_{P_\tau} e_1(x\xi) = \tau e_1 x \xi = (\tau e_1)(x\xi)$, and

$$e_1 e_{P_\tau} e_1(x e_1 y \xi) = e_1(E_{P_\tau}(e_1 E_{A''_{3,\infty}}(x) y)\xi) = \tau e_1 E_{A''_{3,\infty}}(x) y \xi = \tau e_1(x e_1 y \xi).$$

These two relations imply that e_1 and e_{P_τ} are equivalent in $\langle P, e_{P_\tau}\rangle$. Now by (iii) of 3.1.5, e_{P_τ} is a finite projection and e_1 is in a II_1 factor so that $\langle P, e_{P_\tau}\rangle$ is necessarily a finite factor, i.e. $[P : P_\tau] < \infty$. But we know that $\text{tr}(e_1) = \tau$ so $[P : P_\tau] = \text{tr}(e_{P_\tau})^{-1} = \tau^{-1}$. □

4.2. *Restrictions on τ*

We shall now prove (iii) of 4.1.1. We keep the notation of Sec. 4.1. Also define $s_n = e_1 \vee e_2 \vee \ldots \vee e_n$.

Lemma 4.2.1. *If* $1 - s_n \neq 0$ *then it is a minimal projection in* A_n *which belongs to* $Z(A_n)$.

Proof. If w is a word on $e_1, e_2, \ldots, e_n$ then

$$(1 - e_1 \vee e_2 \vee \ldots \vee e_n) w = 0.$$ □

Lemma 4.2.2. *If* $s_n \neq 1$ *then* $e_{n+1} \wedge s_n = e_{n+1}s_{n-1}$.

Proof. By 4.1.3(ii), $e_{n+1}s_ne_{n+1} = E_{A_{n-1}}(s_n)e_{n+1}$. But by 4.2.1 and the bimodule property of $E_{A_{n-1}}$, $E_{A_{n-1}}(s_n) \in Z(A_{n-1})$. Let $p_0, p_1, \ldots, p_k$ be the minimal projections in $Z(A_{n-1})$ with $p_0 = 1 - s_{n-1}$, and write $E_{A_{n-1}}(s_n) = \sum_{i=0}^{k} \lambda_i p_i$. Then since $e_{n+1} \wedge s_n \geqq e_{n+1}s_{n-1}$ and $\lim_{m\to\infty} (e_{n+1}s_ne_{n+1})^m = e_{n+1} \wedge s_n$, and by (iii) of 4.1.3, $\lambda_1 = \lambda_2 = \cdots = \lambda_k = 1$. Since $s_n \neq 1$, $E_{A_n}(s_n) \neq 1$ so by (iii) of 4.1.3, $\lambda_k < 1$. Thus $e_{n+1} \wedge s_n = e_{n+1}s_{n-1}$. □

Corollary 4.2.3. *If* $\mathrm{tr}(1 - s_k) > 0$, $\mathrm{tr}(1 - s_{k+1}) = \mathrm{tr}(1 - s_k) - \tau\, \mathrm{tr}(1 - s_{k-1})$. *And* $\mathrm{tr}(1 - s_1) = 1 - \tau$, $\mathrm{tr}(1 - s_2) = 1 - 2\tau$.

Proof. The trace tr satisfies $\mathrm{tr}(p \vee q) = \mathrm{tr}(p) + \mathrm{tr}(q) - \mathrm{tr}(p \wedge q)$. So by 4.2.2 and (c) of 4.1.1, $\mathrm{tr}(1 - s_{k+1}) = \mathrm{tr}(1 - s_k) - \tau\, \mathrm{tr}(1 - s_{k-1})$. □

For this reason we define the polynomials $P_n(x)$ by $P_0(x) = 0$, $P_1(x) = 1$ and $P_{n+1} = P_n - xP_{n-1}$, so that if $P_{k+2}(\tau) > 0\ \forall k \leqq n$ then $P_{k+2}(\tau) = \mathrm{tr}(1 - s_k)$.

Lemma 4.2.4. *Let* $\sigma = \frac{1+\sqrt{1-4x}}{2}$, $\widetilde{\sigma} = \frac{1-\sqrt{1-4x}}{2}$. *Then*

(i) $P_n(x) = \frac{\sigma^n - \widetilde{\sigma}^n}{\sigma - \widetilde{\sigma}}$

(ii) $P_n(\frac{1}{4}\sec^2\theta) = \sin n\theta / (2^{n-1}\cos^{n-1}\theta \sin\theta)$

(iii) $\deg P_n = \left[\frac{n-1}{2}\right]$.

Proof. (i) The general solution to the difference equation is $P_n = A\sigma^n + B\tilde{\sigma}^n$. The initial conditions give $A + B = 0$, $A(\sigma - \widetilde{\sigma}) = 1$.

(ii) Putting $\sigma = re^{i\theta}$, $\widetilde{\sigma} = re^{-i\theta}$, $x = \frac{1}{4}\sec^2\theta$, $r = \frac{1}{2}\sec\theta$, $\sigma^n - \widetilde{\sigma}^n = 2\,ir^n \sin n\theta, \sigma - \widetilde{\sigma} = 2\,ir\sin\theta$.

(iii) Follows easily by induction from the difference equation. □

Corollary 4.2.5. (i) *The smallest root of* P_n *is* $\frac{1}{4}\sec^2\frac{\pi}{n}$.

(ii) $P_n(\tau) > 0$ *for* $\tau < \frac{1}{4}\sec^2\frac{\pi}{n}$.

(iii) $P_{n+1}(\tau) < 0$ *for* τ *between* $\frac{1}{4}\sec^2\frac{\pi}{n+1}$ *and* $\frac{1}{4}\sec^2\frac{\pi}{n}$.

Proof. (i) By counting the number of distinct values of $\frac{1}{4}\sec^2\frac{m\pi}{n}$ (which are roots of P_n by 4.2.4) we find that all roots of P_n are real and they are the numbers $\frac{1}{4}\sec^2\frac{m\pi}{n}$ with $\frac{m\pi}{n} < \frac{\pi}{2}$. The smallest is $\frac{1}{4}\sec^2\frac{\pi}{n}$.

(ii) By induction the coefficient of $x^{[(n-1)/2]}$ in $P_n(x)$ is positive when $\left[\frac{n-1}{2}\right]$ is even and negative when $\left[\frac{n-1}{2}\right]$ is odd.

(iii) $P_{n+1}(\tau)$ must be negative between its first and second real roots, and $\sec^2 \pi/(n+1) < \sec^2 \pi/n < \sec^2 2\pi/(n+1)$. □

Proof of (iii) of 4.1.1. Suppose $\tau > 1/4$ and $\tau \neq \frac{1}{4}\sec^2 \frac{\pi}{n}$, $n = 3, 4, 5, \ldots$. Then there is a $k \geqq 3$ with $\frac{1}{4}\sec^2 \pi/(k+1) < \tau < \frac{1}{4}\sec^2 \pi/k$. But then $P_n(\tau) > 0$ for all $n \leqq k$ so $P_{n+1}(\tau) = \mathrm{tr}(1 - s_{n-1})$ by 4.2.3. But by (iii) of 4.2.5, $P_{n+1}(\tau) < 0$ which is impossible since $1 - s_{n-1}$ is a projection. □

4.3. *Values of the index*

Theorem 4.3.1. *If N is a subfactor of the II_1 factor M then either $[M : N] \geqq 4$ or $[M : N] = 4\cos^2 \pi/n$ for some $n \geqq 3$.*

Proof. If $[M : N] < \infty$, define the increasing sequence M_i, $i = 0, 1, 2, \ldots$ of II_1 factors by the relations $M_0 = M$, $M_1 = \langle M, e_n \rangle$, $M_{i+1} = \langle M_i, e_{M_{i-1}} \rangle$ for $i \geqq 1$. The inductive limit becomes a II_1 factor with faithful normal normalized trace tr (by uniqueness of the trace — see also [19]). Moreover if $\tau = [M : N]^{-1}$ and $e_i = e_{M_i}$ then the e_i satisfy the conditions of 4.1.1 by 3.4.2 and 3.1.7. By Theorem 4.1.1 either $[M : N] \geqq 4$ or $[M : N] = 4\cos^2 \pi/n$ for some $n \in \mathbb{Z}$, $n \geqq 3$. □

Theorem 4.3.2. *For each $n = 3, 4, \ldots$ there is a subfactor P_τ of R with $[R : P_\tau] = 4\cos^2 \pi/n$ and $P' \cap R = \mathbb{C}$. For each $r \geqq 4$, $r \in \mathbb{R}$, there is a subfactor P of R with $[R : P] = r$.*

Proof. The existence of subfactors with index $r \geqq 4$ was shown in 2.2.5 and the assertion about the relative commutant when $[R : P] = 4\cos^2 \pi/n$ follows from 2.2.4. Thus we only need to construct subfactors with index $4\cos^2 \pi/n$.

To do this note that the conditions of 3.3.2 for finite dimensional C^*-algebras $N \subseteq M$ together with (ii) of 3.3.1 show, by interchanging A and A^T, $\vec{s}$ and $\vec{t}$, that if there is a (τ, M) trace on $\langle M, e_N \rangle$ then there is a $(\tau, \langle M, e_N \rangle)$ trace on $\langle \langle M, e_N \rangle, e_M \rangle$ and so on. Thus we may iterate the basic construction once we have started it. Once the construction has been iterated, the inductive limit has a faithful normal normalized trace on it and the e_i's resulting from the iteration satisfy the conditions of 4.1.1 so by the result of 4.1.1 we may choose P_τ as the subfactor of R.

Thus it suffices to find $N \subseteq M$, finite dimensional C^*-algebras with inclusion matrix A and positive vectors $\vec{s}$ and $\vec{t}$ with $A^T A \vec{t} = (4\cos^2 \pi/n)\vec{t}$ and $AA^T \vec{s} = (4\cos^2 \pi/n)\vec{s}$. In fact the matrix A is enough since the subalgebra N can then be taken as the direct sum of as many copies of $\mathbb{C}$ as there are rows in the matrix.

Let A be the square $n \times n$ matrix (a_{ij}) with $a_{ij} = 1$ if $|i-j| = 1$ and 0 otherwise, e.g. $\begin{pmatrix} 0&1&0 \\ 1&0&1 \\ 0&1&0 \end{pmatrix}$. We leave it to the reader to check the linear algebra. This choice of A was suggested by F. Goodman. □

Remark 4.3.3. If we had started with any $m \times n$ non-negative integer valued matrix we could have made the construction of 4.3.2. Thus 4.1.1 implies that for such a matrix, if $\|A\| \leqq 2$ then $\|A\| = 2\cos\pi/n,\ n = 3, 4, \ldots$. This can also be proved using the well-known result of Kronecker which asserts that if z is an algebraic integer all of whose conjugates have absolute value equal to 1, then z is a root of unity.

5. The Bratteli Diagram; the Relative Commutant

5.1. *The Bratteli Diagram when* $\tau \leqq 1/4$

Let $\{e_1, e_2, \ldots\}$, M and tr be as in Sec. 4.1. Let $A_n = \{e_1, e_2, \ldots, e_n\}''$ and B_n be the algebra generated by $\{e_1, e_2, \ldots, e_n\}$ (without 1). We know that A_n and B_n are finite dimensional C^*-algebras. We want to determine the Bratteli diagram for them and the value of tr on minimal projections. In this section we suppose $\tau \leqq 1/4$. This means that

$$P_{n+2}(\tau) = \mathrm{tr}(1 - e_1 \vee e_2 \vee \ldots \vee e_n) > 0 \quad \text{for all } n \geqq 1$$

(where P_n is as in Sec. 4.2).

Let $\left\{ \begin{matrix} n \\ b \end{matrix} \right\} = \binom{n}{b} - \binom{n}{b-1}$ (ordinary binomial symbols with the convention $\binom{n}{-1} = 0$). We shall show by induction that

(a) $A_n = \bigoplus_{k=0}^{\left[\frac{n+1}{2}\right]} Q_k^n$ where $Q_k^n \cong M_{\left\{ \begin{smallmatrix} n+1 \\ k \end{smallmatrix} \right\}}(\mathbb{C})$.

(b) $Q_0^n = (1 - e_1 \vee e_2 \vee \ldots \vee e_n)\mathbb{C}$ so that

(c) $B_n = \bigoplus_{k=1}^{\left[\frac{n+1}{2}\right]} Q_k^n$

(d) The trace of a minimal projection in Q_k^n is $\tau^k P_{n+2-2k}(\tau)$ for $k = 0, 1, \ldots, \left[\frac{n+1}{2}\right]$.

(e) The inclusion matrix of A_{n-1} in A_n is

(i) When n is even

$$A = (a_{ij}) \text{ with } a_{ij} = \begin{cases} 1 \text{ if } j = i \text{ or } i+1 \\ 0 \text{ otherwise,} \end{cases}$$

$$i, j = 0, 1, \ldots, \left[\frac{n+1}{2}\right]$$

(here the indices i and j refer to the subscript of Q).

(ii) when n is odd

$$A = (a_{ij})\ a_{ij} = \begin{cases} 1 \text{ if } j = i \text{ or } i+1 \\ 0 = \ \text{otherwise} \end{cases}$$

$$i = 0,., \dots, (n+1)/2;\ j = 0, 1, \dots, (n+3)/2.$$

All this information is summed up by the following diagram (which also appears on p. 118 of [31]).

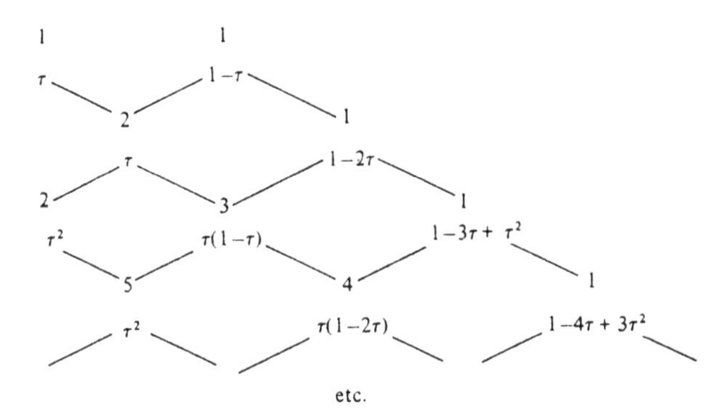

The numbers are the $\left\{ n+1 \atop k \right\}$ and the polynomials, which give the trace on minimal projections in the corresponding matrix algebra, change by multiplication by τ each step down a vertical column.

Proof. The proof will use the basic construction for the inclusion $A_{n-1} \subseteq A_n$ to obtain an almost faithful representation of A_{n+1}.

The truth of assertions $(a) \to (e)$ for $n = 2$ follows immediately from Sec. 3.4. For the inductive step we shall treat only the case $n = 2m$, the odd case being essentially identical. Suppose $(a) \to (e)$ are true for all $k \leqq n$ and apply the basic construction of Sec. 3.1 with $M = A_n$, $N = A_{n-1}$ with respect to tr. Let $E = E_{A_{n-1}}$, $e_{A_{n-1}} = e$. By induction and (ii) of 3.3.1 we know that we have the following Bratteli diagram:

A_{n-1} $\quad\cdots\quad$ $\left\{ n \atop m \right\}$ $\quad$ $\left\{ n \atop m-1 \right\}$ $\quad\cdots\quad$ $\left\{ n \atop 0 \right\}$

$\cap|$

A_n $\quad\cdots\quad$ $\left\{ n+1 \atop m \right\}$ $\quad$ $\left\{ n+1 \atop m-1 \right\}$ $\cdots$ $\left\{ n+1 \atop 1 \right\}$ $\quad$ $\left\{ n+1 \atop 0 \right\}$

$\cap|$

$\langle A_n, e\rangle$ $\quad\cdots\quad$ $\left\{ n+2 \atop m+1 \right\}$ $\quad$ $\left\{ n+2 \atop m \right\}$ $\quad\cdots\quad$ $\left\{ n+2 \atop 1 \right\}$

The numbers on the bottom line follow from 3.7.1, (ii) of 3.3.1 and the identities

$$\begin{Bmatrix} a \\ b \end{Bmatrix} + \begin{Bmatrix} a \\ b-1 \end{Bmatrix} = \begin{Bmatrix} a+1 \\ b \end{Bmatrix}, \quad \begin{Bmatrix} 2m+2 \\ m+1 \end{Bmatrix} = \begin{Bmatrix} 2m+1 \\ m \end{Bmatrix}.$$

By assertion (c), the algebra generated by $\{e_1, e_2, \ldots, e_n\}$ is the direct sum of the first m terms on the A_n line so if we define the faithful (non-normalized) trace Tr on $\langle A_n, e\rangle$ by the rule $\mathrm{Tr}(ep_k) = \tau\ \mathrm{tr}(p_k)$ where p_k is a minimal projection in Q_{k-1}^{n-1} $(k = 1, 2, \ldots, m+1)$, the identity $P_j = P_{j+1} + \tau P_{j-1}$ ensures that Tr agrees with tr on B_n. Moreover as in 3.3.2, uniqueness of the trace and linearity show that $\mathrm{Tr}(ex) = \tau\ \mathrm{tr}(x)$ for all x in A_{n-1} and hence all x in A_n. Also Tr is a faithful positive trace since all the polynomials $P_n(\tau)$ are positive.

But now consider the algebras B_{n+1} and $\langle A_n, e\rangle$. B_{n+1} is generated by B_n and e_{n+1} so that any element can be written $a_0 + \sum_{i=1}^{i} a_i e_{n+1} b_i$ with $a_0 \in B_n$, a_i, $b_i \in A_n$ for $i = 1, 2, \ldots, k$. Multiplication is defined by $e_{n+1} x e_{n+1} = E(x) e_{n+1}$ for $x \in A_n$ (see 4.1.3) and the faithful trace tr defined by $\mathrm{tr}(xe_{n+1}) = \tau\ \mathrm{tr}(x)$ for $x \in A_n$. In $\langle A_n, e\rangle$, sums of the form $a_0 + \sum_{i=1}^{j} a_i e b_i$ with $a_0 \in B_n$, a_i, $b_i \in A_n$ for $i \geqq 1$, form a 2-sided ideal. Since the central support of e is 1 ((iv) of 3.1.5), any element of $\langle A_n, e\rangle$ can be written in this form. Also $exe = E(x)e$ for $x \in A_n$ and the faithful trace Tr on $\langle A_n, e\rangle$ satisfies $\mathrm{Tr}(aeb) = \mathrm{tr}(aeb)$ for a, $b \in A_n$ and $\mathrm{Tr}(b) = \mathrm{tr}(b)$ for $b \in B_n$. Thus we may define a map from B_{n+1} to $\langle A_n, e\rangle$ by $a_0 + \sum_i a_i e_{n+1} b_i \mapsto a_0 + \sum_i a_i e b_i$ which is a surjective isometry for the definite hermitian scalar products defined by tr and Tr (and hence is well defined).

At this stage (†) we have obtained assertion (c) for $n+1$ and the values of tr on the minimal projections in $Q_1^{n+1}, Q_2^{n+1}, \ldots, Q_{m+1}^{n+1}$. But A_{n+1} is just $\{B_{n+1} \cup \{1\}\}''$ and since $\mathrm{tr}(1 - e_1 \vee \ldots \vee e_{n+1}) = P_{n+3}(\tau) > 0$, A_{n+1} is $B_{n+1} \oplus (1 - e_1 \vee e_2 \vee \ldots \vee e_{n+1})\mathbb{C}$. Moreover $x(1 - e_1 \vee e_2 \vee \ldots \vee e_{n+1}) = 0$ for any $x \in B_n$ and $(1 - e_1 \vee \ldots \vee e_{n+1})(1 - e_1 \vee \ldots \vee e_n) \neq 0$ so the Bratteli diagram for $A_n \subseteq A_{n+1}$ is forced to be

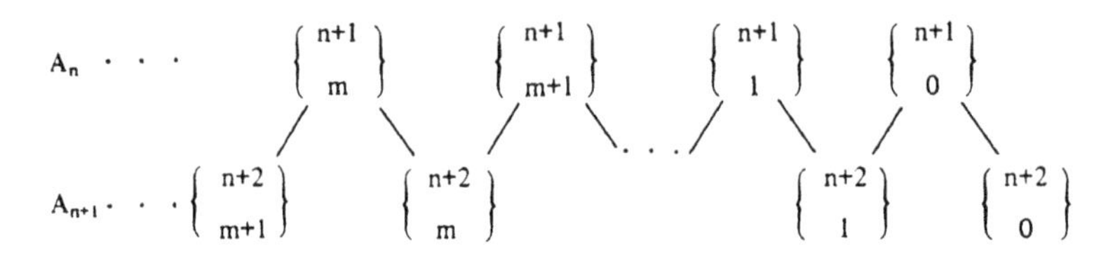

This proves assertions (a) and (e) for $n+1$, and assertion (d) follows from $\mathrm{tr}(1 - e_1 \vee \ldots \vee e_{n+1}) = P_{n+3}(\tau)$. This ends the proof. □

Aside 5.1.1. The binomial identity

$$\sum_{i=0}^{\left[\frac{n+1}{2}\right]} \binom{n+1}{i}^2 = \frac{1}{n+2}\binom{2(n+1)}{n+1}$$

follows from [13, p. 63]. This shows that $\dim A_n$ is the same as the number of ordered reduced words which are thus linearly independent. See 4.1.4.

5.2. *The Bratteli Diagrams* $\tau = \frac{1}{4}\sec^2\frac{\pi}{n}$

If $\tau = \frac{1}{4}\sec^2 \pi/(n+2)$ then $P_k(\tau) > 0$ for all $k \leqq n+1$ so the inductive argument of Sec. 5.1 goes through until the point marked (†) for assertions $(a) \to (e)$ up to step n. At this stage we find that $e_1 \vee e_2 \vee \ldots \vee e_n = 1$ so $B_n = A_n$. From this point on the basic construction for the pair $A_k \subseteq A_{k+1}$ will give an isometric (so faithful) surjective representation of A_{k+2} and the Bratteli diagram will not grow any wider.

To convince the reader of these assertions without boring him with the details, we treat the case $n = 4$. The argument of Sec. 5.1 shows that we have the following diagram

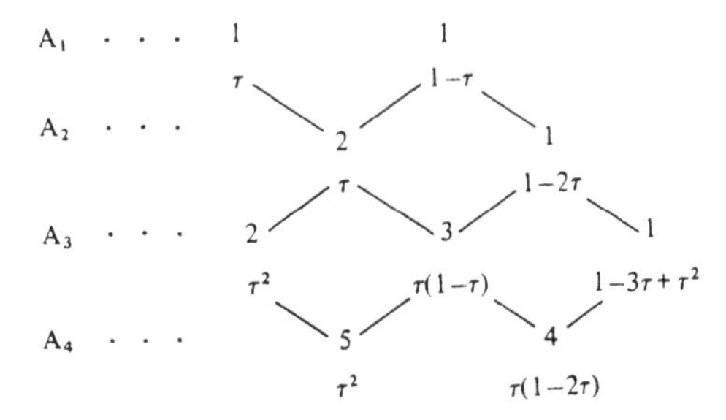

Now if $\vec{s} = \binom{\tau}{1-2\tau}$, $\vec{t} = \begin{pmatrix}\tau^2\\ \tau(1-\tau)\\ 1-3\tau+\tau^2\end{pmatrix}$, and $A = \begin{pmatrix}1&0\\1&1\\0&1\end{pmatrix}$, $\vec{s}$ is an eigenvector for A^TA with eigenvalue $3 = \tau^{-1} = 4\cos^2\frac{\pi}{6}$ and $\vec{t}$ is an eigenvector for AA^T with the same eigenvalue. They give normalized traces on A_2 and A_3 so by Sec. 3.3 the basic construction will continue to give faithful representations of the A_i by the same argument as Sec. 5.1, since $A_n = B_n$ for $n \geqq 4$.

The Bratteli diagrams will be:

For $n = 2 (\tau = 1/2)$

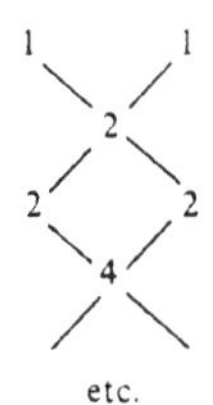

Note that this is the same as the complex Clifford algebras (see [16, p. 148]).
For $n = 3(\tau = 1/\varphi^2,\ \varphi$=golden ratio)

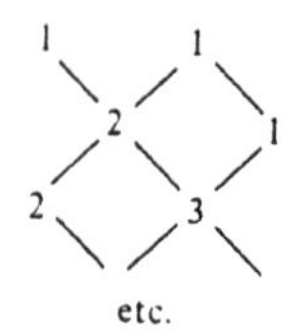

This diagram already appears in [4], [11].
For $n = 4$

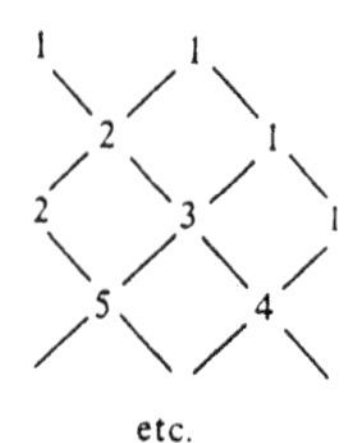

The pattern is now clear. The traces on minimal projections are the same as for the case $\tau < 1/4$ where this makes sense.

5.3. *The relative commutant when* $\tau \leqq 1/4$

One of the main motivations of this paper was to decide whether there is a continuum of values of the index realized by subfactors with trivial relative commutant (see problem 1). It was originally thought that the subfactors P_τ, $\tau \leqq 1/4$, had trivial relative commutant. In this section we shall show that this is only true when $\tau = 1/4$. The calculations were first finished by A. Wassermann to whom the author is grateful. The difficulty is the absence of an orthonormal basis which makes it impossible to exhibit an element of the relative commutant.

The case $\tau < 1/4$. Let us adopt the notation of Sec. 4.1 so R is presented as the algebra generated by the e_i's. We intend to show that $E_{R\cap P'_\tau}(e_1)$ is not a scalar, i.e. $E_{R\cap P'_\tau}(e_1) \neq \tau$. For this we will show that $\|E_{R\cap P'_\tau}(e_1) - \tau\|_1 \neq 0$ (remember $\|x\|_1 = \mathrm{tr}(|x|)$). We shall need some lemmas.

Lemma 5.3.1. $E_{R\cap P'_\tau}(x) = s - \lim\limits_{n\to\infty} E_{R\cap A'_{2,n}}(x)$ *for* $x \in R$.

Proof. Since $P_\tau = \left(\bigcup\limits_{n=1}^{\infty} A_{2,n}\right)''$, the algebras $R\cap A'_{2,n}$ are a decreasing sequence of von Neumann algebras with intersection $P'_\tau \cap R$. The rest is well known (e.g. it is true in $L^2(R, \mathrm{tr})$). $\square$

Lemma 5.3.2. $\|E_{A'_{2,n+1}\cap R}(e_1) - \tau\|_1 = \|E_{A'_n\cap A_{n+1}}(e_{n+1}) - \tau\|_1$.

Proof. Let du denote Haar measure on the unitary group of a finite dimensional C^*-algebra. Then

$$\|E_{A'_{2,n+1}\cap R}(e_1 - \tau)\|_1 = \left\| \int_{U(A_{2,n+1})} (ue_1u^* - \tau)\, du \right\|_1 .$$

Applying the isomorphism σ_{n+1} of 4.1.11 we find that this expression equals

$$\left\| \int_{U(A_n)} (ue_{n+1}u^* - \tau)\, du \right\|_1$$

which equals $\|E_{A'_n\cap A_{n+1}}(e_{n+1}) - \tau\|_1$. □

Thus to show that $P'_\tau \cap R \neq C$, it suffices to show that

$$\lim_{n\to\infty} \|E_{A'_n\cap A_{n+1}}(e_{n+1}) - \tau\|_1 \neq 0.$$

Since we know that the limit exists, it suffices to consider n odd; say $n = 2m - 1$. From Sec. 5.1 let $p_0, p_1, \ldots, p_{m-1}$ be the minimal central projections in A_{n-1} corresponding to Q_0^{n-1}, Q_1^{n-1}, $\ldots$, Q_{m-1}^{n-1}, similarly q_0, q_1, $\ldots$, q_m for A_n and r_0, r_1, $\ldots$, r_m for A_{n+1}. So $p_0 = 1 - e_1 \vee e_2 \vee \ldots \vee e_{n-1}$, $q_0 = 1 - e_1 \vee e_2 \vee \ldots \vee e_n$ and $r_0 = 1 - e_1 \vee e_2 \vee \ldots \vee e_{n+1}$. We also know from Sec. 5.1 and (iii) of 3.3.1 that

$$e_{n+1}r_i = e_{n+1}\, p_{i-1} \quad \text{for } i \geqq 1. \tag{5.3.3}$$

Since all the embeddings on the Bratteli diagram are of multiplicity at most one, the relative commutant of A_n in A_{n+1} is the abelian algebra generated by the mutually orthogonal projections $q_0r_0, q_0r_1, q_1r_0, \ldots, q_mr_m$. Thus

$$E_{A'_n\cap A_{n+1}}(e_1) = \frac{\operatorname{tr}(q_0r_0e_{n+1})}{\operatorname{tr}(q_0r_0)} q_0r_0 + \frac{\operatorname{tr}(q_0r_1e_{n+1})}{\operatorname{tr}(q_0r_1)} q_0r_0 + \cdots$$

$$+ \frac{\operatorname{tr}(q_mr_me_{n+1})}{\operatorname{tr}(q_mr_m)} q_mr_m$$

so that

$$\|E_{A'_n\cap A_{n+1}}(e_{n+1}) - \tau\|_1 = \sum_{i=0}^{m} |\operatorname{tr}(q_ir_ie_{n+1}) - \tau \operatorname{tr}(q_ir_i)|$$

$$+ \sum_{i=0}^{m-1} |\operatorname{tr}(q_ir_{i+1}e_{n+1} - \tau \operatorname{tr}(q_ir_i)|$$

$$\geqq \sum_{i=1}^{m} \tau |\operatorname{tr}(q_ip_{i-1}) - \operatorname{tr}(q_ir_i)| \quad \text{by } 5.3.3$$

$$\geqq \tau \left| \sum_{i=1}^{m} \operatorname{tr}(q_ip_{i-1}) - \operatorname{tr}(q_ir_i) \right|.$$

But from the Bratteli diagram, $q_i p_{i-1}$ is a projection of rank $\left\{ \begin{matrix} 2m-1 \\ i-1 \end{matrix} \right\}$ in the matrix algebra $Q_i^{2m-1}(= q_i A_n)$ and $q_i r_i$ is of rank $\left\{ \begin{matrix} 2m \\ i \end{matrix} \right\}$ in $Q_i^{2m}(= r_i A_{n+1})$. So the last sum may be written

$$L = \tau \left| \sum_{i=1}^{m} \left(\left\{ \begin{matrix} 2m-1 \\ i-1 \end{matrix} \right\} \tau^i \, P_{2(m-i)+1}(\tau) - \left\{ \begin{matrix} 2m \\ i \end{matrix} \right\} \tau^i \, P_{2(m-i)+2}(\tau) \right) \right| .$$

We saw in 4.2.4 that $P_n(\tau) = (\sigma^n - \widetilde{\sigma}^{\,n})/(\sigma - \widetilde{\sigma})$ where $\sigma = (1+\sqrt{1-4\tau})/2$ and $\widetilde{\sigma} = (1-\sqrt{1-4\tau})/2$. Note that for $\tau < 1/4$, $\sigma > 1/2$, $\widetilde{\sigma} < 1/2$ and $\widetilde{\sigma} + (\tau/\widetilde{\sigma}) = 1$, $\sigma + (\tau/\sigma) = 1$. Let us now calculate the relevant limits.

Lemma 5.3.6. (a) $\lim_{m\to\infty} \left(\sum_{i=1}^{m} \left\{ \begin{matrix} 2m-1 \\ i-1 \end{matrix} \right\} \widetilde{\sigma}^{\,2(m-i)+1} \tau^i \right) = 0$

(b) $\lim_{m\to\infty} \left(\sum_{i=1}^{m} \left\{ \begin{matrix} 2m \\ i \end{matrix} \right\} \widetilde{\sigma}^{\,2(m-i)+2} \tau^i \right) = 0$

(c) $\lim_{m\to\infty} \left(\sum_{i=1}^{m} \left\{ \begin{matrix} 2m-1 \\ i-1 \end{matrix} \right\} \sigma^{2(m-i)+1} \tau^i \right) = \tau(1-\tau/\sigma^2)$

(d) $\lim_{m\to\infty} \left(\sum_{i=1}^{m} \left\{ \begin{matrix} 2m \\ i \end{matrix} \right\} \sigma^{2(m-i)+2} \tau^i \right) = \sigma^2(1-\tau/\sigma^2)$.

Proof. (a) and (c). Note that $\sigma^{2(m-i)+1}\tau^i = \tau(\tau/\sigma)^{i-1}\sigma^{2m-1}$ so if $x = \sigma$ or $\widetilde{\sigma}$, in both cases we have to eveluate

$$\lim_{m\to\infty} \tau \left(\sum_{i=1}^{m} \binom{2m-1}{i-1} \left(\frac{\tau}{x}\right)^{i-1} x^{2m-1} - \frac{\tau}{x^2} \sum_{i=1}^{m} \binom{2m-1}{i-2} \left(\frac{\tau}{x}\right)^{i-2} x^{2m-i+1} \right).$$

Since $x + \tau/x = 1$, we recognize the probability of $\geqq m$ successes in $2m-1$ Bernoulli trials. If $x = \sigma$, the probability of success is $> 1/2$ so by the de Moivre–Laplace central limit theorem the limit is $\tau(1-\tau/\sigma^2)$ and if $x = \widetilde{\sigma}$, the probability of success is $< 1/2$ so the limit is 0.

(b) and (d) are proved in the same way with the substitution $x^{2(m-i)+2}\tau^i = x^2(\tau/x)^i x^{2m-i}$. □

Expanding L and using 4.3.5 we see that

$$(\sigma - \widetilde{\sigma})/L = \tau|(1-\tau/\sigma^2)(\tau - \sigma^2)| \neq 0 \quad \text{for } \tau \neq 1/4.$$

This shows that $P_\tau' \cap R \neq \mathbb{C}$ for $\tau < 1/4$.

The case $\tau = 1/4$. In this case we contend that $P'_\tau \cap R = \mathbb{C}$. Luckily we can use another model of $P_{1/4}$. Let R be realized as the closure of the Fermion algebra $\bigotimes_{i=1}^{\infty}(M_2(\mathbb{C}))_i$ with respect to the trace tr. The fixed point algebra of the obvious infinite product action of $U(2)$ is generated by the representation of S_∞ coming from interchanging the tensor product components. The transpositions between successive components may be written $2e_i - 1$ with $\mathrm{tr}(e_i) = 1/4$ and it is a matter of calculation to show that the e_i's satisfy $e_ie_{i\pm1}e_i = \frac{1}{4}e_i$, $e_ie_j = e_je_i$ for $|i-j| \geqq 2$. Thus the e_i algebra is $R^{U(2)}$ and the subfactor $P_{1/4}$ is $M_2(C)'_1 \cap R^{U(2)}$. But it is shown in [31] that $(R^{U(2)})' \cap R = \mathbb{C}$ so that $P'_{1/4} \cap R = M_2(\mathbb{C})_1$ and $M_2(\mathbb{C})_1 \cap R^{U(2)} = \mathbb{C}$. Thus $P'_{1/4} \cap R^{SU(2)} = \mathbb{C}$. For $R^{U(2)}$ see also [12], [15], [31], [2]. □

References

1. Aubert, P. L.: Théorie de Galois pour une W^*-algèbre. Comment. Math. Helv. **39**, 411–433 (1976).
2. Baker, B., Powers, R.: Product states and C^* dynamical systems of product type. Preprint, University of Pennsylvania.
3. Bourbaki, N.: Algèbre, Ch. 8. Hermann, 1959.
4. Bratteli, O.: Inductive limits of finite dimensional C^*-algebras. Trans. A.M.S. **171**, 195–234 (1972).
5. Christensen, E.: Subalgebras of a finite algebra. Math. Ann. **243**, 17–29 (1979).
6. Connes, A.: Periodic automorphisms of the hyperfinite factor of type II_1. Acta Sci. Math. **39**, 39–66 (1977).
7. Connes, A.: Classification of injective factors. Ann. Math. **104**, 73–115 (1976).
8. Connes, A.: Sur la théorie non-commutative de l'intégration. In: Algèbres d'opérateurs. Lecture Notes in Mathematics, Vol. 725. Berlin-Heidelberg-New York: Springer 1978.
9. Connes, A.: A II_1 factor with countable fundamental group. J. Operator Theory **4**, 151–153 (1980).
10. Dixmier, J.: Les algèbres d'opérateurs dans l'espace Hilbertien. Deuxieme édition. Gauthier Villars, 1969.
11. Effros, E., Shen, C.-L.: *A-F* C^*-algebras and continued fractions. Indiana Math. J. **29**, 191–204 (1980).
12. Enomoto, M., Watatani, Y.: $C^*(\mathrm{II})$ and Cuntz algebras from Young diagrams. Preprint.
13. Feller, W.: Probability theory and its applications. I. Wiley and Sons, Inc. 1957.
14. Goldman, M.: On subfactors of factors of type II_1. Mich. Math. J. **7**, 167–172 (1960).
15. Handelman, D.: Notes on ordered K-theory. Vancouver conference on operator algebras (1982).
16. Husemoller, D.: Fibre bundles. Grad. Texts in Math. Vol. 20. Berlin-Heidelberg-New York: Springer 1975.

17. Jones, V.: Sur la conjugaison des sous-facteurs des facteur de type II_1. C.R. Ac. Sci. Paris t. **286**, 597–598 (1977).
18. Jones, V.: L'indice d'un sous-facteur d'un facteur de type II, Comptes Rendus de l'Acad. Sci. Paris **294**, 391–394 (1982).
19. Kadison, R., Fugledge, B.: On a conjecture of Murray and von Neumann. Proc. Natl. Acad. Sci. U.S. **37**, 420–425 (1951).
20. Murray, F., von Neumann, J.: On rings of operators, II. Trans. A.M.S. **41**, 208–248 (1937).
21. Murray, F., von Neumann, J.: On rings of operators. IV. Ann. Math. **44**, 716–808 (1943).
22. Ocneanu, A.: Actions of discrete amenable groups on factors. To appear, Springer Lecture notes in Math. Berlin-Heidelberg-New York: Springer.
23. Popa, S.: On a problem of R. V. Kadison on maximal abelian $*$-subalgebras in factors. Invent. Math. **65**, 269–281 (1981).
24. Skau, C.: Finite subalgebras of a von Neumann algebra. J. Funct. Anal. **25**, 211–235 (1977).
25. Suzuki, N.: Crossed products of rings of operators. Tohoku Math. J. **7**, 167–172 (1960).
26. Suzuki, N.: Extensions of rings of operators in Hilbert spaces. Tohoku Math. J. **14**, 217–232 (1962).
27. Stratila, S., Voiculescu, D.: Representations of AF algebras and of the group $U(\infty)$. Springer Lecture notes in Math.
28. Takesaki, M.: Theory of operator algebras I. Berlin-Heidelberg-New York: Springer 1979.
29. Topping, D.: Lectures on von Neumann algebras. Nostrand Reinhold Math. Studies, 1969.
30. Umegaki, H.: Conditional expectation in an operator algebra I. Tohoku Math. J. **6**, 358–362 (1954).
31. Wasserman, A.: Thesis, University of Pennsylvania, 1981.

Reprinted from Proc. Int. Congr. Math., 1990

THE WORK OF SHIGEFUMI MORI

by
HEISUKE HIRONAKA
Department of Mathematics, Harvard University, Cambridge, MA 02138, USA

The most profound and exciting development in algebraic geometry during the last decade or so was the *Minimal Model Program* or *Mori's Program* in connection with the classification problems of algebraic varieties of dimension three. Shigefumi Mori initiated the program with a decisively new and powerful technique, guided the general research direction with some good collaborators along the way, and finally finished up the program by himself overcoming the last difficulty. The program was constructive and the end result was more than an existence theorem of minimal models. Even just the existence theorem by itself was the most fundamental result toward the classification of general algebraic varieties in dimension 3 up to birational transformations. The constructive nature of the program, moreover, provided a way of factoring a general birational transformation of threefolds into elementary transformations (divisorial contractions, flips and flops) that could be explicitly describable in principle. Mori's theorems on algebraic threefolds were stunning and beautiful by the totally new features unimaginable by those algebraic geometers who had been working, probably very hard too, only in the traditional world of algebraic or complex-analytic surfaces. Three in dimension was in fact a quantum jump from two in algebraic geometry.

Historically, to classify algebraic varieties has always been a fundamental problem of algebraic geometry and even an ultimate dream of algebraic geometers. During the early decades of this century, many new discoveries were made on the new features in classifying algebraic surfaces, unimaginable from the case of curves. They were mainly done by the so-called Italian school of algebraic geometers, such as Guido Castelnuovo (1865–1952), Federigo Enriques (1871–1946), Francesco Severi (1889–1961) and many others. Since then, there have seen several important modernization, precision, reconstruction with rigor, extensions, in the theory of surfaces. The most notable among those were the works by Oscar Zariski during the late 1950s (especially, Castelnuovo's criterion for rational surfaces and minimal models of surfaces) and those by Kunihiko Kodaira during the early 1960s (especially, detailed study on elliptic surfaces and complex-analytic extensions, especially non-algebraic).

In particular, Kodaira and his younger colleague S. Iitaka have produced many talented followers and collaborators. Up to and after the "Algebraic Surfaces" by I. R. Shafarevich, Yu I. Manin, B. G. Moishezon, et al. (1965), the Russian school of algebraic geometers advanced the study of some important algebraic surfaces and their deformations, especially the study of Torelli problems. As for the extensions to positive characteristics, outstanding and remarkable were the works of D. Mumford and E. Bombieri, and their followers during the late 1960s and 1970s. There have been so many other important contributors to the theory of surfaces that I would not try here to list all the important names and works in the theory of surfaces. There were also several unique works on classification problems of non-complete surfaces or complete surfaces with specified divisors, whose studies were instigated as being "2.5 dimensional" by S. Iitaka around 1977.

As for the higher dimensional algebraic varieties, the decade of 1970s saw three lines of new progress, just to name a few. One was the discovery, in the early 1970s, of the gap between the *rationality* and *unirationality*. The second was an attempt to classify *Fano 3-folds*, of which the birational classification was begun by Yu. I. Manin (1971), V. A. Iskovskih (1971, 1977–78), and followed by V. V. Shokurov (1979). However the biregular classification of Fano 3-folds in case $B_2 > 1$, with existence and number of moduli, had to wait a few more years to be achieved by Mori, jointly with S. Mukai, after a new powerful technique was invented by Mori in 1980. (The biregular classification in case $B_2 = 1$ was recently completed by S. Mukai using moduli of vector bundles on $K3$ surfaces.) The third line of progress was an early version of serious attempts toward *higher-dimensional classification* problems, largely inspired by S. Iitaka's bold conjectures proposed around 1970, and especially pressed hard during the latter half of the 1970s, by the Tokyo school of complex-algebraic geometers. The essence of their results was reported by H. Esnault in her Bourbaki talk, exp. 568 (1981).

Notably, K. Ueno produced some structure theorems in higher dimensions during 1977–79, which appeared at that time to boost the program of Iitaka et al. However, the scope of their achievements were limited toward the original goal of classifying general algebraic 3-folds in a style of extending the beautiful old theory of classifying surfaces. Their crucial limitation was the lack of a good higher-dimensional analog to Zariski's theory of minimal models. Technically, as it later became clear, the drawback was the absence of Mori's stunning success in analyzing birational transformations from the point of view of extremal contractions in dimensions higher than 2. In short, a new insight had to be injected into the classification program and it came from the technique of extremal rays inaugurated by Mori which inspired M. Reid, J. Kollár, Y. Kawamata et al. to begin working in earnest in search of minimal models around the turn of the decade.

Early in 1979, Mori brought to algebraic geometry a completely new excitement, that was his proof of *Hartshorne's conjecture*, proposed in 1970, which said that the projective spaces are the only smooth complete algebraic varieties with ample tangent bundles. It was also exciting news to differential geometers, such as Y.-T. Siu and S.-T. Yau who subsequently found an independent proof for Frankel's conjecture of 1961, which was implied by Hartshorne's. It is not clear that Hartshorne's conjecture can actually be proven by the differential-geometric method of Siu and Yau. In all approaches, the most important step was to show the existence of *rational curves* in the manifold in question. Mori's idea was simple and natural as good ones were always so, while the proof was not. The idea was that, under some numerical conditions, rational curves should be obtained by deforming a given curve inside the manifold and degenerating it into a bunch of curves of lower genera. But the difficulty was proving such plentifulness of deformations. There, Mori's ingenuity was to overcome this difficulty first in the cases of *positive characteristics*, where the Frobenius maps did a miracle, and then deduce the complex case from them.

Mori extended and reformulated his new and powerful technique of finding rational curves, which was referred to as *extremel rays* in the *cone of curves* by himself in another monumental paper of his on "Threefolds whose canonical bundles are not nef" (nef=numerically effective). The cone of curves in a projective 3-fold was defined to be the closure of the real convex hull generated by the numerical equivalence classes (or classes in the second homology group over the real numbers) of irreducible algebraic curves, and an extremal ray was to be an edge of the cone within the region where the canonical divisor takes negative intersection number. Mori's discovery was that the cone was locally finitely generated in the canonically negative region and that each extremal ray was generated by a *rational curve*. In the above "not nef" paper, moreover, Mori established an outstanding theorem which, in dimension three, completely classified the geometric structures of the family of rational curves corresponding to an extremal ray. The union of such curves was either an irreducible divisor (either a smooth projective line bundle, or a projective plane, or a quadric in a projective 3-space) or the entire 3-fold. The former led to a birational blowing down while the latter to a fibration map (either conic bundle, or Del Pezzo fiber space or Fano 3-fold). In each case, the description was precise and the target variety (blown-down birational model or the base of fibration) was again *projective*. This result was absolutely stunning to anybody who have had experiences with non-projective or even non-algebraic examples of birational transformations, and it was so beautiful as to encourage many algebraic geometers once again to look into the birational problems in dimension 3. Subsequently, Mori completed the *classification of Fano 3-folds*, jointly with S. Mukai in 1981, and he established a *criterion for uniruledness*, jointly with Y. Miyaoka in 1985. The Fano 3-folds and uniruled 3-folds were those to be investigated separately in all details and there are still left open some important problems about those special 3-folds. Excluding these, the most important general problem was to prove the *existence of minimal models* in which the canonical bundles were nef.

Mori's "not nef" paper showed, for the first time in the history of algebraic geometry, that a general birational transformation between smooth algebraic 3-folds was not completely untouchable and in fact "finitely manageable" in some sense. Naturally the dimension 3 was far more complicated than the dimension 2, and moreover some sort of singularities (seemingly finitely classifiable after Mori) were created in the process of factoring it into elementary transformations. Excited with Mori's discoveries, several algebraic geometers began clustering to him, this time with much clearer vision and better hope than ever, to work on factorization of birational transformations, and more importantly to work on defining what *minimal models* should be and how to obtain them, notably M. Reid, V. V. Shokurov, Y. Kawamata, J. Kollár and of course S. Mori himself. *Mori's Program*, named by J. Kollár and meant to imply the process of obtaining minimal models, was as follows: Start from any projective smooth variety X, which is not birationally uniruled, and find a finite succession of "elementary" birational transformations by which X is transformed to a "minimal model". Of course, the central questions here were the meanings of "elementary transformation" and "minimal model". Firstly, the inverse of a blowing-up with a smooth center in a smooth 3-fold (even in a 3-fold with a "mildly" singular point) was definitely elementary. It was called a *divisorial contraction*. This certainly decreased the Néron–Severi number and hence it should make the variety closer to its minimal model. Suppose we could not make a divisorial contraction any more and the Néron–Severi number reached its minimum. Is the variety then good enough to be called a minimal model? The answer was clearly *no*, although the singularities created by those divisorial contractions were quite acceptable. Unlike the 2-dimensional cases, the resulting 3-fold can still have *extremal rays* which cause unpleasant behaviors of the canonical and pluricanonical bundles. The importance of these bundles had been clearly recognized even from the time of Castelnuovo and Enriques in the theory of classification problems as well as in the theory of deformations. Thus, to *eliminate an extremal ray*, a completely new type of elementary transformation was needed. This was the one later called *flip*, which was, roughly speaking, to take out extremal (canonically negative, or with negative intersection number with the canonical divisor) rational curves and put some rational curves back in with a new imbedding type so that new curves are canonically positive. Such a surgery type operation is unique if it exists. The *existence*, however, turned out to be extremely delicate and hard to prove.

As early as in 1981, encouraged by Mori's "not nef" paper, Miles Reid made fairly clear the idea of what minimal models should be and what elementary transformations would be, by publishing a paper on "Minimal models of canonical 3-folds", which was an expansion of his lecture "Canonical 3-folds" in the Journées de Géométrie Algébrique d'Angers 1980. It then became absolutely clear (mildly suggested by Mori's work before) that some special kind of singularities must be permitted in the good notion of minimal models. M. Reid introduced the notions of *canonical singularity* and *terminal singularity*, which turned out to be very useful in the minimal model program of which he was the first to conjecture in literature. (See Miles Reid "Decomposition of Toric morphisms" 1983.) The former was, as the name suggests, the kind of singularity that appeared in the canonical models and could be quite horrible algebraically and geometrically. The latter was the kind that behaved much better than the former and which people hoped to be the only kind to appear in the minimal models. In any case, Reid's definition was simple and clear and some basic theorems were proved by himself, in which he showed that terminal singularities were closely related to the deformations of the classic singularities of Du Val.

Historically, P. Du Val in 1934 systematically studied the singularities that did not affect the condition of adjunction, that is, in the language of Reid, the canonical singularities in dimension 2. In 1966, M. Artin extended and modernized the classification of what he called *rational singularities*. Du Val singularities were exactly rational singularities of multiplicity 2, or *rational double points*. After M. Reid's introduction of canonical and terminal singularities, rational singularities and their deformations were studied once again. Extensive and direct studies on terminal singularities followed M. Reid's works and by 1987 a complete classification of terminal singularities in dimension 3 was achieved with technically useful lemmas by a combination of works by several mathematicians, V. I. Danilov, D. Morrison, G. Stevens, S. Mori, J. Kollár, N. Shepherd–Barron, and others. Some experimental works were done about higher dimensional terminal singularities, such as the one in dimension 4 by S. Mori, D. Morrison and I. Morrison which seemed to indicate far more complexity than the case of dimension 3.

Back to the minimal model problem, Mori's technique of *extremal ray contraction* had to be generalized to singular varieties, at least to the varieties with only terminal singularities, which were even "rationally factorial". It does not look completely hopeless that eventually Mori's all characteristics method can be modified and extended to such singular cases. (See J. Kollár work that is to appear in the proceedings of the Algebraic Geometry Satellite Conference at Tokyo Metropolitan University, 1990.) However it then was not done in Mori's way. Instead, the generalization was obtained first by generalizing Kodaira's Vanishing Theorem and then by making an ingenious use of this generalization. Here, and subsequently, the contributions of Y. Kawamata were big to the minimal model program.

The celebrated *Vanishing Theorem* was proven by K. Kodaira in 1953. Numerous generalizations, in special cases or in general, were done especially during the 1970s and 80s mostly in an effort to investigate the nicety of the structure of the canonical ring of a projective variety. Here the *canonical ring* means the graded algebra generated by sections of pluricanonical bundles (say, on a desingularized model). As for the nicety, they looked for the property of *being finitely generated* of the canonical algebra or "rationality or better" properties of the singularities of the canonical model. Y. Kawamata had a very clear objective in generalizing and using the Vanishing Theorem, that was to generalize Mori's theory of extremal ray contractions and then verify the minimal model program.

The operation called *flip* is clearly directed, i.e., it changes "canonically negative" to "canonically positive". There is a similar operation for a rational curve with zero intersection number with the canonical divisor, which is called a *flop*. Unlike flips, the inverse of flops are again flops. A flop is symmetric in this sense. A flip changes a variety into another, birational to and better than the original, better in singularities, better in terms of pluricanonical bundles, and so on. Historically, there had been known many examples of flops. The simplest flop, between smooth 3-folds, was known and used even by earlier algebraic geometers. Some flops were shown to be useful in studying degeneration of $K3$ surfaces by V. Kulikov in 1977. In contrast, examples of flips are not so easy to find because 3-folds had to be singular in order to have singularities improved. P. Francia gave an explicit example in his paper published in 1980. Other examples were seen in a paper of M. Reid published in 1983, here with prototypes of minimal models. In 1983, V. V. Shokurov published the *Non-vanishing Theorem*, which was proven by using the Vanishing Theorem. His theorem implied that flips cannot be done infinitely many times.

For Mori's Program, therefore, not only the divisorial contractions are finite but also the flips are finite in any sequence. The very final problem was hence to prove the *existence* of flips for given extremal rational curves, after all the previous works by S. Mori, M. Reid, J. Kollár, V. V. Shokurov and Y. Kawamata, just to name some of the most important contributors to the Program. At any rate, Y. Kawamata reduced the problem of the flip to the existence of a "nice" doubly anticanonical divisor *globally* in a neighborhood of a given extremal rational curve, while the existence had been only proven by M. Reid *locally* about each point of the curve. This seemingly small gap between *global* and *local* was actually enormous. Mori finally overcame this gap by checking cases after cases with very delicate and intricate investigations and established the final existence theorem of algebraic flips by reducing them to sequences of simpler analytic flips and smaller contractions. This monumental paper of Mori was published in the very first issue of the new Journal of the American Mathematical Society, establishing a constructive existence theorem of minimal models which had been shown to have many important consequences.

We need much more of Mori's originality to break the stubborn prejudice that it is a Herculean task to extend the classical classification theorems to all dimensions.

References

1. *Algebraic Manifolds with Ample Tangent Bundles*

[1-1] (with H. Sumihiro) On Hartshorne's conjecture. J. Math. Kyoto Univ. **18** (1978) 223–238.

[1-2] Projective manifolds with ample tangent bundles. Ann. Math. **110** (1979) 593–606.

[1-3] Hartshorne's conjecture and extremal ray (in Japanese). Sugaku **35**, no. 3 (1983) 193–209 [English translation: Hartshorne conjecture and extremal ray. Sugaku Expositions. Amer. Math. Soc. **1**, no. 1 (1988) 15–37].

2. *Threefolds with Non-nef Canonical Bundles*

[2-1] Threefolds whose canonical bundles are not numerically effective. Proc. Nat. Acad. Sci. USA **77** (1980) 3125–3126.

[2-2] Threefolds whose canonical bundles are not numerically effective. Ann. Math. **116** (1982) 133–176.

[2-3] Threefolds whose canonical bundles are not numerically effective. In: Algebraic Threefolds. Proc. of C. I. M. E. Session, Varenna, 1981 (A. Conte, ed.). Lecture Notes in Mathematics, vol. 947. Springer, Berlin Heidelberg New York 1982, pp. 155–189.

3. *Classification of Fano 3-folds*

[3-1] (with S. Mukai) Classification of Fano 3-folds with $B_2 \geq 2$. Manuscripta Math. **36** (1981) 147–162.

[3-2] (with S. Mukai) On Fano 3-folds with $B_2 \geq 2$. In: Algebraic Varieties and Analytic Varieties (S. Iitaka, ed.). Advanced Stud. in Pure Math., vol. 1. Kinokuniya and North-Holland, 1983, pp. 101–129.

[3-3] Cone of curves and Fano 3-folds, Proc. of ICM'82 in Warsaw 1983, pp. 127–132.

[3-4] (with S. Mukai) Classification of Fano 3-folds with $B_2 \geq 2$, I. In: Algebraic and Topological Theories — to the memory of Dr. Takehiko Miyata (M. Nagata et al., eds.). Kinokuniya, Tokyo, 1985, pp. 496–545.

4. *Characterization of Uniruledness*

[4-1] (with Y. Miyaoka) A numerical criterion of uniruledness. Ann. Math. **124** (1986) 65–69.

5. *Completion of the Minimal Model Program, or Mori's Program*

[5-1] On 3-dimensional terminal singularities. Nagoya Math. J. **98** (1985) 43–66.

[5-2] Classification of higher dimensional varieties. In: Algebraic Geometry Bowdoin 1985 (S. J. Bloch, ed.), Proc. Symp. in Pure Math., vol. 46, part 1, 1987, pp. 269–331.

[5-3] Flip theorem and the existence of minimal models for 3-folds. J. Amer. Math. Soc. **1** (1988) 117–253.

[5-4] (with C. Clemens and J. Kollár) Higher Dimensional Complex Geometry, A Summer Seminar at the Univ. of Utah, Salt Lake City, 1987, Astérisque **166**, Soc. Math. France (1988).

[5-5] (announcement of joint work with J. Kollár) Birational classification of algebraic 3-folds. In: Algebraic Analysis, Geometry, and Number Theory, Proc. of JAMI Inaugural Conf. (J. -I. Igusa, ed.), Supplement to Amer. J. Math., Johns Hopkins Univ. Press, 1989, pp. 307–311.

Shigefumi Mori

SHIGEFUMI MORI

Oct. 8, 1996

Personal Data

Birthday:	February 23, 1951
Birthplace:	Nagoya, Japan
Citizenship:	Japanese

Education:

1978 Dr. Sci., Kyoto University, Japan

Professional Experience:

1975	Assistant, Kyoto University
1988	Professor, Nagoya University
1990–present	Professor, RIMS, Kyoto University

Visiting Positions:

1977–80	Assistant Professor,	Harvard University
1985–87	Visiting Professor,	Columbia University
1991–92	Visiting Professor,	University of Utah

Academic Honors:

1990 Japan Academy Prize,
Fields Prize,
Japanese Government Prize (Person of Cultural Merits)

Reprinted from Proc. Int. Congr. Math., 1990

BIRATIONAL CLASSIFICATION OF ALGEBRAIC THREEFOLDS

by

Shigefumi Mori

Research Institute for Mathematical Sciences, Kyoto University, Kyoto 606, Japan

1. Introduction

Let us begin by explaining the background of the birational classification. We will work over the field $\mathbb{C}$ of complex numbers unless otherwise mentioned.

Let X be a non-singular projective variety of dimension r. The canonical divisor class K_X is the only divisor class (up to multiples) naturally defined on an arbitrary X. Its sheaf $\mathcal{O}_X(K_X)$ is the sheaf of holomorphic r-forms. An alternative description is $K_X = -c_1(X)$, where $c_1(X)$ is the first Chern class of X. Therefore it is natural to expect some role of K_X in the classification of algebraic varieties.

The classification of non-singular projective curves C is classical, and summarized in the following table, where $g(C)$ is the genus (the number of holes) of $C, H = \{z \in \mathbb{C} \mid \text{Im}z > 0\}$ and Γ is a subgroup of $SL_2(\mathbb{R})$:

(1.1)

g(C)	0	1	≥ 2
degK_C	-2	0	2g(C) $-$ 2
C	$\mathbb{P}^1$	$\mathbb{C}$/(lattice)	H/Γ

Here we see three different situations. For instance, everything is explicit if $g(C) = 0$; the moduli (to parametrize curves) is the main interest if $g(C) \geq 2$.

Our interest is in generalizing this to higher dimensions. The first difficulty which arises in surface case is that there are too many varieties for genuine classification (*biregular classification*).

(1.2) For a non-singular projective surface X and an arbitrary point $x \in X$, there is a birational morphism $\pi : B_xX \to X$ from a non-singular projective surface B_xX such that $E = \pi^{-1}(x)$ is isomorphic to $\mathbb{P}^1$ (E is called a (-1)-*curve*) and π induces an isomorphism $B_xX - E \simeq X - x$.

In view of (1.2), it is impractical to distinguish X from $B_xX, B_yB_xX, \ldots$ if we want a reasonable classification list. More generally, we say that two algebraic varieties X and Y are *birationally equivalent* and we write $X \sim Y$ if there is a birational mapping $X \cdots\to Y$ or equivalently if their rational function fields $\mathbb{C}(X)$ and $\mathbb{C}(Y)$ are isomorphic function fields over $\mathbb{C}$. We did not face this phenomenon in curve case, since $X \simeq Y$ iff $X \sim Y$ for curves X and Y.

In view of the list (1.1) for curves, we need to divide the varieties into several classes to formulate more precise problems. This is why the *Kodaira dimension* $\kappa(X)$ of a non-singular projective variety X was introduced by [Iitaka1] and [Moishezon].

(1.3) Let $H^0(X, \mathcal{O}(\nu K_X))$ be the space of global ν-ple holomorphic r-forms ($\nu \geq 0, r = \dim X$), and $\phi_0, \ldots, \phi_N$ be its basis. If $N \geq 0$, then

$$\Phi_{\nu K_X} : X \cdots\to \mathbb{P}^N \text{ given by } \Phi_{\nu K}(x) = (\phi_0(x) : \cdots : \phi_N(x))$$

is a rational map. We set $P_\nu(X) = N + 1$. It is important that $H^0(X, \mathcal{O}(\nu K_X))$ and $\Phi_{\nu K}$ are birational invariants, that is $X \sim Y$ induces $H^0(X, \mathcal{O}(\nu K_X)) = H^0(Y, \mathcal{O}(\nu K_Y))$ for $\nu > 0$. We set $\kappa(X) = -\infty$ if $P_\nu(X) = 0$ for all $\nu > 0$. If $P_e(X) > 0$ for some $e > 0$, then

$$\kappa(X) := \operatorname{Max}\{\dim \Phi_{\nu K_X}(X) \mid \nu > 0\}.$$

In particular, $P_\nu(X)$ and $\kappa(X)$ are birational invariants of X.

We remark that $\kappa(X) \in \{-\infty, 0, 1, \ldots, \dim X\}$, and that X with $\kappa(X) = \dim X$ is said to be *of general type*. We have the following table for curves.

(1.4)

$g(C)$	0	1	≥ 2
$\kappa(C)$	$-\infty$	0	1

To have some idea on higher dimension, we can use the easy result $\kappa(X \times Y) = \kappa(X) + \kappa(Y)$. In particular,

(1.5) case ($\kappa(X) = -\infty$) $\quad \kappa(\mathbb{P}^1 \times Y) = -\infty$,

(1.6) case ($0 < \kappa(X) < \dim X$)

$$\kappa(\underbrace{E \times \cdots \times E}_{a \text{ times}} \times \underbrace{C \times \cdots \times C}_{b \text{ times}}) = b \text{ if } g(E) = 1 \text{ and } g(C) \geq 2.$$

The case $0 < \kappa(X) < \dim X$ is studied by the Iitaka fibration.

(1.7) **Iitaka Fibering Theorem** [Iitaka2]. *Let X be a non-singular projective variety with $0 < \kappa(X) < \dim X$. Then there is a morphism $f : X' \to Y'$ of non-singular projective varieties with connected fibers such that $X' \sim X$, $\dim Y' = \kappa(X)$ and $\kappa(f^{-1}(y)) = 0$ for a sufficiently general point $y \in Y'$.*

In (1.7), we cannot expect $\kappa(Y') = \dim Y'$ or even $\kappa(Y') \geq 0$. Therefore X' is not so simple as (1.6). Nevertheless (1.7) reduces the case $0 < \kappa(X) < \dim X$ to the cases $\kappa(X) = -\infty, 0, \dim X$. Thus we can explain the birational classification as in (1.1) for higher dimensions.

2. Birational Classification

For a non-singular projective variety X, we define a graded ring (called the *canonical ring*)

$$R(X) = \oplus_{\nu \geq 0} H^0(X, \mathcal{O}(\nu K_X)).$$

If $\kappa(X) \geq 0$, the ν-canonical image $\Phi_{\nu K}(X)$ is a birational invariant of X. The existence of stable canonical image is interpreted in terms of $R(X)$ by the following easy proposition.

(2.1) **Proposition.** *Let X be a non-singular projective variety of $\kappa(X) \geq 0$. Then $\Phi_{\nu K}(X)$ for sufficiently divisible $\nu > 0$ are all naturally isomorphic iff $R(X)$ is a finitely generated $\mathbb{C}$-algebra.*

(2.2) For X of general type, constructing moduli spaces is one of our main interests. One standard way is to try to find a uniform ν such that $\Phi_{\nu K} : X \cdot\cdot \to \Phi_{\nu K}(X)$ is birational for all X and classify the image. One can expect nice properties of the image (canonical model to be explained later) if there is a stable canonical image. Therefore we would like to ask whether the canonical ring is finitely generated for X of general type (2.1).

(2.3) The case $\kappa(X) = \dim X$ suggests to reduce the birational classification of all varieties to the biregular classification of standard models (like $\Phi_{\nu K}(X)$ for sufficiently divisible ν). However, when $\kappa(X) < \dim X$, there are no obvious candidates for the standard models. For $0 \leq \kappa(X)$, we can ask to find some "standard" models.

We only say the following for $\kappa(X) \leq 0$ at this point.

(2.4) For X with $\kappa(X) = 0$, we would like to find some "standard" model $Y \sim X$ and to classify all such Y.

(2.5) For many X with $\kappa(X) = -\infty$, there exist infinitely many "standard" models $\sim X$. To study the relation among these models is a role of birational geometry. We would like to have a structure theorem of such models. One general problem is to see if all such X are *uniruled*, i.e. there exists a rational curve through an arbitrary point of X, or equivalently there is a dominating rational map $\mathbb{P}^1 \times Y \cdot\cdot \to X$ for some Y of dimension $n-1$. (It is easy to see that uniruled varieties have $\kappa = -\infty$ as in (1.5).)

Since we use the formulation by Iitaka and Moishezon, one basic problem will be the deformation invariance of κ.

(2.6) **Conjecture** [Iitaka1, Moishezon]. *Let $f : X \to Y$ be a smooth projective morphism with connected fibers and connected Y. Then $\kappa(f^{-1}(y))$ and $P_\nu(f^{-1}(y))$ $(\nu \geq 1)$ are independent of $y \in Y$.*

3. Surface Case

We review a few classical results on surfaces which may help the reader to understand the results for 3-folds.

The basic result is the inverse process of (1.2).

(3.1) **Castelnuovo-Enriques.** *Let E be a curve on a non-singular projective surface X'. Then E is a (-1)-curve (i.e. $X' = B_x X$ and E is the inverse image of x for some non-singular projective surface X and $x \in X$) iff $E \simeq \mathbb{P}^1$ and $(E \cdot K_{X'}) = -1$. We write $\mathrm{cont}_E : X' \to X$ and call it the contraction of the (-1)-curve E.*

Finding a (-1)-curve in every exceptional set, we have the following:

(3.2) **Factorization of Birational Morphisms.** *Let $f : X \to Y$ be a birational morphism of non-singular projective surfaces. Then f is a composition of a finite number of contractions of (-1)-curves.*

Starting with a non-singular projective surface X, we can keep contracting (-1)-curves if there are any. After a finite number of contractions, we get a non-singular projective surface $Y(\sim X)$ with no (-1)-curves. Depending on whether K_Y is *nef* ($(K_Y \cdot C) \geq 0$ for all curves C), $\kappa(X)$ takes different values.

(3.3) Case where K_Y is nef. Then Y is the only non-singular projective surface $\sim X$ with no (-1)-curves. To be precise, if Y' is a such surface, then the composite $Y \cdots\to X \cdots\to Y'$ is an isomorphism. This Y is called the *minimal model* of X and denoted by $X_{\min}$. In this case, we have $\kappa(X) \geq 0$.

(3.4) Case where K_Y is not nef. Then an arbitrary Y' (including Y) which is birational to X and has no (-1)-curves is isomorphic to either $\mathbb{P}^2$ or a $\mathbb{P}^1$-bundle over some non-singular curve. In this case, X has no minimal models and we have $\kappa(X) = -\infty$ by (1.5).

The above (3.3) together with (3.4) says that the birational classification of X with $\kappa \geq 0$ is equivalent to the biregular classification of minimal models.

Based on (3.3) and (3.4), the canonical model is defined.

(3.5) Let X be a non-singular projective surface of general type. Then there exists exactly one normal projective surface $Z(\sim X)$ such that Z has only Du Val (rational double) point and K_Z is ample, where Du Val points are defined by one of the following list.

$$A_n : xy + z^{n+1} = 0 \ (n \geq 0),$$
$$D_n : x^2 + y^2 z + z^{n-1} = 0 \ (n \geq 4),$$
$$E_6 : x^2 + y^3 + z^4 = 0,$$
$$E_7 : x^2 + y^3 + yz^3 = 0,$$
$$E_8 : x^2 + y^3 + z^5 = 0.$$

Such Z is called the *canonical model* of X and denoted by X_{can}. The natural map $X_{\text{min}} \cdots \to X_{\text{can}}$ is a morphism which contracts all the rational curves C with $(C \cdot K_{X_{\text{min}}}) = 0$ into Du Val points and is isomorphic elsewhere.

(3.5.1) **Remark.** This X_{can} can also be obtained as $\Phi_{\nu K}(X_{\text{min}}) = \Phi_{\nu K}(X)$ for an arbitrary $\nu \geq 5$ (Bombieri).

(3.6) Let X be a non-singular projective minimal surface with $\kappa = 0$. Thus X has torsion K_X, i.e. some non-zero multiple of it is trivial. There is a precise classification of all such X.

(3.7) The deformation invariance of $\kappa(x)$ and $P_\nu(X)$ was done by [Iitaka3] using the classification of surfaces. [Levine] gave a simple proof without using classification.

4. The Extremal Ray Theory (The Minimal Model Theory)

The first problem in generalizing the results in Sec. 3 to higher dimensions is to find some class of varieties in which there is a reasonable contraction theorem because there is no immediate generalization of (3.1) to 3-folds, since the contraction process inevitably introduces singularities [Mori1]. To define the necessary class of singularities, the first important step was taken by Reid [Reid1,3].

(4.1) **Definition** [Reid3]. Let (X, P) be a normal germ of an algebraic variety (or an analytic space) which is normal. We say that (X, P) has *terminal singularities* (resp. *canonical singularities*) iff

(i) K_X is a $\mathbb{Q}$-*Cartier* divisor, i.e. rK_X is Cartier for some positive integer r (minimal such r is called the *index* of (X, P)), and
(ii) for some (or equivalently, every) resolution $\pi : Y \to (X, P)$, we have $a_i > 0$ (resp. $a_i \geq 0$) for all i in the expression:
$$rK_Y = \pi^*(rK_Y) + \sum a_i E_i,$$
where E_i are all the exceptional divisors and $a_i \in \mathbb{Z}$.

For surfaces, a terminal (resp. canonical) singularity is smooth (resp. a Du Val point). We note that, for projective varieties X with only canonical singularities, the same definitions of $P_\nu(X)$, $\Phi_{\nu K}$ and $\kappa(X)$ work and these are still birational invariants. We can also talk about the ampleness of K_X and the intersection number $(K_X \cdot C) \in \mathbb{Q}$ for such X.

The idea of the cone of curves which is the core of the extremal ray theory was first introduced in Hironaka's thesis [Hironaka].

(4.2) **Definition.** Let X be a projective n-fold. A 1-cycle $\sum a_C C$ is a formal finite sum of irreducible curves C on X with coefficients $a_C \in \mathbb{Z}$. For a 1-cycle Z and a $\mathbb{Q}$-Cartier divisor D, the intersection number $(Z \cdot D) \in \mathbb{Q}$ is defined. Then

$$N_1(X)_{\mathbb{Z}} = \{ \text{ 1-cycles } \}/\{ \text{ 1-cycles } Z \mid (Z \cdot D) = 0 \text{ for all } D\}$$

is a free abelian group of finite rank $\rho(X) < \infty$. Thus $N_1(X) = N_1(X)_{\mathbb{Z}} \otimes_{\mathbb{Z}} \mathbb{R}$ is a finite dimensional Euclidean space. The classes $[C]$ of all the irreducible curves C span a convex cone $NE(X)$ in $N_1(X)$. Taking the closure for the metric topology, we have a closed convex cone $\overline{NE}(X)$. Then

(4.3) **Cone Theorem.** *If X has only canonical singularities, then there exist countably many half lines $R_i \subset NE(X)$ such that*

(i) $\overline{NE}(X) = \sum_i R_i + \{z \in \overline{NE}(X) \mid (z \cdot K_X) \geq 0\}$,
(ii) *for an arbitrary ample divisor H of X and arbitrary $\varepsilon > 0$, there are only finitely many R_i's contained in*

$$\{z \in \overline{NE}(X) \mid (z \cdot K_X) \leq -\varepsilon(z \cdot H)\}.$$

Such an R_i is called an *extremal ray* of X if it cannot be ommitted in (i) of (4.3). We note that an extremal ray exists on X iff K_X is not nef. Each extremal ray R_i defines a contraction of X.

(4.4) **Contraction Theorem.** *Let R be an extremal ray of a projective n-fold X with only canonical singularities. Then there exists a morphism $f : X \to Y$ to a projective variety Y (unique up to isomorphism) such that $f_*\mathcal{O}_X = \mathcal{O}_Y$ and an irreducible curve $C \subset X$ is sent to a point by f iff $[C] \in R$. Furthermore* $\mathrm{Pic}Y = \mathrm{Ker}[(C\cdot) : \mathrm{Pic}X \to \mathbb{Z}]$ *for such a contracted curve C. This f is called the contraction of R and denoted by* cont_R.

The contraction of an extremal ray is not always birational.

(4.5) *Let X be a smooth projective surface with an extremal ray R. Then* cont_R *is one of the following.*

(i) *the contraction of a (-1)-curve,*
(ii) *a $\mathbb{P}^1$-bundle structure $X \to C$ over a non-singular curve,*
(iii) *a morphism to one point, when $X \simeq \mathbb{P}^2$.*

The description of all the possible contractions for a nonsingular projective 3-fold X is given in [Mori1]. Here we only remark that $\mathrm{cont}_R X$ can have a terminal singularity $\mathbb{C}^3/ < \sigma >$ of index 2, where σ is an involution $\sigma(x, y, z) = (-x, -y, -z)$.

(4.6) The category of varieties in which we play the game of the minimal model program is the category $\mathcal{C}$ of projective varieties with only terminal singularities which are $\mathbb{Q}$ *-factorial* (i.e. every Weil divisor is $\mathbb{Q}$-Cartier). The goal of the game is to get a *minimal* (resp. *canonical*) model, i.e. a projective n-fold X with only terminal (resp. canonical) singularities such that K_X is nef (resp. ample). Let us first state the minimal model program which involves two conjectures.

(4.7) Let X be an n-fold $\in \mathcal{C}$. If K_X is nef, then X is a minimal model and we are done. Otherwise, X has an extremal ray R. Then $\mathrm{cont}_R : X \to X'$ satisfies one of the following.

(4.7.1) Case where $\dim X' < \dim X$. Then cont_R is a surjective morphism with connected fibers of dimension > 0 and relatively ample $-K_X$ (like $\mathbb{P}^1$-bundle), and X is uniruled ([Miyaoka-Mori]). This is the case where we can never get a minimal model, and we stop the game since we have the global structure of X, $\mathrm{cont}_R : X \to X'$.

(4.7.2) Case where $\mathrm{cont}_R : X \to X'$ is birational and contracts a divisor. This cont_R is called a *divisorial contraction*. In this case $X' \in \mathcal{C}$ and $\rho(X') < \rho(X)$. Therefore we can work on X' instead of X.

(4.7.3) Case where $\mathrm{cont}_R : X \to X'$ is birational and contracts no divisors. In this case, $K_{X'}$ is not $\mathbb{Q}$-Cartier and $X' \notin \mathcal{C}$. So we cannot continue the game with X'. This is the new phenomenon in dimension ≥ 3.

To get around the trouble in (4.7.3) and to continue the game, Reid proposed the following.

(4.8) **Conjecture (Existence of Flips).** *In the situation of* (4.7.3), *there is an n-fold $X^+ \in \mathcal{C}$ with a birational morphism $f^+ : X^+ \to X'$ which contracts no divisors and such that K_{X^+} is f^+-ample. The map $X \cdots \to X^+$ is called a flip.*

Since $\rho(X^+) = \rho(X)$ in (4.8), the divisorial contraction will not occur for infinitely many times. Therefore the following will guarantee that the game will be over after finitely many steps.

(4.9) **Conjecture (Termination of Flips).** *There does not exist an infinite sequence of flips $X_1 \cdots \to X_2 \cdots \to \cdots$.*

Therefore the minimal model program is completed only when the conjectures (4.8) and (4.9) are settled affirmatively.

The conjecture (4.9) was settled affirmatively by [Shokurov1] for 3-folds and by Kawamata-Matsuda-Matsuki [KMM] for 4-folds. (4.8) was first done by [Tsunoda], [Shokurov2], [Mori3] and [Kawamata6] in a special but important case. Finally (4.8) was done for 3-folds by [Mori5] using the work of [Kawamata6] mentioned above.

(4.10) Thus for 3-folds, we can operate divisorial contractions and flips for a finite number of times and get either a minimal model $\in \mathcal{C}$ or an $X \in \mathcal{C}$ which has an extremal ray R of type (4.7.1). Thus we can get 3-fold analogues of results in Sec. 3.

(4.11) For simplicity of the exposition, we did not state the results in the strongest form and we even omitted various results. Therefore we would like to mention names and give a quick review.

After the prototype of the extremal ray theory was given in [Mori1], the theory has been generalized to the relative setting with a larger class of singularities (toward the conjectures of Reid [Reid3,4]) by Kawamata, Benveniste, Reid, Shokurov and Kollár (in the historical order) and perhaps some others. First through the works of [Benveniste] and [Kawamata2], Kawamata introduced a technique [Kawamata3] which was an ingenious application of the Kawamata-Viehweg vanishing ([Kawamata1] and [Viehweg2]). Based on the works by [Shokurov1] (Non-vanishing theorem) and [Reid2] (Rationality theorem), [Kawamata4] developed the technique to prove Base point freeness theorem (and others) in arbitrary dimensions. The discreteness of the extremal rays was later done by [Kollár1]. As for this section, we refer the reader to the talk of Kawamata.

5. Applications of the Minimal Model Program (MMP) to 3-Folds

Considering MMP in relative setting, one has the factorization generalizing (3.2):

(5.1) **Theorem.** *Let $f : X \to Y$ be a birational morphism of projective 3-folds with only $\mathbb{Q}$-factorial terminal singularities. Then f is a composition of divisorial contractions and flips.*

Since minimal 3-folds have $\kappa \geq 0$ by the hard result of Miyaoka [Miyaoka1-3], one has the following (cf. (3.3) and (3.4)).

(5.2) **Thoerem.** *A 3-fold X has a minimal model iff $\kappa(X) \geq 0$.*

Unlike the surface case, the minimal model of a 3-fold X is not unique; it is unique only in codimension 1. If we are given a $\mathbb{Q}$-factorial minimal model $X_{\min}$, every other $\mathbb{Q}$-factorial minimal models of X are obtained from $X_{\min}$ by operating a simple operation called a *flop* for a finite number of times ([Kawamata6], [Kollár4]). Many important invariants computed by minimal models do not depend on the choice of the minimal model. We refer the reader to the talk of Kollár.

(5.3) **Theorem.** *For a 3-fold X, the following are equivalent.*

(i) $\kappa(X) = -\infty$,
(ii) *X is uniruled,*
(iii) *X is birational to a projective 3-fold Y with only $\mathbb{Q}$-factorial terminal singularities which has an extremal ray of type* (4.7.1).

It will be an important but difficult problem to classify all the possible Y in (iii) of (5.3). There are only finitely many families of such Y with $\rho(Y) = 1$ ([Kawamata7]).

Since a canonical model exists if a minimal model does ([Benveniste] and [Kawamata2]), one has the following (cf. (3.5)).

(5.4) **Theorem.** *If X is a 3-fold of general type, then X has a canonical model and the canonical ring $R(X)$ is a finitely generated $\mathbb{C}$-algebra.*

The argument for (5.4) can be considered as a generalization of the argument for (3.5.1). However the effective part "$\nu \geq 5$" of (3.5.1) has not yet been generalized to dimension ≥ 3.

To study varieties X with $\kappa \geq 0$, [Kawamata4] posed the following.

(5.5) **Conjecture (Abundance Conjecture).** *If X is a minimal variety, then rK_X is base point free for some $r > 0$.*

For 3-folds, there are works by [Kawamata4] and [Miyaoka4] (cf. [KMM]). However the torsionness of K for minimal 3-folds with $\kappa = 0$ is the unsolved, and it remains to prove:

(5.6) **Problem.** *Let X be a minimal 3-fold with H an ample divisor such that $(K_X^3) = 0$ and $(K_X^2 \cdot H) > 0$. Then prove that $\kappa(X) = 2$.*

(5.7) **Remark** ($\kappa = 0$). The 3-folds X with $\kappa(X) = 0$ and $H^1(X, \mathcal{O}_X) \neq 0$ were classified by [Viehweg1] and (5.5) holds for these. This was based on Viehweg's solution of the addition conjecture for 3-folds, and we refer the reader to [Iitaka4]. However not much is known about the 3-folds X with $\kappa(X) = 0$ (or even K_X torsion) and $H^1(X, \mathcal{O}_X) = 0$: so far many examples have been constructed and it is not known if there are only finitely many families. There is a conjecture of [Reid6] in this direction.

By studying the flips more closely, [Kollár-Mori] proved the deformation invariance of κ and P_ν (cf. (3.7)):

(5.8) **Theorem.** *Let $f : X \to \Delta$ (unit disk) be a projective morphism whose fibers are connected 3-folds with only $\mathbb{Q}$-factorial terminal singularities. Then*

(i) *$\kappa(X_t)$ is independent of $t \in \Delta$, where $X_t = f^{-1}(t)$,*
(ii) *$P_\nu(X_t)$ is independent of $t \in \Delta$ for all $\nu \geq 0$ if $\kappa(X_0) \neq 0$.*

Indeed for such a family X/Δ, the simultaneous minimal model program is proved and the (modified) work of [Levine] is used to prove (5.8). We cannot drop the condition "$\kappa(X_0) \neq 0$" at present since the abundance conjecture is not completely solved for 3-folds.

As for other applications (e.g. addition conjecture, deformation space of quotient surface singularities, birational moduli), we refer the reader to [KMM] and [Kollár6].

6. Comments on the Proofs for 3-Folds

Many results on 3-folds are proved by using only the formal definitions of terminal singularities. However some results on 3-folds rely on the classification of 3-fold terminal singularities [Reid3], [Danilov], [Morrison-Stevens], [Mori2] and [KSB] (cf. Reid's survey [Reid5] and [Stevens].) The existence of flips and flops rely heavily on it. Thus generalizing their proofs to higher dimension seems hopeless. At present, there is no evidence for the existence of flips in higher dimensions except that they fit in the MMP beautifully. I myself would accept them as working hypotheses. A more practical problem will be to complete the log-version of the minimal model program for 3-folds [KMM]. This is related to the birational classification of open 3-folds and n-folds with $\kappa = 3$. Since log-terminal singularities have no explicit classification, this might be a good place to get some idea on higher dimension. Shokurov made some progress in this direction [Shokurov3].

There are two other results relying on the classification.

(6.1) **Theorem** [Mori4]. *Every* 3*-dimensional termal singularity deforms to a finite sum of cyclic quotient terminal singularities* (*i.e. points of the form* $\mathbb{C}^3/\mathbb{Z}_r(1,-1,a)$ *for some relatively prime positive integers* a *and* r).

This was used in the Barlow-Fletcher-Reid plurigenus formula for 3-folds [Fletcher] and [Reid5] (cf. also [Kawamata5]). Given a 3-fold X with only terminal singularities, each singularity of X can be deformed to a sum of cyclic quotient singularities $\mathbb{C}^3/\mathbb{Z}_r(1,-1,a)$. Let $S(X)$ be the set of all such (counted with multiplicity). For each $P = \mathbb{C}^3/\mathbb{Z}_r(1,-1,a) \in S(X)$, we let

$$\phi_P(m) = (m - \{m\}_r)\frac{r^2-1}{12r} + \sum_{j=0}^{\{m\}_r-1} \frac{\{aj\}_r(r - \{aj\}_r)}{2r},$$

where $\{m\}_r$ is the integer $s \in [0, r-1]$ such that $s \equiv m (\text{mod } r)$. For a line bundle L on X, let $\chi(L) = \sum_j (-1)^j \dim H^j(X, L)$ and let $c_2(X)$ be the second Chern class of X, which is well-defined since X has only isolated singularities. Then the formula is stated as the following.

(6.2) **The Barlow-Fletcher-Reid Plurigenus Formula.**

$$\chi(\mathcal{O}_X(mK_X)) = \frac{m(m-1)(2m-1)}{12}(K_X^3) + (1-2m)\chi(\mathcal{O}_X) + \sum_{P \in S(X)} \phi_P(m),$$

$$\chi(\mathcal{O}_X) = -\frac{1}{24}(K_X \cdot c_2(X)) + \sum_{P \in S(X)} \frac{r^2-1}{24r}.$$

This is important for effective results on 3-folds (cf. Sec. 7).

(6.3) **Theorem** ([KSB]). *A small deformation of a* 3*-dimensional terminal singularity is terminal.*

This is indispensable in the construction of birational moduli. An open problem in this direction is

(6.4) **Problem.** *Is every small deformation of a* 3*-dimensional canonical singularity canonical?*

Since this remains unsolved, we cannot put an algebraic structure on

$$\{\text{canonical 3-folds}\}/\text{isomorphisms}.$$

7. Related Results

I would like to list some of the directions, which I could not mention in the previous sections. This is by no means exhaustive. For instance, I could not mention the birational automorphism groups (cf. [Iskovskih] for the works before 1983) due to my lack of knowledge.

(7.1) *Effective Classification.* The Kodaira dimension κ is not a simple invariant. For instance, we know that $\kappa(X) = -\infty$ iff $P_\nu(X) = 0(\forall \nu > 0)$. Therefore $P_{12}(X) = 0$ was an effective criterion for a surface X to be ruled, while $\kappa(X) = -\infty$ was not. The 3-dimensional analogue is not known yet.

There are results by Kollár [Kollár2] in the case dim $H^1(X, \mathcal{O}_X) \geq 3$ (cf. [Mori4]). The Barlow-Fletcher-Reid plurigenus formula (6.2) is applied for instance to get $aK_X \sim 0$ with some effectively given $a > 0$ for 3-folds X with numerically trivial K_X by [Kawamata5] and [Morrison], and to get $P_{12}(X) > 0$ for canonical 3-folds X with $\chi(\mathcal{O}_X) \leq 1$ by [Fletcher].

(7.2) *Differential Geometry.* As shown by [Yau], there are differential geometric results (especially when K is positive) which seem out of reach of algebraic geometry. Therefore we welcome differential geometric approaches. In this direction is Tsuji's construction of Kähler-Einstein metrics on canonical 3-folds [Tsuji].

(7.3) *Characteristic p.* [Kollár5] generalized [Mori1] (extremal rays of smooth projective 3-folds over $\mathbb{C}$) to char p. This suggests the possibility of little use of vanishing theorems in MMP for 3-folds. A goal will be the MMP for 3-folds in char p. However even the classification of terminal singularities is open.

(7.4) *Mixed Characteristic Case.* One can ask about the extremal rays (and so on) for arithmetic 3-folds X/S. The methods of [Shokurov2] and [Tsunoda] might work, if X/S is semistable. In the general case, I do not know any results in this direction.

(7.5) *Analytic or Non-projective* 3*-Folds.* Studying analytic or non-projective 3-folds will require a substitute for the cone of curves modulo numerical equivalence. However analytic or non-projective minimal 3-folds can be handled by the flop [Kollár4]. There is a work of [Kollár6].

References

[Benveniste] X. Benveniste: Sur l'anneau canonique de certaines variétés de dimension 3. Invent. Math. **73** (1983) 157–164.

[Danilov] V. I. Danilov: Birational geometry of toric 3-folds. Math. USSR Izv. **21** (1983) 269–279.

[Fletcher] A. Fletcher: Contributions to Riemann-Roch on projective 3-folds with only canonical singularities and applications, In: Algebraic Geometry, Bowdoin 1985. Proc. Symp. Pure Math. **46** (1987) 221–232.

[Hironaka] H. Hironaka: On the theory of birational blowing-up. Thesis, Harvard University 1960.

[Iitaka1] S. Iitaka: Genera and classification of algebraic varieties. 1 (in Japanese). Sugaku **24** (1972) 14–27.

[Iitaka2] S. Iitaka: On D-dimensions of algebraic varieties. J. Math. Soc. Japan **23** (1971) 356–373.

[Iitaka3] S. Iitaka: Deformations of complex surfaces. II J. Math. Soc. Japan **22** (1971) 274–261.

[Iitaka4] S. Iitaka: Birational geometry of algebraic varieties. Proc. ICM83, Warsaw 1984, pp 727–732.

[Iskovskih] V. A. Iskovskih: Algebraic threefolds with special regard to the problem of rationality. Proc. ICM83, Warsaw 1984, pp 733–746.

[Kawamata1] Y. Kawamata: A generalisation of Kodaira-Ramanujam's vanishing theorem. Math. Ann. **261** (1982) 43–46.

[Kawamata2] Y. Kawamata: On the finiteness of generators of the pluri-canonical ring for a threefold of general type. Amer. J. Math. **106** (1984) 1503–1512.

[Kawamata3] Y. Kawamata: Elementary contractions of algebraic 3-folds. Ann. Math. **119** (1984) 95–110.

[Kawamata4] Y. Kawamata: The cone of curves of algebraic varieties. Ann. Math. **119** (1984) 603–633.

[Kawamata5] Y. Kawamata: On the plurigenera of minimal algebraic 3-folds with $K \approx 0$. Math. Ann. **275** (1986) 539–546.

[Kawamata6] Y. Kawamata: The crepant blowing-up of 3-dimensional canonical singularities and its applications to the degeneration of surfaces. Ann. Math. **127** (1988) 93–163.

[Kawamata7] Y. Kawamata: Boundedness of $\mathbb{Q}$-Fano threefolds. Preprint (1990).

[KMM] Y. Kawamata, K. Matsuda, K. Matsuki: Introduction to the minimal model problem. In: Algebraic Geometry, Sendai 1985. Adv. Stud. Pure Math. **10** (1987) 283–360.

[Kollár1] J. Kollár: The cone theorem. Ann. Math. **120** (1984) 1–5.

[Kollár2] J. Kollár: Higher direct images of dualizing sheaves. Ann. Math. (2) **123** (1986) 11–42.

[Kollár3] J. Kollár: Higher direct images of dualizing sheaves. II. Ann. Math. (2) **124** (1986) 171–202.

[Kollár4] J. Kollár: Flops. Nagoya Math. J. **113** (1989) 14–36.

[Kollár5] J. Kollár: Extremal rays on smooth treefolds. Ann. Sci. ENS (to appear).

[Kollár6] J. Kollár: Flips, flops, minimal models etc. Proc. Diff. Geom. Symposium at Harvard, May 1990.

[Kollár-Mori] J. Kollár, S. Mori: To be written up. (1991).

[KSB] J. Kollár, N. Shepherd–Barron: Threefolds and deformations of surface singularities. Invent. Math. **91** (1988) 299–338.

[Levine] M. Levine: Pluri-canonical divisors on Kähler manifolds. Invent. Math. **74** (1983) 293–303.

[Miyaoka1] Y. Miyaoka: Deformations of a morphism along a foliation. In: Algebraic Geometry, Bowdoin 1985. Proc. Symp. Pure Math. **46** (1987) 245–268.

[Miyaoka2] Y. Miyaoka: The Chern classes and Kodaira dimension of a minimal variety. In: Algebraic Geometry, Sendai 1985. Adv. Stud. Pure Math. **10** (1987) 449–476.

[Miyaoka3] Y. Miyaoka: On the Kodaira dimension of minimal threefolds. Math. Ann. **281** (1988) 325–332.

[Miyaoka4] Y. Miyaoka: Abundance conjecture for threefolds: $\nu = 1$ case. Comp. Math. **68** (1988) 203–220.

[Miyaoka-Mori] Y. Miyaoka, S. Mori: A numerical criterion for uniruledness. Ann. Math. **124** (1986) 65–69.

[Moishezon] B. G. Moishezon: Algebraic varieties and compact complex spaces. ICM70, Nice 1971, pp. 643–648.

[Mori1] S. Mori: Threefolds whose canonical bundles are not numerically effective. Ann. Math. **116** (1982) 133–176.

[Mori2] S. Mori: On 3-dimensional terminal singularities. Nagoya Math. J. **98** (1985) 43–66.

[Mori3] S. Mori: Minimal models for semistable degenerations of surfaces. Lectures at Columbia University 1985, unpublished.

[Mori4] S. Mori: Classification of higher-dimensional varieties. In: Algebraic Geometry, Bowdoin 1985. Proc. Symp. Pure Math. **46** (1987) 269–332.

[Mori5] S. Mori: Flip theorem and the existence of minimal models for 3-folds. J. Amer. Math. Soc. **1** (1988) 117–253.

[Morrison] D. Morrison: A Remark on Kawamata's paper "On the plurigenera of minimal algebraic 3-folds with $K \approx 0$". Math. Ann. **275** (1986) 547–553.

[Morrison-Stevens] D. Morrison, G. Stevens: Terminal quotient singularities in dimensions three and four. Proc. AMS **90** (1984) 15–20.

[Nakamura-Ueno] I. Nakamura, K. Ueno: An addition formula for Kodaira dimensions of analytic fibre bundles whose fibres are Moishezon manifolds. J. Math. Soc. Japan **25** (1973) 363–371.

[Reid1] M. Reid: Canonical threefolds. In: Géometrie Algébrique Angers. Sijthoff & Noordhoff (1980) 273–310.

[Reid2] M. Reid: Projective morphisms according to Kawamata. Preprint, Univ. of Warwick (1983).

[Reid3] M. Reid: Minimal models of canonical threefolds. In: Algebraic and Analytic Varieties. Adv. Stud. Pure Math. **1** (1983) 131–180.

[Reid4] M. Reid: Decomposition of toric morphisms. Progress in Math. **36** (1983) 395–418.

[Reid5] M. Reid: Young person's guide to canonical singularities. In: Algebraic Geometry, Bowdoin 1985. Proc. Symp. Pure Math. **46** (1987) 345–416.

[Reid6] M. Reid: The moduli space of 3-folds with $K = 0$ may nevertheless be irreducible. Math. Ann. **278** (1987) 329–334.

[Shokurov1] V. Shokurov: The nonvanishing theorem. Izv. Akad. Nauk SSSR Ser. Mat. **49** (1985) 635–651.

[Shokurov2] V. Shokurov: Letter to M. Reid, 1985.

[Shokurov3] V. Shokurov: Special 3-dimensional flips. Preprint, MPI 1989.

[Stevens] J. Stevens: On canonical singularities as total spaces of deformations. Abh. Math. Sem. Univ. Hamburg **58** (1988) 275–283.

[Tsuji] H. Tsuji: Existence and degeneration of Kähler-Einstein metric for minimal algebraic varieties of general type. Math. Ann. **281** (1988) 123–133.

[Tsunoda] S. Tsunoda: Degenerations of Surfaces. In: Algebraic Geometry, Sendai 1985. Adv. Stud. Pure Math. **10** (1987) 755–764.

[Viehweg1] E. Viehweg: Klassifikationstheorie algebraischer Varietäten der Dimension drei. Comp. Math. **41** (1981) 361–400.

[Viehweg2] E. Viehweg: Vanishing theorems. J. Reine Angew. Math. **335** (1982) 1–8.

[Yau] S. -T. Yau: On the Ricci curvature of a compact Kähler manifold and the complex Monge-Ampère equation I. Comm. Pure Appl. Math. **31** (1978) 339–411.

Reprinted form Proc. Int. Congr. Math., 1990

THE WORK OF E. WITTEN

by

LUDWIG D. FADDEEV

Steklov Mathematical Institute, Leningrad 191011, USSR

It is a duty of the chairman of the Fields Medal Committee to appoint the speakers, who describe the work of the winners at this session. Professor M. Atiyah was asked by me to speak about Witten. He told me that he would not be able to come but was ready to prepare a written address. So it was decided that I shall make an exposition of his address adding my own comments. The full text of Atiyah's address is published separately.

Let me begin by the statement that Witten's award is in the field of Mathematical Physics.

Physics was always a source of stimulus and inspiration for Mathematics so that Mathematical Physics is a legitimate part of Mathematics. In classical time its connection with Pure Mathematics was mostly via Analysis, in particular through Partial Differential Equations. However quantum era gradually brought a new life. Now Algebra, Geometry and Topology, Complex Analysis and Algebraic Geometry enter naturally into Mathematical Physics and get new insights from it. And[1]

In all this large and exciting field, which involves many of the leading physicists and mathematicians in the world, Edward Witten stands out clearly as the most influential and dominating figure. Although he is definitely a physicist (as his list of publications clearly shows) his command of mathematics is rivalled by few mathematicians, and his ability to interpret physical ideas in mathematical form is quite unique. Time and again he has surprised the mathematical community by a brilliant application of physical insight leading to new and deep mathematical theorems.

Now I come to description of the main achievements of Witten. In Atiyah's text many references are given to Feynman Integral, so that I begin with a short and rather schematic reminding of this object.

[1]Small print type here and after refers to Atiyah's text.

In quantum physics the exact answers for dynamical problems can be expressed in a formal way as follows:

$$Z = \int e^{iA} \prod_x d\mu$$

where A is an action functional of local fields — functions of time and space variables x, running through some manifold M. The integration measure is a product of local measures for values of fields in a point x over all M. The result of integration Z could be a number or function of parameters defining the problem — coupling constants, boundary or asymptotical conditions, etc.

In spite of being an ill-defined object from the point of view of rigorous mathematics, Feynman functional integral proved to be a powerful tool in quantum physics. It was gradually realized that it is also a very convenient mathematical means. Indeed the geometrical objects such as loops, connections, metrics are natural candidates for local fields and geometry produces for them interesting action functionals. The Feynman integral then leads to important geometrical or topological invariants.

Although this point of view was expressed and exemplified by several people (e.g., A. Schvarz used 1-forms ω on a three-dimensional manifold with action $A = \int \omega d\omega$ to describe the Ray–Singer torsion) it was Witten who elaborated this idea to a full extent and showed the flexibility and universality of Feynman integral.

Now let me follow Atiyah in description of the main achievements of Witten in this direction.

1. Morse Theory

His paper [2] on supersymmetry and Morse theory is obligatory reading for geometers interested in understanding modern quantum field theory. It also contains a brilliant proof of the classic Morse inequalities, relating critical points to homology. The main point is that homology is defined via Hodge's harmonic forms and critical points enter via stationary phase approximation to quantum mechanics. Witten explains that "supersymmetric quantum mechanics" is just Hodge–de Rham theory. The real aim of the paper is however to prepare the ground for supersymmetric quantum field theory as the Hodge–de Rham theory of infinite-dimensional manifolds. It is a measure of Witten's mastery of the field that he has been able to make intelligent and skilful use of this difficult point of view in much of his subsequent work.

Even the purely classical part of his paper has been very influential and has led to new results in parallel fields, such as complex analysis and number theory.

2. Index Theorem

One of Witten's best known ideas is that the index theorem for the Dirac operator on compact manifolds should emerge by a formally exact functional integral on the loop space. This idea (very much in the spirit of his Morse theory paper) stimulated an extensive development by Alvarez–Gaumé, Getzler, Bismut and others which amply justified Witten's view-point.

3. Rigidity Theorems

Witten [7] produced an infinite sequence of such equations which arise naturally in the physics of string theories, for which the Feynman path integral provides a heuristic explanation of rigidity. As usual Witten's work, which was very precise and detailed in its formal aspects, stimulated great activity in this area, culminating in rigorous proofs of these new rigidity theorems by Bott and Taubes [1]. A noteworthy aspect of these proofs is that they involve elliptic function theory and deal with the infinite sequence of operators simultaneously rather than term by term. This is entirely natural from Witten's view-point, based on the Feynman integral.

4. Knots

Witten has shown that the Jones invariants of knots can be interpreted as Feynman integrals for a 3-dimensional gauge theory [11]. As Lagrangian, Witten uses the Chern–Simons function, which is well-known in this subject but had previously been used as an addition to the standard Yang–Mills Lagrangian. Witten's theory is a major breakthrough, since it is the only intrinsically 3-dimensional interpretation of the Jones invariants: all previous definitions employ a presentation of a knot by a plane diagram or by a braid.

Although the Feynman integral is at present only a heuristic tool it does lead, in this case, to a rigorous development from the Hamiltonian point of view. Moreover, Witten's approach immediately shows how to extend the Jones theory from knots in the 3-sphere to knots in arbitrary 3-manifolds. This generalization (which includes as a specially interesting case the empty knot) had previously eluded all other efforts, and Witten's formulas have now been taken as a basis for a rigorous algorithmic definition, on general 3–manifolds, by Reshetikin and Turaev.

Now I turn to another beautiful result of Witten — proof of positivity of energy in Einstein's Theory of Gravitation.

Hamiltonian approach to this theory proposed by Dirac in the beginning of the fifties and developed further by many people has led to a natural definition of energy. In this approach a metric γ and external curvature h on a space-like initial surface $S^{(3)}$ embedded in space-time $M^{(4)}$ are used as parameters in the corresponding phase space. These data are not independent. They satisfy Gauss–Codazzi constraints — highly nonlinear PDE. The energy H in the asymptotically flat case is given as an integral of indefinite quadratic form of $\nabla\gamma$ and h. Thus it is not manifestly positive. The important statement that it is nevertheless positive may be proved only by taking into account the constraints — a formidable problem solved by Yau and Schoen in the late seventies and as Atiyah mentions, "leading in part to Yau's Fields Medal at the Warsaw Congress".

Witten proposed an alternative expression for energy in terms of solution of a linear PDE with the coefficients expressed through γ and h. This equation is

$$\mathcal{D}^{(3)}\psi = 0$$

where $\mathcal{D}^{(3)}$ is the Dirac operator induced on $S^{(3)}$ by the full Dirac operator on $M^{(4)}$. Witten's formula somewhat schematically can be written as follows:

$$H(\psi_0, \psi_0) = \int (|\nabla\psi|^2 + \psi^* G\psi) dS$$

where ψ_0 is the asymptotic boundary value for ψ and G is proportional to the Einstein tensor $R_{ik} - \frac{1}{2}g_{ik}R$. Due to the equation of motion $G = T$, where T is the energy-momentum tensor of matter and thus manifestly positive. So the positivity of H follows.

This unexpected and simple proof shows another ability of Witten — to solve a concrete difficult problem by specific elegant means.

Reprinted from Proc. Int. Congr. Math., 1990

THE WORK OF EDWARD WITTEN

by

MICHAEL F. ATIYAH

Trinity College, Cambridge CB2 ITQ, England

1. General

The past decade has seen a remarkable renaissance in the interaction between mathematics and physics. This has been mainly due to the increasingly sophisticated mathematical models employed by elementary particle physicists, and the consequent need to use the appropriate mathematical machinery. In particular, because of the strongly non-linear nature of the theories involved, topological ideas and methods have played a prominent part.

The mathematical community has benefited from this interaction in two ways. First, and more conventionally, mathematicians have been spurred into learning some of the relevant physics and collaborating with colleagues in theoretical physics. Second, and more surprisingly, many of the ideas emanating from physics have led to significant new insights in purely mathematical problems, and remarkable discoveries have been made in consequence. The main input from physics has come from quantum field theory. While the analytical foundations of quantum field theory have been intensively studied by mathematicians for many years the new stimulus has involved the more formal (algebraic, geometric, topological) aspects.

In all this large and exciting field, which involves many of the leading physicists and mathematicians in the world, Edward Witten stands out clearly as the most influential and dominating figure. Although he is definitely a physicist (as his list of publications clearly shows) his command of mathematics is rivalled by few mathematicians, and his ability to interpret physical ideas in mathematical form is quite unique. Time and again he has surprised the mathematical community by a brilliant application of physical insight leading to new and deep mathematical theorems.

Witten's output is remarkable both for its quantity and quality. His list of over 120 publications indicates the scope of his research and it should be noted that many of these papers are substantial works indeed.

In what follows I shall ignore the bulk of his publications, which deal with specifically physical topics. This will give a very one-sided view of his contribution, but it is the side which is relevant for the Fields Medal. Witten's standing as a physicist is for others to assess.

Let me begin by trying to describe some of Witten's more influential ideas and papers before moving on to describe three specific mathematical achievements.

2. Influential Papers

His paper [2] on supersymmetry and Morse theory is obligatory reading for geometers interested in understanding modern quantum field theory. It also contains a brilliant proof of the classic Morse inequalities, relating critical points to homology. The main point is that homology is defined via Hodge's harmonic forms and critical points enter via stationary phase approximation to quantum mechanics. Witten explains that "supersymmetric quantum mechanics" is just Hodge–de Rham theory. The real aim of the paper is however to prepare the ground for supersymmetric quantum field theory as the Hodge–de Rham theory of infinite-dimensional manifolds. It is a measure of Witten's mastery of the field that he has been able to make intelligent and skilful use of this difficult point of view in much of his subsequent work.

Even the purely classical part of this paper has been very influential and has led to new results in parallel fields, such as complex analysis and number theory.

Many of Witten's papers deal with the topic of "Anomalies". This refers to classical symmetries or conservation laws which are violated at the quantum level. Their investigation is of fundamental importance for physical models and the mathematical aspects are also extremely interesting. The topic has been extensively written about (mainly by physicists) but Witten's contributions have been deep and incisive. For example, he pointed out and investigated "global" anomalies [3], which cannot be studied in the traditional perturbative manner. He also made the important observation that the η-invariant of Dirac operators (introduced by Atiyah, Patodi and Singer) is related to the adiabatic limit of a certain anomaly [4]. This was subsequently given a rigorous proof by Bismut and Freed.

One of Witten's best known ideas is that the index theorem for the Dirac operator on compact manifolds should emerge by a formally exact functional integral on the loop space. This idea (very much in the spirit of his Morse theory paper) stimulated an extensive development by Alvarez–Gaumé, Getzler, Bismut and others which amply justified Witten's view-point.

Also concerned with the Dirac operator is a beautiful joint paper with Vafa [5] which is remarkable for the fact that it produces sharp uniform bounds for eigenvalues by an essentially topological argument. For the Dirac operator on an odd-dimensional compact manifold, coupled to a background gauge potential, Witten and Vafa prove that there is a constant C (depending on the metric, but independent of the potential) such that *every interval of length C contains an* *eigenvalue*. This is not true for Laplace operators or in even dimensions, and is a very refined and unusual result.

3. The Positive Mass Conjecture

In General Relativity the positive mass conjecture asserts that (under appropriate hypotheses) the total energy of a gravitating system is positive and can only be zero for flat Minkowski space. It implies that Minkowski space is a stable ground state. The conjecture has attracted much attention over the years and was established in various special cases before being finally proved by Schoen and Yau in 1979. The proof involved non-linear P. D. E. through the use of minimal surfaces and was a major achievement (leading in part to Yau's Fields Medal at the Warsaw Congress). It was therefore a considerable surprise when Witten outlined in [6] a much simpler proof of the positive mass conjecture based on linear P. D. E. Specifically Witten introduced spinors and studied the Dirac operator. His approach had its origin in some earlier ideas of supergravity and it is typical of Witten's insight and technical skill that he eventually emerged with a simple and quite classical proof. Witten's paper stimulated both mathematicians and physicists in various directions, demonstrating the fruitfulness of his ideas.

4. Rigidity Theorems

The space of solutions of an elliptic differential equation on a compact manifold is naturally acted on by any group of symmetries of the equation. All representations of compact connected Lie groups occur this way. However, for very special equations, these representations are trivial. Notably this happens for the spaces of harmonic forms, since these represent cohomology (which is homotopy invariant). A less obvious case arises from harmonic spinors (solutions of the Dirac equation), although the relevant space here is the "index" (virtual difference of solutions of D and D^*). This was proved by Atiyah and Hirzebruch in 1970. Witten raised the question whether such "rigidity theorems" might be true for other equations of interest in mathematical physics, notably the Rarita–Schwinger equation. This stimulated Landweber and Stong to investigate the question topologically and eventually Witten [7] produced an infinite sequence of such equations which arise naturally in the physics of string theories, for which the Feynman path integral provides a heuristic explanation of rigidity. As usual Witten's work, which was very precise and detailed in its formal aspects, stimulated great activity in this area, culminating in rigorous proofs of these new rigidity theorems by Bott and Taubes [1]. A noteworthy aspect of these proofs is that they involve elliptic function theory and deal with the infinite sequence of operators simultaneously rather than term by term. This is entirely natural from Witten's view-point, based on the Feynman integral.

5. Topological Quantum Field Theories

One of the remarkable aspects of the Geometry/Physics interaction of recent years has been the impact of quantum field theory on low-dimensional geometry (of 2, 3 and 4 dimensions). Witten has systematized this whole areas by showing that there are, in these dimensions, interesting *topological* quantum field theories [8], [9], [10]. These theories have all the formal structure of quantum field theories but they are purely topological and have no dynamics (i.e. the Hamiltonian is zero). Typically the Hilbert spaces are finite-dimensional and various traces give well-defined invariants. For example, the Donaldson theory in 4 dimensions fits into this framework, showing how rich such structures can be.

A more recent example, and in some ways a more surprising one, is the theory of Vaughan Jones related to knot invariants, which has just been reported on by Joan Birman. Witten has shown that the Jones invariants of knots can be interpreted as Feynman integrals for a 3-dimensional gauge theory [11]. As Lagrangian, Witten uses the Chern–Simons function, which is well-known in this subject but had previously been used as an addition to the standard Yang–Mills Lagrangian. Witten's theory is a major breakthrough, since it is the only intrinsically 3-dimensional interpretation of the Jones invariants: all previous defintions employ a presentation of a knot by a plane diagram or by a braid.

Although the Feynman integral is at present only a heuristic tool it does lead, in this case, to a rigorous development from the Hamiltonian point of view. Moreover, Witten's approach immediately shows how to extend the Jones theory from knots in the 3-sphere to knots in arbitrary 3-manifolds. This generalization (which includes as a specially interesting case the empty knot) had previously eluded all other efforts, and Witten's formulas have now been taken as a basis for a rigorous algorithmic definition, on general 3-manifolds, by Reshetikin and Turaev.

Moreover, Witten's approach is extremely powerful and flexible, suggesting a number of important generalizations of the theory which are currently being studied and may prove to be important.

One of the most exciting recent developments in theoretical physics in the past year has been the theory of 2-dimensional quantum gravity. Remarkably this theory appears to have close relations with the topological quantum field theories that have been developed by Witten [12]. Detailed reports on these recent ideas will probably be presented by various speakers at this congress.

6. Conclusion

From this very brief summary of Witten's achievements it should be clear that he has made a profound impact on contemporary mathematics. In his hands physics is once again providing a rich source of inspiration and insight in mathematics. Of course physical insight does not always lead to immediately rigorous mathematical proofs but it frequently leads one in the right direction, and technically correct proofs can then hopefully be found. This is the case with Witten's work. So far his insight has never let him down and rigorous proofs, of the standard we mathe- maticians rightly expect, have always been forthcoming. There is therefore no doubt that contributions to mathematics of this order are fully worthy of a Fields Medal.

References

1. R. Bott and C. H. Taubes: On the rigidity theorems of Witten. J. Amer. Math. Soc. **2** (1989) 137.
2. E. Witten: Supersymmetry and Morse theory. J. Diff. Geom. **17** (1984) 661.
3. E. Witten: An $SU(2)$ anomaly. Phys. Lett. **117 B** (1982) 324.
4. E. Witten: Global anomalies in string theory. Proc. Argonne–Chicago Symposium on Geometry, Anomalies and Topology (1985).
5. E. Witten and C. Vafa: Eigenvalue inequalities for Fermions in gauge theories. Comm. Math. Phys. **95** (1984) 257.
6. E. Witten: A new proof of the positive energy theorem. Comm. Math. Phys. **80** (1981) 381.
7. E. Witten: Elliptic genera and quantum field theory. Comm. Math. Phys. **109** (1987) 525.
8. E. Witten: Topological quantum field theory. Comm. Math. Phys. **117** (1988) 353.
9. E. Witten: Topological gravity. Phys. Lett. B **206** (1988) 601.
10. E. Witten: Topological sigma models. Comm. Math. Phys. **118** (1988) 411.
11. E. Witten: Quantum field theory and the Jones polynomial. Comm. Math. Phys. **121** (1989) 351.
12. E. Witten: On the structure of the topological phase of two-dimensional gravity. Nuclear Phys. B **340** (1990) 281.

Edward Witten

EDWARD WITTEN

Born:	August, 1951	
Education:	BA Brandeis University	1971
	MA Princeton University	1974
	PhD Princeton University	1976
Employment:	Postdoctoral Fellow Harvard University	Sept. 1976–Aug. 1977
	Junior Fellow, Harvard Society of Fellows	Sept. 1977–Aug. 1980
	Professor of Physics, Princeton University	Sept. 1980–Aug. 1987
	Professor, School of Natural Sciences, Institute for Advanced Study	Sept. 1987–present
Honors and Awards:	MacArthur Fellowship	1982
	Einstein Medal, Einstein Society of Berne, Switzerland	1985
	Award For Physical and Mathematical Sciences, New York Academy of Sciences	1985
	Dirac Medal, International Center For Theoretical Physics	1985
	Alan T. Waterman Award, National Science Foundation	1986
	Fields Medal, International Union of Mathematicians	1990
	Madison Medal, Princeton University	1992
	Member of the Board, Americans For Peace Now	Feb. 1992–present
	Fellow, American Academy of Arts and Sciences	1984
	Fellow, American Physical Society	1984
	Fellow, National Academy of Sciences	1988
	Honorary Ph.D., Brandeis University	1988

	Honorary Ph.D., The Hebrew University of Jerusalem	1993
	Honorary Ph.D., Columbia University, New York	1996
	New Jersey Pride Award	1996

Publications: Author of 193 scientific papers.

Coauthor with M. B. Green and J. H. Schwarz of "Superstring Theory" Volumes 1 and 2, Cambridge University Press, 1987.

Reprinted from Proc. AMS Centennial Symposium, 1988

GEOMETRY AND QUANTUM FIELD THEORY

by
EDWARD WITTEN

1. Introduction. First of all, I would like to thank the American Mathematical Society for inviting me to lecture here on this occasion, and to thank the organizers for arranging such a stimulating meeting. And I would like to echo the sentiments of some previous speakers, who expressed the wish that we will all meet here in good health on the 150th anniversary of the American Mathematical Society, to hear the younger mathematicians explain the solutions of some of the unsolved problems posed this week.

It is a challenge to try to speak about the relation of quantum field theory to geometry in just one hour, because there are certainly many things that one might wish to say. The relationship between theoretical physics and geometry is in many ways very different today than it was just ten or fifteen years ago. It used to be that when one thought of geometry in physics, one thought chiefly of classical physics — and in particular of general relativity — rather than quantum physics. Geometrical ideas seemed (except perhaps to some visionaries) to be far removed from quantum physics — that is, from the bulk of contemporary physics. Of course, quantum physics had from the beginning a marked influence in many areas of mathematics — functional analysis and representation theory, just to mention two. But it would probably be fair to say that twenty years ago the day to day preoccupations of most practicing theoretical elementary particle physicists were far removed from considerations of geometry.

Several important influences have brought about a change in this situation. One of the principal influences was the recognition — clearly established by the middle 1970s — of the central role of nonabelian gauge theory in elementary particle physics. The other main influence came from the emerging study of supersymmetry and string theory. Of course, these different influences are inter-related, since nonabelian gauge theories have elegant supersymmetric generalizations, and in string theory these appear in a fascinating new light. Bit by bit, the study of

1980 *Mathematics Subject Classification* (1985 *Revision*). Primary 81E13, 81E99. Research supported in part by NSF Grant 86-20266 and NSF Waterman Grant 88-17521.

nonabelian gauge theories, supersymmetry, and string theory have brought new questions to the fore, and encouraged new ways of thinking.

An important early development in this process came in the period 1976–77 with the recognition that the Atiyah–Singer index theorem was the proper context for understanding some then current developments in the theory of strong interactions. (In particular, the solution by Gerard't Hooft [**1**] of the "U(1) problem," a notorious paradox in strong interaction theory, involved Yang–Mills instantons, originally introduced in [**2**], and "fermion zero modes" whose proper elucidation involves the index theorem.) Influenced by this and related developments, physicists gradually learned to think about quantum field theory in more geometrical terms. As a bonus, ideas coming at least in part from physics shed new light on some mathematical problems. In the first stage of this process, the purely mathematical problems that arose (at least, those that had motivations independent of quantum field theory, and in which progress could be made) involved "classical" mathematical concepts — partial differential equations, index theory, etc. — where physical considerations suggested new questions or a new point of view.

In the talk just before mine, Karen Uhlenbeck described some purely mathematical developments that at least roughly might be classified in this area. She described advances in geometry that have been achieved through the study of systems of nonlinear partial differential equations. Among other things, she sketched some aspects of Simon Donaldson's work on the geometry of four-manifolds [**3**], in which dramatic advances have been made by studying the moduli spaces of instantons — solutions, that is, of a certain nonlinear system of partial differential equations, the self-dual Yang–Mills equations, which were originally introduced by physicists in the context of quantum field theory [**2**].

If "classical" objects (such as instantons) that arise in quantum field theory could be so interesting mathematically, one might well suspect that mathematicians will soon find the quantum field theories themselves, and not only the "classical" objects that they give rise to, to be of interest. Such a question was indeed raised by Karen Uhlenbeck at the end of her talk, and is much in line with the perspective offered by Michael Atiyah in [**4**], which was the starting point for many of my own efforts.

I will talk today about three areas of recent interest where quantum field theory seems to be the right framework for thinking about a problem in geometry:

(1) Our first problem will be to explain the unexpected occurrence of modular forms in the theory of affine Lie algebras. This problem, which was described the other day by Victor Kac, has two close cousins — to explain "monstrous moonshine" in the theory of the Fischer–Griess monster group [**5, 6**], and to account for the surprising role of modular forms in algebraic topology [**7**], about which Raoul Bott spoke briefly at the end of his talk. Quantum field theory supplies a more or less common explanation for these three phenomena, but the first requires the least preliminary explanation, and it is the one that I will focus on.

(2) The second problem is to give a geometrical definition of the new knot polynomials — the Jones polynomial and its generalizations — that have been discovered in recent years. The essential properties of the Jones polynomial have been described to us the other day by Vaughn Jones.

(3) The third problem is to get a more general insight into Donaldson theory of four-manifolds — which was sketched in the last hour by Karen Uhlenbeck — and the closely related Floer groups of three-manifolds. Here again there are lower dimensional cousins, namely the Casson invariant of three-manifolds, Gromov's theory of maps of a Riemann surface to a symplectic manifold, and Floer's closely related work on fixed points of symplectic diffeomorphisms. But among these formally rather analogous subjects, I will concentrate on Donaldson/Floer theory.

2. Physical Hilbert Spaces and Transition Amplitudes. Let us sketch these three problems in a little more detail. In the first problem, one considers the group $\mathcal{L}G$ of maps $\mathbf{S}^1 \to G$, where G is a finite-dimensional compact Lie group, and $\mathbf{S}^1$ is the ordinary circle. The representations of $\mathcal{L}G$ with "good" properties, analogous to the representations of compact finite-dimensional groups, are the so-called integrable highest weight representations (see [**8, 9**] for introductions.) These representations are rigid (no infinitesimal deformations). From this it follows that any connected group of outer automorphisms of $\mathcal{L}G$ must act at least projectively on any integrable highest weight representation $\mathcal{R}$ of $\mathcal{L}G$. In fact, the group diff $\mathbf{S}^1$ of diffeomorphisms of $\mathbf{S}^1$ acts on $\mathcal{L}G$ by outer automorphisms and acts projectively on the integrable highest weight representations. Thus, in particular, the vector field $d/d\theta$ that generates an ordinary rotation of $\mathbf{S}^1$ is represented on $\mathcal{R}$ by some operator H.

One computes in such a representation the "character"

$$F_{\mathcal{R}}(q) = \mathrm{Tr}_{\mathcal{R}}\, q^H \tag{2.1}$$

(here q is a complex number with $|q| < 1$), and one finds this to be a modular function with a simple transformation law under a suitable congruence subgroup of the modular group. Setting $q = \exp(2\pi i \tau)$, the modular group is of course the group $\mathrm{PSL}(2,\mathbb{Z})$ of fractional linear transformations

$$\tau \longrightarrow \frac{a\tau + b}{c\tau + d}, \tag{2.2}$$

of the upper half-plane, with $a, b, c, d \in \mathbb{Z}$ and $ad - bc = 1$. I will not enter here into the complicated question of exactly what kind of modular functions the characters (2.1) are. (One simple, general statement, which from one point of view is the statement that comes most directly from quantum field theory, is that the $F_{\mathcal{R}}(q)$, with $\mathcal{R}$ running over all highest weight representations of fixed "level," transform as a unitary representation of the full modular group $\mathrm{PSL}(2,\mathbb{Z})$.)

To understand the significance of the modularity of the characters $F_{\mathcal{R}}(q)$, let us recall that the group SL(2,$\mathbb{Z}$) has a natural interpretation as the (orientation preserving) mapping class group of a two-dimensional torus $\mathbf{T}^2$. Thus, we interpret $\mathbf{T}^2$ as the quotient of the $x-y$ plane by the equivalence relations $(x,y) \sim (x+1,y)$ and $(x,y) \sim (x,y+1)$. Clearly, if a,b,c and d are integers such that $ad-bc=1$, the formula $(x,y) \to (ax+by, cx+dy)$ gives a diffeomorphism of $\mathbf{T}^2$ to itself, and every orientation preserving diffeomorphism of $\mathbf{T}^2$ is isotopic to a unique one of these. Thus, $\mathbf{SL}(2,\mathbb{Z})$ can be considered in this sense to arise as a group of diffeomorphisms of a two-dimensional surface.

Thus, while it is natural that the one-dimensional symmetry group diff $\mathbf{S}^1$ plays a role in the representation theory of the loop group $\mathcal{L}G$, the appearance of SL(2,$\mathbb{Z}$) means that in fact a kind of *two-dimensional symmetry* appears in this theory. Our first problem — modular moonshine in the theory of affine Lie algebras — is the problem of explaining the origin of this two-dimensional symmetry.

Now we move on to our second problem. A *braid* is a time dependent history of n points in $\mathbb{R}^2$, which are required, up to a permutation, to end where they begin (Figure 1(a)). Braids with n strands form a group, the Artin braid group $\mathcal{B}_n$, with an evident law of composition, sketched in Figure 1(b). From a braid, one can make a knot (or in general a link) by gluing together the top and bottom as in Figure 1(c). Although every braid gives in this way a unique link, the converse is not so; the same link may arise from many different braids. The crucial difference between braids and links is the following. Braids are classified up to time dependent diffeomorphisms of $\mathbb{R}^2$ (that is, up to diffeomorphisms of $\mathbb{R}^3$ that leave fixed one of the coordinates, the "time" t), while links are classified up to full three-dimensional diffeomorphisms.

If one is given a representation $\mathcal{S}$ of the braid group $\mathcal{B}_n$, and a braid $B \in \mathcal{B}_n$, then $\mathrm{Tr}_{\mathcal{S}} B$ (the trace of the matrix that represents B in the representation $\mathcal{S}$) is an invariant of the *braid* B (and depends in fact only on its conjugacy class in $\mathcal{B}_n$),

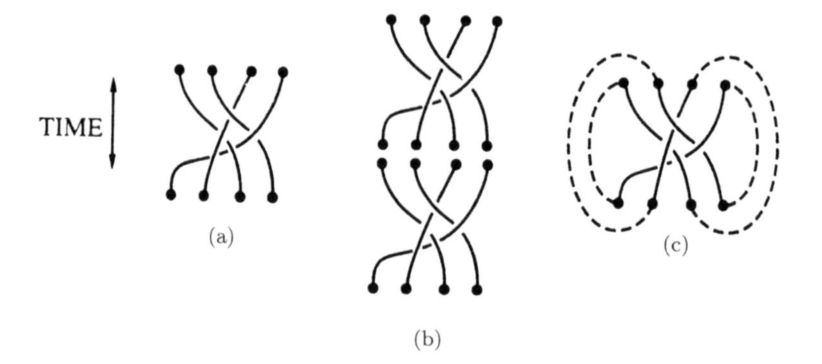

Figure 1. (a) A braid; (b) composition of two braids; (c) making a braid into a link.

but there is no reason for it to be an invariant of the *link* that is obtained by joining the ends of the braid B according to the recipe in the figure.

Nevertheless, Vaughn Jones found a special class of representations of the braid group with the magic property that suitable linear combinations of the braid traces are in fact knot invariants and not just braid invariants. These knot invariants can be combined into the Jones polynomial some of whose remarkable properties were described in Jones's lecture the other day. The discovery of the Jones polynomial stimulated in short order the discovery of some related knot polynomials — the HOMFLY and Kauffman polynomials — whose logical status is rather similar. The challenge of understanding the Jones polynomial is to explain why the Jones braid representations, which obviously have two-dimensional symmetry, should really have three-dimensional symmetry.

Thus, we have two examples where one studies a *group representation* that obviously has d-dimensional symmetry, for some d, but turns out to have $(d+1) = D$-dimensional symmetry, for reasons that might look mysterious. In our first example, the group is $\mathcal{L}G$, $d = 1$, and $D = 2$. In the second example, the group is the braid group, $d = 2$, and $D = 3$.

Our third example, Donaldson/Floer theory, is of a somewhat different nature. In this case, $d = 3$ and $D = 4$, but unlike our previous examples, Donaldson/Floer theory began historically not in the lower dimension but in the upper dimension. The mathematical theory begins in this case with Donaldson's invariants of a closed, oriented four-manifold M. In Donaldson's original considerations, it was important that the boundary of M should vanish. The attempt to generalize the Donaldson invariants to the case that $\partial M = Y \neq \phi$ led to the introduction of the "Floer homology groups" which are vector spaces $\mathrm{HF}^*(Y)$ canonically associated with an oriented three-manifold Y. Though these vector spaces did not originate as group representations, their formal role is just like that of the group representations that entered in our first two examples.

In our first two problems of understanding modular moonshine and the Jones polynomial, the crucial question is to explain why $(d+1)$-dimensional symmetry is present in a construction that appears to only have d-dimensional symmetry. At least from a historical point of view, Donaldson theory is of a completely different nature, since the four-dimensional symmetry has been built in from the beginning. Nevertheless, the logical structure of Donaldson/Floer theory is of a similar nature to that of the first two examples.

In each of our three examples, a pair of dimensions, d and $D = d+1$, plays a key role. With the lower dimension we associate a vector space (the representations of $\mathcal{L}G$ or $\mathcal{B}_n$ or the Floer groups) and with the upper dimension we associate an invariant (the characters $\mathrm{Tr}_{\mathcal{R}} q^H$, the knot polynomials, or the Donaldson invariants of four-manifolds). The facts are summarized in Table 1.

3. Axioms of Quantum Field Theory. Let us now formalize the precise relationship between the vector spaces that appear in dimension d and the invariants

Table 1

Theory	Dimensions	Vector space in lower dimension	Invariant in upper dimensions
Modular moonshine	1, 2	Representations of loop groups	Modular forms $\mathrm{Tr}\, q^H$
Jones polynomial	2, 3	Jones representations of braid group	Invariants of knots and three-manifolds
Donaldson/Floer theory	3, 4	Floer Homology groups	Donaldson invariants of four-manifolds

in dimension $d+1$. (In the physical context, d is called the dimension of space, and $d+1$ is the dimension of space-time.) In formalizing this relationship, we will follow axioms originally proposed (in the context of conformal field theory, essentially our first example) by Graeme Segal [**10**]. (In addition, Michael Atiyah has adapted those axioms for the topological context that is relevant to our second and third examples [**11**], with considerably more precision than I will attempt here.)

So we will consider quantum field theory in space-time dimension $D = d + 1$. The manifolds that we consider will be smooth manifolds possibly endowed with some additional structure. The type of additional structure considered will be characteristic of the theory. For instance, in the case of modular moonshine, this additional structure is a conformal structure; quantum field theories requiring such a structure (but not requiring a choice of Riemannian metric) are called conformal field theories. In the case of Donaldson/Floer theory, the extra structure consists of an orientation; in the case of the Jones polynomial, one requires an orientation and "framing" of tangent bundles (in a suitable stable sense). Theories that require structure of such a purely topological kind may be called topological quantum field theories. The "ordinary" quantum field theories most extensively studied by physicists require metrics on all manifolds considered.

The first notion is that to every d-dimensional manifold X, without boundary, and perhaps with some additional structure characteristic of the particular theory, one associates a vector space $\mathcal{H}_X$. A quantum field theory is said to be "unitary" if these vector spaces actually carry a Hilbert space structure; this is so in the theories of modular moonshine and the Jones polynomial, but not in the case of Donaldson/Floer theory. In the case of the Jones polynomial and Donaldson/Floer theory, the vector spaces $\mathcal{H}_X$ are finite dimensional, and a morphism of vector spaces is taken to mean an arbitrary linear transformation (preserving the unitary structure in the case of the Jones polynomial); in the theory of modular moonshine, the $\mathcal{H}_X$ are infinite dimensional, and it is necessary to be more precise about what is meant by a morphism among these spaces.

In Segal's language, the association $X \to \mathcal{H}_X$ is to be a functor from the category of d-dimensional manifolds with additional structure (and diffeomorphisms preserving the specified structures) to the category of vector spaces (and linear transformations of the appropriate kind).

Certain additional restrictions are imposed. The empty d-manifold ϕ is permitted, and one requires that $\mathcal{H}_\phi = \mathbb{C}$ ($\mathbb{C}$ here being a one-dimensional vector space with a preferred generator which we call "1"). If $X \amalg Y$ denotes the disjoint union of two d-dimensional manifolds X and Y, then one requires $\mathcal{H}_{X \amalg Y} = \mathcal{H}_X \otimes \mathcal{H}_Y$. If $-X$ is X with opposite orientation, and $*$ denotes the dual of a vector space, one requires $\mathcal{H}_{-X} = \mathcal{H}_X^*$.

Since the late 1920s, the spaces $\mathcal{H}_X$ have been known to physicists as the "physical Hilbert spaces" (of the particular quantum field theory under consideration). The association $X \to \mathcal{H}_X$ is roughly half of the basic structure considered in quantum field theory. The second half corresponds in physical terminology to the "transition amplitudes."

To introduce the transition amplitudes, we consider (Figure 2(a) on the next page) a cobordism of oriented (and possibly disconnected or empty) d-dimensional manifolds. Such a cobordism is defined by an oriented $(d+1)$-dimensional manifold W whose boundary is, say, $\partial W = X \cup (-Y)$, where X and Y are oriented d-dimensional manifolds (whose orientations respectively agree or disagree with that induced from W). It is required that whatever structure (conformal structure, framing, metric, etc.) has been introduced on X and Y is extended over W. Such a cobordism is regarded as a morphism from X to Y. To every such morphism of manifolds, a quantum field theory associates a morphism of vector spaces

$$\mathbf{\Phi}_W : \mathcal{H}_X \to \mathcal{H}_Y. \tag{3.1}$$

Of course, this association $W \to \mathbf{\Phi}_W$ should be natural, invariant under any diffeomorphism of W that preserves the relevant structures. Regarding $-W$ as a morphism from $-Y$ to $-X$, one requires that $\mathbf{\Phi}_{(-W)} : \mathcal{H}_{(-Y)} \to \mathcal{H}_{(-X)}$ should be

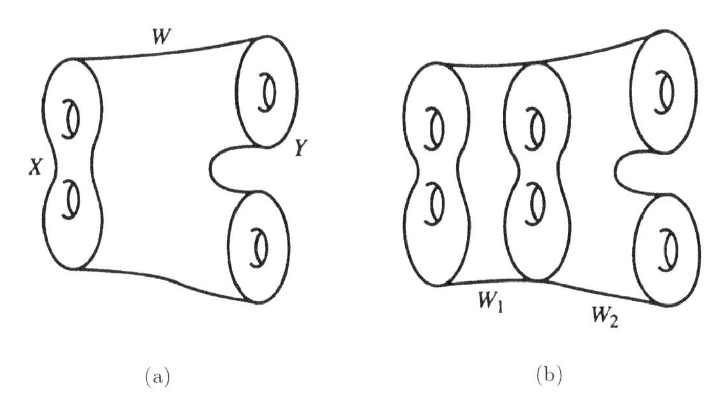

Figure 2. (a) A cobordism of oriented d-dimensional manifolds; (b) a composition of of such cobordisms.

the dual linear transformation to $\mathbf{\Phi}_W$. And if $W = W_1 \cup W_2$ is a composition of cobordisms (Figure 2(b) on the next page), one requires that

$$\mathbf{\Phi}_W = \mathbf{\Phi}_{W_2} \circ \mathbf{\Phi}_{W_1}. \tag{3.2}$$

These requirements correspond physically to relativity, locality, and causality.

A very important special case of this is the case in which W is a closed $D = (d+1)$-dimensional manifold without boundary. Such a W can be regarded as a morphism from the empty d-dimensional manifold ϕ to itself. Since $\mathcal{H}_\phi = \mathbb{C}$, the associated morphism $\mathbf{\Phi}_W : \mathcal{H}_\phi \to \mathcal{H}_\phi$ is simply a number, which for physicists is often called the partition function of W and denoted $Z(W)$. This partition function is the fundamental invariant in quantum field theory; for different choices of theory, one gets the invariants of D-dimensional manifolds indicated in the last column in Table 1.

For $Z(W)$ to be defined in a given quantum field theory, W must of course be endowed with the structure appropriate to the particular quantum field theory in question. For instance, in the case of modular moonshine, W must be a Riemann surface with a conformal structure. In genus one, this means that W is an elliptic curve, which can be represented by a point in the upper half-plane subject to the action of the mapping class group. The naturality of the association $W \to Z(W)$ means that $Z(W)$ can depend only on the equivalence class of the conformal structure of W, and it is this which leads to modular forms. In our other two examples, no metric or conformal structure is present, so we are dealing with topological invariants. In our second example of the Jones polynomial, the invariant $Z(W)$ is an invariant of oriented three-manifolds which is an analog for three-manifolds of the Jones polynomials for knots in $\mathbf{S}^3$. (The actual knot invariants can be obtained by an elaboration of the quantum field theory structure.) In our third example of Donaldson theory, the invariant $Z(W)$ is the prototype of the invariants that appear in the celebrated Donaldson polynomials of oriented four-manifolds.

It is built into the axioms of quantum field theory that the fundamental invariants $Z(W)$ can be computed from a decomposition of the type that is known in the case of three-manifolds as a Heegaard splitting. This means a realization of W as $W = W_1 \cup W_2$, where W_1 and W_2 are D-manifolds joined together along their common boundary Σ. In this case the morphism W from the empty manifold ϕ to itself factorizes as a morphism W_2 from ϕ to Σ composed with a morphism W_1 from Σ to ϕ, i.e.,

$$\mathbf{\Phi}_W = \mathbf{\Phi}_{W_1} \circ \mathbf{\Phi}_{W_2}. \tag{3.3}$$

If 1 is the canonical generator of $\mathcal{H}_\phi$, we then have

$$Z(W) = (1, \mathbf{\Phi}_W \cdot 1) = (1, \mathbf{\Phi}_{W_1} \circ \mathbf{\Phi}_{W_2} \cdot 1). \tag{3.4}$$

Let $v \in \mathcal{H}_\Sigma$ be the vector $v = \mathbf{\Phi}_{W_2}(1)$. Also, think of $-W_1$ as a morphism from ϕ to $-\Sigma$, and let and $w \in \mathcal{H}_{-\Sigma}$ be the vector $w = \mathbf{\Phi}_{-W_1}(1)$. Then (3.4) amounts to

$$Z(W) = (w, v). \tag{3.5}$$

This ability to calculate via Heegaard splittings is part of the conventional *definition* of the Casson invariant (which has a quantum field theory interpretation analogous to that of Donaldson theory), and is essential in the calculability of the three-manifold invariants that are related to the Jones polynomial. Likewise, in the case of modular moonshine, the decomposition (3.5) is the key to the fact that the partition function $Z(W)$ can be written as the character $\mathrm{Tr}_{\mathcal{R}} q^H$ of equation (2.1).

4. Construction of Quantum Field Theories. The question arises, of course, of how these quantum field theories are to be constructed. About this enormous subject it is possible only to say a few words here.

The starting point is always the choice of an appropriate *Lagrangian*, which is the integral of a local functional of appropriate fields. For instance, if one is interested in understanding the Jones polynomial, one picks a finite-dimensional compact simple group G and one considers a connection A on a G-bundle E over a three-manifold M. Let $F = dA + A \wedge A$ denote the curvature of this connection. On the Lie algebra $\mathcal{G}$ of a compact group G, there is an invariant quadratic form which we denote by the symbol Tr (that is, we write $(a,b) = \mathrm{Tr}(ab)$).[1] For the Lagrangian, we take the Chern–Simons invariant of the connection A:

$$\mathcal{L} = \frac{k}{4\pi} \int_M \mathrm{Tr}\left(A \wedge dA + \frac{2}{3} A \wedge A \wedge A \right). \tag{4.1}$$

(Here k is a positive integer, a fact that is required so that the argument $e^{i\mathcal{L}}$ in the Feynman path integral is gauge invariant.) To construct a quantum field theory from this Lagrangian, there are two basic requirements. First, we must construct a functor from Riemann surfaces Σ to Hilbert spaces $\mathcal{H}_\Sigma$; and second, for every cobordism W from Σ to Σ', we must construct a morphism $\mathbf{\Phi}_W : \mathcal{H}_\Sigma \to \mathcal{H}_{\Sigma'}$.

For the first step, one proceeds as follows. Given the surface Σ, we consider the Lagrangian (4.1) on the three-manifold $\Sigma \times \mathbb{R}$. The space of critical points of the Lagrangian, up to gauge transformations, is known in classical mechanics as the "phase space" of the system under investigation. Let us call this phase space $\mathcal{M}_\Sigma$. In the case at hand, the Euler–Lagrange equation for a critical point of the Lagrangian (4.1) is the equation $F = 0$, where $F = dA + A \wedge A$ is the curvature of the connection A. (That is, (4.1) is invariant to first order under variations of the connection of compact support if and only if $F = 0$.) A flat connection on $\Sigma \times \mathbb{R}$ defines a homomorphism of the fundamental group $\pi_1(\Sigma \times \mathbb{R})$ into G. Of course, this is the same as a homomorphism of $\pi_1(\Sigma)$ into G. The classical phase space $\mathcal{M}_\Sigma$

[1]The quadratic form is to be normalized so that the characteristic class $\frac{1}{4\pi}\mathrm{Tr}F \wedge F$ has periods that are multiples of 2π.

associated with the Lagrangian (4.1) is simply the moduli space of homomorphisms of $\pi_1(\Sigma) \to G$, up to conjugation by G.

Now, it is a general fact in the calculus of variations that the phase space associated with a Lagrangian such as (4.1) is always endowed with a canonical symplectic structure ω. Indeed, this is how symplectic structures originally appeared in classical mechanics, and as such it was the starting point of symplectic geometry as a mathematical subject also. In the case at hand, the symplectic structure thus obtained on $\mathcal{M}_\Sigma$ is known [**13**], and has been studied very fruitfully from the point of view of two-dimensional gauge theory [**14**], but my point is that this symplectic structure on $\mathcal{M}_\Sigma$ can be considered to arise from a *three-dimensional* variational problem. This elementary fact is an important starting point for understanding the mysterious three-dimensionality of the Jones polynomial.

Once the appropriate phase space $\mathcal{M}_\Sigma$ is identified, the association $\Sigma \to \mathcal{H}_\Sigma$ is made by "quantizing" $\mathcal{M}_\Sigma$ to obtain a Hilbert space $\mathcal{H}_\Sigma$. Geometric quantization is not sufficiently well developed to make quantization straightforward in general (or perhaps this is actually impossible in general), but in the case at hand quantization can be carried out by choosing a complex polarization of $\mathcal{M}$. This is accomplished by picking a complex structure J on Σ and using the Narasimhan–Seshadri theorem to identify $\mathcal{M}$ with the moduli space of stable holomorphic $G_{\mathbb{C}}$ bundles over Σ. This moduli space is then quantized by defining $\mathcal{H}_\Sigma$ to be the space of holomorphic sections of a certain line bundle over $\mathcal{M}$. This space is independent of J (up to a projective factor) because of its interpretation in terms of quantization of the underlying classical phase space $\mathcal{M}$. The association $\Sigma \to \mathcal{H}_\Sigma$ is the geometric origin of the Jones braid representations (or rather their analog for the mapping class group in genus g).

Once the association $\Sigma \to \mathcal{H}_\Sigma$ is understood, it remains to define for every cobordism W from Σ to Σ', a corresponding morphism $\mathbf{\Phi}_W : \mathcal{H}_\Sigma \to \mathcal{H}_{\Sigma'}$. The key notion here is the "Feynman path integral," that is, Feynman's concept of integration over the whole function space $\mathcal{V}$ of connections on (the given G-bundle E over) W. Roughly speaking, the function space integral with prescribed boundary conditions gives the kernel of the morphism $\mathbf{\Phi}_W$. I have tried to explain heuristically the role of functional integrals in [**12**], and I will not repeat those observations here.

In conclusion, let me point out that if G is a *compact* group, then, as I have argued in [**15**], the quantum field theory associated with the Lagrangian (4.1) is related to the Jones polynomial and its generalizations. However, (4.1) makes sense for any gauge group G with an invariant quadratic form "Tr" on the Lie algebra. It is natural to ask what mathematical constructions are related to the quantum field theories so obtained. One case that can be conveniently analyzed is the case in which one replaces the compact group G by a group $TG = \mathcal{G} \ltimes G$; here $\mathcal{G} \ltimes G$ denotes the semidirect product of G with its own Lie algebra $\mathcal{G}$, the latter regarded as an abelian group acted on by G. It turns out [**16**] that with this choice of gauge group, the quantum field theory derived from (4.1) (with a certain choice of "Tr") is related to recent work of D. Johnson on Reidemeister torsion [**17**], while if instead one considers a certain super-group whose bosonic part is TG then (4.1) (again with a certain choice of "Tr") is related to the Casson invariant of three-manifolds. It is also very interesting to take G to be a semisimple but noncompact Lie group. The corresponding quantum field theories are very little understood, but there are indications that they should be very rich. In fact, it appears [**18**] that the theories based on SL(2,$\mathbb{R}$) and especially SL(2,$\mathbb{C}$) must be intimately connected with the theory of hyperbolic structures on three-manifolds, as surveyed the other day by Thurston.

5. Conclusion. In attending this meeting, I have found it striking how many of the lectures were concerned with questions that are associated with quantum field theory — and in many cases questions that might be characterized as questions about quantum field theory. In time we will hopefully gain a clearer picture of the scope of some of these newly emerging relations between geometry and physics. It is not too much to anticipate that many important constructions relating quantum field theory to topology and differential geometry remain to be discovered. Harder to foresee is whether — by the time of the one hundred fiftieth anniversary of the American Mathematical Society — the influence of quantum field theory will also extend to other areas of mathematics, such as algebraic geometry and number theory, which superficially might appear to be comparatively immune. Let me recall that in his lecture earlier this week, Dick Gross concluded by urging physicists and mathematicians to find a quantum field theory explanation for the appearance of modular forms in the study of the L-functions of elliptic curves. (The question was, of course, motivated by the relation of quantum field theory to the different kinds of modular moonshine.) Perhaps this challenge, or analogous ones about which one might speculate, will be met. Hints today concerning quantum field theoretic insights about number theory are probably no more compelling than hints of quantum field theoretic insight about differential geometry were ten years ago.

What significance might the emerging links between quantum field theory and geometry have for physics? It is very noticeable that the aspects of quantum field theory that are most useful in understanding the geometrical problems that I have been talking about are pretty close to the slightly specialized aspects of quantum field theory that appear in string theory. Modular invariance in the theory of affine Lie algebras is certainly a familiar story to string theorists. The Jones polynomial and its generalizations are related to the "rational conformal field theories" which are one of our main tools for finding exact classical solutions in string theory. The constructions that enter in formulating Donaldson theory as a quantum field theoryneg are also very similar to what string theorists are accustomed to (in the use of world-sheet BRST operators).

Apart from being at least loosely connected with all of the geometrical problems that we have been discussing, string theory seems to be the center of some geometrical questions of central physical interest. The towering puzzle in contemporary theoretical physics is — at least from my standpoint — the puzzle of finding the geometrical context in which string theory should be properly formulated and understood. I am sure many physicists would share this judgment. With our present concepts, this problem (to which I attempted a thumbnail introduction in [**12**]) seems well out of reach. Perhaps it is not too far-fetched to hope that some insight in this central mystery can be obtained from the further study of geometrical questions arising in quantum field theory.

References

1. Gerard't Hooft, *Computation of the quantum effects due to a two-dimensional pseudoparticle*, Phys. Rev. D **14** (1976), 3432.
2. A. Belavin, A. M. Polyakov, A. Schwarz, and Yu. S. Tjupkin, Phys. Lett. B **59** (1975), 85.
3. S. Donaldson, *An applicaton of gauge theory to the topology of four-manifolds*, J. Differential Geom. **18** (1983), 269; *Polynomial invariants for smooth four-manifolds*, preprint, Oxford University.
4. M. F. Atiyah, *New invariants of three and four-dimensional manifolds*, The Mathematical Heritage Of Hermann Weyl (R. Wells, ed.), Amer. Math. Soc., Providence, RI, 1988.
5. J. H. Conway and S. P. Norton, *Monstrous moonshine*, Bull. London Math. Soc. **11** (1979), 308.
6. I. B. Frenkel, A. Meurman, and J. Lepowsky, *A moonshine module for the monster*, Vertex Operators in Mathematics And Physics (J. Lepowsky, S. Mandelstam, and I. M. Singer, eds.), Springer–Verlag.
7. P. Landweber (ed.), *Elliptic curves and modular forms in algebraic topolog*, Lecture Notes in Math., vol. 1326, Springer–Verlag, 1988.
8. V. Kac, *Infinite Dimensional Lie Algebras*, Cambridge Univ. Press, 1985.
9. A. Pressley and G. Segal, *Loop groups*, Oxford Univ. Press, 1987.
10. G. Segal, preprint, Oxford Unversity (to appear).
11. M. F. Atiyah, *Topological quantum field theories*, René Thom Festschrift (to appear).
12. E. Witten, *Geometry and physics*, Proc. Internat. Congr. Math., Berkeley, California, 1986.
13. W. Goldman, *Invariant functions on Lie groups and Hamiltonian flows of surface group representations*, Invent. Math. **85** (1986), 263.
14. M. F. Atiyah and R. Bott, *Yang–Mills equations over Riemann surfaces*, Philos. Trans. Roy. Soc. London. Ser. A **308** (1982), 523.
15. E. Witten, *Quantum field theory and the Jones polynomial*, Comm. Math. Phys. **121** (1989), 351.
16. ———, *Topology-changing amplitudes in* $2+1$ *dimensional gravity*, Nuclear Phys. **B323** (1989), 113.

17. D. Johnson, *A geometric form of Casson's invariant and its connection to Reidemeister torsion*, unpublished lecture notes.
18. E. Witten, 2+1 *dimensional gravity as an exactly soluble system*, Nuclear Phys. **B311** (1988/9), 46.

School of Natural Sciences
Institute for Advanced Study
Princeton
New Jersey 08540

Reprinted from Proc. Int. Congr. Math., 1994

THE WORK OF JEAN BOURGAIN

by
LUIS CAFFARELLI

Institute for Advanced Study, Princeton,
NJ 08540, USA

Introduction

Bourgain's work touches on several central topics of mathematical analysis: the geometry of Banach spaces, convexity in high dimensions, harmonic analysis, ergodic theory, and finally, nonlinear partial differential equations (P.D.E.'s) from mathematical physics. In all of these areas, he made spectacular inroads into questions where progress had been blocked for a long time.

This he did by simultaneously bringing into play different areas of mathematics: number theory, combinatorics, probability, and showing their relevance to the problem in a previously unforeseen fashion.

To give a flavor of his work, I have concentrated on his recent research, of about the last ten years.

The solution of the Λ_p problem

A great part of the work of Bourgain, in the study of the geometry of Banach spaces, concentrated on the question: Given a Banach space of finite dimension n, how large a section can we find that resembles a Hilbert subspace?

Maybe his most relevant paper in this field is his solution of the $\Lambda(p)$ problem: Given a subset Λ of the set of characters of a compact Abelian group. Λ is a p-set ($p > 2$) if the L^p and L^2 norms are equivalent in the subspace of $L^p(G)$ generated by Λ.

The longstanding question was: Do $\Lambda(p)$ and $\Lambda(q)$ sets coincide?

Bourgain answers this problem in the negative with the following sharp estimates:

Among n given characters there is a subset of optimal size $[n^{2/p}]$ for which

$$\left\|\sum a_i\varphi_i\right\|_{L^p} \leq C(p)\left(\sum |a_i|^2\right)^{1/2}. \qquad (*)$$

Through a lacunary argument, one can construct a $\Lambda(p)$ set, which is not $\Lambda(q)$ for any $q > p$.

The converse of Santalo's inequality

Another product of Bourgain's studies is his proof of Santalo's inequality:

Given K the unit ball of a norm on R^n and K^* its dual, Bourgain and Milman prove

$$\text{vol}(K)\text{vol}(K^*) \geq C^n |B|^2$$

for some absolute constant $0 < c < 1$.

This has applications to number theory and computer sciences.

Ergodic theory

In ergodic theory, Bourgain developed a completely new theory, where averages under very sparse (polynomial) families of iterations are studied (and shown to converge).

The basic theorem, from which the general setting follows by a well-known transformation, due to Calderon, is the maximal theorem for $\ell^2(\mathbb{Z})$.

Theorem. *Let $f \in \ell^2(\mathbb{Z})$, and ℓ be a positive integer; let*

$$Mf(n) = \sup_{N>0} \left| \frac{1}{N} \sum_{k=1}^{N} f(n+k^\ell) \right| .$$

Then

$$\|Mf\|_{\ell^2} \leq C\|f\|_{\ell^2}$$

i.e., the maximal function of the partial averages of the k^ℓ iterations of the 1-translation is bounded in ℓ^2.

Oscillatory integrals

An important family of ideas introduced by Stein in harmonic analysis concerns the study of the restriction of classical operators (maximal functions, Hilbert and Fourier transforms) to curves in space (parabolics, circles) that have special relevance to the study of partial differential equations (singular integrals of parabolic type, spherical averages related to the wave equation, etc.).

In this area I'll mention two fundamental contributions of Bourgain:

The circle maximal function

$$Mf(y) = \sup_r \frac{1}{A(S_r)} \int_{x \in S_r(0)} f(y+x)dA$$

was shown by Stein to be a bounded operator in L^p for some range $p(n)$ for $n \geq 3$.

The two-dimensional case ($p \geq 2$) remained open for a long time until Bourgain closed the gap.

As the "solid" maximal function in any dimension can be written as an average of spherical maximal functions in a fixed low dimension, this allows us in particular to prove bounds for the "solid" maximal function independent of dimensions.

The second contribution refers to the restriction of Fourier transforms to spheres, or, related to it, the properties of the characteristic function of the ball as a Fourier multiplier (a natural generalization of taking partial sums of Fourier series).

In two dimensions, these problems were well understood (C. Fefferman).

In higher dimensions some range of continuity is expected around L^2, and a series of results was obtained by Tomas and Stein.

Bourgain considerably sharpened these results, but what is more important than the exact ranges, is the fact that, in doing so, he introduced completely new techniques, where instead of relying on L^2 theory (i.e., decomposing functions and operators in "L^2 pieces") he sharpened the geometric understanding of Besicovitch-type maximal operators and Kakeya sets.

Nonlinear partial differential equations

Bourgain's contributions to nonlinear partial differential equations are very recent, and it is somewhat difficult to decide where to stop in this presentation because results from him and many others in the field (Kenig, Klainerman, Machedon, Ponce, Vega) have been pouring in, in good part thanks to the revitalization of the field brought in by Bourgain's approach.

Let us say that he obtained very sharp results for the well-posedness of the nonlinear Schrödinger equation,

$$iu_t + \Delta u + u|u|^\alpha = 0$$

for non smooth data.

Previous to his work, there was mainly local well-posedness in H^s for large enough s.

By introducing new, suitable space-time functional spaces, Bourgain started a new, more deep and elegant way of treating dispersive equations.

In closing, let me reiterate that some of the outstanding qualities of Bourgain are his power to use whatever it takes — number theory, probabilistic methods, covering techniques, sharp decompositions — to understand the problem at hand, and his versatility, which allowed him to deeply touch so many areas in such a short period of time.

Jean Bourgain

JEAN BOURGAIN

Born February 28, 1954 in Ostende, Belgium

Education:

1977 – Ph.D., Free University of Brussels
1979 – Habilitation Degree, Free University of Brussels

Appointments:

Research Fellowship in Belgium NSF (NFWO), 1975–1981
Professor at Free University of Brussels, 1981–1985
J. L. Doob Professor of Mathematics, University of Illinois, 1985–present
Professor at IHES (France), 1985–1995
Lady Davis Professor of Mathematics, Hebrew University of Jerusalem, 1988
Fairchild Distinguished Professor, Caltech, 1991
Professor IAS, 1994–present

Honors:

Alumni Prize, Belgium NSF, 1979
Empain Prize, Belgium NSF, 1983
Salem Prize, 1983
Damry-Deleeuw-Bourlart Prize (awarded by Belgian NSF), 1985
(quintesimal Belgian Science Prize)
Langevin Prize (French Academy), 1985
E. Cartan Prize (French Academy), 1990 (quintesimal)
Dr. H. C. Hebrew University, 1991
Ostrowski Prize, Ostrowski Foundation (Basel-Switzerland), 1991 (biannual)
Fields Medal, ICM Zurich, 1994
Dr. H. C. Université Marne-la-Vallee (France), 1994
Dr. H. C. Free University of Brussels (Belgium), 1995

Invited speaker:

ICM, Warsaw (1983), Berkeley (1986), Zurich (1994)
ICMP, Paris (1994)

Editorial Board:

- Publications Mathématiques de l'IHES
- Duke Math. J.
- International Mathematical Research Notices
- Geometrical and Functional Analysis (GAFA)
- Journal d'Analyse de Jérusalem

HAMILTONIAN METHODS IN NONLINEAR EVOLUTION EQUATIONS

by
JEAN BOURGAIN*

1. Introduction

One of the main problems in the theory of nonlinear Hamiltonian evolution equations is the behavior of the solutions for time $t \to \infty$. We have in mind the general (not integrable) case, with only few conserved quantities at our disposal. Here one may nevertheless obtain certain information on the longtime behavior of the flows, using the Hamiltonian structure of the equation. Recently, there has been progress along these lines in various directions and we do not intend to try to give a complete survey. In this report, we will discuss the following aspects

(i) Use of methods of statistical mechanics and symplectic geometry
(ii) Use of methods of dynamical systems

The main theme, regarding the first topic, is to establish the existence of a well defined global flow on the support of the Gibbs measure $e^{-\beta H(\phi)} \prod d\phi$ and show its invariance under this flow. This leads thus to the existence of an invariant measure. This measure lives on functions with poor regularity or fields. Proving well-posedness of the Cauchy problem for such data turns out to require a rather delicate analysis of an independent interest. The well-posedness is usually only local in time and a global result is obtained using the Gibbs-measure as a substitute for a conservation law. In this way, global solutions are constructed for data of a regularity considerably below what may be shown by purely PDE methods. We will carry out our discussion here for the nonlinear Schrödinger equation (NLS) in dimensions 1 and 2 and some related equations, completing a line of investigation initiated in the paper of [L-R-S]. Other symplectic invariants, called symplectic capacities, originating from M. Gromov's pioneering work [Grom], allow us to study "squeezing properties" and energy transitions in the symplectically normalized phase space. These normalizations are however such that the resulting theories deal with low regularity solutions and consequently, as a first step, require again to study the flow for such data.

The second topic concerns the persistency of periodic or quasi-periodic (in time) solutions of linear or integrable equations under small Hamiltonian perturbation. A natural approach to such questions is to try to adapt the standard KAM

*IAS and IHES and University of Illinois

technology from classical mechanics. The problem is thus the persistency of finite dimensional tori in an infinite dimensional phase space. In this direction, important contributions are due to S. Kuksin [Kuk$_1$]. His work gives satisfactory results for $1D$ problems with Dirichlet boundary conditions. The shortcoming of the standard KAM approach is the fact that it seems unable to deal with multiplicities or clusters in the normal frequencies; those appear in $1D$ under periodic boundary conditions and in dimension ≥ 2. A different approach, based on the Liapounov–Schmidt decomposition scheme avoids this limitation of the KAM method. It has been elaborated by W. Craig, E. Wayne [C-W$_{1,2}$] for $1D$ time-periodic solutions and more recently by the author. The method is more flexible than KAM and depends less on Hamiltonian structure. Research on its applications is still in progress. Presently, we may deal with time periodic problems in any dimension and time quasi-periodic solutions in $1D$, $2D$. The second part of the report is devoted to a sketch of the underlying ideas. We mainly restrict ourselves to nonlinear perturbations of linear equations here. The included reference list is strictly for the purpose of this exposition.

2. Nonlinear Schrödinger Equations and Invariant Gibbs Measures

We consider the nonlinear Schrödinger equation (NLS)

$$iu_t + \Delta u \pm u|u|^{p-2} = 0 \tag{2.1}$$

with periodic boundary conditions. Thus u is a complex function on $\mathbf{T}^d \times I$ (local) or $\mathbf{T}^d \times \mathbf{R}$ (global). The equation may be rewritten in Hamiltonian format as

$$u_t = i\frac{\partial H}{\partial \,\overline{u}} \tag{2.2}$$

where $H(\phi) = \frac{1}{2}\int_{\mathbf{T}^d} |\nabla\phi|^2 \mp \frac{1}{p}\int_{\mathbf{T}^d} |\phi|^p$. Both the Hamiltonian $H(\phi)$ and the L^2-norm $\int |\phi|^2$ are preserved under the flow. The $1D$ case $p = 4$ is special ($1D$ cubic NLS) because integrable and there are many invariants of motion. This aspect will however play no role in the present discussion. The possible sign choice $\pm$ in (2.1) corresponds to the focusing (resp. defocusing) case. In the focusing case, the Hamiltonian may be unbounded from below and blowup phenomena may occur (for $p \geq 2 + \frac{4}{d}$). The canonical coordinates are $(\mathrm{Re}\,\phi, \mathrm{Im}\phi)$ or alternatively $(\mathrm{Re}\widehat{\phi}, \mathrm{Im}\widehat{\phi})$. The formal Gibbs measure on this infinite dimensional phase space is given by

$$d\gamma_\beta = e^{-\beta H(\phi)} \prod_x d\phi(x) = e^{\pm\frac{\beta}{p}\int|\phi|^p} \cdot e^{-\frac{\beta}{2}\int|\nabla\phi|^2} \prod_x d\phi(x) \tag{2.3}$$

($\beta > 0$ is the reciprocal temperature and we may take $\beta = 1$ in this discussion). From Liouville's theorem, (2.3) defines an invariant measure for the flow of (2.1). Making this statement precise requires one to clarify the following two issues

(i) The rigorous construction (normalization) of the measure (2.3).
(ii) The existence problem for the flow of (2.1) on the support of the measure.

The first issue is well understood in the defocusing case. The case $D = 1$ is trivial, the case $D = 2$, p even integer is based on the Wick-ordering procedure (see [G-J]) and the normalization for $D = 3$, $p = 4$ is due to Jaffe [Ja]. In the focusing case, only the case $D = 1$ is understood [L-R-S] and normalization of the measure is possible for $p \leq 6$, restricting ϕ to an appropriate ball in $L^2(\mathbf{T})$.

The construction of a flow is clearly a PDE issue. The author succeeded in this in the $D = 1$ and $D = 2$, $p = 4$ cases ([B$_1$], [B$_2$]). For $D = 2$, $p = 4$ there is a natural PDE counterpart of the Wick-ordering procedure and equation (2.1) has to be suitably modified (this modification seems physically inessential however). We may summarize the results as follows.

Theorem 2.4. ($D = 1$)

(i) *In the defocusing case, the measure* (2.3) *appears as a weighted Wiener measure, the density being given by the first factor. The same statement is true in the focusing case for* $p \leq 6$, *provided one restricts the measure to an* L^2*-ball* $[\|\phi\|_2 \leq B]$. *The choice of* B *is arbitrary for* $p < 6$ *and* B *has to be sufficiently small if* $p = 6$.

(ii) *Assuming the measure exists, the corresponding* $1D$ *equation* (2.1) *is globally well-posed on a* K_σ *set* A *of data,* $A \subset \bigcap_{s<\frac{1}{2}} H^s(\mathbf{T})$, *carrying the Gibbs measure* γ_β. *The set* A *and the Gibbs measure* γ_β *are invariant under the flow.*

(Recall that the first part of the theorem is due to [L-R-S]).

Remarks.

(i) In dimension 1, the L^2-restriction is acceptable, since L^2 is a conserved quantity and a typical ϕ in the support of the Wiener measure is a function in $H^s(\mathbf{T})$, for all $s < \frac{1}{2}$. Instead of restricting to an L^2-ball, one may alternatively multiply with a weight function with a suitable exponential decay in $\|\phi\|_2$.

(ii) Let for each $N = 1, 2, \ldots$

$$P_N\phi = \phi_N = \sum_{|n|\leq N} \widehat{\phi}(n)\, e^{i\langle n,x\rangle} \tag{2.5}$$

be the restriction operator to the N first Fourier modes. Finite dimensional versions of the PDE model are obtained considering "truncated" equations

$$\begin{cases} iu_t^N + u_{xx}^N \pm P_N\left(u^N|u^N|^{p-2}\right) = 0 \\ u^N(0) = P_N\phi. \end{cases} \tag{2.6}$$

It is proved that for typical ϕ, the solutions u^N of (3.6) converge in the space $C_{H^s(\mathbf{T})}[0, T]$ for all time T and $s < \frac{1}{2}$ to a (strong) solution of

$$\begin{cases} iu_t + u_{xx} \pm P\left(u|u|^{p-2}\right) = 0 \\ u(0) = \phi. \end{cases} \tag{2.7}$$

Theorem 2.8. ($D = 2$, $p = 4$)

(i) *Denote $\widetilde{H}_N$ to be the Wick-ordered Hamiltonians, obtained by replacing*

$$|\phi_N|^4 \text{ by } |\phi_N|^4 - 4a_N|\phi_N|^2 + 2|a_N|^2 \quad \left(a_N = \sum_{|n|\leq N} \frac{1}{|n|^2+\rho} \sim \log N \right).$$

The corresponding measures $e^{-\beta \widetilde{H}_N(\phi)} \prod d\phi$ converge for $N \to \infty$ to a weighted 2-dimensional Wiener measure whose density belongs to all L^p-spaces. Denote $\widetilde{\gamma}_\beta$ to be this "Wick-ordered" Gibbs measure.

(ii) *The measure $\widetilde{\gamma}_\beta$ is invariant under the flow of the "Wick-ordered" equation*

$$iu_t + \Delta u - \left(u|u|^2 - 2u \int |u|^2 \right) = 0 \tag{2.9}$$

which is well-defined. More precisely, denoting u^N the solutions of

$$\begin{cases} iu_t^N + \Delta u^N - P_N \left(u^N|u^N|^2 - 2u^N \int |u^N|^2 \right) = 0 \\ u^N(0) = P_N\phi \end{cases} \tag{2.10}$$

the sequence

$$u^N(t) - \sum_{|n|\leq N} \widehat{\phi}(n)\ e^{i(\langle n,x\rangle + |n|^2 t)} \tag{2.11}$$

converges for typical ϕ in $C_{H^s(\mathbf{T}^2)}[0,T]$ for some $s > 0$, all time T, to

$$u(t) - \sum \widehat{\phi}(n)\ e^{i(\langle n,x\rangle + |n|^2 t)}. \tag{2.12}$$

Remarks.

(i) We repeat that the novelty of Theorem 2.8 lies in the second statement on the existence of a flow. The first statement is a classical result (see G-J]).

(ii) The second terms in (2.11), (2.12) are the solutions to the linear problem

$$\begin{cases} iu_t + \Delta u = 0 \\ u(0) = \phi. \end{cases} \tag{2.13}$$

Here a typical ϕ is a distribution, not a function. However the difference (2.12) between solutions of linear and nonlinear equation is an H^s-function for some $s > 0$, which is a rather remarkable fact.

(iii) The failure in $D = 2$ of typical ϕ to be an L^2-function makes the [L-R-S] construction for $D = 1$ inadequate to deal with the $D = 2$ focusing case. Some recent work on this issue is due to A. Jaffe, but for cubic nonlinearities in the Hamiltonian only. The problem for $D=2$, $p=4$ in the focusing case is open and intimately related to blowup phenomena ($p=4$ is critical in $2D$). Observe that in the preceding the invariance of the (Wick-ordered) Gibbs-measure implies the "quasi-invariance" of the free measure $\prod d^2\phi$. I. Gelfand* proposed to investigate this fact directly. Such approach would avoid normalization problems. For instance, in $1D$, one may consider the focusing equation $iu_t+u_{xx}+u|u|^{p-2}=0$ with $p\geq 6$ and prove this quasi-invariance local in time.

(iv) One may prove an analogue of Theorem 2.8 for $2D$ wave equations, with defocusing polynomial nonlinearity

$$u_{tt}-\Delta u+\int u+f'(u)=0 \quad (u \text{ real } \int>0), \tag{2.14}$$

$$f(u)=u^{2k}=0+(\text{lower order}) \tag{2.15}$$

and replacing $f(u)$ by its Wick ordering: $f(u)$: (cf. [G-J]). However, in this case there is no conservation of the L^2-norm and the result needs to be formulated in terms of the truncated equation. Observe that the (optimal) PDE result on local well-posedness deals with data $\phi\in H^{1/4}$ (see [L-S]).

The $1D$ cubic NLSE appears as the limit of the $1D$ Zakharov model (ZE)

$$\begin{cases} iu_t=-u_{xx}+nu \\ n_{tt}-c^2n_{xx}=c^2\left(|u|^2\right)_{xx} \end{cases} \tag{2.16}$$

when $c\to\infty$. The physical meaning of u,n,c are resp. the electrostatic envelope field, the ion density fluctuation field and the ion sound speed. This model is discussed in [L-R-S]. Defining an auxiliary field $V(x,t)$ by

$$\begin{cases} n_t=-c^2\, V_x \\ V_t=-n_x-|u|^2_x \end{cases} \tag{2.17}$$

we may write (2.16) in a Hamiltonian way, where

$$H=\tfrac{1}{2}\int\left[|u_x|^2+\tfrac{1}{2}\,(n^2+c^2\,V^2)+n|u|^2\right]\,dx \tag{2.18}$$

and $(\operatorname{Re}u,\ \operatorname{Im}u)$, $\left(\widetilde{n},\widetilde{V}\right)$ with $\widetilde{n}=2^{-1/2}n$, $\widetilde{V}=2^{-1/2}\int^x V$ as pairs of conjugate variables. Considering the associated Gibbs measure

$$e^{-\beta H}\cdot\chi_{\{\int|u|^2dx\leq B\}}\prod_x d^2u(x)\;d\widetilde{n}(x)\;d\widetilde{V}(x) \tag{2.19}$$

one gets the $1D$ cubic NLS Gibbs measure as marginal distribution of the u-field.

Theorem 2.20. [B3] *The $1D$ (ZE) is globally well-posed for almost all data $(u_0,\widetilde{n}_0,\widetilde{V}_0)$ in the support of the Gibbs measure which is invariant under the resulting flow.*

*Private communication

Remarks.

(i) In the study of invariant Gibbs measures, it suffices to establish local well-posedness of the IVP for typical data in the support of the measure. One may then exploit the invariance of the measure as a conservation law and generate a global flow. For instance, for the $1D$ NLS $iu_t + u_{xx} \pm u\,|u|^{p-2} = 0$, there is for $p = 4$ a global well-posedness result for L^2-data (L^2 is conserved). However, for $p > 4$, we only dispose presently of a local result (in the periodic case) for data ϕ satisfying

$$\begin{cases} \phi \in H^s \,, \; s > 0 & (p \leq 6) \\ \phi \in H^s \,, \; s > s_* \,, \quad p = 2 + \frac{4}{1-2s_*} & (p > 6) \end{cases} \tag{2.21}$$

and a global flow is established from the invariant measure considerations.

(ii) There has been other investigations in $1D$ on invariant measures, mostly by more probabilistic arguments. In this respect, we mention the works of McKean–Vaninski and in particular Mckean [McK] on the $1D$ cubic NLS. These methods are more general but give less information on the flow.

3. Symplectic Capabilities, Squeezing and Growth of Higher Derivatives

The works of Gromov and Ekeland, Hofer, Zehnder, Viterbo lead to new finite dimensional symplectic invariants, different from Liouville measure on the phase space. Let us recall the following construction of a symplectic capacity for open domains O in $\mathbf{R}^n \times \mathbf{R}^n$, $dp \wedge dq$. Call a smooth function f m-admissible ($m > 0$) if $f = 1$ on a neighborhood of O and $f = 0$ on a nonempty subdomain of O. Denote V_f the associated Hamiltonian vector field $\left(\frac{\partial f}{\partial p}\,,\, -\frac{\partial f}{\partial q}\right)$. Define the symplectic invariant

$$\begin{aligned} c_{2n}(O) = \inf\{m > 0 \mid V_f \text{ has nontrivial periodic orbit of period } \leq 1\,, \\ \text{whenever } f \text{ is } m\text{-admissible for } O\}. \end{aligned} \tag{3.1}$$

Then $c_{2n}(\cdot)$ is monotonic and translation invariant and scales as $c_{2n}(\tau O) = \tau^2 c_{2n}(O)$. The main property is that

$$c_{2n}(B_\rho) = \pi\rho^2 = c_{2n}\left(\Pi_\rho\right) \tag{3.2}$$

where B_ρ is the ball $B_\rho = \left\{|p|^2 + |q|^2 < \rho^2\right\}$ and $\prod_\rho$ a cylinder, say $\prod_\rho = \{p_1^2 + q_1^2 < \rho^2\}$. As a corollary, there is no symplectic squeezing of a ρ-ball in a cylinder of width ρ', $\rho' < \rho$.

Exploiting such invariant in Hamiltonian PDE requires an infinite dimensional setting. Notice that although the theory described above is finite dimensional, a conclusion such as (3.2) is dimension free. An appropriate "finite dimensional approximation" appears to be possible if the flow S_t of the considered equation is of the form

$$\text{linear operator } + \text{ “smooth compact operator”} \tag{3.3}$$

or, more generally, if the evolution of individual Fourier modes on a finite time interval is approximately the same as in a truncated model $\dot{v} = J\nabla H(v,x,t)$, $v = P_N v$. Here the cutoff N should only depend on the required approximation, the time interval $[0,T]$ and the size of the initial data in phase space. Here and also in (3.3), the phase space has to be defined in a specific way, corresponding to the finite dimensional normalizations. Hence, the flow properties derived this way relate to a specific "symplectic Hilbert space", for instance

L^2 for nonlinear Schrödinger equations (in any dimension)
$H^{1/2} \times H^{1/2}$ for nonlinear wave equations (in any dimension)
$H^{-1/2}$ for KdV type equations

and "non-squeezing" refers to that particular space.

Theorem 3.4. ([B_6], [Kuk_2])

There is non-squeezing of balls in cylinders of smaller width

(i) *for nonlinear wave equations $u_{tt} = \nabla u + p(u;t,x)$ with smooth nonlinearity of arbitrary polynomial growth in u in dimension 1 and polynomial in u of degree ≤ 4 (resp. ≤ 2) in dimension 2 (resp. 3, 4).*
(ii) *for certain 1D nonlinear Schrödinger equations.*

The interest of the squeezing or non-squeezing properties lies in its relevance to the energy transition to higher modes, more precisely whether for instance part of the energy may leave a given Fourier mode, which would correspond to squeezing in a small cylinder. The non-squeezing implies also the lack of uniform asymptotic stability of bounded solutions, i.e. $\operatorname{diam} S_t(B_\rho)$ does not tend to 0 for $t \to \infty$ if $\rho > 0$.

The drawback of those results is that they do not relate to properties of the flow in a classical sense, because of the phase space topology. On the other hand, S. Kuksin showed recently that in fact certain squeezing of balls in cylinders may occur in spaces of higher smoothness, if one considers for instance a nonlinear wave equation $u_{tt} = \rho\Delta u + p(u)$ where ρ is a small parameter (small dispersion) ([Kuk_3]). The squeezing phenomena appear in some finite time and are stronger when $\rho \to 0$.

As far as the behaviour of individual smooth solutions are concerned, some examples are obtained in [B_6] of Hamiltonian PDE (in NLS or KdV form) defined as a smooth perturbation of a linear equation, showing in particular that higher derivatives of solutions $u(t)$ for smooth data $u(0) = \phi$ need not be bounded in time. For instance

Proposition 3.5. *There is a Hamiltonian* NLSE *with smooth and local nonlinearity such that* $S_t(B^s(\delta)), t > 0$ *is not a bounded subset of* H^{s_0}, *for any* $s < \infty, \delta > 0$. *Here* $B^s(\delta)$ *denote* $\{\varphi \in H^s \mid \|\varphi\|_s < \delta\}$ *and* s_0 *is some fixed number.*

Another example, closely related to the discussion in the next section is the following. Considering a linear Schrödinger equation

$$-iu_t = -u_{xx} + V(x)u \tag{3.6}$$

where $V(x)$ is a real smooth periodic potential and the periodic spectrum $\{\lambda_k\}$ of $-\frac{d^2}{dx^2} + V$ satisfies a "near resonnance" property

$$\text{dist } (\lambda_{n_j}, \mathbf{Z}\lambda_{n_0}) \to 0 \quad \text{rapidly for} \quad j \to \infty \tag{3.7}$$

for some subsequence $\{n_j\}$. We construct a Hamiltonian perturbation $\Gamma(u) = \frac{\partial}{\partial\,\bar{u}} G$ such that the solution $u_{\varepsilon,q}$ of the IVP

$$\begin{cases} -iu_t = -u_{xx} + V(x)u + \varepsilon\,\Gamma(u) \\ u(0) = q \end{cases} \tag{3.8}$$

satisfies

$$\inf_{q \in O} \sup_t \|u_{\varepsilon,q}(t)\|_{H^{s_0}} \to \infty \quad \text{for} \quad \varepsilon \to 0. \tag{3.9}$$

Here s_0 is again some positive integer and O is some nonempty open subset of $H^{s_0}(\mathbf{T})$.

The general procedure of constructing global solutions piecing together local solutions, using conserved quantities of low smoothness, permit to bound higher derivatives as $C^{|t|}$ for $|t| \to \infty$. The exponential estimate was more recently improved to a polynomial one, provided the Cauchy problem may be solved subcritically with respect to the H^1-norm, for which we assume an *a priori* bound from conservation of the Hamiltonian. The question what is, generically speaking, the "true" order of growth seems an important open problem.

4. Persistency of Periodic and Quasi-periodic Solutions Under Perturbation

One of the most exciting recent developments in nonlinear PDE is the use of the classical KAM-type techniques to construct time quasi-periodic solutions of Hamiltonian equations obtained by perturbation of a linear or integrable PDE. This subject is rapidly developing.

In this brief discussion, we only consider perturbations of linear equations. We work in the real analytic category. Important contributions are due to S. Kuksin [Kuk$_1$], using the standard KAM scheme and more precisely infinite dimensional versions of Melnikov's theorem on the persistency of n-dimensional tori in systems with $N > n$ degrees of freedom. His work yields a rather general theory and we mention only some typical examples of applications to $1D$ nonlinear wave or Schrödinger equations

$$w_{tt} = \left(\frac{\partial^2}{\partial x^2} - V(x;a)\right) w - \varepsilon \frac{\partial \varphi}{\partial w} (x, w; a) \tag{4.1}$$

$$-iu_t = -u_{xx} + V(x,a)u + \varepsilon \frac{\partial \varphi}{\partial |u|^2} \left(x, |u|^2; a\right) u. \tag{4.2}$$

Here $V(x,a)$ is a real periodic smooth potential, depending on n outer parameters $a = (a_1, \ldots, a_n)$. Denote $\{\lambda_j(a)\}$ the Dirichlet spectrum of the Sturm–Liouville operator $-\frac{d^2}{dx^2} + V(x,a)$. Thus $\lambda_j(a) = \pi^2 j^2 + 0(1)$ and we assume the following nondegeneracy condition

$$\det \{\partial \lambda_j(a)/\partial a_k \mid 1 \leq j, k \leq n\} \neq 0 \tag{4.3}$$

(this condition is a substitute for the classical "twist" condition). Denoting $\{\varphi_j\}$ the corresponding eigenfunctions, the $2n$-dimensional linear space

$$Z^0 = \text{span } \{\varphi_j, i\varphi_j \mid 1 \leq j \leq n\} \tag{4.4}$$

is invariant under the flow of equation (4.2) for $\varepsilon = 0$ and foliated into invariant n-tori

$$T^n(I) = \left\{ \sum_{j=1}^{n} (x_j^+ + ix_j^-)\varphi_j \mid (x_j^+)^2 + (x_j^-)^2 = 2I_j, \ j = 1, \ldots, n \right\} \tag{4.5}$$

which are filled with quasi-periodic solutions of (4.2) for $\varepsilon = 0$. A typical result from [Kuk$_1$] is that under assumption (4.3), for most parameter values of a there is an invariant torus $\sum_{a,I}^{\varepsilon}(\mathbf{T}^n)$ nearby the unperturbed torus $\sum_{a,I}^{0}$ given by (4.5) and filled with quasi-periodic solutions of (4.2). The frequency vector ω_ε of a perturbed solution will be $c\varepsilon$ close to $\omega = (\lambda_1, \ldots, \lambda_n)$ of the unperturbed one.

The methods in [K$_1$] leave out the case of periodic boundary conditions, because of certain limitations of the KAM method (second Melnikov condition) excluding multiplicities in the normal frequencies. A different approach has been recently used by W. Craig and C. E. Wayne [C-W$_{1,2}$], based on the Liapounov–Schmidt decomposition and leading to time periodic solutions of perturbed equations under periodic boundary conditions. This method consists in splitting the problem in a (finite dimensional) resonnant part (Q-equation) and an infinite dimensional non-resonnant part (P-equation). In the PDE-case (contrary to the case of a finite dimensional phase space), small divisor problems appear when solving the P-equation by a Newton iteration method, also in the time periodic case. Writing u in the form

$$u = \sum_{m,k} \widehat{u}(m,k)\, e^{im\lambda t}\, \varphi_k(x) \tag{4.6}$$

and letting the linearized operator act on the Fourier coefficients $\widehat{u}(m,k)$, one gets operators of the form

$$(m\lambda - \lambda_k) + \varepsilon\, T \tag{4.7}$$

where the first term is diagonal and T is essentially given by Toeplitz operators with exponentially decreasing matrix elements. The main task is then to obtain reasonable bounds on their inverses. The problem is closely related to a line of research around localization in the Anderson model with quasi-periodic potential. In this case, the operator T in (4.7) is replaced by $-\Delta$, $\Delta =$ lattice Laplacian, and the first term plays the role of the potential. Of primary importance is the structure of "singular sites", i.e. the pairs $(m,\, k)$ such that $|m\lambda - \lambda_k|$ is small. If we think of λ_k as k^2, it is clear that these sites have separation $\to \infty$ for $k \to \infty$. Let us consider the higher dimensional problem for time periodic solutions. We assume the potential $V(x)$ of the form

$$V(x) = V_1(x_1) + \cdots + V_d(x_d) \tag{4.8}$$

in order to avoid certain problems (the dependence of spectrum and eigenfunctions on V) appearing for $d \geq 2$, other than the small divisor issue that is our primary concern here. Alternatively, one may replace in (4.1), (4.2) the term $V.u$ by a Fourier multiplier to introduce a frequency parameter. The eigenfunctions are then simply exponentials. Replacing λ_k by $|k|^2 = k_1^2 + \cdots + k_d^2$, it turns out that the singular sites may still be partitioned into distant clusters, of diameter and mutual distance $\to \infty$ for $|k| \to \infty$. This enables one to large extent to reproduce the arguments from [C-W$_{1,2}$] to deal with that case also.

We now pass to the quasi-periodic problem. The singular sites are now the pairs $(m,\, k)$ satisfying

$$|\langle m,\, \lambda\rangle - \lambda_k| < 1 \text{ where } m \in \mathbb{Z}^b,\, b > 1. \tag{4.9}$$

The structure of those is clearly different here and already in case of a finite dimensional phase space (i.e. finitely many k's), new arguments are needed. There is resemblance with the works of Fröhlich, Spencer, Wittwer [F-S-W] and Surace [Sur] on localization for a quasi-periodic potential. In particular, one relies on a multi-scale analysis. The first step in these investigations is to recover the KAM and Melnikov results using this Liapounov–Schmidt technique. Next, one considers the PDE applications, where the phase space is infinite dimensional. In [B$_4$], we discuss NLS and NLW in $1D$ obtained by perturbing a linear equation. In [B$_5$], $2D$-problems are considered. The presence of spectral clusters of unbounded size leads to certain difficulties we may deal with in $d = 2$ but so far not for larger dimension $d > 2$ (except for time periodic solutions). One may essentially formulate the main model result as follows.

Theorem 4.10. (see [B$_5$]) *Consider a perturbed Schrödinger equation*

$$iu_t - \Delta u + V(x)u + \varepsilon \frac{\partial H}{\partial \overline{u}}(u, \overline{u}, x) = 0 \tag{4.11}$$

or

$$iu_t - \Delta u + (u * V) + \varepsilon \frac{\partial H}{\partial \overline{u}}(u, \overline{u}, x) = 0 \tag{4.12}$$

where $x \in \prod^2, V(x) = V_1(x_1) + V_2(x_2)$ *a real periodic potential in* (4.11) *and* $u * V$ *a real Fourier multiplier in* (4.12). *Denote* $\{\mu_n\}$ *the periodic spectrum of* $-\Delta + V$ *and* $\{\varphi_n\}$ *the corresponding eigenfunctions. Fix indices* $n_1, n_2, \dots, n_b$ *and consider* $\lambda_1 = \mu_{n_1}, \dots, \lambda_b = \mu_{n_b}$ *as parameters (letting* V *vary in an appropriate way). We assume following non-resonnance property* (1^e *Melnikov condition,* cf. [Kuk$_1$])

$$|\langle m, \lambda\rangle + \mu_n| > N_0^{-c} \tag{4.13}$$

for $|m| \leq N_0$, $|n| \leq N_0$ *and* $n \notin \{n_1, \dots, n_b\}$.

(The μ_n *may have a weak dependence on* λ*).*

Consider the solution to the unperturbed equation $\varepsilon = 0$

$$u_0(x, t) = \sum_{j=1}^{b} a_j e^{i\lambda_j t} \varphi_{n_j}(x) \tag{4.14}$$

with $\{a_j\}$ *fixed. For* ε *small enough (depending in particular on* N_0*), there is a (Cantor) set* Δ *of frequencies* λ *(depending on* $|a_j|$ $(j = 1, \dots, b)$*) with* $\text{mes}(C\Delta) \to 0$ for $\varepsilon \to 0$ *in the parameter set, such that if* $\lambda \in \Delta$, *there is a perturbed frequency*

$$|\lambda' - \lambda| < C\varepsilon \tag{4.15}$$

and a perturbed solution

$$u_\varepsilon(x, t) = \sum_n \hat{u}_\varepsilon(n)(t)\varphi_n(x) \tag{4.16}$$

$$\hat{u}_\varepsilon(n)(t) = \sum_m \hat{u}_\varepsilon(n, m)e^{i\langle\lambda', m\rangle t} \tag{4.17}$$

with

$$\hat{u}_\varepsilon(n_j, e_j) = a_j \qquad (e_j = j \text{ unit vector in } \mathbb{Z}^b) \tag{4.18}$$

$$\sum_{(n, m)\notin S} e^{(|m|+|n|)^c} |\hat{u}_\varepsilon(n, m)| < \varepsilon^{1/2} \tag{4.19}$$

with

$$S = \{(n_j, e_j) | j = 1, \dots, b\}.$$

Equations (4.11), (4.12) are parameter dependent, since $V = V(x, \lambda)$. This may sometimes be avoided, relying on amplitude-frequency modulation based on the nonlinear term satisfying an appropriate twist condition (see [C-$W_{1,2}$]). It is our aim to investigate also perturbations of Birkhoff integrable systems, following the same approach. Written in an appropriate normal form, the unperturbed Hamiltonian may be given the form

$$H(p, q, y) = h(p) + \frac{1}{2}\langle A(p)y, y\rangle + 0(|y|^3) \tag{4.20}$$

(cf. [Kuk_1]). Here the phase-space appears as $Z = Z_0 \oplus Y$ and $(p_1, \ldots, p_n, q_1, \ldots, q_n)$ are action-angle variables for Z_0. Specifying $p = a$, write $h(p) = h(a) + \langle \frac{\partial h}{\partial p}(a), p - a\rangle + 0(|p-a|^2)$. Then the λ-frequency vector is given by $\frac{\partial h}{\partial p}(a)$. Thus the usual twist condition

$$\det(\mathrm{Hess}\, H) \neq 0 \tag{4.21}$$

corresponds to (4.3). At this stage, the problem appears essentially as perturbation of a linear system.

References

[B_1]. J. Bourgain, *Periodic nonlinear Schrödinger equations and invariant measures*, preprint IHES (1993), Comm. Math. Phys. **166** (1994), 1–26.

[B_2]. J. Bourgain, *Invariant measures for the 2D-defocusing nonlinear Schrödinger equation*, preprint IHES (1994), Comm. Math. Phys. **176** (1996), 421–445.

[B_3]. J. Bourgain, *On the Cauchy and invariant measure problem for the periodic Zakharov system*, Duke Math. J. **76**, no. 1 (1994), 175–202.

[B_4]. J. Bourgain, *Construction of quasi-periodic solutions for Hamiltonian perturbations of linear equations and applications to nonlinear PDE*, International Math. Research Notices, 1994, Vol. 11, 475–497.

[B_5]. J. Bourgain, *Quasi-periodic solutions of Hamiltonian perturbations of 2D linear Schrödinger equations*, preprint IHES (1995), to appear in Annals. of Math.

[B_6]. J. Bourgain, *Aspects of long time behaviour of solutions of nonlinear Hamiltonian evolution equations*, Geometric Functional Analysis, Vol. 5 no. 2 (1995), 105–140.

[C-W_1]. W. Craig and C. Wayne, *Newton's method and periodic solutions of nonlinear wave equations*, Comm. Pure and Applied Math. **46** (1993), 1409–1501.

[$C-W_2$]. W. Craig and C. Wayne, *Periodic solutions of Nonlinear Schrödinger equations and the Nash–Moser method*, Preprint.

[E-H]. I. Ekeland and H. Hofer, *Symplectic Topology and Hamiltonian Dynamics ll*, Math. Z. **203** (1990), 553–567.

[F-S-W]. J. Frölich, T. Spencer and P. Wittwer, *Localization for a class of one dimensional quasi-periodic Schrödinger operators*, Comm. Math. Phys. **132**, no. 1 (1990), 5–25.

[G-J]. J. Glimm and A. Jaffe, *Quantum Physics*, Springer-Verlag (1987).

[Grom]. M. Gromov, *Pseudo holomorphic curves on almost complex manifolds*, Invent. Math. **82** (1985), 307–347.

[Ja]. A. Jaffe, *Private Notes.*

[Kuk$_1$]. S. Kuksin, *Nearly Integrable Infinite-Dimensional Hamiltonian Systems*, LNM **1556**, Springer-Verlag, (1993).

[Kuk$_2$]. S. Kuksin, *Infinite-dimensional symplectic capacities and a squeezing theorem for Hamiltonian PDE's, Comm. Math. Phys.* **167** (1995), 531–552.

[Kuk$_3$]. S. Kuksin, *On squeezing and flow of energy for nonlinear wave equations*, GAFA, Vol. 5, no. 4 (1995), 668–701.

[L-R-S]. J. Lebowitz, R. Rose and E. Speer, *Statistical mechanics of the nonlinear Schrödinger equation*, J. Stat. Phys. **50** (1988), 657–687.

[L-S]. H. Lindbladt and C. Sogge, *On existence and scattering with minimal regularity for semi-linear wave equations*, preprint.

[McK]. H. McKean, Preprint.

[Sur]. S. Surace, *The Schrödinger equation with a quasi-periodic potential*, Trans. AMS **320**, no. 1 (1990), 321–370.

Reprinted from Proc. Int. Congr. Math., 1994

THE WORK OF PIERRE-LOUIS LIONS

by

S. R. S. VARADHAN

Courant Institute, New York University
251 Mercer Street, New York, NY 10012, USA

Pierre-Louis Lions has made unique contributions over the last fifteen years to mathematics. His contributions cover a variety of areas, from probability theory to partial differential equations (PDEs). Within the PDE area he has done several beautiful things in nonlinear equations. The choice of his problems has always been motivated by applications. Many of the problems in physics, engineering and economics when formulated in mathematical terms lead to nonlinear PDEs; these problems are often very hard. The nonlinearity makes each equation different. The work of Lions is important because he has developed techniques that, with variations, can be applied to classes of such problems. To say that something is nonlinear does not mean much; in fact it could even be linear. The entire class of nonlinear PDEs is therefore very extensive and one does not expect an all-inclusive theory. On the other hand, one does not want to treat each example differently and have a collection of unrelated techniques. It is thus extremely important to identify large classes that admit a unified treatment.

In dealing with nonlinear PDEs one has to allow for nonclassical or nonsmooth solutions. Unlike the linear case one cannot use the theory of distributions to define the notion of a weak solution. One has to invent the appropriate notion of a generalized solution and hope that this will cover a wide class and be sufficient to yield a complete theory of existence, uniqueness, and stability for the class.

Due to the very limited time that is available, I shall focus on three areas within nonlinear PDE where Lions has made major contributions. The first is the so called "viscosity method". This development is a long story that started with some work in collaboration with Crandall. Over many years, in partial collaboration with others (besides Crandall, Evans and Ishii), Lions has developed the method, which is applicable to the large class of nonlinear PDEs known as fully nonlinear second order degenerate elliptic PDEs. The class contains very many important subclasses that arise in different contexts.

By solving a nonlinear PDE one is trying to solve an equation involving an unknown function and its derivatives. Let u be a function in a region G in some R^n and let $Du, D^2u, \ldots, D^k u$ be its derivatives of order up to k. A nonlinear PDE is an equation of the form

$$F[x, u(x), (Du)(x), (D^2u)(x), \ldots, (D^k(u)(x))] = 0 \text{ in } G$$

with some boundary conditions on ∂G. Such a PDE is said to be nonlinear and of order k. The viscosity method applies in cases where $k = 2$ and $F(x, u, p, H)$ has certain monotonicity properties in the arguments u and H. More precisely, it is nondecreasing in u and nonincreasing in H. Here u is a scalar and H is a symmetric matrix of size $n \times n$ with the natural partial ordering for symmetric matrices.

Some of the many examples of such functions are described below.

Linear elliptic equations:

$$-\sum_{i,j} a^{ij}(x) \frac{\partial^2 u}{\partial x_i \partial x_j}(x) + f(x) = 0$$

where the matrix $a^{ij}(x)$ is uniformly positive definite.

In this case the function F is given by

$$F(x, u, p, H) = -\text{Trace } [a(x)H] + f(x).$$

First order equations:

$$f(x, u(x),\ (Du)(x)) = 0$$

These include Hamilton–Jacobi equations where it all started. One added a term of the form $\epsilon\Delta$ to the equation and constructed the solution in the limit as ϵ went to zero. The theory owes its name to its early origins.

If one has a family F_α of such functions one can generate a new one by defining

$$F = \sup_\alpha F_\alpha .$$

If one has a two-parameter family $F_{\alpha\beta}$ of such functions one can generate a new one by defining

$$F = \sup_\alpha \inf_\beta F_{\alpha\beta} .$$

Such examples arise naturally in control theory and game theory and are referred to as Hamilton–Jacobi–Bellman and Isaacs equations.

In order to understand the notion of a generalized solution it is convenient to talk about supersolutions and subsolutions. Suppose u is a subsolution, i.e.

$$F(x, u(x),\ (Du)(x),\ (D^2u(x))) \leq 0$$

and we have another function ϕ, which is smooth, such that $u - \phi$ has a maximum at some point $\hat{x}$. Then by calculus $Du(\hat{x}) = D\phi(\hat{x})$ and $D^2(u)(\hat{x}) \leq D^2(\phi)(\hat{x})$. From the monotonicity properties of F it follows that

$$F(\hat{x}, u(\hat{x}),\ (Du)(\hat{x}),\ (D^2u(\hat{x}))) \geq F(\hat{x}, u(\hat{x}),\ (D\phi)(\hat{x}),\ (D^2\phi(\hat{x}))) .$$

Therefore

$$F(\hat{x}, u(\hat{x}),\ (D\phi)(\hat{x}),\ (D^2\phi(\hat{x}))) \leq 0.$$

The last inequality makes sense without any smoothness assumption on u. We can try to define a nonsmooth subsolution as a u that satisfies the above for arbitrary smooth ϕ and $\hat{x}$ provided $u-\phi$ has a maximum at $\hat{x}$. The definition of a supersolution is similar, and a solution is one that is simultaneosly a super and a subsolution.

Let us consider first a Dirichlet boundary value problem where we want to find a u that solves our equation and has boundary value zero.

The main step is to establish the key comparison theorem (with a long history that began with the work of Crandall and Lions and saw an important contribution from Jensen) that if u is a subsolution and if v is a supersolution in a bounded domain G and if $u \leq v$ on the boundary ∂G then $u \leq v$ in $G \cup \partial G$. This requires some mild regularity conditions on F as well as some nondegeneracy conditions. After all, we have not ruled out $F \equiv 0$. Once such conditions are imposed one can establish the key comparison theorem. From this point on, the theory proceeds in a way similar to the classical Perron's method for solving the Dirichlet problem. Assuming that there is at least one subsolution $\bar{u}$ and at least one supersolution $\bar{v}$ with the given boundary value, one establishes that

$$W(x) = \sup\ \{w(x) : \bar{u} \leq w \leq \bar{v},\ w \text{ is a subsolution}\}$$

is a solution. The comparison theorem is of course enough to guarantee uniqueness. The constructibility of $\bar{u}$ and $\bar{v}$ depends on the circumstances and is relatively easy to establish.

The richness of the theory is in its flexibility. One can prove stability results of various kinds and the validity of various approximation schemes. One can modify the theory to include Neumann boundary conditions. This is tricky because one has to interpret the normal derivative suitably for a function that has no smoothness requirements and the boundary condition can be nonlinear as well. Treating parabolic equations is not any different. One can just consider t as another variable.

I would suggest the survey article by Crandall, Ishii, and Lions that appeared in the Bulletin of the American Mathematical Society in 1992 for those who want to read more about this area.

The second body of work that I want to discuss has to do with the Boltzmann equation and similar equations. During the last six or seven years Pierre-Louis Lions has played the central role in new developments in the theory of the Boltzmann equation and similar transport equations. These are important in kinetic theory and arise in a wide variety of physical applications. We will for simplicity stay within the context of the Boltzmann equation. In R^3 we have a collection of particles moving along and interacting through "collisions" among themselves. As we do not want to keep track of the positions and velocities of the particles individually, we abstract the situation by the density $f(x, v)$ of particles that are at position x with velocity

v. Even if there is no interaction, the density $f(x, v)$ will change in time due to uniform motion of the particles. The time-dependent density $f(t, x, v)$ will satisfy the equation

$$\frac{\partial f}{\partial t} + v.\nabla_x f = 0.$$

The collisions will change this equation to

$$\frac{\partial f}{\partial t} + v.\nabla_x f = Q(f, f).$$

Here Q is a quadratic quantity that represents binary collisions and its precise form depends on the nature of the interaction. Generally it looks like

$$Q(f, f) = \int\int_{R^3 \times s^2} dv_* d\omega B(v - v_*, \omega) \{f' f'_* - f f_*\}.$$

The notation here is standard: v and v_* are the incoming velocities and v' and v'_* are the outgoing velocities. B is the collision kernel. For given incoming velocities v and v_*, ω on the sphere S^2 parametrizes all the outgoing velocities compatible with the conservation of energy and momenta.

$$v' = v - (v - v_*, \omega)\omega, \quad v'_* = v + (v - v_*, \omega)\omega$$

and f', f_*, f'_* are $f(t, x, v)$ with v replaced by the correspondingly changed v', v_*, and v'_*.

This problem of course has a long history. Smooth and unique solutions had been obtained for small time or globally for initial data close to equilibrium. Carleman had studied spatially homogeneous solutions. But a general global existence theorem had never been proved. The work of Lions (in collaboration with DiPerna) is a breakthrough for this and many other related transport problems of great physical interest.

Let me spend a few minutes giving you some idea of the method as developed by Lions and others (mostly his collaborators).

Although the nonlinearity looks somewhat benign it causes a serious problem in trying to establish any existence results. The collision term is quadratic and involves both positive and negative terms. To carry out any analysis one must control each piece separately. One gets certain a priori estimates from the conservation of mass and energy. The Boltzmann H-theorem gives an important additional control if one starts with an initial data with finite entropy. If we denote by Q^+ and Q^- the positive and negative terms in the collision term with considerable effort one is able to obtain only local L_1 bounds on $(1+f)^{-1}Q^{\pm}(f, f)$. The weak solutions are therefore formulated in terms of $\log(1 + f)$. As there are no smoothness estimates in x one has to show that the velocity integrals contained in Q provide the compactification needed to make the weak limit behave properly.

This idea of "velocity averaging", which is central to these methods, is easy to state in a simple context. Suppose we have a function $g(x, v)$ in $R^N \times R^N$ and for some reasonable function $a(v)$ we have a local L_p estimate on $a(v).\nabla_x g(x, v)$. Then for a good test function ψ the velocity integral

$$\int_{R^N} \psi(v) g(x, v)\, dv$$

is in a suitable Sobolev space. Another important step that is needed in dealing with the Vlasov equation is the ability to integrate vector fields with minimal regularity. In nonlinear problems you have to learn to live with the regularity that the problem gives you. The writeup by Lions in the Proceedings of the last ICM (Kyoto 1990) provides a survey with references.

The third and final topic that I would like to touch on is the contribution Lions has made to a class of variational problems. There are many nonlinear PDEs that are Euler equations for variational problems. The first step in solving such equations by the variational method is to show that the extremum is attained. This requires some coercivity or compactness. If the quantity to be minimized has an "energy"-like term involving derivatives, then one has control on local regularity along a minimizing sequence. This usually works if the domain is compact. If the domain is noncompact the situation is far from clear. Take for instance the problem of minimizing

$$\int_{R^N} |(\nabla f)(x)|^2 dx - \int\int V(x - y) f^2(x) f^2(y)\, dx\, dy$$

over functions f with L_2 norm λ (fixed positive number). Here $V(.)$ is a reasonable function decaying at ∞. Because of translation invariance, the minimizing sequence must be centered properly in order to have a chance of converging. The key idea is that, in some complicated but precise sense, if the minimizing sequence cannot be centered, then any member of the sequence can be thought of as two functions with supports very far away from each other. If we denote the infimum by $\sigma(\lambda)$, then along such sequences the infimum will be $\sigma(\lambda_1) + \sigma(\lambda_2)$ with $\lambda_1 + \lambda_2 = \lambda 0 < \lambda_1, \lambda_2 < \lambda$ rather than $\sigma(\lambda)$. If independently one can show that $\sigma(\lambda)$ is strictly subadditive, then one can prove the existence of a minimizer. This idea has been developed fully and applied successfully by Lions to many important and interesting problems.

See the papers in Annales de l'Institut Henri Poincaré, Analyse Non Linéaire 1984 by Lions for many examples where this point of view is successfully used.

References

[1] Michael G. Crandall, Hitoshi Ishi, and Pierre-Louis Lions, *Users Guide to viscosity solutions of second order partial differential equations*, Bull. Amer. Math. Soc. (New Series) **27**, no. 1 (1992), 1–67.

[2] Pierre-Louis Lions, *On Kinetic Equations*, Proceedings of the ICM, Kyoto, 1990, vol. II, 1173–1185.

[3] ———, *The concentration-compactness principle in the calculus of variations. The locally compact case*, Parts 1 and 2, Annales de l'IHP, Analyse non-linéaire, vol. I (1984), 109–145 and 224–283.

PIERRE–LOUIS LIONS

Né le 11.8.1956 à Grasse (France)
Marié, un enfant
Adresse personnelle: 57 rue Cambronne, Paris 75015 — Tél: 44.49.93.16
Adresse professionnelle: CEREMADE, Université Paris-Dauphine, Place de Lattre de Tassigny, 75775 Paris Cedex 16 — Tél: 44.05.46.74.

ETUDES

Ecole Normale Supérieure (Ulm), 1975–1979.
Thèse d'Etat (Directeur de thèse: H. Brézis), Université Pierre et Marie Curie, 1979.

ACTIVITES PROFESSIONNELLES

Attaché de Recherche au C.N.R.S., 1979–1981.
Professeur à l'Université Paris-Dauphine, depuis 1981 (promotion en classe exceptionnelle en 1989). Détachement au CNRS (Directeur de Recherche, classe exceptionnelle) depuis 1995.
Professeur de Mathématiques Appliquées (à temps partiel) à l'Ecole Polytechnique, depuis 1992.
Administrateur du groupe Alcatel-Alsthom, depuis 1996.
Conseil: en Contrôle Optimal au Laboratoire d'Automatique de l'Ecole des Mines de Paris, de 1981 à 1984; en Analyse Numérique et Modélisation à CISI Ingénierie, de 1981 à 1997; en Physique Mathématique au CEA, de 1986 à 1996; en Traitement d'Images à Cognitech Inc. (Santa-Monica), de 1992 à 1996; en Calcul Scientifique à CRS4 (Cagliari) de 1994 à 1997; en techniques quantitatives en Finance et Gestion du Risque à la Caisse Autonome de Refinancement (CDC) de 1995 à 1997; en Modélisation et Calcul Scientifique à Péchiney, de 1995 à 1997; Aérospatiale Espace et Défense, depuis 1996.
Evaluations scientifiques: Morpho Systèmes (1992), Symah Vision (1995).

DISTINCTIONS

Cours Peccot, Collège de France, 1983–1984.
Chaire de la Miller Foundation, Université de Berkeley, 1987.
Conférencier invité au Congrès International des Mathématiciens, Varsovie 1983.
Conférencier invité au Congrès International des Mathématiciens, Kyoto 1990.
Conférencier plénier au Congrès ICIAM, Washington, 1991.
Conférencier plénier au Congrès Européen des Mathématiciens, Paris, 1992.
Conférencier plénier au Congrès de la Société Mathématique Japonaise, Tokyo, 1993.
Sackler Distinguished Lectures in Mathematics, Tel Aviv, 1994.

Conférencier plénier au Congrès International des Mathématiciens, Zürich, 1994.
Conférence "Progress in Mathematics", AMS, Minneapolis, 1994.
Landau Lectures, Jerusalem, 1994.
Andrejewski Lectures, Berlin-Leipzig, 1995.
Conférencier plénier au Congrès de la Société Mathématique Allemande, Ulm, 1995.
Chaire Galilée, Pise, 1996–1997.

Lezione Leonardo da Vinci, Milan, 1997.
Conférencier plénier au Congrès de la Société de Mathématiques Appliquées Espagnole, Vigo, 1997.
Conférencier plénier au Congrès de la Société de Mathématiques Appliquées Italienne, Giardini Nexos, 1998.
Conférencier plénier au Congrès du GAMM, Metz, 1999.
Conférencier plénier au Congrès de la Société Mathématique Italienne, Naples, 1999.
Conférencier plénier au Congrès de la SMAI, Vieux-Boucau 2000.
Conférence "Claude Bernard", Royal Society, Londres, 2000.

Prix de la Fondation Doistau-Blutet, Académie des Sciences, 1986.
Prix IBM, Paris, 1987.
Prix d'équipe Philip Morris pour la Science, Paris, 1991.
Prix Ampère, Académie des Sciences, 1992.
Médaille Fields, Zürich 1994.

Membre de l'Académie des Sciences, Paris.
Membre de l'Accademia de Scienze, Lettere e Arti di Napoli.
Membre de l'Academia Europea.

Docteur Honoris Causa : Heriot-Watt University, Edimbourg, 1995 ; City University of Hong-Kong, 1999.

Chevalier de la Légion d'Honneur.

RESPONSABILITES COLLECTIVES

Responsabilités administratives

Directeur de l'Unité d'Enseignement et de Recherche "Mathématiques de la Décision" de 1986 à 1989.
Membre du Conseil d'Administration de l'Université Paris-Dauphine de 1988 à 1991.
Directeur du Ceremade de 1991 à 1996.
Membre du Conseil Scientifique de l'Université Paris-Dauphine de 1992 à 1996.
Responsable de divers programmes de coopération internationale, (Actions Intégrées, Programme International de Coopération Scientifique...) et de contrats de recherche industriels.
Membre du Comité Exécutif de la Société Mathématique Européenne (1990–1994) et de la Société de Mathématiques Appliquées et Industrielles (1989–1993).
Membre du Comité d'Orientation Stratégique de la Recherche depuis 1995.

Membre du Conseil d'Orientation de l'Université de Cergy-Pontoise de 1995 à 1996.
Membre du Conseil Scientifique de la Défense depuis 1997.

Comité de rédaction de "séries"

Monographs in Mathematics (Birkhaüser).
Partial Differential Equations (Birkhaüser).
Advances in Applications for Applied Sciences (World Scientific).
Ergebnisse der Mathematik (Springer).

Rédacteur en chef des Annales de l'Institut Henri Poincaré, Analyse Non Linéaire.

Editeur de Series in Applied Mathematics, North Holland.
Membre des comités de rédaction de:

Archives for Rational Mechanics and Analysis, Communications in Partial Differential Equations, Proceedings of the Royal Society of Edinburgh, Revista Matematica Iberoamericana, Calculus of Variations and Partial Differential Equations, Journal of Differential and Integral Equations, Numerische Mathematik, Applied Mathematics Letters, Dynamic Systems and Applications, Mathematical Models and Methods for Applied Sciences, Nonlinear Analysis — Theory Methods and Applications, Asymptotic Theory, RAIRO Modélisation Mathématique et Analyse Numérique, Communications in Nonlinear Analysis and Applications, Advances in Differential Equations, Numerical Algorithms, Journal de Mathématiques Pures et Appliquées, Comptes-Rendus de l'Académie des Sciences, Discrete and Continuous Dynamical Systems, Mathematical Research Letters, Journal of Inequalities and Applications, Zeitschrift Analysis Anwendungen, Ricercha Matematica, Interfaces and Free Boundaries, Nonlinear Analysis Forum, International Journal of Differential Equations and Applications, Chinese Annals of Mathematics, Methods and Applications of Analysis, COSMOS (Journal of Science and Technology), Annali di Matematica Pura ed Applicata, Foundations of Computational Mathematics, Computer Methods in Applied Mechanics and Engineering, International Journal of Pure and Applied Mathematics.

Organisateur de colloques

Organisateur ou coorganisateur de 20 colloques internationaux.
Comité organisateur de l'année "Stochastic Differential Equations and their applications", Institute of Mathematics and Applications, Minneapolis.
Responsable des sessions nonlinéaires des écoles CEA–EDF–INRIA de 1987 à 1992.

Comités scientifiques

Président de la Commssion d'Evaluation de l'INRIA depuis 1995, Président du Conseil Scientifique d'EDF depuis 2001, Président du Conseil Scientifique du CEA-DAM depuis 2001.
Conseil Scientifique du CERMICS, Conseil Scientifique de CISI, Conseil Scientifique du Groupement de Recherche sur le "transport de particules chargées" (CNRS) de

1993 à 1995, Panel "Mathematics and Information Sciences" pour le programme "Human Capital and Mobility" (CEE) de 1992 à 1994, Conseil Scientifique du PMT (ENS-Lyon), Comité d'Evaluation du Laboratoire d'Analyse Numérique (Université P.et M. Curie), Comité d'Evaluation du CEA–CESTA (1993 et 1994), Conseil Scientifique du CEA depuis 1995 à 2000, Conseil Scientifique du DMI (ENS-Ulm) de 1996 à 1999, Conseil Scientifique de la Défense depuis 1998, Scientific Advisory Board du Max-Planck Institüt à Leipzig depuis 1999, Scientific Advisory Board de l'Institut IPAM (NSF) à Los Angeles de 1999 à 2000, Comité des programmes scientifiques du CNES depuis 2000, Advisory Commitee of the Engineering Studies de l'Université Pompeu fabra (Barcelone) depuis 2000, Visiting Committee du CEA depuis 2001.

Comités de prix

Cours Peccot (Collège de France) (1988–1994), Prix IBM-France (1988–1994), Prix CISI-Ingénierie (1994–1998), Prix Science et Défense depuis 1990.

ENCADREMENT

Thèses d'Etat ou Habilitations: B. Perthame, Professeur à l'Université Pierre et Marie Curie, Cours Peccot 1989–1990, Prix de la meilleure thèse 1988, Prix Blaise Pascal 1992, Prix CISI Ingénierie 1992, Conférencier invité à l'ICM Zürich 1994, Médaille d'argent du CNRS 1994; G. Barles, Professeur à l'Université de Tours, Prix de la meilleure thèse 1989; M. J. Esteban, Directeur de Recherche au CNRS; C. Le Bris, Directeur de Recherche au CERMICS, Prix Blaise Pascal 1999; B. Desjardins, Ingénieur-Chercheur au CEA, Prix de la meilleure thèse 1999; A. Sayah, Professeur à l'Université de Rabat; C. Villani, Professeur à l'ENS-Lyon; J. Dolbeault, Changé de Recherche au CNRS; E. Rouy, Professeur à l'INSA-Lyon; N. Masmoudi, Professeur au Courant Institute (New York).
25 thèses de 3ème cycle ou nouvelles thèses (4 prix de thèses: B. Perthame, G. Barles, I. Catto et B. Desjardins); 2 thèses en préparation.

Douze chercheurs étrangers post-doctorants.

Reprinted from Proc. Int. Congr. Math., 1994

ON SOME RECENT METHODS FOR NONLINEAR PARTIAL DIFFERENTIAL EQUATIONS

by

PIERRE-LOUIS LIONS

CEREMADE, Université Paris-Dauphine,
Place de Lattre de Tassigny, F-75775 Paris Cedex 16, France

Dedicated to the memory of Ron DiPerna (1947–1989)

1. Introduction

We wish to present here some aspects of a few general methods that have been introduced recently in order to solve nonlinear partial differential equations and related problems in nonlinear analysis.

As is well known, nonlinear partial differential equations have become a rather vast subject with a long history of deep and fruitful connections with many other areas of mathematics and various sciences like physics, mechanics, chemistry, engineering sciences, etc. And we shall not pretend to make any attempt at surveying all recent activities in that field. Also, we shall concentrate on rather theoretical issues leaving completely aside more applied issues such as mathematical modelling, numerical questions that go hand in hand in a fundamental way with the theories. For a discussion of the interaction between nonlinear analysis and modern applied mathematics, we refer the reader to the report by Majda [56] in the preceding Congress.

We shall mainly discuss here recent methods that have been developed recently for the analysis of the major mathematical models of gas dynamics (and compressible fluid mechanics), namely the Boltzmann equation and compressible Euler and Navier–Stokes equations (essentially in the so-called "isentropic regime"). These methods include velocity averaging, regularization by collisions that we shall apply to the solution of the Boltzmann equation (Section 2 below), and compactness via commutators and in particular compensated compactness, which we illustrate on isentropic compressible Euler and Navier–Stokes equations.

This selection of topics (equations and methods) is by no means an exhaustive treatment of all the exciting progresses that have taken place recently in nonlinear partial differential equations: many more important problems have been investigated — see for instance the various reports in this Congress related to Nonlinear Partial Differential Equations — and other methods and theories have been developed. We briefly mention a few in Section 4. And even for the methods that we describe here, much more could be said in particular about applications to other relevant problems.

We only hope that our selection will serve as a good illustration of recent activities. It will also emphasize some current trends that go far beyond the material discussed here. The first one is the analysis of the qualitative behavior of solutions (regularity, compactness, classification of possible behaviors, etc.). The second one, related to the preceding one, concerns the structure of specific nonlinearities and its interplay with the behavior (or possible behaviors) of solutions. Finally, this requires theories and methods that are connected with many branches of mathematics and analysis in particular.

2. Boltzmann Equation

2.1. *Existence and compactness results*

The Boltzmann equation is given by

$$\frac{\partial f}{\partial t} + v.\nabla_x f = Q(f,f) \qquad (x,v) \in \mathbb{R}^{2N},\ t \geq 0 \tag{1}$$

where the unknown f is a nonnegative function on $\mathbb{R}^{2N} \times [0,\infty)$, $N \geq 2$, ∇_x denotes the gradient with respect to x, and we denote by $x \cdot y$ or (x,y) the scalar product in $\mathbb{R}^N$. The nonlinear operator Q can be written as

$$Q(f,f) = Q^+(f,f) - Q^-(f,f) \tag{2}$$

$$Q^+(f,f) = \int_{\mathbb{R}^N} dv_* \int_{S^{N-1}} d\omega\, B(v - v_*, \omega) f' f'_* \tag{3}$$

$$Q^-(f,f) = \int_{\mathbb{R}^N} dv_* \int_{S^{N-1}} d\omega\, B(v - v_*, \omega) f f_* = f L(f), \quad L(f) = f_v^* A \tag{4}$$

where $f_* = f(x, v_*, t)$, $f' = f(x, v', t)$, $f'_* = f(x, v'_*, t)$, $A(z) = \int_{S^{N-1}} B(z,\omega)\, d\omega$, and $B = B(z,\omega)$ is a given nonnegative function of $|z|$ and $|(z,\omega)|$, is called the scattering cross-section or the collision kernel, which depends on the physical interactions of the gas particles (or molecules) and

$$v' = v - (v - v_*, \omega)\omega, \quad v'_* = v_* + (v - v_*, \omega)\omega. \tag{5}$$

A typical example (the so-called hard spheres case) of B is given by; $B = |(z,\omega)|$. We always assume that $A \in L^1_{\text{loc}}(\mathbb{R}^N)$ and $(1 + |z|^2)^{-1} \cdot \int_{|\xi|<R} A(z - \xi) d\xi \to 0$ as $|z| \to \infty$, for all $R \in (0,\infty)$.

Of course, we wish to solve (1) given an initial condition that is the values of f at $t = 0$

$$f|_{t=0} = f_0 \qquad \text{in} \quad \mathbb{R}^{2N}. \tag{6}$$

The initial value problem (1), (6) is a deceivingly simple-looking first-order partial differential equation with nonlinear (quadratic) nonlocal terms. It is a relevant model for the study of a rarefied gas and is currently used for flights in the upper layers of the atmosphere (Mach 20–24, altitude of 70–120 km). The statistical description of a gas in terms of the evolution of the density f of molecules was originally obtained by Boltzmann [6] (see also Maxwell [57], [58]). There is a long history of important mathematical contributions to the study of (1) by Hilbert [31], Carleman [8], [9] etc. Further details on the derivation of (1) and references to earlier mathematical contributions can be found in Grad [28], Cercignani [10], and DiPerna and Lions [18].

The major mathematical difficulty of (1), (6) is the lack of a priori estimates on solutions: only bounds on f in L^1 (with weights) and on $f \log f$ in L^1 are known! Nevertheless, the following result, taken from [18], [20], holds:

Theorem 2.1. *Let $f_0 \geq 0$ satisfy:*

$$\int_{\mathbb{R}^{2N}} f_0(1 + |x|^2 + |v|^2 + |\log \ f_0|)\, dx\, dv < \infty.$$

Then there exists a global weak solution of (1), (6)$f \in C([0,\infty);\ L^1(\mathbb{R}^{2N}_{x,v}))$ *satisfying*

$$\sup_{t\in[0,\infty)} \int_{\mathbb{R}^{2n}} f(t)(1 + |x - vt|^2 + |v|^2 + |\log \ f(t)|)\, dx\, dv < \infty$$

and the following entropy inequality for all $t \geq 0$

$$\int_{\mathbb{R}^{2N}} f(t) \ \log \ f(t)\, dx\, dv + \frac{1}{4}\int_0^t ds \int_{\mathbb{R}^N} dx\, D[f] \leq \int_{\mathbb{R}^{2N}} f_0 \ \log \ f_0\, dx\, dv \tag{7}$$

where $D[f] = \iint_{\mathbb{R}^{2N}} dv\, dv_* \int_{S^{N-1}} B\, d\omega (f'f'_* - ff_*) \log \frac{f'f'_*}{ff_*}$.

Remarks 2.1. (i) We do not want to give here the precise definition of global weak solution as it is a bit too technical. Let us mention that the notion introduced in [18], [20] is modified in Lions [48] (additional properties are imposed on f in [48]).

(ii) Further regularity properties of solutions are an outstanding open problem. It is only known that the regularity of solutions is not "created by the evolution" and has to come from the initial condition f_0. It is tempting to think, in view of the results shown in [48] (see sections 2.2, 2.3 below), that, at least in the model case when $B = \varphi\left(|z|, \frac{|(z,\omega)|}{|z|}\right)$ with $\varphi \in C_0^\infty((0,\infty) \times (0,1))$, f is smooth if f_0 is smooth. Related to the regularity issue is the uniqueness question: uniqueness of weak solutions is not known (it is shown in [48] that any weak solution is equal to a solution with improved bounds assuming that the latter exists!).

(iii) The assumption made upon B corresponds to the so-called angular cut-off.

(iv) Boundary conditions for Boltzmann's equation can be treated: see Hamdache [29] for an analogue of the above result in that case. Realistic boundary conditions require some new a priori estimates and are treated in Lions [46].

(v) Other kinetic models of physical and mathematical interest can be studied by the methods of proof of Theorem 2.1: see for instance DiPerna and Lions [19], Arkeryd and Cercignani [2], Esteban and Perthame [22], and Lions [48].

The strategy of proof for Theorem 2.1 is a classical one, which is almost always the one followed for the proofs of *global existence results*: one *approximates* the problem by a sequence of simpler problems having the same structure (and the same a priori bounds) for which one shows easily the existence of global solutions, and then one tries to *pass to the limit*. This strategy is also useful for the mathematical analysis of numerical methods because one can view numerical solutions as approximated solutions or solutions of approximated problems. This is why the main mathematical problem behind the proof of Theorem 2.2 is the analysis of the behavior of *sequences of solutions* (we could as well consider approximated solutions ...) and in particular of passage to the limit in the equation. This is a delicate question because the available a priori bounds only yield *weak convergences* that are not enough to *pass to the limit in nonlinear terms*. This theme will be recurrent in this report (as it was already in Majda's report [56]).

We thus consider a sequence of (weak or even smooth) nonnegative solutions f^n of (1) corresponding to initial conditions (6) with f_0 replaced by f_0^n and we assume

$$\sup_{n\geq 1}\int_{\mathbb{R}^{2N}} f_0^n(1+|x|^2+|v|^2+|\log\, f_0^n|)\,dx\,dv<\infty \tag{8}$$

$$\sup_{n\geq 1}\sup_{t\geq 0}\int_{\mathbb{R}^{2N}} f^n(t)(1+|x-vt|^2+|v|^2+|\log\, f^n|)\,dx\,dv<\infty \tag{9}$$

$$\sup_{n\geq 1}\int_0^\infty dt\int_{\mathbb{R}^N} dv\, D[f^n]<\infty. \tag{10}$$

Without loss of generality — extracting subsequences if necessary — we may assume that f_0^n, f^n converge weakly in $L^1(\mathbb{R}^{2N})$, $L^1(\mathbb{R}^{2N}\times(0,T))(\forall\, T\in(0,\infty))$ respectively to f_0, f.

Theorem 2.2. *We have for all* $\psi\in C_0^\infty(\mathbb{R}_v^N)$, $T,R\in(0,\infty)$

$$\int_{\mathbb{R}^N} f^n\psi\,dv \underset{n}{\longrightarrow} \int_{\mathbb{R}^N} f\psi\,dv \qquad in\quad L^1(\mathbb{R}_x^N\times(0,T)), \tag{11}$$

$$\left.\begin{aligned} &\int_{\mathbb{R}^N} Q^+(f^n,f^n)\psi\,dv \underset{n}{\longrightarrow} \int_{\mathbb{R}^N} Q^+(f,f)\psi\,dv,\\ &\int_{\mathbb{R}^N} Q^-(f^n,f^n)\psi\,dv \underset{n}{\longrightarrow} \int_{\mathbb{R}^N} Q^-(f,f)\psi\,dv\\ &\qquad\qquad in\ measure\ for\ |x|<R,\ t\in(0,T), \end{aligned}\right\} \tag{12}$$

and f is a global weak solution of (1), (6).

Theorem 2.3. (1) *We have for all* $R, T \in (0, \infty)$

$$Q^+(f^n, f^n) \underset{n}{\longrightarrow} Q^+(f, f) in\ measure\ for\ |x| < R,\ |v| < R,\ t \in (0, T). \qquad (13)$$

(2) *If* f_o^n *converges in* $L^1(\mathbb{R}^{2N})$ *to* f_0, *then* f^n *converges to* f *in* $C([0, T];\ L^1(\mathbb{R}^{2N}_{x,v}))$ *for all* $T \in (0, \infty)$.

Remarks 2.2. (i) Theorem 2.2 is shown in [18] — a simplification of the proof of the passage to the limit (using (13)) is given in [48]. Theorem 2.3 is taken from [48].

(ii) The heart of the matter in Theorem 2.2 is (11), which is a consequence of the velocity averaging phenomenon detailed in Section 2.2 below. The proof of Theorem 2.3 relies upon the results of Section 2.3 below.

(iii) It is shown in Lions [48] that the conclusion in (2) of Theorem 2.3 implies that f_0^n converges to f_0 in $L^1(\mathbb{R}^{2N})$: in other words, no compactification and in particular no regularization is taking place for $t > 0$. As indicated in [47] (see also the recent result of Desvillettes [16]) this fact might be related to the angular cut-off assumption because grazing collisions seem to generate some compactification ("nonlinear hypoelliptic features" in the model studied in [47]).

2.2. *Velocity averaging*

A typical example of the so-called velocity averaging results is the following

Theorem 2.4. *Let* $m \geq 0$, *let* $\theta \in [0, 1)$, *and let* $f, g \in L^p(\mathbb{R}^N_x \times \mathbb{R}^N_v \times \mathbb{R}_t)$ *with* $1 < p \leq 2$. *We assume*

$$\frac{\partial f}{\partial t} + v \cdot \nabla_x f = (-\Delta_{x,t} + 1)^{\theta/2}(-\Delta_v + 1)^{m/2} g \qquad in \quad \mathcal{D}'(\mathbb{R}^{2N+1}). \qquad (14)$$

Then, for all $\psi \in C_0^\infty(\mathbb{R}^N)$, $\int_{\mathbb{R}^N} f(x, v, t)\psi(v)\, dv$ *belongs to the (Besov) space* $B_2^{s,p}(\mathbb{R}^N \times \mathbb{R})$ *— and thus to* $H^{s',p}(\mathbb{R}^N)$ *for all* $0 < s' < s$ *— where* $s = (1 - \theta)\frac{p-1}{p}(1 + m)^{-1}$.

Remarks 2.3. (i) If $m = 0$, $\int_{\mathbb{R}^N} f\psi\, dv \in H^{s,p}$ with $s = \frac{p-1}{p}$. The above exponent s is optimal in general (this is shown in a work to appear by the author). Similar results are available if $2 < p < \infty$ or in more general settings: we refer the reader to DiPerna, Lions, and Meyer [21].

(ii) Such velocity averages are known in statistical physics (or mechanics) as macroscopic quantities. The above result shows that transport equations induce some improved partial regularity on velocity averages (by some kind of dispersive effect).

(iii) The first results in this direction were obtained in Golse, Perthame, and Sentis [27], Golse, Lions, Perthame, and Sentis [26] (where the case $m = 0$ is considered). The case $m \geq 0$, $p = 2$, was treated in DiPerna and Lions [19] while the general case is due to DiPerna, Lions, and Meyer [21] — a slight improvement of the Besov space can be found in Bézard [5]. Two related strategies of proof are proposed in

[21] that both rely on Fourier analysis, one uses some harmonic analysis, namely product Hardy spaces and interpolation theory, while the second one uses classical multipliers theory and careful Littlewood–Paley dyadic decompositions. However, the main idea is rather elementary and described below in extremely rough terms.

As indicated in the preceding remark, we give a caricatural (but accurate!) explanation of the phenomena illustrated by Theorem 2.4. If we Fourier transform (14) in (x,t), we see that we gain decay (=regularity) in (ξ,τ) — dual variables of (x,t) — provided $|\tau + v\cdot\xi| \geq \delta|v\cdot\xi|$ for some $\delta > 0$. On the other hand, the set of v on which we do not gain that regularity, namely $\{v \in \text{Supp }(\psi)/|\tau + v\cdot\xi| < \delta|(\tau,\xi)|\}$, has a measure of order δ, and hence contributes little to the integral $(\int_{\mathbb{R}^n} \hat{f}(\xi,v,\tau)\psi(v)\,dv)$. Balancing the two contributions, we obtain some (fractional) regularity.

Of course, such improved regularity yields local compactness (in (x,t)) of the velocity averages and leads (after some work) to (11).

2.3. *Gain terms and Radon transforms*

We set for $f,g \in C_0^\infty(\mathbb{R}_v^N)$

$$Q^+(f,g) = \int_{\mathbb{R}^N} dv_* \int_{S^{N-1}} d\omega\, B(v - v_*,\omega)\, f' g'_* \tag{15}$$

and we assume (to simplify the presentation) that B satisfies:
$B(z,\omega) = \varphi(|z|, \frac{|(z,\omega)|}{|z|})$ (this is always the case in the context of (1)) and $\varphi \in C_0^\infty((0,\infty)\times(0,1))$. We denote by $\mathcal{M}(\mathbb{R}^N)$ the space of bounded measures on $\mathbb{R}^N$.

Theorem 2.5. *The operator Q^+ from $\mathcal{M}(\mathbb{R}^N)\times H^s(\mathbb{R}^N)$ and $H^s(\mathbb{R}^N)\times\mathcal{M}(\mathbb{R}^N)$ into $H^{s+\frac{N-1}{2}}(\mathbb{R}^N)$ is bounded for all $s \in \mathbb{R}$.*

Remark 2.4. This result is taken from Lions [48] using generalized Radon transforms; a variant of this proof making direct connection with the classical Radon transform has been recently given by Wennberg [72] (this proof, contrarily to the one in [48], does not extend to more general situations such as collision models for mixtures or relativistic models — this case is treated in Andréasson [1]).

The above gain of regularity ($\frac{N-1}{2}$ derivatives) can be shown by writing Q^+ or its adjoint as a "linear combination" of translates of some Radon-like transforms given by

$$R\psi(v) = \int_{S^{N-1}} B(v,\omega)\psi((v,\omega)\omega)\,d\omega \quad , \quad \forall\psi \in C_0^\infty(\mathbb{R}^N) \tag{16}$$

or

$$R\psi(v) = \int_{S^{N-1}} B(v,\omega)\psi(v - (v,\omega)\omega)\,d\omega \quad , \quad \forall\psi \in C_0^\infty(\mathbb{R}^N). \tag{17}$$

In both cases, one integrates φ over the set $\{(v,\omega)\omega \mid \omega \in S^{N-1}\} = \{v - (v,\omega)\omega | \omega \in S^{N-1}\}$, which is the sphere centered at $\frac{v}{2}$ and of radius $\frac{|v|}{2}$. These operators are rather special Fourier integral operators often called generalized Radon transforms (see for instance Phong and Stein [62], Stein [66]). The crucial fact is that the set over which ψ is integrated "movies" with v — except that all these spheres go through 0, but this does not create difficulties because B vanishes if $(v,\omega)\omega = 0$ or if $v - (v,\omega)\omega = 0$. This is the main reason why one can prove that R is bounded from $H^s(\mathbb{R}^N)$ into $H^{s+\frac{N-1}{2}}(\mathbb{R}^N)$ for all $s \in \mathbb{R}(\frac{N-1}{2}$ comes from the stationary phase principle...).

3. Compressible Euler and Navier–Stokes Equations

The compressible Euler and Navier–Stokes equations are the basic models for the evolution of a compressible gas. In the case of aeronautical applications, the main difference between the domains of validity of the Boltzmann equation and the Euler–Navier–Stokes systems is the altitude of the aircraft. This indicates that there should be a transition from the Boltzmann model to those mentioned here. Mathematically, this corresponds to replacing B by $\frac{1}{\varepsilon}$ B in (1) and letting ε go to 0 (at least formally): as is well known, one recovers, taking velocity averages of the limit f(i.e. $\rho = \int_{\mathbb{R}^N} f\,dv$, $\rho u = \int_{\mathbb{R}^N} fv\,dv$, $\rho E = \int_{\mathbb{R}^N} f|v|^2\,dv$), the compressible Euler equations (with $\gamma = \frac{N+2}{N}$) — see Cercignani [10] for more details. This heuristic limit (and related limits) remains completely open from a mathematical viewpoint: partial results can be found in Nishida [61], and Ukaï and Asano [71], and recent progress based upon the material described in Section 2 above is due to Bardos, Golse, and Levermore [4]. Related problems are described in Varadhan's report in this Congress.

The compressible Euler and Navier–Stokes equations take the following form:

$$\frac{\partial p}{\partial t} + \operatorname{div}\,(\rho u) = 0 \qquad x \in \mathbb{R}^N\,,\, t \geq 0 \tag{18}$$

$$\frac{\partial}{\partial t}(\rho u) + \operatorname{div}\,(\rho u \otimes u) - \lambda\Delta u - (\lambda+\mu)\nabla\operatorname{div} u + \nabla p = 0 \quad x \in \mathbb{R}^N\,, t \geq\; 0 \tag{19}$$

and an equation for the pressure p (or equivalently for the total energy or the temperature) that we do not wish to write for reasons explained below. The unknowns ρ, u correspond respectively to the density of the gas ($\rho \geq 0$) and its velocity u (where $u(x,t) \in \mathbb{R}^N$). The constants λ, μ are the viscosity coefficients of the fluid: if $\lambda = \mu = 0$, the above system is called the compressible Euler equations, whereas if $\lambda > 0$, $2\lambda + \mu > 0$, it is called the compressible Navier–Stokes equations. Despite the long history of these problems, the global existence of solutions "in the large" is still open for the full (i.e. with the temperature equation) systems except in the case of compressible Navier–Stokes equations when $N = 1$: in that case, general existence and uniqueness results can be found in Kazhikov and Shelukhin [37], Kazhikov [36], Serre [64], [63], and Hoff [34]. This is why we shall restrict ourselves here to the so-called "isentropic" (or barotropic) case where one postulates that p

is a function of ρ only, and in order to fix ideas we take

$$p = a\rho^\gamma, \quad a > 0, \quad \gamma > 1. \tag{20}$$

This condition is a severe restriction from the mechanical viewpoint (in the Navier–Stokes case, it essentially means considering the adiabatic case and neglecting the viscous heating). Mathematically, it leads to an interesting model problem that is supposed to capture some of the difficulties of the exact systems.

Of course we complement (18)–(19) with initial conditions

$$\rho|_{t=0} = \rho_0, \quad \rho u|_{t=0} = m_0, \quad \text{in} \quad \mathbb{R}^N \tag{21}$$

where $\rho_0 \geq 0$, m_0 are given function on $\mathbb{R}^N$.

We study the case of compressible isentropic Euler equations in Section 3.1. The analogous problem for Navier–Stokes equations is considered in Section 3.3.

3.1. *1D isentropic gas dynamics*

We thus consider the following system

$$\frac{\partial \rho}{\partial t} + \frac{\partial(\rho u)}{\partial x} = 0 \quad , \quad \frac{\partial(\rho u)}{\partial t} + \frac{\partial}{\partial x}(\rho u^2 + a\rho^\gamma) = 0 \qquad x \in \mathbb{R},\ t > 0 \tag{22}$$

where $\rho \geq 0$, and $a > 0$, $\gamma > 1$ are given constants. Without loss of generality (by a simple scaling) we can take $a = \frac{(\gamma-1)^2}{4\gamma}$ (to simplify some of the constants below).

As is well known for such systems of nonlinear hyperbolic (first-order) equations, singularities develop in finite time: even if $\rho_0 = \bar{\rho} + \rho_1 > 0$ on $\mathbb{R}$ with $\bar{\rho} \in \mathbb{R}$, $\bar{\rho} > 0$, $\rho_1, u_0 \in C_0^\infty(\mathbb{R})$, then u_x and ρ_x become infinite in finite time (see Lax [38], [39], [41], and Majda [54], [55] for more details). In addition, bounded solution of (22), (21) are not unique and additional requirements known as (Lax) entropy conditions on the solutions are needed (Lax [41], [40], see also the report by Dafermos in this Congress).

In the case of (22), these requirements take the following form (see Diperna [17], Chen [11], and Lions, Perthame, and Tadmor [53]):

$$\frac{\partial}{\partial t}[\varphi(\rho, \rho u)] + \frac{\partial}{\partial x}[\psi(\rho, \rho u)] \leq 0 \qquad \text{in} \quad \mathcal{D}'(\mathbb{R} \times (0, \infty)) \tag{23}$$

and φ, ψ are given by

$$\begin{cases} \varphi = \displaystyle\int_{\mathbb{R}} dv\, \omega(v)(\rho^{\gamma-1} - (v-u)^2)_+^\lambda, \\ \psi = \displaystyle\int_{\mathbb{R}} dv\, [\theta v + (1-\theta)u]\omega(v)(\rho^{\gamma-1} - (v-u)^2)_+^\lambda \end{cases} \tag{24}$$

where ω is an arbitrary convex function on $\mathbb{R}$ such that ω'' is bounded on $\mathbb{R}$, $\lambda = \frac{3-\gamma}{2(\gamma-1)}$, $\theta = \frac{\gamma-1}{2}$.

Theorem 3.1. *Let* $\rho_0, m_0 \in L^\infty(\mathbb{R})$ *satisfy:* $\rho_0 \geq 0$, $|m_0| \leq C\rho_0$ *a.e. in* $\mathbb{R}$ *for some* $C \geq 0$. *Then there exists* $(\rho, u) \in L^\infty(\mathbb{R} \times (0,\infty))$ $(\rho \geq 0)$ *solution of* (21) – (22) *satisfying* (23).

As explained in Section 2.2, the proof of the existence results depends very much upon the stability and compactness results shown below (in fact one approximates (22) by the vanishing viscosity method; i.e., adding $-\varepsilon\frac{\partial^2\rho}{\partial x^2}, -\varepsilon\frac{\partial^2(\rho u)}{\partial x^2}$ in the equations respectively satisfied by $\rho, \rho u$ where $\varepsilon > 0$, and one lets ε go to 0). We thus consider a sequence (ρ^n, u^n) of solutions of (22) satisfying (23) and we assume that (ρ^n, u^n) is bounded uniformly in n in $L^\infty(\mathbb{R} \times (0,\infty))$ $(\rho^n \geq 0$ a.e.). Without loss of generality, we may assume that (ρ^n, u^n) converges weakly in $L^\infty(\mathbb{R} \times (0,\infty))$–weak* to some $(\rho, u) \in L^\infty(\mathbb{R} \times (0,\infty))$ $(\rho \geq 0$ a.e.). The main mathematical difficulty is the lack of any a priori estimate (except for $\gamma = 3$, the so-called monoatomic case) that would ensure the pointwise compactness needed to pass to the limit in $\rho^n(u^n)^2$ or $(\rho^n)^\gamma$.

Theorem 3.2. $\rho^n, \rho^n u^n$ *converge in measure on* $(-R, R) \times (0, T)$ *(for all* $0 < R$, $T < \infty$*) to* $\rho, \rho u$ *respectively. And* (ρ, u) *is a solution of* (22) *satisfying* (23).

Remarks 3.1. (i) This result shows that the hyperbolic system (22) has compactifying properties because initially at $t = 0$ we did not require that ρ^n or $\rho^n u^n$ converge in measure.

(ii) Theorem 3.2 is essentially due to DiPerna [17] if $\gamma = \frac{2k+3}{2k+1}(k \geq 1)$, Chen [11] if $1 < \gamma \leq \frac{5}{3}$. It is shown in Lions, Perthame, and Tadmor [53] if $\gamma \geq 3$ and in Lions, Perthame, and Souganidis [51] if $1 < \gamma < 3$. The existence result (Theorem 3.1) for $1 < \gamma < \infty$ is taken from [51].

(iii) The proofs in [53], [51] use two main tools: the method introduced by Tartar [69] (and developed by DiPerna [17]) which combines the compensated-compactness theory of Tartar [68], [69], Murat [60] and the entropy inequalities (23), and the kinetic formulation of (22) introduced in [52], [53] where one adds a new "velocity" variable, and writes the unknowns $(\rho, \rho u)$ in terms of macroscopic quantities (velocity averages) associated with a density $f(x, v, t)$ that has a fixed "profile" in v (a "pseudo-maxwellian"). This formulation connects the Boltzmann theory as described in Section 2 and the study of compressible hydrodynamic (or gas dynamics) macroscopic models. More details on this new approach are to be found in Perthame's report in this Congress. In the next section, we present some aspects of the compensated-compactness theory.

3.2. *Compensated compactness and Hardy spaces*

One important point in the compensated-compactness theory developed by Tartar [68], [69] and F. Murat [60] is the systematic detection of nonlinear quantities that enjoy "weak compactness" properties. A typical example known as the div-curl example — it is precisely the one used in the proof of Theorem 3.2 — is given by the following result taken from [60].

Theorem 3.3. *Let* (E^n, B^n) *converge weakly to* (E, B) *in* $L^p(\mathbb{R}^N)^N \times L^q(\mathbb{R}^N)^N$ *with* $1 < p < \infty$, $\frac{1}{q} + \frac{1}{p} = 1$, $N \geq 2$. *We assume that* curl E^n, div B^n *are relatively compact in* $W^{-1,p}(\mathbb{R}^N)$, $W^{-1,q}(\mathbb{R}^N)$ *respectively. Then,* $E^n \cdot B^n$ *converges weakly (in the sense of measures or in distributions sense) to* $E \cdot B$.

Remark 3.2. Let us sketch a proof. We write: $E^n = \nabla\pi^n + \tilde{E}^n$ where div $\tilde{E}^n = 0$, $\tilde{E}^n$ is compact in $L^p(\mathbb{R}^N)$ (Hodge–De Rham decompositions), $\pi^n \in L^p_{\text{loc}}(\mathbb{R}^N)$, $\nabla\pi^n \in L^p(\mathbb{R}^N)$. Then, we only have to pass to the limit in $B^n \cdot \nabla\pi^n = \text{div}\,(\pi^n B^n) - \pi^n \text{div}\, B^n$. The first term passes to the limit because π^n is compact in $L^p_{\text{loc}}(\mathbb{R}^N)$ (Rellich–Kondrakov theorem) while the second term also does because div B^n is relatively compact in $W^{-1,q}(\mathbb{R}^N)$ and $\nabla\pi^n$ is bounded in $L^p(\mathbb{R}^N)$.

As shown in Coifman, Lions, Meyer, and Semmes [12], the above nonlinear phenomenon is intimately connected with some general results in harmonic analysis associated with the (multi-dimensional) Hardy spaces denoted here by $H_p(\mathbb{R}^N)(0 < p \leq 1)$: see Stein and Weiss [67], Fefferman and Stein [23], and Coifman and Weiss [14] for more details on Hardy spaces.

In particular, the following result holds.

Theorem 3.4. *Let* $E \in L^p(\mathbb{R}^N)$ *satisfy* curl $E = 0$ *in* $\mathcal{D}'(\mathbb{R}^N)$, *let* $B \in L^q(\mathbb{R}^N)$ *satisfy* div $B = 0$ *in* $\mathcal{D}'(\mathbb{R}^N)$ *with* $1 < p, q < \infty$, $\frac{1}{r} = \frac{1}{p} + \frac{1}{q} < 1 + \frac{1}{N}$. *Then* $E \cdot B \in H_r(\mathbb{R}^N)$.

Remarks 3.3. (i) This result is taken from [12] (and was inspired by a surprising observation due to Müller [59]).

(ii) The relations between the weak compactness result (Theorem 3.3) and the regularity result (Theorem 3.4) are made clear in [12] and follow from some general considerations on dilation and translation invariant multilinear forms that enjoy a crucial cancellation property ($\int_{\mathbb{R}^N} E \cdot B\, dx = 0$ in Theorem 3.4 above).

(iii) Theorem 3.4 is one of the tools used in the proof by Hélein [30] of the regularity of two-dimensional harmonic maps.

(iv) It is shown in [12] that any element of $H_1(\mathbb{R}^N)$ can be decomposed in a series $\sum_{n\geq 1} \lambda_n E_n \cdot B_n$ where $\|E_n\|_{L^2} = \|B_n\|_{L^2} = 1$, div $B_n =$ curl $E_n = 0$, $\sum_{n\geq 1} |\lambda_n| < \infty$.

If we denote by R_k the Riesz transform $(= \partial_k(-\Delta)^{-1/2})$, then, under the conditions of Theorem 3.4, there exists $\hat{\pi} \in L^p(\mathbb{R}^N)$ such that $E = R\hat{\pi}$. And $E \cdot B = B \cdot R\hat{\pi} = B \cdot R\hat{\pi} + (R \cdot B)\hat{\pi}$ because $R \cdot B = (-\Delta)^{-1/2}\text{div}\, B = 0$. Then we can recover the case $r = 1$ in Theorem 3.4 using the H_1–BMO duality and the result on commutators due to Coifman, Rochberg, and Weiss [13]: indeed, we then obtain $f(R_k g) + (R_k f)g \in H_1(\mathbb{R}^N)$ for each $k \geq 1$, $f \in L^p(\mathbb{R}^N)$, $g \in L^q(\mathbb{R}^N), 1 < p < \infty$, $\frac{1}{p} + \frac{1}{q} = 1$.

3.3. *Isentropic Navier–Stokes equations*

We now consider the system

$$\begin{cases} \dfrac{\partial \rho}{\partial t} + \text{div}\,(\rho u) = 0, \\ \dfrac{\partial \rho u}{\partial t} + \text{div}\,(\rho u \otimes u) - \lambda \Delta u - (\lambda + \mu)\nabla \text{div}\, u + a \nabla \rho^\gamma = 0, \\ \qquad\qquad x \in \mathbb{R}^N, t > 0, \end{cases} \tag{25}$$

where $a > 0,\ 1 < \gamma < \infty, \lambda > 0,\ 2\lambda + \mu > 0,\ \rho(x,t) \geq 0$ on $\mathbb{R}^N \times (0,\infty)$, with the initial conditions (21) that are required to satisfy

$$\begin{cases} \rho_0 \in L^1(\mathbb{R}^N) \cap L^\gamma(\mathbb{R}^N), \quad \rho_0 \geq 0, \\ m_0 = \sqrt{\rho_0}v_0 \quad \text{a.e. with} \quad v_0 \in L^2(\mathbb{R}^N). \end{cases} \tag{26}$$

Theorem 3.5. *We assume* (26) *and* $\gamma \geq \frac{3}{2}$ *if* $N = 2$, $\gamma \geq \frac{9}{5}$ *if* $N = 3$, $\gamma > \frac{N}{2}$ *if* $N \geq 4$. *Then there exists a solution* $(\rho, u) \in L^\infty(0,\infty; L^\gamma(\mathbb{R}^N)) \cap L^2(0,T; H^1(B_R))(\forall\, R,T \in (0,\infty))$ *of* (25), (21) *satisfying in addition:* $\rho \in C([0,\infty); L^p(\mathbb{R}^N))$ *if* $1 \leq p < \gamma$, $\rho|u|^2 \in L^\infty(0,\infty; L^1(\Omega))$, $\rho \in L^q(\mathbb{R}^N \times (0,T))$ *for* $1 \leq q \leq \gamma + \frac{2\gamma}{N} - 1$ *if* $N \geq 2$.

$$\begin{cases} \displaystyle\int_\Omega \frac{1}{2}\rho(t)|u(t)|^2 + \frac{a}{\gamma - 1}\rho(t)^\gamma\, dx + \int_0^t ds \int_\Omega \lambda|\nabla u|^2 + (\lambda+\mu)(\text{div}\, u)^2\, dx \\ \displaystyle\leq \int_\Omega \frac{1}{2}\frac{|m_0|^2}{\rho_0} + \frac{a}{\gamma-1}\rho_0^\gamma\, dx \end{cases} \tag{27}$$

for almost all $t \geq 0$.

Remarks 3.4. (i) This result is taken from Lions [45] (see also [50]). If $N = 1$, more general results are available and we refer to Serre [63], Hoff [32], [33].

(ii) If $N \geq 2$, the uniqueness and further regularity of solutions are completely open as is the case of a general $\gamma > 1$. The case $\gamma = 1$ is also an interesting mathematical problem (see [49]).

(iii) Of course, the equations contained in (25) hold in the sense of distributions.

(iv) The preceding result is rather similar to the results obtained by Leray [42], [43], [44] on the global existence of weak solutions of three-dimensional incompressible Navier–Stokes equations satisfying an energy inequality like (27). Despite many important contributions (like the partial regularity results obtained by Caffarelli, Kohn, and Nirenberg [7]), the uniqueness and regularity of solutions are still open questions.

As explained in the previous sections, the above existence result is based upon a convergence result for sequences of solutions ρ^n, u^n satisfying uniformly in n the properties mentioned in the above result. Hence, without loss of generality, we may assume that (ρ^n, u^n) converge weakly to (ρ, u) in $L^\gamma(\mathbb{R}^N \times (0,T)) \times L^2(0,T; H^1(B_R))$ $(\forall R, T \in (0,\infty))$. Then it is shown in [45], [49] that if $\rho_0^n (= \rho^n|_{t=0})$ converges in $L^1(\mathbb{R}^N)$, then ρ^n converges in $C([0,T]; L^p(\mathbb{R}^N)) \cap L^q(\mathbb{R}^N \times (0,T))$ for all $T \in (0,\infty)$, $1 \leq p < \gamma$, $1 \leq q < \gamma + \frac{2\gamma}{N} - 1$. And (ρ, u) is a solution of (25) with the properties listed in the preceding result. It is also shown in [45], [49] (see also Serre [65]) that the analogue of Theorem 3.2 for the system (25) *does not hold*: in other words, the compactification that took place for the hyperbolic system (22) *is lost* when we add viscous terms while we could expect (from a linear-linearized inspection) that the introduction of viscous terms regularizes the problem! These delicate and surprising phenomena depend in a subtle way on the nonlinearities of the systems we consider. Let us also mention that the proof of the above convergence result is rather delicate and uses in particular the structure of the convective derivatives $(\frac{\partial}{\partial t} + u \cdot \nabla_x)$ that lead with the analysis detailed in [45], [49] to terms like

$$\rho^n R_i R_j (\rho^n u_i^n u_j^n) - \rho^n u_i^n R_i R_j (\rho^n u_j^n),$$

which are shown to converge weakly to $\rho R_i R_j(\rho u_i u_j) - \rho u_i R_i, R_j(\rho u_j)$ under the sole weak convergence stated above on ρ^n, u^n. This weak continuity follows from regularizing properties of the commutators $[u_i^n, R_i R_j]$. It is worth noting that the incompressible limit of such compressible models yields $\rho^n =$ cst (say 1), div $u^n = 0$, in which case the above term reduces to $R_i R_j (u_i^n u_j^n)$ and $R_j(u_i^n u_j^n) = (-\Delta)^{-1/2}\{u^n \cdot \nabla u^n\}$ because div $u^n = 0$. Obviously, curl $(\nabla u^n) = 0$, div $u^n = 0$, and $u^n \cdot \nabla u^n$ is precisely a nonlinear expression for which the compensated-compactness theory applies (see Section 3.2).

4. Perspectives, Trends, Problems and Methods

Let us immediately emphasize that this brief section will select topics in a biased way that reflects the author's tastes.

First of all, we have mentioned above some of the progress made recently and many remaining open questions in gas dynamics and fluid mechanics. There is much more to say and in particular we have not touched here the incompressible models (Euler and Navier–Stokes equations) — see the reports by Beale, Chemin, Constantin, and Avellaneda in this Congress and Majda [56]. Even if many fundamental questions are left open, progress is being made (step by step).

We should also make clear that the topics covered here do not reflect fully the scope of nonlinear partial differential equations and in particular those arising from applications, the variety of mathematical problems and methods developed recently, and their relationships with other fields of mathematics. Let us briefly mention a few more examples of themes covering several related areas that all have important scientific and technological implications: (i) propagation of fronts and interfaces, geometric equations, viscosity solutions, image processing (see the reports by Spruck, Souganidis, and Osher in this Congress), (ii) quantum chemistry, N-body problems, density-dependent and meanfield models, binding, thermodynamic limits, (iii) twinning and defects in solids and crystals, phase transitions, Young measures (see for instance Ball and James [3], James and Kinderlehrer [35], and the report by Sverak).

However, we wish to emphasize that the trends mentioned in the Introduction can also be found in the above themes.

Finally, it is important to develop at the same time the methods — some of which have been briefly presented in this paper — which are certainly interesting by themselves, and we would like to conclude with a few examples of such developments: (i) H-measures of Tartar [70], Gérard [24] (and the related Wigner measures by Lions and Paul [50], Gérard [25]), (ii) nonlinear partial differential equations in infinite dimensions (and in particular the viscosity solutions approach of Crandall and Lions [15]).

References

[1] H. Andréasson, *A regularity property and strong L^1 convergence to equilibrium for the relativistic Boltzmann equation*, preprint **21**, Chalmers Univ., Göteborg, 1994.

[2] L. Arkeryd and C. Cercignani, *On the convergence of solutions of Enskog equations to solutions of the Boltzmann equation*, Comm. Partial Differential Equations, **14** (1989), 1071–1090.

[3] J. Ball and R. D. James, *Fine phase mixtures as minimizers of energy*, Arch. Rational Mech. Anal., **100** (1987), 13–52.

[4] C. Bardos, F. Golse, and D. Levermore, *Fluid dynamics limits of kinetic equations*, I, J. Statist. Phys., **63** (1991), 323–344; II, Comm. Pure Appl. Math., **46** (1993), 667–753.

[5] M. Bézard, *Régularité L^p précisée des moyennes dans les équations de transport*, preprint.

[6] L. Boltzmann, *Weitere Studien über das Wärmegleichgewicht unter Gasmolekülen*. Sitzungsberichte der Akademie der Wisssenschaften, Vienna, **66** (1972), 275–370. (Trans.: Further studies on the thermal equilibrium of gas molecules, in Kinetic Theory, vol. 2 (S. G. Brush, ed.), Pergamon, Oxford (1966), 88–174).

[7] L. Caffarelli, R. V. Kohn, and L. Nirenberg, *On the regularity of the solution of Navier–Stokes equations*, Comm. Pure Appl. Math., **35** (1982), 771–832.

[8] T. Carleman, Acta Math., **60** (1933), 91.

[9] T. Carleman, Problèmes mathématiques dans la théorie cinétique des gaz. Notes written by Carleson and Frostman, Uppsala, Almqvist and Wikselles, 1957.

[10] C. Cercignani, The Boltzmann Equation and its Applications. Springer Verlag, Berlin and New York, 1988.

[11] G. Q. Chen, *The theory of compensated compactness and the system of isentropic gas dynamics*, preprint.

[12] R. Coifman, P. L. Lions, Y. Meyer, and S. Semmes, *Compensated compactness and Hardy spaces*, J. Math. Pures Appl. (9), **72** (1993), 247–286.

[13] R. Coifman, R. Rochberg, and G. Weiss, Ann. of Math. (2), **103** (1976), 611–635.

[14] R. Coifman and G. Weiss, *Extensions of Hardy spaces and their use in analysis*, Bull. Amer. Math. Soc., **83** (1977), 579–945.

[15] M. G. Crandall and P. L. Lions, *Hamilton-Jacobi equations in infinite dimensions, I*, J. Funct. Anal., **63** (1985), 379–396; *II*, J. Funct. Anal., **65** (1985), 308–400; *III*, J. Funct. Anal., **68** (1986), 214–247; *IV*, J. Funct. Anal., **90** (1990), 237–283; *V*, J. Funct. Anal., **97** (1991), 417–465; *VI*, in Evolution Equations, Control Theory and Biomathematics (Ph. Clément and G. Lumer, eds.), Lecture Notes in Pure and Appl. Math. **155**, Dekker, New York, 1994: *VII*, to appear in J. Funct. Anal.

[16] L. Desvillettes, *About the regularizing properties of the non cut-off Kac equation*, preprint, 1994.

[17] R. J. DiPerna, *Convergence of the viscosity method for isentropic gas dynamics*, Comm. Math. Phys., **91** (1983), 27–70.

[18] R. J. DiPerna and P. L. Lions, *On the Cauchy problem for Boltzmann equations: global existence and weak stability*, Ann. of Math. (2), **130** (1989), 312–366.

[19] R. J. DiPerna and P. L. Lions, *Global weak solutions of Vlasov-Maxwell systems*, Comm. Pure Appl. Math., **62** (1989), 729–757.

[20] R. J. DiPerna and P. L. Lions, *Global solutions of Boltzmann's equation and the entropy inequality*, Arch. Rational Mech. Anal., **114** (1991), 47–55.

[21] R. J. DiPerna, P. L. Lions, and Y. Meyer, *L^p-regularity of velocity averages*, Ann. Inst. H. Poincaré Anal. Nonlinéaire, **8** (1991), 271–287.

[22] M. J. Esteban and B. Perthame, *On the modified Enskog equation with elastic or inelastic collisions; Models with spin*, Ann. Inst. H. Poincaré Anal. Nonlinéaire, **8** (1991), 289–398.

[23] C. Fefferman and E. Stein, *H^p spaces of several variables*, Acta Math., **228** (1972), 137–193.

[24] P. Gérard, *Microlocal defect measures*, Comm. Partial Differential Equations, **16** (1991), 1761–1794.

[25] P. Gérard, *Mesures semi-classiques et ondes de Bloch*, in Séminaire EDP, 1990–1991, Ecole Polytechnique, Palaiseau, 1991.

[26] F. Golse, P. L. Lions, B. Perthame, and R. Sentis, *Regularity of the moments of the solutions of a transport equation*, J. Funct. Anal., **76** (1988), 110–125.

[27] F. Golse, B. Perthame, and R. Sentis, *Un résultat pour les équations de transport et application au calcul de la limite de la valeur propre principale d'un opérateur de transport*, C. R. Acad. Sci. Paris Série I, **301** (1985), 341–344.

[28] H. Grad, *Principles of the kinetic theory of gases*, Handbuch der Physik, **12**, Springer Verlag, Berlin (1958), 205–294.

[29] K. Hamdache, *Global existence for weak solutions for the initial boundary value problems of Boltzmann equation*, Arch. Rational Mech. Anal., **119** (1992), 309–353.

[30] F. Hélein, *Régularité des applications faiblement harmoniques entre une surface et une variété riemanienne*, C. R. Acad. Sci. Paris Série I, **312** (1990), 591–596.

[31] D. Hilbert, *Begründung der kinetischen Gastheorie*, Math. Ann. **72** (1912), 562–577.

[32] D. Hoff, *Construction of solutions for compressible, isentropic Navier–Stokes equations in one space dimension with non smooth initial data*, Proc. Roy. Soc. Edinburgh Sect. A, **103** (1986), 301–315.

[33] D. Hoff, *Global existence for 1D compressible, isentropic Navier–Stokes equations with large initial data*, Trans. Amer. Math. Soc., **303** (1987), 169–181.

[34] D. Hoff, *Global well-posedness of the Cauchy problem for non-isentropic gas dynamics with discontinuous initial data*, J. Differential Equations, **95** (1992), 33–73.

[35] R. D. James and D. Kinderlehrer, *Theory of diffusionless phase transformations*, Lecture Notes in Phys. **344**, (M. Rascle, D. Serre, and M. Slemod, eds.), Springer Verlag, Berlin and New York, 1989.

[36] A. V. Kazhikov, *Cauchy problem for viscous gas equations*, Sibirsk. Mat. Zh., **23** (1982), 60–64.

[37] A. V. Kazhikov and V. V. Shelukhin, *Unique global solution with respect to time of the initial boundary value problems for one-dimensional equations of a viscous gas*, J. Appl. Math. Mech., **41** (1977), 273–282.

[38] P. D. Lax, *Hyperbolic systems of conservation laws, II*, Comm. Pure Appl. Math., **10** (1957), 537–566.

[39] P. D. Lax, *Development of singularities of solutions of nonlinear hyperbolic partial differential equations*, J. Math. Phys., **5** (1964), 611–613.

[40] P. D. Lax, *Shock waves and entropy*, in Contributions to Nonlinear Functional Analysis (Zarantonello, ed.), Academic Press, New York, (1973), 603–634.

[41] P. D. Lax, Hyperbolic systems of conservation laws and the mathematical theory of shock waves. CBMS-NSF Regional Conferences Series in Applied Mathematics, **11**, 1973.

[42] J. Leray, *Etude de diverses équations intégrales nonlinéaires et de quelques problèmes que pose l'hydrodynamique*, J. Math. Pures Appl. (9), **12** (1933), 1–82.

[43] J. Leray, *Essai sur les mouvements plans d'un liquide visqueux que limitent des parois*, J. Math. Pures Appl. (9), **13** (1934), 331–418.

[44] J. Leray, *Essai sur le mouvements d'un liquide visqueux emplissant l'espace*, Acta Math., **63** (1934), 193–248.

[45] P. L. Lions, *Existence globale de solutions pour les équations de Navier–Stokes compressibles isentropiques*, C. R. Acad. Sci. Paris Série I, **316** (1993), 1335–1340. *Compacité des solutions des équations de Navier–Stokes compressibles isentropiques*, C. R. Acad. Sci. Paris Série I, **317** (1993), 115–120. *Limites incompressibles et acoustique pour des fluides visqueux compressibles isentropiques*, C. R. Acad. Sci. Paris Série I, **317** (1993), 1197–1202.

[46] P. L. Lions, *Conditions at infinity for Boltzmann's equation*, Comm. Partial Differential Equations, **19** (1994), 335–367.

[47] P. L. Lions, *On Boltzmann and Landau equations*, Phil. Trans. Roy. Soc. London Ser. A, **346** (1994), 191–204.

[48] P. L. Lions, *Compactness in Boltzmann's equation via Fourier integral operators and applications. Parts I–III*, to appear in J. Math. Kyoto Univ., 1994.

[49] P. L. Lions, Mathematical Topics in Fluid Mechanics., to appear in Oxford Univ. Press.

[50] P. L. Lions and T. Paul, *Sur les mesures de Wigner*, Rev. Mat. Iberoamericana, **9** (1993), 553–618.

[51] P. L. Lions, B. Perthame, and P. E. Souganidis, *Existence and compactness of entropy solutions for the one-dimensional isentropic gas dynamics systems*, to appear in Comm. Pure Appl. Math.
[52] P. L. Lions, B. Perthame, and E. Tadmor, *A kinetic formulation of multidimensional scalar conservation laws and related equations*, J. Amer. Math. Soc., **7** (1994), 169–191.
[53] P. L. Lions, B. Perthame, and E. Tadmor, *Kinetic formulation of the isentropic gas dynamics and p-systems*, to appear in Comm. Math. Phys.
[54] A. Majda, Compressible Fluid Flow and Systems of Conservation Laws in Several Space Variables, Springer, Berlin and New York, 1984.
[55] A. Majda, *Mathematical fluid dynamics: The interaction of nonlinear analysis and modern applied mathematics*, in Proceedings of the AMS Centennial Symposium, August 8–12, 1988.
[56] A. Majda, *The interaction of Nonlinear Analysis and Modern Applied Mathematics*, in Proc. Internat. Congress Math., Kyoto, 1990, vol. I, Springer, Berlin and New York, 1991.
[57] J. C. Maxwell, Scientific papers. Vol. 2, Cambridge Univ. Press, Cambridge, 1880 (Reprinted by Dover Publications, New York, 1965).
[58] J. C. Maxwell, *On the dynamical theory of gases*, Phil. Trans. Roy. Soc. London Ser. A, **157** (1886), 49–88.
[59] S. Müller, *A surprising higher integrability property of mappings with positive determinant*, Proc. Amer. Math. Soc., **21** (1989), 245–248.
[60] F. Murat, *Compacité par compensation*, Ann. Scuola Norm. Sup. Pisa Cl. Sci (4), **5** (1978), 489–507; *II*, in Proceedings of the International Meeting on Recent Methods on Nonlinear Analysis (E. De Giorgi, E. Magenes, and U. Mosco, eds.), Pitagora, Bologna, 1979; *III*, Ann. Scuola Norm. Sup. Pisa Cl. Sci (4), **8** (1981), 69–102.
[61] T. Nishida, *Fluid dynamical limit of the nonlinear Boltzmann equation to the level of the compressible equation*, Comm. Math. Phys., **61** (1978), 119–148.
[62] D. H. Phong and E. Stein, *Hilbert integrals, singular integrals and Radon transforms*, Ann. of Math. (2).
[63] D. Serre, *Solutions faibles globales des équations de Navier–Stokes pour un fluide compressible*, C. R. Acad. Sci. Paris Série I, **303** (1986), 629–642.
[64] D. Serre, *Sur l'équation monodimensionnelle d'un fluide visqueux, compressible et conducteur de chaleur*, C. R. Acad. Sci. Paris Série I, **303** (1986), 703–706.
[65] D. Serre, *Variations de grande amplitude pour la densité d'un fluide visqueux compressible*, Phys. D, **48** (1991), 113–128.
[66] E. Stein, *Oscillatory integrals in Fourier analysis*, in Beijing Lectures in Harmonic Analysis (E. Stein, ed.), Princeton Univ. Press, Princeton, NJ (1986), 307–355.
[67] E. Stein and G. Weiss, *On the theory of H^p spaces*, Acta Math., **103** (1960), 25–62.
[68] L. Tartar, *Compensated compactness and applications to partial differential equations*, in Nonlinear Analysis and Mechanics: Heriot-Watt Symposium, vol. 4 (R. J. Knops, ed.), Research Notes in Math., Pitman, London, 1979.
[69] L. Tartar, *The compensated compactness method applied to systems of conservation laws*, in Systems of Nonlinear Partial Differential Equations (J. M. Ball, ed.), NATO ASI Series C III, Reidel, New York, 1983.

[70] L. Tartar, *H-measures, a new approach for studying homogenization, oscillations and concentration effects in partial differential equations*, Proc. Roy. Soc. Edinburgh Sect. A, **115** (1990), 193–230.

[71] S. Ukaï and K. Asano, *The Euler limit and initial layer of the nonlinear Boltzmann equation*, Hokkaido Math. J., **12** (1983), 311–332.

[72] B. Wennberg, *Regularity estimates for the Boltzmann equation*, preprint **2**, 1994, Chalmers Univ., Göteborg.

PRESENTATION OF JEAN-CHRISTOPHE YOCCOZ*

by
ADRIEN DOUADY

Ecole Normale Supérieure
45, rue d'Ulm, F-75230 Paris Cedex, France

1. Curriculum

Jean-Christophe Yoccoz is a pure product, and the best vintage, of the French system. A former student of the Ecole Normale Supérieure where he was placed 1st in 1975, placed 1st at the Ecole Polytechnique the same year, tied in 1st place for the Agrégation de Mathématiques in 1977, he defended his doctoral thesis in 1985 and was invited to teach the Cours Peccot at the Collège de France in 1987. Today, at age 37, he is Professor at the Université de Paris-Sud (Orsay), a member of the I.U.F. (Institut Universitaire de France) and the URA[1] "Topology and Dynamics" of the CNRS[2] at Orsay.

He did his National Service in Cooperation at the IMPA[3] in Rio de Janeiro, and that had a profound influence on him. He visits Brazil regularly, but also the International Center at Trieste. By the way, his wife is Brazilian.

Yoccoz is a student of Michel Herman, and thus he became perhaps the best specialist of the Theory of Dynamical Systems.

2. The Theory of Dynamical Systems

This theory seeks to describe *the long-term evolution* of a system when the law of elementary evolution is known. The time can be continuous or discrete.

In the case of *continuous time*, the law of infinitesimal evolution is interpreted as a differential equation, which is given by a *vector field*, and the problem is to understand the long-term evolution of the solutions. Sometimes strange attractors are obtained.

A typical example is the problem of the *stability of the Solar System*, which has led Poincaré to create the theory at the turn of the century.

*Translated from French to English by Chee Whye Chin.
[1]Unité de Recherche Associée.
[2]Centre National de la Recherche Scientifique.
[3]Instituto Nacional de Matemática Pura e Aplicada.

In the case of *discrete time*, the elementary evolution is given by a map f, which gives the state of the system at time $n+1$ as a function of the state at time n. It is then a question of iterating f a large number of times.

When two maps f and g describe the same phenomenon in different representations, they are *conjugated* by the map h which expresses the change of representation. Any conjugation can be interpreted in this manner. Two conjugate maps thus have the same dynamical properties. Therefore, the classification of maps up to conjugation is a central problem in the theory.

3. C^∞ Conjugation to Rotation

The simplest example is where the state space is a circle, and the iteration map as well as its inverse are infinitely differentiable, that is to say a C^∞ *diffeomorphism*. For such a map f, Poincaré has defined the *rotation number* $\alpha = \mathrm{Rot}(f) \in \mathbf{T} = \mathbf{R}/\mathbf{Z}$. The question then is: when will f be C^∞-conjugated to the rotation $\mathcal{R}_\alpha : t \mapsto t+\alpha$?

If α is rational, say $\alpha = \frac{p}{q}$, this requires that one has $f^q = I$, which essentially never occurs. The interesting case is thus that where α is irrational. It has been investigated by Denjoy — who has shown that f is always topologically conjugated to $\mathcal{R}_\alpha$ —, Birkhoff, Arnold, Herman and many others, and of course Yoccoz. They have all stressed the importance of the *arithmetic properties* of α.

For rational α, resonances occur. If α is irrational, compensations almost always occur and one observes certain regularities. But if α, while being irrational, is extremely close to the rationals with moderately large denominators, it happens that a resonance begins and before it dies out another takes over, and one can get a very complicated situation. What matters is therefore the distance $\delta_q(\alpha)$ of α from the rationals with a denominator bounded by q, and the way in which this distance tends to 0 when q tends to infinity.

4. Diophantine Conditions

We say that α is diophantine if $\delta_q(\alpha)$ is bounded below by an expression of the form $\frac{c}{q^\beta}$.

For a C^∞ diffeomorphism of the circle of rotation number α, Herman has shown that f is necessarily C^∞-conjugated to the rotation $\mathcal{R}_\alpha$ if α is diophantine of exponent 2. This result was an important breakthrough. In fact Herman had proven a stronger theorem: the same result holds under a weaker hypothesis, satisfied for almost all values of α.

In his thesis, Yoccoz improved upon the theorem of Herman: he gave a simpler proof and obtained the result under the hypothesis that α is diophantine without any restriction on the exponent — a hypothesis weaker than that of Herman.

Counter-examples of Herman show that this result is optimal.

5. The R-analytic Case

The same question can be asked in the **R**-analytic setting.

Yoccoz has proven in his thesis that an **R**-analytic diffeomorphism of the circle with a diophantine rotation number α is necessarily **R**-analytically conjugated to the rotation $\mathcal{R}_\alpha$. Recently, he has given a description of the exact set of the rotation numbers α having this property. It is a complicated set: whereas the sets that one defines in this manner are generally of type F_σ, the one here is only $F_{\sigma\delta}$.

The C^∞ functions and the **R**-analytic functions have a different consistency. When one works in the **R**-analytic setting, the first thing to do is to extend the maps to the complex values of the variable. A map $f : \mathbf{T} \to \mathbf{T}$ is thus extended to an annular neighbourhood Ω of $\mathbf{T}$ in the cylinder $\mathbf{C}/\mathbf{Z}$, and if f is **R**-analytically conjugated to $\mathcal{R}_\alpha$ there is an annulus A in Ω which is invariant under f. For $z \in A$, the closure of the orbit of A is an **R**-analytic curve, corresponding to a circle parallel to the equator $\mathbf{T}$ of $\mathbf{C}/\mathbf{Z}$. The minimum thickness of A, its modulus, which occurs in the neighbourhood of its boundary are some of the properties which the geometric reasoning has taken on.

6. Converse of the Theorem of Bruno

The question of local linearizability of holomorphic diffeomorphisms in the neighbourhood of a fixed point is closely related to the above. It is the following:

Is a function $f : z \mapsto a_1 z + a_2 z^2 + a_3 z^3 + \cdots$ holomorphically conjugated in the neighbourhood of 0 to its linear part $z \mapsto a_1 z$?

The result is easy if $|a_1| \neq 1$ (Schröder, Böttcher); the interesting case is that where a_1 is of the form $e^{2i\pi\alpha}$. It was studied by Fatou — who dealt with the case where α is rational, Cremer — who gave examples of non-linearizability, Siegel — who showed that f is linearizable whenever α is Diophantine (and regardless of the tail $a_2 z^2 + \cdots$), Bruno — who improved upon the theorem of Siegel by proving the result under the weaker hypothesis $\sum \frac{\operatorname{Log} q_{n+1}}{q_n} < \infty$ (where the $\frac{p_n}{q_n}$ are the reduced terms of the expansion of α in continued fractions), and finally Yoccoz who proved the converse of the theorem of Bruno.

Siegel and Bruno worked by force, solving the problem formally and bounding the coefficients of the conjugant. Yoccoz has a more geometric and finer approach. There is a construction called renormalization, which associates, to a map f of angle α, a map f_1 having an angle α_1 whose expansion in continued fractions is the same as that of α but shifted with the loss of the first term. By a thorough quantitative study of the properties of this operation and its iterates, Yoccoz obtained a very enlightening proof of the theorem of Bruno, and he was able to prove the converse: that for any α not satisfying the condition of Bruno one can choose the tail in such a way that f has periodic points arbitrarily close to 0, which excludes linearizability. In fact the simplest tail ($f = a_1 z + z^2$) does the trick.

There remained a question: Is the non-linearizability always due to the presence of small cycles? The examples constructed by Cremer and Yoccoz could lead one to believe so. The question has been solved in the negative by Perez-Marco, a student of Yoccoz who has again refined his method. It curiously involves the condition $\sum \frac{\operatorname{Log}\operatorname{Log} q_{n+1}}{q_n} < \infty$, weaker than that of Bruno.

7. Game of Settings

Geometry and Analysis are involved in all parts of the Theory of Dynamical Systems. But they have a particular way of interacting in Complex Dynamical Systems, thanks to the inequalities of Schwarz, Koebe, Groetzsch etc., powerful inequalities that can be applied under purely topological hypotheses. It is a method that Yoccoz has developed enormously.

8. *MLC*: the Size of Limbs

Most of what we know about the dynamical properties of the family of complex quadratic polynomials is concentrated on the topological properties of its locus of connectedness M, known by the name of *Mandelbrot set*. Its combinatorial properties are now well understood, and Thurston has proposed a synthetic model of it. But to know that M is actually homeomorphic to its model, one piece of information is lacking: that M is locally connected.

The set M contains copies of itself. Yoccoz showed that M is locally connected at every point which is "not infinitely renormalizable", that is to say which is not contained in the intersection of a decreasing sequence of copies of M. To prove the full *MLC* conjecture, it remains now to show that the intersection of a decreasing sequence of copies of M is reduced to a point.

The first case consists of the points of the cardioid: the set M is formed by a large cardioid Γ, filled, and by *limbs* attached to the points of Γ with rational internal arguments (internal arguments define the natural parameterization of Γ).

Yoccoz showed that the diameter of a limb $M_{p/q}$ attached to the point of internal argument p/q is bounded above by an expression of the form $\frac{c}{q}$ where c is a constant. This result is certainly not optimal (according to Hubbard one might hope for $\frac{\operatorname{Log} q}{q^2}$), but it is sufficient for showing that M is locally connected at the points of Γ. By the same method, he obtained the local connectedness at any point which is on the boundary of a hyperbolic component.

9. *MLC*: Yoccoz Puzzles

To show that M is locally connected at the point c which is neither infinitely renormalizable nor on the boundary of a hyperbolic component, Yoccoz uses the method called "Yoccoz Puzzles". According to a principle which I stand by in Complex Dynamics:

We plough in the dynamical plane.

We harvest in the plane of parameters.

There are indeed figures in the dynamical plane which are reproduced more or less faithfully in the plane of parameters.

The starting point is an article by Branner-Hubbard, which deals with a certain family of cubic polynomials (see the presentation by Lyubich in this Congress). They have to show that certain sets in the plane of parameters, which are expected to be reduced to a point, are actually so. According to the inequality of Grötzsch, it suffices to enclose such a set in a sequence of nested annuli whose sum of moduli is infinite. That is what they do, but first in the dynamical plan where the annuli considered are coverings of one another, so that up to a constant the moduli are reciprocals of integers and the divergence follows from a very thorough combinatorial study.

In their case, the transfer to the plane of parameters is easy, because the annuli considered are reproduced there conformally.

The *MLC* Conjecture is in a sense also a statement of the form "the points are actually points". In the plane of parameters which contains M, one can define the *pieces* bounded by the external rays and the equipotential arcs. Such a piece cuts out a connected set from M, and to prove *MLC* at a point c it suffices to prove that the intersection of the pieces which are neighbourhoods of c is reduced to c.

The situation is similar to that of Branner-Hubbard, and the proof in the dynamical plane can be carried out along the same lines. Most of the difficulties lie in the passage to the plane of parameters, and there Yoccoz produced a tour de force of Analysis. Indeed, away from M and K, there is a conformal correspondence between the dynamical plane and the plane of parameters, but on these sets there is no longer any correspondence (within M there are small copies of M which are not found in the dynamical plane), the annuli do not have the same modulus and it is necessary to carry out fine Analysis to show that the ratio of the moduli is bounded, and that the divergence is thus preserved.

Yoccoz did not type up his manuscript, but one can read a proof in the redactions which Milnor has made in a preprint, and Hubbard (Three theorems of Yoccoz) in the book dedicated to Milnor.

10. C^∞ Conjugation

I have dwelt at length on Complex Dynamics, because that is what I understand best, but the works of Yoccoz in Real Dynamics are equally important. Most are in collaboration.

Palis and Yoccoz have obtained a complete system of C^∞ conjugation invariants for the Morse-Smale diffeomorphisms:

- local invariants which describe the normal forms at the attractive or repulsive points;
- global invariants which compare the coordinates adapted to these normal forms where the basins overlap.

The case of a North-South dynamics on S^n is easy: the second invariant is the change of charts. But in the general case there are saddle points. Palis and Yoccoz show that these saddle points do not produce new invariants by virtue of a theorem of inessential singularities: If two Morse-Smale diffeomorphisms are C^∞-conjugate over the union of the attractive and repulsive basins, the conjugation extends in a C^∞ manner to the whole manifold.

11. Other Works with Palis

Yoccoz has written at least three other articles with Palis. One on the centralizers of diffeomorphisms, where they show that, under certain fairly general conditions, starting from a hyperbolic diffeomorphism f one can obtain by an arbitrarily small perturbation a diffeomorphism f_1 which commutes only with itself and its iterates, and such that any diffeomorphism f_2 sufficiently close to f_1 has the same property.

Another article on homoclinic bifurcations completes a result of Newhouse and establishes a converse to a result of Palis-Takens: if the Hausdorff dimension of the hyperbolic set creating this bifurcation is larger than one, the structurally stable maps are not prevalent in the neighbourhood, destroying an old dream of Thom.

To believe Michel Serres, in such a collaboration, there is always a fox who hunts and a boar who digs. How many times have Palis and Yoccoz exchanged roles?

12. Works with Le Calvez and Raphael Douady

With Le Calvez, another student of Herman, Yoccoz has proved that there is no minimal homeomorphism on the annulus $S^1 \times \mathbf{R}$. In other words, there is no homeomorphism of the sphere S^2 preserving the two poles, and such that any other point has a dense orbit.

The methods are those from 2-dimensional topology. The central lemma is that, in the neighbourhood of a fixed point which is neither attractive nor repulsive, and for which there is a neighbourhood containing no complete orbit, the map behaves, from the point of view of the index, as $z \mapsto e^{2\pi ip/q}z(1 - \overline{z}^{rq})$ for some integers r and q.

I also want to mention an article with Raphael Douady. For a diffeomorphism f of the circle conjugated to $\mathcal{R}_\alpha$ by a map h of class C^1, the measure μ_s of density $(h')^{1-s}$ satisfies $f_*((f')^s\mu_s) = \mu_s$. Douady and Yoccoz show that there exists a unique measure satisfying this property whenever f, a diffeomorphism of class C^2, has an irrational rotation number, even if the conjugant is only a homeomorphism.

13. Cocorico

This return to diffeomorphisms of the circle ends our brief guided tour of the works of Yoccoz.

With two Fields medals for France, even if it amounts to a coincidence, we can celebrate. But on an occasion like this, one should recall a proverb from our gardeners:

If the rose is pretty, it's because the manure is fat.

It is the responsibility of each national community, in Mathematics, to ensure that the quality of education in mathematics, in particular at the secondary level, is preserved. For us French, at a time when draconian reductions in working hours threaten, this task will be tough.

Jean-Christophe Yoccoz

Reprinted from Proc. Int. Congr. Math., 1994

RECENT DEVELOPMENTS IN DYNAMICS

by
JEAN-CHRISTOPHE YOCCOZ

Lab. Math., Bâtiment 425
Université Paris–Sud Orsay
F-91405 Orsay Cedex, France

1. Introduction

1.1 Broadly speaking, the goal of the theory of dynamical systems is, as it should be, to understand *most* of the dynamics of *most* systems.

The dynamical systems that we will consider in this survey are smooth maps f from a smooth manifold M to itself; the time variable then runs amongst non-negative integers.

Frequently, we will also assume that the map f is a diffeomorphism, allowing the time variable to take all integer values. We could also consider smooth flows on M, with a real time variable: the ideas and concepts are pretty much the same in this case.

Given two dynamical systems $f : M \to M$ and $g : N \to N$, a morphism from f to g is a smooth map $h : M \to N$ such that $g \circ h = h \circ f$, in other words the diagram

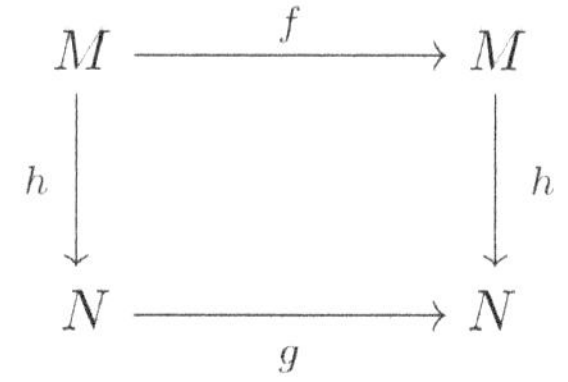

is commutative.

When h is a diffeomorphism, we will say that f and g are (smoothly) *conjugated.* When h is an embedding, f is a *subsystem* of g. When h is a submersion, f is an *extension* of g, and g is a *factor* of f.

The ultimate goal of the theory should be to classify dynamical systems up to conjugacy. This can be achieved for some classes of simple systems [PY1]; but even for (say) smooth diffeomorphisms of the two-dimensional torus, such a goal is totally unrealistic. Hence we have to settle to the more limited, but still formidable, task to understand most of the dynamics of most systems.

The word "most" in the last sentence may assume both a topological and metrical meaning. From a topological point of view, it means open and dense, or more frequently G_δ-dense (i.e. countable intersection of open and dense); from a metrical point of view, we would like to understand the trajectories of Lebesgue for almost every point of the system; when considering a smoothly parametrized family of maps or diffeomorphisms, we would also like to understand the dynamics for almost all values of the parameter.

1.2 The dynamical features that we are able to understand fall into two classes, hyperbolic dynamics and quasiperiodic dynamics; it may very well happen, especially in the conservative case, that a system exhibits both hyperbolic and quasiperiodic features.

I will not try to give a precise definition of what is hyperbolic or quasiperiodic: actually, we seek to extend these concepts, keeping a reasonable understanding of the dynamics, in order to account for as many systems as we can. The big question is then: Are these concepts sufficient to understand most systems?

1.3 The prototype of a quasiperiodic dynamical system is a translation T in a compact abelian group G; typically, G is the n-dimensional torus $\mathbf{T}^n = \mathbf{R}^n/\mathbf{Z}^n$, but the additive group $\mathbf{Z}_p$ of p-adic integers (or more generally any profinite abelian group) is also relevant.

Every translation commutes with T, hence is a symmetry of the dynamics of T: this makes the dynamics homogeneous, with a group of symmetries acting transitively. Another significant feature is that the family of iterates of T is equicontinuous; the topological entropy of T is zero.

Finally, the Haar measure on G is invariant under T, and the unitary operator $\varphi \mapsto \varphi \circ T$ of $L^2(G)$ induced by T has a discrete spectrum.

1.4 As prototypes of hyperbolic dynamical systems, we will consider two examples.

The first one is the Bernoulli shift σ on the profinite abelian group $\Sigma = \{0,1\}^{\mathbf{Z}}$, defined by

$$\sigma((x_i)_{i\in\mathbf{Z}}) = (y_i)_{i\in\mathbf{Z}}, \quad y_i = x_{i+1}.$$

For the second one, we consider a matrix $A \in \mathrm{GL}(n, \mathbf{Z})$ which is hyperbolic, i.e. no eigenvalue has modulus one. Such a matrix induces an automorphism of $\mathbf{T}^n$, which is a typical example of Anosov diffeomorphism.

Let us consider some significant features of the dynamics (in both examples).

Perhaps the most important is the *shadowing property*: define an ε-pseudo orbit as a sequence $(z_i)\_i \in \mathbf{Z}$ such that $d(f(z_i), z_{i+1}) < \varepsilon$ for all i; then, for given $\delta > 0$, there exists $\varepsilon > 0$ such that every ε-pseudo orbit $(z_i)\_i \in \mathbf{Z}$ is "shadowed" by a true orbit $(w_i)\_i \in \mathbf{Z}$ in the sense that $d(w_i, z_i) < \delta$ for all i.

A counterpart of the shadowing property is the *expansivity property*: there exists $\varepsilon_o > 0$ such that

$$\operatorname*{Sup}_n \, d(f^n x, f^n y) \geq \varepsilon_o$$

for all distinct x, y: this makes the shadowing orbit unique (for δ small enough) and is in contrast with the equicontinuity of iterates of the quasi-periodic case.

In both examples, the topological entropy is strictly positive. As automorphisms of compact abelian groups, the two examples preserve the Haar measure; the corresponding unitary operators have a Lebesgue spectrum.

2. Quasiperiodic Dynamics

2.1 Before giving some specific results, let us begin with a broad overview.

There are three approaches to quasiperiodic dynamics that have been very fruitful.

The first one is the function-theoretical approach, dealing with the stability of diophantine quasiperiodic motions. This includes the so-called KAM-theory, and techniques where functional equations are solved via Newton's method (combined with smoothing operators) or implicit function theorems in Fréchet spaces (which are conceptual analogues of Newton's method). In several special but important contexts. Herman [H2] has been able to solve the functional equations via the Schauder–Tichonoff fixed point theorem.

Finally, Rüssmann [Ru2] has announced the proof of several KAM-theorems relying only on the standard fixed point theorem.

In the symplectic context, the variational approach has also been quite successful; there is a huge number of results related to the existence of periodic orbits. We will present briefly the pioneering work of Mather on quasiperiodic dynamics in this context.

The last approach to quasiperiodic phenomena is more geometric, and frequently coined as "renormalization". Roughly speaking, the combinatorics of the recurrence are unravelled in an infinite sequence of simple successive steps, each of them involving a change of scales both in time and space. Typically, for a circle diffeomorphism f, with irrational rotation number α having convergents $(p_n/q_n)_{-}n \geq 0$, two successive iterates f^{q_n}, $f^{q_{n+1}}$ give rise to a circle diffeomorphism f_n, which is the "nth-renormalization" of f (Herman, Yoccoz). Sullivan has developed this approach when the recurrence is combinatorially described as a translation in a profinite abelian group.

2.2 Let us consider a holomorphic germ $f(z) = \lambda z + O(z^2)$, $\lambda \in \mathbf{C}^*$, in one complex variable.

We are interested in the dynamics near the fixed point 0, when the eigenvalue λ has modulus 1 but is not a root of unity; we write $\lambda = e^{2\pi i\alpha}$, with irrational $\alpha \in (0, 1)$.

It is convenient to assume some normalization on f; we will consider the class S_α of germs as above that are defined and univalent in the unit disk $\mathbf{D}$.

The germ f is always formally linearizable: there exists a unique formal power series $h_f(z) = z + O(z^2)$ satisfying $h_f \circ R_\lambda = f \circ h_f$, where $R_\lambda : z \to \lambda z$ is the linear part of f.

Consider $V_f = \text{int}\left(\bigcap_{n\geq 0} f^{-n}(\mathbf{D})\right)$; it is easy to see that $0 \in V_f$ if and only if h_f is convergent, and that in this case there exists $r_f > 0$ such that the restriction of h_f to $\{|z| < r_f\}$ is a conformal representation of the component U_f of 0 in V_f. Actually, when $\overline{U}_f \subset \mathbf{D}$, r_f is the radius of convergence of h_f.

Let us define

$$r(\alpha) = \inf_{S_\alpha} r_f,$$

and denote by $(p_n/q_n)\_n \geq 0$ the convergents of α.

Siegel [Si] proved in 1942 that $r(\alpha) > 0$ as soon as the diophantine condition $\text{Log}\, q_{n+1} = O(\text{Log}\, q_n)$ holds; he achieved this first breakthrough through small divisors problems by a direct estimation of the coefficients of h_f. Later, Brjuno [Br], through a refinement of Siegel's estimates, proved that if

$$\text{(B)} \qquad \Phi(\alpha) = \sum_{n\geq 0} q_n^{-1}\, \text{Log}\, q_{n+1} < +\infty,$$

then $r(\alpha) > 0$ and even $\text{Log}\, r(\alpha) > 2\Phi(\alpha) - c$ (for some $c > 0$ independent of α). See also [C].

Using a "renormalization" approach based on a geometric construction, I gave a new proof of the Siegel–Brjuno theorem and proved the converse ([Y5], [Y4]).

Theorem. (1) *If* $\Phi(\alpha) < +\infty$, *then*

$$|\text{Log}\, r(\alpha) + \Phi(\alpha)| < c,$$

for some $c > 0$ *independent of* α.

(2) *If* $\Phi(\alpha) = +\infty$, *the quadratic polynomial* $P_\lambda(z) = \lambda z + z^2$ *is not linearizable: every neighborhood of* 0 *contains a periodic orbit, distinct from* 0.

Actually, one first constructs a nonlinearizable germ with this property, and then shows that the same holds for the quadratic polynomial, via Douady–Hubbard's theory of quadratic-like maps.

Significant progress has been achieved by Perez–Marco [PM2], [PM3] in the understandings of the dynamics in the nonlinearizable case. He first showed that for a germ $f \in S_\alpha$ that is not linearizable and has no periodic orbit in $\mathbf{D}$ (except for 0) to exist, it is necessary and sufficient that

$$\sum_{n\geq 0} q_n^{-1}\, \text{Log} \quad \text{Log}\, q_{n+1} = +\infty.$$

He also defines "degenerate" Siegel disks as follows: assuming f to be univalent in a neighborhood of $\overline{\mathbf{D}}$, the connected component K_f of 0 in $\bigcap_{\mathbf{Z}} f^{-n}(\overline{\mathbf{D}})$ is a full, compact, connected, invariant subset of $\overline{\mathbf{D}}$ that meets S^1. When α satisfies the diophantine condition (H) (see 2.3), one has just $K_f = \overline{U}_f$.

These invariant sets provide a rich connection with the theory of analytic circle diffeomorphisms; if $k : \mathbf{H}/\mathbf{Z} \to \mathbf{C} - K_f$ is a conformal representation, the map $g = k^{-1} f\ k$ is defined in some strip $\{0 < \text{Im } z < \delta\}$ and extends by Schwarz's reflection principle to a circle diffeomorphism with the same rotation number as f.

2.3 Let us now consider analytic circle diffeomorphisms. For $\delta > 0$, define $B_\delta = \{z \in \mathbf{C}/\mathbf{Z}, |\text{ Im } z| < \delta\}$. For irrational $\alpha \in \mathbf{R}/\mathbf{Z}$, let $S_\alpha(\delta)$ be the set of orientation preserving analytic diffeomorphisms f of $\mathbf{R}/\mathbf{Z}$ with rotation number α that extend to a univalent map from B_δ to $\mathbf{C}/\mathbf{Z}$.

By Denjoy's theorem, f is conjugated to the translation $R_\alpha : z \mapsto z + \alpha$ on the circle by a homeomorphism h_f of $\mathbf{R}/\mathbf{Z}$ (uniquely defined if we require $h_f(0) = 0$). As for germs, h_f is analytic if and only if the circle $\mathbf{R}/\mathbf{Z}$ is contained in the interior of $\bigcap_{n \geq 0} f^{-n}(B_\delta)$. There are two kinds of results, depending on whether we assume or not that f is near the translation R_α; the breakthroughs (under more restrictive arithmetic conditions) are respectively due to Arnold (1960) and Herman (1976). We state the results in their final form before some comments.

Theorem 1. (Arnold [A], Rüssmann, Yoccoz [Y6]) *Assume that* Σq_n^{-1} Log $q_{n+1} < +\infty$. *There exists* $\varepsilon = \varepsilon(\alpha, \delta)$ *such that, if*

$$\|f - R_\alpha\|_{C^o(B_\delta)} < \varepsilon(\alpha, \delta),$$

then h_f *is analytic. Moreover, the diophantine condition is optimal.*

Theorem 2. (Herman, Yoccoz) *Assume that the rotation number satisfies the diophantine condition* (*H*) *below. Then* h_f *is analytic. Moreover, the diophantine condition is optimal.*

The Arithmetic Condition (H)

Assume that $0 < \alpha < 1$ and define $\alpha_o = \alpha, \alpha_n = \{\alpha_{n-1}^{-1}\}$ for $n \geq 1$. For $m \geq n \geq 0$, define inductively $\Delta(m, n)$ as follows:

$$\Delta(n, n) = 0, \quad \forall n \geq 0$$

$$\Delta(m+1, n) = \begin{cases} \exp \Delta(m, n) & \text{if} \quad \Delta(m, n) \leq \text{Log } \alpha_m^{-1} \\ \alpha_m^{-1}(\Delta(m, n) - \text{Log } \alpha_m^{-1} + 1) & \text{if} \quad \Delta(m, n) \geq \text{Log } \alpha_m^{-1} \end{cases}$$

Then α satisfies (H) if for every $n \geq 0$ we have $\Delta(m, n) \geq \text{Log } \alpha_m^{-1}$ for $m \geq m_o = m_o(n)$.

The set of numbers α satisfying (H) is a $F_{\sigma\delta}$ set (a countable intersection of F_σ sets) but neither a F_σ or a G_δ set (this explains why the definition has to be complicated). Numbers α such that

$$\text{Log } q_{n+1} = 0((\text{Log } q_b)^c), \quad \text{for some } c > 0$$

satisfy (H). On the other hand, condition (H) is strictly stronger than condition (B). Indeed, for numbers $\alpha = 1/(a_1 + 1/(a_2 + \cdots$ such that

$$a_i \leq a_{i+1} \leq \exp(a_i)$$

condition (B) is always fulfilled; on the other hand, defining $b_o = 0$, $b_n = \exp(b_{n-1})$, the number α satisfies (H) if and only if, for any $k \geq 0$, we have $a_{m+k} \leq b_m$ for m large enough; for instance, if $a_{i+1} \geq \exp(a_i^\theta)$, for some $\theta \in (0, 1)$, α does not satisfy (H).

Conditions (B) and (H) are closely related: let $\mathcal{H}_o$ be the set of irrational α such that $\Delta(m, 0) \geq \text{Log } \alpha_m^{-1}$ for large m; then α satisfies (H) if and only if its orbit under $\text{GL}(2, \mathbf{Z})$ is contained in $\mathcal{H}_o$, whereas it satisfies (B) if and only if its orbit meets $\mathcal{H}_o$.

Remarks. (1) The fact that the optimal arithmetic condition is not the same in the local and global conjugacy theorems is in strong contrast with the smooth (C^∞) case; then in both theorems, the optimal arithmetic condition is the standard one

$$\text{Log } q_{n+1} = O(\text{Log } q_n)$$

(Moser [Mo2], Herman [H1], Yoccoz [Y2]).

(2) Another important difference between the smooth and analytic cases is that the effect of good rational approximations is cumulative in the analytic case, but not in the smooth case. Another way to see this difference is to observe that the arithmetic condition in the smooth case is given by the linearized equation, whereas both conditions (H) and (B) do not appear naturally when looking at linear difference equations.

(3) All known proofs ([H1], [Y2], [KO1], [KO2], [KS]) of *global* conjugacy theorems (smooth or analytic) are based on a renormalization scheme that relies in an essential way on the relationship between the good rational approximations of the rotation number (given by the continued fraction).

2.4 When several frequencies are involved, KAM techniques are available, but they do not give as much geometric insight as we would like to have. One would like to have some geometric renormalization scheme as above, but the problem, of a purely arithmetical nature, is then to understand thoroughly the relationships between good rational approximations.

Here is a test case. Consider the following two theorems.

Theorem 1. (Arnold [A], Moser [Mo2]) *Let $\alpha = (\alpha_1, \ldots, \alpha_n) \in \mathbf{T}^n$ satisfy the diophantine condition:* $\exists_\gamma > 0,\ \tau \geq 0$ *s.t.*

$$|\langle k, \alpha\rangle + k_o| \geq \gamma \|k\|^{-n-\tau}$$

for all $(k_o, k_1 \cdots k_n) \in \mathbf{Z}^{n+1} - \{0\}$.

There exists $\varepsilon = \varepsilon(\alpha)$ *and* $k = k(\tau)$ *such that if* f *is a smooth diffeomorphism of* $\mathbf{T}^n$ *satisfying*

$$\|f - R_\alpha\|_{C^k} < \varepsilon,$$

then there exists a (small) translation R_λ *and a smooth diffeomorphism* h *such that*

$$f = R_\lambda \circ h \circ R_\alpha \circ h^{-1}.$$

Theorem 2. (Moser [Mo3]) *Let* $(\alpha_1, \ldots, \alpha_n) \in \mathbf{T}^n$ *satisfy the diophantine condition:* $\exists_\gamma > 0,\ \tau \geq 0$ *s.t.*

$$|k_o\alpha_i - k_i| \geq \gamma \|k\|^{-\frac{1}{n}-\tau}, \quad 1 \leq i \leq n$$

for all $(k_o, \ldots, k_n) \in \mathbf{Z}^{n+1} - \{0\}$.

There exists $\varepsilon = \varepsilon(\alpha)$ *and* $k = k(\tau)$ *such that if* $f_1, \ldots, f_n$ *are smooth commuting diffeomorphisms of* $\mathbf{T}^1$ *satisfying*

$$\|f_i - R_{\alpha_i}\|_{C^k} < \varepsilon, \quad \rho(f_i) = \alpha_i$$

then there exists a smooth diffeomorphism h *such that*

$$f_i = h\, R_{\alpha_i} h^{-1}, \quad 1 \leq i \leq n.$$

Problem 1: Prove Theorem 2 without assuming that f_i is close to R_{α_i}.

Problem 2: Find the *optimal* arithmetical conditions in Theorem 1 and Theorem 2 in the *analytic* case.

The first problem should be easier: diophantine conditions in smooth small divisors problems tend to be more "stable" than in analytic ones.

2.5 Codimension 1 invariant tori

The fundamental result of Moser [Mo1] on the existence of invariant curves for near integrable area-preserving twist diffeomorphisms of the annulus was first genera- lized by Rüssmann as a "translated curve" theorem (removing the area-preserving hypothesis) [Ru1], [H3]. This has recently been further generalized to higher dimensions as follows (by Cheng–Sun [CS] and Herman [H6]).

Let L be a smooth orientation preserving diffeomorphism of $\mathbf{T}^n \times \mathbf{R}$ such that $L(\mathbf{T}^n \times \{0\}) = \mathbf{T}^n \times \{0\}$ and the restriction of L to $\mathbf{T}^n \times \{0\}$ is a translation; let also $\alpha \in \mathbf{T}^n$ satisfy the standard diophantine condition (see Theorem 1 in 2.4).

Then, if F is a smooth diffeomorphism of $\mathbf{T}^n \times \mathbf{R}$ close enough to L, there exists a translation R in $\mathbf{T}^n \times \mathbf{R}$ such that $R \circ F$ leaves invariant a codimension one torus T, going through the origin, C^∞-close to $\mathbf{T}^n \times \{0\}$, and $R \circ F/T$ is smoothly conjugated to the given translation R_α.

Herman has derived important consequences of this result.

The first is the failure of the quasi-ergodic hypothesis. The ergodic (resp. quasi-ergodic) hypothesis states that the generic Hamiltonian flow is ergodic (resp. has a dense orbit) on the generic (compact, connected) energy surface. The classical KAM-theorems provide for open sets of Hamiltonian flows a set of positive measure (on each energy surface) made of diophantine invariant tori; hence the ergodic hypothesis fails. Herman has discovered a rigidity property of the rotation number in the symplectic context that guarantees a similar phenomenon: there exist a nonempty open set of Hamiltonian flows and energy values for which the energy surface contains a Cantor set of *codimension one* diophantine invariant tori; the orbits "between" the tori are thus constrained to stay there.

Another important consequence is the failure of a conjecture of Pesin: Herman shows that on any manifold M (of dimension ≥ 3) there exists a nonempty open set of volume-preserving diffeomorphisms whose Lyapunov exponents are all 0 on a set of positive volume. In dimension 2, this follows from Moser's twist theorem.

2.6 In the symplectic context, Mather has been pioneering the study of quasiperiodic motions through a variational approach. In one degree of freedom, we now have (due to Aubry [AD], Mather [Ma1], Le Calvez [LeC1], ...) a fairly satisfactory theory of Aubry–Mather Cantor sets. In more degrees of freedom, Mather has obtained a yet partial generalization that seems quite promising in understanding somewhat Arnold diffusion ([Ma3], [Ma2]).

Let me explain this very roughly for discrete time (diffeomorphisms). Consider an integrable diffeomorphism L of $\mathbf{T}^n \times \mathbf{R}^n = T^*\mathbf{T}^n$:

$$L(\theta, r) = (\theta + \nabla\ell(r), r),$$

with strictly convex ℓ superlinear at ∞.

To each invariant lagrangian torus $\{r = r_o\}$, we can associate the cohomology class $r_o \in H^1(\mathbf{T}^n, \mathbf{R})$ and the rotation number $\alpha = \nabla\ell(r_o)$ that belongs in a natural way to $H_1(\mathbf{T}^n, \mathbf{R})$.

Let now $F : (\theta, r) \mapsto (\Theta, R)$ be an exact symplectic diffeomorphism close to L. Writing

$$\Sigma R_i \, d\Theta_i - \Sigma r_i \, d\theta_i = dH \; (\theta, \Theta)$$

we obtain the generating function H of F, defined on $\mathbf{R}^n \times \mathbf{R}^n$ and satisfying

$$H(\theta + k, \Theta + k) = H(\theta, \Theta), \quad k \in \mathbf{Z}^n.$$

Given an invariant measure μ with compact support, we transport it via $(\theta, r) \mapsto (\theta, \Theta)$ to the diagonal quotient $(\mathbf{R}^n \times \mathbf{R}^n)/\mathbf{Z}^n$ and consider, for

$\omega \in \mathbf{R}^n = H^1(\mathbf{T}^n, \mathbf{R})$, the ω-action:

$$A_\omega(\mu) = \int [H(\theta, \Theta) - \langle \omega, \Theta - \theta \rangle] \, d\mu.$$

The invariant measure is *minimal* if it minimizes the ω-action (amongst invariant measures, or equivalently amongst all measures) for some cohomology class ω.

On the other hand, to any invariant measure, one can associate a rotation number

$$\alpha(\mu) = \int (\Theta - \theta) \, d\mu \in \mathbf{R}^n = H_1(\mathbf{T}^n, \mathbf{R}).$$

Then μ is minimal if and only if it minimizes the action A_o amongst all invariant measures with the same rotation number.

For any ω, there exist ω-minimal measures; there also exist minimal measures with any given rotation number α. The correspondence between α and ω is realized by Legendre transform (in a nonsmooth, nonstrictly convex context).

The support of such a minimal measure is an invariant torus in the integrable case and shares in the general case some properties of Aubry–Master sets: in particular, it is the *graph* of the restriction to a closed subset of a *Lipschitz* map from $\mathbf{T}^n$ to $\mathbf{R}^n$.

The key point for further progresses is to understand the "shadowing" properties of these minimal measures. With one degree of freedom, Mather has proved (see also Le Calvez) that there are no obstructions except for the obvious ones: if $(\Lambda_n)_{-}n \in \mathbf{Z}$ is a sequence of Aubry–Mather sets, *not separated by an invariant curve*, there exists an orbit coming successively (in the prescribed order) close to each of the Λ_i. In more degrees of freedom, invariant tori do not separate and there is no obvious obstruction preventing an orbit to come successively close to the supports of any given sequence of minimal measures (for a generic diffeomorphsim). Mather has a partial result in this direction.

2.7 Renormalization theory for quadratic polynomials

The Aubry–Mather sets and minimal measures we have just discussed are important generalizations of the classical KAM quasiperiodic motions. Another nonstandard "generalization" is provided by the dynamics of infinitely renormalizable quadratic polynomials.

The key tool is the Douady–Hubbard theory of quadratic-like maps [DH2], i.e. ramified covering $f : U \to U'$ of degree 2, with U, U' simply connected and $U \subset\subset U'$. Such a map is quasiconformally conjugated to a quadratic polynomial, its filled-in Julia set is $K_f = \bigcap_{n \geq 0} f^{-n}(U')$.

An integer $n \geq 2$ is a renormalization period for the quadratic polynomial $P_c : z \mapsto z^2 + c$ if there exist open neighborhoods $U_n \subset\subset U'_n$ of 0 such that $P_c^n : U_n \to U'_n$ is quadratic-like with connected filled-in Julia set. The quadratic polynomial is

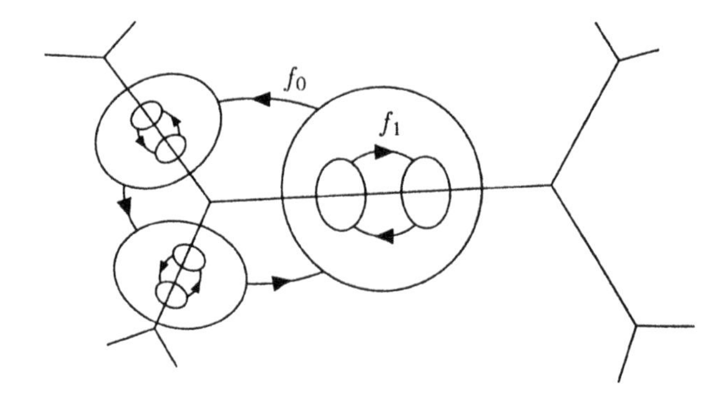

Figure 1

infinitely renormalizable if the set $\mathcal{N} = \{n_1 < n_2 < \cdots\}$ of its renormalization periods is infinite; then n_k divides n_{k+1} (we write $n_{k+1} = p_{k+1}n_k$, $p_1 = n_1$); let $f_o = P_c$ and $f_k = P_c^{n_k}/U_{n_k}$. Then f_{k+1} is the restriction of f_k^{pk} to the smaller domain $U_{n_{k+1}}$; it is called the renormalization of f_k. (See Figure 1).
In the study of the dynamics of rational maps, a key point is to understand the geometry and the dynamics of the post-critical set:

$$P(f) = \overline{\{f^n(c), n \geq 1, c \text{ critical value}\}}.$$

In our case, $P(f_o)$ is contained in the fully invariant compact set

$$K_\infty = \bigcap_{k\geq 0} \bigcup_{0\leq j<n_k} f_o^j(K_{f_k}).$$

We have a natural continuous surjective map

$$K_\infty \xrightarrow{p} \mathbf{Z}_{\mathcal{N}} = \varprojlim \mathbf{Z}/n_k\mathbf{Z}$$

onto the profinite abelian group of $\mathcal{N}$-adic integers, the dynamics on $\mathbf{Z}_{\mathcal{N}}$ being translation by 1; the post-critical set is sent to the set of positive integers. The map p is known to be a homeomorphism for real c (Sullivan), but it is also known that it is not always injective.

Problem: Find a necessary and sufficient condition on the combinatorics for p to be a homeomorphism.

Very little is known in the general case (with complex c). On the other hand, a beautiful approach pioneered by Sullivan [Su1], [Su2] has been fruitful in important particular cases. The general strategy is the following; one first constructs, from an infinitely renormalizable quadratic-like map, a geometric object that is a compact set laminated by Riemann surfaces (Riemann lamination). The dynamical properties of the initial quadratic-like map correspond to some properties of the complex geometry of this Riemann lamination. Such laminations have, as usual Riemann surfaces, a Teichmüller space; the renormalization operator (from f_k to f_{k+1}) corresponds to a map between such Teichmüller spaces and we are led to study the *dynamics* of this new map (at the "parameter" level). This map does not increase the Teichmüller distance, and the central problem is to understand to which extent it is contracting. There are partial results in this direction by Sullivan (for real c, with $(p_k)\_k \geq 1$ bounded) and McMullen (under a potentially more general geometric assumption) [McM].

3. Hyperbolic Dynamics

3.1 Before we discuss some recent developments, we recall some "classical" hyperbolic dynamics, as developed by Anosov, Sinaï, Smale, Palis, ... in the 1960s [Bo], [Sm], [Sh], [Y9].

The central concept is that of a *basic set* : if f is a smooth diffeomorphism of a manifold M, a basic set of f is a *compact*, *invariant* subset K of M that is *transitive* (f/K has a dense orbit), *locally maximal* (K is the maximal invariant set in an open neighborhood), and *hyperbolic* : the tanget bundle $E = TM/K$ admits an invariant splitting $E = E^s \oplus E^u$ in a stable subbundle E^s uniformly contracted by Tf and an unstable subbundle E^u uniformly contracted by Tf^{-1}.

The dynamics on a basic set are fairly well understood (and completely so when $\dim M = 2$); in particular, the existence of Markov partitions allows us to reduce the study of periodic orbits, invariant measures, ... to the same problems in symbolic dynamics, i.e. subshifts of a finite type on a finite alphabet.

The existence of a basic set K for a diffeomorphism is a semilocal property: it only involves the dynamics of f near K. One gets to more global properties (Anosov diffeomorphisms, Axiom A diffeomorphisms, ...) if one asks that some big invariant subset, carrying "most" of the dynamical properties of f, is hyperbolic.

For instance the *chain recurrent set* $C(f)$ of a smooth diffeomorphism of a compact manifold is the locus of points that are periodic for some arbitrarily small C^o-perturbation of f. Let us say that f is *uniformly hyperbolic* if $C(f)$ is hyperbolic.

It can be proven that $C(f)$ is then a finite union of disjoint basic sets. Uniformly hyperbolic diffeomorphisms form an open subset of $\mathrm{Diff}^\infty(M)$ (it is even open in the C^1-topology), and they are *stable* [R], [Ro]: two C^1-close uniformly hyperbolic diffeomorphisms are topologically conjugated on a neighborhood of their respective chain recurrent sets. Actually, a deep theorem of Mañé (extended by Palis) states that the converse is also true: a diffeomorphism that is C^1 stable in this sense is uniformly hyperbolic [M3].

It was hoped at some point that such globally hyperbolic diffeomorphisms could account, at least in the dissipative case, for most diffeomorphisms. This was shown to be too optimistic when Newhouse [N1], [N2] discovered in the 1970s that there exist open sets of diffeomorphisms that exhibit generically infinitely many attractive periodic orbits (this is not compatible to any global uniform hyperbolic behaviour).

Nevertheless, uniformly hyperbolic diffeomorphisms still constitute a good starting point from which one can bifurcate and study more complicated diffeomorphisms. Also, there are many important classes of diffeomorphisms that are not uniformly hyperbolic, but that admit many basic sets that together should carry a lot of information on the dynamics.

3.2 The conceptual apparatus to study weaker forms of hyperbolicity is based on Oseledets' theorem (1968) [O] and Pesin's theory (1976) [Pe1], [Pe2], [FHY]. Oseledets theorem, itself based on a subadditive ergodic theorem, asserts the existence of Lyapunov exponents of a diffeomorphisms on a (Borel) set of points that has full measure with respect to all invariant measures. From this starting point, Pesin then constructed the stable and unstable "foliations" associated to the nonzero exponents and proved the crucial fact that they are absolutely continuous.

How frequently are all (or some) of the Lyapunov exponents of a non-uniformly hyperbolic diffeomorphism different from 0? I have mentioned above Herman's theorem (see 2.5), which indicates that we cannot be too optimistic. On the other hand, there have been several breakthroughs showing that it tends to happen with positive measure in the parameter space.

3.3 The first crucial step in this direction is Jakobson's theorem (1981) [J]. He considers real quadratic polynomials $P_c(x) = x^2 + c$, for c in some subset $A_\varepsilon \subset [-2, -2+\varepsilon]$, whose relative measure tends to 1 as ε goes to zero. For such a parameter, let α be the negative fixed point of P_c and $I = (\alpha, -\alpha)$; he constructs a countable partition $I = \bigcup I_i$ (mod 0) into disjoint open intervals and a map $T : \bigcup\limits_i \to I$ that is *uniformly expanding* (with bounded distortion) and whose restriction to each I_i is an iterate $P_c^{k_i}$ realizing a diffeomorphism onto I (see Figure 2).

An important point is that although the k_i are not bounded, the measures of the I_i for which $k_i \geq k$ is exponentially small with k. From the existence of such a map T, it is easy to deduce that P_c has an ergodic invariant measure that is equivalent

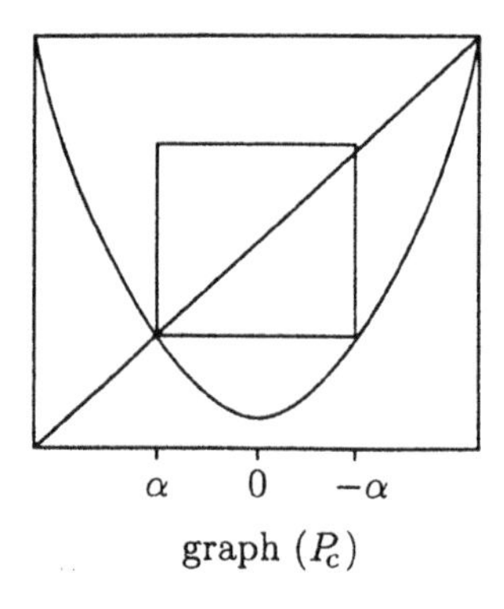

graph (P_c)

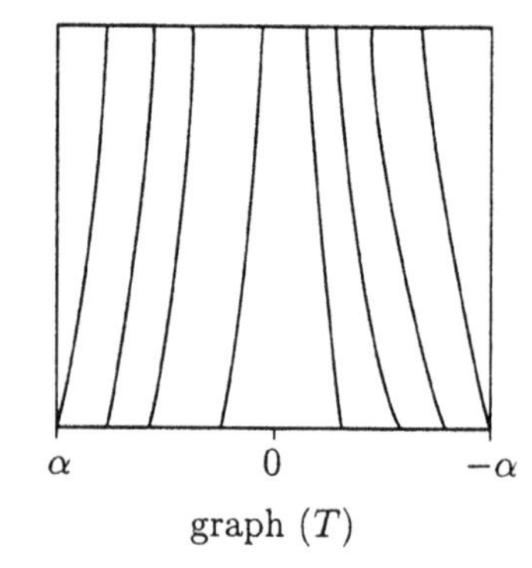

graph (T)

Figure 2

to Lebesgue measure on the (real) Julia set, and that the corresponding Lyapunov exponent is positive.

This kind of result has since been extended in several directions. Rees [Re] has proved that a similar statement holds for an holomorphic family of rational maps (see also [Be]). Jakobson and Swiatek have extended the set of values of c for which the map T is constructed (putting no restriction on the k_i's) [JS].

For *complex* quadratic polynomials whose all periodic orbits are repulsive and that are not infinitely renormalizable, I proved that the dynamics are still sufficiently expanding to guarantee that the Julia sets are locally connected, as a consequence of a (very weak) self-similarity property [Y10]. Lyubich [Ly1] then went on to prove that such a Julia set has measure 0. These results are related to previous work of Branner–Hubbard [BH] and McMullen [McM] on complex cubic polynomials.

3.4 Another very important breakthrough, going to higher dimensions, was achieved by Benedicks–Carleson (1989). They consider Hénon's family [He] of polynomial diffeomorphisms of the plane:

$$H_{a,b}(x,y) = (x^2 + a - y,\ bx).$$

The parameter b is the constant value of the Jacobian; it is fixed and very small. The parameter a belongs to a subset $A_\varepsilon \subset [-2+\varepsilon,\ -2+2\varepsilon]$ of relative large measure, with $0 < |b| \ll \varepsilon \ll 1$. For such parameters, the rectangle $U = \{(x,y), |x| < 2 - \frac{3}{4}\varepsilon, |y| < 3b\}$ satisfies $H_{a,b}(U) \subset\subset U$, and one wants to describe the "attractor"

$$\Lambda_{a,b} = \bigcap_{n \geq 0} H^n_{a,b}(U).$$

What emerges from Benedicks–Carleson's study [BC2], together with more recent work of Benedicks–Young [BY] and Jakobson–Newhouse [JN] is the following structure (see Figure 3): one can construct an open subrectangle $V \subset U$, a countable family of disjoint subrectangles $V_i \subset V$ such that $\bigcup_i V_i$ "essentially" covers $V \cap \Lambda$, and a map $T : \bigcup_i V_i \to V$, whose restriction to V_i is some iterate H^{k_i} of H, and that is *uniformly* hyperbolic; the k_i's are not bounded, but they take big values on very small sets.

From there, one constructs a nice Sinaï–Bowen–Ruelle invariant measure on Λ; it describes the asymptotics of a positive *Lebesgue* measure set of orbits in U, and the Lyapunov exponents with respect to that measure are nonzero. One also recovers many classical properties of uniformly hyperbolic attractors (it is easy to see that Λ cannot be uniformly hyperbolic; in fact, there is a dense subset of Λ where the stable and unstable manifolds are tangent).

One should note that the admissible set A_ε of values of the parameter a has an empty interior; also, two distinct values of a give rise to attractors that admit the same qualitative description, but are definitely not conjugated. This is in strong contrast to the uniformly hyperbolic case.

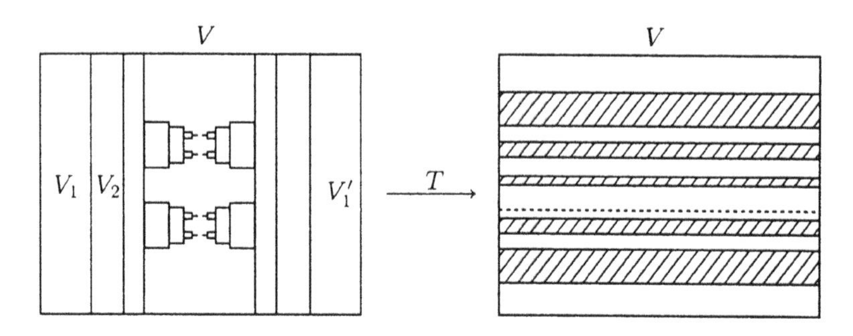

Figure 3

The kind of phenomenon that we have tried to describe is not particular to the Hénon family. A first extension of these results, extremely important for applications (see below), was given by Mora–Viana [MV], who introduced the concept of "Hénon-like" families. More recently, Viana [V] proved similar results for some families in higher dimensions, for instance skew products:

$$T : \mathbf{T}^2 \times \mathbf{R}^2 \to \mathbf{T}^2 \times \mathbf{R}^2$$

$$T(\theta, (x, y)) = (A(\theta),\ H_{a(\theta),b}(x, y)),$$

where A is an Anosov diffeomorphisms of $\mathbf{T}^2$ (for instance linear hyperbolic) and a is a Morse function on $\mathbf{T}^2$ (subjected to some conditions). In this context, because of the uniform hyperbolicity on the base, it is no more necessary to exclude parameters.

3.5 What we would like to do in the next few years is to obtain a conceptual theory of "weakly hyperbolic basic sets" (including of course the striking examples considered above). For a smooth diffeomorphisms f of a manifold M, such a "weakly hyperbolic basic set" should again be a *compact*, *invariant*, *transitive*, *locally maximal* subset K of M satisfying moreover some kind (?) of *weak hyperbolicity* condition. Let me speculate, based on the examples above, on what could be some aspects of this theory.

(1) One would be able to cover "most" of K with a countable family of disjoint open sets V_i and define on $\bigcup V_i$ a *uniformly hyperbolic* map T whose restriction to V_i is some iterate f^{k_i} of f;

(2) One would thus obtain K as the limit (for the Hausdoff distance on compact sets) of an increasing sequence of (uniformly hyperbolic) basic sets

$$K_n = \bigcap_{m \in \mathbf{Z}} f^{-m}\left(\overline{\bigcup_{k_i \leq n} V_i}\right);$$

(3) One should be able to construct some kind of Sinaï–Bowen–Ruelle invariant measure, whose "restriction" to most unstable manifolds would be absolutely continuous with respect to some Hausdorff measure on the unstable manifold. The Lyapunov exponents with respect to this measure would be nonzero;

(4) There would exist some "infinite Markov partition" (as the V_i above) allowing a description by symbolic dynamics (with an infinite alphabet).

4. Parameter Space

4.1 I want to discuss now "how many" dynamical systems we are able to understand.

Let us start with a "test case", the family of quadratic (real or complex) polynomials $P_c : x \mapsto x^2 + c$, where only one (real or complex) parameter c is involved.

If the critical point 0 escapes under iteration to infinity or converges to some attractive periodic orbit, the dynamics on J_c are *uniformly hyperbolic* (expanding) and stable; such parameters c form an open set U_{hyp}.

If there is a periodic orbit with eigenvalue λ of modulus 1, the parameter c is determined (algebraically) by the period and λ (up to a finite number of choices): the dynamics on the Julia set is of "weak-hyperbolic" type when λ is a root of unity, whereas quasiperiodic features are dominant when λ is not a root of unity (see 2.2).

We are left with the case where all periodic orbits are repulsive, but the critical point belongs to the Julia set J_c (preventing it to be uniformly hyperbolic): it is here natural to discuss separately the cases where P_c is infinitely renormalizable and it is not.

In the non-infinitely renormalizable case, the dynamics on the Julia set still exhibit some (very weak) form of hyperbolicity (see 3.3). As consequence, I proved that such parameters are *rigid*, i.e. they are determined by the combinatorics of the Julia set [Y10].

In the infinitely renormalizable case, the dynamics on the Julia set exhibit many quasiperiodic features (see 2.7). Swiatek [Sw] and Lyubic [Ly2] have proved that for *real* quadratic polynomials, these parameters are rigid (in the real sense). As a consequence, the open set $U_{\text{hyp}} \cap \mathbf{R}$ of (uniform) hyperbolicity is dense in $\mathbf{R}$. Actually, one would expect that even the following stronger statement should be true: for almost all $c \in \mathbf{R}$, either $c \in U_{\text{hyp}}$ or there exists (as in Jakobson's theorem) an invariant measure on $J_c \cap \mathbf{R}$, absolutely continuous with respect to Lebesgue measure, for which the Lyapunov exponent is nonzero.

On the other hand, although there is a partial result by Lyubich, it is not known whether infinitely renormalizable parameters are rigid in the complex sense. This is the missing step in the Douady–Hubbard's conjecture that the Mandelbrot set is locally connected; actually, assuming this rigidity, we would have from Douady–Hubbard [DH1] and Thurston a complete *topological* description of the Mandelbrot set (whereas we have only a *combinatorial* one at the moment).

4.2 In higher dimensions, such as for instance for diffeomorphisms of surfaces, we are still very far from having a near complete understanding of the "parameter space".

Nevertheless, the results that we have discussed above and others in the same line have led to a change of point of view in looking at these problems.

The classical, uniformly hyperbolic, basic sets have a strong stability property known as hyperbolic continuation: a nearby diffeomorphism admits basic set close to the original one and the dynamics on the two basic sets are conjugated by a homeomorphisms close to the identity. The "parameter space", for instance the space of all smooth diffeomorphisms of the given compact manifold, was in the 1960s and 1970s mostly considered from a topological point of view; one was looking for dynamical features appearing on some open set, or some G_δ set (dense into some open set).

Although this point of view remains important, properties of "weakly hyperbolic basic set" such as Hénon-like attractors have given a strong impetus on the measure-theoretic point of view of the parameter space: typically, in generic parameter family of diffeomorphisms, on expects to meet these weakly hyperbolic features on a F_σ subset of the parameter space (because one needs to exclude parameters), but one that has positive Lebesgue measure.

4.3 I would like here to emphasize the importance of the Hénon family, or rather of the Hénon-like families introduced by Mora and Viana (see 3.4), for the study of surface diffeomorphisms.

Consider a smooth diffeomorphism f_o of a surface M that exhibits a homoclinic tangency: this means that f_o has a fixed saddle point p such that the stable manifold $W^s(p)$ and the unstable manifold $W^u(p)$ are tangent along a homoclinic orbit $(f^n(q))_{n \in Z}$. This is a codimension one phenomenon.

Assume that the tangency is quadratic and consider a one-parameter family of diffeomorphisms $(f_t)_{-|t|} < \varepsilon$ unfolding generically the tangency.

One would like to understand the orbits under f_t, $|t|$ small, which remain in an appropriately small neighborhood of the orbit of q under f_o. To do this, the first step is to compute the return map R_t into some small neighborhood V of q; it is a disjoint union $V = \bigcup_{n \geq n_o} V_n$; R_t is equal to f_t^n on V_n and is a Hénon-like family of approximately constant Jacobian. The striking conclusion is that every dynamical feature exhibited by the Hénon family, or Hénon-like families, actually occurs in a generic one-parameter family near homoclinic bifurcations. For instance, if the fixed point p is dissipative, the Jacobian of these Hénon-like families will be very small and we will get for a positive measure set of parameters Hénon-like attractors. Let me recall in this context an older result of Newhouse: there are arbitrarily close to 0 intervals of parameter values in which for generic t the diffeomorphisms f_t has *infinitely many* periodic orbits; still we suspect (but we don't know) that this only happens for a set of t of Lebesgue measure zero.

Homoclinic bifurcations are by no means the only codimension one bifurcations where Hénon-like families occur; another such example, studied by Diaz, Rocha, and Viana, is the critical saddle-node bifurcation [DRV1], [DRV2].

4.4 Palis has proposed a general program to study the dynamics of (non-uniformly hyperbolic) diffeomorphisms (of compact surfaces, to begin with). He suggests that one should look first to a *dense* subset in the space of non-uniformly hyperbolic diffeomorphisms for which we have at least some grasp of the dynamics; he conjectures actually that homoclinic tangencies could be such a subset. Then one should consider generic parametrized families through these "simple" diffeomorphisms, and study which dynamical features are "persistent" in the measure theoretical sense, i.e. they occur on sets of parameters of positive measure or even relative positive density near the initial diffeomorphism.

Starting with Newhouse–Palis–Takens [NPT] and Palis–Takens ([PT1], [PT2]), there has been a great deal of effort and results related to the study of homoclinic bifurcations, which give rise to an extremely rich number of complicated phenomena; still we are quite far from a satisfactory understanding of f. One very interesting feature of these results, when the fixed saddle point belongs to some (uniformly hyperbolic) basic set, is the subtle relationship between the geometry of the basic set (Hausdorff dimension, thickness, ...) and the dynamics near the bifurcating parameter [PY2].

4.5 Still there is a very central difficulty in carrying out Palis's program, a difficulty that occurs at very many places in dynamical systems, the so-called closing-lemma problem. Pugh's closing lemma [Pu] asserts that if p is a recurrent point for a smooth diffeomorphism f, then one can perturb f *in the C^1-topology* in order for p to become periodic. See also [M1].

We still have no idea whether it is possible to achieve the same goal by a C^2 (or even $C^{1+\varepsilon}$) perturbation. In particular, we still don't know whether the C^2 diffeomorphisms of $\mathbf{T}^2$ that have a periodic orbit form a C^2-dense subset of $\mathrm{Diff}^2(\mathbf{T}^2)$! Guttiercz [G] has constructed an example (on the noncompact surface $\mathbf{T}^2 - \{0\}$ that indicates that the localized perturbations used by Pugh in the C^1-case cannot be sufficient in the C^2-case. Also Herman [H4], [H5] has constructed a Hamiltonian flow on a compact sympletic manifold, for which no periodic orbit can be created by smooth perturbations of the Hamiltonian (because of KAM-theory and a sympletic rigidity of the rotation number).

In a similar vein, recent results of Herman suggest that it would be very interesting to know the answer to the following question: Let $\mathcal{C}$ be the set of smooth diffeomorphisms (of a compact manifold M) that are of finite order on a nonempty open set (depending on the diffeomorphism); what is the closure of $\mathcal{C}$ in the C^∞-topology?

References

[A] V. I. Arnold, *Small denominators I: on the mapping of a circle into itself*, Izv. Akad. Nauk., serie Math. **25**, 21–86 (1961) = Trans. Am. Math. Soc., 2nd series, **46**, 213–284.

[AD] S. Aubry and P. Y. Le Daeron, *The discrete Frenkel–Kantorova model and its extensions*, Phys. D, **8**, 381–422 (1983).

[BC1] M. Benedicks and L. Carleson, *On iterations of* $1-ax^2$ *on* $(-1,1)$, Ann. of Math. (2), **122**, 1–25 (1985).

[BC2] M. Benedicks and L. Carleson, *The dynamics of the Hénon map*, Ann. of Math. (2) **133**, 73–169 (1991).

[BY] M. Benedicks and L. S. Young, *SBR-measures for certain Hénon maps*, Invent. Math., 112–3, 541–576 (1993).

[Be] J. Bernard, *Sur la dynamique des exemples de Lattès*, thèse Univ. Paris-Sud, (1994).

[Bo] R. Bowen, Equilibrium states and the ergodic theory of Anosov diffeomorphisms, Lecture Notes in Math. 470, (1975), Springer–Verlag, Berlin and New York.

[BH] B. Branner and J. H. Hubbard, *The iteration of cubic polynomials*, Acta Math., **160**, 143–206 (1988); **169** 229–325 (1992).

[Br] A. D. Brjuno, *Analytical form of differential equations*, Trans. Moscow Math. Soc. **25**, 131–288 (1971); **26**, 199–239 (1972).

[CS] C. Q. Cheng and Y. S. Sun, *Existence of invariant tori in three-dimensional measure preserving mappings*, Celestial Mech. **47**, 275–293 (1990).

[C] T. M. Cherry, *A singular case of iteration of an analytic function; a contribution to the small divisor problem*, in Nonlinear Problems of Engineering, ed. W. Ames, Acad. Press, New York and San Diego, CA, 29–50 (1964).

[DRV1] L. J. Diaz, J. Rocha, and M. Viana, *Saddle-node cycles and prevalence of strange attractors*, preprint IMPA 1993.

[DRV2] L. J. Diaz, J. Rocha, and M. Viana, *Global strange attractors for dissipative maps of the annulus*, in preparation.

[DH1] A. Douady and J. H. Hubbard, *Étude dynamique des polynômes complexes*, Publ. Math. Orsay 82–84 (1984).

[DH2] A. Douady and J. H. Hubbard, *On the dynamics of polynomial-like mappings*, Ann. Sci. École Norm. Sup. (4), **18**, 287–343 (1985).

[FHY] A. Fathi, M. Herman, and J. C. Yoccoz, *A proof of Pesin's stable manifold theorem dans Geometric dynamics*, Proceedings of the Symposium in Rio de Janeiro 1981, Springer Lecture Notes no. 1007.

[G] C. Gutierrez, *a counter-example to a* C^2 *closing lemma*, Ergodic Theory Dynamical Systems **7**, 509–530 (1987).

[He] M. Hénon, *A two-dimensional mapping with a strange attractor*, Comm. Math. Phys. **50**, 69–77 (1976).

[H1] M. R. Herman, *Sur la conjugaison differentiable des difféomorphisms du cercle à des rotations*, Publ. Math. I.H.E.S. **49**, 5–233 (1979).

[H2] M. R. Herman, *Simple proofs of local conjugacy theorems for diffeomorphisms of the circle with almost every rotation number*, Bol. Soc. Brasil. Mat., **16**, 45–83 (1985).

[H3] M. R. Herman, *Sur les courbes invariantes par les difféomorphisms de l'anneau*, Astérisque, vol. **1**, 103–104 (1983), vol. **2**, 144 (1986).

[H4] M. R. Herman, *Exemples de flots hamiltoniens dont aucune perturbation en topologie* C^∞ *n'a d'orbites périodiques sur un ouvert de surfaces d'énergie*, C. R. Acad. Sci. Paris, t. 321, Série I, 989–994 (1991).

[H5] M. R. Herman, *Differentiabilité optimale et contre-exemples à la fermeture en topologie* C^∞ *des orbites récurrentes des flots hamiltoniens*, C. R. Acad. Sci. Paris, t. 313, Série I, 49–51 (1991).

[H6] M. R. Herman *Théorème des tores translatés et quelques applications à la stabilité topologique des systèmes dynamiques conservatifs*, en preparation.

[J] M. V. Jakobson, *Absolutely continuous invariant measures for one-parameter families of one-dimensional maps*, Comm. Math. Ph. **81**, 39–88 (1981).

[JN] M. V. Jakobson and S. Newhouse, *Strange attractors in strongly dissipative surface diffeomorphisms*, in preparation.

[JS] M. V. Jakobson and G. Swiatek, *Metric properties of non-renormalizable S-unimodal maps*, preprint (1994).

[KO1] Y. Katznelson and D. Ornstein, *The differentiability of conjugation of certain diffeomorphisms of the circle*, Ergodic Theory Dynamical Systems **9**, 643–680 (1989).

[KO2] Y. Katznelson and D. Ornstein, *The absolute continuity of conjugation of certain diffeomorphisms of the circle*, Ergodic Theory Dynamical Systems **9**, 681–690 (1989).

[KS] K. M. Khanin and Ya. G. Sinai, *A new proof of M. Herman's theorem*, Comm. Math. Phys. **112**, 89–101 (1987).

[LeC1] P. Le Calvez, *Propriétés dynamiques des zones d'instabilité*, Ann. Sci. École Norm. Sup. (4), **20**, 443–464 (1987).

[LeC2] P. Le Calvez, *Étude topologique des applications deviant la verticale*, Ensaios Mat., vol. **2**, 1990.

[Ly1] M. Lyubich, *On the Lebesgue measure of the Julia set of a quadratic polynomial*, preprint Stonybrook (1991).

[Ly2] M. Lyubich, *Geometry of quadratic polynomials: Moduli, rigidity and local connectivity*, preprint Stonybrook (1993).

[M1] R. Mañé, *An ergodic closing lemma*, Ann. of Math. (2) **116**, 503–540 (1982).

[M2] R. Mañé, Ergodic Theory and Differentiable Dynamics, Springer–Verlag, Berlin and New York, 1987.

[M3] R. Mañé, *a proof of the C^1 stability conjecture*, Publ. Math. I.H.E.S. **66**, 161–210 (1988).

[Ma1] J. Mather, *Existence of quasiperiodic orbits for twist diffeomorphisms of the annulus*, Topology, **21**, 457–467 (1982).

[Ma2] J. Mather, *Minimal measures*, Comm. Math. Helv., **64**, 375–394 (1989).

[Ma3] J. Mather, *Action minimizing invariant measures for positive definite Lagrangian systems*, Math. Z., **207**, 169–207 (1991).

[McM] C. McMullen, *Renormalization and 3-manifolds which fiber over the circle*, preprint (1994).

[dMvS] W. de Melo and S. van Strien, One Dimensional Dynamics, Springer–Verlag, Berlin and New York, 1993.

[MV] L. Mora and M. Viana, *Abundance of strange attractors*, Acta Math. **171**, 1–71 (1993).

[Mo1] J. Moser, *A rapidly convergent iteration method, part II*, Ann. Scuola Norm. Sup. Pisa Cl. Sci. (3), **20**, 499–535 (1996).

[Mo2] J. Moser, *On invariant curves of area preserving mappings of an annulus*, Nachr. Akad. Wiss. Göttingen, Math.-Phys. Kl., 1–20 (1962).

[Mo3] J. Moser, *On commuting circle mappings and simultaneous diophantine approximations*, Math. Z. **205** 105–121 (1990).

[N1] S. Newhouse, *Diffeomorphisms with infinitely many sinks*, Topology, **13**, 9–18 (1974).

[N2] S. Newhouse, *The abundance of wild hyperbolic sets and non smooth stable sets of diffeomorphisms*, Publ. Math. I.H.E.S. **50**, 101–151 (1979).

[NPT] S. Newhouse, J. Palis, and F. Takens, *Bifurcations and stability of families of diffeomorphisms*, Publ. Math. I.H.E.S. **57**, 5–71 (1983).

[O] V. Oseledec, *A multiplicative ergodic theorem: Lyapunov characteristic numbers for dynamical systems*, Trans. Moscow Math. Soc. **19**, 197–231 (1968).

[P] J. Palis, *On the C^1 — Ω-stability conjecture*, Publ. Math. I.H.E.S. **66**, 211–215 (1988).

[PdM] J. Palis and W. de Melo, Geometric Theory of Dynamical Systems, Springer–Verlag, Berlin and New York, 1982.

[PT1] J. Palis and F. Takens, *Cycles and measure of bifurcation sets for two-dimensional diffeomorphisms*, Invent, Math. **82**, 397–422 (1985).

[PT2] J. Palis and F. Takens, *Hyperbolicity and the creation of homoclinic orbits*, Ann. of Math. (2) **125**, 337–374 (1987).

[PT3] J. Palis and F. Takens, Hyperbolicity and Sensitive Chaotic Dynamics, Cambridge Univ. Press, London, 1993.

[PY1] J. Palis and J. C. Yoccoz, *Differentiable conjugacy of Morse Smale diffeomorphisms*, Bol. Soc. Brasil. Mat. **25**, 25–48 (1990).

[PY2] J. Palis and J. C. Yoccoz, *Homoclinic tangencies for hyperbolic sets of large Hausdorff dimension*, Acta Math. **172**, 92–136 (1994).

[PM1] R. Perez-Marco, *Solution complète au problème Siegel de linearisation d'une application holomorphe au voisinage d'un point fixe*, Sem. Bourbaki no. 753, Astérisque, 206 (1992).

[PM2] R. Perez-Marco, *Sur les dynamiques holomorphes non linéarisables et une conjecture de V. I. Arnold*, Ann. Sci. École Norm. Sup. (4), **26**, 565–644 (1993).

[PM3] R. Perez-Marco, *Non-linearizable dynamics having an noncountable number of symmetries*, to appear in Invent. Math.

[PM4] R. Perez-Marco, *Fixed point and circle maps*, in préparation.

[Pe1] Y. Pesin, *Families of invariant manifolds corresponding to non-zero characteristic exponents*, Math. USSR-Izv. **10**, 1261–1305 (1976).

[Pe2] Y. Pesin, *Characteristic Lyapunov exponents and ergodic theory*, Russian Math. Surveys **32**, 55–114 (1977).

[Pu] C. Pugh, *The closing lemma*, Amer. J. Math. **89**, 956–1009 (1967).

[Re] M. Rees, *Positive measure sets of ergodic rational maps*, Ann. Sci. École Norm. Sup. (4), **19**, 383–407 (1986).

[R] J. Robbin, *A structural stability theorem*, Ann. of Math. (2), **94**, 447–493 (1971).

[Ro] C. Robinson, *Structural stability of vector fields*, Ann. of Math. (2) **99**, 154–175 (1974).

[Ru1] H. Rüssmann, *Kleine Nenner I; Über invariante Kurven differenzierbarer Abbildungen eines Kreisringes*, Nachr. Akad. Wiss. Göttingen Math.-Phys. Kl., 67–105 (1970).

[Ru2] H. Rüssmann, *Non-degeneracy in the perturbation theory of integrable dynamical systems*, in Number Theory and Dynamical Systems, Dodson–Vickers (eds), L.N. 134, Lond. Math. Soc., 1–18 (1989).

[Sh] M. Shub, *Stabilité globale des systèmes dynamiques*, Astérisque 50 (1978).

[Si] C. L. Siegel, *Iteration of analytic functions*, Ann. of Math. (2) **43**, 807–812 (1942).

[Sm] S. Smale, *Differentiable dynamical systems*, Bull. Amer. Math. Soc. **73**, 747–817 (1967).

[Su1] D. Sullivan, *Bounded structure of infinitely renormalizable mappings*, in Universality and Chaos, ed. P. Cvitanovic, 2nd edition, Adam Hilger, 1989.

[Su2] D. Sullivan, *Bounds, quadratic differentials and renormalization conjectures*, in AMS Centennial Publications, vol. **2**, Mathematics in the 21st Century, 1992.

[Sw] G. Swiatek, *Hyperbolicity is dense in the real quadratic family*, preprint Stonybrook, 1992.

[T] F. Takens, *Homoclinic bifurcations*, Proc. Internat. Congress Math., Berkeley (1986), 1229–1236.

[V] M. Viana, *Strange attractors exhibiting multidimensional expansion*, in preparation.

[Y1] J. C. Yoccoz, *Il n'y a pas de contre-exemple de Denjoy analytique*, C. R. Acad. Sci. Paris Série I, t. 298 (1984), 141–144.

[Y2] J. C. Yoccoz, *Conjugaison différentiable des difféomorphismes du cercle dont le nombre de rotation vérifie une condition diophantienne*, Ann. Sci. École Norm. Sup. (4), t. 17 (1984), 333–359.

[Y3] J. C. Yoccoz, *Centralisateurs et conjugaison différentiable des difféomorphismes du cercle*, Thèse d'Etat, Juin 1985, Université de Paris-Sud.

[Y4] J. C. Yoccoz, *Théorème de Siegel, nombres de Brjuno et polynômes quadratiques*, manuscrit (1987).

[Y5] J. C. Yoccoz, *Linéarisation des germes de difféomorphismes holomorphes de* ($\mathbf{C}$, 0), C. R. Acad. Sci. Paris Série I, t. 306 (1988), 55–58.

[Y6] J. C. Yoccoz, *Conjugaison analytique des difféomorphismes du cercle*, manuscrit (1989).

[Y7] J. C. Yoccoz, *Polynômes quadratiques et attracteur de Hénon*, Sém. Bourbaki no. 734, Astérisque (1990).

[Y8] J. C. Yoccoz, *Travaux de Herman sur les tores invariants*, Sém. Bourbaki no. 754, fevrier 1992.

[Y9] J. C. Yoccoz, *Hyperbolic dynamics*, cours donné à Hilleröd 1993, basé sur des notes de Workshop on Dynamical Systems, Int. Center Theor. Physics, Trieste, 1991, à paraître chez Kluwer.

[Y10] J. C. Yoccoz, *Vers la connexité locale du lieu de connexité des polynômes quadratiques*, en préparation.

Reprinted from Proc. Int. Congr. Math., 1994

THE WORK OF EFIM ZELMANOV

by
WALTER FEIT

Department of Mathematics, Yale University, New Haven, CT 06511, USA

0. Introduction

Efim Zelmanov has received a Fields Medal for the solution of the restricted Burnside problem. This problem in group theory had long been known to be related to the theory of Lie algebras. In fact, to a large extent *it is* the problem in Lie algebras. A precise statement of it can be found in Section 2 below.

In proving the necessary properties of Lie algebras, Zelmanov built on the work of many others, though he went far beyond what had previously been done in this direction. For instance, he greatly simplified Kostrikin's results [K] which settled the case of prime exponent and then extended these methods to handle the prime power case.

However, while the case of exponent 2 is trivial, the case of exponent 2^k for arbitrary k is the most difficult case that needed to be addressed. The results from Lie algebras that work for exponent p^k with p an odd prime are not adequate for exponent 2^k. This indicated that a new approach was necessary here. Zelmanov was the first to realize that in the case of groups of exponent 2^k the theory of Jordan algebras is of great significance. Even though Vaughan-Lee later removed the need for Jordan algebras [V], it seems probable that this proof could not have been discovered without them, as the ideas used arise most naturally from Jordan algebras.

Zelmanov had earlier made fundamental contributions to Jordan algebras and was an expert in this area, thus he was uniquely qualified to attack the restricted Burnside problem.

Below the background from the theory of Jordan algebras and some of Zelmanov's contributions to this theory are first discussed (I am grateful to McCrimmon and Jacobson for much of this material). See [J1] and [J2] for the general theory of Jordan algebras. Then the Burnside problems are described and some of the things that were earlier known about them are listed. Section 4 contains some consequences of the restricted Burnside problem. Finally, some relevant results from Lie and Jordan algebras are mentioned (in a necessarily sketchy manner).

Zelmanov himself has written a set of expository notes on these topics [Z11]. It contains all the appropriate definitions and some of the material used in the proof of the restricted Burnside problem. It also includes material on several related questions, such as the Kurosh–Levitzky problem. Of course, ultimately, the details are the heart of the matter, and for these the reader should consult [Z8] and [Z9], or [V].

This circle of ideas illustrates the unity of mathematics once again. Although many formal identities are used to settle the restricted Burnside problem, it seems unlikely that they could have been discovered without the conceptual framework provided by the seemingly unrelated and diverse fields of Lie and Jordan algebras.

1. Jordan Algebras

Jordan algebras were introduced in the 1930s by the physicist P. Jordan in an attempt to find an algebraic setting for quantum mechanics, essentially different from the standard setting of hermitian matrices. Hermitian matrices or operators are not closed under the associative product xy, but are closed under the symmetric products $xy + yx$, xyx, x^n. An empirical investigation indicated that the basic operation was the Jordan product

$$x \cdot y = \frac{1}{2}(xy + yx),$$

and that all other properties flowed from the commutative law $x \cdot y = y \cdot x$ and the Jordan identity $(x^2 \cdot y) \cdot x = x^2 \cdot (y \cdot x)$. (For example, the Jordan triple product $\{xyz\} = \frac{1}{2}(xyz + zyx)$ can be expressed as $x \cdot (y \cdot z) + (x \cdot y) \cdot z - (x \cdot z) \cdot y$, though the tetrad

$$\{xyzw\} = \frac{1}{2}(xyzw + wzyx)$$

cannot be expressed in terms of the Jordan product.) Jordan took these as axioms for the variety of Jordan algebras. Algebras resulting from the Jordan product in an associative algebra were called special, so the physicists were seeking algebras that were exceptional (= nonspecial). In a fundamental paper [JNW] Jordan, von Neumann, and Wigner classified all finite-dimensional formally-real Jordan algebras. These are direct sums of five types of simple algebras: algebras determined by a quadratic form on a vector space (a special subalgebra of the Clifford algebra of the quadratic form) and four types of algebras of hermitian $(n \times n)$-matrices over the four composition algebras (the reals, complexes, quaternions, and octonions). The algebra of hermitian matrices over the octonions is Jordan only for $n \leq 3$, and is exceptional if $n = 3$, so there was only one exceptional simple algebra in their list (now known as the Albert algebra, of dimension 27). At the end of their paper Jordan, von Neumann, and Wigner expressed the hope that by dropping the assumption of finite dimensionality one might obtain exceptional simple algebras other than Albert algebras.

Algebraists developed a rich structure theory of Jordan algebras over fields of characteristic $\neq 2$. First, the analogue of Wedderburn's theory of finite-dimensional associative algebras was obtained by Albert. Next this was extended by Jacobson to an analogue of the Wedderburn–Artin theory of semisimple rings with minimum condition on left or right ideals. In this, the role of the one-sided ideals was played by inner deals, defined as subspaces B such that $U_b x$ is in B for all x in the algebra A and all b in B where $U_a = 2L_a^2 - L_{a^2}$ and L_a is the left multiplication by a in the Jordan algebra A. If A is the Jordan subalgebra of an associative algebra the $U_b x = bxb$ in the associative product. Using the definition of semi-simplicity (called nondegeneracy) that A contains no $z \neq 0$ such that $U_z = 0$, Jacobson showed that every nondegenerate Jordan algebra with d.c.c. on inner ideals is the direct sum of simple algebras that are of classical type (analogues of those found in [JNW]: (Type I) Jordan algebras of nondegenerate quadratic forms; (Type II) algebras $H(A, *)$ of hermitian elements in a $*$-simple artinian associative algebra $A((n \times n)$-matrices over a division algebra with involution, or over a direct sum of a division algebra and its opposite under the exchange involution, or matrices over a split quaternion algebra with standard involution); (Type III) 27-dimensional exceptional Albert algebras; (Type IV) Jordan division algebras, defined by the condition that U_a is invertible for every $a \neq 0$.

Up to this point, the structure theory treated only algebras with finiteness conditions because the primary tool was the use of primitive idempotents to introduce coordinates. In 1975 Alfsen, Schultz, and Stömer obtained a Gelfand–Naimark theorem for Jordan C^*-algebras, and once again the basic dimensional structure theorem, but here again it was crucial that the hypotheses guaranteed a rich supply of idempotents.

In three papers [Z1], [Z2], [Z3], Zelmanov revolutionized the structure theory of Jordan algebras. These deal with prime Jordan algebras, where A is called prime if $U_B C = 0$ for ideals B and C in A implies that either B or $C = 0$. In [Z1] Zelmanov proved the remarkable result that a prime Jordan algebra without nil ideals (improved in [Z3] to prime and nondegenerate) is either i-special (a homomorphic image of a special Jordan algebra) or is a form of the 27-dimensional exceptional algebra. This applied in particular to simple algebras. The proof required the introduction of a host of novel concepts and techniques as well as sharpening of earlier methods, e.g. the coordinatization theorem of Jacobson and analogues of results on radicals due to Amitsur.

The paper [Z3] is devoted to the study of i-special Jordan algebras. Zelmanov showed that a prime nondegenerate i-special algebra is special, and he determined their structure as either of hermitian type or of Clifford type. Paper [Z2], which preceded [Z3] obtained these results for Jordan division algebras.

The principal tool in both papers is the study of the free associative algebra $\Phi\langle X\rangle$ on $X = \{x_1, x_2, \ldots\}$. This becomes a Jordan algebra $\Phi\langle X\rangle^+$ by replacing the given associative multiplication ab by $a \cdot b = \frac{1}{2}(ab + ba)$. The subalgebra $SJ\langle X \rangle$ of $\Phi\langle X \rangle^+$ generated by X is called the free special Jordan algebra.

We also have the subalgebra $H\langle X \rangle$ of $\Phi\langle X \rangle^+$ of symmetric elements $(a^* = a)$ under the involution in $\Phi\langle X \rangle$ fixing the elements of X. It was shown by Paul Cohn in 1954 that $SJ\langle X \rangle \subsetneqq H\langle X \rangle$ and $H\langle X \rangle$ is the subalgebra of $\Phi\langle X \rangle^+$ generated by X and all the tetrads $x_i x_j x_k x_l$ with $i < j < k < l$. Zelmanov has obtained a completely unanticipated supplement to Cohn's theorem: the existence of elements f in $SJ\langle X \rangle$ such that if $I(f)$ denotes the ideal generated by f then

$$\{I(f), p, q, r\} \in SJ\langle X \rangle$$

for p, q, r in $SJ\langle X \rangle$. This is used to sort out the two types of i-special algebras: Clifford types characterized by the identity $f \equiv 0$ and hermitian types by the nonidentity $f \not\equiv 0$.

One of the consequences of Zelmanov's theory is that the only exceptional simple Jordan algebras, even including infinite-dimensional ones, are the forms of the 27-dimensional Albert algebras. This laid to rest the hope that had been raised by Jordan, von Neumann, and Wigner in [JNW]. Another consequence of Zelmanov's results is that the free Jordan algebra in three or more generators has zero divisors (elements a such that U_a is not injective). This is in sharp contrast to the theorem of Malcev and Neumann that any free associative algebra can be imbedded in a division algebra.

Motivated by applications to analysis and differential geometry, Koecher, Loos, and Myberg extended the structure theory of Jordan algebras to triple systems and Jordan pairs. Zelmanov applied his methods to obtain new results on these.

Lie methods were used in these papers based on the Tits–Koecher construction. The final work in this line of investigation was [Z4] in which Zelmanov applied the theory of Jordan triple systems to study graded Lie algebras with finite gradings in which the homogeneous parts could be infinite dimensional.

To encompass characteristic 2 (which is essential for applications to the restricted Burnside problem) it is necessary to deal with quadratic Jordan algebras [JM]. These were introduced by McCrimmon in [Mc] as the natural extension of Jordan algebras to algebras over any commutative ring. This amounted to replacing the product $a \cdot b = \frac{1}{2}(ab + ba)$ in an associative algebra by the product $U_a b$.

In the joint paper with McCrimmon [ZM], the results of [Z3] were extended to quadratic Jordan algebras.

2. Burnside Problems

We begin with some definitions and notation.

A group is *locally finite* if every finite subset generates a finite group. In 1902 Burnside [B1] studied torsion groups and asked when such groups are locally finite. The most general form of the question is the Generalized Burnside Problem (GBP).

(GBP) Is a torsion group necessarily locally finite?

Equivalently

(GBP)′ Is every finitely generated torsion group finite?

A group G has a *finite exponent* e if $x^e = 1$ for all x in G and e is the smallest natural number with this property. Clearly a group with a finite exponent is a torsion group. A more restricted version of GBP, which already occurs in Burnside's work, is the ordinary Burnside Problem (BP).

(BP) Is every group that has a finite exponent locally finite?

There is a universal object $B(r, e)$, (the Burnside group of exponent e on r generators), which is the quotient of the free group on r generators by the subgroup generated by all eth powers. BP is equivalent to

(BP)′ Is $B(r, e)$ finite for all natural numbers e and r?

Burnside proved that groups of exponent 2 (trivial) and exponent 3 are locally finite. In 1905 Burnside [B2] showed that a subgroup of $GL(n, \mathbf{C})$ of finite exponent is finite. Schur in 1911 [Sc] proved that a finitely generated torsion subgroup of $GL(n, \mathbf{C})$ has finite exponent, and hence a torsion subgroup of $GL(n, \mathbf{C})$ is locally finite. This was very important as it showed that answers to BP or GBP would necessarily involve groups not describable in terms of linear transformations over $\mathbf{C}$. Other methods were required. In handling groups of exponent 3 Burnside had used only the multiplication table of a group. However, his methods were totally inadequate to handle, for instance, groups of prime exponent greater than 3.

During the 1930s people began to study finite quotients of $B(r, e)$ and considered the following statement.

(RBP) $B(r, e)$ has only finitely many finite quotients.

This is equivalent to

(RBP)′ $B(r, e)$ has a unique maximal finite quotient $RB(r, e)$.

W. Magnus called the question of the truth or falsity of RBP *the restricted Burnside problem.* If such a unique maximal finite quotient $RB(r, e)$ exists for some e and r, then necessarily every finite group on r generators and exponent e is a homomorphic image of $RB(r, e)$. If $RB(r, e)$ exists for some e and all r we say that RBP is true for e.

3. Results

In 1964 Golod [G] constructed infinite groups for every prime p, which are generated by 2 elements and in which every element has order a power of p, thus giving a negative answer to GBP. A few years later in 1968 Adian and Novikov [AN] showed that $B(2,e)$ is infinite for e odd and $e > 4380$, thus giving a negative answer to BP. The bound has been improved since then as $B(r,e)$ is finite for $e = 2,3,4$, or 6, but in no other case with $r > 1$ is it known to be finite.

In a seminal paper Hall and Higman [HH] in 1956 proved a series of results concerning RBP. Let π be a set of primes. Consider the following two statements.

(1) There are only finitely many finite simple π-groups of any given exponent.
(2) The Schreier conjecture is true for π-groups, i.e. for any finite simple π-group G, $\mathrm{Aut}(G)/G$ is solvable.

A special case of one of their results is the following.

Theorem [HH]. *Suppose that statements* (1) *and* (2) *are true for the set* π. *Then if for every prime* p *in* π, *and natural numbers* m *and* r, $RB(r,p^m)$ *exists; then RBP is true for any exponent* e *that is a* π*-number.*

The classification of the finite simple groups shows that (1) and (2) are true for any set of primes π. Hence the truth of RBP will follow once it is proved that $RB(r,p^m)$ exists for all primes p and all natural numbers m and r.

In 1959 Kostrikin announced that $RB(r,p)$ exists for p a prime and any natural number r. Kostrikin's original argument had some difficulties. He published a corrected and updated version of his proof in his book [K], which contains numerous references to Zelmanov.

In 1989 Zelmanov announced that RBP is true for all exponents p^m with p any prime, and hence for all exponents by the remarks above. The proof appeared in 1990–91 in Russian. The English translation appeared in [Z8] and [Z9].

It should be mentioned that analogous questions have been raised for associative, Lie, and Jordan algebras. Golod's work was actually motivated by the associative algebra question and the counterexamples for groups arose as corollaries. The questions for Lie and Jordan algebras will be discussed below.

4. Some Consequences

This section contains some consequences of RBP. The ideas used in the proof, in addition to the actual result, have also been applied widely.

The next three results were proved by Zelmanov [Z10] as direct consequences of RBP.

Theorem 1. *Every periodic pro-p-group is locally finite.*

Corollary 2. *Every infinite compact (Hausdorff) group contains an infinite abelian subgroup.*

Theorem 3. *Every periodic compact (Hausdorff) group is locally finite.*

Theorem 3 was conjectured by Platonov [Ko].

Shalev showed that RBP implies the following.

Theorem 4 [Sh]. *A pro-p-group is p-adic analytic if and only if there exists a natural number n such that the wreath product $Z_p \wr Z_{p^n}$ is not a homomorphic image of any subgroup of G.*

The "only if" part of Theorem 4 is elementary, but the converse is equivalent to RBP. Since then, Zelmanov jointly with others, has made several further contributions to the study of pro-p-groups, see e.g. [ZS], [ZW].

5. Lie Algebras

Let G be a finite group of exponent p^k, p a prime. Let $G = G_0$ and $G_{i+1} = [G, G_i]$ for all i. Choose s with $G_s \neq \langle 1 \rangle$, $G_{s+1} = \langle 1 \rangle$. Then

$$G = G_0 > \cdots > G_{s+1} = \langle 1 \rangle$$

is the lower central series of G. Define

$$L(G) = \sum_{i=0}^{s} G_i/G_{i+1}$$

as abelian groups. Then $L(G)$ becomes a Lie ring with $[a_i G_i, a_j G_j] = [a_i, a_j]G_{i+j+1}$, and $L(G)$ has the same nilpotency class as G. Furthermore $L(G)/pL(G)$ is a Lie algebra over Z_p.

Let L be a Lie algebra.

L satisfies the *Engel identity* (E_n) if $\mathrm{ad}(x)^n = 0$ for all x in L.

An element x in L is *nilpotent* if $\mathrm{ad}(x)^n = 0$ for some n.

If G has exponent p then $L(G)$ is a Lie algebra over Z_p that satisfies (E_{p-1}). Kostrikin proved

Theorem 5 [K]. *If L is a Lie algebra over Z_p that satisfies (E_{p-1}) then L is locally nilpotent.*

Theorem 1 implies the existence of $RB(r, p)$ and so yields RBP for prime exponent. Observe that for prime exponent $e = p$, the case $p = 2$ is trivial, so that it may be assumed that $p > 2$. This is in sharp contrast to prime power exponents $e = p^k$, where $p = 2$ is the most complicated case.

An element a of L is a *sandwich* if $[[L, a], a] = 0$ and $[[[L, a], L], a] = 0$. L is a *sandwich algebra* if it is generated by finitely many sandwiches. This concept was introduced by Kostrikin and is of fundamental importance for the proof of RBP. A first critical result is

Theorem 6 [ZK]. *Every sandwich Lie algebra is locally nilpotent.*

Theorem 6 is essential for the proof of Theorem 5.

The main result in [Z8] is rather technical but it has the following consequence.

Theorem 7 [Z8]. *Every Lie ring satisfying an Engel condition is locally nilpotent.*

More importantly, it implies

Theorem 8. *$RB(r, p^k)$ exists for p an odd prime.*

Once again an essential part of the proof requires Theorem 2. Let L be a Lie algebra over an infinite field of characteristic p that satisfies an Engel condition. They way to apply Theorem 2 is to construct a polynomial $f(x_1, \ldots, x_t)$ that is not identically zero, such that every element in $f(L)$ is a sandwich in L. Actually such a polynomial is not constructed but its existence for $p > 2$ follows only after a very complicated series of arguments, which constitute the bulk of the paper [Z8]. This of course settles RBP for odd exponent. (It might be mentioned that the classification of finite simple groups is not required here, only that groups of odd order are solvable.)

6. The Case of Exponent 2^k

The outline of the proof of RBP for exponent 2^k is similar to that for exponent p^k with $p > 2$ described in the previous section. However, the construction of the function f is vastly more complicated. It is here that quadratic Jordan algebras play an essential role, most especially the results of [ZM]. The details are extremely technical and cannot be presented here. The reader should consult [Z9] for a complete proof.

General References

[AN] S. I. Adian and P. S. Novikov, *On infinite periodic groups, I, II, III*, Math. USSR-lzv. **2** (1968), 209–236, 241–479, 665–685.

[B1] W. Burnside, *On an unsettled question in the theory of discontinuous groups*, Quart. J. Math. **33** (1902), 230–238.

[B2] ———, *On criteria for the finiteness of the order of a group of linear substitutions*, Proc. London Math. Soc. (2) **3** (1905), 435–440.

[G] E. Golod, *On nil algebras and residually finite p-groups*, Math. USSR-lzv. **28** (1964), 273–276.

[HH] P. Hall and G. Higman, *On the p-length of p-soluble groups and reduction theorems for Burnside's problem*, Proc. London Math. Soc. **6** (1956), 1–40.

[J1] N. Jacobson, Structure and representation of Jordan algebras, Amer. Math. Soc., Providence, RI 1968.

[J2] ———, Structure theory of Jordan algebras, Univ. of Arkansas Lecture Notes **5** (1981).

[JM] N. Jacobson and K. McCrimmon, *Quadratic Jordan algebras of quadratic forms with base points*, J. Indian Math. Soc. **35** (1971), 1–45.

[JNW] P. Jordan, J. von Neumann, and F. Wigner, *On an algebraic generalization of the quantum mechanical formalism*, Ann. of Math. (2) **35** (1934), 29–73.

[K] A. I. Kostrikin, Around Burnside, (Russian) MR (89d, 20032).

[Ko] Kourovka Notebook, 12th ed., Inst. Mat. SO AN SSSR Novosibirsk, 1992. Problem 3.41.

[Mc] K. McCrimmon, A *general theory of Jordan rings*, PNAS **56** (1966), 1072–1079.

[Sc] I Schur. Über Gruppen periodischer Substitutionen, Collected Works **I**, 442–450, Springer, Berlin, Heidelberg, and New York 1973.

[Sh] A. Shalev, *Characterization of p-adic analytic groups in terms of wreath products*, J. Algebra **145** (1992), 204–208.

[V] M. R. Vaughan–Lee, The restricted Burnside problem, 2nd ed., Oxford University Press, Oxford, 1993.

References to Zelmanov's Work

[Z1] E. Zelmanov, *Primary Jordan algebras*, Algebra and Logic **18** (1979), 103–111.

[Z2] ———, *Jordan division algebras*, Algebra and Logic **18** (1979), 175–190.

[Z3] ———, *On prime Jordan algebras II*, Siberian Math. J. **24** (1983), 73–85.

[Z4] ———, *Lie algebras with a finite grading*, Math. USSR-Sb. **52** (1985), 347–385.

[Z5] ———, *On Engel Lie Algebras*, Soviet Math. Dokl. **35** (1987), 44–47.

[Z6] ———, *Engelian Lie Algebras*, Siberian Math. J. **29** (1989), 777–781.

[Z7] ———, *Weakened Burnside problem*, Siberian Math. J. **30** (1989), 885–891.

[Z8] ———, *Solution of the restricted Burnside problem for groups of odd exponent*, Math. USSR-lzv. **36** (1991), 41–60.

[Z9] ———, *A solution of the restricted Burnside problem for 2-groups*, Math. USSR-Sb. **72** (1992), 543–564.

[Z10] ———, *On periodic compact groups*, Israel J. Math. **77** (1992), 83–95.

[Z11] ———, Nil rings and periodic groups, Korean Math. Soc. Lecture Notes 1992, Korean Math. Soc. Seoul, South Korea.

[ZK] ———and A. I. Kostrikin, *A theorem on sandwich algebras*, Proc. Steklov Inst. Math. **183** issue **4** (1991), 121–126.

[ZM] ———and K. McCrimmon, *The structure of strongly prime quadratic Jordan algebras*, Adv. in Math. **69** (1988), 113–122.

[ZS] ———and Shalev, *Pro-p-groups of finite coclass*, Math. Proc. Cambridge Philos. Soc. **111** (1992), 417–421.

[ZW] ———and J. S. Wilson, *Identities of Lie algebras for pro-p-groups*, J. Pure Appl. Algebra **81** (1992), 103–109.

Efim I. Zelmanov

EFIM I. ZELMANOV

Date of Birth: September 7, 1955

Place of Birth: Former USSR

Education:
Dr: Leningrad State University, 1985
Ph.D.: Novosibirsk State University, 1980
M.S.: Novosibirsk State University, 1977

Academic Positions:
Yale University, Professor, 1995
University of Chicago, Professor 1994–95
University of Wisconsin–Madison, Professor, 1990–94
Institute of Math., Academy of Sciences of the USSR, Novosibirsk, Leading Research, 1986–present
Institute of Mathematics in Novosibirsk, Special Lecture series, 1985–87
Institute of Math., Academy of Sciences of the USSR, Novosibirsk, Senior Research, 1985–86
Institute of Math., Academy of Sciences of the USSR, Novosibirsk, Junior Research, 1980–85
Novosibirsk State Univ., Teacher, 1977–81

Other Scientific Activities:
Editor, Journal of Algebra
Editor, Communications in Algebra
Editor, Nova Journal of Algebra and Geometry
Editor, International Journal of Algebra and Computation
Editor, Transactions of the American Mathematical Society
Editor, Algebra i Logica
Editor, Research Announcements

Awards:
Andre Aizenstadt Prize, May 1996
Fields Medal, August 1994
Medal, College de France, January 1992
NSF Research Grant, 1992–94
H.I. Romnes Faculty Fellowship, 1993
NSF Research Grant, 1994–97

Invited Addresses:

Valentine Lecture, Kansas State University, April 1996
Owen Lecture, Detroit, Michigan, March 1996
Russel Marker Lectures, Penn. State University, March 1996
Brazilian Colloquium, Rio de Janeiro, July 1995
Sir Henry Cooper Fellow, Auckland, New Zealand, July 1995
Leonardo da Vinci Lecture, Milan Italy, June 1995
Colloquiums at Jerusalem and Tel Aviv, June 1995
Joint AMS-Israel Math. Union Meeting, Jerusalem, May 1995
Blyth Lectures at the Univ. Toronto, March 1995
Colloquiums at Yale, Univ. Chicago, Univ. Illinois–Urbana, Univ. Illinois–Chicago, Northwestern, Vanderbilt, University of Michigan, Michigan State Univ., Joint Boston Colloquium, Binghampton, CUNY, Granada, Oviedo, Ottawa, Waterloo, Canberra, Rome, 1994–95
Hans Zassenhaus Memorial Lectures, Columbus, October 1995
International Congress of Mathematicians, Zurich, Switzerland, August 1994
Conference on Group Theory, Galway, Ireland, 1993
Conference on Non-associative Algebras, Oviedo, Spain, 1993
Conference on Combinatorial Group Theory, Edinburgh, 1993
Adrian A. Albert Memorial Lectures, Univ. of Chicago, 1992
Conference on Lie Algebras, New Haven, 1992
Colloquiums at Oxford, Cambridge, London, Manchester, Warwick, Glasgow, Edinburgh, Jerusalem, Tel–Aviv, Haifa, Zaragoza, Malaga, Canberra, 1991
London, Mathematical Society Meeting, 1991
Colloquium at College de France, Paris, 1991
Joint Meeting of Math. Societies of Ontario, and Quebec, Canada, 1991
Distinguished Lecture Series, University of Iowa, 1990
International Congress of Mathematicians, Kyoto, Japan, August 1990
Colloquium at Max Planck Institute, Bonn, Germany, 1990
Colloquiums and Seminars at Yale, Harvard, Columbia, Univ. of Chicago, Univ. of Virginia, Univ. of Michigan, Univ. of Wisconsin, Ohio State Univ., 1989
Conference on Algebra, Novosibirsk, USSR, 1989
International Congress of Mathematicians, Warsaw, Poland, August 1983.

ON THE RESTRICTED BURNSIDE PROBLEM

by
EFIM I. ZELMANOV

Yale University
Department of Mathematics
10 Hillhouse Avenue
New Haven, CT 06520-8283 USA

In 1902 W. Burnside formulated his famous problems on periodic groups.

A group G is said to be *periodic* if for an arbitrary element $g \in G$ there exists a natural number $n = n(g)$ depending on g such that $g = 1$. A group G is said to be *periodic of bounded exponent* if there exists $n \geq 1$ such that for an arbitrary element $g \in G$ there holds $g^n = 1$. The minimal n with this property is called the exponent of G.

The General Burnside Problem: Is it true that a finitely generated periodic group is finite?

The Burnside Problem: Is it true that a finitely generated group of bounded exponent is finite?

After many unsuccessful attempts to solve the problems in positive the following weaker version of The Burnside Problem was formulated.

The Restricted Burnside Problem: Is it true that there are only finitely many *finite* m generated groups of exponent n?

Let F_m be the free group of rank m. Let F_m^n be the subgroup of F_m generated by all n-th powers $g^n, g \in F_m$. The Burnside Problem asks whether the group $B(m,n) = F_m/F_m^n$ is finite. The Restricted Burnside Problem asks whether the group $B(m,n)$, even if infinite, still has only finitely many subgroups of finite index. If yes, then factoring out the intersection of all subgroups of finite index in $B(m,n)$ we'll get $B_0(m,n)$, the universal finite m generated group of exponent n having all other finite m generated groups of exponent n its homomorphic images.

In 1964 Golod and Shafarevich constructed counter examples to the General Burnside Problem. Since then new infinite finitely generated periodic groups were constructed by S. V. Alyoshin [2], R. I. Grigorchuk [13], N. Gupta and S. Sidki [16], V. I. Sushchansky [47].

In 1968 P. S. Novikov and S. I. Adian [40] were able to construct counter examples to The Burnside Problem for groups of odd exponent $n \geq 4381$ (later cut to $n \geq 665$ by S. I. Adian still under assumption that n is odd).

Only very recently S. Ivanov [20] proved that the group $B(2, 2^k)$, where k is sufficiently big is also infinite. Thus The Burnside Problem has negative solution for all sufficiently big exponents whether even or odd.

Tarsky Monsters (see A. Yu. Olshansky [41]) and other Monsters brought to life by geometric methods show how wild infinite periodic groups can be. In a sense, this is the strongest form of a negative solution of The Burnside Problem.

Speaking of what is known on the positive side, the General Burnside Problem has positive solution for linear groups (W. Burnside, I. Schur) and The Burnside Problem has positive solution for groups of exponent 3 (W. Burnside), exponent 4 (I. N. Sanov) and exponent 6 (M. Hall). For all other exponents the problem is either open or solved in negative.

At the same time there were two major reasons to believe that the Restricted Burnside Problem would have a positive solution. One of these reasons was the Reduction Theorem obtained by P. Hall and G. Higman [17]. Let $n = p_1^{k_1} \cdots p_r^{k_r}$, where p_i are distinct prime numbers, $k_i \geq 1$, and assume that (a) the Restricted Burnside Problem for groups of exponents $p_i^{k_i}$ has a positive solution, (b) there are finitely many finite simple groups of exponent n, (c) the group of outer automorphisms $\mathrm{Out}(G) = \mathrm{Aut}(G)/\mathrm{Inn}(G)$ is solvable for any finite simple group of exponent n. Then the Restricted Burnside Problem for groups of exponent n also has positive solution.

The announced classification of finite simple groups (see [11]) implies the conditions (b) and (c). Remark that if the exponent n is odd or $n = p^\alpha q^\beta$, where p, q are prime numbers, then by the celebrated theorems of Feit and Thompson [8] and W. Burnside there are no finite simple groups of exponent n and the conditions (b), (c) are satisfied automatically.

Another reason was the close relation of the problem to Lie algebras. Suppose that $n = p^k$, where p is a prime number. A finite group of exponent p^k is nilpotent. Let G be an m generated finite group of exponent $n = p^k$. To solve the Restricted Burnside Problem is to find an upper bound for the order $|G|$ that depends only on m and n. It is easy to see, however, that it is sufficient to find an upper bound $f(m, n)$ for the class of nilpotency of G.

Consider the Zassenhaus filtration of G : $G = G_1 > G_2 > \cdots > G_s = (1)$. The subgroup G_k is generated by p-powers of left-normed group commutators $(a_1, \ldots, a_i)^{p^j}$, where

$$i \cdot p^j \geq k, (a_1, a_2) = a_1^{-1} a_2^{-1} a_1 a_2, (a_1, \ldots, a_i) = (\cdots (a_1, a_2), a_3), \ldots, a_i)$$

and $a_1, a_2, \ldots$ are arbitrary elements from G. Factors G_i/G_{i+1} are elementary abelian p-groups and thus can be viewed as vector spaces over the field $\mathbb{Z}/p\mathbb{Z}$. Consider the direct sum

$$\tilde{L}(G) = \bigoplus_{i \geq 1} G_i/G_{i+1}.$$

The brackets $[a_iG_{i+1}, b_jG_{j+1}] = (a_i, b_j)G_{i+j+1}$, where $a_i \in G_i, b_j \in G_j$, defines a structure of a Lie algebra on $\tilde{L}(G)$.

Suppose that the group G is generated by elements $g_1, \ldots, g_m$ and let $L(G)$ be the subalgebra of $\tilde{L}(G)$ generated by cosets $g_1G_2, \ldots, g_mG_2$. If $L(G)^c = (0)$, that is, if $[[\ldots[a_1, a_2], a_3], \ldots, a_c] = 0$ for arbitrary elements $a_1, a_2, \ldots, a_c \in L(G)$, then an arbitrary group commutator in $g_1, \ldots, g_m$ of length c is a product of elements $\rho_k^{p^{m_k}}$, where ρ_k is a commutator in $g_1, \ldots, g_m$ of length i and $i \cdot p^{m_k} > c$. Let $\rho_1, \rho_2, \ldots, \rho_r$ be all left-normed group commutators in $g_1, \ldots, g_m$ of length $\leq c$, $r \leq m^c$. Then an arbitrary element $g \in G$ can be represented as $g = \rho_1^{k_1} \cdots \rho_r^{k_r}$, $0 \leq k_i < n$. Hence,

$$|G| \leq n^{m^c}.$$

Suppose that $n = p$ is a prime number. Then Zassenhous filtration of G coincides with the lower central series, $G_{i+1} = (G_i, G)$. W. Magnus [36] proved that in this case the Lie algebra $L(G)$ satisfies Engel's identity

$$[\ldots[y, \underbrace{x], x], \ldots, x}_{p-1}] = 0 \qquad (E_{p-1}).$$

Thus the problem for groups of prime exponent p has been reduced to the following problem in Lie algebras:

Is it true that a finitely generated Lie algebra over $\mathbb{Z}/p\mathbb{Z}$ that satisfies Engel's identity E_{p-1} is nilpotent?

If the answer is "yes" then the degree of nilpotency can depend only on p and on the number of generators. This problem has been successfully solved by A. I. Kostrikin [28, 29] who solved in this way the Restricted Burnside Problem for groups of prime exponent.

If G is a finite group of prime power exponent $p^k, k > 1$, then the Lie algebra $L(G)$ does not necessarily satisfy Engel's identity E_{p^k-1} ([15, 18]). But

(1) $L(G)$ satisfies the linearized Engel's identity E_{p^k-1}, that is, for arbitrary elements $a_1, a_{p^k-1} \in L(G)$ we have

$$\sum \operatorname{ad}(a_{\sigma(1)}) \cdots \operatorname{ad}(a_{\sigma(p^k-1)}) = 0,$$

where σ runs over the whole symmetric group S_{p^k-1} (G. Higman, [19]),

(2) for an arbitrary commutator ρ on the generators $g_iG_2, 1 \leq i \leq m$ of $L(G)$ we have

$$\operatorname{ad}(\rho)^{p^k} = 0.$$

(I. N. Sanov [43]).

Now let us turn to what was happening in associative and Lie nil algebras.

An associative algebra A is said to be *nil* if for an arbitrary element $a \in A$ there exists a natural number $n = n(a)$ such that $a^{n(a)} = 0$. The algebra A is said to be *nil of bounded degree* if all numbers $n(a), a \in A$ are uniformly bounded from above.

A. G. Kurosh [32] and J. Levitzky (see [3]) formulated two problems for nil algebras that were similar to Burnside's problems.

The General Kurosh–Levitzky Problem: Is every finitely generated nil algebra nilpotent?

The Kurosh–Levitzky Problem: Is every finitely generated nil algebra of bounded degree nilpotent?

In fact, it was a counter example to the General Kurosh–Levitzky Problem that was constructed by the method of E. S. Golod and I. R. Shafarevich. Then this counter example was used to construct the first counter example to the General Burnside Problem. We have already mentioned above that since 1964 there appeared many new examples of finitely generated infinite periodic groups. But so far the example of E. S. Golod and I. R. Shafarevich remains the only example of a nonnilpotent finitely generated nil ring.

For nil algebras of bounded degree the situation is quite different. Unlike its group theoretic counterpart The Kurosh–Levitzky Problem has only positive solutions in all important classes of algebras.

The most general class of algebras where the General Kurosh–Levitzky Problem has positive solution is the class of PI-algebra. Let $f(x_1, \ldots, x_k)$ be a nonzero element of the free associative algebra on the free generators $x_1, \ldots, x_k$. We say that an associative algebra A satisfies the polynomial identity $f = 0$ (and thus is a PI-algebra) if $f(a_1, \ldots, a_k) = 0$ for arbitrary elements $a_1, \ldots, a_k \in A$.

Theorem (I. Kaplansky [25]). *A finitely generated nil PI-algebra is nilpotent.*

In 1956 A. I. Shirshov suggested another purely combinatorial direct approach to Kurosh–Levitzky problems.

Theorem (A. I. Shirshov [46]). *Suppose that an associative algebra A is generated by elements $a_1, \ldots, a_m$ and assume that (1) A satisfies a polynomial identity of degree n, (2) every product of a_i's of length $\leq n$ is a nilpotent element. Then the algebra A is nilpotent.*

It is very important that the nilpotency assumption is imposed here not on every element of A but only on words in generators (even on finitely many of them).

Now let us go back to Lie algebras.

It is natural to call an element $a \in L$ nilpotent of the generator $\mathrm{ad}(a)$ is nilpotent. With this definition both Kurosh–Levitzky problems become meaningful for Lie algebras. Moreover, by the results of G. Higman and I. N. Sanov (see above) the Lie algebra of a finite group G of exponent p^n satisfies the assumptions of The Kurosh–Levitzky Problem in the form of A. I. Shirshov (the role of words is played by commutators).

In [56, 57] we solved this problem for Lie algebras satisfying a linearized Engel's identity E_n.

Theorem 1. *Suppose that a Lie algebra L is generated by elements $a_1, \dots, a_m$ and assume that there exist integers $n \geq 1, m \geq 1$ such that* (1) *L satisfies the linearized Engel's identity E_n,* (2) *for an arbitrary commutator ρ in a_i's we have* $\mathrm{ad}(\rho)^m = 0$. *Then L is nilpotent.*

Corollary. *A Lie ring that satisfies E_n is locally nilpotent.*

From Theorem 1 we derive:

Theorem 2. *The Restricted Burnside Problem has a positive solution for groups of exponent p^k.*

In view of the Reduction Theorem of P. Hall and G. Higman and the announced classification of finite simple groups this implies that the Restricted Burnside Problem has positive solution for groups of an arbitrary exponent.

Now we shall try to explain briefly the idea of the proof of Theorem 1.

Recall that an algebra is said to be locally nilpotent if every finitely generated subalgebra of it is nilpotent.

In [54] we showed that to prove Theorems 1, 2 it suffices to prove that a Lie algebra over an infinite field which satisfies an Engel's identity is locally nilpotent.

An element a of a Lie algebra L is called a *sandwich* if

$$[[L, a], a] = (0), [[[L, a], L], a] = (0),$$

(see A. I. Kostrikin, [29]). In case of algebras of odd characteristics the second equality easily follows from the first one. However if char $= 2$ both conditions are necessary. We call a Lie algebra a *sandwich algebra* if it is generated by a finite collection of sandwiches. The following theorem is due to A. I. Kostrikin and the author.

Theorem about Sandwich Algebras [32]. *A sandwich Lie algebra is nilpotent.*

This theorem suggests the following plan of attack on Theorem 1 (which has been outlined in [54]).

Assume that there exists a nonzero Lie algebra $\mathcal{L}$ over an infinite field K which satisfies an Engel's identity but isn't locally nilpotent. Factoring out the locally nilpotent radical of $\mathcal{L}$ (see [27, 42]) we may assume that $\mathcal{L}$ does not contain any nonzero locally nilpotent ideals.

Suppose we managed to construct a Lie polynomial $f(x_1, \dots, x_r)$ (an element of the free Lie algebra) such that f is not identically zero on $\mathcal{L}$ and for arbitrary elements $a_1, \dots, a_r \in \mathcal{L}$ the value $f(a_1, \dots, a_r)$ is a sandwich of $\mathcal{L}$. The K-linear span of $f(\mathcal{L}) = \{f(a_1, \dots, a_r) \mid a_1, \dots, a_r \in \mathcal{L}\}$ is an ideal in $\mathcal{L}$. By the theorem about sandwich algebras the ideal $Kf(L)$ is locally nilpotent, which contradicts our assumption.

A year of effort that followed [54] did not bring us a desired sandwich-valued polynomial (its existence a posteriori followed from Theorem 1). Instead in

November 1988 we constructed an even sandwich-valued superpolynomial f. This means that for a Lie superalgebra $L = L_0 + L_1$ satisfying a superization of E_n every value of f on L is a sandwich of L_0. It turned out to be a good substitute of sandwich-valued polynomials. The sketch of this rather complicated construction appeared in [55]. Unfortunately it worked only for characteristic $\neq 2, 3$.

In January of 1989 we constructed another "generalized" nonzero sandwich-valued polynomial (this time involving "divided powers" of ad-generators). The full linearization of a "generalized polynomial" is an ordinary polynomial and every value of such a linearization is a linear combination of sandwiches. This approach worked for an arbitrary characteristic ([56, 57]).

Some lengthy computations from the proof (which are really hard to read) may be explained better within the framework of Jordan Algebra Theory (see [21, 22]). We shall demonstrate the idea for the simpler case $p \neq 2, 3$.

The first (less computational) part of the construction of a sandwich-valued polynomial is a construction of a polynomial f such that $f(\mathcal{L}) \neq (0)$ and for an arbitrary element $a \in f(\mathcal{L})$ we have $\mathrm{ad}(a)^3 = 0$.

Choose arbitrary elements $a, b \in f(\mathcal{L})$ and consider the subspaces $\mathcal{L}^+ = \mathcal{L}\,\mathrm{ad}(a)^2$, $\mathcal{L}^- = \mathcal{L}\,\mathrm{ad}(b)^2$. Then for an arbitrary element $c \in \mathcal{L}^+$ the operation $x \circ_c y = [[x, c], y]$; $x, y \in \mathcal{L}^-$, defines the structure of a Jordan algebra on $\mathcal{L}^-$ (see [6, 24, 26]). The pair of subspaces $(\mathcal{L}^-, \mathcal{L}^+)$ is a so-called Jordan Pair (see [34, 39]).

For $p = 2$ or $p = 3$ we define $\mathcal{L}^-$ and $\mathcal{L}^+$ with the divided squares of adjoint operators and apply Kevin McCrimman's theory of Quadratic Jordan Algebras [22, 37, 38].

For odd p we managed to translate Jordan arguments into the language of elementary computations in [56]. Later M. Vaughan–Lee (see [48]) was able to do this even for $p = 2$. In both cases, however, there was a price to pay: lengthy computations.

The question about upper bounds for orders (or classes of nilpotency) of finite m generated groups of exponent n seems to be rather complicated. Exact orders of universal finite Burnside groups $B_0(m, n)$ are known for $n = 2, 3, 4, 5$ (see [48]). G. Higman found an upper bound for the class of nilpotency of $B_0(m, 5)$ that grows linearly in m. M. Vaughan–Lee found an upper bound for the class of $B_0(m, 7)$ which is polynomial in m.

The difficulty of the problem is related to the fact that the Burnside Problem is a Ramsey-type problem and, conversely, many Ramsey-type problems can be reformulated as Burnside-type problems. It has been an open problem for many years whether or not there exists a primitive recursive upper bound in van der Waerden's theorem and in other major Ramsey-type theorems. This problem was solved by S. Shelah [12, 45], whose bound lies in the fifth class of the Grzegorchyk hierarchy.

Our original proof in [56] yielded an upper bound that is a variant of Ackermann's function and thus grows faster than any primitive recursive function.

The primitive recursive upper bound for $|B_0(m,n)|$ was found in [49]. Let $T(m,1) = m$, $T(m,n+1) = m^{T(m,n)}$.

Theorem 3. *Let G be a finite m generated groups of exponent n. Then $|G| \leq T(m, n^{n^n})$.*

S. I. Adian and N. N. Repin [1] found a lower bound for the class of $B_0(2,p)$, p is prime, which is exponential in p.

Remark, that A. Belov [5] proved that there exists a constant $\alpha > 0$ such that an arbitrary m generated associative ring satisfying the identity $x^n = 0$ is nilpotent of class $\leq m^{\alpha n}$.

The following generalization of Theorem 1 solves The General Kurosh–Levitsky Problem (in Shirshov's form) in the class of Lie PI-algebras.

Theorem 4. *There exists a function $h(m,n)$ with the following property. Suppose that Lie algebra L is generated by m elements $a_1, \ldots, a_m$ and (1) L satisfies a polynomial identity of degree n, (2) for an arbitrary commutator ρ in $a_1, \ldots, a_m$ of length $\leq h(m,n)$ the operator* ad(ρ) *is nilpotent. Then the algebra L is nilpotent.*

It is known that an infinite finitely generated p-group can be residually finite (such are the examples of E. S. Golod, R. I. Grigorchuk, N. Gupta-S. Sidki). However, Theorem 4 implies

Theorem 5. *A finitely generated residually finite p-group G whose Lie algebra $L(G)$ is PI, is finite.*

In particular, the Lie algebra $L(G)$ is PI if the pro-p completion $G_{\hat{p}}$ does not contain an abstract free subgroup of rank 2 (see [52]).

V. P. Platonov conjectured that periodic compact groups are locally finite. J. S. Wilson [51] proved that (under the assumption that there are finitely many simple sporadic groups) it suffices to prove the conjecture for pro-p groups. A periodic pro-p group G clearly does not contain a free abstract subgroup, thus the Lie algebra $L(G)$ is PI.

Theorem 6 [58]. *A periodic pro-p group is locally finite.*

A. Shalev [44] proved that the positive solution of The Restricted Burnside Problem implies (and in fact, is equivalent to) the following criterion of analyticity of a pro-p group. A finite p-group P is said to be a section of a pro-p group G if there exist an open subgroup K of G and a closed normal subgroup H in K such that $K/H \cong P$. Let C_m denote the cyclic group of order m.

Theorem 7 (A. Shalev, [44]). *A finitely generated pro-p group G is not analytic if and only if for any n the wreath product C_p wr C_{p^n} is a section of G.*

References

1. S. I. Adian and N. N. Repin, *On exponential lower bound for class of nilpotency of Engel Lie algebras*, Matem. Zametki **39, no. 3** (1986), 444–452.
2. S. V. Alyoshin, *Finite automata and the Burnside problem on periodic groups*, Matem. Zametki **11, no. 3** (1972), 319–328.
3. S. A. Amitsur, *Jacob Levitzki 1904–1956*, Israel J. Math. **19** (1974), 1–2.
4. R. Baer, *The higher commutator subgroups of a group*, Bull. Amer. Math. Soc. **50** (1944), 143–160.
5. A. Belov, *Some estimates for nilpotency classes of nil algebras over a field of an arbitrary characteristic and the Height theorem*, Commun. in Algebra **20** (1992), 2919–2922.
6. G. Benkart *Inner ideals and the structure of Lie algebras*, Dissertation, Yale Univ., 1974.
7. W. Burnside, *On an unsettled question in the theory of discontinuous groups*, Quart. J. Pure Appl. Math. **33** (1902), 230–238.
8. W. Feit and J. Thompson, *Solvability of groups of odd order*, Pacific J. Math. **13** (1963), 755–1029.
9. E. S. Golod, *On nil algebras and residually finite p-groups*, Izv. Akad. Nauk SSSR **28, no. 2** (1964), 273–276.
10. E. S. Golod and I. R. Shafarevich, *On towers of class fields*, Izv. Akad. Nauk SSSR **28, no. 2** (1964), 261–272.
11. D. Gorenstein, *Finite Simple Groups*, New York, 1882.
12. R. L. Graham, B. L. Rothshild and J. H. Spencer, *Ramsey Theory*, Wiley–Interscience, New York, 1990.
13. R. I. Grigorchuk, *On the Burnside problem for periodic groups*, Funct. Anal. Appl. **14, no. 1** (1980), 53–54.
14. O. Grün, *Zusammenhang zwischen Potenz bildung und Kommutator bildung*, J. Reine Angew. Mth. **182** (1940), 158–177.
15. F. J. Grunewald, G. Havas, J. L. Mennicke and M. Neumann, *Groups of exponent eight*, Bull. Austral. Math. Soc. **20** (1979), 7–16.
16. N. Gupta and S. Sidki, *On the Burnside problem for periodic groups*, Math. Z **182** (1983), 385–386.
17. P. Hall and G. Higman, *On the p-length of p-soluble groups and reduction theorems for Burnside's problem*, Proc. London Math. Soc. **6** (1956), 1–42.
18. G. Havas and M. F. Newman, *Application of computers to questions like those of Burnside*, In Lecture Notes in Math., Springer–Verlag, 1980, p. 806.
19. G. Higman, *Lie ring methods in the theory of finite nilpotent groups*, Proc. Intern. Congr. Math. Edinburgh (1958), 307–312.
20. S. Ivanov, *The free Burnside groups of sufficiently large exponents*, Intern. J. Algebra and Computation **4, no. 1 & 2** (1994), 1–308.
21. N. Jacobson, *Structure and Representations of Jordan Algebras*, AMS, Providence, R.I., 1969.
22. ———, *Lectures on quadratic Jordan algebras*, Tata Inst. of Fund. Research, Bombay, 1069.

23. S. A. Jennings, *The structure of the group ring of a p-group over a modular field*, TAMS **50** (1941), 175–185.
24. I. L. Kantor, *Classification of irreducibly transitively differential groups*, Doklady AN SSSR **5** (1964), 1404–1407.
25. I. Kaplansky, *Rings with a polynomial identity*, Bull. Amer. Math. Soc. **54** (1948), 575–580.
26. M. Koecher, *Imbedding of Jordan algebras into Lie algebras*, Amer. J. Math. **89** (1967), 787–816.
27. A. I. Kostrikin, *On Lie rings with Engel's condition*, Doklady AN SSSR, **108, no. 4** (1956), 580–582.
28. ———, *On the Burnside Problem*, Izvestia AN SSSR **23, no. 1** (1959), 3–34.
29. ———, *Sandwiches in Lie algebras*, Matem. Sbornik **110** (1979), 3–12.
30. ———, *Around Burnside*, Nauka, Moscow, 1986.
31. A. I. Kostrikin and E. I. Zelmanov, *A theorem on sandwich algebras*, Proc. V.A. Steklov Math. Inst. **183** (1988), 142–149.
32. A. G. Kurosh, *Problems in ring theory which are related to the Burnside Problem on periodic groups*, Izvestia AN SSSR **5, no. 3** (1941), 233–240.
33. M. Lazard, *Sur les groupes nilpotents et les anneaux de Lie*, Ann. Sci. Ecole Norm. Sup. **71, no. 3** (1954), 101–190.
34. O. Loos, *Jordan Pairs*, Springer–Verlag, 1975.
35. W. Magnus, *Über Gruppen und zugeordnete Liesche Ringe*, J. Reine Angew Math. **182** (1940), 142–159.
36. ———, *A connection between the Baker–Hausdorff formula and a problem of Burnside*, Ann. Math. **52** (1950), 11–26.
37. K. McCrimmon, *A general theory of Jordan rings*, Proc. Nat. Acad. Sci. USA **56** (1966), 1072–1079.
38. K. McCrimmon and E. Zelmanov, *The structure of strongly prime quadratic Jordan algebras*, Adv. Math. **69** (1988), 113–222.
39. K. Meyberg, *Lectures on algebras and triple systems*, The Univ. of Virginia, Charlottesville, 1972.
40. P. S. Novikov and S. I. Adian, *On infinite periodic groups. I, II, III*, Izvestia AN SSSR **32, no. 1** (1968), 212–244; **no. 2**, 251–254; **no. 3**, 709–731.
41. A. Yu. Olshansky, *Geometry of defining relations in groups*, Nauka, Moscow 1989, pp. 133–138.
42. B. I. Plotkin, *Algebraic sets of elements in groups and Lie algebras*, Uspekhi Mat. Nauk **13, no. 6** (1958), 133–138.
43. I. N. Sanov, *On a certain system of relations in periodic groups of prime power exponent*, Izvestia AN SSSR **15** (1951), 477–502.
44. A. Shalev, *Characterization of p-adic analytic groups in terms of wreath products*, J. Algebra **145** (1992), 204–208.
45. S. Shelah, *Primitive recursive bounds for van der Waerden Numbers*, J. Amer. Math. Soc. (1988), 683–688.
46. A. I. Shirshov, *On rings with identical relations*, Matem. Sbornik **43** (1957), 277–283.
47. V. I. Sushchansky, *Periodic p-groups of permutations and the General Burnside Problem*, Doklady AN SSSR **247, no. 3** (1979), 447–461.

48. M. Vaughan–Lee, *The Restricted Burnside Problem*, Second edition, Oxford Univ. Press, 1993, 107–145.
49. M. Vaughan–Lee and E. I. Zelmanov, *Upper bounds in the Restricted Burnside Problem*, J. Algebra **162** (1992), 107–145.
50. G. E. Wall, *On the Lie ring of a group of prime exponent*, Proc. 2nd Intern. Conf. Theory of Groups, Canberra, 1973, pp. 667–690.
51. J. S. Wilson, *On the structure of compact torsion groups*, Monatsh. Math. **96** (1983), 404–410.
52. J. S. Wilson and E.I. Zelmanov, *Identities for Lie algebras of pro-p groups*, J. Pure and Appl. Algebra **81** (1992), 103–109.
53. H. Zassenhaus, *Ein verfahren, jeder endlichen p-Gruppe einem Lie–Ring mit der Charakteristik p zuzuordnen*, Abh. Math. Sem. Univ. Hamburg **13** (1940), 200–207.
54. E. I. Zelmanov, *On some problems in the theory of groups and Lie algebras*, Matem. Sbornik **180, no. 2** (1989), 159–167.
55. ———, *On the restricted Burnside problem*, Siber. Math. J. **30, no. 6** (1989), 68–74.
56. ———, *The solution of the restricted Burnside problem for groups of odd exponent*, Math. USSR Izv. **36** (1991), 41–60.
57. ———, *The solution of the restricted Burnside problem for 2-groups*, Matem. Sbornik **182** (1991), 568–592.
58. ———, *On periodic compact groups*, Israel J. Math **77** (1992), 83–95.

Reprinted from Proc. Int. Congr. Math., 1998

THE WORK OF RICHARD EWEN BORCHERDS

by
PETER GODDARD
St John's College
Cambridge CB2 1TP

1. Introduction

Richard Borcherds has used the study of certain exceptional and exotic algebraic structures to motivate the introduction of important new algebraic concepts: vertex algebras and generalized Kac-Moody algebras, and he has demonstrated their power by using them to prove the "moonshine conjectures" of Conway and Norton about the Monster Group and to find whole new families of automorphic forms.

A central thread in his research has been a particular Lie algebra, now known as the Fake Monster Lie algebra, which is, in a certain sense, the simplest known example of a generalized Kac-Moody algebra which is not finite-dimensional or affine (or a sum of such algebras). As the name might suggest, this algebra *appears* to have something to do with the Monster group, *i.e.* the largest sporadic finite simple group.

The story starts with the observation that the Leech lattice can be interpreted as the Dynkin diagram for a Kac-Moody algebra, $\mathcal{L}_\infty$. But $\mathcal{L}_\infty$ is difficult to handle; its root multiplicities are not known explicitly. Borcherds showed how to enlarge it to obtain the more amenable Fake Monster Lie algebra. In order to construct this algebra, Borcherds introduced the concept of a vertex algebra, in the process establishing a comprehensive algebraic approach to (two-dimensional) conformal field theory, a subject of major importance in theoretical physics in the last thirty years.

To provide a general context for the Fake Monster Lie algebra, Borcherds has developed the theory of generalized Kac-Moody algebras, proving, in particular, generalizations of the Kac-Weyl character and denominator formulae. The denominator formula for the Fake Monster Lie algebra motivated Borcherds to construct a "real" Monster Lie algebra, which he used to prove the moonshine conjectures. The results for the Fake Monster Lie algebra also motivated Borcherds to explore the properties of the denominator formula for other generalized Kac-Moody algebras, obtaining remarkable product expressions for modular functions, results on the moduli spaces of certain complex surfaces and much else besides.

2. The Leech Lattice and the Kac-Moody Algebra $\mathcal{L}_\infty$

We start by recalling that a finite-dimensional simple complex Lie algebra, $\mathcal{L}$, singular invariant bilinear form (,) on $\mathcal{L}$ which induces such a form on the rank dimensional space spanned by the roots of $\mathcal{L}$. Suppose $\{\alpha_i : 1 \leq i \leq \operatorname{rank}\mathcal{L}\}$ is a basis of simple roots for $\mathcal{L}$. Then the numbers $a_{ij} = (\alpha_i, \alpha_j)$ have the following properties:

$$a_{ii} > 0, \tag{1}$$

$$a_{ij} = a_{ji}, \tag{2}$$

$$a_{ij} \leq 0 \quad \text{if } i \neq j, \tag{3}$$

$$2a_{ij}/a_{ii} \in \mathbb{Z}. \tag{4}$$

The symmetric matrix $A = (a_{ij})$ obtained in this way is positive definite.

The algebra $\mathcal{L}$ can be reconstructed from the matrix A by the system of generators and relations used to define $\mathcal{L}_\infty$,

$$[e_i, f_i] = h_i, \qquad [e_i, f_i] = 0 \quad \text{for } i \neq j, \tag{5}$$

$$[h_i, e_j] = a_{ij} e_j, \qquad [h_i, f_j] = -a_{ij} f_j, \tag{6}$$

$$\mathrm{Ad}(e_i)^{n_{ij}}(e_j) = \mathrm{Ad}(f_i)^{n_{ij}}(f_j) = 0, \quad \text{for } n_{ij} = 1 - 2a_{ij}/a_{ii}. \tag{7}$$

These relations can be used to define a Lie algebra, $\mathcal{L}_A$, for any matrix A satisfying the conditions (1-4). $\mathcal{L}_A$ is called a (symmetrizable) Kac-Moody algebra. If A is positive definite, $\mathcal{L}_A$ is semi-simple and, if A is positive semi-definite, $\mathcal{L}_A$ is a sum of affine and finite-dimensional algebras.

Although Kac and Moody only explicitly considered the situation in which the number of simple roots was finite, the theory of Kac-Moody algebras applies to algebras which have a infinite number of simple roots. Borcherds and others [1] showed how to construct such an algebra with simple roots labelled by the points of the Leech lattice, Λ_L. We can conveniently describe Λ_L as a subset of the unique even self-dual lattice, $II_{25,1}$, in 26-dimensional Lorentzian space, $\mathbb{R}^{25,1}$. $II_{25,1}$ is the set of points whose coordinates are all either integers or half odd integers which have integral inner product with the vector $(\frac{1}{2}, \ldots, \frac{1}{2}; \frac{1}{2}) \in \mathbb{R}^{25,1}$, where the norm of $x = (x_1, x_2, \ldots, x_{25}; x_0)$ is $x^2 = x_1^2 + x_2^2 + \cdots + x_{25}^2 - x_0^2$.

The vector $\rho = (0, 1, 2, \ldots, 24; 70) \in II_{25,1}$ has zero norm, $\rho^2 = 0$; the Leech lattice can be shown to be isomorphic to the set $\{x \in II_{25,1} : x \cdot \rho = -1\}$ modulo displacements by ρ. We can take the representative points for the Leech lattice to have norm 2 and so obtain an isometric correspondence between Λ_L and

$$\{r \in II_{25,1} : r \cdot \rho = -1, r^2 = 2\}. \tag{8}$$

Then, with each point r of the Leech lattice, we can associate a reflection $x \mapsto \sigma_r(x) = x - (r \cdot x)r$ which is an automorphism of $II_{25,1}$. Indeed these reflections σ_r generate a Weyl group, W, and the whole automorphism group of $II_{25,1}$ is the semi-direct product of W and the automorphism group of the affine Leech lattice, which is the Dynkin/Coxeter diagram of the Weyl group W. To this Dynkin diagram can be associated an infinite-dimensional Kac-Moody algebra, $\mathcal{L}_\infty$, generated by elements $\{e_r, f_r, h_r : r \in \Lambda_L\}$ subject to the relations (5–7). Dividing by the linear combinations of the h_r which are in the centre reduces its rank to 26.

The point about Kac-Moody algebras is that they share many of the properties enjoyed by semi-simple Lie algebras. In particular, we can define a Weyl group, W, and for suitable (*i.e.* lowest weight) representations, there is a straightforward generalization of the Weyl character formula. For a representation with lowest weight λ, this generalization, the Weyl-Kac character formula, states

$$\chi\lambda = \sum_{w \in W} \det(w) w(e^{\rho+\lambda})/e^{\rho} \prod_{\alpha>0} (1 - e^{\alpha})^{m_\alpha}, \tag{9}$$

where ρ is the Weyl vector, with $\rho \cdot r = -r^2/2$ for all simple roots r, m_α is the multiplicity of the root α, the sum is over the elements w of the Weyl group W, and the product is over positive roots α, that is roots which can be expressed as the sum of a subset of the simple roots with positive integral coefficients.

Considering even just the trivial representation, for which $\lambda = 0$ and $\chi_0 = 1$, yields a potentially interesting relation from (9),

$$\sum_{w \in W} \det(w) w(e^{\rho}) = e^{\rho} \prod_{\alpha>0} (1 - e^{\alpha})^{m_\alpha}. \tag{10}$$

Kac showed that this denominator identity produces the Macdonald identities in the affine case. Kac-Moody algebras, other than the finite-dimensional and affine ones, would seem to offer the prospect of new identities generalizing these but the problem is that in other cases of Kac-Moody algebras, although the simple roots are known (as for $\mathcal{L}_\infty$), which effectively enables the sum over the Weyl group to be evaluated, the root multiplicities, m_α, are not known, so that the product over positive roots cannot be evaluated.

No general simple explicit formula is known for the root multiplicities of $\mathcal{L}_\infty$ but, using the "no-ghost" theorem of string theory, I. Frenkel established the bound

$$m_\alpha \leq p_{24}\left(1 - \frac{1}{2}\alpha^2\right), \tag{11}$$

where $p_k(n)$ is the number of partitions of n using k colours. This bound is saturated for some of the roots of $\mathcal{L}_\infty$ and, where it is not, there is the impression that is because something is missing. What seems to be missing are some simple roots of zero or negative norm. In Kac-Moody algebras all the simple roots are specified by

(1) to be of positive norm, even though some of the other roots they generate may not be.

3. Vertex Algebras

Motivated by Frenkel's work, Borcherds introduced in [3] the definition of a vertex algebra, which could in turn be used to define Lie algebras with root multiplicities which are explicitly calculable. A vertex algebra is a graded complex vector space, $V = \otimes_{n\in\mathbb{Z}} V_n$, together with a "vertex operator", $a(z)$, for each $a \in V$, which is a formal power series in the complex variable z,

$$a(z) = \sum_{m\in\mathbb{Z}} a_m z^{-m-n}, \quad \text{for } a \in V_n, \tag{12}$$

where the operators a_m map $V_n \to V_{n-m}$ and satisfy the following properties:

1. $a_n b = 0$ for $n > N$ for some integer N dependent on a and b;
2. there is an operator (derivation) $D : V \to V$ such that $[D, a(z)] = \frac{d}{dz} a(z)$;
3. there is a vector $\mathbf{1} \in V_0$ such that $\mathbf{1}(z) = 1$, $D\mathbf{1} = 0$;
4. $a(0)\mathbf{1} = a$;
5. $(z - \zeta)^N (a(z)b(\zeta) - b(\zeta)a(z)) = 0$ for some integer N dependent on a and b.

[We may define vertex operators over other fields or over the integers with more effort if we wish but the essential features are brought out in the complex case.]

The motivation for these axioms comes from string theory, where the vertex operators describe the interactions of "strings" (which are to be interpreted as models for elementary particles). Condition (5) states that $a(z)$ and $b(\zeta)$ commute apart from a possible pole at $z = \zeta$, *i.e.* they are local fields in the sense of quantum field theory. A key result is that, in an appropriate sense,

$$(a(z-\zeta)b)(\zeta) = a(z)b(\zeta) = b(\zeta)a(z). \tag{13}$$

More precisely

$$\int_0 d\zeta \int_\zeta dz (a(z-\zeta)b)(\zeta) f = \int_0 dz \int_0 d\zeta a(z)b(\zeta) f - \int_0 d\zeta \int_0 dz b(\zeta)a(z) f, \tag{14}$$

where f is a polynomial in z, ζ, $z - \zeta$ and their inverses, and the integral over z is a circle about ζ in the first integral, one about ζ and the origin in the second integral and a circle about the origin excluding the ζ in the third integral. The axioms originally proposed by Borcherds [2] were somewhat more complicated in from and follow from those given here from the conditions generated by (14).

We can associate a vertex algebra to any even lattice Λ, the space V then having the structure of the tensor product of the complex group ring $\mathbb{C}(\Lambda)$ with the symmetric algebra of a sum $\bigoplus_{n>0} \Lambda_n$ of copies Λ_n, $n \in \mathbb{Z}$, of Λ. In terms of string theory, this is the Fock space describing the (chiral) states of a string moving in a space-time compactified into a torus by imposing perodicity under displacements by the lattice Λ.

The first triumph of vertex algebras was to provide a natural setting for the Monster group, M. M acts on a graded infinite-dimensional space $V^\natural$, constructed by Frenkel, Lepowsky and Meurman, where $V^\natural = \oplus_{n\geq -1} V_n^\natural$, and the dimensions of dim $V_n^\natural$ is the coefficent, $c(n)$ of q^n in the elliptic modular function,

$$j(\tau) - 744 = \sum_{n=-1}^{\infty} c(n) q^n$$
$$= q^{-1} + 196884q + 21493760q^2 + \cdots, \quad q = e^{2\pi i \tau}. \tag{15}$$

A first thought might have been that the Monster group should be related to the space V_{Λ_L}, the vertex algebra directly associated with the Leech lattice, but V_{Λ_L} has a grade 0 piece of dimension 24 and the lowest non-trivial representation of the Monster is of dimension 196883. $V^\natural$ is related to V_{Λ_L} but is a sort of twisted version of it; in string theory terms it corresponds to the string moving on an orbifold rather than a torus.

The Monster group is precisely the group of automorphisms of the vertex algebra $V^\natural$,

$$ga(z)g^{-1} = (ga)(z), \quad g \in M. \tag{16}$$

This characterizes M in a way similar to the way that two other sporadic simple finite groups, Conway's C_{o_1} and the Mathieu group M_{24}, can be characterized as the automorphism groups of the Leech lattice (modulo -1) and the Golay Code, respectively.

4. Generalized Kac-Moody Algebras

In their famous moonshine conjectures, Conway and Norton went far beyond the existence of the graded representation $V^\natural$ with dimension given by j. Their main conjecture was that, for each element $g \in M$, the Thompson series

$$T_g(q) = \sum_{n=-1}^{\infty} \mathrm{Trace}(g|V_n^\natural) a^n \tag{17}$$

is a Hauptmodul for some genus zero subgroup, G, of $\mathrm{SL}_2(\mathbb{R})$, *i.e.*, if

$$H = \{\tau : \mathrm{Im}(\tau) > 0\} \tag{18}$$

denotes the upper half complex plane, G is such that the closure of H/G is a compact Riemann surface, $\overline{H/G}$, of genus zero with a finite number of points removed and $T_g(q)$ defines an isomorphism of $\overline{H/G}$ onto the Riemann sphere.

To attack the moonshine conjectures it is necessary to introduce some Lie algebraic structure. For any vertex algebra, V, we can introduce [2, 4] a Lie algebra of operators

$$L(a) = \frac{1}{2\pi i}\oint a(z)dz = a_{-h+1}, \quad a \in V_h. \tag{19}$$

Closure $[L(a), L(b)] = L(L(a)b)$ follows from (14), but this does not define a Lie algebra structure directly on V because $L(a)b$ is not itself antisymmetric in a and b. However, DV is in the kernel of the map $a \mapsto L(a)$ and $L(a)b = -L(b)a$ in V/DV, so it does define a Lie algebra $\mathcal{L}^0(V)$ on this quotient [2], but this is not the most interesting Lie algebra associated with V.

Vertex algebras of interest come with an additional structure, an action of the Virasoro algebra, a central extension of the Lie algebra of polynomial vector fields on the circle, spanned by L_n, $n \in \mathbb{Z}$ and 1,

$$[L_m, L_n] = (m-n)L_{m+n} + \frac{c}{12}m(m^2-1)\delta_{m,-n}, \quad [L_n, c] = 0, \tag{20}$$

with $L_{-1} = D$ and $L_0 a = ha$ for $a \in V_h$. For V_Λ, $c = \dim\Lambda$, and for $V^\natural$, $c = 24$. The Virasoro algebra plays a central role in string theory. The space of "physical states" of the string is defined by the Virasoro conditions: let

$$P^k(V) = \{a \in V : L_0 a = ka; L_n a = 0, n > 0\}, \tag{21}$$

the space of physical states in $P^1(V)$. The space $P^1(V)/L_{-1}P^0(V)$ has a Lie algebra structure defined on it (because $L_{-1}V \cap P^1(V) \subset L_{-1}P^0(V)$). This can be reduced in size further using a contravariant form (which it possesses naturally for lattice theories). The "no-ghost" theorem states that the space of physical states $P^1(V)$ has lots of null states and is positive semi-definite for V_Λ, where Λ is a Lorentzian lattice with $\dim\Lambda \leq 26$. So we can quotient $P^1(V)/L_{-1}P^0(V)$ further by its null space with the respect to the contravariant form to obtain a Lie algebra $\mathcal{L}(V)$.

The results of factoring by the null space are most dramatic when $c = 26$. The vertex algebra V_L has a natural grading by the lattice L and the "no-ghost" theorem states that the dimension of the subspace of $\mathcal{L}(V)$ of non-zero grade α is $p_{24}(1-\frac{1}{2}\alpha^2)$ if Λ is a Lorentzian lattice of dimension 26 but $p_{k-1}(1-\alpha^2/2) - p_{k-1}(\alpha^2/2)$ if $\dim\Lambda = k \neq 26$, $k > 2$. Thus the algebra $\mathcal{L}'_M = \mathcal{L}(V_{II_{25,1}})$ saturates Frenkel's bound, and Borcherds initially named it the "Monster Lie algebra" because it appeared to be directly connected to the Monster; it is now known as the "Fake Monster Lie algebra."

Borcherds [4] had the great insight not only to construct the Fake Monster Lie algebra, but also to see how to generalize the definition of a Kac-Moody algebra effectively in order to bring $\mathcal{L}'_M$ within the fold. What was required was to relax the condition (1), requiring roots to have positive norm, and to allow them to be either zero or negative norm. The condition (4) then needs modification to apply only in the space-like case $a_{ii} > 0$ and the same applies to the condition (7) on the generators. The only condition which needs to be added is that

$$[e_i, e_j] = [f_i, f_j] = 0 \quad \text{if } a_{ij} = 0. \tag{22}$$

The closeness of these conditions to those for Kac-Moody algebras means that most of the important structural results carry over; in particular there is a generalization of the Weyl-Kac character formula for representations with highest weight λ,

$$\chi\lambda = \sum_{w\in W} \det(w) w \left(e^\rho \sum_\mu \epsilon_\lambda(\mu) e^{\mu+\lambda} \right) e^\rho \prod_{\alpha>0} (1-e^\alpha)^{m_\alpha}, \tag{23}$$

where the second sum in the numerator is over vectors μ and $\epsilon_\lambda(\mu) = (-1)^n$ if μ can be expressed as the sum of n pairwise orthogonal simple roots with non-positive norm, all orthogonal to λ, and 0 otherwise. Of course, putting $\lambda = 0$ and $\chi_\lambda = 1$ again gives a denominator formula.

The description of generalized Kac-Moody algebras in terms of generators and relations enables the theory to be taken over rather simply from that of Kac-Moody algebras but it is not so convenient as a method of recognising them in practice, *e.g.* from amongst the algebras $\mathcal{L}(V)$ previously constructed by Borcherds. But Borcherds [3] gave an alternative characterization of them as graded algebras with an "almost posititive definite" contravariant bilinear form. More precisely, he showed that a graded Lie algebra, $\mathcal{L} = \bigoplus_{n\in Z} \mathcal{L}_n$, is a generalized Kac-Moody algebra if the following conditions are satisfied:

1. $\mathcal{L}_0$ is abelian and $\dim \mathcal{L}_n$ is finite if $n \neq 0$;
2. $\mathcal{L}$ possesses an invariant bilinear form such that $(\mathcal{L}_m, \mathcal{L}_n) = 0$ if $m \neq n$;
3. $\mathcal{L}$ possesses an involution ω which is -1 on $\mathcal{L}_0$ and such that $\omega(\mathcal{L}_m) \subset \mathcal{L}_{-m}$;
4. the contravariant bilinear form $\langle L, M\rangle = -(L, \omega(M))$ is positive definite on $\mathcal{L}_m$ for $m \neq 0$;
5. $\mathcal{L}_0 \subset [\mathcal{L}, \mathcal{L}]$.

This characterization shows that the Fake Monster Lie algebra, $\mathcal{L}'_M$, is a generalised Kac-Moody algebra, and its root multiplicities are known to be given by $p_{24}(1 - \frac{1}{2}\alpha^2)$, but Borcherds' theorem establishing the equivalence of his two definitions does not give a constructive method of finding the simple roots. As we remarked in the context of Kac-Moody algebras, if we knew both the root multiplicities and the simple roots, the denominator formula

$$\sum_{w\in W} \det(w) w \left(e^\rho \sum_\mu \epsilon_\mu(\alpha) e^\mu \right) = e^\rho \prod_{\alpha>0} (1-e^\alpha)^{m_\alpha} \tag{24}$$

might provide an interesting identity. Borcherds solved [4] the problem of finding the simple roots, or rather proving that the obvious ones were all that there were, by inverting this argument. The positive norm simple roots can be identified with the Leech lattice as for $\mathcal{L}_\infty$. Writing $\mathrm{II}_{25,1} = \Lambda_L \oplus \mathrm{II}_{1,1}$, which follows by uniqueness or the earlier comments, the 'real' or space-like simple roots are $\{(\lambda, 1, \frac{1}{2}\lambda^2 - 1)$: $\lambda \in \Lambda_L\}$. (Here we are using we are writing $\mathrm{II}_{1,1} = \{(m, n) : m, n \in \mathbb{Z}\}$ with (m, n) having norm $-2mn$.) Light-like simple roots are quite easily seen to be $n\rho$, where n is a positive integer and $\rho = (0, 0, 1)$. The denominator identity is then used to prove that there are no other light-like and that there are no time-like simple roots.

The denominator identity provides a remarkable relation between modular functions (apparently already known to some of the experts in the subject) which is the precursor of other even more remarkable identities. If we restrict attention to vectors $(0, \sigma, \tau) \in II_{25,1} \otimes \mathbb{C}$, with $\mathrm{Im}(\sigma) > 0$, $\mathrm{Im}(\tau) > 0$, it reads

$$p^{-1} \prod_{m>0, n\in\mathbb{Z}} (1 - p^m q_n)^{c'(mn)} = \Delta(\sigma)\Delta(\tau)(j(\sigma) - j(\tau)) \tag{25}$$

where $c'(0) = 24$, $c'(n) = c(n)$ if $n \neq 0$, $p = e^{2\pi i\tau}$, and

$$\Delta(\tau)^{-1} = q^{-1} \prod_{m>1} (1 - q^n)^{-24} = \sum_{n\geq 0} p_{24}(n) q^{n-1}. \tag{26}$$

5. Moonshine, the Monster Lie Algebra and Automorphic Forms

The presence of $j(\sigma)$ in (25) suggests a relationship to the moonshine conjectures and Borcherds used [5, 6] this as motivation to construct the "real" Monster Lie Algebra, $\mathcal{L}_M$ as one with denominator identity obtained by multiplying each side of (25) by $\Delta(\sigma)\Delta(\tau)$, to obtain the simpler formula

$$p^{-1} \prod_{m>0, n\in\mathbb{Z}} (1 - p^m q^n)^{c(mn)} = j(\sigma) - j(\tau). \tag{27}$$

This looks like the denominator formula for a generalized Kac-Moody algebra which is graded by $\mathrm{II}_{1,1}$ and is such that the dimension of the subspace of grade $(m, n) \neq (0, 0)$ is $c(mn)$, the dimension of $V^\natural_{mn}$. It is not difficult to see that this can be constructed by using the vertex algebra which is the tensor product $V^\natural \otimes V_{\mathrm{II}_{1,1}}$ and defining $\mathcal{L}_M$ to be the generalised Lie algebra, $\mathcal{L}(V^\natural \otimes V_{\mathrm{II}_{1,1}})$, constructed from the physical states.

Borcherds used [5, 6] twisted form of the denominator identity for $\mathcal{L}_M$ to prove the moonshine conjectures. The action of M on $V^\natural$ provides an action on $V = V^\natural \otimes V_{\mathrm{II}_{1,1}}$ induces an action on the physical state space $P^1(V)$ and on its quotient, $\mathcal{L}_M = \mathcal{L}(V)$, by its null space. The "no-ghost" theorem implies that the part of $\mathcal{L}_M$ of grade (m, n), $(\mathcal{L}_M)_{(m,n)}$, is isomorphic to $V^\natural_{mn}$ as an M module. Borcherds adapted the argument he used to establish the denominator identity to prove the twisted relation

$$p^{-1} \exp\left(- \sum_{N>0} \sum_{m>0, n\in\mathbb{Z}} \mathrm{Tr}(g^N | V^\natural_{mn} p^{mN} q^{nN} / N \right)$$

$$= \sum_{m\in\mathbb{Z}} \mathrm{Tr}(g | V^\natural_m) p^m - \sum_{n\in\mathbb{Z}} \mathrm{Tr}(g | V^\natural_n) q^n. \tag{28}$$

These relations on the Thompson series are sufficient to determine them from their first few terms and to establish that they are modular functions of genus 0.

Returning to the Fake Monster Lie Algebra, the denominator formula given in (25) was restricted to vectors of the form $v = (0, \sigma, \tau)$ but we consider it for more general $v \in II_{25,1} \otimes \mathbb{C}$, giving the denominator function

$$\Phi(v) = \sum_{w \in W} \det(w) e^{2\pi i (w(\rho), v)} \prod_{n>0} (1 - e^{2\pi i n (w(\rho), v)})^{24}. \tag{29}$$

This expression converges for $\mathrm{Im}(v)$ inside a certain cone (the positive light cone). Using the explicit form for $\Phi(v)$ when $v = (0, \sigma, \tau)$, the known properties of j and Δ and the fact that $\Phi(v)$ manifestly satisfies the wave equation, Borcherds [6, 7, 9] establishes that $\Phi(v)$ satisfies the functional equation

$$\Phi(2v/(v, v)) = -((v, v)/2)^{12} \Phi(v). \tag{30}$$

It also has the properties that

$$\Phi(v + \lambda) = \Phi(v) \quad \text{for } \lambda \in \mathrm{II}_{25,1} \tag{31}$$

and

$$\Phi(w(v)) = \det(w)\Phi(v) \quad \text{for } w \in \mathrm{Aut}(\mathrm{II}_{25,1})^+, \tag{32}$$

the group of automorphisms of the lattice $\mathrm{II}_{25,1}$ which preserve the time direction. These transformations generate a discrete subgroup of the group of conformal transformations on $\mathbb{R}^{25,1}$, which is itself isomorphic to $O_{26,2}(\mathbb{R})$; in fact the discrete group is isomorphic to $\mathrm{Aut}(\mathrm{II}_{26,2})^+$. The denominator function for the Fake Monster Lie algebra defines in this way an automorphic form of weight 12 for the discrete subgroup $\mathrm{Aut}(\mathrm{II}_{26,2})^+$ of $O_{26,2}(\mathbb{R})^+$. This result once obtained is seen not to depend essentially on the dimension 26 and Borcherds has developed this approach of obtaining representations of modular functions as infinite products from denominator formulae for generalized Kac-Moody algebras to obtain a plethora of beautiful formulae [7, 9, 11], *e.g.*

$$j(\tau) = q^{-1} \prod_{n>0} (1 - q)^{-744} (1 - q^2)^{80256} (1 - q^3)^{-12288744} \ldots, \tag{33}$$

where $f_0(\tau) = \sum_n c_0(n) q^n$ is the unique modular form of weight $\frac{1}{2}$ for the group $\Gamma_0(4)$ which is such that $f_0(\tau) = 3q^{-3} + \mathcal{O}(q)$ at $q = 0$ and $c_0(n) = 0$ if $n \equiv 2$ or 3 mod 4. he has also used these denominator functions to establish results about the moduli spaces of Enriques surfaces and families of K3 surfaces [8, 10].

Displaying penetrating insight, formidable technique and brilliant originality, Richard Borcherds has used the beautiful properties of some exceptional structures to motiviate new algebraic theories of great power with profound connections with other areas of mathematics and physics. He has used them to establish outstanding conjectures and to find new deep results in classical areas of mathematics. This is surely just the beginning of what we have to learn from what he has created.

References

[1] R. E. Borcherds, J. H. Conway, L. Queen and N. J. A. Sloane, *A monster Lie algebra?* Adv. Math. 53 (1984) 75–79.
[2] R. E. Borcherds, *Vertex algebras, Kac-Moody algebras and the monster*, Proc. Nat. Acad. Sci. U.S.A. 83b (1986) 3068–3071.
[3] R. E. Borcherds, *Generalized Kac-Moody algebras*, J. Alg. 115 (1988) 501–512.
[4] R. E. Borcherds, *The monster Lie algebra*, Adv. Math. 83 (1990) 30–47.
[5] R. E. Borcherds, *Monstrous moonshine and monstrous Lie algebras*, Invent. Math. 109 (1992) 405–444.
[6] R. E. Borcherds, *Sporadic groups and string theory*, in *Proceedings of the First European Congress of Mathematics, Paris July 1992*, ed A. Joseph *et al.*, Vol. 1, Birkhauser (1994) pp. 411–421.
[7] R. E. Borcherds, *Automorphic forms on* $O_{s+2,2}(R)$ *and infinite products*, Invent. Math. 120 (1995) 161–213.
[8] R. E. Borcherds, *The moduli space of Enriques surfaces and the fake monster Lie superalgebra*, Topology 35 (1996) 699–710.
[9] R. E. Borcherds, *Automorphic forms and Lie algebras*, in Current developments in mathematics, International Press (1996).
[10] R. E. Borcherds, L. Katzarkov, T. Pantev and N. I. Shepherd-Barron, *Families of K3 surfaces*, J. Algebraic Geometry 7 (1998) 183–193.
[11] R. E. Borcherds, *Automorphic forms with singularities on Grassmannians*, Invent. Math. 132 (1998) 491–562.

Richard E. Borcherds

RICHARD E. BORCHERDS

Full name: Richard Ewen Borcherds

Date of birth: November 29, 1959

Place of birth: Cape Town, South Africa

Education: Undergraduate, Cambridge University (1978–1981).
Ph.D., Cambridge University (1981–1983).

Appointments: Research Fellow, Trinity College, Cambridge (1983–1987);
Morrey Assistant Professor, University of California, Berkeley (1987–1988);
Royal Society University Research Fellow, Cambridge University (1988–1992);
Lecturer, Cambridge University (1992–1993);
Professor, University of California, Berkeley (1993–1996);
Royal Society Research Professor , Cambridge University (1996–1999);
Professor, University of California at Berkeley (1999–present).

Prize: Fields Medal, International Mathematical Union, 1998.

Reprinted from Proc. Int. Congr. Math., 1998

WHAT IS MOONSHINE?

by

RICHARD E. BORCHERDS

Department of Pure Mathematics and
Mathematical Sciences Cambridge University
Cambridge CB2 1SB
England

This is an informal write up of my talk at the I.C.M. in Berlin. It gives some background to Goddard's talk [Go] about the moonshine conjectures. For other survey talks about similar topics see [B94], [B98], [LZ], [J], [Ge], [Y].

The classification of finite simple groups shows that every finite simple group either fits into one of about 20 infinite families, or is one of 26 exceptions, called sporadic simple groups. The monster simple group is the largest of the sporadic finite simple groups, and was discovered by Fischer and Griess [G]. Its order is

$$8080, 17424, 79451, 28758, 86459, 90496, 17107, 57005, 75436, 80000, 00000$$

$$= 2^{46} \cdot 3^{20} \cdot 5^{9} \cdot 7^{6} \cdot 11^{2} \cdot 13^{3} \cdot 17 \cdot 19 \cdot 23 \cdot 29 \cdot 31 \cdot 41 \cdot 47 \cdot 59 \cdot 71$$

(which is roughly the number of elementary particles in the earth). The smallest irreducible representations have dimensions $1, 196883, 21296876, \ldots$. The elliptic modular function $j(\tau)$ has the power series expansion

$$j(\tau) = q^{-1} + 744 + 196884q + 21493760q^2 + \cdots,$$

where $q = e^{2\pi i\tau}$, and is in some sense the simplest nonconstant function satisfying the functional equations $j(\tau) = j(\tau+1) = j(-1/\tau)$. John McKay noticed some rather weird relations between coefficients of the elliptic modular function and the representations of the monster as follows:

$$1 = 1,$$

$$196884 = 196883 + 1,$$

$$21493760 = 21296876 + 196883 + 1,$$

where the numbers on the left are coefficients of $j(\tau)$ and the numbers on the right are dimensions of irreducible representations of the monster. At the time he discovered these relations, several people thought it so unlikely that there could be a relation between the monster and the elliptic modular function that they politely told McKay that he was talking nonsense. The term "monstrous moonshine" (coined by Conway) refers to various extensions of McKay's observation, and in particular to relations between sporadic simple groups and modular functions.

For the benefit of readers who are not native English speakers, I had better point out that "moonshine" is not a poetic terms referring to light from the moon. It means foolish or crazy ideas. (Quatsch in German.) A typical example of its use is the following quotation from E. Rutherford (the discoverer of the nucleus of the atom): "The energy produced by the breaking down of the atom is a very poor kind of thing. Anyone who expects a source of power from the transformations of these atoms is talking moonshine." (Moonshine is also a name for corn whiskey, especially if it has been smuggled or distilled illegally.)

We recall the definition of the elliptic modular function $j(\tau)$. The group $SL_2(\mathbf{Z})$ acts on the upper half plane H by

$$\begin{pmatrix} a & b \\ c & d \end{pmatrix}(\tau) = \frac{a\tau + b}{c\tau + d}.$$

A modular function (of level 1) is a function f on H such that $f((a\tau+b)/(c\tau+d)) = f(\tau)$ for all $\left(\begin{smallmatrix} a & b \\ c & d \end{smallmatrix}\right) \in SL_2(\mathbf{Z})$. It is sufficient to assume that f is invariant under the generators $\tau \mapsto \tau + 1$ and $\tau \mapsto -1/\tau$ of $SL_2(\mathbf{Z})$. The elliptic modular function j is the simplest nonconstant example, in the sense that any other modular function can be written as a function of j. It can be defined as follows:

$$\begin{aligned} j(\tau) &= \frac{E_4(\tau)^3}{\Delta(\tau)} \\ &= q^{-1} + 744 + 196884q + 21493760q^2 + \cdots, \\ E_4(\tau) &= 1 + 240 \sum_{n>0} \sigma_3(n) q^n \\ &= 1 + 240q + 2160q^2 + \cdots \\ &\quad \times (\sigma_3(n) = \sum_{d|n} d^3), \\ \Delta(\tau) &= q \prod_{n>0} (1 - q^n)^{24} \\ &= q - 24q + 252q^2 + \cdots. \end{aligned}$$

A modular form of weight k is a holomorphic function

$$f(\tau) = \sum_{n \geq 0} c(n) q^n$$

on the upper half plane satisfying the functional equation $f((a\tau+b)/(c\tau+d)) = (c\tau+d)^k f(\tau)$ for all $\begin{pmatrix} a & b \\ c & d \end{pmatrix} \in SL_2(\mathbf{Z})$. The function $E_4(\tau)$ is an Eisenstein series and is a modular form of weight 4, while $\Delta(\tau)$ is a modular form of weight 12.

The function $j(\tau)$ is an isomorphism from the quotient $SL_2(\mathbf{Z})\backslash H$ to $\mathbf{C}$, and is uniquely defined by this up to multiplication by a constant or addition of a constant. In particular any other modular function is a function of j, so j is in some sense the simplest nonconstant modular function.

An amusing property of j (which so far seems to have no relation with moonshine) is that $j(\tau)$ is an algebraic integer whenever τ is an imaginary quadratic irrational number. A well known consequence of this is that

$$\exp(\pi\sqrt{163}) = 262537412640768743.99999999999925\ldots$$

is very nearly an integer. The explanation of this is that $j((1+i\sqrt{163})/2)$ is exactly the integer

$$-262537412640768000 = -2^{18}3^3 5^3 23^3 29^3,$$

and

$$\begin{aligned} j((1+i\sqrt{163})/2) &= q^{-1} + 744 + 196884q + \cdots \\ &= -e^{\pi\sqrt{163}} + 744 + (\text{something very small}). \end{aligned}$$

McKay and Thompson suggested that there should be a graded representation $V = \oplus_{n\in\mathbf{Z}} V_n$ of the monster, such that $\dim(V_n) = c(n-1)$, where $j(\tau) - 744 = \sum_n c(n)q^n = q^{-1} + 196884q + \cdots$. Obviously this is a vacuous statement if interpreted literally, as we could for example just take each V_n to be a trivial representation. To characterize V, Thompson suggested looking at the McKay-Thompson series

$$T_g(\tau) = \sum_n Tr(g|V_n)q^{n-1}$$

for each element g of the monster. For example, $T_1(\tau)$ should be the elliptic modular function. Conway and Norton [C-N] calculated the first few terms of each McKay-Thomson series by making a reasonable guess for the decomposition of the first few V_n's into irreducible representations of the monster. They discovered the astonishing fact that all the McKay-Thomson series appeared to be Hauptmoduls for certain genus 0 subgroups of $SL_2(\mathbf{R})$. (A Hauptmodul for a subgroup Γ is an isomorphism from $\Gamma\backslash H$ to $\mathbf{C}$, normalized so that its Fourier series expansion starts off $q^{-1}+O(1)$.)

As an example of some Hauptmoduls of elements of the monster, we will look at the elements of order 2. There are 2 conjugacy classes of elements of order 2, usually called the elements of types $2A$ and $2B$. The corresponding McKay-Thompson series start off

$$T_{2B}(\tau) = q^{-1} + 276q - 2048q^2 + \cdots \qquad \text{Hauptmodul for } \Gamma_0(2),$$

$$T_{2A}(\tau) = q^{-1} + 4372q + 96256q^2 + \cdots \qquad \text{Hauptmodul for } \Gamma_0(2).$$

The group $\Gamma_0(2)$ is $\{\left(\begin{smallmatrix} a & b \\ c & d \end{smallmatrix}\right) \in SL_2(\mathbf{Z}) \in SL_2(\mathbf{Z}) | c \text{ is even}\}$, and the group $\Gamma_0(2)+$ is the normalizer of $\Gamma_0(2)$ in $SL_2(\mathbf{R})$. Ogg had earlier commented on the fact that the full normalizer $\Gamma_0(p)+$ of $\Gamma_0(p)$ for p prime is a genus 0 group if and only if p is one of the primes 2, 3, 5, 7, 11, 13, 17, 19, 23, 29, 31, 41, 47, 59, or 71 dividing the order of the monster.

Conway and Norton's conjectures were soon proved by A. O. L. Atkin, P. Fong, and S. D. Smith. The point is that to prove something is a virtual character of a finite group it is only necessary to prove a finite number of congruences. In the case of the moonshine module V, proving the existence of an infinite dimensional representation of the monster whose McKay-Thompson series are given Hauptmoduls requires checking a finite number of congruences and positivity conditions for modular functions, which can be done by computer.

This does not give an explicit construction of V, or an explanation about why the conjectures are true. Frenkel, Lepowsky, and Meurman managed to find an explicit construction of a monster representation $V = \oplus V_n$, such that $\dim(V_n) = c(n-1)$, and this module had the advantage that it came with some extra algebraic structure preserved by the monster. However it was not obvious that V satisfied the Conway-Norton conjectures. So the main problem in moonshine was to show that the monster modules constructed by Frenkel, Lepowsky and Meurman on the one hand, and by Atkin, Fong, and Smith on the other hand, were in fact the same representation of the monster.

Peter Goddard [Go] has given a description of the proof of this in his talk in this volume, so I will only give a quick sketch of this. The main steps of the proof are as follows:

- 1. The module V constructed by Frenkel, Lepowsky, and Meurman has an algebraic structure making it into a "vertex algebra". A detailed proof of this is given in [?].
- 2. Use the vertex algebra structure on V and the Goddard-Thorn no-ghost theorem [?] from string theory to construct a Lie algebra acted on by the monster, called the monster Lie algebra.
- 3. The monster Lie algebra is a "generalized Kac-Moody algebra" ([K90]); use the (twisted) Weyl-Kac denominator formula to show that $T_g(\tau)$ is a "completely replicable function".
- 4. Y. Martin [M], C. Cummins, and T. Gannon [C-G] proved several theorems showing that completely replicable functions were modular functions of Hauptmoduls for genus 0 groups. By using these theorems it follows that T_g is a Hauptmodul for a genus 0 subgroup of $SL_2(\mathbf{Z})$, and hence V satisfies the moonshine conjectures. (The original proof used an earlier result by Koike [Ko] showing that the appropriate Hauptmoduls were completely replicable, together with a boring case by case check and the fact that a completely replicable function is characterized by its first few coefficients.)

We will now give a brief description of some of the terms above, starting with vertex algebras. The best reference for finding out more about vertex algebras is Kac's book [Ko]. In this paragraph we give a rather vague description. Suppose that V is a commutative ring acted on by a group G. We can form expressions like

$$u(x)v(y)w(z)$$

where $u, v, w \in V$ and $x, y, z \in G$, and the action of $x \in G$ on $u \in V$ is denoted rather confusingly by $u(x)$. (This is not a misprint for $x(u)$; the reason for this strange notation is to make the formulas compatible with those in quantum field theory, where u would be a quantum field and x a point of space-time.) For each fixed $u, v, \ldots \in V$, we can think of $u(x)v(y)\cdots$ as a function from G^n to V. We can rewrite the axioms for a commutative ring acted on by G in terms of these functions. We can now think of a vertex algebra roughly as follows: we are given lots of functions from G^n to V satisfying the axioms mentioned above, with the difference that these functions are allowed to have certain sorts of singularities. In other words a vertex algebra is a sort of commutative ring acted on by a group G, except that the multiplication is not defined everywhere but has singularities. In particular we cannot recover an underlying ring by defining the product of u and v to be $u(0)v(0)$, because the function $u(x)v(y)$ might happen to have a singularity at $u = v = 0$.

It is easy to write down examples of vertex algebras: any commutative ring acted on by a group G is an example. (Actually this is not quite correct: for technical reasons we should use a formal group G instead of a group G.) Conversely any vertex algebra "without singularities" can be constructed in this way. Unfortunately there are no easy examples of vertex algebras that are not really commutative rings. One reason for this is that nontrivial vertex algebras must be infinite dimensional; the point is that if a vertex algebra has a nontrivial singularity, then by differentiating it we can make the singularity worse and worse, so we must have an infinite dimensional space of singularities. This is only possible if the vertex algebra is infinite dimensional. However there are plenty of important infinite dimensional examples; see for example Kac's book for a construction of the most important examples, and [?] for a construction of the monster vertex algebra.

Next we give a brief description of generalized Kac-Moody algebras. The best way to think of these is as infinite dimensional Lie algebras which have most of the good properties of finite dimensional reductive Lie algebras. Consider a typical finite dimensional reductive Lie algebra G, (for example the Lie algebra $G = M_n(\mathbf{R})$ of $n \times n$ real matrices). This has the following properties:

- 1. G has an invariant symmetric bilinear form $(,)$ (for example $(a, b) = -Tr(a, b)$).
- 2. G has a (Cartan) involution ω (for example, $\omega(a) = -a^t$).
- 3. G is graded as $G = \oplus_{n \in \mathbf{Z}} G_n$ with G_n finite dimensional and with ω acting as -1 on the "Cartan subalgebra" G_0. (For example, we could put the basis element $e_{i,j}$ of $M_n(\mathbf{R})$ in G_{i-j}.)

- 4. $(a, \omega(a)) > 0$ if $g \in G_n$, $g \neq 0$.

Conversely any Lie algebra satisfying the conditions above is essentially a sum of finite dimensional and affine Lie algebras. Generalized Kac-Moody algebras are defined by the same conditions with one small change: we replace condition 4 by

- 4'. $(a, \omega(a)) > 0$ if $g \in G_n$, $g \neq 0$ and $n \neq 0$.

This has the effect of allowing an enormous number of new examples, such as all Kac-Moody algebras and the Heisenberg Lie algebra (which behaves like a sort of degenerate affine Lie algebra). Generalized Kac-Moody algebras have many of the properties of finite dimensional semisimple Lie algebras, and in particular they have an analogue of the Weyl character formula for some of their representations, and an analogue of the Weyl denominator formula. An example of the Weyl-Kac denominator formula for the algebra $G = SL_2[z, z^{-1}]$ is

$$\prod_{n>0}(1-q^{2n})(1-q^{2n-1}z)(1-q^{2n-1}z^{-1}) = \sum_{n\in\mathbf{Z}}(-1)^n q^{n^2} z^n.$$

This is the Jacobi triple product identity, and is also the Macdonald identity for the affine Lie root system corresponding to A_1.

Dyson described Macdonald's discovery of the Macdonald identities in [D]. Dyson found identities for $\eta(\tau)^m = q^{m/24}\prod_{n>0}(1-q^n)^m$ for the following values of m:

$$3, 8, 10, 14, 15, 21, 24, 26, 28, \ldots$$

and wondered where this strange sequence of numbers came from. (The case $m = 3$ is just the Jacobi triple product identity with $z = 1$.) Macdonald found his identities corresponding to affine root systems, which gave an explanation for the sequence above: with one exception, the numbers are the dimensions of simple finite dimensional complex Lie algebras. The exception is the number 26 (found by Atkin), which as far as I know has not been explained in terms of Lie algebras. It seems possible that it is somehow related to the fake monster Lie algebra and the special dimension 26 in string theory.

Next we give a quick explanation of "completely replicable" functions. A function is called completely replicable if its coefficients satisfy certain relations. As an example of a completely replicable function, we will look at the elliptic modular function $j(\tau) - 744 = \sum c(n)q^n$. This satisfies the identity

$$j(\sigma) - j(\tau) = p^{-1}\prod_{\substack{m>0\\ n\in\mathbf{Z}}}(1-p^m q^n)^{c(mn)},$$

where $p = e^{2\pi i\sigma}$, $q = e^{2\pi i\tau}$. (This formula was proved independently in the 80's by Koike, Norton, and Zagier, none of whom seem to have published their proofs.) Comparing coefficients of $p^m q^n$ on both sides gives many relations between the coefficients of j whenever we have a solution of $m_1 n_1 = m_2 n_2$ in positive integers, which are more or less the relations needed to show that j is completely replicable. For example, from the relation $2 \times 2 = 1 \times 4$ we get the relation

$$c(4) = c(3) + \frac{c(1)^2 - c(1)}{2}$$

or equivalently

$$20245856256 = 864299970 + \frac{196884^2 - 196884}{2}.$$

In the rest of this paper we will discuss various extensions of the original moonshine conjectures, some of which are still unproved. The first are Norton's "generalized moonshine" conjectures [N]. If we look at the Hauptmodul $T_{2A}(\tau) = q^{-1} + 4372q + \cdots$ we notice that one of the coefficients is almost the same as the dimension 4371 of the smallest non-trivial irreducible representation of the baby monster simple group, and the centralizer of an element of type 2A in the monster is a double cover of the baby monster. Similar things happen for other elements of the monster, suggesting that for each element g of the monster there should be some sort of graded moonshine module $V_g = \oplus_n V_{g,n}$ acted on by a central extension of the centralizer $Z_M(g)$. In particular we would get series $T_{g,h}(\tau) = \sum_n Tr(h|V_{g,n})q^n$ satisfying certain conditions. Some progress has been made on this by Dong, Li, and Mason [?], who proved the generalized moonshine conjectures in the case when g and h generate a cyclic group by reducing to the case when $g = 1$ (the ordinary moonshine conjectures). G. Höhn [H] has made some progress in the harder case when g and h do not generate a cyclic group by constructing the required modules for the baby monster (when g is of type $2A$). It seems likely that his methods would also work for the Fischer group Fi_{24}, but it is not clear how to go further than this. There might be some relation to elliptic cohomology (see [Hi] for more discussion of this), as this also involves pairs of commuting elements in a finite group and modular forms.

The space V_g mentioned above does not always have an invariant vertex algebra structure on it. Ryba discovered that a vertex algebra structure sometimes magically reappears when we reduce V_g modulo the prime p equal to the order of g. In fact V_g/pV_g can often be described as the Tate cohomology group $\hat{H}^0(g, V)$ for a suitable integral form V of the monster vertex algebra. This gives natural examples of vertex algebras over finite fields which do not lift naturally to characteristic 0. (Note that most books and papers on vertex algebras make the assumption that we work over a field of characteristic 0; this assumption is often unnecessary and excludes many interesting examples such as the one above.)

We will finish by describing some more of McKay's observations about the monster, which so far are completely unexplained. The monster has 9 conjugacy classes of elements that can be written as the product of two involutions of type $2A$, and their orders are 1, 2, 3, 4, 5, 6, 2, 3, 4. McKay pointed out that these are exactly the numbers appearing on an affine E_8 Dynkin diagram giving the linear relation between the simple roots. They are also the degrees of the irreducible representations of the binary icosahedral group. A similar thing happens for the baby monster: this time there are 5 classes of elements that are the product of two involutions of type $2A$ and their orders are 2, 4, 3, 2, 1. (This is connected with the fact that the baby monster is a "3,4-transposition group".) These are the numbers on an affine F_4 Dynkin diagram, and if we take the "double cover" of an F_4 Dynkin diagram we get an E_7 Dynkin diagram. The number on an E_7 Dynkin diagram are 1, 1, 2, 2, 3, 3, 4, 2 which are the dimensions of the irreducible representations of the binary octahedral group. The double cover of the baby monster is the centralizer of an element of order 2 in the monster. Finally a similar thing happens for $Fi_{24}.2$: this time there are 3 classes of elements that are the product of two involutions of type $2A$ and their orders are 2, 3, 1. (This is connected with the fact that $F_{24}.2$ is a "3-transposition group".) These are the numbers on an affine G_2 Dynkin diagram, and if we take the "triple cover" of an G_2 Dynkin diagram we get an E_6 Dynkin diagram. The number on an E_6 Dynkin diagram are 1, 1, 1, 2, 2, 2, 3, which are the dimensions of the irreducible representations of the binary tetrahedral group. The triple cover of $Fi_{24}.2$ is the centralizer of an element of order 3 in the monster.

The connection between Dynkin diagrams and 3-dimensional rotation groups is well understood (and is called the McKay correspondence), but there is no known explanation for the connection with the monster.

References

[B94] R. E. Borcherds, *Simple groups and string theory*, First European congress of mathematics, Paris July 1992, Ed. A. Joseph and others, Vol. 1, p. 411–421, Birkhauser 1994.

[B98] R. E. Borcherds, *Automorphic forms and Lie algebras*, Current developments in mathematics 1996, to be published by international press, 1998. Also available from www.dpmms.cam.ac.uk/~reb.

[CG] C. J. Cummins and T. Gannon, *Modular equations and the genus zero property of moonshine functions*, Invent. Math. 129 (1997), no. 3, 413–443.

[CN] J. H. Conway and S. Norton, *Monstrous moonshine*, Bull. London. Math. Soc. 11 (1979) 308–339.

[D] F. J. Dyson, *Missed opportunities*, Bull. Amer. Math. Soc. 78 (1972), 635–652.

[DLM] C. Dong, H. Li and G. Mason, *Modular invariance of trace functions in orbifold theory*, preprint q-alg/9703016.

[FLM] I. B. Frenkel, J. Lepowsky and A. Meurman, *Vertex operator algebras and the monster*, Academic press 1988. (Also see the announcement *A natural representation of the Fischer-Griess monster with the modular function J as character*, Proc. Natl. Acad. Sci. USA 81 (1984), 3256–3260.)

[Ge] R. W. Gebert, *Introduction to vertex algebras, Borcherds algebras and the monster Lie algebra*, Internat. J. Modern Physics A8 (1993), no 31, 5441–5503.

[Go] *The Work of R.E. Borcherds*, preprint math/9808136.

[GT] P. Goddard and C. B. Thorn, *Compatibility of the dual Pomeron with unitarity and the absence of ghosts in the dual resonance model*, Phys. Lett., B 40, No. 2 (1972), 235–238.

[G] R. L. Griess, *The friendly giant*, Invent. Math. 69 (1982), no. 1, 1–102.

[Hi] F. Hirzebruch, T. Berger and R. Jung, *Manifolds and modular forms*, Aspects of Mathematics, E20. Friedr. Vieweg & Sohn, Braunschweig, 1992. ISBN: 3-528-06414-5.

[H] G. Höhn, *Selbstduale Vertexoperatorsuperalgebren und das Babymonster* (Self-dual vertex-operator superalgebras and the Baby Monster), Dissertation, Rheinische Friedrich-Wilhelms-Universität Bonn, Bonn, 1995. Bonner Mathematische Schriften 286. Universität Bonn, Mathematisches Institut, Bonn, 1996.

[J] E. Jurisich, *Generalized Kac-Moody Lie algebras, free Lie algebras and the structure of the Monster Lie algebra*, J. Pure Appl. Algebra 126 (1998), no. 1-3, 233–266.

[K90] V. G. Kac, *Infinite dimensional Lie algebras*, third edition, Cambridge University Press, 1990. (The first and second editions (Birkhauser, Basel, 1983, and C.U.P., 1985) do not contain the material on generalized Kac-Moody algebras.)

[Ko] M. Koike, *On Replication Formula and Hecke Operators*, Nagoya University preprint.

[LZ] B. Lian and G. Zuckerman, *New perspectives on the BRST-algebraic structure in string theory*, hepth/921107, Communications in Mathematical Physics 154, (1993) 613–64, and Moonshine cohomology, q-alg/ 950101 Finite groups and Vertex Operator Algebras, RIMS publication (1995) 87–11.

[M] Y. Martin, *On modular invariance of completely replicable functions*, in: *Moonshine, the Monster, and related topics* (South Hadley, MA, 1994), 263–286, Contemp. Math., 193, Amer. Math. Soc., Providence, RI, 1996.

[N] S. P. Norton, Appendix to G. Mason's paper in *The Arcata Conference on Representations of Finite Groups* (Arcata, Calif., 1986), 181–210, Proc. Sympos. Pure Math., 47, Part 1, Amer. Math. Soc., Providence, RI, 1987.

Reprinted from Proc. Int. Congr. Math., 1998

THE WORK OF WILLIAM TIMOTHY GOWERS

by

BÉLA BOLLOBÁS

Dept. of Math. Sciences
University of Memphis
Memphis TN 38152, USA
and
Trinity College, Cambridge CB2 1TQ
England

It gives me great pleasure to report on the beautiful mathematics of William Timothy Gowers that earned him a Fields Medal at ICM'98.

Gowers has made spectacular contributions to the theory of Banach spaces, pure combinatorics, and combinatorial number theory. His hallmark is his exceptional ability to attack difficult and fundamental problems the *right* way: a way that with hindsight is very natural but a priori is novel and extremely daring.

In functional analysis Gowers has solved many of the best-known and most important problems, several of which originated with Banach in the early 1930s. The shock-waves from these results will reverberate for many years to come, and will dramatically change the theory of Banach spaces. The great success of Gowers is due to his exceptional talent for combining techniques of analysis with involved and ingenious combinatorial arguments.

In combinatorics, Gowers has made fundamental contributions to the study of randomness: his tower type lower bound for Szemerédi's lemma is a tour de force. In combinatorial number theory, he has worked on the notoriously difficult problem of finding arithmetic progressions in sparse sets of integers. The ultimate aim is to prove Szemerédi's theorem with the optimal bound on the density that suffices to ensure long arithmetic progressions. Gowers proved a deep result for progressions of length four, thereby hugely improving the previous bound. The difficult and beautiful proof, which greatly extends Roth's argument, and makes clever use of Freiman's theorem, amply demonstrates Gowers' amazing mathematical power.

1. Banach Spaces

A major aim of functional analysis is to understand the connection between the geometry of a Banach space X and the algebra $\mathcal{L}(X)$ of bounded linear operators from the space X into itself. In particular, what conditions imply that a space X contains 'nice' subspaces, and that $\mathcal{L}(X)$ has a rich structure?

In order to start this global project, over the past sixty years numerous major concrete questions had to be answered. As Hilbert said almost one hundred years ago, "Wie überhaupt jedes menschliche Unternehmen Ziele verfolgt, so braucht die mathematische Forschung Probleme. Durch die Lösung von Problemen stählt sich die Kraft des Forschers; er findet neue Methoden und Ausblicke, er gewinnt einen weiteren und freieren Horizont."

In this spirit, the theory of Banach spaces has been driven by a handful of fundamental problems, like the basis problem, the unconditional basic sequence problem, Banach's hyperplane problem, the invariant subspace problem, the distortion problem, and the Schröder-Bernstein problem. For over half a century, progress with these major problems had been very slow: it is due to Gowers more than to anybody else that a few years ago the floodgates opened, and with the solutions of many of these problems the subject now has a 'spacious, free horizon'.

If a space (infinite-dimensional separable Banch space) X can be represented as a sequence space then an operator $T \in \mathcal{L}(X)$ is simply given by an infinite matrix, so it is desirable to find a basis of the space. A *Schauder basis* or simply *basis* of a space X is a sequence $(e_n)_{n=1}^{\infty} \subset X$ such that every vector $x \in X$ has a *unique* representation as a norm-convergent sum $x = \sum_{n=1}^{\infty} a_n e_n$. In 1973, solving a forty year old problem, Enflo [4] proved that not every separable Banach space has a basis, so our operators cannot always be given in this siple way. On the other hand, it is almost trivial that every Banach space contains a *basic sequence*: a sequence $(x_n)_{n=1}^{\infty}$ that is a basis of its closed linear span.

The relationship between an operator $T \in \mathcal{L}(X)$ and closed subspaces of X can also be very involved. In the 1980s Enflo [5] and Read [22] solved in the negative the invariant subspace problem for Banach spaces, and a little later Read [23] showed that this phenomenon can arise on a 'nice' space as well: he constructed a bounded linear operator on ℓ_1 that has only trivial invariant subspaces.

Although a basis $(e_n)_{n=1}^{\infty}$ of a space X leads to a representation of the operators on X as matrices, it does not guarantee that $\mathcal{L}(X)$ has a rich structure. For example, it does not guarantee that $\mathcal{L}(X)$ contains many non-trivial projections. Thus, if $x = \sum_{n=1}^{\infty} a_n e_n$ and $\epsilon_n = 0, 1$, then $\sum_{n=1}^{\infty} \epsilon_n a_n e_n$ need not even converge. Similarly, a permutation of a basis need not be a basis, and if $\sum_{n=1}^{\infty} a_n e_n$ is convergent and $\pi : \mathbb{N} \to \mathbb{N}$ is a permutation then $\sum_{n=1}^{\infty} a_{\pi(n)} e_{\pi(n)}$ need not converge. A basis is said to be *unconditional* if it *does* have these very pleasant properties; equivalently, a basis $(e_n)_{n=1}^{\infty}$ is unconditional if there is a constant $C > 0$ such that, if $(a_n)_{n=1}^{m}$

and $(\lambda_n)_{n=1}^m$ are scalar sequences with $|\lambda_n| \leq 1$ for all n, then

$$\left\| \sum_{n=1}^{m} \lambda_n a_n e_n \right\| \leq C \left\| \sum_{n=1}^{m} a_n e_n \right\|.$$

Also, a sequence $(x_n)_{n=1}^\infty$ is an *unconditional basic sequence* if it is an unconditional basis of its closed linear span. The standard bases of c_0 and ℓ_p, $1 \leq p < \infty$, are all unconditional (and symmetric).

An unconditional basis guarantees much more structure than a basis, so it is not surprising that even classical spaces like $C([0,1])$ and L_1 fail to have unconditional bases. However, the fundamental question of whether every space has a subspace with an unconditional basis (or, equivalently, whether every space contains an unconditional basic sequence) was open for many years, even after Enflo's result.

The search for a subspace with an unconditional basis is closely related to the search for other 'nice' subspaces. For example, it is trivial that not every space contains a Hilbert space, but it is far from clear whether every space contains c_0 or ℓ_p for some $1 \leq p < \infty$. Indeed, this question was answered only in 1974, when Tsirelson [28] constructed a counterexample by a clever inductive procedure. This development greatly enhanced the prominence of the unconditional basic sequence problem.

The breakthrough came in the summer of 1991, when Gowers and Maurey [17] independently constructed spaces without unconditional basic sequences. As the constructions and proofs were almost identical, they joined forces to simplify the proofs and to exploit the consequences of the result. The Gowers-Maurey space X_{GM} is based on a construction of Schlumprecht [25] that eventually enabled Odell and Schlumprecht [21] to solve the famous *distortion problem.* Odell and Schlumprecht constructed a space isomorphic to ℓ_2 that contains no subspace almost isometric to ℓ_2. The main difficulty Gowers and Maurey had to overcome in order to make use of Schlumprecht's space X_S was that X_S itself had an unconditional basis.

Johnson observed that the proofs could be modified to show that the Gowers-Maurey space not only has no unconditional basic sequence, but it does not even have a *decomposable subspace* either: so subspace of X_{GM} can be written as a topological direct sum of two (infinite-dimensional) subspaces. Thus the space X_{GM} is not only the first example of a *non-decomposable* infinite-dimensional space, but it is also *hereditarily indecomposable.* Equivalently, every closed subspace Y of X_{GM} is such that every projection in $\mathcal{L}(Y)$ is essentially trivial: either its rank or its corank is finite. To appreciate how exotic a hereditarily indecomposable space is, note that a space X is hereditarily indecomposable if and only if the distance between the unit spheres of any two infinite-dimensional subspaces is 0: if Y and Z are infinite-dimensional subspaces then

$$\inf\{\|y - z\| : \ y \in Y, \ z \in Z, \ \|y\| = \|z\| = 1\} = 0.$$

In fact, Gowers and Maurey [16] showed that if X is a complex hereditarily indecomposable space then the algebra $\mathcal{L}(X)$ is rather small. An operator $S \in \mathcal{L}(X)$ is said to be *strictly singular* if there is no subspace $Y \subset X$ such that the restriction of S to Y is an isomorphism. Equivalently, $S \in \mathcal{L}(X)$ is strictly singular if for every (infinite-dimensional) subspace $Y \subset X$ and every $\epsilon > 0$ there is a vector $y \in Y$ with $\|S_y\| < \epsilon\|y\|$.

Theorem 1. *Let X be a complex hereditarily indecomposable space. Then every operator $T \in \mathcal{L}(X)$ is a linear combination of the identity and a strictly singular operator.*

Gowers [9] was the first to solve *Banach's hyperplane problem* when he constructed a space with an unconditional basis that is not isomorphic to any of its hyperplanes or even proper subspaces. The theorem above implies that every complex hereditarily indecomposable space answers Banach's hyperplane problem since it is not isomorphic to *any* of its proper subspaces. In fact, Ferenczi [7] showed that a complex Banach space X is hereditarily indecomposable if and only if for every subspace $Y \subset X$, every bounded linear operator from Y into X is a linear combination of the inclusion map and a strictly singular operator. Recently, Argyros and Felouzis [1] showed that every Banach space contains either ℓ_1 or a subspace that is a quotient of a hereditarily indecomposable space.

It was not by chance that in order to construct a space without an unconditional basis, Gowers and Maurey constructed a hereditarily indecomposable space. As shown by the following stunning *dichotomy theorem* of Gowers [12], having an unconditional basis or being hereditarily indecomposable are the only two 'pure states' for a space.

Theorem 2. *Every infinite-dimensional Banach space contains an infinite-dimensional subspace that either has an unconditional basis or is hereditarily indecomposable.*

Gowers based his proof of the dichotomy theorem on a combinatorial game played on sequences and subspaces. In order to describe this game, we need some definitions. Given a space X with a basis $(e_n)_{n=1}^\infty$, the *support* of a vector $a = \sum_{n=1}^\infty a_n e_n \in X$ is $\text{supp}(a) = \{n : a_n \neq 0\}$. A vector $a = \sum_{n=1}^\infty a_n e_n$ *precedes* a vector $b = \sum_{n=1}^\infty b_n e_n$ if $n < m$ for all $n \in \text{supp}(a)$ and $m \in \text{supp}(b)$. A *block basis* is a sequence $x_1 < x_2 < \dots$ of non-zero vectors, and a *block subspace* is the closed linear span of a block basis. For a subspace $Y \subset X$, write $\sum(Y)$ for the set of all sequences $(x_i)_1^n$ of non-zero vectors of norm at most 1 in Y with $x_1 < \cdots < x_n$. Call a set $\sigma \subset \sum(X)$ *large* if $\sigma \cap \sum(Y) \neq \emptyset$ for every (infinite-dimensional) block subspace Y. For a set $\sigma \subset \sum(X)$ and a sequence $\Delta = (\delta_i)_{i=1}^\infty$ of positive reals, the *enlargement of σ by Δ* is

$$\sigma_\Delta = \left\{(x_i)_1^n \in \sum(X) : \ \|x_i - y_i\| < \delta_i, \ 1 \leq i \leq n, \text{ for some } (y_i)_1^n \in \sigma\right\}.$$

And now for the two-player game (σ, Y) defined by a set $\sigma \subset \sum(X)$ and a block subspace $Y \subset X$. The first player, *Hider*, chooses a block subspace $Y_1 \subset Y$; the second player, *Seeker*, replies by picking a finitely supported vector $y_1 \in Y_1$. Then Hider chooses a block subspace $Y_2 \subset Y$, and Seeker picks a finitely supported vector $y_2 \in Y_2$. Proceeding in this way, Seeker wins the (σ, Y)-game if, at any stage, the sequence $(y_i)_1^n$ is in σ. Hider wins if he manages to make the game go on for ever. Clearly, Seeker has a winning strategy for the (σ, Y) game if σ is big when measured by Y.

The combinatorial foundation of Gowers' dichotomy theorem is then the following result [12].

Theorem 3. *Let X be a Banach space with a basis and let $\sigma \subset \sum(X)$ be large. Then for every positive sequence Δ there is a block subspace $Y \subset X$ such that Seeker has a winning strategy for the (σ_Δ, Y)-game.*

The beautiful proof of this result bears some resemblence to arguments of Galvin and Prikry [8] and Ellentuck [3] concerning Ramsey-type results for sequences.

Gowers' dichotomy theorem has been the starting point of much new research on Banach spaces. For example, it can be used to tackle the still open problem of classifying minimal Banach spaces. A Banach space is *minimal* if it embeds into all of its infinite-dimensional subspaces. Casazza et al [2] used the dichotomy theorem to show that every minimal Banach space embeds into a minimal Banach space with an unconditional basis. Hence, a minimal space is either reflexive or embeds into c_0 or ℓ_1.

The *Schröder-Bernstein problem* asks whether two Banach spaces are necessarily isomorphic if each is a complemented subspace of the other. In [13] Gowers gaver the first counterexample, and later with Maurey [16] constucted the following further examples with even stronger paradoxical properties.

Theorem 4. *For every $n \geq 1$ there is a Banach space X_n such that two finite-codimensional subspaces of X_n are isomorphic if and only if they have the same codimension modulo n. Also, there is a Banach space Z_n such that two product spaces Z_n^r and Z_n^s are isomorphic if and only if r and s are equal modulo n.*

For $n \geq 2$, the space Z_n can be used to solve the Schröder-Bernstein problem; even more, with $X = Z_3$ and $Y = Z_3 \oplus Z_3$ we have $Y \oplus Y = Z_3^4 \cong Z_3 = X$. Thus not only are X and Y complemented subspaces of each other, but $X \cong Y \oplus Y$ and $Y \cong X \oplus X$. However, $X = Z_3$ and $Y = Z_3^2$ are not isomorphic.

The last result we shall discuss here is Gowers' solution of Banach's homogeneous spaces problem. A space is *homogeneous* if it is isomorphic to all of its subspaces. Banach asked whether there were any examples other than ℓ_2. Gowers proved the striking result that homogeneity, in fact, characterizes Hilbert space [12].

Theorem 5. *The Hilbert space ℓ_2 is the only homogeneous space.*

To prove this, Gowers could make use of results of Szankowski [25], and Komorowski and Tomczak-Jaegermann [19] that imply that a homogeneous space with an unconditional basis is isomorphic to ℓ_2. What happens if X is homogeneous but does not have an unconditional basis? By the dichotomy theorem, X has a subspace Y that either has an unconditional basis or is hereditarily indecomposable. Since $X \cong Y$ and X does not have an unconditional basis, Y is hereditarily indecomposable. But this is impossible, since a hereditarily indecomposable space is not isomorphic to any of its proper subspaces, let alone all of them!

2. Arithmetic Progressions

In 1936 Erdős and Turán [6] conjectured that, for every positive integer k and $\delta > 0$, there is an integer N such that every subset of $\{1, \ldots, N\}$ of size at least δN numbers contains an arithmetic progression of length k. In 1953 Roth [24] used exponential sums to prove the conjecture in the special case $k = 3$: this was one of the results Davenport highlighted in 1958 when Roth was awarded a Fields Medal. In 1969 Szemerédi found an entirely combinatorial proof for the case $k = 4$, and six years later he proved the full Erdős-Turán conjecture. Szemerédi's theorem trivially implies van der Waerden's theorem.

In 1977 Fürstenberg [7] used techniques of ergodic theory to prove not only the full theorem of Szemerédi, but also a number of substantial extensions of it. This proof revolutionized ergodic theory.

In spite of these beautiful results, there is still much work to be done on the Erdős-Turán problem. Write $f(k, \delta)$ for the minimal value of N that will do in Szemerédi's theorem. The proofs of Szemerédi and Fürstenberg give extremely weak bounds for $f(k, \delta)$, even in the case $k = 4$. In order to improve these bounds, and to make it possible to attack some considerable extensions of Szemerédi's theorem, it would be desirable to use exponential sums to prove the general case.

Recently, Gowers [15] set out to do exactly this. He introduced a new notion of *pseudorahdomness*, called *quadratic uniformity* and, using techniques of harmonic analysis, showed that a quadratically uniform set contains about the expected number of arithmetic progressions of length four. In order to find arithmetic progressions in a set that is not quadratically uniform, Gowers avoided the use of Szemerédi's uniformity lemma or van der Waerden's theorem, and instead made use of Weyl's inequality and, more importantly, Freiman's theorem. This theorem states that if for some finite set $A \subset \mathbb{Z}$ the sum $A + A = \{a + b : a, b \in A\}$ is not much larger than A then A is not far from a generalized arithmetic progression. By ingenious and involved arguments Gowers proved the following result [14].

Theorem 6. *There is an absolute constant C such that*

$$f(4, \delta) \leq \exp \exp \exp((1/\delta)^C).$$

In other words, if $A \subset \{1, \ldots, N\}$ has size at least $|A| = \delta N > 0$ and $N \geq \exp\exp\exp((1/\delta)^C)$, then A contains an arithmetic progression of length 4.

The bound in this theorem is imcomparably better than the previous best bounds.

The entirely new approach of Gowers raises the hope that one could prove the full theorem of Szemerédi with good bounds on $f(k, \delta)$. In fact, there is even hope that Gowers' method could lead to a proof of the Erdős conjecture that if $A \subset \mathbb{N}$ is such that $\sum_{a \in A} 1/a = \infty$ then A contains arbitrarily long arithmetic progressions. The most famous special case of this conjecture is that the primes contain arbitrarily long arithmetic progressions.

3. Combinatorics

The basis of Szemerédi's original proof of his theorem on arithmetic progressions was a deep lemma that has become an extremely important tool in the study of the structure of graphs. This, result, *Szemerédi's uniformity lemma*, states that the vertex set of every graph can be partitioned into boundedly many pieces $V_1, \ldots, V_k$ such that 'most' pairs (V_i, V_j) are 'uniform'. In order to state this lemma precisely, recall that, for a graph $G = (V, E)$, and sets U, $W \subset V$, the *density* $d(U, W)$ is the proportion of the elements (u, w) of $U \times W$ such that uw is an edge of G. For ϵ, $\delta > 0$ a pair (U, W) is called (ϵ, δ)-*uniform* if for any $U' \subset U$ and $W' \subset W$ with $|U'| \geq \delta|U|$ and $|W'| \geq \delta|W|$, the densities $d(U', W')$ and $d(U, W)$ differ by at most $\epsilon/2$.

Szemerédi's uniformity lemma [27] claims that for all ϵ, δ, $\eta > 0$ there is a $K = K(\epsilon, \delta, \eta)$ such that the vertex set of any graph G can be partitioned into at most K sets $U_1, \ldots, U_k$ of sizes differing by at most 1, such that at least $(1 - \eta)k^2$ of the pairs (U_i, U_j) are (ϵ, δ)-uniform.

Loosely speaking, a 'Szemerédi partition' $V(G) = \bigcup_{i=1}^k U_i$ is one such that for most pairs (U_i, U_j) there are constants α_{ij} such that if $U_i' \subset U_i$ and $U_j' \subset U_j$ are not too small then G contains about $\alpha_{ij}|U_i'||U_j'|$ edges from U_i' to U_j'. In some sense, Szemerédi's uniformity lemma gives a classification of all graphs. The main drawback of the lemma is that the bound $K(\epsilon, \delta, \eta)$ is extremely large: in the case $\epsilon = \delta = \eta$, all we know about $K(\epsilon, \epsilon, \epsilon)$ is that it is at most a tower of 2*s* of height proportional to ϵ^{-5}. This is an enormous bound, and in many applications a smaller bound, say of the type $e^{\epsilon^{-100}}$ would be significantly more useful. As the lemma is rather easy to prove, it was not unreasonable to expect a bound like this.

It was a great surprise when Gowers [14] proved the deep result that $K(\epsilon, \delta, \eta)$ is of tower type in $1/\delta$, even if ϵ and η are kept large.

Theorem 7. *There are constants c_0, $\delta_0 > 0$ such that for $0 < \delta < \delta_0$ there is a graph G that does not have a $(1/2, \delta, 1/2)$-uniform partition into K sets, where K is a tower of 2s of height at most $c_0\delta^{-1/16}$.*

It is well known that even exponential lower bounds are hard to come by, let alone tower type lower bounds, so this is a stunning result indeed! The proof, which makes use of clever random choices to construct graphs whose small sets of vertices do *not* behave like subsets of random graphs, goes some way towards clarifying the nature of randomness. It also indicates that any proof of an upper bound for $K(\epsilon, \delta, \eta)$ must involve a long sequence of refinements of partitions, each exponentially larger than the previous one.

This sketch has been all too brief, and a deeper study of Gowers' work would be needed to properly appreciate his clarity of thought and mastery of elaborate structures. However, I hope that enough has been said to give some taste of his remarkable mathematical achievements. In the theory of Banach spaces, not only has he solved many of *the* main classical problems of the century, but he has also opened up exciting new directions. In combinatorics, too, he has tackled some of the most notorious questions, bringing about their solution with the same exceptional blend of combinatorial power and technical skill. Hilbert would surely agree that Gowers has given us wider and freer horizons.

References

[1] S. A. Argyros and V. Felouzis, Interpolating H. I. Banach spaces, to appear.

[2] P. G. Casazza, N. J. Kalton, D. Kutzerova and M. Mastyło, Complex interpolation and complementably minimal spaces, in *Ineraction between Functional Analysis, Harmonic Analysis, and Probability*, Lecture Notes in Pure and Applied Math., Dekker, N.Y., 1996, pp. 135–143.

[3] E. Ellentuck, A new proof that analytic sets are Ramsey, *J. Symbolic Logic* 39 (1974), 163–165.

[4] P. Enflo, A counterexample to the approximation property in Banach spaces, *Acta Math.* 130 (1973), 309–317.

[5] P. Enflo, On the invariant-subspace problem in Banach spaces, *Acta Math.* 158 (1987), 213–313.

[6] P. Erdős and P. Turán, On some sequences of integers, *J. London Math. Soc.* 11 (1936), 261–264.

[7] V. Ferenczi, Operators on subspaces of hereditarily indecomposable Banach spaces, *Bull. London Math. Soc.* 29 (1997), 338–344.

[8] F. Galvin and K. Prikry, Borel sets and Ramsey's theorem, *J. Symbolic Logic* 38 (1973), 193–198.

[9] W. T. Gowers, A solution to Banach's hyperplane problem, *Bull. London Math. Soc.* 26 (1994), 523–530.

[10] W. T. Gowers, A Banach space not containing c_0, ℓ_1 or a reflexive subspace, *Trans. Amer. Math. Soc.* 344 (1994), 407–420.

[11] W. T. Gowers, A hereditarily indecomposable space with an asymptotic unconditional basis, *Oper. Theory: Adv. Appl.* 77 (1995), 111–120.

[12] W. T. Gowers, A new dichotomy for Banach spaces, *Geom. Funct. Anal.* 6 (1996), 1083–1093.

[13] W. T. Gowers, A solution to the Schroeder-Bernstein problem for Banach spaces, *Bull. London Math. Soc.* 28 (1996), 297–304.

[14] W. T. Gowers, Lower bounds of tower type for Szemerédi's uniformity lemma, *Geom. Funct. Anal.* 7 (1997), 322–337.

[15] W. T. Gowers, A new proof of Szemerédi's theorem for arithmetic progressions of length four, *Geom. Funct. Anal.* 8 (1998), 529–551.

[16] W. T. Gowers and B. Maurey, The unconditional basis sequence problem, *J. Amer. Math. Soc.* 6 (1993), 851–874.

[17] W. T. Gowers and B. Maurey, Banach spaces with small spaces of operators, *Math. Ann.* 307 (1997), 543–568.

[18] R. C. James, Bases and reflexivity of Banach spaces, *Ann. of Math.* (2) 52 (1950), 518–527.

[19] R. Komorowski and N. Tomczak-Jaegermann, Banach spaces without local unconditional structure, *Israel J. Math.* 89 (1995), 205–226.

[20] J. Lindenstrauss and L. Tzafriri, Classical Banach Spaces I: Sequence Spaces, Springer-Verlag, Berlin and New York 1977.

[21] E. Odell and T. Schlumprecht, The distortion problem, *Acta. Math.* 173 (1994), 259–281.

[22] C. J. Read, A solution to the Invariant Subspace Problem, *Bull. London Math. Soc.* 16 (1984), 337–401.

[23] C. J. Read, A solution to the Invariant Subspace Problem on the space ℓ_1, *Bull. London Math. Soc.* 17 (1985), 305–317.

[24] K. F. Roth, On certain sets of integers, *J. London Math. Soc.* 28 (1953) 245–252.

[25] T. Schlumprecht, An arbitrarily distortable Banach space, *Israel J. Math.* 76 (1991), 81–95.

[26] E. Szemerédi, On sets of integers containing no k elements in arithmetic progression, *Acta Arith.* 27 (1975), 199–245.

[27] E. Szemerédi, Regular partitions of graphs, in "Proc. Colloque Inter. CNRS" (J. C. Bermond, J.-C. Fournier, M. Las Vergnas, D. Sotteau, eds.), 1978, pp. 399–401.

[28] B. S. Tsirelson, Not every Banach space contains ℓ_p or c_0, *Functional Anal. Appl.* 8 (1974), 139–141.

William T. Gowers

AUTOBIOGRAPHY OF WILLIAM TIMOTHY GOWERS

I was born in 1963 in Marlborough, England, but grew up in London. My parents were both musicians - my father a composer and my mother a piano teacher. Music was always an important part of my life as well, and at the age of nine I went to boarding school in Cambridge, where I sang in the King's College Choir. From there I went, with an academic scholarship, to Eton College. Though I enjoyed other subjects, I knew from about the age of 11 or 12 that mathematics was the one I liked best, if only because it involved the least rote-learning, and in my last year at Eton I was selected for the team that represented the UK at the International Mathematical Olympiad in Washington DC. It was at about this point that I began to take seriously the idea that I might become a mathematician, though I still had very little idea what this meant.

From Eton I went to Trinity College Cambridge, where I first met Béla Bollobás, who was later to become my research supervisor. I was fortunate to have, throughout my childhood, a succession of unconventional, enlightened and inspiring mathematics teachers - Mrs Gazzard when I was five or six, Mrs Briggs at King's and Dr Norman Routledge at Eton. All three encouraged intellectual curiosity, independent thought and a love of mathematics. Bollobás was a worthy successor, to put it mildly. I could not live up to his standards as an undergraduate, but I worked harder than I would have for anybody else and had a far more interesting mathematical life, thanks to his "exercises for the enthusiast" (one couldn't exactly go to a supervision and claim not to be enthusiastic) and wonderful lectures.

In my fourth year at Cambridge I took Part III of the Mathematical Tripos, and decided that I wanted to do research under Bollobás in the geometry of Banach spaces. This was partly because a good friend of mine (Imre Leader) had been one of his students for a year and was clearly enjoying it, and partly because Bollobás lectured on the subject that year, and set me a very interesting dissertation topic on it as well (about type and cotype). He agreed to take me on, and turned out to be as good a research supervisor as he had been a director of studies. In particular, he would get his research students to read recent papers and explain them to him and his other research students, in what he called mini-seminars. This gave us a very good training in giving talks (notes absolutely not allowed) and was an efficient way to become acquainted with the literature. Apart from this, however, he did not encourage us to read papers, maintaining that it is better to think hard about a problem for yourself. Of course, this principle must be qualified, as he was well aware, but it has remained a principle of mine and I recommend it to others, at least in the areas of mathematics where I have worked.

My research took a little while to get off the ground, but after about eighteen months I had a fairly hard result, and this result suggested further questions that could be realistically tackled, so it was the basis for a fruitful research project. At the end of my second year I married Emily Thomas, who had been in my year at Trinity as an undergraduate and was a research student in Latin literature (in her third year - we were out of step because of Part III). A day or two after we returned from our honeymoon, she was elected to a research fellowship at Trinity. At the end of my three years I wrote up what I had done for the next year's research fellowship competition, and was also awarded a fellowship.

Trinity operates in an unusual way, in that the successful candidates are formally admitted as research fellows less than 24 hours after they learn that they have been successful. So suddenly I went from being a graduate student, who looked up at the High Table where the fellows, including my wife, were enjoying better food and (one imagined) high-powered conversation, to being on the other side. This experience, coming on the heels of my very happy time as an undergraduate and graduate, implanted in me a love of Trinity, and sense of belonging there, that people who have not been through the Oxbridge system, and even some who have, find hard to understand.

After a little over a year of this life, free of all formal duties, I was invited to apply for a lectureship at University College London. I did, and got it, and a couple of months later, a lectureship came up there in Classics, which went to my wife.

The summer before I was due to leave Cambridge for London, I went to a conference in Jerusalem that was to change my life. During my time as a research fellow, I had thought about an old problem in Banach spaces called the unconditional basic sequence problem, and sketched out an argument for a potential counterexample to a stronger, but still very interesting, statement. In order to build such a counterexample, I needed a space with a strong "distortability" property, but conversations I had had with Ted Odell, of the Univeristy of Texas at Austin, suggested that such a space might well exist. In Jerusalem, Thomas Schlumprecht, who was later to prove a string of spectacular theorems with Odell, gave a talk in which he outlined the very interesting properties of a new space that he had constructed, and these properties were suspiciously similar to what I needed for my approach to the problem. I suddenly felt as though a famous problem was within reach and resolved to devote all my energies to it. Fairly quickly I produced the counterexample to the strengthened statement, and was then faced with serious difficulties in strengthening this example to deal with the original statement. But after about a month I knew I had got there, and after another month or so I had a write-up that was ready to circulate.

Not long after that I received an email message from Bernard Maurey, telling me that he had also been to Schlumprecht's talk, and had also found a counterexample to the unconditional basic sequence problem. He had my preprint, and could tell that our examples were very similar. I was devastated, and desperately tried to take my mind off it by going to watch an afternoon showing of Terminator II. Much later I learned that he had had a similar reaction on hearing of my work, though how he dealt with it I do not know.

However, I soon got used to the situation, and if I had had the benefit of hindsight I would not only not have minded, but would have been positively pleased. Maurey and I wrote up the result jointly, and went on to write another paper, developing the first, which gave counterexamples to several other old questions in the subject. My blood runs a little cold when I wonder how different life might have been if I had taken a month off after the Jerusalem conference, or thought about a different problem for a while, or already had children by 1991. But as it is, I lost nothing from having to share the result, and gained a great deal, as Maurey was a wonderful collaborator. (I would send him vague, ill-formed ideas by email, and within hours he would explain why they didn't work, but how, with the help of an amazingly clever argument, he could prove such and such a modification, which might perhaps have been what I was getting at?)

There were many kindred spirits at UCL and my duties were not particularly onerous, with the result that my time there was one of my most fruitful mathematically. I found a sequence of further counterexamples to Banach-space problems, and eventually also a theorem, which, when combined with a result of Ryszard Komorowski and Nicole Tomczak-Jaegermann, showed that every Banach space that is isomorphic to all its infinite-dimensional subspaces is isomorphic to a Hilbert space. This answered a problem of Banach, and was very good from the public-relations point of view, as one does not want to be labelled as a mere producer of counterexamples, however unfair such an implied criticism may be.

Just before the Christmas of 1992, my first child, John, was born, and life began to get more complicated. In 1993 I received a very flattering invitation from I.H.E.S., to go and visit there for two years. David Larman, my unusually enlightened Head of Department at UCL, was happy to let me go there, but two years seemed a very long time and in the end we decided to go for a shorter visit of a few months.

This was a blissful time, during which John learnt to walk and started to talk. Though I would never want to give up teaching, I certainly enjoyed having an extended period devoted to research (at the time, I was writing up some of my results and trying with absolutely no success to prove that P does not equal NP), not to mention French cheese, wine, bread, weather, architecture, countryside and so on. During this time, I applied for, but did not get, a lectureship back at Cambridge.

When we returned to England, Emily was pregnant again, and Richard was born just after Christmas. I applied again to Cambridge, and this time I was appointed. Better still, there was room for me back at Trinity (university and college appointments are made independently at Cambridge), so in the summer of 1995, I returned there after four very happy years at UCL.

By this time I felt the need of a change, mathematically speaking, and was working on combinatorial problems - as was natural, given that Bollobás is a combinatorialist - and this led, in a rather curious way, to my next main direction of research. I had, and still have, a collection of about half a dozen problems that I very much liked and would come back to every so often. Amongst them was Szemerédi's theorem, which I regarded as a problem because I could not believe that such a simple statement was so hard to prove. In 1996, I made what I thought was genuine progress on the problem by discovering a proof for progressions of length three. I later realized that the proof I had found was essentially the same as Roth's 1953 proof, but by that time I was thinking hard about progressions of length four, and understood what I was doing much better than I would have if I had started by familiarizing myself with all the literature on the subject. Not long afterwards I found a proof for progressions of length four, which generalized easily to the entire theorem, wrote it up, and presented it to the Cambridge combinatorics seminar over two sessions.

My excitement was short-lived, however, because the "proof" I had discovered was seriously flawed. While looking over the write-up I became suspicious of one small lemma, where an inequality seemed to be at the wrong extreme. It gradually became apparent that the lemma was false, as was the higher-level lemma I needed it for, as was the result *that* was needed for. In fact, I found a counterexample to the main statement on which my argument depended.

I did not realize it at the time (and it is a good thing that I didn't), but I had not discovered anything essentially new - part of my argument was still correct, but that part was more or less the same as part of Furstenberg's ergodic-theoretic proof of Szemerédi's theorem. Nevertheless, the counterexample was a suggestive one, and I eventually did manage to find a new proof of the theorem, by showing that all counterexamples to my proposed intermediate statement were of a similar form. But this came only after another two years of extraordinarily hard slog. Looking back on it now, I cannot imagine that I would have been prepared to put in that much work if it had not been for the intense disappointment when the earlier argument collapsed. Since then, I have never minded finding counterexamples to my cherished mathematical beliefs (which seems to happen often).

1997 was memorable for three reasons. Early on in the year I obtained a proof of Szemerédi's theorem for progressions of length four. The two decisive moments in this both felt like accidents. I stumbled on Proposition 3.1 (as presented in the article in this volume) after hours of unsuccessful formal manipulations of a similar kind, and I chanced on a reference to Freiman's theorem while browsing in a book on combinatorial geometry. I had thought about the problem enough to recognise that both these results were highly significant to me, and from then on there was no turning back. Whatever further difficulties I might encounter, it was obvious that there *was* a proof along these lines.

In October, our daughter Madeline was born. After two boys, we were thrilled to have a girl - not for the first time I felt that life was working out almost tastelessly well. And that feeling was only increased when I started to get hints that I was being considered for a Fields medal (nothing explicit - just urgent instructions to write up any results I had, sudden unexplained demands for CVs etc.). Up to that point I had assumed that my area of mathematics was too unfashionable for there to be the slightest chance of this. In January 1998 I had an extraordinary three days in which I was told first that I was to be the next Rouse Ball Professor of Mathematics at Cambridge (a chair first held by Littlewood) and then that, along with my colleague Richard Borcherds, I would indeed be receiving a Fields medal. There followed several months of trying not to think about it and to sound casual when colleagues asked whether I would be going to the ICM in Berlin that summer.

The main difference the Fields medal has made to my life so far is that it led, indirectly, to a visit to Princeton, with Emily visiting the Classics department. This was initially to have been for one year but we extended it to two, towards the end of which a vacancy in Latin literature arose unexpectedly in Cambridge. Emily applied for and got it, so that, at the time of writing, we are once again at the same institution, though Emily decided not to take this too far and went to St John's College rather than Trinity.

Returning to Cambridge a second time was almost like starting a new job, partly because two years was long enough to get thoroughly used to Princeton, a town that feels in many ways like the centre of the mathematical universe, and partly because in my absence the Cambridge mathematics department had moved to brand new, purpose-built buildings. The new Centre for Mathematical Sciences gives the impression of a department that is looking decades into the future. It is difficult to conclude this sketch properly, since I hope that a conclusion is inappropriate and that I will have a part to play in that future.

A NEW PROOF OF SZEMERÉDI'S THEOREM FOR ARITHMETIC PROGRESSIONS OF LENGTH FOUR*

by

W. T. GOWERS

University of Cambridge
Cambridge CB3 OWB, UK

1. Introduction

The starting point for this paper is the following famous theorem of Szemerédi, published in 1975.

Theorem 1.1. For every positive integer k and every real number $\delta > 0$, there exists a positive integer N such that every subset of $\{1, 2, \ldots, N\}$ of cardinality at least δN contains an arithmetic progression of length k.

By the time it was proved, it had become a renowned and long-standing conjecture of Erdős and Turán [ET]. Szemerédi's theorem immediately implies another well-known, but significantly easier theorem, due to van der Waerden.

Theorem 1.2. For every pair of positive integers k and r there exists a positive integer N such that, however the set $\{1, 2, \ldots, N\}$ is partitioned into r sets, one of those sets contains an arithmetic progression of length k.

The first progress towards the theorem was due to Roth [R1], who proved the result in the special case $k = 3$, using exponential sums. Szemerédi later found a different, more combinatorial proof of this case, which he was able to extend to prove the result first for $k = 4$ [S1] and then eventually in the general case [S2]. There was then a further breakthrough due to Furstenberg [Fu], who showed that techniques of ergodic theory could be used to prove many Ramsey-theoretic results, including Szemerédi's theorem and certain extensions of Szemerédi's theorem that were previously unknown.

These results left an obvious avenue unexplored: can Roth's proof for $k = 3$ be generalized to prove the whole theorem? The answer was shown to be yes in the two papers [G2], which dealt with progressions of length four, and [G3], which tackled

*Adapted and modified from *A new proof of Szemerédi's theorem for arithmatic progressions of length four*, Geometric and Functional Analysis **8** (1998) 529–551. ©Birkhäuser 1998 (ISSN 1016-443X)

the general case. The motivation for generalizing Roth's argument is twofold. First, his argument is very natural and beautiful, and it is curious that it should not have an obvious generalization (though there are good reasons for this, as will become clear). Second, the bounds arising from the known proofs of Szemerédi's theorem are very weak, and in general for this sort of problem all the best bounds tend come from the use of exponential sums. For example, Roth shows that when $k = 3$ one can take N to be $\exp\exp(C/\delta)$ for some absolute constant C, which is far better than the bound given by any known combinatorial argument. This estimate has been reduced by Szemerédi [S3] and Heath-Brown [H-B] to $\exp\big((1/\delta)^C\big)$, also using exponential sums. More recently, Bourgain [Bo] has reduced the value of C to $2+\epsilon$.

With our new approach, it is possible to show that there is an absolute constant $c > 0$ such that every subset of $\{1, 2, \ldots, N\}$ of size at least $N(\log\log N)^{-c}$ contains an arithmetic progression of length four. Equivalently, there is an absolute constant C such that any subset of $\{1, 2, \ldots, N\}$ of size at least δN contains an arithmetic progression of length four, as long as $N \geq \exp\exp\big((1/\delta)^C\big)$.

Although a bound of this type may seem weak (and is almost certainly far from best possible) it is nevertheless a significant improvement on what went before. Even to state the earlier bounds needs some effort. Let us define the tower function T inductively by $T(1) = 2$ and $T(n+1) = 2^{T(n)}$. Next, define a function W inductively by $W(1) = 2$ and $W(n+1) = T(W(n))$. The previous best known bound for N has not been carefully calculated, but is at least as bad as $W(1/\delta)$. Even the bounds for van der Waerden's theorem are weak: to show that any r-colouring of $\{1, 2, \ldots, N\}$ gives a monochromatic arithmetic progression of length four, the proofs need N to be at least as large as $T(T(r))$.

These earlier estimates rely on van der Waerden's theorem in its full generality, for which the best known bounds, due to Shelah [Sh], involve functions of the same type as the function W above. An important feature of our proof is that we avoid using van der Waerden's theorem, and also have no need for Szemerédi's uniformity lemma, which is known to require a bound similar to the function T [G1]. Instead, our main tools are a well known consequence of Weyl's inequality and a deep theorem of Freiman.

It should be mentioned that Roth himself did find a proof for $k = 4$ [R2] which used analytic methods, but these were combined with certain combinatorial arguments of Szemerédi and the proof still used van der Waerden's theorem. The argument of this paper is quite different and more purely analytic, which is why it gives a better bound.

2. Quadratically Uniform Sets

In this section, we shall reduce Szemerédi's theorem for progressions of length four to a question that looks somewhat different. The rough idea is to define a notion of pseudorandomness, which we shall call quadratic uniformity, and show that every pseudorandom set, in the appropriate sense, contains about the same number of arithmetic progressions of length four as a random set of the same size. In later sections, we shall then prove that a set which *fails* to be pseudorandom can be restricted to a large arithmetic progression where its density increases noticeably. These two facts then easily imply the result.

In order to define quadratic uniformity, we shall need to introduce some notation. Given a positive integer N, we shall write $\mathbb{Z}_N$ for the group of integers mod N. When N is clear from the context (which will be always) we shall write ω for the number $\exp(2\pi i/N)$. From time to time we shall need elements of $\mathbb{Z}_N$ to have multiplicative inverses, so for convenience we shall take N to be a prime for the rest of the paper. Given any function $f:\mathbb{Z}_N\to\mathbb{C}$, we shall define its rth Fourier coefficient $\hat{f}(r)$ to be $\sum_{s\in\mathbb{Z}_N} f(s)\omega^{-rs}$. It would be more standard to write $\sum_{s\in\mathbb{Z}_N} f(s)e(-rs/N)$, where $e(x)$ is the function $\exp(2\pi ix)$. However, we have found the less standard notation convenient.

The three main properties of the discrete Fourier transform that we shall use are the following, which are discrete forms of Parseval's identity, the convolution identity and the inversion formula respectively.

$$\sum_{r\in\mathbb{Z}_N}|\hat{f}(r)|^2 = N\sum_{s\in\mathbb{Z}_N}|f(s)|^2, \tag{1}$$

$$(f*g)^\wedge(r) = \hat{f}(r)\hat{g}(r) \qquad (r\in\mathbb{Z}_N), \tag{2}$$

$$f(s) = N^{-1}\sum_r \hat{f}(r)\omega^{rs}. \tag{3}$$

We shall also use the simple fact that if f is a real-valued function then $\hat{f}(-r)=\overline{\hat{f}(r)}$ for any r.

There are two classes of functions to which we shall apply Fourier techniques. The first is what we call *balanced* functions associated with subsets $A\subset\mathbb{Z}_N$. Given such a set A, of size δN, we define its balanced function $f=f_A$ by

$$f(s)=\begin{cases}1-\delta & s\in A,\\ -\delta & s\notin A.\end{cases}$$

This is the characteristic function of A minus the constant function $\delta\mathbf{1}$. Note that $\sum_{s\in\mathbb{Z}_N} f_A(s)=\hat{f}_A(0)=0$ and that $\hat{f}_A(r)=\hat{A}(r)$ for $r\neq 0$. (Here, we have identified A with its characteristic function. We shall continue to do this.) The second class of functions that will interest us is functions of the form

$$g(s)=\begin{cases}\omega^{\phi(s)} & s\in B,\\ 0 & s\notin B,\end{cases}$$

where B is a subset of $\mathbb{Z}_N$ and $\phi:B\to\mathbb{Z}_N$.

Another convention we shall adopt from now on is that any sum is over $\mathbb{Z}_N$ if it is not specified as being over another set. The next lemma contains some well known facts about functions on $\mathbb{Z}_N$ with small Fourier coefficients. When we say below that one statement with constant c_i implies another with constant c_j, we mean that the second statement follows from the first provided that $c_j \geq \gamma(c_i)$, for some function γ which tends to zero at zero. In fact, $\gamma(c_i)$ will always be some power of c_i.

Lemma 2.1. *Let f be a function from $\mathbb{Z}_N$ to the unit disc in $\mathbb{C}$. The following are equivalent.*

(i) $\sum_r |\hat{f}(r)|^4 \leq c_1 N^4$.
(ii) $\max_r |\hat{f}(r)| \leq c_2 N$.
(iii) $\sum_k \left|\sum_s f(s)\overline{f(s-k)}\right|^2 \leq c_3 N^3$.
(iv) $\sum_k \left|\sum_s f(s)\overline{g(s-k)}\right|^2 \leq c_4 N^2 \|g\|_2^2$ *for every function* $g : \mathbb{Z}_N \to \mathbb{C}$.

Proof. Using identities (2) and (1) above, we have

$$\begin{aligned}\sum_k \Big|\sum_s f(s)\overline{g(s-k)}\Big|^2 &= \sum_{s-t=u-v} f(s)\overline{g(t)f(u)}g(v)\\ &= \sum_{s+v=u+t} f(s)g(v)\overline{f(u)g(t)}\\ &= \sum_x |f * g(x)|^2\\ &= N^{-1}\sum_r |(f*g)^\wedge(r)|^2\\ &= N^{-1}\sum_r |\hat{f}(r)|^2|\hat{g}(r)|^2\\ &\leq \Big(\sum_r |\hat{f}(r)|^4\Big)^{1/2}\Big(\sum_r |\hat{g}(r)|^4\Big)^{1/2}\end{aligned}$$

by the Cauchy-Schwarz inequality. If $f = g$, then equality holds above, which gives the equivalence between (i) and (iii) with $c_1 = c_3$. It is obvious that (iv) implies (iii) if $c_3 \geqslant c_4$. Using the additional inequality

$$\Big(\sum_r |\hat{g}(r)|^4\Big)^{1/2} \leqslant \sum_r |\hat{g}(r)|^2,$$

we can deduce (iv) from (i) if $c_4 \geqslant c_1^{1/2}$.

Since $\max_r |\hat{f}(r)| \le \left(\sum_r |\hat{f}(r)|^4\right)^{1/4}$, one can see that (ii) follows from (i) if $c_2 \geqslant c_1^{1/4}$. For the reverse implication, we use the fact that

$$\sum_r |\hat{f}(r)|^4 \le \max_r |\hat{f}(r)|^2 \sum_r |\hat{f}(r)|^2 .$$

By identity (1) and the restriction on the image of f, we have the estimate $\sum_r |\hat{f}(r)|^2 \leqslant N^2$, so that (i) follows from (ii) if $c_1 \geqslant c_2^2$. □

If f satisfies condition (i) with $c_1 = \alpha$, then we shall say that f is *α-uniform*. If f is the balanced function of a set A, we shall say also that A is α-uniform. (This definition coincides with the definition made by Chung and Graham of a *quasirandom* subset of $\mathbb{Z}_N$ [CG].)

Roth's proof can be presented as follows. Let A be a subset of $\mathbb{Z}_N$ of size δN. If A is α-uniform for a suitable α (a power of δ, where $|A| = \delta N$) then A contains roughly the expected number of arithmetic progressions of length three. (This follows easily from Lemma 2.3 below.) If not, then some non-zero Fourier coefficient of the characteristic function of A is a large fraction of N. It follows easily that there is a subset $I = \{a+d, a+2d, \dots, a+md\} \subset \mathbb{Z}_N$ such that m is a substantial fraction of N and $|A \cap I| \geqslant (\delta+\epsilon)m$ for some $\epsilon > 0$ which is also a power of δ. It can be shown quite easily (see for example Lemma 6.2 of this paper) that I can be partitioned into genuine arithmetic progressions (that is, when considered as subsets of $\mathbb{Z}$) of size about $m^{1/2}$. Hence, there is an arithmetic progression P of about this size such that $|A \cap P| \geqslant (\delta + \epsilon)|P|$. Now repeat the argument for P. The number of times it can be repeated depends only on δ, so, provided N is large enough, there must be an arithmetic progression of size three in A.

It turns out that, even if α is extremely small, an α-uniform set need not contain roughly the expected number of arithmetic progressions of length four. (An example is presented in [G3].) For this reason, if we wish to have an approach similar to the above one, but for progressions of length four, then we need a stronger notion of pseudorandomness. Given a function $f : \mathbb{Z}_N \to \mathbb{Z}_N$ and $k \in \mathbb{Z}_N$, define a function $\Delta(f;k)$ by $\Delta(f;k)(s) = f(s)\overline{f(s-k)}$. Notice that if $f(s) = \omega^{\phi(s)}$ for some function $\phi : \mathbb{Z}_N \to \mathbb{Z}_N$, then $\Delta(f;k)(s) = \omega^{\phi(s)-\phi(s-k)}$.

Lemma 2.2. *Let f be a function from $\mathbb{Z}_N$ to the closed unit disc in $\mathbb{C}$. Then the following are equivalent.*

(i) $\sum_u \sum_v \left|\sum_s f(s)\overline{f(s-u)f(s-v)}f(s-u-v)\right|^2 \leqslant c_1 N^4$.
(ii) $\sum_k \sum_r \left|\Delta(f;k)^\wedge(r)\right|^4 \leqslant c_2 N^5$.
(iii) $|\Delta(f;k)^\wedge(r)| \geqslant c_3 N$ *for at most $c_3^2 N$ pairs (k,r).*
(iv) *For all but $c_4 N$ values of k the function $\Delta(f;k)$ is c_4-uniform.*

Proof. The equivalence of (i) and (ii) with $c_1 = c_2$ follows, as in the proof of the equivalence of (i) and (iii) in Lemma 2.1, by expanding. Alternatively, it can be deduced by applying that result to each function $\Delta(f;k)$ and adding. If $|\Delta(f;k)^\wedge(r)| \geqslant c_3N$ for more than c_3^2N pairs (k,r) then obviously

$$\sum_k \sum_r |\Delta(f;k)^\wedge(r)|^4 > c_3^6 N^5,$$

so (ii) implies (iii) provided that $c_2 \leqslant c_3^6$. If (ii) does not hold, then there are more than $c_2N/2$ values of k such that $\sum_r \left|\Delta(f;k)^\wedge(r)\right|^4 \leq c_2N^4/2$. By the implication of (i) from (ii) in Lemma 1 this implies that there are more than $c_2N/2$ values of k such that $\max_r |\Delta(f;k)^\wedge(r)| \geqslant (c_2/2)^{1/2}N$, and hence (iii) implies (ii) as long as $c_2 \geqslant 2c_3^2$. Finally, it is easy to see that (iv) implies (ii) if $c_2 \geqslant 2c_4$ and (ii) implies (iv) if $c_2 \leqslant c_4^2$. □

A function satisfying property (i) above with $c_1 = \alpha$ will be called *quadratically* α-uniform. A set will be called quadratically α-uniform if its balanced function is. Let us define a *square* and a *cube in* $\mathbb{Z}_N$ to be sequences of the form $(s, s+a, s+b, s+a+b)$ and $(s, s+a, s+b, s+c, s+a+b, s+a+c, s+b+c, s+a+b+c)$ respectively. The number of squares in a set A is easily seen to be $N^{-1}\sum_r |\hat{A}(r)|^4$. It follows that if A has cardinality δN, then it contains at least δ^4N^3 squares and is α-uniform if and only if it contains at most $(\delta^4+\alpha)N^3$ squares. It is not hard to show that A contains at least δ^8N^4 cubes, and that A is quadratically uniform if and only if it contains at most $\delta^8(1+\epsilon)N^4$ cubes for some small ϵ. However, we shall not need this result. The aim of the rest of this section is to show that a quadratically uniform set contains roughly the expected number of arithmetic progressions of length four. This we do in a sequence of lemmas, all individually straightforward.

Lemma 2.3. *Let f, g and h be three functions from $\mathbb{Z}_N$ to $\mathbb{C}$ all taking values of modulus at most 1. Let $\alpha > 0$ and suppose also that h is α-uniform. Then for any distinct non-zero elements a, $b \in \mathbb{Z}_N$, we have the inequality*

$$\Big|\sum_{x,d} f(x)g(x+ad)h(x+bd)\Big| \leqslant \alpha^{1/4}N^2.$$

Proof. The triples $(x, x+ad, x+bd)$ are exactly those triples (x,y,z) for which $(a-b)x + by - az = 0$. So let us rename the numbers $a-b$, b and a as a, b and c and examine the sum $\sum_{ax+by+cz=0} f(x)g(y)h(z)$, assuming merely that a, b and c are non-zero. We have

$$\sum_{ax+by+cz=0} f(x)g(y)h(z) = N^{-1}\sum_{x,y,z} f(x)g(y)h(z)\sum_r \omega^{-(ax+by+cz)r}$$

$$= N^{-1}\sum_r \hat{f}(ar)\hat{g}(br)\hat{h}(cr).$$

By the Cauchy-Schwarz inequality, the magnitude of this last sum is at most

$$N^{-1}\Big(\sum_r |\hat{f}(ar)|^2\Big)^{1/2}\Big(\sum_r |\hat{g}(br)|^2\Big)^{1/2}\max_r |\hat{h}(cr)|.$$

Since N is assumed to be prime and a, b and c are non-zero, this is simply $N^{-1}\|\hat{f}\|_2\|\hat{g}\|_2 \max_r |\hat{h}(r)|$. By Parseval's identity and the ℓ_∞ bounds for f and g, this is at most $N \max_r |\hat{h}(r)|$. Since h is α-uniform, $\max_r |\hat{h}(r)| \leqslant \alpha^{1/4}N$, by the trivial implication of (ii) from (i) in Lemma 2.1, and the result is proved. □

The next lemma shows that quadratic uniformity is a stronger concept than uniformity. (Functions such as $f(x) = \omega^{x^2}$ show that it is strictly stronger, as can easily be checked.)

Lemma 2.4. *Let f be a quadratically α-uniform function from $\mathbb{Z}_N$ to $\mathbb{C}$ taking values of modulus at most 1. Then f is $\alpha^{1/2}$-uniform.*

Proof. This is a simple application of the Cauchy-Schwarz inequality. We shall use the equivalent form (iii) of the definition of α-uniformity. Then

$$\begin{aligned}\Big(\sum_k\Big|\sum_s f(s)\overline{f(s-k)}\Big|^2\Big)^2 &= \Big(\sum_k\sum_{s,t} f(s)\overline{f(s-k)f(t)}f(t-k)\Big)^2\\ &= \Big(\sum_{u,v}\sum_s f(s)\overline{f(s-u)f(s-v)}f(s-u-v)\Big)^2\\ &\leqslant N^2\sum_{u,v}\Big|\sum_s f(s)\overline{f(s-u)f(s-v)}f(s-u-v)\Big|^2\\ &\leqslant \alpha N^6,\end{aligned}$$

by the assumption that f is quadratically α-uniform. This proves the lemma. □

We now come to the main lemma, where the connection between quadratic uniformity and progressions of length four is first made. After this lemma, there will remain the straightforward task of moving back from balanced functions to characteristic functions of sets.

Lemma 2.5. *Let f_1, f_2, f_3 and f_4 be functions from $\mathbb{Z}_N$ to $\mathbb{C}$ taking values of modulus at most 1. Suppose also that f_4 is quadratically α-uniform. Then*

$$\Big|\sum_{s,d} f_1(s)f_2(s+d)f_3(s+2d)f_4(s+3d)\Big| \leqslant \alpha^{1/8}N^2\,.$$

Proof. Let $\alpha : \mathbb{Z}_N \to [0,1]$ be a function with the property that $\Delta(f_4;r)$ is $\alpha(r)$-uniform for every $r \in \mathbb{Z}_N$. By hypothesis (and Lemmas 2.1 and 2.2), such a function can be found with $\sum_r \alpha(r) \leqslant \alpha N$. Then

$$\begin{aligned}&\Big|\sum_{s,d} f_1(s)f_2(s+d)f_3(s+2d)f_4(s+3d)\Big|^2\\ &\qquad\leqslant N\sum_s\Big|\sum_d f_1(s)f_2(s+d)f_3(s+2d)f_4(s+3d)\Big|^2\end{aligned}$$

$$\leqslant N\sum_s\Big|\sum_d f_2(s+d)f_3(s+2d)f_4(s+3d)\Big|^2$$

$$=N\sum_s\sum_{c,d} f_2(s+c)\overline{f_2(s+d)}f_3(s+2c)\overline{f_3(s+2d)}f_4(s+3c)\overline{f_4(s+3d)}$$

$$=N\sum_s\sum_{c,u}\Delta(f_2;u)(s+c)\Delta(f_3;2u)(s+2c)\Delta(f_4;3u)(s+3d),$$

where in the last line we made the substitution $u=c-d$.

It follows from Lemma 2.3 that for any fixed value of u the last expression (when summed over s and c only) has size at most $\alpha(3u)^{1/4}N^2$. But $\sum_u \alpha(3u)^{1/4}\leqslant \alpha^{1/4}N$, by Jensen's inequality, so the whole sum is at most $\alpha^{1/4}N^3$. The result follows after multiplying by the factor N and taking square roots. □

Lemma 2.6. *Let A_1, A_2, A_3 and A_4 be subsets of $\mathbb{Z}_N$ of sizes δ_1, δ_2, δ_3 and δ_4 respectively. Suppose also that A_3 is α^4-uniform and A_4 is quadratically α^8-uniform. Let T be the number of quadruples of the form $(s,s+d,s+2d,s+3d)$ in $A_1\times A_2\times A_3\times A_4$. Then $|T-\delta_1\delta_2\delta_3\delta_4N^2|\leqslant 5\alpha N^2$.*

Proof. For each i let f_i be the balanced function of A_i. Then the quantity T we wish to estimate is

$$\sum_{s,d}(\delta_1+f_1(s))(\delta_2+f_2(s+d))(\delta_3+f_3(s+2d))(\delta_4+f_4(s+3d)).$$

If we expand out the product of four brackets, we obtain a sum of 16 terms which we can consider separately. The term that takes δ_i from each bracket sums to $\delta_1\delta_2\delta_3\delta_4N^2$. Any term that involves three δ_i and an f_j sums to zero (since $\sum_s f_j(s)=0$). Any term that involves two δ_i and two f_j also sums to zero: for example taking δ_1, δ_2, f_3 and f_4 results, after a change of variables, in the sum $\delta_1\delta_2\Big(\sum_s f_3(s)\Big)\Big(\sum_t f_4(t)\Big)$. Since both f_3 and f_4 are α^4-uniform, by hypothesis and Lemma 2.4, any term that involves one δ_i and three f_j sums to a quantity of modulus at most αN^2, by Lemma 2.3. Finally, Lemma 2.5 shows that the term with no δ_is sums to a quantity of modulus at most αN^2 as well. The result follows. □

For technical reasons we need one further simple lemma before we come to the main theorem of this section.

Lemma 2.7. *Let A be an α^4-uniform subset of $\mathbb{Z}_N$ of cardinality δN, and let P be an interval of the form $\{a+1,\dots,a+M\}$, where $M=\beta N$. Then $\big||A\cap P|-\beta\delta N\big|\leqslant\alpha N$.*

Proof. First, we can easily estimate the Fourier coefficients of the set P. Indeed,

$$|\hat{P}(r)|=\Big|\sum_{s=1}^{M}\omega^{-r(a+s)}\Big|$$

$$=|(1-\omega^{rM})/(1-\omega^r)|\leqslant N/2r\,.$$

(We also know that it is at most M, but will not need to use this fact.) This estimate implies that $\sum_{r\neq 0}|\hat{P}(r)|^{4/3}\leqslant N^{4/3}$. Therefore,

$$\begin{aligned}\big||A\cap P|-\beta\delta N\big| &= N^{-1}\Big|\sum_{r\neq 0}\hat{A}(r)\hat{P}(r)\Big|\\ &\leqslant N^{-1}\Big(\sum_{r\neq 0}|\hat{A}(r)|^4\Big)^{1/4}\Big(\sum_{r\neq 0}|\hat{P}(r)|^{4/3}\Big)^{3/4}\\ &\leqslant \Big(\sum_{r\neq 0}|\hat{A}(r)|^4\Big)^{1/4}\leq \alpha N\end{aligned}$$

as claimed. □

We are now in a position to prove that every sufficiently quadratically uniform set contains an arithmetic progression of length four.

Theorem 2.8. *Let $\delta>0$, let $\alpha=\delta^4/1000$, let $N>200\delta^{-3}$ and let A be a subset of $\mathbb{Z}_N$ of size at least δN that is quadratically α^8-uniform. Then A contains an arithmetic progression, even when considered as a subset of $\{1,2,\ldots,N\}$.*

Proof. Let $A_3=A_4=A$ and let A_1 and A_2 be the set of all elements of A that lie between $2N/5$ and $3N/5$. Since A is α^4-uniform, by Lemma 2.4, A_1 and A_2 have cardinality at least $(\delta/5-\alpha)N$, by Lemma 2.7. Since $\alpha\leq\delta/10$, this is at least $\delta N/10$.

Lemma 2.6 now tells us that the number of quadruples $(s,s+d,s+2d,s+3d)$ in $A_1\times A_2\times A_3\times A_4$ is at least $\delta^4N^2/200$, since $5\alpha=\delta^4/200$. At most δN of these can have $d=0$, and by our lower bound on N these do not exhaust all of them. The rest correspond to genuine progressions in $\{1,2,\ldots,N\}$, by our choice of A_1 and A_2. The result is proved. □

3. Finding Many Additive Quadruples

We have just seen that a quadratically uniform set must contain an arithmetic progression of length four. We now begin an argument of several steps, which will eventually show that if A is a subset of $\mathbb{Z}_N$ of cardinality δN which *fails* to be quadratically α-uniform (note that for convenience we have replaced α^8 by α in our statements), then there is an arithmetic progression $P\subset\mathbb{Z}_N$ (which is still an arithmetic progression when regarded as a subset of $\{1,2,\ldots,N\}$) of size N^β such that $|A\cap P|\geqslant(\delta+\epsilon)|P|$, where β and ϵ depend on α and δ only.

If A fails to be quadratically α-uniform, then so does its balanced function f (by definition). This tells us that there are many values of k for which the function $\Delta(f;k)$ has a large (meaning proportional to N) Fourier coefficient r. In the next result, we shall show that the set of pairs (k,r) for which $\Delta(f;k)^\wedge(r)$ is large is far from arbitrary.

Proposition 3.1. *Let $\alpha > 0$, let $f : \mathbb{Z}_N \to D$, let $B \subset \mathbb{Z}_N$ and let $\phi : B \to \mathbb{Z}_N$ be a function such that*

$$\sum_{k\in B} \left|\Delta(f;k)^\wedge(\phi(k))\right|^2 \geqslant \alpha N^3.$$

Then there are at least $\alpha^4 N^3$ quadruples $(a,b,c,d) \in B^4$ such that $a+b=c+d$ and $\phi(a)+\phi(b)=\phi(c)+\phi(d)$.

Proof. Expanding the left hand side of the inequality we are assuming gives us the inequality

$$\sum_{k\in B}\sum_{s,t} f(s)\overline{f(s-k)f(t)}f(t-k)\omega^{-\phi(k)(s-t)} \geqslant \alpha N^3.$$

If we now introduce the variable $u = s-t$ we can rewrite this as

$$\sum_{k\in B}\sum_{s,u} f(s)\overline{f(s-k)f(s-u)}f(s-k-u)\omega^{-\phi(k)u} \geqslant \alpha N^3.$$

Since $|f(x)| \leqslant 1$ for every x, it follows that

$$\sum_u \sum_s \left| \sum_{k\in B} \overline{f(s-k)}f(s-k-u)\omega^{-\phi(k)u}\right| \geqslant \alpha N^3$$

which implies that

$$\sum_u \sum_s \left| \sum_{k\in B} \overline{f(s-k)}f(s-k-u)\omega^{-\phi(k)u}\right|^2 \geqslant \alpha^2 N^4.$$

For each u and x let $f_u(x) = \overline{f(-x)}f(-x-u)$ and let $g_u(x) = B(x)\omega^{\phi(x)u}$. The above inequality can be rewritten

$$\sum_u \sum_s \left| \sum_k f_u(k-s)\overline{g_u(k)}\right|^2 \geqslant \alpha^2 N^4.$$

By Lemma 2.1, we can rewrite it again as

$$\sum_u \sum_r |\hat{f}_u(r)|^2 |\hat{g}_u(r)|^2 \geqslant \alpha^2 N^5.$$

Since $\sum_r |\hat{f}(r)|^4 \leqslant N^4$, the Cauchy-Schwarz inequality now implies that

$$\sum_u \Big(\sum_r |\hat{g}_u(r)|^4\Big)^{1/2} \geqslant \alpha^2 N^3.$$

Applying the Cauchy-Schwarz inequality again, we can deduce that

$$\sum_{u,r} |\hat{g}_u(r)|^4 = \sum_{u,r}\left|\sum_{k\in B} \omega^{\phi(s)u-rs}\right|^4 \geqslant \alpha^4 N^5.$$

Expanding the left hand side of this inequality we find that

$$\sum_{u,r}\sum_{a,b,c,d\in B}\omega^{u(\phi(a)+\phi(b)-\phi(c)-\phi(d))}\omega^{-r(a+b-c-d)}\geqslant\alpha^4N^5.$$

But now the left hand side is exactly N^2 times the number of quadruples $(a,b,c,d)\in B^4$ for which $a+b=c+d$ and $\phi(a)+\phi(b)=\phi(c)+\phi(d)$. This proves the proposition. □

We shall call a quadruple with the above property *additive*. In the next section, we shall show that functions with many additive quadruples have a very interesting structure.

4. An Application of Freiman's Theorem

There is a wonderful theorem due to Freiman about the structure of finite sets $A\subset\mathbb{Z}$ with the property that $A+A=\{x+y: x,y\in A\}$ is not much larger than A. Let us define a d-dimensional arithmetic progression to be a set of the form $P_1+\dots+P_d$, where the P_i are ordinary arithmetic progressions. It is not hard to see that if $|A|=m$ and A is a subset of a d-dimensional arithmetic progression of size Cm, then $|A+A|\leqslant 2^dCm$. Freiman's theorem [F1, 2] tells us that these are the *only* examples of sets with small sumset.

Theorem 4.1. *For every real number C there are constants d and K, depending only on C, such that, whenever A is a subset of $\mathbb{Z}$ with $|A|=m$ and $|A+A|\leqslant Cm$, there exists an arithmetic progression Q of dimension at most d such that $|Q|\leqslant Km$ and $A\subset Q$.*

Freiman's proof of his theorem did not give a bound for d and K, but recently an extremely elegant proof was discovered by Ruzsa which gives quite a good bound [Ru]. As it happens, we do not need the full strength of Freiman's theorem, and can obtain a better bound for Szemerédi's theorem with an appropriate weakening of it. This has the added advantage that not all of Ruzsa's proof is needed. Here we shall prove the weakening in outline. Some results that we quote can be found presented in full in [N].

The sets to which the result will be applied are graphs of functions with many additive quadruples, such as are produced by Proposition 3.1. Obviously, such a set is not a subset of $\mathbb{Z}$, but it is easy to embed it "isomorphically" into $\mathbb{Z}$, in a sense that we shall make precise later. If γ is such a graph, then we can regard γ as a subset of $\mathbb{Z}^2$. To every additive quadruple we can associate a quadruple of points $(x,y,z,w)\in\gamma$ such that $x+y=z+w$, where the addition is in $\mathbb{Z}^2$. It turns out to be convenient to consider instead quadruples with $x-y=z-w$ but they are clearly in one-to-one correspondence with the other kind.

The assumption that A is a subset of $\mathbb{Z}^2$ containing many quadruples (x,y,z,w) with $x-y=z-w$ tells us virtually nothing about the size of $A+A$, since half of A might be very nice and the remainder arbitrary. Even the stronger property that all large subsets of A contain many such quadruples (which comes out of Proposition 11) is not enough. For example, A could be the union of a horizontal line and a vertical line. What we shall show is that A has a reasonably large *subset* B such that $|B+B|$ is reasonably small. We will then be able to apply Freiman's theorem to the set B. This result, in its qualitative form, is due to Balog and Szemerédi [BS]. However, they use Szemerédi's uniformity lemma, which, as we mentioned in the introduction, produces a very weak bound. We therefore need a different argument, which will be the main task of this section. We begin with a combinatorial lemma.

Lemma 4.2. *Let X be a set of size m, let $\delta > 0$ and let $A_1, \ldots, A_n$ be subsets of X such that $\sum_{x=1}^{n} \sum_{y=1}^{n} |A_x \cap A_y| \geqslant \delta^2 mn^2$. There is a subset $K \subset [n]$ of cardinality at least $2^{-1/2}\delta^5 n$ such that for at least 90% of the pairs $(x, y) \in K^2$ the intersection $A_x \cap A_y$ has cardinality at least $\delta^2 m/2$. In particular, the result holds if $|A_x| \geqslant \delta m$ for every x.*

Proof. For every $j \leq m$ let $B_j = \{i : j \in A_i\}$ and let $E_j = B_j^2$. Choose five numbers $j_1, \ldots, j_5 \leqslant m$ at random (uniformly and independently), and let $X = E_{j_1} \cap \cdots \cap E_{j_5}$. The probability p_{xy} that a given pair $(x, y) \in [n]^2$ belongs to E_{j_r} is $m^{-1}|A_x \cap A_y|$, so the probability that it belongs to X is p_{xy}^5. By our assumption we have that $\sum_{x,y=1}^{n} p_{xy} \geqslant \delta^2 n^2$, which implies (by Hölder's inequality) that $\sum_{x,y=1}^{n} p_{xy}^5 \geqslant \delta^{10} n^2$. In other words, the expected size of X is at least $\delta^{10} n^2$.

Let Y be the set of pairs $(x, y) \in X$ such that $|A_x \cap A_y| < \delta^2 m/2$, or equivalently $p_{xy} < \delta^2/2$. Because of the bound on p_{xy}, the probability that $(x, y) \in Y$ is at most $(\delta^2/2)^5$, so the expected size of Y is at most $\delta^{10} n^2/32$.

It follows that the expectation of $|X| - 16|Y|$ is at least $\delta^{10} n^2/2$. Hence, there exist $j_1, \ldots, j_5$ such that $|X| \geqslant 16|Y|$ and $|X| \geqslant \delta^{10} n^2/2$. This proves the lemma, with $X = K^2$ (so $K = B_{j_1} \cap \cdots \cap B_{j_5}$). □

Let A be a subset of $\mathbb{Z}^D$ and identify A with its characteristic function. Then $A * A(x)$ is the number of pairs $(y, z) \in A^2$ such that $y - z = x$. (Recall that we have a non-standard use for the symbol "$*$".) Hence, the number of quadruples $(x, y, z, w) \in A^4$ with $x - y = z - w$ is $\|A * A\|_2^2$. The next result is a precise statement of the Balog-Szemerédi theorem, but, as we have mentioned, the bounds obtained in the proof are new.

Proposition 4.3. *Let A be a subset of $\mathbb{Z}^D$ of cardinality m such that $\|A * A\|_2^2 \geqslant c_0 m^3$. There are constants c and C depending only on c_0 and a subset $A'' \subset A$ of cardinality at least cm such that $|A'' - A''| \leqslant Cm$.*

Proof. The function $f(x) = A * A(x)$ (from $\mathbb{Z}^D$ to $\mathbb{Z}$)is non-negative and satisfies $\|f\|_\infty \leqslant m$, $\|f\|_2^2 \geqslant c_0 m^3$ and $\|f\|_1 = m^2$. This implies that $f(x) \geqslant c_0 m/2$ for at least $c_0 m/2$ values of x, since otherwise we would have

$$\|f\|_2^2 < (c_0/2)m.m^2 + (c_0 m/2).m^2 = c_0 m^3.$$

Let us call a value of x for which $f(x) \geqslant c_0 m/2$ a *popular difference* and let us define a graph G with vertex set A by joining a to b if $b-a$ (and hence $a-b$) is a popular difference. The average degree in G is at least $c_0^2 m/4$, so there must be at least $c_0^2 m/8$ vertices of degree at least $c_0^2 m/8$. Let $\delta = c_0^2/8$, let $a_1, \dots, a_n$ be vertices of degree at least $c_0^2 m/8$, with $n \geqslant \delta m$, and let $A_1, \dots, A_n$ be the neighbourhoods of the vertices $a_1, \dots, a_n$. By Lemma 4.2 we can find a subset $A' \subset \{a_1, \dots, a_n\}$ of cardinality at least $\delta^5 n/\sqrt{2}$ such that at least 90% of the intersections $A_i \cap A_j$ with $a_i, a_j \in A'$ are of size at least $\delta^2 m/2$. Set $\alpha = \delta^6/\sqrt{2}$ so that $|A'| \geqslant \alpha m$.

Now define a graph H with vertex set A', joining a_i to a_j if and only if $|A_i \cap A_j| \geqslant \delta^2 m/2$. The average degree of the vertices in H is at least $(9/10)|A'|$, so at least $(4/5)|A'|$ vertices have degree at least $(4/5)|A'|$. Define A'' to be the set of all such vertices.

We claim now that A'' has a small difference set. To see this, consider any two elements $a_i, a_j \in A''$. Since the degrees of a_i and a_j are at least $(4/5)|A'|$ in H, there are at least $(3/5)|A'|$ points $a_k \in A'$ joined to both a_i and a_j. For every such k we have $|A_i \cap A_k|$ and $|A_j \cap A_k|$ both of size at least $\delta^2 m/2$. If $b \in A_i \cap A_k$, then both $a_i - b$ and $a_k - b$ are popular differences. It follows that there are at least $c_0^2 m^2/4$ ways of writing $a_i - a_k$ as $(p-q)-(r-s)$, where $p, q, r, s \in A$, $p - q = a_i - b$ and $r - s = a_k - b$. Summing over all $b \in A_i \cap A_k$, we find that there are at least $\delta^2 c_0^2 m^3/8$ ways of writing $a_i - a_k$ as $(p-q)-(r-s)$ with $p, q, r, s \in A$. The same is true of $a_j - a_k$. Finally, summing over all k such that a_k is joined in H to both a_i and a_j, we find that there are at least $(3/5)|A'|\delta^4 c_0^4 m^6/64 \geqslant \alpha\delta^4 c_0^4 m^7/120$ ways of writing $a_i - a_j$ in the form $(p-q)-(r-s)-\big((t-u)-(v-w)\big)$ with $p, q, \dots, w \in A$.

Since there are at most m^8 elements in A^8, the number of differences of elements of A'' is at most $120m/\alpha\delta^4 c_0^4 \leqslant 2^{38} m/c_0^{24}$. Note also that the cardinality of A'' is at least $(4/5)\alpha m \geqslant c_0^{12} m/2^{19}$. The proposition is proved. □

Combining Theorem 4.1 and Proposition 4.3 gives us the following consequence of Freiman's theorem.

Corollary 4.4. *Let A be a subset of $\mathbb{Z}^D$ of cardinality m such that $\|A*A\|_2^2 \geqslant c_0 m^3$. Then there is an arithmetic progression Q of cardinality at most Cm and dimension at most d such that $|A \cap Q| \geqslant cm$, where C, d and c are constants depending only on c_0.* □

It follows by simple averaging arguments that there is a one-dimensional arithmetic progression of length around $m^{1/d}$ that intersects substantially with A. It is this consequence that we shall now indicate how to prove more directly.

One of Freiman's ideas was the following very useful definition. Let A and B be subsets of Abelian groups. A function $\phi : A \to B$ is said to be a *(Freiman) homomorphism of order k* if, whenever $a_1, \dots, a_{2k}$ are elements of A such that

$$a_1 + \cdots + a_k = a_{k+1} + \cdots + a_{2k}$$

we have also

$$\phi(a_1) + \cdots + \phi(a_k) = \phi(a_{k+1}) + \cdots + \phi(a_{2k}).$$

If in addition ϕ is a bijection and the reverse implication is also true, then ϕ is said to be not just a homomorphism but an isomorphism. The rough idea is that a Freiman isomorphism preserves much of the additive structure of a set but not necessarily the full linear structure. So the set $\{1,2,25\} \subset \mathbb{Z}$ is 5-isomorphic to the set $\{(0,0),(1,0),(0,1)\} \subset \mathbb{Z}^2$ via, for example, the map $1 \mapsto (0,0)$, $2 \mapsto (1,0)$ and $25 \mapsto (0,1)$, because any group-theoretic fact that distinguishes the two sets must concern sums of more than five (not necessarily distinct) elements. They are not 25-isomorphic, because, for example, $1+24\times 2 = 24\times 1+25$ while $(0,0)+24\times(1,0) \neq 24\times(0,0)+(0,1)$.

Amongst the simple properties that we shall need of Freiman isomorphism are the following.

(1) If ϕ is an isomorphism of order 2 from A to B and $P \subset A$ is an arithmetic progression, then $\phi(P)$ is an arithmetic progression.

(2) If ϕ is an isomorphism of order 2 from A to B then it induces a well-defined bijection between $A+A$ and $B+B$ by the formula $\phi(x+y) = \phi(x)+\phi(y)$.

(3) If A is a finite subset of $\mathbb{Z}^d$ and k is any positive integer then A is isomorphic of order k to some subset of $\mathbb{Z}$.

These and similar facts are easy exercises and we shall assume some of them without giving proofs.

Two of the key lemmas in Ruzsa's argument are the following, of which the first is due to Plünnecke and the second to Ruzsa. The notation $kA - lA$ stands for the set of all sums of the form $a_1 + \cdots + a_k - a_{k+1} - \cdots - a_{k+l}$ with all a_i in A.

Lemma 4.5. *Let A be a subset of an Abelian group and suppose that $|A-A| \leqslant C|A|$. Then $|kA - lA| \leqslant C^{k+l}|A|$.*

Lemma 4.6. *Let A be a subset of $\mathbb{Z}$ and suppose that $|kA-kA| \leqslant C|A|$. Then there is a prime number $N \leqslant 8C|A|$ and a subset $A' \subset A$ of size at least $|A|/k$ which is isomorphic of order k to a subset of $\mathbb{Z}_N$.*

Broadly speaking, if we have a set of integers with small sumset, Lemmas 4.5 and 4.6 allow us to replace it (or at least a large part of it) with a subset of $\mathbb{Z}_N$ of size proportional to N, while preserving the relevant additive structure. It follows that the next lemma, which is about large subsets of $\mathbb{Z}_N$, has an immediate consequence for sets with small sumsets. Before stating it we need a further definition. Given an element x of $\mathbb{Z}_N$, define $|x|$ to be the distance from x to 0 - that is, y if x is congruent to $y \in \{0,1,\ldots,\lfloor N/2\rfloor\}$ and $N-y$ if it is congruent to $y \in \{\lfloor N/2\rfloor+1,\ldots,N-1\}$. If K is a subset of $\mathbb{Z}_N$ and $\gamma > 0$, then the *Bohr neighbourhood* $B(K,\gamma) \subset \mathbb{Z}_N$ is the set of all $x \in \mathbb{Z}_N$ such that $|rx| \leqslant \delta N$ for every $r \in K$.

The A and δ of the next lemma should not be confused with the A and δ of the main result - it is just helpful to reuse those letters.

Lemma 4.7. *Let $A \subset \mathbb{Z}_N$ be a set of size at least δN. Then there is a set K of size at most $2\delta^{-2}$ such that every point in the Bohr neighbourhood $B(K,1/4)$ can be written in at least $\delta^4 N^3/2$ ways as $a+b-c-d$, with a, b, c and d in A.*

Proof. Let K be the set $\{r \in \mathbb{Z}_N : r \neq 0, |\hat{A}(r)| \geqslant \delta^{3/2}N/\sqrt{2}\}$, and write $g(x)$ for $A * A * (-A) * (-A)(x)$. The function g counts the number of ways of writing x as $a+b-c-d$ with a, b, c and d in A, so we want to show that g is large everywhere on the Bohr neighbourhood $B(K, 1/4)$. To do this, we use the convolution identity and inversion formula to rephrase the problem in terms of Fourier transforms. Indeed, the Fourier transform of g is $|\hat{A}|^4$, so

$$\begin{aligned} Ng(x) &= \sum_r |\hat{A}(r)|^4 \omega^{rx} \\ &= |\hat{A}(0)|^4 + \sum_{r \in K} |\hat{A}(r)|^4 \omega^{rx} + \sum_{r \notin K \cup \{0\}} |\hat{A}(r)|^4 \omega^{rx}. \end{aligned}$$

The first of these terms is $\delta^4 N^4$. If x belongs to $B(K, \delta)$, then the real part of ω^{rx} is non-negative for every $r \in K$, so the real part of the second term is non-negative. As for the last term, we can bound it above in modulus by

$$\max_{r \notin K \cup \{0\}} |\hat{A}(r)|^2 \sum_r |\hat{A}(r)|^2$$

which, by Parseval's identity and the definition of K, is at most $(\delta^3/2)N^2 . \delta N^2 = \delta^4 N^4/2$. It follows that the real part of $g(x)$, and hence also the modulus, is at least $\delta^4 N^3/2$, as claimed. All that remains is to remark that K has cardinality at most $2\delta^{-2}$, or else the sum $\sum_r |\hat{A}(r)|^2$ would be larger than is allowed by Parseval's identity and the size of A. □

The next lemma is a very standard application of the pigeonhole principle.

Lemma 4.8. *Let $K \subset \mathbb{Z}_N$ be a set of size k. Then the Bohr neighbourhood $B(K, 1/4)$ contains an arithmetic progression P, centred on 0, of length at least $N^{1/k}/4$.*

Proof. Let $K = \{r_1, \ldots, r_k\}$ and consider the N points of the form $(r_1 x, r_2 x, \ldots, r_k x) \in \mathbb{Z}_N^k$. Let m be the largest integer less than $N^{1/k}$ and for $1 \leqslant i \leqslant m$ let $P_i \subset \mathbb{Z}_N$ be the interval $\{x \in \mathbb{Z}_N : (i-1)N/m < x \leqslant iN/m\}$. Since $m < N^{1/k}$ we can partition $\mathbb{Z}_N^k$ into fewer than N sets of the form $P_{i_1} \times \cdots \times P_{i_k}$. By the pigeonhole principle one of these sets must contain at least two points $(r_1 x, r_2 x, \ldots, r_k x)$ and $(r_1 y, r_2 y, \ldots, r_k y)$. Let $z = x - y$. By the way the sets are constructed, we find that $|r_i z| < N/m$ for every i. But then for any integer t we may also say that $|r_i t z| < tN/m$. Hence, tz belongs to $B(K, 1/4)$ whenever $|t| \leqslant m/4$. Since we may certainly choose m to be greater than $N^{1/k}/2$, the result follows. □

Corollary 4.9. *Let C be a constant and let A be a subset of $\mathbb{Z}^2$ of size n such that $|A + A| \leqslant Cn$. Then there is an arithmetic progression P of size at least n^{c_1} such that $|A \cap P| \geqslant c_2|P|$, where c_1 and c_2 are constants that have a power-type dependence on C.*

Proof. Property (3) of isomorphisms tells us that A is 8-isomorphic to a subset A_1 of $\mathbb{Z}$. This implies, with room to spare, that $|A_1| = n$ and $|A_1+A_1| \leqslant Cn$. Lemma 4.5 implies that $|8A_1-8A_1| \leqslant C^{16}n$ and Lemma 4.6 now implies that A_1 has a subset A_2 of size at least $C^{16}n/8$ that is 8-isomorphic to a subset $A_3 \subset \mathbb{Z}_N$, where N is some prime less than $8C^{16}n$. Writing δ for $|A_3|/N$, we can now apply Lemmas 4.7 and 4.8 to give us an arithmetic progression Q, centred at 0, of length at least $N^{\delta^2/2}/2$, such that every $x \in Q$ can be written in at least $\delta^4 N^3/2$ ways as $a+b-c-d$ with a, b, c, d belonging to A_3. Elementary properties of Freiman isomorphisms now tell us that similar statements holds for A_2 and the corresponding subsets of A_1 and A. In particular, there is an arithmetic progression $P \subset \mathbb{Z}^2$ (meaning a set of the form $\{a, a+d, \ldots, a+hd\}$ with a, $d \in \mathbb{Z}^2$ and $d \neq 0$), of the same size as Q, such that every $x \in P$ can be written in at least $\delta^4 N^3/2$ ways as $a+b-c-d$ with a, b, c and d in A.

Turning to characteristic functions, we may therefore say that

$$\sum_{a,b,c,d\in A} P(a+b-c-d) \geqslant \delta^4 N^3 |P|/2.$$

Hence, by averaging, there must exist b, c and d such that

$$\sum_{a\in A} P(a+b-c-d) = |A \cap (P-b+c+d)| \geqslant \delta^4 N^3 n^{-3}|P|/2 \geqslant \delta^4 |P|/2.$$

Since δ has a power-type dependence on C, the lemma is proved. □

Corollary 4.10. *Let $B \subset \mathbb{Z}_N$ be a set of cardinality βN, and let $\phi : B \to \mathbb{Z}_N$ be a function with at least $c_0 N^3$ additive quadruples. Then there are constants γ and η with power-type dependence on β and c_0 only, a mod-N arithmetic progression $P \subset \mathbb{Z}_N$ of cardinality at least N^γ and a linear function $\psi : P \to \mathbb{Z}_N$ such that $\phi(s)$ is defined and equal to $\psi(s)$ for at least $\eta|P|$ values of $s \in P$.*

Proof. (Sketch) Let γ be the graph of ϕ, embedded in the obvious (but unnatural) way into $\mathbb{Z}^2$. It is a straightforward exercise to show that even after the embedding, which destroys some additive quadruples, there are still plenty left. (Partition $\mathbb{Z}_N^2$ into the four parts according to whether the two coordinates are less than or greater than $N/2$. Easy Fourier analysis shows that many of the additive quadruples come from entirely within some part, and these ones are not destroyed by the embedding.)

Lemma 4.3 gives us a large subset of γ with small sumset and Corollary 4.9 shows that this subset has a large intersection with some reasonably large arithmetic progression Q in $\mathbb{Z}_2$. Since γ is the graph of a function, the common difference $d = (d_1, d_2)$ of this arithmetic progression must have $d_1 \neq 0$. Hence, Q is the graph of a linear function defined on some arithmetic progression P with common difference d_1. The result follows, and it is easy to check that the dependences are of the kind claimed. □

5. Obtaining Quadratic Bias

Let $A \subset \mathbb{Z}_N$ be a set which fails to be quadratically α-uniform and let f be the balanced function of A. Then there is a subset $B \subset \mathbb{Z}_N$ of cardinality at least αN, and a function $\phi : B \to \mathbb{Z}_N$ such that $|\Delta(f;k)^\wedge(\phi(k))| \geqslant \alpha N$ for every $k \in B$. From Section 3 we know that B contains at least $\alpha^{12}N^3$ additive quadruples for the function ϕ. The last section then implies that ϕ can be restricted to a large arithmetic progression P where it often agrees with a linear function $s \mapsto as + b$. We shall now use this fact to show that $\mathbb{Z}_N$ can be uniformly covered by large arithmetic progressions $P_1, \ldots, P_N$ such that, for every s we can choose a quadratic function $\psi_s : P_s \to \mathbb{Z}_N$ such that $\sum_{z \in P_s} f(z)\omega^{-\psi_s(z)}$ is on average large in modulus (meaning an appreciable fraction of $|P_s|$). In the next section we shall use this result to find an arithmetic progression where the density of A increases.

Proposition 5.1. *Let $A \subset \mathbb{Z}_N$ have balanced function f. Let P be an arithmetic progression (in $\mathbb{Z}_N$) of cardinality T. Suppose that there exist λ and μ such that $\sum_{k \in P} |\Delta(f;k)^\wedge(\lambda k + \mu)|^2 \geqslant \beta N^2 T$. Then there exist quadratic polynomials ψ_0, $\psi_1, \ldots, \psi_{N-1}$ such that*

$$\sum_s \Big| \sum_{z \in P+s} f(z)\omega^{-\psi_s(z)} \Big| \geqslant \beta NT/\sqrt{2}.$$

Proof. Expanding the assumption we are given, we obtain the inequality

$$\sum_{k \in P} \sum_{s,t} f(s)f(s-k)f(t)f(t-k)\omega^{-(\lambda k+\mu)(s-t)} \geqslant \beta N^2 T.$$

Substituting $u = s - t$, we deduce that

$$\sum_{k \in P} \sum_{s,u} f(s)f(s-k)f(s-u)f(s-k-u)\omega^{-(\lambda k+\mu)u} \geqslant \beta N^2 T.$$

Let $P = \{x+d, x+2d, \ldots, x+td\}$. Then we can rewrite the above inequality as

$$\sum_{i=1}^{T} \sum_{s,u} f(s)f(s-x-id)f(s-u)f(s-k-id-u)\omega^{-(\lambda x+\lambda id+\mu)u} \geqslant \beta N^2 T. \quad (*)$$

Since there are exactly T ways of writing $u = y + jd$ with $y \in \mathbb{Z}_N$ and $1 \leqslant j \leqslant T$, we can rewrite the left-hand side above as

$$\frac{1}{T} \sum_s \sum_{i=1}^{T} \sum_y \sum_{j=1}^{T} f(s)f(s-x-id)f(s-y-jd)f(s-x-id-y-jd)$$
$$\times\, \omega^{-(\lambda x+id+\mu)(y+jd)}.$$

Let us define $\gamma(s,y)$ by the equation

$$\left|\sum_{i=1}^{T}\sum_{j=1}^{T} f(s-x-id)f(s-y-jd)f(s-x-id-y-jd)\omega^{-(\phi(x)+i\mu)(y+jd)}\right| = \gamma(s,y)T^2.$$

Since $|f(s)| \leqslant 1$, $(*)$ tells us that the average value of $\gamma(s,y)$ is at least β.

In general, suppose we have real functions f_1, f_2 and f_3 such that

$$\left|\sum_{i=1}^{T}\sum_{j=1}^{T} f_1(i)f_2(j)f_3(i+j)\omega^{-(ai+bj-2cij)}\right| \geqslant cT^2.$$

Since $2cij = c((i+j)^2 - i^2 - j^2)$, we can rewrite this as

$$\left|\sum_{i=1}^{T}\sum_{j=1}^{T} f_1(i)\omega^{-(ai+ci^2)} f_2(i)\omega^{-(bj+cj^2)} f_3(i+j)\omega^{c(i+j)^2}\right| \geqslant cT^2$$

and then replace the left hand side by

$$\frac{1}{N}\left|\sum_{r}\sum_{i=1}^{T}\sum_{j=1}^{T}\sum_{k=1}^{2T} f_1(i)\omega^{-(ai+ci^2)} f_2(j)\omega^{-(bj+cj^2)} f_3(k)\omega^{ck^2}\omega^{-r(i+j-k)}\right|.$$

If we now set $g_1(r) = \sum_{i=1}^{T} f_1(i)\omega^{-(ai+ci^2)}\omega^{-ri}$, $g_2(r) = \sum_{j=1}^{T} f_2(j)\omega^{-(bj+cj^2)}\omega^{-rj}$ and $g_3(r) = \sum_{k=1}^{2T} f_3(k)\omega^{-ck^2}\omega^{-rk}$, then we have

$$\left|\sum_{r} g_1(r)g_2(r)g_3(r)\right| \geqslant cT^2N,$$

which implies, by the Cauchy-Schwarz inequality, that $\|g_1\|_\infty\|g_2\|_2\|g_3\|_2 \geqslant cT^2N$. Since $\|g_2\|_2^2 \leqslant NT$ and $\|g_3\|_2^2 \leqslant 2NT$ (by identity (1) of Section 2), this tells us that $|g_1(r)| \geqslant cT/\sqrt{2}$ for some r. In particular, there exists a quadratic polynomial ψ such that $\left|\sum_{i=1}^{T} f_1(i)\omega^{-\psi(i)}\right| \geqslant cT/\sqrt{2}$.

Let us apply this general fact to the functions $f_1(i) = f(x-s-id)$, $f_2(j) = f(s-y-jd)$ and $f_3(k) = f(s-x-y-kd)$. It gives us a quadratic polynomial $\psi_{s,y}$ such that

$$\left|\sum_{i=1}^{T} f(s-x-id)\omega^{-\psi_{s,y}(i)}\right| \geqslant \gamma(s,y)T/\sqrt{2}.$$

Let $\gamma(s)$ be the average of $\gamma(s,y)$, and choose ψ_s to be one of the $\psi_{s,y}$ in such a way that

$$\left|\sum_{i=1}^{T} f(s-x-id)\omega^{-\psi_s(i)}\right| \geqslant \gamma(s)T/\sqrt{2}.$$

If we now sum over s, we have the required statement (after a small change to the definition of the ψ_s). □

Combining the above result with the results of the previous section, we obtain a statement of the following kind. If A fails to be quadratically uniform, then $\mathbb{Z}_N$ can be uniformly covered by large arithmetic progressions, on each of which the balanced function of A exhibits "quadratic bias". It is not immediately obvious that this should enable us to find a progression where the restriction of A has an increased density. That is a task for the next section.

6. An Application of Weyl's Inequality

A famous result of Weyl asserts that, if α is an irrational number and k is an integer, then the sequence $\alpha, 2^k\alpha, 3^k\alpha, \ldots$ is equidistributed mod 1. As an immediate consequence, if α is any real number and $\epsilon > 0$, then there exists n such that the distance from $n^2\alpha$ to the nearest integer is at most ϵ. This is the result we need to finish the proof. For the purposes of a bound, we need an estimate for n in terms of ϵ. It is not particularly easy to find an appropriate statement in the literature. In the longer paper to come, we shall give full details of the deduction of the statement we need, with estimates, from Weyl's inequality. Here we shall merely state the result in a convenient form, almost certainly not with the best known bound.

Theorem 6.1. *Let N be sufficiently large and let $a \in \mathbb{Z}_N$. For any $t \leqslant N$ there exists $p \leqslant t$ such that $|p^2a| \leqslant Ct^{-1/8}N$, where C is an absolute constant.*

Before we apply Theorem 6.1, we need a standard lemma (essentially due to Dirichlet).

Lemma 6.2. *Let $\phi : \mathbb{Z}_N \to \mathbb{Z}_N$ be linear (i.e., of the form $\phi(x) = ax + b$) and let r, $s \leqslant N$. For some $m \leqslant (2rN/s)^{1/2}$ the set $\{0, 1, 2, \ldots, r-1\}$ can be partitioned into arithmetic progressions $P_1, \ldots, P_m$ such that the diameter of $\phi(P_j)$ is at most s for every j. Moreover, the sizes of the P_j differ by at most 1.*

Proof. Let t be an integer greater than or equal to $(2rN/s)^{1/2}$ and note that this is at least $r^{1/2}$. Of the numbers $\phi(0), \phi(1), \ldots, \phi(t)$, at least two must be within N/t and hence there exists $u \leqslant t$ such that $|\phi(u) - \phi(0)| \leqslant N/t$. Split $\{0, 1, \ldots, r-1\}$ into u congruence classes mod u, each of size at most $\lceil r/u \rceil$. Each congruence class is an arithmetic progression. If P is a set of at most st/N consecutive elements of a congruence class, then P is an arithmetic progression with $\phi(P)$ of diameter at most s. Hence, each congruence class can be divided into at most $2rN/ust$ subprogressions P with $\phi(P)$ of diameter at most s and with different Ps differing in size by at most 1. Since the congruence classes themselves differ in size by at most 1, it is not too hard to see that the whole of $\{0, 1, \ldots, r\}$ can be thus partitioned. Hence, the total number of subprogressions is at most $2rN/st \leqslant (2rN/s)^{1/2}$. (Note that we cannot make t larger because we needed the estimate $r/u \geqslant st/N$ above.) □

Proposition 6.3. *There is an absolute constant C with the following property. Let $\psi : \mathbb{Z}_N \to \mathbb{Z}_N$ be any quadratic polynomial and let $r \in \mathbb{N}$. For some $m \leqslant Cr^{1-1/128}$ the set $\{0, 1, 2, \dots, r-1\}$ can be partitioned into arithmetic progressions $P_1, \dots, P_m$ such that the diameter of $\psi(P_j)$ is at most $Cr^{-1/128}N$ for every j. The lengths of any two P_j differ by at most 1.*

Proof. Let us write $\psi(x) = ax^2 + bx + c$. By Theorem 6.1 we can find $p \leqslant r^{1/2}$ such that $|ap^2| \leqslant C_1 r^{-1/8}N$ for some absolute constant C_1. Then for any s we have

$$\begin{aligned}\psi(x+sp) &= a(x+sp)^2 + b(x+sp) + c\\ &= s^2(ap^2) + \theta(x,p)\end{aligned}$$

where θ is a bilinear function of x and p. (Throughout this paper, we use the word "linear" where "affine" is, strictly speaking, more accurate.)

For any u, the diameter of the set $\{s^2(ap^2) : 0 \leqslant s < u\}$ is at most $u^2|ap^2| \leqslant C_1u^2r^{-1/8}N$. Therefore, for any $u \leqslant r^{1/4}$, we can partition the set $\{0, 1, \dots, r-1\}$ into arithmetic progressions of the form

$$Q_j = \{x_j, x_j + p, \dots, x_j + (u_j - 1)p\},$$

such that, for every j, $u - 1 \leqslant u_j \leqslant u$ and there exists a linear function ϕ_j such that, for any subset $P \subset Q_j$,

$$\operatorname{diam}(\psi(P)) \leqslant C_1u^2r^{-1/8}N + \operatorname{diam}(\phi_j(P))\,.$$

Let us choose $u = r^{1/64}$, with the result that $u^2r^{-1/16} = r^{-1/32}$. By Lemma 6.2, if $v \leqslant u^{1/2}/2$, then every Q_j can be partitioned into arithmetic progressions P_{jt} of length $v-1$ or v in such a way that $\operatorname{diam}(\phi_j(P_{jt})) \leqslant 2u^{-1/2}N$ for every t. This, with our choice of u above, gives us the result. □

Corollary 6.4. *Let $\psi : \mathbb{Z}_N \to \mathbb{Z}_N$ be a quadratic polynomial and let $r \leqslant N$. There exists $m \leqslant Cr^{1-1/128}$ (where C is an absolute constant) and a partition of the set $\{0, 1, \dots, r-1\}$ into arithmetic progressions $P_1, \dots, P_m$ such that the sizes of the P_j differ by at most one, and if $f : \mathbb{Z}_N \to D$ is any function such that*

$$\left|\sum_{x=0}^{r-1} f(x)\omega^{-\psi(x)}\right| \geqslant \alpha r,$$

then

$$\sum_{j=1}^{m}\left|\sum_{x\in P_j} f(x)\right| \geqslant \alpha r/2\,.$$

Proof. By Proposition 6.3 we can choose $P_1, \dots, P_m$ such that $\operatorname{diam}(\phi(P_j)) \leqslant CNr^{-1/128}$ for every j. For sufficiently large r this is at most $\alpha N/4\pi$. By the triangle inequality,

$$\sum_{j=1}^{m}\Big|\sum_{x\in P_j} f(x)\omega^{-\psi(x)}\Big| \geqslant \alpha r.$$

Let $x_j \in P_j$. The estimate on the diameter of $\psi(P_j)$ implies that $|\omega^{-\psi(x)} - \omega^{-\psi(x_j)}|$ is at most $\alpha/2$ for every $x \in P_j$. Therefore

$$\begin{aligned}\sum_{j=1}^{m}\Big|\sum_{x\in P_j} f(x)\Big| &= \sum_{j=1}^{m}\Big|\sum_{x\in P_j} f(x)\omega^{-\psi(x_j)}\Big| \\ &\geqslant \sum_{j=1}^{m}\Big|\sum_{x\in P_j} f(x)\omega^{-\psi(x)}\Big| - \sum_{j=1}^{m}(\alpha/2)|P_j| \\ &\geqslant \alpha r/2.\end{aligned}$$

The statement about the sizes of the P_j follows easily from our construction. □

7. Putting Everything Together

Theorem 7.1. *There is an absolute constant C with the following property. Let A be a subset of $\mathbb{Z}_N$ with cardinality δN. If $N \geqslant \exp\exp\big((1/\delta)^C\big)$, then A contains an arithmetic progression of length four.*

Proof. Suppose that the result is false. Then Theorem 2.8 implies that A is not quadratically $10^{-24}\delta^{32}$-uniform. Let $\alpha = 10^{-24}\delta^{32}$ and let f be the balanced function of A. The implication of (ii) from (iii) in Lemma 2.2 then implies that there is a set $B \subset \mathbb{Z}_N$ of cardinality at least $\alpha N/2$ together with a function $\phi : B \to \mathbb{Z}_N$, such that $|\Delta(f;k)^{\sim}(\phi(k))| \geqslant (\alpha/2)^{1/2}N$ for every $k \in B$. In particular,

$$\sum_{k\in B} |\Delta(f;k)^{\wedge}(\phi(k))|^2 \geqslant (\alpha/2)^2 N^3.$$

Hence, by Proposition 3.1, ϕ has at least $(\alpha/2)^8 N^3$ additive quadruples. Corollary 4.10 then implies the existence of an arithmetic progression P of size at least N^{γ} and a linear function $\psi : P \to \mathbb{Z}_N$ such that $\phi(x) = \psi(x)$ for at least $\eta|P|$ values of $x \in P$. Here, γ and η depend in a power-type way (that can be effectively calculated) on α and hence on δ.

This gives us the hypotheses we need to apply Proposition 5.1. This gives us quadratic polynomials $\psi_0, \psi_1, \ldots, \psi_{N-1}$ such that

$$\sum_{s}\Big|\sum_{z\in P+s} f(z)\omega^{-\psi_s(z)}\Big| \geqslant \beta NT/\sqrt{2}$$

with $\beta = \alpha\eta/2$ and $T = |P| = N^\gamma$. Corollary 6.4 implies that we can partition each $P + s$ into further progressions $P_{s1}, \ldots, P_{sm}$ (in the mod-N sense) of cardinalities differing by at most one and all at least $cT^{1/128}$, where c is an absolute constant, such that

$$\sum_s \sum_{j=1}^m \Big| \sum_{x \in P_{sj}} f(x) \Big| \geqslant \beta NT/2\sqrt{2}.$$

It is an easy consequence of Lemma 6.2 that we can also insist that the P_{sm} are genuine arithmetic progressions (in $\{0, 1, \ldots, N-1\}$ and not just in $\mathbb{Z}_N$), except that now the condition on the sizes is that the average length of a P_{sj} is $cT^{1/256}$ (for a slightly different c) and no P_{sj} has more than twice this length. With such a choice of P_{sj}, let p_{sj} equal $\sum_{x \in P_{sj}} f(x)$, and let q_{sj} be p_{sj} if this is positive, and zero otherwise. Then $\sum_s \sum_{j=1}^m p_{sj} = T \sum_x f(x) = 0$, which implies that $\sum_s \sum_{j=1}^m q_{sj} \geqslant \beta NT/4\sqrt{2}$. Hence, there exists a choice of s and j such that $\sum_{x \in P_{sj}} f(x) \geqslant \beta T/4m\sqrt{2} = c_1 \beta T^{1/256}$, where c_1 is another absolute constant. Then $|P_{sj}|$ is at least $c_1 \beta T^{1/256}$ and $|A \cap P_{sj}|$ is at least $(\delta + c_2\beta)|P_{sj}|$.

We now repeat the argument, replacing A and $\{0, 1, 2, \ldots, N\}$ by $A \cap P_{sj}$ and P_{sj}. We are now operating with a new density of at least $\delta + \delta^C$, for some absolute constant C, so it is clear that the number of iterations we will need to make is bounded above by δ^{-C}. At each iteration we replace the current N with a new one which is at least $N^\gamma = N^{\delta^K}$, where K is another absolute constant. This tells us that the theorem is proved, provided that $N^{\delta^{K\delta^{-C}}}$ is sufficiently large. The restriction comes in Theorem 2.8, which tells us that we must have $N^{\delta^{K\delta^{-C}}} \geqslant 200\delta^{-3}$. A small calculation now gives the result stated. □

An alternative formulation of the condition on N and δ is that δ should be at least $(\log\log N)^{-c}$ for some absolute constant $c > 0$. We have the following immediate corollary.

Corollary 7.2. *There is an absolute constant $c > 0$ with the following property. If the set $\{1, 2, \ldots, N\}$ is coloured with at most $(\log\log N)^c$ colours, then there is a monochromatic arithmetic progression of length four.* □

8. Concluding Remarks

Most of the above proof generalizes reasonably easily, with the result that it is not hard to guess the basic outline of a proof of Szemerédi's complete theorem. To be more precise, the results of Sections 2 and 6 have straightforward generalizations, and the result of Section 5 can also be generalized appropriately, although not in quite as obvious a manner. The main difficulty with the general case is in proving a suitable generalization of Corollary 4.10. What is needed, which is the main result of [G3], is a generalization to functions defined on subsets of $\mathbb{Z}_N^k$ with a suitable arithmetical structure. For such functions, one must find a large arithmetic progression $P \subset \mathbb{Z}_N$ and a set of the form $Q = (P + r_1) \times \ldots \times (P + r_k)$ such that ϕ agrees with a multilinear function γ for many points in Q. Even the case $k = 2$ is not at all easy.

The bounds obtained for Theorem 7.1 and Corollary 7.2 improve enormously on any that were previously known. However, they are most unlikely to be best possible. It is tempting to try to improve them to bounds of the form $\delta \geqslant (\log N)^{-c}$ by using ideas from the papers of Szemerédi [S3] and Heath-Brown [H-B], but this idea raises difficulties that appear to be serious. It would also be nice to combine the ideas of this paper with techniques for proving additive statements about the primes in order to show that the primes contain infinitely many progressions of length four. However, the argument of Lemma 4.7 leads to bounds that are much too weak to be of use with sets of density $(\log N)^{-1}$. On the other hand, it ought to be possible to improve them significantly and such an improvement might well help to solve some old problems about the primes.

References

[BS] A. Balog and E. Szemerédi, *A Statistical Theorem of Set Addition*, Combinatorica **14** (1994), 263–268.

[Bo] J. Bourgain, *On triples in arithmetic progression*, Geom. Funct. Anal. **9** (1999), 968–984.

[CGW] F. R. K. Chung and R. L. Graham, *Quasi-random subsets of* $\mathbb{Z}_n$, J. Combin. Theory Ser. A **61** (1992), 64–86.

[ET] P. Erdős and P. Turán, *On some sequences of integers*, J. London Math. Soc. **11** (1936), 261–264.

[F1] G. R. Freiman, Foundations of a Structural Theory of Set Addition, (in Russian), Kazan Gos. Ped. Inst., Kazan (1966).

[F2] G. R. Freiman, Foundations of a Structural Theory of Set Addition, Translations of Mathematical Monographs **37**, Amer. Math. Soc., Providence, R. I., USA.

[Fu] H. Furstenberg, *Ergodic behaviour of diagonal measures and a theorem of Szemerédi on arithmetic progressions*, J. Analyse Math. **31** (1977), 204–256.

[G1] W. T. Gowers, *Lower Bounds of Tower Type for Szemerédi's Uniformity Lemma*, Geometric and Functional Analysis **7** (1997), 322–337.

[G2] W. T. Gowers, *A new proof of Szemerédi's theorem for arithmetic progressions of length four*, Geometric and Functional Analysis **8** (1998), 529–551.

[G3] W. T. Gowers, *A new proof of Szemerédi's theorem*, Geometric and Functional Analysis **11** (2001), 465–588.

[H-B] D. R. Heath-Brown, *Integer sets containing no arithmetic progressions*, J. London Math. Soc. (2) **35** (1987), 385–394.

[N] M. B. Nathanson, Additive Number Theory: Inverse Problems and the Geometry of Sumsets, Graduate Texts in Mathematics 165, Springer-Verlag 1996.

[R1] K. F. Roth, *On certain sets of integers*, J. London Math. Soc. **28** (1953), 245–252.

[R2] K. F. Roth, *Irregularities of sequences relative to arithmetic progressions, IV*, Period. math. Hungar. **2** (1972), 301–326.

[Ru] I. Ruzsa, *Generalized arithmetic progressions and sumsets*, Acta Math. Hungar. **65** (1994), 379–388.

[Sh] S. Shelah, *Primitive Recursive Bounds for van der Waerden Numbers*, J. Amer. Math. Soc. **1** (1988), 683–697.

[S1] E. Szemerédi, *On sets of integers containing no four elements in arithmetic progression*, Acta Math. Acad. Sci. Hungar. **20** (1969), 89–104.

[S2] E. Szemerédi, *On sets of integers containing no k elements in arithmetic progression*, Acta Arith. **27** (1975), 299–345.

[S3] E. Szemerédi, *Integer sets containing no arithmetic progressions*, Acta Math. Hungar. **56** (1990), 155–158.

[W] H. Weyl, *Über die Gleichverteilung von Zahlen mod Eins*, Math. Annalen **77** (1913), 313–352.

Reprinted from Proc. Int. Congr. Math., 1988

THE WORK OF MAXIM KONTSEVICH

by
CLIFFORD HENRY TAUBES
Harvard University
Cambridge, MA 02138, USA

Maxim Kontsevich is known principally for his work on four major problems in geometry. In each case, it is fair to say that Kontsevich's work and his view of the issues has been tremendously influential to subsequent developments. These four problems are:

- Kontsevich presented a proof of a conjecture of Witten to the effect that a certain, natural formal power series whose coefficients are intersection numbers of moduli spaces of complex curves satisfies the Korteweg-de Vries hierarchy of ordinary, differential equations.
- Kontsevich gave a construction for the universal Vassiliev invariant for knots in 3-space, and generalized this construction to give a definition of pertubative Chern-Simons invariants for three dimensional manifolds. In so doing, he introduced the notion of Graph Cohomology which succinctly summarizes the algebraic side of the invariants. His constructions also vastly simplified the analytic aspects of the definitions.
- Kontsevich used the notion of stable maps of complex curves with marked points to compute the number of rational, algebraic curves of a given degree in various complex projective varieties. Moreover, Kontsevich's techniques here have greatly affected this branch of algebraic geometry. Kontsevich's formulation with Manin of the related Mirror Conjecture about Calabi-Yau 3-folds has also proved to be highly influential.
- Kontsevich proved that every Poisson structure can be formally quantized by exhibiting an explicit formula for the quantization.

What follows is a brief introduction for the non-expert to these four areas of Kontsevich's work. Here, I focus almost solely on the contributions of Kontsevich to the essential exclusion of many other; and I ask to be pardonned for my many and glaring omissions.

1. Intersection Theory on the Moduli Space of Curves and the Matrix Airy Function [1]

To start the story, fix integers $g \geq 0$ and $n > 0$ which are constrained so $2g + n \geq 2$. That is, the compact surface of genus g with n punctures has negative Euler characteristic. Introduce the moduli space $M_{g,n}$ of smooth, compact, complex curves of genus g with n distinct marked points. This is to say that a point in $M_{g,n}$ consists of an equivalence class of tuple consisting of a complex structure j on a compact surface C of genus g, together with an ordered set $\Lambda \equiv \{x_1, \ldots, x_n\} \subset C$ of n points. The equivalence is under the action of the diffeomorphism group of the surface. This $M_{g,n}$ has a natural compactification (known as the Deligne-Mumford compactification) which will not be notationally distinguished. Suffice it to say that the compactification has a natural fundamental class, as well as an n-tuple of distinguished, complex line bundles. Here, the i'th such line bundle, L_i, at the point $(j, \Lambda) \in M_{g,n}$ is the holomorphic cotangent space at $x_i \in \Lambda$.

With the preceding understood, note that when $\{d_1, \ldots, d_n\}$ are non-negative integers which sum to the dimension of $M_{g,n}$ (which is $3g - 3 + n$). Then, a number is obtained by pairing the cohomology class

$$\prod_{1 \leq i \leq n} c_i(L_i)^{d_i}$$

with the afore-mentioned fundamental class of $M_{g,n}$. (Think of representing these Chern classes by closed 2-forms and then integrating the appropriate wedge product over the smooth part of $M_{g,n}$.) Using Poincaré duality, such numbers can be viewed as intersection numbers of varieties on $M_{g,n}$ and hence the use of this term in the title of Kontsevich's article.

As g, n and the integers $\{d_1, \ldots, d_n\}$ vary, one obtains in this way a slew of intersection numbers from the set of spaces $\{M_{g,n}\}$. In this regard, it proved convenient to keep track of all these numbers with a generating functional. The latter is a formal power series in indeterminants $t_0, t_1, \ldots$ which is written schematically as

$$F(t_0, t_1, \ldots) = \sum_{(k)} \langle \tau_0^{k_0} \tau_1^{k_1} \cdots \rangle \prod_{i \geq 0} \frac{t_i^{k_i}}{k_i!}, \tag{1}$$

where, (k) signifies the multi-index $(k_0, k_1, \ldots)$ consisting of non-negative integers where only finitely many are non-zero. Here, the expression $\langle \tau_0^{k_0} \tau_1^{k_1} \cdots \rangle$ is the number which is obtained as follows: Let

$$n = k_1 + k_2 + \cdots, \quad \text{and} \quad g = \frac{1}{3}(2(k_1 + 2k_2 + 3k_3 + \cdots) - n) + 1.$$

If g is not a positive integer, set $\langle \tau_0^{k_0} \tau_1^{k_1} \cdots \rangle = 0$. If g is a positive integer, construct on $M_{g,n}$ the product of $c_1(L_j)$ for $1 \leq j \leq k_1$ times the product of $c_1(L_j)^2$ for $k_1 + 1 \leq j \leq k_1 + k_2$ times ... etc.; and thus construct a form whose dimension is

$3g - 3 + n$, which is that of $M_{g,n}$. Finally, pair this class on the fundamental class of $M_{g,n}$ to obtain $\langle \tau_0^{k_0} \tau_1^{k_1} \cdots \rangle$.

By comparing formal properties of two hypothetical quantum field theories, E. Witten was led to conjecture that the formal series $U \equiv \partial^2 F/\partial t_0^2$ obeys the classical KdV equation,

$$\frac{\partial U}{\partial t_1} = U \frac{\partial U}{\partial t_0} + \frac{1}{12} \frac{\partial^3 U}{\partial t_0^3}. \tag{2}$$

(As U is a formal power series, this last formula can be viewed as a conjectural set of relations among the intersection numbers which appear in the definition of F in (1).)

Kontsevich gave the proof that U obeys this KdV equation. His proof of Equation (2) is remarkable if nothing else then for the fact that he gives what is essentially an explicit calculation of the intersection numbers $\{\langle \tau_0^{k_0} \tau_1^{k_1} \cdots \rangle\}$. To this end, Kontsevich first introduces a model for $M_{g,n}$ based on what he calls ribbon graphs with metrics. (A ribbon graph is obtained from a 3-valent graph by more or less thickening the edges to bands. They are related to Riemann surfaces through the classical theory of quadratic differentials.) With an explicit, almost combinatorial model for $M_{g,n}$ in hand, Kontsevich proceeds to identify the classes $c_1(L_j)$ directly in terms of his model. Moreover, this identification is sufficiently direct to allow for the explicit computation of the integrals for $\{\langle \tau_0^{k_0} \tau_1^{k_1} \cdots \rangle\}$. It should be stressed here that this last step involves some extremely high powered combinatorics. Indeed, many of the steps in this proof exhibit Kontsevich's unique talent for combinatorial calculations. In any event, once the coefficients of U are obtained, the proof ends with an identification of the expression for U with a novel expansion for certain functions which arises in the KdV story. (These are the matrix Airy functions referred to at the very start of this section.)

2. Feynman Diagrams and Low Dimensional Topology [2]

From formal quantum field theory arguments, E. Witten suggested that there should exist a family of knot invariants and three manifolds invariants which can be computed via multiple integrals over configuration spaces. Kontsevich gave an essentially complete mathematical definition of these invariants, and his ideas have profoundly affected subsequent developments.

In order to explain, it proves useful to first digress to introduce some basic terminology. First of all, the three dimensional manifolds here will be all taken to be smooth, compact and oriented, or else Euclidean space. A knot in a three manifold is a connected, 1-dimensional submanifold, which is to say, the embedded image of the circle. A link is a finite, disjoint collection of knots. A knot or link invariant is an assignment of some algebraic data to each knot or link (for example, a real number), where the assignments to a pair of knots (or links) agree when one member of the pair is the image of the other under a diffeomorphism of the ambient manifold. (One might also restrict to diffeomorphisms which can be connected by a path of diffeomorphisms to the identity map.)

A simple example is provided by the Gauss linking number an invariant of links with two components which can be computed as follows: Label the components as K_1 and K_2. A point in K_1 together with one in K_2 provides the directed vector from the former to the latter, and thus a point in the 2-sphere. Since both K_1 and K_2 are copies of the circle, this construction provides a map from the 2-torus (the product of two circles) to the 2-sphere. The Gauss linking number is the degree of this map. (The invariance of the degree under homotopies implies that this number is an invariant of the link.) Alternately, one can introduce the standard, oriented volume form ω on the 2-sphere, and then the Gauss linking number is the integral over the $K_1 \times K_2$ of the pull-back of the form ω.

With conjectured the existence of a vast number of knot, link and 3-manifold invariants of a form which generalizes this last formula for the Gauss linking number. Independently of Kontsevich, significant work towards constructing these invariants for knots and links had been carried out by Bar-Natan, Birman, Garoufalidis, Lin, and Guadagnini-Martinelli-Mintchev. Meanwhile, Axelrod and Singer had developed a formulation of the three-manifold invariants.

In any event, what follows is a three step sketch of Kontsevich's formulation for an invariant of a three-manifold M with vanishing first Betti number.

Step 1: The invariants in question will land in a certain graded, abelian group which is constructed from graphs. Kontsevich calls these groups "graph cohomology groups." To describe the groups, introduce the set G_0 of pairs consisting of a compact graph Γ with only three-valent vertices and a certain kind of orientation o for Γ. To be precise, o is an orientation for

$$\left(\bigoplus_{\text{edges}(\Gamma)} \mathbb{R}\right) \oplus H^1(\Gamma).$$

Note that isomorphisms between such graphs pull back the given o. Thus, one can think of G_0 as a set of isomorphism classes. Next, think of the elements of G_0 as defining a basis for a vector space over $\mathbb{Z}$ where consistency forces the identification of $(\Gamma, -o)$ with $-(\Gamma, o)$.

One can make a similar definition for graphs where all vertices are three valent save for one four valent vertex. The resulting $\mathbb{Z}$-module is called G_1. In fact, for each $n \geq 0$ there is a $\mathbb{Z}$-module G_n which is constructed from graphs with all vertices being at least 3-valent, and with the sum over the vertices of (valence -3) equal to n.

With the set $\{G_n\}_{n\geq 0}$ more or less understood, remark that there are natural homomorphisms $\partial : G_n \to G_{n+1}$ which obey $\partial^2 = 0$. Indeed, ∂ is defined schematically as follows:

$$\partial(\Gamma, o) = \sum_{e \in \text{edges}(\Gamma)} (\Gamma/e, \text{ induced orientation from } o).$$

Here, Γ/e is the graph which is obtained from Γ by contracting e to a point. The induced orientation is quite natural and left to the reader to work out. In any event, with ∂ in hand, the modules $\{G_n\}$ define a differential complex, whose cohomology groups are

$$GC_* \equiv \text{kernel}(\partial : G_* \to G_{*+1})/\text{Image}(\partial : G_{*-1} \to G_*). \tag{3}$$

This is 'graph cohomology'. For the purpose of defining 3-manifold invariants, only GC_0 is required.

Step 2: Fix a point $p \in M$ and introduce in $M \times M$ the subvariety

$$\Sigma = (p \times M) \cup (M \times p) \cup \Delta,$$

where Δ denotes the diagonal. A simple Meyer-Vietoris argument finds closed 2-forms on $M \times M - \Sigma$ which integrate to 1 on any linking 2-sphere of any of the three components of Σ. Moreover, there is such a form ω with $\omega \wedge \omega = 0$ near Σ. In fact, near Σ, this ω can be specified almost canonically with the choice of a framing for the tangent bundle of M. (The tangent bundle of an oriented 3-manifold can always be framed. Furthermore, Atiyah essentially determind a canonical frame for TM.) Away from Σ, the precise details of ω are immaterial. In any event, fix ω using the canonical framing for TM.

With ω chosen, consider a pair (Γ, o) from G_0. Associate to each vertex of Γ a copy of M, and to each oriented edge e of Γ, the copy of $M \times M$ where the first factor of M is labeled by the staring vertex of e, and the second factor by the ending vertex. Associate to this copy of $M \times M$ the form ω, and in this way, the edge e labels a (singular) 2-form ω_e on $\times_{\text{vertices}(\Gamma)} M$.

Step 3: At least away from all versions of the subvariety Σ, the forms $\{\omega_e\}_{e \in \text{edges}(\Gamma)}$ can be wedged together to give a top dimensional form $\Pi_{e \in \text{edges}(\Gamma)} \omega_e$, on $\times_{v \in \text{vertices}(\Gamma)} M$. It is a non-trivial task to prove that this form is integrable. In any event, the assignment of this integral to the pair (Γ, o) gives a $\mathbb{Z}$-linear map from G_0 to $\mathbb{R}$. The latter map does not define an invariant of M from the pair (Γ, o) as there are choices involved in the definition of ω, and these choices effect the value of the integral. However, Kontsevich found a Stokes theorem argument which shows that this map from G_0 to $\mathbb{R}$ descends to the kernel of ∂ as an invariant of M. That is, these graph-parameterized integrals define a 3-manifold invariant with values in the dual space $(GC_0)^*$. (A recent paper by Bott and Cattaneo has an exceptionally elegant discussion of these points.)

Kontsevich's construction of 3-manifold invariants completely separates the analytic issues from the algebraic ones. Indeed, the module GC_0 encapsulates all of the algebra; while the analysis, as it were, is confined to issues which surround the integrals over products of M. In particular, much is known about GC_0; for example, it is known to be highly non-trivial.

Kontsevich has a similar story for knots which involves integrals over configuration spaces that consist of points on the knot and points in the ambient space. Here, there is a somewhat more complicated analog of graph cohomology. In the case of knots in 3-sphere, Kontsevich's construction is now known to give all Vassiliev invariant of knots.

In closing this section, it should be said that Kontsevich has a deep understanding of these and related graph cohomology in terms of certain infinite dimensional algebras [3].

3. Enumeration of Rational Curves Via Torus Actions [4]

The general problem here is as follows: Suppose X is a compact, complex algebraic variety in some complex projective space. Fix a 2-dimensional homology class on X and 'count' the number of holomorphic maps from the projective line $\mathbb{P}^1$ into X which represent the given homology class. To make this a well posed problem, maps should be identified when they have the same image in X. The use of quotes around the word count signifies that further restrictions are typically necessary in order to make the problem well posed. For example, a common additional restriction fixes some finite number of points in X and requires the maps in question to hit the given points.

These algebro-geometric enumeration problems were considered very difficult. Indeed, for the case where $X = \mathbb{P}^2$, the answer was well understood prior to knotsevich's work only for the lowest multiples of the generator of $\mathrm{H}_2(\mathbb{P}^2; \mathbb{Z})$. Kontsevich synthesized an approach to this counting problem which has been quickly adopted by algebraic geometers as the method of choice. Of particular interest are the counts made by Kontsevich for the simplest case of $X = \mathbb{P}^2$ and for the case where X is the zero locus in $\mathbb{P}^4$ of a homogeneous, degree 5 polynomial. (The latter has trivial canonical class which is the characterization of a Calabi-Yau manifold.)

There are two parts to Kontsevich's approach to the counting problem. The first is fairly general and is roughly as follows: Let V be a compact, algebraic variety and let β denote a 2-dimentional homology class on V. Knotsevich introduces a certain space M of triples (C, x, f) where C is a connected, compact, reduced complex curve, while $x = (x_1, \ldots, x_k)$ is a k-tuple of pairwise distinct points on C and $f : C \to V$ is a holomorphic map which sends the fundamental class of C to β. Moreover, the associated automomorphism group of f is suitably constrained. (Here, k could be zero.) This space M is designed so that its compactification is a reasonable, complex algebraic space with a well defined fundamental class. (This compactification covers, in a sense, the oft used Deligne-Mumford compactification of the space of complex curves with marked points.) The utilization of this space M with its compactification is one key to Kontsevich's approach. In particular, suppose $X \subset V$ is an algebraic subvariety. Under certain circumstances, the problem of counting holomorphic maps from C into X can be computed by translating the latter problem into that of evaluating the pairing of M's fundamental class with certain products of Chern classes on M. The point here is that the condition that a map $f : C \to V$ lie in X can be reinterpreted as the condition that the corresponding points in M lie in the zero locus of a certain section of a certain bundle over M.

With these last points understood, Part 2 of Kontsevich's approach exploits the observation that $V = \mathbb{P}^n$ has a non-trivial torus action. Such an action induces one on M and its compactification. Then, in the manner of Ellingsrud and Stromme, Kontsevich uses one of Bott's fixed point formulas to obtain a formula for the appropriate Chern numbers in various interesting examples.

4. Deformation Quantization of Poisson Manifolds

This last subject comes from very recent work of Kontsevich, so the discussion here will necessarily be brief. A 'Poisson structure' on a manifold X can be thought of as a bilinear map

$$B_1 : C^\infty(X) \otimes C^\infty(X) \to C^\infty(X)$$

which gives a Lie algebra structure to $C^\infty(X)$. In particular, B_1 sends a pair (f, g) to $\langle \alpha, \mathrm{d}f \wedge \mathrm{d}g \rangle$ where α is an non-degenerate section of $\Lambda^2 TX$ which satisfies a certain quadratic differential constraint. The problem of quantizing such a Poisson structure can be phrased as follows: Let h be a formal parameter (think Planck's constant). Find a set of bi-differential operators B_2, $B_3, \ldots$ so that

$$f * g \equiv fg + h \cdot B_1(f, g) + h^2 \cdot B_2(f, g) + \cdots$$

defines an associative product taking pairs of functions on X and returning a formal power series with $C^\infty(X)$ valued coefficients. (A bi-differential operator acts as a differential operator on each entry separately.) Kontsevich solves this problem by providing a formula for $\{B_{2,3}, \ldots\}$ in terms of B_1. The solution has the following remarkable form

$$f * g = \sum_{0 \le n \le \infty} h^n \sum_{\Gamma \in G[n]} \omega_\Gamma B_{\Gamma,\alpha}(f, g),$$

where

- $G[n]$ is a certain set of $(n(n+1))^n$ labeled graphs with $n+2$ vertices and n edges.
- $B_{\Gamma,\alpha}$ is a bi-differential operator whose coefficients are constructed from multiple order derivatives of the given α by a rules which come from the graph Γ.

- ω_Γ is a number which is obtained from Γ by integrating a certain Γ-dependent differential form over the configuration space of n distinct points in the upper half plane.

The details can be found in [5].

References

[1] *Intersection theory on the moduli space of curves and the matrix Airy Function*, Commun. Math. Phys. 147 (1992) 1–23.

[2] *Feynman diagrams and low-dimensional topology*, in: First European Congress of Mathematics (Paris 1992), Vol. II, Progress in Mathematics 120, Birkhauser (1994) 97–121.

[3] *Formal, (non)-commutative symplectic geometry*, in: Proceedings of the I. M. Gelfand Seminar, 1990–1992, edited by L. Corwin, I. Gelfand & Lepowsky, Birkhauser (1993), 173–188.

[4] *Enumeration of rational curve via torus actions*, in: The Moduli Space of Curves, edited by R. Dijkgraaf, C. Faber & G. Van der Geer, Progress in Mathematics 129, Birkhauser (1995) 120–139.

[5] *Deformation quantization of Poisson manifolds*, I, preprint 1997.

Maxim Kontsevich

MAXIM KONTSEVICH

BIOGRAPHICAL SKETCH

Maxim Kontsevich, born on August 25, 1964 in Klimki near Moscow graduated from Moscow State University in 1985, and after that worked until 1990 as a junior researcher in the Institute for Problems of Information Transmission. After being a visitor in several prestigious institutions such as Max-Planck-Institut für Mathematik in Bonn, Harvard University and Institute for Advanced Study in the United States, he hold a professor position at the University of California at Berkeley from July 1993 to August 1995. Since then, he is permanent professor at the IHÉS where he was hired at the age of 31.

RESEARCH ACTIVITY

Maxim Kontsevich works mainly on mathematical structures related with the modern theoretical physics.

In late 80-ies he proposed simultaneously with G.Segal a rigorous mathematical formulation of two-dimensional Conformal Field Theory.

In his thesis he proved a remarkable conjecture of E.Witten relating characteristic classes of moduli spaces of stable curves with integrable systems. The proof includes essentially the first use in mathematics of the technique of Feynman diagrams. Then he discovered a field-theoretic proof of the main result of the theory of so-called finite type invariants (Vassiliev invariants) of knots in 3-dimensional space.

Later he found a deep connection between operads, Lie algebra cohomology, Feynman graphs and topology, summarized in his talk on the first European Congress of Mathematicians (Paris 1992). In his paper with Yuri Manin he proposed the mathematical formulation of the topological sigma-model, together with an important new notion of a stable map, giving foundations for the theory of quantum cohomology. In his talk on the International Congress of Mathematicians (Berlin, 1998) he proposed a new viewpoint on Mirror Symmetry as an equivalence between two different triangulated categories, the derived category of coherent sheaves on a complex algebraic variety and so-called Fukaya category associated with a dual symplectic manifold. This viewpoint was later confirmed in theoretical physics after the discovery of D-branes. Also the homological mirror conjecture of Kontsevich has many deep consequences for the pure algebra and deformation theory.

In his paper "Deformation quantization of Poisson manifolds" he proved a long standing conjecture on the existence of a formal noncommutative deformation of the algebra of functions on a smooth manifold in the direction of a given Poisson bracket. He not only proved the existence but explicitly constructed the deformation using methods of the perturbative quantum field theory and a delicate analysis of possible counterterms. Surprisingly there is an action of the absolute motivic Galois group on the space of possible universal formulas.

With Yan Soibelman he worked on the geometric description of collapsing Calabi-Yau manifolds in mirror symmetry. This work led him to a description of singular Ricci-nonnegative spaces in probabilistic terms, conjectural compactification of the moduli space of conformal field theories, and the nonarchimedean Kahler geometry.

Also recently he worked on new foundations of the theory of qunatum fields, and on a rigorous approach to renormalization.

PRIZES and DISTINCTIONS

Prix Otto Hahn, Max-Planck Gesellschaft, 1992.

Prize of Mairie of Paris, First European Congress of Mathematicians, Paris, 1992.

Iagolnitzer Prize of International Association of Mathematical Physics, Brisbane (Australia), 1997.

Fields Medal, International Congress of Mathematicians, Berlin, 1998.

MAIN PUBLICATIONS

[1] Intersection theory on the moduli spaces of curves and the matrix Airy function, *Comm. Math. Phys.* **147** (1992) 1–23.

[2] Vassiliev's knot invariants, *Adv. Soviet Math.* **16** (Part 2) (1993) 137–150.

[3] Formal (non)-commutative symplectic geometry, in *The Gelfand Mathematical Seminars, 1990–1992*, eds. L. Corwin, I. Gelfand and J. Lepowsky (Birkhäuser 1993) 173–187.

[4] Feynman diagrams and low-dimensional topology, in *First European Congress of Mathematics, 1992, Paris, Volume II*, Progress in Mathematics **120** (Birkhäuser 1994) 97–121.

[5] Gromov-Witten classes, quantum cohomology, and enumerative geometry, (with Yu. Manin), *Comm. Math. Phys.* **164** (3) (1994) 525–562; and *Mirror Symmetry II*, eds. B. Greene and S.-T. Yau (AMS and International Press, 1997) 607–654.

[6] Homological algebra of Mirror Symmetry, in *Proceedings of the International Congress of Mathematicians, Zuerich 1994*, Vol. **I** (Birkhäuser 1995) 120–139.

[7] Lyapunov exponents and Hodge structures, in: *The Mathematical Beauty of Physics: In Memory of Claude Itzykson: Saclay, 5-7 June 1996*, eds. J. M. Drouffe and

J. B. Zuber, Advanced Series in Mathematical Physics **24** (World Scientific, 1997) 318–332.

[8] Deformation quantization of Poisson manifolds, I, Preprint IHÉS/M/97/72 (1997).

[9] Operads and motives in deformation quantization, *Lett. Math. Phys.* **48** (1999) 35–72.

[10] Homological mirror symmetry and torus fibrations (with Y. Soibelman), in: *Symplectic geometry and mirror symmetry*, eds. K. Fukaya, Y.-G. Oh., K. Ono and G. Tian (World Scientific, 2001) 203–263.

Reprinted from The Gelfand Mathematical Seminars, 1990–1992

FORMAL (NON)-COMMUTATIVE SYMPLECTIC GEOMETRY

by

MAXIM KONTSEVICH

Some time ago B. Feigin, V. Retakh and I had tried to understand a remark of J. Stasheff [15] on open string theory and higher associative algebras [16]. Then I found a strange construction of cohomology classes of mapping class groups using as initial data any differential graded algebra with finite-dimensional cohomology and a kind of Poincaré duality.

Later generalizations to the commutative and Lie cases appeared. In attempts to formulate all this I have developed a kind of (non)-commutative calculus. The commutative version has fruitful applications in topology of smooth manifolds in dimensions ≥ 3. The beginnings of applications are perturbative Chern-Simons theory (S. Axelrod and I.M. Singer [1] and myself), V. Vassiliev's theory of knot invariants and discriminants (see [19], new results in [2]) and V. Drinfeld's works on quasi-Hopf algebras (see [6]), also containing elements of Lie calculus.

Here I present the formal aspects of the story. Theorem 1.1 is the main motivation for my interest in non-commutative symplectic geometry. Towards the end the exposition becomes a bit more vague and informal. Nevertheless, I hope that I will convince the reader that non-commutative calculus has every right to exist.

I have benefited very much from conversations with B. Feigin, V. Retakh, J. Stasheff, R. Bott, D. Kazhdan, G. Segal, I.M. Gelfand, I. Zakharevich, J. Cuntz, Yu. Manin, V. Ginzburg, M. Kapranov and many others.

1. Three Infinite-Dimensional Lie Algebras

Let us define three Lie algebras. The first one, denoted by ℓ_n, is a certain Lie subalgebra of derivations of the free Lie algebra generated by $2n$ elements $p_1, \ldots, p_n, q_1, \ldots, q_n$.

By definition, ℓ_n consists of the derivations acting trivially on the element $\Sigma[p_i, q_i]$.

The second Lie algebra a_n is defined in the same way for the free associative algebra without unit generated by $p_1, \ldots, p_n, q_1, \ldots, q_n$.

The third Lie algebra c_n is the Lie algebra of polynomials

$$F \in \mathbf{Q}[p_1, \dots, p_n, q_1, \dots, q_n]$$

such that $F(0) = F'(0) = 0$, with respect to the usual Poisson bracket

$$\{F, G\} = \sum \left(\frac{\partial F}{\partial p_i} \frac{\partial G}{\partial q_i} - \frac{\partial F}{\partial q_i} \frac{\partial G}{\partial p_i} \right).$$

One can define c_n also as the Lie algebra of derivations of a free polynomial algebra $\mathbf{Q}[p_*, q_*]$ preserving the form $\Sigma\, dp_i \wedge dq_i$ and the codimension one ideal $(p_1, \dots, p_n, q_1, \dots, q_n)$.

In Section 4 we shall give an interpretation of the algebras ℓ_n, a_n as Poisson algebras in some versions of non-commutative geometry.

Our aim is a computation of the stable homology (with trivial coefficients) of these Lie algebras. The spirit of the (quite simple) computations is somewhere between Gelfand-Fuks computations (see [8] and [7]) and cyclic homology.

It is well known that all classical series of locally-transitive infinite-dimensional Lie algebras (formal vector fields, hamiltonian fields, contact fields, ...) have trivial or uninteresting stable (co)homology (see [11]). Our algebra h_n is a subalgebra of the algebra of hamiltonian vector fields, consisting of the vector fields preserving a point. Applying the Shapiro lemma one can relate its cohomology with the cohomology of the algebra of all polynomial (or formal) hamiltonian vector fields with coefficients in the adjoint representation. We want to mention here the recent work of I.M. Gelfand and O. Mathieu (see [9]) where some nonstable classes for the Lie algebra of all formal hamiltonian vector fields were constructed using cylic homology and non-commutative deformations.

If we denote by h_n one of these three series of algebras, then we have a sequence of natural embeddings $h_1 \subset h_2 \subset \cdots \subset h_\infty$ where the last algebra corresponds to the case of a countable infinite number of generators. Of course, $H_*(h_\infty) = \varinjlim H_*(h_n)$.

Let us denote by $\hat{h}_n$ the completion of h_n with respect to the natural grading on it. Then the continuous cohomology of $\hat{h}_n$ is in a sense dual to $H_*(h_n)$. More precisely, the grading on h_n induces a grading on its homology, $H_k(h_n) = \bigoplus_i H_k^{(i)}(h_n)$.

$$H^k_{\text{cont}}(\hat{h}_n) = \bigoplus_i (H_k^{(i)}(h_n))^*.$$

For the limit algebras h_∞ we have a structure of Hopf algebra on its homology (as is usual in K-theory). The multiplication comes from the homomorphism $h_\infty \oplus h_\infty \to h_\infty$ and the comultiplication is dual to the multiplication in cohomology.

This Hopf algebra is commutative and cocommutative. Thus $H_*(h_\infty)$ is a free polynomial algebra (in the $\mathbf{Z}/2\mathbf{Z}$-graded sense) generated by the subspace $PH_*(h_\infty)$ of primitive elements.

In all three cases we have an evident subalgebra $sp(2n) \subset h_n$ consisting of linear derivations. The primitive homology of $sp(2\infty)$ is well-known:

$$PH_k(sp(2\infty), \mathbf{Q}) = \left\{ \begin{array}{ll} \mathbf{Q}, & k = 3 (\mathrm{mod}\, 4) \\ 0, & k \neq 3 (\mathrm{mod}\, 4) \end{array} \right\}.$$

Now we can state our main result:

Theorem 1.1. *$PH_k(h_\infty)$ is equal to the direct sum of $PH_k(sp(\infty))$ for all three cases and*

(1) *(for the case ℓ_∞)*

$$\bigoplus_{n \geq 2} H^{2n-2-k}(\mathit{OutFree}(n), \mathbf{Q}),$$

where $\mathit{OutFree}(n)$ denotes the group of outer automorphisms of a free group with n generators,

(2) *(for the case a_∞)*

$$\bigoplus_{m>0, 2-2g-m<0} H^{4g-4+2m-k}(\mathcal{M}_{g,m}/\Sigma_m, \mathbf{Q}),$$

where $\mathcal{M}_{g,m}/\Sigma_m$ denotes the (coarse) moduli space of smooth complex algebraic curves of genus g with m punctures, (the quotient space modulo the action of the symmetric group is equal to the moduli space of curves with unlabeled punctures),

(3) *(for the case c_∞)*

$$\bigoplus_{n \geq 2} (\textit{Graph homology})_k^{(n)}$$

(see the definition below).

The grading on the homology groups arising from the natural grading on h_∞ is equal to $(2n-2)$, $(4g-4+2m)$ and $(2n-2)$, respectively.

2. Hamiltonian Vector Fields in the Ordinary Sense

Before starting the proof of the third case of Theorem 1.1, we define the graph complex. By a *graph* we mean a finite 1-dimensional CW-complex. Let us call an *orientation* of the graph Γ a choice of orientation of the real vector space $\mathbf{R}^{\{\text{edges of } \Gamma\}} \oplus H^1(\Gamma, \mathbf{R})$. For $n \geq 2$, $k \geq 1$ denote by $G_k^{(n)}$ the vector space over $\mathbf{Q}$ generated by the equivalence classes of pairs (Γ, or) where Γ is a connected nonempty graph with Euler characteristic $1-n$ and k vertices, such that degrees of all vertices are greater than or equal to 3 and (or) is an orientation of Γ. We impose the relation $(\Gamma, -\text{or}) = -(\Gamma, \text{or})$.

It follows that $(\Gamma, \text{or}) = 0$ for every graph Γ containing a simple loop (i.e., an edge attached by both ends to one vertex). The reason is that such graphs have automorphisms reversing orientation in our sense.

It is easy to see that $G_k^{(n)}$ is finite-dimensional for all k, n.

Define a differential on the vector space $G_*^{(n)} = \oplus_k G_k^{(n)}$ by the formula (for Γ without simple loops):

$$d(\Gamma, \mathrm{or}) = \sum_{e \in \{\text{edges of } \Gamma\}} (\Gamma/e, \text{induced orientation}).$$

Here Γ/e denotes the result of the contraction of the edge e, the "induced orientation" is the product of the natural orientation on the codimension-1 co-oriented subspace $\mathbf{R}^{\{\text{edges of } \Gamma\}/e} \subset \mathbf{R}^{\{\text{edges of } \Gamma\}}$ and the orientation on $H^1(\Gamma/e, \mathbf{R}) \simeq H^1(\Gamma, \mathbf{R})$.

One can easily check that $d^2 = 0$. Hence we have an infinite sequence of finite-dimensional complexes $G_*^{(n)}$. Define the graph complex without specification as the direct sum of the complexes $G_*^{(n)}$.

The homology groups of graph complexes have important topological applications. In a sense they are universal characteristic classes for diffeomorphism groups of manifolds in odd dimensions ≥ 3. The idea of this relation comes from perturbative Chern-Simons theory. We shall describe this somewhere later.

Proof of the third case of Theorem 1.1. Recall that our Lie algebras c_n are $\mathbf{Z}_{\geq 0}$-graded. Thus the standard chain complex $\bigwedge^*(c_n)$ is graded. We consider the case when n is much larger than the grading degree.

It is well known that every Lie algebra acts (through the adjoint representation) trivially on its homology. The algebra $sp(2n) \subset c_n$ acts reductively on $\bigwedge^*(c_n)$. Hence the chain complex is canonically quasi-isomorphic to the subcomplex of $sp(2n)$-invariants.

The underlying vector space of the Lie algebra c_n as a representation of $sp(2n)$ is equal to

$$\bigoplus_{j \geq 2} S^j(V),$$

where $V = \mathbf{Q}\langle p_1, \ldots, p_n, q_1, \ldots, q_n \rangle$ is the defining $2n$-dimensional representation of $sp(2n)$.

Thus our chain complex as a representation of $sp(2n)$ is equal to the sum

$$\bigoplus_{k_2 \geq 0, k_3 \geq 0, \ldots} (\wedge^{k_2}(S^2(V) \otimes \wedge^{k_3}(S^3(V) \otimes \ldots).$$

Every summand is a space of tensors on V satisfying some symmetry conditions.

We can construct $(N-1)!! = 1 \cdot 3 \cdot \ldots \cdot (N-1)$ invariant elements in $V^{\otimes N}$ for every even N. Namely, each decomposition of the finite set $\{1, \ldots, N\}$ into pairs $(i_1, j_1), \ldots, (i_{N/2}, j_{N/2})$ where $i_1 < j_1, \ldots, i_{N/2} < j_{N/2}$; $i_1 < \cdots < i_{N/2}$ gives the tensor $\omega_{i_1 j_1} \ldots \omega_{i_{N/2} j_{N/2}}$, where ω_{ij} denotes the tensor of the standard skew-symmetric product on V^*.

By the Main Theorem of Invariant Theory these tensors will form a base of the space $(V^{\otimes N})^{sp(2n)}$ and there are no nonzero invariants for odd N if $2n = \dim(V)$ is sufficiently large.

Let us consider the space $\wedge^{k_2}(S^2(V) \otimes \wedge^{k_3}(S^3(V) \otimes \ldots$ as a quotient space of $(V^{\otimes 2})^{\otimes k_2} \otimes (V^{\otimes 3})^{\otimes k_3} \otimes \ldots$ Every pairing on the set $\{1, 2, \ldots, 2k_2 + 3k_3 + \ldots\}$ gives a graph with labeled vertices and edges in the following way: we consider $\{1, 2, \ldots, k_2 + k_3 + \ldots\}$ as a set of vertices, and the set of pairs as a set of edges. One can also provide in a canonical way these graphs with orientations. The passing to the quotient spaces modulo the action of symmetric groups corresponds to the consideration of graphs *without* labelings.

It is easy to see that we obtain a vector space analogous to our graph complex with two differences: 1) we consider now graphs not necessarily empty or connected, 2) vertices have degrees greater than or equal to 2. The differential in the new complex can be described in the same way as for the graph complex. The homology of the new complex is equal to the homology of c_∞.

One can check that the multiplication in the stable homology can be identified with the operation of disjoint union of graphs. Thus the primitive part arises from the subcomplex corresponding to the nonempty connected graphs.

There is a direct summand subcomplex of the last complex, consisting of "polygons", i.e. connected graphs with degrees of all vertices equal 2. It is easy to see that for $k \neq 3 (\mathrm{mod}\, 4)$ there exists an automorphism of k-gon reversing the orientation. Hence we obtain the trivial $sp(2\infty)$-part of primitive stable homology.

Let us consider now the subcomplex consisting of connected nonempty graphs containing at least one vertex of degree ≥ 3. We can associate with such a graph a new graph with degrees of all vertices greater than or equal to 3. The new graph is just the old graph with removed vertices of degree 2.

One can introduce a partial order on the set of equivalence classes of graphs by the possibility of obtaining one graph from another by a sequence of edge contractions.

In such a way we define a certain filtration on the bigger complex by the ordered set of graphs with degrees ≥ 3. For any such graph Γ the corresponding graded subquotient complex is the quotient complex of tensor products over the set of edges of Γ of some standard complexes modulo the action of the finite group $\mathrm{Aut}\,(\Gamma)$.

The standard complex for an edge has dimension 1 in each degree $k \geq 0$, because there exists unique up to an isomorphism way to put k points to the interior of the standard interval (edge). The differential in this standard complex kills all classes in positive degrees. Hence it has only one nontrivial homology in degree 0.

We see that the spectral sequence associated with the filtration by graphs with degrees ≥ 3 collapses at the first term to the graph complex. It is clear that Euler characteristic of a graph is preserved under any edge contraction. Thus the graph complex is the direct sum of its subcomplexes over all possible Euler characteristics.

The degree in the sense of the natural grading on $c_{2\infty}$ of a cycle associated with a graph Γ is equal to $-2\chi(\Gamma)$. $\square$

There are a lot of nontrivial classes in the graph complex. For example, any finite-dimensional Lie algebra $\mathfrak{g}$ with fixed nondegenerate invariant scalar product on it defines a sequence of classes of graph homology in positive even degrees.

One can choose an orthogonal base in $\mathfrak{g}$ with respect to the scalar product. The tensor of structure constants will be skew-symmetric 3-tensor. Each 3-valent graph defines up to a sign a way to contract indices in some tensor power of the tensor of structure constants. It is easy to see that we obtain a function $\Phi_{\mathfrak{g}}(\Gamma, \text{or})$ on the set of equivalence classes of 3-valent graphs with orientation.

The immediate consequence of the Jacobi identity is the fact that

$$\sum_{\text{equiv. classes of } (\Gamma,\text{or})} \frac{\Phi_{\mathfrak{g}}(\Gamma, \text{or})}{\#\,\text{Aut}(\Gamma)} (\Gamma, \text{or})$$

gives closed chains in all $G_*^{(n)}$. Thus one can construct some classes for every simple Lie algebra using the Killing scalar product.

3. Moduli Spaces of Graphs

It will be useful for us to describe graph homology as a kind of homology of topological spaces.

Denote by $\mathcal{G}^{(n)}$ for $n \geq 2$ the set of equivalence classes of pairs (Γ, metric) where Γ is a nonempty connected graph with Euler characteristic equalling $(1-n)$ and degrees of all vertices greater than or equal to 3, (metric) is a map from the set of edges to the set of positive real numbers $\mathbf{R}_{>0}$. One can introduce a topology on $\mathcal{G}^{(n)}$ using Hausdorff distance between metrized spaces associated in the evident way with pairs (Γ, metric). It is better to consider $\mathcal{G}^{(n)}$ not as an ordinary space, but as an orbispace (i.e. don't forget automorphism groups). Mention here that $\mathcal{G}^{(n)}$ is a non-compact and non-smooth locally polyhedral space. It has a finite stratification by combinatorial types of graphs with strata equal to some quotient spaces of Euclidean spaces modulo actions of finite groups.

A fundamental fact on the topology of $\mathcal{G}^{(n)}$ is the following theorem of M. Culler, K. Vogtmann (see [4]):

Theorem 3.1. *$\mathcal{G}^{(n)}$ is a classifying space of the group OutFree(n) of outer automorphisms of a free group with n generators.*

The virtual cohomological dimension of $\mathcal{G}^{(n)}$ is equal to $2n-3$ and the actual dimension is equal to $3n-3$.

Any representation of the group OutFree(n) gives a local system on $\mathcal{G}^{(n)}$. We can define (co)homology, also homology with closed support and cohomology with compact support of $\mathcal{G}^{(n)}$ with coefficients in any local system.

Let us denote by ϵ the 1-dimensional local system with the fiber over (Γ, metric) equal to $\wedge^n(H^1(\Gamma, \mathbf{Q}))$. A simple check shows that the chain complex computing $H_*^{\text{closed}}(\mathcal{G}^{(n)}, \epsilon)$ arising from the stratification above coincides with the shifted graph complex $G^{(n)}_{*+n-1}$.

Now we have a geometric realization of homology arising in the first and the third cases of Theorem 1.1.

Define a *ribbon graph* (or a *fatgraph* in other terms) as a graph with fixed cyclic orders on the sets of half-edges attached to each vertex. One can associate an oriented surface with boundary to each ribbon graph by replacing edges by thin oriented rectangles (ribbons) and glueing them together at all vertices according to the chosen cyclic order.

Denote by $\mathcal{R}^{(g,m)}$ the moduli space of connected ribbon graphs with metric, such that degrees of all vertices greater than or equal to 3 and the corresponding surface has genus g and m boundary components.

Theorem 3.2. *$\mathcal{R}^{(g,m)}$ is canonically isomorphic as an orbispace to $\mathcal{M}_{g,m} \times \mathbf{R}^m/\Sigma_m$, (and, hence is a classifying space of some mapping class group).*

This theorem follows from results of K. Strebel and/or R. Penner (see [17], [13] or an exposition in [12]).

The space $\mathcal{R}^{(g,m)}$ is a non-compact but smooth orbispace (orbifold), so there is a rational Poincaré duality. We want to mention here that due to the factor $\mathbf{R}^m$ and to the action of the symmetric group the orbifold $\mathcal{R}^{(g,m)}$ is *not* oriented for $m > 1$.

The virtual cohomological dimension of $\mathcal{R}^{(g,m)}$ is equal to $4g - 4 + m$ for $g \geq 1$ and to $m - 3$ for $g = 0$ (see [12]), the actual dimension is equal to $6g - 6 + 3m$.

Thus vector spaces arising in Theorem 1.1 could be written as

$$\bigoplus_{n\geq 2} H^{2n-2-k}(\mathcal{G}^{(n)}, \mathbf{Q}), \quad \bigoplus_{m>0, 2-2g-m<0} H^{4g-4+2m-k}(\mathcal{R}^{(g,m)}, \mathbf{Q}), \bigoplus_{n} \geq 2$$

respectively.

The evident forgetful map $\mathcal{R}^{(g,m)} \to \mathcal{G}^{(2g+m-1)}$ is proper. The orientation sheaf on $\mathcal{R}^{(g,m)}$ coincides with the pullback of ϵ under this map. Hence we obtain a sequence of linear maps

$$H^{2n-2-k}(\mathcal{G}^{(n)}, \mathbf{Q}) \to \bigoplus_{g,m:2g+m-1=n} H^{4g-4+2m-k}(\mathcal{R}^{(g,m)}, \mathbf{Q})$$

$$\simeq \bigoplus_{g,m:2g+m-1=n} H^{\text{closed}}_{2g+m-1-k}(\mathcal{R}^{(g,m)}, \epsilon) \to H^{\text{closed}}_{k+n-1}(\mathcal{G}^{(n)}, \epsilon).$$

We shall see in Section 5 that the composition map is zero.

4. (Non)-Commutative Symplectic Geometry

We shall describe here a (non)-commutative formalism surprisingly parallel to the usual calculus of differential forms and Poisson brackets. Almost everything will work literally at the same way in three possible worlds: Lie algebras, associative algebras and commutative algebras. Our formalism could be extended to the case of "Koszul dual pairs of quadratic operads" (see [10]) including Poisson algebras and, probably, operator algebras etc.

Let us fix a wor(l)d $\mathcal{A} \in \{\text{Lie, associative, commutative}\}$.

Definitions. *A formal $\mathcal{A}$-supermanifold is a complete free finitely generated $\mathbf{Z}/2\mathbf{Z}$-graded $\mathcal{A}$-algebra (nonunital in the case $\mathcal{A} \in \{Lie, associative\}$).*

A local coordinate system on a manifold is a choice of generators of the corresponding algebra.

A (formal) diffeomorphism between two manifolds is a continuous isomorphism between graded algebras.

A vector field is a continuous derivation.

Submanifold is a free quotient algebra.

The tangent space at zero is dual to the space of generators of algebra (= the quotient space of algebra by the maximal proper ideal).

All definitions above are just general categorical nonsense.

In ordinary calculus we can consider differential forms as functions on the odd tangent bundle to the manifold. We can define this object without difficulties in our situation:

Definition. *For supermanifold X, "the total space of the odd tangent bundle" ΠTX is the free differential envelope of X.*

For example, if X has coordinates $x_1, \ldots, x_n$ then ΠTX has coordinates $x_1, \ldots, x_n, dx_1, \ldots, dx_n$. The algebra corresponding to ΠTX is $\mathbf{Z}_{\geq 0}$-graded by the number of differentials. In other words, there is a canonical action of the multiplicative group scheme $\mathbf{G}_m$ on ΠTX. The presence of differential on the free differential envelope means that there is a canonical action of the odd affine group scheme $\mathbf{G}_a^{0|1}$ on ΠTX.

Now we are coming to the delicate point: what is the notion of function? We propose the following strange definition (where $\cdot$ denotes the operation in A):

Definition. *For $\mathcal{A}$-algebra A the space of* 0*-forms $F(A)$ is the quotient space*

$$A \otimes A/(\textit{subspace generated by } a \otimes b - b \otimes a \textit{ and } a \otimes (b \cdot c) - (a \cdot b) \otimes c)\,.$$

Of course, in the supercase one has to make appropriate sign corrections.

Functor $F(A)$ coincides with A^2 in the commutative case, with $A^2/[A, A]$ in the associative case (for unital A, $F(A)^*$ is equal to the space of traces on A), and with the functor considered by Drinfeld (see [6]) for the Lie case.

By functoriality we obtain the action of $\mathbf{G}_m$ and $\mathbf{G}_a^{0|1}$ on $F(\Pi TX)$ for any X. In other words, $F(\Pi TX)$ is a $\mathbf{Z}_{\geq 0}$-graded complex. We shall call it the de Rham complex of the manifold X.

Notations. *$F^i(X)$ for $i \geq 0$ is the i-th homogeneous comportent of $F(\Pi TX)$, d is the differential $F^i(X) \to F^{i+1}(X)$.*

It is clear that $F^0(X)$ coincides with the functor F applied to the algebra corresponding to the manifold.

One can define for a vector field ξ on a manifold two vector fields L_ξ, i_ξ on ΠTX by formulas

$$L_\xi(a) = \xi(a),\ L_\xi(da) = d(\xi(a)),\ i_\xi(a) = 0,\ i_\xi(da) = \xi(a)$$

for every $a \in A$. The following commutator relations hold:

$$L_\xi = i_\xi d + d i_\xi,\ i_\xi i_\eta + i_\eta i_\xi = 0,\ [L_\xi, i_\eta] = i_{[\xi,\eta]},\ [L_\xi, L_\eta] = L_{[\xi,\eta]}.$$

By functoriality we have analogous operations on the de Rham complex.

Using these formulas we can prove easily that for local manifolds the de Rham complex is exact. It follows from the fact that $L_e = [i_e, d]$ is an invertible operator on $F^*(X)$ where e denotes the Euler vector field $x_1 \frac{\partial}{\partial x_1} + \cdots + x_n \frac{\partial}{\partial x_n}$ on a manifold with coordinates $x_1, \ldots, x_n$.

The mini-theory developed above works well for many other functors (algebras) $\to$ (vector spaces) instead of $F(A)$. The advantage of our definition is the existence of symplectic theory.

It follows easily from the definitions that any 2-form on a manifold defined a skew-symmetric bilinear form on the tangent space at 0 (through the first coefficient in its Taylor expansion).

Definition. *Symplectic supermanifold is a pair (X, ω) where ω is closed even 2-form on X with nondegenerate restriction to T_0X.*

One can check that for any nondegenerate 2-form ω the operator $\xi \to i_\xi\omega$ is an isomorphism between the space of vector fields and the space of 1-forms. Thus by usual arguments vector fields preserving symplectic structure are in one-to-one correspondence with 0-forms. In fact, the Lie algebra of hamiltonian vector fields depends (up to inner automorphism) only on the dimension of the symplectic manifold.

Theorem 4.1. (Darboux theorem) *A symplectic manifold is isomorphic to the flat manifold, i.e. with*

$$\omega = \sum c_{\alpha\beta} dt_\alpha \otimes dt_\beta.$$

We shall not use here this theorem, so the proof will be omitted.

For the case of associative or Lie manifolds there exists a simple description of closed 2-forms:

Theorem 4.2. *For (associative or Lie) free algebra A there exists a canonical isomorphism $F^2_{\text{closed}}(A) \simeq [A, A]$.*

Proof. First of all, we define a map $t : F^1(A) \to [A, A]$ by the formula $t(a \otimes db) = [a, b]$. It is clear that this map is onto and it vanishes on $dF^0(A)$. Thus for the associative case we obtain the short sequence

$$A \to A^2/[A, A] \xrightarrow{d} F^1(A) \xrightarrow{t} [A, A] \to 0$$

exact everywhere, but the middle term. If we choose coordinates then we obtain a grading on all terms of this sequence. Simple dimension count shows that Euler characteristics of all graded components are zero (we know generating function of $F^1(A)$ because there exists an isomorphism $F^1(A) \simeq \{\text{derivations of } A\}$). Thus the sequence above is exact and coincides with the exact sequence

$$0 \to F^0(A) \xrightarrow{d} F^1(A) \xrightarrow{d} F^2_{\text{closed}}(A) \to 0.$$

Analogous but more lengthy arguments work for Lie algebras too. (For another approach see [6]). □

5. Sketch of the Proof of Theorem 1.1 for the Associative and the Lie Cases

Now we can combine all facts together.

It follows from Theorem 4.2 that algebras ℓ_n, a_n are algebras of hamiltonian vector fields on flat symplectic manifolds in non-commutative geometries. Thus they are canonically equivalent as vector spaces to 0-forms.

In the associative case the vector space of 0-forms on a flat manifold with the cotangent space V at zero as $GL(V)$-module is equal to

$$\bigoplus_{n \geq 2} (V^{\otimes n})^{\mathbf{Z}/n\mathbf{Z}}$$

where cyclic group $\mathbf{Z}/n\mathbf{Z}$ acts by permutations of factors in $V^{\otimes n}$. The same arguments as in the commutative case lead to the ribbon version of the graph complex. By Theorem 3.2 we obtain at the end *all* cohomology groups of all moduli spaces of complex curves with unlabeled punctures.

In the case of Lie algebras the situation is a bit more complicated. We say (without proof) that the space of 0-forms is now equal to

$$\bigoplus_{n \geq 2} (V^{\otimes n} \otimes L_n)^{\Sigma_n}$$

where L_n is a certain $(n-2)!$-dimensional representation of the symmetric group Σ_n. Again using the same strategy as in Section 2 we obtain the Lie version of the graph complex. As the vector space it will be the direct sum over equivalence classes of graphs of some vector spaces. The vector space associated with graph Γ will be the subspace of $\mathrm{Aut}(\Gamma)$-invariants in the tensor product of natural (degree of vertex

$-2)!$-dimensional vector spaces over all vertices twisted with the 1-dimensional local system ϵ.

On the other hand, we can construct a finite cell-complex homotopy equivalent to $B\mathrm{OutFree}(n)$ passing from the natural stratification of $\mathcal{G}^{(n)}$ to its baricentric subdivision. The corresponding cochain complex carries some filtration by graphs (by the minimal graph corresponding to strata attached to the cell). Computations show that the spectral sequence associated with this filtration collapses at the second term to the Lie version of graph complex. □

Two evident functors

$$\{\text{commutative algebras}\} \to \{\text{associative algebras}\} \to \{\text{Lie algebras}\}$$

lift to correspondences between 3 types of calculus, in particular, to homomorphisms of Lie algebras

$$c_\infty \to a_\infty \to \ell_\infty.$$

The composite map goes through sub- and quotient algebra $sp(2\infty)$, so it is zero on the nontrivial part of primitive homology. One can identify arising maps with geometric maps from Section 3.

The entire story above has an odd analogue. One has to consider superalgebras and odd symplectic structures. Then the stable homology will be described in the same way but with the twisted by ϵ coefficients.

The odd version of (commutative) graph homology plays the same role for smooth even-dimensional manifolds ($\dim \geq 4$) as the even version for odd dimensions.

6. Poisson Brackets: Formulas and Interpretations

In all 3 worlds the space of 1-forms on the flat manifold with coordinates $x_1, \ldots, x_n$ can be identified with the direct sum of n copies of the corresponding free algebra A:

$$(a_1, \ldots, a_n) \leftrightarrow \sum a_i \otimes dx_i.$$

Thus we can define linear operators $\frac{\partial}{\partial x_i} : F(A) \to A$ by formula $dH = \sum \frac{\partial H}{\partial x_i} \otimes dx_i$.

In the associative case 0-forms are linear combinations of cyclic words (of length ≥ 2) in alphabet $x_1, \ldots, x_n$. For example,

$$\frac{\partial(xxyxz)}{\partial x} = xyxz + yxzx + zxxy, \quad \frac{\partial(xxyxz)}{\partial y} = xzxx,$$

where $xxyxz$ is considered as a cyclic word.

The following basic identity holds in all 3 cases:

$$\sum \left[x_i, \frac{\partial H}{\partial x_i}\right] = 0.$$

It is just nothing in the commutative world. In the associative world one can prove this identity immediately using the description of F^0 above. The Lie case follows from the associative case by embeddings of free Lie algebras into free associative algebras.

There are universal formulas for Poisson brackets:

$$\{G, H\} = \sum \left(\frac{\partial G}{\partial p_i} \otimes \frac{\partial H}{\partial q_i} - \frac{\partial G}{\partial q_i} \otimes \frac{\partial H}{\partial p_i}\right),$$

hamiltonian vector field, corresponding to H is $\dot{p}_i = \frac{\partial H}{\partial q_i}$, $\dot{q}_i = \frac{\partial H}{\partial p_i}$. The invariance of $\Sigma[p_i, q_i]$ is equivalent to the identity above.

V. Drinfeld in [6] used another Poisson bracket on $F(A)$ (for the Lie case):

$$\{G, H\} = \sum_{i=1}^{n} x_i \otimes \left[\frac{\partial G}{\partial x_i}, \frac{\partial H}{\partial x_i}\right],$$

$$H \mapsto \text{vector field } \dot{x}_i = \left[x_i, \frac{\partial H}{\partial x_i}\right].$$

Later we shall give an interpretation of this bracket as a Kirillov bracket on the dual space to the "Lie" algebra $\mathbf{C}\oplus\cdots\oplus\mathbf{C}$ (n summands). In the Lie world "Lie" algebras are commutative algebras. As an abstract Lie algebra $F(A)$ with this bracket is a trivial central extension by $\langle x_1 \otimes x_1, \ldots, x_n \otimes x_n\rangle$ of the Lie algebra of derivations D of the free Lie algebra generated by x_i such that

$$\forall i\ \exists y_i \text{ such that } D(x_i) = [x_i, y_i], \quad D\left(\sum x_i\right) = 0.$$

Recall, that

(1) Teichmüller group $T_{g,1}$ is the group of automorphisms of the free group generated by $\{p_1, \ldots, p_g, q_1, \ldots, q_g\}$ preserving the element $\Pi p_i q_i p_i^{-1} q_i^{-1}$,
(2) pure braid group with n strings is the group of automorphisms of the free group generated by $\{x_1, \ldots, x_n\}$ preserving conjugacy classes of x_i and the element $x_1 x_2 \ldots x_n$.

Thus we see that in the Lie world Poisson algebra is an analogue of the Teichmüller group for flat symplectic manifolds, and an analogue of the pure braid group for Kirillov brackets.

Also, if K is a subfield of $\mathbf{C}$ containing all roots of unity and C is a smooth algebraic curve defined over K of genus g with one puncture or of genus 0 with $n+1$ punctures then the Galois group $\mathrm{Gal}(\bar{K}/K)$ acts on the ℓ-adic completion of the fundamental group of $C(\mathbf{C})$ through the ℓ-adic pro-nilpotent group with the corresponding Poisson algebra as the Lie algebra (ℓ is an arbitrary prime).

7. How Big are Stable Homologies?

We collect here some attempts to understand the "size" of stable homologies of Poisson algebras. The situation is not clear because different approaches give contradictory hints.

7.1. *Explicit constructions of stable classes*

There is a generalization of the construction mentioned at the end of Section 2 for the commutative case.

Let us start from some general remark. If $\mathfrak{G}$ is a Lie superalgebra and $D \in \mathfrak{G}^1$ is an odd element such that $[D, D] = 0$ then one can associate with D for any $k \geq 0$ some homology class of $\mathfrak{G}$ in degree k. The reason is that D produces a homomorphism from $0|1$-dimensional commutative algebra $\mathbf{A}^{0|1}$ to $\mathfrak{G}$ and $\dim H_k(\mathbf{A}^{0|1}) = (1|0)$ for k even and $(0|1)$ for k odd.

We can construct superanalogs of algebras ℓ_n, a_n, c_n starting from flat symplectic supermanifolds in the sense of the previous section. One can see that the stable homology are the same as in the pure even case because Main Theorem of Invariant Theory works with appropriate corrections also in the supercase.

Thus any odd Hamiltonian with vanishing Poisson bracket with itself produces stable classes in even degrees. For example, a finite-dimensional Lie algebra $\mathfrak{g}$ with a nondegenerate scalar product on it gives: 1) a pure odd symplectic (in ordinary super-commutative sense) manifold $X = \Pi\mathfrak{g}$, 2) an odd cubic polynomial H on X arising from the structure constants. Jacobi identity implies $[H, H] = 0$.

Formally the same construction works in all three cases. Define duality between types of algebras as

$$\text{Lie} \leftrightarrow \text{commutative}, \ \text{associative} \leftrightarrow \text{associative}.$$

For any finite-dimensional $\mathcal{A}$-algebra V with nondegenerate invariant scalar product on it, the odd vector space ΠV considered as a manifold of the dual type carries a symplectic structure and an odd hamiltonian vector field with square equal to 0.

If we restrict ourselves to finite-dimensional simple algebras over the field of complex numbers, then we obtain many examples (Dynkin diagrams) for the case of Lie algebras, essentially one example (matrix algebra) for the associative case, and *no* nontrivial examples in the commutative case (one-dimensional commutative algebra gives zero classes in nontrivial part of $PH_*(\ell_\infty)$).

We have tried to deform these examples. It turns out that there are many new classes in the case c_∞, some classes for a_∞ and no classes for ℓ_∞.

The basic example for the associative case is the following: symplectic manifold X is $0|1$-dimensional with the coordinate x and symplectic structure $\omega = dx \otimes dx$, odd hamiltonian H is arbitrary linear combination of $x \otimes x^{2k}$, $k \geq 0$. One can prove that the linear span of all stable classes via isomorphism of Theorem 1.1 is equal to the space of all polynomials in Morita-Miller-Mumford classes on moduli spaces of curves.

Thus the conclusion from this approach is that the nontrivial primitive part of stable homology of Poisson algebras looks big for the commutative case, moderate for the associative case and small or zero for the Lie case.

7.2. *Euler characteristics of (generalized) graph complexes*

The absolute value of Euler characteristic of a finite complex of vector spaces gives an estimate from below for the total dimension of its homology. It is much easier to compute Euler characteristics for our generalized graph complexes in an "orbifold" sense (see [3]). The last adjective means that we count each graph Γ with the weight equal to $1/\#\,\mathrm{Aut}\,(\Gamma)$. It is reasonable to expect that the "most" part of graphs has no nontrivial automorphisms.

The generating function

$$\sum_{k\geq 1} t^k \times (\text{orbifold Euler characteristic of the subcomplex of graphs with } \chi = -k)$$

in all cases is an asymptotic expansion for $t \to 0$ of

$$\log\left(\frac{\int_{\text{near }0} \exp(-F(x)/t)dx}{\sqrt{2\pi t}}\right)$$

where $F(x)$ is a series in x equal to

(1) $\sum\limits_{n\geq 2} \frac{x^n}{n(n-1)}$ for the Lie case,

(2) $\sum\limits_{n\geq 2} \frac{x^n}{n}$ for the associative case,

(3) $\sum\limits_{n\geq 2} \frac{x^n}{n!}$ for the commutative case.

These formulas follows from Feynman rules. The second and the third integral coincide! (It is a simple exercise in calculus.)

Thus we obtain quite big but the same numbers (Bernoulli numbers) for the second and the third cases and some bigger numbers for the first (Lie) case. It is absolutely different from the previous picture.

7.3. *Conjecture*

Computations show that $\dim(H_2(c_\infty)) = 1$. The unique up to a factor class corresponds to the ordinary "quantization", i.e. deformation of the Lie algebra structure on ordinary hamiltonian vector fields. (Recall that by the Shapiro lemma $H^*(c_n)$ is more or less equal to the deformation cohomology of hamiltonian fields). The graph representing this class has 2 vertices and 3 edges connecting both vertices.

We conjecture that for all 3 cases (or 6, if one takes into account also odd versions) stable homology of Poisson algebras are finite-dimensional.

This conjecture has a non-trivial consequence that the difference between the virtual and the actual rational Euler characteristic for moduli spaces of open curves tends to $+\infty$ when genus tends to $+\infty$.

7.4. *Poisson world*

As we mentioned before, the whole story can be told for some more general classes of algebras with a set of basic binary operations and quadratic relations between these operations (like Jacobi identity or associativity). One example of such situation is the case of Poisson algebras, i.e. vector spaces V with structures of commutative and Lie algebra on it satisfying condition

$$[a, bc] = b[a, c] + c[a, b].$$

As in Section 4 one can define "Poisson algebras in the Poisson world".

Poisson world is a degenerated relative of the associative world. For example, like in the associative case, there are $n!$ linearly independent polylinear monomials in n indeterminates $x_1, \dots, x_n$ for any $n \geq 1$.

On the other hand, generalized graph complex for the Poisson world contains as a direct summands graph complexes for commutative and Lie cases. We expect that better understanding of the underlying geometry of Poisson graph complexes gives more clear picture in three classical cases.

8. Duality

Here we shall be very concise.

First of all, any differential graded $\mathcal{A}$-algebra V defines a manifold ΠV^* in the dual world with the action of $\mathbf{A}^{0|1}$ (ΠV^* could be infinite-dimensional). It is just the usual (co)bar construction. Applying the bar construction twice we obtain a differential graded algebra quasi-isomorphic to the initial one. Hence suitably defined homotopy categories for dual types of algebras are dual (see [14]).

Define *strong homotopy* $\mathcal{A}$-algebra as a manifold in the dual world with the action of $\mathbf{A}^{0|1}$. Homotopy theories of differential graded algebras and strong homotopy algebras coincide. The advantage of strong homotopy algebras is that their homotopy types are in one-to-one correspondence with equivalence types of so called *minimal* strong homotopy algebras, i.e. manifolds with odd vector fields with square equal zero and with the vanishing first Taylor coefficient at zero point (see [18]).

Another quite different aspect of duality is a kind of Lie theory. On the tensor product $V \otimes U$ of $\mathcal{A}$-algebra V and $\mathcal{A}^{\text{dual}}$-algebra U there is a canonical structure of Lie algebra. Category of $\mathcal{A}^{\text{dual}}$-algebras is (more or less) equivalent to the category of functors

$$\{\mathcal{A}\text{-algebras}\} \to \{\text{Lie algebras}\}$$

preserving limits. At the moment we don't understand why the homotopy theory gives the same duality as the Lie theory.

In general, we expect 4 constructions. If V is an $\mathcal{A}^{\text{dual}}$-algebra, then as formal flat $\mathcal{A}$-manifolds

(1) V is a group-like object in the category of $\mathcal{A}$-manifolds (Lie theory),
(2) there is an odd vector field of homogeneity degree 1 with the square equal to 0 on ΠV (bar construction),
(3) there is an even Poisson bracket of homogeneity degree 1 on V^* (Kirillov bracket),
(4) there is an odd Poisson bracket of homogeneity degree 1 on ΠV^* (odd Kirillov bracket).

In the last three cases Taylor coefficients of corresponding structures are structure constants of algebra V. To construct group law we use a) the structure of Lie algebra on $V \otimes U$ for arbitrary $\mathcal{A}$-algebra U and b) Campbell-Dynkin-Hausdorff formula.

9. Towards a global geometry

J. Cuntz and D. Quillen ([5]), following A. Grothendieck, define *smooth* non-commutative associative algebra (with or without unit) as an algebra having the lifting property with respect to the nilpotent extensions. They proved that this property is equivalent to the existence of "connection with zero torsion on the tangent bundle". The last notion means the following: starting from an algebra A one can construct a new algebra TA adding *even* symbols da, $a \in A$ satisfying the Leibniz rule $d(a \cdot b) = a \cdot db + da \cdot b$. There is a $\mathbf{Z}_{\geq 0}$-grading on TA by the number of differentials. Connection with zero torsion on the tangent bundle is a derivation (= vector field) D of TA of homogeneity degree 1 such that $Da = da$ for $a \in A$.

It seems that both definitions of smoothness are equivalent in other cases.

It follows from results of J. Cuntz and D. Quillen that for smooth algebras co-homology of the de Rham complex (which is Karoubi-de Rham complex in the associative unital case) gives the "right" cohomology, i.e. cyclic homology of algebras.

We propose the following picture:

Let V be a finite-dimensional $\mathcal{A}$-algebra. Then V defines a functor

$$\{\text{finitely generated } \mathcal{A}\text{-algebras}\} \to \{\text{affine schemes over} \mathbf{C} f\}$$

by associating with $\mathcal{A}$-algebra A the scheme $A(V)$ of its homomorphisms to V. We expect that if A is smooth then $A(V)$ is smooth, vector fields on A go to vector fields on $A(V)$, and if V carries a nondegenerate invariant scalar product then differential forms for A go to differential forms on $A(V)$ and symplectic structures go to symplectic structures.

Examples.

$\mathbf{G}_m = \langle x, y : xy = 1\rangle$ is smooth unital associative algebra. Its de Rham cohomologies are 1-dimensional in degrees $1, 3, 5, \ldots$ and zero in even degrees (compare with the cohomology of the representation space $\mathbf{G}_m(\mathrm{Mat}_N(\mathbf{C})) = GL(N, \mathbf{C})$).

$P = \langle p : p^2 = p\rangle$ is 1-dimensional smooth nonunital associative algebra. And what is more, this non-commutative manifold is symplectic. De Rham cohomologies are 1-dimensional in degrees $0, 2, 4, \ldots$ and zero in odd degrees. Representation spaces are symplectic manifolds homotopy equivalent to the disjoint union of complex Grassmanians.

References

[1] S. Axelrod and I. M. Singer, *Chern-Simons perturbation theory*, M. I. T. preprint (October 1991).

[2] D. Bar-Natan, *On the Vassiliev knot invariants*, Harvard preprint (August 1992).

[3] K. S. Brown, *Cohomology of Group*, Graduate Texts in Mathematics **87** (Springer, Berlin, Heidelberg, New York, 1982).

[4] M. Culler and K. Vogtmann, *Moduli of graphs and automorphisms of free groups*, Invent. Math. **84** (1986) 91–119.

[5] J. Cuntz, D. Quillen, *Algebra extensions and nonsingularity*, to appear.

[6] V. G. Drinfel'd, *On quasitriangular Quasi-Hopf algebras and a group closely connected with Gal*($\bar{\mathbf{Q}}/\mathbf{Q}$), Leningrad Math. J. **2** (1991) 829–860.

[7] D. B. Fuks, *Cohomology of Infinite-Dimensional Lie Algebras* (Consultants Bureau, New York and London, 1986).

[8] I. M. Gelfand and D. B. Fuks, *Cohomology of Lie algebras of formal vector fields*, Izv. Akad. Nauk SSSR, Ser. Mat. **34** (2) (1970) 322–337.

[9] I. M. Gelfand and O. Mathieu, *On the Cohomology of the Lie algebra of Hamiltonian vector fields*, preprint RIMS-781 (1991).

[10] V. Ginzburg and M. Kapranov, *Koszul duality for quadratic operads*, to appear.

[11] V. W. Guillemin and S. D. Shnider, *Some stable results on the cohomology of classical infinite-dimensional Lie algebras*, Trans. Am. Math. Soc. **179** (1973) 275–280.

[12] J. Harer, *The cohomology of the moduli space of curves*, LNM **1337** (Springer, 1988) 138–221.

[13] R. C. Penner, *The decorated Teichmüller space of punctured surfaces*, Commun. Math. Phys. **113** (1987) 299–339.

[14] D. Quillen, *Rational Homotopy Theory*, Ann. Math. **90** (1969) 205–295.

[15] J. Stasheff, *An almost groupoid structure for the space of (open) strings and implications for string field theory*, in: Advances in Homotopy Theory: Proceedings of a Conference in Honour of the 60th Birthday of I. M. James (Cortona, June 1988)

LMS Lecture Notes Series **139**, (eds.) S. Salamon, B. Steev and W. Sutherland (Cambridge University Press, 1989) 165–172.
[16] J. Stasheff, *On the homotopy associativity of H-Spaces I, II*, Trans. AMS **108** (1963) 275–312.
[17] K. Strebel, *Quadratic Differentials* (Springer, Berlin, Heidelberg, New York, 1984).
[18] D. Sullivan, *Infinitesimal computations in topology*, Publ. I.H.E.S. **47** (1978) 269–331.
[19] V. Vassiliev, *Complements to Discriminants of Smooth Maps: Topology and Applications* (Amer. Math. Soc. Press, 1992).

COMMENTS ON "FORMAL (NON)-COMMUTATIVE SYMPLECTIC GEOMETRY"

by
MAXIM KONTSEVICH

Here are directions to some new references to "Formal non-commutative symplectic geometry".

One can find a complete proof of the main theorem in [4], in [3] new cohomological operations on (commutative) graph cohomology are introduced. In [8], [5] and [6] a natural general framework is developed in which versions of graph complexes can be defined, see also [9], [15] and [16]. Applications to topology and other stuff are discussed in [11], [12] and [17]. Some calculations of (commutative) graph cohomology are in [1]. We recommend [10] and [7] for the different approaches to a "global" noncommutative geometry. The main idea is to see non-commutative definitions (for the associative operad) via spaces of matrix representations, see [13], [2] and [14].

References

[1] D. Bar-Natan and B. McKay, *Graph cohomology – An overview and some computations*, draft available at http://www.math.toronto. edu/ drorbn/papers/GCOC.

[2] R. Bocklandt and L. Le Bruyn, *Necklace Lie algebras and noncommutative symplectic geometry*, Math. Z. **240** (1) (2002) 141–167, and e-print: math.AG/0010030.

[3] J. Conant and K. Vogtmann, *Infinitesimal operations on graphs and graph homology*, e-print: math.QA/0111198.

[4] J. Conant and K. Vogtmann, *On a theorem of Kontsevich*, e-print: math.QA/0208169.

[5] E. Getzler and M. Kapranov, *Cyclic operads and cyclic homology*, Geometry, Topology & Physics for Raonl Bott, Conference Proceedings and Lecture Notes in Geometry and Topology, **IV** (Internat. Press, Cambridge, MA, 1995) 161–201.

[6] E.Getzler and M. Kapranov, *Modular operads*, Comp. Math. **110** (1) (1998) 65–126, and e-print: dg-ga/9408003.

[7] V. Ginzburg, *Non-commutative symplectic geometry, quiver varieties, and operads*, Math. Res. Lett. **8** (3) (2001) 377–400, and e-print: math.QA/0005165.

[8] V. Ginzburg and M. Kapranov, *Koszul duality for operads*, Duke Math. J. **76** (1) (1994) 203–272.

[9] V. Hinich and A. Vaintrob, *Cyclic operads and algebra of chord diagrams*, Selecta Math. (N.S.) **8** (2) (2002) 237–282, and e-print: math.QA/0005197.

[10] M. Kapranov, *Non-commutative geometry based on commutator expansions*, J. Reine Angew. Math. **505** (1998) 73–118, and e-print: math.AG/9802041.

[11] M. Kontsevich, *Feynman diagrams and low-dimensional topology*, First European Congress of Mathematics, Vol. II (Paris 1992), Progr. Math. **120** (Birkhäuser, Basel, 1994) 97–121.

[12] M. Kontsevich, *Rozansky-Witten invariants via formal geometry*, Comp. Math. **115** (1) (1999) 115–127, and e-print: dg-ga/9704009.

[13] M. Kontsevich, *Non-commutative smooth spaces*, talk on Arbeitstagung 1999, available at http://www.mpim-bonn.mpg.de/html/preprints.

[14] L. Le Bruyn, *Noncommutative geometry*, e-print: math.AG/9904171 and a book in preparation on his home page.

[15] M. Markl, *Cyclic operads and homology of graph complexes*, e-print: math.QA/9801095.

[16] M. Movshev, *A definition of graph homology and graph K-theory of algebras*, e-print: math.KT/991111.

[17] M. Penkava, *Infinity algebras and homology of graph complexes*, e-print: q-alg/9601018.

Reprinted from Notices of the Amer. Math. Soc. 1999

THE WORK OF CURTIS T. McMULLEN

by
JOHN MILNOR
Institute for Mathematical Science
State University of New York
Stony Brook

Curt McMullen has made important contributions to the study of Kleinian groups, hyperbolic 3-manifolds, and holomorphic dynamics. Indeed, following the lead of Dennis Sullivan, he clearly regards these three areas as different facets of one unified branch of mathematics. Following are descriptions of a few selected topics. I hope that these will illustrate the variety and depth of his work. However, by all means the reader should look at the original papers, since he is a master expositor. See especially his two books and his Berlin lecture [Mc10].

Solving the Quintic. His first work was on Smale's theory of purely iterative algorithms. By definition, these are numerical algorithms which can be carried out by iterating a single rational function, without allowing any "if $\cdots$ then $\cdots$" branching. In [Mc1] he showed that the roots of a polynomial of degree n can be computed by a generally convergent purely iterative algorithm if and only if $n \leq 3$. With Peter Doyle [DMc] he showed that these roots can be computed by a tower of finitely many such algorithms if and only if $n \leq 5$.

A Fat Julia Set. The Julia set J of a rational map f from the Riemann sphere $\widehat{\mathbb{C}} = \mathbb{C} \cup \{\infty\}$ to itself can be described roughly as the compact set consisting of all points $z \in \widehat{\mathbb{C}}$ such that the iterates of f, restricted to any neighborhood of z, behave chaotically. It is not known whether such a Julia set can have positive area, without being the entire Riemann sphere. However, McMullen [Mc2] produced very simple examples for the more general question in which we replace the rational map by a transcendental function, such as the map $z \mapsto \sin(z)$ of Figure 1. (A different and more complicated example was given in [EL].)

The Kra Conjecture. To any Riemann surface X one can associate the Banach space $Q(X)$ consisting of all holomorphic quadratic differentials $\Phi = \phi(z)\, dz^2$ for which the norm $\|\Phi\| = \int |\phi|\, dz\, d\overline{z}$ is finite. Any covering map $f : X \to Y$ induces a push-forward operation $f_* : Q(X) \to Q(Y)$, where the image differential at a point y is obtained by summing over the points of $f^{-1}(y)$. This operation can never increase

Figure 1. The Julia set for $z \mapsto \sin(z)$, shown in black, has positive area but no interior points.

norms, $\|f_*(\Phi)\| \leq \|\Phi\|$. In the special case of the universal covering $f : \widetilde{Y} \to Y$ of a hyperbolic surface of finite area, Irwin Kra conjectured in 1972 that there is always some definite amount of cancellation between the different preimages of a point of Y, so that $\|f_*\|$ is strictly less than 1. This was proved by McMullen [Mc3, Mc5]. In fact McMullen considered a completely arbitrary covering map $f : X \to Y$, showing that $\|f_*\| < 1$ if and only if this covering is nonamenable.

Cusps in the boundary of Teichmüller space. Next a problem in Kleinian groups. In 1970, Lipman Bers compactified the Teichmüller space of complex structures on a hyperbolic Riemann surface of finite area by adding an ideal boundary consisting of algebraic limits of associated Kleinian groups. He conjectured that the "cusps", corresponding to ideal limits in which some simple closed curve has been pinched to a point, are everywhere dense in this boundary. (Compare Figure 2.) This was proved by McMullen [Mc4] in 1991, using a careful estimate for the change in the associated group representation $\pi_1(S) \to \mathrm{PSL}_2(\mathbb{C})$ as some simple closed curve in the surface S shrinks to a point.

Thurston Geometrization. The still unproved Thurston Geometrization Conjecture asserts that every compact 3-manifold can be cut up along spheres and tori into pieces, each of which admits a simple geometric structure. Here eight possible geometries must be allowed. For six of these eight geometries, the problem is now well understood, but difficulties remain in the hyperbolic case, while the spherical

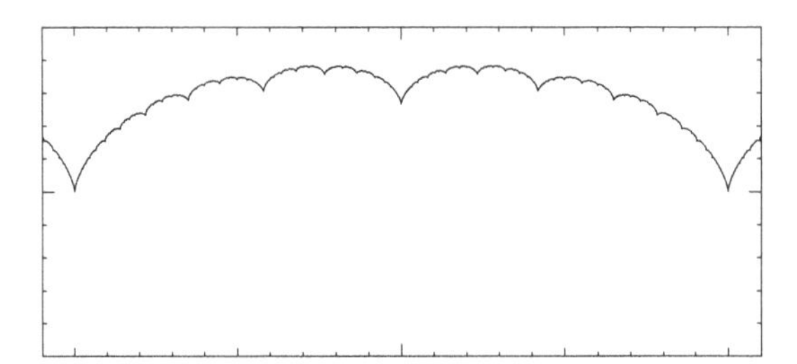

Figure 2. (courtesy of David Wright). Dense cusps in the boundary of Teichmüller space for a punctured torus, using the Maskit embedding. Teichmüller space is the region underneath this boundary curve.

case, including the classical Poincaré conjecture, is still intractable. For references, compare [Mi2].

Thurston outlined proofs that a 3-manifold admits a hyperbolic structure in two important special cases. First suppose that M is a *Haken manifold*, that is, suppose that M is S^2-irreducible and can be built up inductively from 3-balls by gluing together submanifolds of the boundary, taking care that no essential simple closed curve in this submanifold bounds a disk in the manifold. Thurston showed that M can be given a hyperbolic structure if and only if its fundamental group is infinite, and every $\mathbb{Z} \oplus \mathbb{Z}$ in its fundamental group comes from a boundary torus. McMullen [Mc6] used his work on the Kra conjecture to give a new and simpler proof of this theorem. (The details are still quite complicated.) The second case handled by Thurston concerned 3-manifolds which fiber over the circle. Again McMullen gave a new proof, which will be discussed below.

Renormalization. Let f be a smooth even map from the closed interval $I = [-1, 1]$ into itself with a nondegenerate critical point at the origin, and with no other critical points. We will say that f is *renormalizable* if there is an integer $n \geq 2$ so that the n fold iterate $g = f^{\circ n}$ maps the subinterval $\{x\,;\,|x| \leq |g(0)|\}$ into itself with only one nondegenerate critical point. If we rescale by setting $\widehat{f}(x) = g(\alpha x)/\alpha$ where $\alpha = g(0)$, then $\widehat{f}$ will be a new map from the interval I into itself satisfying the original hypothesis. This $\widehat{f}$ is called the *renormalization* $\mathcal{R}_n(f)$. In 1978 Mitchell Feigenbaum, and independently Pierre Coullet and Charles Tresser, considered the special case $n = 2$, and studied maps f which are *infinitely renormalizable*, so that we can form a sequence of iterated renormalizations f, $\mathcal{R}_2 f$, $\mathcal{R}_2^{\circ 2} f$, ..., each mapping I to itself with one critical point. They observed empirically that this sequence of maps always seems to converge to a fixed smooth limit map. Their ideas, motivated by renormalization ideas from statistical mechanics, and by attempts to understand the onset of turbulence in fluid mechanics, now occupy a central role in one-dimensional dynamics, since the infinitely renormalizable maps are the most difficult ones to understand.

This construction was extended to the complex case by Douady and Hubbard, using the idea of a *quadratic-like map*, that is, a proper holomorphic map $f : U \to V$

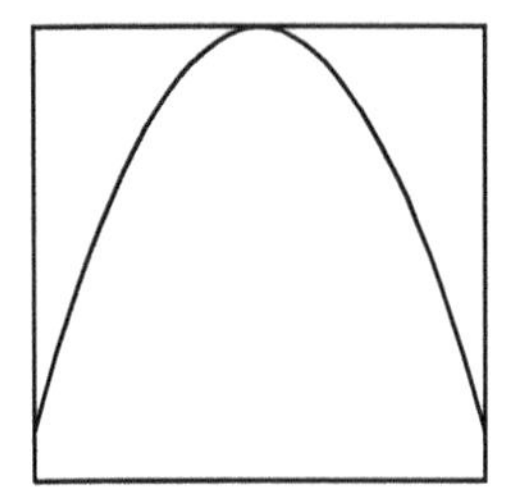
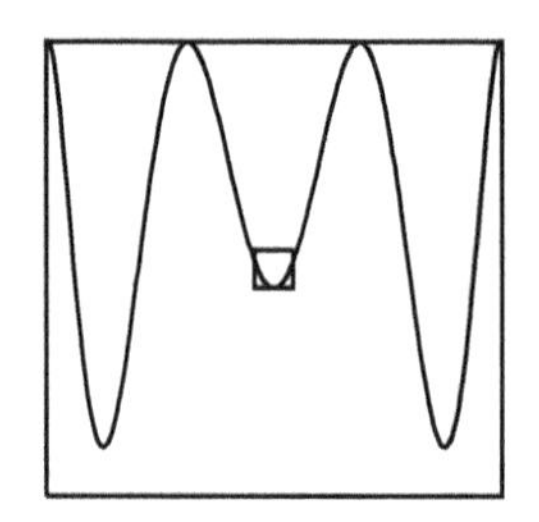
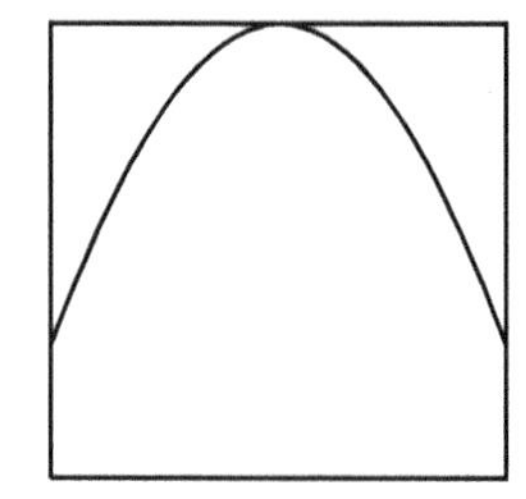

Figure 3. A quadratic map f, its iterate $f^{\circ 3}$, and its renormalization $\mathcal{R}_3(f)$. The right hand box in obtained from the small box in the middle by magnifying and rotating 180°.

of degree two, where U and V are simply connected open sets in $\mathbb{C}$ and $\overline{U}$ is a compact subset of V. This has led to important work by mathematicians such as Dennis Sullivan, Curt McMullen, and Mikhail Lyubich. Most subsequent progress in understanding the real case has been based on complex methods. (One exception is Martens [Ma].) McMullen's book [Mc7] provided the first careful presentation of the foundations of renormalization. As one example, he was the first to notice the possibility of an aberrant form of "crossed renormalization" in the complex case, which does not fit into the usual pattern. He used his work on renormalization to obtain partial results on the generic hyperbolicity conjecture for real quadratic maps; that is, the conjecture that every such map can be approximated by one with an attracting periodic orbit. For example he showed that every component of the interior of the Mandelbrot set which meets the real axis is hyperbolic. The full conjecture was later proved by Lyubich [L1] and by Graczyk and Świątek [GS].

McMullen's second book [Mc8] developed renormalization theory further and tied it up with Mostow rigidity, and also with Thurston geometrization. (See also [Mc10].) He introduced the concept of a "deep point" in a fractal subset $X \subset \mathbb{C}$. By definition, p is *deep* if there are positive constants ϵ and c so that the distance from an arbitrary point q to X is at most $c|p-q|^{1+\epsilon}$. Taking $p = 0$ for convenience, we can understand this concept by zooming in on the origin so as to magnify the set X by some large constant λ. Replacing X by the magnified copy λX, we can replace the constant c by c/λ^{ϵ}, which tends to zero as $\lambda \to \infty$. In other words, these magnified images will fill out the complex plane more and more densely, with gaps which become smaller and smaller, as λ becomes large. (Compare [Mi1].)

Now consider a hyperbolic 3-manifold which may have infinite volume. The *convex core* of such a manifold M can be described as the smallest geodesically convex subset which is a strong deformation retract of M. Assuming both upper and lower bounds for the injectivity radius at points of the convex core, McMullen's inflexibility theorem asserts that two such manifolds M and M' which are "pseudo-isometric" must actually be related by a diffeomorphism which becomes exponentially close to an isometry as we penetrate deeper into the convex core. Closely related is the statement that actions of $\pi_1(M)$ and $\pi_1(M')$ on the 2-sphere at infinity for hyperbolic 3-space are quasiconformally conjugate, and that this quasiconformal conjugacy is actually conformal at every deep point of the limit set for this action.

As an application, he gave a new proof of the second Thurston geometrization theorem. To any surface diffeomorphism $\psi : S \to S$ we can associate the *mapping torus* T_ψ, that is the quotient of $S \times \mathbb{R}$ under the $\mathbb{Z}$ action which is generated by $(x, t) \mapsto (\psi(x), t + 1)$. If S has genus two or more and ψ is pseudo-Anosov, then Thurston showed that T_ψ is a hyperbolic 3-manifold. McMullen proved this by using his inflexibility result to construct a hyperbolic structure on $S \times \mathbb{R}$ which is invariant under the given $\mathbb{Z}$ action.

Next he applied these ideas to renormalization. One basic result is a rigidity theorem for bi-infinite "towers" of renormalizations. We can think of such a tower as a bi-infinite sequence $(\ldots, q_{-1}, q_0, q_1, q_2, \ldots)$ of quadratic-like maps $q_j : U_j \to V_j$, where each q_{j+1} is a renormalization $\mathcal{R}_{n_j}(q_j)$. If the renormalization periods n_j are bounded, and if the annuli $V_j \smallsetminus \overline{U}_j$ have modulus bounded away from zero, then he showed that the entire tower is uniquely determined up to a suitable isomorphism relation by its quasiconformal conjugacy class.

Consider an infinitely renormalizable real quadratic map f, with periodic combinatorics. Using the complex theory, McMullen showed that the successive renormalizations converge exponentially fast to a map which is periodic under renormalization. Closely related is the statement that the critical point is a deep point for the Julia set of f. (Figure 4.)

Figure 4. The critical point (at the center of symmetry) is a deep point of the Julia set for the Feigenbaum infinitely renormalizable map $z \mapsto 1 - az^2$, where $a = 1.401155189 \cdots$. This Julia set has no interior points.

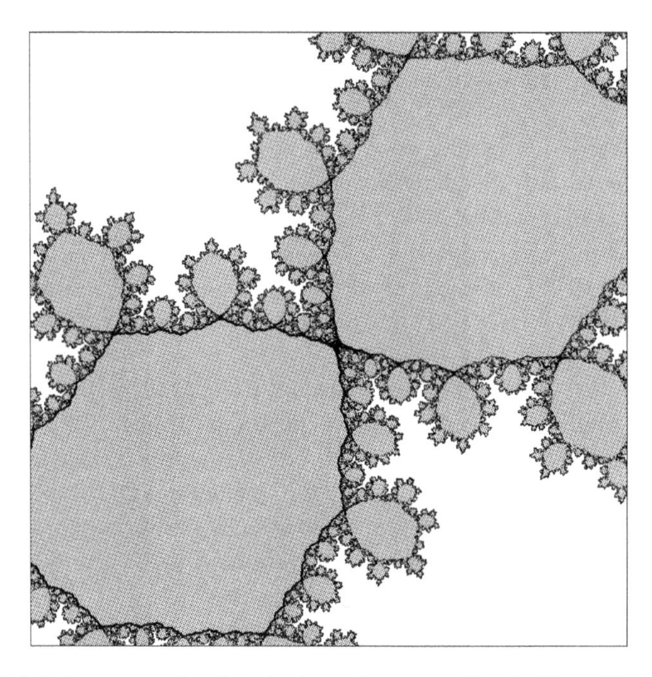

Figure 5. Filled Julia set associated with the golden mean Siegel disk, with rotation number $\rho = 1/(1+1/(1+1/(1+\cdots)))$. The Siegel disk is the large region to the lower left.

Now consider a quadratic map $f(z) = z^2 + c$ which has a Siegel disk of rotation number ρ. That is, choose the constant c so that the derivative of f at one of its two fixed points is equal to $e^{2\pi i \rho}$, where $\rho \in \mathbb{R} \smallsetminus \mathbb{Q}$ satisfies a suitable Diophantine condition. If the continued fraction expansion of ρ is periodic, then McMullen used similar ideas in [Mc9] to show that this Siegel disk is "self-similar" about the critical point 0 (the central point in Figure 5). In fact, his argument can be used to show that the entire Julia set J of f is *asymptotically self-similar* in the following sense: There is a scale factor λ with $|\lambda| > 1$, so that the magnified images $\lambda^n J$ converge to a well defined limit set $\widehat{J} = \lambda \widehat{J} \subset \widehat{\mathbb{C}}$ as $n \to \infty$, using the Hausdorff topology for compact subsets of the Riemann sphere. (In this particular example, $\lambda = 1.8166\cdots$ is real.) The corresponding limit for the boundary of the Siegel disk is a quasicircle contained in $\widehat{J}$, while the corresponding limit for the filled Julia set $K(f)$ (the union of bounded orbits for f) is the entire sphere $\widehat{\mathbb{C}}$.

There has been very significant subsequent work in renormalization, based in part on McMullen's ideas. Compare the discussion in [Me]. Note in particular [L2], which implies that the boundary of the Mandelbrot set is asymptotically self-similar about the Feigenbaum point, and [L3] which proves existence of a full horseshoe structure for the real renormalization operator, and uses it, together with work of Martens and Nowicki, to prove that every real quadratic map outside a set of measure zero has either a periodic attractor or an absolutely continuous asymptotic measure.

References

[DMc] P. Doyle and C. McMullen, *Solving the quintic by iteration*, Acta Math. **163** (1989), 151–180.

[EL] A. Eremenko and M. Lyubich, *Examples of entire functions with pathological dynamics*, Journal London Math. Soc. **36** (1987), 458–468.

[GS] J. Graczyk and G. Świątek, *Generic hyperbolicity in the logistic family*, Annals of Math. **146** (1997), 399–482.

[L1] M. Lyubich, *Dynamics of complex polynomials, parts I, II*, Acta Math. **178** (1997), 185–297; see also: *Geometry of quadratic polynomials: moduli, rigidity, and local connectivity*, Stony Brook I.M.S. preprint 1993#9.

[L2] ———, *Feigenbaum-Coullet-Tresser universality and Milnor's hairiness conjecture*, Annals of Math., to appear.

[L3] ———, *Regular and stochastic dynamics in the real quadratic family*, to appear, Proc. Nat. Acad. Sci. U.S.A.; see also: *Almost every real quadratic map is either regular or stochastic*, Stony Brook I.M.S. preprint 1997#8.

[Ma] M. Martens, *The periodic points of renormalization*, Stony Brook I.M.S. preprint 1995#3.

[Mc1] C. McMullen, *Families of rational maps and iterative root finding algorithms*, Annals of Math. **125** (1987), 467–493.

[Mc2] ———, *Area and Hausdorff dimension of Julia sets of entire functions*, Trans. Amer. Math. Soc. **300** (1987), 329–342.

[Mc3] ———, *Amenability, Poincaré series and quasiconformal maps*, Invent. Math. **97** (1989), 95–127.

[Mc4] ———, *Cusps are dense*, Annals of Math. **133** (1991), 425–454.

[Mc5] ———, *Amenable coverings of complex manifolds and holomorphic probability measures*, Invent. Math. **110** (1992), 29–37.

[Mc6] ———, *Riemann surfaces and the geometrization of 3-manifolds*, Bull. Amer. Math. Soc. **27** (1992), 207–216.

[Mc7] ———, *Complex Dynamics and Renormalization*, Annals Math. Studies, vol. 135, Princeton University Press, Princeton, NJ, 1994.

[Mc8] ———, *Renormalization and 3-Manifolds which Fiber over the Circle*, Annals Math. Studies, Vol. 142, Princeton University Press, Princeton, NJ, 1996.

[Mc9] ———, *Self-similarity of Siegel disks and the Hausdorff dimension of Julia sets*, Acta. Math. **180** (1998), 247–292.

[Mc10] ———, *Rigidity and inflexibility in conformal dynamics*, Proc. Int. Cong. Math. Berlin, Vol. 2, 1998, pp. 841–855.

[Me] W. de Melo, *Rigidity and renormalization in one dimensional dynamical systems*, Proc. Int. Cong. Math. Berlin, Vol. 2, 1998, pp. 765–778.

[Mi1] J. Milnor, *Self-similarity and hairiness in the Mandelbrot set*, Computers in Geometry and Topology, M. Tangora, editor, Lect. Notes Pure Appl. Math., Dekker, 1989, pp. 211–259.

[Mi2] ———, *Collected Papers*, The Fundamental Group, Vol. II, Publish or Perish, 1995. See pp. 93–95; but note that unpublished claims in the spherical case have not been substantiated.

Curtis T. McMullen

CURTIS T. McMULLEN

ACADEMIC POSITIONS

Harvard University, Cambridge, MA.
Professor, 1998–present.

University of California, Berkeley, CA.
Professor, 1990–1998.

Princeton University, Princeton, NJ.
Professor, 1990–92. Assistant Professor, 1987–1990.

Massachusetts Institute of Technology, Cambridge, MA.
C.L.E. Moore Instructor in Mathematics, 1985–86.

EDUCATION

Harvard University Cambridge, MA.
Ph.D. in Mathematics, June 1985.
Advisor: Prof. Dennis Sullivan, CUNY and IHES.

Cambridge University Cambridge, England.
Herchel Smith Fellow, 1980–81.

Williams College Williamstown, MA.
B.A. 1980. Valedictorian.

ACADEMIC RECOGNITIONS

Fields Medal, 1998.
Invited Speaker, ICM 1990 (Kyoto) and 1998 (Berlin).
Fellow of the American Academy of Arts and Sciences, 1998.
Salem Prize, 1991.
Alfred P. Sloan Fellowship, 1988.

PERSONAL DATA

Born 21 May 1958, in Berkeley, California.

Reprinted from Proc. Int. Congr. Math., 1998

RIGIDITY AND INFLEXIBILITY IN CONFORMAL DYNAMICS

by
CURTIS T. McMULLEN*
Mathematics Department, Harvard University

1. Introduction

This paper presents a connection between the rigidity of hyperbolic 3-manifolds and universal scaling phenomena in dynamics.

We begin by stating an inflexibility theorem for 3-manifolds of infinite volume, generalizing Mostow rigidity (§2). We then connect this inflexibility to dynamics and discuss:

- The geometrization of 3-manifolds which fiber over the circle (§2);
- The renormalization of unimodal maps $f : [0, 1] \to [0, 1]$ (§4),
- Real-analytic circle homeomorphisms with critical points (§5), and
- The self-similarity of Siegel disks (§6).

Chaotic sets for these four examples are shown in Figure 1. The snowflake in the first frame is the limit set Λ of a Kleinian group Γ acting on the Riemann sphere $S^2_\infty = \partial\mathbb{H}^3$. Its center c is a *deep point* of Λ, meaning the limit set is very dense at microscopic scales near c. Because of the inflexibility and combinatorial periodicity of $M = \mathbb{H}^3/\Gamma$, the limit set is also self-similar at c with a universal scaling factor.

The remaining three frames show deep points of the (filled) Julia set for other conformal dynamical systems: the Feigenbaum polynomial, a critical circle map and the golden ratio Siegel disk. Our goal is to explain an inflexibility theory that leads to universal scaling factors and convergence of renormalization for these examples as well.

The qualitative theory of dynamical systems, initiated by Poincaré in his study of celestial mechanics, seeks to model and classify stable regimes, where the topological

*Research supported in part by the NSF.

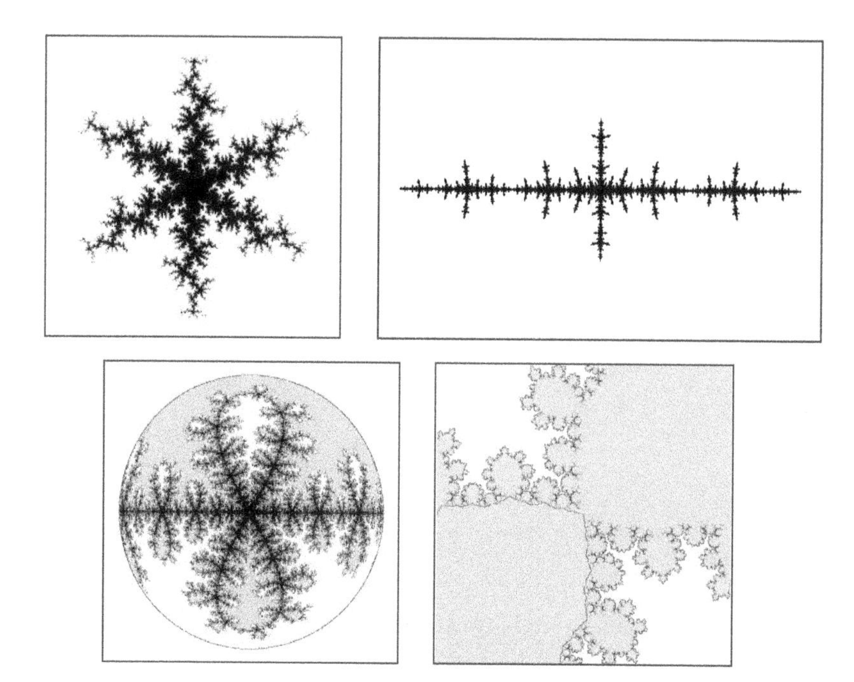

Figure 1. Dynamical systems with deep points: a totally degenerate Kleinian group, the Feigenbaum polynomial, a critical circle map and the golden mean Siegel disk.

form of the dynamics is locally constant. In the late 1970s physicists discovered a rich, universal structure in the onset of instability. One-dimensional dynamical systems emerged as elementary models for critical phenomena, phase transitions and renormalization.

In pure mathematics, Mostow and others have developed a rigidity theory for compact manifolds M^n of constant negative curvature, $n \geq 3$, and other quotients of symmetric spaces. This theory shows M is determined up to isometry by $\pi_1(M)$ as an abstract, finitely-presented group. Remarkably, rigidity of M is established via the ergodic theory of $\pi_1(M)$ acting on the boundary of the universal cover of M.

In our case, $M = \mathbb{H}^3/\Gamma$ is a hyperbolic 3-manifold, the boundary of its universal cover $\mathbb{H}^3$ is isomorphic to S^2, and the action of $\pi_1(M) \subset \mathrm{Isom}^+(\mathbb{H}^3) = PSL_2(\mathbb{C})$ on S^2 is conformal. Similarly, upon complexification, 1-dimensional dynamical systems give rise to holomorphic maps on the Riemann sphere $\hat{\mathbb{C}} \cong S^2$. Hyperbolic space $\mathbb{H}^3$ enters the dynamical picture as a means to organize *geometric limits* under rescaling (§3). The universality observed by physicists can then be understood, as in the case of 3-manifolds, in terms of rigidity of these geometric limits.

We conclude with progress towards the classification of hyperbolic manifolds (§7), where geometric limits also play a central role.

2. Hyperbolic 3-Manifolds and Fibrations

A *hyperbolic manifold* is a complete Riemannian manifold with a metric of constant curvature -1. Mostow rigidity states that any two closed, homotopy equivalent hyperbolic 3-manifolds are actually isometric.

In this section we discuss a remnant of rigidity for *open* manifolds. Let $\mathrm{core}(M) \subset M$ denote the *convex core* of M, defined as the closure of the set of geodesic loops in M. The manifold M satisfies $[r, R]$-*injectivity bounds*, $r > 0$, if for any $p \in \mathrm{core}(M)$, the largest embedded ball $B(p, s) \subset M$ has radius $s \in [r, R]$.

Let $f : M \to N$ be a homotopy equivalence between a pair of hyperbolic 3-manifolds. Then f is a K-*quasi-isometry* if, when lifted to the universal covers,

$$\mathrm{diam}(\tilde{f}(B)) \leq K(\mathrm{diam}\, B + 1) \quad \forall B \subset \tilde{M},$$

and

$$\mathrm{diam}(\tilde{f}^{-1}(B)) \leq K(\mathrm{diam}\, B + 1) \quad \forall B \subset \tilde{N}.$$

A diffeomorphism $f : M \to N$ is an *asymptotic isometry* if f is exponentially close to an isometry deep in the convex core. That is, there is an $A > 1$ such that for any nonzero vector $v \in T_p M$, $p \in \mathrm{core}(M)$, we have

$$\left| \log \frac{|Df(v)|}{|v|} \right| \leq C A^{-d(p, \partial \mathrm{core}(M))}.$$

In [Mc2] we show:

Theorem 2.1 (Geometric Inflexibility). *Let M and N be quasi-isometric hyperbolic 3-manifolds with injectivity bounds. Then M and N are asymptotically isometric.*

Mostow rigidity is a special case: if M and N are closed, then any homotopy equivalence $M \sim N$ is a quasi-isometry, injectivity bounds are automatic, and $\partial\mathrm{core}(M) = \emptyset$, so an asymptotic isometry is an isometry.

To sketch the proof of Theorem 2.1, recall any hyperbolic 3-manifold M determines a conformal dynamical system, namely the action of its fundamental group $\pi_1(M)$ on the sphere at infinity $S^2_\infty = \partial \mathbb{H}^3$ for the universal cover $\tilde{M} \cong \mathbb{H}^3$. The *limit set* $\Lambda \subset S^2_\infty$ is the chaotic locus for this action; its convex hull covers the core of M. The action is properly discontinuously on the rest of the sphere, and the quotient $\partial M = (S^2_\infty - \Lambda)/\pi_1(M)$ gives a natural Riemann surface at infinity for M.

A quasi-isometric deformation of M determines a quasiconformal deformation v of ∂M, which in turn admits a (harmonic) visual extension V to an equivalent deformation of M. The strain $SV(p)$ is the average of the ellipse field $Sv = \bar{\partial} v$ over all visual rays γ from p to ∂M. By our injectivity bounds, γ corkscrews chaotically before exiting the convex core. Thus the ellipses of Sv on ∂M appear in random orientations as seen from p (Figure 2). This randomness provides abundant

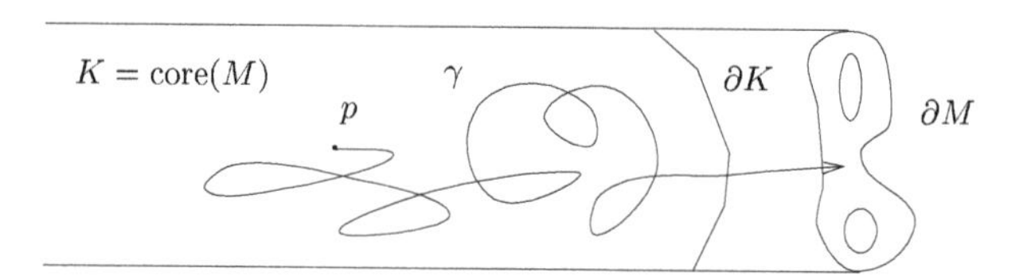

Figure 2. An observer deep in the convex core sees a kaleidoscopic view of ∂M.

cancellation in the visual average, and we find the metric distortion $\|SV(p)\|$ is exponentially small compared to $\|Sv\|_\infty$. Thus V is an infinitesimal asymptotic isometry.

In dimension 3, any two quasi-isometric hyperbolic manifolds are connected by a smooth path in the deformation space, so the global theorem follows from the infinitesimal version.

Inflexibility is also manifest on the sphere at infinity. Let us say a local homeomorphism ϕ on $S^2_\infty \cong \hat{\mathbb{C}}$ is $C^{1+\alpha}$*-conformal* at z if the complex derivative $\phi'(z)$ exists and

$$\phi(z+t) = \phi(z) + \phi'(z) \cdot t + O(|t|^{1+\alpha}).$$

We say $x \in \Lambda \subset S^2_\infty$ is a *deep point* if Λ is so dense at x that for some $\beta > 0$,

$$B(y,s) \subset B(x,r) - \Lambda \implies s = O(r^{1+\beta}).$$

It is easy to see that a geodesic ray $\gamma \subset \mathbb{H}^3$ terminating at a deep point in the limit set penetrates the convex hull of Λ at a linear rate. From the inflexibility theorem we find:

Corollary 2.2. *Let M and N satisfy injectivity bounds, and let $\phi : S^2_\infty \to S^2_\infty$ be a quasiconformal conjugacy between $\pi_1(M)$ and $\pi_1(N)$. Then ϕ is $C^{1+\alpha}$-conformal at every deep point of the limit set of $\pi_1(M)$.*

The inflexibility theorem is motivated by the following application to 3-manifolds that fiber over the circle. Let S be a closed surface of genus $g \geq 2$ and let $\psi \in \mathrm{Mod}(S)$ be a pseudo-Anosov mapping class. Let

$$T_\psi = S \times [0,1]/\{(x,0) \sim (\psi(x),1)\}$$

be the 3-manifold fibering over the circle with fiber S and monodromy ψ. By a deep theorem of Thurston, T_ψ is hyperbolic. To find its hyperbolic structure, let $V(S)$ denote the variety of representations $\rho : \pi_1(S) \to \mathrm{Isom}(\mathbb{H}^3)$, and define

$$\mathcal{R} : V(S) \to V(S)$$

by $\mathcal{R}(\rho) = \rho \circ \psi_*^{-1}$. We refer to $\mathcal{R}$ as a *renormalization operator*, because it does not change the group action on $\mathbb{H}^3$, only its marking by $\pi_1(S)$.

Let $QF(S) \overset{Q}{\cong} \text{Teich}(S) \times \text{Teich}(\bar{S}) \subset V(S)$ denote the space of quasifuchsian groups, and define

$$M(X, \psi) = \lim_{n \to \infty} Q(X, \psi^{-n} Y), \quad \text{for any } (X, Y) \in \text{Teich}(S) \times \text{Teich}(\bar{S}).$$

Then $M = M(X, \psi)$ has injectivity bounds, its convex core is homeomorphic to $S \times [0, \infty)$, and the manifolds M and $\mathcal{R}(M)$ are quasi-isometric. By the inflexibility theorem there is an asymptotic isometry $\Psi : M \to M$ in the homotopy class of ψ, so the convex core of M is asymptotically periodic. As n tends to ∞, the marking of $\mathcal{R}^n(M)$ moves into the convex core at a linear rate, and we find:

Theorem 2.3. *The renormalizations $\mathcal{R}^n(M(X, \psi))$ converge exponentially fast to a fixed-point M_ψ of $\mathcal{R}$.*

Since $\mathcal{R}(M_\psi) = M_\psi$, the map ψ is realized by an isometry α on M_ψ, and the quotient $T_\psi = M_\psi / \langle \alpha \rangle$ gives the desired hyperbolic structure on the mapping cylinder of ψ.

This iterative construction of T_ψ hints at a dynamical theory of the action of $\text{Mod}(S)$ on the variety $V(S)$, as does the following result [Kap]:

Theorem 2.4 (Kapovich). *The derivative $D\mathcal{R}_\psi$ is hyperbolic on the tangent space to $V(S)$ at M_ψ for all pseudo-Anosov mapping classes on closed surfaces.*

The snowflake in the first frame of Figure 1 is a concrete example of the limit set Λ for a Kleinian group $\Gamma = \pi_1(M(X, \psi))$ as above. In this example S is a torus, made hyperbolic by introducing a single orbifold point $p \in S$ of order 3; and $\psi = \left(\begin{smallmatrix} 1 & 1 \\ 1 & 2 \end{smallmatrix}\right) \in SL_2(\mathbb{Z}) \cong \text{Mod}(S)$ is the simplest pseudo-Anosov map. The suspension of $p \in S$ gives a singular geodesic $\gamma \subset T_\psi$ forming the orbifold locus of the mapping torus of ψ.

The picture is centered at a deep point $c \in \Lambda$ fixed by an elliptic element of order 3 in Γ. The limit set Λ is a nowhere dense but very furry tree, with six limbs meeting at c. By general results, Λ is a locally connected dendrite, with Hausdorff dimension two but measure zero ([CT], [Th1, Ch. 8], [Sul1], [BJ1]); in fact by [BJ2] we have $0 < \mu_h(\Lambda) < \infty$ for the gauge function $h(r) = r^2 |\log r \log\log\log r|^{1/2}$.

One can easily construct a quasiconformal automorphism ϕ of Γ, with $\phi(c) = c$ and $\phi \circ \gamma = \psi_*(\gamma) \circ \phi$ for all $\gamma \in \Gamma$. By Corollary 2.2, ϕ is $C^{1+\alpha}$-conformal at c, and we find:

Theorem 2.5. *The limit set Λ is self-similar at each elliptic fixed-point in Λ, with scaling factor $\phi'(c) = e^L$. Here L is the complex length of the singular geodesic γ on T_ψ.*

In particular the self-similarity factor e^L is inherited from the geometry of the rigid manifold T_ψ, and it is universal across all manifolds $M(X, \psi)$ attracted to M_ψ under renormalization.

3. Geometric Limits in Dynamics

In this section we extend the inflexibility of Kleinian groups and their limit sets to certain other conformal dynamical systems $\mathcal{F}$ and their Julia sets J, where we will find:

> *The conformal structure at the deep points of J is determined by the topological dynamics of $\mathcal{F}$.*

Consider the space $\mathcal{H}$ of all holomorphic maps $f : U(f) \to V(f)$ between domains in $\hat{\mathbb{C}}$. Introduce a (non-Hausdorff) topology on $\mathcal{H}$ such that $f_n \to f$ if for any compact $K \subset U(f)$, we have $K \subset U(f_n)$ for all $n \gg 0$ and $f_n|K \to f|K$ uniformly.

A *holomorphic dynamical system* is a subset $\mathcal{F} \subset \mathcal{H}$. Given a sequence of dynamical systems $\mathcal{F}_n \subset \mathcal{H}$, the *geometric limit* $\mathcal{F} = \limsup \mathcal{F}_n$ consists of all maps $f = \lim f_{n_i}$ obtained as limits of subsequences $f_{n_i} \in \mathcal{F}_{n_i}$.

To bring hyperbolic space into the picture, identify $\hat{\mathbb{C}}$ with the boundary of the Poincaré ball model for $\mathbb{H}^3$, let $\mathrm{F}\mathbb{H}^3$ be its frame bundle, and let $\omega_0 \in \mathrm{F}\mathbb{H}^3$ be a standard frame at the center of the ball. Given any other $\omega \in \mathrm{F}\mathbb{H}^3$, there is a unique Möbius transformation g sending ω_0 to ω, and we define

$$(\mathcal{F}, \omega) = g^*(\mathcal{F}) = \{g^{-1} \circ f \circ g : f \in \mathcal{F}\}.$$

In other words, $(\mathcal{F}, \omega)$ is $\mathcal{F}$ as 'seen from' ω.

We say $\mathcal{F}$ is *twisting* if it is essentially nonlinear — for example, if there exists an $f \in \mathcal{F}$ with a critical point, or if $\mathcal{F}$ contains a free group of Möbius transformations.

Given a closed set $J \subset \hat{\mathbb{C}}$, we say $(\mathcal{F}, J)$ is *uniformly twisting* if $\limsup(\mathcal{F}, \omega_n)$ is twisting for any sequence $\omega_n \in \mathrm{F}(\mathrm{hull}(J))$, the frame bundle over the convex hull of J in $\mathbb{H}^3$. Informally, uniform twisting means $\mathcal{F}$ is quite nonlinear at every scale around every point of J.

For a Kleinian group, the pair $(\Gamma, \Lambda(\Gamma))$ is uniformly twisting iff $M = \mathbb{H}^3/\Gamma$ has injectivity bounds. Thus geometric inflexibility, Corollary 2.2, is a special case of [Mc2]:

Theorem 3.1 (Dynamic Inflexibility). *Let $(\mathcal{F}, J)$ be uniformly twisting, and let ϕ be a quasiconformal conjugacy from $\mathcal{F}$ to another holomorphic dynamical system $\mathcal{F}'$. Then ϕ is $C^{1+\alpha}$-conformal at all deep points of J.*

The next three sections illustrate how such inflexibility helps explain universal scaling in dynamics.

4. Renormalization of Interval Maps

Let $f : I \to I$ be a real-analytic map on an interval. The map f is *quadratic-like* if $f(\partial I) \subset \partial I$ and f has a single quadratic critical point $c_0(f) \in \mathrm{int}(I)$. The basic example is $f(x) = x^2 + c$ on $[-a, a]$ with $f(a) = a$. We implicitly identify maps that are linearly conjugate.

If an iterate $f^p|L$ is also quadratic-like for some interval L, with $c_0(f) \in L \subset I$, then we can take the least such $p > 1$ and define the *renormalization* of f by

$$\mathcal{R}(f) = f^p|L.$$

The order of the intervals $L, f(L), \ldots, f^p(L) = L \subset I$ determines a permutation $\sigma(f)$ on p symbols.

The map f is *infinitely renormalizable* if the sequence $\mathcal{R}^n(f)$ is defined for all $n > 0$. The *combinatorics* of f is then recorded by the sequence of permutations $\tau(f) = \langle \sigma(\mathcal{R}^n(f)) \rangle$. We say f has *bounded combinatorics* if only finitely many permutations occur, and *periodic combinatorics* if $\tau(\mathcal{R}^q f) = \tau(f)$ for some $q \geq 1$.

Theorem 4.1. *Let $f : I \to I$ be infinitely renormalizable, with combinatorics of period q. Then $\mathcal{R}^{qn}(f) \to F$ exponentially fast as $n \to \infty$, where F is the unique fixed-point of the renormalization operator $\mathcal{R}^q$ with the same combinatorics as f.*

For example, the Feigenbaum polynomial $f(x) = x^2 - 1.4101155\ldots$, arising at the end of the cascade of period doublings in the quadratic family, has $\tau(f) = \langle (12), (12), (12), \ldots \rangle$. Under renormalization, $\mathcal{R}^n(f)$ converges exponentially fast to a solution of the functional equation

$$F \circ F(x) = \alpha^{-1} F(\alpha x).$$

To formulate the speed of convergence more completely, extend $f : I \to I$ to a complex analytic map on a neighborhood of $I \subset \mathbb{C}$, and let $F : W \to \mathbb{C}$ denote the maximal analytic continuation of the renormalization fixed-point. Then we find there is an $A > 1$ such that for any compact $K \subset W$, we have

$$\sup_{z \in K} |\mathcal{R}^n(f)(z) - F(z)| = O(A^{-n}),$$

where $\mathcal{R}^n(f)$ is suitably rescaled.

Now suppose only that f has bounded combinatorics. Under iteration of f, all but countably many points in I are attracted to the postcritical Cantor set

$$P(f) = \overline{\bigcup_{n>0} f^n(c_0(f))} \subset I.$$

Theorem 4.2. *Let f and g be infinitely renormalizable maps with the same bounded combinatorics. Then $f|P(f)$ and $g|P(g)$ are $C^{1+\alpha}$-conjugate.*

Thus quantitative features of the attractor $P(f)$ (such as its Hausdorff dimension) are determined by the combinatorics $\tau(f)$.

These universal properties of quadratic-like maps were observed experimentally and linked to renormalization by Feigenbaum and Coullet-Tresser in the late 1970s. A program for applying complex quadratic-like maps to renormalization was formulated by Douady and Hubbard in the early 1980s. Sullivan introduced a wealth of new ideas and established the convergence $\mathcal{R}^{nq}(f) \to F$ ([Sul3], [Sul4]). The inflexibility theory gives a new proof yielding, in addition, exponential speed of convergence and $C^{1+\alpha}$-smoothness of conjugacies.

Our approach to renormalization is via *towers* [Mc2]. For simplicity we treat the case of the Feigenbaum polynomial f. By Sullivan's *a priori* bounds, the sequence of renormalizations $\langle \mathcal{R}^n(f) \rangle$ is compact, and all limits are complex quadratic-like maps with definite moduli. Passing to a subsequence we can arrange that $\mathcal{R}^{n+i}(f) \to f_i$ and obtain a tower

$$\tau = \langle f_i : i \in \mathbb{Z} \rangle \quad \text{such that } f_{i+1} = f_i \circ f_i \; \forall i.$$

The Julia set $J(\tau) = \bigcup J(f_i)$ is dense in $\mathbb{C}$, and we deduce that τ is rigid — it admits no quasiconformal deformations. Convergence of renormalization, $\mathcal{R}^n(f) \to F$, then easily follows.

The rapid speed of convergence of renormalization comes from inflexibility of the one-sided tower $\tau = \langle f, f^2, f^4, \ldots \rangle$. To establish this inflexibility, we first show the full dynamical system $\mathcal{F}(f) = \{f^{-i} \circ f^j\}$ contains copies of f^{2^n} near every $z \in J(f)$ and at every scale. Thus $(\mathcal{F}(f), J(f))$ is *uniformly twisting.* Next we use expansion in the hyperbolic metric on $\mathbb{C} - P(f)$ to show $c_0(f)$ is a *deep point* of $J(f)$. Finally by Theorem 3.1, a quasiconformal conjugacy ϕ from f to $\mathcal{R}(f) = f \circ f$ is actually $C^{1+\alpha}$-conformal at the critical point. At small scales ϕ provides a nearly linear conjugacy from $\mathcal{R}^n(f)$ to $\mathcal{R}^{n+1}(f)$, and exponential convergence follows.

The second frame of Figure 1 depicts the Julia set of the infinitely renormalizable Feigenbaum polynomial f, centered at its critical point. The Julia set $J(f)$ is locally connected [JH], [LS]; it is still unknown if $\text{area}(J(f)) = 0$.

Milnor has observed that the Mandelbrot set M is quite dense at the Feigenbaum point $c = -1.4101155\cdots \in \partial M$ and at other fixed-points of tuning [Mil], and it is reasonable to expect that c is a deep point of M. Lyubich has recently given an elegant proof of the hyperbolicity of renormalization at its fixed-points, including a new proof of exponential convergence of $\mathcal{R}^n(f)$ via the Banach space Schwarz lemma, and a proof of Milnor's conjecture that blowups of M around the Feigenbaum point converge to the whole plane in the Hausdorff topology [Lyu].

5. Critical Circle Maps

A *critical circle map* $f : S^1 \to S^1$ is a real-analytic homeomorphism with a single cubic critical point $c_0(f) \in S^1$. A typical example is the *standard map*

$$f(x) = x + \Omega + K \sin(x), \quad x \in \mathbb{R}/2\pi\mathbb{Z}, \quad \Omega \in \mathbb{R}$$

with $K = -1$ and $c_0 = 0$. These maps arise in KAM theory and model the disappearance of invariant circles ([FKS], [Lan], [Rand], [Mak], [DGK]). Another class of examples are the rational maps

$$f(z) = \lambda z^2 \frac{z-3}{1-3z}, \quad |\lambda| = 1, \tag{1}$$

acting on $S^1 = \{z : |z| = 1\}$ with $c_0(f) = 1$.

If $f : S^1 \to S^1$ has no periodic points, then it is topologically conjugate to a rigid rotation by angle $2\pi\rho(f)$, where the *rotation number* $\rho(f)$ is irrational [Y]. The behavior of f is strongly influenced by the continued fraction of its rotation number,

$$\rho(f) = 1/(a_1 + 1/(a_2 + 1/(a_3 + \cdots))), \quad a_i \in \mathbb{N}.$$

By truncating the continued fraction we obtain rational numbers $p_n/q_n \to \rho(f)$. We say $\rho(f)$ is of *bounded type* if $\sup a_i < \infty$.

Theorem 5.1 (De Faria-de Melo). *Let f_1, f_2 be two critical circle maps with equal irrational rotation numbers of bounded type. Then f_1 and f_2 are $C^{1+\alpha}$-conjugate.*

We sketch the proof from [dFdM]. Consider a complex analytic extension of $f(z)$ to a neighborhood of S^1. Let the *Julia set* $J(f)$ be the closure of the set of periodic points of f. As for maps of the interval, one finds the critical point $c_0(f)$ is a *deep point* of $J(f)$, and the full dynamical system $(\mathcal{F}(f), J(f))$ is *uniformly twisting.* Because of the good arithmetic of $\rho(f)$, the forward orbit of the critical point is spread evenly along S^1, so in fact the Julia set is deep at *every* point on the circle. To complete the proof, one constructs a quasiconformal conjugacy between f_1 and f_2, and then applies the inflexibility Theorem 3.1 to deduce that $\phi|S^1$ is $C^{1+\alpha}$.

To bring renormalization into the picture, it is useful to work on the universal cover $\mathbb{R}$ of $S^1 = \mathbb{R}/2\pi\mathbb{Z}$. One can then treat the lifted map $f : \mathbb{R} \to \mathbb{R}$ and the deck transformation $g(x) = x + 2\pi$ on an equal footing. The maps (f, g) form a basis for a subgroup $\mathbb{Z}^2 \subset \mathrm{Diff}(\mathbb{R})$, and any matrix $\left(\begin{smallmatrix} a & b \\ c & d \end{smallmatrix}\right) \in GL_2(\mathbb{Z})$ determines a *renormalization operator* by

$$\mathcal{R}(f, g) = (f^a g^b, f^c g^d).$$

When the continued fraction of $\rho(f)$ is periodic, one can choose $\mathcal{R}$ such that $\mathcal{R}^n(f, g)$ converges exponentially fast to a fixed-point of renormalization (F, G). For the more general case where $\rho(f)$ is of bounded type, a finite number of renormalization operators suffice to relate any two adjacent levels of the tower $\tau = \langle f^{q_n} \rangle$.

The third frame in Figure 1 depicts the Julia set of the rational map $f(z)$ given by Eq. (5.1), with $\lambda \approx -0.7557 - 0.6549i$ chosen so $\rho(f)$ is the golden ratio. The picture is centered at the deep point $c_0(f) \in J(f)$. Petersen has shown $J(f)$ is locally connected [Pet]; it is an open problem to determine if $\mathrm{area}(J(f)) = 0$.

Levin has proposed a similar theory for critical circle *endomorphisms* such as $f(z) = \lambda z^3(z-2)/(1-2z)$ [Lev].

6. The Golden-Ratio Siegel Disk

Let $f(z) = \lambda z + z^2$, where $\lambda = e^{2\pi i\theta}$.

Siegel showed that f is analytically conjugate to the rotation $z \mapsto \lambda z$ on a neighborhood of the origin when θ is Diophantine ($|\theta - p/q| > C/q^n$). The *Siegel disk* D for f is the maximal domain on which f can be linearized. For θ of bounded type, Herman and Świątek proved that ∂D is a quasicircle passing through the critical point $c_0(f) = -\lambda/2$ ([Dou1], [Sw]). In particular, the critical point provides the only obstruction to linearization.

Now suppose θ is a quadratic rational such as the golden ratio:

$$\theta = \frac{\sqrt{5}-1}{2} = 1/(1+1/(1+1/(1+\cdots))).$$

Then the continued fraction of θ is preperiodic; there is an $s > 0$ such that $a_{n+s} = a_n$ for all $n \gg 0$. Experimentally, a universal structure emerges at the transition from linear to nonlinear behavior at ∂D [MN] [Wid]. In [Mc4] we prove:

Theorem 6.1. *If θ is a quadratic irrational, then the boundary of the Siegel disk D for f is self-similar about the critical point $c_0(f) \in \partial D$.*

More precisely, there is a map $\phi : (\bar{D}, c_0) \to (\bar{D}, c_0)$ which is a $C^{1+\alpha}$-conformal contraction at the critical point, and locally conjugates f^{q_n} to $f^{q_{n+s}}$.

Theorem 6.2. *Let f and g be quadratic-like maps with Siegel disks having the same rotation number of bounded type. Then $f|\bar{D}_f$ and $g|\bar{D}_g$ are $C^{1+\alpha}$ conjugate.*

For instance, let D_a be the Siegel disk for $f_a(z) = \lambda z + z^2 + az^3$. Then the Hausdorff dimension of ∂D_a is *constant* for small values of a. As for the Julia set we have:

Theorem 6.3. *If θ has bounded type, then the Hausdorff dimension of the Julia set of $f(z) = e^{2\pi i\theta}z + z^2$ is strictly less than two.*

A blowup of the golden ratio Siegel disk, centered at the critical point $c_0(f) \in \partial D$, is shown in the final frame of Figure 1. The picture is self-similar with a universal scaling factor $1.8166\ldots$ depending only on the rotation number. The Julia set of f is locally connected [Pet]. Recently Buff and Henriksen have shown that the golden Siegel disk contains a Euclidean triangle with vertex resting on the critical point [BH]; empirically, an angle of approximately 120° will fit.

The mechanism of rigidity for Siegel disks is visible in the geometry of the *filled* Julia set $K(f) = \{z : f^n(z) \text{ remains bounded for all } n > 0\}$. Under iteration, every point in the interior of $K(f)$ eventually lands in the Siegel disk, and $\partial K(f) = J(f)$. The gray cauliflower forming the interior of $K(f)$ in Figure 1 is visibly dense at the critical point. In fact $c_0(f)$ is a *measurable* deep point of $K(f)$, meaning

$$\frac{\text{area}(K(f) \cap B(c_0, r))}{\text{area}(B(c_0, r))} = 1 - O(r^\beta), \quad \beta > 0. \tag{2}$$

For the proof of Theorem 6.2, one starts with a quasiconformal conjugacy ϕ from f to g furnished by the theory of polynomial-like maps [DH]. Since f and g have the same linearization on their Siegel disks, we can assume ϕ is conformal on D_f. But then ϕ is conformal throughout int $K(f)$. By (2) the conformal behavior dominates near $c_0(f)$, and we conclude ϕ is $C^{1+\alpha}$-conformal at the critical point. This smoothness is spread to all points of ∂D_f using the good arithmetic of θ.

The self-similarity of ∂D is established similarly, using a conjugacy from f^{q_n} to $f^{q_{n+s}}$.

The dictionary. Table 3 summarizes the parallels which emerge between hyperbolic manifolds, quadratic-like maps on the interval, critical circle maps and Siegel disks. This table can be seen as a contribution to Sullivan's dictionary between conformal dynamical systems ([Sul2], [Mc1]).

7. Surface Groups and Their Geometric Limits

For a complete classification of conformal dynamical systems, one must go beyond the bounded geometry of the preceding examples, and confront short geodesics, unbounded renormalization periods and Liouville rotation numbers. We conclude with an example of such a complete classification in the setting of hyperbolic geometry.

Let S be the compact surface obtained by removing a disk from a torus. Let $AH(S) \subset V(S)$ be the set of discrete faithful representations such that $\rho(\pi_1(\partial S))$ is parabolic. A representation $\rho : \pi_1(S) \to \Gamma$ in $AH(S)$ gives a hyperbolic manifold $M = \mathbb{H}^3/\Gamma$ homeomorphic to $\text{int}(S) \times \mathbb{R}$. To each end of M one can associate an *end invariant*

$$E^\pm(M) = \begin{cases} \partial^\pm(M) \in \text{Teich}(S) & \text{or} \\ \epsilon^\pm(M) \in \mathbb{P}\mathcal{L}(S). \end{cases}$$

In the first case the end is naturally completed by a hyperbolic punctured torus $\partial^\pm(M)$; in the second case the end is pinched along a simple curve or lamination $\epsilon^\pm(M)$.

Identifying $\text{Teich}(S) \cup \mathcal{ML}(S)$ with $\bar{\mathbb{H}} = \mathbb{H} \cup \mathbb{R} \cup \infty$, we may now state:

Theorem 7.1 (Minsky). *The pair of end invariants establishes a bijection*

$$E : AH(S) \to \bar{\mathbb{H}} \times \bar{\mathbb{H}} - \bar{\mathbb{R}} \times \bar{\mathbb{R}}$$

with E^{-1} continuous.

<table>
<tr><th>Hyperbolic Manifolds</th><th>Interval Maps</th><th>Siegel Disks/
Circle Maps</th></tr>
<tr><td>Discrete surface group
$\Gamma \subset PSL_2(\mathbb{C})$
$M = \mathbb{H}^3/\Gamma$</td><td>$\mathbb{R}$-quadratic polynomial
$f(z) = z^2 + c$</td><td>Nonlinear rotation
$f(z) = \lambda z + z^2$ or
$\lambda z^2(z-3)/(1-3z)$</td></tr>
<tr><td>Representation
$\rho : \pi_1(S) \to \Gamma$</td><td>Quadratic-like map
$f : U \to V$</td><td>Holomorphic commuting
pair (f, g)</td></tr>
<tr><td>Ending lamination
$\epsilon(M) \in \mathcal{GL}(S)$</td><td>Tuning invariant
$\tau(f) = \langle \sigma(\mathcal{R}^n(f)) \rangle$</td><td>continued fraction
$\theta = [a_1, a_2, \ldots]$, $\lambda = e^{2\pi i\theta}$</td></tr>
<tr><td>Inj. radius $(M) > r > 0$</td><td>Bounded combinatorics</td><td>Bounded type</td></tr>
<tr><td>Cut points in Γ
$= \cup_1^\infty$ (Cantor sets)</td><td colspan="2">Postcritical set $P(f) = \overline{\cup f^n(c)}$, $f'(c) = 0$
=(Cantor set)   = (circle or quasi-circle)</td></tr>
<tr><td>($\mathbb{R}$-tree of $\epsilon(M)$, $\pi_1(S)$)</td><td>$(\underleftarrow{\lim}\, \mathbb{Z}/p_i, x \mapsto x+1)$</td><td>$(\mathbb{R}/\mathbb{Z}, x \mapsto x+\theta)$</td></tr>
<tr><td>$\Lambda(\Gamma)$ is locally connected</td><td>$J(f)$ is locally connected</td><td>$J(f)$ is locally connected</td></tr>
<tr><td>area $\Lambda(\Gamma) = 0$</td><td colspan="2">area $(J(f) = 0$?</td></tr>
<tr><td>Inj. radius $\in [r, R]$ in core (M)</td><td colspan="2">$(\mathcal{F}(f), J(f))$ is uniformly twisting</td></tr>
<tr><td>Mapping class $\psi \in \mathrm{Mod}(S)$</td><td>Kneading permutation</td><td>Automorphism $\left(\begin{smallmatrix} a & b \\ c & d \end{smallmatrix}\right)$ of $\mathbb{Z}^2$</td></tr>
<tr><td colspan="3">Renormalization Operators</td></tr>
<tr><td>$\mathcal{R}(\rho) = \rho \circ \phi^{-1}$</td><td>$\mathcal{R}(f) = f^p(z)$</td><td>$\mathcal{R}(f, g) = (f^a g^b, f^c g^d)$</td></tr>
<tr><td colspan="3">Stable Manifold of Renormalization</td></tr>
<tr><td>M = asymptotic fiber</td><td>f = limit of doublings</td><td>θ = golden ratio</td></tr>
<tr><td>Elliptic points deep in $\Lambda(\Gamma)$</td><td colspan="2">Critical point $c_0(f)$ deep in $J(f)$ or $K(f)$</td></tr>
<tr><td>$\rho \circ \psi^{-n}$, $n = 1, 2, 3 \ldots$</td><td>f^n, $n = 1, 2, 4, 8, 16, \ldots$</td><td>$f^n$, $n = 1, 2, 3, 5, 8, \ldots$</td></tr>
<tr><td>Geometric limit of $\mathcal{R}^n(\rho)$</td><td>Quadratic-like tower
$\langle f_i : i \in \mathbb{Z} \rangle$; $f_{i+1} = f_i \circ f_i$</td><td>Tower of commuting pairs</td></tr>
<tr><td>Hyperbolic 3-manifold $S \times [0,1]/\psi$
fibering over the circle</td><td colspan="2">Fixed-points of
Renormalization</td></tr>
<tr><td colspan="3">Conformal structure is $C^{1+\alpha}$-rigid at deep point $\Longrightarrow$
Renormalization converges exponentially fast</td></tr>
<tr><td>M is asymptotically rigid</td><td colspan="2">$J(f)$ is self-similar at the critical point $c_0(f)$</td></tr>
</table>

Table 3.

Corollary 7.2. *Each Bers' slice of $AH(S)$ is bounded by a Jordan curve naturally parameterized by $\mathbb{R} \cup \infty$, with rational points corresponding to cusps.*

Corollary 7.3. *Geometrically finite manifolds are dense in $AH(S)$.*

Theorem 7.1 establishes a special case of Thurston's *ending lamination conjecture* [Mc1, §4]. We remark that E is *not* a homeomorphism, and indeed $AH(S)$ is not even a topological manifold with boundary [Mc3, Appendix].

The proof of Theorem 7.1 from [Min] can be illustrated in the case $E(M) = (\tau, \lambda)$, with $\tau \in \mathbb{H}$ and $\lambda \in \mathbb{R}$ an irrational number with continued fraction $[a_1, a_2, \ldots]$. By rigidity of manifolds in $\partial AH(S)$, it suffices to construct a quasi-isometry

$$\phi : M \to M(a_1, a_2, \ldots)$$

from M to a model Riemannian manifold explicitly constructed from the ending invariant. The quasi-isometry is constructed piece by piece, over blocks M_i of M corresponding to terms a_i in the continued fraction.

The construction yields a description not only of manifolds in $AH(S)$, but also of their geometric limits, which we formulate as follows.

Theorem 7.4. *Every geometric limit $M = \lim M_n$, $M_n \in AH(S)$, is determined up to isometry by a sequence $\langle a_i, i \in I\rangle$, where*

- *$I \subset \mathbb{Z}$ is a possibly infinite interval,*
- *$a_i \in \mathrm{Teich}(S) \cup \{*\}$ if i is an endpoint of I; and*
- *$a_i \in \{1, 2, 3, \ldots, \infty\}$ otherwise.*

Here $\langle a_i \rangle$ should be thought of as a generalized continued fraction, augmented by Riemann surface data for the geometrically finite ends of M. (The special point $\{*\}$ is used for the triply-punctured sphere.)

For example, the sequence $\langle a_i \rangle = \langle \ldots, \infty, \infty, \infty, \ldots \rangle$ determines the periodic manifold

$$M_\infty \cong \mathrm{int}(S) \times \mathbb{R} - \left(\bigcup_{\mathbb{Z}} \gamma_i \times \{i\} \right),$$

where $\gamma_i \subset S$ are simple closed curves and $i(\gamma_i, \gamma_{i+1}) = 1$. These curves enumerate the rank two cusps of M_∞. Geometrically, M_∞ is obtained from the Borromean rings complement $S^3 - B$ (itself a hyperbolic manifold) by taking the $\mathbb{Z}$-covering induced by the linking number with one component of B.

In general the coefficients $\langle a_i \rangle$ in Theorem 7.4 specify how to obtain M by Dehn filling the cusps of M_∞. Compare [Th2, §7].

Corollaries 7.2 and 7.3 are reminiscent of two open conjectures in dynamics: the local connectivity of the Mandelbrot set, and the density of hyperbolicity for complex quadratic polynomials.

Quadratic polynomials, however, present an infinite variety of parabolic bifurcations, in contrast to the single basic type occurring for punctured tori. This extra diversity is reflected in the topological complexity of the boundary of the Mandelbrot set, versus the simple Jordan curve bounding a Bers slice.

Parabolic bifurcations can be analyzed by Ecalle cylinders [Dou2] and parabolic towers [Hin], both instances of geometric limits as in §3. A complete understanding of complex quadratic polynomials will likely entail a classification of all their geometric limits as well.

References

[BJ1] C. J. Bishop and P. W. Jones, Hausdorff dimension and Kleinian groups, *Acta Math.* **179** (1997), 1–39.

[BJ2] C. J. Bishop and P. W. Jones, The law of the iterated logarithm for Kleinian groups, In *Lipa's Legacy (New York, 1995)*, Vol. 211 of *Contemp. Math.*, pp. 17–50. Amer. Math. Soc., 1997.

[BH] X. Buff and C. Henriksen, Scaling ratios and triangles in Siegel disks. Preprint, Cornell, 1998.

[CT] J. W. Cannon and W. P. Thurston, Group invariant Peano curves. Preprint, Princeton, 1985.

[dFdM] E. de Faria and W. de Melo, Rigidity of critical circle mappings I, II. SUNY IMS Preprints 1997/16,17.

[DGK] T. W. Dixon, T. Gherghetta, and B. G. Kenny, Universality in the quasiperiodic route to chaos, *Chaos* **6** (1996), 32–42.

[Dou1] A. Douady, Disques de Siegel et anneaux de Herman, In *Séminaire Bourbaki, 1986/87*, pages 151–172. Astérisque, volume 152–153, 1987.

[Dou2] A. Douady, Does a Julia set depend continuously on the polynomial? In R. Devaney, editor, *Complex Analytic Dynamics* Amer. Math. Soc., 1994.

[DH] A. Douady and J. Hubbard, On the dynamics of polynomial-like mappings, *Ann. Sci. Éc. Norm. Sup.* **18** (1985), 287–344.

[Hin] B. Hinkle, Parabolic limits of renormalization, SUNY IMS Preprint 1997/7.

[Lan] O. E. Lanford III, Renormalization group methods for circle mappings with general rotation number, In *VIIIth International Congress on Mathematics Physics* (Marseille, 1986), pp. 532–536. World Scientific, 1987.

[JH] Y. Jiang and J. Hu, The Julia set of the Feigenbaum quadratic polynomial is locally connected, Preprint, 1993.

[Kap] M. Kapovich, On dynamics of pseudo-Anosov homeomorphisms on the representation variety of surface groups, *Ann. Acad. Sci. Fenn. Math.* **23** (1998), 83–100.

[Lev] G. Levin, Bounds for maps of an interval with one reflecting critical point. I, To appear, Fund. Math.

[LS] G. Levin and S. van Strien, Local connectivity of the Julia set of real polynomials, *Ann. of Math.* **147** (1998), 471–541.

[Lyu] M. Lyubich, Feigenbaum-Coullet-Tresser universality and Milnor's Hairiness Conjecture, *Annals of Math.* **149** (1999), 319–420.

[FKS] L. P. Kadanoff, M. J. Feigenbaum and S. J. Shenker, Quasiperiodicity in dissipative systems: a renormalization group analysis, *Phys. D* **5** (1982), 370–386.

[Mak] R. S. MacKay, *Renormalisation in Area-Preserving Maps*, World Scientific, 1993.

[MN] N. S. Manton and M. Nauenberg, Universal scaling behavior for iterated maps in the complex plane, *Commun. Math. Phys.* **89** (1983), 555–570.

[Mc1] C. McMullen, The classification of conformal dynamical systems, in *Current Developments in Mathematics, 1995*, pp. 323–360, International Press, 1995.

[Mc2] C. McMullen, *Renormalization and 3-Manifolds which Fiber over the Circle*, Vol. 142 of *Annals of Math. Studies*, Princeton University Press, 1996.

[Mc3] C. McMullen, Complex earthquakes and Teichmüller theory, *J. Amer. Math. Soc.* **11** (1998), 283–320.

[Mc4] C. McMullen, Self-similarity of Siegel disks and the Hausdorff dimension of Julia sets, *Acta Math.* **180** (1998), 247–292.

[Mil] J. Milnor, Self-similarity and hairiness in the Mandelbrot se, In M. C. Tangora, editor, *Computers in Geometry and Topology*, Lect. Notes Pure Appl. Math., pp. 211–259. Dekker, 1989.

[Min] Y. Minsky, The classification of punctured torus groups, *Annals of Math.* **149** (1999), 559–626.

[Pet] C. L. Petersen, Local connectivity of some Julia sets containing a circle with an irrational rotation, *Acta Math.* **177** (1996), 163–224.

[Rand] D. Rand, Universality and renormalisation in dynamical systems, In *New Directions in Dynamical Systems*, pp. 1–56. Cambridge University Press, 1988.

[Sul1] D. Sullivan, Growth of positive harmonic functions and Kleinian group limit sets of zero planar measure and Hausdorff dimension two, In *Geometry at Utrecht*, volume 894 of *Lecture Notes in Mathematics*, pp. 127–144, Springer-Verlag, 1981.

[Sul2] D. Sullivan, Conformal dynamical systems, In *Geometric Dynamics*, volume 1007 of *Lecture Notes in Mathematics*, pp. 725–752, Springer-Verlag, 1983.

[Sul3] D. Sullivan, Quasiconformal homeomorphisms in dynamics, topology and geometry, In *Proceedings of the International Conference of Mathematicians*, pp. 1216–1228, Amer. Math. Soc., 1986.

[Sul4] D. Sullivan, Bounds, quadratic differentials and renormalization conjectures, In F. Browder, editor, *Mathematics into the Twenty-first Century: 1988 Centennial Symposium, August 8–12*, pp. 417–466, Amer. Math. Soc., 1992.

[Sw] G. Świątek, On critical circle homeomorphisms, *Bol. Soc. Bras. Mat.* **29** (1998), 329–351.

[Th1] W. P. Thurston, *Geometry and Topology of Three-Manifolds*, Lecture Notes, Princeton University, 1979.

[Th2] W. P. Thurston, Hyperbolic structures on 3-manifolds II: Surface groups and 3-manifolds which fiber over the circle, Preprint, 1986.

[Wid] M. Widom, Renormalization group analysis of quasi-periodicity in analytic maps, *Commun. Math. Phys.* **92** (1983), 121–136.

[Y] J.-C. Yoccoz, Il n'y a pas de contre-exemple de Denjoy analytique, *C. R. Acad. Sci. Paris Sér. I Math.* **298** (1984), 141–144.

Reprinted from Proc. Int. Congr. Math., 2002

THE WORK OF LAURENT LAFFORGUE

by
GÉRARD LAUMON
CNRS and Université Paris-Sud
UMR 8628, Mathématique
F-91405 Orsay Cedex, France

Laurent Lafforgue has been awarded the Fields Medal for his proof of the Langlands correspondence for the full linear groups GL_r $(r \geq 1)$ over function fields.

What follows is a brief introduction to the Langlands correspondence and to Lafforgue's theorem.

1. The Langlands Correspondence

A global field is either a number field, i.e. a finite extension of $\mathbb{Q}$, or a function field of characteristic $p > 0$ for some prime number p, i.e. a finite extension of $\mathbb{F}_p(t)$ where $\mathbb{F}_p$ is the finite field with p elements. The global fields constitute a primary object of study in number theory and arithmetic algebraic geometry.

The conjectural Langlands correspondence, which was first formulated by Robert Langlands in 1967 in a letter to André Weil, relates two fundamental objects which are naturally attached to a global field F:

- its Galois group $\mathrm{Gal}(\bar{F}/F)$, where $\bar{F}$ is an algebraic closure of F, or more accurately its motivic Galois group of F which is by definition the tannakian group of the tensor categroy of Grothendieck motives over F,
- the ring $\mathbb{A}$ adèles of F, or more precisely the collection of Hilbert spaces $L^2(G(F)\backslash G(\mathbb{A}))$ for all reductive groups G over F.

Roughly speaking, for any (connected) reductive group G over F, Langlands introduced a dual group ${}^{\mathrm{L}}G = \hat{G} \rtimes \mathrm{Gal}(\bar{F}/F)$, the connected component $\hat{G}$ of which is the complex reductive group whose roots are the co-roots of G and vice versa. And he predicted that a large part of the spectral decomposition of the Hilbert space $L^2(G(F)\backslash G(\mathbb{A}))$, equipped with the action by right translations of $G(\mathbb{A})$, is governed by representations of the motivic Galois group of F with values in ${}^{\mathrm{L}}G$.

Of special importance is the group $G = \mathrm{GL}_r$, the Langlands dual of which is simply the direct product ${}^{\mathrm{L}}\mathrm{GL}_r = \mathrm{GL}_r(\mathbb{C}) \times \mathrm{Gal}(\bar{F}/F)$. Indeed, any complex reductive group $\hat{G}$ may be embedded into $\mathrm{GL}_r(\mathbb{C})$ for some r.

The particular case $G = \mathrm{GL}_1$ of the Langlands correspondence is the abelian class field theory of Teiji Takagi and Emil Artin which was developed in the 1920s as a wide extension of the quadratic reciprocity law.

The Langlands correspondence embodies a large part of number theory, arithmetic algebraic geometry and representation theory of Lie groups. Small progress made towards this conjectural correspondence had already amazing consequences, the most striking of them being the proof of Fermat's last theorem by Andrew Wiles. Famous conjectures, such as the Artin conjecture on L-functions and the Ramanujan-Petersson conjecture, whould follow from the Langlands correspondence.

2. Lafforgue's Main Theorem

Over number fields, the Langlands correspondence in its full generality seems still to be out of reach. Even its precise formulation is very involved. In the function field case the situation is much better. Thanks to Lafforgue, the Langlands correspondence for $G = \mathrm{GL}_r$ is now completely understood.

From now on, F is a function field of characteristic $p > 0$. We also fix some auxiliary prime number $\ell \neq p$.

As Alexandre Grothendieck showed, any algebraic variety over F gives rise to ℓ-adic representations of $\mathrm{Gal}(\bar{F}/F)$ on its étale cohomology groups and the irreducible ℓ-adic representations of $\mathrm{Gal}(\bar{F}/F)$ are good substitutes for irreducible motives over F. Therefore, the Langlands correspondence may be nicely formulated using ℓ-adic representations.

Let r be a positive integer. On the one hand, we have the set $\mathcal{G}_r$ of isomorphism classes of rank r irreducible ℓ-adic representations of $\mathrm{Gal}(\bar{F}/F)$ the determinant of which is of finite order. To each $\sigma \in \mathcal{G}_r$, Grothendieck attached an Eulerian product $L(\sigma, s) = \Pi_x L_x(\sigma, s)$ over all the places x of F, which is in fact a rational function of p^{-s} and which satisfies a functional equation of the form $L(\sigma, s) = \varepsilon(\sigma, s)L(\sigma^\vee, 1-s)$ where $\sigma^\vee$ is the contragredient representation of σ and $\varepsilon(\sigma, s)$ is some monomial in p^{-s}. If σ is unramified at a place x, we have

$$L_x(\sigma, s) = \prod_{i=1}^{r} \frac{1}{1 - z_i p^{-s\deg(x)}}$$

where $z_1, \ldots, z_r$ are the *Frobenius eigenvalues* of σ at x and $\deg(x)$ is the degree of the place x.

One the other hand, we have the set $\mathcal{A}_r$ of isomorphism classes of cuspidal automorphic representations of $\mathrm{GL}_r(\mathbb{A})$ the central character of which is of finite order. Thanks to Langlands' theory of Eisenstein series, they are the building blocks of the spectral decomposition of $L^2(\mathrm{GL}_r(F)\backslash\mathrm{GL}_r(\mathbb{A}))$. To each $\pi \in \mathcal{A}_r$, Roger Godement and Hervé Jacquet attached an Eulerian product $L(\pi, s) = \Pi_x L_x(\pi, s)$ over all the places x of F, which is again a rational function of p^{-s}, satisfying a functional equation $L(\pi, s) = \varepsilon(\pi, s)L(\pi^\vee, 1-s)$. If π is unramified at a place x, we have

$$L_x(\pi, s) = \prod_{i=1}^{r} \frac{1}{1 - z_i p^{-s\deg(x)}}$$

where $z_1, \ldots, z_r$ are called the *Hecke eigenvalues* of π at x.

Theorem (i) (The Langlands Conjecture) *There is a unique bijective correspondence $\pi \to \sigma(\pi)$, preserving L-functions in the sence that $L_x(\sigma(\pi), s) = L_x(\pi, s)$ for every place x, between $\mathcal{A}_r$ and $\mathcal{G}_r$.*

(ii) (The Ramanujan-Petersson Conjecture) *For any $\pi \in \mathcal{A}_r$ and for any place x of F where π is unramified, the Hecke eigenvalues $z_1, \ldots, z_r \in \mathbb{C}^\times$ of π at x are all of absolute value* 1.

(iii) (The Deligne Conjecture) *Any $\sigma \in \mathcal{G}_r$ is pure of weight zero, i.e. for any place x of F where σ is unramified, and for any field embedding $\iota : \bar{\mathbb{Q}}_\ell \hookrightarrow \mathbb{C}$, the images $\iota(z_1), \ldots, \iota(z_r)$ of the Frobenius eigenvalues of σ at x are all of absolute value* 1.

As I said earlier, in rank $r = 1$, the theorem is a reformulation of the abelian class field theory in the function field case. Indeed, the reciprocity law may be viewed as an injective homomorphism with dense image

$$F^\times\backslash\mathbb{A}^\times \to \mathrm{Gal}(\bar{F}/F)^{\mathrm{ab}}$$

from the idèle class group to the maximal abelian quotient of the Galois group.

In higher ranks r, the first breakthrough was made by Vladimir Drinfeld in the 1970s. Introducting the fundamental concept of shtuka, he proved the rank $r = 2$ case. It is a masterpiece for which, among others works, he was awarded the Fields Medal in 1990.

3. The Strategy

The strategy that Lafforgue is following, and most of the geometric objects that he is using, are due to Drinfeld. However, the gap between the rank two case and the general case was so big that it took more than twenty years to fill it.

Lafforgue considers the ℓ-adic cohomology of the moduli stack of rank r Drinfeld shtukas (see the next section) as a representation of $\mathrm{GL}_r(\mathbb{A}) \times \mathrm{Gal}(\bar{F}/F) \times \mathrm{Gal}(\bar{F}/F)$. By comparing the Grothendieck-Lefschetz trace formula (for Hecke operators twisted by powers of Frobenius endomorphisms) with the Arthur-Selberg trace formula, he tries to isolate inside this representation a subquotient which decomposes as

$$\bigoplus_{\pi \in \mathcal{A}_r} \pi \otimes \sigma(\pi)^\vee \otimes \sigma(\pi)\,.$$

Such a comparison of trace formulas was first made by Yasutaka Ihara in 1967 for modular curves over $\mathbb{Q}$. Since, it has been extensively used for Shimura varieties and Drinfeld modular varieties by Langlands, Robert Kottwitz and many others. There are two main difficulties to overcome to complete the comparison:

- to prove suitable cases of a combinatorial conjecture of Langlands and Diana Shelstad, which is known as the *Fundamental Lemma*,
- to compare the contribution of the "fixed points at infinity" in the Grothendieck-Lefschetz trace formula with the weighted orbital integrals of James Arthur which occur in the geometric side of the Arthur-Selberg trace formula.

For the moduli space of shtukas, the required cases of the Fundamental Lemma were proved by Drinfeld in the 1970s. So, only the second difficulty was remaining after Drinfeld had completed his proof of the rank 2 case. This is precisely the problem that Lafforgue has solved after seven years of very hard work. The proof has been published in three papers totalling about 600 pages.

4. Drinfeld Shtukas

Let X be "the" smooth, projective and connected curve over $\mathbb{F}_p$ whose field of rational functions is F. It plays the role of the ring of integers of a number field. Its closed points are the places of F. For any such point x we have the completion F_x of F at x and its ring of integers $\mathcal{O}_x \subset F_x$.

Let $\mathcal{O} = \Pi_x \mathcal{O}_x \subset \mathbb{A}$ be the maximal compact subring of the ring of adèles. Weil showed that the double coset space

$$\mathrm{GL}_r(F)\backslash\mathrm{GL}_r(\mathbb{A})\backslash\mathrm{GL}_r(\mathcal{O})$$

can be naturally identified with the set of isomorphism classes of rank r vector bundles on X.

Starting from this observation, with the goal of realizing a congruence relation between Hecke operators and Frobenius endomorphisms, Drinfeld defined a rank r *shtuka* over an arbitrary field k of characteristic p as a diagram

$$ {}^\tau\mathcal{E} \underset{\varphi}{\xrightarrow{\sim}} \overbrace{\mathcal{E}'' \hookrightarrow \mathcal{E}' \hookleftarrow \mathcal{E}}^{\text{Hecke}}$$

where $\mathcal{E}$, $\mathcal{E}'$ and $\mathcal{E}'$ are rank r vector bundles on the curve X_k deduced from X by extending the scalars to k, where $\mathcal{E} \hookrightarrow \mathcal{E}'$ is an elementary upper modification of ε at some k-rational point of X which is called the *pole* of the shtuka, where $\mathcal{E}'' \hookrightarrow \mathcal{E}'$ is an elementary lower modification of $\mathcal{E}'$ at some k-rational point of X which is called the *zero* of the shtuka, and where ${}^{\tau}\mathcal{E}$ is the pull-back of $\mathcal{E}$ by the endomorphism of X_k which is the identity on X and the Frobenius endomorphism on k.

Drinfeld proved that the above shtukas are the k-rational points of an algebraic stack over $\mathbb{F}_p$ which is equipped with a projection onto $X \times X$ given by the pole and the zero. More generally, he introduced level structures on rank r shtukas and he constructed an algebraic stack Sht_r parametrizing rank r shtukas equipped with a compatible system of level structures. This last algebraic stack is endowed with an algebraic action of $\mathrm{GL}_r(\mathbb{A})$ through the Hecke operators.

5. Iterated Shtukas

The geometry at infinity of the moduli stack Sht_r is amazingly complicated. The algebraic stack Sht_r is not of finite type and one needs to *truncate* it to obtain manageable geometric objects. Bounding the Harder-Narasimhan polygon of a shtuka, Lafforgue defines a family of open substacks $(\mathrm{Sht}_r^{\leq P})_P$ which are all of finite type and whose union is the whole moduli stack. But in doing so, he loses the action of the Hecke operators which do not stabilize those open substacks.

In order to recover the action of the hecke operators, Lafforgue enlarges Sht_r by allowing specific degenerations of shtukas that he has called *iterated shtukas*.

More precisely, Lafforgue lets the isomorphism $\varphi : {}^{\tau}\mathcal{E} \xrightarrow{\sim} \mathcal{E}''$, appeaing in the definition of a shtuka, degenerate to a *complete homomorphism* ${}^{\tau}\mathcal{E} \Rightarrow \mathcal{E}''$ i.e. a continuous family of complet homomorphisms between the stalks of the vector bundles ${}^{\tau}\mathcal{E}$ and $\mathcal{E}''$.

Let me recall that a complete homomorphism $V \Rightarrow W$ between two vector spaces of the same dimension r is a point of the partial compactification $\widetilde{\mathrm{Hom}}(V,W)$ of $\mathrm{Isom}(V,W)$ which is obtained by successively blowing up the quasi-affine variety $\mathrm{Hom}(V,W) - \{0\}$ along its closed subsets

$$\{f \in \mathrm{Hom}(V,W) - \{0\} | \operatorname{rank}(f) \leq i\}$$

for $i = 1, \ldots, r-1$. If $V = W$ is the standard vector space of dimension r, the quotient of $\widetilde{\mathrm{Hom}}(V,W)$ by the action of the homotheties is the Procesi-De Concini compactification of PGL_r.

In particular, Lafforgue obtains a smooth compactification, with a normal crossing divisor at infinity, of any truncated moduli stack of shtukas without level structure.

6. One Key of the Proof

Lafforgue proves his main theorem by an elaborate induction on r. Compared to Drinfeld's proof of the rank 2 case, a very simple but crucial novelty in Lafforgue's proof is the distinction in the ℓ-adic cohomology of Sht_r between the r-negligible part (the part where all the irreducible constituents as Galois modules are of dimension < 1) and the r-essential part (the rest). Lafforgue shows that the difference between the cohomology of Sht_r and the cohomology of any truncated stack $\mathrm{Sht}_r^{\leq P}$ is r-negligible. He also shows that the cohomology of the boundary of $\mathrm{Sht}_r^{\leq P}$ is r-negligible. Therefore, the r-essential part, which is defined purely by considering the Galois action and which is naturally endowed with an action of the Hecke operators, occurs in the ℓ-adic cohomology of any truncated moduli stack $\mathrm{Sht}_r^{\leq P}$ and also in their compactifications.

At this point, Lafforgue makes an extensive use of the proofs by Richard Pink and Kazuhiro Fujiwara of a conjecture of Deligne on the Grothendieck-Lefschetz trace formula.

7. Compactification of Thin Schubert Cells

In proving the Langlands conjecture for functions fields, Lafforgue tried to construct nice compactifications of the truncated moduli stacks of shtukas with arbitrary level structures. A natural way to do that is to start with some nice compactifications of the quotients of $\mathrm{PGL}_r^{n+1}/\mathrm{PGL}_r$ for all integers $n \geq 1$, and to apply a procedure similar to the one which leads to iterated shtukas.

Lafforgue constructed natural compactifications of $\mathrm{PGL}_r^{n+1}/\mathrm{PGL}_r$. In fact, he remarked that $\mathrm{PGL}_r^{n+1}/\mathrm{PGL}_r$ is the quotient of $\mathrm{GL}_r^{n+1}/\mathrm{GL}_r$ by the obvious free action of the torus $\mathbb{G}_\mathrm{m}^{n+1}/\mathbb{G}_\mathrm{m}$ and that $\mathrm{GL}_r^{n+1}/\mathrm{GL}_r$ may be viewed as a thin Schubert cell in the Grassmanian variety of r-planes in a $r(n+1)$-dimensional vector space. And, more generally, he constructed natural compactifications of all similar quotients of thin Schubert cells in the Grassmannian variety of r-planes in a finite-dimensional vector space.

Let me recall that thin Schubert cells are by definition intersection of Schubert varieties and that Israel Gelfand, mark Goresky, Robert MacPherson and Vera Serganova constructed natural bijections between thin Schubert cells, matroids and certain convex polyhedra which are called polytope matroids.

For $n = 1$ and arbitrary r, Lafforgue's compactification of $\mathrm{PGL}_r^2/\mathrm{PGL}_r$ coincides with the Procesi-De Concini compactification of PGL_r. It is smooth with a normal crossing divisor at infinity.

For $n = 2$ and arbitrary r, Lafforgue proves that his compactification of $\mathrm{PGL}_r^3/\mathrm{PGL}_r$ is smooth over a toric stack, and thus can be desingularized.

For $n \geq 3$ and $r \geq 3$, the geometry of Lafforgue's compactifications is rather mysterious and not completely understood.

Gerd Faltings linked the search of good local models for Shimura varieties in bed characteristics to the search of smooth compactifications of G^{n+1}/G for a reductive group G. He gave another construction of Lafforgue's compactifications of $\mathrm{PGL}_r^{n+1}/\mathrm{PGL}_r$ and he succeded in proving that Lafforgue's compactification of $\mathrm{PGL}_r^{n+1}/\mathrm{PGL}_r$ is smooth for $r = 2$ and arbitrary n.

8. Conclusion

I hope that I gave you some idea of the depth and the technical strength of Lafforgue's work on the Langlands correspondence for which we are now honoring him with the Fields Medal.

L. Lafforgue (left) and G. Laumon

LAURENT LAFFORGUE

I.H.É.S., Bures-sur-Yvette, France
Né le 6 novembre 1966 à Antony (Hauts-de-Seine), France
Nationalité française

1986–1990	Élève à l'École Normale Supérieure de Paris
1988–1991	Étudiant en géométrie algébrique (et en théorie d'Arakelov avec Christophe Soulé)
1990–1991	Chargé de recherche au C.N.R.S. dans l'équipe puis "Arithmétique et Géométrie Algébrique" de
1992–2000	l'Université paris-Sud
1991–1992	Service militarire à l'École Spéciale Militaire de Saint-Cry-Coëtquidan
1993–1994	Thèse sur les D-chtoucas de Drinfeld sous la direction de Gérard Laumon
1994–2000	Suite de l'étude des chtoucas
2000	Professeur à l'Institut des hautes Études Scientifiques

Reprinted from Proc. Int. Congr. Math., 2002

THE WORK OF VLADIMIR VOEVODSKY

by
CHRISTOPHE SOULÉ
CNRS and Institute des Hautes Etudes Scientifiques
35, route de Chartres
91440 Bures-sur-Yvette, France

Vladimir Voevodsky was born in 1966. He studied at Moscow State University and Harvard University. He is now Professor at the Institute for Advanced Study in Princeton.

Among his main achievements are the following: he defined and developed motivic cohomology and the $\mathbf{A}^1$-homotopy theory of algebraic varieties; he proved the Milnor conjectures on the K-theory of fields.

Let us state the first Milnor conjecture. Let F be a field and n a positive integer. The *Milnor K-group* of F is the abelian group $K_n^M(F)$ defined by the following generators and relations. The generators are sequences $\{a_1, \ldots, a_n\}$ of n units $a_i \in F^*$. The relations are

$$\begin{aligned}&\{a_1, \ldots, a_{k-1}, xy, a_{k+1}, \ldots, a_n\}\\ &\quad = \{a_1, \ldots, a_{k-1}, x, a_{k+1}, \ldots, a_n\} + \{a_1, \ldots, a_{k-1}, y, a_{k+1}, \ldots, a_n\}\end{aligned}$$

for all a_i, x, $y \subset F^*$, $1 \leq k \leq n$, and the *Steinberg relation*

$$\{a_1, \ldots, x, \ldots, 1-x, \ldots, a_n\} = 0$$

for all $a_i \in F^*$ and $x \in F - \{0, 1\}$.

On the other hand, let $\overline{F}$ be an algebraic closure of F and $G = \mathrm{Gal}(\overline{F}/F)$ the absolute Galois group of F, with its profinite topology. The *Galois cohomology* of F with $\mathbf{Z}/2$ coefficients is, by definition,

$$H^n(F, \mathbf{Z}/2) = H^n_{\text{continuous}}(G, \mathbf{Z}/2).$$

Theorem 2.1. (Voevodsky 1996 [5]) *Assume* $1/2 \in F$ *and* $n \geq 1$. *The Galois symbol*

$$h_n : K_n^M(F)/2K_n^M(F) \to H^n(F, \mathbf{Z}/2)$$

is an isomorphism.

This was conjectured by Milnor in 1970 [1]. When $n = 2$, Theorem 1 was proved by Merkurjev in 1983. The case $n = 3$ was then solved independently by Merkurjev-Suslin and Rost.

There exists also a Galois symbol on $K_n^M(F)/pK_n^M(F)$ for any prime p invertible in F. When $n = 2$ and F is a number field, Tate proved that it is an isomorphism. In 1983 Merkurjev and Suslin proved that it is an isomorphism when $n = 2$ and F is any field. Both Voevodsky and Rost have made a lot of progress towards proving that, for any F, any $n > 0$ and any p invertible in F, the Galois symbol is an isomorphism.

The map h_n in Theorem 1 is defined as follows. When $n = 1$, we have $K_1^M(F) = F^*$ and $H^1(F, \mathbf{Z}/2) = \mathrm{Hom}(G, \mathbf{Z}/2)$. The map

$$h_1 : F^*/(F^*)^2 \to \mathrm{Hom}(G, \mathbf{Z}/2)$$

maps $a \in F^*$ to the quadratic character χ_a defined by

$$\chi_a(g) = g(\sqrt{a})/\sqrt{a} = \pm 1$$

for any $g \in G$ and any square root $\sqrt{a}$ of a in $\bar{F}$. That h_1 is bijective is a special case of Kummer theory. When $n \geq 2$, we just need to define h_n on the generators $\{a_1, \ldots, a_n\}$ of $K_n^M(F)$. It is given by a cup-product:

$$h_n(\{a_1, \ldots, a_n\}) = \chi_{a_1} \cup \cdots \cup \chi_{a_n}.$$

The fact that h_n is compatible with the Steinberg relation was first noticed by Bass and Tate.

Theorem 1 says that $H^n(F, \mathbf{Z}/2)$ has a very explicit description. In particular, an immediate consequence of Theorem 1 and the definition of h_n is the following

Corollary 1. *The graded $\mathbf{Z}/2$-algebra $\bigoplus_{n \geq 0} H^n(F, \mathbf{Z}/2)$ is spanned by elements of degree one.*

This means that absolute Galois groups are very special groups. Indeed, it is seldom seen that the cohomology of a group or a topological space is spanned in degree one.

Corollary 2. (Bloch) *Let X be a complex algebraic variety and $\alpha \in H^n(X(\mathbf{C}), \mathbf{Z})$ a class in its singular cohomology. Assume that $2\alpha = 0$. Then, there exists a nonempty Zariski open subset $U \subset X$ such that the restriction of α to U vanishes.*

If Theorem 1 was extended to $K_n^M(F)/pK_n^M(F)$ for all n and p, Corollary 2 would say that any torsion class in the integral singular cohomology of X is supported on some hypersurface. (Hodge seems to have believed that such a torsion class should be Poincaré dual to an analytic cycle, but this is not always true.)

With Orlov and Vishik, Voevodsky proved a second conjecture of Milnor relating the Witt group of quadratic forms over F to its Milnor K-theory [3].

A very serious difficulty that Voevodsky had to overcome to prove Theorem 1 was that, when $n = 2$, Merkurjev made use of the algebraic K-theory of conics over F, but, when $n \geq 2$, one needed to study special quadric hypersurfaces of dimension $2^{n-1} - 1$. And it is quite hard to compute the algebraic K-theory of varieties of such a high dimension. Although Rost had obtained crucial information about the K-theory of these quadrics, this was not enough to conclude the proof when $n > 3$. Instead of algebraic K-theory, Voevodsky used *motivic cohomology*, which turned out to be more computable.

Given an algebraic variety X over F and two integers p, $q \in \mathbf{Z}$, Voevodsky defined an abelian group $H^{p,q}(X, \mathbf{Z})$, called motivic cohomology. These groups are analogs of the singular cohomology of CW-complexes. They satisfy a long list of properties, which had been anticipated by Beilinson and Lichtenbaum. For example, when n is a positive integer and X is smooth, the group

$$H^{2n,n}(X, \mathbf{Z}) = \mathrm{CH}^n(X)$$

is the Chow group of codimension n algebraic cycles on X modulo linear equivalence. And when X is a point we have

$$H^{n,n}(\text{point}) = K_n^M(F).$$

It is also possible to compute Quillen's algebraic K-theory from motivic cohomology. Earlier constructions of motivic cohomology are due to Bloch (at the end of the seventies) and, later, to Suslin. The way Suslin modified Bloch's definition was crucial to Voevodsky's approach and, as a matter of fact, several important papers on this topic were written jointly by Suslin and Voevodsky [4, 7]. There exist also two very different definitions of $H^{p,q}(X, \mathbf{Z})$, due to Levine and Hanamura; according to the experts they lead to the same groups. But it seems fair to say that Voevodsky's approach to motivic cohomology is the most complete and satisfactory one.

A larger context in which Voevodsky developed motivic cohomology is the $\mathbf{A}^1$-homotopy of algebraic manifolds [6], which is a theory of "algebraic varieties up to deformations", developed jointly with Morel [2]. Starting with the category of smooth manifolds (over a fixed field F), they first embed this category into the category of *Nisnevich sheaves*, by sending a given manifold to the sheaf it represents. A Nisnevich sheaf is a sheaf of sets on the category of smooth manifolds for the Nisnevich topology, a topology which is finer (resp. coarser) than the Zariski (resp. étale) topology. Then Morel and Voevodsky define a homotopy theory of Nisnevich sheaves in much the same way the homotopy theory of CW-complexes is defined. The parameter space of deformations is the affine line $\mathbf{A}^1$ instead of the real unit interval $[0, 1]$. Note that, in this theory there are *two circles* (corresponding to the two degrees p and q for motivic cohomology)! The first circle is the sheaf represented by the smooth manifold $\mathbf{A}^1 - \{0\}$ (indeed, $\mathbf{C} - \{0\}$ has the homotopy type of a circle). The second circle is $\mathbf{A}^1/\{0, 1\}$ (note that $\mathbf{R}/\{0, 1\}$ is a loop). The latter is not represented by a smooth manifold. But, if we identify 0 and 1 in the sheaf of sets represented by $\mathbf{A}^1$ we get a presheaf of sets, and $\mathbf{A}^1/\{0, 1\}$ can be defined as the sheaf attached to this presheaf. This example shows why it was useful to embed the category of algebraic manifolds into a category of sheaves.

It is quite extraordinary that such a homotopy theory of algebraic manifolds exists at all. In the fifties and sixties, interesting invariants of differentiable manifolds were introduced using algebraic topology. But very few mathematicians anticipated that these "soft" methods would ever be successful for algebraic manifolds. It seems now that any notion in algebraic topology will find a partner in algebraic geometry. This has long been the case with Quillen's algebraic K-theory, which is precisely analogous to topological K-theory. We mentioned that motivic cohomology is an algebraic analog of singular cohomology. Voevodsky also computed the algebraic analog of the Steenrod algebra, i.e. cohomological operations on motivic cohomology (this played a decisive role in the proof of Theorem 1). Morel and Voevodsky developed the (stable) $\mathbf{A}^1$-homotopy theory of algebraic manifolds. Voevodsky defined *algebraic cobordism* as homotopy classes of maps from the suspensions of an algebraic manifold to the classifying space MGL. There is also a direct geometric definition of algebraic cobordism, due to Levine and Morel (see Levine's talk in these proceedings), which should compare well with Voevodsky's definition. And the list is growing: Morava K-theories, stable homotopy groups of spheres, etc...

Vladimir Voevodsky is an amazing mathematician. He has demonstrated an exceptional talent for creating new abstract theories, about which he proved highly nontrivial theorems. He was able to use these theories to solve several of the main long standing problems in algebraic K-theory. The field is completely different after his work. He opened large new avenues and, to use the same word as Laumon, he is leading us closer to the world of *motives* that Grothendieck was dreaming about in the sixties.

References

[1] J. Milnor, Algebraic K-theory and quadratic forms, *Inv. Math.* **9** (1970) 318–344.
[2] F. Morel and V. Voevodsky, $\mathbf{A}^1$-homotopy theory of schemes, *Publ. Math. IHES*, **90** (1999) 45–143.
[3] D. Orlov, A. Vishik and V. Voevodsky, An exact sequence for Milnor's K-theory with applications to quadratic forms (2000), to appear.
[4] A. Suslin and V. Voevodsky, Singular homology of abstract algebraic varieties, *Inv. Math.* **123** (1996) 61–94.
[5] V. Voevodsky, On 2-torsion in motivic cohomology (2001), to appear.
[6] V. Voevodsky, The $\mathbf{A}^1$-homotopy theory, in *Proc. of the Int. Congr. Math.*, Volume **1** (Berlin, 1998) 579–604.
[7] V. Voevodsky, A. Suslin and E. Friedlander, Cycles, transfers and motivic homology theories, *Annals of Maths. Studies* **143** (Princeton University Press, 2000).

Vladimir Voevodsky

VLADIMIR VOEVODSKY

Work
School of Mathematics
Institute for Advanced Study
Princeton, NJ 08540
USA
Email: Vladimir@math.ias.edu
Tel: (609) 734-8042

Residence
22 Earle Lane
Princeton, NJ 08540
USA
Tel: (609) 279-2835

PERSONAL

Date of birth June 4, 1966
Place of birth Moscow, Russia

EDUCATION

Harvard University

Ph.D. in Mathematics June 1992

"Homology of schemes and covariant motives" (advisor: David Kazhdan)

Moscow University

B.S. in Mathematics June 1989

ACADEMIC POSITIONS

Institute for Advanced Study Jan. 2002–present
Professor
Institute for Advanced Study Sept. 1998–Dec. 2001
Member
Northwestern University Sept. 1996–June 1999
Associate Professor
Max-Planck Institute Sept. 1996–Aug. 1997
Visiting Scholar
Harvard University Sept. 1996–June 1997
Visiting Scholar
Harvard University July 1993–July 1996
Junior Fellow of Harvard Society of Fellows

Institute for Advanced Study Sept. 1992–May 1993
Member

GRANTS AND FELLOWSHIPS

Sloan Fellowship 1996–1998
NSF Grant "Motivic Homology with Finite Coefficients" 1995–1998
NSF Grant "A^1-homology theory" 1999–2002
Clay Prize Fellowship 1999, 2000, 2001

PRIZES

Fields Medal 2002

Papers in Refereed Journals

[1] G. B Shabat and V. Voevodsky, Equilateral triangulations of Riemann surfaces and curves over algebraic number fields, *Soviet Math. Kokl.* **39** (1) (1989) 38–41.
[2] M. Kapranov and V. Voevodsky, ∞-Groupoids as a model for a homotopy category, *Russian Math. Surveys* **45** (5) (1990) 239–240.
[3] M. Kapranov and V. Voevodsky, Combinatorial-Geometric aspects of polycategory theory, *Cahiers Top. et Geom. Diff.* **32** (1) (1991) 11–27.
[4] M. Kapranov and V. Voevodsky, The ∞-groupoids and homotopy types, *Cahiers Top. et Geom. Diff.* **32** (1) (1991) 29–46.
[5] M. Kapranov and V. Voevodsky, The free n-category generated by a cube, oriented matroids and higher Bruhat orders, *Functional Anal. Appl.* **25** (1) (1991) 50–52.
[6] V. Voevodsky, Galois groups of functional fields of finite type over Q, *Russian Math. Surveys* **46** (5) (1991) 202–203.
[7] V. Voevodsky, Etale topologies of schemes over fields of finite type over Q, *Math. USSR Izv.* **37** (3) (1991) 511–523.
[8] V. Voevodsky, Galois representations connected with hyperbolic curves, *Math. USSR Izv.* **39** (3) (1992) 1281–1291.
[9] M. Kapranov and V. Voevodsky, Braided monoidal 2-categories and Manin-Schechtman higher braid groups, *J. of Pure and Appl. Algebra* **92** (3) (1994) 241–267.
[10] V. Voevodsky, A nilpotence theorem for cycles algebraically equivalent to zero, *International Math. Research Notes* **4** (1995) 187–199.
[11] V. Voevodsky, Homology of schemes, *Selecta Mathematica, New Series* **2** (1) (1996) 111–153.
[12] A. Suslin and V. Voevodsky, Singular homology of abstract algebraic varieties, *Inv. Math.* **123** (1996) 61–94.
[13] F. Morel and V. Voevodsky, $\mathbf{A}^1$-Homotopy theory of schemes, *Publications Math. IHES* **90** (1999) 45–143.
[14] V. Voevodsky, Motivic cohomology are isomorphic to higher Chow groups, *IMRN* **7** (2002) 351–356.

Proceedings of Refereed Conferences

[1] M. Kapranov and V. Voevodsky, 2-Categories and Zamolodchikov's Tetrahedra equations, *Proc. of Symp. in Pure Math.* **56** (Part 2) (1994) 177–259.

[2] A. Suslin and V. Voevodsky, Motivic cohomology with finite coefficients and Bloch-Kato conjecture, in: *Arithmetic and Geometry of Algebraic Cycles* (Kluwer, 2000) 117–189.

[3] V. Voevodsky, A possible new approach to the motivic spectral sequence for algebraic K-theory, *Contemp. Math.* **293** (2002) 371–379.

[4] V. Voevodsky, Open problems in the motivic stable homotopy theory, in: *Motives, polylogarithms and Hodge Theory I* (International Press, 2002) 3–35.

Other Major Publications

[1] V. Voevodsky, Galois group $Gal(\overline{Q} = Q)$ and Teihmuller modular groups, in: *Proc. Conf. Constr. Methods and Alg. Number Theory* (Minsk, 1989).

[2] M. Kapranov and V. Voevodsky, Multidimensional categories (in Russian), *Proc. Conf. of Young Scientists* (Moscow Univ. Press, 1990).

[3] G. B. Shabat and V. Voevodsky, Drawing curves over number fields, *Grothendieck Festchrift* **III** (Birkhäuser, Boston, 1990) 199–227.

[4] V. Voevodsky, Triangulations of oriented manifolds and ramified coverings of sphere (in Russian), *Proc. Conf. of Young Scientists* (Moscow Univ. Press, 1990).

[5] V. Voevodsky, Triangulated categories of motives over a field, in: *Cycles, Transfers and Motivic Homology Theories*, Ann. of Math. Studies **143** (Princeton Univ. Press, 2000) 188–238.

[6] E. Friedlander and V. Voevodsky, Bivariant cycle cohomology, in: *Cycles, Transfers and Motivic Homology Theories*, Ann. of Math. Studies **143** (Princeton Univ. Press, 2000) 138–187.

[7] A. Suslin and V. Voevodsky, Relative equidimensional cycles and Chow sheaves, in: *Cycles, Transfers and Motivic Homology Theories*, Ann. of Math. Studies **143** (Princeton Univ. Press, 2000) 10–86.

[8] V. Voevodsky, Cohomological theory of presheaves with transfers, in: *Cycles, Transfers and Motivic Homology Theories*, Ann. of Math. Studies **143** (Princeton Univ. Press, 2000) 87–137.

Internal Memoranda and Progress Reports

[1] G. B. Shabat and V. Voevodsky, Piece-Wise Euclidean approximation of Jacobians of algebraic curves, *CSTARCI Math.*, Preprint 01-90 (Moscow, 1990).

[2] V. Voevodsky, Flags and Grothendieck cartographical group in higher dimensions, *CSTARCI Math.*, Preprint 05-90 (Moscow, 1990).

[3] V. Voevodsky, Algebraic Morava K-theories and the Bloch-Kato conjecture for $\mathbf{Z}/2$-coefficients, Preprint (June 1995).

[4] V. Voevodsky, The Milnor conjecture, MPIM Preprint (Dec. 1996).

[5] V. Voevodsky, Δ-closed classes, Preprint (2000). http://www.math.uiuc.edu/K-theory/0442

[6] V. Voevodsky, Homotopy theory of simplicial sheaves in completely decomposable topologies, Preprint (2000). http://www.math.uiuc.edu/K-theory/0443
[7] V. Voevodsky, Unstable homotopy categories in Nisnevich and cdh-topologies, Preprint (2000). http://www.math.uiuc.edu/K-theory/0444
[8] D. Orlov, A. Vishik and V. Voevodsky, An exact sequence for Milnor's K-theory with applications to quadratic forms, Preprint (2000), submitted to *Annals of Math.* http://www.math.uiuc.edu/K- theory/0454
[9] V. Voevodsky, C. Mazza and C. Weibel, Lectures on motivic cohomology, Preprint (2001). http://www.math.uiuc.edu/K-theory/0486
[10] V. Voevodsky, Reduced power operations in motivic cohomology, Preprint (2001), submitted to *Pub. IHES.* http://www.math.uiuc.edu/K-theory/0487
[11] V. Voevodsky, On 2-torsion in motivic cohomology, Preprint (2001), submitted to *Pub. IHES.*
[12] V. Voevodsky, Lectures on motivic cohomology 2000/2001 (written by P. Deligne), Preprint (2001). http://www.math.uiuc.edu/K-theory/0527.
[13] V. Voevodsky Cancellation theorem, Preprint (2001). http://www.math.uiuc.edu/K-theory/0541

Reprinted from Motives, Polylogarithms and Hodge Theory

OPEN PROBLEMS IN THE MOTIVIC STABLE HOMOTOPY THEORY, I

by

VLADIMIR VOEVODSKY[a]
School of Mathematics
Institute for Advanced Study
Princeton NJ, USA

Contents

1 Introduction **857**

2 Slice Filtration **859**

3 Main Conjectures **861**

3.1 *Motivic Eilenberg-MacLane spectra* 862
3.2 *Motivic Thom spectrum* 864
3.3 *Algebraic K-theory spectrum* 865
3.4 *Sphere spectrum* 865

4 Slice-Wise Cellular Spectra **866**

5 Reformulations in Terms of Rigid Homotopy Groups **868**

6 Rigid Homology and Rigid Adams Spectral Sequence **870**

7 Slice Spectral Sequence and Convergence Problems **874**

8 Possible Strategies of the Proof **880**

1. Introduction

In this paper we discuss a number of conjectures concentrated around the notion of the slice filtration and the related notion of the rigid homotopy groups. Many of

[a]Supported by the NSF grants DMS-97-29992 and DMS-9901219, Sloan Research Fellowship and Veblen Fund.

the ideas discussed below are in greater or lesser degree the result of conversations I had with Fabien Morel, Mike Hopkins and, more recently, Charles Rezk.

In topology there is a direct connection between the homotopy groups of a spectrum and the Postnikoff tower which describes how one can build this spectrum from the topological Eilenberg-MacLane spectra. On the level of cohomology theories this results in the existence of a spectral sequence which starts from cohomology with coefficients in the homotopy groups of a spectrum and converges to the cohomology theory represented by the spectrum. The connection exists because the Eilenberg-MacLane spectrum corresponding to an abelian group A has only one non trivial homotopy group which equals A.

The motivic Eilenberg-MacLane spectrum corresponding to an abelian group A has many non trivial motivic homotopy groups. As a result, for a motivic spectrum E, one can not recover a Postnikoff tower describing how to build E out of motivic Eilenberg-MacLane spectra by looking at the motivic homotopy groups of E. There is a spectral sequence which starts with cohomology with coefficients in the sheaves of motivic homotopy groups of E and converges to the theory represented by E but the cohomology with coefficients in the sheaves of homotopy groups are not ordinary cohomology theories in the sense of the motivic homotopy theory. In particular the trace maps p_* defined by a finite field extension E/F in these theories fail to satisfy the condition $p_*p^* = \deg(E/F)Id$ which holds for ordinary motivic cohomology. The problem of constructing the "right" spectral sequence received a lot of attention in the particular case of algebraic K-theory. Recently S. Bloch and S. Lichtenbaum gave a construction which works for fields in [1] and E.M. Friedlander and A. Suslin generalized it to varieties over a field in [3].

In the first section of this paper we define for any spectrum E a canonical Postnikoff tower (1) which we call the slice tower of E. The main conjecture of this paper (Conjecture 10) implies that for any spectrum E its slices $s_i(E)$ have unique and natural module structures over the motivic Eilenberg-MacLane spectrum $H_{\mathbf{Z}}$ and therefore represent ordinary cohomology theories. The main theme of all the conjectures presented here is that the slices $s_i(E)$ play the same role in the motivic homotopy theory as objects of the form $\Sigma^i H_{\pi_i(E)}$ play in topology. In Section 3 we formulate conjectures providing explicit description of the slices of the motivic Eilenberg-MacLane spectrum, motivic Thom spectrum, algebraic K-theory spectrum and the sphere spectrum. The most surprising here is the description of the slices of the sphere spectrum which was first suggested by Charles Rezk.

In Section 4 we introduce a class of *slice-wise cellular* spectra whose slices are the motivic Eilenberg-MacLane spectra corresponding to complexes of abelian groups (as opposed to complexes of sheaves with transfers which may appear for a general E). Modulo the conjectures of Section 3 we show that it contains all the standard spectra mentioned above. For spectra of this class the slices are determined by actual abelian groups which we call the rigid homotopy groups. An important property of rigid homotopy groups is that for the standard spectra they are expected to be finitely generated abelian groups which do not depend on the base scheme (as long as it is normal and connected). In Section 5 we show (again modulo the conjectures) that rigid homotopy groups have a number of properties which are similar to the properties of the usual stable homotopy groups. In particular rigid homotopy groups are finitely generated if rigid homology are finitely generated and rationally rigid homotopy groups are isomorphic to rigid homology.

The slice tower defines a slice spectral sequence which, for a slice-wise cellular spectrum, starts with motivic cohomology with coefficients in the rigid homotopy groups of the spectrum and tries to converge to the motivic homotopy groups of the spectrum. We conjecture that for the algebraic K-theory the slice spectral sequence coincides with the spectral sequence constructed in [1] and [3] but the precise relation of the two approaches remains to be understood. In general it seems to be hard to figure out whether or not the slice spectral sequence converges. In Section 7 we formulate some conjectures about the convergence of the slice spectral sequence and show how they are related to the convergence problem for the motivic Adams spectral sequence. Unlike all the rest of conjectures of this paper for which a clear strategy exists at least for varieties over a field of characteristic zero the convergence conjectures are simply guesses.

Three other groups of conjectures in motivic homotopy theory, not included in to this paper, seem to be slowly crystallizing. One group describes the behavior of slice filtration with respect to the functors $f_*, f^*, f_!, f^!$ for morphisms of different types. In view of Conjecture 10 one should probably include Conjecture 17 into that group. The second group describes a theory of operadic description of T-loop spaces. The third one concerns explicit constructions of the slice filtration. It seems that something like the construction used by E. M. Friedlander and A. Suslin to get the spectral sequence for algebraic K-theory can be used to produce explicit models for the spectra $f_n E$ for any E. Somehow the third and the second group should be related and should in particular provide a proof of Conjecture 16 but it is still all very murky.

This paper was written during my stay at the Institute for Advanced Study in Princeton. It is a very special place and I am very grateful to all people who make it to be what it is.

2. Slice Filtration

Let S be a Noetherian scheme and $SH(S)$ the stable motivic homotopy category defined in [14, §5]. Recall that we denote by $\Sigma_T^\infty(X,x)$ the suspension spectrum of a pointed smooth scheme X over S. The T-desuspensions

$$\Sigma^{\infty-q}(X,x) = \Sigma_T^{-q}\Sigma_T^\infty(X,x)$$

of the suspension spectra for all smooth schemes over S and all $q \geq 0$ form a set of generators of $SH(S)$ i.e. the smallest triangulated subcategory in $SH(S)$ which is closed under direct sums and contains objects of the form $\Sigma^{\infty-q}(X,x)$ coincides with the whole $SH(S)$. Let $SH^{\text{eff}}(S)$ be the smallest triangulated subcategory in $SH(S)$ which is closed under direct sums and contains suspension spectra of spaces but not their T-desuspensions. The categories $\Sigma_T^q SH^{\text{eff}}(S)$ for $q \in \mathbf{Z}$ form a filtration of $SH(S)$ in the sense that we have a sequence of full embeddings

$$\cdots \subset \Sigma_T^{q+1} SH^{\text{eff}}(S) \subset \Sigma_T^q SH^{\text{eff}}(S) \subset \Sigma_T^{q-1} SH^{\text{eff}}(S) \subset \cdots$$

and the smallest triangulated subcategory which contains $\Sigma_T^q SH^{\text{eff}}(S)$ for all q and is closed under direct sums coincides with $SH(S)$. This filtration is called the *slice filtration.*

Remark 2.1. The intersection of $\Sigma_T^q SH^{\text{eff}}(S)$ for all q is non-zero. As an example suppose that $S = \text{Spec}(k)$ where k is a field and choose a prime number l not equal to the characteristic of k. Consider the sequence of morphisms between the motivic Eilenberg-MacLane spectra

$$\Sigma^{0,n(l-1)} H_{\mathbf{Z}/l} \to \Sigma^{0,(n+1)(l-1)} H_{\mathbf{Z}/l}$$

given by multiplication with the motivic cohomology class

$$\tau_l \in H^{0,l-1}(\text{Spec}(k), \mathbf{Z}/l).$$

Let $H_{\text{et},\mathbf{Z}/l}$ be the homotopy colimit of this sequence. It is clear from the definition that there is a canonical isomorphism $\Sigma^{0,l-1} H_{\text{et},\mathbf{Z}/l} = H_{\text{et},\mathbf{Z}/l}$. Therefore, if this object belongs to $\Sigma_T^q SH^{\text{eff}}$ for at least one q then it belongs to the intersection of these subcategories for all q. In fact $H_{\text{et},\mathbf{Z}/l}$ belongs to SH^{eff} since $H_{\mathbf{Z}/l}$ is an effective spectrum (see Conjecture 1 below) and SH^{eff} is closed under formation of homotopy colimits. This example is particularly important because the spectrum $H_{\text{et},\mathbf{Z}/l}$, at least for varieties over a field, represents the etale cohomology with $\mathbf{Z}/l$ coefficients

$$H^{p,q}_{\text{et},\mathbf{Z}/l}(X_+) = H^p_{\text{et}}(X, \mu_l^{\otimes q}).$$

Since all the triangulated categories we consider have arbitrary direct sums and sets of compact generators a theorem of Neeman [8, Theorem 4.1] implies that he inclusions $i_q : \Sigma_T^n SH^{\text{eff}} \to SH(S)$ have right adjoints r_q. Since i_q is a full embedding the adjunction $Id \to r_q i_q$ is an isomorphism. Define f_q as $i_q \circ r_q$. Note that we have canonical morphisms $f_{q+1} = f_{q+1} f_q \to f_q$. A standard argument implies the following result.

Theorem 2.2. *There exist unique up to a canonical isomorphism triangulated functors $s_q : SH(S) \to SH(S)$ and natural transformations*

$$\pi_q : Id \to s_q$$

$$\sigma_q : s_q \to \Sigma^{1,0} f_{q+1}$$

satisfying the following conditions:

(1) *for any E the sequence*

$$f_{q+1}E \to f_qE \stackrel{\pi_q}{\to} s_q(E) \stackrel{\sigma_q}{\to} \Sigma^{1,0} f_{q+1}E \tag{1}$$

is a distinguished triangle

(2) *for any E the object $s_q(E)$ belongs to $\Sigma_T^q SH^{\text{eff}}$*

(3) *for any E the object $s_q(E)$ is right orthogonal to $\Sigma_T^{q+1} SH^{\text{eff}}$ i.e. for any object X in $\Sigma_T^{q+1} SH^{\text{eff}}$ we have $Hom(X, s_q(E)) = 0$.*

For any E in $SH(S)$ the sequence of distinguished triangles (1) is called the slice tower of E. The direct sum s_* of functors s_q for $q \in \mathbf{Z}$ is a triangulated functor from SH to SH which commutes with direct sums. This functor does not commute with smash products but for any E and F there is a canonical morphism $s_*(E) \wedge s_*(F) \to s_*(E \wedge F)$. In the following section we will see that in many ways the functor s_* reminds of the functor H_{π_*} from the topological stable homotopy category to itself which takes a spectrum E to $\oplus_{i \in \mathbf{Z}} \Sigma^i H_{\pi_i(E)}$. The main difference between s_* and H_{π_*} is that the former is a triangulated functor while the later is not.

3. Main Conjectures

This section contains the main conjectures predicting the structure of the slices of four standard spectra. The first three are the spectra described in [14, §6]. The Eilenberg-MacLane spectra representing motivic cohomology are considered in the first section, the algebraic Thom spectrum representing algebraic cobordisms in the second and the spectrum representing algebraic K-theory in the third. In the last section we consider the sphere spectrum representing the motivic stable (co-) homotopy groups.

In the standard topological approach one associates to a ring spectrum E a graded Hopf algebroid whose ring of objects is the ring of homotopy groups of E and the ring of morphisms is a ring of homotopy groups of $E \wedge E$. This algebroid can then be used to compute the Adams spectral sequence build on E and other interesting things. Unfortunately this approach only works for nice enough E which is usually reflected by some "flatness" condition. Already in the case of the ordinary Eilenberg-MacLane spectrum corresponding to integral cohomology it does not work very well. Instead we are going to consider directly the cosimplicial spectrum $N(E)$ with terms of the form $N^i(E) = E^{\wedge(i+1)}$, cofaces given by unit morphisms and codegeneracies given by the multiplication morphisms. Since our goal here is to present some conjectures describing the structure of the motivic stable homotopy category we do not discuss the definition of the homotopy category of cosimplicial spectra of which $N(E)$ is an object. In most examples we deal with below it will be enough to think of $N(E)$ as of a cosimplicial object in SH.

3.1. *Motivic Eilenberg-MacLane spectra*

Our first group of conjectures describes the slices of the motivic Eilenberg-MacLane spectrum $H_{\mathbf{Z}}$ and of the associated standard cosimplicial spectrum $N(H_{\mathbf{Z}})$.

Conjecture 1.

$$s_q(H_{\mathbf{Z}}) = \begin{cases} H_{\mathbf{Z}} & \text{for } q = 0, \\ 0 & \text{for } q \neq 0. \end{cases} \tag{2}$$

This conjecture is equivalent to the combination of two statements. One is that $H_{\mathbf{Z}}$ is an effective spectrum. This seems to be easy enough to prove by showing that the motivic Eilenberg-MacLane spaces $K(\mathbf{Z}(n), 2n)$ can be build out of n-fold T-suspensions. Another one is that $H^{p,q}(X, \mathbf{Z}) = 0$ for $q < 0$. This is currently known for regular schemes S over a field through the comparison of motivic cohomology with the higher Chow groups.

Our next goal is to describe $s_*(N(H_{\mathbf{Z}}))$. Unfortunately we do not know how to formulate the expected answer in one coherent conjecture. Instead we formulate a rather imprecise conjecture about the structure of $s_*(H_{\mathbf{Z}}^{\wedge n})$ and conjectures giving explicit descriptions for $s_*(N(H_k))$ where $k = \mathbf{Q}$ or $k = \mathbf{Z}/l$.

Conjecture 2. *The objects $s_q(H_{\mathbf{Z}} \wedge H_{\mathbf{Z}})$ are isomorphic to direct sums of the form $\oplus_{p\geq 0}\Sigma^{p,q}H_{X_{p,q}}$ where for q or p non zero $X_{p,q}$ is a finite abelian group of the form $\oplus\mathbf{Z}/l_j$ where l_j are prime numbers and $X_{0,0} = \mathbf{Z}$.*

This conjecture is known for $S = \mathrm{Spec}(k)$ where k is a field of characteristic zero (see Section 8). Conjecture 2 clearly implies a similar result for all smash powers of $H_{\mathbf{Z}}$ and H_k for $k = \mathbf{Q}$ or $k = \mathbf{Z}/l$ and in combination with Conjecture 1 it implies that all the terms of the cosimplicial spectra $s_q(N(H_k))$ are direct sums of finitely many copies of $\Sigma^{p,q}H_k$.

To describe $s_*(N(H_k))$ explicitly define a Hopf algebra $A^{\mathrm{rig}}_{*,*}(k)$ which we call the rigid Steenrod algebra (over k) as follows. For $k = \mathbf{Q}$ we set $A^{\mathrm{rig}}_{*,*}(k) = \mathbf{Q}$. Denote by $S_k[x_1, \ldots, x_n]$ and $\Lambda_k[x_1, \ldots, x_n]$ the symmetric and exterior algebras in variables $x_1, \ldots, x_n$ over a field k. For $k = \mathbf{Z}/l$ we set

$$A^{\mathrm{rig}}_{*,*}(k) = S_k[\xi_1, \ldots, \xi_n, \ldots] \otimes \Lambda_k[\tau_0, \ldots, \tau_n, \ldots]$$

where the bidegree of ξ_i is $(2(l^i - 1), l^i - 1)$, the bidegree of τ_i is $(2l^i - 1, l^i - 1)$ and the comultiplication is given by

$$\Delta(\xi_n) = \xi_n \otimes 1 + \sum_{i=1}^{n-1} \xi_{n-i}^{l^i} \otimes \xi_i + 1 \otimes \xi_n,$$

$$\Delta(\tau_n) = \tau_n \otimes 1 + \sum_{i=0}^{n-1} \xi_{n-i}^{l^i} \otimes \tau_i + 1 \otimes \tau_n.$$

For $l \neq 2$ the rigid Steenrod algebra is the usual (dual) Steenrod algebra of topology (see [4, Theorem 3]). Any Hopf algebra A over a field k defines a cosimplicial algebra $N(A)$ with terms $N(A)^i = A^{\otimes i}$, coface maps given by the unit and comultiplication and codegeneracy maps given by the counit. For a bigraded abelian group $A_{*,*}$ denote by $H_{A_{*,*}}$ the spectrum $\oplus_{p,q}\Sigma^{p,q}H_{A_{p,q}}$. A graded Hopf algebra $A_{*,*}$ defines a cosimplicial spectrum $H_{N(A_{*,*})}$ whose terms are spectra of the form $H_{A_{*,*}^{\otimes i}}$.

Conjecture 3. *For $k = \mathbf{Q}$ or $k = \mathbf{Z}/l$ there is an isomorphism of cosimplicial spectra*

$$s_*(N(H_k)) = H_{N(A^{\mathrm{rig}}_{*,*}(k))}$$

such that for any $q \in \mathbf{Z}$

$$s_q(N(H_k)) = \Sigma^{0,q}H_{N(A^{\mathrm{rig}}_{*,q}(k))}.$$

Using the standard elements in the motivic homology and cohomology of the lens spaces

$$K(\mathbf{Z}/l(1),1) = (\mathbf{A}^\infty - \{0\})/\mu_l$$

one can assign elements in $\pi_{p,q}(H_{\mathbf{Z}/l} \wedge H_{\mathbf{Z}/l})$ to the generators ξ_n and τ_m and then use the multiplicative structure on $H_{\mathbf{Z}/l}$ to define a homomorphism from $A^{\mathrm{rig}}_{*,*}(\mathbf{Z}/l)$ considered as a bigraded vector space generated by the monomials in ξ_n and τ_m to $\pi_{*,*}(H_{\mathbf{Z}/l} \wedge H_{\mathbf{Z}/l})$. The $H_{\mathbf{Z}/l}$-module structure on $H_{\mathbf{Z}/l} \wedge H_{\mathbf{Z}/l}$ allows one to extend any element of the homotopy group $\pi_{p,q}(H_{\mathbf{Z}/l} \wedge H_{\mathbf{Z}/l})$ to a morphism from $\Sigma^{p,q}H_{\mathbf{Z}/l}$ to $H_{\mathbf{Z}/l} \wedge H_{\mathbf{Z}/l}$ and thus gives a morphism

$$H_{A^{\mathrm{rig}}_{*,*}(\mathbf{Z}/l)} \to H_{\mathbf{Z}/l} \wedge H_{\mathbf{Z}/l}. \tag{3}$$

Conjecture 4. *The morphism (3) is an isomorphism.*

We know how to prove this conjecture for $S = \mathrm{Spec}(k)$ where k is a field of characteristic zero (see [11] and further papers of these series). It is one of the elements of the computation of the algebra of cohomological operations in in motivic cohomology needed for the proof of the Milnor conjecture given in [13]. Doing the same thing with the higher smash powers of H_k one can define morphisms

$$H_{(A^{\mathrm{rig}}_{*,*}(\mathbf{Z}/l))^{\otimes i}} \to H_{\mathbf{Z}/l}^{\wedge(i+1)}. \tag{4}$$

and Conjecture 4 implies that they are also isomorphisms. The morphisms (4) do not commute with the coface and codegeneracy morphisms and thus do not give a morphism $H_{N(A^{\mathrm{rig}}_{*,*}(\mathbf{Z}/l))} \to N(H_{\mathbf{Z}/l})$. For example the two morphisms $H_{\mathbf{Z}/l} \to H_{\mathbf{Z}/l} \wedge H_{\mathbf{Z}/l}$ defined by the unit $\mathbf{1} \to H_{\mathbf{Z}/l}$ which are the zero dimensional coface morphisms in $N(H_{\mathbf{Z}/l})$ do not coincide while the coface morphisms $H_{\mathbf{Z}/l} \to H_{A^{\mathrm{rig}}_{*,*}(\mathbf{Z}/l)}$ coincide by construction. When we pass to slices this problem should disappear and explicit computations confirm it. However we do not know an explicit construction of a morphism in either direction required by Conjecture 3 for $k = \mathbf{Z}/l$.

3.2. *Motivic Thom spectrum*

Let MGL be the motivic Thom spectrum representing the algebraic cobordism. An analog of the standard argument from topology should be sufficient to show that

$$MGL^{*,*}((\mathbf{P}^\infty)^n) = MGL^{*,*}(\mathbf{1})[[t_1, \ldots, t_n]]. \tag{5}$$

Together with the obvious properties of the morphism $\mathbf{P}^\infty \times \mathbf{P}^\infty \to \mathbf{P}^\infty$, this formula implies that the image of t_1 under the induced map on algebraic cobordisms is a formal group law. It gives a homomorphism

$$MU_* \to MGL_{*,*}(\mathbf{1}) \tag{6}$$

from the Lazard ring MU_* to $MGL_{*,*}(\mathbf{1})$ which sends MU_{2q} to $MGL_{2q,q}(\mathbf{1})$.

Conjecture 5. *There exists an isomorphism*

$$s_q(MGL) = \Sigma_T^q H_{MU_{2q}} \tag{7}$$

compatible with the homomorphism (6).

The compatibility condition in this conjecture means the following. An isomorphism of the form (7) defines in particular a homomorphism of abelian groups $MU_{2q} \to \pi_{2q,q}(s_q(MGL))$. On the other hand the definition of s_q's shows that we have a canonical homomorphisms $\pi_{p,q}(E) \to \pi_{p,q}(s_q(E))$ and, therefore, (6) also defines a homomorphism $MU_{2q} \to \pi_{2q,q}(s_q(MGL))$. The condition requires the two homomorphisms to be the same. Conjecture 10 discussed in the following section implies that that there exists a unique morphism $H_{MU_{2q}} \to s_q(MGL)$ satisfying the compatibility condition.

Consider the graded cosimplicial ring $\pi_*(N(MU))$ where MU is the complex cobordisms spectrum and π_* refer to topological homotopy groups. Alternatively it can be defined as the ring of functions on the simplicial scheme which represents the functor sending an affine scheme Spec(R) to the nerve of the groupoid whose objects are the formal group laws of dimension one over R and morphisms are changes of the generator.

Conjecture 6. *There is an isomorphism*

$$s_q(N(MGL)) = \Sigma_T^q H_{\pi_{2q}(N(MU))}$$

which coincides with the isomorphism (7) *on the zero term.*

It seems to be possible to repeat the argument used to compute oriented cohomology of classifying spaces for algebraic cobordism which leads to a canonical isomorphism

$$MGL \wedge MGL = MGL[b_1, \ldots, b_n, \ldots]$$

where

$$E[x_1, \ldots, x_n] = \oplus_{i_1, \ldots, i_n \geq 0} E.$$

In particular it seems that Conjecture 6 is relatively simple modulo Conjecture 5. Thus the situation here is different from the case of Eilenberg-MacLane spectra where Conjecture 1 is relatively easy while Conjecture 3 is hard.

3.3. *Algebraic K-theory spectrum*

In this paper we denote the algebraic K-theory spectrum by KGL to distinguish it from the space BGL. It is (2,1)-periodic that is we have a canonical isomorphism $T \wedge KGL = KGL$. This immediately implies that $s_q(KGL) = \Sigma_T^q s_0(KGL)$ for all q.

Conjecture 7.

$$s_0(KGL) = H_{\mathbf{Z}}.$$

The slices of the standard cosimplicial spectrum associated with the algebraic K-theory are described by the following analog of Conjecture 6.

Conjecture 8.

$$s_q(N(KGL)) = \Sigma_T^q H_{\pi_{2q}(N(KU))}.$$

3.4. *Sphere spectrum*

Conjectures 1 and 6 lead to a complete computation of the slices of the sphere spectrum. Observe first that the cone of the unit morphism $\mathbf{1} \to MGL$ belongs to $\Sigma_T^1 SH^{\mathrm{eff}}$. It can be seen from the fact that this cone is built out of the n-fold T-desuspensions of the suspension spectra of the Thom spaces $MGL(n)$ and each $MGL(n)$ can be built out of the n-fold T-suspensions of open subsets of $BGL(n)$. This observations implies that for $n > q$ the morphisms $s_q(\cos k_{n+1} N(MGL)) \to s_q(\mathbf{1})$ are isomorphisms and in particular that $s_q(\mathbf{1}) = \mathrm{Tot}(s_q(N(MGL)))$. The right hand side can be computed from Conjecture 6. The n-th term of $s_q(N(MGL))$ is just $\Sigma_T^q H_{\pi_{2q}(N^n(MU))}$. Correspondingly the total object $\mathrm{Tot}(s_q(N(MGL)))$ is nothing but the Eilenberg-MacLane spectrum of the form $\Sigma_T^q H_{\pi_{2q} N(MU)}$ where $H_{\pi_{2q} N(MU)}$ is the Eilenberg-MacLane spectrum corresponding to the complex of abelian groups associated with the cosimplicial abelian group $\pi_{2q}(N(MU))$. The cohomology groups of $\pi_*(N(MU))$ are denoted in topology by

$$H^n(\pi_{2q}(N(MU))) = \mathrm{Ext}^n_{MU_*(MU)}(MU_*, MU_*)_{2q}.$$

They form the E_2-term of the Adams-Novikov spectral sequence (see [10]). Summarizing we have.

Conjecture 9.

$$s_q(\mathbf{1}) = \Sigma_T^q H_{\pi_{2q}(N(MU))}.$$

The particular case of this conjecture for $q = 0$ looks as follows.

Conjecture 10.

$$s_0(\mathbf{1}) = H_{\mathbf{Z}}.$$

Consider the canonical morphism $\mathbf{1} \to H_{\mathbf{Z}}$ and let $\bar{H}_{\mathbf{Z}}$ be its fiber (the desuspension of its cone). Conjecture 10 is equivalent to the combination of two statements. One is that $\bar{H}_{\mathbf{Z}}$ belongs to $\Sigma_T^1 SH^{\mathrm{eff}}$ and another one is that $H_{\mathbf{Z}}$ is right orthogonal to $\Sigma_T^1 SH^{\mathrm{eff}}$. Let us call the first one the divisibility conjecture and the second one the T-rigidity conjecture. T-rigidity conjecture is also a part of Conjecture 1 and was discussed there. The divisibility part so far is unknown even over a field of characteristic zero. This conjecture seems to be very important and more fundamental that the rest of the conjectures of this paper. In particular it provides the only way we know to characterize the Eilenberg-MacLane spectra without giving an explicit definition. One of the implications of Conjecture 10 is that for any spectrum its slices have unique and natural module structures over the Eilenberg-MacLane spectrum which explains that all our conjectures predict that different objects of the form $s_*(-)$ are generalized Eilenberg-MacLane spectra.

4. Slice-Wise Cellular Spectra

Many important spectra including the algebraic cobordism spectrum and the algebraic K-theory spectrum are T-cellular that is they belong to the smallest triangulated subcategory closed under direct sums which contains the spheres T^i for $i \in \mathbf{Z}$. Unfortunately we do not know whether or not the Eilenberg-MacLane spectrum is T-cellular. In this section we will describe another class of spectra which contains $H_{\mathbf{Z}}$ and such that its objects have many of the nice properties of T-cellular spectra. Using Conjecture 10 and Conjecture 2 we will show that it contains T-cellular spectra.

Definition 4.1. An object E of $SH(S)$ is called slice-wise cellular if for any $q \in \mathbf{Z}$ the slice $s_q(E)$ of E belongs to the smallest triangulated subcategory of $SH(S)$ closed under direct sums which contains the Eilenberg-MacLane spectrum $\Sigma_T^q H_{\mathbf{Z}}$.

Our definition immediately implies that the subcategory of slice-wise cellular objects is a triangulated subcategory closed under direct sums and direct summands. Conjecture 1 implies that the Eilenberg-MacLane spectrum $H_{\mathbf{Z}}$ is slice-wise cellular.

Lemma 4.2. *The subcategory of slice-wise cellular spectra is closed under smash product.*

Proof. The proof is modulo Conjecture 2. Let E and F be slice-wise cellular spectra. We need to show that $s_q(E\wedge F)$ is in the smallest triangulated subcategory of $SH(S)$ closed under direct sums which contains the Eilenberg-MacLane spectrum $\Sigma^q_T H_{\mathbf{Z}}$. Replacing E or F by an appropriate T-suspension we may assume that $q=0$. For any F we have $F=\mathrm{hocolim} f_{-n}F$ and both the smash product and the functor s_0 commute with homotopy colimits. Thus $s_0(E\wedge F)=\mathrm{hocolim} s_0(f_{-n}E\wedge f_{-n}F)$ and it is enough to prove that $s_0(f_{-n}E\wedge f_{-n}F)$ is of the required form for any n. For any $m>n$ the smash product $f_mE\wedge f_{-n}F$ is in $\Sigma^1_T SH^{\mathrm{eff}}$. Thus the slice tower of E gives a finite sequence of distinguished triangles of the form

$$0\to s_0(f_nE\wedge f_{-n}F)\cong s_0(s_nE\wedge f_{-n}F),$$

$$s_0(f_nE\wedge f_{-n}F)\to s_0(f_{n-1}E\wedge f_{-n}F)\to s_0(s_{n-1}E\wedge f_{-n}F),$$

$$\dots$$

$$s_0(f_{1-n}E\wedge f_{-n}F)\to s_0(f_{-n}E\wedge f_{-n}F)\to s_0(s_{-n}E\wedge f_{-n}F).$$

It is enough to check now that $s_0(s_{-m}E\wedge f_{-n}F)$ are in the smallest triangulated subcategory closed under direct sums which contains $H_{\mathbf{Z}}$. Our assumption on E implies that it is enough to check that $s_0(\Sigma^{p,q}H_{\mathbf{Z}}\wedge f_{-n}F)$ is of the required form. Using the slice tower of F we reduce the problem to the case of $s_0(\Sigma^{p,q}H_{\mathbf{Z}}\wedge H_{\mathbf{Z}})$ where our result follows from Conjectures 2 and 1. □

Definition 4.3. An object is called T-connective or just connective if it belongs to $\Sigma^q_T SH^{\mathrm{eff}}$ for some q.

Proposition 4.4. *A connective spectrum E is slice-wise cellular if and only if $E\wedge H_{\mathbf{Z}}$ is slice-wise cellular.*

Proof. The proof requires Conjectures 10, 1 and 2. We will only prove the "only if" part. We may clearly assume that E belongs to SH^{eff}. Consider the Adams tower for the Eilenberg-MacLane spectrum

$$\begin{array}{ccccc} & & \dots & & \\ \bar{H}_{\mathbf{Z}}^{\wedge n} & \to & \bar{H}_{\mathbf{Z}}^{\wedge n-1} & \to & \bar{H}_{\mathbf{Z}}^{\wedge n-1}\wedge H_{\mathbf{Z}}, \\ & & \dots & & \\ \bar{H}_{\mathbf{Z}}^{\wedge 2} & \to & \bar{H}_{\mathbf{Z}} & \to & \bar{H}_{\mathbf{Z}}\wedge H_{\mathbf{Z}}, \\ \bar{H}_{\mathbf{Z}} & \to & \mathbf{1} & \to & H_{\mathbf{Z}}. \end{array} \tag{8}$$

By the divisibility part of Conjecture 10, $\bar{H}_{\mathbf{Z}}^{\wedge n}$ belongs to $\Sigma^n_T SH^{\mathrm{eff}}$ and since E is assumed to be effective so does $E\wedge\bar{H}_{\mathbf{Z}}^{\wedge n}$. Therefore applying the functor s_q to the tower (8) smashed with E we get a finite sequence of distinguished triangles of the form

$$\begin{array}{ccccc} 0 & \to & s_q(E \wedge \bar{H}_{\mathbf{Z}}^{\wedge q}) & \cong & s_q(E \wedge \bar{H}_{\mathbf{Z}}^{\wedge q} \wedge H_{\mathbf{Z}}), \\ & & \dots & & \\ s_q(E \wedge \bar{H}_{\mathbf{Z}}^{\wedge 2}) & \to & s_q(E \wedge \bar{H}_{\mathbf{Z}}) & \to & s_q(E \wedge \bar{H}_{\mathbf{Z}} \wedge H_{\mathbf{Z}}), \\ s_q(E \wedge \bar{H}_{\mathbf{Z}}) & \to & s_q(E) & \to & s_q(E \wedge H_{\mathbf{Z}}). \end{array} \tag{9}$$

Therefore it is enough to show that objects $s_q(E \wedge \bar{H}_{\mathbf{Z}}^{\wedge n} \wedge H_{\mathbf{Z}})$ are of the required form. The multiplication morphism allows one to make $\bar{H}_{\mathbf{Z}}^{\wedge n} \wedge H_{\mathbf{Z}}$ into a direct summand of $\Sigma^{-n,0} H_{\mathbf{Z}}^{\wedge(n+1)}$. Our result follows now from Conjectures 2 and 1. □

Corollary 4.5. *The category of slice-wise cellular spectra contains the sphere spectrum* **1** *and therefore all T-cellular spectra.*

The algebraic cobordism spectrum MGL is a slice-wise cellular spectrum. There are two ways to see it. One is to use the fact that

$$MGL \wedge H_{\mathbf{Z}} = H_{\mathbf{Z}[b_1,\dots,b_n,\dots]}$$

where b_i are of bidegree $(2i, i)$ and Proposition 4.4. Another one is to note that the Thom spaces $MGL(n)$ out of which the motivic Thom spectrum is built can be built in turn out of spheres and then use Corollary 4.5. Similarly the spectrum KGL is a slice-wise cellular spectrum. We can not use Proposition 4.4 directly to prove it since it is not connective but we can use the fact that it is built out of suspension spectra of the space BGL and these spaces are T-cellular.

5. Reformulations in Terms of Rigid Homotopy Groups

Definition 5.1. The rigid homotopy groups of an object E in $SH(S)$ are given by

$$\pi_{p,q}^{\mathrm{rig}}(E) = \pi_{p,q}(s_q(E)).$$

More generally we define the presheaves of rigid homotopy groups setting

$$\underline{\pi}_{p,q}^{\mathrm{rig}}(E) : U/S \mapsto \mathrm{Hom}(\Sigma^{p,q}\Sigma_T^{\infty} U_+, s_q(E)).$$

Note that by definition the rigid homotopy groups are the values of the presheaves of rigid homotopy groups on the base scheme S. Conjecture 10 implies that a slice of any spectrum has a unique structure of a module over the Eilenberg-MacLane spectrum. This leads to the following conjecture (for the definition of a presheaf with transfers over a general base scheme S see [11]).

Conjecture 11. *The presheaves of rigid homotopy groups have canonical structures of presheaves with transfers.*

We have canonical homomorphisms from the motivic stable homotopy groups $\pi_{p,q}(E) = \mathrm{Hom}(\Sigma^{p,q}\mathbf{1}, E)$ to the rigid homotopy groups. For any E the group $\pi_{*,*}(E)$ is a module over the ring of motivic homotopy groups of spheres $\pi_{*,*}(\mathbf{1})$ and one can easily see that the submodule $\pi_{*,<0}(\mathbf{1})\pi_{*,*}(E)$ goes to zero in the rigid homotopy groups. In general it seems that nothing else can be said about this homomorphism.

Conjectures of the previous sections allow us to compute the rigid homotopy groups of the standard spectra explicitly provided we know homotopy groups of the form $\pi_{p,0}H_{\mathbf{Z}}$ i.e. the motivic cohomology of weight zero.

Conjecture 12. *For a normal connected scheme S one has*

$$\pi_{p,0}H_{\mathbf{Z}} = \begin{cases} \mathbf{Z} \text{ for } p = 0, \\ 0 \text{ for } p \neq 0. \end{cases} \tag{10}$$

For a regular S over a field this conjecture is known. It follows from the comparison of motivic cohomology with the higher Chow groups. For any S of characteristic zero it can be proved using resolution of singularities and the blow-up long exact sequence in generalized cohomology established in [12]. In this case one can prove more namely that for any Noetherian S of characteristic zero which is of finite dimension and any $p \in \mathbf{Z}$ one has

$$\pi_{p,0}(H_{\mathbf{Z}}) = H^{-p}_{cdh}(S, \mathbf{Z}).$$

The same should be true for all Noetherian S of finite dimension.

Conjecture 10 implies that the map

$$\mathrm{Hom}(H_{\mathbf{Z}}, \Sigma^{p,0}H_{\mathbf{Z}}) \to \pi_{p,0}(H_{\mathbf{Z}})$$

defined by the unit map $\mathbf{1} \to H_{\mathbf{Z}}$ is an isomorphism for any $n \in \mathbf{Z}$. Together with Conjecture 12 it implies that for a normal connected scheme S the functor

$$D(Ab) \to SH(S) \tag{11}$$

which sends a complex of abelian groups C to the Eilenberg-MacLane spectrum H_C is a full embedding and its image coincides with the smallest triangulated subcategory of $SH(S)$ closed under direct sums which contains the Eilenberg-MacLane spectrum $H_{\mathbf{Z}}$. Thus in the case of a normal connected scheme S a spectrum E is slice-wise cellular if and only if there exist complexes of abelian groups $\Pi_q(E)$ such that $s_qE = \Sigma^{0,q}H_{\Pi_q(E)}$.

Combining Conjecture 12 with Conjectures 1, 5, 7 and 9 we get that for a normal connected S the rigid homotopy groups of the standard spectra are given by the following formulae:

$$\pi^{\mathrm{rig}}_{p,q}(H_{\mathbf{Z}}) = \begin{cases} \mathbf{Z} \text{ for } p = q = 0, \\ 0 \quad \text{otherwise;} \end{cases} \tag{12}$$

$$\pi^{\mathrm{rig}}_{p,q}(MGL) = \begin{cases} MU_{2q} \text{ for } p = 2q, \\ \quad 0 \quad \text{ for } p \neq 2q; \end{cases} \tag{13}$$

$$\pi^{\mathrm{rig}}_{p,q}(KGL) = \begin{cases} \mathbf{Z} \text{ for } p = 2q, \\ 0 \text{ for } p \neq 2q; \end{cases} \tag{14}$$

$$\pi^{\mathrm{rig}}_{p,q}(\mathbf{1}) = Ext^{2q-p}_{MU_*(MU)}(MU_*, MU_*)_{2q}. \tag{15}$$

In particular these groups do not depend on S which is one of the reasons we call them "rigid". Note that in the first three cases the rigid homotopy groups of a motivic spectrum coincide with the homotopy groups of its topological counterpart but in the last case they do not.

6. Rigid Homology and Rigid Adams Spectral Sequence

Define the rigid homology of a spectrum with coefficients in a commutative ring R setting

$$H^{\mathrm{rig}}_{p,q}(E, R) = \pi^{\mathrm{rig}}_{p,q}(E \wedge H_R).$$

The unit map $\mathbf{1} \to H_R$ defines the rigid analog of the Hurewicz map

$$\pi^{\mathrm{rig}}_{p,q}(E) \to H^{\mathrm{rig}}_{p,q}(E, R).$$

These homomorphisms have a number of useful properties analogous to the properties of the usual topological Hurewicz map which are missing for the motivic Hurewicz homomorphisms $\pi_{p,q} \to H_{p,q}$.

Lemma 6.1. *Let E be a connective spectrum such that the groups $H^{\mathrm{rig}}_{p,q}(E, \mathbf{Z})$ are finitely generated. Then the groups $\pi^{\mathrm{rig}}_{p,q}(E)$ are finitely generated.*

Proof. Conjecture 10 implies that smashing the Adams tower (8) with E and applying s_q one gets a finite Postnikoff tower for $s_q(E)$ whose quotients are direct summands of objects of the form $s_q(E \wedge H_{\mathbf{Z}}^{\wedge n})$. Conjecture 2 implies that they are finite direct sums of objects of the form $s_q(E \wedge H_A)$ for finitely generated abelian groups A. Together with our condition on E it implies that $\pi^{\mathrm{rig}}_{p,q}(E) = \pi_{p,q}(s_q(E))$ are finitely generated. □

Lemma 6.2. *For any spectrum E the homomorphism*

$$\pi^{\mathrm{rig}}_{p,q}(E) \otimes \mathbf{Q} \to H^{\mathrm{rig}}_{p,q}(E, \mathbf{Q})$$

is an isomorphism.

Proof. Note first that both sides as functors in E take filtered homotopy colimits to filtered colimits. Thus since any spectrum E is a filtered homotopy colimit of its connective parts $f_{-n}E$ we may assume that E is connective and thus that it is in SH^{eff}. The rational version of Conjecture 10 implies that smashing the Adams tower with the rational Moore spectrum $\mathbf{1}_{\mathbf{Q}}$ and with E and applying functor s_q we get a finite Postnikoff tower for $s_q(E \wedge \mathbf{1}_{\mathbf{Q}})$ and the rational version of Conjecture 3 implies that this tower degenerates providing an isomorphism $s_q(E \wedge \mathbf{1}_{\mathbf{Q}}) \to s_q(E \wedge H_{\mathbf{Q}})$. Finally since $\mathbf{1}_{\mathbf{Q}}$ is a filtered homotopy colimit of sphere spectra we have $s_q(E \wedge \mathbf{1}_{\mathbf{Q}}) = s_q(E) \wedge \mathbf{1}_{\mathbf{Q}}$ and the homotopy groups of this spectrum are isomorphic to $\pi^{\mathrm{rig}}_{p,q}(E) \otimes \mathbf{Q}$. □

Remark 6.3. The statement of Lemma 6.2 does not hold for motivic homotopy groups. In particular while $s_q(\mathbf{1}_{\mathbf{Q}})$ should be isomorphic to $s_q(H_{\mathbf{Q}})$ that is should be zero for $q \neq 0$ and $H_{\mathbf{Q}}$ for $q = 0$ the morphism $\mathbf{1}_{\mathbf{Q}} \to H_{\mathbf{Q}}$ is not an isomorphism at least in some cases.

Multiplication on H_R defines for any E and F a homomorphism

$$H^{\mathrm{rig}}_{*,*}(E,R) \otimes_R H^{\mathrm{rig}}_{*,*}(F,R) \to H^{\mathrm{rig}}_{*,*}(E \wedge F, R).$$

Charles Rezk pointed out that using Conjectures 2, 1 and 10 one can prove the following "Künneth Theorem".

Theorem 6.4. *Let E be a slice-wise cellular spectrum, F any spectrum and k a field. Then the multiplication homomorphism*

$$H^{\mathrm{rig}}_{*,*}(E,k) \otimes_k H^{\mathrm{rig}}_{*,*}(F,k) \to H^{\mathrm{rig}}_{*,*}(E \wedge F, k)$$

is an isomorphism.

For any commutative ring R and any spectrum E we can apply the functor s_q to the Adams tower (8) based on H_R smashed with E. This gives a sequence of distinguished triangles of the form

$$\begin{array}{ccccc} & & \cdots & & \\ s_q(E \wedge \bar{H}_R^{\wedge(n+1)}) & \to & s_q(E \wedge \bar{H}_R^{\wedge n}) & \to & s_q(E \wedge \bar{H}_R^{\wedge n} \wedge H_R), \\ & & \cdots & & \\ s_q(E \wedge \bar{H}_R^{\wedge 2}) & \to & s_q(E \wedge \bar{H}_R) & \to & s_q(E \wedge \bar{H}_R \wedge H_R), \\ s_q(E \wedge \bar{H}_R) & \to & s_q(E) & \to & s_q(E \wedge H_R), \end{array} \tag{16}$$

which defines a spectral sequence whose E_1-term consists of the rigid homology groups

$$E^1_{p,n} = H^{\mathrm{rig}}_{p,q}(E \wedge \bar{H}_R^{\wedge n}) = \pi_{p,q}(s_q(E \wedge \bar{H}_R^{\wedge n} \wedge H_R))$$

and the r-th differential is of the form

$$H^{\mathrm{rig}}_{p,q}(E \wedge \bar{H}_R^{\wedge n}) \to H^{\mathrm{rig}}_{p-1,q}(E \wedge \bar{H}_R^{\wedge(n+r)}).$$

This spectral sequence is called the rigid Adams spectral sequence with coefficients in R. The complexes

$$E \wedge \bar{H}_R \to \Sigma^1_s(E \wedge \bar{H}_R \wedge H_R) \to \Sigma^2_s(E \wedge \bar{H}_R^{\wedge 2} \wedge H_R) \to \dots$$

which define the E_1-term of the Adams spectral sequence are isomorphic to the normalized chain complexes of the cosimplicial objects $E \wedge N(H_R)$ (this is actually true for the Adams spectral sequence based on any commutative ring spectrum). Thus the E_1-term of the rigid Adams spectral sequence with coefficients in R can be identified with the collection of complexes of abelian groups of the form $\pi^{\mathrm{rig}}_{p,q}(E \wedge N(H_R))$. If k is a field we have by Theorem 6.4

$$\pi^{\mathrm{rig}}_{p,q}(E \wedge H_k^{\wedge(n+1)}) = H^{\mathrm{rig}}_{p,q}(E \wedge H_k^{\wedge n}) = (H^{\mathrm{rig}}_{*,*}(E) \otimes H^{\mathrm{rig}}_{*,*}(H_k)^{\otimes n})_{p,q}.$$

Together with Conjecture 3 this implies that the rigid homology with coefficients in $\mathbf{Z}/l$ are comodules over the rigid Steenrod algebra $A^{\mathrm{rig}}_{*,*}(k)$ and that the E_2-term of the rigid Adams spectral sequence with coefficients in $\mathbf{Z}/l$ can be identified with Ext-groups from k to $H^{\mathrm{rig}}_{*,*}(E, k)$ in the category of comodules over $A^{\mathrm{rig}}_{*,*}$.

According to Conjecture 10 for any connective spectrum E and any q there exists n such that $s_q(E \wedge \bar{H}^{\wedge(n+1)}) = 0$. Thus if $R = \mathbf{Z}$ and E is connective the triangles (16) give a finite Postnikoff tower for $s_q(E)$ and we conclude that the rigid Adams spectral sequence with integral coefficients converges for any connective spectrum E. Unfortunately this spectral sequence is not very convenient in practice since we do not know how to describe its E_2-term.

To describe the convergence properties of the rigid Adams spectral sequence with coefficients in $\mathbf{Z}/l$ denote the associated filtration on rigid homotopy groups by

$$a_l^i \pi^{\mathrm{rig}}_{p,q}(E) = \mathrm{Im}(\pi^{\mathrm{rig}}_{p,q}(E \wedge \bar{H}^{\wedge i}_{\mathbf{Z}/l}) \to \pi^{\mathrm{rig}}_{p,q}(E)).$$

Proposition 6.5. *Let E be an effective spectrum. Then one has*

$$a_l^{q+1} \pi^{\mathrm{rig}}_{p,q}(E) \subset l\pi^{\mathrm{rig}}_{p,q}(E) \tag{17}$$

Proof. Denote by M_l the Moore spectrum $cone(\mathbf{1} \xrightarrow{l} \mathbf{1})$. To show that the inclusion (17) holds we have to show that the composition

$$\pi^{\mathrm{rig}}_{p,q}(E \wedge \bar{H}^{\wedge(q+1)}_{\mathbf{Z}/l}) \to \pi^{\mathrm{rig}}_{p,q}(E) \to \pi^{\mathrm{rig}}_{p,q}(E \wedge M_l)$$

is zero. This composition factors through the morphism

$$\pi^{\mathrm{rig}}_{p,q}(E \wedge M_l \wedge \bar{H}^{\wedge(q+1)}_{\mathbf{Z}/l}) \to \pi^{\mathrm{rig}}_{p,q}(E \wedge M_l)$$

which by Lemma 6.6 factors through the morphism

$$\pi^{\text{rig}}_{p,q}(E \wedge M_l \wedge \bar{H}_{\mathbf{Z}}^{\wedge(q+1)}) \to \pi^{\text{rig}}_{p,q}(E \wedge M_l).$$

Since E is effective so is $E \wedge M_l$ and thus by Conjecture 10 we have

$$\pi^{\text{rig}}_{p,q}(E \wedge M_l \wedge \bar{H}_{\mathbf{Z}}^{\wedge(q+1)}) = \pi_{p,q}(s_q(E \wedge M_l \wedge \bar{H}_{\mathbf{Z}}^{\wedge(q+1)})) = 0. \qquad \square$$

Lemma 6.6. *For any $i \geq 0$ there exists a morphism*

$$M_l \wedge \bar{H}_{\mathbf{Z}/l}^{\wedge i} \to M_l \wedge \bar{H}_{\mathbf{Z}}^{\wedge i} \tag{18}$$

such that the diagram

$$\begin{array}{ccc} M_l \wedge \bar{H}_{\mathbf{Z}/l}^{\wedge i} & \to & M_l \wedge \bar{H}_{\mathbf{Z}}^{\wedge i} \\ \downarrow & & \downarrow \\ M_l & = & M_l \end{array}$$

commutes.

Proof. Proceed by induction on i. To prove the lemma for $i = 1$ it is sufficient to show that the composition

$$M_l \wedge \bar{H}_{\mathbf{Z}/l} \to M_l \to M_l \wedge H_{\mathbf{Z}} \tag{19}$$

is zero. Consider the morphism of distinguished squares

$$\begin{array}{ccccc} M_l \wedge \bar{H}_{\mathbf{Z}} & \to M_l \to & M_l \wedge H_{\mathbf{Z}} \\ \downarrow & \downarrow & \downarrow \\ M_l \wedge \bar{H}_{\mathbf{Z}/l} & \to M_l \to & M_l \wedge H_{\mathbf{Z}/l}. \end{array}$$

The composition of (19) with the right vertical arrow is the composition of the two lower arrows which is zero. On the other hand the right vertical arrow is a split monomorphism. Therefore (19) is zero. For $i > 1$ one defines the morphism (18) inductively as the composition

$$M_l \wedge \bar{H}_{\mathbf{Z}/l}^{\wedge i} = M_l \wedge \bar{H}_{\mathbf{Z}/l}^{\wedge(i-1)} \wedge \bar{H}_{\mathbf{Z}/l} \to M_l \wedge \bar{H}_{\mathbf{Z}}^{\wedge(i-1)} \wedge \bar{H}_{\mathbf{Z}/l} \to$$

$$M_l \wedge \bar{H}_{\mathbf{Z}}^{\wedge(i-1)} \wedge \bar{H}_{\mathbf{Z}} = M_l \wedge \bar{H}_{\mathbf{Z}}^{\wedge i}. \qquad \square$$

Proposition 6.5 easily implies the following convergence result.

Proposition 6.7. *Let E be a connective spectrum and l a prime number such that the rigid homology groups $H^{\text{rig}}_{p,q}(E, \mathbf{Z}_{(l)})$ are finitely generated. Then the rigid Adams spectral sequence for E with coefficients in $\mathbf{Z}/l$ converges to $\pi^{\text{rig}}_{p,q}(E) \otimes \mathbf{Z}_{(l)}$, that is*

$$\cap_{i \geq 0} a_l^i \pi^{\text{rig}}_{*,*}(E) \otimes \mathbf{Z}_{(l)} = 0$$

and the canonical homomorphisms

$$(a_l^i \pi_{p,q}^{\mathrm{rig}}(E)/a_l^{i+1}\pi_{p,q}^{\mathrm{rig}}(E)) \otimes \mathbf{Z}_{(l)} \to H_{p,q}^{\mathrm{rig}}(E \wedge \bar{H}_{\mathbf{Z}/l}^{\wedge i})_\infty$$

where the subscript ∞ denotes the infinite term of the spectral sequence, are isomorphisms.

Proof. A four term exact sequence similar to the one used in the proof of Lemma 7.2 implies that it is sufficient to prove that for any p, q and n one has

$$\cap_{i\geq 0}\mathrm{Im}(\pi_{p,q}^{\mathrm{rig}}(E \wedge \bar{H}_{\mathbf{Z}/l}^{\wedge(n+i)}) \to \pi_{p,q}^{\mathrm{rig}}(E \wedge \bar{H}_{\mathbf{Z}/l}^{\wedge n})) = 0. \tag{20}$$

We can invert all the primes but l and assume that the rigid homology of E are finitely generated $\mathbf{Z}_{(l)}$-modules. The same argument as the one used in the proof of Lemma 6.1 shows that then the rigid homotopy groups of E are finitely generated $\mathbf{Z}_{(l)}$-modules. Conjecture 2 implies that the same holds for the spectra $E \wedge \bar{H}_{\mathbf{Z}/l}^{\wedge n}$ and thus (20) follows from Proposition 6.5. □

The E_2-term of the rigid Adams spectral sequence with $\mathbf{Z}/l$-coefficients for $E = \mathbf{1}$ consists of the Ext-groups $\mathrm{Ext}_{A_*(l)}(\mathbf{Z}/l, \mathbf{Z}/l)$ in the category of comodules over the rigid Steenrod algebra and according to (15) and Proposition 6.7 the E_∞-term gives the quotients of a filtration on the groups $\mathrm{Ext}_{MU_*(MU)}(MU_*, MU_*) \otimes \mathbf{Z}_{(l)}$. For $l > 2$ the rigid Steenrod algebra coincides with the topological Steenrod algebra and therefore, as was pointed out by Charles Rezk, the rigid Adams spectral sequence in this case looks exactly like the algebraic Adams-Novikov spectral sequence [9, 10]. There is no doubt that these two spectral sequences are indeed isomorphic.

7. Slice Spectral Sequence and Convergence Problems

For any $n \in \mathbf{Z}$ the slice tower (1) of a spectrum E defines in the usual way a spectral sequence of abelian groups which starts with groups of the form $\pi_{p,n}(s_q(E))$. We call these spectral sequences the slice spectral sequences for E. The r-th differential in the n-th slice spectral sequence goes from $\pi_{p,n}(s_q(E))$ to $\pi_{p-1,n}(s_{q+r}(E))$ which suggests that one can visualize it in the same way as one visualizes the Adams spectral sequence in topology. One considers p as the horizontal and q as the vertical index. The differentials then go from a given column ("stem") to the previous one reaching higher and higher in the vertical direction. For any n one has $\pi_{p,n}(s_q(E)) = 0$ for $q < n$ therefore each of this spectral sequences is zero below the horizontal line $n = q$. In particular for each term there are only finitely many incoming differentials. Let $f_q\pi_{p,n}(E)$ be the image of $\pi_{p,n}(f_qE)$ in $\pi_{p,n}(E)$. These subgroups form a filtration on $\pi_{p,n}(E)$ and one verifies in the standard manner that one gets canonical monomorphisms

$$f_q\pi_{p,n}(E)/f_{q+1}\pi_{p,n}(E) \to \pi_{p,n}(s_q(E))_\infty \tag{21}$$

where the subscript ∞ indicates that we consider the infinite term of the spectral sequence.

Definition 7.1. A spectrum E is called convergent with respect to the slice filtration if for any $p, n, q \in \mathbf{Z}$ one has

$$\cap_{i\geq 0} f_{q+i}\pi_{p,n}(f_q E) = 0. \tag{22}$$

Lemma 7.2. *Let E be a spectrum convergent with respect to the slice filtration. Then the homomorphisms* (21) *are isomorphisms.*

Proof. A standard argument shows that the homomorphisms (21) fit into exact sequences of the form

$$0 \to f_q\pi_{p,n}(E)/f_{q+1}\pi_{p,n}(E) \to \pi_{p,n}(s_q(E))_\infty \to$$

$$\to \cap_{i\geq 1} f_{q+i}\pi_{p-1,n}(f_{q+1}E) \to \cap_{i\geq 0} f_{q+i}\pi_{p-1,n}(E)$$

which implies the statement of the lemma. □

For E as in Lemma 7.1 $f_q\pi_{p,n}(E)$ is a nondegenerate filtration on $\pi_{p,n}(E)$ and its quotients are subquotients of the groups $\pi_{p,n}(s_q(E))$. We say that E is bounded with respect to the slice filtration if for any p, n there exists q such that $\pi_{p,n}(f_{q+i}E) = 0$ for $i > 0$. Any bounded E is clearly convergent. If in a distinguished triangle two out of three terms are bounded then so is the third. If one term is bounded and another one is convergent then the third one is convergent. Any direct sum of convergent spectra is convergent. Conjecture 1 implies that any E of the form $\Sigma^{p,q}H_C$ where C is a complex of abelian groups or more generally any spectrum which belongs to the smallest triangulated subcategory which contains objects of this form is bounded. We will see below that the intersection (22) is non zero for the rational Moore spectrum $\mathbf{1_Q}$ and thus one can not expect any spectrum to be convergent. We say that E is a finite spectrum if it belongs to the smallest triangulated subcategory which contains T-desuspensions of suspension spectra of smooth schemes over S and is closed under direct summands.

Conjecture 13. *Any finite spectrum is convergent with respect to the slice filtration.*

The convergence property for the slice spectral sequence is closely related to the convergence property for the motivic analog of the classical Adams spectral sequence which is defined by taking motivic homotopy groups of the Adams tower (8) smashed with E. We will show this modulo the following conjecture.

Conjecture 14. *Let (X, x) be a smooth pointed scheme of dimension d over S. Then*

$$\pi_{p,q}(\Sigma_T^\infty(X, x) \wedge H_{\mathbf{Z}}) = 0$$

for $q > d$.

The existing techniques imply this conjecture for $S = \mathrm{Spec}(k)$ where k is a field of characteristic zero. Using resolution of singularities and the blow-up long exact sequence in generalized homology (see [12]) one reduces the problem to a smooth proper X of pure dimension $d' \leq d$. The Spanier-Whitehead duality (loc.cit) and the Thom isomorphism for motivic cohomology imply that for such X one has

$$\pi_{p,q}(\Sigma_T^\infty X \wedge H_{\mathbf{Z}}) = H^{2d'-p,d'-q}(X, \mathbf{Z}).$$

The right hand side is zero for $d' < q$ by the rigidity part of Conjecture 1 which is known for varieties over a field.

Assuming Conjecture 14 we immediately see that for any finite spectrum E the spectrum $E \wedge H_{\mathbf{Z}}$ is bounded and in particular convergent. Conjectures 2 and 4 imply that $H_{\mathbf{Z}} \wedge H_{\mathbf{Z}}$ splits into a direct sum of the form $\Sigma^{0,q} H_{C_q}$ where C_q is a complex of abelian groups which is bounded in both directions and has no homology groups in dimensions less than $2q$. Together with the fact that for any smooth scheme X one has

$$\pi_{p,q}(\Sigma_T^\infty X \wedge H_{\mathbf{Z}}) = 0$$

for $2q - p > \dim(S)$ it implies that for a finite E only finitely many summands of $H_{\mathbf{Z}}^{\wedge i}$ will contribute to $\pi_{p,n}(f_q(E \wedge H_{\mathbf{Z}}^{\wedge i}))$ for any given p. This implies in turn that for a finite spectrum E and any $i \geq 0$ the spectrum $E \wedge H_{\mathbf{Z}}^{\wedge i}$ is bounded. Thus a finite spectrum E is convergent with respect to the slice filtration if and only if all the spectra $E \wedge \bar{H}^{\wedge i}$ forming the Adams tower for E are.

Lemma 7.3. *If E is a finite spectrum convergent with respect to the slice filtration then E is convergent with respect to the Adams filtration i.e. for any p, n and q one has*

$$\cap_{i \geq 0} \mathrm{Im}(\pi_{p,n}(E \wedge \bar{H}_{\mathbf{Z}}^{\wedge(q+i)}) \to \pi_{p,n}(E \wedge \bar{H}_{\mathbf{Z}}^{\wedge q})) = 0. \tag{23}$$

Proof. We may assume that E is effective. If E is convergent with respect to the slice filtration then as was shown above the spectra $E \wedge \bar{H}_{\mathbf{Z}}^{\wedge q}$ are convergent with respect to the slice filtration. On the other hand Conjecture 10 implies that the morphism

$$E \wedge \bar{H}_{\mathbf{Z}}^{\wedge(q+i)} \to E \wedge \bar{H}_{\mathbf{Z}}^{\wedge q}$$

factors through the morphism

$$f_i(E \wedge \bar{H}_{\mathbf{Z}}^{\wedge q}) \to E \wedge \bar{H}_{\mathbf{Z}}^{\wedge q}.$$

Combining we conclude that (23) holds. □

Lemma 7.4. *Let E be a connective spectrum convergent with respect to the Adams filtration (in the sense of Lemma 7.3) and such that $E \wedge H_{\mathbf{Z}}$ is bounded with respect to the slice filtration. Then E is convergent with respect to the slice filtration.*

Proof. Conjecture 10 implies that for any E and any q the morphism $s_q(E) \to s_q(E) \wedge H_{\mathbf{Z}}$ is a split monomorphism and thus the morphism $\bar{H}_{\mathbf{Z}} \wedge s_q(E) \to s_q(E)$ is zero. It implies easily that a connective spectrum E is convergent with respect to the Adams filtration if and only if f_qE are convergent with respect to the Adams filtration for all q. Therefore, it is sufficient to show that for E satisfying the conditions of the lemma the intersection $\cap_{q\geq 0} f_q\pi_{p,n}(E)$ is zero. Our assumption on E together with Conjecture 2 implies that all the spectra $E \wedge \bar{H}_{\mathbf{Z}}^{\wedge i} \wedge H_{\mathbf{Z}}$ are bounded with respect to the slice filtration. An inductive argument shows now that any element in this intersection will lie in the image of the homomorphism

$$\pi_{p,n}(E \wedge \bar{H}_{\mathbf{Z}}^{\wedge(i+1)}) \to \pi_{p,n}(E)$$

for any i and thus is zero by the assumption. □

Combining Lemmas 7.3 and 7.4 we see that, modulo the rest of the conjectures, Conjecture 13 is equivalent to the conjecture predicting that finite spectra are convergent with respect to the Adams filtration.

Consider the slice spectral sequence for a slice-wise cellular spectrum E. For simplicity let us assume that S is normal and connected such that $s_q(E) = \Sigma^{0,q}H_{\Pi_q(E)}$ where Π_q is a complex of abelian groups whose homology are the rigid homotopy groups of E and thus

$$\pi_{p,n}(s_q(E)) = H^{-p,q-n}(S, \Pi_q(E)).$$

Remark 7.5. If E is not a slice-wise cellular spectrum the groups $\pi_{p,n}(s_q(E))$ should still be isomorphic to motivic cohomology with coefficients in some complexes of sheaves with transfers whose cohomology presheaves are the presheaves of rigid homotopy groups of E. Thus for an arbitrary E the slice spectral sequence starts with motivic cohomology and approximates the motivic homotopy groups of the spectrum.

The canonical filtration

$$\cdots \subset \tau^{\geq 2}\Pi_q(E) \subset \tau^{\geq 1}\Pi_q(E) \subset \tau^{\geq 0}\Pi_q(E) \subset \cdots$$

on the complex $\Pi_q(E)$ gives a sequence of distinguished triangles of the form

$$\Sigma^{0,q}H_{\tau^{\geq(p+1)}\Pi_q(E)} \to \Sigma^{0,q}H_{\tau^{\geq p}\Pi_q(E)} \to \Sigma^{p,q}H_{H^p(\Pi_q(E))} \to \Sigma^{1,q}H_{\tau^{\geq(p+1)}\Pi_q(E)} \quad (24)$$

and since $s_q(E) = \mathrm{hocolim}_{p<0}\Sigma^{0,q}H_{\tau^{\geq p}\Pi_q(E)}$ this sequence defines a spectral sequence which starts with the groups $H^{p-p',q-n}(S, \pi^{\mathrm{rig}}_{p,q}(E))$ and converges to $\pi_{p',n}$ $(s_q(E))$ (the fact that it really converges requires some extra work based on the vanishing of $H^{i,j}(S, A)$ for $i > j + \mathrm{dim}S$). We see that in the case of a slice-wise cellular spectrum the combination of the slice spectral sequence with the spectral sequences generated by the towers (24) provides an "approximation" of the motivic homotopy groups of E by motivic cohomology of S with coefficients in the rigid homotopy groups of E.

Let us look in more detail on the slice spectral sequences for MGL, KGL and $\mathbf{1}$. Conjecture 5 predicts that $s_q(MGL) = \Sigma^{2q,q} H_{MU_{2q}}$ and in particular the complex Π_q in this case has only one nontrivial cohomology group in dimension $-2q$. Thus the tower (24) degenerates and we have

$$\pi_{p,n}(s_q MGL) = H^{2q-p,q-n}(S, MU_{2q}). \tag{25}$$

The slice spectral sequence starts with groups $H^{2q-p,q-n}(S, MU_{2q})$ and the r-th differential is of the form

$$H^{2q-p,q-n}(S, MU_{2q}) \to H^{2q-p+2r+1,q+r-n}(S, MU_{2(q+r)}).$$

If we reindex it we get the "motivic Atiyah-Hirzebruch spectral sequence" for MGL.

Conjecture 15. *The spectrum MGL is convergent with respect to the slice filtration.*

For $q > \dim(S) + p - n$ the group (25) is zero and thus Conjecture 15 implies that MGL is bounded with respect to the slice filtration. If S is regular and local over a field one has $H^{p,q}(S, \mathbf{Z}) = 0$ for $p > 2q$ or $p = 2q$ and $q \neq 0$ and $H^{0,0}(S, \mathbf{Z}) = \mathbf{Z}$. The same is expected to be true for any regular local S. Thus in this case the only nontrivial groups in the slice spectral sequence contributing to $\pi_{2q,q} MGL$ are $H^{0,0}(S, MU_{2q}) = MU_{2q}$ and all outgoing differentials are zero. Together with the convergence conjecture it implies that the homomorphism (6) maps MU_{2q} surjectively to $MGL_{2q,q}$ which implies the surjectivity part of [14, Conjecture 1, p. 601]. Since MU_* has no torsion the injectivity part can be proved by considering the rational coefficient case where everything splits.

For the algebraic K-theory spectrum Conjecture 7 implies that the slice spectral sequence has the same form as the spectral sequence constructed in [1] and [3]. We expect that these two spectral sequences are isomorphic.

Consider now the case of the sphere spectrum $\mathbf{1}$. By Conjecture 9 $s_q(\mathbf{1})$ is of the form $\Sigma_T^q H_{N_{2q}}$ where N_{2q} is a complex of abelian groups whose cohomology groups are given by

$$H^r(N_{2q}) = \mathrm{Ext}^r_{MU_*(MU)}(MU_*, MU_*)_{2q}. \tag{26}$$

These groups are zero for $r < 0$ and $r > q$ and thus the sequences (24) in this case give finite Postnikoff towers for $s_q(\mathbf{1})$ and we have a strongly convergent spectral sequence which starts with the groups

$$H^{r-p,q-n}(S, \pi^{\mathrm{rig}}_{r,q}(\mathbf{1})) = H^{r-p,q-n}(S, \mathrm{Ext}^{2q-r}_{MU_*(MU)}(MU_*, MU_*)_{2q}) \tag{27}$$

and converges to $\pi_{p,n}(s_q(\mathbf{1}))$. Let us consider two particular cases. First assume that $S = \mathrm{Spec}(k)$ where k is an algebraically closed field of characteristic zero. Then for a torsion abelian group A we have $H^{i,j}(k, A) = 0$ for $i \neq 0$ or $j < 0$ and $H^{0,j}(k, A) = A(j)$ for $j \geq 0$ where $A(j)$ denotes the twisting by the j-th power of roots of unity. For $q \neq 0$ the groups (26) are torsion and thus the spectral sequence computing $\pi_{p,n}(s_q(\mathbf{1}))$ degenerates and for $q \neq 0$ we get

$$\begin{aligned}\pi_{p,n}(s_q(\mathbf{1})) &= H^{0,q-n}(S, \pi^{\mathrm{rig}}_{p,q}(\mathbf{1})) \\ &= H^{0,q-n}(S, \mathrm{Ext}^{2q-p}_{MU_*(MU)}(MU_*, MU_*)_{2q}) \\ &= \mathrm{Ext}^{2q-p}_{MU_*(MU)}(MU_*, MU_*)_{2q}(q-n).\end{aligned}$$

For $n = 0$ the slice spectral sequence in this case becomes isomorphic to the usual Adams-Novikov spectral sequence in topology which shows that our conjectures predict that for $S = \mathrm{Spec}(k)$ as above one has

$$\pi_{p,0}(\mathbf{1}) = \pi^s_p(S^0).$$

Let now k be any field and consider the part of the slice spectral sequence which contributes to $\pi_{0,0}(\mathbf{1})$. The picture one gets here is very similar to the picture obtained by Fabien Morel in his work in [6] and [5] on $\pi_{0,0}$ based on the motivic Adams spectral sequence. The groups which contribute to $\pi_{0,0}(\mathbf{1})$ are $\pi_{0,0}(s_q(\mathbf{1}))$ and the groups which contribute to $\pi_{0,0}(s_q(\mathbf{1}))$ in the spectral sequence, defined by the tower (24), are

$$H^{r,q}(S, \pi^{\mathrm{rig}}_{r,q}(\mathbf{1})) = H^{r,q}(S, \mathrm{Ext}^{2q-r}_{MU_*(MU)}(MU_*, MU_*)_{2q}).$$

For any field k we have

$$H^{i,j}(\mathrm{Spec}(k), A) = 0$$

for $i > j$ thus the only nontrivial contributions come from cohomology with coefficients in $\mathrm{Ext}^{2q-r}_{MU_*(MU)}(MU_*, MU_*)_{2q}$ for $r \leq q$. All such groups are zero except for the ones with $r = q$ which are equal to $\mathbf{Z}$ for $r = q = 0$ and to $\mathbf{Z}/2$ for $r = q > 0$ ([10]). This is consistent with the conjecture of Fabien Morel predicting that $\pi_{0,0}(\mathbf{1})$ for any field k is isomorphic to its Grothendieck-Witt ring of quadratic forms. In terms of the Grothendieck-Witt ring the f_q filtration is then expected to coincide with the filtration by the powers of the ideal of forms of even dimension and the Milnor conjecture becomes a degeneracy result for the spectral sequence in this range. The results of [6] and [5] imply that in this case the slice filtration $f_n\pi$ coincides with the Adams filtration. In general it is not so since the Adams filtration on any spectrum which is a module over the Eilenberg-MacLane spectrum is trivial while the slice filtration may be not.

Finally consider the slice spectral sequence for the rational Moore spectrum $\mathbf{1}_{\mathbf{Q}}$ such that $\pi_{p,q}(\mathbf{1}_{\mathbf{Q}}) = \pi_{p,q}(\mathbf{1}) \otimes \mathbf{Q}$. We have $s_q(\mathbf{1}_{\mathbf{Q}}) = 0$ for $q \neq 0$ and $s_0(\mathbf{1}_{\mathbf{Q}}) = H_{\mathbf{Q}}$.

Thus

$$\pi_{p,n}(s_q(\mathbf{1_Q})) = \begin{cases} H^{-p,-n}(S,\mathbf{Q}) & \text{for } q=0,\\ 0 & \text{for } q \neq 0.\end{cases}$$

The slice spectral sequence in this case degenerates and the intersection of all the terms of the filtration $f_q\pi_{p,n}(\mathbf{1_Q})$ equals to the kernel of the motivic Hurewicz homomorphism

$$\pi_{p,n}(\mathbf{1_Q}) \to H^{-p,-n}(S,\mathbf{Q}).$$

If $k = \mathbf{Q}$ the group $\pi_{0,0}(\mathbf{1}) \otimes \mathbf{Q}$ contains at least two linearly independent elements while $H^{0,0}(\mathrm{Spec}(\mathbf{Q}),\mathbf{Q}) = \mathbf{Q}$ which implies that in this case the intersection (22) is not zero.

8. Possible Strategies of the Proof

We know of two strategies which can be used to prove the conjectures of this paper. The first one looks as follows. One starts with Conjecture 1. The existing techniques are sufficient to prove it in the case when S is a regular scheme over a field. For any such S one has $p^*(H_\mathbf{Z}) = H_\mathbf{Z}$ where $p : S \to \mathrm{Spec}(k)$ is the canonical morphism. An unstable version of this fact is proved in [11]. A stable version which easily follows will be done in one of the later papers of the series. The inverse image functor p^* obviously takes effective objects to effective objects and, for p of the form we consider here, it also takes rigid objects to rigid objects (see [12]). This argument show that it is enough to consider the case $S = \mathrm{Spec}(k)$. The effectiveness part of the conjecture in this case should follows from the description of the Eilenberg-MacLane spaces in terms of effective cycles given in [11] and will be considered in the next paper of the series. The rigidity part follows from the comparison between motivic cohomology defined in terms of motivic complexes in [17] and motivic cohomology defined in terms of SH since for the former one has $H^{p,q} = 0$ for $q < 0$ by definition. The s-stable form of this comparison result is proved in [11]. The T-stable form will be proved in one of the later papers of the series. It requires the cancellation theorem proved through comparison with higher Chow groups in [15]. The same argument proves Conjecture 12 for regular schemes S over a field.

The next step in this strategy is to prove Conjecture 2. The existing techniques are sufficient to do it for a regular scheme S of characteristic zero. As before one first reduces the problem to $S = \mathrm{Spec}(\mathbf{Q})$ by showing that $p^*(H_\mathbf{Z}) = H_\mathbf{Z}$. The Eilenberg-MacLane spectrum is built out of the suspension spectra of the Eilenberg-MacLane spaces $K(\mathbf{Z}(q), 2q)$ and thus it is sufficient to prove an analog of Conjecture 2 for $\Sigma_T^\infty K(\mathbf{Z}(q), 2q) \wedge H_\mathbf{Z}$ and show that the corresponding direct sum decompositions are compatible with the assembly morphisms of the Eilenberg-MacLane spectrum. To do it one constructs a functor ρ from DM, the stable version of the category DM_-^{eff} of [17], to SH right adjoint to the functor $\lambda : SH \to DM$ which takes the suspension spectrum of a smooth scheme to its "motive". The unstable version of this construction is described in [11]. For any E one gets a natural morphism of the form

$$E \wedge H_{\mathbf{Z}} \to \rho\lambda(E). \tag{28}$$

The morphism (28) is an isomorphism if E is a sphere. Using the Spanier-Whitehead duality established in [12] one can show that it is an isomorphism if E is the suspension spectrum of a scheme which is smooth and proper over S. If S is the spectrum of a field of characteristic zero the resolution of singularities implies that the suspension spectra of smooth projective varieties generate SH and thus (28) is an isomorphism for any E. In particular one concludes that

$$\Sigma_T^\infty K(\mathbf{Z}(q), 2q) \wedge H_{\mathbf{Z}} \cong \rho\tilde{M}(K(\mathbf{Z}(q), 2q))$$

where $\tilde{M}$ is the functor which takes a pointed space to its "motive" in DM. The spaces $K(\mathbf{Z}(q), 2q)$ for $q > 0$ can be represented by spaces of effective cycles (see [11]) on T^q and thus in the case of characteristic zero by infinite symmetric powers of T^q. In the second paper of the series started by [11] we analyze the structure of the motives of infinite symmetric powers. We first show that they admit an analog of the Steenrod splitting such that the motive of an infinite symmetric product splits as a direct sum of motives of "reduced" finite symmetric products. For any prime l the motive of each finite symmetric product localized in l can be represented as a direct summand of the iterated circle powers (the μ_n-version of the cyclic power). This construction uses the fact that motives are functorial with respect to "correspondences" given by finite relative cycles. Finally the motives of iterated circle powers of spheres are explicitly computed and showed to be isomorphic to direct sums of Tate motives of the form required by Conjecture 2.

The next step which is again possible to do over a general base scheme is to construct reduced power operations and use them to prove Conjecture 4. Together with explicit computations of the action of reduced power operations in motivic cohomology of the lens spaces this should lead to a proof of Conjecture 3.

The next step is to prove Conjecture 10. We do not know how to deal with the divisibility part of it yet but one may hope to prove it by looking at the geometry of the symmetric products of spheres. Over a field of characteristic zero $K(\mathbf{Z}(q), 2q)$ is built out of q-fold T-suspensions of $\Sigma_s^1(\mathbf{A}^{(m-1)q} - \{0\})/S_m$ where S_m is the symmetric group and $\mathbf{A}^{(m-1)q}$ is identified with the q-th power of the subspace V of $\mathbf{A}^m$, on which S_m acts by permutation of coordinates, given by the equation $\sum_{i=1}^m x_i = 0$. What one has to show is that the suspension spectrum of $(\mathbf{A}^{(m-1)q} - \{0\})/S_m$ considered as a pointed space belongs to $\Sigma_T^1 SH^{\mathrm{eff}}$. It can probably be done by explicit computation using some resolution of singularities for these spaces.

Assuming Conjecture 10 one can try to prove Conjecture 5 as follows. First note that Conjecture 10 shows that there exists a unique homomorphism $MU_* \to \pi_{*,*}(MGL)$ compatible with the homomorphism (6). To verify that it is an isomorphism it is sufficient to check that its analogs with coefficients in $\mathbf{Q}$ and $\mathbf{Z}/l$ are isomorphisms. The usual approach to the computation of the homology of MU as a module over the Steenrod algebra should work with no problem in the rigid setting which together with this identification of the E_2-term of the rigid Adams spectral sequence and the convergence theorem 6.7 should lead to a proof of Conjecture 5. Conjecture 6 seem to be easy modulo Conjecture 5 and how one proves Conjecture 9 using Conjecture 6 is explained in Section 3.2.

Conjectures 7 and 8 about algebraic K-theory can not be proved by means of the Adams spectral sequence since KGL is not a connective spectrum. There are several things one can do about it. One is to try to prove an analog of Conner and Floyd formula [2]

$$K^{*,*}(X) = MGL^{*,*}(X) \otimes_{MGL^{*,*}(S)} K^{*,*}(S)$$

and use it together with Conjecture 5 to get Conjecture 7. Alternatively, one can use the Adams spectral sequence to approximate the slices of the suspension spectrum of BGL and use the fact that $KGL = \mathrm{hocolim}_{n\geq 0}\Sigma_T^{-n}\Sigma_T^\infty BGL$. One can also use the following approach.[a] Define the s-stable homotopy category $SH_s(S)$ starting from the unstable $\mathbf{A}^1$-homotopy category and inverting S_s^1 but not S_t^1. We get the s-suspension spectrum functor

$$\Sigma_s^\infty : \mathcal{H}_{\mathbf{A}^1,\bullet}(S) \to SH_s(S)$$

and the t-suspension spectrum functor

$$\Sigma_t^\infty : SH_s(S) \to SH(S)$$

such that $\Sigma_T^\infty = \Sigma_t^\infty \Sigma_s^\infty$. The functor Σ_t^∞ has a right adjoint which we denote by Ω_t^∞. The definition of the slice filtration can be given in the context of SH_s in the same way as we did for $SH(S)$ except that all s-spectra are effective.

Conjecture 16.

$$\Omega_t^\infty(\Sigma_T^n SH^{\mathrm{eff}}(S)) \subset \Sigma_T^n SH_s(S)$$

This conjecture says that for a space (X, x) the space $\Omega_T^\infty\Sigma_T^\infty(\Sigma_T^n(X, x))$ can be built, at least s-stably, from n-fold T-suspensions. It connects the theme of this paper to another bunch of conjectures describing the hypothetical theory of operadic description of T-loop spaces. Any such theory should provide a model for $\Omega_T^\infty\Sigma_T^\infty$ which could then be used to prove Conjecture 16.

Let us show how Conjecture 16 can be used to prove Conjecture 7. The unit $\mathbf{1} \to KGL$ of the ring structure of KGL defines by Conjecture 10 a morphism $H_{\mathbf{Z}} \to$

[a]This approach is further elaborated in [16].

$s_0(KGL)$. The functor Ω_t^∞ clearly reflects isomorphisms between effective spectra. Thus to prove Conjecture 7 it is sufficient to show that

$$\Omega_t^\infty H_{\mathbf{Z}} \to \Omega_t^\infty s_0(KGL) \tag{29}$$

is an isomorphism. Consider the distinguished triangle

$$f_1 KGL \to f_0 KGL \to s_0(KGL).$$

Applying to it the functor Ω_t^∞ we get a triangle of the form

$$\Omega_t^\infty f_1 KGL \to \Omega_t^\infty f_0 KGL \to \Omega_t^\infty s_0(KGL). \tag{30}$$

It is easy to see that $\Omega_t^\infty f_0(KGL) = \Omega_t^\infty(KGL)$ and that $\Omega_t^\infty s_0(KGL)$ is orthogonal to $\Sigma_T^1 SH_s$. Conjecture 16 implies that $\Omega_t^\infty f_1 KGL$ belongs to $\Sigma_T^1 SH_s$ and therefore the triangle (30) is isomorphic to the triangle

$$f_1 \Omega_t^\infty KGL \to \Omega_t^\infty KGL \to s_0 \Omega_t^\infty(KGL).$$

Since the space $BGL \times \mathbf{Z}$ represents algebraic K-theory in the unstable category at least for a regular S (see [7]) the s-spectrum $\Omega_t^\infty KGL$ can be represented by the sequence of spaces which consists of s-deloopings of $BGL \times \mathbf{Z}$. Since BGL is build out of spheres in a rather explicit way it seems to be easy to show that $s_0 \Omega_t^\infty(BGL) = H_{\mathbf{Z}}$ where $H_{\mathbf{Z}}$ is considered as the Eilenberg-MacLane s-spectrum i.e. just the sequence of usual simplicial $K(\mathbf{Z}, n)$'s. Finally Conjectures 1 and 12 imply that for a normal connected S one has $\Omega_t^\infty(H_{\mathbf{Z}}) = H_{\mathbf{Z}}$. Thus we see that (29) is an endomorphism of $H_{\mathbf{Z}}$ and one verifies immediately that it takes unit to the unit which in turn implies that it is an identity.

This is how the first strategy looks like. It can be extended to prove the conjectures for all schemes S of characteristic zero but we have no idea how to extend it to positive or mixed characteristic. The bottleneck of this approach is the method used to prove Conjecture 2. One problem is that in positive characteristic the spaces of effective cycles are not representable by symmetric products and the argument used to establish the fact that their motives are direct sums of Tate motives does not work. The other problem is that without resolution of singularities I do not know how to prove that the morphism (28) is an isomorphism for $E = H_{\mathbf{Z}}$.

The second strategy is much less detailed than the first one but it may offer a way to prove conjectures of this paper in their full generality. I learned the ideas on which it is based from Mike Hopkins, Fabien Morel and Markus Rost. This strategy takes Conjecture 5 and closely related to it Conjecture 6 as the starting point. Over a general base scheme it is much easier to work with cobordisms than with motivic cohomology since the Thom spectrum is built directly from the suspension spectra of smooth varieties while the Eilenberg-MacLane spectrum is not. One can then attempt to prove the torsion part of Conjecture 6 by some analog of Quillen's argument. We do not know what to do with the rational part but may be one can prove Lemma 6.2 directly. This would imply in particular that $s_0(MGL) = H_{\mathbf{Z}}$ and thus we get Conjecture 10 since $s_0(MGL) = s_0(\mathbf{1})$ for simple geometric reasons. On the other hand knowing Conjecture 6 it seems possible to show that $H_{\mathbf{Z}}$ is the homotopy colimit of an explicit diagram built out of many copies of suspended MGL's and since the rigid homology of MGL are easy to compute we can compute rigid homology of $H_{\mathbf{Z}}$ thus solving Conjecture 3. This approach would also imply another result which deserves a separate formulation.

Conjecture 17. *For any morphism of schemes $f : S' \to S$ the natural morphism $f^* H_{\mathbf{Z}} \to H_{\mathbf{Z}}$ is an isomorphism.*

References

[1] Spencer Bloch and Stephen Lichtenbaum, A spectral sequence for motivic cohomology, www.math.uiuc.edu/K-theory/062 (1994).

[2] Pierre E. Conner and Edwin E. Floyd, *The Relation of Cobordism to K-theory*, Lecture Notes in Math. **28** (Springer, Heidelberg, 1966).

[3] Eric M. Friedlander and Andrei Suslin, The spectral sequence relating algebraic K-theory to motivic cohomology, www.math.uiuc.edu/K-theory/432 (2000).

[4] John Milnor, The Steenrod algebra and its dual, *Annals of Math.* **67** (1) (1958) 150–171.

[5] Fabien Morel, Suite spectrale d'Adams et conjectures de Milnor, www.math.jussieu.fr/morel (1999).

[6] Fabien Morel, Suite spectrale d'Adams et invariants cohomologiques des formes quadratiques, *C.R.Acad.Sci.Paris* **328** (1999) 963–968.

[7] Fabien Morel and Vladimir Voevodsky, $\mathbf{A}^1$-homotopy theory of schemes, *Publ. Math. IHES* (90) (1999) 45–143.

[8] Amnon Neeman, The Grothendieck duality theorem via Bousfield's techniques and Brown representability, *J. Amer. Math. Soc.* **9** (1) (1996) 205–236.

[9] Sergei P. Novikov, Methods of algebraic topology from the point of view of cobordism theory (Russian), *Izv. AN SSSR* **31** (1967) 855–951.

[10] Douglas C. Ravenel, *Complex Cobordism and Stable Homotopy Groups of Spheres* (Academic Press, Inc., 1986).

[11] Vladimir Voevodsky, $\mathbf{A}^1$-homotopy theory and finite correspondences, In preparation.

[12] Vladimir Voevodsky, Functoriality of the motivic stable homotopy categories, In preparation.

[13] Vladimir Voevodsky, The Milnor Conjecture, www.math.uiuc.edu/K-theory/170 (1996).

[14] Vladimir Voevodsky, The $\mathbf{A}^1$-homotopy theory, in: *Proceedings of the International Congress of Mathematicians*, Volume 1 (Berlin, 1998) 579–604.

[15] Vladimir Voevodsky, Motivic cohomology are isomorphic to higher Chow groups, www.math.uiuc.edu/K-theory/378 (1999).

[16] Vladimir Voevodsky, A possible new approach to the motivic spectral sequence, www.math.uiuc.edu/K-theory/469 (2001).

[17] Vladimir Voevodsky, Eric M. Friedlander and Andrei Suslin, *Cycles, Transfers and Motivic Homology Theories* (Princeton University Press, 2000).

Proceedings of the International Congress
of Mathematicians, Madrid, Spain, 2006

THE WORK OF ANDREI OKOUNKOV

by
GIOVANNI FELDER
Department of Mathematics, ETH Zurich, 8092 Zurich, Switzerland
E-mail: felder@math.ethz.ch

Andrei Okounkov's initial area of research is group representation theory, with particular emphasis on combinatorial and asymptotic aspects. He used this subject as a starting point to obtain spectacular results in many different areas of mathematics and mathematical physics, from complex and real algebraic geometry to statistical mechanics, dynamical systems, probability theory and topological string theory. The research of Okounkov has its roots in very basic notions such as partitions, which form a recurrent theme in his work. A partition λ of a natural number n is a non-increasing sequence of integers $\lambda_1 \geq \lambda_2 \geq \cdots \geq 0$ adding up to n. Partitions are a basic combinatorial notion at the heart of the representation theory. Okounkov started his career in this field in Moscow where he worked with G. Olshanski, through whom he came in contact with A. Vershik and his school in St. Petersburg, in particular S. Kerov. The research programme of these mathematicians, to which Okounkov made substantial contributions, has at its core the idea that partitions and other notions of representation theory should be considered as random objects with respect to natural probability measures. This idea was further developed by Okounkov, who showed that, together with insights from geometry and ideas of high energy physics, it can be applied to the most diverse areas of mathematics.

This is an account of some of the highlights mostly of his recent research.

I am grateful to Enrico Arbarello for explanations and for providing me with very useful notes on Okounkov's work in algebraic geometry and its context.

1. Gromov–Witten Invariants

The context of several results of Okounkov and collaborators is the theory of Gromov–Witten (GW) invariants. This section is a short account of this theory. GW invariants originate from classical questions of enumerative geometry, such as: how many rational curves of degree d in the plane go through $3d-1$ points in general position? A completely new point of view on this kind of problems appeared at the end of the eighties, when string theorists, working on the idea that space-time is the product of four-dimensional Minkowski space with a Ricci-flat compact complex three-fold, came up with a prediction for the number of rational curves of given degree in the quintic $x_1^5 + \cdots + x_5^5 = 0$ in $\mathbb{C}P^4$. Roughly speaking, physics gives predictions for differential equations obeyed by generating functions of numbers of curves. Solving these equations in power series gives recursion relations for the numbers. In particular a recursion relation of Kontsevich gave a complete answer to the above question on rational curves in the plane.

In general, Gromov–Witten theory deals with intersection numbers on moduli spaces of maps from curves to complex manifolds. Let V be a nonsingular projective variety over the complex numbers. Following Kontsevich, the compact moduli space $\overline{M}_{g,n}(V, \beta)$ (a Deligne–Mumford stack) of *stable maps* of class $\beta \in H_2(V)$ is the space of isomorphism classes of data $(C, p_1, \ldots, p_n, f)$ where C is a complex projective connected nodal curve of genus g with n marked smooth points $p_1, \ldots, p_n$ and $f : C \to V$ is a stable map such that $[f(C)] = \beta$. Stable means that if f maps an irreducible component to a point then this component should have a finite automorphism group. For each $j = 1, \ldots, n$ two natural sets of cohomology classes can be defined on these moduli space: 1) pull-backs $\mathrm{ev}_j^* \alpha \in H^*(\overline{M}_{g,n}(V, \beta))$ of cohomology classes $\alpha \in H^*(V)$ on the target V by the evaluation map $\mathrm{ev}_j : (C, p_1, \ldots, p_n, f) \mapsto f(p_j)$; 2) the powers of the first Chern class $\psi_j = c_1(L_j) \in H^2(\overline{M}_{g,n}(V, \beta))$ of the line bundle L_j whose fiber at $(C, p_1, \ldots, p_n, f)$ is the cotangent space $T^*_{pj}C$ to C at p_j. The *Gromov–Witten invariants* of V are the intersection numbers

$$\langle \tau_{k_1}(\alpha_1) \cdots \tau_{k_n}(\alpha_n) \rangle_{\beta,g}^V = \int_{\overline{M}_{g,n}(V,\beta)} \prod \psi_j^{k_j} \mathrm{ev}_j^* \alpha_j.$$

If all k_i are zero and the α_i are Poincaré duals of subvarieties, the Gromov–Witten invariants have the interpretation of counting the number of curves intersecting these subvarieties. As indicated by Kontsevich, to define the integral one needs to construct a *virtual fundamental class,* a homology class of degree equal to the "expected dimension"

$$\mathrm{vir\,dim}\ \overline{M}_{g,n}(V, \beta) = -\beta \cdot K_V + (g-1)(3 - \dim V) + n, \tag{1}$$

where K_V is the canonical class of V. This class was constructed in works of Behrend–Fantechi and Li–Tian.

The theory of Gromov–Witten invariants is already non-trivial and deep in the case where V is a point. In this case $\overline{M}_{g,n} = \overline{M}_{g,n}(\{\mathrm{pt}\})$ is the Deligne–Mumford moduli space of stable curves of genus g with n marked points. Witten conjectured and Kontsevich proved that the generating function

$$F(t_0, t_1, \ldots) = \sum_{n=0}^{\infty} \frac{1}{n!} \sum_{k_1 + \cdots + k_n = 3g-3+n} \langle \tau_{k_1} \cdots \tau_{k_n} \rangle_g^{\mathrm{pt}} \prod_{j=1}^{n} t_{k_j},$$

involving simultaneously all genera and numbers of marked points, obeys an infinite set of partial differential equations (it is a tau-function of the Korteweg–de Vries integrable hierarchy obeying the "string equation") which are sufficient to compute all the intersection numbers explicitly. One way to write the equations is as Virasoro conditions

$$L_k(e^F) = 0, \quad k = -1, 0, 1, 2 \ldots,$$

for certain differential operators L_k of order at most 2 obeying the commutation relations $[L_j, L_k] = (j-k)L_{j+k}$ of the Lie algebra of polynomial vector fields.

Before Okounkov few results were available for general projective varieties V and they were mostly restricted to genus $g = 0$ Gromov–Witten invariants (quantum cohomology). For our purpose the conjecture of Eguchi, Hori and Xiong is relevant here. Again, Gromov–Witten invariants of V can be encoded into a generating function F_V depending on variables $t_{j,a}$ where a labels a basis of the cohomology of V. Eguchi, Hori and Xiong extended Witten's definition of the differential operators L_k and conjectured that F_V obeys the Virasoro conditions $L_k(e^{F_V}) = 0$ with these operators.

2. Gromov–Witten Invariants of Curves

In a remarkable series of papers ([10], [11], [12]), Okounkov and Pandharipande give an exhaustive description of the Gromov–Witten invariants of curves. They prove the Eguchi–Hori–Xiong conjecture for general projective curves V, give explicit descriptions in the case of genus 0 and 1, show that the generating function for $V = \mathbb{P}^1$ is a tau-function of the Toda hierarchy and consider also in this case the $\mathbb{C}^\times$-equivariant theory, which is shown to be governed by the 2D-Toda hierarchy. They also show that GW invariants of $V = \mathbb{P}^1$ are unexpectedly simple and more basic than the GW invariants of a point, in the sense that the latter can be obtained as a limit, giving thus a more transparent proof of Kontsevich's theorem.

A key ingredient is the *Gromov–Witten/Hurwitz correspondence* relating GW invariants of a curve V to *Hurwitz numbers*, the numbers of branched covering of V with given ramification type at given points. A basic beautiful formula of Okounkov and Pandharipande is the formula for the *stationary* GW invariants of a curve V of genus $g(V)$, namely those for the Poincaré dual ω of a point:

$$\langle \tau_{k_1}(\omega) \cdots \tau_{k_n}(\omega) \rangle^{\bullet V}_{\beta = d\cdot[V], g} = \sum_{|\lambda| = d} \left(\frac{\dim \lambda}{d!} \right)^{2-2g(V)} \prod_{i=1}^{n} \frac{pk_i + 1(\lambda)}{(k_i+1)!}. \tag{2}$$

The (finite!) summation is over all partitions λ of the degree d and $\dim \lambda$ is the dimension of the corresponding irreducible representation of S_d. The genus g of the domain is fixed by the condition that the cohomological degree of the integrand is equal to the dimension of the virtual fundamental class. It is convenient here to include also stable maps with possibly disconnected domains and this is indicated by the bullet. The functions $p_k(\lambda)$ on partitions are described below.

Hurwitz numbers can be computed combinatorially and are given in terms of representation theory of the symmetric group by an explicit formula of Burnside. If the covering map at the ith point looks like $z \to z^{k_i+1}$, i.e., if the monodromy at the ith point is a cycle of length $k_i + 1$, the formula is

$$H_d^V(k_1+1,\ldots,k_n+1) = \sum_{|\lambda|=d} \left(\frac{\dim\lambda}{d!}\right)^{2-2g(V)} \prod_{i=1}^{n} fk_i + 1(\lambda).$$

Thus in this case the GW/Hurwitz correspondence is given by the substitution rule $f_{k+1}(\lambda) \to p_{k+1}(\lambda)/(k+1)!$. The functions f_k and p_k are basic examples of *shifted symmetric functions,* a theory initiated by Kerov and Olshanski, and the results of Okounkov and Pandharipande offer a geometric realization of this theory. A shifted symmetric polynomial of n variables $\lambda_1,\ldots,\lambda_n$ is a polynomial invariant under the action of the symmetric group given by permuting $\lambda_j - j$. A shifted symmetric function is a function of infinitely many variables $\lambda_1, \lambda_2 \ldots$, restricting for each n to a shifted symmetric polynomial of n variables if all but the first n variables are set to zero. Shifted symmetric functions form an algebra $\Lambda^* = \mathbb{Q}[p_1, p_2, \ldots]$ freely generated by the *regularized shifted power sums*, appearing in the GW invariants:

$$p_k(\lambda) = \sum_j \left(\left(\lambda_j - j + \frac{1}{2}\right)^k - \left(-j + \frac{1}{2}\right)^{-k}\right)(1-2^{-k})\zeta(-k)$$

The second term and the Riemann zeta value "cancel out" in the spirit of Ramanujan's second letter to Hardy: $1+2+3+\cdots = -\frac{1}{12}$. The shifted symmetric functions $f_k(\lambda)$ appearing in the Hurwitz numbers are central characters of the symmetric groups S_n: $f_1 = |\lambda| = \sum \lambda_i$ and the sum of the elements of the conjugacy class of a cycle of length $k \geq 2$ in the symmetric group S_n is a central element acting as $f_k(\lambda)$ times the identity in the irreducible representation corresponding to λ. The functions p_k and f_k are two natural shifted versions of Newton power sums.

In the case of genus $g(V) = 0$, 1, Okounkov and Pandharipande reformulate (2) in terms of expectation values and traces in fermionic Fock spaces and get more explicit descriptions and recursion relations. In particular if $V = E$ is an elliptic curve the generating function of GW invariants reduces to the formula of Bloch and Okounkov [1] for the character of the infinite wedge projective representation of the algebra of polynomial differential operators, which is expressed in terms of Jacobi theta functions. As a corollary, one obtains that

$$\sum_d q^d \langle \tau k_1(\omega) \cdots \tau k_n(\omega)\rangle_d^{\bullet E}$$

belongs to the ring $\mathbb{Q}[E_2, E_4, E_6]$ of *quasimodular forms.*

As shown by Eskin and Okounkov [2], one can use quasimodularity to compute the asymptotics as $d \to \infty$ of the number of connected ramified degree d coverings of a torus with given monodromy at the ramification points. By a theorem of Kontsevich–Zorich and Eskin–Mazur, this asymptotics gives the volume of the moduli space of holomorphic differentials on a curve with given orders of zeros, which is in turn related to the dynamics of billiards in rational polygons. Eskin and Okounkov give explicit formulae for these volumes and prove in particular the Kontsevich–Zorich conjecture that they belong to $\pi^{-2g}\mathbb{Q}$ for curves of genus g.

3. Donaldson–Thomas Invariants

As is clear from the dimension formula (1) the case of three-dimensional varieties V plays a very special role. In this case, which in the Calabi–Yau case $K_V = 0$ is the original context studied in string theory, it is possible to define invariants counting curves by describing curves by equations rather than in parametric form. Curves in V of genus g and class $\beta \in H_2(V)$ given by equations are parametrized by Grothendieck's Hilbert schemes Hilb $(V; \beta, \mathcal{X})$ of subschemes of V with given Hilbert polynomial of degree 1. The invariants $\beta, \mathcal{X} = 2 - g$ are encoded in the coefficients of the Hilbert polynomial. R. Thomas constructed a virtual fundamental class of Hilb$(V; \beta, \mathcal{X})$ for three-folds V of dimension $-\beta.\ K_V$, the same as the dimension of $\overline{M}_{g,0}(V, \beta)$. Thus one can define Donaldson–Thomas (DT) invariants as intersection numbers on this Hilbert scheme. There is no direct geometric relation between Hilb$(V; \beta, \chi)$ and $\overline{M}_{g,0}(V, \beta)$, and indeed the (conjectural) relation between Gromov–Witten invariants and Donaldson–Thomas invariants is quite subtle. In its simplest form, it relates the GW invariants $\int_{\overline{M}^{\bullet}_{g,n}(V,\beta)} \prod \mathrm{ev}_i^* \gamma^i$ to the DT invariants $\int_{\mathrm{Hilb}(V;\beta,\mathcal{X})} \prod c2(\gamma_i)$. The class $c_2(\gamma)$ is the coefficient of $\gamma \in H^*(V)$ in the Künneth decomposition of the second Chern class of the ideal sheaf of the universal family $\nu \subset \mathrm{Hilb}(V; \beta, \chi) \times V$.

The conjecture of Maulik, Nekrasov, Okounkov and Pandharipande [6], [7], inspired by ideas of string theory [14] states that suitably normalized generating functions $Z'_{GW}(\gamma; u)_\beta, Z'_{DT}(\gamma; q)_\beta$ are essentially related by a coordinate transformation:

$$(-iu)^{-d} Z'_{GW}(\gamma_1, \ldots, \gamma_n; u)_\beta = q^{-d/2} Z'_{DT}(\gamma_1, \ldots, \gamma_n; q)_\beta, \quad \text{if } q = -e^{iu},\ \beta \neq 0.$$

Here $d = -\beta \cdot K_V$ is the virtual dimension. Moreover these authors conjecture that there $Z'_{DT}(\gamma; q)_\beta$ is a *rational* function of q. This has the important consequence that all (infinitely many) GW invariants are determined in principle by finitely many DT invariants. Versions of these conjectures are proven for local curves and the total space of the canonical bundle of a toric surface. The GW/DT correspondence can be viewed as a far-reaching generalization of formula (2), to which it reduces in the case where V is the product of a curve with $\mathbb{C}^2$.

4. Other Uses of Partitions

Here is a short account of other results of Okounkov based on the occurrence of partitions.

One early result of Okounkov [9] is his first proof of the Baik–Deift–Johansson conjecture (two further different proofs followed, one by Borodin, Okounkov and Olshanski and one by Johansson). This conjecture states that, as $n \to \infty$, the joint distribution of the first few rows of a random partition of n with the Plancherel measure $P(\lambda) = (\dim \lambda)^2/|\lambda|!$, natural from representation theory, is the same, after proper shift and rescaling, as the distribution of the first few eigenvalues of a Gaussian random hermitian matrix of size n. The proof involves comparing random surfaces given by Feynman diagrams and by ramified coverings and contains many ideas that anticipate Okounkov's later work on Gromov–Witten invariants.

Random partitions also play a key role in the work [8] of Nekrasov and Okounkov on $N = 2$ supersymmetric gauge theory in four dimensions. Seiberg and Witten gave a formula for the effective "prepotential", postulating a duality with a theory of monopoles. The Seiberg–Witten formula is given in terms of periods on a family of algebraic curves, closely connected with classical integrable systems. Nekrasov showed how to rigorously define the prepotential of the gauge theory as a regularized instanton sum given by a localization integral on the moduli space of antiselfdual connections on $\mathbb{R}^4$. Nekrasov and Okounkov show that this localization integral can be written in terms of a measure on partitions with periodic potential and identify the Seiberg–Witten prepotential with the surface tension of the limit shape.

Partitions of n also label $(\mathbb{C}^\times)^2$-invariant ideals of codimension n in $\mathbb{C}[x, y]$ and thus appear in localization integrals on the Hilbert scheme of points in the plane. Okounkov and Pandharipande [13] describe the ring structure of the equivariant quantum cohomology (genus zero GW invariants) of this Hilbert scheme in terms of a time-dependent version of the Calogero–Moser operator from integrable systems.

5. Dimers

Dimers are a much studied classical subject in statistical mechanics and graph combinatorics. Recent spectacular progress in this subject is due to the discovery by Okounkov and collaborators of a close connection of planar dimer models with real algebraic geometry.

A *dimer configuration* (or perfect matching) on a bipartite graph G is a subset of the set of edges of G meeting every vertex exactly once. For example if G is a square grid we may visualize a dimer configuration as a tiling of a checkerboard by dominoes. In statistical mechanics one assigns positive weights (Boltzmann weights) to edges of G and defines the weight of a dimer configuration as the product of the weights of its edges. The basic tool is the Kasteleyn matrix of G, which is up to certain signs the weighted adjacency matrix of G. For finite G Kasteleyn proved that the partition function (i.e., the sum of the weights of all dimer configurations) is the absolute value of the determinant of the Kasteleyn matrix.

Kenyon, Okounkov and Sheffield consider a doubly periodic bipartite graph G embedded in the plane with doubly periodic weights. For each natural number n one then has a probability measure on dimer configurations on $G_n = G/n\mathbb{Z}^2$ and statistical mechanics of dimers is essentially the study of the asymptotics of these probability measures in the thermodynamic limit $n \to \infty$. One key observation is the Kasteleyn matrix on G_1 can be twisted by a character $(z, w) \in (\mathbb{C}^\times)^2$ of $\mathbb{Z}^2$ and thus one defines the *spectral curve* as the zero set $P(z, w) = 0$ of the determinant of the twisted Kasteleyn matrix $P(z, w) = \det K(z, w)$. This determinant is a polynomial in $z^{\pm 1}, w^{\pm 1}$ with real coefficients and thus defines a real plane curve.

The main observation of Kenyon, Okounkov and Sheffield [3] is that the spectral curve belongs to the very special class of (simple) Harnack curves, which were studied in the 19th century and have reappeared recently in real algebraic geometry. Kenyon, Okounkov and Sheffield show that in the thermodynamic limit, three different *phases* (called gaseous, liquid and frozen) arise. These phases are characterized by qualitatively different long-distance behaviour of pair correlation functions. One can see these phases by varying two real parameters (B_1, B_2) (the "magnetic field") in the weights, so that the spectral curve varies by rescaling the variables. The regions in the (B_1, B_2)-plane corresponding to different phases are described in terms of the *amoeba* of the spectral curve, namely the image of the curve by the map Log: $(z, w) \mapsto (\log |z|, \log |w|)$. The amoeba of a curve is a closed subset of the plane which looks a bit like the microorganism with the same name. The amoeba itself corresponds to the liquid phase, the bounded components of its complement to the gaseous phase and the unbounded components to the frozen phase. This insight has a lot of consequences for the statistics of dimer models and lead Okounkov and collaborators to beautiful results on interfaces with various boundary conditions [3], [5].

Such a precise and complete description of phase diagrams and shapes of interfaces is unprecedented in statistical mechanics.

6. Random Surfaces

One useful interpretation of dimers is as models for random surfaces in three-dimensional space. In the simplest case one considers a model for a melting or dissolving cubic crystal in which at a corner some atoms are missing (see figure).

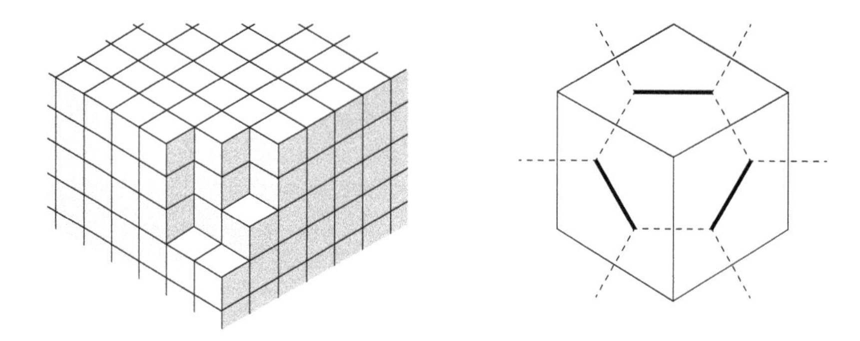

A melting crystal corner (left) and the relation between tilings and dimers (right).

Viewing the corner from the (1,1,1) direction one sees a tiling of the plane by 60° rhombi, which is the same as a dimer configuration on a honeycomb lattice (each tile covers one dimer of the dimer configuration). In this simple model, one gives the same probability for every configuration with given missing volume. If one lets the size of the cubes go to zero keeping the missing volume fixed, the probability measure concentrates on an a surface, the limit shape. More generally, every planar dimer model can be rephrased as a random surface model and limit shapes for more general crystal corner geometries can be defined. Kenyon, Okounkov and Sheffield show that the limit shape is given by the graph of (minus) the *Ronkin function* $R(x,y) = (2\pi i)^{-2} \int_{|z|=|w|=1} \log(P(e^x z, e^y w)) dz dw / zw$ of the spectral curve (in the case of the honeycomb lattice with equal weights, $P(z,w) = z+w+1$). This function is affine on the complement of the amoeba and strictly convex on the amoeba. So the connected components of the complement of the amoeba are the projections of the facets of the melting crystal.

In addition to this surprising connection with real algebraic geometry, random surfaces of this type are essential in the GW/DT correspondence, see Section 3, as they arise in localization integrals for DT invariants of toric varieties.

7. The Moduli Space of Harnack Curves

The notions used by Okounkov and collaborators in their study of dimer models arose in an independent recent development in real algebraic geometry. Their result bring a new probabilistic point of view in this classical subject.

In real algebraic geometry, unsurmountable difficulties already appear when one consider curves. The basic open question is the first part of Hilbert's 16th problem: what are the possible topological types of a smooth curve in the plane given by a polynomial equation $P(z,w) = 0$ of degree d? Topological types up to degree 7 are known but very few general results are available. In a recent development in real algebraic geometry in the context of toric varieties the class of Harnack curves plays an important role and can be characterized in many equivalent way. In one definition, due to Mikhalkin, a Harnack curve is a curve such that the map to its amoeba is 2:1 over the interior, except at possible nodal points; equivalently, by a theorem of Mikhalkin and Rullgård, a Harnack curve is a curve whose amoeba has area equal to the area of the Newton polygon of the polynomial P. These equivalent properties determine the topological type completely.

Kenyon and Okounkov prove [4] that *every* Harnack curve is the spectral curve of some dimer model. They obtain an explicit parametrization of the moduli space of Harnack curves with fixed Newton polygon by weights of dimer models, and deduce in particular that the moduli spaces are connected.

8. Concluding Remarks

Andrei Okounkov is a highly creative mathematician with both an exceptional breadth and a sense of unity of mathematics, allowing him to use and develop, with perfect ease, techniques and ideas from all branches of mathematics to reach his research objectives. His results not only settle important questions and open new avenues of research in several fields of mathematics, but they have the distinctive feature of mathematics of the very best quality: they give simple complete answers to important natural questions, they reveal hidden structures and new connections between mathematical objects and they involve new ideas and techniques with wide applicability.

Moreover, in addition to obtaining several results of this quality representing significant progress in different fields, Okounokov is able to create the ground, made of visions, intuitive ideas and techniques, where new mathematics appears. A striking example for this concerns the relation to physics: many important developments in mathematics of the last few decades have been inspired by high energy physics, whose intuition is based on notions often inaccessible to mathematics. Okounkov's way of proceeding is to develop a mathematical intuition alternative to the intuition of high energy physics, allowing him and his collaborators to go beyond the mere verification of predictions of physicists. Thus, for example, in approaching the topological vertex of string theory, instead of stacks of D-branes and low energy effective actions we find mathematically more familiar notions such as localization and asymptotics of probability measures. As a consequence, the scope of Okounkov's research programme goes beyond the context suggested by physics: for example the Maulik–Nekrasov–Okounkov–Pandharipande conjecture is formulated (and proved in many cases) in a setting which is much more general than the Calabi–Yau case arising in string theory.

References

[1] Bloch, Spencer, and Okounkov, Andrei, The character of the infinite wedge representation. *Adv. Math.* **149** (1) (2000), 1–60.

[2] Eskin, Alex, and Okounkov, Andrei, Asymptotics of numbers of branched coverings of a torus and volumes of moduli spaces of holomorphic differentials. *Invent. Math.* **145** (1) (2001), 59–103.

[3] Kenyon, Richard, Okounkov, Andrei, and Sheffield, Scott, Dimers and amoebae. *Ann. of Math.* (2) **163** (3) (2006), 1019–1056.

[4] Kenyon, Richard, and Okounkov, Andrei, Planar dimers and Harnack curves. *Duke Math. J.* **131** (3) (2006), 499–524.

[5] Kenyon, Richard, and Okounkov, Andrei, Limit shapes and the complex burgers equation. Preprint; arXiv:math-ph/0507007.

[6] Maulik, D., Nekrasov, N., Okounkov, A., and Pandharipande, R., Gromov-Witten theory and Donaldson-Thomas theory, I. Preprint; arXiv:math.AG/0312059.

[7] Maulik, D., Nekrasov, N., Okounkov, A., and Pandharipande, R., Gromov-Witten theory and Donaldson-Thomas theory, II. Preprint; arXiv:math. AG/0406092.

[8] Nekrasov, Nikita A., and Okounkov, Andrei, Seiberg-Witten theory and random partitions. In *The unity of mathematics,* Progr. Math. 244, Birkhäuser, Boston, MA, 2006, 525–596.

[9] Okounkov, Andrei, Random matrices and random permutations. *Internat. Math. Res. Notices* **2000** (20) (2000), 1043–1095.

[10] Okounkov, Andrei, and Pandharipande, Rahul, Gromov-Witten theory, Hurwitz theory, and completed cycles. *Ann. of Math.* (2) **163** (2) (2006), 517–560.

[11] Okounkov, Andrei, and Pandharipande, Rahul, The equivariant Gromov-Witten theory of $\mathbf{P}^1$. *Ann. of Math.* (2) **163** (2) (2006), 561–605.

[12] Okounkov, Andrei, and Pandharipande, Rahul, Virasoro constraints for target curves. *Invent. Math.* **163** (1) (2006), 47–108.

[13] Okounkov, Andrei, and Pandharipande, Rahul, Quantum cohomology of the Hilbert scheme of points in the plane. Preprint; arXiv:math.AG/0411210.

[14] Okounkov, Andrei, Reshetikhin, Nikolai, and Vafa, Cumrun, Quantum Calabi-Yau and classical crystals. In *The unity of mathematics,* Progr. Math. 244, Birkhäuser, Boston, MA, 2006, 597–618.

Andrei Okounkov

ANDREI OKOUNKOV

B.S. in Mathematics, Moscow State University, 1993

Ph.D. in Mathematics, Moscow State University, 1995

Positions until today

1994–1995	Research Fellow in the Dobrushin Mathematical Laboratory at the Institute for Problems of Information Transmission, Russian Academy of Sciences
1996	Member in the Institute of Advanced Study in Princeton
1997	Member in the Mathematical Sciences Research Institute in Berkeley, on leave from University of Chicago
1996–1999	L. E. Dickson instructor in Mathematics, University of Chicago
1998–2001	Assistant Professor of Mathematics, University of California at Berkeley
2001–2002	Professor of Mathematics, University of California at Berkeley
Present position	Professor of Mathematics, Princeton University

Proceedings of the International Congress
of Mathematicians, Madrid, Spain, 2006

THE WORK OF TERENCE TAO

by
CHARLES FEFFERMAN

Mathematics at the highest level has several flavors. On seeing it, one might say:

(A) What amazing technical power!
(B) What a grand synthesis!
(C) How could anyone not have seen this before?
(D) Where on earth did this come from?

The work of Terence Tao encompasses all of the above. One cannot hope to capture its extraordinary range in a few pages. My goal here is simply to exhibit a few contributions by Tao and his collaborators, sufficient to produce all the reactions (A)... (D). I shall discuss the Kakeya problem, non-linear Schrodinger equations and arithmetic progressions of primes.

Let me start with a vignette from Tao's work on the Kakeya problem, a beautiful and fundamental question at the intersection of geometry and combinatorics. I shall state the problem, comment briefly on its significance and history, and then single out my own personal favorite result, by Nets Katz and Tao.

The original Kakeya problem was to determine the least possible area of a plane region inside which a needle of length 1 can be turned a full 360 degrees. Besicovitch and Pál showed that the area can be taken arbitrarily small.

In its modern form, the Kakeya problem is to estimate the fractal dimension of a "Besicovitch set" $E \subset \mathbb{R}^n$, i.e., a set containing line segments of length 1 in all directions.

There are several relevant notions of "fractal dimension". Here, let us use the Minkowski dimension, defined in terms of coverings of E by small balls of a fixed radius δ. The Minkowski dimension is the infimum of all β such that, for small δ, E can be covered by $\delta^{-\beta}$ balls of radius δ. We want to prove that any Besicovitch set $E \subset \mathbb{R}^n$ has Minkowski dimension at least $\beta(n)$, with $\beta(n)$ as large as possible. (Perhaps $\beta(n) = n$.)

Regarding the central importance of this problem, perhaps it is enough to say that it is intimately connected with the multiplier problem for Fourier transforms, and with the restriction of Fourier transforms to hypersurfaces; these in turn are closely connected with non-linear PDE via Strichartz estimates and their variants. There are also connections with other hard, interesting problems in combinatorics.

Let me sketch some of the history of the problem over the last 30 years. The basic result of the 1970s is that $\beta(2) = 2$. (This is due to Davies, and is closely related to the early work of A. Córdoba. See [7], [8].) In the 1980s, Drury [9] showed that $\beta(n) \geq \frac{n+1}{2}$ for $n \geq 3$. (See also Christ et al [4].)

Then, about 1990, J. Bourgain and, shortly afterwards, T. Wolff discovered that Besicovitch sets of small fractal dimension have geometric structure (they contain "bouquets" and "hairbrushes"). During the 1990s, Bourgain also discovered a connection between the Kakeya problem and Gowers' work on the Balog-Szemerédi theorem from combinatorics. These insights led to small, hard-won improvements in the value of $\beta(n)$. The work looks deep and forbidding. See [1], [2], [24].

The connection with Gowers' work arises in the following result. (We write $\#(S)$ for the number of elements of a set S.)

Deep Theorem (Bourgain, using ideas from Gowers' improvement of Balog–Szemerédi). *Let A, B be subsets of an abelian group, and let $G \subset A \times B$. Assume that $\#(A)$, $\#(B)$, and $\#\{a + b : (a,b) \in G\}$ are at most N. Then $\#\{a - b : (a,b) \in G\} \geq CN^{2-1/13}$, for a universal constant C.*

The point is that one improves on the trivial bound N^2. From the Deep Theorem, one quickly obtains a result on $\beta(n)$ by slicing the set E with three parallel hyperplanes H, H', H'', with H'' halfway between H and H'.

Enter Nets Katz and Terence Tao, who proved the following result in 1999.

Little Lemma. *Under the assumptions of the Deep Theorem, we have $\#\{a - b : (a,b) \in G\} \geq C\,N^{2-1/6}$, for a universal constant C.*

Note that the Little Lemma is strictly sharper than the Deep Theorem, Nevertheless, its proof takes only a few pages, and can be understood by a bright high-school student. After reading the proof, one has not the faintest clue where the idea came from (see (D)).

The Little Lemma and its refinements led to the estimate $\beta(n) \geq \frac{4n+3}{7}$ for the Kakeya problem, which at the time was the best result known for $n > 8$. In high dimensions, the high-school accessible paper [18] went further than all the deep, forbidding work that came before it. Since then, there has been further progress, with Nets Katz, Izabella Laba, and Terence Tao playing a leading rôle. The subject still looks deep and forbidding. In particular, regarding (A), let me refer the reader to the tour-de-force [17] by these authors.

Unfortunately, the complete solution to the Kakeya problem still seems far away.

Next, I shall discuss "interaction Morawetz estimates". This simple idea, with profound consequences for PDE, was discovered by the "I-Team": J. Colliander, M. Keel, G. Staffilani, H. Takaoka, and Terence Tao. Let me start with the 3D non-linear Schrodinger equation (NLS):

$$i\partial_t u + \Delta_x u = \pm|u|^{p-1}u, \quad u(x,0) = u_0(x) \text{ given}, \tag{1}$$

where u is a complex-valued function of $(x,t) \in \mathbb{R}^3 \times \mathbb{R}$, and $p > 1$ is given.

This equation is important in physics and engineering. For instance, it describes the propagation of light in a fiber-optic cable. The behavior of solutions of (1) depends strongly on the $\pm$ sign and on the value of p. In particular, the minus sign is "focussing", and we may expect solutions of (1) to develop singularities; while the plus sign is "defocussing", and we expect solutions of (1) to spread out over large regions of space, as $t \longrightarrow \pm\infty$. In the defocussing case, the non-linear term in (1) should eventually become negligibly small, and the solution of (1) ought to behave like a solution of the (linear) free Schrödinger equation $(i\partial_t + \Delta_x)u = 0$. From now on, we restrict attention to the defocussing case.

We note two obvious conserved quantities for (1): the "mass" $\int_{\mathbb{R}^3} |u(x,t)|^2 dx$, and the energy,

$$E = \frac{1}{2}\int_{\mathbb{R}^3} |\Delta_x u(x,t)|^2\, dx + \frac{1}{p+1}\int_{\mathbb{R}^3} |u(x,t)|^{p+1}\, dx.$$

How can we prove that solutions of (1) spread out for large time? A fundamental tool is the Morawetz estimate. C. Morawetz first discovered this wonderful, simple idea for the non-linear Klein-Gordon equation. Let me describe it here for cubic 3D NLS, i.e., for equation (1) with $p = 3$. There, the Morawetz estimate asserts that

$$\int_0^T \int_{\mathbb{R}^3} \frac{|u(x,t)|^4}{|x|} dx\, dt \le C \sup_{0\le t\le T} \|(-\Delta_x)^{1/4} u(\cdot,t)\|^2_{L^2(\mathbb{R}^3)}, \tag{2}$$

for any $T > 0$ and any solution of (1).

The good news is that (2) instantly shows that u must eventually become small in any given bounded region of space. (If not, then the left-hand side of (2) grows linearly in T as $T \longrightarrow \infty$, while the right-hand side of (2) remains bounded, thanks to conservation of mass and energy.)

The bad news is that (2) does not rule out a scenario in which $u(x,t)$ remains concentrated near a moving center $x = x_0(t)$. The trouble is that the weight function $\frac{1}{|x|}$ is concentrated near $x = 0$, whereas $u(\cdot, t)$ may be concentrated somewhere else.

The I-Team found an amazingly simple and straightforward way to overcome the bad news. Let me sketch the idea, starting with the classic proof of (2). To derive (2), we start with the quantity

$$M_0(t) = \operatorname{Im} \int_{\mathbb{R}^3} \bar{u}(x,t) \cdot \left[\frac{x}{|x|} \cdot \nabla_x u(x,t)\right] dx. \tag{3}$$

On one hand, $M_0(t)$ is controlled by the right-hand side of (2), when $0 \le t \le T$. On the other hand, a computation using (1) shows that

$$\frac{d}{dt} M_0(t) = 4\pi^2 |u(0,t)|^2 + 2\int_{\mathbb{R}^3} |\nabla_\Omega u(x,t)|^2 \frac{dx}{|x|} + \int_{\mathbb{R}^3} \frac{|u(x,t)|^4}{|x|} dx, \tag{4}$$

where ∇_Ω denotes the angular part of the gradient.

The Morawetz estimate (2) follows at once.

The I-Team simply replaced $M_0(t)$ by a weighted average of translates,

$$M(t) = \int_{\mathbb{R}^3} M_y(t)|u(y,t)|^2\, dy,$$

where

$$M_y(t) = \operatorname{Im} \int_{\mathbb{R}^3} \bar{u}(x,t) \cdot \left[\frac{x-y}{|x-y|} \cdot \nabla_x\, u(x,t) \right] dx.$$

This puts the greatest weight on those $y \in \mathbb{R}^3$ where $u(y,t)$ lives — an eminently sensible idea.

Starting with $M(t)$ and proceeding more or less as in the proof of the Morawetz estimate, one obtains easily the

Interaction Morawetz Estimate.

$$\int_0^T \int_{\mathbb{R}^3} |u(x,t)|^4\, dx\, dt \leq C\|u(\cdot,0)\|^2_{L^2(\mathbb{R}^3)} \cdot \sup_{0\leq t\leq T} \|(-\Delta_x)^{1/4}\, u(\cdot,t)\|^2_{L^2(\mathbb{R}^3)}. \tag{5}$$

Again, the right-hand side is bounded for large T, thanks to conservation of mass and energy. This time however, the left-hand side grows linearly in T, even if our solution is concentrated in a moving ball. The I-Team has overcome the bad news. They made it look effortless. Why did no one think of it before? (See (C).)

Observe that the right-hand side of (5) is much weaker than the energy; we need only half an x-derivative, as opposed to a full gradient. The original purpose of the interaction Morawetz estimate was to derive global existence for cubic defocussing 3D NLS in Sobolev spaces in which the energy may be infinite. That is a big achievement (see [6]), but I will not discuss it further here, except to point out that the proof involves additional ideas and formidable work.

Instead, let me say a few words about the defocussing quintic 3D NLS, i.e., the case $p = 5$ of equation (1). This equation is particularly natural and deep, because it is critical for the energy. One knows that finite-energy initial data lead to solutions for a short time, and that small-energy initial data lead to global solutions. The challenge is to prove global existence for initial data with large, finite energy.

To appreciate the difficulty of the problem, we have only to turn to the tour-de-force [3] by J. Bourgain, solving the problem in the radially symmetric case. The general case is an order of magnitude harder; a singularity can form only at the origin in the radial case, but it may form anywhere in the general case. The I-Team settled the general case using a version of the interaction Morawetz estimate for quintic NLS (with cutoffs, which unfortunately greatly complicate the analysis). This is natural, since in a sense one must overcome the same bad news as before. Their result [5] is as follows.

Theorem. *Take* $p = 5$ *in the defocussing case in* (1). *Then, for any finite-energy initial data* u_0, *there is a global solution* $u(x, t)$ *of NLS. If* u_0 *belongs to* H^s *with* $s > 1$, *then $u(\cdot, t)$ also belongs to H^s for all t. Moreover, there exist solutions $u_\pm$ of the free Schrödinger equation, such that*

$$\int_{\mathbb{R}^3} |\nabla_x(u(x,t) - u_\pm(x,t))|^2 \, dx$$

tends to zero as t tends to $\pm\infty$.

I will not try to describe their proof, except to say that they use an interaction Morawetz estimate with cutoffs, along with ideas from Bourgain [3], especially the "induction on energy", as well as other ideas that I cannot begin to describe here. The details are highly formidable; see (A).

We come now to Tao's great joint paper ([16]; see also Green [15]) with Ben Green, in which they prove the following result. Here again, $\#(S)$ denotes the number of elements of a set S.

Theorem GT. *There exist arbitrarily long arithmetic progressions of primes. More precisely, given $k \geq 3$, there exist constants $c(k) > 0$ and $N_0(k) \geq 1$, such that for any $N > N_0(k)$, we have $\#\{k$-term arithmetic progressions among the primes less* than $N\} > \frac{c(k)N^2}{(\log N)^k}$.

The lower bound here agrees in order of magnitude with a natural guess. (Green and Tao are currently working on a more precise result, with an optimal $c(k)$.)

To convey something of the range and depth of the ideas in the proof, let me start with the classic theorem of Szemerédi on sets of positive density. Here, $\mathbb{Z}_N$ denotes the cyclic group of order N.

Theorem Sz 1. *Given k and δ, we have for large enough N that any subset $E \subset \mathbb{Z}_N$ with $\#(E) > \delta \cdot N$ contains an arithmetic progression of length k.*

Szemerédi's theorem also gives a lower bound for the number of k-term progressions in E. (See [23].) It is convenient to speak of functions f rather than sets E. (One obtains Theorem Sz 1 from Theorem Sz 2 below, simply by taking f to be the indicator function of E.) Thus, Szemerédi's theorem may be rephrased as follows.

Theorem Sz 2. *Given k, δ, the following holds for large enough N. Let $f : \mathbb{Z}_N \to \mathbb{R}$, with $0 \leq f(x) \leq 1$ for all x, and with*

(1) $Av_{x \in \mathbb{Z}_N} f(x) > \delta$.

Then

(2) $Av_{x,r \in \mathbb{Z}_N} \{f(x) \cdot f(x+r) \ldots f(x+(k-1)r)\} \geq c(k,\delta) > 0$, *where* $c(k,\delta)$ depends only on k, δ *(and not on N or f).*

(In (1), (2) and similar formulas, "*Av*" denotes the mean.)

In Theorems Sz 1 and 2, δ stays fixed as N grows. If instead we could take $\delta \sim 1/\log N$, then the Green–Tao theorem would follow. However, such an improvement of Theorems Sz 1, 2 seems utterly out of reach, and may be false.

There are three very different proofs of Theorems Sz 1, 2; they are due to Szemerédi [21], Furstenberg [10], and Gowers [14]. Without doing justice to the remarkable ideas in these arguments, let me just say that Szemerédi used combinatorics, Furstenberg used ergodic theory, and Gowers used (non-linear) Fourier analysis. It is hard to see anything in common in these three proofs. In a sense, the Green–Tao paper synthesizes them all, by quoting Theorem Sz 2 and using ideas that go back to the proofs of Furstenberg and Gowers. See (B).

Green and Tao prove a powerful extension of Theorem Sz 2, in which the hypothesis $0 \leq f(x) \leq 1$ is replaced by $0 \leq f(x) \leq v(x)$ for a suitable non-negative weight function $v(x)$. The function $v(x)$ is assumed to satisfy three conditions, which we describe crudely here.

- $Av_{x \in \mathbb{Z}_N}\, v(x) = 1$.
- We assume an upper bound on the quantity

$$Av_{\vec{x}} = (x_1, \ldots, x_t) \in (\mathbb{Z}_N)^t \left\{ \prod_{i=1}^{m} v(\lambda_i(\vec{x})) \right\}$$

  for certain affine functions $\lambda_1, \ldots, \lambda_m$: $(\mathbb{Z}_N)^{\pm} \longrightarrow \mathbb{Z}_N$.
- For any $h_1, \ldots, h_m \in \mathbb{Z}_N$, we assume that

$$Av_{x \in \mathbb{Z}_N} \{ v(x+h_1) \ldots v(x+h_m) \} \leq \sum_{i \neq j} \tau_m(h_i - h_j),$$

  for a function $\tau_m : \mathbb{Z}_N \longrightarrow \mathbb{R}$ that satisfies

$$Av_{h \in \mathbb{Z}N} \{ (\tau_m(h)^q \} \geq C(m,q)$$

  for any q.

Such a function $v(x)$ is called a "pseudo-random measure" by Green and Tao. Their extension of Szemerédi's theorem is as follows.

Theorem GTS (Green–Tao–Szemerédi). *Let $k, \delta > 0$, suppose N is large enough and let v be a pseudo-random measure. Let $f : \mathbb{Z}_N \longrightarrow \mathbb{R}$, with $0 \leq f(x) \leq v(x)$, and with $Av_{x \in \mathbb{Z}_N} f(x) \geq \delta$.*

Then $Av_{x,r \in \mathbb{Z}_N} \{ f(x) \cdot f(x+r) \ldots f(x+(k-1)r) \} \geq c(k,\delta) > 0$, where $c(k,\delta)$ depends on k, δ, but not on N or f.

The point is that there are pseudo-random measures $v(x)$ that are large on sparse subsets of $\mathbb{Z}_N$ (e.g., the primes up to N). We will return to this point.

Let me say a few words about the proof of Theorem GTS, and then afterwards describe how it applies to the primes.

It is in the proof of Theorem GTS that Szemerédi's theorem is combined with ideas from Furstenberg and Gowers.

Green and Tao break up the function f into a "uniform" and an "anti-uniform" part, $f = f_U + f_{U^\perp}$.

They expand out $Av_{x,r\in\mathbb{Z}_N}\{f(x)\cdot f(x+r)\dots f(x+(k-1)r)\}$ into a sum of terms

(3) $Av_{x,r\in\mathbb{Z}_N}\{f_0(x)\cdot f_1(x+r)\dots f_{k-1}(x+(k-1)r)\}$, where each f_i is either f_U or $f_{U^\perp}$.

The terms (3) that contain any factor f_U are $o(1)$, thanks to ideas that go back to Gowers' proof.

This leaves us with the term (3) in which each f_i is $f_{U^\perp}$. Let us call this the "critical term".

To control that term, Green and Tao partition $\mathbb{Z}_N$ into subsets $E_1, E_2, \dots, E_A$, and then define a function $\bar{f}_{U^\perp}$ on $\mathbb{Z}_N$ by averaging $f_{U^\perp}$ over each E_α. Green and Tao then prove that

(4) replacing $f_{U^\perp}$ by $\bar{f}_{U^\perp}$ makes a difference $o(1)$ in the critical term,

and moreover,

(5) $0 \le \bar{f}_{U^\perp} \le 1$ and $Av_{x\in\mathbb{Z}_N}\ \bar{f}_{U^\perp}(x) \ge \delta$.

Consequently, the classic Szemerédi theorem (Theorem Sz 2) applies to $\bar{f}_{U^\perp}$, completing the proof of the Green–Tao–Szemerédi theorem.

The proof of (4) and (5) is based on ideas that go back to Furstenberg's proof of Szemerédi's theorem.

Once the Green–Tao–Szemerédi theorem is established, one can take $f(x) = \log x$ for x prime, $f(x) = 0$ otherwise. If we can find a pseudo-random measure v such that

(6) $0 \le f(x) \le v(x)$ for all x,

then Theorem GTS applies, and it yields arbitrarily long arithmetic progressions of primes as in Theorem GT. A first guess for $v(x)$ is the standard Von Mangoldt function $\Lambda(x) = \log p$ for $x = p^k$, p prime; $\Lambda(x) = 0$ otherwise. Λ may indeed be a pseudo-random measure, but that would be very hard to prove. Fortunately, another function v can be seen to be a pseudo-random measure satisfying (6), thanks to important work of Goldston–Yildirim [11], [12], [13], using not-so-hard analytic number theory.

Thus, in the end, a great theorem on the prime numbers is proven without hard analytic number theory. The difficulty lies elsewhere.

I have repeatedly used the phrase "tour-de-force"; I promise that I am not exaggerating.

There are additional first-rate achievements by Tao that I have not mentioned at all. For instance, he has set forth a program [22] for proving the global existence and regularity of wave maps, by using the heat flow for harmonic maps. This has an excellent chance to work, and it may well have important applications in general relativity. I should also mention Tao's joint work with Knutson [19] on the saturation conjecture in representation theory. It is most unusual for an analyst to solve an outstanding problem in algebra.

Tao seems to be getting stronger year by year. It is hard to imagine what can top the work he has already done, but we await Tao's future contributions with eager anticipation.

References

[1] Bourgain, J., Besicovitch-type maximal operators and applications to Fourier analysis. *Geom. Funct. Anal.* **1** (2) (1991), 147–187.

[2] Bourgain, J., On the dimension of Kakeya sets and related maximal inequalities. *Geom. Funct. Anal.* **9** (2) (1999), 256–282.

[3] Bourgain, J., Global well-posedness of defocusing 3D critical NLS in the radial case. *J. Amer. Math. Soc.* **12** (1999), 145–171.

[4] Christ, M., Duoandikoetxea, J., Rubio de Francia, J. L., Maximal operators associated to the Radon transform and the Calderón-Zygmund method of rotations. *Duke Math. J.* **53** (1986), 189–209.

[5] Colliander, J., Keel, M., Staffilani, G., Takaoka, H., Tao, T., Global well-posedness and scattering for the energy-critical nonlinear Schrödinger equation in $\mathbb{R}^3$. *Ann. of Math.*, to appear.

[6] Colliander, J., Keel, M., Staffilani, G., Takaoka, H., Tao, T., Global existence and scattering for rough solutions of a nonlinear Schrodinger equation in $\mathbb{R}^3$. *Comm. Pure Appl. Math.* **57** (8) (2004), 987–1014.

[7] Córdoba, A., The Kakeya maximal function and the spherical summation multipliers. *Amer. J. Math.* **99** (1977), 1–22.

[8] Davies, R., Some remarks on the Kakeya problem. *Proc. Cambridge Philos. Soc.* **69** (1971), 417–421.

[9] Drury, S., L^p estimates for the x-ray transform. *Illinois J. Math.* **27** (1983), 125–129.

[10] Furstenberg, H., Ergodic behavior of diagonal measures and a theorem of Szemerédi on arithmetic progressions. *J. Analyse Math.* **31** (1977), 204–256.

[11] Goldston, D., and Yildirim, C. Y., Higher correlations of divisor sums related to primes, I: Triple correlations. *Integers* **3** (2003), A5 (electronic).

[12] Goldston, D., and Yildirim, C. Y., Higher correlations of divisor sums related to primes, III: k-correlations. Preprint, September 2002; arXiv: math.NT/0209102.

[13] Goldston, D., and Yildirim, C. Y., Small gaps between primes, I. Preprint, April 2005; arXiv:math.NT/0504336.

[14] Gowers, T., A new proof of Szemerédi's theorem. *Geom. Funct. Anal* **11** (2001), 465–588.

[15] Green, B., Roth's theorem in the primes. *Ann. of Math.* **161** (2005), 1609–1636.

[16] Green, B., and Tao, T., The primes contain arbitrarily long arithmetic progressions. *Ann. of Math.*, to appear.

[17] Katz, N. H., Łaba, I., Tao, T., An improved bound on the Minkowski dimension of Besicovitch sets in $\mathbb{R}^3$. *Ann. of Math.* **152** (2000), 383–446.

[18] Katz, N. H., and Tao, T., Bounds on arithmetic projections, and applications to the Kakeya conjecture. *Math. Res. Lett.* **6** (1999), 625–630.
[19] Knutson, A., and Tao, T., The honeycomb model of GL_n ($\mathbb{C}$) tensor products I: Proof of the saturation conjecture. *J. Amer. Math. Soc.* **12** (4) (1999), 1055–1090.
[20] Morawetz, C., Time decay for the nonliner Klein-Gordon equation. *Proc. Roy. Soc. Ser. A* **306** (1968), 291–296.
[21] Szemerédi, E., On sets of integers containing no k elements in arithmetic progression. *Acta Arith.* **27** (1975), 299–345.
[22] Tao, T., Geometric renormalization of large energy wave maps. Preprint, November 2004; arXiv:math.AP/0411354.
[23] Varnavides, P., On certain sets of positive density. *J. London Math. Soc.* **34** (1959), 358–360.
[24] Wolff, T., Recent work connected with the Kakeya problem. In *Prospects in mathematics* (Princeton, NJ, 1996), Amer. Math. Soc., Providence, RI, 1999, 129–162.

Department of Mathematics, Fine Hall, Washington Road, Princeton, NJ 08544-1000, U.S.A.
E-mail: cf@math.princeton.edu

Terence Tao

TERENCE TAO

B.Sc. Flinders University, December 1991

Ph.D. Princeton University, June 1996

Positions until today

1996–present	University of California, Los Angeles (first as assistant professor, now full professor)
1999 and 2000	University of New South Wales
2001–2003	Clay Mathematical Institute
2001–2003	Australian National University

Proceedings of the International Congress
of Mathematicians, Madrid, Spain, 2006

THE DICHOTOMY BETWEEN STRUCTURE AND RANDOMNESS, ARITHMETIC PROGRESSIONS, AND THE PRIMES

by
TERENCE TAO*

ABSTRACT. A famous theorem of Szemerédi asserts that all subsets of the integers with positive upper density will contain arbitrarily long arithmetic progressions. There are many different proofs of this deep theorem, but they are all based on a fundamental dichotomy between structure and randomness, which in turn leads (roughly speaking) to a decomposition of any object into a structured (low-complexity) component and a random (discorrelated) component. Important examples of these types of decompositions include the Furstenberg structure theorem and the Szemerédi regularity lemma. One recent application of this dichotomy is the result of Green and Tao establishing that the prime numbers contain arbitrarily long arithmetic progressions (despite having density zero in the integers). The power of this dichotomy is evidenced by the fact that the Green–Tao theorem requires surprisingly little technology from analytic number theory, relying instead almost exclusively on manifestations of this dichotomy such as Szemerédi's theorem. In this paper we survey various manifestations of this dichotomy in combinatorics, harmonic analysis, ergodic theory, and number theory. As we hope to emphasize here, the underlying themes in these arguments are remarkably similar even though the contexts are radically different.

Mathematics Subject Classification (2000). Primary 11P32, 37A45, 05C65, 05C75, 42A99.

Keywords. Szemerédi's theorem, ergodic theory, graph theory, hypergraph theory, arithmetic combinatorics, arithmetic progressions, prime numbers

1. Introduction

In 1975, Szemerédi [53] proved the following deep and enormously influential theorem:

Theorem 1.1 (Szemerédi's theorem). *Let A be a subset of the integers $\mathbb{Z}$ of positive upper density, thus $\limsup_{N\to\infty} \frac{|A\cap[-N,N]|}{|[-N,N]|} > 0$. Here $|A|$ denotes the cardinality of*

*The author is supported by a grant from the Packard foundation.

a set A, and $[-N, N]$ denotes the integers between $-N$ and N. Then for any $k \geq 3$, A contains infinitely many arithmetic progressions of length k.

Several proofs of this theorem are now known. The original proof of Szemerédi [53] was combinatorial. A later proof of Furstenberg [11], [13] used ergodic theory and has led to many extensions. A more quantitative proof of Gowers [19], [20] was based on Fourier analysis and arithmetic combinatorics (extending a much older argument of Roth [50] handling the $k = 3$ case). A fourth proof by Gowers [21] and Rödl, Nagle, Schacht, and Skokan [46], [47], [48], [49] relied on the structural theory of hypergraphs. These proofs are superficially all very different (with each having their own strengths and weaknesses), but have a surprising number of features in common. The main difficulty in all of the proofs is that one *a priori* has no control on the behaviour of the set A other than a lower bound on its density; A could range from being a very random set, to a very structured set, to something in between. In each of these cases, A will contain many arithmetic progressions — but the *reason* for having these progressions varies from case to case. Let us illustrate this by informally discussing some representative examples:

- (Random sets) Let $0 < \delta < 1$, and let A be a random subset of $\mathbb{Z}$, which each integer n lying in A with an independent probability of δ. Then A almost surely has upper density δ, and it is easy to establish that A almost surely has infinitely many arithmetic progressions of length k, basically because each progression of length k in $\mathbb{Z}$ has a probability of δ^k of also lying in A. A more refined version of this argument also applies when A is *pseudorandom* rather than random — thus we allow A to be deterministic, but require that a suitable number of correlations (e.g. pair correlations, or higher order correlations) of A are negligible. The argument also extends to sparse random sets, for instance one where $\mathbf{P}(n \in A) \sim 1/\log n$.
- (Linearly structured sets) Consider a quasiperiodic set such as $A := \{n : \{\alpha n\} \leq \delta\}$, where $0 < \delta < 1$ is fixed, α is a real number (e.g. $\alpha = \sqrt{2}$) and $\{x\}$ denotes the fractional part of x. Such sets are "almost periodic" because there is a strong correlation between the events $n \in A$ and $n + L \in A$, thanks to the identity $\{\alpha(n+L)\} - \{\alpha n\} = \{\alpha L\} \mod 1$. An easy application of the Dirichlet approximation theorem (to locate an approximate period L with $\{\alpha L\}$ small) shows that such sets still have infinitely many progressions of any given length k. Note that this argument works regardless of whether α is rational or irrational.
- (Quadratically structured sets) Consider a "quadratically quasiperiodic" set of the form $A := \{n : \{\alpha n^2\} \leq \delta\}$. If α is irrational, then this set has upper density δ, thanks to Weyl's theorem on equidistribution of polynomials. (If α is rational, one can still obtain some lower bound on the upper density.) It is not linearly structured (there is no asymptotic correlation between the events $n \in A$ and $n + L \in A$ as $n \to \infty$ for any fixed non-zero L), however it has quadratic structure in the sense that there is a strong correlation between the events $n \in A$,

$n + L \in A$, $n + 2L \in A$, thanks to the identity

$$\{\alpha n^2\} - 2\{\alpha(n+L)^2\} + \{\alpha(n+2L)^2\} = 2\{\alpha L^2\} \mod 1.$$

In particular A does not behave like a random set. Nevertheless, the quadratic structure still ensures that A contains infinitely many arithmetic progressions of any length k, as one first locates a "quadratic period" L with $\{\alpha L^2\}$ small, and then for suitable $n \in A$ one locates a much smaller "linear period" M with $\{\alpha LMn\}$ small. If this is done correctly, the progression $n, n + LM, \ldots$, $n + (k-1)LM$ will be completely contained in A. The same arguments also extend to a more general class of quadratically structured sets, such as the "2-step nilperiodic" set $A = \{n : \{\lfloor\sqrt{2}n\rfloor\sqrt{3}n \leq \delta\}$, where $\lfloor x \rfloor$ is the greatest integer function.

- (Random subsets of structured sets) Continuing the previous example $A := \{n : \{\alpha n^2\} \leq \delta\}$, let A' be a random subset of A with each $n \in A$ lying in A' with an independent probability of δ' for some $0 < \delta' < 1$. Then this set A' almost surely has a positive density of $\delta\delta'$ if α is irrational. The set A' almost surely has infinitely many progressions of length k, since A already starts with infinitely many such progressions, and each such progression as a probability of $(\delta')^k$ of also lying in A'. One can generalize this example to random sets $\tilde{A}$ where the events $n \in \tilde{A}$ are independent as n varies, and the probability $\mathbf{P}(n \in \tilde{A})$ is a "quadratically almost periodic" function of n such as $\mathbf{P}(n \in \tilde{A}) = F(\{\alpha n^2\})$ for some nice (e.g. piecewise continuous) function F taking values between 0 and 1; the preceding example is the case where $F(x) := \delta' 1_{x<\delta}$. It is also possible to adapt this argument to (possibly sparse) pseudorandom subsets of structured sets, though one needs to take some care in defining exactly what "pseudorandom" means here.
- (Sets containing random subsets of structured sets) Let A'' be any set which contains the set A' (or $\tilde{A}$) of the previous example. Since A' contains infinitely many progressions of length k, it is trivial that A'' does also.

As the above examples should make clear, the reason for the truth of Szemerédi's theorem is very different in the cases when A is random, and when A is structured. These two cases can then be combined to handle the case when A is (or contains) a large (pseudo-)random subset of a structured set. Each of the proofs of Szemerédi's theorem now hinge on a *structure theorem* which, very roughly speaking, asserts that *every* set of positive density is (or contains) a large pseudorandom subset of a structured set; each of the four proofs obtains a structure theorem of this sort in a different way (and in a very different language). These remarkable structural results — which include the Furstenberg structure theorem and the Szemerédi regularity lemma as examples — are of independent interest (beyond their immediate applications to arithmetic progressions), and have led to many further developments and insights. For instance, in [27] a "weighted" structure theorem (which was in some sense a hybrid of the Furstenberg structure theorem and the Szemerédi regularity lemma) was the primary new ingredient in proving that the primes $P := \{2, 3, 5, 7, \ldots\}$ contained arbitrarily long arithmetic progressions. While that latter claim is ostensibly a number-theoretical result, the method of proof in fact uses surprisingly little from number theory, being much closer in spirit to the proofs of Szemerédi's theorem (and in fact Szemerédi's theorem is a crucial ingredient in the proof). This can be seen from the fact that the argument in [27] in fact proves the following stronger result:

Theorem 1.2 (Szemerédi's theorem in the primes [27]). *Let A be a subset of the primes P of positive* relative *upper density, thus* $\limsup_{N\to\infty} \frac{|A\cap[-N,N]|}{|P\cap[-N,N]|} > 0$. *Then for any $k \geq 3$, A contains infinitely many arithmetic progressions of length k.*

This result was first established in the $k = 3$ case by Green [22], the key step again being a (Fourier-analytic) structure theorem, this time for subsets of the primes. The arguments used to prove this theorem do not directly address the important question of whether the primes P (or any subset thereof) have any pseudorandomness properties (but see Section 5 below). However, the structure theorem does allow one to (essentially) describe any dense subset of the primes as a (sparse) pseudorandom subset of some unspecified dense set, which turns out to be sufficient (thanks to Szemerédi's theorem) for the purpose of establishing the existence of arithmetic progressions.

There are now several expositions of Theorem 1.2; see for instance [42], [25], [55], [56], [37]. Rather than give another exposition of this result, we have chosen to take a broader view, surveying the collection of structural theorems which underlie the proof of such results as Theorem 1.1 and Theorem 1.2. These theorems have remarkably varied contexts — measure theory, ergodic theory, graph theory, hypergraph theory, probability theory, information theory, and Fourier analysis — and can be either qualitative (infinitary) or quantitative (finitary) in nature. However, their *proofs* tend to share a number of common features, and thus serve as a kind of "Rosetta stone" connecting these various fields. Firstly, for a given class of objects, one quantifies what it means for an object to be "(pseudo-)random" and an object to be "structured". Then, one establishes a *dichotomy between randomness and structure*, which typically looks something like this:

> *If an object is not (pseudo-)random, then it (or some non-trivial component of it) correlates with a structured object.*

One can then iterate this dichotomy repeatedly (e.g. via a stopping time argument, or by Zorn's lemma), to extract out all the correlations with structured objects, to obtain a *weak structure theorem* which typically looks as follows:

> *If A is an arbitrary object, then A (or some non-trivial component of A) splits as the sum of a structured object, plus a pseudorandom error.*

In many circumstances, we need to improve this result to a *strong structure theorem*:

*If A is an arbitrary object, then A (or some non-trivial component of A) splits as the sum of a structured object, plus a small error, plus a **very** pseudorandom error.*

When one is working in an infinitary (qualitative) setting rather than a finitary (quantitative) one — which is for instance the case in the ergodic theory approach — one works instead with an *asymptotic structure theorem*:

*If A is an arbitrary object, then A (or some non-trivial component of A) splits as the sum of a "compact" object (the limit of structured objects), plus an **infinitely** pseudorandom error.*

The reason for the terminology "compact" to describe the limit of structured objects is in analogy to how a compact operator can be viewed as the limit of finite rank operators; see [12] for further discussion.

In many applications, the small or pseudorandom errors in these structure theorems are negligible, and one then reduces to the study of structured objects. One then exploits the structure of these objects to conclude the desired application.

Our focus here is on the structure theorems related to Szemerédi's theorem and related results such as Theorem 1.2; we will not have space to describe all the generalizations and refinements of these results here. However, these types of structural theorems appear in other contexts also, for instance the Komlós subsequence principle [40] in probability theory. The Lebesgue decomposition of a spectral measure into pure point, singular continuous, and absolutely continuous spectral components can also be viewed as a structure theorem of the above type. Also, the stopping time arguments which underlie the structural theorems here are also widely used in harmonic analysis, in particular obtaining fundamental decompositions such as the Calderón–Zygmund decomposition or the atomic decomposition of Hardy spaces (see e.g. [52]), as well as the tree selection arguments used in multilinear harmonic analysis (see e.g. [43]). It may be worth investigating whether there are any concrete connections between these disparate structural theorems.

2. Ergodic Theory

We now illustrate the above general strategy in a number of contexts, beginning with the ergodic theory approach to Szemerédi's theorem, where the dichotomy between structure and randomness is particularly clean and explicit Informally speaking, the ergodic theory approach seeks to understand the set A of integers by analyzing the asymptotic correlations of the shifts $A+n := \{a+n : a \in A\}$ (or of various asymptotic averages of these shifts), and treating these shifts as occurring on an abstract measure space. More formally, let X be a measure space with probability measure $d\mu$, and let $T : X \to X$ be a bijection such that T and T^{-1} are both measure-preserving maps. The associated shift operator $T : f \mapsto f \circ T^{-1}$ is thus a unitary operator on the Hilbert space $L^2(X)$ of complex-valued square-integrable functions with the usual inner product $\langle f, g\rangle := \int_X f\overline{g}\, d\mu$. A famous transference result known as the *Furstenberg correspondence principle*[1] (see [11], [13], [12]) shows that Szemerédi's theorem is then equivalent to

Theorem 2.1 (Furstenberg recurrence theorem [11]). *Let X and T be as above, and let $f \in L^\infty(X)$ be any bounded non-negative function with $\int_X f\, d\mu > 0$. Then for any $k \geq 1$ we have*

$$\liminf_{N\to\infty} \mathbf{E}_{1\leq n\leq N} \int_X f T^n f \dots T^{(k-1)n} f\, d\mu > 0.$$

Here and in the sequel we use $\mathbf{E}_{n\in I} a_n$ as a shorthand for the average $\frac{1}{|I|}\sum_{n\in I} a_n$.

When $k = 2$ this is essentially the Poincaré recurrence theorem; by using the von Neumann ergodic theorem one can also show that the limit exists (thus the lim inf can be replaced with a lim). The $k = 3$ case can be proved by the following argument, as observed in [12]. We need to show that

$$\liminf_{N\to\infty} \mathbf{E}_{1\leq n\leq N} \int_X f T^n f T^{2n} f\, d\mu > 0 \tag{1}$$

whenever f is bounded, non-negative, and has positive integral.

The first key observation is that any sufficiently pseudorandom component of f will give a negligible contribution to (1) and can be dropped. More precisely, let us call f is *linearly pseudorandom* (or *weakly mixing*) with respect to the shift T if we have

$$\lim_{N\to\infty} \mathbf{E}_{1\leq n\leq N} |\langle T^n f, f\rangle|^2 = 0. \tag{2}$$

Such functions are negligible for the purpose of computing averages such as those in (1); indeed, if at least one of $f, g, h \in L^\infty(X)$ is linearly pseudorandom, then an easy application of van der Corput's lemma (which in turn is an application of Cauchy–Schwarz) shows that

$$\lim_{N\to\infty} \mathbf{E}_{1\leq n\leq N} \int_X f T^n g T^{2n} h\, d\mu = 0.$$

We shall refer to these types of results — that pseudorandom functions are negligible when averaged against other functions — as *generalized von Neumann theorems*.

In view of this generalized von Neumann theorem, one is now tempted to "quotient out" all the pseudorandom functions and work with a reduced class of

[1]Morally speaking, to deduce Szemerédi's theorem from Furstenberg's theorem, one takes X to be the integers $\mathbb{Z}$, T to be the standard shift $n \mapsto n+1$, and μ to be the density $\mu(A) = \lim_{N\to\infty} \frac{|A\cap[-N,N]|}{|[-N,N]|}$. This does not quite work because not all sets A have a well-defined density, however additional arguments (e.g. using the Hahn-Banach theorem) can fix this problem.

"structured" functions. In this particular case, it turns out that the correct notion of structure is that of a *linearly almost periodic function*, which are in turn generated by the *linear eigenfunctions* of T. To make this more precise, we need the following dichotomy:

Lemma 2.2 (Dichotomy between randomness and structure). *Suppose that $f \in L^\infty(X)$ is not linearly pseudorandom. Then there exists an linear eigenfunction $g \in L^\infty(X)$ of T (thus $Tg = \lambda g$ for some $\lambda \in \mathbb{C}$) such that $\langle f, g \rangle \neq 0$.*

Remark 2.3 Observe that if g is a linear eigenfunction of T with $Tg = \lambda g$, then $|\lambda| = 1$ and $\lim_{N\to\infty} \mathbf{E}_{1\leq n\leq N} \int_X gT^n\overline{g}^2T^{2n}g\ d\mu = \int_X |g|^4$. Thus linear eigenfunctions can and do give nontrivial contributions to the expression in (1). One can view Lemma 2.2 as a converse to this observation.

The proof of this lemma follows easily from spectral theory and is omitted here. It has the following consequence. Let $\mathcal{Z}_1$ be the σ-algebra generated by all the eigenfunctions of T, this is known as the *Kronecker factor* of X, and roughly speaking encapsulates all the "linear structure" in the measure preserving system. Given every function $f \in L^2(X)$, we have the decomposition $f = f_{U^\perp} + f_U$, where $f_{U^\perp} := \mathbf{E}(f|\mathcal{Z}_1)$ is the conditional expectation of f with respect to the σ-algebra $\mathcal{Z}_1$ (i.e. the orthogonal projection from $L^2(X)$ to the $\mathcal{Z}_1$-measurable functions). By construction, $f_U := f - \mathbf{E}(f|\mathcal{Z}_1)$ is orthogonal to every eigenfunction of T, and is hence linearly pseudorandom by Lemma 2.2. In particular, we have established

Proposition 2.4 (Asymptotic structure theorem). *Let f be bounded and non-negative, with positive integral. Then we can split*[2] *$f = f_{U^\perp} + f_U$, where $f_{U^\perp}$ is bounded, non-negative, and $\mathcal{Z}_1$-measurable (and thus approximable in L^2 to arbitrary accuracy by finite linear combinations of linear eigenfunctions), with positive integral, and f_U is linearly pseudorandom.*

This result is closely related to the Koopman–von Neumann theorem in ergodic theory. In the language of the introduction, it asserts (very roughly speaking) that any set A of integers can be viewed as a (linearly) pseudorandom set where the "probability" $f_{U^\perp}(n)$ that a given element n lies in A is a (linearly) almost periodic function of n.

Note that the linearly pseudorandom component f_U of f gives no contribution to (1), thanks to the generalized von Neumann theorem. Thus we may freely replace f by $f_{U^\perp}$ if desired; in other words, for the purposes of proving (1) we may assume without loss of generality that f is measurable with respect to the Kronecker

[2]The notation is from [27]; the subscript U stands for "Gowers uniform" (pseudorandom), and $U^\perp$ for "Gowers anti-uniform" (structured).

factor $\mathcal{Z}_1$. In the notation of [14], we have just shown that the Kronecker factor is a *characteristic factor* for the recurrence in (1). (In fact it is essentially the universal factor for this recurrence, see [64], [39] for further discussion.)

We have reduced the proof of (1) to the case when f is structured, in the sense of being measurable in $\mathcal{Z}_2$. There are two ways to obtain the desired "structured recurrence" result. Firstly there is a "soft" approach, in which one observes that every $\mathcal{Z}_1$-measurable square-integrable function f is *almost periodic*, in the sense that for any $\varepsilon > 0$ there exists a set of integers n of positive density such that $T^n f$ is within ε of f in $L^2(X)$; from this it is easy to show that $\int_X f T^n f T^{2n} f \, d\mu$ is close to $\int_X f^3$ for a set of integers n of positive density, which implies (1). This almost periodicity can be verified by first checking it for polynomial combinations of linear eigenfunctions, and then extending by density arguments. There is also a "hard" approach, in which one obtains algebraic and topological control on the Kronecker factor $\mathcal{Z}_1$. In fact, from a spectral analysis of T one can show that $\mathcal{Z}_1$ is the inverse limit of a sequence of σ-algebras, on each of which the shift T is isomorphic to a shift $x \mapsto x + \alpha$ on a compact abelian Lie group G. This gives a very concrete description of the functions f which are measurable in the Kronecker factor, and one can establish (1) by a direct argument similar to that used in in the introduction for linearly structured sets. This "hard" approach gives a bit more information; for instance, it can be used to show that the limit in (1) actually converges, so one can replace the lim inf by a lim.

It turns out that these arguments extend (with some non-trivial effort) to the case of higher k. For sake of exposition let us just discuss the $k = 4$ case, though most of the assertions here extend to higher k. We wish to prove that

$$\liminf_{N\to\infty} \mathbf{E}_{1\leq n\leq N} \int_X f T^n f T^{2n} f T^{3n} f \, d\mu > 0 \tag{3}$$

whenever f is bounded, non-negative, and has positive integral. Here, it turns out that we must strengthen the notion of pseudorandomness (and hence generalize the notion of structure); linear pseudorandomness is no longer sufficient to imply negligibility. For instance, let f be a *quadratic eigenfunction*, in the sense that $Tf = \lambda f$, where λ is no longer constant but is itself a linear eigenfunction, thus $T\lambda = c\lambda$ for some constant c. As an example, if $X = (\mathbb{R}/\mathbb{Z})^2$ with the skew shift $T(x, y) = (x + \alpha,\ y + x)$ for some fixed number α, then the function $f(x, y) = e^{2\pi i y}$ is a quadratic eigenfunction but not a linear one. Typically such quadratic eigenfunctions will be linearly pseudorandom, but if $|\lambda| = |c| = 1$ (which is often the case) then we have the identity

$$\mathbf{E}_{1\leq n\leq N} \int_X f T^n \overline{f}^3 T^{2n} f^3 T^{3n} \overline{f} \, d\mu = \int_X |f|^8 \, d\mu \tag{4}$$

and so we see that these functions can give non-trivial contributions to expressions such as (1). The correct notion of pseudorandomness is now *quadratic pseudorandomness*, by which we mean that

$$\lim_{H\to\infty}\lim_{N\to\infty}\mathbf{E}_{1\leq n\leq N}\mathbf{E}_{1\leq h\leq H}|\langle T^h f\overline{f}, T^n(T^h f\overline{f})\rangle|^2 = 0.$$

In other words, f is quadratically pseudorandom if and only if $T^h f\overline{f}$ is asymptotically linearly pseudorandom on the average as $h \to \infty$. Several applications of van der Corput's lemma give a generalized von Neumann theorem, asserting that

$$\lim_{N\to\infty}\mathbf{E}_{1\leq n\leq N}\int_X f_0 T^n f_1 T^{2n} f_2 T^{3n} f_3\, d\mu = 0$$

whenever f_0, f_1, f_2, f_3 are bounded functions with at least one function quadratically pseudorandom.

One would now like to construct a factor $\mathcal{Z}_2$ (presumably larger than the Kronecker factor $\mathcal{Z}_1$) which will play the role of the Kronecker factor for the average (3); in particular, we would like a statement of the form

Lemma 2.5 (Dichotomy between randomness and structure). *Suppose that $f \in L^\infty(X)$ is not linearly pseudorandom. Then there exists a $\mathcal{Z}_2$-measurable function $g \in L^\infty(X)$ such that $\langle f, g\rangle \neq 0$.*

which would imply[3]

Proposition 2.6 (Asymptotic structure theorem). *Let f be bounded and non-negative, with positive integral. Then we can split $f = f_{U^\perp} + f_U$, where $f_{U^\perp}$ is bounded, non-negative, and $\mathcal{Z}_2$-measurable, with positive integral, and f_U is quadratically pseudorandom.*

This reduces the proof of (3) to that of $\mathcal{Z}_2$-measurable f. The existence of such a factor $\mathcal{Z}_2$ (which would be a *characteristic factor* for this average) is trivial to construct, as we could just take $\mathcal{Z}_2$ to be the entire σ-algebra, and it is in fact easy (via Zorn's lemma) to show the existence of a "best" such factor, which embed into all other characteristic factors for this average (see [64]). Of course, for the concept of characteristic factor to be useful we would like $\mathcal{Z}_2$ to be as small as possible, and furthermore to have some concrete structural description of the factor. An obvious guess for $\mathcal{Z}_2$ would be the σ-algebra generated by all the linear and quadratic eigenfunctions, but this factor turns out to be a bit too small (see [14];

[3]One can generalize this structure theorem to obtain similar characteristic factors $\mathcal{Z}_3$, $\mathcal{Z}_4$ for cubic pseudorandomness, quartic pseudorandomness, etc. Applying Zorn's lemma, one eventually obtains the *Furstenberg structure theorem*, which decomposes any measure preserving system as a weakly mixing extension of a distal system, and thus decomposes any function as a distal function plus an "infinitely pseudorandom" error; see [13]. However this decomposition is not the most "efficient" way to prove Szemerédi's theorem, as the notion of pseudorandomness is too strong, and hence the notion of structure too general. It does illustrate however that one does have considerable flexibility in where to draw the line between randomness and structure.

this is related to the example of the 2-step nilperiodic set in the introduction). A more effective candidate for $\mathcal{Z}_2$, analogous to the "soft" description of the Kronecker factor, is the space of all "quadratically almost periodic functions". This concept is a bit tricky to define rigorously (see e.g. [13], [12], [54]), but roughly speaking, a function f is linearly almost periodic if the orbit $\{T^n f : n \in \mathbb{Z}\}$ is precompact in $L^2(X)$ viewed as a Hilbert space, while a function f is quadratically almost periodic if the orbit is precompact in $L^2(X)$ viewed as a Hilbert *module* over the Kronecker factor $L^\infty(\mathcal{Z}_1)$; this can be viewed as a matrix-valued (or more precisely compact operator-valued) extension of the concept of a quadratic eigenfunction. Another rough definition is as follows: a function f is linearly almost periodic if $T^n f(x)$ is close to $f(x)$ for many constants n, whereas a function f is quadratically almost periodic if $T^{n(x)} f(x)$ is close to $f(x)$ for a function $n(x)$ which is itself linearly almost periodic. It turns out that with this "soft" proposal for $\mathcal{Z}_2$, it is easy to prove Lemma 2.5 and hence Proposition 2.6, essentially by obtaining a "relative" version of the proof of Lemma 2.2. The derivation of (3) in this soft factor is slightly tricky though, requiring either van der Waerden's theorem, or the color focusing argument used to prove van der Waerden's theorem; see [11], [13], [12], [54].

More recently, a more efficient "hard" factor $\mathcal{Z}_2$ was constructed by Conze–Lesigne [7], Furstenberg–Weiss [14], and Host–Kra [38]; the analogous factors for higher k are more difficult to construct, but this was achieved by Host–Kra in [39], and also subsequently by Ziegler [64]. This factor yields more precise information, including convergence of the limit in (3). Here, the concept of a 2-*step nilsystem* is used to define structure. A 2-step nilsystem is a compact symmetric space G/Γ, with G a 2-step nilpotent Lie group and Γ is a closed subgroup, together with a shift element $\alpha \in G$, which generates a shift $T(x\Gamma) := \alpha x\Gamma$. The factor $\mathcal{Z}_2$ constructed in these papers is then the inverse limit of a sequence of σ-algebras, on which the shift is equivalent to a 2-step nilsystem. This should be compared with the "hard" description of the Kronecker factor, which is the 1-step analogue of the above result. Establishing the bound (3) then reduces to the problem of understanding the structure of arithmetic progressions $x\Gamma$, $\alpha x\Gamma$, $\alpha^2 x\Gamma$, $\alpha^3 x\Gamma$ on the nilsystem, which can be handled by algebraic arguments, for instance using the machinery of Hall–Petresco sequences [44].

The ergodic methods, while non-elementary and non-quantitative (though see [54]), have proven to be the most powerful and flexible approach to Szemerédi's theorem, leading to many generalizations and refinements. However, it seems that a purely "soft" ergodic approach is not quite capable by itself of extending to the primes as in Theorem 1.2, though it comes tantalizingly close. In particular, one can use Theorem 2.1 and a variant of the Furstenberg correspondence principle to establish Theorem 1.2 when the set of primes P is replaced by a random subset $\tilde{P}$ of the positive integers, with $n \in \tilde{P}$ with independent probability $1/\log n$ for $n > 1$; see [60]. Roughly speaking, if A is a subset of $\tilde{P}$, the idea is to construct an abstract measure-preserving system generated by a set $\tilde{A}$, in which $\mu(T^{n_1}\tilde{A} \cap \ldots \cap T^{n_k}\tilde{A})$ is the normalized density of $(A+n_1)\cap\ldots\cap(A+n_k)$ for any $n_1,\ldots,n_k$. Unfortunately, this approach requires the ambient space $\tilde{P}$ to be extremely pseudorandom and does not seem to extend easily to the primes.

3. Fourier Analysis

We now turn to a more quantitative approach to Szemerédi's theorem, based primarily on Fourier analysis and arithmetic combinatorics. Here, one analyzes a set of integers A finitarily, truncating to a finite setting such as the discrete integral $\{1,\ldots,N\}$ or the cyclic group $\mathbb{Z}/N\mathbb{Z}$, and then testing the correlations of A with linear phases such as $n\mapsto e^{2\pi ikn/N}$, quadratic phases $n\mapsto e^{2\pi ikn^2/N}$, or similar objects. This approach has lead to the best known bounds on Szemerédi's theorem, though it has not yet been able to handle many of the generalizations of this theorem that can be treated by ergodic or graph-theoretic methods. In analogy with the ergodic arguments, the $k=3$ case of Szemerédi's theorem can be handled by linear Fourier analysis (as was done by Roth [50]), while the $k=4$ case requires quadratic Fourier analysis (as was done by Gowers [19]), and so forth for higher order k (see [20]). The Fourier analytic approach seems to be closely related to the theory of the "hard" characteristic factors discovered in the ergodic theory arguments, although the precise nature of this relationship is still being understood.

It is convenient to work in a cyclic group $\mathbb{Z}/N\mathbb{Z}$ of prime order. It can be shown via averaging arguments (see [63]) that Szemerédi's theorem is equivalent to the following quantitative version:

Theorem 3.1 (Szemerédi's theorem, quantitative version). *Let $N>1$ be a large prime, let $k\geq 3$, and let $0<\delta<1$. Let $f:\mathbb{Z}/N\mathbb{Z}\to\mathbb{R}$ be a function with $0\leq f(x)\leq 1$ for all $x\in\mathbb{Z}/N\mathbb{Z}$ and $\mathbf{E}_{x\in\mathbb{Z}/N\mathbb{Z}}f(x)\geq\delta$. Then we have*

$$\mathbf{E}_{x,r\in\mathbb{Z}/N\mathbb{Z}}f(x)T^r f(x)\ldots T^{(k-1)r}f(x)\geq c(k,\delta)$$

for some $c(k,\delta)>0$ depending only on k and δ, where $T^r f(x):=f(x+r)$ is the shift operator on $\mathbb{Z}/N\mathbb{Z}$.

We remark that the Fourier-analytic arguments in Gowers [20] give the best known lower bounds on $c(k,\delta)$, namely $c(k,\delta)>2^{-2^{1/\delta^{c_k}}}$ where $c_k:=2^{2^{k+9}}$. In the $k=3$ case it is known that $c(3,\delta)\geq\delta^{C/\delta^2}$ for some absolute constant C, see [5]. A conjecture of Erdős and Turán [8] is roughly equivalent to asserting that $c(k,\delta)>e^{-C_k/\delta}$ for some C_k. In the converse direction, an example of Behrend shows that $c(3,\delta)$ cannot exceed $e^{c\log^2(1/\delta)}$ for some small absolute constant c, with similar results for higher values of k; in particular, $c(k,\delta)$ cannot be as large as any fixed power of δ. This already rules out a number of elementary approaches to Szemerédi's theorem and suggests that any proof must involve some sort of iterative argument.

Let us first describe (in more "modern" language) Roth's original proof [50] of Szemerédi's theorem in the $k = 3$ case. We need to establish a bound of the form

$$\mathbf{E}_{x,r\in\mathbb{Z}/N\mathbb{Z}} f(x)T^r f(x)T^{2r} f(x) \geq c(3,\delta) > 0 \tag{5}$$

when f takes values between 0 and 1 and has mean at least δ. As in the ergodic argument, we first look for a notion of pseudorandomness which will ensure that the average in (5) is negligible. It is convenient to introduce the *Gowers* $U^2(\mathbb{Z}/N\mathbb{Z})$ *uniformity norm* by the formula

$$\|f\|_{U^2(\mathbb{Z}/N\mathbb{Z})}^4 := \mathbf{E}_{n\in\mathbb{Z}/N\mathbb{Z}}|\mathbf{E}_{x\in\mathbb{Z}/N\mathbb{Z}} T^n f(x)\overline{f(x)}|^2,$$

and informally refer to f as *linearly pseudorandom* (or *linearly Gowers-uniform*) if its U^2 norm is small; compare this with (2). The U^2 norm is indeed a norm; this can be verified either by several applications of the Cauchy–Schwarz inequality, or via the Fourier identity

$$\|f\|_{U^2(\mathbb{Z}/N\mathbb{Z})}^4 = \sum_{\xi\in\mathbb{Z}/N\mathbb{Z}} |\hat{f}(\xi)|^4, \tag{6}$$

where $\hat{f}(\xi) := \mathbf{E}_{x\in\mathbb{Z}/N\mathbb{Z}} f(x)e^{-2\pi i x\xi/N}$ is the usual Fourier transform. Some further applications of Cauchy–Schwarz (or Plancherel's theorem and Hölder's inequality) yields the generalized von Neumann theorem

$$|\mathbf{E}_{x,r\in\mathbb{Z}/N\mathbb{Z}} f_0(x)T^r f_1(x)T^{2r} f_2(x)| \leq \min_{j=0,1,2} \|f_j\|_{U^2(\mathbb{Z}/N\mathbb{Z})} \tag{7}$$

whenever f_0, f_1, f_2 are bounded in magnitude by 1. Thus, as before, linearly pseudorandom functions give a small contribution to the average in (5), though now that we are in a finitary setting the contribution does not vanish completely.

The next step is to establish a dichotomy between linear pseudorandomness and some sort of usable structure. From (6) and Plancherel's theorem we easily obtain the following analogue of Lemma 2.2:

Lemma 3.2 (Dichotomy between randomness and structure). *Suppose that* $f : \mathbb{Z}/N\mathbb{Z} \to \mathbb{C}$ *is bounded in magnitude by* 1 *with* $\|f\|_{U^2(\mathbb{Z}/N\mathbb{Z})} \geq \eta$ *for some* $0 < \eta < 1$. *Then there exists a linear phase function* $\phi : \mathbb{Z}/N\mathbb{Z} \to \mathbb{R}/\mathbb{Z}$ *(thus* $\phi(x) = \xi x/N + c$ *for some* $\xi \in \mathbb{Z}/N\mathbb{Z}$ *and* $c \in \mathbb{R}/\mathbb{Z}$*) such that* $|\mathbf{E}_{x\in\mathbb{Z}/N\mathbb{Z}} f(x)e^{-2\pi i\phi(x)}| \geq \eta^2$.

The next step is to iterate this lemma to obtain a suitable structure theorem. There are two slightly different ways to do this. Firstly there is the original *density increment argument* approach of Roth [50], which we sketch as follows. It is convenient to work on a discrete interval $[1, N/3]$, which we identify with a subset of $\mathbb{Z}/N\mathbb{Z}$ in the obvious manner. Let $f : [1, N/3] \to \mathbb{R}$ be a non-negative function bounded in magnitude by 1, and let η be a parameter to be chosen later. If $f - \mathbf{E}_{1\leq x\leq N/3} f(x)$ is not linearly pseudorandom, in the sense that $\|f - \mathbf{E}_{1\leq x\leq N/3} f(x)\|_{U^2(\mathbb{Z}/N\mathbb{Z})} \geq \eta$, then we apply Lemma 3.2 to obtain a correlation with a linear phase ϕ. An easy application of the Dirichlet approximation theorem then shows that one can partition $[1, N/3]$ into arithmetic progressions (of length roughly $\eta^2\sqrt{N}$) on which ϕ is essentially constant (fluctuating by at most $\eta^2/100$, say). A pigeonhole argument (exploiting the fact that $f - \mathbf{E}_{1\leq x\leq N/3}f(x)$ has mean zero) then shows that on one of these progressions, say P, f has significantly higher density than on the average, in the sense that $\mathbf{E}_{x\in P}f(x) \geq \mathbf{E}_{x\in\mathbb{Z}/N\mathbb{Z}}f(x) + \eta^2/100$. One can then apply an affine transformation to convert this progression P into another discrete interval $\{1,\ldots,N'/3\}$, where N' is essentially the square root of N. One then iterates this argument until linear pseudorandomness is obtained (using the fact that the density of f cannot increase beyond 1), and one eventually obtains

Theorem 3.3 (Local Structure theorem). *Let $f : [1, N/3] \to \mathbb{R}$ be a non-negative function bounded by* 1, *and let $\eta > 0$. Then there exists a progression P in $[1, N/3]$ of length at least $c(\eta)N^{c(\eta)}$ for some $c(\eta) > 0$, on which we have the splitting $f = f_{U^\perp} + f_U$, where $f_U^\perp := \mathbf{E}_{x\in P}f(x) \geq \mathbf{E}_{1\leq x\leq N/3}f(x)$ is the mean of f on P, and f_U is linearly pseudorandom in the sense that*

$$\|f_U\|_{U^2(\mathbb{Z}/M\mathbb{Z})} \leq \eta$$

where we identify P with a subset of a cyclic group $\mathbb{Z}/M\mathbb{Z}$ of cardinality $M \approx 3|P|$ in the usual manner.

More informally, any function will contain an arithmetic progression P of significant size on which f can be decomposed into a non-trivial structured component $f_{U^\perp}$ and a pseudorandom component f_U. In the language of the introduction, it is essentially saying that any dense set A of integers will contain components which are dense pseudorandom subsets of long progressions. Once one has this theorem, it is an easy matter to establish Szemerédi's theorem in the $k = 3$ case. Indeed, if $A \subseteq \mathbb{Z}$ has upper density greater than δ, then we can find arbitrarily large primes N such that $|A \cap [1, N/3]| \geq \delta N/3$. Applying Theorem 3.3 with $\eta := \delta^3/100$, and f equal to the indicator function of $A \cap [1, N/3]$, we can find a progression P in $\{1,\ldots,N/3\}$ of length at least $c(\delta)N^{c(\delta)}$ on which $\mathbf{E}_{x\in P}f(x) \geq \delta$ and $f - \mathbf{E}_{x\in P}f(x)$ is linearly pseudorandom in the sense of Theorem 3.3. It is then an easy matter to apply the generalized von Neumann theorem to show that $A \cap P$ contains many arithmetic progressions of length three (in fact it contains $\gg \delta^3|P|^3$ such progressions). Letting N (and hence $|P|$) tend to infinity we obtain Szemerédi's theorem in the $k = 3$ case. An averaging argument of Varnavides [63] then yields the more quantitative version in Theorem 3.1 (but with a moderately bad bound for $c(3,\delta)$, namely $c(3,\delta) = 2^{-2^{C/\delta^C}}$ for some absolute constant C).

A more refined structure theorem was given in [23] (see also [35]), which was termed an "arithmetic regularity lemma" in analogy with the Szemerédi regularity lemma which we discuss in the next section. That theorem has similar hypotheses to Theorem 3.3, but instead of constructing a single progression on P on which one has pseudorandomness, one partitions $[1, N/3]$ into *many* long progressions[4], where on most of which the function f becomes linearly pseudorandom (after subtracting the mean). A related structure theorem (with a more "ergodic" perspective) was also given in [56]. Here we give an alternate approach based on Fourier expansion and the pigeonhole principle. Observe that for any $f : \mathbb{Z}/N\mathbb{Z} \to \mathbb{C}$ and any threshold λ we have the Fourier decomposition $f = f_{U^\perp} + f_U$, where the "structured" component $f_{U^\perp} := \sum_{\xi:|\hat{f}(\xi)|\geq\lambda} \hat{f}(\xi)e^{2\pi i x\xi/N}$ contains all the significant Fourier coefficients, and the "pseudorandom" component $f_U := \sum_{\xi:|\hat{f}(\xi)|\leq\lambda} \hat{f}(\xi)e^{2\pi i x\xi/N}$ contains all the small Fourier coefficients. Using Plancherel's theorem one can easily establish

Theorem 3.4 (Weak structure theorem). *Let $f : \mathbb{Z}/N\mathbb{Z} \to \mathbb{C}$ be a function bounded in magnitude by* 1, *and let* $0 < \lambda < 1$. *Then we can split $f = f_{U^\perp} + f_U$, where $f_{U^\perp}$ is the linear combination of at most $O(1/\lambda^2)$ linear phase functions $x \mapsto e^{2\pi i x\xi/N}$, and f_U is linearly pseudorandom in the sense that $\|f_U\|_{U^2(\mathbb{Z}/N\mathbb{Z})} \leq \lambda$.*

This theorem asserts that an arbitrary bounded function only has a bounded amount of significant linear Fourier-analytic structure; after removing this bounded amount of structure, the remainder is linearly pseudorandom.

This theorem, while simple to state and prove, has two weaknesses which make it unsuitable for such tasks as counting progressions of length three. Firstly, even though f is bounded by 1, the components $f_{U^\perp}, f_U$ need not be. Related to this, if f is non-negative, there is no reason why $f_{U^\perp}$ should be non-negative also. Secondly, the pseudorandomness control on f_U is not very good when compared against the complexity of $f_{U^\perp}$ (i.e. the number of linear exponentials needed to describe $f_{U^\perp}$). In practice, this means that any control one obtains on the structured component of f will be dominated by the errors one has to concede from the pseudorandom component. Fortunately, both of these defects can be repaired, the former by a Fejér summation argument, and the latter by a pigeonhole argument (which introduces a second error term f_S, which is small in L^2 norm). More precisely, we have

Theorem 3.5 (Strong structure theorem). *Let $f : \mathbb{Z}/N\mathbb{Z} \to \mathbb{R}$ be a non-negative function bounded by* 1, *and let* $0 < \varepsilon < 1$. *Let $F : \mathbb{N} \to \mathbb{N}$ be an arbitrary increasing function (e.g. $F(n) = 2^{2^n}$). Then there exists an integer $T = O_{F,\varepsilon}(1)$ and a decomposition $f = f_{U^\perp} + f_S + f_U$, where $f_{U^\perp}$ is the linear combination of at most T linear phase functions, f_U is linearly pseudorandom in the sense that $\|f_U\|_{U^2(\mathbb{Z}/N\mathbb{Z})} = O(1/F(T))$, and f_S is small in the sense that $\|f_S\|_{L^2(\mathbb{Z}/N\mathbb{Z})} := (\mathbf{E}_{n\in\mathbb{Z}/N\mathbb{Z}}|f_S(n)|^2)^{1/2} = O(\varepsilon)$. Furthermore, $f_{U^\perp}, f_U$ are bounded in magnitude by* 1. *Also, $f_{U^\perp}$ and $f_{U^\perp} + f_S$ are non-negative with the same mean as f.*

This theorem can be proven by adapting arguments from [26], [35], or [56]; we omit the details. Note that we have the freedom to set the growth function F

[4] Actually, for technical reasons it is more efficient to replace the notion of an arithmetic progression by a slightly different object known as a *Bohr set*; see [23], [35] for details.

arbitrarily fast in the above proposition; this corresponds roughly speaking to the fact that in the ergodic counterpart to this structure theorem (Proposition 2.4) the pseudorandom error f_U has asymptotically *vanishing* Gowers U^2 norm. One can view $f_{U^\perp}$ as a "coarse" Fourier approximation to f, and $f_{U^\perp} + f_S$ as a "fine" Fourier approximation to f; this perspective links this proposition with the graph regularity lemmas that we discuss in the next section.

Theorem 3.5 can be used to deduce the structure theorems in [23], [56], [35], while a closely related result was also established in [4]. It can also be used to directly derive the $k = 3$ case of Theorem 3.1, as follows. Let f be as in that proposition, and let $\varepsilon := \delta^3/100$. We apply Theorem 3.5 to decompose $f = f_{U^\perp} + f_S + f_U$. Because $f_{U^\perp}$ has only T Fourier exponentials, it is easy to see that $f_{U^\perp}$ is almost periodic, in the sense that $\|T^n f_{U^\perp} - f_{U^\perp}\|_{L^2(\mathbb{Z}/N\mathbb{Z})} \leq \varepsilon$ for at least $c(\varepsilon, T)N$ values of $n \in \mathbb{Z}/N\mathbb{Z}$, for some $c(\varepsilon, T) > 0$. For such values of n, one can easily verify that

$$\mathbf{E}_{x\in\mathbb{Z}/N\mathbb{Z}} f_{U^\perp}(x) T^n f_{U^\perp}(x) T^{2n} f_{U^\perp}(x) \geq \delta^3/2.$$

Because f_S is small, we can also deduce that

$$\mathbf{E}_{x\in\mathbb{Z}/N\mathbb{Z}} (f_{U^\perp} + f_S)(x) T^n (f_{U^\perp} + f_S)(x) T^{2n} (f_{U^\perp} + f_S)(x) \geq \delta^3/4$$

for these values of n. Averaging in n (and taking advantage of the non-negativity of $f_{U^\perp} + f_S$) we conclude that

$$\mathbf{E}_{x,n\in\mathbb{Z}/N\mathbb{Z}} (f_{U^\perp} + f_S)(x) T^n (f_{U^\perp} + f_S)(x) T^{2n} (f_{U^\perp} + f_S)(x) \geq \delta^3 c(\varepsilon, T)/4.$$

Adding in the pseudorandom error f_U using the generalized von Neumann theorem (7), we conclude that

$$\mathbf{E}_{x,n\in\mathbb{Z}/N\mathbb{Z}} f(x) T^n f(x) T^{2n} f(x) \geq \delta^3 c(\varepsilon, T)/4 - O(1/F(T)).$$

If we choose F to be sufficiently rapidly growing depending on δ and ε, we can absorb the error term in the main term and conclude that

$$\mathbf{E}_{x,n\in\mathbb{Z}/N\mathbb{Z}} f(x) T^n f(x) T^{2n} f(x) \geq \delta^3 c(\varepsilon, T)/8.$$

Since $T = O_{F,\varepsilon}(1) = O_\delta(1)$, we obtain the $k = 3$ case of Theorem 3.1 as desired.

Roth's original Fourier-analytic argument was published in 1953. But the extension of this Fourier argument to the $k > 3$ case was not achieved until the work of Gowers [19], [20] in 1998. For simplicity we once again restrict attention to the $k = 4$ case, where the theory is more complete. Our objective is to show

$$\mathbf{E}_{x,r\in\mathbb{Z}/N\mathbb{Z}} f(x) T^r f(x) T^{2r} f(x) T^{3r} f(x) \geq c(4, \delta) > 0 \quad (8)$$

whenever f is non-negative, bounded by 1, and has mean at least δ. There are some significant differences between this case and the $k = 3$ case (5). Firstly, linear pseudorandomness is not enough to guarantee that a contribution to (8) is negligible: for instance, if $f(x) := e^{2\pi i \xi x^2/N}$, then

$$\mathbf{E}_{x,r\in\mathbb{Z}/N\mathbb{Z}} f(x) T^r \overline{f}^3(x) T^{2r} f^3(x) T^{3r} \overline{f}(x) = 1$$

despite f being very linearly pseudorandom (the U^2 norm of f is $N^{-1/4}$); compare this example with (4). One must now utilize some sort of "quadratic Fourier analysis" in order to capture the correct concept of pseudorandomness and structure. Secondly, the Fourier-analytic arguments must now be supplemented by some results from arithmetic combinatorics (notably the Balog–Szemerédi theorem, and results related to Freiman's inverse sumset theorem) in order to obtain a usable notion of quadratic structure. Finally, as in the ergodic case, one cannot rely purely on quadratic phase functions such as $e^{2\pi i(\xi x^2+\eta x)/N}$ to generate all the relevant structured objects, and must also consider generalized quadratic objects such as locally quadratic phase functions, 2-step nilsequences (see below), or bracket quadratic phases such as $e^{2\pi i\lfloor\sqrt{2}n\rfloor\sqrt{3}n}$.

Let us now briefly sketch how the theory works in the $k=4$ case. The correct notion of pseudorandomness is now given by the *Gowers* U^3 *uniformity norm*, defined by

$$\|f\|_{U^3(\mathbb{Z}/N\mathbb{Z})}^8 := \mathbf{E}_{n\in\mathbb{Z}/N\mathbb{Z}}\|T^n f\overline{f}\|_{U^2(\mathbb{Z}/N\mathbb{Z})}^4.$$

This norm measures the extent to which f behaves quadratically; for instance, if $f=e^{2\pi iP(x)/N}$ for some polynomial P of degree k in the finite field $\mathbb{Z}/N\mathbb{Z}$, then one can verify that $\|f\|_{U^3(\mathbb{Z}/N\mathbb{Z})}=1$ if P has degree at most 2, but (using the Weil estimates) we have $\|f\|_{U^3(\mathbb{Z}/N\mathbb{Z})}=O_k(N^{-1/16})$ if P has degree $k>2$. Repeated application of Cauchy–Schwarz then yields the generalized von Neumann theorem

$$|\mathbf{E}_{x,r\in\mathbb{Z}/N\mathbb{Z}}f_0(x)T^r f_1(x)T^{2r}f_2(x)T^{3r}f_3(x)| \le \min_{0\le j\le 3}\|f_j\|_{U^3(\mathbb{Z}/N\mathbb{Z})} \qquad (9)$$

whenever f_0, f_1, f_2, f_3 are bounded in magnitude by 1. The next step is to establish a dichotomy between quadratic structure and quadratic pseudorandomness in the spirit of Lemma 3.2. In the original work of Gowers [19], it was shown that a function which was not quadratically pseudorandom had local correlation with quadratic phases on medium-length arithmetic progressions. This result (when combined with the density increment argument of Roth) was already enough to prove (8) with a reasonable bound on $c(4,\delta)$ (basically of the form $1/\exp(\exp(\delta^{-C}))$); see [19], [20]. Building upon this work, a stronger dichotomy, similar in spirit to Lemma 2.5, was established in [29]. Here, a number of essentially equivalent formulations of quadratic structure were established, but the easiest to state (and the one which generalizes most easily to higher k) is that of a *(basic) 2-step nilsequence*, which can be viewed as a notion of "quadratic almost periodicity" for sequences. More precisely, a 2-step nilsequence a sequence of the form $n\mapsto F(T^n x\Gamma)$, where F is a Lipschitz function on a 2-step nilmanifold G/Γ, $x\Gamma$ is a point in this nilmanifold, and T is a shift operator $T: x\Gamma\mapsto \alpha x\Gamma$ for some fixed group element $\alpha\in G$. We remark that quadratic phase sequences such as $n\mapsto e^{2\pi i\alpha n^2}$ are examples of 2-step nilsequences, and generalized quadratics such as $n\mapsto e^{2\pi i\lfloor\sqrt{2}n\rfloor\sqrt{3}n}$ can also be written (outside of sets of arbitrarily small density) as 2-step nilsequences.

Lemma 3.6 (Dichotomy between randomness and structure [29]). *Suppose that* $f : \mathbb{Z}/N\mathbb{Z} \to \mathbb{C}$ *is bounded in magnitude by* 1 *with* $\|f\|_{U^3(\mathbb{Z}/N\mathbb{Z})} \geq \eta$ *for some* $0 < \eta < 1$. *Then there exists a 2-step nilsequence* $n \mapsto F(T^n x\Gamma)$, *where* G/Γ *is a nilmanifold of dimension* $O_\eta(1)$, *and* F *is a bounded Lipschitz function* G/Γ *with Lipschitz constant* $O_\eta(1)$, *such that* $|\mathbf{E}_{1 \leq x \leq N} f(x)\overline{F(T^n x\Gamma)}| \geq c(\eta)$ *for some* $c(\eta) > 1$. (*We identify the integers from* 1 *to* N *with* $\mathbb{Z}/N\mathbb{Z}$ *in the usual manner.*)

In fact the nilmanifold G/Γ constructed in [29] is of a very explicit form, being the direct sum of at most $O_\eta(1)$ circles (which are one-dimensional), skew shifts (which are two-dimensional), and Heisenberg nilmanifolds (which are three-dimensional). The dimension $O_\eta(1)$ is in fact known to be polynomial in η, but the best bounds for $c(\eta)$ are currently only exponential in nature. See [29] for further details and discussion.

The proof of Lemma 3.6 is rather lengthy but can be summarized as follows. If f has large U^3 norm, then by definition $T^n f\overline{f}$ has large U^2 norm for many n. Applying Lemma 3.2, this shows that for many n, $T^n f\overline{f}$ correlates with a linear phase function of some frequency $\xi(n)$ (which can be viewed as a kind of "derivative" of the phase of f in the "direction" n). Some manipulations involving the Cauchy–Schwarz inequality then show that $\xi(n)$ contains some additive structure (in that there are many quadruples n_1, n_2, n_3, n_4 with $n_1 + n_2 = n_3 + n_4$ and $\xi(n_1) + \xi(n_2) = \xi(n_3) + \xi(n_4)$). Methods from additive combinatorics (notably the Balog–Szemerédi(-Gowers) theorem and Freiman's theorem, see e.g. [61]) are then used to "linearize" ξ, in the sense that $\xi(n)$ agrees with a (generalized) linear function of n on a large (generalized) arithmetic progression. One then "integrates" this fact to conclude that f itself correlates with a certain "anti-derivative" of $\xi(n)$, which is a (generalized) quadratic function on this progression. This in turn can be approximated by a 2-step nilsequence. For full details, see [29].

Thus, quadratic nilsequences are the only obstruction to a function being quadratically pseudorandom. This can be iterated to obtain structural results. The following "weak" structural theorem is already quite useful:

Theorem 3.7 (Weak structure theorem [35]). *Let* $f : \mathbb{Z}/N\mathbb{Z} \to \mathbf{C}$ *be a function bounded in magnitude by* 1, *and let* $0 < \lambda < 1$. *Then we can split* $f = f_{U^\perp} + f_U$, *where* $f_{U^\perp}$ *is a 2-step nilsequence given by a nilmanifold of dimension* $O_\lambda(1)$ *and by a bounded Lipschitz function* F *with Lipschitz constant* $O_\lambda(1)$, *and* f_U *is quadratically pseudorandom in the sense that* $\|f_U\|_{U^3(\mathbb{Z}/N\mathbb{Z})} \leq \lambda$. *Furthermore,* $f_{U^\perp}$ *is non-negative, bounded by* 1, *and has the same mean as* f.

This is an analogue of Theorem 3.4, and asserts that any bounded function has only a bounded amount of quadratic structure, with the function becoming quadratically pseudorandom once this structure is subtracted. It cannot be proven in quite the same way as in Theorem 3.4, because we have no "quadratic Fourier inversion formula" that decomposes a function neatly into quadratic components (the problem being that there are so many quadratic objects that such a formula is necessarily overdetermined). However, one can proceed by a finitary analogue of the ergodic theory approach, known as an "energy increment argument". In the ergodic setting, one uses all the quadratic objects to create a σ-algebra $\mathcal{Z}_2$, and sets $f_{U^\perp}$ to be the conditional expectation of f with respect to that σ-algebra. In the finitary setting, it turns out to be too expensive to try to use *all* the 2-step nilsequences to create a σ-algebra. However, by adopting a more adaptive approach, selecting only those 2-step nilsequences which have some significant correlation with f (or some component of f), one can obtain the above theorem; we omit the details.

It is likely that quantitative versions of this structure theorem will improve the known bounds on Szemerédi's theorem in the $k = 4$ case; see [32], [33], [34]. A closely related version of this argument was also essential in establishing Theorem 1.2, see Section 5 below.

4. Graph Theory

We now turn to the third major line of attack to Szemerédi's theorem, based on graph theory (and hypergraph theory), and which is perhaps the purest embodiment of the strategy of exploiting the dichotomy between randomness and structure. For graphs, the relevant structure theorem is the *Szemerédi regularity lemma*, which was developed in [53] in the original proof of Szemerédi's theorem, and has since proven to have many further applications in graph theory and computer science; see [41] for a survey. More recently, the analogous regularity lemma for hypergraphs have been developed in [21], [46], [47], [48], [49], [58]. Roughly speaking, these very useful lemmas assert that any graph (binary relation) or hypergraph (higher order relation), no matter how complex, can be modelled effectively as a pseudorandom sub(hyper)graph of a finite complexity (hyper)graph. Returning to the setting of the introduction, the graph regularity lemma would assert that there exists a colouring of the integers into finitely many colours such that relations such as $x - y \in A$ can be viewed approximately as pseudorandom relations, with the "probability" of the event $x - y \in A$ depending only on the colour of x and y.

The strategy of the graph theory approach is to abstract away the arithmetic structure in Szemerédi's theorem, converting the problem to one of finding solutions to an abstract set of equations, which can be modeled by graphs or hypergraphs. As before, we first illustrate this with the simple case of the $k = 3$ case of Szemerédi's theorem, which we will take in the form of Theorem 3.1. For simplicity we specialize to the case when f is the indicator function of a set A (which thus has density at least δ in $\mathbb{Z}/N\mathbb{Z}$); it is easy to see (e.g. by probabilistic arguments) that this special case in fact implies the general case. The key observation is that the problem of locating an arithmetic progression of length three can be recast as the problem of solving three constraints in three unknowns, where each constraint only involves two of the unknowns. Specifically, if $x, y, z \in \mathbb{Z}/N\mathbb{Z}$ solve the system of constraints

$$\begin{array}{rrrl} & y & +2z & \in A \\ -x & & +z & \in A \\ -2x & -y & & \in A \end{array} \qquad (10)$$

then $y+2z, -x+z, -2x-y$ is an arithmetic progression of length three in A. Conversely, each such progression comes from exactly N solutions to (10). Thus, it will suffice to show that there are at least $c(3,\delta)N^3$ solutions to (10). Note that we already can construct at least δN^2 "trivial solutions" to (10), in which $y+2z = -x+z = -2x+y$ is an element of A. Furthermore, these trivial solutions (x, y, z) are "edge-disjoint" in the sense that no two of these solutions share more than one value in common (i.e. if (x, y, z) and (x', y', z') are distinct trivial solutions then at most one of $x = x'$, $y = y'$, $z = z'$ are true). It turns out that these trivial solutions automatically generate a large number of non-trivial solutions to (10) — without using any further arithmetic structure present in these constraints. Indeed, the claim now follows from the following graph-theoretical statement.

Lemma 4.1 (Triangle removal lemma [51]). *For every $0 < \delta < 1$ there exists $0 < \sigma < 1$ with the following property. Let $G = (V, E)$ be an (undirected) graph with $|V| = N$ vertices which contains fewer than σN^3 triangles. Then it is possible to remove $O(\delta N^2)$ edges from G to create a graph G' which contains no triangles whatsoever.*

To see how the triangle removal lemma implies the claim, consider a vertex set V which consists of three copies V_1, V_2, V_3 of $\mathbb{Z}/N\mathbb{Z}$ (so $|V| = 3N$), and consider the tripartite graph $G = (V, E)$ whose edges are of the form

$$E = \{(y,z) \in V_2 \times V_3 : y + 2z \in A\} \cup \{(x,z) \in V_1 \times V_3 : -x + z \in A\}$$
$$\cup \{(x,y) \in V_1 \times V_2 : -2x - y \in A\}.$$

One can think of G as a variant of the Cayley graph for A. Observe that solutions to (10) are in one-to-one correspondence with triangles in G. Furthermore, the δN^2 trivial solutions to (10) correspond to δN^2 edge-disjoint triangles in G. Thus to delete all the triangles one needs to remove at least δN^2 edges. Applying Lemma 4.1 in the contrapositive (adjusting N, δ, σ by constants such as 3 if necessary), we see that G contains at least σN^3 triangles for some $\sigma = \sigma(\delta) > 0$, and the claim follows.

The only known proof of the triangle removal lemma proceeds by a structure theorem for graphs known as the *Szemerédi regularity lemma.* In order to emphasize the similarities between this approach and the previously discussed approaches, we shall not use the standard formulation of this lemma, but instead use a more recent formulation from [57], [58] (see also [1], [45]), which replaces graphs with functions, and then obtains a structure theorem decomposing such functions into a structured (finite complexity) component, a small component, and a pseudorandom (regular) component. More precisely, we work with functions $f : V \times V \to \mathbb{R}$; this can be thought of as a weighted, directed generalization of a graph on V in which every edge (x, y) is assigned a real-valued weight $f(x, y)$. The first step is to define a notion of pseudorandomness. For graphs, this concept is well understood. There are many equivalent formulations of this concept (see [6]), but we shall adopt one particularly close to the analogous concepts in previous sections, by introducing the *Gowers* $\square^2$ *cube norm* as

$$\|f\|_{\square^2}^4 := \mathbf{E}_{x,y,x',y' \in V} f(x,y) f(x,y') f(x',y) f(x',y');$$

when f is the incidence function of a graph, the right-hand side essentially counts the number of 4-cycles in that graph. Again, one can use the Cauchy–Schwarz inequality to establish that the $\square^2$ norm is indeed a norm; alternatively, one can use spectral theory and observe that the $\square^2$ norm is essentially the Schatten-von Neumann p-norm of f with $p = 4$. We refer to f as *pseudorandom* if its $\square^2$ norm is small. By two applications of Cauchy–Schwarz we have the generalized von Neumann inequality

$$|\mathbf{E}_{x,y,z \in V} f(x,y) g(y,z) h(z,x)| \leq \min(\|f\|_{\square^2}, \|g\|_{\square^2}, \|h\|_{\square^2}) \tag{11}$$

whenever f, g, h are bounded in magnitude by 1 (note that this generalizes (5)).

The next step, as before, is to establish a dichotomy between pseudorandomness and structure. The analogue of Lemma 2.2 or Lemma 3.2 is

Lemma 4.2 (Dichotomy between randomness and structure). *Suppose that* $f : V \times V \to \mathbb{R}$ *is bounded in magnitude by* 1 *with* $\|f\|_{\square^2(\mathbb{Z}/N\mathbb{Z})} \geq \eta$ *for some* $0 < \eta < 1$. *Then there exists sets* $A, B \subset V$ *such that* $|\mathbf{E}_{x,y \in V} f(x,y) 1_A(x) 1_B(y)| \geq \eta^4/4$. *Here* $1_A(x)$ *denotes the indicator function of* A *(thus* $1_A(x) = 1$ *if* $x \in A$ *and* $1_A(x) = 0$ *otherwise).*

This lemma follows from an easy application of the pigeonhole principle and is omitted. One can iterate it (by and energy increment argument, as in Theorem 3.7) to obtain a weak version of the Szemerédi regularity lemma:

Theorem 4.1 (Weak structure theorem) [10]. *Let* $f : V \times V \to \mathbb{R}$ *be a non-negative function bounded by* 1, *and let* $\varepsilon > 0$. *Then we can decompose* $f = f_{U^\perp} + f_U$, *where* $f_{U^\perp} = \mathbf{E}(f|\mathcal{Z} \otimes \mathcal{Z})$, $\mathcal{Z}$ *is a* σ*-algebra of* V *generated by at most* $2/\varepsilon$ *sets, and* $\|f_U\|_{\square^2} \leq \varepsilon$.

As with Theorem 3.4, the above theorem is too weak to be of much use, because the control one has on the pseudorandomness of f_U is fairly poor compared to the control on the complexity of $f_{U^\perp}$. The following strong version of the regularity lemma is far more useful (compare with Theorem 3.5):

Theorem 4.2 (Strong structure theorem [57]). *Let* $f : V \times V \to \mathbb{R}$ *be a non-negative function bounded by* 1, *and let* $\varepsilon > 0$. *Let* $F : \mathbb{N} \to \mathbb{N}$ *be an arbitrary*

increasing function (e.g. $F(n) = 2^{2^n}$*). Then there exists an integer* $T = O_{F,\varepsilon}(1)$ *and a decomposition* $f = f_{U^\perp} + f_S + f_U$*, where* $f_{U^\perp} = \mathbf{E}(f|\mathcal{Z} \otimes \mathcal{Z})$*,* $\mathcal{Z}$ *is generated by at most* T *sets in* V*,* f_U *is pseudorandom in the sense that* $\|f_U\|_{\square^2} = O(1/F(T))$*, and* f_S *is small in the sense that* $\|f_S\|_{L^2(V \times V)} := (\mathbf{E}_{x,y \in V}|f_S(x,y)|^2)^{1/2} = O(\varepsilon)$*. Furthermore,* $f_{U^\perp}, f_U$ *are bounded in magnitude by* 1*. Also,* $f_{U^\perp}$ *and* $f_{U^\perp} + f_S$ *are non-negative and bounded by* 1*.*

One can view $f_{U^\perp}$ as a "coarse" approximation to f, as it is measurable with respect to a fairly low-complexity σ-algebra, and $f_{U^\perp} + f_S = \mathbf{E}(f|\mathcal{Z}^{(n')} \otimes \mathcal{Z}^{(n')})$ as a "fine" approximation to f, which is considerably more complex but is also a far better approximation to f, in fact the accuracy of the fine approximation exceeds the complexity of the coarse approximation by any specified growth function F. Also the difference between the coarse and fine approximations is controlled by an arbitrarily small constant ε.

Theorem 4.2 already easily implies the Szemerédi regularity lemma in its traditional formulation; see [57]. It also implies Lemma 4.1, similar to how Theorem 3.5 implies the $k = 3$ version of Szemerédi's theorem; we omit the standard details.

As in the other two approaches, the above arguments extend (with some additional difficulties) to higher values of k. Again we restrict attention to the $k = 4$ case for simplicity. To locate a progression of length four in a set $A \subset \mathbb{Z}/N\mathbb{Z}$ is now equivalent to solving the system of constraints

$$\begin{array}{lllll} & y & +2z & +3w & \in A \\ -x & & +z & +2w & \in A \\ -2x & -y & & +w & \in A \\ -3x & -2y & -z & & \in A. \end{array} \qquad (12)$$

This in turn follows from a hypergraph analogue of the triangle removal lemma. Define a 3-*uniform hypergraph* to be a pair $H = (V, E)$ where V is a finite set of vertices and E is a finite set of unordered triplets (x, y, z) in V, which we refer to as the *edges* of H. Define a *tetrahedron* in H to be a quadruple (x, y, z, w) of vertices such that all four triplets $(x, y, z), (y, z, w), (z, w, x), (w, x, y)$ are edges of H.

Lemma 4.5 (Tetrahedron removal lemma) [9]. *For every* $0 < \delta < 1$ *there exists* $0 < \sigma < 1$ *with the following property. Let* $H = (V, E)$ *be a* 3*-uniform hypergraph graph with* $|V| = N$ *vertices which contains fewer than* σN^4 *tetrahedra. Then it is possible to remove* $O(\delta N^3)$ *edges from* H *to create a hypergraph* H' *which contains no tetrahedra whatsoever.*

Letting f be the indicator function of H, we now have a situation where

$$\mathbf{E}_{x,y,z,w \in V} f(x, y, z) f(y, z, w) f(z, w, x) f(w, x, y) \leq \sigma$$

and we need to remove some small components from f so that this average now vanishes completely. Again, the key step here is to obtain a structure theorem that decomposes f into structured parts, small errors, and pseudorandom errors.

The notion of pseudorandomness is now captured by the Gowers $\Box^3$ cube norm, defined by

$$\|f\|_{\Box^3}^8 := \mathbf{E}_{x,y,z,x',y',z' \in V} f(x,y,z) \ldots f(x',y',z')$$

where the product is over the eight values of f with first co-ordinate x or x', second co-ordinate y or y', and third co-ordinate z or z'. In the case when f indicator function of a hypergraph H, this norm essentially counts the number of octahedra present in H. One can obtain a strong structure theorem analogous to Theorem 4.4, but with one significant difference. In Theorem 4.4, the structured component $f_{U^\perp}(x,y)$ can be broken up into a small number of components which are of the form $1_A(x)1_B(y)$. In the 3-uniform hypergraph analogue of Theorem 4.4, the structured component $f_{U^\perp}(x,y,z)$ will be broken up into a small number of components of the form $1_A(x,y)1_B(y,z)1_C(z,x)$. It turns out that in order to conclude the proof of Lemma 4.5, this structural decomposition is not sufficient by itself; one must also turn to the functions $1_A(x,y)$, $1_B(y,z)$, $1_C(z,x)$ generated by this structure theorem and decompose them further, essentially by invoking Theorem 4.4. This leads to some technical complications in the argument, although this approach to Szemerédi's theorem is still the most elementary and self-contained. See [21], [46], [47], [48], [49], [58] for details.

5. The Primes

Having surveyed the three major approaches to Szemerédi's theorem, we now turn to the question of counting progressions in the primes (or in dense subsets of the primes). The major new difficulty here, of course, is that the primes have asymptotically zero density rather than positive density, and even the most recent quantitative bounds on Szemerédi's theorem (see the discussion after Theorem 3.1) are not strong enough by themselves to overcome the "thinness" of the primes. However, it turns out that the primes (and functions supported on the primes) are still within the range of applicability of structure theorems. For instance, to oversimplify dramatically, the structure theorem in [27] essentially[5] represents the primes (or any dense subset of the primes) as a (sparse) pseudorandom subset of a set of positive density. Since sets of positive density already contain many progressions thanks to Szemerédi's theorem, it turns out that enough of these

[5]This is a gross oversimplification. The precise statement is that after eliminating obvious irregularities in the primes caused by small residue classes, and excluding a small and technical exceptional set, a normalized counting function on the primes can be decomposed as a bounded function (which is thus spread out over a set of positive density), plus a pseudorandom error. Ignoring the initial elimination of obvious irregularities and the exceptional set, and pretending the bounded function was the indicator function of a positive density set A, one recovers the interpretation of the primes as a sparse pseudorandom subset of A.

progressions survive when passing to a pseudorandom subset that one can conclude Theorem 3.1.

Interestingly, Theorem 1.2 can be tackled by (quantitative) ergodic methods, by Fourier-analytic methods, and by graph-theoretic methods, with the three approaches leading to slightly different results. For instance, the establishment of infinitely many progressions of length three in the primes by van der Corput [62] was Fourier-analytic, as was the corresponding statement for dense subsets of the primes (i.e. the $k = 3$ case of Theorem 1.2), proven 76 years later by Green [22]. The argument in [27] which proves Theorem 1.2 in full combines ideas from all three approaches, but is closest in spirit to the ergodic approach, albeit set in the finitary context of a cyclic group $\mathbb{Z}/N\mathbb{Z}$ rather than on an infinitary measure space. The argument in [59], which shows that the Gaussian primes (or any dense subset thereof) contains infinitely many constellations of any prescribed shape, and can be viewed as a two-dimensional analogue of Theorem 1.2, was proven via the (hyper)graph-theoretical approach. Finally, a more recent argument in [30], [31], in which precise asymptotics for the number of progressions of length four in the primes are obtained, as well as a "quadratic pseudorandomness" estimate on a renormalized counting function for the primes, proceeds by returning back to the original Fourier-analytic approach, but now using quadratic Fourier-analytic tools (Lemma 3.6 and Theorem 3.7) rather than linear ones.

As mentioned in the introduction, these results are discussed in other surveys [42], [25], [55], [56], [37], and we will only sketch some highlights here. In all the results, the strategy is to try to isolate the "structured" component of the primes from the "pseudorandom" component. There is some obvious structure present in the primes; for instance, they are almost all odd, they are almost all coprime to three, and so forth. This obvious structure can be normalized away fairly easily. For instance, to remove the bias the primes have towards being odd, one can replace the primes $P = \{2, 3, 5, \ldots\}$ with the renormalized set $P_{2,1} := \{n : 2n + 1 \text{ prime}\} = \{1, 2, 3, 5, \ldots\}$. Each arithmetic progression in $P_{2,1}$ clearly induces a corresponding progression in P, but the set $P_{2,1}$ has no bias modulo 2. More generally, to reduce all the bias present in residue classes mod p for all $p < w$ (where w is a medium-sized parameter to be chosen later), one can work with a set $P_{W,b} := \{n : Wn + b \text{ prime}\}$, where W is the product of all the primes less than w and $1 \leq b < W$ is a number coprime to W. This "W-trick" allows for some technical simplifications.

Next, it is convenient not to work with the primes as a set, but rather as a renormalized counting function. One convenient choice is the von Mangoldt function $\Lambda(n)$, defined as $\log p$ if n is a power of a prime p and 0 otherwise. Actually, because of the W-trick, it is better to consider a renormalized von Mangoldt function such as $\Lambda_{W,b}(n) := \frac{W}{\phi(W)}\Lambda(Wn + b)$, where $\phi(W)$ is the Euler totient function of W. The prime number theorem in arithmetic progressions asserts that the asymptotic average value of $\Lambda_{W,b}(n)$ is equal to 1. To establish progressions of length k in the primes, it suffices to obtain a nontrivial lower bound for the asymptotic value of the average

$$\mathbf{E}_{1\leq n,r\leq N}\Lambda_{W,b}(n)\Lambda_{W,b}(n+r)\ldots\Lambda_{W,b}(n+(k-1)r). \tag{13}$$

In fact this quantity is conjectured to asymptotically equal 1 as $W, N \to \infty$, with W growing much slower than N (a special case of the Hardy–Littlewood prime tuples conjecture); the intuition is that by removing all the bias present in the small residue classes, we have eliminated all the "obvious" structure in the primes, and the renormalized function $\Lambda_{W,b}$ should now fluctuate pseudorandomly around its mean value 1. However, this conjecture has only been verified in the cases $k = 3, 4$ (leading to an asymptotic count for the number of progressions of primes of length k less than a large number N); for the cases $k > 4$ we only have a lower bound of $c(k)$ for some small $c(k) > 0$.

Let us cheat slightly by pretending that $\Lambda_{W,b}$ is a function on the cyclic group $\mathbb{Z}/N\mathbb{Z}$ rather than on the integers $\mathbb{Z}$; there are some minor technical truncation issues that need to be addressed to pass from one to the other but we shall ignore them here. In order to show that (13) is close to 1, an obvious way to proceed would be to establish some kind of pseudorandomness control on the deviation $\Lambda_{W,b} - 1$ from the mean, and then some sort of generalized von Neumann theorem to show that this deviation is negligible. Based on the experience with Szemerédi's theorem, one would expect linear pseudorandomness to be the correct notion for $k = 3$, quadratic pseudorandomness for $k = 4$, and so forth. In the $k = 3$ case it is indeed a standard computation (using Vinogradov's method, or a modern variant of that method such as the one based on Vaughan's identity) to show that $\Lambda_{W,b} - 1$ is has small Fourier coefficients, which is a reasonable proxy for linear pseudorandomness; the point being that the W-trick has eliminated all the "major arcs" which would otherwise destroy the pseudorandomness. It then remains to obtain a generalized von Neumann theorem, similar to (7). In preceding sections, one was working with functions that were bounded (and hence square integrable), and one could obtain these theorems easily from Plancherel's theorem. In the current setting, the L^2 estimates on $\Lambda_{W,b}$ are unfavourable, and what one needs instead is some sort of l^p bound on the Fourier coefficients of $\Lambda_{W,b}$ for some $2 < p < 3$. This can be done by a more careful application of Vinogradov's method, but can also be achieved using harmonic analysis methods arising from restriction theory; see [22], [28]. The key new insight here is that while the Fourier coefficients of $\Lambda_{W,b}$ are difficult to understand directly, one can *majorize* $\Lambda_{W,b}$ pointwise by (a constant multiple of) a much better behaved function ν of comparable size, whose Fourier coefficients are much easier to obtain bounds for (indeed ν is essentially linearly pseudorandom once one subtracts off its mean, which is essentially 1). This "enveloping sieve" ν is essentially the Selberg upper bound sieve, and can be viewed as a "smoothed out" version[6] of $\Lambda_{W,b}$. Restriction theory (related to the method of the large sieve) is then used to pass from Fourier control of ν to Fourier control of $\Lambda_{W,b}$.

A similar idea was used in [22], [28] to establish the $k = 3$ case of Theorem 1.2; we sketch the argument from [28] here as follows. The main objective is to establish a lower bound for expressions such as

$$\mathbf{E}_{x,r\in\mathbb{Z}/N\mathbb{Z}}\Lambda_{W,b}1_A(x)\Lambda_{W,b}1_A(x+r)\Lambda_{W,b}1_A(x+2r) \tag{14}$$

for large sets A. Restriction theory still allows us to obtain good l^p upper bound for the Fourier coefficients of $\Lambda_{W,b}1_A$. This functions as a substitute for Plancherel's theorem (which is not favourable here), and one can now obtain structure theorems such as Theorem 3.4 (and with some more effort, Theorem 3.5). This decomposes $\Lambda_{W,b}1_A$ into some structured component $f_{U^\perp}$ and a linearly pseudorandom component f_U. The generalized von Neumann theorem lets us dispose the contribution of f_U to (14), so let us focus on $f_{U^\perp}$. One can try to use the complexity bound on $f_{U^\perp}$ (controlling the number of linear phases that comprise $f_{U^\perp}$) to get some lower bound here, but this would require developing a strong structure theorem analogous to Theorem 3.5. It turns out that one can argue more cheaply, using a weaker structure theorem analogous to Theorem 3.4. The key observation is that because $\Lambda_{W,b}1_A$ is dominated (up to a constant) by the enveloping sieve ν, the structured component of $\Lambda_{W,b}1_A$ (which is essentially a convolution of $\Lambda_{W,b}1_A$ with a Fejér-like kernel) is pointwise dominated (up to a constant) by a corresponding structured component of ν. But since ν is linearly pseudorandom after subtracting off its mean, the structured component of ν turns out to essentially be just the mean of ν, which is bounded. We conclude that $f_{U^\perp}$ is bounded, at which point one can just apply Szemerédi's theorem (Theorem 3.1) directly to obtain a good lower bound on this contribution to (14), and one can now conclude the $k = 3$ case of Theorem 1.2.

The proof of Theorem 1.2 for general k in [27] follows the same general strategy, but it is convenient to abandon the Fourier framework (which becomes quite complicated for $k > 3$) and instead take an approach which borrows ingredients from all three approaches, especially the ergodic theory approach. From the Fourier approach one borrows the Gowers uniformity norms $U^{k-2}(\mathbb{Z}/N\mathbb{Z})$, which are a convenient way to define the appropriate notion of pseudorandomness for counting progressions of length k. One still needs an enveloping sieve ν, but instead of using a Selberg-type sieve that enjoys good Fourier coefficient control, it turns out to be more convenient to use an enveloping sieve[7] of Goldston and Yıldırım [15], [16], [17]

[6]What is essentially happening here is that we are viewing the primes not as a zero density subset of the integers, but as a positive density subset of a set of "almost primes" which can be controlled efficiently via sieve theory.

[7]A related enveloping sieve was also used in the recent establishment of narrow gaps in the primes [18].

which has good control on k-point correlations (indeed, it behaves pseudorandomly after subtracting off its mean, which is essentially 1).

The next step is a generalized von Neumann theorem to show that the contribution of pseudorandom functions are negligible. The fact that the functions involved are no longer bounded by 1, but are instead dominated by ν, makes this theorem somewhat trickier to establish, however it can still be achieved by a number of applications of the Cauchy–Schwarz and taking advantage of the pseudorandomness properties of $\nu - 1$. This type of argument is inspired by certain "sparse counting lemmas" arising from the hypergraph approach, particuarly from [21].

The main step, as in previous sections, is a structure theorem which decomposes $\Lambda_{W,b}$ (or $\Lambda_{W,b}1_A$) into a structured component and a pseudorandom component. In principle one could use higher order Fourier analysis (or the precise characteristic factors achieved in [39], [64] to obtain this decomposition, but this looks rather difficult technically, though progress has been made in the $k = 4$ case. Fortunately, there is a "softer" approach in which one defines structure purely by duality; to oversimplify substantially, one defines a function to be structured if it is approximately orthogonal to all pseudorandom functions. One can then obtain a soft structural theorem in which the structural component is essentially a conditional expectation of the original function to a certain σ-algebra generated by certain special structured functions which are called "dual functions" in [27]. This σ-algebra (the finitary analogue of a characteristic factor) is not too tractable to work with, but somewhat miraculously, one can utilize the pseudorandomness properties of ν and a large number of applications of the Cauchy–Schwarz inequality to show that the conditional expectation of ν with respect to this σ-algebra remains bounded (outside of a small exceptional set, which turns out to have a negligible impact). Since $\Lambda_{W,b}1_A$ is pointwise dominated by a constant multiple of ν, the structured component of $\Lambda_{W,b}1_A$ is similarly bounded and can thus be controlled using Szemerédi's theorem. Combining this with the generalized von Neumann theorem to handle the pseudorandom component, one obtains Theorem 1.2. The result for the Gaussian prime constellations is similar, but uses the Gowers cube norms $\Box^{k-2}$ instead of the uniformity norms, and replaces Szemerédi's theorem by a hypergraph removal lemma similar to Lemma 4.1 and Lemma 4.5; see [58], [59].

The arguments used to prove Theorem 1.2 give a lower bound for the expression (13), but do not compute its asymptotic value (which should be 1). As mentioned earlier, for $k = 3$ this can be achieved by the circle method. More recently, the $k = 4$ case has been carried out in [30], [31]; the same method in fact allows one to asymptotically count the number of solutions to any two linear homogeneous equations in four prime unknowns. The key point is to show that $\Lambda_{W,b} - 1$ is quadratically pseudorandom, as the generalized von Neumann theorem will then allow one to control (13) satisfactorily. It turns out that a variant of Lemma 3.6 applies here, and reduces matters to showing that $\Lambda_{W,b} - 1$ does not correlate significantly with any 2-step nilsequences. This task is attackable by Vinogradov's method, although it is rather lengthy and it turns out to be simpler to first replace $\Lambda_{W,b} - 1$ with the closely related Möbius function.

References

[1] Alon, N., Shapira, A., A characterization of the (natural) Graph properties testable with one-sided error. In *46th Symposium on Foundations of Computer Science*, IEEE Computer Soc. Press, Los Alamitos, CA, 2005, 429–438.

[2] Behrend, F. A., On sets of integers which contain no three terms in arithmetic progression. *Proc. Nat. Acad. Sci.* **32** (1946), 331–332.

[3] Bergelson, V., Host, B., Kra, B., Multiple recurrence and nilsequences. *Invent. Math.* **160** (2) (2005), 261–316.

[4] Bourgain, J., A Szemerédi type theorem for sets of positive density in $\mathbb{R}^k$. *Israel J. Math.* **54** (3), (1986), 307–316.

[5] Bourgain, J., On triples in arithmetic progression. *Geom. Func. Anal.* **9** (1999), 968–984.

[6] Chung, F., Graham, R., Wilson, R. M., Quasi-random graphs, *Combinatorica* **9** (1989), 345–362.

[7] Conze, J. P., Lesigne, E. Sur un théorème ergodique pour les mesures diagonales. In *Probabilités*, 1–31, Publ. Inst. Rech. Math. Rennes, 1987-1, Univ. Rennes I, Rennes, 1988, 1–31.

[8] Erdős, P., Turán, P., On some sequences of integers. *J. London Math. Soc.* **11** (1936), 261–264.

[9] Frankl, P., Rödl, V., Extremal problems on set systems. *Random Structures and Algorithms* **20** (2), (2002), 131–164.

[10] Frieze, A., Kannan, R., Quick approximation to matrices and applications. *Combinatorica* **19** (2) (1999), 175–220.

[11] Furstenberg, H., Ergodic behavior of diagonal measures and a theorem of Szemerédi on arithmetic progressions. *J. Analyse Math.*, **31** (1977), 204–256.

[12] Furstenberg, H., *Recurrence in Ergodic theory and Combinatorial Number Theory.* Princeton University Press, Princeton NJ 1981.

[13] Furstenberg, H., Katznelson, Y., Ornstein, D., The ergodic-theoretical proof of Szemerédi's theorem. *Bull. Amer. Math. Soc.* **7** (1982), 527–552.

[14] Furstenberg, H., Weiss, B., A mean ergodic theorem for $1/N \sum_{n=1}^{N} f(T^n x) g(T^{n^2} x)$, In *Convergence in ergodic theory and probability* (Columbus OH 1993), Ohio State Univ. Math. Res. Inst. Publ. 5, Walter de Gruyter, Berlin, 1996. 193–227.

[15] Goldston, D., Yıldırım, C. Y., Higher correlations of divisor sums related to primes, I: Triple correlations. *Integers*, **3** (2003) A5, 66pp. (electronic)

[16] Goldston, D., Yıldırım, C. Y., Higher correlations of divisor sums related to primes, III: k-correlations, preprint.

[17] Goldston, D., Yıldırım, C. Y., Small gaps between primes, I, preprint.

[18] Goldston, D., Motohashi, Y., Pintz, J., Yıldırım, C. Y., Small gaps between primes exist. *Proc. Japan Acad. Ser. A Math. Sci.* **82** (4) (2006), 61–65.

[19] Gowers, T., A new proof of Szemerédi's theorem for arithmetic progressions of length four. *Geom. Func. Anal.* **8** (1998), 529–551.

[20] Gowers, T., A new proof of Szemerédi's theorem, *Geom. Func. Anal.* **11** (2001), 465–588.

[21] Gowers, T., Hypergraph regularity and the multidimensional Szemerédi theorem. Preprint.

[22] Green, B. J., Roth's theorem in the primes. *Ann. of Math.* **161** (3) (2005), 1609–1636.

[23] Green, B. J., A Szemerédi-type regularity lemma in abelian groups. *Geom. Func. Anal.* **15** (2) (2005), 340–376.

[24] Green, B. J., Finite field models in arithmetic combinatorics. In *Surveys in Combinatorics*, London Math. Soc. Lecture Note Ser. 327, (Cambridge University Press, Cambridge 2005, 1–27.

[25] Green, B. J., Long arithmetic progressions of primes. Preprint.

[26] Green, B. J., Konyagin, S., On the Littlewood problem modulo a prime. *Canad. J. Math.*, to appear.

[27] Green, B. J., Tao, T., The primes contain arbitrarily long arithmetic progressions. *Ann. of Math.*, to appear.

[28] Green, B. J., Tao, T., Restriction theory of Selberg's sieve, with applications. *J Théor: Nombres Bordeaux* **18** (2006), 147–182.

[29] Green, B. J., Tao, T., An inverse theorem for the Gowers U^3 norm. *Proc. Edinburgh Math. Soc.*, to appear.

[30] Green, B. J., Tao, T., Quadratic uniformity of the Möbius function. Preprint.

[31] Green, B. J., Tao, T., Two linear equations in four prime unknowns. Preprint.

[32] Green, B. J., Tao, T., New bounds for Szemerédi's theorem, I: Progressions of length 4 in finite field geometries. preprint.

[33] Green, B. J., Tao, T., New bounds for Szemerédi's theorem, II: A new bound for $r_4(N)$. In preparation.

[34] Green, B. J., Tao, T., New bounds for Szemerédi's theorem, III: A polylog bound for $r_4(N)$. In preparation.

[35] Green, B. J., Tao, T., On arithmetic regularity lemmas. In preparation.

[36] Hardy, G. H., Littlewood, J. E. Some problems of "partitio numerorum"; III: On the expression of a number as a sum of primes. *Acta Math.*, **44** (1923), 1–70.

[37] Host, B., Progressions arithmetiques dans les nombres premiers (d'apres B. Green and T. Tao), *Seminaire Bourbaki*, Mars 2005, 57eme annee, 2004–2005, no. 944.

[38] Host, B., Kra, B., Convergence of Conze-Lesigne averages, *Ergodic Theory Dynam. Systems* **21** (2) (2001), 493–509.

[39] Host, B., Kra, B., Non-conventional ergodic averages and nilmanifolds. *Annals of Math.* **161** (1) (2005) 397–488.

[40] Komlós, J., A generalization of a problem of Steinhaus. *Acta Math. Hungar.* **18** (1967), 217–229.

[41] Komlós, J., Simonovits, M., Szemerédi's regularity lemma and its applications in graph theory. In *Combinatorics*, Paul Erdös is eighty, Vol. 2 (Keszthely, 1993), Bolyai Soc. Math. Stud. 2, János Bolyai Math. Soc., Budapest, 1996, 295–352.

[42] Kra, B., The Green–Tao Theorem on arithmetic progressions in the primes: an ergodic point of view. *Bull. Amer. Math. Soc. (N.S)* **43** (1) (2006), 3–23.

[43] Lacey, M., Thiele, C. L^p estimates on the bilinear Hilbert transform for $2 < p < \infty$. *Ann. of Math.* **146** (1997), 693–724.
[44] Leibman, A., Polynomial sequences in groups. *J. Algebra* **201** (1998), 189–206.
[45] Lovász, L., Szegedy, B. Szemerédi's lemma for the analyst. Preprint.
[46] Nagle, B., Rödl, V., Schacht, M., The counting lemma for regular k-uniform hypergraphs. *Random Structures and Algorithms* **28** (2) (2006), 113–179.
[47] Rödl, V., Schacht, M., Regular partitions of hypergraphs. *Combin. Probab. Comput.*, to appear.
[48] Rödl, V., Skokan, J., Regularity lemma for uniform hypergraphs. *Random Structures Algorithms* **25** (1) (2004), 1–42.
[49] Rödl, V., Skokan, J., Applications of the regularity lemma for uniform hypergraphs. *Random Structures Algorithms* **28** (2) (2006), 180–194.
[50] Roth, K. F., On certain sets of integers. *J. London Math. Soc.*, **28** (1953), 245–252.
[51] Ruzsa, I., Szemerédi E., Triple systems with no six points carrying three triangles. *Colloq. Math. Soc. J. Bolyai* **18** (1978), 939–945.
[52] Stein, E., *Harmonic Analysis: Real Variable Methods, Orthogonality, and Oscillatory Integrals*. Princeton University Press, Princeton, NJ, 1993.
[53] Szemerédi, E., On sets of integers containing no k elements in arithmetic progression. *Acta Arith.* **27** (1975), 299–345.
[54] Tao, T., A quantitative ergodic theory proof of Szemerédi's theorem. *Electron. J. Combin.*, to appear.
[55] Tao, T., Obstructions to uniformity, and arithmetic patterns in the primes. *Quarterly J. Pure Appl. Math.* **2** (2006), 199–217.
[56] Tao, T., Arithmetic progressions in the primes. *Collect. Math.* (2006), Vol. Extra., 37–88.
[57] Tao, T., Szemerédi's regularity lemma revisited. *Contrib. Discrete Math.* **1** (1) (2006), 8–28.
[58] Tao, T., A variant of the hypergraph removal lemma. *J. Combin. Theory Ser: A* **113** (7) 1257–1280.
[59] Tao, T., The Gaussian primes contain arbitrarily shaped constellations. *J. Anal. Math.* **99** (2006), 109–176.
[60] Tao, T., An ergodic transference theorem. unpublished.
[61] Tao, T., Vu, V., Additive Combinatorics, book in preparation, Cambridge University Press.
[62] van der Corput, J.G., Über Summen von Primzahlen und Primzahlquadraten, *Math. Ann.* **116** (1939), 1–50.
[63] Varnavides, P., On certain sets of positive density, *J. London Math. Soc.* **34** (1959) 358–360.
[64] Ziegler, T., Universal characteristic factors and Furstenberg averages. *J. Amer. Math. Soc.* **20** (2007), 53–97.

Department of Mathematics, UCLA, Los Angeles CA 90095, U.S.A.
E-mail: tao@math.ucla.edu

Proceedings of the International Congress of Mathematicians
Hyderabad, India, 2010

THE WORK OF NGÔ BAO CHÂU

by
JAMES ARTHUR*

ABSTRACT. Ngô Bao Châu has been awarded a Fields Medal for his proof of the fundamental lemma. I shall try to describe the role of the fundamental lemma in the theory of automorphic forms. I hope that this will make it clear why the result will be a cornerstone of the subject. I will also try to give some sense of Ngô's proof. It is a profound and beautiful argument, built on insights mathematicians have contributed for over thirty years.

Mathematics Subject Classification (2000). Primary 11F55; Secondary 14D23.

Keywords. Fundamental lemma, trace formula, Hitchin fibration, affine Springer fibres, stabilization.

The Formal Statement

Here is the statement of Ngô's primary theorem. It is taken from the beginning of the introduction of his paper [N2].

Théorème 1. *Soient k un corps fini à q éléments, $\mathcal{O}$ un anneau de valuation discrète complet de corps résiduel k et F son corps des fractions. Soit G un schéma en groupes réductifs au-dessus de $\mathcal{O}$ dont le nombre de Coxeter multiplié par deux est plus petit que la caractéristique de k. Soient $(\kappa,\, p_\kappa)$ une donnée endoscopique de G au-dessus de $\mathcal{O}$ et H le schéma en groupes endoscopiques associeé.*

On a l'égalité entre la κ-intégrale orbitale et l'intégrale orbitale stable

$$\Delta_G(a)\mathbf{O}_a^k(1_{\mathfrak{g}}, dt) = \Delta_H(a_H)\mathbf{SO}_{aH}(1_{\mathfrak{h}}, dt) \tag{1}$$

associées aux classes de conjugaison stable semi-simples régulières a et a_H de $\mathfrak{g}(F)$et $\mathfrak{h}(F)$ qui se correspondent, aux fonctions caractéristiques $1_{\mathfrak{g}}$ et $1_{\mathfrak{h}}$ des compacts $\mathfrak{g}(\mathcal{O})$et $\mathfrak{h}(\mathcal{O})$ dans $\mathfrak{g}(F)$et $\mathfrak{h}(F)$ et oú on a noté

$$\Delta_G(a) = q^{-\mathrm{val}(\mathfrak{D}_G(a))/2} et\, \Delta_H(a_H) = q^{-\mathrm{val}(\mathfrak{D}_H(a_H))/2}$$

$\mathfrak{D}_G$ and $\mathfrak{D}_H$ étant les fonctions discriminant de G et de H.

*Department of Mathematics, University of Toronto, Toronto, ON M5S 2E4 Canada. E-mail: arthur@math.toronto.edu

In §1.11 of his paper, Ngô describes the various objects of his assertion in precise terms. At this point we simply note that the "orbital integrals" he refers to are integrals of locally constant functions of compact support. The assertion is therefore an identity of sums taken over two finite sets. Observe however that there is one such identity for every pair (a, a_H) of "regular orbits". As a approaches a singular point, the size of the two finite sets increases without bound, and so therefore does the complexity of the identity. Langlands called it the *fundamental lemma* when he first encountered the problem in the 1970's. It was clearly fundamental, since he saw that it would be an inescapable precondition for any of the serious applications of the trace formula he had in mind. He called it a lemma because it seemed to be simply a family of combinatorial identities, which would soon be proved. Subsequent developments, which culminated in Ngô's proof, have revealed it to be much more. The solution draws on some of the deepest ideas in modern algebraic geometry.

Ngô's theorem is an infinitesimal form of the fundamental lemma, since it applies to the Lie algebras $\mathfrak{g}$ and $\mathfrak{h}$ of the groups G and H. However, Waldspurger had previously used methods of descent to reduce the fundamental lemma for groups to its Lie algebra variant [W3]. Ngô's geometric methods actually apply only to fields of positive characteristic, but again Waldspurger had earlier shown that it suffices to treat this case [W1].[1] Therefore Ngô's theorem does imply the fundamental lemma that has preoccupied mathematicians in automorphic forms since it was first conjectured by Langlands in the 1970's.

I would like to thank Steve Kudla for some helpful suggestions.

Automorphic Forms and the Langlands Programme

To see the importance of the fundamental lemma, we need to recall its place in the theory of automorphic forms. *Automorphic forms* are eigenforms of a commuting family of natural operators attached to reductive algebraic groups. The corresponding eigenvalues are of great arithmetic significance. In fact, the information they contain is believed to represent a unifying force for large parts of number theory and arithmetic geometry. The *Langlands programme* summarizes much of this, in a collection of interlocking conjectures and theorems that govern automorphic forms and their associated eigenvalues. It explains precisely how a theory with roots in harmonic analysis on algebraic groups can characterize some of the deepest objects of arithmetic. There has been substantial progress in the Langlands programme since its origins in a letter from Langlands to Weil in 1967. However, its deepest parts remain elusive.

[1]Another proof of this reduction was subsequently established by Cluckers, Hales and Loeser, by completely different methods of motivic integration.

The operators that act on automorphic forms are differential operators (Laplace-Beltrami operators) and their combinatorial p-adic analogues (Hecke operators). They are best studied implicitly in terms of group representations. One takes G to be a connected reductive algebraic group over a number field F, and R to be the representation of $G(\mathbb{A})$ by right translation on the Hilbert space $L^2(G(F)\backslash G(\mathbb{A}))$. We recall that $G(\mathbb{A})$ is the group of points in G with values in the ring $\mathbb{A} = \mathbb{A}_F$ of adéles of F, a locally compact group in which the diagonal image of $G(F)$ is discrete. Automorphic forms, roughly speaking, are functions on $G(F)\backslash G(A)$ that generate irreducible subrepresentations of R, which are in turn known as *automorphic representations.* Their role is similar to that of the much more elementary functions

$$e^{inx}, \quad n \in \mathbb{Z}, \quad x \in \mathbb{Z}\backslash\mathbb{R},$$

in the theory of Fourier series. We can think of x as a *geometric* variable, which ranges over the underlying domain, and n as a *spectral* variable, whose automorphic analogue contains hidden arithmetic information.

The centre of the Langlands programme is the principle of functoriality. It postulates a reciprocity law for the spectral data in automorphic representations of different groups G and H, for any L-homomorphism $\rho : {}^L H \rightarrow {}^L G$ between their L-groups. We recall that ${}^L G$ is a complex, nonconnected group, whose identity component $\hat{G}$ can be regarded as a complex dual group of G. There is a special case of this that is of independent interest. It occurs when H is an *endoscopic group* for G, which roughly speaking, means that ρ maps $\widehat{H}$ injectively onto the connected centralizer of a semisimple element of $\widehat{G}$. The theory of endoscopy, due also to Langlands, is a separate series of conjectures that includes more than just the special case of functoriality. Its primary role is to describe the internal structure of automorphic representations of G in terms of automorphic representations of its smaller endoscopic groups H. The fundamental lemma arises when one tries to use the trace formula to relate the automorphic representations of G with those of its endoscopic groups.[2]

The Trace Formula and Transfer

The trace formula for G is an identity that relates spectral data with geometric data. The idea, due to Selberg, is to analyze the operator

$$R(f) = \int_{G(\mathbb{A})} f(y)R(y)\mathrm{d}y$$

[2]Endoscopic groups should actually be replaced by *endoscopic data,* objects with slightly more structure, but I will ignore this point.

on $L^2(G(F)\backslash G(\mathbb{A}))$ attached to a variable test function f on $G(\mathbb{A})$. One observes that $R(f)$ is an integral operator, with kernel

$$K(x,y) = \sum_{\gamma \in G(F)} f(x^{-1}\gamma y), \quad x, y \in G(\mathbb{A}).$$

One then tries to obtain an explicit formula by expressing the trace of $R(f)$ as the integral

$$\int_{G(F)\backslash G(\mathbb{A})} \sum_{\gamma \in G(F)} f(x^{-1}\gamma x)\mathrm{d}x$$

of the kernel over the diagonal. The formal outcome is an identity

$$\sum_{\{\gamma\}} \int_{G_\gamma(F)\backslash G(\mathbb{A})} f(x^{-1}\gamma x)\,\mathrm{d}x = \sum_{\pi} tr(\pi(f)), \tag{2}$$

where $\{\gamma\}$ ranges over the conjugacy classes in $G(F)$, $G_\gamma(F)$ is the centralizer of γ in $G(F)$, and π ranges over automorphic representations.

The situation is actually more complicated. Unless $G(F)\backslash G(\mathbb{A})$ is compact, a condition that fails in the most critical cases, $R(f)$ is not of trace class, and neither side converges. One is forced first to truncate the two sides in a consistent way, and then to evaluate the resulting integrals explicitly. It becomes an elaborate process, but one that eventually leads to a rigorous formula with many new terms on each side [A1]. However, the original terms in (2) remain the same in case π occurs in the *discrete* part of the spectral decomposition of R, and γ is *anisotropic* in the strong sense that G_γ is a maximal torus in G with $G_\gamma(F)\backslash G_\gamma(\mathbb{A})$ compact. If γ is anisotropic, and f is a product of functions f_v on the completions $G(F_v)$ of $G(F)$ at valuations v on F, the corresponding integral in (2) can be written

$$\begin{aligned}
&\int_{G\gamma(F)\backslash G(\mathbb{A})} f(x^{-1}\gamma x)\,\mathrm{d}x \\
&\quad = \mathrm{vol}(G_\gamma(F)\backslash G_\gamma(\mathbb{A})) \int_{G_\gamma(\mathbb{A})/G(\mathbb{A})} f(x^{-1}\gamma x)\,\mathrm{d}x \\
&\quad = \mathrm{vol}(G_\gamma(F)\backslash G_\gamma(\mathbb{A})) \prod_v \int_{G_\gamma(F_v)\backslash G(F_v)} f_v(x_v^{-1}\gamma x_v)\,\mathrm{d}x_v.
\end{aligned}$$

The factor

$$\mathbf{O}_\gamma(f_v) = \mathbf{O}_\gamma(f_v, \mathrm{d}t_v) = \int_{G_{\gamma(F_v)}\backslash G(F_v)} f_v(x_v^{-1}\gamma x_v)\,dx_v$$

is the "orbital integral" of f_v over the conjugacy class of γ in $G(F_v)$. It depends on a choice of Haar measure $\mathrm{d}t_v$ on $T(F_v) = G_\gamma(F_v)$, as well as the underlying Haar measure $\mathrm{d}x_v$ on $G(F_v)$, and makes sense if γ is replaced by any element $\gamma_v \in G(F_v)$ that is *strongly regular*, in the sense that $G_{\gamma v}$ is any maximal torus.

The goal is to compare automorphic spectral data on different groups G and H by establishing relations among the geometric terms on the left hand sides of their associated trace formulas. This presupposes the existence of a suitable transfer correspondence $f \to f^H$ of test functions from $G(\mathbb{A})$ to $H(\mathbb{A})$. The idea here is to define the transfer locally at each completion v by asking that the orbital integrals of f_v^H match those of f_v. Test functions are of course smooth functions of compact support, a condition that for the totally disconnected group $G(F_v)$ at a p-adic place v becomes the requirement that f_v be locally constant and compactly supported. The problem is to show for both real and p-adic places v that f_v^H, defined only in terms of conjugacy classes in $H(F_v)$, really is the family of orbital integrals of a smooth function of compact support on $H(F_v)$.

The transfer of functions is a complex matter, which I have had to oversimplify. It is founded on a corresponding transfer mapping $\gamma_{H,v} \to \gamma_v$ of strongly regular conjugacy classes over v from any local endoscopic group H for G to G itself. But this only makes sense for *stable* (strongly regular) conjugacy classes, which in the case of G are defined as the intersections of $G(F_v)$ with conjugacy classes in the group $G(\overline{F}_v)$ over an algebraic closure $\overline{F}_v$. A *stable* orbital integral of f_v is the sum of ordinary orbital integrals over the finite set of conjugacy classes in a stable conjugacy class. Given f_v, H and $\gamma_{H,v}$, Langlands and Shelstad set $\mathbf{SO}_{\gamma H,v}(f_v^H)$ equal to a certain linear combination of orbital integrals of f_v over the finite set of conjugacy classes in the stable image γ_v of $\gamma_{H,v}$. The coefficients are subtle but explicit functions, which they introduce and call transfer factors [LS]. They then conjecture that as the notation suggests, $\{\mathbf{SO}_{\gamma H,v}(f_v^H)\}$ is the set of stable orbital integrals of a smooth, compactly supported function f_v^H on $H(F_v)$.

We can at last say what the fundamental lemma is. For a test function $f = \prod_v f_v$ on $G(\mathbb{A})$ to be *globally* smooth and compactly supported, it must satisfy one further condition. For almost all p-adic places v, f_v must equal the characteristic function 1_{Gv} of an (open) hyperspecial maximal compact subgroup K_v of $G(F_v)$. The fundamental lemma is the natural variant at these places of the Langlands-Shelstad transfer conjecture. It asserts that if f_v equals 1_{G_v}, we can actually take f_v^H to be an associated characteristic function 1_{Hv} on $H(F_v)$. It is in these terms that we understand the identity (1) in Ngô's theorem. We of course have to replace 1_{G_v} and 1_{Hv} by their analogues $1_{\mathfrak{g}v}$ and $1\mathfrak{h}_v$ on the Lie algebras $\mathfrak{g}(F_v)$ and $\mathfrak{h}(F_v)$ of $G(F_v)$ and $H(F_v)$, and the mapping $\gamma_{H,v} \to \gamma_v$ by a corresponding transfer mapping $a_{H,v} \to a_v$ of stable adjoint orbits. The superscript κ on the left hand side of (1) is an index that determines an endoscopic group $H = H^\kappa$ for G over F_v by a well defined procedure. It also determines a corresponding linear combination of orbital integrals (called a κ-orbital integral) on $\mathfrak{g}(F_v)$, indexed by the $G(F_v)$-orbits in the stable orbit a_v. The coefficients depend in a very simple way on κ, and when normalized by the quotient $\Delta_G(\cdot)\Delta_H(\cdot)^{-1}$ of discriminant functions, represent the specialization of the general Langlands-Shelstad transfer factors to the Lie algebra $\mathfrak{g}(F_v)$. The term on the left hand side of (1) is a K-orbital integral of $1_{\mathfrak{g}_v}$, and the term on the right hand side is a corresponding stable orbital integral of $1_{\mathfrak{h}_v}$.

The Hitchin Fibration

We have observed that local information, in the form of the Langlands-Shelstad transfer conjecture and the fundamental lemma, is a requirement for the comparison of global trace formulas. However, it is sometimes also possible to go in the opposite direction, and to deduce local information from global trace formulas. The most important such result is due to Waldspurger. In 1995, he used a special case of the trace formula to prove that the fundamental lemma implies the Langlands-Shelstad transfer conjecture for p-adic places v [W1]. (The archimedean places v had been treated by local means earlier by Shelstad. See [S].) The fundamental lemma would thus yield the full global transfer mapping $f \to f^H$. It is indeed fundamental!

Ngô had a wonderful idea for applying global methods to the fundamental lemma itself. He observed that the Hitchin fibration [H], which Hitchin had introduced for the study of the moduli space of vector bundles on a Riemann surface, was related to the geometric side of the trace formula. His idea applies to the field $F = k(X)$ of rational functions on a (smooth, projective) curve X over a finite field of large characteristic. This is a global field, which combines the arithmetic properties of a number field with the geometric properties of the field of meromorphic functions on a Riemann surface, and for which both the trace formula and the Hitchin fibration have meaning. Ngô takes G to be a quasisplit group scheme over X. His version of the Hitchin fibration also depends on a suitable divisor D of large degree on X.

The total space of the Hitchin fibration $\mathcal{M} \to \mathcal{A}$ is an algebraic (Artin) stack[3] $\mathcal{M}$ over k. To any scheme S over k, it attaches the groupoid $\mathcal{M}(S)$ of Higgs pairs (E, ϕ), where E is a G-torsor over $X \times S$, and $\phi \in H^0(X \times S, \mathrm{Ad}(E) \otimes \mathcal{O}_X(D))$ is a section of the vector bundle $\mathrm{Ad}(E)$ obtained from the adjoint representation of G on its Lie algebra $\mathfrak{g}$, twisted by the line bundle $\mathcal{O}_X(D)$. Ngô observed that in the case $S = Spec(\kappa)$, the definitions lead to a formal identify

$$\sum_{\xi} \left(\sum_{\{a\}} \int_{G^{\xi}_{a}(F) \backslash G^{\xi}(\mathbb{A})} f_D(\mathrm{Ad}(x)^{-1} a)\, dx \right) = |\{\mathcal{M}(k)\}|, \tag{3}$$

whose right hand side equals the number of isomorphism classes in the groupoid $\mathcal{M}(k)[N1, \S 1]$. On the left hand side, ξ ranges over the set $\ker^1(F, G)$ of locally trivial elements in $H^1(F, G)$, a set that frequently equals $\{1\}$, and G^{ξ} is an inner twist of G by ξ, equipped with a trivialization over each local field F_v, with Lie algebra $\mathfrak{g}^{\xi}$. Also, $\{a\}$ ranges over the $G^{\xi}(F)$ orbits in $\mathfrak{g}^{\xi}(F)$, and $G^{\xi}_{a}(F)$ is the stabilizer of a in

[3] I am little uncomfortable discussing objects in which I do not have much experience. I apologize in advance for any inaccuracies.

$G^\xi(F)$, while

$$f_D = \bigotimes_v f_{D,v},$$

where v ranges over the valuations of F (which is to say the closed points of X) and $f_{D,v}$ is the characteristic function in $\mathfrak{g}^\xi(F_v)$ of the open compact subgroup $\varpi_v^{-d_v(D)}\mathfrak{g}^\xi(\mathcal{O}_v)$.

The expression in the brackets in (3) is the analogue for the Lie algebra $\mathfrak{g}^\xi$ of the left hand geometric side of (2). It is to be regarded in the same way as (2), as part of a formal identity between two sums that both diverge. On the other hand, as in (2), the sum over the subset of orbits $\{a\}$ that are anisotropic actually does converge.

The base $\mathcal{A}$ of the Hitchin fibration is an affine space over k. As a functor, it assigns to any S the set

$$\mathcal{A}(S) = \bigotimes_{i=1}^{r} H^0(X \times S, \mathcal{O}_X(e_i D)),$$

where $e_1, \ldots, e_r$ are the degrees of the generators of the polynomial algebra of G-invariant polynomials on $\mathfrak{g}$. Roughly speaking, the set $\mathcal{A}(k)$ attached to $S = \mathrm{Spec}(k)$ parametrizes the stable $G(\mathbb{A})$-orbits in $\mathfrak{g}(\mathbb{A})$ that have representatives in $\mathfrak{g}(F)$, and intersect the support of the function f_D. The Chevalley mapping from $\mathfrak{g}$ to its affine quotient $\mathfrak{g}/G$ determines a morphism h from $\mathcal{M}$ to $\mathcal{A}$ over k. This is the Hitchin fibration. Ngô uses it to isolate the orbital integrals that occur on the left hand side of (3). In particular, he works with the open subscheme $\mathcal{A}^{\mathrm{ani}}$ of $\mathcal{A}$ that represents orbits that are anisotropic over $\bar{k}$. The restriction

$$h^{\mathrm{ani}} : \mathcal{M}^{\mathrm{ani}} \to \mathcal{A}^{\mathrm{ani}}, \quad \mathcal{M}^{\mathrm{ani}} = h^{-1}(\mathcal{A}^{\mathrm{ani}}) = \mathcal{M} \times_{\mathcal{A}} \mathcal{A}^{\mathrm{ani}}, \tag{4}$$

of the morphism h to $\mathcal{A}^{\mathrm{ani}}$ is then proper and smooth, a reflection of the fact that the stabilizer in G of any anisotropic point $a \in \mathfrak{g}(F)$ is an anisotropic torus over the maximal unramified extension of F. (See [N2, §4].)

Affine Springer Fibres

The Hitchin fibration can be regarded as a "geometrization" of a part of the global trace formula. It opens the door to some of the most powerful techniques of algebraic geometry. Ngô uses it in conjunction with another geometrization, which had been introduced earlier, and applies to the fibres $\mathcal{M}_a$ of the Hitchin fibration. This is the interpretation of the local orbital integral

$$\mathbf{O}_{\gamma v}(1_{\mathfrak{g}_v}) = \int_{G_{a_v}(F_v)\backslash G(F_v)} 1_{\mathfrak{g}_v}(\mathrm{Ad}(x_v)^{-1}av)dx_v$$

in terms of affine Springer fibres.

The original Springer fibre of a nilpotent element N in a complex semisimple Lie algebra is the variety of Borel subalgebras (or more generally, of parabolic subalgebras in a given adjoint orbit under the associated group) that contain N. It was used by Springer to classify irreducible representations of Weyl groups. The affine Springer fibre of a topologically unipotent (regular, semisimple) element $a_v \in \mathfrak{g}(F_v)$, relative to the adjoint orbit of the lattice $\mathfrak{g}(\mathcal{O}_v)$, is the set

$$\mathcal{M}_v(a,k) = \{x_v \in G(F_v)/G(\mathcal{O}_v) : \mathrm{Ad}(x_v)^{-1}a_v \in \mathfrak{g}(\mathcal{O}_v)\}$$

of lattices in the orbit that contain a_v. Suppose for example that a_v is anisotropic over F_v, in the strong sense that the centralizer $G_{av}(F_v)$ is compact. If one takes the compact (abelian) groups $G_{a_v}(F_v)$ and $\mathfrak{g}(\mathcal{O}_v)$ to have Haar measure 1, one sees immediately that $\mathbf{O}_{a_v}(1_{\mathfrak{g}v})$ equals the order $|\mathcal{M}_v(a,k)|$ of $\mathcal{M}_v(a,k)$. (Topologically unipotent means that the linear operator $\mathrm{ad}(a_v)^n$ on $\mathfrak{g}(F_v)$ approaches 0 as n approaches infinity. In general, the closer a_v is to 0, the larger is the set $\mathcal{M}_v(a,k)$, and the more complex the orbital integral $\mathbf{O}_{a_v}(1_{\mathfrak{g}_v})$.)

Kazhdan and Lusztig introduced affine Springer fibres in 1988, and established some of their geometric properties [KL]. In particular, they proved that $\mathcal{M}_v(a,k)$ is the set of k-points of an inductive limit $\mathcal{M}_v(a)$ of schemes over k. (It is this ind-scheme that is really called the affine Springer fibre.) Their results also imply that if a_v is anisotropic over the maximal unramified extension of $F_v, \mathcal{M}_v(a)$ is in fact a scheme.

The study of these objects was then taken up by Goresky, Kottwitz and MacPherson. Their strategy was to obtain information about the orbital integral $|M_v(a,k)|$ from some version of the Lefschetz fixed point formula. They realized that relations among orbital integrals could sometimes be extracted from cohomology groups of affine Springer fibres $\mathcal{M}_v(a)$ and $\mathcal{M}_v(a_H)$, for the two different groups G and H. Following this strategy, they were able to establish the identity (1) for certain pairs $(a_v, a_{H,v})$ attached to unramified maximal tori [GKM]. Goresky, Kottwitz and MacPherson actually worked with certain equivariant cohomology groups. Laumon and Ngô later added a deformation argument, which allowed them to prove the fundamental lemma for unitary groups [LN]. However, the equivariant cohomology groups that led to these results are not available in general.

It was Ngô's introduction of the global Hitchin fibration that broke the impasse. He formulated the affine Springer fibre $\mathcal{M}_v(a)$ as a functor of schemes S over k, in order that it be compatible with the relevant Hitchin fibre $\mathcal{M}_a$[N2, §3.2]. He also introduced a third object to mediate between the two kinds of fibre. It is a Picard stack $\mathcal{P} \to \mathcal{A}$, which acts on $\mathcal{M}$, and represents the natural symmetries of the Hitchin fibration. Ngô attached this object to the group scheme J over $\mathcal{A}$ obtained from the G-centralizers of regular elements in $\mathfrak{g}$, and the Kostant section from semisimple conjugacy classes to regular elements.

The stack $\mathcal{P}$ plays a critical role. Ngô used it to formulate the precise relation between the Hitchin fibre $\mathcal{M}_a$ at any $a \in \mathcal{A}^{ani}(\bar{k})$ with the relevant affine Springer fibres $\mathcal{M}_v(a)$ [N2, Proposition 4.15.1]. Perhaps more surprising is the fact that as a group object in the category of stacks, $\mathcal{P}$ governs the stabilization of anisotropic Hitchin fibres $\mathcal{M}_a$. Ngô analyses the characters $\{\kappa\}$ on the abelian groups of connected components $\pi_0(\mathcal{P}_a)$. He shows that they are essentially the geometric analogues of objects that were used to stabilize the anisotropic part of the trace formula.

Stabilization

Could one possibly establish the fundamental lemma from the trace formula? Any such attempts have always foundered on the lack of a transfer of unit functions 1_{G_v} to 1_{H_v} by orbital integrals. In some sense, however, this is exactly what Ngô does. It is not the trace formula for automorphic forms that he uses, but the Grothendieck-Lefschetz trace formula of algebraic geometry. Moreover, it is the "spectral" side of this trace formula that he transfers from $\mathfrak{g}$ to $\mathfrak{h}$ (the Lie algebras of G and H), in the form of data from cohomology, rather than its "geometric" side, in the form of data given by fixed points of Frobenius endomorphisms. This is in keeping with the general strategy of Goresky, Kottwitz and MacPherson. The difference here is that Ngô begins with perverse cohomology attached to the global Hitchin fibration, rather than the ordinary equivariant cohomology of a local affine Springer fibre.

Stabilization refers to the operation of writing the trace formula for G, or rather each of its terms $I(f)$, as a linear combination

$$I(f) = \sum_H \iota(G,H) S^H(f^H) \tag{5}$$

of stable distributions on the endoscopic groups H of G over F. (A stable distribution is a linear form whose values depend only on the stable orbital integrals of the given test function. The resulting identity of stable distributions for any given H, obtained by induction on $\dim(H)$ from (5) and the trace formula for G, is known as the stable trace formula.) The process is most transparent for the anisotropic terms[4]

$$I^{\mathrm{ani}}(f) = \sum_{\{\gamma\}} \mathrm{vol}(G_\gamma(F)\backslash G_\gamma(\mathbb{A})) \cdot \prod_v (\mathbf{O}_\gamma(f_v)), \tag{6}$$

in which $\{\gamma\}$ ranges over the set of anisotropic conjugacy classes in $G(F)$. It was carried out in this case by Langlands [L] and Kottwitz [K2], assuming the existence of the global transfer mapping $f \to f^H$ (which Waldspurger later reduced to the fundamental lemma). This is reviewed by Ngô in the first chapter (§1.13) of his paper [N2].

[4]This expression only makes sense if the split component AG of G is trivial. In general, one must include AG in the volume factors.

The idea for the stabilization of (6) can be described very roughly as follows. One first groups the conjugacy classes $\{\gamma\}$ into stable conjugacy classes $\{\gamma\}_{\mathrm{st}}$ in $G(F)$, for representatives γ attached to anisotropic tori $T = G_\gamma$. The problem is to quantify the obstruction for the contribution of $\{\gamma\}_{\mathrm{st}}$ to be a stable distribution on $G(\mathbb{A})$. For any v, the set of $G(F_v)$-conjugacy classes in the stable conjugacy class of γ in $G(F_v)$ is bijective with the set

$$\ker(H^1(F_v, G) \to H^1(Fv, T))$$

of elements in the finite abelian group $H^1(F_v, T)$ whose image in the Galois cohomology set $H^1(F_v, G)$ is trivial. Let me assume for simplicity in this description that G is simply connected. The set $H^1(F_v, G)$ is then trivial for any p-adic place v, and becomes a concern only when v is archimedean. The obstruction for $\{\gamma\}_{\mathrm{st}}$ is thus closely related to the abelian group

$$\operatorname{coker}\left(H^1(F,T) \longrightarrow \bigoplus_v H^1(F_v,T)\right).$$

The next step is to apply Fourier inversion to this last group. According to Tate-Nakayama duality theory, its dual group of characters κ is isomorphic to $\widehat{T}^\Gamma$, the group of elements in the complex dual torus $\widehat{T}$ that are invariant under the natural action of the global Galois group $\Gamma = \mathrm{Gal}(\overline{F}\backslash F)$. On the other hand, each $\kappa \in \widehat{T}^\Gamma$ maps to a semisimple element in the complex dual group $\widehat{G}$, which can be used to define an endoscopic group $H = H^\kappa$ for G. One accounts for the local archimedean sets $H^1(F_v, G)$ simply by defining the local contribution of a complementary element in $H^1(F_v, T)$ to be 0. In this way, one obtains a global contribution to (6) for any κ. It is a global κ-orbital integral, whose local factor at almost any v appears on the left hand side of the identity in the fundamental lemma.

One completes the stabilization of (6) by grouping the indices (T, κ) into equivalence classes that map to a given H. The corresponding contributions to the right hand side of (6) become the summands of H in (5). Notice that the summands with $\kappa = 1$ correspond to the endoscopic group H with $\widehat{H} = \widehat{G}$ (a quasisplit inner form G^* of G). Like all of the other summands, they are defined directly. This is in contrast to the more exotic parts I(f) of the trace formula [A1, §29], where the contribution of $H = G^*$ (known as the stable part $I_{\mathrm{st}}(f)$ of $I(f)$ in case $G = G^*$ is already quasisplit) can only be constructed from (5) indirectly by induction on $\dim(H)$.

The heart of Ngô's proof is an analogue of the stabilization of (6) for the geometrically anisotropic part (4) of the Hitchin fibration.[5] This does not depend on the transfer of functions, and is therefore unconditional. Ngô formulates it as an identity of the $\{\kappa\}$-component $(\cdot)_\kappa$ of an object attached to G with the stable

[5]Recall that the left hand side of (3) differs from that of (2) in having a supplementary sum over $\xi \in \ker^1(F, G)$. This is part of the structure of the Hitchin fibration. But it also actually leads to a slight simplification of the stabilization of (6) by Langlands and Kottwitz. (See [N2, §1.13].)

component $(\cdot)_{\rm st}$ of a similar object for the corresponding endoscopic group. I will only be able to describe his steps in the most general of terms.

Since $\mathcal{M}^{\rm ani}$ is a smooth Deligne-Mumford stack, the purity theorems of [D] and [BBD] can be applied to the proper morphism $h^{\rm ani}$ in (4). They yield an isomorphism

$$h_*^{\rm ani}\overline{\mathbb{Q}}_\ell \cong \bigoplus_n {}^p H^n(h_*^{ani}\overline{Q}_\ell)[-n], \tag{7}$$

whose left hand side is a priori only an object in the derived category $D_c^b(\mathcal{A})$ of the bounded complexes of sheaves on $\mathcal{A}$ with constructible cohomology, but whose right hand summands are pure objects in the more manageable abelian subcategory of perverse sheaves on $\mathcal{A}$. Ngô then considers the action of the stack $\mathcal{P}^{\rm ani}$ over $\mathcal{A}^{\rm ani}$ on either side. Appealing to a homotopy argument, he observes that this action factors through the quotient $\pi_0(\mathcal{P}^{\rm ani})$ of connected components, a sheaf of finite abelian groups on $\mathcal{A}^{\rm ani}$. As we noted earlier, an analysis of this sheaf then leads him to the dual characters $\{\kappa\}$ that were part of the stabilization of (6), and relative to which one can take equivariant components ${}^pH^n(f_*^{\rm ani}\overline{\mathbb{Q}}_\ell)_\kappa$ of the summands in (7). On the other hand, if H corresponds to κ, we have the morphism ν from $\mathcal{A}_H$ to $\mathcal{A}$ that comes from the embedding $\widehat{H} \subset \widehat{G}$ of two dual groups of equal rank. It provides a pullback mapping of sheaves from $\mathcal{A}$ to $\mathcal{A}_H$. Ngô's stabilization of (4) then takes the form of an isomorphism

$$v^*\left(\bigoplus_n {}^pH^n(h_*^{\rm ani}\overline{\mathbb{Q}})_\kappa[2r](r)\right) \cong \bigoplus_n {}^pH^n(h_{H,*}^{\rm ani}\overline{\mathbb{Q}}_\ell)_{\rm st}, \tag{8}$$

for a degree shift $[2r]$ and Tate twist (r) attached to a certain positive integer $r = r_H^G(D)$. (See [N2, Theorem 6.4.2].)

Ngô's "geometric stabilization" identity (8), whose statement I have oversimplified slightly,[6] is a key theorem. In particular, it leads directly to the fundamental lemma. For it implies a similar identity for the stalks of the sheaves at a point $a_H \in \mathcal{A}_H$ (with image $a \in \mathcal{A}$ under ν). After some further analysis, the application of a theorem of proper base change reduces what is left to an endoscopic identity for the cohomology of affine Springer fibres. This is exactly what Goresky, Kottwitz and MacPherson had been working towards. Once it is available, an application of the Grothendieck-Lefschetz trace formula gives a relation among points on affine Springer fibres, which leads to the fundamental lemma. (See [LN, §3.10] for example.)

However, it is more accurate to say that the (global) stabilization identity (8) is *parallel* to the (local) fundamental lemma. Ngô actually had to prove the two theorems together. In a series of steps, which alternate between local and global arguments, and go back and forth between the two theorems, he treats special cases

[6]The isomorphism is between the semisimplifications of the graded perverse sheaves. Moreover, v, $h^{\rm ani}$ and $h_H^{\rm ani}$ should be replaced by their preimages $\tilde{\nu}$, $\tilde{h}^{\rm ani}$ and $\tilde{h}_H^{\rm ani}$ relative to certain finite morphisms.

that become increasingly more general, until the proof of both theorems is at last complete. Everything of course depends on the original divisor D on X, which in Ngô's argument is allowed to vary in such a way that its degree approaches infinity. The main technical result that goes into the proof of (8) is a theorem on the support of the sheaves on the left hand side. As I understand it, this is highly dependent on the fact that these objects are actually perverse sheaves.

Further Remarks

I should also mention two important generalizations of the fundamental lemma. One is the "twisted fundamental lemma" conjectured by Kottwitz and Shelstad, which will be needed for any endoscopic comparison that includes the twisted trace formula. Waldspurger [W3] had reduced this conjecture to the primary theorem of Ngô, together with a variant [N2, Théorème 2] of (1) that Ngô proves by the same methods. Another is the "weighted fundamental lemma", which applies to the more general geometric terms in the trace formula that are obtained by truncation. It is needed for any endoscopic comparisons that do not impose unsatisfactory local constraints on the automorphic representations. Once again, Waldspurger had reduced the conjectural identity to its analogue for a Lie algebra over a local field of positive characteristic. Chaudouard and Laumon have recently proved the weighted fundamental lemma for Lie algebras by extending the methods of Ngô to other terms in the trace formula [CL]. This has been a serious enterprise, which requires a geometrization of analytic truncation methods in order to deal with the failure of the full Hitchin fibration $\mathcal{M} \to \mathcal{A}$ to be proper. In any case, all forms of the fundamental lemma have now been proved, including the most general "twisted, weighted fundamental lemma".

I have emphasized the role of transfer in the comparison of trace formulas. This is likely to lead to a classification of automorphic representations for many groups G, beginning with orthogonal and symplectic groups [A2], according to Langlands' conjectural theory of endoscopy. The fundamental lemma also has other important applications. For example, its proof fills a longstanding gap in the theory of Shimura varieties. Kottwitz observed some years ago that the key geometric terms in the Grothendieck-Lefschetz formula for a Shimura variety are actually twisted orbital integrals [K1]. The twisted fundamental lemma now allows a comparison of these terms with corresponding terms in the stable trace formula. (See [K3].) This in turn leads to reciprocity laws between the arithmetic data in the cohomology of many such varieties with the spectral data in automorphic forms.

This completes my report. It will be clear that Ngô's proof is deep and difficult. What may be less clear is the enormous scope of his methods. The many diverse geometric objects he introduces are all completely natural. That they so closely reflect objects from the trace formula and local harmonic analysis, and fit together so beautifully in Ngô's proof, is truly remarkable.

References

[A1] J. Arthur, *An introduction to the trace formula,* in *Harmonic Analysis, the Trace Formula, and Shimura Varieties,* Clay Mathematics Proceedings, vol. 4, 2005, 1–263.

[A2] J. Arthur, *The Endoscopic Classification of Representations: Orthogonal and, Symplectic Groups,* in preparation.

[BBD] A. Beilinson, J. Bernstein and P. Deligne, *Faiseaux pervers,* Astérisque **100** (1982).

[CL] P.-H. Chaudouard and G. Laumon, *Le lemme fondamental pondere* II: *Énoncés cohomologiques,* preprint.

[D] P. Deligne, *La conjecture de Weil* II, Publ. Math. de I.H.E.S. **52** (1980), 137–252.

[GKM] M. Goresky, R. Kottwitz, and R. MacPherson, *Homology of affine Springer-fibres in the unramified case,* Duke Math. J. **121** (2004), 509–561.

[H] N. Hitchin, *Stable bundles and integrable connections,* Duke Math. J. **54** (1987), 91–114.

[KL] D. Kazhdan and G. Lusztig, *Fixed point varieties on affine flag manifolds,* Israel J. Math. **62** (1988), 129–168.

[K1] R. Kottwitz, *Shimura varieties and twisted orbital integrals,* Math. Ann. **269** (1984), 287–300.

[K2] R. Kottwitz, *Stable trace formula: elliptic singular terms,* Math. Ann. **275** (1986), 365–399.

[K3] R. Kottwitz, *Shimura varieties and λ-adic representations,* in *Automorphic Forms, Shimura Varieties and L-functions,* vol. I, Academic Press, 1990, 161–209.

[L] R. Langlands, *Les débuts d'une formule des traces stables,* Publ. Math. Univ. Paris VII **13**, 1983.

[LS] R. Langlands and D. Shelstad, *On the definition of transfer factors,* Math. Ann. **278** (1987), 219–271.

[LN] G. Laumon and B.C. Ngô, *Le lemme fondamental pour les groupes unitaires,* Annals of Math. **168** (2008), 477–573.

[N1] B.C. Ngô, *Fibration de Hitchin et endoscopie,* Invent. Math. 164 (2006), 399–453.

[N2] B.C. Ngô, *Le lemme fondamental pour les algébres de Lie,* Publ. Math. I.H.E.S. 111 (2010), 1–169.

[S] D. Shelstad, *Tempered endoscopy for real groups* I: *geometric transfer with canonical factors,* Contemp. Math. **472** (2008), 215–246.

[W1] J.-L. Waldspurger, *Le lemme fondamental implique le transfer,* Compositio Math. **105** (1997), 153–236.

[W2] J.-L. Waldspurger, *Endoscopie et changement de caractéristique,* J. Inst. Math. Jussieu **5** (2006), 423–525.

[W3] J.-L. Waldspurger, *L'endoscopie tordue n'est pas si tordue,* Memoirs of AMS **908** (2008).

Ngô Bao Châu

NGÔ BAO CHÂU

Studied Mathematics at Êcole Normale Supérieure, Paris, France.

Ph.D., Université de Paris-Sud, Orsay, 1997.

Positions held

Researcher, CNRS, Paris 13 University, 1998–2004.

Professor, Universite de Paris-Sud 11, Orsay, 2004–, on leave since 2007.

Member, Institute for Advanced Study, Princeton, 2007–2010.

Professor, University of Chicago, 2010–

Proceedings of the International Congress of Mathematicians
Hyderabad, India, 2010

ENDOSCOPY THEORY OF AUTOMORPHIC FORMS

by
NGÔ BAO CHÂU*

ABSTRACT. Historically, Langlands has introduced the theory of endoscopy in order to measure the failure of automorphic forms from being distinguished by their L-functions as well as the defect of stability in the Arthur-Selberg trace formula and ℓ-adic cohomology of Shimura varieties. However, the number of important achievements in the domain of automorphic forms based on the idea of endoscopy has been growing impressively recently. Among these, we will report on Arthur's classification of automorphic representations of classical groups and recent progress on the determination of ℓ-adic galois representations attached to Shimura varieties originating from Kottwitz's work. These results have now become unconditional; in particular, due to recent progress on local harmonic analysis. Among these developments, we will report on Waldspurger's work on the transfer conjecture and the proof of the fundamental lemma.

Mathematics Subject Classification (2010). Main: 11F70; Secondary: 14K10

Keywords. Automorphic forms, endoscopy, transfer conjecture, fundamental lemma, Hitchin fibration.

1. Langlands' Functoriality Conjecture

This section contains an introduction of the functoriality principle conjectured by Langlands in [39].

1.1. *L-functions of Dirichlet and Artin*

The proof by Dirichlet for the infiniteness of prime numbers in an arithmetic progression of the form $m + Nx$ for some fixed integers m, N with $(m, N) = 1$, was a triumph of the analytic method in elementary number theory, *cf.* [13]. Instead of studying congruence classes modulo N which are prime to N, Dirichlet attached to each character $\chi : (\mathbb{Z}/N\mathbb{Z})^\times \to \mathbb{C}^\times$ of the group $(\mathbb{Z}/N\mathbb{Z})^\times$ of invertible elements in

*Supported by the Institute for Advanced Study, the NSF and the Symonyi foundation. School of Mathematics, Institute for Advanced Study, Princeton NJ 08540 USA.
Département de mathématiques, Université Paris-Sud, 91405 Orsay FRANCE.
E-mail: ngo@ias.edu

$\mathbb{Z}/N\mathbb{Z}$, the Euler product

$$L_N(s,\chi) = \prod_{p\nmid N}(1-\chi(p)p^{-s})^{-1}. \tag{1}$$

This infinite product converges absolutely for all complex numbers s having real part $\Re(s) > 1$ and defines a holomorphic function on this domain of the complex plane. For $N = 1$ and trivial character χ, this function is the Riemann zeta function. As for the Riemann zeta function, general Dirichlet L-function has a meromorphic continuation to the whole complex plane. However, in contrast with the Riemann zeta function that has a simple pole at $s = 1$, the Dirichlet L-function associated with a non trivial character χ admits a holomorphic continuation. This property of holomorphicity was a key point in Dirichlet's proof for the infiniteness of prime numbers in an arithmetic progression. Another important property is the functional equation relating $L(s,\chi)$ and $L(1-s,\bar{\chi})$.

Let $\sigma : \mathrm{Gal}(\bar{\mathbb{Q}}/\mathbb{Q}) \to \mathbb{C}^\times$ be a finite order character of the Galois group of the field of rational numbers $\mathbb{Q}$. For each prime number p, we choose an embedding of the algebraic closure $\bar{\mathbb{Q}}$ of $\mathbb{Q}$ into the algebraic closure $\bar{\mathbb{Q}}_p$ of the field of p-adic numbers $\mathbb{Q}_p$. This choice induces a homomorphism $\mathrm{Gal}(\bar{\mathbb{Q}}_p/\mathbb{Q}_p) \to \mathrm{Gal}(\bar{\mathbb{Q}}/\mathbb{Q})$ from the local Galois group at p to the global Galois group. The Galois group $\mathrm{Gal}(\bar{\mathbb{F}}_p/\mathbb{F}_p)$ of the finite field $\mathbb{F}_p$ is a canonical quotient of $\mathrm{Gal}(\bar{\mathbb{Q}}_p/\mathbb{Q}_p)$. We have the exact sequence

$$1 \to I_p \to \mathrm{Gal}(\bar{\mathbb{Q}}_p/\mathbb{Q}_p) \to \mathrm{Gal}(\bar{\mathbb{F}}_p/\mathbb{F}_p) \to 1 \tag{2}$$

where I_p is the inertia group. Recall that $\mathrm{Gal}(\bar{\mathbb{F}}_p/\mathbb{F}_p)$ is an infinite procyclic group generated by the substitution of Frobenius $x \mapsto x^p$ in $\bar{\mathbb{F}}_p$. Let the inverse of this substitution denote Fr_p.

Let $\sigma : \mathrm{Gal}(\bar{\mathbb{Q}}/\mathbb{Q}) \to \mathbb{C}^\times$ be a character of finite order. For all but finitely many primes p, say for all $p \nmid N$ for some integer N, the restriction of σ to the inertia group I_p is trivial. In that case $\sigma(\mathrm{Fr}_p) \in \mathbb{C}^\times$ is a well defined root of unity. Artin defines the L-function

$$L_N(s,\sigma) = \prod_{p\nmid N}(1-\sigma(\mathrm{Fr}_p)p^{-s})^{-1}. \tag{3}$$

Artin's reciprocity law implies the existence of a Dirichlet character χ such that

$$L_N(s,\chi) = L_N(s,\sigma). \tag{4}$$

As a consequence, the $L_N(s,\sigma)$ satisfies all the properties of the Dirichlet L-functions. In particular, it is holomorphic for nontrivial σ and it satisfies a functional equation with respect to the change of variables $s \leftrightarrow 1-s$.

Finite abelian quotients of $\mathrm{Gal}(\bar{\mathbb{Q}}/\mathbb{Q})$ correspond to finite abelian extensions of $\mathbb{Q}$. According to Kronecker-Weber's theorem, abelian extensions are obtained by adding roots of unity to $\mathbb{Q}$. Since general extensions of $\mathbb{Q}$ are not abelian, it is natural to seek a non abelian generalization of Artin's reciprocity law.

Let $\sigma : \mathrm{Gal}(\bar{\mathbb{Q}}/\mathbb{Q}) \to \mathrm{GL}(n, \mathbb{C})$ be a continuous n-dimensional complex representation. Since Galois groups are profinite groups, the image of σ is a finite subgroup of $\mathrm{GL}(n, \mathbb{C})$. There exists an integer N, such that for every prime $p \nmid N$, the restriction of σ to the inertia group I_p is trivial. In that case, $\sigma(\mathrm{Fr}_p)$ is well defined in $\mathrm{GL}(n, \mathbb{C})$, and its conjugacy class does not depend on the particular choice of embedding $\bar{\mathbb{Q}} \to \bar{\mathbb{Q}}_p$. The Artin L-function attached to σ is the Euler product

$$L_N(s, \sigma) = \prod_{p \nmid N} \det(1 - \sigma(\mathrm{Fr}_p)p^{-s})^{-1}. \tag{5}$$

Again, this infinite product converges absolutely for a complex number s with real part $\Re(s) > 1$ and defines a holomorphic function on this domain of the complex plane. It follows from the Artin-Brauer theory of characters of finite groups that the Artin L-function has meromorphic continuation to the complex plane.

Conjecture 1 (Artin). *If σ is a nontrivial irreducible n-dimensional complex representation of* $\mathrm{Gal}(\bar{\mathbb{Q}}/\mathbb{Q})$, *the L-function $L(s, \sigma)$ admits holomorphic continuation to the complex plane.*

The case $n = 1$ follows from Artin's reciprocity theorem and Dirichlet's theorem. The general case would follow from Langlands's conjectural nonabelian reciprocity law. According to this conjecture, it should be possible to attach to σ as above a cuspidal automorphic representation π of the group $\mathrm{GL}(n)$ with coefficients in the ring of the adeles $\mathbb{A}_{\mathbb{Q}}$ so that the Artin L-function of σ has the same Eulerian development as the principal L-function attached to σ. According to the Tamagawa-Godement-Jacquet theory *cf.* [62, 17], the latter extends to an entire function on complex plane that satisfies a functional equation. In the case $n = 2$, if the image of σ is solvable, the reciprocity law was established by Langlands and Tunnel by means of the solvable base change theory. The case where the image of σ in $\mathrm{PGL}_2(\mathbb{C}) = \mathrm{SO}_3(\mathbb{C})$ is the the nonsolvable group of symmetries of the icosahedron is not known in general, though some progress on this question has been made [64].

1.2. *Elliptic curves*

Algebraic geometry is a generous supply of representations of Galois groups. However, most interesting representations have ℓ-adic coefficients instead of complex coefficients. Any system of polynomial equations with rational coefficients, homogeneous or not, defines an algebraic variety. The groups of ℓ-adic cohomology attached to it are equipped with a continuous action of $\mathrm{Gal}(\bar{\mathbb{Q}}/\mathbb{Q})$. In contrast with complex representations, ℓ-adic representations might not have finite image.

The study of the case of elliptic curves is the most successful so far. Let E be an elliptic curve defined over $\mathbb{Q}$. The first ℓ-adic cohomology group of E is a 2-dimensional $\mathbb{Q}_\ell$-vector space equipped with a continuous action of $\mathrm{Gal}(\bar{\mathbb{Q}}/Q)$. In other words, we have a continuous 2-dimensional ℓ-adic representation

$$\sigma_{E,\ell} : \mathrm{Gal}(\bar{\mathbb{Q}}/\mathbb{Q}) \to \mathrm{GL}(2, \mathbb{Q}_\ell) \tag{6}$$

for every prime ℓ. The $\mathbb{Q}$-elliptic curve E can be extended to a $\mathrm{Spec}(\mathbb{Z}[N^{-1}])$-elliptic curve E_N for some integer N, i.e. E can be defined by homogeneous equation with coefficients in $\mathbb{Z}[N^{-1}]$ such that for every prime $p \nmid N$, the reduction of E_n modulo p is an elliptic curve defined over the finite field $\mathbb{F}_p$. If $p \neq \ell$, this implies that the restriction of $\sigma_{E,\ell}$ to inertia I_p is trivial. It follows that the conjugacy class of $\sigma_{E,\ell}, (\mathrm{Fr}_p)$ in $\mathrm{GL}(2, \mathbb{Q}_\ell)$ is well defined. The number of points on E_N with coefficients in $\mathbb{F}_p$ is given by the Grothendieck-Lefschetz fixed points formula

$$|E_N(\mathbb{F}_p)| = 1 - \mathrm{tr}(\sigma_{E,\ell}(\mathrm{Fr}_p)) + p. \tag{7}$$

It follows that $\mathrm{tr}(\sigma_{E,\ell}(\mathrm{Fr}_p))$ is an integer independent of the prime ℓ. Since it is also known that $\det(\sigma_{E,\ell}(\mathrm{Fr}_p)) = p$, the eigenvalues of $\sigma_E(\mathrm{Fr}_p)$ are conjugate algebraic integers of eigenvalue $p^{1/2}$, independent of ℓ. We can therefore drop the ℓ in the expressions $\mathrm{tr}(\sigma_{E,\ell}(\mathrm{Fr}_p))$ and $\det(\sigma_{E,\ell}(\mathrm{Fr}_p))$ as well as in the characteristic polynomial of $\sigma_{E,\ell}(\mathrm{Fr}_p)$.

The L-function attached to the elliptic curve E is defined by Euler product

$$L_N(s, E) = \prod_{p \nmid N} \det(1 - \sigma_E(\mathrm{Fr}_p)p^{-s})^{-1}. \tag{8}$$

Since the complex eigenvalues of $\sigma_E(\mathrm{Fr}_p)$ are of complex absolute value $p^{1/2}$, the above infinite product is absolute convergent for $\Re(s) > 3/2$ and converges to a homolomorphic function on this domain of the complex plane.

Shimura, Taniyama and Weil conjectured that the there exists a weight two holomorphic modular form f whose L-function $L(s, E)$ has the same Eulerian development as $L_N(s, E)$ at the places $p \nmid N$. It follows, in particular, that $L(s, E)$ has a meromorphic continuation to the complexe plane and it satisfies a functional equation. As it was shown by Frey and Ribet, a more spectacular consequence is the last Fermat's theorem is actually true. The Shimura-Taniyama-Weil conjecture is now a celebrated theorem of Wiles and Taylor [73, 63] in the semistable case. The general case is proved in [7].

The Shimura-Taniyama-Weil conjecture fits well with Langlands's reciprocity conjecture, *cf.* [39]. Though the main drive of Wiles's work consists of the theory of deformation of Galois representations, it needed as input the reciprocity law for solvable Artin representations $\sigma : \mathrm{Gal}(\bar{\mathbb{Q}}/\mathbb{Q}) \to \mathrm{GL}_2(\mathbb{C})$ that was proved by Langlands and Tunnell. The interplay between the p-adic theory of deformations of Galois representations and Langlands's functoriality principle should be a fruitful theme to reflect upon *cf.* [44].

1.3. *The Langlands conjectures*

Let G be a reductive group over a global field F which can be a finite extension of $\mathbb{Q}$ or the field of rational functions of a smooth projective curve over a finite field. For each absolute value v on F, F_v denotes the completion of F with respect to v, and if v is nonarchimedean, $\mathcal{O}_v$ denotes the ring of integers of F_v. Let $\mathbb{A}_F$ denote the ring of adeles attached to F, defined as the restricted product of the F_v with respect to $\mathcal{O}_v$.

By discrete automorphic representation, we mean an irreducible representation of the group $G(\mathbb{A}_F)$, the group of adeles points of G, that occurs as a subrepresentation of

$$L^2(G(F)\backslash G(\mathbb{A}_F))_\chi \tag{9}$$

where χ is a unitary character of the center of G [6]. Such a representation can develop as a completed tensor product $\pi = \hat{\bigotimes}_v \pi_v$ where π_v are irreducible admissible smooth representations of $G(F_v)$ for all nonarchimedean place v. For almost all nonarchimedean place v, π_v has a unique $G(\mathcal{O}_v)$-invariant line l_v. The Hecke algebra $\mathcal{H}_v$ of compactly supported complex valued functions on $G(F_v)$ that are bi-invariant under the action of $G(\mathcal{O}_v)$ acts on that line. Assume that G is unramified at v then $\mathcal{H}_v$ is a commutative algebra whose structure could be described in terms of a duality between reductive groups, [8].

Reductive groups over an algebraically closed field are classified by their root datum $(X^*, X_*, \Phi, \Phi^\wedge)$, where X^* and X_* are the group of characters, respectively cocharacters of a maximal torus and $\Phi \subset X^*, \Phi^\vee \subset X_*$ are, respectively, the finite subset of roots and of coroots, *cf.* [61]. By the exchange of roots and coroots, we have the dual root datum which is the root datum of a complex reductive group $\hat{G}$. The reductive group G is defined over F and becomes split over a Galois extension E of F. The group $\mathrm{Gal}(E/F)$ acts on the root datum of G in fixing a basis. It thus defines an action of $\mathrm{Gal}(E/F)$ on the complex reductive group $\hat{G}$. The semi-direct product ${}^L G = G \rtimes \mathrm{Gal}(E/F)$ was introduced by Langlands and is known as the L-group attached to G, *cf.* [39].

Suppose G unramified at a nonarchimedean place v; in other words, assume that the finite extension E is unramified over v. After a choice of embedding $E \to \bar{F}_v$, the Frobenius element $\mathrm{Fr}_v \in \mathrm{Gal}(\bar{\mathbb{F}}_v/\mathbb{F}_v)$, where F_v denotes the residue field of F_v, defines an element of $\mathrm{Fr}_v \in \mathrm{Gal}(E/F)$. There exists an isomorphism, known as the Satake isomorphism, between the Hecke algebra $\mathcal{H}_v$ and the algebra of $\hat{G}$-invariant polynomial functions on the connected component $\hat{G} \rtimes \{Fr_v\}$ of ${}^L G = G \rtimes \mathrm{Gal}(E/F)$. The line l_v acted on by the Hecke algebra $\mathcal{H}_v$ defines a semisimple element $s_v \in \hat{G} \rtimes \{\mathrm{Fr}_v\}$ up to $\hat{G}$-conjugacy in this component.

Unramified representations of $G(F_v)$ are classified by semisimple $\hat{G}$-conjugacy classes in $\hat{G} \rtimes \{\mathrm{Fr}_v\}$. In order to classify all irreducible admissible smooth representations of $G(F_v)$ for all nonarchimedean v, Langlands introduced the group

$$L_{F_v} = W_{F_v} \times \mathrm{SL}(2, \mathbb{C})$$

where W_{F_v} is the Weil group of F_v. The subgroup W_{F_v} of $\mathrm{Gal}(\bar{F}_v/F_v)$ consists of elements whose image in $\mathrm{Gal}(\bar{\mathbb{F}}_v/\mathbb{F}_v)$ is an integral power of Fr_v.

According to theorems of Laumon, Rapoport, and Stuhler in equal characteristic case, and Harris-Taylor and Henniart in unequal characteristic case, there is a natural bijection between the set of n-dimensional representations of L_{F_v} and the set of irreducible admissible smooth representations of $\mathrm{GL}_n(F_v)$ preserving L-factors and ϵ-factors of pairs, [51, 20, 22, 23].

According to Langlands, there should be also a group L_F attached to the global field F such that automorphic representations of $\mathrm{GL}_n(n, \mathbb{A}_F)$ are classified by n-dimensional complex representations of L_F. The hypothetical group L_F should be equipped with a surjective homomorphism to the Weil group W_F.

When F is the field of rational functions of a curve defined over a finite field $\mathbb{F}_q$, the situation is much better. Instead of complex representations of the hypothetical L-group L_F, one parametrizes automorphic representations by ℓ-adic representations of the Weil group W_F. Recall that in the function field case W_F is the subgroup of $\mathrm{Gal}(\bar{F}/F)$ consisting of elements whose image in $\mathrm{Gal}(\bar{\mathbb{F}}_q/\mathbb{F}_q)$ is an integral power of Fr_q. In a tour de force, Lafforgue proved that there exists a natural bijection between irreducible n-dimensional ℓ-adic representation of the Weil group W_F and cuspidal automorphic representations of $GL_n(\mathbb{A}_F)$ following a strategy initiated by Drinfeld, who settled the case $n = 2$ [14, 46, 47]. In the number fields case, only a part of ℓ-adic representations of W_F coming from motives should correspond to a part of automorphic representations.

Let us come back to the general case where G is a reductive group over a global field that can be either a number field or a function field. According to Langlands, automorphic representations should be partitioned into packets parametrized by conjugacy classes of homomorphisms $L_F \to {}^L\mathrm{G}$ compatible with the projections to W_F. At non-archimedean places, irreducible admissible smooth representations of $G(F_v)$ should also be partitioned into finite packets parametrized by conjugacy of homomorphism $L_{F_v} \to \hat{G} \rtimes W_{F_v}$ compatible with the projections to W_{F_v}. The parametrization of the local component of an automorphic representation should derive from the global parametrization by the homomorphism $L_{F_v} \to L_F$ that is only well defined up to conjugation.

This reciprocity conjecture on global parametrization of automorphic representations seems for the moment out of reach, in particular because of the hypothetical nature of the group L_F. In constrast, Langlands' functoriality conjecture is not dependent on the existence of L_F.

Conjecture 2 (Langlands). *Let H and G be reductive groups over a global field F and let ϕ be a homomorphism between their L-groups ${}^LH \to {}^LG$ compatible with projection to W_F. Then for each automorphic representation π_H of $H(\mathbb{A}_F)$, there exists an automorphic representation π of $G(\mathbb{A}_F)$ such that at each unramified place*

v where π_H is parametrized by a conjugacy class $s_v(\pi_H)$ in $\hat{H} \rtimes \{\mathrm{Fr}_v\}$, the local component of π is also unramified and parametrized by $\pi(s_v(\pi_H))$.

At least in the number field case, the existence of L_F seems to depend upon the validity of the functoriality principle. Some of the most important conjectures in number theory and in the theory of automorphic representations. As explained in [39], Artin conjecture follows from the case of functoriality when $\hat{H}$ is trivial. It is also explained in loc. cit how the generalized Ramanujan conjecture and the generalized Sato-Tate conjecture would also follow from the functoriality conjecture.

The approach based on a combination of the converse theorem of Cogdell and Piateski-Shairo, and the Langlands-Shahidi method was succesful in establishing some startling cases of functoriality beyond endoscopy, *cf.* [26]. However, it suffers obvious limitation as Langlands-Shahidi method is based on the representation of a Levi component of a parabolic group on the Lie algebra of its unipotent radical.

Recently, the p-adic method was also successful in establishing a weak form of the functoriality conjecture. The most spectacular result is the proof of the Sato-Tate conjecture [21] deriving from this weak form. We will not discuss this topic in this survey.

So far, the most successful method in establishing special cases of functoriality is endoscopy. We will discuss this topic in more details in the next section.

2. Endoscopy Theory and Applications

The endoscopy theory is primarily focused in the structure of the packet of representations that have the same conjectural parametrization, either global $L_F \to {}^L G$ or local $L_{F_v} \to \hat{G} \rtimes W_{F_v}$. The existence of the packet is closely related to the lack of stability in the trace formula. As shown in [42], the answer to this question derives from the comparison of trace formulas. It is quite remarkable that the inconvenient unstability in the trace formula turned out to be a possibility. The quest for a stable trace formula bringing the necessity of comparing two trace formulas, turned out to be an efficient tool for establishing particular cases of functoriality.

A good number of known cases of functoriality fits into a general scheme that is nowadays known as the theory of endoscopy and twisted endoscopy: Jacquet-Langlands theory, solvable base change, automorphic induction and the Arthur lift from classical groups to linear groups.

Another source of endoscopic phenomenon was the study of continuous cohomology of Shimura varieties as first recognized by Langlands [40]. The work of Kottwitz has definitely shaped this theory by proposing precise conjecture on the ℓ-adic cohomology of Shimura variety as Galois module [34]. This description has been established in many important cases by means of comparison of the Grothendieck-Lefschetz fixed points formula and the Arthur-Selberg trace formula.

2.1. *Packets of representations*

First intuitions of endoscopy come from the theory of representations of $\mathrm{SL}(2,\mathbb{R})$. The restriction of discrete series representations of $\mathrm{GL}(2,\mathbb{R})$ to $\mathrm{SL}(2,\mathbb{R})$ is reducible. Their irreducible factors having the same Langlands parameter obtained by composition $W(\mathbb{R}) \to \mathrm{GL}(2,\mathbb{C}) \to \mathrm{PGL}(2,\mathbb{C})$ and thus belong to the same packet. Packet of representations is understood to be dual stable conjugacy relation between conjugacy classes. For instance, the rotations of angle θ and $-\theta$ centered at the origin of the plane are not conjugate in $\mathrm{SL}(2,\mathbb{R})$, but become conjugate either in $\mathrm{GL}(2,\mathbb{R})$ or in $\mathrm{SL}(2,\mathbb{C})$.

In general, if G is a quasi-split reductive group over a local field F_v, and $\Pi_v(G)$ is the set of irreducible representations of G, Langlands conjectured that $\Pi(G)$ is a disjoint union of finite sets $\Pi_{v,\phi}(G)$ that are called L-packets and indexed by admissible homomorphisms $\phi_v : L_{F_v} \to {}^L G_v$. The work of Shelstad [59] in the real case suggested the following description of the set $\Pi_v(G)$ in general, *cf.* [42].

Let S_{ϕ_v} denote the centralizer of the image of ϕ_v in $\hat{G}$, and $S^0(\phi_v)$ its neutral component. Let $Z(\hat{G})$ denote the center of $Z(\hat{G})$ and $Z(\hat{G})^\Gamma$ denote the subgroup of invariants under the action of the Galois group Γ. The group $\mathcal{S}_{\phi_v} = S_{\phi_v}/S^0_{\phi_v} Z(\hat{G})^\Gamma$ should control completely the structure of the finite set Π_{ϕ_v} and also the characters of the representations belonging to Π_{ϕ_v}. If we further assume ϕ_v tempered, i.e its image is contained in a relatively compact subset of $\hat{G}$, then there should be a bijection $\pi \mapsto \langle s, \pi\rangle$ from Π_{ϕ_v} onto the set of irreducible characters of $\mathcal{S}_{\phi_v}$. In particular, the cardinal of the finite set Π_{ϕ_v} should equal the number of conjugacy classes of Π_{ϕ_v}.

There is also a conjectural description of multiplicity in the automorphic spectrum of each member of a global L-packet. We can attach any admissible homomorphism $\phi : L_F \to {}^L G$ local parameter $\phi_v : L_{F_v} \to {}^L G_v$. By definition, the global L-packet Π_ϕ is the infinite product of local L-packets Π_{ϕ_v}. For a representation $\pi = \bigotimes_v \pi_v$ with $\pi_v \in \Pi_v$ to appear in the automorphic spectrum, all but finitely many local components must be unramified. For those representations, there is a conjectural description of its automorphic multiplicity $m(\pi,\phi)$ that was made precise by Kottwitz based on the case of SL_2 worked out by Labesse and Langlands *cf.* [38]. In [31], Kottwitz introduced a group $\mathcal{S}_\phi$ equipped with homomorphism $\mathcal{S}_\phi \to \mathcal{S}_{\phi_v}$. The conjectural formula for $m(\pi,\phi)$ is

$$m(\pi,\phi) = |\mathcal{S}_\phi|^{-1} \sum_{\epsilon\in\mathcal{S}_\phi} \Pi_v \langle \epsilon_v, \pi_v\rangle.$$

For each v, ϵ_v denotes the image of ϵ in $\mathcal{S}_{\phi_v}$ and $\langle \epsilon_v, \pi_v\rangle$, the value of the character of $\mathcal{S}_{\phi_v}$ corresponding to π_v evaluated on ϵ_v.

If the above general description has an important advantage of putting the automorphic theory in perspective, it also suffers a considerable inconvenience of being dependent on the hypothetical Langlands group L_F.

For quite a long time, we have known only a few low rank cases including the case of inner forms of SL(2) due to Labesse and Langlands [38], the cyclic base change for GL(2) due to Saito, Shintani and Langlands [41] and the case of $U(3)$ and its base change due to Rogawski [58]. Later, the cyclic base change for GL(n) was established by Arthur and Clozel [3]. Recently, this field has been undergoing spectacular developments. For quasisplit classical groups, Arthur has been able to establish the existence and the description of local packets as well as an automorphic multiplicity formula for global packets [2]. For p-adic groups, the local description becomes unconditional based on the local Langlands conjecture for GL(n) proved by Harris-Taylor and Henniart. Arthur's description of global packet as well as his automorphic multiplicity formula is based on cuspidal automorphic representations of GL(n) instead of the hypothetical group L_F. This description relies on a little bit of intricate combinatorics that goes beyond the scope of this report. The unitary case was also settled by Moeglin [52], the case of inner forms of SL(n) by Hiraga and Saito [24]. The general case of Jacquet-Langlands correspondence has been also established by Badulescu [4].

Most of the above developments were made possible by the formidable machine that is the Arthur trace formula and its stabilization. The comparison of the trace formula for two different groups, one being endoscopic to the other, proved to be a quite fruitful method. Arthur's parametrization of automorphic forms on quasi-split classical groups derives from the possibility of realizing these groups as twisted endoscopic groups of GL(n) and the comparison between the twisted trace formula of GL(n) and the ordinary trace formula for the classical group. This procedure is known as the stabilization of the twisted trace formula. The structure of the L-packets derives from the stabilization of ordinary trace formula for classical groups. For both twisted and untwisted, Arthur needed to assume the validity of certain conjectures on orbital integrals: the transfer and the fundamental lemma.

2.2. *Construction of Galois representations*

Based on indications given in Shimura's work, Langlands proposed a general strategy to constructing Galois representations attached to automorphic representation incorporated in ℓ-adic cohomology of Shimura varieties. This domain also recorded important developments due to Kottwitz, Clozel, Harris, Taylor, Yoshida, Labesse, Morel, Shin and others.

In particular, a non negligible portion of the global Langlands correspondence for number fields is now known. A number field F is of complex multiplication if it is a totally imaginary quadratic extension of a totally real number field F^+. In particular, the complex conjugation induces an automorphism c of F that is independent of complex embedding of F. Let $\Pi = \bigotimes_v \Pi_v$ be a cuspidal automorphic representation of $\mathrm{GL}(n, \mathbb{A}_F)$ such that $\Pi^\wedge \simeq \Pi \circ c$, whose component at infinity Π_∞ has the same infinitesimal character as some irreducible algebraic representation satisfying certain regularity condition. Then for every prime number ℓ, there exists a continuous representation $\sigma : \mathrm{Gal}(\bar{F}/F) \to \mathrm{GL}(n, \bar{\mathbb{Q}}_\ell)$ so that for every prime p of F that does not lie above ℓ, the local component π_v of π corresponds to the ℓ-adic local representation of $\mathrm{Gal}(\bar{F}_v/F_v)$ via the local Langlands correspondence established by Harris-Taylor and Henniart. This important theorem is due to Clozel, Harris, and Labesse [11], Morel [53] and Shin [60] with some difference in the precision.

Under the above assumptions on the number field F and the automorphic representation Π, there exists a unitary group $U(F^+)$ with respect to the quadratic extension F/F^+ that gives rise to a Shimura variety and an automorphic representation π of U whose base change to $\mathrm{GL}(n, F)$ is Π. The base change from the unitary group U to the linear group $\mathrm{GL}(n, F)$ is a case of the theory of twisted endoscopy. It is based on a comparison of the twisted trace formula for $\mathrm{GL}(n, F)$ and the ordinary trace formula for $U(F^+)$. For more details, see [10, 37].

Following the work of Kottwitz on Shimura varieties, it is possible to attach Galois representation to automorphic forms. Algebraic cuspidal automorphic representations of unitary group appears in ℓ-adic cohomology of Shimura variety. In [35], Kottwitz proved a formula for the number of points on certain type of Shimura varieties with values in a finite field at a place of good reduction, and in [34], he showed how to stabilize this formula in a very similar manner to the stabilization of the trace formula. He also needed to assume the validity of the same conjectures on local orbital integrals as in the case of stabilization of the trace formula.

Kottwitz' formula for the number of points allow to show the compatibility with the local correspondence at the unramified places. More recently, Shin proved a formula for fixed points on Igusa varieties that looks formally similar to Kottwitz' formula that allows him to prove the compatibility with the local correspondence at a ramified place [60].

Morel was able to calculate the intersection cohomology of non-compact unitary Shimura varieties when the other authors confined themselves in the compact case [53]. The description of the intersection cohomology has been conjectured by Kottwitz.

We observe the remarkable similarity between Arthur's works on the classification of automorphic representations of classical groups and the construction of Galois representations attached to automorphic representations by Shimura varieties. Both need the stablization of a twisted trace formula and of an ordinary trace formula or similar formula thereof.

3. Stabilization of the Trace Formula

The main focus of the theory of endoscopy is the stabilization of the trace formula. The trace formula allows us to derive properties of automorphic representations from a careful study of orbital integrals. The orbital side of the trace formula is not stable but the defect of stability can be expressed by an endoscopic group. It follows the endoscopic case of the functoriality conjecture.

This section will give more details about the stabilization of the orbital side of the trace formula.

3.1. *Trace formula and orbital integrals*

In order to simplify the exposition, we will consider only semisimple groups G defined over a global field F. The Arthur-Selberg trace formula for G has the following form

$$\sum_{\gamma \in G(F)/\sim} \mathbf{O}_\gamma(f) + \cdots = \sum_n \mathrm{tr}_\pi + \cdots \tag{10}$$

where γ runs over the set of anisotropic conjugacy classes of $G(F)$ and π over the set of discrete automorphic representations. The trace formula contains also more complicated terms related to hyperbolic conjugacy classes on one side and the continuous spectrum on the other side.

The test function f is of the form $f = \bigotimes_v f_v$ where for v, f_v is a smooth compactly supported function on $G(F_v)$ and for almost all nonarchimedean places v, f_v the unit function of the unramified Hecke algebra of $G(F_v)$. The global orbital integral

$$\mathbf{O}_\gamma(f) = \int_{I^\gamma(F)\backslash G(\mathbb{A})} f(g^{-1}\gamma g)dg \tag{11}$$

is convergent for anisotropic conjugacy classes $\gamma \in G(F)$. Here $I_\gamma(F)$ is the discrete group of F-points on the centralizer I_γ of γ. After choosing a Haar measure $dt = \bigotimes dt_v$ on $I_\gamma(\mathbb{A})$, we can express the above global integral as follows

$$\mathbf{O}_\gamma(f) = vol(I_\gamma(F)\backslash I_\gamma(\mathbb{A}), dt) \prod_v \mathbf{O}_\gamma(f_v, dg_v/dt_v). \tag{12}$$

The torus I_γ has an integral form well defined up to finitely many places, and the measure dt is chosen so that $I_\gamma(\mathcal{O}_v)$ has volume one for almost all v. Over a nonarchimedean place, the local orbital integral

$$\mathbf{O}_\gamma(f_v, dg_v/dt_v) = \int_{I_\gamma(F_v)\backslash G(F_v)} f(g^{-1}\gamma g)\frac{dg_v}{dt_v} \tag{13}$$

is defined for every locally constant function $f_v \in C_c^\infty(G(F_v))$ with compact support. Local orbital integral $\mathbf{O}_\gamma(f_v, dg_v/dt_v)$ is convergent for every v and equals 1 for almost all v. The volume term is finite when the global conjugacy class γ is anisotropic.

Arthur introduced a truncation operator to deal with the continuous spectrum in the spectral expansion and hyperbolic conjugacy classes in the geometric expansion.

In the geometric expansion, Arthur has more complicated local integrals that he calls weighted orbital integrals, see [2].

3.2. *Stable orbital integrals*

For $\mathrm{GL}(n)$, two regular semisimple elements in $\mathrm{GL}(n,F)$ are conjugate if and only if they are conjugate in $\mathrm{GL}(n,\bar{F})$, where $\bar{F}$ is an algebraic closure of F and this latter condition is tantamount to request that γ and γ' have the same characteristic polynomial. For a general reductive group G, we also have a characteristic polynomial map $\chi : G \to T/W$ where T is a maximal torus and W is its Weyl group. An element is said to be strongly regular semisimple if its centralizer is a torus. Strongly regular semisimple elements γ, $\gamma' \in G(\bar{F})$ have the same characteristic polynomial if and only if they are $G(\bar{F})$-conjugate. However, there are possibly more than one $G(F)$-conjugacy classes within the set of strongly regular semisimple elements having the same characteristic polynomial in $G(F)$. These conjugacy classes are said to be stably conjugate.

Let $\gamma, \gamma' \in G(F)$ be such that there exist $g \in G(\bar{F})$ with $\gamma' = g\gamma g^{-1}$. For all $\sigma \in \mathrm{Gal}(\bar{F}/F)$, since γ, γ' are defined over $F, \sigma(g)^{-1}g$ belongs to the centralizer of γ. The map

$$\sigma \mapsto \sigma(g)^{-1}g \tag{14}$$

defines a cocycle with values in $I_\gamma(\bar{F})$ whose image in $G(\bar{F})$ is a boundary. For a fixed $\gamma \in G(F)$, assumed strongly regular semisimple, the set of $G(F)$-conjugacy classes in the stable conjugacy class of γ can be identified with the subset A_γ of elements $\mathrm{H}^1(F, I_\gamma)$ whose image in $\mathrm{H}^1(F, G)$ is trivial. For local fields, the group $\mathrm{H}^1(F, I_\gamma)$ is finite but for global field, it can be infinite.

For a local nonarchimedean field F, A_γ is a subgroup of the finite abelian group $\mathrm{H}^1(F, I_\gamma)$. One can form linear combinations of orbital integrals within a stable conjugacy class using characters of A_γ. In particular, the stable orbital integral

$$\mathbf{SO}_\gamma(f) = \sum_{\gamma'} \mathbf{O}_{\gamma'}(f)$$

is the sum over a set of representatives γ' of conjugacy classes within the stable conjugacy class of γ. One needs to choose in a consistent way Haar measures on different centralizers $I'_\gamma(F)$. For strongly regular semisimple γ, the tori $I_{\gamma'}$ for γ' in the stable conjugacy class of γ, are in fact canonically isomorphic, so that we can transfer a Haar measure from $I_\gamma(F)$ to $I_{\gamma'}(F)$. Obviously, the stable orbital integral $\mathbf{SO}_\gamma$ depends only on the characteristic polynomial of γ. If a is the characteristic polynomial of a strongly regular semisimple element γ, we set $\mathbf{SO}_a = \mathbf{SO}_Y$. A stable distribution is an element in the closure of the vector space generated by the distributions of the forms $\mathbf{SO}_a$ with respect to the weak topology.

In some sense, stable conjugacy classes are more natural than conjugacy classes. In order to express the difference between orbital integrals and stable orbital integrals, one needs to introduce other linear combinations of orbital integrals known as κ-orbital integrals. For each character $\kappa : A_\gamma \to \mathbb{C}^\times$, κ-orbital integral is a linear combination

$$\mathbf{O}_\gamma^\kappa(f) = \sum_{\gamma'} \kappa(\mathrm{cl}(\gamma'))\mathbf{O}_{\gamma'}(f)$$

over a set of representatives γ' of conjugacy classes within the stable conjugacy class of γ, $\mathrm{cl}(\gamma')$ being the class of γ' in A_γ. For any γ' in the stable conjugacy class of γ, A_γ and $A_{\gamma'}$ are canonical isomorphic so that the character κ on A_γ defines a character of $A_{\gamma'}$. Now, $\mathbf{O}_\gamma^\kappa$ and $\mathbf{O}_{\gamma'}^\kappa$, are not equal but differ by the scalar $\kappa(cl(\gamma'))$ where $\mathrm{cl}(\gamma')$ is the class of γ' in A_γ. Even though this transformation rule is simple enough, we can't a priori define κ-orbital $\mathbf{O}_a^\kappa$ for a characteristic polynomial a as in the case of stable orbital integral. This is a source of an important technical difficulty in the theory of endoscopy: the transfer factor.

3.3. *Stable distributions and the trace formula*

Test functions for the trace formula are finite combination of functions f on $G(\mathbb{A})$ of the form $f = \bigotimes_{v\in|F|} f_v$ where for all v, f_v is a smooth function with compact support on $G(F_v)$ and for almost all finite place v, f_v is the characteristic function of $G(\mathcal{O}_v)$ with respect to an integral form of G which is well defined almost everywhere.

The trace formula defines a linear form in f. For each v, it induces an invariant linear form in f_v. There exists a Galois theoretical cohomological obstruction that prevents this linear form from being stably invariant. Let $\gamma \in G(F)$ be a strongly regular semisimple element. Let $(\gamma'_v) \in G(\mathbb{A})$ be an adelic element with γ'_v stably conjugate to γ for all v and conjugate for almost all v. There exists a cohomological obstruction that prevents the adelic conjugacy class (γ'_v) from being rational. In fact the map

$$\mathrm{H}^1(F, I_\gamma) \to \bigoplus_v H_1(F_v, I_\gamma) \tag{15}$$

is not surjective in general. Let $\hat{I}_\gamma$ denote the dual complex torus of I_γ equipped with a finite action of the Galois group $\Gamma = \mathrm{Gal}(\bar{F}/F)$. For each place v, the Galois group $\Gamma_v = \mathrm{Gal}(\bar{F}_v/F_v)$ of the local field also acts on $\hat{I}_\gamma$. By local Tate-Nakayama duality as reformulated by Kottwitz, $\mathrm{H}^1(F_v, I_\gamma)$ can be identified with the group of characters of $\pi_0\ (\hat{I}_\gamma^{\Gamma_v})$. By global Tate-Nakayama duality, an adelic class in $\bigotimes_v \mathrm{H}^1(F_v, I_\gamma)$ comes from a rational class in $\mathrm{H}^1(F, I_\gamma)$ if and only if the corresponding characters on $\pi_0(\hat{I}_\gamma^{\Gamma_v})$, after restriction to $\pi_0(\hat{I}_\gamma^{\Gamma})$, sum up to the trivial character. The original problem with conjugacy classes within a stable conjugacy class, complicated by the presence of the strict subset A_γ of $\mathrm{H}^1(F, I_\gamma)$, was solved in Langlands [42] and in a more general setting by Kottwitz [32].

In [42], Langlands outlined a program to derive from the usual trace formula a stable trace formula. The key point is to apply Fourier transform on the finite group $\pi_0(\hat{I}_\gamma^{\Gamma})$ and the part of the trace formula corresponding to the stable conjugacy class of γ becomes a sum over the group of characters of $\pi_0(\hat{I}_\gamma^\Gamma)$. By definition, the term corresponding to the trivial character of $\pi_0(\hat{I}_\gamma^\Gamma)$ is the stable trace formula. The other terms can be expressed as product of κ-orbital integrals.

Langlands conjectured that these κ-orbital integrals can also be expressed in terms of stable orbital integrals of endoscopic groups. The precise constant occuring in these conjectures were worked out in his joint work with Shelstad *cf.* [45]. There are in fact two conjectures: the transfer and the fundamental lemma that we will review in a similar but simpler context of Lie algebras. Admitting these conjectures, Langlands and Kottwitz proved that the correction terms in the elliptic part match with the stable trace formula for endoscopic groups. This equality is known under the name of the stabilization of the elliptic part of the trace formula.

The whole trace formula was eventually stabilized by Arthur under more local assumptions that are the weighted transfer and the weighted fundamental lemma *cf.* [1]. Arthur's classification of automorphic forms of quasisplit classical groups depends upon the stabilization of twisted trace formula. For this purpose, Arthur's local assumptions are more demanding: the twisted weighted transfer and the twisted weighted fundamental lemma.

3.4. *The transfer and the fundamental lemma*

We will state the two conjectures about local orbital integrals known as the transfer conjecture and the fundamental lemma in the case of Lie algebra. The statements in the case of Lie group are very similar but the constant known as the transfer factor more complicated.

Assume for simplicity that G is a split group over a local non-archimedean field F. Let $\hat{G}$ denote the connected complex reductive group whose root system is related to the root system of G by exchanging roots and coroots. Let γ be a regular semisimple F-point on the Lie algebra $\mathfrak{g}$ of G. Its centralizer I_γ is a torus defined over F. By the Tate-Nakayama duality, a character κ of $\mathrm{H}^1(F, I_\gamma)$ corresponds to a semisimple element of $\hat{G}$ that is well defined up to conjugacy. Let $\hat{H}$ be the neutral component of the centralizer of κ in $\hat{G}$. For a given torus I_γ, we can define an action of the Galois group of F on $\hat{H}$ that factors through the component group of the centralizer of κ in $\hat{G}$. By duality, we obtain a quasi-split reductive group H over F which is an endoscopic group of G.

The endoscopic group H is not a subgroup of G in general. Nevertheless, it is possible to transfer stable conjugacy classes from H to G, and from the Lie algebra $\mathfrak{h} = \mathrm{Lie}(H)$ to $\mathfrak{g}$. Assume for simplicity that H is also split. The inclusion $\hat{H} = \hat{G}_\kappa \subset \hat{G}$ induces an inclusion of Weyl groups $W_H \subset W$. It follows that there exists a canonical map $t/W_H \to \mathfrak{t}/W$ that realizes the transfer of stable conjugacy classes from $\mathfrak{h}$ to $\mathfrak{g}$. If $\gamma_H \in \mathfrak{h}(F)$ has characteristic polynomial $a_H \in \mathfrak{t}/W_H(F)$ mapping to the characteristic polynomial a of $\gamma \in G(F)$, we will say that the stable conjugacy class of γ_H transfers to the stable conjugacy class of γ.

Kostant has constructed a section $\mathfrak{t}/W \to \mathfrak{g}$ of the characteristic polynomial morphism $\mathfrak{g} \to \mathfrak{t}/W$ *cf.* [29]. For every $a \in (\mathfrak{t}/W)(F)$, the Kostant section defines a distinguished conjugacy class with the stable conjugacy class of a. As shown by Kottwitz *cf.* [36], the Kostant section provides us a rather simple definition of the Langlands-Shelstad transfer factor in the case of Lie algebra. Let $\Delta(\gamma_H, \gamma)$ be the unique complex function depending on regular semisimple conjugacy classes $\gamma_H \in \mathfrak{h}(F)$ and $\gamma \in \mathfrak{g}(F)$ with the characteristic polynomial $a_H \in (\mathfrak{t}/W_H)(F)$ of γ_H mapping to the characteristic polynomial $a \in (\mathfrak{t}/W)(F)$ of γ and satisfying the following property

- $\Delta(\gamma_H, \gamma)$ depends only on the stable conjugacy class of γ_H,
- if γ and γ' are stably conjugate then $\Delta(\gamma_H, \gamma') = \langle \mathrm{inv}(\gamma, \gamma'), \kappa\rangle \Delta(\gamma_H, \gamma)$ where $inv(\gamma, \gamma')$ is the cohomological invariant lying in $\mathrm{H}^1(F, I_Y)$ defined by the cocyle (14),
- if γ is conjugate to the Kostant section at a, $\Delta(\gamma_H, \gamma) = |\Delta_G(\gamma)^{-1}\Delta_H(\gamma_H)|^{1/2}$ where Δ_G, Δ_H are the usual discriminant functions on $\mathfrak{g}$ and $\mathfrak{h}$ and $|.|$ denotes the standard absolute value of the nonarchimedean field F.

Conjecture 3 (Transfer). *For every $f \in C_c^\infty(G(F))$ there exists $f^H \in C_c^\infty(H(F))$ such that*

$$\mathbf{SO}_{\gamma H}(f^H) = \Delta(\gamma H, \gamma)\mathbf{O}_\gamma^\kappa(f) \tag{16}$$

for all strongly regular semisimple elements γ_H and γ with the characteristic polynomial $a_H \in (\mathfrak{t}/W_H)(F)$ of γ_H mapping to the characteristic polynomial $a \in (\mathfrak{t}/W)(F)$ of γ.

Under the assumption γ_H and γ regular semisimple with the characteristic polynomial $a_H \in (\mathfrak{t}/W_H)(F)$ of γ_H mapping to the characteristic polynomial $a \in (\mathfrak{t}/W)(F)$ of γ, their centralizers in H and G are canonically isomorphic tori. We can therefore transfer Haar measures between those locally compact groups.

Assume that we are in unramified situation i.e. both G and H have reductive models over $\mathcal{O}_F$. Let $1_{\mathfrak{g}(\mathcal{O}_F)}$ be the characteristic function of $\mathfrak{g}(\mathcal{O}_F)$ and $1_{\mathfrak{h}(\mathcal{O}_F)}$ the characteristic function of $\mathfrak{h}(\mathcal{O}_F)$.

Conjecture 4 (Fundamental lemma). *The equality (16) holds for $f = 1_{\mathfrak{g}(\mathcal{O}_F)}$ and $f^H = 1_{\mathfrak{h}(\mathcal{O}F)}$.*

In the case of Lie group instead of Lie algebra, there is a more general version of the fundamental lemma. Let $\mathcal{H}_G$ be the algebra of $G(\mathcal{O}_F)$-biinvariant functions with compact support on $G(F)$ and $\mathcal{H}_H$ the similar algebra for $H(F)$. Using Satake isomorphism, we have a canonical homomorphism $b : \mathcal{H}_G \to \mathcal{H}_H$.

Conjecture 5. *The equality (16) holds for any $f \in \mathcal{H}_G$ and for $f^H = b(f)$.*

In [68], Waldspurger also stated another beautiful conjecture in the same spirit. Let G_1 and G_2 be two semisimple groups with isogeneous root systems i.e. there exists an isomorphism between their maximal tori which maps a root of G_1 on a scalar multiple of a root of G_2 and conversely. In this case, there is an isomorphism $\mathfrak{t}_1/W_1 \simeq \mathfrak{t}_2/W_2$. We can therefore transfer regular semisimple stable conjugacy classes from $\mathfrak{g}_1(F)$ to $\mathfrak{g}_2(F)$ and back.

Conjecture 6 (Nonstandard fundamental lemma). *Let $\gamma_1 \in \mathfrak{g}_1(F)$ and $\gamma_2 \in \mathfrak{g}_2(F)$ be regular semisimple elements having the same characteristic polynomial. Then we have*

$$\mathbf{SO}_{\gamma 1}(1_{\mathfrak{g}1(\mathcal{O}_F)}) = \mathbf{SO}_{\gamma 2}(1_{\mathfrak{g}2(\mathcal{O}_F)}). \tag{17}$$

3.5. *The long march*

Let us remember the long march to the conquest of the transfer conjecture and the fundamental lemma.

The theory of endoscopy for real groups is almost entirely due to Shelstad. She proved, in particular, the transfer conjecture for real groups. The fundamental lemma does not make sense for real groups.

Particular cases of the fundamental lemma were proved in low rank case by Labesse-Langlands for SL(2) [38], Kottwitz for SL(3) [30], Rogawski for U(3) [58], Hales, Schroder and Weissauer for Sp(4). The first case of twisted fundamental lemma was proved by Saito, Shintani and Langlands in the case of base change for GL(2). The conjecture 4 in the case of stable base change was proved by Kottwitz [33] for unit and then 5 by Clozel and Labesse independently for Hecke algebra. Kazhdan [27], and Waldspurger [66] proved 4 for SL(n). More recently, Laumon and myself proved the case U(n) [50] in equal characteristic.

The following result is to a large extent a collective work.

Theorem 7. *The conjectures 3, 4, 5 and 6 are true for p-adic fields.*

In the landmark paper [67], Waldspurger proved that the fundamental lemma implies the transfer conjectures. Due to his and Hales' works, the case of Lie group follows from the case of Lie algebra. Waldspurger also proved that the twisted fundamental lemma follows from the combination of the fundamental lemma with his nonstandard variant [68]. In [19], Hales proved that if we know the fundamental lemma for the unit for almost all places, we know it for the entire Hecke algebra for all places. In particular, if we know the fundamental lemma for the unit element at all but finitely many places, we also know it at the remaining places. More details on Hales' argument can be found in [53].

The problem is reduced to the fundamental lemma for Lie algebra. Following Waldspurger and, independently, Cluckers, Hales and Loeser, it is enough to prove the fundamental lemma for a local field in characteristic p, see [69] and [12].

For local fields of characteristic p, the approach using algebraic geometry was eventually successful. This approach originated in the work of Kazhdan and Lusztig who introduced the affine Springer fiber, *cf.* [28]. In [18], Goresky,

Kottwitz and MacPherson gave an interpretation of the fundamental lemma in terms of the cohomology of the affine Springer. They also introduced the use of the equivariant cohomology and proved the fundamental lemma for unramified elements assuming the purity of cohomology of affine Springer fiber. Later in [49], Laumon proved the fundamental lemma for general element in the Lie algebra of unitary group also by using the equivariant cohomology and admitting the same purity assumption. The conjecture of purity of cohomology of affine Springer fiber is still unproved.

The Hitchin fibration was introduced in this context in [54]. Laumon and I used this approach, combined with [49], to prove the fundamental lemma for unitary group in [50]. The equivariant cohomology is no longer used for effective calculation of cohomology but to prove a qualitative property of the support of simple perverse sheaves occurring in the cohomology of Hitchin fibration. Later, I realized that the equivariant cohomology does not work in general simply due to the lack of toric action. The general case was proved in [56] with essentially the same strategy as in [50] but with a major difference. Since the equivariant cohomology does not provide a general argument for the determination of the support of simple perverse sheaves occurring in the cohomology of Hitchin fibration, an entirely different argument was needed. This new argument is a blend of an observation of Goresky and MacPherson on perverse sheaves and Poincaré duality with some particular geometric properties of algebraic integrable systems *cf.* [57].

4. Affine Springer Fibers and the Hitchin Fibration

In this section, we will describe the geometric approach to the fundamental lemma.

4.1. *Affine Spriger fibers*

Let $k = \mathbb{F}_q$ be a finite field with q elements. Let G be a reductive group over k and $\mathfrak{g}$ its Lie algebra. Let $F = k((\pi))$ and $\mathcal{O}_F = k[[\pi]]$. Let $\gamma \in \mathfrak{g}(F)$ be a regular semisimple element. According to Kazhdan and Lusztig [28], there exists a k-scheme $\mathcal{M}_\gamma$ whose set of k points is

$$\mathcal{M}_\gamma(k) = \{g \in G(F)/G(\mathcal{O}_F) | \mathrm{ad}(g)^{-1}(\gamma) \in \mathfrak{g}(\mathcal{O}_F)\}.$$

They proved that the affine Springer fiber $\mathcal{M}_\gamma$ is finite dimensional and locally of finite type.

The centralizer $I_\gamma(F)$ acts on $\mathcal{M}_\gamma(k)$. The group $I_\gamma(F)$ can be given a structure of infinite dimensional group $\tilde{P}_\gamma$ over k, acting on $\mathcal{M}_\gamma$. There exists a unique quotient $\mathcal{P}_\gamma$ of $\tilde{P}_\gamma$ such that the above action factors through $\mathcal{P}_\gamma$ and there exists an open subvariety of $\mathcal{M}_\gamma$ over which $\mathcal{P}_\gamma$ acts simply transitively.

Here is a simple but important example. Let $G = \mathrm{SL}_2$ and let γ be the diagonal matrix

$$\gamma = \begin{pmatrix} \pi & 0 \\ 0 & -\pi \end{pmatrix}.$$

In this case $\mathcal{M}_\gamma$ is an infinite chain of projective lines with the point ∞ in each copy being identified with the point 0 of the next one. The group $\mathcal{P}_\gamma$ is $\mathbb{G}_m \times \mathbb{Z}$ with $\mathbb{G}_m$ acting on each copy of $\mathbb{P}^1$ by rescaling and the generator of $\mathbb{Z}$ acting by translation from each copy to the next one. The dense open orbit is obtained by removing from $\mathcal{M}_\gamma$ its double points.

We have a cohomological interpretation for stable κ-orbital integrals. Let us fix an isomorphism $\bar{\mathbb{Q}}_\ell \simeq \mathbb{C}$ so that κ can be seen as taking values in $\bar{\mathbb{Q}}_\ell$. Then we have the formula

$$\mathbf{O}_\gamma^\kappa(1_{\mathfrak{g}(\mathcal{O}_F)}) = \sharp\mathcal{P}_\gamma^0(k)^{-1}\mathrm{tr}(\mathrm{Fr}_q, \mathrm{H}^*(\mathcal{M}_\gamma \otimes_k \bar{k}, \bar{\mathbb{Q}}_\ell)_\kappa)$$

where Fr_q denotes the action of the geometric Frobenius on the ℓ-adic cohomology of the affine Springer fiber. In the case where the component group $\pi_0(\mathcal{P}_\gamma)$ is finite, $\mathrm{H}^*(\mathcal{M}_\gamma, \bar{\mathbb{Q}}_\ell)_\kappa$ is the biggest direct summand of $\mathrm{H}^*(\mathcal{M}_\gamma, \bar{\mathbb{Q}}_\ell)$ on which $\mathcal{P}_\gamma$ acts through the character κ. By taking $\kappa = 1$, we obtained a cohomological interpretation of the stable orbital integral

$$\mathbf{SO}_\gamma(1_{\mathfrak{g}(\mathcal{O}_F)}) = \sharp\mathcal{P}_\gamma^0(k)^{-1}\mathrm{tr}(\mathrm{Fr}_q, \mathrm{H}^*(\mathcal{M}_\gamma, \bar{\mathbb{Q}}_ell)_{st})$$

where the index st means the direct summand where acts trivially. When $\pi_0(\mathcal{P}_\gamma)$ is infinite, the definition of $\mathrm{H}^*(\mathcal{M}_\gamma, \bar{\mathbb{Q}}_\ell)_{st}$ and $\mathrm{H}^*(\mathcal{M}_\gamma, \bar{\mathbb{Q}}_\ell)_\kappa$ is a little bit more complicated.

Cohomological interpretation of the fundamental lemma follows from the above cohomological interpretation of stable and K-orbital integrals. In general, it does not seem possible to prove the cohomological fundamental lemma by a direct method because the ℓ-adic cohomology of the affine Springer fiber is as complicated as the orbital integrals. Nevertheless, in the case of unramified conjugacy classes, by using a large torus action of the affine Springer fiber and the Borel-Atiyah-Segal localization theorem for equivariant cohomology, Goresky, Kottwitz and MacPherson proved a formula for the ℓ-adic cohomology of unramified affine fibers in assuming the purity conjecture. It should be noticed however that there may be no torus action on the affine Springer fibers associated to most ramified conjugacy classes.

4.2. *The Hitchin fibration*

The Hitchin fibration appears in a quite remote area from the trace formula and the theory of endoscopy. It is fortunate that the geometry of the Hitchin fibration and the arithmetic of endoscopy happen to be just different smiling faces of Bayon Avalokiteshvara.

In [25], Hitchin constructed a large family of algebraic integrable systems. Let X be a smooth projective complex curve and Bun_G^{st} the moduli space of stable G-principal bundles on X. The cotangent bundle $T{*}\mathrm{Bun}_G^{st}$ is naturally a symplectic variety so that its algebra of analytic functions is equipped with a Poisson bracket {f, g}. It has dimension $2d$ where d is the dimension of Bun_G. Hitchin proves the existence of d Poisson commuting algebraic functions on $T^*\mathrm{Bun}_G$ that are algebraically independent

$$f = (f_1, \ldots, f_d) : T^*\mathrm{Bun}_G^{st} \to \mathbb{C}^d. \tag{18}$$

The Hamiltonian vector fields associated to $f_1, \ldots, f_d$ form d commuting vector fields along the fiber of f. Hitchin proved that generic fibers of f are open subsets of abelian varieties and Hamiltonian vector fields are linear.

To recall the construction of Hitchin, it is best to relax the stability condition and consider the algebraic stack Bun_G of all principal G-bundles instead of its open substack Bun_G^{st} of stable bundles. Following Hitchin, a Higgs bundle is a pair (E, ϕ) where $E \in \mathrm{Bun}_G$ is a principal G-bundle over X and ϕ is a global section of $\mathrm{ad}(E)$ K, K being the canonical bundle of X. Over the stable locus, the moduli space $\mathcal{M}$ of all Higgs bundles coincide with $T^*\mathrm{Bun}_G^{st}$ by Serre's duality.

According to Chevalley and Kostant, the algebra $\mathbb{C}[\mathfrak{g}]^G$ of adjoint invariant function is a polynomial algebra generated by homogeneous functions $a_1, \ldots, a_r$ of degree $e_1 + 1, \ldots, e_r + 1$ where $e_1, \ldots, e_r$ are the exponents of the root system. If (E, ϕ) is a Higgs bundle then $a_i(\phi)$ is well defined as a global section of $K^{\otimes(e_i+1)}$. This defines a morphism $f : \mathcal{M} \to \mathcal{A}$ where $\mathcal{A}$ is the affine space

$$\mathcal{A} = \bigoplus_{i=1}^{r} H^0(X, K^{\otimes(e_i+1)}).$$

whose dimension equals somewhat miraculously the dimension d of $\dim(\mathrm{Bun}_G)$. This construction applies also to a more general situation where K is replaced by an arbitrary line bundle, but of course the symplectic form as well as the equality of dimension are lost. It is not difficult to extend Hitchin's argument to prove that, after passing from the coarse moduli space to the moduli stack, the generic fiber of f is isomorphic to an extension of a finite group by an abelian variety. More canonically, the generic fiber of f is a principal homogeneous space under the action of the extension of a finite group by an abelian variety. On the infinitesimal level, this action is nothing but the action of the Hamiltonian vector fields along the fibers of f. We observe that Hamiltonian vector fields act also on singular fibers of f, and we would like to understand the geometry of those fibers by this action.

In [54], we constructed a smooth Picard stack $g : \mathcal{P} \to \mathcal{A}$ that acts on $f : \mathcal{M} \to \mathcal{A}$. In particular, for every $a \in \mathcal{A}, \mathcal{P}_a$ acts on $\mathcal{M}_a$ in integrating the infinitesimal action of the Hamiltonian vector fields. For generic parameters a, the action of $\mathcal{P}_a$ on $\mathcal{M}_a$ is simply transitive but for degenerate parameters a, it is not.

We observe the important product formula

$$[\mathcal{M}a/\mathcal{P}a] = \prod_{v\in X} [\mathcal{M}_{a,v}/\mathcal{P}_{a,v}] \tag{19}$$

that expresses the quotient $[\mathcal{M}_a/\mathcal{P}_a]$ as an algebraic stack as the product of affine Springer fibers $\mathcal{M}_{a,v}$ by its group of symmetry $\mathcal{P}_{a,v}$. For almost all $v, \mathcal{M}_{a,v}$ is a discrete set acted on simply transitively by $\mathcal{P}_{a,v}$.

In order to get an insight of the product formula, it is best to switch the base field from the field of complex numbers to a finite field k. In this case, it is instructive to count the number of k-points on the Hitchin fiber $\mathcal{M}_a$ as well as on the quotients $[\mathcal{M}_a/\mathcal{P}_a]$. In order to get actual numbers, we assume that the component group $\pi_0(\mathcal{P}_a)$ is finite. This is the case for a in an open subset $\mathcal{A}^{ell}$ of $\mathcal{A}$, called the elliptic part, to which we will restrict ourselves from now on.

More details about the following discussion can be found in [54, 55]. For $a \in \mathcal{A}^{ell}(k)$, the fiber $\mathcal{M}_a$ is a proper Deligne-Mumford stack and the number of its k-points can be expressed as a sum

$$|\mathcal{M}_a(k)| = \sum_{\gamma\in\mathfrak{g}(F)/\simeq,\chi(f)=a} \mathbf{O}_\gamma(1_D) \tag{20}$$

over rational conjugacy classes $\gamma \in \mathfrak{g}(F)/\sim, F$ denoting the function field of X within the stable conjugacy class defined by a, of global orbital integral (11) of certain adelic function 1_D, whose local expression $1_D = \Pi_{v\in|X|} 1_{D_v}$ is given by the choice of a global section of the line bundle $K = \mathcal{O}_X(D)$. The number of k-points on thc quotient $[\mathcal{M}_a/\mathcal{P}_a]$ can be expressed as a product of stable orbital integrals

$$|[\mathcal{M}_a/\mathcal{P}_a](k)| = \prod_{v\in|X|} \mathbf{So}_a(1_{K_v}) \tag{21}$$

We will now look for an expression of the sum of global orbital integrals (20) in terms of stable orbital (21) plus correcting terms as in the stabilization of the trace formula. In our geometric terms, this expression becomes

$$|\mathcal{M}_a(k)| = |\mathcal{P}^0_a(k)| \sum_\kappa \mathbf{O}^\kappa_{\gamma a}(1_D) \tag{22}$$

where $\mathbf{O}^\kappa_{\gamma a}$ are κ-orbital integrals attached to the Kostant conjugacy class γ_a in the stable class a with respect to a Frobenius invariant character $\kappa : \pi_0(\mathcal{P}_a) \to \mathbb{Q}_\ell{}^\times$. The component group $\pi_0(\mathcal{P}_a)$ or the smile of Avalokiteshvara is the origin of endoscopic pain.

The cohomological interpretation of the formula (22) is the decomposition into direct sum of the cohomology of $\mathcal{M}_a$ with respect to the action of $\pi_0(\mathcal{P}_a)$

$$\mathrm{H}^*(\mathcal{M}_a \otimes_k \bar{k}, \mathbb{Q}_\ell) = \bigoplus_{\kappa:\pi_0(\mathcal{P}_a)\to\mathbb{Q}_\ell^\times} \mathrm{H}^*(\mathcal{M}_a \otimes_k \bar{k}, \mathbb{Q}_\ell)_\kappa. \tag{23}$$

It is not obvious to understand how this decomposition depends on a since the component group $\pi_0(\mathcal{P}_a)$ also depends on a. According to a theorem of Grothendieck, the component groups $\pi_0(\mathcal{P}_a)$ for varying a can be interpolated as fiber of a sheaf of abelian groups $\pi_0(\mathcal{P})$ for the étale topology of $\mathcal{A}$. Restricted to the elliptic part $\mathcal{A}^{ell}, \pi_0(\mathcal{P})$ is a sheaf of finite abelian groups. One of the difficulties to understand the decomposition (23) lies in the fact that $\pi_0(P)$ is not a constant sheaf. Nevertheless, the sheaf $\pi_0(\mathcal{P})$ acts on the perverse sheaves of cohomology

$$ {}^p\mathrm{H}^n(f_*\mathbb{Q}_\ell|_{\mathcal{A}}) $$

and decomposes it into a direct sum canonically indexed by a finite set of semisimple conjugacy classes of the dual group $\hat{G}$

$$ {}^p\mathrm{H}^n(f_*\mathbb{Q}_\ell|_{\mathcal{A}^{ell}}) = \bigoplus_{[\kappa]\in\hat{G}/\sim} {}^p\mathrm{H}^n(f_*\mathbb{Q}_\ell|_{\mathcal{A}^{ell}}) $$

This peculiar decomposition reflects the combinatorial complexity of the stabilization of the trace formula, see [54, 55]. Among the direct summand, the main term corresponding to $\kappa = 1$ is called the stable piece. For instance, the surprising appearance of semisimple conjugacy classes of the dual group reflects the presence of the equivalence classes of endoscopic groups in the stabilization of the trace formula.

The stabilization of the trace formula as envisionned by Langlands and Kottwitz suggests that the $[\kappa]$-part in the above decomposition should correspond to the stable part in the similar decomposition for an endoscopic group. This prediction can be realized in a clean geometric formulation after we pass to the étale scheme $\tilde{\mathcal{A}}$ over $\mathcal{A}$ *cf.* [56] which depends on the choice of a point $\infty \in X$. It was constructed in such a way that over $\tilde{\mathcal{A}}, \pi_0(\mathcal{P})$ becomes a quotient of the constant sheaf, whose sections over any connected test scheme are cocharacters of the maximal torus T. Over $\tilde{\mathcal{A}}^{ell}$, we obtain a finer decomposition

$$ {}^p\mathrm{H}^n(f_*\mathbb{Q}_\ell|_{\tilde{\mathcal{A}}^{ell}}) = \bigoplus_{[\kappa]\in\hat{T}} {}^p\mathrm{H}^n(f_*\mathbb{Q}_\ell|_{\tilde{\mathcal{A}}^{ell}})_\kappa $$

indexed by a finite subset of the maximal torus $\hat{T}$ in $\hat{G}$.

Let $\kappa \in \hat{T}$ correspond to a nontrivial piece in the above decomposition. The κ-component of the above direct sum is supported by the locus $\tilde{\mathcal{A}}^{ell}_\kappa$ in $\tilde{\mathcal{A}}^{ell}$ given by the elements $\tilde{a} \in \tilde{\mathcal{A}}^{ell}$ such that $\kappa : \mathbf{X}_*(T) \to \mathbb{Q}_\ell^\times$ factors through $\pi_0(\mathcal{P}_{\tilde{a}})$. This locus is not connected; its connected components are classified by homomorphism $\rho : \pi_1(X, \infty) \to \pi_0(\hat{G}_\kappa)$. Such a homomorphism defines a reductive group scheme H over X whose dual group is $\hat{H}_\rho$ by outer twisting. It can be checked that the connected component of $\tilde{\mathcal{A}}^{ell}_\kappa$ corresponding to ρ is just the Hitchin base $\mathcal{A}_{H_\rho}$ for the reductive group scheme H_ρ. Let $\iota_{\kappa,\rho} : \tilde{\mathcal{A}}_{H_\rho} \to \tilde{\mathcal{A}}$ denote this closed immersion.

Theorem 8. *Let G be a split semisimple group. There exists an isomorphism*

$$ \bigoplus_n {}^p\mathrm{H}^n(f_*\mathbb{Q}_\ell|_{\tilde{\mathcal{A}}^{ell}})_\kappa[2r](r) \sim \bigoplus_\rho (\iota_{\kappa,\rho})_* \bigoplus_n {}^p\mathrm{H}^n(f_{H_\rho,*}\mathbb{Q}_\ell|_{\tilde{\mathcal{A}}^{ell}_{H_\rho}})_\kappa $$

where ρ are homomorphisms $\rho : \pi_1(X, \infty) \to \pi_0(\hat{G}_\kappa)$ and where r is some multiple of deg(K).

Here we stated our theorem in the case of split group, but it is valid for quasi-split group as well. In fact, the theorem was first proved for quasi-split unitary group by Laumon and myself in [50] before the general case was proved in [56]. To be more precise, the above theorem is proved under the assumption that the characteristic of the residue field is at least twice the Coxeter number of G.

The fundamental lemma for Lie algebra in equal characteristic case follows from the above theorem by a local-global argument. The unequal characteristic case follows from the equal characteristic case by theorem of Waldspurger [69] and Cluckers, Hales, Loeser [12]. Waldspurger assumes that p does not divide the order of the Weyl group and Cluckers, Hales, Loeser needs a much stronger lower bound on p. In number field case, these assumptions do not matter as Hales proved that the validity of the fundamental lemma at almost all places implies its validity at the remaining places. Currently, the fundamental lemma for local fields of positive characteristic small with respect to G, is not known.

4.3. *Support theorem*

The main ingredient in the proof of theorem 8 is the determination of the support of simple perverse sheaves that appear as constituent of perverse cohomology of $f_*\mathbb{Q}_\ell$.

Let C be a pure ℓ-adic complex over a scheme S of finite type over a finite field k. Its perverse cohomology ${}^p\mathrm{H}^n(C)$ are then perverse sheaves and geometrically semisimple according to a theorem of Beilinson, Bernstein, Deligne and Gabber *cf.* [5]. According to Goresky and MacPherson, geometrically simple perverse sheaves are of the following form: let Z be a closed irreducible subscheme of $S \otimes_k \bar{k}$ with $i : Z \to S \otimes_k \bar{k}$ denoting the closed immersion, let U be a smooth open subscheme of Z with $j : U \to Z$ denoting the open immersion, let $\mathcal{K}$ be a local system on U, then $K = i_* j_{!*}\mathcal{K}[\dim(Z)]$ is a simple perverse sheaf, $j_{!*}$ being the functor of intermediate extension, and every simple perverse sheaf on $S \otimes_k \bar{k}$ is of this form. In particular, the support $Z = \mathrm{supp}(K)$ of a simple perverse sheaf is well defined. For a pure ℓ-adic complex C over a scheme S, we can ask the question what is the set of supports of simple perverse sheaves occurring as direct factors of the perverse sheaves of cohomology ${}^p\mathrm{H}^n(C)$.

The main topological ingredient in the proof of theorem 8 is the determination of this set of supports. We state only the result in characteristic zero. In characteristic p, we prove a weaker result, more complicated to state but enough for the purposes of the fundamental lemma.

Theorem 9. *Assume the base field k is the field of complex numbers. Then for any simple perverse sheaf K direct factor of ${}^p\mathrm{H}^n(f_*\mathbb{Q}_\ell|_{\tilde{\mathcal{A}}^{ell}})_{st}$, the support of K is $\tilde{\mathcal{A}}^{ell}$.*

Similarly, if K is a direct factor of ${}^p\mathrm{H}^n(f_*\mathbb{Q}_\ell|_{\tilde{\mathcal{A}}^{ell}})_\kappa$, then the support of k is of the form $\iota_\rho(\tilde{\mathcal{A}}_{H_\rho})$ for certain homomorphism $\rho : \pi_1(X,\infty) \to \pi_0(\hat{G}_\kappa)$.

If we know two perverse sheaves having simple constituents of the same support, in order to construct an isomorphism between them, it is enough to construct an isomorphism over an open subset of the support. Over a small enough open subscheme, the isomorphism can be constructed directly.

Let us explain the proof of the nonstandard fundamental lemma conjectured by Waldspurger. Let G_1, G_2 be semisimple groups with isogeneous root systems. Their Hitchin moduli spaces $\mathcal{M}_1$, $\mathcal{M}_2$ map to the same base $\mathcal{A} = \mathcal{A}_1 = \mathcal{A}_2$. Let restrict to the elliptic locus and put $\mathcal{A} = \mathcal{A}^{ell}$. In order to prove $(f_{1*}\mathbb{Q}_\ell)_st \sim (f_{2*}\mathbb{Q}_\ell)_{st}$, it is enough to prove that they are isomorphic over an open subscheme of $\mathcal{A}$, as we know every simple perverse sheaf occurring in either one of these two complexes have support $\mathcal{A}^{ell}$. Over an open subscheme of $\mathcal{A}^{ell}$, $\mathcal{M}_1$ is acted on simply transitively by extension of a finite group by an abelian scheme and so is $\mathcal{M}_2$. The nonstandard fundamental lemma follows now from the fact that the above two abelian schemes are isogeneous and isogeneous abelian varieties have the same cohomology.

4.4. *Weighted fundamental lemma*

According to Waldspurger, the twisted fundamental lemma follows from the usual fundamental lemma and its nonstandard variant. Combining with his theorem that the fundamental lemma implies the transfer, the local results needed to stabilize the elliptic part of the trace formula and the twisted trace formula.

The classification of automorphic forms on quasi-split classical group requires the full power of the stabilization of the entire trace formula. For this purpose, Arthur needs more the twisted weighted fundamental lemma. This conjecture is an identity between twisted weighted orbital integrals.

The weighted fundamental lemma is now a theorem due to Chaudouard and Laumon *cf.* [9]. In the particular case of Sp(4), it was previously proved by Whitehouse *cf.* [72]. They introduced a condition of χ-stability in Higgs bundles such that the restriction of the Hitchin map $f : \mathcal{M} \to \mathcal{A}$ to the open subset $\mathcal{A}^\heartsuit$ of stable conjugacy classes that are generically regular semisimple and to moduli stack of χ-stable bundles $M^\heartsuit_{\chi-st}$

$$f^\heartsuit_{\chi-st} : \mathcal{M}^\heartsuit_{\chi-st} \to \mathcal{A}^\heartsuit$$

is a proper morphism. This is an extension of the proper morphism $f^{ell} : \mathcal{M}^{ell} \to \mathcal{A}^{ell}$ that depends on a stability parameter χ. Chaudouard and Laumon extended the support theorem from f^{ell} to $f^\heartsuit_{\chi-st}$. They also showed that the number of points on a hyperbolic fiber of $\mathcal{A}^\heartsuit$ can be expressed in terms of weighted orbital integrals. The weighted fundamental lemma follows. It is quite remarkable that the moduli space depends on the stability parameter χ, though the number of points and the ℓ-adic complex of cohomology don't.

Finally, Waldspurger showed that the twisted weighted fundamental lemma follows from the weighted fundamental lemma and its nonstandard variant. He also showed that, if these statements are known for a local field of characteristic p, they are also known for a p-adic local field with the same residue field, provided the residual characteristic does not divide the order of the Weyl group.

5. Functoriality Beyond Endoscopy

The unstability of the trace formula has been instrumental in establishing the first cases of the functoriality conjecture. The stable trace formula now fully established by Arthur should be the main tool in our quest for more general functoriality.

In [43], Langlands proposed new insights for the general case of functoriality principle. He observed that we are primarily concerned with the question how to distinguish automorphic representations π of G whose hypothetical parametrization $\sigma : L_F \to {}^L G$ has image contained in a smaller subgroup. Assume π of Ramanujan type (or tempered), the Zariski closure of the image of σ is not far from being determined by the order of the pole at 1 of the L-functions $L(s, \rho, \pi)$ for all representations ρ of ${}^L G$. Though we are not in position to work directly with these L-functions individually, the stable trace formula can be effective in dealing with the sum of L-functions attached to all automorphic representations π or the sum of their logarithmic derivative. Nontempered representations, especially the trivial representation, represent an obstacle to this strategy as they contribute to this sum the dominant term. The subsequent article [15], directly inspired from [43], might have proposed a method to subtract the dominant contribution. Other works [65, 48, 16], more or less inspired from [43], are the first encouraging steps on this new path that might lead us to the general case of functoriality.

Acknowledgment

I would like to express my deep gratitude to Laumon and Kottwitz who have been helping and encouraging me in the long march in pursuit of the lemma. I would like to thank Langlands for his comments on this report and to Moeglin for a useful conversation on its content. I would like to thank Dottie Phares for helping me with English.

References

[1] Arthur, J., *A stable trace formula. III. Proof of the main theorems.* Ann. of Math. (2) 158 (2003), no. 3, 769–873.

[2] Arthur, J., *An introduction to the trace formula.* in Harmonic analysis, the trace formula, and Shimura varieties, 1–263, Clay Math. Proc., 4, Amer. Math. Soc., Providence, RI, 2005.

[3] Arthur, J., Clozel, L., *Simple algebras, base change, and the advanced theory of the trace formula.* Annals of Math. Studies, 120. Princeton University Press, Princeton, NJ, 1989.

[4] Badulescu, I, *Global Jacquet-Langlands correspondence, multiplicity one and classification of automorphic representations.* With an appendix by Neven Grbac. Invent. Math. 172 (2008), no. 2, 383–438.

[5] Beilinson, A., Bernstein, J., Deligne P.: Faisceaux pervers. *Astérisque* 100 (1982).

[6] Borel, A., Jacquet, H., *Automorphic forms and automorphic representations,* in Proc. of Symp. in Pure Math. 33 ed. by A. Borel and W. Casselman, Amer. Math. Soc., 1977.

[7] Breuil, C.; Conrad, B.; Diamond, F.; Taylor, R., *On the modularity of elliptic curves over Q: wild 3-adic exercises.* J. Amer. Math. Soc. 14 (2001), no. 4, 843–939.

[8] Cartier, P. Representation of p-adic groups: A survey, in Proc. of Symp. in Pure Math. 33 ed. by A. Borel and W. Casselman, Amer. Math. Soc., 1977.

[9] Chaudouard, P.-H., Laumon, G., *Le lemme fondamental pondéré,* preprint.

[10] Clozel, L. *Représentations galoisiennes associees aux représentations automorphes autoduales de GL(n),* Publ. IHES 73 (1991), 97145.

[11] Clozel, L., Harris, M., Labesse, J.-P. *Construction of automorphic Galois representations.* Chevaleret's Book Project.

[12] Cluckers, R., Hales, T., Loeser, F, *Transfer Principle for the Fundamental Lemma,* Chevaleret's Book Project.

[13] Dirichlet, G., *Démonstration d'un théorème sur la progression arithmétique,* 1834, *Werke* Bd. I, p. 307.

[14] Drinfeld, V. *Langlands' conjecture for* GL(2) *over functional fields.* Proceedings of the International Congress of Mathematicians (Helsinki, 1978), pp. 565–574, Acad. Sci. Fennica, Helsinki, 1980.

[15] Frenkel, E., Langlands, R., Ngô, B.C. *Formule des traces et fonctorialité: le début d'un programme,* preprint.

[16] Frenkel, E., Ngô, B.C. *Geometrization of trace formulas,* preprint.

[17] Godement, R., Jacquet, H., *Zeta functions of simple algebras.* Lecture Notes in Mathematics, Vol. 260. Springer-Verlag, Berlin-New York, 1972.

[18] Goresky, M., Kottwitz, R., MacPherson, R.: *Homology of affine Springer fiber in the unramified case.* Duke Math. J. 121 (2004) 509–561.

[19] Hales, T., *On the fundamental lemma for standard endoscopy: reduction to unit elements.* Canad. J. Math. 47 (1995), no. 5, 974–994.

[20] Harris, M., Taylor, R., *The geometry and cohomology of some simple Shimura varieties.* With an appendix by Vladimir G. Berkovich. Annals of Mathematics Studies, 151. Princeton University Press, Princeton, NJ, 2001.

[21] Harris, M., Shepherd-Barron, N. and Taylor, R. *A family of Calabi-Yau varieties and potential automorphy.* Annals of Math. 171 (2010), 779–813.

[22] Henniart, G., *Une preuve simple des conjectures de Langlands pour* GL(n) *sur un corps p-adique.* Invent. Math. 139 (2000), no. 2, 439–455.

[23] Henniart, G. *On the local Langlands and Jacquet-Langlands correspondences.* International Congress of Mathematicians. Vol. II, 1171–1182, Eur. Math. Soc., Zurich, 2006.

[24] Hiraga, K., Saito, H., *On L-packets for inner forms of* SL(n), preprint.

[25] Hitchin N.: Stable bundles and integrable systems. *Duke Math. J.* 54 (1987) 91–114.
[26] Kim, H., Shahidi, F. *Symmetric cube L-functions for* GL_2 *are entire.* Ann. of Math. (2) 150 (1999), no. 2, 645–662.
[27] Kazhdan, D., *On lifting* in Lie group representations, II, 209–249, Lecture Notes in Math., 1041, Springer, Berlin, 1984.
[28] Kazhdan, D. Lusztig, G. *Fixed point varieties on affine flag manifolds.* Israel J. Math. 62 (1988), no. 2, 129–168.
[29] Kostant, B., *Lie group representations on polynomial rings.* Amer. J. Math. 85 (1963) 327–404.
[30] Kottwitz, R., *Unstable orbital integrals on* SL(3). Duke Math. J. 48 (1981), no. 3, 649–664.
[31] Kottwitz, R., *Stable trace formula: cuspidal tempered terms.* Duke Math. J. 51 (1984), no. 3, 611–650.
[32] Kottwitz, R. *Stable trace formula: singular elliptic terms.* Math. Ann. 275 (1986), no. 3, 365–399.
[33] Kottwitz, R. *Base change for unit elements of Hecke algebras.* Compositio Math. 60 (1986), no. 2, 237–250.
[34] Kottwitz, R., *Shimura varieties and* λ*-adic representations,* in Automorphic forms, Shimura varieties, and L-functions, Vol. I (Ann Arbor, MI, 1988), 161–209, Perspect. Math., 10, Academic Press, Boston, MA, 1990.
[35] Kottwitz, R., *Points on some Shimura varieties over finite fields.* J. Amer. Math. Soc. 5 (1992), no. 2, 373–444.
[36] Kottwitz, R., *Transfer factors for Lie algebras.* Represent. Theory 3 (1999), 127–138.
[37] Labesse, J.-P. *Changement de base CM et séries discrétes.* Chevaleret's Book Project.
[38] Labesse, J.-P., Langlands, R. *L-indistinguishability for* SL(2), Can. J. Math. 31 (1979), pp. 726–785.
[39] Langlands, R., *Problems in the Theory of Automorphic Forms,* Lecture Notes in Mathematics, Vol. 170. Springer-Verlag, Berlin-New York, 1970.
[40] Langlands, R., Letter to Lang (1970) and Langlands' comments on this letter, available at http://publications.ias.edu/rpl.
[41] Langlands, R. *Base change for* GL(2). Annals of Math. Studies, 96. Princeton University Press, Princeton, N.J., 1980.
[42] Langlands, R., *Les débuts d'une formule des traces stables.* [Mathematical Publications of the University of Paris VII], 13. Université de Paris VII, U.E.R. de Mathématiques, Paris, 1983. v+188 pp.
[43] Langlands R. *Beyond endoscopy,* available at http://publications.ias.edu/rpl.
[44] Langlands, R., *A review of Haruzo Hida's p-adic automorphic forms on Shimura varieties.* Bull. of the A.M.S. 44 (2007) no. 2, 291–308.
[45] Langlands, R., Shelstad, D., *On the definition of transfer factors* Math. Ann. 278 (1987), no. 1–4, 219–271.
[46] Lafforgue, L. *Chtoucas de Drinfeld et correspondance de Langlands.* Invent. Math. 147 (2002), no. 1, 1–241.
[47] Laforgue, L. *Chtoucas de Drinfeld, formule des traces d'Arthur-Selberg et correspondance de Langlands.* Proceedings of the International Congress of Mathematicians, Vol. I (Beijing, 2002), 383–400, Higher Ed. Press, Beijing, 2002.
[48] Lafforgue, L. *Construire des noyaux de la fonctorialité,* preprint.

[49] *Sur le lemme fondamental pour les groupes unitaires,* preprint.
[50] Laumon, G. Ngô, B.-C., *Le lemme fondamental pour les groupes unitaires.* Ann. of Math. (2) 168 (2008), no. 2, 477–573.
[51] Laumon, G., Rapoport, M., Stuhler U. *D-elliptic sheaves and the Langlands correspondence.* Invent. Math. 113 (1993), no. 2, 217–338.
[52] Moeglin, C., *Classification et changement de base pour les séries discrétes des groupes unitaires p-adiques.* Pacific J. Math. 233 (2007), no. 1, 159–204.
[53] Morel, S. *On the cohomology of certain noncompact Shimura varieties.* With an appendix by Robert Kottwitz. Annals of Math. Studies, 173. Princeton University Press, Princeton, NJ, 2010.
[54] Ngô, B.C. *Fibration de Hitchin et endoscopie.* Inv. Math. 164 (2006) 399–453.
[55] Ngô, B.C. *Fibration de Hitchin et structure endoscopique de la formule des traces.* International Congress of Mathematicians. Vol. II, 1213–1225, Eur. Math. Soc., Zurich, 2006.
[56] Ngô, B.C. *Le lemme fondamental pour les algébres de Lie,* to appear in Publ. Math. de l'I.H.É.S.
[57] Ngô, B.C. *Decomposition theorem and abelian fibration,* Chevaleret's book project.
[58] Rogawski, J. *Automorphic Representations of Unitary Groups in Three Variables.* Annals of Math. Studies 123 (1990).
[59] Shelstad, D., *L-Indistinguishability for Real Groups.* Math. Annalen 259 (1982), 385–430.
[60] Shin, S. *Galois representations arising from some compact Shimura varieties.* to appear in Ann. of Math.
[61] Springer, A. *Reductive groups,* in Proc. of Symp. in Pure Math. 33 ed. by A. Borel and W. Casselman, Amer. Math. Soc., 1977.
[62] Tamagawa, T., *On ζ-functions of a division algebra,* Ann. of Math. (2) 77 (1963) 387–405.
[63] Taylor, R.; Wiles, A., *Ring-theoretic properties of certain Hecke algebras.* Ann. of Math. (2) 141 (1995), no. 3, 553–572.
[64] Taylor, R., *On icosahedral Artin representations. II.* Amer. J. Math. 125 (2003), no. 3, 549–566.
[65] Venkatesh, A., *"Beyond endoscopy" and special forms on GL(2).* J. Reine Angew. Math. 577 (2004), 23–80.
[66] Waldspurger, J.-L. *Sur les intégrales orbitales tordues pour les groupes linéaires: un lemme fondamental.* Canad. J. Math. 43 (1991), no. 4, 852–896.
[67] Waldspurger, J.-L. *Le lemme fondamental implique le transfert.* Compositio Math. 105 (1997), no. 2, 153–236.
[68] Waldspurger, J.-L. *L'endoscopie tordue n'est pas si tordue.* Mem. Amer. Math. Soc. 194 (2008), no. 908, x+261 pp.
[69] Waldspurger, J.-L. *Endoscopie et changement de caractéristique.* J. Inst. Math. Jussieu 5 (2006), no. 3, 423–525.
[70] Waldspurger, J.-L., *A propos du lemme fondamental pondéré tordu.* Math. Ann. 343 (2009), no. 1, 103–174.
[71] Waldspurger, J.-L., *Endoscopie et changement de caracteristique: intégrales or-bitales pondérés.* Ann. Inst. Fourier (Grenoble) 59 (2009), no. 5, 1753–1818.

[72] Whitehouse, D., *The twisted weighted fundamental lemma for the transfer of automorphic forms from* GSp(4). *Formes automorphes. II. Le cas du groupe* GSp(4). Astrisque no. 302 (2005), 291–436.
[73] Wiles, A., *Modular elliptic curves and Fermat's last theorem.* Ann. of Math. (2) 141 (1995), no. 3, 443–551.

Proceedings of the International Congress of Mathematicians
Hyderabad, India, 2010

THE WORK OF ELON LINDENSTRAUSS

by
HARRY FURSTENBERG

I've been asked to describe some of the achievements of Elon Lindenstrauss — our Fields medalist. Elon Lindenstrauss's work continues a tradition of interaction between dynamical systems theory and diophantine analysis. This tradition goes back at least to the year 1914 — when Hermann Weyl published a paper entitled "An application of number theory to statistical mechanics and the theory of perturbations." In that paper, Weyl used what we would call Kronecker's Theorem to show the validity of the ergodic hypothesis in certain situations. In the meantime the roles have been reversed, with dynamical systems theory and ergodic theory providing the tools for answering questions in number theory.

The number theoretical issues arising in the work of Lindenstrauss have to do with so-called diophantine approximation — in which one asks whether inequalities having real solutions have integer solutions. In this area we encounter a phenomenon which is reminiscent of ergodic behavior. It can be described crudely by saying that whatever is not excluded for some good reason and can happen in principle, will eventually happen — at least approximately. There is a good reason that

$$-\varepsilon < x^2 - (1+\sqrt{2})^2 y^2 < \varepsilon$$

cannot be solved for small ε (this would imply that $\sqrt{2}$ is well approximable). But this doesn't apply to the three variable inequality:

$$-\varepsilon < x^2 - (1+\sqrt{2})^2 y^2 - \alpha z^2 < \varepsilon \quad (\alpha \neq 0 \text{ arbitrary})$$

and indeed by the relatively recently established Oppenheim conjecture, for any positive ε, this has a solution in integers (x, y, z) not all 0.

An important advance has come about by enlarging the scope of dynamics to include what will be referred to as "homogeneous dynamics". Every since Poincaré dynamical theory had broken out of the shackles of Ordinary Differential Equations and a dynamical system comes about whenever we have a 1-parameter group $\{T_t\}$ — think of t as time — of transformations acting in a space X, which we identify as the phase space of the system. We have *homogeneous* dynamics when X is a homogeneous space of a Lie group; we can write $X = G/\Gamma$. For any 1-parameter subgroup $\{g(t)\} \subset G$ we can set $T_t(g\Gamma) = g(t)g\Gamma$. Homogeneous dynamics allows one further abuse of the term "dynamics", extending the action from a 1-parameter subgroup of G to an arbitrary Lie subgroup $H \subset G$, so that the time parameter can be higher dimensional. This liberalization of viewpoint has been quite fruitful in the recent application of dynamics to number theory.

One particular homogeneous space has been the focus of activity in this work, it is a space that appears implicitly in Minkowski's geometry of numbers. Namely, for a dimension d, we consider the space Ω_d of unimodular lattices spanned by d independent vectors in $\mathbb{R}^d$. The group $\mathrm{SL}(d, \mathbb{R})$ acts transitively on this space in a natural way: $\Omega_d \cong \mathrm{SL(d,\mathbb{R})/SL(d,\mathbb{Z})}$. There is a measure on Ω_d invariant under the action of the group and the measure of Ω_d is finite. Nonetheless the space Ω_d is non-compact in its natural topology. This is important, as is Mahler's criterion for a set $\Sigma \subset \Omega_d$ to have compact closure. Namely, $\bar{\Sigma}$ is compact unless there is a sequence $\{\sigma_n\} \subset \Sigma$ and vectors $v_n \in \sigma_n$ with $||v_n|| \to 0$.

There is a broad spectrum of problems for which this is relevant. Namely, let $\Phi(x_1, x_2, \ldots, x_d)$ be a homogeneous polynomial and we ask if for arbitrarily small $\varepsilon > 0$ one can solve $|\Phi(x_1, x_2, \ldots, x_1)| < \varepsilon$ in integers not all 0. (This would in fact imply that the range of Φ on $\mathbb{Z}^d$ is dense in either $\mathbb{R}^+, \mathbb{R}^-$, or both). Now define the subgroups

$$H_\Phi \subset G = \mathrm{SL(d,\mathbb{R})} \text{ by } \mathrm{H_\Phi = \{h \in G : \Phi(h\bar{v}) = \Phi(\bar{v}) \text{ for all } \bar{v} \in \mathbb{R}^d\}}.$$

In general for a non-compact group H, one expects orbits Hx to be unbounded, and then Mahler's criterion will come into play. If we take $x_0 \in \Omega_d$ to be the lattice $\mathbb{Z}^d$, then if $H_\Phi x_0$ is unbounded, this will imply that there exist $h \in H_\Phi$ and $\vec{v} \in \mathbb{Z}^d$ with $||h\vec{v}||$ arbitrarily small which means that $\Phi(\vec{v})$ is arbitrarily small. This was the strategy leading to the solution of the Oppenheim conjecture in the 80's by Margulis. Here $\Phi(x_1, x_2, x_3) = \alpha x_1^2 - \beta x_2^2 - \gamma x_3^2$ and H_Φ has the property investigated by Marina Ratner motivated by conjectures of Raghunathan and Dani — of being generated by unipotent subgroups. (A linear transformation is unipotent if 1 is its unique eigenvalue.) By this theory one can classify all the closed H_Φ-invariant subsets of Ω_3 and in particular, one sees that an H_Φ-orbit has compact closure only if it is already compact. Margulis shows that this can happen to the orbit of $x_0 = \mathbb{Z}^d$ only if α, β, γ are commensurable. Otherwise this orbit is unbounded which leads to the conclusion that $|\Phi(x_1, x_2, x_3)| < \varepsilon$ has integer solutions.

Another notorious diophantine approximation problem is Littlewood's conjecture: for *all* pairs of real number α, β if for x real we denote by $||x||$ the distance of x to the nearest integer, then

$$\liminf_{n \to \infty} n||n\alpha||\ ||n\beta|| = 0.$$

This fits into the framework just discussed for the polynomial

$$\psi(x_1, x_2, x_3) = x_1(\alpha x_1 - x_2)(\beta x_1 - x_3)$$

where we disallow $x_1 = 0$. A linear transformation carries this to

$$\Theta(X, Y, Z) = XYZ$$

and H_Θ is (locally) just the diagonal subgroup $\left\{\begin{pmatrix} e^{-t-s} & 0 & 0 \\ 0 & e^t & 0 \\ 0 & 0 & e^s \end{pmatrix}\right\}$. This has no non-trivial unipotent subgroups; and the Ratner theory does not apply. Nonetheless, Margulis has conjectured that a bounded orbit for H_θ is necessarily compact and this conjecture, as in the foregoing discussion, has the Littlewood conjecture as a consequence.

We have here a contrast of unipotent homogeneous dynamics with what might be called — with Katok — higher rank hyperbolic dynamics. The former is "tame": neighboring points separate at a polynomial rate, whereas in hyperbolic dynamics they can separate at an exponential rate. Thanks largely to the work of Ratner, the unipotent theory may be said to be largely understood, whereas the hyperbolic theory is in a less satisfactory shape.

The earliest confirmations of Raghunathan's conjectures for unipotent actions came from the case $d = 2$ with results regarding the horocycle flow which corresponds to the subgroup $\left\{\begin{pmatrix} 1 & t \\ 0 & 1 \end{pmatrix}\right\}$. The hyperbolic counterpart, $\left\{\begin{pmatrix} e^t & 0 \\ 0 & e^{-t} \end{pmatrix}\right\}$, leads to the geodesic flow which is the prototypical example of chaotic dynamics. This would lead one to expect that the higher dimensional cases of diagonal group actions can only get worse, thus leaving little hope for a dynamical approach to the Littlewood conjecture.

Among those who spearheaded the initiative to understand the phenomenon of rigidity in the hyperbolic framework was Anatole Katok, who, in a paper with Ralph Spatzier gave conditions for a rigidity result in the hyperbolic setup. In this paper, the importance of the acting group being of rank≥ 2 is underscored. An analogy is drawn to a phenomenon I have studied; namely the paucity of closed subsets of the group $\mathbb{R}/\mathbb{Z}$ invariant under two endomorphisms $x \to px$ (mod 1)and $x \to qx$ (mod 1), provided $\{p^n q^m\}$ is not contained in some $\{r^n\}$. (That is to say $\log p/\log q$ is irrational). The only closed sets are $\mathbb{R}/\mathbb{Z}$ itself and finite sets of rationals. It is an open question whether the only invariant measures are correspondingly the obvious ones: Lebesgue measure and atomic measures supported on rational and combinations of these. This example has been instructive for the following reason. Namely if one adds the condition that one or the other transformation, $x \to px$ or qx (mod 1) has *positive entropy* with respect to the invariant measure in question, then the measure must have a Lebesgue component. This result of Dan Rudolph which partially answers our query regarding $\times p$, $\times q$ suggests that for diagonal homogeneous actions, positive entropy will also play a significant role. This is the case already in the paper of Katok and Spatzier where other hypotheses are necessary. The state-of-the-art theorem in this regard is due to Einsiedler, Katok and Lindenstrauss and it depends heavily on new ideas of Lindenstrauss, requiring only positive entropy along some 1-parameter subgroup to conclude that an invariant measure is of an algebraic character. This theorem provides the crucial step to proving a modified version of Littlewood's conjecture — a version representing the first significant advance on the Littlewood problem: for all but a set of dimension 0 of pairs α, β of real numbers, $\lim \inf_{n\to\infty} n||n\alpha||\ ||n\beta|| = 0$.

One of the seminal contributions of Lindenstrauss to this realm is his broadening of the notion of recurrence of a measure to a wide variety of situations, in particular, to situations where the measure is not invariant under a certain set of transformations. Quoting Lindenstrauss, "the only thing which is really needed is some form of recurrence which produces the complicated orbits which are the life and blood of ergodic theory."

This brings us to what is possibly the most exciting work of Elon Lindenstrauss; namely the solution of the Quantum Unique Ergodicity question in the arithmetic case. From the mathematical standpoint, the issue is whether eigenfunctions of the Laplace operator on a negatively curved manifold tend to be more and more evenly spread over the space as the eigenvalue tends to negative infinity. In the special case of arithmetic hyperbolic surfaces, the so-called Hecke operators come into the picture and they act on the limiting measure arising from such a sequence of eigenfunctions. This action is recurrent and the tools developed by Lindenstrauss become applicable to this situation at hand, and lead elegantly to a solution of the problem.

Solving the so-called arithmetic quantum unique ergodicity conjecture of Rudnick and Sarnak is exciting if for no other reason than that the conjecture has been established provisionally, based on the generalized Riemann hypothesis. While this doesn't bring us closer to a solution of this famous question, this connection does testify to the depth of the mathematics involved.

I close my introductory remarks by mentioning one of the corollaries of Elon Lindenstrauss's handling of the arithmetic QUE conjecture; namely replacing reals by adéles and integers by rationals, we can speak of the adelic analogue of geodesic flow: namely, the action of the diagonal of $SL_2(\mathbb{A})$ on $SL_2(\mathbb{A})/SL_2(\mathbb{Q})$. The striking statement is that the adelic geodesic flow is uniquely ergodic.

I think it is fair to say that there is both power and beauty in the mathematical work of Elon Lindenstrauss.

Elon Lindenstrauss

ELON LINDENSTRAUSS

B.Sc. in Mathematics and Physics, The Hebrew University of Jerusalem, 1991;

M.Sc. in Mathematics, The Hebrew University, 1995;

Ph.D. in Mathematics, The Hebrew University, 1999.

Positions held

Member, Institute for Advanced Study, Princeton, 1999–2001.

Szegö Assistant Professor, Stanford University, 2001–2003.

Visiting Member, Courant Institute, New York University, 2003–2005.

Professor, Princeton University, 2004–2010.

Professor, Hebrew University, 2008–

Proceedings of the International Congress of Mathematicians
Hyderabad, India, 2010

EQUIDISTRIBUTION IN HOMOGENEOUS SPACES AND NUMBER THEORY

by

ELON LINDENSTRAUSS*

ABSTRACT. We survey some aspects of homogeneous dynamics — the study of algebraic group actions on quotient spaces of locally compact groups by discrete subgroups. We give special emphasis to results pertaining to the distribution of orbits of explicitly describable points, especially results valid for the orbits of *all* points, in contrast to results that characterize the behavior of orbits of *typical* points. Such results have many number theoretic applications, a few of which are presented in this note. Quantitative equidistribution results are also discussed.

Mathematics Subject Classification (2010). Primary 37A17; Secondary 37A45, 11J13, 11B30, 11J71

Keywords. invariant measures, homogeneous spaces, geometry of numbers, quantitative equidistribution, arithmetic combinatorics, quantum unique ergodicity, entropy.

1. Introduction

1.1. In this note we discuss a certain very special class of dynamical systems of algebraic origin, in which the space is the quotient of a locally compact group G by a discrete subgroup Γ and the dynamics is given by the action of some closed subgroup $H < G$ on G/Γ by left translations, or more generally by the action of a subgroup of the group of affine transformations on G that descends to an action on G/Γ. There are several natural classes of locally compact groups one may consider — connected Lie groups, linear algebraic groups (over $\mathbb{R}$, or $\mathbb{Q}_p$, or perhaps general local field of arbitrary characteristic), finite products of linear algebraic groups over different fields, or the closely related case of linear algebraic groups over adeles of a global field such as $\mathbb{Q}$.

1.2. Such actions turn out to be of interest for many reasons, but in particular are intimately related to deep number theoretic questions. They are also closely

*The research presented was supported by the NSF (most recently by grants DMS-0554345 and DMS-0800345) and the Israel Science Foundation.

Einstein Institute of Mathematics, The Hebrew University of Jerusalem, Jerusalem, 91904, Israel *and* Princeton University, Princeton NJ 08540 USA. E-mail: elonl@math.princeton.edu connected to another rich area: the spectral theory of such quotient spaces, also known as the theory of automorphic forms, which has so many connections to both analytic and algebraic number theory that they are hard to separate.

From the point of view of these connections between dynamics and number theory, perhaps the most interesting quotient space is the space X_d of lattices in $\mathbb{R}^d$ up to homothety, which is naturally identified with $\mathrm{PGL}(d,\mathbb{R})/\mathrm{PGL}(d,\mathbb{Z})$. There are several historical sources for the use of this space in number theory. One prominent historical source is H. Minkowski's work on Geometry of Numbers c. 1895; and while (like most mathematical research areas) it is hard to draw the precise boundaries of the Geometry of Numbers, certainly at its heart is a systematic use of lattices, and implicitly the space of lattices, to the study of number theoretic problems of independent interest.

The use of tools and techniques of ergodic theory and dynamical systems, and perhaps no less importantly the use of the dynamical point of view, to study these actions has proven to be a remarkably powerful method with applications in several rather diverse areas in number theory and beyond, but in particular for many of the problems considered in the Geometry of Numbers. This is a very active direction of current research sometimes referred to as Flows on Homogeneous Spaces, though the shorter term Homogeneous Dynamics seems to be gaining popularity.

1.3. We present below a Smörgåsbord of topics from the theory. The selection is somewhat arbitrary, and is biased towards aspects that I have personally worked on. A brief overview of the topics discussed in each section is given below:

§2. Actions of unipotent and diagonalizable groups are discussed. Thanks to the deep work of several mathematicians, the actions of unipotent groups are quite well understood (at least on a qualitative level). The actions of diagonalizable groups are much less understood. These diagonalizable actions behave quite differently depending on whether the acting group is one dimensional or of higher dimensions; in the latter case there are several long-standing conjectures and a few partial results toward these conjectures that are powerful enough to have applications of independent interest.

§3. We consider why the rigidity properties of an action of a multiparameter diagonalizable group is harder to understand than actions of unipotent groups (or groups generated by unipotents), and highlight one difference between these two classes of groups: growth rates of the Haar measure of norm-balls in these groups.

§4. Three applications of the measure classification results for multiparameter diagonalizable groups are presented: results regarding Diophantine approximations and Littlewood's Conjecture, Arithmetic Quantum Unique Ergodicity, and an equidistribution result for periodic orbits of the diagonal group in X_3 (a problem considered by Linnik with strong connections to L-functions and automorphic forms).

§5. We present recent progress in the study of actions of another natural class of groups that share with unipotent groups the property of large norm-balls: Zariski dense subgroups of semisimple groups or more generally groups generated by unipotents.

§6. We conclude with a discussion of the quantitative aspects of the density and equidistribution results presented in the previous sections regarding orbits of group actions on homogeneous spaces.

2. Actions of Unipotent and Diagonalizable Groups

2.1. Part of the beauty of the subject is that for a given number theoretic application one is led to consider a very concrete dynamical system. Perhaps the best way to illustrate this point is by example. An important and influential milestone in the theory of flows on homogeneous spaces has been Margulis' proof of the longstanding Oppenheim Conjecture in the mid 1980's [Mar87]. The Oppenheim Conjecture states that if $Q(x_1, \dots, x_d)$ is an indefinite quadratic form in $d \geq 3$ variables, not proportional to a form with integral coefficients, then

$$\inf \{|Q(v)| : v \in \mathbb{Z}^d \setminus \{0\}\} = 0. \tag{2.1}$$

By restricting Q to a suitably chosen rational subspace, it is easy to reduce the conjecture to the case of $d = 3$, and instead of considering the values of an arbitrary indefinite ternary quadratic form on the lattice $\mathbb{Z}^d$ one can equivalently consider the values an arbitrary lattice ξ in $\mathbb{R}^d$ attains on the fixed indefinite ternary quadratic form, say $Q_0(x, y, z) = 2xz - y^2$. The symmetry group

$$\mathrm{SO}(1, 2) = \{h \in \mathrm{SL}(3, \mathbb{R}) : Q_0(v) = Q_0(hv) \text{ for all } v \in \mathbb{R}^3\}$$

is a noncompact semisimple group. By the definition of H, for every $h \in H = \mathrm{SO}(1, 2)$ and $\xi \in X_3$ the set of values Q_0 attains at nonzero vectors of the lattice ξ coincides with the set of values this quadratic form attains at nonzero vectors of the lattice $h.\xi$, i.e. the lattice obtained from ξ by applying the linear map h on each vector. It is now an elementary observation, using Mahler's Compactness Criterion, that for $\xi \in X_3$,

$$\inf \{|Q_0(v)| : v \in \xi \setminus \{0\}\} = 0 \iff \text{the orbit } H.\xi \text{ is unbounded.}$$

G. A. Margulis established the conjecture by showing that any orbit of H on X_3 is either periodic or unbounded (see [DM90a] for a highly accessible account); the lattices corresponding to periodic orbits are easily accounted for, and correspond precisely to indefinite quadratic forms proportional to integral forms. Here and throughout, an orbit of a group H acting on a topological space X is said to be *periodic* if it is closed and supports a finite H-invariant measure.

We note that the homogeneous space approach for studying values of quadratic forms was noted by M.S. Raghunathan who also gave a much more general conjecture in this direction regarding orbit closures of connected unipotent groups in the quotient space G/Γ. In retrospect one can identify a similar approach in the remarkable paper [CSD55] by Cassels and Swinnerton-Dyer.

2.2. This example illustrates an important point: in most cases it is quite easy to understand how a typical orbit behaves, e.g. to deduce from the ergodicity of H acting on X_3 that for almost every ξ the orbit $H.\xi$ is dense in X_3; but for many number theoretical applications one needs to know how orbits of individual points behave — in this case, one needs to understand the orbit $H.\xi$ for **all** $\xi \in X_3$.

2.3. Raghunathan's Conjecture regarding the orbit closures of groups generated by one parameter unipotent subgroups, as well as an analogous conjecture by S.G. Dani regarding measures invariant under such groups [Dan81] have been established in their entirety[1] in a fundamental series of papers by M. Ratner [Ra91a, Ra90a, Ra90b, Ra91b].

Theorem 1 (Ratner). *Let G be a real Lie group, $H < G$ a subgroup generated by one parameter* Ad*-unipotent groups, and Γ a lattice in G. Then:*

(i) *Any H-invariant and ergodic probability measure μ on G/Γ is an L-invariant measure supported on a single periodic L-orbit of some subgroup $L \leq G$ containing H*
(ii) *For any $x \in G/\Gamma$, the orbit closure $\overline{H.x}$ is a periodic orbit of some subgroup $L \leq G$ containing H.*

A measure μ as in (i) above will be said to be *homogeneous.*

This fundamental theorem of Ratner, which in applications is often used in conjunction with the work of Dani and Margulis on nondivergence of unipotent flows [Mar71, Dan86] and related estimates on how long a unipotent trajectory can spend near a periodic trajectory of some other group (e.g. as developed in [DM90b, DM93] or [Ra91b]) give us very good (though non-quantitative) understanding of the behavior of individual orbits of groups H generated by one parameter unipotent subgroups, such as the group SO(1, 2) considered above. It has been extended to algebraic groups over $\mathbb{Q}_p$ and to S-algebraic groups (products $G = \prod_{p \in S} \mathbb{G}_i(\mathbb{Q}_p)$ with the convention that $\mathbb{Q}_\infty = \mathbb{R}$) by Ratner [Ra95] and Margulis-Tomanov [MT94].

2.4. These theorems on unipotent flows have numerous number theoretical applications, much too numerous to list here. A random sample of such applications, to give a flavor of their diverse nature, is the substantial body of work regarding counting of integer and rational points on varieties, e.g. Eskin, Mozes and Shah [EMS96] who give the asymptotic behavior as $T \to \infty$ of the number of elements $\gamma \in \mathrm{SL}(d, \mathbb{Z})$ with

[1] Special cases of Raghunathan's Conjecture were established by Dani and Margulis [DM90b] using a rather different approach.

a given characteristic polynomial satisfying $||\gamma|| < T$ (see also H. Oh's survey [Oh10] for some more recent counting results of interest); Vatsal's proof of a conjecture of Mazur regarding non-vanishing of certain L-functions associated to elliptic curves at the critical point [Vat02]; Elkies and McMullen's study of gaps in the sequence $\sqrt{n}$ mod 1 [EM04]; and Ellenberg and Venkatesh theorems on representing positive definite integral quadratic forms by other forms [EV08].

2.5. The action of one parameter diagonalizable groups on homogeneous spaces, such as the action of $a_t = \begin{pmatrix} e^{t/2} & 0 \\ 0 & e^{-t/2} \end{pmatrix}$ on X_2 is fairly well understood (at least in some aspects), but these $\mathbb{R}$-actions behave in a drastically different way than e.g. one parameter unipotent groups. The case of a_t acting on X_2 is particularly well studied. There is a close collection between this action and the continued fraction expansion of real numbers that has been used already by E. Artin [Art24], and was further elucidated by C. Series [Ser85] and others, that essentially allows one to view this system as a flow over a simple symbolic system. Any ergodic measure preserving flow of sufficiently small entropy can be realized as an invariant measure for the action of a_t on X_2, and there is a wealth of irregular orbit closures. There is certainly also a lot of mystery remaining regarding this action and in particular due to the lack of rigidity it is extremely hard to understand the behavior of specific orbits of the action, e.g.:

Question 1. *Is the orbit of the lattice*

$$\begin{pmatrix} 1 & \sqrt[3]{2} \\ 0 & 1 \end{pmatrix} \mathbb{Z}^2$$

under the semigroup $\{a_t : t \geq 0\}$ *dense in* X_2*?*

Even showing that this orbit is unbounded is already equivalent to the continued fraction expansion of $\sqrt[3]{2}$ being unbounded, a well known and presumably difficult problem. While Artin constructs in [Art24] a point in X_2 which has a dense a_t-orbit in a way that can be said to be explicit, I do not know of any construction of a lattice in X_2 generated by vectors with algebraic entries that is known to have a dense a_t-orbit.

2.6. Actions of higher rank diagonal groups are much more rigid than one parameter diagonal group, though not quite as rigid as the action of groups generated by unipotents. Many of the properties such actions are expected to satisfy are still conjectural, though there are several quite usable partial results that can be used to obtain nontrivial number theoretic consequences. A basic example of such actions is the action of the $(d-1)$-dimensional diagonal group $A < \mathrm{PGL}(d,\mathbb{R})$ on the space of lattices X_d for $d \geq 3$. A similar phenomenon is exhibited in a somewhat more elementary setting by the action of a multiplicative semigroup Σ of integers containing at least two multiplicative independent elements on the 1-torus $\mathbb{T} = \mathbb{R}/\mathbb{Z}$. This surprising additional rigidity of multidimensional diagonalizable groups has been discovered by Furstenberg [Fur67] in the context of multiplicative semigroups acting on $\mathbb{T}$, and is in a certain sense implicit in the work of Cassels and Swinnerton-Dyer [CSD55].

2.7. Actions of diagonalizable groups also appear naturally in many contexts. In the aforementioned paper of Cassels and Swinnerton-Dyer [CSD55] the following conjecture is given:

Conjecture 2. *Let $F(x_1, \ldots, x_d) = \prod_{i=i}^{d}(\sum_{j=i}^{d} g_{ij}x_j)$ be a product of d-linearly independent linear forms in d variables, not proportional to an integral form (as a homogeneous polynomial in d variables), with $d \geq 3$. Then*

$$\inf \{|F(v)| : v \in \mathbb{Z}^d \setminus \{0\}\} = 0. \tag{2.2}$$

This conjecture in shown in [CSD55] to imply Littlewood's Conjecture (see §4.1), and seems to me to be the more fundamental of the two. As pointed out by Margulis, e.g. in [Mar97], Conjecture 2 is equivalent to the following:

Conjecture 2'. *Any A-orbit $A.\xi$ in X_d for $d \geq 3$ is either periodic or unbounded.*

2.8. A somewhat more elementary action with similar features was studied by Furstenberg [Fur67]. Let Σ be the multiplicative semigroup of $\mathbb{N}$ generated by two multiplicative independent integers a, b (i.e. $\log a/\log b \notin \mathbb{Q}$). In stark contrast to cyclic multiplicative semigroups, Furstenberg has shown that any Σ-invariant closed subset $X \subset \mathbb{T} = \mathbb{R}/\mathbb{Z}$ is either finite or $\mathbb{T}$ and gave the following influential conjecture:

Conjecture 3. *Let $\Sigma = \{a^n b^k : n, k \geq 0\}$ be as above. The only Σ-invariant probability measure on $\mathbb{R}/\mathbb{Z}$ with no atoms is the Lebesgue measure.*

This conjecture can be phrased equivalently in terms of measures on G/Γ invariant under left translation by a rank two diagonalizable group H for an appropriate *solvable* group G and lattice $\Gamma < G$; e.g. if a, b are distinct primes, we can take

$$H = \{(s,t,r) : s \in \mathbb{R}^\times, t \in \mathbb{Q}_a^\times, r \in \mathbb{Q}_r^\times, |s| \cdot |t|_a \cdot |r|_b = 1\}$$

$$G = H^\times \ltimes (\mathbb{R} \times \mathbb{Q}_a \times \mathbb{Q}_b)$$

$$\Gamma = \{(s,s,s) : s = a^n b^m, n, m \in \mathbb{Z} \ltimes \{(t,t,t) : t \in \mathbb{Z}[\tfrac{1}{ab}]\}.$$

2.9. Ergodic theoretic entropy is a key invariant in ergodic theory whose introduction in the late 1950s by Kolmogorov and Sinai completely transformed the subject. At first sight it seems quite unrelated to the type of questions considered above. However, it has been brought to the fore in the study of multiparameter diagonalizable actions by D. Rudolph (based on earlier work of R. Lyons [Lyo88]), who established an important partial result towards Furstenberg's Conjecture (Conjecture 3): Rudolph classified such measures under a positive entropy condition [Rud90]. A. Katok and R. Spatzier were the first to extend this type of results to flows on homogeneous spaces [KS96], but due to a subtle question regarding ergodicity of subactions their results do not seem to be applicable in the number theoretic context.

2.10. Some care needs to be taken when stating the expected measure classification result for actions of multiparameter diagonalizable groups on a quotient space G/Γ, even for $G = \mathrm{PGL}(3,\mathbb{R})$ and A the full diagonal group, since as pointed out by M. Rees [Ree82] (see also [EK03, §9]), any such conjecture should take into account possible scenarios where the action essentially degenerates into a one parameter action where no such rigidity occurs. An explicit conjecture regarding measures invariant under multiparameter diagonal flows was given by Margulis in [Mar00, Conjecture 2]; a similar but less explicit conjecture by Katok and Spatzier was given in [KS96], and by Furstenberg (unpublished). For the particular case of the action of the diagonal group A on the space of lattice in X_d such degeneration cannot occur[2] and one has the following conjecture:

Conjecture 4. *Let μ be an A-invariant and ergodic probability measure on X_d for $d \geq 3$ (and $A < \mathrm{PGL}(3,\mathbb{R})$ the group of diagonal matrices). Then μ is homogeneous (cf. §2.3).*

More generally, we quote the following from [EL06]:

Conjecture 5. *Let S be a finite set of places for $\mathbb{Q}$ and for every $v \in S$ let G_v be a linear algebraic group over $\mathbb{Q}_v$. Let $G_S = \prod_{v \in S} G_v$, $G \leq G_S$ closed, and $\Gamma < G$ discrete. For each $v \in S$ let $A_v < G_v$ be a maximal $\mathbb{Q}_v$-split torus, and let $A_S = \prod_{v \in S} A_v$. Let A be a closed subgroup of $A_S \cap G$ with at least two independent elements. Let μ be an A-invariant and ergodic probability measure on G/Γ. Then at least one of the following two possibilities holds:*

(i) *μ is homogeneous, i.e. is the L-invariant measure on a single, finite volume, L-orbit for some closed subgroup $A \leq L \leq G$.*
(ii) *There is some S-algebraic subgroup L_S with $A \leq L_S \leq G_S$, an element $x \in G/\Gamma$, an algebraic homeomorphism $\phi : L_S \to \tilde{L}_S$ onto some S-algebraic group $\tilde{L}_S$, and a closed subgroup $H < \tilde{L}_S$ with $H \geq \phi(\Gamma)$ so that (i) $\mu((L_S \cap G).x_\Gamma) = 1$, (ii) $\phi(A)$ does not contain two independent elements and (iii) the image of μ to L_S/H is not supported on a single point.*

2.11. To obtain a measure classification result in the homogeneous spaces setting with only an entropy assumption and no assumptions regarding ergodicity of subactions (which are nearly impossible to verify in most applications of the type considered here) requires a rather different strategy of proof than [KS96], using two different and complementary methods. The first, known as the *high entropy*

[2]For *probability* measures; there are non-homogeneous A-invariant and ergodic Radon measures on X_d.

method, was developed by M. Einsiedler and Katok [EK03] and utilizes non-commutativity of the unipotent subgroups normalized by the acting group, and e.g. in the case of A acting on X_d for $d \geq 3$ allows one to conclude that any measure of sufficiently high entropy (or positive entropy in "sufficiently many directions") is the uniform measure. The other method, the *low entropy method,* was developed by the author [Lin06] where in particular an analogue to Rudolph's theorem for the action of the maximal $\mathbb{R}$-split torus[3]on $\mathrm{SL}(2,\mathbb{R}) \times \mathrm{SL}(2,\mathbb{R})/\Gamma$ is given. Even though the measure under study is invariant under a diagonalizable group and a priori has no invariance under any unipotent element, ideas from the theory of unipotent flows, particularly from a series of papers of Ratner on the horocycle flow [Ra82a, Ra82b, Ra83], are used in an essential way. These two methods can be combined successfully as was done in a joint paper with Einsiedler and Katok [EKL06] where the following partial result toward Conjecture 4 is established:

Theorem 2 ([EKL06]). *Let A be the group of diagonal matrices as above and $d \geq 3$. Let μ be an A-invariant and ergodic probability measure on X_d. If for some $a \in A$ the entropy $h_\mu(a) > 0$ then μ is homogeneous.*

2.12. The high entropy method was developed further by Einsiedler and Katok in [EK05] and the low entropy method was developed further by Einsiedler and myself in [EL08]; these can be combined to give in particular the following theorem, which we state for simplicity for real algebraic groups but holds in the general S-algebraic setting of Conjecture 5 (see [EL06, §2.1.4] for more details[4]):

Theorem 3. *Let G be a semisimple real algebraic group, $A < G$ the connected component of a maximal $\mathbb{R}$-split torus, and $\Gamma < G$ an irreducible lattice. Let μ be an A-invariant and ergodic probability measure on G/Γ. Assume that:*

(i) *the $\mathbb{R}$-rank of G is ≥ 2*
(ii) *there is no reductive proper subgroup $L < G$ so that μ is supported on a single periodic L-orbit*
(iii) *there is some $a \in A$ for which $h_\mu(a) > 0$.*

Then μ is the uniform measure on G/Γ.

If (ii) does not hold, one can reduce the classification of A-invariant measures μ on this periodic L-orbit to the classification of $A \cap [L,L]$-invariant and ergodic measures μ' on $[L,L]/\wedge$, with $\wedge$ a lattice in $[L,L]$. If $\wedge$ is reducible, up to finite index, $[L,L]/\wedge = \prod_{i=1}^{s} L_i/\wedge_i$ and $\mu = \prod_{i=1}^{s_1} \mu_i'$, with μ_i' an $A \cap L_i$-invariant measure on $L_i/\wedge_i$. As long as there is some L_i with $\mathbb{R}$-rank ≥ 2 and some element $a' \in A \cup L_i$

[3] Which in this case is simply the product of the diagonal group from each factor.
[4] There is a slight inaccuracy in the statement of [EL06, Thm. 2.4]: either one needs to assume to begin with that $h_\mu(a) > 0$ for some $a \in A$ or one needs to allow the trivial group $H = \{e\}$ in the first case listed there.

with $h_{\mu'_i}(a') > 0$, one can apply Theorem 3 recursively to obtain a more explicit, but less concise measure classification result.

New ideas seem to be necessary to extend Theorem 3 to non-maximally split tori; in part this seems to be related to the fact that for non-maximal A much more general groups L, even solvable ones, need to be considered in case (ii).

3. A Remark on Invariant Measures, Individual Orbits, and Size of Groups

3.1. One important difference between a group H generated by unipotent one parameter subgroups (considered as a subgroup of some ambient algebraic group G, which for simplicity we assume in this paragraph to be simple) and diagonalizable groups such as the group A of diagonal matrices in $G = \mathrm{PGL}(d, \mathbb{R})$ is the size of norm-balls in the groups H or A respectively under any nontrivial finite dimensional representation ρ of G (in particular, the adjoint representation): if λ_H and λ_A denote Haar measure on H and A respectively,

$$\lambda_H(\{h \in H : ||\rho(h)|| < T\}) \geq CT^{\alpha} \quad \text{for some } \alpha = \alpha(\rho) > 0 \tag{3.1}$$

while

$$\lambda_A(\{a \in A : ||\rho(a)|| < T\}) \asymp (\log T)^{d-1}. \tag{3.2}$$

We shall loosely refer to groups as in (3.1) for which the volume of norm-balls is polynomial as *thick* in G, and groups where this volume is polylogarithmic as in (3.2) as *thin*.

3.2. Such norm-balls appear naturally when one studies how orbits of nearby points x and y diverge — an important element of Ratner's proof of Theorem 1. Suppose e.g. G is a linear algebraic group over $\mathbb{R}, \Gamma < G$ a lattice and $H < G$ some closed subgroup. If $x = exp(w).y$ for $w \in \mathrm{Lie}(G)$ small, $h.x = exp(\mathrm{Ad}h(w)).h.y$ and these will still be reasonably close for all $h \in H$ with $||Ad(h)|| < ||w||^{-1}$. One can gain in the range of usable elements of H by allowing $h.x$ to be compared with a more carefully chosen point $h'.y \in H.y$, but in any case the range of usable $h \in H$ includes elements of norm bounded at most by a polynomial in $||w||^{-1}$. The entropy condition of Theorems 2 and 3 can be thought of as a partial compensation for the fact that the acting group is thin.

3.3. The size of norm-balls also plays an important role in another important aspect of the dynamics, namely the extent to which the behavior of individual orbits relates to any possible classification of invariant measures. We recall the following definition due to Furstenberg:

Definition 1. Let X be a locally compact space, and H an amenable group acting continuously on X. A point $x \in X$ will be said to be *generic* for an H-invariant measure μ along a Følner sequence[5] $\{F_n\}$ in H (that is usually kept implicit) if for any $f \in C_c(X)$

$$\lim_{n\to\infty} \frac{\int_{F_n} f(h,x)d\lambda_H(h)}{\lambda_H(F_n)} \to \int_X f(y)d\mu(y)$$

where λ_H is the left invariant Haar measure on H.

By the pointwise ergodic theorem (which in this generality can be found in [Lin01]) and separability of $C_c(X)$, if $\{F_n\}$ is a sufficiently nice Følner sequence (e.g. for $H = \mathbb{R}^k$, F_n can be taken to be any increasing sequence of boxes whose shortest dimension $\to \infty$ as $n \to \infty$), and if μ is an H-invariant and ergodic probability measure, then μ almost every $x \in X$ is generic for μ along $\{F_n\}$.

3.4. As is well-known, if X is uniquely ergodic, i.e. there is a *unique* H-invariant probability measure μ on X (which will necessarily be also H-ergodic, as the ergodic measures are the extreme points of the convex set of all H-invariant probability measures) then something much stronger is true: *every* $x \in X$ is generic for μ along *any* Følner sequence (we will also say in this case that the H-orbit of x is μ-equidistributed in X along any Føner sequence).

Even if there are only two H-invariant and ergodic probability measures on X, or even if there is a unique H-invariant and ergodic probability measure on X but X is not compact, individually orbits may behave in somewhat complicated ways, failing to be generic for any measure on X. The most one can say is that if $\{F_n\}$ is Følner sequence, for large n the push forward of $(\lambda_H(F_n))^{-1}\lambda_H|F_n$ restricted to a large Følner set F_n under the map $h \mapsto h.x$ is close to a linear combination (depending on n) of the two H-invariant and ergodic measures in the former case, or to c times the unique H-invariant probability measure in the latter case for some $c \in [0,1]$ (which again may depend on n).

3.5. For unipotent flows, the connection between distribution properties of individual orbits and the ensemble of invariant probability measures is exceptionally sharp. In [Ra91b] Ratner has shown that if u_t is a one parameter unipotent group, G a real Lie group, and $\Gamma < G$ a lattice then any $x \in G/\Gamma$ is generic for some homogeneous measure μ whose support contains x. A uniform version where one is allowed to vary the unipotent group as well as the starting point was given by Dani and Margulis [DM93, Thm. 2]. Another useful result in the same spirit by Mozes and Shah [MS95] classifies limits of sequences of homogeneous probability measures $(m_i)_i$ in G/Γ that are invariant and ergodic under some one parameter unipotent subgroup of G (possibly different for different i); such a limiting measure is also a homogeneous probability measure. Often if the volume of the corresponding sequence of periodic orbits goes to ∞ one can show that these

[5] A sequence of sets $F_n \subset H$ is said to be a Følner sequence if for any compact $K \subset G$ we have that $\lambda_H(F_n \Delta K F_n)/\lambda_H(F_n) \to 0$ as $n \to \infty$; a group H is said to be amenable if it has a Følner sequence.

homogeneous probability measures converge to the uniform measure on G/Γ. In the proof of all these results, the thickness of unipotent groups (and groups generated by unipotents), under the guise of the polynomial nature of unipotent flows, plays a crucially important role.

Even for $G = \mathrm{SL}(2,\mathbb{R})$, the connection between invariant measures and distribution properties of individual orbits for the action of unipotent groups on *infinite* volume quotients is not well understood outside the geometrically finite case, though there is some interesting work in this direction, e.g. [SS08].

3.6. For diagonalizable flows, the connection between invariant measures and behavior of individual orbits is much more tenuous. Certainly if $X = G/\Gamma$ is compact then for any $\xi \in X$ the A-orbit closure $\overline{A.\xi}$ supports an A-invariant measure: but this measure may not be unique, nor does the support of μ have to coincide with $\overline{A.\xi}$. Counterexamples given by Maucourant [Mau10] to the topological counterpart of Conjecture 5 in [Mar00] are of precisely this type: they give an A orbit whose limit set is the support of two (or more) different homogeneous measures. An example in a similar spirit has been given by U. Shapira [Sha10,LS10] for the action of the full diagonal group A on X_3: Here ξ is the lattice

$$\xi = \begin{pmatrix} 1 & a & 0 \\ 0 & 1 & 0 \\ 0 & a & 1 \end{pmatrix} \mathbb{Z}^3$$

which for a typical $a \in \mathbb{R}$ will spiral between two *infinite* homogeneous measures supported on the closed orbits through the standard lattice $\mathbb{Z}^d$ of the groups

$$H_1 = \begin{pmatrix} * & * & 0 \\ * & * & 0 \\ 0 & 0 & * \end{pmatrix} \quad \text{and} \quad H_2 = \begin{pmatrix} * & 0 & 0 \\ 0 & * & * \\ 0 & * & * \end{pmatrix}.$$

3.7. In special cases *isolation results* give a weak substitute for diagonal actions to the "linearization" techniques used in [DM93, MS95, Ra91b] for unipotent flows. An isolation result of this type for the action of A on X_d for $d \geq 3$ by Cassels and Swinnerton-Dyer [CSD55][6] gives in particular that if $\xi, \xi_0 \in X_d$ with

$$A.\xi_0 \subset \overline{A/\xi} \setminus (A.\xi) \quad \text{and} \quad A.\xi_0 \text{ periodic} \tag{3.3}$$

then $A.\xi$ is unbounded; this has been strengthened by Barak Weiss and myself [LW01] to show that under the same assumptions $\overline{A.\xi}$ is a periodic orbit of some closed connected group H with $A \leq H \leq \mathrm{PGL}(d,\mathbb{R})$ (such periodic orbits are easily classified and in particular unless $H = A$ are unbounded). Results of this nature under somewhat less restrictive conditions than (3.3), along with some Diophantine applications, were recently given by U. Shapira and myself [LS10].

[6] In the paper, Cassels and Swinnerton-Dyer treat only the case of $d = 3$, but the general case is similar.

Using the Cassels Swinnerton-Dyer isolation result it is easy to show that Conjecture 4 implies Conjecture 2: indeed, if $A.\xi$ is a bounded orbit in X_d then $\overline{A.\xi}$ supports an A-invariant probability measure, and hence by the ergodic decomposition $\overline{A.\xi}$ supports an A-invariant and ergodic probability measure. Assuming Conjecture 4 this measure will be homogeneous, and by the classification alluded to in the previous paragraph the only compactly supported A-invariant homogeneous probability measures are the probability measures on periodic A-obits. Thus $\overline{A.\xi}$ contains an A-periodic measure, and unless $A.\xi$ is itself periodic we get a contradiction to the Cassels-Swinnerton-Dyer Isolation Theorem.

3.8. The field of arithmetic combinatorics has witnessed dramatic progress over the last few years with remarkable applications. One of the basic results is the following exponential sum estimate by Bourgain, Glibichuk and Konyagin [BGK06]: for any δ there are $c, \varepsilon > 0$ so that if p is prime, $\tilde{H}$ a subgroup of $(\mathbb{Z}/p\mathbb{Z})^\times$ with $|\tilde{H}| > p^\delta$,

$$\max_{b \in (\mathbb{Z}/p\mathbb{Z})^\times} \frac{|\sum_{h \in \tilde{H}} e(bh/p)|}{|\tilde{H}|} < cp^{-\varepsilon}$$

with $e(x) = \exp(2\pi i x)$. Bourgain has proved a similar estimate with p replaced by an arbitrary integer N; this involves considerable technical difficulties since one is interested in a result in which the error term does not depend on the decomposition of N into primes. If $\tilde{H}$ is the reduction modulo N of some multiplicative semigroup $H \subset \mathbb{Z}^\times$, we can interpret this estimate as saying that for any $0 \leq b < N$, the periodic H-orbit $\{\frac{hb}{N} \bmod 1 : h \in \tilde{H}\}$ is close to being equidistributed in $\mathbb{T}$ in a quantitative way provided $|\tilde{H}| > N^\delta$.

3.9. Of particular interest to us is the semigroup $H = \{a^n b^k : n, k \in \mathbb{N}\}$ where a, b are multiplicatively independent integers. For a certain sequence of N_i (relatively prime to ab) it may well happen that $|H \bmod N| > N^\delta$ for a fixed δ, even though H is a thin sequence in the sense of §3.1. For such a sequence N_i and any choice of $b_i \in (\mathbb{Z}/N_i\mathbb{Z})^\times$, the sequence of periodic H-orbits $H.\frac{b_i}{N_i} \bmod 1$ would become equidistributed in a quantitative way as $i \to \infty$ by the theorem of Bourgain quoted above (§3.8). However there are sequences of N for which $|H \bmod N|$ is rather small — $(\log N)^{c \log\log\log N}$ [APR83]. A trivial lower bound on $|H \bmod N|$ is

$$|H \bmod N| \geq (\log_a N)(\log_b N)/2,$$

and if there were infinitely many N_i with $|H \bmod N_i| \ll (\log N_i)^2$ then the orbits $H.\frac{1}{N_i} \bmod 1$ would spend a positive proportion of their mass very close to 0, and hence fail to equidistribute.

Using the Schmidt Subspace Theorem (more precisely, its S-algebraic extension by Schlickewei) in an elegant and surprising way Bugeaud, Corvaja and Zannier [BCZ03] show that

$$\lim_{N \to \infty} \frac{|H \bmod N|}{(\log N)^2} \to \infty$$

giving credence to the following conjecture, presented as a question by Bourgain in [Bou09]:

Conjecture 6. *Let* $H = \{a^n b^k : n, k \in \mathbb{N}\}$, *with* a, b *multiplicatively independent. Then for any sequence* $\{(b_i, N_i)\}$ *with* $N_i \to \infty$ *and* $b_i \in (\mathbb{Z}/N_i\mathbb{Z})^\times$ *the sequence of* H*-periodic orbits* $H.\frac{b_i}{N_i}$ *mod 1 becomes equidistributed as* $i \to \infty$, *i.e. for any* $f \in C(\mathbb{T})$,

$$|H|^{-1} \sum_{h \in H} f\left(h.\frac{b_i}{N_i}\right) \to \int_{\mathbb{T}} f dx.$$

Even if one assumes (or proves) Conjecture 3 regarding H-invariant measures, this conjecture seems challenging due to the absence of a strong connection between individual orbits and invariant measures for diagonalizable group actions (cf. §3.6).

4. Some Applications of the Rigidity Properties of Diagonalizable Group Actions

4.1. The partial measure classification results for actions of diagonalizable groups mentioned above, e.g. Theorems 2 and 3, have several applications. We give below a sample of three theorems, in the proof of which one of the major ingredients is the classification of positive entropy invariant measures. Several other applications are discussed in Einsiedler's notes for his lecture at this ICM [Ein10].[7]

Multiparameter diagonal groups and Diophantine approximations

4.2. Using the variational principle relating topological entropy and ergodic theoretic entropy, together with an averaging argument and use of semicontinuity properties of entropy for measures supported on compact subsets of X_d in [EKL06] the following partial result towards Conjecture 2 was deduced from Theorem 2 (see either [EKL06] or [EL10, §12] for more details):

Theorem 4 (Einsiedler, Katok and L. [EKL06]). *The set of degree* d *homogeneous polynomials* $F(x_1, \ldots, X_d)$ *that can be factored as a product of* d *linearly independent forms in* d *variables that fail to satisfy* (2.2) *have Hausdorff dimension zero.*

By Conjecture 2 above, the set of such F is expected to be countable; the trivial upper bound on the dimension of the set of such F is $d(d-1)$.

[7]Einsiedler and I have worked together for some years on many aspects of the action of diagonalizable groups, and there is some overlap between this paper and Einsiedler's [Ein10], as well as our joint contribution to the proceedings of the previous ICM in Madrid [EL06]. However the selection of topics and style is quite different in these three papers.

4.3. Recall the following well known conjecture of Littlewood regarding simultaneous Diophantine approximations:

Conjecture 7 (Littlewood). *For any* $x, y \in \mathbb{R}^2$,

$$\inf \{n|nx - m||ny - k| : (n, m, k) \in \mathbb{Z}^3,\ n \neq 0\} = 0. \tag{4.1}$$

Similar ideas as in the proof of Theorem 4 allows one to prove that the Hausdorff dimension of the set of exceptional pairs $(x, y) \in \mathbb{R}^2$ that do not satisfy (4.1) is zero. Indeed, one can be a bit more precise: for a sequence of integers $(a_k)_{k\in\mathbb{N}}$ define its combinatorial entropy as

$$h_{comb}((a_k)) = \lim_{n\to\infty} \frac{\log W_n((a_k))}{n}$$

where $W_n((a_k))$ counts the number of possible n-tuples $(a_k, a_{k+1}, \ldots, a_{k+n-1})$ (if (a_k) is unbounded, $W_n((a_k)) = \infty$). Then the techniques of [EKL06] gives the following explicit sufficient criterion for a real number x to satisfy Littlewood's conjecture for all $y \in \mathbb{R}$:

Theorem 5. *Let* $x = a_1 + \frac{1}{a_2 + \frac{1}{a_3 + \cdots}}$ *be the continued fraction expansion of* $x \in \mathbb{R}$. *If* $h_{comb}((ak)) > 0$ *then for every* $y \in \mathbb{R}$ *equation* (4.1) *holds.*

Periodic orbits of diagonal groups

4.4. Unlike the case for groups generated by unipotents, it is not hard to give a sequence of A-periodic orbits $A.x_i$ in X_d (for any $d \geq 2$) so that the associated probability measures $m_{A.x_i}$ fail to converge to the uniform measure (cf. [ELMV09, §7]). Indeed, as pointed out to me by U. Shapira, such an example is implicit already in an old paper by Cassels [Cas52].

4.5. However, when the periodic orbits are appropriately grouped their behavior improves markedly: define for any A-periodic $\xi \in X_d$ an order in the ring D of $d \times d$ (possibly singular) diagonal matrices by

$$\mathcal{O}(\xi) = \{h \in D : h.\bar{\xi} \subseteq \bar{\xi}\}$$

where $\bar{\xi}$ is a lattice representing the homothety equivalence class ξ. This is a discrete subring of D containing 1; $\mathrm{stab}_A(\xi) = \{a \in A : a.\xi = \xi\}$ is precisely the set of invertible elements of $\mathcal{O}(\xi)$ and moreover $\mathbb{Z}[\mathrm{stab}_A(\xi)] \subseteq \mathcal{O}(\xi)$. Since ξ is A-periodic, $\mathrm{stab}_A(\xi)$ contains $d-1$-independent units and $\mathcal{O}(\xi)$ is a lattice in D (considered as an additive group), isomorphic as a ring to an order in a totally real number field K of degree d over $\mathbb{Q}$. For a given order $\mathcal{O} < D$ set

$$C(\mathcal{O}) = \{A.y : \mathcal{O}(y) = \mathcal{O}\};$$

for any A-periodic $\xi \in X_d$ the collection $C(\mathcal{O}(\xi))$ can be shown to be finite.

Theorem 6 (Einsiedler, Michel, Venkatesh and L. [ELMV10]). *Let $A.x_i$ can be a sequence of distinct A-periodic orbits in X_3, and set $C_i = C(\mathcal{O}(x_i))$. Then for any $f \in C_c(X_3)$ we have that*

$$\frac{1}{|C_i| \cdot |A/\mathrm{stab}_A(x_i)|} \sum_{A.y \in C_i} \int_{A/\mathrm{stab}_A(x_i)} f(a.y)da \to \int_{x_3} f.$$

For $d = 2$ the corresponding statement is a theorem of Duke [Duk88] proved using the theory of automorphic forms, with some previous substantial partial results by Linnik and Skubenko (see [Li68]). Weaker results about the distribution of periodic A-orbits for $d \geq 3$ in substantially greater generality were obtained in [ELMV09].

4.6. In the case of periodic A-orbits $A.\xi$ whose corresponding order $\mathcal{O}(\xi)$ is maximal (equivalently, is isomorphic to the full integer ring $\mathcal{O}_K$ of a totally real number field K), $\mathcal{C}(\xi)$ can be identified with the ideal class group of $\mathcal{O}_K$, and in particular has a natural structure of a group. It is quite challenging to make use of the group structure of $\mathcal{C}(\xi)$ in the dynamical context. In particular, it would be of interest to prove equidistribution of the collection of A-orbits corresponding to (possibly quite small) subgroups of the ideal class group.

4.7. We refer the reader to the comprehensive survey [MV06] by Michel and Venkatesh for more details on this and related equidistribution questions.

Diagonal flows and Arithmetic Quantum Unique Ergodicity

4.8 In [RS94], Z. Rudnick and P. Sarnak conjectured the following:

Conjecture 8. *Let M be a compact Riemannian manifold of negative sectional curvature. Let ϕ_i be an orthonormal sequence of eigenfunctions of the Laplacian on M. Then*

$$\int_M f(x)|\phi_i(x)|^2 d\,\mathrm{vol}(x) \to \frac{1}{\mathrm{vol}(M)} \int_M f(x) d\,\mathrm{vol}(x) \quad \forall f \in C^\infty(M). \tag{4.2}$$

There is also a slightly stronger form of this conjecture for test functions in phase space. Both versions of the conjecture are open, and there does not seem to be strong evidence for it in high dimensions. However in the special case of $M = \mathbb{H}/\Gamma$ with Γ an arithmetic lattice of congruence type (either congruence sublattices of PGL(2, $\mathbb{Z}$) or of PGL(1, $\mathcal{O}$) for $\mathcal{O}$ an order in an indefinite quaternion algebra over $\mathbb{Q}$; in the latter case M is compact) we have a lot of extra symmetry that aids the analysis: an infinite commuting ensemble of self-adjoint operators, generated by the Laplacian and, for each prime p outside a possible finite set P of "bad" primes, a corresponding Hecke operators T_p.

Theorem 7 (Brooks and L. [BL10,Lin06]). *Let $M = \mathbb{H}/\Gamma$ be as above, and $p \notin P$, with M compact. Then any orthonormal sequence ϕ of joint eigenfunctions of the Laplacian and T_p on M satisfies* (4.2).

This theorem refines a previous theorem that relied on work by Bourgain and myself [BL03]. When Γ is a congruence subgroup of $\mathrm{SL}(2, \mathbb{Z})$, i.e. M is not compact, there is an extra complication in that one needs to show that no mass escapes to the cusp in the limit. Under the assumption of ϕ_i being joint eigenfunctions of all Hecke operators this has been established by Soundararajan [Sou09].

4.9. The proof of Theorem 7 does not quite use multiparameter diagonalizable flows but rather the following theorem (generalized in [EL08]) of similar but somewhat more general flavor:

Definition 2. Let X be locally compact space, H a locally compact group acting continuously on X, and μ any σ-finite measure on X (not necessarily H invariant). Then μ is H-recurrent if for every set $B \subset X$ with $\mu(B) > 0$ for almost every $x \in X$ the set $\{h \in H : h.x \in B\}$ is unbounded (has noncompact closure).

Theorem 8 ([Lin06]). *Let $G = \mathrm{PGL}(2,\mathbb{R}) \times \mathrm{PGL}(2,\mathbb{Q}_p)$, $H = \mathrm{PGL}(2,\mathbb{Q}_p)$ considered as a subgroup of G, A_1 the diagonal subgroup of $SL(2, \mathbb{R})$ (also considered as a subgroup of G), and $\wedge < G$ an irreducible lattice. Let μ be a probability measure on $G/\wedge$ which is (i) A_1-invariant (ii) H-recurrent (iii) a.e. A_1-ergodic component of μ has positive entropy (with respect to A_1). Then μ is the uniform measure on $G/\wedge$.*

Note that if μ as in Theorem 8 were invariant under any unbounded subgroup of H, by Poincare recurrencé it would be H-recurrent.

The connection to Theorem 7 uses the fact that for $\Gamma < \mathrm{PGL}(2,\mathbb{R})$ of congruence type as above and $p \notin P$, $\mathbb{H}/\wedge$ can be identified with $K\backslash G/\Gamma$ for G as in Theorem 8 and $K < G$ the compact subgroup $\mathrm{PO}(2,\mathbb{R}) \times \mathrm{PGL}(2,\mathbb{Z}_p)$; let $\pi : G/\wedge \to \mathbb{H}/\Gamma$ be the projection corresponding to this identification. The Hecke operator T_p is related to this construction as follows: for $f \in L^2(\mathbb{H}/\Gamma)$ and $\tilde{x} \in G/\wedge$

$$[T_p f](\pi(\tilde{x})) = p^{-1/2} \int_{\mathrm{PGL}(2,\mathbb{Z}_p)\left(\begin{smallmatrix} p & 0 \\ 0 & 1 \end{smallmatrix}\right)\mathrm{PGL}(2,\mathbb{Z}_p)} \tilde{f} \circ \pi((e,h).\tilde{x})\, dh.$$

The crux of both [BL03] and [BL10] is the verification of the entropy assumption (iii) above, which can be rephrased in terms of decay rates of measures of small tubes in $G/\wedge$.

4.10. Note that though Theorems 4 and 5 are clearly partial results, in Theorem 6 and Theorem 7 one essentially obtains unconditionally full equidistribution statements using only the partial measure classification results currently available.

4.11. A more detailed discussion of quantum unique ergodicity in the arithmetic context can be found in Soundararajan's contribution to these proceedings [Sou10], which also include a discussion of some recent exciting results of Holowinsky and Soundararajan [HS09] regarding an analoguous question for holomorphic forms.

5. Zariski Dense Subgroups of Groups Generated by Unipotents

5.1. An important difference between groups generated by unipotent subgroups and diagonalizable groups is the size of norm-balls in these groups. Given a closed subgroup $H < G$ with large norm-balls, i.e. for which

$$\lambda_H(\{h \in H : ||\mathrm{Ad}(h)|| < T\}) \geq CT^\alpha \quad \text{for some } \alpha > 0 \tag{5.1}$$

the discussion in §3 might lead us to hope that we may be able to understand the behavior of individual H-orbits for the action of H on a quotient space G/Γ for a lattice $\Gamma < G$.

5.2. A natural class of groups which satisfy the thickness condition (5.1) are Zariski dense discrete subgroups $\wedge$ of semisimple algebraic groups. For instance, one may look at the action of a subgroup $\wedge < \mathrm{SL}(d,\mathbb{Z})$ with a large Zariski closure on $\mathbb{T}^d$, or at the action of a subgroup $\wedge < G$ with large Zariski closure (in the simplest case, G) on G/Γ where G is a simple real algebraic group. Two substantial papers addressing this question appeared in the same Tata Institute Studies volume by Furstenberg [Fur98] and by N. Shah [Sh98], the latter paper addressing this question when $\wedge$ is generated by unipotent elements.

5.3. In the context of actions of subgroups $\wedge < \mathrm{SL}(d,\mathbb{Z})$ on $\mathbb{T}^d$, under the assumption of strong irreducibility of the $\wedge$-action and that the identity component of the Zariski closure of $\wedge$ is semisimple, Muchnik [Muc05] and Guivarc'h and Starkov [GS04] show that for any $x \in \mathbb{T}^d$ the orbit $\wedge.x$ is either finite or dense, in analogy with theorems of Furstenberg (cf. §2.8) and Berend [Ber84] who address this question in the context of the action of two or more commuting automorphisms of $\mathbb{T}^d$.

5.4. Groups $\wedge$ as above with a large Zariski closure are not amenable, and hence in general there is no reason why the behavior of individual orbits in a continuous action of $\wedge$ on a compact (or locally compact) space X should be governed by $\wedge$-invariant measures, even to the more limited extent manifest by actions of diagonalizable groups. A natural substitute for invariant measures in this context was suggested by Furstenberg (e.g. in [Fur98]): choose an arbitrary auxiliary probability measure ν on $\wedge$ whose support generates $\wedge$, subject to an integrability condition, e.g. the finite moment condition $\int_{||g||} \delta dv(g) < \infty$ for some $\varepsilon > 0$ (if $\wedge$ is finitely generated one can take ν to be finitely supported). A measure μ on X is said to be v-stationary if

$$\nu * \mu := \int /g_* \mu d\nu(g) = \mu.$$

Unlike invariant measures, even in the nonamenable setting, if X is compact then for every $x \in X$ there is a ν-stationary probability measure supported on $\overline{\wedge}.x$.

5.5. In analogy with Conjecture 3, one may conjecture that if ν is a measure on $\mathrm{SL}(d,\mathbb{Z})$ whose support generates a subgroup $\wedge$ acting strongly irreducibly on $\mathbb{T}^d$ and whose Zariski closure is semisimple, in particular if $\wedge$ is Zariski dense in $\mathrm{SL}(d,\mathbb{R})$, any ν-stationary probability measure on $\mathbb{T}^d$ is a linear combination of Lebesgue measure $\lambda_{\mathbb{T}^d}$ and finitely supported measures each on a finite $\wedge$-orbit. In particular, one may hope that any ν-stationary measure is in fact $\wedge$-invariant, a phenomenon Furstenberg calls *stiffness.* Guivarc'h posed the following question, suggesting that a much stronger statement might be true: whether under the conditions above, for any $x \in \mathbb{T}^d$ with at least one irrational component,

$$\nu^{*k} * \delta_x := \underbrace{\nu * \cdots * \nu}_{k} * \delta_x \to \lambda_{\mathbb{T}^d} \quad \text{as } k \to \infty. \tag{5.2}$$

Equation (5.2) clearly implies that if μ is any nonatomic measure, $\nu^{*k} * \nu \to \lambda_{\mathbb{T}^d}$, hence it implies the above classification of ν-stationary measures.

5.6. In joint work with Bourgain, Furman and Mozes, a positive quantitative answer to Guivarc'h question is given under the assumption that $\wedge$ acts totally irreducibly on $\mathbb{T}^d$ and has a proximal element,[8] in particular, if $\wedge$ is Zariski dense in $\mathrm{SL}(d,\mathbb{R})$:

Theorem 9 (Bourgain, Furman, Mozes and L. [BFLM10]). *Let $\wedge < \mathrm{SL}_d(\mathbb{R})$ satisfy the assumptions above, and let ν be a probability measure supported on a set of generators of $\wedge$ satisfying the moment condition of §5.4. Then there are constants $C, c > 0$ so that if for a point $x \in \mathbb{T}^d$ the measure $\mu_n = v^{*n} * \delta_x$ satisfies that for some $a \in \mathbb{Z}^d \setminus \{0\}$*

$$|\widehat{\mu}_n(a)| > t > 0, \quad \textit{with} \quad n > C.\log\left(\frac{2||a||}{t}\right),$$

then x admits a rational approximation p/q for $p \in \mathbb{Z}^d$ and $q \in \mathbb{Z}_+$ satisfying

$$\left\|x - \frac{p}{q}\right\| < e^{-cn} \quad \textit{and} \quad |q| < \left(\frac{2||a||}{t}\right)^C. \tag{5.3}$$

This proof uses in an essential way the techniques of arithmetic combinatorics, particularly a nonstandard projections theorem by Bourgain [Bou10].

[8] An element $g \in \mathrm{SL}(d,\mathbb{R})$ is said to be proximal if it has a simple real eigenvalue strictly larger in absolute value than all other eigenvalues.

5.7. A purely ergodic theoretic approach to classifying Λ-stationary measures, as well as Λ-orbit closures, has been developed by Y. Benoist and J. F. Quint. Their approach has a considerable advantage that it is significantly more general in scope, though the analytic approach of [BFLM10] where applicable gives much more precise and quantitative information. In particular, in [BQ09] the following is proved for homogeneous quotients G/Γ:

Theorem 10 (Benoist and Quint). *Let G be the connected component of a simple real algebraic group, Γ a lattice in G. Let ν be a finitely supported probability measure G whose support generates a Zariski dense subgroup $\Lambda < G$ then*

1. *Any non-atomic ν-stationary measure on G/Γ is the uniform measure on G/Γ.*
2. *For any $x \in G/\Gamma$, the orbit $\Lambda.x$ is either finite or dense. Moreover, in the latter case the Cesaro averages $\frac{1}{n}\Sigma_{k=1}^{n}\nu^{*n} * \delta_x$ converge weak* to the uniform measure on G/Γ.*

It is not known in this case if the sequence $\nu^{*n} * \delta_x$ converges to the uniform measure. A technique introduced by Eskin and Margulis [EM04] to establish nondivergence of the sequence of measures $\nu^{*k} * \delta_x$ on G/Γ and further developed by Benoist and Quint is used crucially in this work, and in particular gives a useful substitute in this context for the linearization techniques for unipotent flows discussed in §3.5. Some of the ideas of Ratner's Measure Classification Theorem (see §2.3) are used in the proof of Theorem 10, as well as the result itself.

6. Quantitative Aspects

6.1. As we have seen, dynamical techniques applied in the context of homogeneous spaces are extremely powerful, and have many applications in number theory and other subjects. However, they have a major deficiency, in that they are quite hard to quantify. For example, Margulis' proof of the Oppenheim conjecture (cf. §2.1) does not give any information about the size of the smallest $v \in \mathbb{Z}^3 \setminus \{0\}$ satisfying $|Q(v)| < \varepsilon$ for a given indefinite ternary quadratic form Q not proportional to a rational one (note that necessarily any quantitative statement of this type needs to be somewhat involved as the qualitative statement fails for integral Q, and any quantitative statement has to take into account how well Q can be approximated by forms proportional to rational forms of a given height.)

Contrast this with the proof by Davenport and Heilbronn [DH46] of the Oppenheim Conjecture for diagonal forms with $d \geq 5$ variables (forms of the type $Q(x_1, \ldots, x_d) = \Sigma_i \lambda_i x_i^2$ where not all λ_i have the same sign) using a variant of the Hardy-Littlewood circle method, from which it can be deduced[9] that the

[9] At least for forms that are not too well approximated by forms proportional to rational ones, though by Meyer's Theorem for $d \geq 5$ rational forms should not cause any significant complication.

shortest vector v with $|Q(v) < \varepsilon$ is $O(\varepsilon^{-C})$, and the much more recent work of Götze and Margulis [GM10] who treat the general $d \geq 5$ case using substantially more elaborate analytic tools and obtain a similar quantitative estimate.

6.2. Overcoming this deficiency is an important direction of research within the theory of flows on homogeneous spaces. There is one general class in which at least in principle it had long been known that fairly sharp quantitative equidistribution statements can be given, and that is for the action of horocyclic groups. Recall that $U < G$ is said to be *horocyclic* if there is some $g \in G$ for which $U = \{u \in G : g^n u g^{-n} \to e \text{ as } n \to \infty\}$; the prototypical example is $U = \begin{pmatrix} 1 & t \\ 0 & 1 \end{pmatrix}$ in SL(2,$\mathbb{R}$). Such quantitative equidistribution results have been given by Sarnak [Sar81] and Burger [Bur90, Thm. 2] and several other authors since. Even in this well-understood case, quantitative equidistribution results have remarkable applications such as in the work of Michel and Venkatesh on subconvex estimates of L-functions [Ven05, MV09].

6.3. Another case which is well understood, particularly thanks to the work of Green and Tao [GT07], is the action of a subgroup of G on $G\backslash\Gamma$ when G is *nilpotent*; these nilsystems appear naturally in the context of combinatorial ergodic theory, and have a different flavor from the type of dynamics we consider here, e.g. when G is a semisimple group or a solvable group of exponential growth.

6.4. We list below several nonhorospherical quantitative equidistribution results closer to the main topics of this note:

(a) Using deep results from the theory of automorphic forms, and under some additional assumptions that are probably not essential, Einsiedler, Margulis and Venkatesh were able to give a quantitative analysis of equidistribution of *periodic* orbits of semisimple groups on homogeneous spaces [EMV09] with a polynomial rate of convergence — a result that I suspect should have many applications.

(b) Let ν be a probability measure on $\mathrm{SL}(d,\mathbb{Z})$ as in §5.6. Theorem 9 quoted above from [BFLM10] gives a quantitative equidistribution statement for successive convolutions $\nu^{*n} * \delta_x$ for $x \in \mathbb{T}^d$, which in particular gives quantitative information on the random walk associated with ν on $(\mathbb{Z}/N\mathbb{Z})^d$ as $N \to \infty$ irrespective of the prime decomposition of N. This has turned out to be useful in the recent work of Bourgain and P. Varjú [BV10] that show that the Cayley graphs of $\mathrm{SL}(d,\mathbb{Z}/N\mathbb{Z})$ with respect to a finite set S of elements in $\mathrm{SL}(d,\mathbb{Z})$ generating a Zariski dense subgroup of $\mathrm{SL}(d,\mathbb{R})$ are a family of expanders as $Nt \to \infty$ as long as N is not divisible by some fixed set of prime numbers depending on S.

(c) In joint work with Margulis we give an effective dynamical proof of the Oppenheim Conjecture, i.e. one that does give bounds on the minimal size of a nonzero integral vector v for which $|Q(v)| < \varepsilon$. The bound obtained is of the form $||v|| \ll \exp(\varepsilon^{-C})$. Nimish Shah has drawn my attention to a paper of Dani [Dan94] which has a proof of the Oppenheim conjecture that in principle is quantifiable, i.e. without the use of minimal sets or the axiom of choice, though it is not immediately apparent what quality of quantification may be obtained from his method.

(d) In work with Bourgain, Michel and Venkatesh [BLMV09] we have given an effective version of Furstenberg's Theorem (cf. §2.8), giving in particular that if a, b are multiplicatively independent integers, for sufficiently large C depending on a, b and some $\theta > 0$, for all $N \in \mathbb{N}$ and m relatively prime to N,

$$\left\{ \frac{a^n b^k m}{N} : 0 \leq n, k \leq C \log N \right\}$$

intersects any interval in $\mathbb{R}/\mathbb{Z}$ of length $\gg$ log log log N^{θ}. This has been generalized by Z. Wang [Wan10] in the context of commuting actions of toral automorphisms.

Clearly, there is ample scope for further research in this direction, particularly regarding the quality of these quantitative results and their level of generality. In particular, I think any improvement on the quality of the estimate obtained in (d) above would be quite interesting.

Acknowledgments

It has been my good fortune to be around many knowledgeable and generous mathematicians, from whom I have learned a lot, and I hope at least some traces of their insight are represented in this note. Of these I particularly want to mention Benjy Weiss, my advisor, Peter Sarnak, who has opened my eyes regarding the connections between homogeneous dynamics and many other areas, in particular quantum unique ergodicity, Manfred Einsiedler, with whom I have had a long-term collaboration that has been very meaningful to me mathematically and otherwise, Akshay Venkatesh, who was good enough to share with me his rare insight into the essence of a large swath of mathematics, particularly number theory, and Jean Bourgain from whom I have learned much of what I know about arithmetic combinatorics. I also learned a lot from my ongoing collaboration with Gregory Margulis. I am grateful to the Clay Mathematics Institute for providing generous support at a crucial time. I am also grateful to Manfred Einsiedler and Michael Larsen for going over these notes at short notice, giving constructive and helpful comments. Last, but certainly not least, I want to thank my family — my wife Abigail, my daughters, my parents, and my sisters for too many things to list here.

References

[APR83] L. M. Adleman, C. Pomerance, and R. S. Rumely, *On distinguishing prime numbers from composite numbers*, Ann. of Math. (2) **117** (1983), no. 1, 173–206. MR683806 (84e:10008)

[Art24] E. Artin, *Ein mechanisches System mit quasiergodischen Bahnen*, Abhandlungen des Mathematischen Seminars **3** (1924), 170–175.

[BQ09] Y. Benoist and J.-F. Quint, *Mesures stationnaires et fermés invariants des espaces homogenes* (2009). Preprint.

[Ber84] D. Berend, *Multi-invariant sets on compact abelian groups*, Trans. Amer. Math. Soc. **286** (1984), no. 2, 505–535. MR760973 (86e:22009)

[Bou09] J. Bourgain, *On the distribution of the residues of small multiplicative subgroups of* $\mathbb{F}_p$, Israel J. Math. **172** (2009), 61–74. MR2534239

[Bou10] J. Bourgain, *The discretized sum product and projection theorems*, J. Anal. Math. (2010). To appear.

[BL03] J. Bourgain and E. Lindenstrauss, *Entropy of quantum limits*, Comm. Math. Phys. **233** (2003), no. 1, 153–171. MR1957735 (2004c:11076)

[BFLM10] J. Bourgain, A. Furman, E. Lindenstrauss, and S. Mozes, *Stationary measures and equidistribution for orbits of non-abelian semigroups on the torus*, J. Amer. Math. Soc. (2010). To appear.

[BLMV09] J. Bourgain, E. Lindenstrauss, P. Michel, and A. Venkatesh, *Some effective results for* × a × b, Ergodic Theory Dynam. Systems **29** (2009), no. 6, 1705–1722. MR2563089

[BGK06] J. Bourgain, A. A. Glibichuk, and S. V. Konyagin, *Estimates for the number of sums and products and for exponential sums in fields of prime order*, J. London Math. Soc. (2) **73** (2006), no. 2, 380–398. MR2225493 (2007e:11092)

[BV10] J. Bourgain and P. P. Varju, *Expansion in* $\mathrm{SL}_d(\mathbb{Z}/q\mathbb{Z})$, *q arbitrary* (2010), available at arXiv:1006.3365.

[BL10] S. Brooks and E. Lindenstrauss, *Graph Eigenfunctions and Quantum Unique Ergodicity* (2010).

[BCZ03] Y. Bugeaud, P. Corvaja, and U. Zannier, *An upper bound for the G.C.D. of* a^n-1 *and* b^n-1, Math. Z. **243** (2003), no. 1, 79–84. MR1953049 (2004a:11064)

[Bur90] M. Burger, *Horocycle flow on geometrically finite surfaces*, Duke Math. J. **61** (1990), no. 3, 779–803. MR1084459 (91k:58102)

[Cas52] J. W. S. Cassels, *The product of n inhomogeneous linear forms in n variables*, J. London Math. Soc. **27** (1952), 485–492. MR0050632 (14,358d)

[CSD55] J. W. S. Cassels and H. P. F. Swinnerton-Dyer, *On the product of three homogeneous linear forms and the indefinite ternary quadratic forms*, Philos. Trans. Roy. Soc. London. Ser. A. **248** (1955), 73–96. MR0070653 (17,14f)

[Dan81] S. G. Dani, *Invariant measures and minimal sets of horospherical flows*, Invent. Math. **64** (1981), no. 2, 357–385. MR629475 (83c:22009)

[Dan86] S. G. Dani, *On orbits of unipotent flows on homogeneous spaces. II*, Ergodic Theory Dynam. Systems **6** (1986), no. 2, 167–182. MR857195 (88e:58052)

[Dan94] S. G. Dani, *A proof of Margulis' theorem on values of quadratic forms, independent of the axiom of choice*, Enseign. Math. (2) **40** (1994), no. 1–2, 49–58. MR1279060 (95e:11075)

[DM90a] S. G. Dani and G. A. Margulis, *Values of quadratic forms at integral points: an elementary approach*, Enseign. Math. (2) **36** (1990), no. 1–2, 143–174. MR1071418 (91k:11053)

[DM90b] S. G. Dani and G. A. Margulis, *Orbit closures of generic unipotent flows on homogeneous spaces of* SL(3, **R**), Math. Ann. **286** (1990), no. 1–3, 101–128. MR1032925 (91k:22026)

[DM93] S. G. Dani and G. A. Margulis, *Limit distributions of orbits of unipotent flows and values of quadratic forms*, I. M. Gel'fand Seminar, Adv. Soviet Math., vol. 16, Amer. Math. Soc., Providence, RI, 1993, pp. 91–137. MR1237827 (95b:22024)

[DH46] H. Davenport and H. Heilbronn, *On indefinite quadratic forms in five variables,* J. London Math. Soc. **21** (1946), 185–193. MR0020578 (8,565e)

[Duk88] W. Duke, *Hyperbolic distribution problems and half-integral weight Mass forms*, Invent. Math. **92** (1988), no. 1, 73–90. MR931205 (89d:11033)

[Ein10] M. Einsiedler, *Applications of measure rigidity of diagonal actions*, International Congress of Mathematicians (Hyderabad, India, 2010), 2010.

[EK03] M. Einsiedler and A. Katok, *Invariant measures on G/Γ for split simple Lie groups G*, Comm. Pure Appl. Math. **56** (2003), no. 8, 1184–1221. MR1989231 (2004e:37042)

[EK05] M. Einsiedler and A. Katok, *Rigidity of measures — the high entropy case and non-commuting foliations*, Israel J. Math. **148** (2005), 169–238. Probability in mathematics. MR2191228 (2007d:37034)

[EKL06] M. Einsiedler, A. Katok, and E. Lindenstrauss, *Invariant measures and the set of exceptions to Littlewood's conjecture*, Ann. of Math. (2) **164** (2006), no. 2, 513–560. MR2247967 (2007j:22032)

[EL06] M. Einsiedler and E. Lindenstrauss, *Diagonalizable flows on locally homogeneous spaces and number theory*, International Congress of Mathematicians. Vol. II, Eur. Math. Soc., Zurich, 2006, pp. 1731–1759. MR2275667 (2009d:37007)

[EL08] M. Einsiedler and E. Lindenstrauss, *On measures invariant under diagonalizable actions: the rank-one case and the general low-entropy method*, J. Mod. Dyn. **2** (2008), no. 1, 83–128. MR2366231 (2009h:37057)

[EL10] M. Einsiedler and E. Lindenstrauss, *Diagonal actions on locally homogeneous spaces*, Proceedings of the 2007 Clay Summer School on Homogeneous Flows, Moduli Spaces and Arithmetic, Amer. Math. Soc., Providence, RI, 2010. To appear.

[ELMV09] M. Einsiedler, E. Lindenstrauss, P. Michel, and A. Venkatesh, *Distribution of periodic torus orbits on homogeneous spaces*, Duke Math. J. **148** (2009), no. 1, 119–174. MR2515103

[ELMV10] M. Einsiedler, E. Lindenstrauss, P. Michel, and A. Venkatesh, *Distribution of periodic torus orbits and Duke's theorem for cubic fields* (2010). to appear, Annals of Math.

[EMV09] M. Einsiedler, G. A. Margulis, and A. Venkatesh, *Effective equidistribution for closed orbits of semisimple groups on homogeneous spaces*, Invent. Math. **177** (2009), no. 1, 137–212. MR2507639

[EM04] N. D. Elkies and C. T. McMullen, *Gaps in $\sqrt{n}$ mod 1 and ergodic theory*, Duke Math. J. **123** (2004), no. 1, 95–139. MR2060024 (2005f:11143)

[EV08] J. S. Ellenberg and A. Venkatesh, *Local-global principles for representations of quadratic forms*, Invent. Math. **171** (2008), no. 2, 257–279. MR2367020 (2008m:11081)

[EM04] A. Eskin and G. Margulis, *Recurrence properties of random walks on finite volume homogeneous manifolds*, Random walks and geometry, Walter de Gruyter GmbH & Co. KG, Berlin, 2004, pp. 431–444. MR2087794 (2005m:22025)

[Fur67] H. Furstenberg, *Disjointness in ergodic theory, minimal sets, and a problem in Diophantine approximation*, Math. Systems Theory **1** (1967), 1–49. MR0213508 (35 #4369)

[EMS96] A. Eskin, S. Mozes, and N. Shah, *Unipotent flows and counting lattice points on homogeneous varieties*, Ann. of Math. (2) **143** (1996), no. 2, 253–299, DOI 10.2307/2118644. MR1381987 (97d:22012)

[Fur98] H. Furstenberg, *Stiffness of group actions*, Lie groups and ergodic theory (Mumbai, 1996), Tata Inst. Fund. Res. Stud. Math., vol. 14, Tata Inst. Fund. Res., Bombay, 1998, pp. 105–117. MR1699360 (2000f:22008)

[GM10] F. Götze and G. Margulis, *Distribution of Values of Quadratic Forms at Integral Points* (Apr. 2010), available at arXiv:1004.5123.

[GT07] B. Green and T. Tao, *The quantitative behaviour of polynomial orbits on nilmanifolds*, Ann. of Math. (2007), available at arxiv:0709.3562. To appear.

[GS04] Y. Guivarc'h and A. N. Starkov, *Orbits of linear group actions, random walks on homogeneous spaces and toral automorphisms*, Ergodic Theory Dynam. Systems **24** (2004), no. 3, 767–802. MR2060998 (2005f:37058)

[HS09] R. Holowinsky and K. Soundararajan, *Mass equidistribution of Hecke eigenforms*, available at arxiv:0809.1636.

[KS96] A. Katok and R. J. Spatzier, *Invariant measures for higher-rank hyperbolic abelian actions*, Ergodic Theory Dynam. Systems **16** (1996), no. 4, 751–778. MR1406432 (97d:58116)

[Lin01] E. Lindenstrauss, *Pointwise theorems for amenable groups*, Invent. Math. **146** (2001), no. 2, 259–295. MR1865397 (2002h:37005)

[Lin06] E. Lindenstrauss, *Invariant measures and arithmetic quantum unique ergodicity*, Ann. of Math. (2) **163** (2006), no. 1, 165–219. MR2195133 (2007b:11072)

[LS10] E. Lindenstrauss and U. Shapira, *Homogeneous orbit closures and applications* (2010). Preprint.

[LW01] E. Lindenstrauss and Barak Weiss, *On sets invariant under the action of the diagonal group*, Ergodic Theory Dynam. Systems **21** (2001), no. 5, 1481–1500. MR1855843 (2002j:22009)

[Li68] Yu. V. Linnik, *Ergodic properties of algebraic fields*, Translated from the Russian by M. S. Keane. Ergebnisse der Mathematik und ihrer Grenzgebiete, Band 45, Springer-Verlag New York Inc., New York, 1968. MR0238801 (39 #165)

[Lyo88] R. Lyons, *On measures simultaneously* 2- *and* 3-*invariant*, Israel J. Math. **61** (1988), no. 2, 219–224. MR941238 (89e:28031)

[Mar71] G. A. Margulis, *The action of unipotent groups in a lattice space*, Mat. Sb. (N.S.) **86(128)** (1971), 552–556 (Russian). MR0291352 (45 #445)

[Mar87] G. A. Margulis, *Formes quadratriques indefinies et flots unipotents sur les espaces homogenes*, C. R. Acad. Sci. Paris Ser. I Math. **304** (1987), no. 10, 249–253 (English, with French summary). MR882782 (88f:11027)

[Mar97] G. A. Margulis, *Oppenheim conjecture*, Fields Medallists' lectures, World Sci. Ser. 20th Century Math., vol. 5, World Sci. Publ., River Edge, NJ, 1997, pp. 272–327. MR1622909 (99e:11046)

[Mar00] G. A. Margulis, *Problems and conjectures in rigidity theory*, Mathematics: frontiers and perspectives, Amer. Math. Soc., Providence, RI, 2000, pp. 161–174. MR1754775 (2001d:22008)

[MT94] G. A. Margulis and G. M. Tomanov, *Invariant measures for actions of unipotent groups over local fields on homogeneous spaces*, Invent. Math. **116** (1994), no. 1–3, 347–392. MR1253197 (95k:22013)

[Mau10] F. Maucourant, *A non-homogeneous orbit closure of a diagonal subgroup*, Ann. of Math. **171** (2010), no. 1, 557–570.

[MV06] Ph. Michel and A. Venkatesh, *Equidistribution, L-functions and ergodic theory: on some problems of Yu. Linnik*, International Congress of Mathematicians. Vol. II, Eur. Math. Soc., Zurich, 2006, pp. 421–457. MR2275604 (2008g:11085)

[MV09] Ph. Michel and A. Venkatesh, *The subconvexity problem for* GL_2, Publ. IHES (2009), available at arxiv:0903.3591. To appear.

[MS95] S. Mozes and N. Shah, *On the space of ergodic invariant measures of unipotent flows*, Ergodic Theory Dynam. Systems **15** (1995), no. 1, 149–159. MR1314973 (95k:58096)

[Muc05] R. Muchnik, *Semigroup actions on* $\mathbb{T}^n$, Geom. Dedicata **110** (2005), 1–47. MR2136018 (2006i:37022)

[0h10] H. Oh, *Orbital counting via mixing and unipotent flows*, Proceedings of the 2007 Clay Summer School on Homogeneous Flows, Moduli Spaces and Arithmetic, Amer. Math. Soc., Providence, RI, 2010. To appear.

[Ra82a] M. Ratner, *Rigidity of horocycle flows*, Ann. of Math. (2) **115** (1982), no. 3, 597–614. MR657240 (84e:58062)

[Ra82b] M. Ratner, *Factors of horocycle flows*, Ergodic Theory Dynam. Systems 2 (1982), no. 3–4, 465–489 (1983). MR721735 (86a:58076)

[Ra83] M. Ratner, *Horocycle flows, joinings and rigidity of products*, Ann. of Math. (2) **118** (1983), no. 2, 277–313. MR717825 (85k:58063)

[Ra90a] M. Ratner, *On measure rigidity of unipotent subgroups of semisimple groups*, Acta Math. **165** (1990), no. 3–4, 229–309. MR1075042 (91m:57031)

[Ra90b] M. Ratner, *Strict measure rigidity for unipotent subgroups of solvable groups*, Invent. Math. **101** (1990), no. 2, 449–482. MR1062971 (92h:22015)

[Ra91a] M. Ratner, *On Raghunathan's measure conjecture*, Ann. of Math. (2) 134 (1991), no. 3, 545–607. MR1135878 (93a:22009)

[Ra91b] M. Ratner, *Raghunathan's topological conjecture and distributions of unipotent flows*, Duke Math. J. **63** (1991), no. 1, 235–280. MR1106945 (93f:22012)

[Ra95] M. Ratner, *Raghunathan's conjectures for Cartesian products of real and p-adic Lie groups*, Duke Math. J. **77** (1995), no. 2, 275–382. MR1321062 (96d:22015)

[Ree82] M. Rees, *Some R^2-anosov flows*, 1982. unpublished.

[RS94] Z. Rudnick and P. Sarnak, *The behaviour of eigenstates of arithmetic hyperbolic manifolds*, Comm. Math. Phys. **161** (1994), no. 1, 195–213. MR1266075 (95m:11052)

[Rud90] D. J. Rudolph, $\times 2$ *and* $\times 3$ *invariant measures and entropy*, Ergodic Theory Dynam. Systems **10** (1990), no. 2, 395–406. MR1062766 (91g:28026)

[SS08] O. Sarig and B. Schapira, *The generic points for the horocycle flow on a class of hyperbolic surfaces with infinite genus*, Int. Math. Res. Not. IMRN (2008), Art. ID rnn 086, 37. MR2439545 (2009k:37075)

[Sar81] P. Sarnak, *Asymptotic behavior of periodic orbits of the horocycle flow and Eisenstein series*, Comm. Pure Appl. Math. **34** (1981), no. 6, 719–739. MR634284 (83m:58060)

[Sch80] W. M. Schmidt, *Diophantine approximation*, Lecture Notes in Mathematics, vol. 785, Springer, Berlin, 1980. MR568710 (81j:10038)

[Ser85] C. Series, *The modular surface and continued fractions*, J. London Math. Soc.

[Sh98] N. A. Shah, *Invariant measures and orbit closures on homogeneous spaces for actions of subgroups generated by unipotent elements*, Lie groups and ergodic theory (Mumbai, 1996), Tata Inst. Fund. Res. Stud. Math., vol. 14, Tata Inst. Fund. Res., Bombay, 1998, pp. 229–271. MR1699367 (2001a:22012)

[Sha10] U. Shapira, *A solution to a problem of Cassels and Diophantine properties of cubic numbers*, Ann. of Math. (2010). To appear.

[Sou09] K. Soundararajan, *Quantum unique ergodicity for* $\mathrm{SL}_2(\mathbb{Z})\backslash\mathbb{H}$ (2009), available at arxiv:0901.4060.

[Sou10] K. Soundararajan, *Quantum unique ergodicity and number theory*, International Congress of Mathematicians (Hyderabad, India, 2010), 2010.

[Vat02] V. Vatsal, *Uniform distribution of Hccgncr points*, Invcnt. Math. 148 (2002), no. 1, 1–46. MR1892842 (2003j:11070)

[Ven05] A. Venkatesh, *Sparse equidistribution problems, period bounds, and subconvexity*, Ann. of Math. (2005), available at arXiv:math/0506224. To appear.

[Wan10] Z. Wang, *Quantitative Density under Higher Rank Abelian Algebraic Toral Actions* (2010). Submitted.

Proceedings of the International Congress of Mathematicians
Hyderabad, India, 2010

THE WORK OF STANISLAV SMIRNOV

by
HARRY KESTEN*

Mathematics Subject Classification (2000). 60F05; 60K35

Keywords. Fields medal, Conformal Invariance, Cardy's formula, crossing probabilities

Stanislav (Stas for short) Smirnov is receiving a Fields medal for his ingenious and astonishing work on the existence and conformal invariance of scaling limits or continuum limits of lattice models in statistical physics.

Like many Fields medalists, Stas demonstrated his mathematical skills at an early age. According to Wikipedia he was born on Sept 3,1970 and was ranked first in the 1986 and 1987 International Mathematical Olympiads. He was an undergraduate at Saint Petersburg State University and obtained his Ph.D. at Caltech in 1996 with Nikolai Makarov as his thesis advisor. Stas has also worked on complex analysis and dynamical systems, but in these notes we shall only discuss his work on limits of lattice models. This work should make statistical physicists happy because it confirms rigorously what so far was only accepted on heuristic grounds. The success of Stas in analyzing lattice models in statistical physics will undoubtedly be a stimulus for further work.

Before I start on the work for which Stas is best known, let me mention a wonderful result of his (together with Hugo Duminil-Copin, [21]) which he announced only two months ago. They succeeded in rigorously verifying that the connective constant of the planar hexagonal lattice is $\sqrt{2+\sqrt{2}}$. The *connective constant* μ of a lattice $\mathcal{L}$ is defined as $\lim_{n\to\infty}[c_n]^{1/n}$, where c_n is the number of self-avoiding paths on $\mathcal{L}$ of length n which start at a fixed vertex v. It is usually easy to show by subadditivity (or better submultiplicativity; $c_{n+m} \le c_n c_m$) that this limit exists and is independent of the choice of v. However, the value of μ is unknown for most $\mathcal{L}$. Thus this result of Stas is another major success in Statistical Physics.

*Malott Hall, Cornell University, Ithaca, NY., 14853, USA.
E-mail: kesten@math.cornell.edu

1. Percolation

Since the result for which Stas is best known deals with percolation, it is appropriate to describe this model first. The first percolation problem appeared in Amer. Math Monthly, vol. 1 (1894), proposed by M.A.C.E. De Volson Wood ([9]). He proposes the following problem: "An equal number of white and black balls of equal size are thrown into a rectangular box, what is the probability that there will be contiguous contact of white balls from one end of the box to the opposite end? "As a special example, suppose there are 30 balls in the length of the box, 10 in the width, and 5 (or 10) layers deep." Apart from an incorrect solution by one person who misunderstands the problem, there is no reaction and we still have no answer. Next there is a hiatus of almost 60 years to 1954 when Broadbent ([1]) asks Hammersley at a symposium on Monte-Carlo methods a question which I interpret as follows: Think of the edges of $\mathbb{Z}^d$ as tubes through which fluid can flow with probability p and are blocked with probability $1-p$. Alternatively we assign the color blue or yellow to the edges or call the edges occupied or vacant.) p is the same for all edges, and the edges are independent of each other. If fluid is pumped in at the origin, how far can it spread? Can it reach infinity? Physicists are interested in the model since it seems to be one of the simplest models which has a phase transition. In fact Broadbent and Hammersley ([1, 2]) proved that there exists a value p_c, strictly between 0 and 1, such that ∞ is reached with probability 0 when $p < p_c$, but can be reached with strictly positive probability for $p > p_c$. p_c is called the *critical probability.* The *percolation probability* $\theta(p)$ is defined as the probability that infinity is reached from the origin (or from any other fixed vertex).

Let E be a set of edges. Say that a point a is *connected* (in E) to a point b if there is an open path (in E) from a to b. One can then define the *open clusters* as maximal connected components of open edges in E. By translation invariance, the Broadbent and Hammersley result shows that on $\mathbb{Z}^d$, for $p < p_c$, with probability 1 all open clusters are finite, while it can be shown for $p > p_c$, that with probability 1 there exists a unique infinite open cluster (see [3] for uniqueness). We can do the same thing when we replace $\mathbb{Z}^d$ by another lattice. We can also have all edges open, but the vertices open with probability p and closed with probability $1-p$. In obvious terminology, we talk about bond and site percolation. Site percolation is more general than bond percolation, in the sense that any bond percolation model is equivalent to a site percolation model on another graph, but not vice versa. For Stas' brilliant result we shall consider exclusively site percolation on the 2-dimensional triangular lattice. See Figure 1.

We would like to have a global (as opposed to microscopic) description of such systems. Can we tell what $\theta(p, \mathcal{L})$ is? And similarly, what is the behavior of the "average cluster size" and some other functions. We have a fair understanding of the system for $p \neq p_c$ fixed. For instance, if $p < p_c$, then (with probability 1) there is a translation invariant system of finite clusters, and the probability that the volume

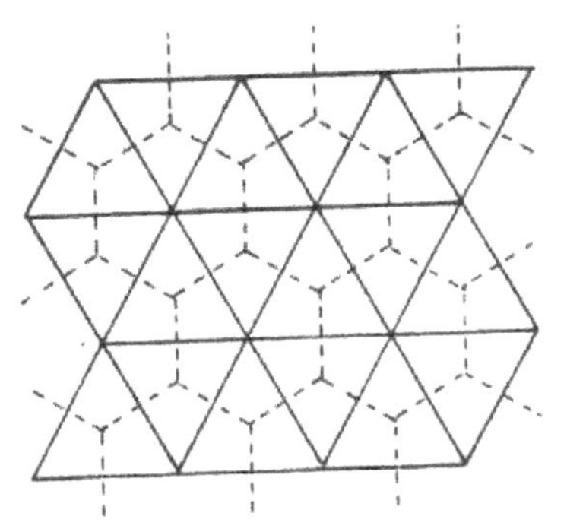

—— $= \mathcal{G}$, the triangular lattice,
- - - $= \mathcal{G}_d$, the hexagonal lattice.

Figure 1.

of the cluster of a fixed site exceeds n decreases exponentially in n (see [10], Theorem 6.75). If $p > p_c$, then there is exactly one infinite open cluster. Also, if $\mathcal{C}$ denotes the open cluster of the origin, then for some constants $0 < c_1(p) < c_2(p) < \infty$,

$$c_1 n^{(d-1)/d} \leq -\log[P_p\{|\mathcal{C}| = n\}] \leq c_2 n^{(d-1)/d}.$$

For $d = 2$ we even know that

$$0 < -\lim_{n\to\infty} n^{-(d-1)/d} \log[P_p\{|\mathcal{C}| = n\}] < \infty,$$

i.e., for some $0 < c(p) < \infty$,

$$P_p\{|\mathcal{C}| = n\} = \exp[-(c + o(1))n^{(d-1)/d}]$$

(see [10], Section 8.6). For these reasons the most interesting behavior can be expected to be for p equal or close to p_c. We have here a system with a function $\theta(p, \mathcal{L})$, which has a phase transition, but, at least in dimension 2, is continuous. I am told that physicists have been successful in analyzing such systems by making an extra assumption, the so-called *scaling hypothesis*: for $p \neq p_c$ there is a single length scale ξ (p), called the *correlation length,* such that for p close to p_c, at distance n the picture of the system looks like a single function of $n/\xi(p)$. More explicitly, it is assumed that many quantities behave like $T(n/\xi(p))$ for some function T which is the same for a class of lattices $\mathcal{L}$. What happens when $p = p_c$ where there is no special length scale singled out (other than the lattice spacing)? The correlation length is assumed to go to ∞ as $p \to p_c$. Therefore, investigating what happens as $p \to p_c$ automatically entails looking at a piece of our system which is many lattice spacings large. For convenience we shall think of looking at our system in a fixed piece of space, but letting the lattice spacing go to 0. We shall call this "taking the scaling limit" or "taking the continuum limit." We shall try to explain Stas' result that this limit exists and is conformally invariant if we consider critical site percolation on the triangular lattice in the plane.

2. The Scaling Limit

What do we expect or hope for? One hopes that at least the cluster distribution and the distribution of the curves separating two adjacent clusters converge in some sense in the scaling limit. Since there is no special scale, one expects scale invariance of the limit. If $\mathcal{L}$ has enough symmetry you can also hope for rotational symmetry of the scaling limit. In dimension two, scale and rotation invariance together should give invariance under holomorphic transformations. If one believes in scale invariance, then one can expect power laws, i.e., that certain functions behave like a power of n or $|p - p_c|$ for n large or p close to p_c. E.g., if we set $R = R(p) =$ the radius of the open cluster of the origin, then scale invariance at $p = p_c$ would give that

$$\frac{P_{pc}\{R \geq xy\}}{P_{pc}\{R \geq y\}} \to g(x) \tag{2.1}$$

for some function $g(x)$, as $y \to \infty$ and $x \geq 1$ fixed. This, in turn, would imply $g(xy) = g(x)g(y)$ and $g(x) = x^{\lambda}$ for some constant λ. Necessarily $\lambda \leq 0$, since (2.1) is less than or equal to 1 for $x \geq 1$. Now let $\varepsilon > 0$ and $(1+\varepsilon)^k \leq t \leq (1+\varepsilon)^{k+1}$. Then

$$P_{pc}\{R \geq t\} \leq P_{pc}\{R \geq (1+\varepsilon)^k\} = P_{pc}\{R \geq 1\} \prod_{j=1}^{k} \frac{P_{pc}\{R \geq (1+\varepsilon)^j\}}{P_{pc}\{R \geq (1+\varepsilon)^{j-1}\}}. \tag{2.2}$$

Since

$$\frac{P_{pc}\{R \geq (1+\varepsilon)^j\}}{P_{pc}\{R \geq (1+\varepsilon)^{(j-1)}\}} \to g(1+\varepsilon) = (1+\varepsilon)^{\lambda} \text{ as } j \to \infty,$$

we obtain

$$P_{pc}\{R \geq t\} \leq t^{\lambda + o(1)} \text{ as } t \to \infty.$$

By replacing k by $k+1$ and reversing the inequality in the lines following (2.2) we see that

$$P_{pc}\{R \geq t\} = t^{\lambda + o(1)} \text{ as } t \to \infty \text{ or } \lim_{t \to \infty} \frac{\log P_{pc}\{R \geq t\}}{\log t} = \lambda. \tag{2.3}$$

Of course we did not prove (2.1) here, nor did we obtain information about λ. The complete proof of (2.3) and evaluation of λ in [15] is much more intricate.

An example of a different but related kind of power law which one may expect says

$$\frac{\log[\theta(p)]}{\log(p - p_c)} \to \beta \text{ as } p \downarrow p_c.$$

Exponents such as λ and β are called *critical exponents*. It is believed that all these exponents can be obtained as algebraic functions of only a small number of independent exponents. Physicists have indeed found (non-rigorously) that various quantities behave as powers. Still on a heuristic basis, they believe that these exponents are *universal,* in the sense that they depend basically on the dimension of the lattice only. In particular they should exist and be the same for the bond and site version on $\mathbb{Z}^2$ and the bond and site version on the triangular lattice. For the planar lattices physicists even predicted values for these exponents.

The pathbreaking work of Stas and Lawler, Schramm, Werner has made it possible to prove some power laws for various processes such as site percolation on the triangular lattice, loop erased random walk, or processes related to the uniform spanning tree. Nevertheless, there still is no proof of universality for percolation, because the percolation results so far are for one lattice only, namely site percolation on the triangular lattice. As stated by Stas in his lecture at the last ICM ([20], p. 1421), "The point which is perhaps still less understood both from mathematics and physics points of view is why there exists a universal conformally equivalent scaling limit." *From now on, all further results tacitly assume that we are dealing with site percolation on the triangular lattice.* As far as I know no other two dimensional percolation results have been proven. For this lattice p_c equals 1/2.

Somehow, the knowledge and guesses about other similar systems convinced people that it would be helpful to prove that the scaling limit for percolation at p_c in two dimensions exists and is conformallly invariant. This is still vague since we did not specify what it means that the scaling limit exists and is conformally invariant. It seems that M. Aizenman (see [13], bottom of p. 556) was the first to express this as a requirement about the scaling limit of *crossing probabilities.*

A crossing probability of a Jordan domain $\mathcal{D}$ with boundary the Jordan curve $\partial\mathcal{D}$ is a probability of the form

$$P\{\exists \text{ an occupied path in } \overline{\mathcal{D}} \text{ from the arc } [a,b] \text{ to the arc } [c,d]\},$$

where $\overline{\mathcal{D}}$ = closure of $\mathcal{D}$, and a, b, c, d are four points on $\partial\mathcal{D}$ such that one successively meets these points as one traverses $\partial\mathcal{D}$ counterclockwise, and the interiors of the four arcs $[a,b],[b,c],[c,d]$ and $[d,a]$ are disjoint. We may also replace "occupied path" by "vacant path" in this definition. It seems reasonable to require that each crossing probability converges to some limit if our percolation configuration converges. As we shall see soon that this is indeed the case in the Stas' development. However, see [6] and [7] for a stricter sense of convergence.

To be more specific, let $\mathcal{D}$ be a Jordan domain in $\mathbb{R}^2$ with a smooth boundary $\partial\mathcal{D}$. Also let $\tau = \exp(2\pi/3)$ and consider three points of $\partial\mathcal{D}$ and label these $A(1), A(\tau), A(\tau^2)$ as one traverses $\partial\mathcal{D}$ counterclockwise. (More general $\mathcal{D}$ should be allowed, but we don't want to discuss technicalities here.) As shown by Stas, there then exist three functions

$$h(A(\alpha), A(\tau\alpha), A(\tau^2\alpha), z), \quad \alpha \in \{1, \tau, \tau^2\},$$

which are the unique harmonic solutions of the mixed Dirichlet-Neumann problem

$$\begin{aligned} &h(A(\alpha), A(\tau\alpha), A(\tau^2\alpha), z) = 1 \text{ at } A(\alpha), \\ &h(A(\alpha), A(\tau\alpha), A(\tau^2\alpha), z) = 0 \text{ on the arc } A(\tau\alpha), A(\tau^2\alpha), \\ &\frac{\partial}{\partial(\tau\nu)} h(A(\alpha), A(\tau\alpha), A(\tau^2\alpha), z) = 0 \text{ on the arc } A(\alpha), A(\tau\alpha), \\ &\frac{\partial}{\partial(-\tau^2\nu)} h(A(\alpha), A(\tau\alpha), A(\tau^2\alpha), z) = 0 \text{ on the arc } A(\tau^2\alpha), A(\alpha), \end{aligned} \tag{2.4}$$

where these functions are regarded as functions of z, and ν is the counterclockwise pointing unit tangent to $\partial\mathcal{D}$. The harmonic solution to these boundary conditions (2.4) is unique, and hence its determination is a conformally invariant problem. More specifically, let Φ be a conformal equivalence between $\mathcal{D}$ and a domain $\tilde{\mathcal{D}}$, and for simplicity assume that the equivalence extends to $\partial\mathcal{D}$. Let $\tilde{h}$ be the harmonic solution of the boundary problem (2.4) with A replaced by $\tilde{A} = \Phi(A)$. Then the uniqueness of the solution implies that

$$h(A(\alpha), A(\tau\alpha), A(\tau^2\alpha), z) = \tilde{h}(\Phi(A(\alpha), \Phi(A(\tau\alpha), \Phi(A(\tau^2\alpha), \Phi(z)).$$

In shorter notation,

$$h = \tilde{h} \circ \Phi. \tag{2.5}$$

Thus, the solution of (2.4) is a conformal invariant of the points $A(1), A(\tau), A(\tau^2), z$ and the domain $\mathcal{D}$. By the Riemann mapping theorem we may choose Φ such that $\tilde{\mathcal{D}}$ has a simple form and then use (2.5) to obtain h on $\mathcal{D}$. Carleson observed that if we take $\tilde{\mathcal{D}}$ to be an equilateral triangle, then the solution $\tilde{h}(A(\alpha), A(\tau\alpha), A(\tau^2\alpha), z)$ is just a linear function which is 1 at the vertex $A(\alpha)$, and 0 on the opposite side $(A(\tau\alpha), A(\tau^2\alpha))$, and similarly when α is replaced by $\tau\alpha$ or $\tau^2\alpha$. For Stas this elegant form made the problem that much more attractive to work on.

Stas achieves his main result by making the following choices: On the triangular lattice, let $A(1) = (2/\sqrt{3}, 0)$, $A(\tau) = (1\sqrt{3}, 1)$, $A(\tau^2) =$ the origin. These are the vertices of an equilateral triangle $\mathcal{D}$ of height 1 and one vertex at the origin. One further takes z on the arc $[A(\tau^2), A(1)] = [(0,0), A(1)]$. Actually we are cheating a bit because the points $A(1), A(\tau), A(\tau^2)$ and z may not lie in $\delta\mathcal{L}$ but we shall ignore this difficulty here and on several places below. For $\alpha \in \{1, \tau, \tau^2\}$, and $z \in [(0,0), A(1)]$, define

$$\begin{aligned} Q_\alpha^\delta(z) = {} &\text{there exists in } \mathcal{D} \text{ a simple, occupied path, from the} \\ &\text{arc } [A(\alpha), A(\tau\alpha)] \text{ to the arc } [A(\tau^2\alpha), A(\alpha)], \text{ and this} \\ &\text{path separates } z \text{ from the arc } [A(\tau\alpha), A(\tau^2\alpha)], \end{aligned} \tag{2.6}$$

and

$$H^\delta(A(\alpha), A(\tau\alpha), A(\tau^2\alpha), z) = P\{Q_\alpha^\delta(z)\}. \tag{2.7}$$

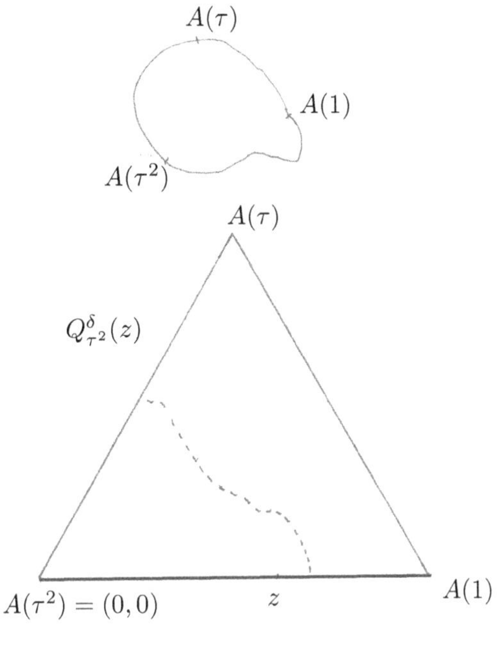

Figure 2.

Stas then formulates his main result as follows: For percolation on $\delta\mathcal{L}$, with $\mathcal{L}$ the triangular lattice, as $\delta \to 0$,

$$H^\delta(A(\alpha), A(\tau\alpha), A(\tau^2\alpha), z) \to h(A(\alpha), A(\tau\alpha), A(\tau^2\alpha), z), \text{ uniformly on } \mathcal{D}. \tag{2.8}$$

The basic structure of the argument is now well known. It is shown that the H^δ are Hölder continuous, so that every sequence $\delta_n \to 0$ has a further subsequence δ_n^* along which the functions $H^{\delta_n^*}$ converge. Moreover the limit along this subsequence has to be harmonic and to satisfy the boundary conditions (2.4). The limit is therefore unique and independent of the choice of the subsequence δ_n^*. Thus $\lim_{\delta\to 0} H^\delta$ exists and is harmonic and conformally invariant (because the solution h to the problem (2.4) is conformally invariant). Note that this proof also yields the convergence of crossing probabilities to a computable limit. Indeed, it follows directly from the definitions that $Q^\delta_{\tau^2}(z)$ is just the event that there exists a crossing in $\mathcal{D}$ from the arc $[A(\tau), A(\tau^2)]$ to the arc $[z, A(1)]$. It then follows from (2.8) that the probability of the existence of such a crossing converges, (as $\delta \to 0$) to $h(A(\tau^2),\, A(1), A(\tau), z_1) = 1 - z_1\sqrt{3}/2$, where $||z - (z_1, 0)|| \to 0$. The value $1 - z_1\sqrt{3}/2$ comes from the fact that $h(z)$ is linear on the segment from $A(\tau^2)$ to $A(1)$ and that $z \to (z_1, 0)$ as $\delta \to 0$.

Stas' proof is quite ingenious. Quite apart from the clever introduction of the variable z, there are steps which one would never expect to work. It uses estimates which rely on quite unexpected cancellations. The principal part of the argument is to show that any subsequential limit (as $\delta \to 0$ through some subsequence δ_n) of the H^δ is harmonic. In turn, this relies on the H^δ being approximations to discrete harmonic functions. Rather than trying to prove harmonicity locally from properties of a second derivative, Stas shows that certain contour integrals of H^δ tend to zero as $\delta \downarrow 0$ and applies Morera's theorem.

Thus these crossing probabilities have limits, which can be computed explicitly. These limits agree with Cardy's formula ([8]). This shows that certain finite collections of crossings of (suitably oriented equilateral) triangles converge weakly and that their probabilities behave as expected, or desired. But much more can be said. [6, 7], and later [4, 23], show that in "the full scaling limit" there is also weak convergence of the occurrence of loops, and loops inside loops or touching other loops, etc. As stated in the abstract of [5]: "These loops do not cross but do touch each other-indeed, any two loops are connected by a finite 'path' of touching loops."

3. Schramm-Loewner Evolutions (SLE)

A short time before Smirnov's paper, Schramm had tried to find out how conformal invariance could be used (if shown to apply) to study also other models than percolation. Loewner introduced his evolutions when he tried to prove Bieberbach's conjecture. Roughly speaking, Loewner represented a family of curves (one for each $z \in \mathbb{H}$) by means of a single function U_t. Here $\mathbb{H}$ is the open upper halfplane, U_t is a given function, and after a reparametrization, g_t is a solution of the initial value problem

$$\frac{\partial}{\partial t} g_t(z) = \frac{2}{g_t(z) - U_t}, \quad g_0(z) = z. \tag{3.1}$$

Let

$$T_z = \sup\{s : \text{solution is well defined for } t \in [0, s) \text{ with } g_s(z) \in \mathbb{H}\}$$

and $H_t := \{z : T_z > t\}$. Then g_t is the unique conformal transformation from H_t onto $\mathbb{H}$ for which $g_t(z) - z \to 0$ as $z \to \infty$ (see [14], Theorem 4.6). The g_t arising in this way are called *Loewner chains* and $\{U_t\}$ the *driving function.* See [14], Theorem 4.6. The original Loewner chains were defined without any probability concepts. In particular the driving function $\{U_t\}$ was deterministic. Reference [16] raised the question whether a random driving function could produce some of the known random curves as Loewner chain $\{g_t\}$. Schramm showed in [16] that if the process $\{g_t\}$ has certain Markov properties, then one can obtain this process as Loewner chain only if the driving function is $\sqrt{\kappa} \times$ Brownian motion, for some $\kappa \geq 0$. The processes which have such a driving function are called SLE's (originally this stood for "Stochastic Loewner Evolution", but is now commonly read as Schramm-Loewner evolution). When a chain is an SLE_κ (in obvious notation) new computations become possible or much simplified. In particular, the existence and explicit values of most of the critical exponents have now been rigorously established (but see questions Q2 and Q4 below). Stas has made major contributions to these determinations in [15, 22]. In particular he provided essential steps for showing that a certain interface between occupied and vacant sites in percolation is an SLE_6 curve.

The SLE calculations confirm predictions of physicists, as well as a conjecture of Mandelbrot. As a result, the literature on SLE_κ has grown by leaps and bounds in the last few years, and the study of properties of SLE is becoming a subfield by itself. SLE_κ processes with different κ can have quite different behavior. A good survey of percolation and SLE is in [17], and [14] is a full length treatment of SLE.

4. Generalization and Some Open Problems

I don't know of any lattice model in physics which has as much independence built in as percolation. It is therefore of great significance that Stas has a way to attack problems concerning the existence and conformal invariance of a scaling limit for some models with dependence between sites, and in particular for the two-dimensional Ising model. This is perhaps the oldest lattice model, and the literature on it is enormous. I am largely ignorant of this literature and have not worked my way through Stas' papers on these models. Nevertheless I am excited by the fact that Stas is seriously attacking such models.

For the people who are new to this, the Ising model again assigns a random variable (usually called a spin) to each site of a lattice $\mathcal{L}$. Denote the spin at a site v by $\sigma(v)$. Again $\sigma(v)$ can take only two values, which are usually taken to be ± 1. The interaction between two sites u and v is $J(u,v)\sigma(u)\sigma(v)$ and in the simplest case

$$J(u,v) = \begin{cases} J \text{ if } u \text{ and } v \text{ are neighbors} \\ 0 \text{ otherwise.} \end{cases}$$

We restrict ourselves to this simplest case, which takes $J \geq 0$ constant. However, in order to discuss boundary conditions we also need another constant, $\tilde{J}$ say. For any *finite* set $\wedge \subset \mathcal{L}$ we consider the probability distribution of the spin configuration on $\wedge$. This configuration is of course the vector $\{\sigma(v)\}_{v\in\wedge}$, and so can also be viewed as a point in $\{-1,1\}^\wedge$. For any fixed $\tilde{\sigma}$ and $\wedge$ we define

$$H(\sigma,\tilde{\sigma}) = H_\wedge(\sigma,\tilde{\sigma}) = -\sum_{u,v\in\wedge} J\sigma(u)\sigma(v) - \sum_{u\in\wedge, v\notin\wedge} \tilde{J}\sigma(u)\tilde{\sigma}(v), \tag{4.1}$$

and the normalizing constant (also called partition function)

$$Z = Z(\wedge,\beta,\tilde{\sigma}) = \sum_{\sigma} \exp[= \beta H_\wedge(\sigma,\tilde{\sigma})].$$

Here the sum over σ runs over $\{-1,1\}^\wedge$, the collection of possible spin configurations on $\wedge.\beta \geq 0$ is a parameter, which is usually called the "inverse temperature."

Let $\tilde{\sigma}$ be fixed outside $\wedge$. Then, given the boundary condition $\sigma(v) = \tilde{\sigma}(v)$ for $v \notin \wedge$, the distribution of the spins in $\wedge$ is given by

$$P\{\sigma(u) = \tau(u) \text{ for } u \in \wedge | \sigma(v) = \tilde{\sigma}(v), v \notin \wedge\}$$
$$= [Z(\wedge, \beta, \tilde{\sigma})]^{-1} \exp[-\beta H_\wedge(\tau, \tilde{\sigma})].$$

This defines a probability measure for the spins in a finite $\wedge$. A probability distribution for all spins simultaneously has to be obtained by taking a limit as $\wedge \uparrow \mathcal{L}$. The second sum in the right hand side of (4.1) shows the influence of boundary conditions. At sufficiently low temperature there can be two extremal states, obtained by taking $\wedge \uparrow \mathcal{L}$ under different boundary conditions. It now becomes unclear how to deal with boundary conditions when one wants to take a continuum limit.

To conclude, here are some problems on percolation. These also have appeared in other lists, (see in particular [17]), but you may like to be challenged again.

Q1 Prove the existence and find the value of critical exponents of percolation on other two-dimensional lattices than the triangular one and establish universality in two dimensions.

This seems to be quite beyond our reach at this time. Probably even moreso is the same question in dimension >2.

Q2 Prove a power law and find a critical exponent for the probability that there are j disjoint occupied paths from the disc $\{z : |z| \le r\}$ to $\{z : |z| \ge R\}$. For $j = 1$ this is the one-arm problem of [15]. For $j \ge 2$, the problem is solved, at least for the triangular lattice, if some of the arms are occupied and some are vacant (see Theorem 4 in [22]), but it seems that there is not even a conjectured exponent for the case when all arms are to be occupied or all vacant.

More specific questions are

Q3 Is the percolation probability (right) continuous at p_c? Equivalently, is there percolation at p_c? This is only a problem for $d > 2$. The answer in $d = 2$ is that there is no percolation at p_c;

Q4 Establish the existence and find the value of a critical exponent for the expected number of clusters per site. This quantity is denoted by

$$\kappa(p) = \sum_{n=1}^{\infty} \frac{1}{n} P_p\{|C| = n\}$$

in [10], p. 23. The answer is still unknown, even for critical percolation on the two-dimensional triangular lattice. It is known that $\kappa(p)$ is twice differentiable on [0,1], but it is believed that the third derivative at p_c fails to exist; see [12], Chapter 9. This problem is mainly of historical interest, because there was an attempt to prove that p_c for bond percolation on $\mathbb{Z}^2$ equals 1/2, by showing that $\kappa(p)$ has only one singularity in (0,1).

5. Conclusion

I have been amazed and greatly pleased by the progress which Stas Smirnov and coworkers have made in a decade. They have totally changed the fields of random planar curves and of two dimensional lattice models. Stas has shown that he has the talent and insight to produce surprising results, and his work has been a major stimulus for the explosion in the last 15 years or so of probabilistic results about random planar curves.

As some of the listed problems here show, there still are fundamental, and probably difficult, issues to be settled. I wish Stas a long and creative career, and that we all may enjoy his mathematics.

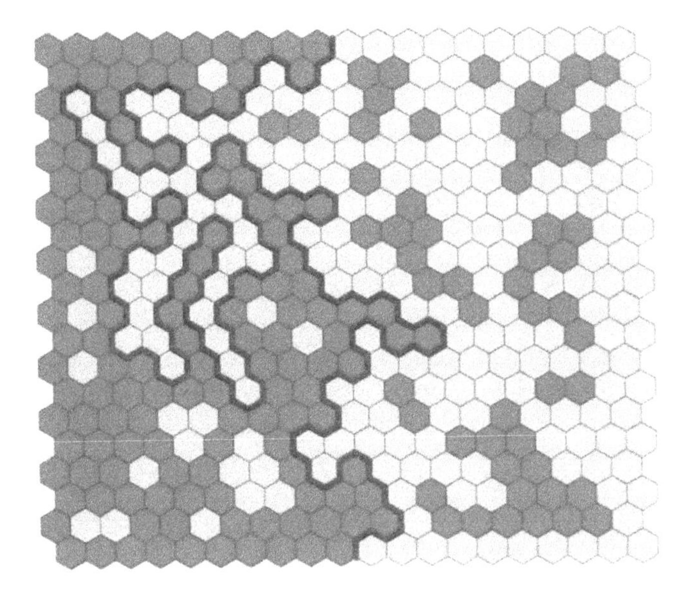

Figure 3.

References

[1] S.R. Broadbent, *in discusssion of Symposium on Monte Carlo methods*, J. Roy. Stat. Soc. (B) **16** (1954) 68.

[2] S.R. Broadbent and J.M. Hammersley, *Percolation processes,* Proc. Cambr. Phil. Soc. **53** (1957) 629–641 and 642–645.

[3] R.M. Burton and M. Keane, *Density and uniqueness in percolation*, Comm. Math. Phys. **121** (1989) 501–505.

[4] F. Camia, *Scaling limits of two-dimensional percolation: an overview*, arXiv:0810.1002v1.

[5] F. Camia and C.M. Newman, *Continuum nonsimple loops and 2D critical percolation,* J. Stat. Phys. **116** (2004) 157–173.
[6] F. Camia and C.M. Newman, *Two-dimensional critical percolation: the full scaling limit,* Comm. Math. Phys. **268** (2006) 1–38.
[7] F. Camia and C.M. Newman, *Critical percolation exploration path and* SLE_6*: a proof of convergence,* Probab. Theory and Relat. Fields **139** (2007) 473–519.
[8] J.L. Cardy, *Critical percolation in finite geometries,* J. Phys. A, **25** (1992) L 201–L 206.
[9] M.A.C.E. De Volson Wood, Problem 5, Amer. Math. Monthly, **1** (1894) 211.
[10] G. Grimmett,*Percolation, second edition,* 1999, Springer-Verlag.
[11] G. Grimmett, *The random cluster model,* 2006, Springer-Verlag.
[12] H. Kesten, *Percolation theory for mathematicians,* 1982, Birkhäuser.
[13] R.P. Langlands, C. Piget, Ph. Pouliot, and Y. Saint-Aubin, *On the universality of crossing probabilities in two-dimensional percolation,* J. Stat. Phys. **67** (1992) 553–574.
[14] G.F. Lawler, *Conformally invariant processes in the plane,* Amer. Math. Soc., Math. Surveys and Monographs, vol. **114** (2005).
[15] G.F. Lawler, O. Schramm and W. Werner, *One-arm exponent for critical 2D percolation,* Elec. J. Probab. **7** (2002) Paper # 2.
[16] O. Schramm, *Scaling limits of loop-erased random walks and uniform spanning trees,* Israel J. Math. **118** (2000) 221–288.
[17] O. Schramm, *Conformally invariant scaling limits: an overview and a collection of problems,* pp. 513–543 in Proc. Internat. Congress of Mathematicians, Madrid,2006.
[18] S. Smirnov I, *Critical percolation in the plane. Conformal invariance, Cardy's formula, scaling limits,* C.R. Acad. Sci. Paris Sér.I Math. **333** (2001) 239–244.
[19] S. Smirnov II, *long version of I,* Preprint, 2001. arXiv: 0909.4499.
[20] S. Smirnov, *Towards conformal invariance of 2D lattice models,* Proc. Internat. Congress Mathematiciens **II,** Madrid, Spain (2006) 1421.
[21] S. Smirnov and H. Duminil-Copin, *The connective constant of the honeycomb lattice equals* $\sqrt{2+\sqrt{2}}$, Ann. Math. **175** (2012) 1653–1665.
[22] S. Smirnov and W. Werner, *Critical exponents for two-dimensional percolation,* Math. Res. Letters **8** (2001) 729–744.
[23] N. Sun, *Conformally invariant scaling limits in planar critical percolation,* arXiv:0911. 0063v1.
[24] W. Werner, *Random planar curves and Schramm-Loewner evolutions,* Lecture Notes in Math, vol. **1840,** J. Picard ed., Springer-Verlag 2004.

Stanislav Smirnov

STANISLAV SMIRNOV

Graduation from St. Petersburg State University, 1992;

Ph.D., California Institute of Technology, 1996.

Positions held

Gibbs Instructor at the Yale University, 1996–1999;

Short-term positions at MPI Bonn and IAS Princeton, 1997–1998;

Researcher at the Royal Institute of Technology in Stockholm, 1998–2001;

Professor at the Royal Institute of Technology in Stockholm, from 2001;

Researcher at the Swedish Royal Academy of Sciences 2001–2004;

Professor at the University of Geneva, 2003–.

Proceedings of the International Congress of Mathematicians
Hyderabad, India, 2010

DISCRETE COMPLEX ANALYSIS AND PROBABILITY

by
STANISLAV SMIRNOV*

ABSTRACT. We discuss possible discretizations of complex analysis and some of their applications to probability and mathematical physics, following our recent work with Dmitry Chelkak, Hugo Duminil-Copin and Clément Hongler.

Mathematics Subject Classification (2010). Primary 30G25; Secondary 05C81, 60K35, 81T40, 82B20.

Keywords. Discrete complex analysis, discrete analytic function, Ising model, self-avoiding walk, conformal invariance

1. Introduction

The goal of this note is to discuss some of the applications of discrete complex analysis to problems in probability and statistical physics. It is not an exhaustive survey, and it lacks many references. Forgoing completeness, we try to give a taste of the subject through examples, concentrating on a few of our recent papers with Dmitry Chelkak, Hugo Duminil-Copin and Clément Hongler [CS08, CS09, CS10, DCS10, HS10]. There are certainly other interesting developments in discrete complex analysis, and it would be a worthy goal to write an extensive exposition with an all-encompassing bibliography, which we do not attempt here for lack of space.

Complex analysis (we restrict ourselves to the case of one complex or equivalently two real dimensions) studies analytic functions on (subdomains of) the complex plane, or more generally analytic structures on two dimensional manifolds. Several things are special about the (real) dimension two, and we won't discuss an interesting and often debated question, why exactly complex analysis is so nice

*This research was supported by the European Research Council AG "CONFRA" and by the Swiss National Science Foundation. We would like to thank Dmitry Chelkak for comments on the preliminary version of this paper.
Section de Mathematiques, Université de Genéve. 2-4 rue du Liévre, CP 64, 1211 Geneve 4, SUISSE. E-mail: stanislav.smirnov@unige.ch

and elegant. In particular, several definitions lead to identical class of analytic functions, and historically different adjectives (regular, analytic, holomorphic, monogenic) were used, depending on the context. For example, an *analytic* function has a local power series expansion around every point, while a *holomorphic* function has a complex derivative at every point. Equivalence of these definitions is a major theorem in complex analysis, and there are many other equivalent definitions in terms of Cauchy-Riemann equations, contour integrals, primitive functions, hydrodynamic interpretation, etc. Holomorphic functions have many nice properties, and hundreds of books were devoted to their study.

Consider now a discretized version of the complex plane: some graph embedded into it, say a square or triangular lattice (more generally one can speak of discretizations of Riemann surfaces). Can one define analytic functions on such a graph? Some of the definitions do not admit a straightforward discretization: e.g. local power series expansions do not make sense on a lattice, so we cannot really speak of discrete analyticity. On the other hand, as soon as we define discrete derivatives, we can ask for the holomorphicity condition. Thus it is philosophically more correct to speak of *discrete holomorphic,* rather than discrete analytic functions. We will use the term *preholomorphic* introduced by Ferrand [Fer44], as we prefer it to the term *monodiffric* used by Isaacs in the original papers [Isa41, Isa52] (a play on the term monogenic used by Cauchy for continuous analytic functions).

Though the preholomorphic functions are easy to define, there is a lack of expository literature about them. We see two main reasons: firstly, there is no canonical preholomorphicity definition, and one can argue which of the competing approaches is better (the answer probably depends on potential applications). Secondly, it is straightforward to transfer to the discrete case beginnings of the usual complex analysis (a nice topic for an undergraduate research project), but the easy life ends when it becomes necessary to multiply preholomorphic functions. There is no easy and natural way to proceed and the difficulty is addressed depending on the problem at hand.

As there seems to be no canonical discretization of the complex analysis, we would rather adopt a utilitarian approach, working with definitions corresponding to interesting objects of probabilistic origin, and allowing for a passage to the scaling limit. We want to emphasize, that we are concerned with the following triplets:

1. A planar graph,
2. Its embedding into the complex plane,
3. Discrete Cauchy-Riemann equations.

We are interested in triplets such that the discrete complex analysis approximates the continuous one. Note that one can start with only a few elements of the triplet, which gives some freedom. For example, given an embedded graph, one can ask which discrete difference equations have solutions close to holomorphic functions.

Or, given a planar graph and a notion of preholomorphicity, one can look for an appropriate embedding.

The ultimate goal is to find lattice models of statistical physics with preholomorphic observables. Since those observables would approximate holomorphic functions, some information about the original model could be subsequently deduced.

Below we start with several possible definitions of the preholomorphic functions along with historical remarks. Then we discuss some of their recent applications in probability and statistical physics.

2. Discrete Holomorphic Functions

For a given planar graph, there are several ways to define *preholomorphic* functions, and it is not always clear which way is preferable. A much better known class is that of *discrete harmonic* (or *preharmonic)* functions, which can be defined on any graph (not necessarily planar), and also in more than one way. However, one definition stands out as the simplest: a function on the vertices of graph is said to be *preharmonic* at a vertex v, if its discrete Laplacian vanishes:

$$0 = \Delta H(u) := \sum_{v:\ \text{neighbor of}\ u} (H(v) - H(u)). \tag{1}$$

More generally, one can put weights on the edges, which would amount to taking different resistances in the electric interpretation below. Preharmonic functions on planar graphs are closely related to discrete holomorphicity: for example, their gradients defined on the oriented edges by

$$F(\vec{uv}) := H(v) - H(u), \tag{2}$$

are preholomorphic. Note that the edge function above is antisymmetric, i.e. $F(\vec{uv}) = -F(\vec{vu})$.

Both classes with the definitions as above are implicit already in the 1847 work of Kirchhoff [Kir47], who interpreted a function defined on oriented edges as an electric current flowing through the graph. If we assume that all edges have unit resistance, than the sum of currents flowing from a vertex is zero by the first Kirchhoff law:

$$\sum_{u:\ \text{neighbor of}\ v} F(\vec{uv}) = 0, \tag{3}$$

and the sum of the currents around any oriented closed contour γ (for the planar graphs it is sufficient to consider contours around faces) face is zero by the second Kirchhoff law:

$$\sum_{\vec{uv} \in \gamma} F(\vec{uv}) = 0. \tag{4}$$

The two laws are equivalent to saying that F is given by the gradient of a potential function H as in (2), and the latter function is preharmonic (1). One can equivalently think of a hydrodynamic interpretation, with F representing the flow of liquid. Then conditions (3) and (4) mean that the flow is divergence- and curl-free correspondingly. Note that in the continuous setting similarly defined gradients of harmonic functions on planar domains coincide up to complex conjugation with holomorphic functions. And in higher dimensions harmonic gradients were proposed as one of their possible generalizations.

There are many other ways to introduce discrete structures on graphs, which can be developed in parallel to the usual complex analysis. We have in mind mostly such discretizations that restrictions of holomorphic (or harmonic) functions become *approximately* preholomorphic (or preharmonic). Thus we speak about graphs embedded into the complex plane or a Riemann surface, and the choice of embedding plays an important role. Moreover, the applications we are after require passages to the scaling limit (as mesh of the lattice tends to zero), so we want to deal with discrete structures which converge to the usual complex analysis as we take finer and finer graphs.

Preharmonic functions satisfying (1) on the square lattices with decreasing mesh fit well into this philosophy, and were studied in a number of papers in early twentieth century (see e.g. [PW23, Bou26, Lus26]), culminating in the seminal work of Courant, Friedrichs and Lewy. It was shown in [CFL28] that solution to the Dirichlet problem for a discretization of an elliptic operator converges to the solution of the analogous continuous problem as the mesh of the lattice tends to zero. In particular, a preharmonic function with given boundary values converges in the scaling limit to a harmonic function with the same boundary values in a rather strong sense, including convergence of all partial derivatives.

Preholomorphic functions distinctively appeared for the first time in the papers [Isa41, Isa52] of Isaacs, where he proposed two definitions (and called such functions "monodiffric"). A few papers of his and others followed, studying the first definition (5), which is asymmetric on the square lattice. More recently, the first definition was studied by Dynnikov and Novikov [DN03] in the triangular lattice context, where it becomes symmetric (the triangular lattice is obtained from the square lattice by adding all the diagonals in one direction).

The second, symmetric, definition was reintroduced by Ferrand, who also discussed the passage to the scaling limit [Fer44, LF55]. This was followed by extensive studies of Duffin and others, starting with [Duf56].

Both definitions ask for a discrete version of the Cauchy-Riemann equations $\partial_{i\alpha}F = i\partial_\alpha F$ or equivalently that z-derivative is independent of direction. Consider a subregion Ω_ε of the mesh ε square lattice $\varepsilon\mathbb{Z}^2 \subset \mathbb{C}$ and define a function on its vertices. Isaacs proposed the following two definitions, replacing the derivatives by discrete differences. His "monodiffric functions of the first kind" are required to satisfy inside Ω_ε the following identity:

$$F(z+i\varepsilon) - F(z) = i(F(z+\varepsilon) - F(z)), \tag{5}$$

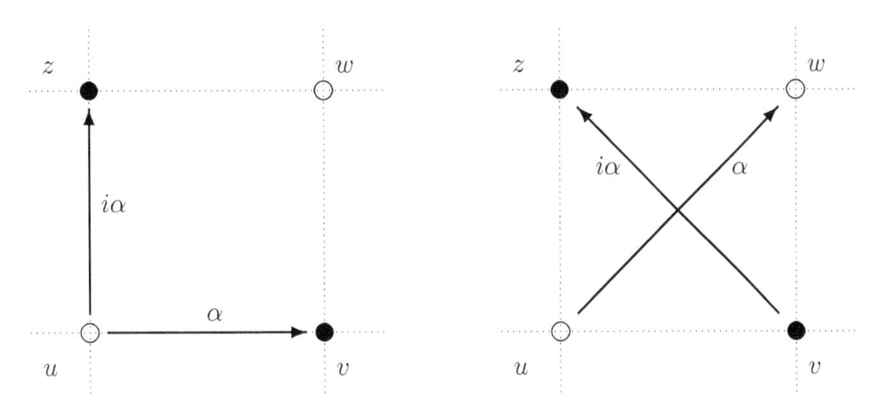

Figure 1. The first and the second Isaacs' definitions of discrete holomorphic functions: multiplied by i difference along the vector α is equal to the difference along the rotated vector $i\alpha$. Note that the second definition (on the right) is symmetric with respect to lattice rotations, while the first one is not.

which can be rewritten as

$$\frac{F(z+i\varepsilon)-F(z)}{(z+i\varepsilon)-z}=\frac{F(z+\varepsilon)-F(z)}{(z+\varepsilon)-z}.$$

We will be working with his second definition, which is more symmetric and also appears naturally in probabilistic context (but otherwise the theories based on two definitions are almost the same). We say that a function is *preholomorphic*, if inside Ω_ε it satisfies the following identity, illustrated in Figure 1:

$$F(z+i\varepsilon)-F(z+\varepsilon)=i(F(z+\varepsilon(1+i))-F(z)), \tag{6}$$

which can also be rewritten as

$$\frac{F(z+i\varepsilon)-F(z+\varepsilon)}{(z+i\varepsilon)-(z+\varepsilon)}=\frac{F(z+\varepsilon(1+\varepsilon))-F(z)}{(z+\varepsilon(1+i))-z}.$$

It is easy to see that restrictions of continuous holomorphic functions to the mesh ε square lattice satisfy this identity up to $O(\varepsilon^3)$. Note also that if we color the lattice in the chess-board fashion, the complex identity (6) can be written as two real identities (its real and imaginary parts), one involving the real part of F at black vertices and the imaginary part of F at white vertices, the other one — vice versa. So unless we have special boundary conditions, F splits into two "demi-functions" (real at white and imaginary at black vs. imaginary at black and real at white vertices), and some prefer to consider just one of those, i.e. ask F to be purely real at black vertices and purely imaginary at white ones.

The theory of so defined preholomorphic functions starts much like the usual complex analysis. It is easy to check, that for preholomorphic functions sums are also preholomorphic, discrete contour integrals vanish, primitive (in a simply-connected domain) and derivative are well-defined and are preholomorphic functions on the dual square lattice, real and imaginary parts are preharmonic on their respective black and white sublattices, etc. Unfortunately, the product of two preholomorphic functions is no longer preholomorphic: e.g., while restrictions of 1, z, and z^2 to the square lattice are preholomorphic, the higher powers are only approximately so.

Situation with other possible definitions is similar, with much of the linear complex analysis being easy to reproduce, and problems appearing when one has to multiply preholomorphic functions. Pointwise multiplication cannot be consistently defined, and though one can introduce convolution-type multiplication, the possible constructions are non-local and cumbersome. Sometimes, for different graphs and definitions, problems appear even earlier, with the first derivative not being preholomorphic.

Our main reason for choosing the definition (6) is that it naturally appears in probabilistic context. It was also noticed by Duffin that (6) nicely generalizes to a larger family of *rhombic lattices,* where all the faces are rhombi. Equivalently, one can speak of *isoradial graphs,* where all faces are inscribed into circles of the same radius — an isoradial graph together with its dual forms a rhombic lattice.

There are two main reasons to study this particular family. First, this is perhaps the largest family of graphs for which the Cauchy-Riemann operator admits a nice discretization. Indeed, restrictions of holomorphic functions to such graphs are preholomorphic to higher orders. This was the reason for the introduction of complex analysis on rhombic lattices by Duffin [Duf68] in late sixties. More recently, the complex analysis on such graphs was studied for the sake of probabilistic applications [Mer01, Ken02, CS08].

On the other hand, this seems to be the largest family where certain lattice models, including the Ising model, have nice integrability properties. In particular, the critical point can be defined with weights depending only on the local structure, and the star-triangle relation works out nicely. It seems that the first appearance of related family of graphs in the probabilistic context was in the work of Baxter [Bax78], where the eight vertex and Ising models were considered on Z-invariant graphs, arising from planar line arrangements. These graphs are topologically the same as the isoradial ones, and though they are embedded differently into the plane, by [KS05] they always admit isoradial embeddings. In [Bax78] Baxter was not passing to the scaling limit, and so the actual choice of embedding was immaterial for his results. However, his choice of weights in the models would suggest an isoradial embedding, and the Ising model was so considered by Mercat [Mer01], Boutilier and de Tiliére [BdT08, BdT09], Chelkak and the author [CS09]. Additionally, the dimer and the uniform spanning tree models on such graphs also have nice properties, see e.g. [Ken02].

We would also like to remark that rhombic lattices form a rather large family of graphs. While not every topological quadrangulation (graph all of whose faces are quadrangles) admits a rhombic embedding, Kenyon and Schlenker [KS05] gave a simple topological condition necessary and sufficient for its existence.

So this seems to be the most general family of graphs appropriate for our subject, and most of what we discuss below generalizes to it (though for simplicity we speak of the square and hexagonal lattices only).

3. Applications of Preholomorphic Functions

Besides being interesting in themselves, preholomorphic functions found several diverse applications in combinatorics, analysis, geometry, probability and physics.

After the original work of Kirchhoff, the first notable application was perhaps the famous article [BSST40] of Brooks, Smith, Stone and Tutte, who used preholomorphic functions to construct tilings of rectangles by squares.

Several applications to analysis followed, starting with a new proof of the Riemann uniformization theorem by Ferrand [LF55]. Solving the discrete version of the usual minimization problem, it is immediate to establish the existence of the minimizer and its properties, and then one shows that it has a scaling limit, which is the desired uniformization. Duffin and his co-authors found a number of similar applications, including construction of the Bergman kernel by Dieter and Mastin [DM71]. There were also studies of discrete versions of the multi-dimensional complex analysis, see e.g. Kiselman's [Kis05].

In [Thu86] Thurston proposed *circle packings* as another discretization of complex analysis. They found some beautiful applications, including yet another proof of the Riemann uniformization theorem by Rodin and Sullivan [RS87]. More interestingly, they were used by He and Schramm [HS93] in the best result so far on the Koebe uniformization conjecture, stating that any domain can be conformally uniformized to a domain bounded by circles and points. In particular, they established the conjecture for domains with countably many boundary components. More about circle packings can be learned form Stephenson's book [Ste05]. Note that unlike the discretizations discussed above, the circle packings lead to nonlinear versions of the Cauchy-Riemann equations, see e.g. the discussion in [BMS05].

There are other interesting applications to geometry, analysis, combinatorics, probability, and we refer the interested reader to the expositions by Lovász [Lov04], Stephenson [Ste05], Mercat [Mer07], Bobenko and Suris [BS08].

In this note we are interested in applications to probability and statistical physics. Already the Kirchhoff's paper [Kir47] makes connection between the Uniform Spanning Tree and preharmonic (and so preholomorphic) functions.

Connection of Random Walk to preharmonic functions was certainly known to many researchers in early twentieth century, and figured implicitly in many papers. It is explicitly discussed by Courant, Friedrichs and Lewy in [CFL28], with preharmonic functions appearing as Green's functions and exit probabilities for the Random Walk.

More recently, Kenyon found preholomorphic functions in the dimer model (and in the Uniform Spanning Tree in a way different from the original considerations of Kirchhoff). He was able to obtain many beautiful results about statistics of the dimer tilings, and in particular, showed that those have a conformally invariant scaling limit, described by the Gaussian Free Field, see [Ken00, Ken01]. More about Kenyon's results can be found in his expositions [Ken04, Ken09]. An approximately preholomorphic function was found by the author in the critical site percolation on the triangular lattice, allowing to prove the Cardy's formula for crossing probabilities [Smi01b, Smi01a].

Finally, we remark that various other discrete relations were observed in many integrable two dimensional models of statistical physics, but usually no explicit connection was made with complex analysis, and no scaling limit was considered. Here we are interested in applications of integrability parallel to that for the Random Walk and the dimer model above. Namely, once a preholomorphic function is observed in some probabilistic model, we can pass to the scaling limit, obtaining a holomorphic function. Thus, the preholomorphic observable is approximately equal to the limiting holomorphic function, providing some knowledge about the model at hand. Below we discuss applications of this philosophy, starting with the Ising model.

4. The Ising Model

In this Section, we discuss some of the ways how preholomorphic functions appear in the Ising model at criticality. The observable below was proposed in [Smi06] for the hexagonal lattice, along with a possible generalization to $O(N)$ model. Similar objects appeared earlier in Kadanoff and Ceva [KC71] and in Mercat [Mer01], though boundary values and conformal covariance, which are central to us, were never discussed.

The scaling limit and properties of our observable on isoradial graphs were worked out by Chelkak and the author in [CS09]. It is more appropriate to consider it as a *fermion* or a *spinor,* by writing $F(z)\sqrt{dz}$, and with more general setup one has to proceed in this way.

Earlier we constructed a similar fermion for the random cluster representation of the Ising model, see [Smi06, Smi10] and our joint work with Chelkak [CS09] for generalization to isoradial graphs (and also independent work of Riva and Cardy [RC06] for its physical connections). It has a simpler probabilistic interpretation than the fermion in the spin representation, as it can be written as the probability of the interface between two marked boundary points passing through a point inside, corrected by a complex weight depending on the winding.

The fermion for the spin representation is more difficult to construct. Below we describe it in terms of contour collections with distinguished points. Alternatively it corresponds to the partition function of the Ising model with a $\sqrt{z}$ monodromy at a given edge, corrected by a complex weight; or to a product of order and disorder operators at neighboring site and dual site.

We will consider the Ising model on the mesh ε square lattice. Let Ω_ε be a discretization of some bounded domain $\Omega \subset \mathbb{C}$. The Ising model on Ω_ε has configurations σ which assign ± 1 (or simply $\pm$) spins $\sigma(v)$ to vertices $v \in \Omega_\varepsilon$ and Hamiltonian defined (in the absence of an external magnetic field) by

$$H(\sigma) = -\sum_{\langle u,v\rangle} \sigma(u)\sigma(v),$$

where the sum is taken over all edges $\langle u, v\rangle$ inside Ω_ε. Then the partition function is given by

$$Z = \sum_{\sigma} \exp(-\beta H(\sigma)),$$

and probability of a given spin configuration becomes

$$\mathbb{P}(\sigma) = \exp(-\beta H(\sigma))/Z.$$

Here $\beta \geq 0$ is the temperature parameter (behaving like the reciprocal of the actual temperature), and Kramers and Wannier have established [KW41] that its critical value is given by $\beta_C = \log(\sqrt{2}+1)/2$.

Now represent the spin configurations graphically by a collection of interfaces — contours on the dual lattice, separating plus spins from minus spins, the so-called *low-temperature expansion*, see Figure 2. A contour collection is a set of edges, such that an even number emanates from every vertex. In such case the contours can be represented as a union of loops (possibly in a non-unique way, but we do not

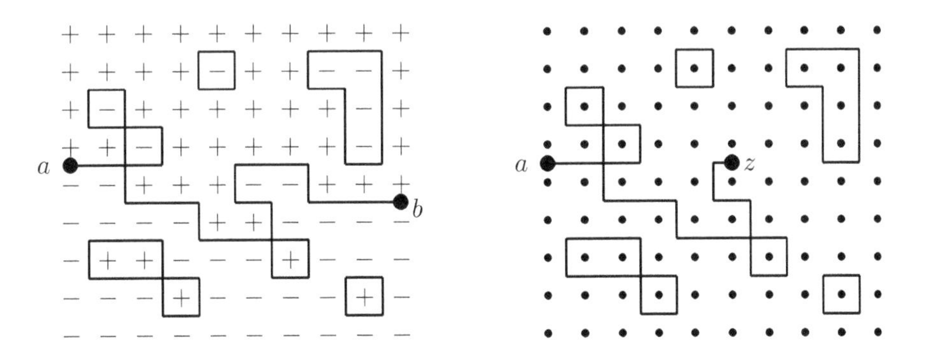

Figure 2. Left: configuration of spins in the Ising model with Dobrushin boundary conditions, its contour representation, and an interface between two boundary points. Right: an example of a configuration considered for the Fermionic observable: a number of loops and a contour connecting a to z. It can be represented as a spin configuration with a monodromy at z.

distinguish between different representations). Note that each contour collection corresponds to two spin collections which are negatives of each other, or to one if we fix the spin value at some vertex. The partition function of the Ising model can be rewritten in terms of the contour configurations ω as

$$Z = \sum_{\omega} x^{\text{length of contours}}.$$

Each neighboring pair of opposite spins contributes an edge to the contours, and so a factor of $x = \exp(-2,\beta)$ to the partition function. Note that the critical value is $x_c = \exp(-2\beta_c) = \sqrt{2} - 1$.

We now want to define a preholomorphic observable. To this effect we need to distinguish at least one point (so that the domain has a non-trivial conformal modulus). One of the possible applications lies in relating interfaces to Schramm's SLE curves, in the simplest setup running between two boundary points. To obtain a discrete interface between two boundary points a and b, we introduce Dobrushin boundary conditions: + on one boundary arc and − on another, see Figure 2. Then those become unique points with an odd number of contour edges emanating from them.

Now to define our fermion, we allow the second endpoint of the interface to move inside the domain. Namely, take an edge center z inside Ω_ε, and define

$$F_\varepsilon(z) := \sum_{\omega(a\to z)} x^{\text{length of contours}} \mathcal{W}(\omega(a \to z)), \tag{7}$$

where the sum is taken over all contour configurations $\omega = \omega(a \to z)$ which have two exceptional points: a on the boundary and z inside. So the contour collection can be represented (perhaps non-uniquely) as a collection of loops plus an interface between a and z.

Furthermore, the sum is corrected by a *Fermionic* complex weight, depending on the configuration:

$$\mathcal{W}(\omega(a \to z)) := \exp(-i\ s \ \text{winding}(\gamma, a \to z)).$$

Here the winding is the total turn of the interface γ connecting a to z, counted in radians, and the *spin* s is equal to 1/2 (it should not be confused with the Ising spins ± 1). For some collections the interface can be chosen in more than one way, and then we trace it by taking a left turn whenever an ambiguity arises. Another choice might lead to a different value of winding, but if the loops and the interface have no "transversal" self-intersections, then the difference will be a multiple of 4π and so the complex weight $\mathcal{W}$ is well-defined. Equivalently we can write

$$\mathcal{W}(\omega(a \to z)) = \lambda^{\#\ \text{signed turns of}\ \gamma}, \quad \lambda := \exp\left(-is\frac{\pi}{2}\right),$$

see Figure 3 for weights corresponding to different windings.

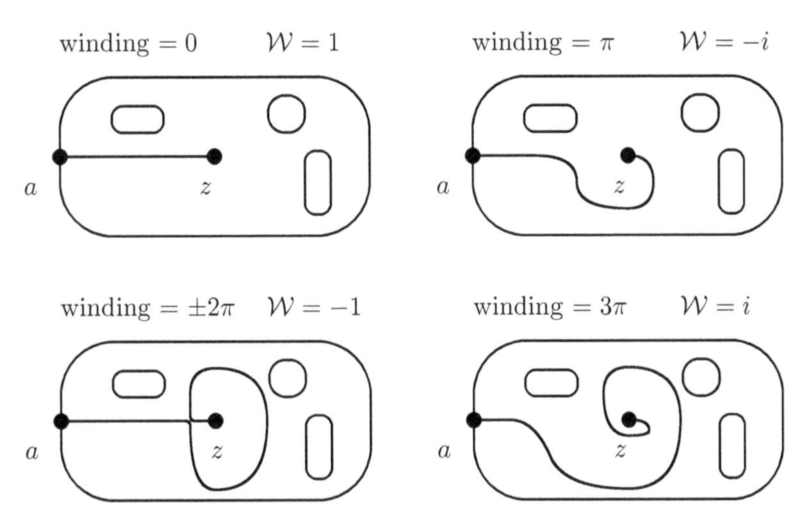

Figure 3. Examples of Fermionic weights one obtains depending on the winding of the interface. Note that in the bottom left example there are two ways to trace the interface from a to z without self-intersections, which give different windings $\pm 2\pi$, but the same complex weight $\mathcal{W} = -1$.

Remark 1. Removing complex weight $\mathcal{W}$ one retrieves the correlation of spins on the dual lattice at the dual temperature x^*, a corollary of the Kramers- Wannier duality.

Remark 2. While such contour collections cannot be directly represented by spin configurations, one can obtain them by creating a *disorder operator*, i.e. a *monodromy* at z: when one goes one time around z, spins change their signs.

Our first theorem is the following, which is proved for general isoradial graphs in [CS09], with a shorter proof for the square lattice given in [Smi10]:

Theorem 1 (Chelkak, Smirnov). *For Ising model at criticality, F is a preholomorphic solution of a Riemann boundary value problem. When mesh $\varepsilon \to 0$,*

$$F_\varepsilon(z)/\sqrt{\varepsilon} \rightrightarrows \sqrt{P'(z)} \text{ inside } \Omega,$$

where P is the complex Poisson kernel at a: a conformal map $\Omega \to \mathbb{C}_+$ such that $a \mapsto \infty$. Here both sides should be normalized in the same chart around b.

Remark 3. For non-critical values of x observable F becomes *massive preholomorphic,* satisfying the discrete analogue of the massive Cauchy-Riemann equations: $\bar{\partial} F = im(x - x_c)\bar{F}$, cf. [MS09].

Remark 4. Ising model can be represented as a dimer model on the Fisher graph. For example, on the square lattice, one first represents the spin configuration as above — by the collection of contours on the dual lattice, separating + and − spins. Then the dual lattice is modified with every vertex replaced by a "city" of

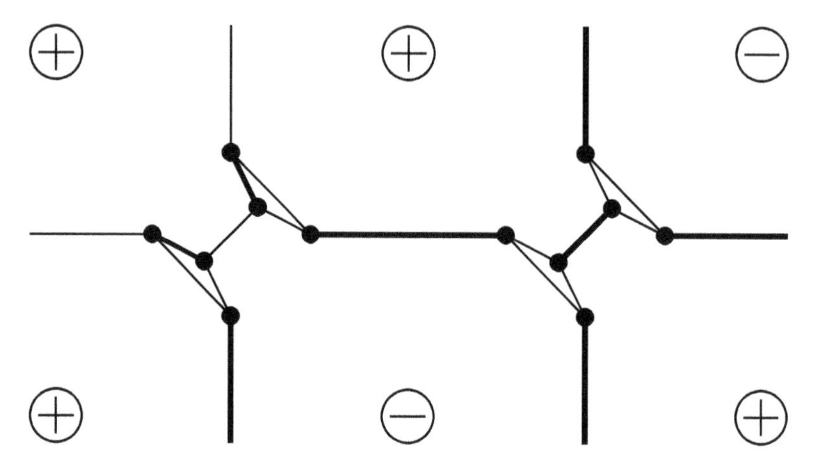

Figure 4. Fisher graph for a region of the square lattice, a spin configuration and a corresponding dimer configuration, with dimers represented by the bold edges.

six vertices, see Figure 4. It is easy to see that there is a natural bijection between contour configurations on the dual square lattice and dimer configuration on its Fisher graph.

Then, similarly to the work of Kenyon for the square lattice, the coupling function for the Fisher lattice will satisfy difference equations, which upon examination turn out to be another discretization of Cauchy-Riemann equations, with different projections of the preholomorphic function assigned to six vertices in a "city". One can then reinterpret the coupling function in terms of the Ising model, and this is the approach taken by Boutilier and de Tiliere [BdT08, BdT09].

This is also how the author found the observable discussed in this Section, observing jointly with Kenyon in 2002 that it has the potential to imply the convergence of the interfaces to the Schramm's SLE curve.

The key to establishing Theorem 1 is the observation that the function F is preholomorphic. Moreover, it turns out that F satisfies a stronger form of preholomorphicity, which implies the usual one, but is better adapted to fermions.

Consider the function F on the centers of edges. We say that F is *strongly* (or *spin*) preholomorphic if for every centers u and v of two neighboring edges emanating from a vertex w, we have

$$\mathrm{Proj}(F(v), 1/\sqrt{\alpha}) = \mathrm{Proj}(F(u), 1/\sqrt{\alpha}),$$

where α is the unit bisector of the angle uwv, and $\mathrm{Proj}(p, q)$ denotes the orthogonal projection of the vector p on the vector q. Equivalently we can write

$$F(v) + \bar{\alpha}\overline{F(v)} = F(u) + \bar{\alpha}\overline{F(u)}. \tag{8}$$

This definition implies the classical one for the square lattice, and it also easily adapts to the isoradial graphs. Note that for convenience we assume that the

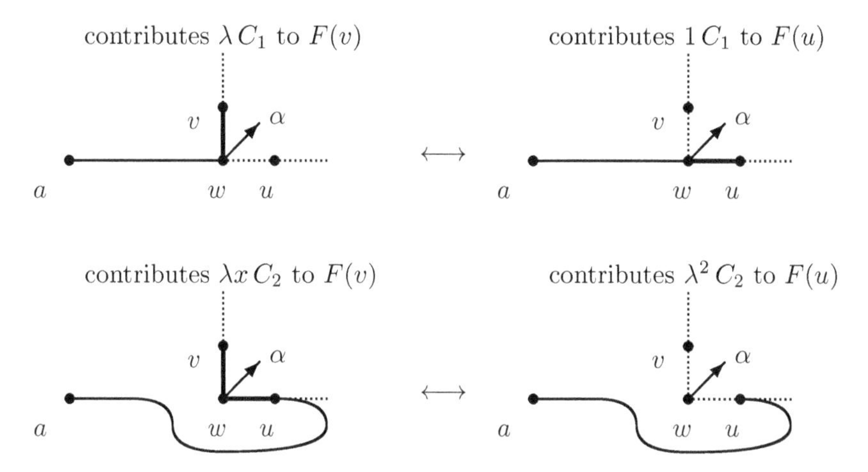

Figure 5. Involution on the Ising model configurations, which adds or erases half-edges vw and uw. There are more pairs, but their relative contributions are always easy to calculate and each pair taken together satisfies the discrete Cauchy-Riemann equations. Note that with the chosen orientation constants C_1 and C_2 above are real.

interface starts from a in the positive real direction as in Figure 2, which slightly changes weights compared to the convention in [CS09].

The strong preholomorphicity of the Ising model fermion is proved by constructing a bijection between configurations included into $F(v)$ and $F(u)$. Indeed, erasing or adding half-edges wu and wv gives a *bijection* $\omega \leftrightarrow \tilde{\omega}$ between configuration collections $\{\omega(u)\}$ and $\{\omega(v)\}$, as illustrated in Figure 5. To check (8), it is sufficient to check that the sum of contributions from ω and $\tilde{\omega}$ satisfies it. Several possible configurations can be found, but essentially all boil down to the two illustrated in Figure 5.

Plugging the contributions from Figure 5 into the equation (8), we are left to check the following two identities:

$$\lambda + \lambda\bar{\lambda} = 1 + \lambda\bar{1}, \quad \lambda x + \lambda\overline{\lambda x} = \lambda^2 + \lambda\bar{\lambda}^2. \tag{9}$$

The first identity always holds, while the second one is easy to verify when $x = x_c = \sqrt{2} - 1$ and $\lambda = \exp(-\pi i/4)$. Note that in our setup on the square lattice λ (or the spin s) is already fixed by the requirement that the complex weight is well-defined, and so the second equation in (9) uniquely fixes the allowed value of x. In the next Section we will discuss a more general setup, allowing for different values of the spin, corresponding to other lattice models.

To determine F using its preholomorphicity, we need to understand its behavior on the boundary. When $z \in \partial\Omega_\varepsilon$, the *winding* of the interface connecting a to z inside Ω_ε is uniquely determined, and coincides with the winding of the boundary itself. This amounts to knowing $\text{Arg}(F)$ on the boundary, which would be sufficient to determine F knowing the singularity at a or the normalization at b.

In the continuous setting the condition obtained is equivalent to the Riemann Boundary Value Problem (a homogeneous version of the Riemann-Hilbert-Privalov BVP)

$$\operatorname{Im}(F(z)\cdot(\text{tangent to } \partial\Omega)^{1/2}) = 0, \tag{10}$$

with the square root appearing because of the Fermionic weight. Note that the homogeneous BVP above has conformally covariant solutions (as $\sqrt{dz}$-forms), and so is well defined even in domains with fractal boundaries. The Riemann BVP (10) is clearly solved by the function $\sqrt{P'_\alpha(z)}$, where P is the Schwarz kernel at a (the complex version of the Poisson kernel), i.e. a conformal map

$$P : \Omega \to \mathbb{C}+, \quad a \mapsto \infty.$$

Showing that on the lattice F_ε satisfies a discretization of the Riemann BVP (10) and converges to its continuous counterpart is highly non-trivial and a priori not guaranteed — there exist "logical" discretizations of the Boundary Value Problems, whose solutions have degenerate or no scaling limits. We establish convergence in [CS09] by considering the primitive $\int_{z_0}^{z} F^2(u)du$, which satisfies the Dirichlet BVP even in the discrete setting. The big technical problem is that in the discrete case F^2 is no longer preholomorphic, so its primitive is a priori not preholomorphic or even well-defined. Fortunately, in our setting the imaginary part is still well-defined, so we can set

$$H_\varepsilon(z) := \frac{1}{2\varepsilon}\operatorname{Im}\int^{z} F(z)^2 dz.$$

While the function H is not exactly preharmonic, it is approximately so, vanishes exactly on the boundary, and is positive inside the domain. This allows to complete the (at times quite involved) proof. A number of non-trivial discrete estimates is called for, and the situation is especially difficult for general isoradial graphs. We provide the needed tools in a separate paper [CS08].

Though Theorem 1 establishes convergence of but one observable, the latter (when normalized at b) is well behaved with respect to the interface traced from a. So it can be used to establish the following, see [CDHKS14]:

Corollary 1. *As mesh of the lattice tends to zero, the critical Ising interface in the discretization of the domain Ω with Dobrushin boundary conditions converges to the Schramm's SLE(3) curve.*

Convergence is almost immediate in the topology of (probability measures on the space of) Loewner driving functions, but upgrading to convergence of curves requires extra estimates, cf. [KS09, DCHN09, CDHKS14]. Once interfaces are related to SLE curves, many more properties can be established, including values of dimensions and scaling exponents.

But even without appealing to SLE, one can use preholomorphic functions to a stronger effect. In a joint paper with Hongler [HS10] we study a similar observable, when both ends of the interface are allowed to be inside the domain. It turns out to be preholomorphic in both variables, except for the diagonal, and so its scaling limit can be identified with the Green's function solving the Riemann BVP. On the other hand, when two arguments are taken to be nearby, one retrieves the probability of an edge being present in the contour representation, or that the nearby spins are different. This allows to establish conformal invariance of the energy field in the scaling limit:

Theorem 2 (Hongler, Smirnov). *Let $a \in \Omega$ and $\langle x^\varepsilon, y^\varepsilon \rangle$ be the closest edge from $a \in \Omega_\varepsilon$. Then, as $\varepsilon \to 0$, we have*

$$\mathbb{E}_+[\sigma_x^\varepsilon \sigma_y^\varepsilon] = \frac{\sqrt{2}}{2} + \frac{l_\Omega(a)}{\pi} \cdot \varepsilon + o(\varepsilon),$$

$$\mathbb{E}_{\text{free}}[\sigma_x^\varepsilon \sigma_y^\varepsilon] = \frac{\sqrt{2}}{2} + \frac{l_\Omega(a)}{\pi} \cdot \varepsilon + o(\varepsilon),$$

where the subscripts + and free *denote the boundary conditions and l_Ω is the element of the hyperbolic metric on Ω.*

This confirms the Conformal Field Theory predictions and, as far as we know, for the first time provides the multiplicative constant in front of the hyperbolic metric.

These techniques were taken further by Hongler in [Hon10], where he showed that the (discrete) energy field in the critical Ising model on the square lattice has a conformally covariant scaling limit, which can be then identified with the corresponding Conformal Field Theory. This was accomplished by showing convergence of the discrete energy correlations in domains with a variety of boundary conditions to their continuous counterparts; the resulting limits are conformally covariant and are determined exactly. Similar result was obtained for the scaling limit of the spin field on the domain boundary.

5. The $O(N)$ Model

The Ising preholomorphic function was introduced in [Smi06] in the setting of general $O(N)$ models on the hexagonal lattice. It can be further generalized to a variety of lattice models, see the work of Cardy, Ikhlef, Rajabpour [RC07, IC09]. Unfortunately, the observable seems only partially preholomorphic (satisfying only some of the Cauchy-Riemann equations) except for the Ising case. One can make an analogy with divergence-free vector fields, which are not a priori curl-free.

The argument in the previous Section was adapted to the Ising case, and some properties remain hidden behind the notion of the strong holomorphicity. Below we present its version generalized to the $O(N)$ model, following our joint work [DCS10] with Duminil-Copin. While for $N \neq 1$ we only prove that our observable is divergence-free, it still turns out to be enough to deduce some global information, establishing the Nienhuis conjecture on the exact value of the connective constant for the hexagonal lattice:

Theorem 3 (Duminil-Copin, Smirnov). *On the hexagonal lattice the number $C(k)$ of distinct simple length k curves from the origin satisfies*

$$\lim_{k\to\infty}\frac{1}{k}\log C(k) = \log\sqrt{2+\sqrt{2}}. \tag{11}$$

Self-avoiding walks on a lattice (those without self-intersections) were proposed by chemist Flory [Flo53] as a model for polymer chains, and turned out to be an interesting and extensively studied object, see the monograph [MS93].

Using Coulomb gas formalism, physicist Nienhuis argued that the connective constant of the hexagonal lattice is equal to $\sqrt{2+\sqrt{2}}$, meaning that (11) holds. He even proposed better description of the asymptotic behavior:

$$C(k) \approx \left(\sqrt{2+\sqrt{2}}\right)^k k^{11/32},\ k\to\infty. \tag{12}$$

Note that while the exponential term with the connectivity constant is lattice-dependent, the power law correction is supposed to be universal.

Our proof is partially motivated by Nienhuis' arguments, and also starts with considering the self-avoiding walk as a special case of $O(N)$ model at $N=0$. While a "half-preholomorphic" observable we construct does not seem sufficient to imply conformal invariance in the scaling limit, it can be used to establish the critical temperature, which gives the connective constant.

The general $O(N)$ model is defined for positive integer values of N, and is a generalization of the Ising model (to which it specializes for $N=1$), with ± 1 spins replaced by points on a sphere in the N-dimensional space. We work with the graphical representation, which is obtained using the *high-temperature expansion,* and makes the model well defined for all non-negative values of N.

We concentrate on the hexagonal lattice in part because it is trivalent and so at most one contour can pass through a vertex, creating no ambiguities. This simplifies the reasoning, though general graphs can also be addressed by introducing additional weights for multiple visits of vertices. We consider configurations ω of disjoint simple loops on the mesh ε hexagonal lattice inside domain Ω_ε, and two parameters: loop-weight $N\ge 0$ and (temperature-like) edge-weight $x\ge 0$. Partition function is then given by

$$Z = \sum_\omega N^{\#\text{ loops}}\, x^{\text{length of contours}}.$$

A typical configuration is pictured in Figure 6, where we introduced Dobrushin boundary conditions: besides loops, there is an interface γ joining two fixed boundary points a and b. It was conjectured by Kager and Nienhuis [KN04] that in the interval $N\in[0,2]$ the model has conformally invariant scaling limits for $x = x_c(N) := 1/\sqrt{2+\sqrt{2-N}}$ and $x\in(x_c(N),+\infty)$, The two different limits correspond to dilute/dense regimes, with the interface γ conjecturally converging

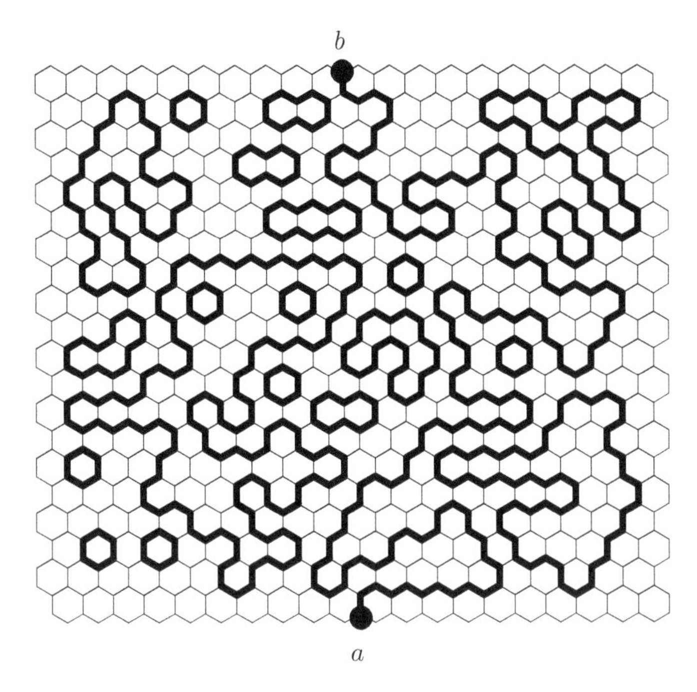

Figure 6. The high-temperature expansion of the $O(N)$ model leads to a gas of disjoint simple loops. Probability of a configuration is proportional to $N^{\#\text{ loops}} x^{\text{length}}$. We study it with Dobrushin boundary conditions: besides loops, there is an interface between two boundary points a and b.

to the Schramm's SLE curves for an appropriate value of $\kappa \in [8/3, 4]$ and $\kappa \in [4, 8]$ correspondingly. The scaling limit for low temperatures $x \in (0, x_c)$ is not conformally invariant.

Note that for $N = 1$ we do not count the loops, thus obtaining the low-temperature expansion of the Ising model on the dual triangular lattice. In particular, the critical Ising corresponds to $x = 1/\sqrt{3}$ by the work [Wan50] of Wannier, in agreement with Nienhuis predictions. And for $x = 1$ one obtains the critical site percolation on triangular lattice (or equivalently the Ising model at infinite temperature). The latter is conformally invariant in the scaling limit by [Smi01b, Smi01a].

Note also that the Dobrushin boundary conditions make the model well-defined for $N = 0$: then we have only one interface, and no loops. In the dilute regime this model is expected to be in the universality class of the self-avoiding walk.

Analogously to the Ising case, we define an observable (which is now a *parafermion* of fractional spin) by moving one of the ends of the interface inside the domain. Namely, for an edge center z we set

$$F_\varepsilon(z) := \sum_{\omega(a \to z)} x^{\text{length of contours}} \mathcal{W}(\omega(a \to z)), \tag{13}$$

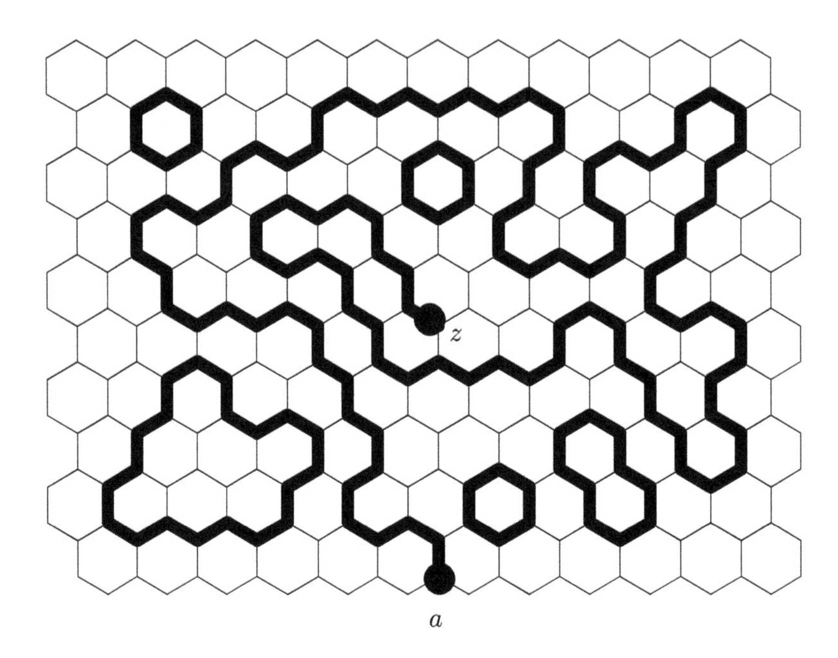

Figure 7. To obtain the parafermionic observable in the $O(N)$ model we consider configurations with an interface joining a boundary point z to an interior point z and weight them by a complex weight depending on the winding of the interface.

where the sum is taken over all configurations $\omega = \omega(a \to z)$ which have disjoint simple contours: a number of loops and an interface γ joining two exceptional points, a on the boundary and z inside. As before, the sum is corrected by a complex weight with the *spin* $s \in \mathbb{R}$:

$$\mathcal{W}(\omega(a \to z)) := \exp(-i\, s \text{ winding}(\gamma, a \to z)),$$

equivalently we can write

$$\mathcal{W}(\omega(a \to z)) := \lambda^{\#\text{ signed turns of }\gamma}, \quad \lambda := \exp\left(-is\frac{\pi}{3}\right).$$

Note that on hexagonal lattice one turn corresponds to $\pi/3$, hence the difference in the definition of λ.

Our key observation is the following

Lemma 4. *For $N \in [0,2]$, set $2\cos(\theta) = N$ with parameter $\theta \in [0, \pi/2]$. Then for*

$$s = \frac{\pi - 3\theta}{4\pi}, \quad x^{-1} = 2\cos\left(\frac{\pi+\theta}{4}\right) = \sqrt{2 - \sqrt{2-N}}, \quad \text{or} \tag{14}$$

$$s = \frac{\pi + 3\theta}{4\pi}, \quad x^{-1} = 2\cos\left(\frac{\pi-\theta}{4}\right) = \sqrt{2 + \sqrt{2-N}}, \tag{15}$$

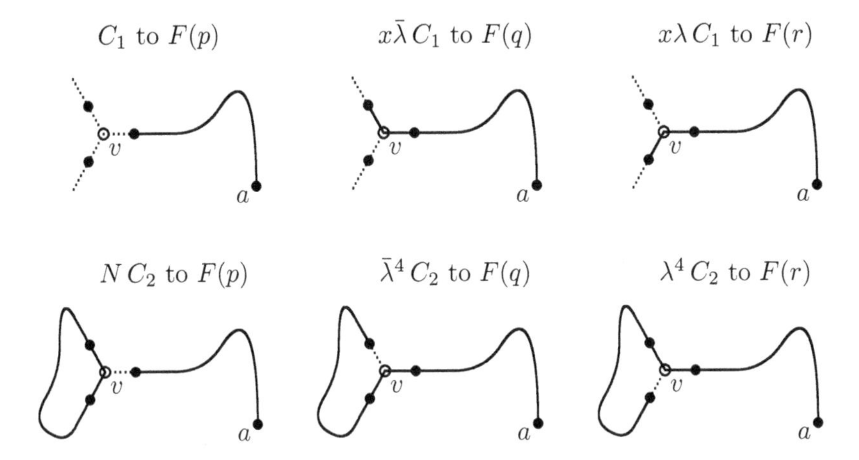

Figure 8. Configurations with the interface ending at one of the three neighbors of v are grouped into triplets by adding or removing half-edges around v. Two essential examples of triplets are pictured above, along with their relative contributions to the identity (13).

the observable F satisfies the following relation for every vertex v inside Ω_ε :

$$(p-v)F(p)+(q-v)F(q)+(r-v)F(r)=0, \tag{16}$$

where p, q, r are the mid-edges of the three edges adjacent to v.

Above solution (14) corresponds to the dense, and (15) — to the dilute regime. Note that identity (16) is a form of the first Kirchhoff's law, but apart from the Ising case $N=1$ we cannot verify the second one.

To prove Lemma 4, we note that configurations with an interface arriving at p, q or r can be grouped in triplets, so that three configurations differ only in immediate vicinity of v, see Figure 8. It is enough then to check that contributions of three configurations to (16) sum up to zero. But the relative weights of configurations in a triplet are easy to write down as shown in Figure 8, and the coefficients in the identity (16) are proportional to the three cube roots of unity: 1, $\tau := \exp(i2\pi/3)$, $\bar{\tau}$ (if the neighbors of v are taken in the counterclockwise order). Therefore we have to check just two identities:

$$N+\tau\bar{\lambda}^4+\bar{\tau}\lambda^4=0,$$
$$1+\tau x\bar{\lambda}+\bar{\tau}x\lambda=0.$$

Recalling that $\lambda=\exp(-is\pi/3)$, the equations above can be recast as

$$-\frac{2\pi}{3}-4s\frac{\pi}{3}=\pm(\pi-\theta)+2\pi k, \quad k\in\mathbb{Z},$$
$$x=-1\Big/\left(2\cos\left(\frac{(2+s)\pi}{3}\right)\right).$$

The first equation implies that

$$s = \pm\left(-\frac{3}{4} + \frac{3\theta}{4\pi}\right) - \frac{1}{2} - \frac{3}{2}k, \quad k \in \mathbb{Z}, \tag{17}$$

and the second equation then determines the allowed value of x uniquely. Most of the solutions of (17) lead to observables symmetric to the two main ones, which are provided by solutions to the equations (14) and (15).

When we set $N = 0$, there are no loops, and configurations contain just an interface from a to z, weighted by x^{length}. This corresponds to taking $\theta = \pi/2$ and one of the solutions is given by $s = 5/8$ and $x_c = 1/\sqrt{2+\sqrt{2}}$, as predicted by Nienhuis. To prove his prediction, we observe that summing the identity (16) over all interior vertices implies that

$$\sum_{z \in \partial\Omega_\varepsilon} F(z)\eta(z) = 0,$$

where the sum taken over the centers z of oriented edges $\eta(z)$ emanating from the discrete domain Ω_ε into its exterior. Since $F(a) = 1$ by definition, we conclude that F for other boundary points sums up to 1. As in the Ising model, the winding on the boundary is uniquely determined, and (for this particular critical value of x), one observes that considering the real part of F we can get rid of the complex weights, replacing them by explicit positive constants (depending on the slope of the boundary). Thus we obtain an equation

$$\sum_{z \in \partial\Omega_\varepsilon/\{a\}} \sum_{\omega(a\to z)} x_c^{\text{length of contours}} \asymp 1,$$

regardless of the size of the domain Ω_ε. A simple counting argument then shows that the series

$$\sum_k C(k)x^k = \sum_{\text{simple walks from } a \text{ inside } \mathbb{C}} x^{\mathrm{length}},$$

converges when $x < x_c$ and diverges when $x > x_c$, clearly implying the conjecture.

Note that establishing the holomorphicity of our observable in the scaling limit would allow to relate self-avoiding walk to the Schramm's SLE with $\kappa = 8/3$ and together with the work [LSW04] of Lawler, Schramm and Werner to establish the more precise form (12) of the Nienhuis prediction.

6. What's Next

Below we present a list of open questions. As before, we do not aim for completeness, rather we highlight a few directions we find particularly intriguing.

Question 1. As was discussed, discrete complex analysis is well developed for isoradial graphs (or rhombic lattices), see [Duf68, Mer01, Ken02, CS08]. Is there a more general discrete setup where one can get similar estimates, in particular convergence of preholomorphic functions to the holomorphic ones in the scaling limit? Since not every topological quadrangulation admits a rhombic embedding [KS05], can we always find another embedding with a sufficiently nice version of discrete complex analysis? Same question can be posed for triangulations, with variations of the first definition by Isaacs (5), like the ones in the work of Dynnikov and Novikov [DN03] being promising candidates.

Question 2. Variants of the Ising observable were used by Hongler and Kytölä to connect interfaces in domains with more general boundary conditions to more advanced variants of SLE curves, see [HK09]. Can one use some version of this observable to describe the spin Ising loop soup by a collection of branching interfaces, which converge to a branching SLE tree in the scaling limit? Similar argument is possible for the random cluster representation of the Ising model, see [KS10]. Can one construct the energy field more explicitly than in [Hon10], e.g. in the distributional sense? Can one construct other Ising fields?

Question 3. So far "half-preholomorphic" parafermions similar to ones discussed in this paper have been found in a number of models, see [Smi06, RC06, RC07, IC09], but they seem fully preholomorphic only in the Ising case. Can we find the other half of the Cauchy-Riemann equations, perhaps for some modified definition? Note that it seems unlikely that one can establish conformal invariance of the scaling limit operating with only half of the Cauchy-Riemann equations, since there is no conformal structure present.

Question 4. In the case of the self-avoiding walk, an observable satisfying only a half of the Cauchy-Riemann equations turned out to be enough to derive the value of the connectivity constant [DCS10]. Since similar observables are available for all other $O(N)$ models, can we use them to establish the critical temperature values predicted by Nienhuis? Our proof cannot be directly transfered, since some counting estimates use the absence of loops. Similar question can be asked for other models.

Question 5. If we cannot establish the preholomorphicity of our observables exactly, can we try to establish it approximately? With appropriate estimates that would allow to obtain holomorphic functions in the scaling limit and hence prove conformal invariance of the models concerned. Note that such more general approach worked for the critical site percolation on the triangular lattice [Smi01b, Smi01a], though approximate preholomorphicity was a consequence of exact identities for quantities similar to discrete derivatives.

Question 6. Can we find other preholomorphic observables besides ones mentioned here and in [Smi06]? It is also peculiar that all the models where preholomorphic observables were found so far (the dimer model, the uniform spanning tree, the Ising model, percolation, etc.) can be represented as dimer models. Are there any models in other universality classes, admitting a dimer representation? Can then

Kenyon's techniques [Ken04, Ken09] be used to find preholomorphic observables by considering the Kasteleyn's matrix and the coupling function?

Question 7. Throughout this paper we were concerned with linear discretizations of the Cauchy-Riemann equations. Those seem more natural in the probabilistic context, in particular they might be easier to relate to the SLE martingales, cf. [Smi06]. However, there are also well-known nonlinear versions of the Cauchy-Riemann equations. For example, the following version of the Hirota equation for a complex-valued function F arises in the context of the circle packings, see e.g. [BMS05]:

$$\frac{(F(z+i\varepsilon)-F(z-\varepsilon))(F(z-i\varepsilon)-F(z+\varepsilon))}{(F(z+i\varepsilon)-F(z+\varepsilon))(F(z-i\varepsilon)-F(z-\varepsilon))}=-1. \tag{18}$$

Can we observe this or a similar equation in the probabilistic context and use it to establish conformal invariance of some model? Note that plugging into the equation (18) a smooth function, we conclude that to satisfy it approximately it must obey the identity

$$(\partial_x F(z))^2+(\partial_y F(z))^2=0.$$

So in the scaling limit (18) can be factored into the Cauchy-Riemann equations and their complex conjugate, thus being in some sense linear. It does not seem possible to obtain "essential" non-linearity using just four points, but using five points one can create one, as in the next question.

Question 8. A number of non-linear identities was discovered for the correlation functions in the Ising model, starting with the work of Groeneveld, Boel and Kasteleyn [GBK78, BK78]. We do not want to analyze the extensive literature to-date, but rather pose a question: can any of these relations be used to define discrete complex structures and pass to the scaling limit? In two of the early papers by McCoy, Wu and Perk [MW80, Per80], a quadratic difference relation was observed in the full plane Ising model first on the square lattice, and then on a general graph. To better adapt to our setup, we rephrase this relation for the correlation $C(z)$ of two spins (one at the origin and another at z) in the Ising model at criticality on the mesh ε square lattice. In the full plane, one has

$$C(z+i\varepsilon)C(z-i\varepsilon)+C(z+\varepsilon)C(z-\varepsilon)=2C(z)^2. \tag{19}$$

Note that C is a *real-valued* function, and the equation (19) is a discrete form of the identity

$$C(z)\Delta C(z)+|\nabla C(z)|^2=0.$$

The latter is conformally invariant, and is solved by moduli of analytic functions. Can one write an analogous to (19) identity in domains with boundary, perhaps approximately? Can one deduce conformally invariant scaling limit of the spin correlations in that way?

Question 9. Recently there was a surge of interest in random planar graphs and their scaling limits, see e.g. [DS09, LGP08]. Can one find observables on random planar graphs (weighted by the partition function of some lattice model) which after an appropriate embedding (e.g. via a circle packing or a piecewise-linear Riemann surface) are preholomorphic? This would help to show that planar maps converge to the Liouville Quantum Gravity in the scaling limit.

Question 10. Approach to the two-dimensional integrable models described here is in several aspects similar to the older approaches based on the Yang- Baxter relations [Bax89]. Some similarities are discussed in Cardy's paper [Car09]. Can one find a direct link between the two approaches? It would also be interesting to find a link to the three-dimensional consistency relations as discussed in [BMS09].

Question 11. Recently Kenyon investigated the Laplacian on the vector bundles over graphs in relation to the spanning trees [Ken10]. Similar setup seems natural for the Ising observable we discuss. Can one obtain more information about the Ising and other models by studying difference operators on vector bundles over the corresponding graphs?

Question 12. Can anything similar be done for the three-dimensional models? While preholomorphic functions do not exist here, preharmonic vector fields are well-defined and appear naturally for the Uniform Spanning Tree and the Loop Erased Random Walk. To what extent can they be used? Can one find any other difference equations in three-dimensional lattice models?

References

[Bax78] R. J. Baxter. Solvable eight-vertex model on an arbitrary planar lattice. *Philos. Trans. Roy. Soc. London Ser. A*, 289(1359):315–346, 1978.

[Bax89] R. J. Baxter. *Exactly solved models in statistical mechanics.* Academic Press Inc. [Harcourt Brace Jovanovich Publishers], London, 1989. Reprint of the 1982 original.

[BdT08] Cédric Boutillier and Béatrice de Tiliére. The critical Z-invariant Ising model via dimers: the periodic case. Probab. Theory Related Fields, 147: 379–413, 2010.

[BdT09] Cédric Boutillier and Béatrice de Tiliére. The critical Z-invariant Ising model via dimers: locality property. Comm. Math. Physics, 301(2) 473–516, 2011.

[BK78] R. J. Boel and P. W. Kasteleyn. Correlation-function identities and inequalities for Ising models with pair interactions. *Comm. Math. Phys.*, 61(3):191–208, 1978.

[BMS05] Alexander I. Bobenko, Christian Mercat, and Yuri B. Suris. Linear and nonlinear theories of discrete analytic functions. Integrable structure and isomonodromic Green's function. *J. Reine Angew. Math.*, 583:117–161, 2005.

[BMS09] Vladimir V. Bazhanov, Vladimir V. Mangazeev, and Sergey M. Sergeev. Quantum geometry of 3-dimensional lattices and tetrahedron equation. In XVIth *International Congress on Mathematical Physics*, 23–44. World Sci. Publ., Hackensack, NJ.

[Bou26] George Bouligand. Sur le probleme de Dirichlet. *Ann. Soc. Pol. Math.*, 4:59–112, 1926.

[BS08] Alexander I. Bobenko and Yuri B. Suris. *Discrete differential geometry*, volume 98 of *Graduate Studies in Mathematics*. American Mathematical Society, Providence, RI, 2008.

[BSST40] R. L. Brooks, C. A. B. Smith, A. H. Stone, and W. T. Tutte. The dissection of rectangles into squares. *Duke Math. J.*, 7:312–340, 1940.

[Car09] John Cardy. Discrete holomorphicity at two-dimensional critical points. *J. Stat. Phys.*, 137(5–6):814–824, 2009.

[CFL28] R. Courant, K. Friedrichs, and H. Lewy. Über die partiellen Differenzengleichungen der mathematischen Physik. *Math. Ann.*, 100:32–74, 1928.

[CS08] Dmitry Chelkak and Stanislav Smirnov. Discrete complex analysis on isoradial graphs. Adv. in Math., 228: 1590–1630, 2011.

[CS09] Dmitry Chelkak and Stanislav Smirnov. Universality in the 2D Ising model and conformal invariance of fermionic observables. Invent. Math., 189: 515–580, 2012.

[CDHKS14] Dmitry Chelkak, Hugo Duminil-Copin, Clément Hongler, Antti Kemppainen and Stanislav Smirnov. Convergence of Ising interfaces to Schramm's SLE curves. C. R. Acad. Sci. Paris Sér. I Math., 352: 157–161, 2014.

[DCHN09] Hugo Duminil-Copin, Clément Hongler, and Pierre Nolin. Connection probabilities and RSW-type bounds for the FK Ising model. Communications on Pure and Applied Mathematics, 64(9): 1165–1198, 2011.

[DCS10] Hugo Duminil-Copin and Stanislav Smirnov. The connective constant of the honeycomb lattice equals $\sqrt{2+\sqrt{2}}$. Ann. Math., 175: 1653–1666, 2012.

[DM71] Charles R. Deeter and C. Wayne Mastin. The discrete analog of a minimum problem in conformal mapping. *Indiana Univ. Math. J.*, 20:355–367, 1970/1971.

[DN03] I. A. Dynnikov and S. P. Novikov. Geometry of the triangle equation on two-manifolds. *Moscow Math.* J., 3:419–438, 2003.

[DS09] Bertrand Duplantier and Scott Sheffield. Duality and the Knizhnik-Polyakov-Zamolodchikov relation in Liouville quantum gravity. *Phys. Rev. Lett.*, 102(15):150603, 4, 2009.

[Duf56] R. J. Duffin. Basic properties of discrete analytic functions. *Duke Math.* J., 23:335–363, 1956.

[Duf68] R. J. Duffin. Potential theory on a rhombic lattice. *J. Combinatorial Theory*, 5:258–272, 1968.

[Fer44] Jacqueline Ferrand. Fonctions préharmoniques et fonctions préholomorphes. *Bull. Sci. Math. (2)*, 68:152–180, 1944.

[Flo53] P. Flory. *Principles of Polymer Chemistry*. Cornell University Press, 1953.

[GBK78] J. Groeneveld, R. J. Boel, and P. W. Kasteleyn. Correlation-function identities for general planar Ising systems. *Physica A: Statistical and Theoretical Physics*, 93(1–2):138–154, 1978.

[HK09] Clément Hongler and Kalle Kytölä. Dipolar SLE in Ising model with plus-minus-free boundary conditions. Preprint, 2009.

[Hon10] Clément Hongler. Conformal invariance of the Ising model correlations. Ph.D. thesis, Universté de Genéve, 2010.

[HS93] Zheng-Xu He and Oded Schramm. Fixed points, Koebe uniformization and circle packings. *Ann. of Math. (2)*, 137(2):369–406, 1993.

[HS10] Clément Hongler and Stanislav Smirnov. Energy density in the 2D Ising model. *Probab. Theory Relat. Fields*, 151:735–756, 2011.

[IC09] Yacine Ikhlef and John Cardy. Discretely holomorphic parafermions and integrable loop models. *J. Phys.* A, 42(10):102001, 11, 2009.

[Isa41] Rufus Philip Isaacs. A finite difference function theory. *Univ. Nac. Tucuman. Revista A.*, 2:177–201, 1941.

[Isa52] Rufus Isaacs. Monodiffric functions. Construction and applications of conformal maps. In *Proceedings of a symposium,* National Bureau of Standards, Appl. Math. Ser., No. 18, pages 257–266, Washington, D. C., 1952. U. S. Government Printing Office.

[KC71] Leo P. Kadanoff and Horacio Ceva. Determination of an operator algebra for the two-dimensional Ising model. *Phys. Rev. B (3),* 3:3918–3939, 1971.

[Ken00] Richard Kenyon. Conformal invariance of domino tiling. *Ann. Probab.,* 28(2):759–795, 2000.

[Ken01] Richard Kenyon. Dominos and the Gaussian free field. *Ann. Probab.,* 29(3):1128–1137, 2001.

[Ken02] Richard Kenyon. The Laplacian and Dirac operators on critical planar graphs. *Invent. Math.,* 150(2):409–439, 2002.

[Ken04] Richard Kenyon. An introduction to the dimer model. In *School and Conference on Probability Theory,* ICTP Lect. Notes, XVII, pages 267–304 (electronic). Abdus Salam Int. Cent. Theoret. Phys., Trieste, 2004.

[Ken09] Richard Kenyon. Lectures on dimers. In *Statistical mechanics,* volume 16 of *IAS/Park City Math. Ser.,* pages 191–230. Amer. Math. Soc., Providence, RI, 2009.

[Ken10] Richard Kenyon. Spanning forests and the vector bundle Laplacian. *Ann. Probab.*, 39(5):1983–2017, 2011.

[Kir47] G. Kirchhoff. Ueber die Auflösung der Gleichungen, auf welche man bei der Untersuchung der linearen Vertheilung galvanischer Ströme geführt wird. *Annalen der Physik und Chemie,* 148(12):497–508, 1847.

[Kis05] Christer O. Kiselman. Functions on discrete sets holomorphic in the sense of Isaacs, or monodiffric functions of the first kind. *Sci. China Ser. A,* 48(suppl.):86–96, 2005.

[KN04] Wouter Kager and Bernard Nienhuis. A guide to stochastic Löwner evolution and its applications. *J. Statist. Phys.,* 115(5–6):1149–1229, 2004.

[KS05] Richard Kenyon and Jean-Marc Schlenker. Rhombic embeddings of planar quad-graphs. *Trans. Amer. Math. Soc.,* 357(9):3443–3458 (electronic), 2005.

[KS09] Antti Kemppainen and Stanislav Smirnov. Random curves, scaling limits and Loewner evolutions. Preprint, University of Helsinki, 2009, arXiv: 1212.6215.

[KS10] Antti Kemppainen and Stanislav Smirnov. Conformal invariance in random cluster models. II. Full scaling limit. In preparation, 2010.

[KW41] H.A. Kramers and G.H. Wannier. Statistics of the two-dimensional ferromagnet. Pt. 1. *Phys. Rev., II. Ser.*, 60:252–262, 1941.

[LF55] Jacqueline Lelong-Ferrand. *Représentation conforme et transformations á intégrale de Dirichlet bornée.* Gauthier-Villars, Paris, 1955.

[LGP08] Jean-Francois Le Gall and Frédéric Paulin. Scaling limits of bipartite planar maps are homeomorphic to the 2-sphere. *Geom. Funct. Anal.*, 18(3):893–918, 2008.

[Lov04] László Lovász. Discrete analytic functions: an exposition. In *Surveys in differential geometry. Vol. IX,* Surv. Differ. Geom., IX, pages 241–273. Int. Press, Somerville, MA, 2004.

[LSW04] Gregory F. Lawler, Oded Schramm, and Wendelin Werner. On the scaling limit of planar self-avoiding walk. In *Fractal geometry and applications: a jubilee of Benoît Mandelbrot, Part 2,* volume 72 of *Proc. Sympos. Pure Math.*, pages 339–364. Amer. Math. Soc., Providence, RI, 2004.

[Lus26] L. Lusternik. Uber einege Anwendungen der direkten Methoden in Variationsrechnung. *Recueil de la Société Mathématique de Moscou,* pages 173201, 1926.

[Mer01] Christian Mercat. Discrete Riemann surfaces and the Ising model. *Comm.. Math. Phys.*, 218(1):177–216, 2001.

[Mer07] Christian Mercat. Discrete Riemann surfaces. In *Handbook of Teichmüller theory. Vol. I,* volume 11 of *IRMA Lect. Math. Theor. Phys.*, pages 541–575. Eur. Math. Soc., Zürich, 2007.

[MS93] Neal Madras and Gordon Slade. *The self-avoiding walk.* Probability and its Applications. Birkhäuser Boston Inc., Boston, MA, 1993.

[MS09] Nikolai Makarov and Stanislav Smirnov. Off-critical lattice models and massive SLEs. In XVIth *International Congress on Mathematical Physics,* pages 362–371. World Sci. Publ., Hackensack, NJ, 2009.

[MW80] Barry M. McCoy and Tai Tsun Wu. Non-linear partial difference equations for the two-dimensional Ising model. *Physics Review Letters,* 45:675–678, 1980.

[Per80] J. H. H. Perk. Quadratic identities for Ising model correlations. *Physics Letters*, 79A:3–5, 1980.

[PW23] H. B. Phillips and N. Wiener. Nets and the Dirichlet problem. *Mass. J. of Math.*, 2:105–124, 1923.

[RC06] V. Riva and J. Cardy. Holomorphic parafermions in the Potts model and stochastic Loewner evolution. *J. Stat. Mech. Theory Exp.*, (12):P12001, 19 pp. (electronic), 2006.

[RC07] M. A. Rajabpour and J. Cardy. Discretely holomorphic parafermions in lattice Z_N models. *J. Phys.* A, 40(49):14703–14713, 2007.

[RS87] Burt Rodin and Dennis Sullivan. The convergence of circle packings to the Riemann mapping. *J. Differential Geom.*, 26(2):349–360, 1987.

[Smi01a] Stanislav Smirnov. Critical percolation in the plane. Preprint, arXiv: 0909.4499, 2001.

[Smi01b] Stanislav Smirnov. Critical percolation in the plane: Conformal invariance, Cardy's formula, scaling limits. *C. R. Math. Acad. Sci. Paris,* 333(3): 239–244, 2001.

[Smi06] Stanislav Smirnov. Towards conformal invariance of 2D lattice models. Sanz-Solé, Marta (ed.) *et al.*, Proceedings of the international congress of mathematicians (ICM), Madrid, Spain, August 22–30, 2006. Volume II: Invited lectures, 1421–1451. Zurich: European Mathematical Society (EMS), 2006.

[Smi10] Stanislav Smirnov. Conformal invariance in random cluster models. I. Holomorphic spin structures in the Ising model. *Ann. of Math. (2),* 172:101–133, 2010.

[Ste05] Kenneth Stephenson. *Introduction to circle packing. The theory of discrete analytic functions.* Cambridge University Press, Cambridge, 2005.

[Thu86] William P. Thurston. Zippers and univalent functions. In *The Bieberbach conjecture (West Lafayette, Ind., 1985),* volume 21 of *Math. Surveys Monogr.,* pages 185–197. Amer. Math. Soc., Providence, RI, 1986.

[Wan50] G. H. Wannier. Antiferromagnetism. The triangular Ising net. *Phys. Rev.,* 79(2):357–364, Jul 1950.

Proceedings of the International Congress of Mathematicians
Hyderabad, India, 2010

THE WORK OF CÉDRIC VILLANI

by

HORNG-TZER YAU*

1. Introduction

The starting point of Cédric Villani's work goes back to the introduction of entropy in the nineteenth century by L. Carnot and R. Clausius. At the time, entropy was a vague concept and its rigorous definition had to wait until the fundamental work of L. Boltzmann who introduced nonequilibrium statistical physics and the famous H functional. Boltzmann's work, though a fundamental breakthrough, did not resolve the question concerning the nature of entropy and time arrow; the debate on this central question continued for a century until today. J. von Neumann, in recommending C. Shannon to use entropy for his uncertainty function, quipped that entropy is a good name because "nobody knows what entropy really is, so in a debate you will always have the advantage".

The first result of Villani I will report on concerns the fundamental connection between entropy and its dissipation. In this work, we will see that rigorous mathematical analysis is not just a display of powerful analytic skill, but also leads to deep insights into nature. Based on this work, Villani has developed a general theory, hypercoercivity, which applies to broad systems of equations. In a separate direction, entropy was used by Villani as a fundamental tool in optimal transport and the study of curvature in metric spaces. Finally, I will describe Villani's work on Landau damping, which predicts a very surprising decay (and thus the word damping) of the electric field in a plasma *without* particle collisions, and therefore without entropy increase. This is in sharp contrast with Boltzmann's picture that the time irreversibility comes from collision processes.

*Partially supported by NSF grants DMS-0757425, 0804279.
Department of Mathematics, Harvard University, Cambridge MA 02138.
E-mail: htyau@math.harvard.edu

2. Boltzmann Equation

The Boltzmann equation was derived by L. Boltzmann in 1873 based on his physical intuition of collision processes. The most striking feature of the Boltzmann equation, the time irreversibility, contradicts the reversibility of the Newton equations. This fact is most concisely expressed via the Boltzmann H-theorem stating that the entropy

$$S = -\iint f \log f \, dv \, dx$$

is always nondecreasing. Furthermore, the entropy production vanishes if and only if the state is spatially homogeneous and Maxwellian in the velocity variable. The Boltzmann H-theorem is semi-rigorous in the sense that if the solution to the Boltzmann equation is sufficiently smooth then the original proof of Boltzmann is rigorous. The mathematical study of the Boltzmann equation started perhaps from T. Carleman and H. Grad in the middle of the last century. Despite decades of intensive research, most fundamental questions concerning the Boltzmann equation remain open, e.g., 1. Are solutions of the Boltzmann equation smooth if the initial data are sufficiently smooth? 2. The Boltzmann H-theorem states that the entropy increases, but what is the rate? Or, more generally, how fast do solutions to the Boltzmann equation approach the equilibrium (Maxwellian) states?

The first question, the regularity of the Boltzmann equation, is only understood for small perturbation of equilibrium measures. There is a general framework of renormalized solutions developed by R. DiPerna and P.-L. Lions [16], but precise estimates on the solutions remain elusive. The second question, the decay to equilibrium for the Boltzmann equation, is where Villani made his fundamental contribution. Before we describe Villani's work in some detail, several important recent results concerning the Boltzmann equation should be mentioned here. This incomplete list includes the well-posedness and the approach to equilibrium for small perturbation data by Y. Guo [24], and the recent extension of this approach to long-range interactions and soft potentials by P. Gressman and R. Strain [23], the weak shock solutions by S.-H. Yu [52], the derivation of incompressible Navier-Stokes equations from the Boltzmann equation by C. Bardos, F. Golse, D. Levermore and L. Saint-Raymond [5, 17].

The Boltzmann equation is given by

$$\partial_t f + v \nabla_x f = Q(f, f)$$

where $f(t, x, v)$ is the probability density in the phase space at the time t. The nonlinear term Q is the collision operator

$$Q(f, f) = \iint [f(v')f(v'_*) - f(v)f(v_*)] B(v - v_*, \sigma) dv_* d\sigma,$$

where v, v_* are the incoming velocities, $v', v,$ the outgoing velocities, σ the collision angle and B is the scattering kernel depending on the details of the microscopic interactions. The Boltzmann H-functional (negative of the entropy) is defined by

$$H(f) = \int_{\mathbb{R}^3} \int_{\mathbb{R}^3} f(x,v) \log f(x,v) dx\, dv$$

and the Boltzmann H-theorem states that

$$\partial_t H(f(t)) = -D(f(t)) \leq 0, \quad D(f) = \frac{1}{4} \int\int\int [f(v)f(v_*) \\ -f(v')f(v'_*)] \log \frac{f(v)f(v_*)}{f(v')f(v'_*)} d\sigma\, dv\, dv_*. \tag{2.1}$$

The dissipation D vanishes if and only if the state is Maxwellian:

$$D(f) = 0 \quad \text{if and only if} \quad f = M_{\rho,\mu,T} := \rho(x) \frac{e^{-\frac{|v-\mu(x)|^2}{2T(x)}}}{(2\pi T(x))^{3/2}} \tag{2.2}$$

where M is any local equilibrium state of the Boltzmann equation with density ρ, velocity u and temperature T which can depend on the space variable. If ρ, μ, T are independent of the space variables, M is called a global equilibrium.

To understand the approach to equilibrium via the Boltzmann H-theorem, we first consider the spatially homogeneous case, i.e., the function $f(x,v)$ depends only on v. C. Cercignani in 1983 [11] conjectured that, under suitable assumptions on the collision kernel B, there is a constant K such that

$$D(f) \geq K(f)H(f|M), \quad H(f|M) := \iint dx\, dv\, f \log \frac{f}{M} \tag{2.3}$$

where M is a global equilibrium and $H(f|M)$ is the entropy of f relative to a global equilibrium M. The Cercignani conjecture is similar to the logarithmic Sobolev inequality for the diffusion process, but the dissipation operator D is nonlinear in the function f. If the Cercignani conjecture holds, then the decay to global equilibrium would be exponentially fast. Through counterexamples, A.V. Bobylev and C. Cercignani [7] proved that this conjecture is false if the constant K depends on the function f only through finite Sobolev norms and moments. On the other hand, it was shown by E. Carlen and M. Carvalho [12] that $D(f) \geq \Theta_f(H(f|M))$, where the function Θ is not explicit but depends on f only through its moments and some derivatives. The conjecture was finally settled in a joint work by G. Toscani and Villani [45] and the subsequent work by Villani [47]. The conclusion is as surprising as it is beautiful: The Cercignani conjecture is in general false, but it is always almost correct in the following sense.

Theorem 2.1. *For a reasonable physical scattering kernel B, if f is smooth and with certain decay property in high momentum regime, then for any $\varepsilon > 0$ we have*

$$D(f) \geq K_\varepsilon(f)H(f|M)^{1+\varepsilon}$$

where $K_\varepsilon(f)$ depends on the smoothness and moments of f.

This inequality then implies that the entropic convergence rate is faster than $C_\varepsilon(f_0)t^{-1/\varepsilon}$ for any initial smooth data f_0. This is a much deeper inequality than the logarithmic Soblev inequality, as the operator D is nonlinear and the inequality fails for $\varepsilon = 0$ except in certain nonphysical situations, such as when the collision kernel is quadratic at large velocities.

Villani's next project in this direction is the very ambitious extension of this theorem to the spatially inhomogeneous case. One's immediate reaction to this question is that this is beyond reach since there is no global existence theory for the Boltzmann equation. The key physical question, however, is to understand the mechanism that leads to relaxation in the space variable. If we assume that good smooth solutions are given, the intrinsic difficulties are immediately visible: The identity $D(f) = 0$ implies that f is a local Maxwellian, but not necessary a global one, i.e., the density, temperature, and velocity parameters in the Maxwellian (2.2) depend on the space variable. Therefore, the relaxation to the global Maxwellian requires an additional mechanism different from the consideration of the entropy production. The only control on the space variable in the Boltzmann equation is the first order operator $v \cdot \nabla_x$. Now we have a formidable problem: It is analogue of a hypoelliptic problem, but the elliptic part is a nonlinear integral operator! Numerically, the entropy does decay very fast in the spatially inhomogeneous case, but the entropy production is far from monotonic. The main result in this direction is the following theorem by L. Desvillettes and Villani [15].

Theorem 2.2. *Suppose that $f_t(x, v)$ is a regular solution to the Boltzmann equation and f_t satisfies some lower bound estimate in the large velocity region. Under some assumptions on the collision kernel B, for any $\varepsilon > 0$ there is a C_ε such that*

$$H(f_t|M) \leq C_\varepsilon(f_0)t^{-1/\varepsilon}$$

where f_0 is the initial value of the Boltzmann equation.

This result assumes that the regularity of $f_t(x, v)$ is given, but is a large data theorem in the sense that there is no smallness condition on the initial function f_0. With a smallness condition, i.e., if the initial data is near a global Maxwellian, the assumptions of Theorem 2.2 can be verified, see, e.g., [24, 25]. Furthermore, significant progress was made in this direction for soft and long range potentials [23] and the decay rates can be exponentially fast for certain collision operators [22]. The method introduced to prove Theorem 2.2 is a very powerful one; Villani later developed a general theory, hypocoercivity [50], to estimate the large time asymptotics of a general class of hypoelliptic operators.

This program was also continued by younger mathematicians, in particular in the series of papers by C. Mouhot, C. Baranger, R. Strain, M. Gualdani and S. Mischler, on the spectral gap for the linearized Boltzmann operator [6, 34] and on the matching of the nonlinear convergence to equilibrium with the linearized theory, in a homogeneous setting [33] and in an inhomogeneous hypocoercive setting [19].

Finally, we mention that Villani's other work related to the Boltzmann equation includes a series of papers on the influence of grazing collisions, mainly with L. Desvillettes and R. Alexandre: existence of renormalized solutions (with defect measure) for the Boltzmann equation without cutoff [2], the rigorous derivation of the Fokker–Planck–Landau equation from the Boltzmann equation in the grazing collision limit [3, 46], and sharp regularity bounds associated with entropy production [1].

3. Optimal Transportation and Curvature

The optimal transport problem, also known as Monge-Kantorovich problem, is an ancient engineering problem seeking to minimize the cost to transport mass. For our purpose, the initial and final mass distributions are given by two probability measures μ and ν on a compact measurable metric space X. The goal is to find a measurable map $T : X \to X$ with $T_{\#}\mu = \nu$ to minimize the transportation cost

$$\int c(x, T(x)) d\mu(x). \tag{3.1}$$

The square root of the minimal transportation cost with the squared distance transportation cost function $c(x, y) = d(x, y)^2$ is called the 2-Wasserstein distance, W_2, between these two measures. The minimizer T is called the optimal transport map. The existence and uniqueness of the optimal transport map was proved in the Euclidean space by Y. Brenier [9] and in the Riemannian manifolds by R. McCann [32].

The probability measures on the metric space X with the Wasserstein distance constitute a compact metric space, called the *Wasserstein space* $(P(X), W_2) =: P_2(X)$ on X. We now take X to be a compact manifold M with metric tensor g which in turns generates a geodesic distance d and the normalized volume measure $\nu = \mathrm{dvol}_M/\mathrm{vol}_\mathrm{M}$. The information (negative of the entropy) $H(\mu)$ of a measure $\mu = \rho v$ absolutely continuous w.r.t. ν is defined by

$$H(\mu) = \int \rho \log \rho \, d\nu.$$

In a study of nonlinear heat equations, F. Otto [38] defined a formal Riemannian structure on $P_2(M)$ and interpreted these equations as gradient flows on the Wasserstein space $P_2(M)$ with this formal Riemannian structure. Subsequently, F. Otto and Villani [39] found the remarkable property that the entropy, viewed as a functional on the Wasserstein space $P_2(M)$, is concave if the Ricci curvature of the manifold M is nonnegative. This provided the first link between the concavity of entropy on the Wasserstein space and the Ricci curvature of the underlying manifold. This relation was subsequently established rigorously in [13], partly motivated by the earlier work [31]. F. Otto and Villani [39] also argued that the converse should hold, and it was rigorously established in [41].

If we replace the convexity of entropy by a lower bound K on the Hessian of entropy on $P_2(M)$, the corresponding condition on the Ricci curvature becomes the lower bound

$$\mathrm{Ric} \geq Kg.$$

Furthermore, the volume measure can be generalized to the weighted volume measure $e^{-\Phi}\mathrm{dvol}$ provided the Ricci curvature is replaced by the Bakry-Émery tensor

$$\mathrm{Ric}_\infty := \mathrm{Ric} + \mathrm{Hess}\ (\Phi) \geq Kg.$$

Using this heuristic idea, F. Otto and Villani [39] then provided a unified approach to a wide range of inequalities in analysis and geometry including the logarithmic Sobolev inequality and Talagrand's concentration inequality.

If $P_2(M)$ is a regular Riemannian manifold, a lower bound K on the Hessian of the entropy functional is equivalent to the displacement convexity inequality

$$H(\mu_t) \leq (1-t)H(\mu_0) + tH(\mu_1) - K\frac{t(1-t)}{2}W_2(\mu_0,\mu_1)^2 \tag{3.2}$$

for any Wasserstein geodesic μ_t. Notice that this definition depends only on the concept of geodesic on the Wasserstein space which can be defined on any metric space. There is no need for a Riemannian structure on M if we take (3.2) as the definition that the Ricci curvature on a metric space X is bounded below by K. With this definition of a lower bound on the Ricci curvature, J. Lott and Villani [36] proved the fundamental stability result that the lower bound on the Ricci curvature is stable under the Gromov-Hausdorff convergence. A closely related definition of Ricci curvature lower bounds, and similar stability results were obtained independently by K.-T. Sturm [43, 44]. The main statement of Lott-Villani's results can be stated as follows.

Theorem 3.1. *Let $\{(X_i, d_i, \nu_i)\}$ be a sequence of compact measured length spaces and $\lim_{i\to\infty}(X_i, d_i, \nu_i) = (X, d, \nu)$ in the measured Gromov-Hausdorff topology. If the Ricci curvature of (X_i, d_i, ν_i) is bounded below by K then the Ricci curvature of (X, d, ν) is also bounded below by K.*

This theorem demonstrates the robustness of this definition of Ricci curvature lower bounds. On the other hand, the definition is also a very effective notion since it allows one to generalize many theorems in Riemannian geometry to the setting of metric spaces, including the Bishop-Gromov theorem, logarithmic Sobolev inequality and Bonnet-Myers theorem. Moreover, the definition can easily be discretized [8]. We note that there are other notions and definitions of curvatures on metric spaces or graphs. This includes the work of Y. Ollivier [37] and F. R. Chung and S.-T. Yau's definitions of curvatures on graphs [18, 28].

To summarize, Villani has brought the tools of entropy and its time-evolution from the study of convergence to equilibrium in the Boltzmann equation to a geometric setting involving the Wasserstein space. In addition to the Ricci curvature, Villani has explored connections with other geometric or analytic problems, such as the Sobolev inequality, for which he has provided a new proof based on optimal transport and entropy-type functionals, in collaboration with D. Cordero-Erausquin and B. Nazaret [14]. Like [36], this paper was a starting point for other developments, including the work of A. Figalli, F. Maggi and A. Pratelli [21] on quantitative anisotropic isoperimetric inequalities. Villani also wrote a series of papers with A. Figalli, G. Loeper and L. Rifford relating the smoothness of optimal transport with the shape of the cut locus in Riemannian geometry [20, 30].

4. Landau Damping

The last theorem of Villani I will describe is a rigorous proof of Landau damping in the nonlinear setting. The fundamental equation governing plasma dynamics is the Vlasov-Poisson equation, which, for periodic data is given by

$$\partial_t f + v\nabla_x f + E\nabla_v f = 0, \quad f(t,x,v) \geq 0$$

where $f(t,x,v)$ is the density of charged particles with velocity $v \in \mathbb{R}^3$ at $x \in \mathbb{T}^3$, the unit torus in $\mathbb{R}^3$. The electric field E is related to the density of charged particles $\rho(t,x) = \int f(t,x,v)dv$ via the Poisson equation

$$E := E[\rho] := -\nabla\phi, \quad -\Delta\phi = \rho(x) - 1$$

where the constant 1 is the density of background charges normalized to be one. This equation describes the dynamics of galaxies if we make a sign change in the electric field due to the sign difference between the Coulomb and gravitational forces. The result I will describe is valid with both signs, but I will use the language of plasma physics.

The Vlasov-Poisson equation describes collisionless dynamics and is time reversible. It is well known that dissipative dynamics often approach equilibrium exponentially fast, but for reversible dynamics the state f_t at any given time carries the same information as the initial data and decay to equilibrium can only be valid after certain averaging. On the other hand, fast relaxation to equilibrium in nature is ubiquitous even for systems governed by Newtonian dynamics. The common explanation has been that the relaxation is due to collision processes which produce dissipation. In 1946, L. Landau [27] revolutionized this common belief by arguing that the electric field in the Vlasov-Poisson equation, which is a collisionless equation, decays exponentially fast. He computed this rate of convergence for the linearized Vlasov-Poisson equation. This astonishing discovery is thus termed Landau damping. Despite intensive studies, the understanding of Landau damping for the Vlasov-Poisson equation is very limited.

The Vlasov-Poisson equation has infinitely many stationary solutions. In fact, any density $g(v)$ satisfying the normalization condition $Jg(v)dv = 1$ is a stationary solution. The stability analysis of the linearized Vlasov-Poisson equation was mainly due to the work of O. Penrose [40] in the sixties. It states that f^0 is stable if for any $\sigma \in S^2$ and $f_\sigma(v) = \int_{v\sigma+\sigma^\perp} f^0(z)dz$, then for any w such that $f'_\sigma(w) = 0$ one has

$$\int \frac{f'_\sigma(v)dv}{v-w} dv < 1.$$

The Landau damping for the linearized Vlasov-Poisson was already understood in the sixties by the work of A. Saenz [42]; for the quasi-linear case only nonrigorous results were available. On the other hand, it was pointed out by G. Backus [4] that the linear approximation is not expected to be valid for the full nonlinear equation in the large time regime. For the nonlinear Landau damping, the only partial results available were examples of solution to the Vlasov-Poisson equation that exhibit Landau damping [10, 26]. Last year Mouhot and Villani [35] proved that, for any analytic data near an analytic linearly stable stationary state, the electric field decay exponentially fast. Notice that the analyticity assumption is not an artifact of the proof, it is in fact necessary [29]. This resolves the long standing problem of Landau damping. We will not be able to state their theorem precisely nor in its general form, but the following limited version in the physical dimension $d = 3$ gives a flavor of the depth of the full theorem.

Theorem 4.1. (nonlinear Landau damping for general interaction). *There is an analytic norm $\|\cdot\|_a$ on functions of the phase space such that the following holds: Let $f^0 : \mathbb{R}^d \to \mathbb{R}_+$ be an analytic stationary state satisfying Penrose's linear stability criterion. Suppose the initial profile $f_i \geq 0$ is near the analytic stationary state in the sense that*

$$\|f_i - f^0\|_a \leq \varepsilon, \tag{4.1}$$

for some ε sufficiently small. Then there are analytic profiles $f_{+\infty}(v)$, $f_{-\infty}(v)$ such that

$$f(t,\cdot) \stackrel{t\to\pm\infty}{\longrightarrow} f_{\pm\infty} \quad weakly$$

exponentially fast. Furthermore, the marginal density of the unique solution of the nonlinear Vlasov equation with initial value $f(0,\cdot) = f_i$ converges exponentially fast as time $t \to \pm\infty$, i.e., there exists ρ_∞ and $\lambda > 0$ such that for all integer r we have

$$\|\rho(t,\cdot) - \rho_\infty\|_{C^r(\mathbb{T}^3)} \leq Ce^{\lambda|t|}, \quad t \to \pm\infty, \tag{4.2}$$

where C^r denotes the L_∞ norm of the derivatives up to order r.

Since the Vlasov-Poisson equation is time reversible, the profiles f_t keep the memory of the initial datum for all time. The fast relaxations in Mouhot-Villani's theorem only refer to averaged quantities such as density in the position space or in the weak sense. This is due to the fact that weak convergence preserves only the information of low frequency modes; the information at low frequencies was transferred to high frequencies to "maintain the constant total information for all times". Although f_t carries all information of the initial data for all time, more and more information is stored at high frequency modes. Hence if we only look at low frequency modes (such as weak convergence), there is a loss of information and this is responsible for the fast relaxation of various averaged quantities. This resembles the phenomena in turbulence and it requires very precise understanding of this transfer of information to yield a mathematical proof. Mouhot-Villani's theorem is the first rigorous result to establish a fast decay to equilibrium, a time irreversible behavior, in confined collisionless time-reversible dynamics.

5. Conclusion

In Villani's work, we have seen not only rigorous mathematical analysis providing deep insights into physical behavior, but also important new mathematics emerging from the study of natural phenomena, in the spirit of Maxwell and Boltzmann. Besides his research articles, Villani has written extensive surveys and books [48, 49, 50, 51], and, through these, as well as the insights of his work, he has inspired a generation of young mathematicians with deep, rich, physically motivated mathematical questions. We are witnessing the beginning of Villani's spectacular career and influence in shaping the directions of analysis and mathematics.

References

[1] Alexandre, R., Desvillettes, L., Villani, C. and Wennberg, B.: Entropy dissipation and long-range interactions. *Arch. Ration. Mech. Anal.* **152**, no. 4, 327–355 (2000).

[2] Alexandre, R. and Villani, C.: On the Boltzmann equation for long-range interactions. *Comm. Pure Appl. Math.* **55**, no. 1, 30–70 (2002).

[3] Alexandre, R. and Villani, C.: On the Landau approximation in plasma physics. *Ann. Inst. H. Poincar Anal. Non Linaire* **21**, no. 1, 61–95 (2004).

[4] Backus, G: Linearized plasma oscillations in arbitrary electron velocity distributions. *J. Mathematical Phys.* **1**, 178–191 (1960); erratum, 559.

[5] Bardos, C., Golse, F., Levermore, C. D.: Fluid dynamic limits of kinetic equations. II. Convergence proofs for the Boltzmann equation. *Comm. Pure Appl. Math.* **46**, no. 5, 667–753 (1993).

[6] Baranger, C., Mouhot, C.: Explicit spectral gap estimates for the linearized Boltzmann and Landau operators with hard potentials. *Rev. Mat. Iberoamericana* **21**, no. 3, 819–841 (2005).

[7] Bobylev, A. V. and Cercignani, C.: On the rate of entropy production for the Boltzmann equation. *J. Stat. Phys.* **94**, 603–618 (1999).
[8] Bonciocat, A. I. and Sturm, K. S.: Mass transportation and rough curvature bounds for discrete spaces. *J. Funct. Anal.* **256**, no. 9, 2944–2966 (2009).
[9] Brenier, Y.: Polar factorization and monotone rearrangement of vector-valued functions. *Comm. Pure Appl. Math.*, **44**, no. 4, 375–417 (1991).
[10] Caglioti, E. and Maffei, C.: Time asymptotics for solutions of Vlasov–Poisson equation in a circle. *J. Statist. Phys.* **92**, 1–2, 301–323 (1998).
[11] Cercignani, C.: H-theorem and trend to equilibrium in the kinetic theory of gases. *Arch. Mech.* **34**, 231–241 (1982).
[12] Carlen, E. and Carvalho, M.: Strict entropy production bounds and stability of the rate of convergence to equilibrium for the Boltzmann equation. *J. Stat. Phys.* **67**, 575–608 (1992).
[13] Cordero-Erausquin, D., R. McCann and Schmuckenschläger, M.: A Riemannian interpolation inequality à la Borell, Brascamp and Lieb. *Invent. Math.* **146**, no. 2, 219–257 (2001).
[14] Cordero-Erausquin, D., Nazaret, B. and Villani, C.: A mass-transportation approach to sharp Sobolev and Gagliardo-Nirenberg inequalities. *Adv. Math.* **182**, no. 2, 307–332 (2004).
[15] Desvillettes, L. and Villani, C.: On the trend to global equilibrium for spatially inhomogeneous kinetic systems: the Boltzmann equation. *Invent. Math.* **159**, 245–316 (2005).
[16] DiPerna, R. J. and Lions, P.-L.: On the Cauchy problem for Boltzmann equations: global existence and weak stability. *Ann. of Math.* (2) **130**, 2 321–366 (1989).
[17] Golse, F. and Saint-Raymond, L.: The Navier-Stokes limit of the Boltzmann equation for bounded collision kernels. *Invent. Math.* **155**, no. 1, 81 161 (2004).
[18] Chung, F. R. K. and Yau, S.T.: Logarithmic Harnack inequalities. *Math. Res. Lett.* **3**, no. 6, 793–812 (1996).
[19] Gualdani, M., Mischler S., Mouhot C.: Factorization for non-symmetric operators and exponential H-theorem, arXiv:1006.5523.
[20] Figalli, A., Rifford, L. and Villani, C.: Nearly round spheres look convex. Preprint (2009).
[21] Figalli, A., Maggi, F., Pratelli, A.: A mass transportation approach to quantitative isoperimetric inequalities, to appear in *Invent. Math.*
[22] Gressman, P. and Strain, R.: Global strong solutions of the Boltzmann equation without angular cut-off, arXiv:0912.0888v1.
[23] Gressman, P. and Strain, R.: Global Classical Solutions of the Boltzmann Equation with Long-Range Interactions and Soft Potentials, arXiv:1002.3639.
[24] Guo, Y.: Classical solutions to the Boltzmann equation for molecules with an angular cutoff. *Arch. Ration. Mech. Anal.* **169**, no. 4, 305–353 (2003).
[25] Guo, Y.: The Landau equation in a periodic box. *Comm. Math. Phys.* **231**, no. 3, 391–434 (2002).
[26] Hwang, J.-H. and Velazquez, J.: On the existence of exponentially decreasing solutions of the nonlinear Landau damping problem. *Indiana Univ. Math. J.* **58**, no. 6, 2623–2660 (2009).
[27] Landau, L.: On the vibration of the electronic plasma. *J. Phys. USSR 10* **25**, (1946).

[28] Lin, Y. and Yau, S.T.: Ricci curvature and eigenvalue estimate on locally finite graphs. *Math. Res. Lett.* **17**, no. 2, 343–356 (2010).
[29] Lin, Z. W., Zeng, C. C.: Small BGK waves and nonlinear Landau damping, arXiv:1003.3005.
[30] Loeper, G. and Villani, C.: Regularity of optimal transport in curved geometry: the nonfocal case. To appear in *Duke Math. J.*
[31] McCann, R. J.: A convexity principle for interacting gases. *Adv. Math.* **128**, no. 1, 153–179 (1997).
[32] McCann, R. J.: Polar factorization of maps on Riemannian manifolds. *Geom. Fund. Anal.* **11**, no. 3, 589–608 (2001).
[33] Mouhot, C.: Rate of convergence to equilibrium for the spatially homogeneous Boltzmann equation with hard potentials. *Comm. Math. Phys.* **261**, no. 3, 629–672 (2006).
[34] Mouhot, C., Strain, R.: Spectral gap and coercivity estimates for linearized Boltzmann collision operators without angular cutoff. *J. Math. Pures Appl.* **87**, no. 5, 515–535 (2007).
[35] Mouhot, C. and Villani, C.: On Landau damping. arXiv:0904.2760, (2009).
[36] Lott, J. and Villani, C.: Ricci curvature via optimal transport. *Ann. Math.* **169**, 903–991 (2009).
[37] Ollivier, Y.: Ricci curvature of Markov chains on metric spaces. *J. Funct. Anal.* **256**, no. 3, 810–864 (2009).
[38] Otto, F.: The geometry of dissipative evolution equations: the porous medium equation. *Comm. Partial Differential Equations* **26**, no. 1–2, 101–174 (2001).
[39] Otto, F. and Villani, C.: Generalization of an inequality by Talagrand and links with the logarithmic Sobolev inequality. *J. Funct. Anal.* **173**, no. 2, 361–400 (2000).
[40] Penrose, O.: Electrostatic instability of a non-Maxwellian plasma. *Phys. Fluids* **3**, 258–265 (1960).
[41] von Renesse, M.-K. and Sturm, K.-T.: Transport inequalities, gradient estimates, entropy, and Ricci curvature. *Comm. Pure Appl. Math.* **58**, no. 7, 923–940 (2005).
[42] Saenz, A. W.: Long-time behavior of the electric potential and stability in the linearized Vlasov theory. *J. Mathematical Phys.* **6**, 859 875 (1965).
[43] Sturm, K.-T.: On the geometry of metric measure spaces. I. *Acta Math.* **196**, no. 1, 65–131 (2006).
[44] Sturm, K.-T.: On the geometry of metric measure spaces. II. *Acta Math.* **196**, no. 1, 133–177 (2006).
[45] Toscani, G. and Villani, C.: Sharp entropy dissipation bounds and explicit rate of trend to equilibrium for the spatially homogeneous Boltzmann equation. *Commun. Math. Phys.* **203**, 667–706 (1999).
[46] Villani, C.: On a new class of weak solutions to the spatially homogeneous Boltzmann and Landau equations. *Arch. Rational Mech. Anal.* **143**, no. 3, 273–307 (1998).
[47] Villani, C.: Cercignani's conjecture is sometimes true and always almost true. *Comm. Math. Phys.* **234**, 3, 455–490 (2003).
[48] Villani, C.: A review of mathematical topics in collisional kinetic theory. In Handbook of mathematical fluid dynamics, Vol. I. North-Holland, Amsterdam, 2002, pp. 71–305.

[49] Villani, C.: Topics in optimal transportation, vol. 58 of Graduate Studies in Mathematics. American Mathematical Society, Providence, RI, 2003.
[50] Villani, C.: Hypocoercivity. Mem. Amer. Math. Soc. **202**, no. 950 (2009).
[51] Villani, C.: Optimal transport, vol. 338 of Grundlehren der Mathematischen Wissenschaften. Springer-Verlag, Berlin, 2009.
[52] Yu, S. H.: Nonlinear wave propagations over a Boltzmann shock profile, *J. Amer. Math. Soc.*, (2010).

Cédric Villani

CÉDRIC VILLANI

Studied Mathematics at École Normale Supéreure in Paris from 1992 to 1996;

Ph.D., Université Paris-Dauphine, 1998.

Positions held

Assistant Professor at École Normale Supéreure in Paris, 1996–2000.

Professor at École Normale Supériure de Lyon, 2000-

Semester-long visiting positions in Georgia Tech, Atlanta (1999),

Miller Institute, Berkeley (2004) and Institute for Advanced Study, Princeton (2009)

Director of the Institut Henri Poincaré (IHP) in Paris, 2009-

Part-time visitor of the Institut des Hautes Études Scientifiques (IHES), 2009-

Proceedings of the International Congress of Mathematicians
Hyderabad, India, 2010

LANDAU DAMPING

by
CÉDRIC VILLANI*

ABSTRACT. In this note I describe the solution of a longstanding problem in mathematical physics: the extension of the Landau damping from the linearized to the nonlinear Vlasov equation.

Mathematics Subject Classification (2010). 82C40 (82D10, 70F45)

Keywords. Kinetic equations, galactic and plasma dynamics, collisionless relaxation

In 1936, Lev Landau introduced the basic collisional kinetic model for plasma physics, now commonly called the Landau–Fokker–Planck equation. With this model he imported in plasma physics Boltzmann's notion of relaxation by increase of entropy, or equivalently loss of information.

In 1946, Landau came back to this field with a much more daring concept: relaxation without entropy increase, with preservation of information, even when collisions are neglected. This notion led to the extremely influential idea that conservative partial differential equations may exhibit irreversible features.

Landau's analysis was not directly based on the relevant kinetic model in plasma physics, the Vlasov–Poisson equation, but only on a linearized approximation. The validity of this approximation in large time has been questioned. A recent work in collaboration with Mouhot [22] fills this gap and thus demonstrates that relaxation is possible in confined reversible systems, without entropy increase nor radiation. In this note I shall describe the main results and the main insights brought by the proof.

1. Mean-field Approximation

Large particle systems interacting via long-range collective interactions occur in many situations in physics. Consider the most fundamental situation of classical

*In memory of Paul Malliavin and Michelle Schatzman, who passed away just too soon to attend this conference — with admiration for their mathematical talent, and gratitude for their renewed support.

École Normale Supérieure de Lyon & Institut Henri Poincaré, 11 rue Pierre et Marie Curie, 75231 Paris Cedex 05, FRANCE. E-mail: villani@ihp.jussieu.fr particles interacting via Newton's equations in $\mathbb{R}^d$:

$$m_i \ddot{x}_i(t) = \sum_j F_{j\to i}(t),$$

where m_i is the mass of particle $i, x_i(t) \in \mathbb{R}^d$ its position at time $t, \ddot{x}_i(t)$ its acceleration, and $F_{j\to i}$ is the force exerted by particle j on particle i. If all masses are equal and the force derives from an interaction potential, in a dimensional units we obtain, after proper time-rescaling,

$$\ddot{x}_i(t) = -\frac{1}{N}\sum_j \nabla W(x_i(t) - x_j(t)).$$

In applications N can be of the order of 10^{20}, and then such a large system of equations is hopeless. The mean-field limit $N \to \infty$ transforms this system of many simple equations in just one (complicated) equation. To perform the limit, first rewrite the equations in terms of the empirical measure $\widehat{\mu}_t^N(dx\,dv) = N^{-1}\sum \delta_{(x_i(t),\dot{x}_i(t))}$:

$$\frac{\partial \hat{\mu}^N}{\partial t} + v\cdot\nabla_x \hat{\mu}^N + F^N(t,x)\cdot\nabla_v\hat{\mu}^N = 0, \quad F^N = -(\nabla W *_{x,v} \hat{\mu}^N);$$

then take the limit $N \to \infty$ to get an equation for the limit measure $\mu_t(dx\,dv)$. Assuming that $\mu_t(dx\,dv) = f(t,x,v)dx\,dv$, we can formally simplify by the invariant measure $dx\,dv$; the result is the **nonlinear Vlasov equation** with interaction potential W. In this model the unknown $f = f(t,x,v)$ is a time-dependent density distribution in phase space (position, velocity), and the equation is

$$\frac{\partial f}{\partial t} + v\cdot\nabla_x f + F(t,x)\cdot\nabla_v f = 0 \tag{1}$$

$$F = -\nabla W *_x \rho \quad \rho(t,x) = \int f(t,x,v)dv, \tag{2}$$

To escape the discussion of boundary conditions and to avoid dispersion effects at infinity, I shall only consider periodic data, that is, $x \in \mathbb{T}^d = \mathbb{R}^d/\mathbb{Z}^d$.

The most important case of application is the Vlasov–Poisson equation in plasma physics, where heavy ions are treated as a fixed background, $f(t,x,v)$ is the density of electrons, treated as a continuum, and the interaction potential is the Coulomb potential (the fundamental solution of $-\Delta$). Another archetypal interaction is the Newton potential, which differs from the Coulomb potential only by a change of sign and units; the resulting equation is the gravitational Vlasov–Poisson equation, of considerable importance in astrophysics.

Before going on, let me notice that while the mean-field limit is well-understood for smooth interactions [4, 8, 23], it has never been put on rigorous footing for Coulomb or Newton interactions. For singular potentials the only available results are those of Hauray and Jabin [11], which miss the Coulomb singularity by (a little bit more than) one order, and assume very stringent conditions of uniform interparticle separation at initial time.

2. Qualitative Behavior of the Vlasov Equation

The Vlasov equation (1) is a time-reversible transport equation; it is in some sense Hamiltonian [1] [15, Section 6]. In contrast with the Boltzmann equation, it keeps the value of Boltzmann's entropy, $-\iint f \log f \, dv \, dx$, constant in time. Its invariances are well-known: preservation of the energy (kinetic energy + potential energy), and preservation of all integrals of f, that is, all functionals of the form $\iint A(f) dv \, dx$.

This equation admits many, many equilibria: first, any spatially homogeneous function $f^0 = f^0(v)$ is an equilibrium; next, there is a general recipe to construct inhomogeneous equilibria on $\mathbb{T}^d \times \mathbb{R}^d$, known as BGK (Bernstein–Greene–Kruskal) waves [3]; the theory of these equilibria is still in their infancy in spite of wide speculation.

The long-time behavior of the Vlasov–Poisson equation has been the object of much debate and speculation: does the kinetic distribution converge to an equilibrium by means of conservative phenomena? Which equilibria are stable and which ones are not? Is there a recipe to predict the "most likely" asymptotic equilibria? Is there an invariant measure on solutions of the Vlasov–Poisson equations? These questions are of great interest in particular in astrophysics, since the apparent approximate homogeneity of galaxies cannot be explained by means of the very slow entropy production mechanisms. At the end of the sixties, Lynden-Bell introduced the mysterious notion of (collisionless) **violent relaxation** to solve this paradox [16, 17]; this is still the object of much debate.

In this ocean of conjectures and mysteries about collisionless relaxation, the only little island on which we can set foot, so far, is the **Landau damping**, which holds in the neighborhood of stable homogeneous equilibria, as I shall now explain.

3. Linearization

In the sequel I will use Fourier transform in both x and v variables, writing

$$\hat{f}(k,v) = \int f(x,v) e^{-2i\pi k\cdot x} \, dx, \; \tilde{f}(k,\eta) = \iint f(x,v) e^{-2i\pi k\cdot x} e^{-2i\pi k\eta\cdot v} dv \, dx.$$

Let $f^0 = f^0(v)$ be a homogeneous equilibrium. Let us write $f(t,x,v) = f^0(v) + h(t,x,v)$, assume $\|h\| \ll 1$ in some sense, and accordingly neglect the quadratic term in (1). The result is the **linearized Vlasov equation**:

$$\frac{\partial h}{\partial t} + v \cdot \nabla_x h + F[h](t,x) \cdot \nabla_v f^0 = 0, \quad F[h] = -\nabla_x W *_{x,v} h; \tag{3}$$

here $F[h]$ is the force field generated by the distribution h.

Applying the Duhamel formula (considering $F[\cdot]\cdot\nabla_v f^0$ as perturbation of $v\cdot\nabla_x$), then taking the Fourier transform in x and integrating in v yields the *closed equation* on $\rho^1(t,x) = \int h(t,x,v)dv$:

$$\hat{\rho}^1(t,k) = \tilde{h}_i(k,kt) + \int_0^t K^0(t-\tau,k)\hat{\rho}^1(\tau,k)d\tau, \tag{4}$$

where

$$K^0(t,k) = -4\pi^2\hat{W}(k)\tilde{f^0}(kt)|k|^2 t. \tag{5}$$

Appreciate the miracle: the Fourier modes $\tilde{\rho}^1(k)$, $k \in \mathbb{Z}$, evolve in time independently of each other, and satisfy a convolution equation — a simple instance of **Volterra equation**. In a way this expresses a property of complete integrability, which can actually be made more formal [21].

The convolution equation (4) can be studied by means of Fourier–Laplace transform. If (a) $\tilde{f^0}(\eta) = O(e^{-2\pi\lambda_0|\eta|})$, (b) the Laplace transform (in time) of the kernel K^0 does not approach 1 in a strip $\{0 \leq \mathcal{R}e\xi \leq \lambda_L|k|\}$, and (c) $\tilde{h}_i(k,\eta) = O(e^{-2\pi\lambda|\eta|})$, then (4) implies exponential time-decay of nonzero modes of the density perturbation: $\hat{\rho}_1(t,k) = O(e^{-2\pi\lambda'|k|t})$, for any rate $\lambda' < \min(\lambda,\lambda_0,\lambda_L)$. As a consequence the force $F[h]$ decays exponentially fast. This phenomenon discovered in [14] is called **Landau damping**. (Physics textbooks usually focus on λ_L, the Landau damping rate, as dictating the relaxation rate; but λ_0 and λ should not be forgotten in case f^0 and h_i are not entire functions.)

The existence of a positive decay rate λ_L is guaranteed by the analyticity of f^0 and the *Penrose stability condition*, which in dimension $d = 1$ reads

$$\forall\omega \in \mathbb{R}, \quad (f^0)'(\omega) = 0 \Rightarrow \hat{W}(k)\int \frac{(f^0)'(v)}{v-w}dv < 1. \tag{6}$$

The multidimensional version is that for any $k \in \mathbb{Z}^d$, the one-dimensional marginal of f^0 along the axis k satisfies this criterion. For instance, this stability criterion always holds true for Coulomb interaction in dimension 3 if f^0 is a radially symmetric distribution. On the contrary, for Newton interaction, even the Gaussian distribution may be stable or unstable, depending on the temperature (or equivalently, up to rescaling, on the size of the periodic box); this is the phenomenon of *Jeans instability*, which qualitatively explains the fact that stars tend to cluster in galaxies rather than spread around uniformly.

The following theorem summarizes the situation; since Landau's original work it has proven and reproven by many authors in various formalisms and with various degrees of precision, generality and rigor [2, 7, 12, 20, 22, 25, 27, 28].

Theorem 3.1 (Landau's damping theorem). *Let $f^0 = f^0(v)$ be an analytic homogeneous equilibrium, with $|\tilde{f^0}(\eta)| = O(e^{-2\pi\lambda_0|n|})$, and let W be an interaction potential such that $\nabla W \in L^1(\mathbb{T}^d)$. Let K^0 be defined in* (5)*; assume that there is $\lambda_L > 0$ such that the Laplace transform $(K^0)^L(\xi,k)$ of $K^0(t,k)$ stays away from the value 1 when $0 \leq \mathcal{R}e\xi \lesssim \lambda_L|k|$. Let further $h_i = h_i(x,v)$ be an analytic initial perturbation such that $\tilde{h}_i(k,\eta) = O(e^{-2\pi\lambda|n|})$. Then if h solves the linearized Vlasov* *equation* (3) *with initial datum* h_i, *one has exponential decay of the force field: for any* $k \neq 0$, *and any* $\lambda' < \min(\lambda_0, \lambda_L, \lambda)$,

$$\hat{F}[h](t,k) = O(e^{-2\pi\lambda'|k|t}).$$

In particular, $F[h]$ *converges to 0 exponentially fast as* $t \to \infty$.

Moreover, Penrose's stability condition (6) *guarantees the existence of* $\lambda_L > 0$.

Note that high modes decay faster, low modes decay slower. The infrared cutoff imposed by the periodic boundary conditions implies a uniform lower bound on the decay rate of the various modes; if the problem is set in the whole space, the very slow decay of very low spatial frequencies prevents the exponential decay of the force [9, 10].

4. Nonlinear Landau Damping

The impact of Landau's discovery cannot be overestimated: Landau damping nowadays is one of the cornerstones of classical plasma physics [26].

However, half a century ago, Backus [2] raised a serious objection against Landau's reasoning. He argued that the linearization approximation is not justified in large times for the Vlasov equation, because the amplitude of $\nabla_v h(t,x,v)$ grows at least linearly in time, due to the appearance of fast oscillations as $t \to \infty$; so even if $\nabla_v h$ is initially of size $O(\varepsilon)$, after time $O(1/\varepsilon)$ it will be of size $O(1)$.

O'Neil [24] further predicted that the linearization approximation anyway breaks down on time scales $O(1/\sqrt{\varepsilon})$, where ε is the size of the perturbation; this is well checked on numerical schemes. So if one is interested in larger time scales, the question naturally arises whether damping does hold for the nonlinear Vlasov equation, at least in the perturbative regime near a stable spatially homogeneous equilibrium.

This question seems to pose formidable difficulties: the nonlinear equation does not have the beautiful structure of the linearized equation; moreover the density $f(t,x,v)$ develops fast oscillations which prevent any uniform smoothness bound, a fortiori analytic regularity. Numerical simulations on such long time scales are not fully reliable, and sometimes subject to controversy. Isichenko [13] argued that the convergence to equilibrium in the nonlinear case should be very slow. However, Caglioti and Maffei [5] proved the existence of some exponentially damped solutions, leaving open the question of their genericity.

The following recent theorem by Mouhot and myself ends the debate:

Theorem 4.1 (nonlinear Landau damping). *Let* f^0 *be an analytic profile satisfying the Penrose linear stability condition. Further assume that the interaction* *potential W satisfies*

$$\widehat{W} = O\left(\frac{1}{|k|^2}\right). \tag{7}$$

Then one has nonlinear stability and nonlinear damping close to f^0. More precisely, there is $\varepsilon > 0$ such that if f_i is an initial datum satisfying

$$|\tilde{f} - \tilde{f}_0|(k,\eta) \le \varepsilon e^{-2\pi\mu|k|} e^{-2\pi\lambda|\eta|}, \qquad \iint |f_i(x,v) - f^0(v)| e^{2\pi\beta|v|} dx\, dv \le \epsilon$$

and $f(t,x,v)$ is the solution of the nonlinear Vlasov equation (1) *with interaction potential W and initial datum f_i, then $F[f](t,\cdot)$ converges exponentially fast to 0 as $t \to +\infty$. Moreover, $f(t,\cdot)$ converges weakly to an analytic homogeneous equilibrium $f_\infty = f_\infty(v)$.*

This theorem is perturbative, and in fact there is convincing numerical evidence that the conclusion should not hold for large perturbations of equilibrium. Nevertheless the strength of Theorem 4.1 is that it theoretically demonstrates the possibility of relaxation to equilibrium without any dissipation or randomness in a non-radiating, time-reversible, entropy-preserving system.

Theorem 4.1 also predicts the existence of a limit distribution $f_\infty(v)$. The constructive nature of the proof of Theorem 4.1 provides a natural approximation scheme for that limit. By time-reversibility there is also a limit distribution in negative times, and one may check that in general it differs from f_∞, implying that the limit distribution does not depend only on the invariants of the equation.

There is no contradiction between the reversibility of the Vlasov equation and the effective irreversibility of the behavior expressed by Theorem 4.1: the explanation is that although information is preserved for all times, it becomes stored in high-frequency variations of the distribution function in the kinetic variable. This transfer of information from low to high modes acts like a cascade in phase space, which we like to interpret in terms of regularity: the regularity deteriorates in the velocity variable because of the fast oscillations, but at the same time the regularity of the force in the position variable improves with time.

The rest of this text is devoted to a sketchy presentation of the main tools underlying the proof of Theorem 4.1.

5. Gliding Analytic Regularity

To estimate solutions of the nonlinear Vlasov equation in analytic regularity, let us look for an analytic norm which behaves well under composition (because the solution of a linear transport equation is obtained by composition with the trajectories of particles); and which does not fear the fast oscillations.

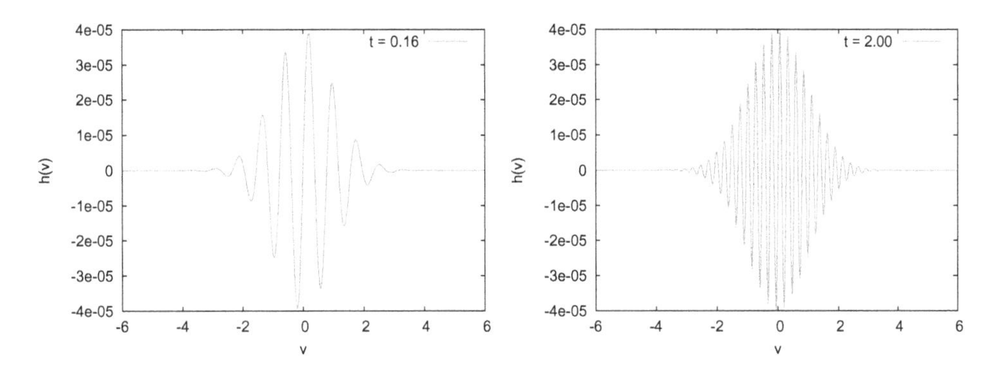

Figure 1. A slice of the distribution function (relative to a homogeneous equilibrium) for gravitational Landau damping, at two different times; notice the fast oscillations of the distribution function, which are very difficult to capture by an experiment. Image courtesy of Francis Filbet.

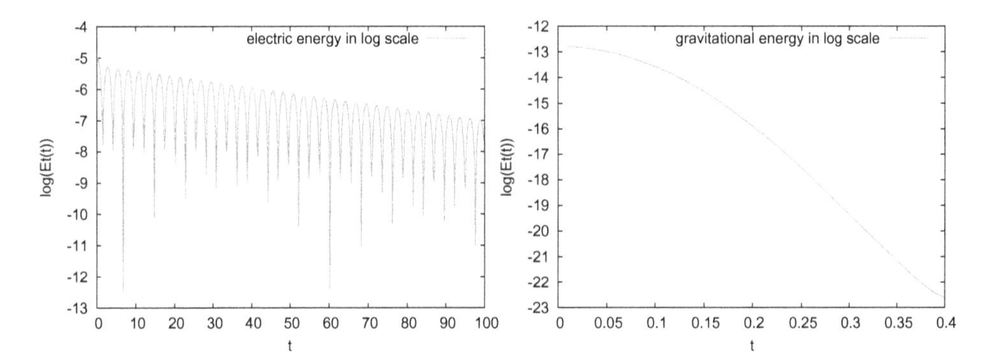

Figure 2. Time-evolution of the norm of the field, for electrostatic (on the left) and gravitational (on the right) interactions. In the electrostatic case, the fast time-oscillations are called Langmuir oscillations, and should not be mistaken with the velocity oscillations.

The first problem (composition) is solved by using algebra norms, well-known in certain areas of mathematics: two such norms are defined (say in dimension 1) by

$$\|f\|_{\mathcal{F}^\lambda} = \sum_{k\in\mathbb{Z}} e^{2\pi\lambda|k|}|\hat{f}(k)| \quad \|f\|_{\mathcal{C}^\lambda} = \sum_{n\in\mathbb{N}_0} \frac{\lambda^n}{n!}\|f^{(n)}\|_{L^\infty}, \tag{8}$$

where $f^{(n)}$ stands for the derivative of order n of f, and $\mathbb{N} = \{0, 1, 2, \ldots\}$. The first norm (as it is written) makes sense only for periodic functions, while the second one makes sense for any smooth function on $\mathbb{R}$. Both satisfy $\|fg\| \leq \|f\|\|g\|$, and as a consequence also satisfy nice formulas for the composition: with obvious notation,

$$\|f \circ (\mathrm{Id} + G\|_\lambda \leq \|f\|_\nu, \quad \nu = \lambda + \|G\|_\lambda.$$

The second problem (fast oscillations) is resolved by taking away the contribution of the free transport. This is like a scattering philosophy: to estimate the solution of the perturbed kinetic equation at time t, first evolve it backwards from time t to time 0, using the (reversed) free transport equation.

So the smoothness scale is devised by comparison with the solution of the free transport, and there is an information cascade from low to high modes, as in weak turbulence theory. We call this the **gliding regularity**.

All in all, we introduce a functional norm which mixes the two recipes appearing in (8) (one recipe for the position variable, another one for the velocity variable), and kills fast oscillations by replacing differentiation along ∇_v by differentiation along $\nabla_v + t\nabla_x$:

$$\|f\|_{\mathcal{Z}_\tau^{\lambda,(\mu,\gamma);p}} = \sum_{k\in\mathbb{Z}^d}\sum_{n\in\mathbb{N}_0^d} e^{2\pi\mu|k|}(1+|k|)^\gamma \frac{\lambda^n}{n!}\left\|(\nabla_v + 2i\pi\tau k)^n \hat{f}(k,v)\right\|_{L_p(dv)}. \tag{9}$$

By default, $\tau = 0, \gamma = 0$ and $p = \infty$.

The five indices might seem a burden, however they provide a lot of flexibility. The parameter τ can be adjusted as one wishes, but should not be too far from the physical time. Please note that the $\mathcal{Z}$ spaces are ordered with respect to the parameters λ, μ, γ and (cheating a bit) p, but not with respect to the parameter τ, at least not with uniform constants.

One can work out nice properties of the $\mathcal{Z}$ norms with respect to product, composition, differentiation, inversion. For instance:

$$\|f(x + X(x,v), v + V(x,v))\|_{\mathcal{Z}_\tau^{\lambda,\mu;p}} \le \|f\|_{\mathcal{Z}_\sigma^{\alpha,\beta;p}},$$

where $\alpha = \lambda + \|V\|_{\mathcal{Z}_\tau^{\lambda,\mu}}, \beta = \mu + \lambda|\tau - \sigma| + \|X - \sigma V\|_{\mathcal{Z}_\tau^{\lambda,\mu}}$.

Finally, as soon as one has a good decay in velocity space, one may embed the complicated $\mathcal{Z}$ spaces into more naive functional spaces, such as

$$\|f\|_{\mathcal{Y}_\tau^{\lambda,\mu}} := \sup_{k,\eta} |\tilde{f}(k,\eta)| e^{2\pi\lambda|\eta + k\tau|} e^{2\pi\mu|k|}.$$

6. Characteristics

Let us consider the linear Vlasov problem, where particles move in a given force field F, satisfying the same estimates as if it was induced by a solution $\bar{f}$ of the free transport. Then the regularity of F *improves with time*: if $\bar{f}$ is analytic, then

$$\|F(t,\cdot)\|_{\mathcal{F}^{\lambda t+\mu}} = O(1)$$

for some $\lambda, \mu > 0$.

Then the solution f is given by the composition of the initial datum, f_i, by the characteristic equations (trajectories). So, to understand the solution of the equation, it is sufficient to understand these characteristics; we define

$$S_{t,\tau}(x,v) = (X_{t,\tau}(x,v), V_{t,\tau}(x,v))$$

as the position and velocity at time τ of particles which are transported by the force field F and which at time t will have position x and velocity v. To compare S with the free transport evolution $S^0_{t,\tau}(x,v) = (x - (t-\tau)v, v)$, introduce

$$\Omega_{t,\tau} = S_{t,\tau} \circ S^0_{\tau,t}. \tag{10}$$

That is, start from time τ, evolve by the free dynamics up to time t, and then evolve it backwards by the perturbed dynamics to time τ. As $t \to \infty$, $\Omega_{t,\tau}$ converges to what is usually called a scattering transform.

A fixed point argument shows the following: if $\lambda' < \lambda, \mu' < \mu$ and

$$\|F(t,\cdot)\|_{\mathcal{F}^{\lambda t+\mu}} \le \frac{\varepsilon(\mu-\mu')(\lambda-\lambda')^2}{C},$$

for C large enough, then

$$\|\Omega_{t,\tau} - \mathrm{Id}\|_{\mathcal{Z}^{\lambda',\mu'}_\tau} \le C\varepsilon e^{-2\pi(\lambda-\lambda')\tau} \min\left(t-\tau, \frac{1}{\lambda-\lambda'}\right).$$

This estimate shows that the dynamics asymptotically looks like free transport; it is good because it is (a) uniform as $t \to \infty$; (b) small as $\tau \to t$; (c) exponentially small as $\tau \to \infty$.

The loss of regularity index is roughly of order $O(\varepsilon^{1/3})$; we shall see later how to improve this by playing on the parameters of the norm.

7. Reaction

Now let us consider the force as the unknown, and let the force act on a given time-dependent distribution $\bar{f}(t,x,v) = f^0(v) + h(t,x,v)$. Then the equation is

$$\frac{\partial f}{\partial t} + v \cdot \nabla_x f + F[f](t,x) \cdot \nabla_v \bar{f}(t,x,v) = 0, \tag{11}$$

which formally describes the evolution of a gas of particles which acts by forcing the distribution $\bar{f}$, such that there is a flux of particles from distribution f to distribution $\bar{f}$, which exactly reacts to the effect of the force and guarantees that $\bar{f}$ is unaffected.

Let us set artificially $f^0 = 0$ to focus on the effect of the nonlinearity. Applying the Duhamel principle, Fourier transform and integration in velocity, we obtain

$$\hat{\rho}(t,k) = \tilde{f}_i(k,kt) + \int_0^t \sum_\ell \widehat{\nabla W}(k-\ell)\hat{\rho}(\tau,k-\ell)(\widetilde{\nabla_v f})(\tau,\ell,k(t-\tau))d\tau. \tag{12}$$

Of course all modes of the density are now coupled; to bound them all together, let's use the norm $\|\rho\|_{\mathcal{F}^{\lambda t+\mu}}$. We assume that $\widehat{W}(k) = O(1/|k|^{1+\gamma})$ as $k \to \infty$, and that $\tilde{f}$ satisfies the same estimates as a solution of free transport:

$$\widetilde{\nabla_v f}(\tau, \ell, k(t-\tau)) \leq C|k|(t-\tau)e^{-2\pi\bar{\lambda}|k(t-\tau)+\ell\tau|}e^{-2\pi\bar{\mu}|k|}.$$

Plugging this bound in (12) leads to an integral equation replacing (4):

$$\|\rho(t)\|_{\mathcal{F}^{\lambda t+\mu}} \leq A(t) + C\int_0^t K(t,\tau)\|\rho(\tau)\|_{\mathcal{F}^{\lambda\tau+\mu}}\,d\tau, \tag{13}$$

where $A(t) = \sum_k e^{2\pi(\lambda t+\mu)|k|}|\tilde{h}_i(k, kt)|$ remains bounded if λ is small enough, and

$$K(t,\tau) = \sup_{k,\ell}\left(\frac{|k|(t-\tau)e^{-2\pi(\bar{\lambda}-\lambda)|k(t-\tau)+\ell\tau|}e^{-2\pi(\bar{\mu}-\mu)|\ell|}}{1+|k-\ell|^\gamma}\right). \tag{14}$$

Note that the argument inside the supremum is not uniformly small for large k and large t: a **resonance** phenomenon occurs for

$$k(t-\tau) + \ell\tau = 0,$$

similar to the celebrated *echo experiment* performed by Malmberg and collaborators in the sixties [18, 19].

The bad news about kernel (14) is that it grows linearly with time: $K(t,\tau)$ is in general not better than $O(\tau)$, and $\int_0^t K(t,\tau)d\tau = O(t)$, suggesting a potential superexponential instability. But the good news is that the interaction comes with an important delay. To appreciate this, compare the integral equations $\varphi(t) \leq \int_{t-1}^t \tau\varphi(\tau)d\tau$ (allowing superexponential growth) and $\varphi(t) \leq t\varphi(t/2)$ (imposing subexponential growth).

Also the influence of the singularity of the interaction potential W is seen on (14): the more singular it is, the slower the decay as $|k-\ell| \to \infty$, the stronger the coupling between different modes.

To estimate solutions of (13) one can use exponential moment estimates. The idea is that

$$\int_0^t e^{-\varepsilon t}K(t,\tau)e^{\varepsilon\tau}\,d\tau \tag{15}$$

will be smaller if K favors large values of $t-\tau$. In the present case,

$$\int_0^t e^{-\varepsilon t}K(t,\tau)e^{\varepsilon\tau}d\tau \leq \frac{C}{\varepsilon^r t^{\gamma-1}}, \tag{16}$$

for some constants $C > 0, r > 0$, and ε arbitrarily small. The important fact is that the bound on the right-hand side of (16) decays as $t \to \infty$, at least for $\gamma > 1$. One can use this information to show that solutions of (13) cannot grow faster than $O(e^{\varepsilon t})$, where ε is as small as desired; stated otherwise, there is an arbitrarily small loss on the decay rate.

This method accommodates with the presence of f^0, at the price of technical estimates involving further information on $K(t,\tau)$:

$$\left(\int e^{-2\varepsilon t}K(t,\tau)^2 e^{2\varepsilon\tau}\,d\tau\right)^{1/2} \leq \frac{C}{\varepsilon^\gamma t^{\gamma-1/2}}, \quad \sup_{\tau\geq 0}\int_\tau^\infty e^{\varepsilon\tau}K(t,\tau)e^{-\varepsilon t}\,dt \leq \frac{C}{\varepsilon^r}.$$

As $\gamma \to 1$, the coupling becomes so strong that the previous method no longer works; instead one can work out a more complicated scheme where all modes are estimated separately, rather than within a single norm. The resulting infinite system of inequalities also provides an arbitrarily small loss on the exponential decay rate.

8. Newton's Scheme

To overcome the loss of decay rate observed in the solution of the linearized problem, we adapt to the present setting the classical Newton algorithm, thus constructing the solution of the nonlinear Vlasov equation as a superposition of solutions of linear equations: $f = \lim_{n\to\infty} f^n$, $f^n = f^n(t,x,v)$ being defined as

$$f^n = f^0 + h^1 + \cdots + h^n,$$

where

- $f^0 = f^0(v)$ is the homogeneous equilibrium;
- h^1 solves the linearized Vlasov equation around f^0, starting from $f_i - f^0$;
- for any $n \geq 1, h^{n+1}$ solves the linear equation

$$\frac{\partial h^{n+1}}{\partial t} + v\cdot\nabla_x h^{n+1} + F[f^n]\cdot\nabla_v h^{n+1} + F[h^{n+1}]\cdot\nabla_v f^n$$
$$= -F[h^n]\cdot\nabla_v h^n \tag{17}$$

with initial datum $h^{n+1}(0,\cdot) = 0$. The fact that h^n appears quadratically in the right-hand side of (17) formally guarantees that the convergence of the scheme is extremely fast (almost like δ^{2^n}).

The analytic regularity of the solution of this system of equations is first estimated for short times, as in Cauchy–Kowalevskaya theory.

Large time estimates are much more tricky and involve all the ingredients from sections 5 to 7. First one composes by the characteristics induced by$F[f^n]$, in order to get rid of the term $F[f^n]\cdot\nabla_v h^{n+1}$. This does not harm much if we can show that these trajectories are asymptotic to free transport in a suitable sense. Then the reaction analysis and echo control provide the decay of the force, with an arbitrarily small loss on the rate of decay. The overall goal is to set up a virtuous circle: if $F[f^n]$ decays fast, the trajectories will be close to free transport trajectories, and in particular will induce a good mixing of h^{n+1}; and in turn this will imply a fast decay of $F[h^{n+1}]$.

The implementation of these ideas is particularly technical. Let $\rho_t^n(x,v) = \int h^n(t,x,v)\,dv$, and $\Omega_{t,\tau}^n = S_{t,\tau}^n \circ S_{t,\tau}^0$, where S^n are the characteristics induced by $F[f^n]$. Two key estimates which are propagated along the scheme are:

$$\sup_{\tau\geq 0} \|\rho_\tau^n\|_{\mathcal{F}^{\lambda_n\tau+\mu_n}} \leq \delta_n, \qquad \sup_{\tau\geq\tau\geq 0} \|h_\tau^n \circ \Omega_{t,\tau}^{n-1}\|_{\mathcal{Z}_{\tau-\frac{bt}{1+bv}}^{\lambda_n(1+b),\mu_n;1}} \leq \delta_n, \tag{18}$$

where $b(t) = B/(1+t)$ for some well-chosen parameter $B > 0$. Notice the shift in the indices of the norm of h^n, where the regularity is modulated depending on the final time t: this trick, combined with the decay of the force field, allows to circumvent the fixed loss of regularity due to the composition by the characteristics.

A number of auxiliary estimates are propagated: schematically,

- $\Omega^n \simeq \mathrm{Id}, \nabla\Omega^n \simeq I$;
- $\Omega^n - \Omega^k$ is small and $(\Omega^k)^{-1} \circ \Omega^n \simeq \mathrm{Id}$ as $k \to \infty$, uniformly in n;
- $h^k \circ \Omega^n$, $\nabla h^k \circ \Omega^n$, $V^2 h^k \circ \Omega^n$ are small as $k \to \infty$, uniformly in n;
- $(\nabla h^{n+1}) \circ \Omega^n \simeq \nabla(h^{n+1} \circ \Omega^n)$

A key step is a self-consistent estimate on $\rho^{n+1} = \int h^{n+1} dv$: among other ingredients, the assumption $\widehat{W}(k) = O(1/|k|^2)$ is used there to ensure that $\|\nabla F^{n+1}\| \leq C\|\rho^{n+1}\|$, so that

$$\|F^{n+1} \circ \Omega^n - F^{n+1}\| \leq \|\nabla F^{n+1}\| \|\Omega^n - \mathrm{Id}\| \leq \|\rho^{n+1}\| - \|\Omega^n - \mathrm{Id}\|,$$

with the same norm on the left-hand and right-hand sides.

The implementation of the scheme is done in a number of steps at each stage, each of which involves a small loss on the gliding regularity, and large constants. But the latter are all eventually wiped out by the extraordinarily fast convergence of the Newton scheme:

$$\delta_n = O(\delta^{a^n}), \quad 1 < a < 2.$$

In the end remains the *uniform bound*

$$\sup_{n\in\mathbb{N}} \sup_{t\geq 0} (\|f^n(t,\cdot) - f^0\|_{\mathcal{Z}_t^{\lambda,\mu;1}} + \|F^n(t,\cdot)\|_{\mathcal{F}^{\lambda t+\mu}}) = O(\|f_i - f^0\|). \tag{19}$$

From this follows a uniform bound on the solution f, and the exponential decay on the force $F(t,\cdot)$, which in turn implies that $f(t,x+vt,v)$ converges to some distribution function $g(x,v)$. Then $f(t,x,v)$ is asymptotic to $g(x-vt,v)$, and the existence of the asymptotic profile $f_\infty(v)$ follows by the homogenization properties of the free transport.

9. Conclusions

Theorem 4.1 establishes that Landau damping survives nonlinearity: this solves a controversial problem posed half a century ago. The proof of this result is technical and complex, but constructive and based on elementary tools. It provides a hands-on approach of the long-time behavior of the nonlinear Vlasov equation, and singles out the mechanism and the important ingredients behind Landau damping: confinement, mixing, and the Riemann–Lebesgue lemma.

The construction bears several similarities with the Kolmogorov–Arnold–Moser theory [6]. Indeed, the linearized Vlasov equation is completely integrable in some sense, the nonlinearity acts as a perturbation, and the loss of regularity occurring in the solution of the linearized Landau problem can be overcome by a Newton scheme. In our case, the most severe reason for the loss of regularity is the formation of echoes due to the oscillatory nature of solutions. In this sense the proof provides an unexpected bridge between three of the most famous paradoxical statements from classical mechanics of the twentieth century: Landau damping, KAM theory, and the echo experiment. This is all the more remarkable that this bridge only appears in the treatment of the nonlinear Vlasov equation, while Landau was dealing specifically with the linearized equation.

However, in contrast with classical KAM theory, the solution of the linearized Vlasov equation implies a loss of infinitely many derivatives; in Fourier space, this is like mutiplication by $e^{|\xi|^\alpha}$ with $0 < \alpha < 1$. This high loss of regularity is one of the main reasons why we are unable to run a classical Nash–Moser regularization scheme and get results in C^k regularity. Instead, we are only able to work in Gevrey regularity, and formulate a guess for the critical regularity: Gevrey-3 (that is, derivatives growing like $n!^3$).

After this theorem, many new problems can be formulated: extension to other models, to inhomogeneous equilibria, long-time behavior of less smooth data, mean-field limit in the perturbative regime... A number of old problems also remain wide open, such as the understanding of the statistical theory of the Vlasov equation. When addressing these issues, just as the problem which motivated Theorem 4.1, we must bear in mind that the goal of mathematical physics is not to rigorously prove what physicists already know, but rather through mathematics to get new insights about the physics, and from physics to identify new mathematical problems.

References

[1] Ambrosio, L., and Gangbo, W. Hamiltonian ODE's in the Wasserstein space of probability measures. *Comm. Pure Appl. Math. 51* (2007), 18–53.

[2] Backus, G. Linearized plasma oscillations in arbitrary electron distributions. *J. Math. Phys. 1* (1960), 178–191, 559.

[3] Bernstein, I.B., Greene, J.M. and Kruskal, M.D. Exact nonlinear plasma oscillations. *Phys. Rev. 108,* 3 (1957), 546–550.

[4] Braun, W., and Hepp, K. The Vlasov dynamics and its fluctuations in the $1/N$ limit of interacting classical particles. *Commun. Math. Phys. 56* (1977), 125–146.

[5] Caglioti, E., and Maffei, C. Time asymptotics for solutions of Vlasov–Poisson equation in a circle. *J. Statist. Phys. 92,* 1–2 (1998), 301–323.

[6] Chierchia, L. A. N. Kolmogorov's 1954 paper on nearly-integrable Hamiltonian systems. A comment on: "On conservation of conditionally periodic motions for a small change in Hamilton's function" [Dokl. Akad. Nauk SSSR (N.S.) **98** (1954), 527–530]. *Regul. Chaotic Dyn. 13,* 2 (2008), 130–139.

[7] Degond, P. Spectral theory of the linearized Vlasov–Poisson equation. *Trans. Amer. Math. Soc. 294,* 2 (1986), 435–453.

[8] Dobrušin, R. L. Vlasov equations. *Funktsional. Anal. i Prilozhen. 13,* 2 (1979), 48–58, 96.

[9] Glassey, R., and Schaeffer, J. Time decay for solutions to the linearized Vlasov equation. *Transport Theory Statist. Phys. 23,* 4 (1994), 411–453.

[10] Glassey, R., and Schaeffer, J. On time decay rates in Landau damping. *Comm. Partial Differential Equations 20,* 3–4 (1995), 647–676.

[11] Hauray, M., and Jabin, P.-E. N-particles approximation of the Vlasov equations with singular potential. *Arch. Ration. Mech. Anal. 183,* 3 (2007), 489–524.

[12] Hayes, J. N. On non-Landau damped solutions to the linearized Vlasov equation. *Nuovo Cimento (10) 30* (1963), 1048–1063.

[13] Isichenko, M. Nonlinear Landau damping in collisionless plasma and inviscid fluid. *Phys. Rev. Lett. 78,* 12 (1997), 2369–2372.

[14] Landau, L.D. On the vibration of the electronic plasma. *J. Phys. USSR 10* (1946), 25. English translation in *JETP 16,* 574. Reproduced in *Collected papers of L.D. Landau,* edited and with an introduction by D. ter Haar, Pergamon Press, 1965, pp. 445–460; and in *Men of Physics: L.D. Landau,* Vol. 2, Pergamon Press, D. ter Haar, ed. (1965).

[15] Lott, J. Some geometric calculations on Wasserstein space. *Comm. Math. Phys. 277,* 2 (2008), 423–437.

[16] Lynden-Bell, D. The stability and vibrations of a gas of stars. *Mon. Not. R. astr. Soc. 124,* 4 (1962), 279–296.

[17] Lynden-Bell, D. Statistical mechanics of violent relaxation in stellar systems. *Mon. Not. R. astr. Soc. 13,* 6 (1967), 101–121.

[18] Malmberg, J., and Wharton, C. Collisionless damping of electrostatic plasma waves. *Phys. Rev. Lett. 13,6* (1964), 184–186.

[19] Malmberg, J., Wharton, C., Gould, R., and O'Neil, T. Plasma wave echo experiment. *Phys. Rev. Letters 20,* 3 (1968), 95–97.

[20] Maslov, V. P., and Fedoryuk, M. V. The linear theory of Landau damping. *Mat. Sb. (N.S.) 127(169),* 4 (1985), 445–475, 559.

[21] Morrison, P.J. Hamiltonian description of Vlasov dynamics: Action-angle variables for the continuous spectrum. *Transp. Theory Statist. Phys. 29,* 3–5 (2000), 397–414.

[22] Mouhot, C., and Villani, C. On Landau damping. Preprint, 2009. Available online at arXiv:0904.2760

[23] Neunzert, H. An introduction to the nonlinear Boltzmann–Vlasov equation. In *Kinetic theories and the Boltzmann equation,* C. Cercignani, Ed., vol. 1048 of *Lecture Notes in Math.,* Springer, Berlin, Heidelberg, 1984, pp. 60–110.

[24] O'Neil, T. Collisionless damping of nonlinear plasma oscillations. *Phys. Fluids 8,* 12 (1965), 2255–2262.

[25] Penrose, O. Electrostatic instability of a non-Maxwellian plasma. *Phys. Fluids 3* (1960), 258–265.

[26] Ryutov, D. D. Landau damping: half a century with the great discovery. *Plasma Phys. Control. Fusion 41* (1999), A1–A12.

[27] Sáenz, A. W. Long-time behavior of the electric potential and stability in the linearized Vlasov theory. *J. Mathematical Phys. 6* (1965), 859–875.

[28] van Kampen, N. On the theory of stationary waves in plasma. *Physica 21* (1955), 949–963.

CPSIA information can be obtained
at www.ICGtesting.com
Printed in the USA
LVHW021436130323
741520LV00005B/283

9 789814 696180